Illustrated Dictionary of MYCOLOGY

by

Miguel Ulloa
Department of Botany
Institute of Biology
Universidad Nacional Autónoma de México

and

Richard T. Hanlin
Department of Plant Pathology
University of Georgia

with the assistance of

Samuel Aguilar and **Elvira Aguirre Acosta**
Universidad Nacional Autónoma de México

APS PRESS

The American Phytopathological Society
St. Paul, Minnesota

Cover art by M. Ulloa

This book has been reproduced directly from electronic files submitted in final form to APS Press by the authors. No editing or proofreading has been done by the Press.

Reference in this publication to a trademark, proprietary product, or company name by personnel of the U.S. Department of Agriculture or anyone else is intended for explicit description only and does not imply approval or recommendation to the exclusion of others that may be suitable.

Library of Congress Catalog Card Number: 00-103517
International Standard Book Number: 0-89054-257-0

Printed in the United States of America on acid-free paper

The American Phytopathological Society
3340 Pilot Knob Road
St. Paul, Minnesota 55121-2097, USA

This book is dedicated to my wife Carmen
and to my academic colleagues

M. Ulloa

Table of contents

Preface

When the *Diccionario Ilustrado de Micología* first appeared in 1991, it provided, for the first time, a reference which Spanish-speaking students of mycology could consult in their own language. The *Diccionario* contained some 2,000 technical terms, many of which were illustrated by drawings and/or photographs that were integrated into the text. These illustrations served to assist the student in understanding and interpreting the various terms they depicted. In addition, the etymology was given for each entry term, thus enabling the student to learn the derivation of these terms. That the *Diccionario* filled a need is demonstrated by the fact that it was soon out of print.

After reviewing the *Diccionario* soon after its publication, it was apparent that English-speaking/reading students also could benefit from its unique features, but few possess the language skills necessary to do so. Because of our earlier collaboration on mycological works, it was decided to undertake the laborious task of translating it into English. As the translation slowly progressed, however, new terms appeared in the literature, and it was felt desirable to include them to make the dictionary as current as possible. This also led us to include some older terms that had been omitted from the Spanish edition. The result is that while this work is nominally based on the Spanish version, it is much expanded, with nearly twice as many terms and some 80% more illustrations.

The text of the dictionary was initially entered into the computer in Word Perfect 6.0, but as the text increased greatly in size, it was migrated to Adobe Page Maker 6.5 for ease of handling. The color transparencies were digitized with a Polaroid Sprintscan slide scanner at 1350 dpi resolution and the images, obtained in black and white, were optimized with Photoshop 5.0. The same procedure was followed for the black and white transparencies of the drawings. Once the text was completed, the illustrations were then inserted near the terms they represent. The completed manuscript was furnished as film-ready copy on diskette for printing.

Although all of the drawings and the majority of the photographs were made by the authors, we are grateful to various individuals for permission to include additional illustrations belonging to them; they are acknowledged at the end of the book. Special mention must be made of two colleagues, Samuel Aguilar and Elvira Aguirre, of the Instituto de Biología, Universidad Nacional Autónoma de México. Both spent many months assisting with various phases of the project. The former contributed greatly in editing the book. Without their help, the project would have taken much longer to complete. We also thank María C. González, who helped calculate the scales of many of the illustrations, and María Ángeles Herrera-Campos, who provided some data on lichens, both of the Instituto de Biología. Finally, we acknowledge the technical advice of Antonio Bolívar, the Managing Editor of Redacta, S. A., of Mexico City.

We are indebted to Héctor Hernández, Director of the Instituto de Biología, for his continuous support, and to APS Press for their willingness to publish the present work. Their encouragement and support as we learned the details of computer publishing eased the task of preparing the final manuscript.

Miguel Ulloa
Mexico City, Mexico

Richard T. Hanlin
Athens, Georgia

December 1999

Preface to the Spanish edition

The idea of doing this *Diccionario ilustrado de micología* occurred to me in July, 1984, while I was working on the book *El reino de los hongos, micología básica y aplicada*, and was due principally to the uneasiness I frequently felt on finding in the mycological bibliography written in Spanish many technical terms that were poorly translated or inappropriately applied. With this in mind, and taking into account the convenience and necessity of assembling a mycological lexicon of broad coverage, and complementing it with a considerable number of high quality illustrations, I dedicated myself to undertaking the first work of this type written in Spanish, with the object of contributing to filling a vacuum that had existed in this aspect of mycology, in Mexico as well as in other Spanish speaking countries.

The dictionary brings together around 2,000 terms, the majority of them mycological, which I have considered as of major usage in the instruction and investigation of fungi. Many terms are not strictly mycological, since they are employed in the nomenclature of various biological sciences, but I have included them because they frequently are employed in various specialities of mycology.

For each term is given the equivalent in English, the Greek (transliterated into Spanish) or Latin etymology (sometimes both), the definition, one or more examples of fungi (generally a genus or species) to which the term is applied, the taxonomic order to which the examples belong, one or more figures, when possible, and when appropriate, the opposite or related term, which can be consulted under the corresponding entries.

Since the major part of the mycological bibliography that is consulted in our field is in English it is understandable that students of the fungi are more or less familiar with their lexicology in this language, and that at times they may encounter certain difficulties on trying to use them translated into Spanish; for them, a list of the terms in English is included at the end of the dictionary, followed by their equivalents in Spanish. When the ending of the words in English changes in its plural form, this is given in parentheses, e.g., *perithecium* (*perithecia*).

In case the user of this book needs to know the taxonomic position of the fungi used to illustrate various concepts, he/she can consult the alphabetical list of taxonomic orders and classes to which the genera belong, as well as the taxonomic scheme, included at the end of the work. The correct placement of the taxa will permit the reader to refer to other works to obtain additional information about the terms consulted.

With the objective of making the work more useful, didactic, and attractive, I have placed special interest on including a considerable quantity of figures (766); whenever it was convenient, or possible, I illustrated the fungal structures and the biological processes or phenomena in which the fungi are involved, and which I considered most important.

Of the total of 611 photographs, more than half (326) are mine; all of the drawings (158) are drawings in black and white made by me (many are based on or adapted from drawings or photographs already published, and which are found in the works cited in the bibliography). The initials of the authors of the photographs are indicated in the captions, and the names corresponding to these initials, as well as the institutions in which these authors work, are presented in alphabetical order in the credits of the photographs at the end of the work. The scale at which the figures are printed also is given in the caption of each figure.

Although the review I made of the mycological bibliography did not pretend to be exhaustive, I am sure that it was sufficiently extensive to attain an important coverage of the subject. The orthography of the terms in Spanish and their Greek or Latin etymologies is based principally on the *Diccionario de botanica* by P. Font Quer (1963), although the selection and editing of the definitions derived from the analysis of various possible interpretations, principaly of those that are considered in the specialized dictionaries of mycology written in English (which are noted below). Inasmuch as certain information was obtained from Font Quer and from two other works done in Mexico, which are the *Glosario para Spermatophyta, Flora mesoamericana*, by M. Sousa and S. Zárate (1983), and the *Glosario botánico ilustrado*, by N. P. Moreno (1984), and the last two are designed to be utilized in the study of higher plants, some data were useful in the preparation of the present mycological dictionary. The orthography of the terms in English and the major part of the information that was utiulized in the editing of the definitions was derived from a complete and careful review of the following works: *A Glossary of Mycology*, by W. H. Snell and E. A. Dick (1957), *Collegiate Dictionary of Botany*, by D. Swartz (1971), and *Ainsworth and Bisby´s Dictionary of the Fungi*, by D. L. Hawksworth, B. C. Sutton and G. C. Ainsworth (1983), as well as from the other works cited in the bibliography. In the majority of cases, the definitions given by the different authors agree in essence and they served as the basis for editing them according to my own criteria, trying always to include representative examples. In the available botanical and mycological dictionaries, the inclusion of examples is not common. It is appropriate to make clear that the present book is basically a compilation of terms relative to the macro- and micromorphology, reproduction, physiology and ecology of fungi, since terms referring to metabolic products (such as organic acids, antibiotics, pigments and others) and of properties related to the color, taste or odor of these products, are deliberately excluded, terms that, e.g., are described in Snell and Dick. Except for the names of classes of fungi (such as Chytridiomycetes, Zygomycetes, etc.), which for practical reasons are explained, the names of genera and other suprageneric taxonomic categories are also excluded, entries which are considered in Hawksworth *et al.* In the appendix are given some terms common in taxonomy.

The principal purposes of this *Diccionario ilustrado de micología* are to collect the major part of the basic words utilized in the exercise of mycology and related sciences, to help in the translation and interpretation of the nomenclature that is utilized in English, and to standardize the usage of the terminology in Spanish and English, trying to present in a uniform manner the modern use of a lexicology as extensive and specific as that employed in the study of the fungi, and enriching the information with the aid of a good number of pertinent illustrations. The book represents more a case of the graphic fucus that I have always followed in the exercise of my profession.

I am grateful for the collaboration of Ricardo Valenzuela, who besides providing various photographs, also supplied me with herbarium specimens for use in taking various slides; of Patricia Lappe, who helped me prepare the list of terms in English and in assembling the measurements of the fungal structures illustrated; of Calixto Benavides, who with great care collaborated with me in the reviewing and arranging of the figures and their legends, as well as their corresponding scales, besides the meticulous task of reviewing the text proofs of the dictionary; of Laura Estrada Cuéllar and Magdalena Reyes Téllez, for painstaikingly typing the text; and of Evangelina Pérez Silva and Teófilo Herrera, for their suggestions of species of fungi used to represent various terms. I also express my gratitude, in a special way, to Elvira Aguirre Acosta, for her constant interest in the progress of the dictionary, and her valuable assistance involved in various aspects of the work, particularly in the search for herbarium specimens to be photographed, and in the preparation of the list of taxonomic orders and classes to which the genera of fungi utilized as examples of the included terms belong. Finally, I recognize the important assistance of Antonio Lot Helgueras, Director of the Institute of Biology of the Universidad Nacional Autónoma de Mexico (UNAM), as well as of the personnal of Redacta, S.A. for their painstaking and professional participation in composing this work.

Miguel Ulloa

Mexico City, Mexico, 1991

Introduction

An essential aspect in mastering any specialized or professional subject is to acquire a thorough knowledge of the technical terms used in that field. Once learned, these terms enable the student to communicate with other professionals in the area and to comprehend technical publications. Like other areas of biology and of science in general, mycology has an extensive terminology that has developed over many decades. The beginning student of mycology is introduced to the most important of these terms in introductory courses and textbooks, but as he/she progresses to advanced studies, new and unfamiliar terms are encountered, especially in the literature. It is at this point that recourse to a mycological dictionary is necessary.

The *Illustrated Dictionary of Mycology* is designed to provide mycology students at all levels with a complete, up-to-date reference they may consult. It contains over 3,800 terms for which definitions are given, along with the etymological derivations of these terms. The terms are illustrated with 774 black and white photographs and 548 drawings. In addition, one or more species of fungi to which a particular term applies is included with the definition. Although most of the entries are strictly mycological in nature, a number of terms of broader biological usage that are frequently encountered in the mycological literature are included for convenience.

Translating the text from Spanish into English was done in a stepwise manner. An initial "rough" translation was made and checked against the original text to assure accuracy of meaning. The next step was to realphabetize the text according to the English alphabet. This was a major task due to differences in the alphabets of the two languages; e.g., many terms that begin with "e" in Spanish begin with "s" in English, and certain letters in the Spanish alphabet do not occur in English. This meant that numerous terms had to be moved individually, a time-consuming process. Once this was accomplished, the text was divided into the 26 letters of the alphabet for ease in handling as new terms were added and the translated text was rewritten where necessary to achieve a smoother presentation. As this process progressed, additional illustrations were added to illustrate more of the terms in the dictionary. The entire project was further slowed by the authors being located in different countries. Several visits were made to each others' laboratories to assure continuity of thought and to resolve various issues that arose. After the text appeared complete and ready for insertion of the illustrations, the final formatting was achieved.

Because of the additional terms and illustrations, the present volume is twice as large and more complete than the Spanish edition on which it is nominally based. All of the drawings are originals made for the dictionary by the senior author, and the majority of the photographs likewise were taken by the authors. Additional photographs, however, were solicited from various individuals for inclusion; they are gratefully acknowledged at the end of the book.

Although the dictionary is not taxonomically oriented, it was nonetheless considered useful to provide a taxonomic outline of the taxa mentioned as examples of the different terms in the text, so that students unfamiliar with a particular genus can locate the group to which it belongs. Of the multitude of taxonomic schemes available, it was decided to utilize that of Alexopoulos *et al.*, in their book *Introductory Mycology*, 4th ed. (1996), with some modifications included from the classification system of the *Ainsworth and Bisby's Dictionary of the Fungi*, 8th ed.

by Hawksworth *et al.*, 1995, for those fungal genera that were not considered by Alexopoulos *et al.* As for the Ascomycetes, the most recent, and more complete, classification scheme of Eriksson and Winka (1997) was included as an appendix to the classification scheme utilized in the dictionary. Not only does the latter incorporate the most recent molecular data, but it also avoids the gaps present in other recently published schemes.

There are, of course, other dictionaries of mycology available to the student, all of which can provide basic definitions of mycological terms. The main feature of the present volume that we hope will be of especial benefit to the student is the large number of illustrations (1 322) used to interpret many of the terms. These illustrations are inserted into the text near the term(s) to which they refer for ease of reference. In addition, one or more species of fungi to which a particular refers is given for each entry.

Orismology is the science of defining the technical terms of a particular field of study. The word is derived from the Greek *horismós*, and this from *horízein*, meaning to delimit or make a boundary. The presence of complete etymologies for all of the terms included in the dictionary provides the more scholarly inclined student with a unique opportunity to delve more deeply into the origin and derivation of the mycological terms he/she uses regularly in the study of fungi. It is our hope that the presence of these etymologies will encourage some students to broaden their horizons and achieve a new level of knowledge as they learn about the exciting world of the fungi.

Abbreviations and symbols

Abb.- abbreviation

adj.- adjective

AF.- Anglo-French

Ar.- Arabic

Azt.- Aztequism (from Nahuatl)

ca.- about, approximately (< L. *circa*)

Cf.- compare (< L. *confer*)

Chin.- Chinese

cm - centimeter

diam.- diameter

dim.- diminutive

E.- English

ed.- edition

e.g.- for example (< L. *exempli gratia*)

et al.- and others (< L. *et alii.*, *aliorum*)

etc.- and so forth (< L. *et cetera*)

F.- French

f.- feminine

G.- German

Ga.- Galic

gen.- genus(singular) and genera (plural)

genit.- genitive

Gr.- Greek

Hind.- Hindi

IE.- Indo-European

i.e.- that is (< L. *id est*)

It.- Italian

L.- Latin

lit.- literally

LL.- Late Latin

m - meter (< L. *metrum*, measure)

Ma.- Malay

MD.- Middle Dutch

ME.- Middle English

med.- medical

MF.- Middle French

MHG.- Middle High German

ML.- Middle Latin

mm - millimeter

µm - micrometer

mycol.- mycology

NL.- NeoLatin

nm -	nanometer
OE.-	Old English
OF.-	Old French
OHG.-	Old High German
OIt.-	Old Italian
ON.-	Old Norse (Norwegian)
ONF.-	Old North French
OPr.-	Old provençal
pl.-	plural
pp.-	past participle
pref.-	prefix
prob.-	probably
prp.-	present participle
Scand.-	Scandinavian
sing.-	singular
Sp.-	Spanish
sp.-	species
spp.-	species (plural)
subsp.-	subspecies
suf.-	suffix
syn.-	synonym
var.-	variation, variety
VL.-	Vulgar Latin
< -	derived from
> -	giving rise to
= -	equals, equivalent to
# -	number
°C -	degrees centigrade

ILLUSTRATED DICTIONARY
OF
MYCOLOGY

a

abaxial (NL. *abaxialis* < L. *ab-*, away from, from + *axialis*, axial < *axis*, axis, axle + *-alis* > E. suf. *-al*, relating to or belonging to): facing away from the main or central axis; e.g., the abaxial side of a basidiospore is the ventral or convex side that faces away from the longitudinal axis of the basidium. Cf. **adaxial**.

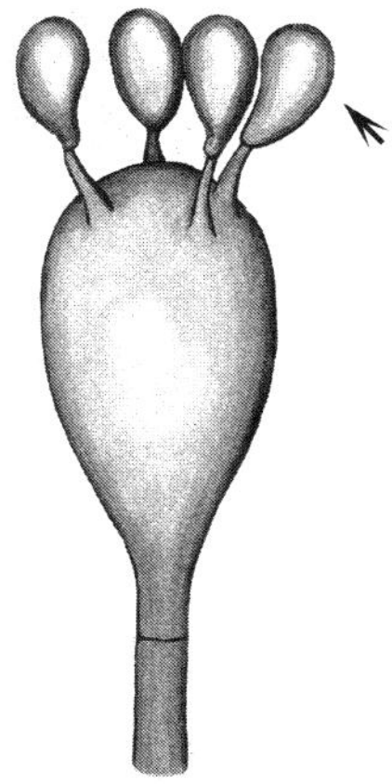

Abaxial side of the basidiospore of *Amanita* sp., x 1 100.

aberrant (L. *aberrans*, genit. *aberrantis*, wandering < *aberrare*, to deviate from + suf. *-ant* > E. *-ant*, one that performs, being): differing in minor or major respects from the common fundamental characters of a taxon; cultural or structural variation from the norm.

abhymenial (L. *ab-*, away from + *hymenium* < Gr. *hyménion*, dim. of *hymén*, membrane + L. suf. *-alis* > E. *-al*, relating to or belonging to): opposite the hymenial or spore-bearing surface.

abjection (ME. *abjectioun* < L. *abjectus*, cast away, discharged, set free + E. suf. *-tion*, result of an action, state of): the active (forcible) separation of a spore from a sporogenous cell (conidiophore, sterigma, sporophore, etc.); e.g., in the basidiomycetes the discharge of a basidiospore from the sterigma of a basidium is a process of abjection, in which a microscopic drop of liquid is produced between the hilum of the spore and the apex of the sterigma, as is observed in *Calocera cornea* (Dacrymycetales). Also, the ballistoconidia or ballistospores of various species of yeasts (such as *Sporobolomyces roseus* and *Itersonilia perplexans*, sporobolomycetaceous asexual yeasts) are discharged by abjection.

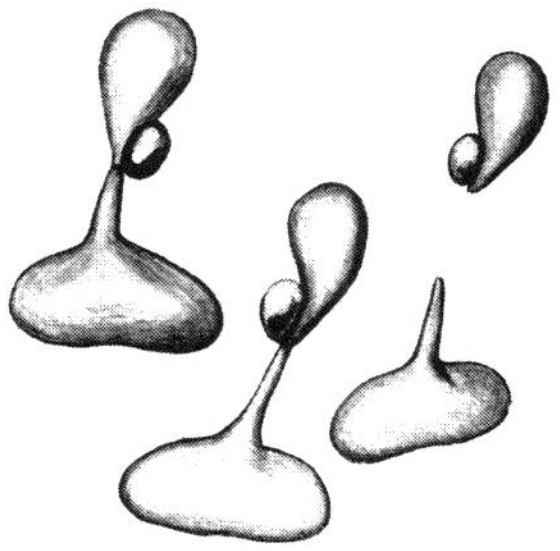

Abjection of the ballistospore of *Itersonilia perplexans*, x 3 160.

abjunction (L. *abjungere*, to separate, disarticulate < *ab-*, off, without, away from + *junctus*, joined, united + E. suf. *-tion*, result of an action, state of): the delimitation and cutting off of spores from portions of hyphae by formation of transverse septa, such as occurs in the development of arthrospores (arthroconidia) in *Geotrichum* and in the blastospores of *Monilia* (moniliaceous asexual fungi). Regarded by some as a syn. of **abstriction**.

Abjunction and **abscission** (successive stages of **abstriction**) of the arthrospores of *Geotrichum candidum*, x 1 800.

ablastic, **ablastous** (Gr. *ablastóo*, to not germinate, germinate poorly + suf. *-íkos* > L. *-icus* > E. *-ic*, belonging to, relating to; or + L. *-osus* > OF. *-ous*, *-eus* > E. *-ous*, possessing the qualities of): the incapacity of spores or other fungal propagules to germinate.

aboospore (L. *ab-*, away from + Gr. *oón*, egg + *sporá*, spore): an oospore that results from parthenogenetic development of an oosphere, as occurs in some members of the Saprolegniales.

abortive (L. *abortivus*, prematurely born < *abortare*, to miscarry < *ab-*, away from, without + *ortus*, birth + *-ivus* > E. *-ive*, quality or tendency, fitness): incompletely developed; an organ or structure whose development remains in a rudimentary state and does not reach maturity.

abraded (L. *abradere*, to scrape off, to rub or wear away, erode): *Lichens*. Having the thallus surface worn, eroded.

abrupt (L. *abruptus*, broken off < *abrumpere*, to break off < *ab-*, off, away from + *rumpere*, to burst): terminating suddenly, not gradually; appearing to have been cut off transversely; e.g., the stipe of some basidiomycetes in the gen. *Hohenbuehelia* (Agaricales) has this characteristic. Syn. of **truncate**.

abscission (L. *abscission* < *abscissio*, genit. *abscissionis*, the act of cutting-off, a breaking-off < *abscindere*, to tear off, separate + *-io*, *-ionis*, state, result of > E. *-ion*): separation, such as the detachment of a spore from the sporogenous cell.

absorption (L. *absorption* < *absorptio*, genit. *absorptionis*, the act of absorbing or swallowing < *absorptus*, ptp. of *absorbere*, to swallow, engulf + *-io*, *-ionis* > E. suf. *-ion*, result of an action, state of): the action of absorbing, which is the passage of solid, liquid, or gaseous materials (usually nutritive) dissolved in water into or through living cells, by means of osmosis. The fungi typically acquire their nutrients by absorption through the cell wall and/or plasma membrane.

abstriction (NL. *abstrictio*, genit. *abstrictionis*, the act of binding or pressing together < *strictus*, drawn tight + *-io*, *-ionis* > E. suf. *-ion*, result of an action, state of): *Sporogenesis*. The production of spores in a sporogenous filament through the successive formation of transverse walls, so that the spores remain grouped in short chains, delimited by more or less distinct constrictions, as occurs, e.g., in *Geotrichum* and *Monilia* (moniliaceous asexual fungi). In abstriction, **abjunction** occurs first, and then **abscission**, especially by constriction.

abyssal (Gr. *ábyssos*, abyss, a deep, immeasurable space + L. suf. *-alis* > E. *-al*, relating to or belonging to): pertaining to an abyss, especially with reference to organisms that inhabit bodies of water at great depths; e.g., *Periconia abyssa* (dematiaceous asexual fungi), which is found on wood submerged at depths of 3,975-5,315 m., where the hydrostatic pressure is 50 atmospheres or greater.

acanthophysis, pl. **acanthophyses** (Gr. *ákantha*, spine + *phýsis*, action of springing forth, giving birth, a growth): *Basidiomycetes*. A clavate, coralloid, or botryose paraphysis, or a cystidium, that is covered with adornments or spines. Such paraphyses are present in the basidiocarps of some Aphyllophorales, such as *Aleurodiscus mirabilis*, *Fistulina radicata* and *Vararia* spp. In some Agaricales acanthophysoid cystidia are present, such as the cheilocystidia and pileocystidia of *Mycena brownii* and some other species of *Mycena*.

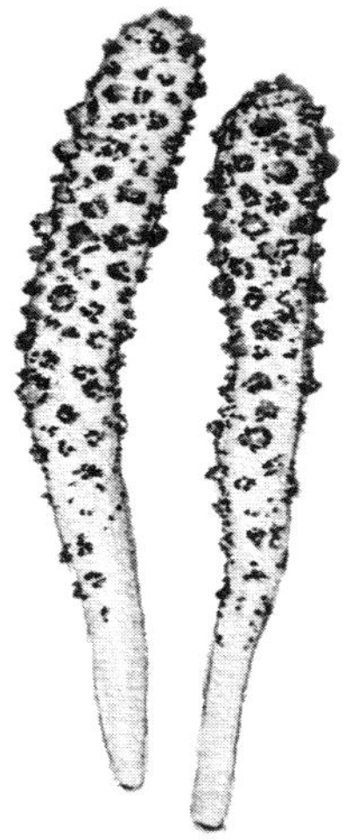

Acanthophyses from the hymenophore of *Fistulina radicata*, x 350.

acaryallagic (Gr. *a-*, without + *káryon*, nucleus + *allagé*, permutation, commutation + suf. *-íkos* > L. *-icus* > E. *-ic*, belonging to, relating to): see **akaryallagic**.

acaryote (Gr. *a-*, without + *káryon*, nut, nucleus): see **akaryote**.

acaudate (L. *acaudatus*, without a tail < *a-*, without + *cauda*, tail + suf. *-atus* > E. *-ate*, provided with or likeness): without a tail. Cf. **caudate**.

acaulescent, **acauline**, **acaulose** (L. *acaulis*, without a stem < *a-*, without + *caulis*, stem + *-escens*, genit. *-escentis*, that which turns, beginning to, slightly > E. *-escent*; or + L. suf. *-inus* > E. *-ine*, of or pertaining to; or + L. *-osus*, full of, augmented, prone to > ME. *-ose*): without a stalk or pedicel, sessile; Myxomycetes whose sporangia lack a pedicle or peduncle are referred to as acauline (especially referring to the species that have aethalia, such as *Lycogala*, Liceales).

accumbent (L. *accumbens*, genit. *accumbentis*, reclining < *accumbere*, to recline, lay down + *-entem* > E. *-ent*, being): reclining, lying against something; e.g., like the basidiocarps of *Poria*, *Corticium* and other gen. of Aphyllophorales.

Acaulescent sporangium of *Lycogala epidendrum*, x 8 (*MU*).

-aceae (L. *-aceae*, fem. pl. adjectival suf.): ending of taxonomic families, e.g., Mucoraceae, Saccharomycetaceae, etc.

-aceous (L. suf. *-aceus* > E. *-aceous*, of or pertaining to, with the nature of): a suf. used to indicate a relation to or having the nature of; e.g., micaceous and crustaceous mean similar to mica and crust, respectively. Other examples are pucciniaceous, ustilaginaceous and agaricaceous, for fungi having the fundamental characteristics of the families Pucciniaceae, Ustilaginaceae and Agaricaceae, respectively.

acephalous (Gr. *aképhalos*, headless < *a-*, without + *kephalé*, head + L. *-osus* > OF. *-ous*, *-eus* > E. *-ous*, having, possessing the qualities of): without a head; applied to structures that do not have a cap or enlargement on the upper end. Cf. **capitate**.

acerose, **acerous** (L. *acerosus*, needle-shaped < *acer*, sharp + *-osus*, full of, augmented, prone to > ME. *-ose*; or + L. *-osus* > OF. *-ous*, *-eus* > E. *-ous*, having, possessing the qualities of): having the shape of a needle, ending in a point, like pine leaves. E.g., the phialides of *Penicillium verruculosum*, *P. purpurogenum* and of *Paecilomyces fumosoroseus* (moniliaceous asexual fungi) are acerose.

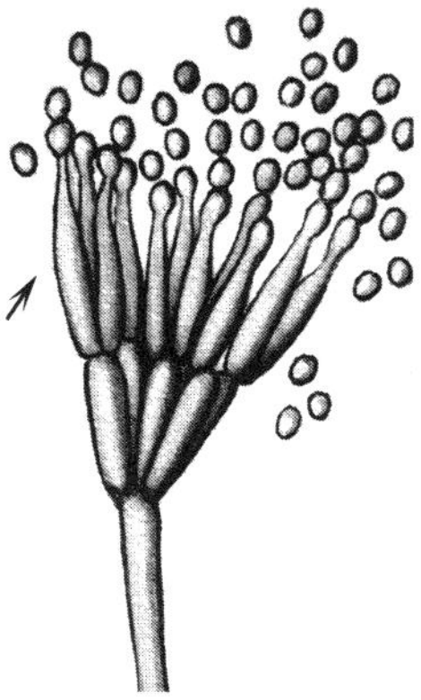

Acerose phialides of *Penicillium purpurogenum*, x 700.

acervate, **acervatus** (L. *acervatus*, heaped up, piled < *acervare*, to heap up, pile + suf. *-atus* > E. *-ate*, provided with or likeness): forming cushion-like piles or masses.

acervulus, pl. **acervuli** (NL. *acervulus* < dim. of L. *acervus*, heap): an aggregation of hyphae in the form of a small cushion, which erupts through the epidermis of parasitized host plants, and on which are formed short, crowded conidiophores. The acervulus is the asexual sporophore characteristic of the melanconiaceous asexual fungi, e.g., *Colletotrichum*.

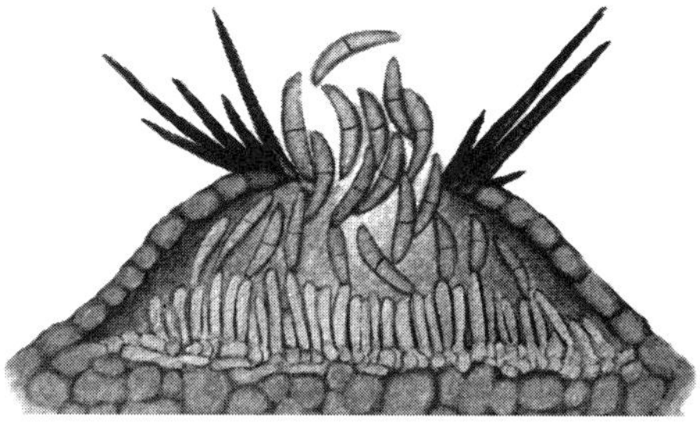

Acervulus of *Colletotrichum falcatum*, x 1 100.

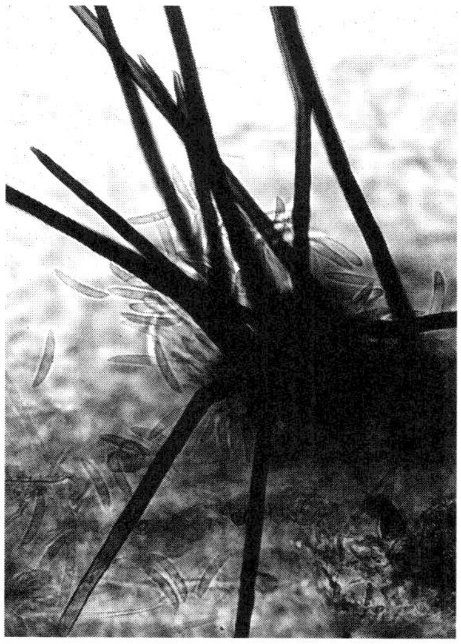

Acervulus of *Colletotrichum circinans*, x 350 (*MU*).

acetabulate (L. *acetabulum*, receptacle; wide, shallow goblet, orig. for vinegar + suf. *-atus* > E. *-ate*, provided with or likeness): syn. of **acetabuliform**.

acetabuliform (NL. *acetabuliformis* < L. *acetabulum*, shallow, wide cup, orig. for vinegar + *-formis* < *forma*, shape): with the shape of a shallow, wide cup, like the apothecia of various lichens and Pezizales e.g., *Cookeina sulcipes* (Pezizales).

acetabulum, pl. **acetabula** (L. *acetabulum*, a vinegar cup, receptacle; a wide, shallow goblet, orig. for vinegar < *acetum*, vinegar + dim. suf. *-ulum*): a shallow, saucer-shaped structure, such as the apothecium of certain fungi (Discomycetes); e.g., *Peziza* and *Helvella* (Pezizales).

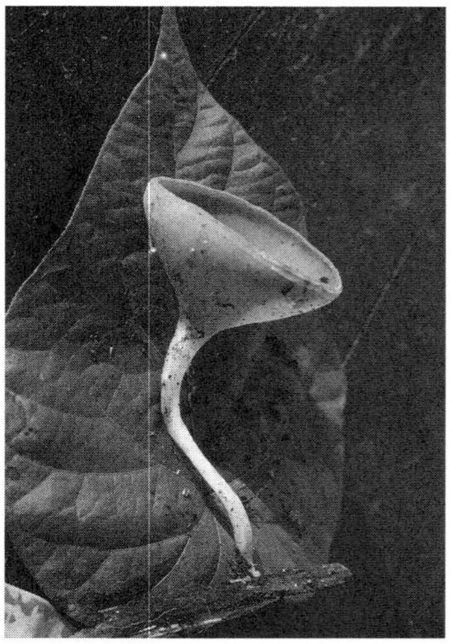

Acetabuliform apothecium of *Cookeina sulcipes*, x 1 (*EPS*).

Acetabuliform apothecia of *Cookeina sulcipes*, x 1 (*TH*).

achlorophyllous (Gr. *a-*, without + *chlorós*, green, the color of grass + *phýllon*, leaf + L. *-osus* > OF. *-ous*, *-eus* > E. *-ous*, having, possessing the qualities of): lacking chlorophyll, the photosynthetic pigment of green plants.

achroic, **achrous** (Gr. *a-*, without + *chrōs* or *chroía*, the superficial appearance or color of a thing + suf. *-íkos* > L. *-icus* > E. *-ic*, belonging to, relating to; or + L. *-osus* > OF. *-ous*, *-eus* > E. *-ous*, having, possessing the qualities of): having no color or pigment.

acicular (L. *acicularis*, narrow, stiff and pointed < *acicula*, dim. of *acus*, a point, needle + suf. *-aris* > E. *-ar*, like, pertaining to): a long, very slender, sharp-pointed structure; needle-shaped. Certain types of cystidia, as in *Pouzarella nodospora* (Agaricales), are acicular, as are the points of the appendages of the cheilocystidia in *Mycena longiseta* (Agaricales).

aciculate (NL. *aciculatus* < L. *acicula*, dim. of *acus*, a point, needle + suf. *-atus* > E. *-ate*, provided with or likeness): **1.** Having needle-like structures. **2.** Having fine grooves on the surface, as though made by a needle. A structure in the shape of a needle is **acicular**, as the ascospores of *Lophodermium juniperinum* (Rhytismatales).

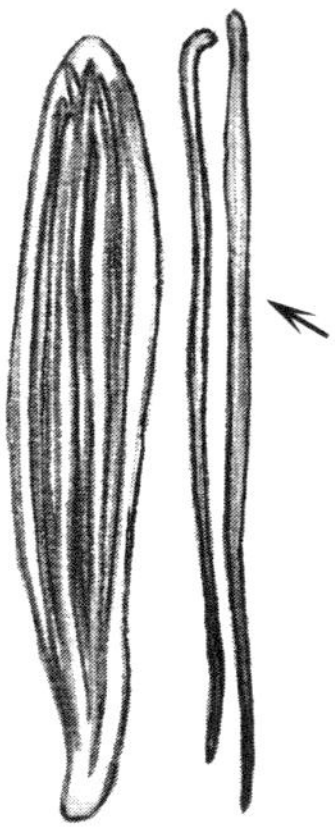

Aciculate ascospores of *Lophodermium juniperinum*, x 590.

aciculiform (L. *acicula*, dim. of *acus*, a point, needle + *-formis* < *forma*, shape): having the shape of a needle. Syn. of **acicular**.

acidophilous (L. *acidus*, sour, acid + Gr. *phílos*, have an affinity for + L. *-osus* > OF. *-ous*, *-eus* > E. *-ous*, possessing the qualities of): applied to organisms that grow in acid substrates; in particular, it refers to certain lichens that grow in peaty soils. Cf. **basophil**.

acinaciform (L. *acinaciformis* < *acinaces*, sabre, scimitar, curved sword < Gr. *akinákes*, short sword + L. *-formis* < *forma*, shape): in the shape of a sword, like the appendages that cover the zygosporangium of *Absidia spinosa* (Mucorales).

Acinaciform appendages of *Absidia spinosa*, x 180 (*EAA*).

acinose, **acinosus**, **acinous** (L. *acinosus*, like grapes < *acinus*, grape seed, berry + *-osus*, full of, augmented, prone to > ME. *-ose*; or + L. *-osus* > OF. *-ous*, *-eus* >

E. *-ous*, having, possessing the qualities of): consisting of a cluster of seeds or berries, or with a granular structure similar to a cluster of grapes. See **botryose**.

acolumellate (Gr. *a-*, without + L. *columna*, column + dim. suf. *-ella* + suf. *-atus* > E. *-ate*, provided with or likeness): lacking a columella. *Gasteromycetes*. Refers to a fruiting body in which the gleba lacks a basal columella, e.g. in *Astraeus* (Sclerodermatales).

acrasids (Gr. *akrásia*, bad mixture, or *akrátos*, immisible + E. *-id*, having the quality of, that which): the cellular slime fungi, in which the myxamoebae (cellular vegetative phase) retain their individuality when they unite to form a **pseudoplasmodium**. They are classified in the phylum Acrasiomycota of the kingdom Protista.

acrasin (Gr. *akrásia*, bad mixture, or *akrátos*, immisible + NL. *-in*, suf. used in chemistry to denote an activator or compound): a chemotactic compound produced by *Dictyostelium* and other gen. of Dictyosteliomycota, and which controls the aggregation of the myxamoebae to form the pseudoplasmodia.

Acrasiomycetes (Gr. *akrásia*, bad mixture, or *akrátos*, immisible + L. *-mycetes*, ending of class < Gr. *mýkes*, genit. *mýketos*, fungus): a group of slime fungi with myxamoebae that aggregate to form pseudoplasmodia which are transformed into fruiting bodies called sorocarps. According to the classification system followed in this dictionary, this group is not recognized as a class of the kingdom Fungi but as the phylum Acrasiomycota of the kingdom Protista.

acrocont, **acrokont**, **acrocontous**, **acrokontous** (Gr. *ákros*, apex + *kontós*, oar, flagellum; or + L. *-osus* > OF. *-ous*, *-eus* > E. *-ous*, having, possessing the qualities of): a motile cell which has the flagellum (swimming organ) in the anterior or front part, as occurs in the zoospores of the aquatic fungi of the phylum Hyphochytriomycota, of the kingdom Stramenopila, e.g., *Rhizidiomyces* (Hyphochytriales). Cf. **opisthocont** and **pleurocont**.

acrogenous (Gr. *ákros*, apex + *génos*, origin, birth + L. *-osus* > OF. *-ous*, *-eus* > E. *-ous*, possessing the qualities of): formed at the apex, like the spores of certain fungi, e.g., as in *Monilia* and *Cladosporium* (moniliaceous and dematiaceous asexual fungi, respectively).

acrochroic (Gr. *ákros*, apex + *chrõs*, color + suf. *-íkos* > L. *-icus* > E. *-ic*, belonging to, relating to): having pigment concentrated principally in the apex, as is observed in hyphae of the Clavariaceae.

acronema, pl. **acronemata** (Gr. *ákros*, apex + *nêma*, genit. *nêmatos*, filament): the tip of the flagellum consisting only of the extension of the two central microtubules but without the nine peripheral triplets of microtubules.

acropetal (Gr. *ákros*, apex + L. *petere*, to direct toward, grow toward + L. suf. *-alis* > E. *-al*, relating to or belonging to): refers to a structure that develops from the base toward the apex; e.g., in a chain of conidia with acropetal development the conidium at the tip is the youngest and of most recent formation. *Monilia* and *Cladosporium* (moniliaceous and dematiaceous asexual fungi, respectively) are examples of gen. of fungi that produce conidia in this manner. Cf. **basipetal**.

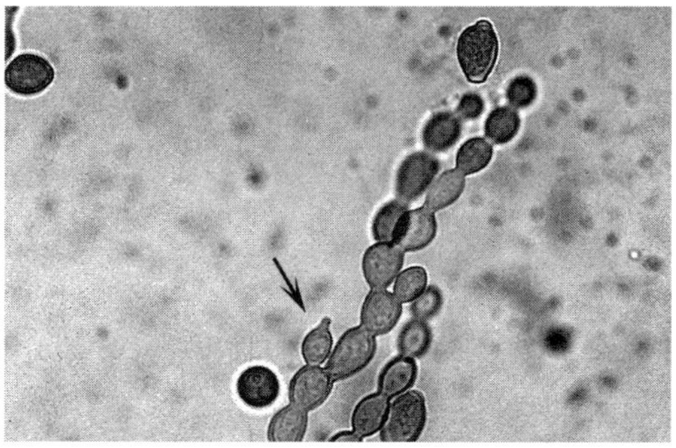

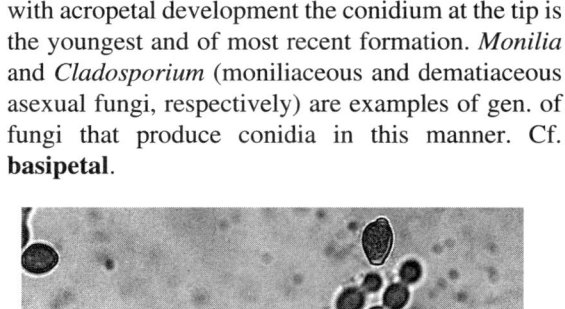

Acropetal succession of conidia of *Monilia fructicola*, x 320 (*MU*).

acropleurogenous (Gr. *ákros*, apex + *pleurá*, side, flank + *génos*, origin, birth + L. *-osus* > OF. *-ous*, *-eus* > E. *-ous*, having, possessing the qualities of): that which originates at the apex and later moves toward the side due to indeterminate growth of the apex; e.g., the conidia of *Trichothecium* (moniliaceous asexual fungi) are acropleurogenous. Also called **pleuroacrogenous**.

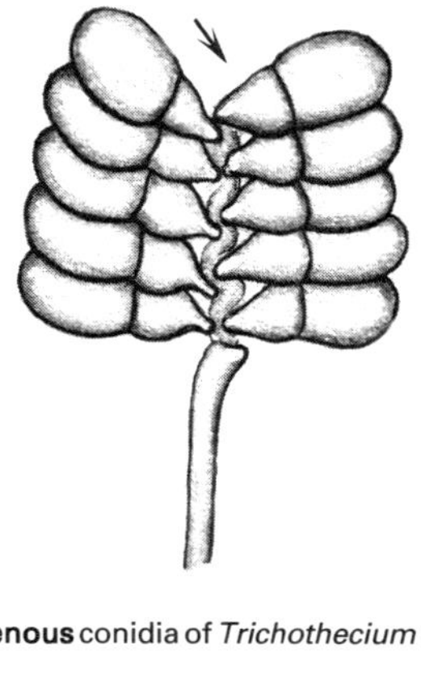

Acropleurogenous conidia of *Trichothecium roseum*, x 1 260.

acrosporous (Gr. *ákros*, apex + *sporá*, spore + L. *-osus* > OF. *-ous*, *-eus* > E. *-ous*, having, possessing the qualities of): a type of maturation in which a spore (conidium, spermatium, etc.) forms on the end of a mother cell or sporophore; i.e., the cells continue delimitation and maturation in sequence from base to apex, with the youngest spore at the apex.

acroton (Gr. *akrotónos*, strained to the utmost, fully extended < *ákros*, apex + *tónos*, tension): *Lichens*. A needle-like structure with lateral branches.

actinogyrose (Gr. *aktîs*, genit. *aktînos*, radius, ray + *gýros*, circular, round + L. *-osus*, full of, augmented, prone to > ME. *-ose*): a lichen apothecium of the gyrodisk type (with concentric circles on the upper surface) but with radial furrows or folds and without a proper margin, like that of *Umbilicaria cylindrica* (Lecanorales).

actinolichen (Gr. *aktîs*, genit. *aktînos*, radius, ray + *leichén*, lichen): a lichenized association between an actinomycete (*Streptomyces*) and an alga (*Chlorella xanthella*).

actinomycete (Gr. *aktîs*, genit. *aktînos*, radius, ray + L. *-mycetes*, ending of class < Gr. *mýkes*, genit. *mýketos*, fungus): a filamentous gram positive bacterium. The Actinomycetes have sometimes been erroneously classified as filamentous conidial fungi. They comprise saprobes and animal and human pathogens. These prokaryotic microorganisms are the producers of some important antibiotics, such as amphotericin, cycloheximide, nystatin and streptomycin.

aculeate (L. *aculeatus*, provided with spines or prickles < *aculeus*, spur, spine, sting < *acus*, needle + suf. *-atus* > E. *-ate*, provided with or likeness): **1.** Having spurs or spines; e.g., the columella of the sporangium of *Mucor spinosus* (Mucorales) and the basidiospores of *Laccaria laccata* (Agaricales). **2.** Shaped like a needle or spine, like the synnemata of *Doratomyces stemonitis* (stilbellaceous asexual fungi), and the hymenophore of the hydnaceous basidiomycetes (*Hydnum*, *Sarcodon*, *Hymenochaete* and other gen. of Aphyllophorales). Sometimes the term **trichiform** is applied to the aculeate caulocystidia of *Laccaria laccata* and other Agaricales.

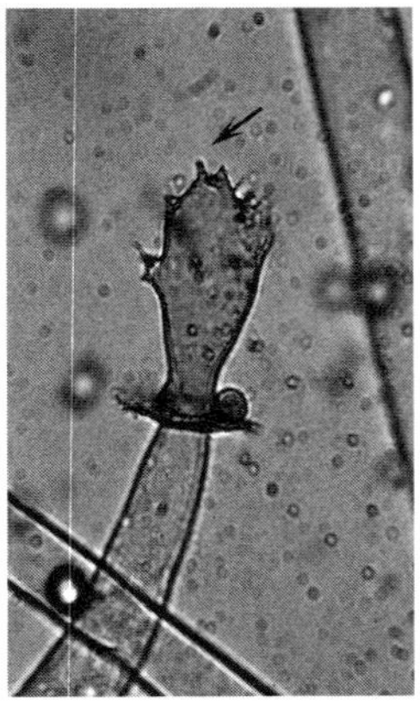

Aculeate columella of the sporangium of *Mucor spinosus*, x 800 (*MU*).

Aculeate synnemata of *Doratomyces stemonitis*, x 27 (*MU*).

aculeus, pl. **aculei** (L. *aculeus*, spur, spine, sting < *acus*, needle + E. suf. *-ule* < L. dim. suf. *-ulus*): each of the sharp spines or teeth of the hymenophore of *Hydnum* and *Sarcodon*, among other Hydnaceae (Aphyllophorales).

acuminate (L. *acuminatus*, taper-pointed, ptp, of *acuminare*, to sharpen + suf. *-atus* > E. *-ate*, provided with or likeness): tapering gradually to a point, pointed. E.g., the conidia of *Alatospora acuminata* (moniliaceous asexual fungi), the apex of the perithecial stroma of *Cordyceps stylophora* (Hypocreales), the cystidia of *Psathyrella squamosa*, *Mycena alcalina* and *Macrocystidia cucumis* (Agaricales), and the pileus apex in *Weraroa* (Hymenogastrales) are all acuminate.

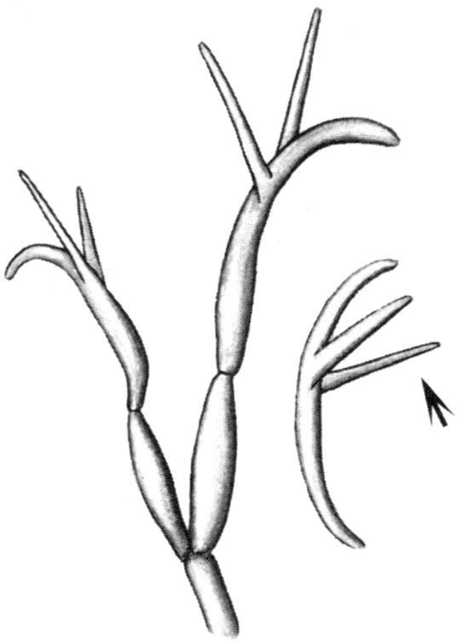

Acuminate conidia of *Alatospora acuminata*, x 1 250.

Acuminate apex of the perithecial stroma of *Cordyceps stylophora*, x 4 (*MU*).

acute (L. *acutus*, sharpened < *acuere*, to sharpen < *acus*, needle): sharp at the apex, ending in a point, pointed; narrowed gradually with an angle of less than 90°. E.g., the distal ends of the aciculate, aculeate, or accuminate cystidia of various Agaricales are acute.

acyanophilic (Gr. *a-*, without + *kýanos*, dark blue + *phílos*, have an affinity for + suf. *-íkos* > L. *-icus* > E. *-ic*, belonging to, relating to): not staining blue; refers to the negative reaction of cell or spore walls when placed in 1% cotton blue in lactic acid. Cf. **cyanophilic**.

adaxial (NL. *adaxialis* < L. *ad-*, toward + *axialis*, axial < *axis*, axis, axle + L. suf. *-alis* > E. *-al*, relating to or belonging to): the structure or side that is closest to an axis; e.g., the adaxial side of a basidiospore is the dorsal or concave side that is nearest the longitudinal axis of the basidium. Cf. **abaxial**.

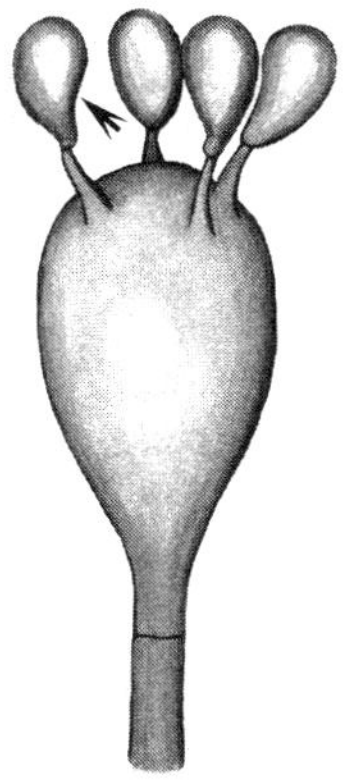

Adaxial side of the basidiospore of *Amanita* sp., x 1 170.

adelomycete (Gr. *adélos*, hidden, not known + *mýkes*, genit. *mýketos*, fungus): a fungus whose sexual state is not known.

adelphogamy (Gr. *adelphós*, brother, close relative + *gámos*, sexual union + *-y*, E. suf. of concrete nouns): a particular type of pseudomixis, which consists of the copulation of a mother cell with an independent daughter cell, as happens, e.g., in yeasts of the gen. *Zygosaccharomyces* (Saccharomycetales). Cf. **pedogamy**.

adelphotaxis, pl. **adelphotaxes** (Gr. *adelphós*, brother + *táxis*, movement): a tactic phenomenon of mutual attraction exhibited by the zoospores of *Achlya* (Saprolegniales), which may result in the auto-aggregation of zoospores. This chemotactic process is triggered by certain exogenous attractants and is amplified by signals from aggregated zoospores.

adenose (Gr. *adén*, genit. *adénos*, a gland + L. *-osus*, full of, augmented, prone to > ME. *-ose*): having glands; gland-like.

adherent (ME. < ML. *adherent* < L. *adhaerent* < *adhaerens* < *ad-*, near, toward + *haerere*, to stick + *-entem* > E. *-ent*, being): syn. of **adnate**.

adhesorium, pl. **adhesoria** (L. *adhesus*, adhesion + suf. *-orium*, the place of a thing): a tubular, bulbous extension produced by the germinating cyst of *Plasmodiophora brassicae* (Plasmodiophorales), from which the **Stachel** is discharged through the cell wall and into the cytoplasm of a root hair of the host plant (cabbage); in a few seconds the cytoplasm of the parasitic fungus enters through the hole made by the Stachel, where it is caught up by the cytoplasmic currents of the host and subsequently develops into a plasmodium that continues the infection process in the plant. In *Polymyxa betae*, also in the Plasmodiophorales, a similar infection mechanism is found, with an adhesorium derived from the evagination of the sheath (**Rohr**), which contains the Stachel; the Stachel is discharged from the Rohr, perforates the host cell wall, and in a period of one second an infective, amoeboid unit is injected. The total time that elapses from formation of the adhesorium to penetration of the host is ca. one minute.

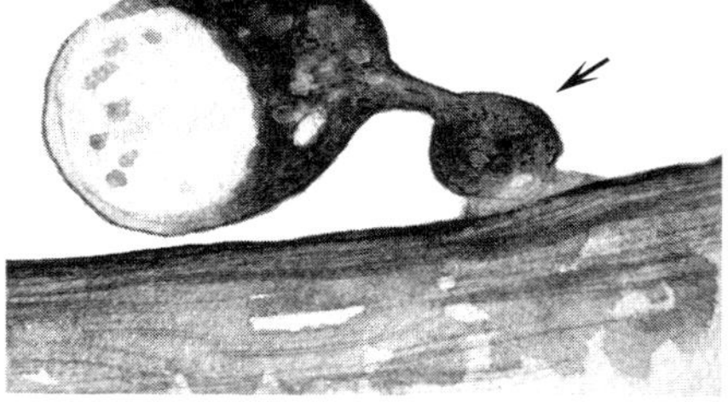

Adhesorium of *Plasmodiophora brassicae*. Drawing based on a transmission electron micrograph published by Beckett *et al.* in their *Atlas of Fungal Ultrastructure*, 1974, x 13 470.

adiaspore (Gr. *a-*, without + *diá*, through, separating, dividing + *sporá*, spore): a spore that increases notably in size but which does not divide. *Med. Mycol.* The large spherical chlamydospore (adiaspore) that is derived from a conidium of *Chrysosporium* (=*Emmonsia*) *pruinosa* (moniliaceous asexual fungi), which increases greatly in size on being inhaled and introduced into the bronchial tubes and lungs of rodents and other small wild, warm-blooded animals, where it causes the disease known as **adiasporomycosis**. Adiaspores also can be formed from conidia incubated *in vitro* at temperatures near 37° C.

adiasporomycosis, pl. **adiasporomycoses** (Gr. *adiaspore* < *a-*, without + *diá*, through, separating, dividing + *sporá*, spore + *mýkes* > L. *myces*, fungus + suf. *-osis*, disease-causing): a mycotic infection of warm-blooded animals caused by adiaspores of *Chrysosporium* (=*Emmonsia*) *pruinosa* (moniliaceous asexual fungi) that enter the lungs and bronchial tubes of these animals.

adnate (L. *adnatus*, broadly attached < *ad-*, near, toward + *natus*, born): *Agaricology*. The condition in which the small gills or lamellae of agarics reach the foot or stipe and adhere to it throughout their width, as in the gen. *Laccaria* (Agaricales). Also called **adherent**, **attached** or **concrescent**, which also is used to designate a film or certain scales that are strongly adhered to the surface of the pileus. Cf. **free**.

Adnate lamellae of *Laccaria proxima*, x 3 (*MU*).

adpressed (L. *adpressus*, pressed closely against something < *apprimere*, to press): see **appressed**.

adspersed (L. *ad-*, towards + *sparsus*, spread out < *spargere*, to scatter): scattered; of wide distribution.

adunc, **aduncate**, **aduncous** (L. *aduncus*, hooked < *ad-*, toward + *uncus*, hook + suf. *-atus* > E. *-ate*, provided with or likeness; or + L. *-osus* > OF. *-ous*, *-eus* > E. *-ous*, having, possessing the qualities of): curved inward like a hook. See **unciform**.

Adpressed fibrous scales of the pileus of *Agaricus placomyces*, x 1 (*MRO*).

Adpressed scales of the pileus of *Macrolepiota procera*, x 5 (*MU*).

adventitious (L. *adventicius*, added from another source and not inherent or innate, supplemental; arising or occurring sporadically or in other than the usual location < ML. *adventus*, coming from outside, with the E. suf. *-ous*, having, possessing the qualities of < OF. *-ous*, *-eus* < L. *-osus*): appearing in an abnormal or unusual position; accidental, unexpected, supplementary; e.g., an adventitious septum is one that forms independently or in the absence of nuclear division, particularly in relation to the movement of cytoplasm from one part to the other of the hypha of the fungus. In contrast is the **primary septum**, which forms in direct association with nuclear division. See also **adventitious ramification**.

adventitious ramification (L. *adventicius*, added from another source and not inherent or innate, supplemental; arising or occurring sporadically or in other than the usual location < ML. *adventus*, coming from outside, with the E. suf. *-ous*, having, possessing the qualities of < OF. *-ous*, *-eus* < L. *-osus*); L. *ramificatio*, production of branches from an axis or lower branch < *ramus*, branch, pl. *rami* + *-ationem*, action, state or condition, or result > E. suf. *-ation*): branching that occurs outside the normal pattern in the branches of the lichen *Cladonia* (Lecanorales), as a regeneration following damage.

adventitious septum, pl. **septa** (L. *adventicius*, added from another source and not inherent or innate, supplemental; arising or occurring sporadically or in other than the usual location < ML. *adventus*, coming from outside, with the E. suf. *-ous*, having, possessing the qualities of < OF. *-ous*, *-eus* < L. *-osus*); L. *septum*, barrier, partition): see **septum**.

aecidiole, **aecidiolum**, pl. **aecidiola** (NL. *aecidiolum*, dim. of *aecidium* < Gr. *aikídion* < *aikía*, lesion, injury, wound, alluding to the symptoms on infected host plants + dim. suf. *-ídion* > L. *-idium*; the L. dim. suf. *-olum* > E. *-ole*): *Ascolichens*. Syn. of **spermogonium**.

aecidium, pl. **aecidia** (L. *aecidium* < Gr. *aikídion* < *aikía*, lesion, injury, wound, alluding to the symptoms on the leaves and stems of plants infected by these fungi + dim. suf. *-ídion* > L. *-idium*): an alternative name given to the **aecium**, which is the preferred term.

aeciospore (L. *aecium* < Gr. *aikía*, wound, damage + *sporá*, spore): a spore produced in an **aecium**.

aecium, pl. **aecia** (L. *aecium* < Gr. *aikía*, wound, lesion + dim. suf. *-íon* > L. *-ium*): also called **aecidium**; *Uredinales* (the plant pathogenic fungi commonly known as rusts). A goblet- or cup-shaped sorus (stage I of the demicyclic or macrocyclic rusts) in which originate the binucleate aeciospores (aecidiospores) capable of infecting a plant of a different species; in *Puccinia graminis tritici*, the aecia form on the lower surface of the leaves of barberry, and in *Aecidium* sp. they are produced on the leaves of *Diospyros digyna*.

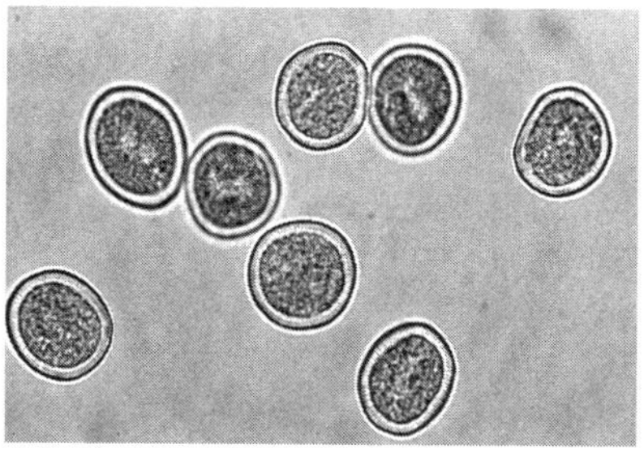

Aeciospores of *Gymnosporangium clavipes*, x 740 (*CB*).

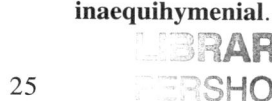

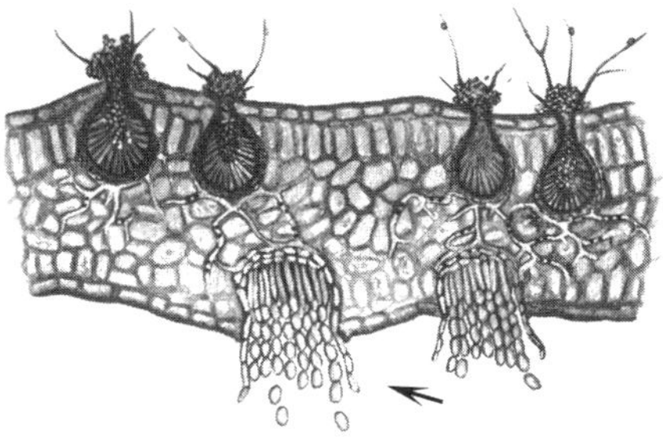

Aecium and aeciospores of *Puccinia graminis tritici*. Drawing made from a longitudinal section of a barberry leaf, x 80.

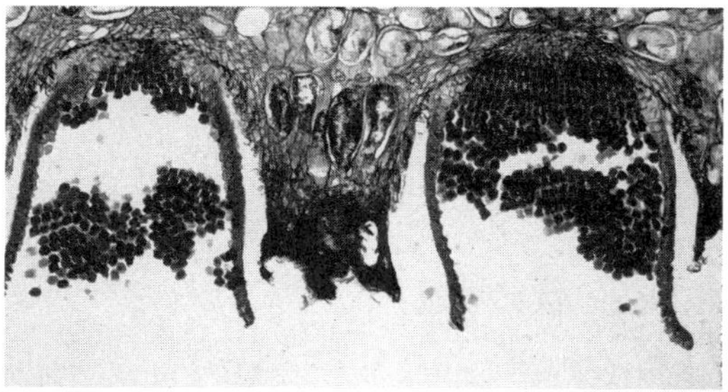

Aecium and aeciospores of *Puccinia graminis tritici*, seen in longitudinal section of a barberry leaf, x 100 (*MU*).

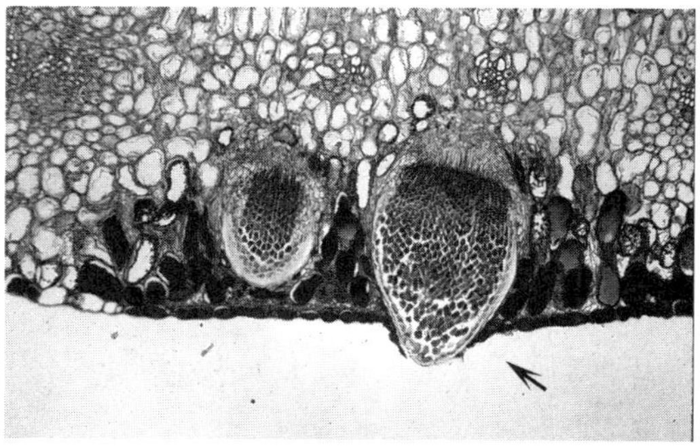

Aecium and aeciospores of *Puccinia graminis tritici*, seen in longitudinal section of a barberry leaf, x 125 (*MU*).

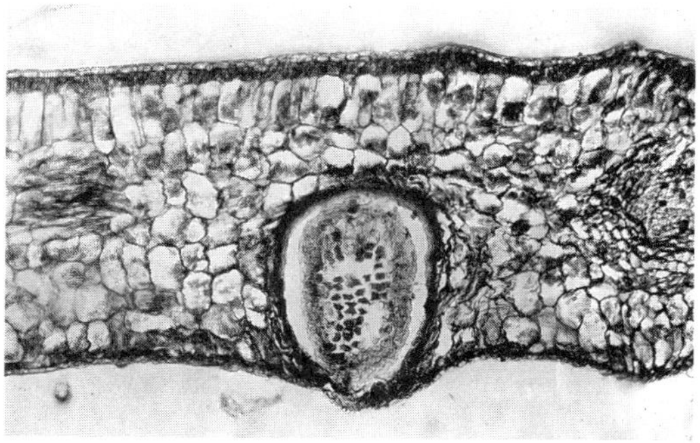

Aecium and aeciospores of *Aecidium* sp., seen in longitudinal section of a host leaf (*Diospyros digyna*), x 160 (*MU*).

aequihymenial (L. *aequus*, equal + *hymenium*, membrane + L. suf. *-alis* > E. *-al*, relating to or belonging to): see **aequihymeniiferous**.

aequihymeniiferous (L. *aequus*, equal + *hymenium*, membrane + *-ferous*, bearer < *ferre*, to bear, carry + L. *-osus* > OF. *-ous*, *-eus* > E. *-ous*, having, possessing the qualities of): bearer of a hymenium that develops in a uniform manner, in which the basidia mature and liberate their basidiospores almost simultaneously over all the surface of the gills. This type of hymenial development also is called **aequihymenial**, and it is characteristic of all of the gen. of Agaricales except *Coprinus*. Cf. **inaequihymeniiferous** and **inaequihymenial**.

aeroallergen (L. *aer*, air + *allergen* < Gr. *állos*, another + *érgon*, work + *génos*, origin < *gennáo* to engender, produce): a particle, viable or not, present in the atmosphere, that produces an allergy in humans and higher animals, generally respiratory, such as asthma and rhinitis. The spores of many fungi, e.g., the conidia of *Penicillium* and *Aspergillus* (moniliaceous asexual fungi) and ascospores, basidiospores and other types of spores and fungal propagules, along with pollen grains, are among the most common aeroallergens.

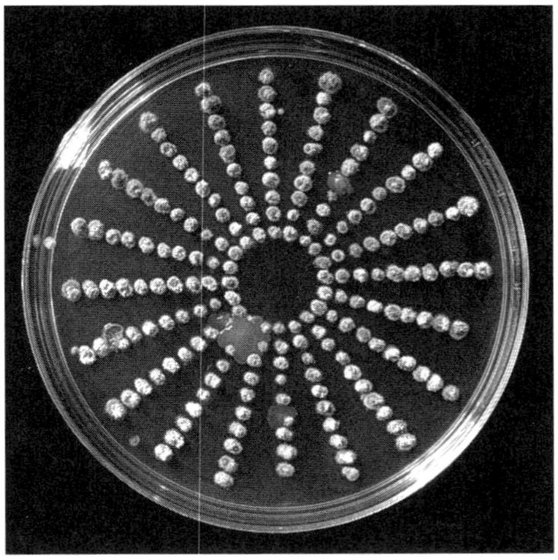

Colonies of *Penicillium aurantiogriseum* growing on an agar plate; their radial arrangement corresponds to the position of the holes in one of the metallic covers used in a six-stage Andersen air sampler. The mold colonies developed from the spores trapped from the air by the sampler and which, after passing through the holes, were impacted on the agar surface. These types of spores are among the most common **aeroallergens**, x 0.6 (*MU*).

aeroaquatic (L. *aer*, air + E. *aquatic*, growing or living in or frequenting water < L. *aqua*, water + Gr. *-tikós* > L. *-ticus* > E. *-tic*, relation, fitness, inclination or ability): aquatic hyphomycetes found in still and stagnant fresh, non-marine, and occasionally marine waters. Usually, aeroaquatic fungi survive in water and mud of low oxygen content, but they apparently need higher oxygen levels for colonization of new substrates. The conidia of these fungi often have three-dimensional shapes such as coils, spirals and many-celled cage-like structures, formed at the air-water interface, the attribute for which they are named. Some examples of this type of fungi are *Helicoon* and *Varicosporina* (moniliaceous asexual fungi).

aerobe (Gr. *aér*, *aéros*, air + *bíos*, life): an organism that requires free oxygen to live. Most fungi are aerobes and obtain their energy from respiration maintained by the oxygen in the air. Cf. **anaerobe**.

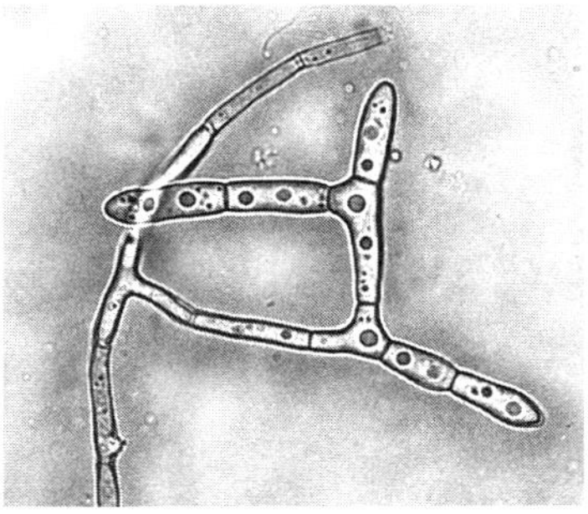

Aeroaquatic tridimensional conidium of *Varicosporina ramulosa*, x 980 (*MCG*).

aerogenic (Gr. *aér*, *aéros*, air + *génos*, origin < *gennáo*, to engender, produce + suf. *-íkos* > L. *-icus* > E. *-ic*, belonging to, relating to): an organism that produces detectable gas (CO_2) during the breakdown of carbohydrates, as yeasts and molds can do.

aerohygrophilous (Gr. *aér*, *aéros*, air + *hygrós*, humid, wet + *phílos*, have an affinity for + L. *-osus* > OF. *-ous*, *-eus* > E. *-ous*, possessing the qualities of): refers to species of lichens whose thallus has a hydrophobic surface and which inhabit ecological niches where water is only available as vapor or in liquid form in the substrate, such as *Ramalina maciformis* (Lecanorales), a species that lives in hot deserts and which can absorb sufficient water from the air to remain physiologically active.

aeromycology (L. *aer*, air + Gr. *mýkes*, fungus + *lógos*, study, treatise + *-y*, E. suf. of concrete nouns): the branch of aerobiology that treats of the study of fungi in the air, in various aspects related to the biology of the dispersal of fungal propagules by the air and its implications in agriculture, medicine and industry, basic as well as applied.

A six-stage Andersen air sampler, one of the spore trappers most commonly used in **aeromycology**. Each stage includes a plate with 200 holes of uniform diameter (which decreases at each stage to simulate the human respiratory tract) through which air is drawn at 28 liters/min to impact on petri dishes containing agar media immediately below, x 0.1 (*CC*).

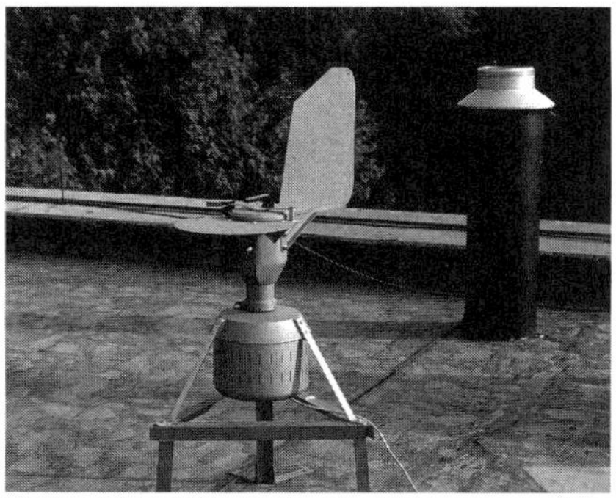

A Burkard sampler, a device used to trap spores and other fungal propagules, as well as pollen grains and other viable and nonviable particles, from the air. Sampling periods may be 24 hours or several days, and the particles are trapped on a moving tape covered with glycerine/hexane. This type of sampler is used to monitor the airspora in agricultural fields and other environments studied in **aeromycology**, and in general in aerobiology, x 0.04 (*CC*).

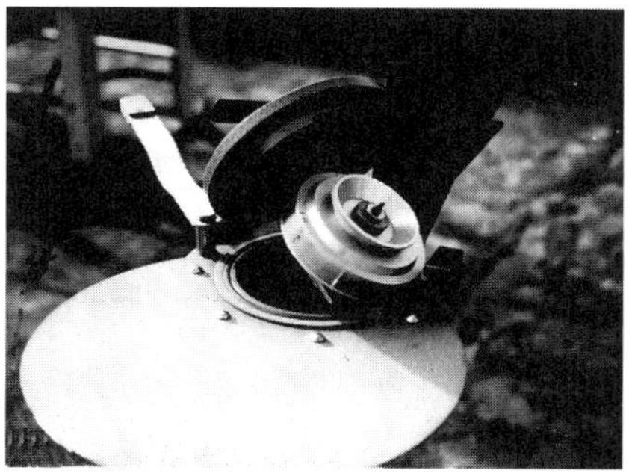

A close-up of the same Burkard sampler; the tape used to trap spores and other particles is attached to a rotating drum, x 0.1 (*CC*).

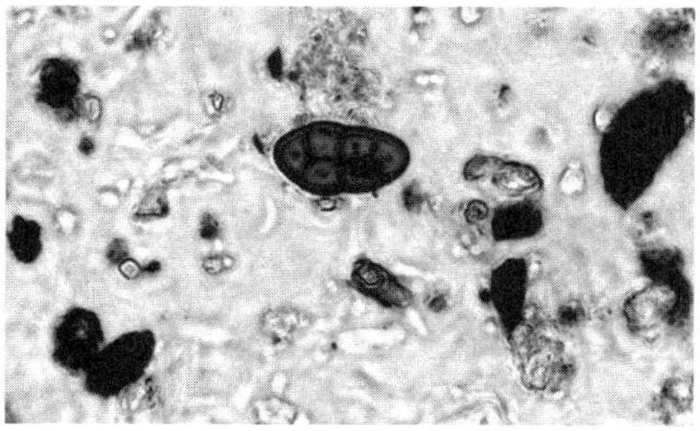

A fungal spore, along with nonviable particles, trapped by the Burkard sampler, x 430 (*CC*).

aerotaxis, pl. **aerotaxes** (Gr. *aér*, *aéros*, air + *táxis*, disposition): a chemotactic phenomenon in which the air, oxygen or other gas acts as an excitant. This is shown in the zoospores of *Rhizophydium pollinis-pini* (Chytridiales), which are extremely sensitive to the presence of free oxygen; when mounted in water beneath a cover glass they immediately move toward the edge of the drop or toward air bubbles in the preparation.

aerotropism (Gr. *aér*, *aéros*, air + *trópos*, a turn, change in manner < *tropé*, a turning < *trépo*, to revolve, turn towards + *-ismós* > L. *-ismus* > E. *-ism*, state, phase, tendency, action): movement stimulated by air, oxygen or another gas. The sporangiophores of *Phycomyces nitens* (Mucorales) turn toward small bits of iron placed near them, apparently in response to small quantities of ozone liberated by the oxidation of the iron. Aluminum foil or thin slivers of glass also induce curvature of the sporangiophores, due to the presence of ozone adsorbed on these materials. See **tropism**.

aethalium, pl. **aethalia** (Gr. *aíthalos*, soot + dim. suf. *-íon* > L. *-ium*): a semiglobose, pulvinate or irregular fructification, relatively large, that results from the transformation of all or most of a plasmodial mass, in which various sporangia remain partially or totally enveloped by a common peridium; typical of some Myxomycetes, such as *Lycogala* (Liceales) and *Fuligo* (Physarales).

Aethalia of *Lycogala epidendrum*, x 6 (*MU*).

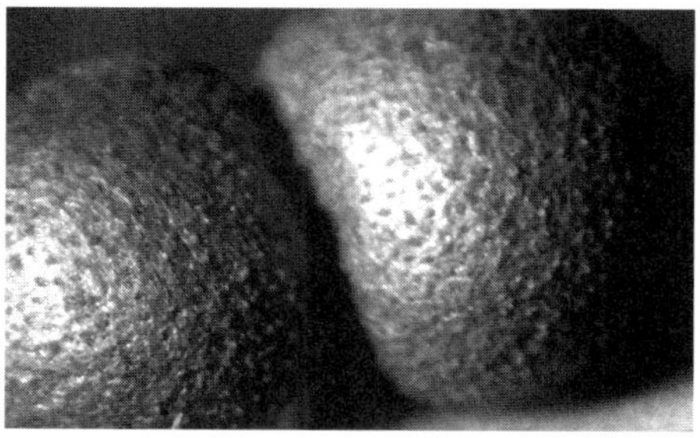

Aethalia of *Lycogala epidendrum* on dead wood, x 9 (*MU*).

Aethalium of *Fuligo cinerea*, x 0.3.

aetiology (ML. *aetiologia*, statement of causes < Gr. *aitíologia* < *aitía*, cause + *lógos*, word, discourse, treatise + *-y*, E. suf. of concrete nouns): see **etiology**.

agamic, **agamous** (Gr. *ágamous*, unwed < *a*-, without + *gámos*, sexual union + suf. *-íkos* > L. *-icus* > E. *-ic*, belonging to, relating to; or + L. *-osus* > OF. *-ous*, *-eus* > E. *-ous*, having, possessing the qualities of): asexual, without sexual union, not reproducing sexually.

agamogenesis, pl. **agamogeneses** (Gr. *ágamous*, unwed < *a*-, without + *gámos*, sexual union + OF. suf. *-ous*, *-eus* < L. *-osus*, possessing the qualities of + Gr. *génesis*, engendering, origin): agamic or asexual generation, i.e., vegetative, which occurs without nuclear fusion.

agar (Ma. *agaragar*, seaweed from which a gelatin is obtained): mucilaginous substance (galactan) obtained from *Gracilaria lichenoides*, various species of the gen. *Gelidium*, and other marine algae, principally from the Asiatic seas. It is frequently employed as a gel for various culture media for fungi, bacteria, and other organisms.

agaric (L. *agaricum*, a fungus < Gr. *agarikón*, mushroom, excrescence, with the E. ending *-ic*, which denotes relation): one of the Agaricales, a very large group of basidiomycetes (ca. 6000 spp.), terrestrial or lignicolous, saprobic, mycorrhizal, rarely pathogenic, with edible, hallucinogenic, and poisonous species.

agaricoid (< gen. *Agaricus* < Gr. *agarikón*, mushroom, excrescence + L. suf. *-oide* < Gr. *-oeídes*, similar to): *Agaricology*. Like the fruiting body of the gen. *Agaricus* (Agaricales). A fruiting body (mushroom) with the hymenophore arranged in lamellae, but with strong evidence of an affinity to a particular family of the Agaricales. See **lepiotoid**, which is the type of fruiting body that includes *Agaricus*, in modern agaric taxonomy. Cf. **gasteroid**.

agarolytic (Ma. *agaragar*, seaweed from which a gelatin is obtained, agar + Gr. *lytikós*, able to loosen < *lýtos*, dissolvable, broken + suf. *-íkos* > L. *-icus* > E. *-ic*, belonging to, relating to): capable of causing lysis and liquefaction of agar, a solidifying agent of many culture media. The production of agarolytic enzymes, although a relatively rare phenomenon among microorganisms, has been reported from bacteria (*Pseudomonas*, *Alginomonas*, *Agarbacterium*, *Nocardia*, and others), mostly from marine habitats. Recently, an agarolytic activity of the plasmodium of *Fuligo septica* (Physarales) has been reported, apparently being the first one in eukaryotic microorganisms.

agglutinate (L. *agglutinatus* < *agglutinare*, to stick < *ag-* < *ad-*, to + *gluten*, glue + suf. *-atus* > E. *-ate*, provided with or likeness): to unite, clump together. Something that is united but not fused. E.g., the gloeospores of *Acremonium*, *Rhinocladiella* and *Fusarium* (moniliaceous, dematiaceous and tuberculariaceous asexual fungi, respectively) are agglutinated by a mucilage, but they can be separated without altering their integrity.

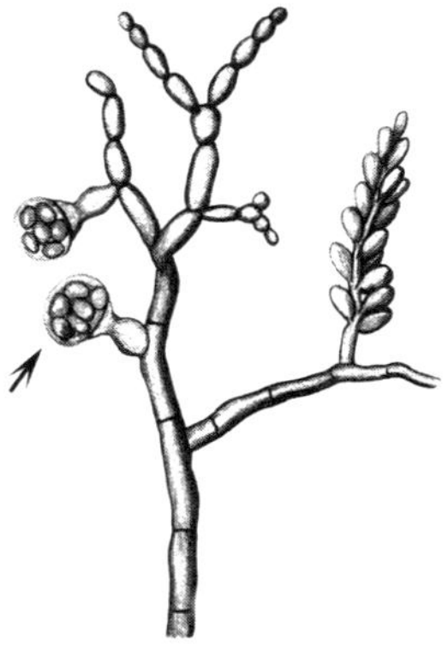

Agglutinate conidia of the *Phialophora* type in *Rhinocladiella* (=*Fonsecaea*) *pedrosoi*, x 500.

agnotobiotic (Gr. *a*-, without + *gnótos*, well known + *bioté*, life, manner of living + suf. *-íkos* > L. *-icus* > E. *-ic*, belonging to, relating to): a crude culture or mixture whose specific identity is unknown. Cf. **axenic**, **gnotobiotic** and **monoxenic**.

akaryallagic (Gr. *a*-, without + *káryon*, nucleus + *allagé*, permutation, commutation + E. suf. *-ic* < L. *-icus* < Gr. *-íkos*, belonging to, relating to): reproduction in which the nucleus remains unchanged, i.e., asexual or vegetative multiplication. Cf. **karyallagic**.

akaryote (Gr. *a*-, without + *káryon*, nut, nucleus): lacking a nucleus. *Plasmodiophoromycetes*. A stage (prophase) in nuclear division during which the nucleus has very little chromatin, for which reason it does not stain and consequently is not evident.

akinete (Gr. *akínetos*, motionless): a thick-walled, nonmotile spore of cyanobacteria in lichen thalli.

akinetospore (Gr. *akínetos*, motionless + *sporá*, spore): a motionless spore, devoid of flagella, that goes through a rest period before germinating.

alate (L. *alatus*, winged <*ala*, wing + suf. *-atus* > E. *-ate*, provided with or likeness): having wings, or wing-like projections. E.g., the stipe of *Boletellus russellii* (Agaricales), has broken and velvety extensions, called wings, in the edges or reticulations. Also, the basidiospores of this species are winged, as are those of *Lactarius pterosporus* (Agaricales) because their edges are so large that they give the spores a winged appearance. Also called **pterate**.

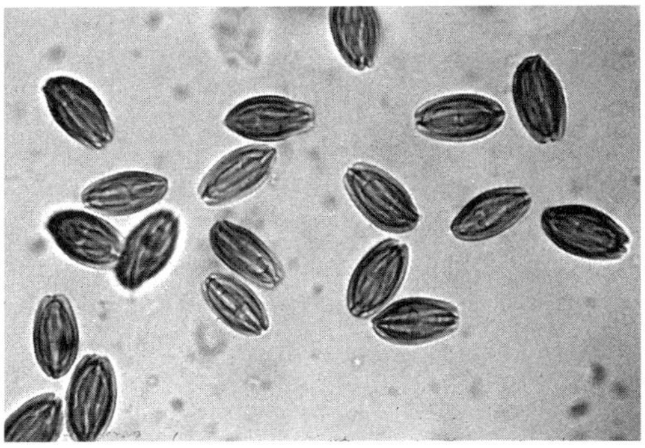

Alate basidiospores of *Boletus russellii*, x 450 (*MU*).

alepidote (Gr. *a-*, without + *lepidotós*, covered with scales): without scales, entirely smooth. Cf. **squamulose** and **lepidote**.

-ales (L. *-alis*, pl. *-ales*, fem. suf.): ending of taxonomic orders, e.g., Blastocladiales, Entomophthorales, etc.

aleuriospore (Gr. *áleuron*, wheat flour, meal + *sporá*, spore): an asexually reproduced spore that originates by swelling of a terminal or lateral cell of a hypha; characteristic of some conidial fungi, such as the gen. *Chalara*, *Humicola* and *Nigrospora* (dematiaceous asexual fungi).

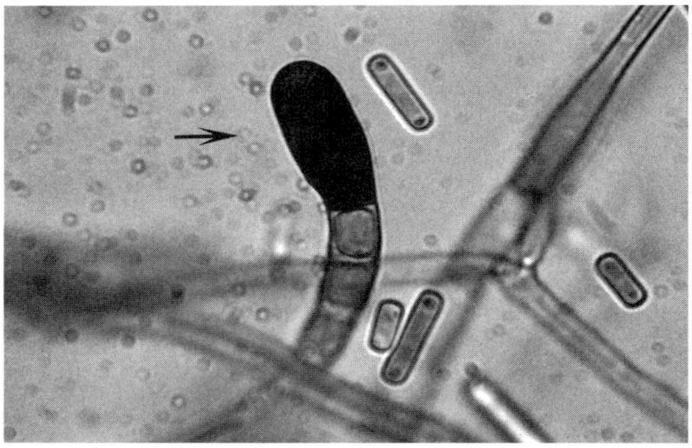

Aleuriospore and conidia of *Chalara* (=*Thielaviopsis*) *basicola*, x 600 (*MU*).

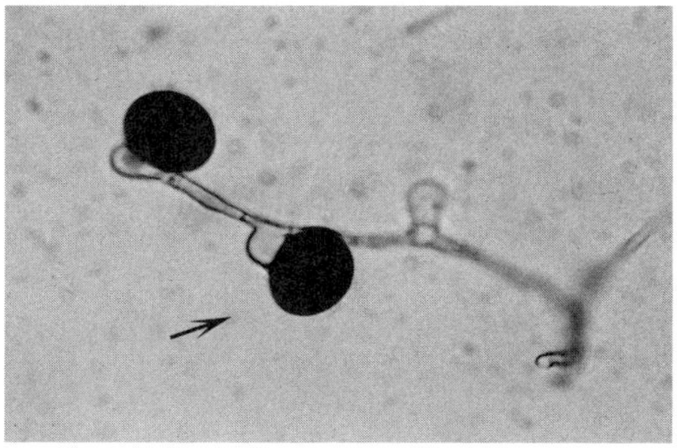

Aleuriospores of *Nigrospora* sp., x 800 (*MU*).

algicolous (L. *alga*, alga, in general, any pigmented, photosynthetic thallophyte + *-cola*, inhabitant + *-osus* > OF. *-ous*, *-eus* > E. *-ous*, possessing the qualities of): living or growing on algae, whether as a parasite or symbiont; e.g., *Phlyctidium anatropum* (Chytridiales) that parasitizes the thallus of the alga *Stigeoclonium* sp., *Spathulospora phycophila* (Spathulosporales), a laboulbeniomycete that lives as a parasite on the thallus of *Ballia callitricha* and *B. scoparia*, marine rhodophyceous algae, and *Mycosphaerella ascophylli* (Dothideales), a loculoascomycete that has established a symbiotic mode of living that qualifies it as mycophycobiosis, i.e., an obligate symbiotic association between a systemic marine fungus and a macroalga (*Ascophyllum nodosum* of the Phaeophyceae).

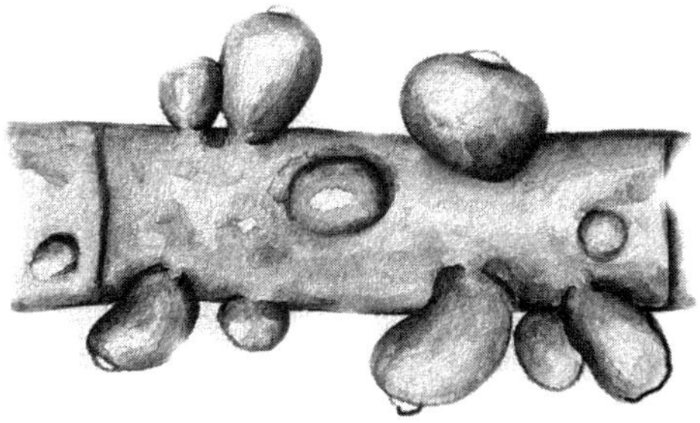

Algicolous thalli of *Phlyctidium anatropum* parasitizing a thallus of the alga *Stigeoclonium* sp., x 1 000.

aliform (L. *aliformis*, shaped like a wing < *ala*, wing + *-formis* < *forma*, shape): in the shape of a wing; e.g., like the stellate projections of the ascospore wall of *Emericella variecolor* (Eurotiales).

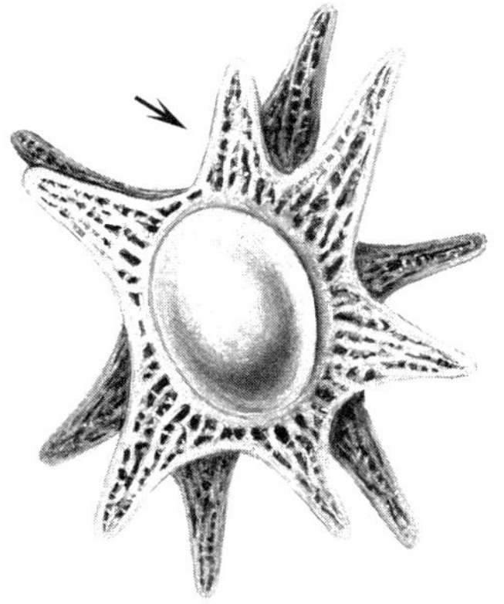

Aliform projections of the ascospore wall of *Emericella variecolor*, x 4 250.

allantoid (L. *allantoideus*, sausage-shaped < Gr. *allantoeidés*, similar to a sausage < *allás*, genit. *allantós*, sausage + *-oeídes*, similar to): with the shape of a sausage; e.g., like the Hülle cells of *Aspergillus puniceus* (moniliaceous asexual fungi) and the ascospores of the Diatrypaceae (Xylariales).

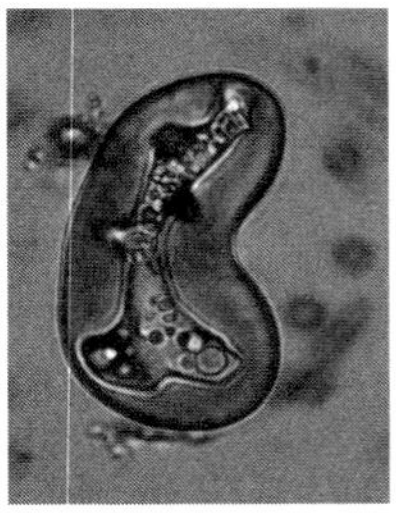

Allantoid Hülle cell of *Aspergillus puniceus*, x 700 (*MU*).

allantospore, **allantoid spore** (Gr. *allantoeidés*, similar to a sausage < *allás*, genit. *allantós*, sausage + *sporá*, spore + *-oeídes*, similar to): a spore that is sausage-shaped, slightly arcuate, with rounded ends; a Saccardoan term applied originally to conidial fungi to designate sausage-shaped spores, such as the conidia of *Dendrodochium* (moniliaceous asexual fungi), the asexual state of *Nectria magnusiana* (Hypocreales). The basidiospores of *Pleurotus* spp. and of *Panellus stypticus* (Agaricales) also are allantoid.

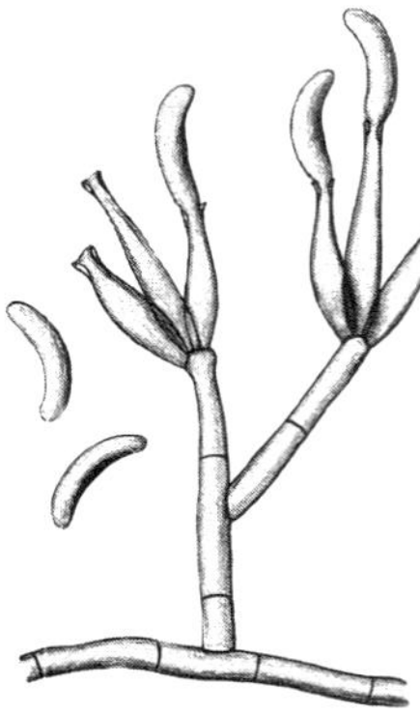

Allantospores of *Dendrodochium* sp., x 600.

allelo- (Gr. *allélo* < *allélon*, of/to one another, reciprocally): *Genetics*. Also called **allelomorph**; a factor or gene located on a site or locus of a chromosome, and which together with the other equal allele, located on the same locus of the other homologous chromosome, forms a pair of genes that control the same type of characters; in fungi with bipolar sexual compatability, the latter is controlled by a pair of allelic genes or allelomorphs; in those that have tetrapolar compatability, two pairs of alleles determine sexual union.

allelomorph (Gr. *allélo* < *allélon*, of/to one another + *morphé*, form): see **allelo-**.

allochronic (Gr. *állos*, other + *chronikós*, of or concerning time < *chrónos*, time + suf. *-íkos* > L. *-icus* > E. *-ic*, belonging to, relating to): occurring at different time periods, e.g., contemporary and fossil species of fungi.

allochrous, **allochroous** (Gr. *allóchroos*, that which takes another color < *állos*, other + *chrõs*, *chroíakos*, colored + L. *-osus* > OF. *-ous*, *-eus* > E. *-ous*, having, possessing the qualities of): changing from one color to another; e.g., the basidiocarps of *Psilocybe* (Agaricales), which on being rubbed or damaged, change from brown to bluish or greenish tones; in this particular case they are said to be **caerulescent**.

allochthonous (Gr. *állos*, other + *thón*, *thonós*, country + L. *-osus* > OF. *-ous*, *-eus* > E. *-ous*, possessing the qualities of): refers to an organism that is not native to the country in which it grows; also called **exotic**. E.g., *Amanita phalloides* (Agaricales) is considered allochthonous as it was introduced into the United States from Europe. Cf. **autochthonous**.

allocyst (Gr. *állos*, other + *kýstis*, bladder; here vessel, cell): *Agaricales*. A cystidium formed in the mycelium; an obovoid, claviform or pyriform cell with a thin or slightly thickened wall, that is formed individually or in a chain in the terminal part of a submerged, haploid or dikaryotic mycelium, of *Pholiota gummosa* (Agaricales). Germination of allocysts has not been observed, for which reason they are considered a form of food reserve; they appear to correspond to what have been called **chlamydospores**.

allopatric (Gr. *állos*, other + L. *patris*, a father, one's country + Gr. suf. *-íkos* > L. *-icus* > E. *-ic*, belonging to, relating to): occurring in different geographical regions. Cf. **sympatric**.

alpha-conidium, **α-conidium**, pl. **conidia** (*alpha*, α, first letter of the Gr. alphabet; NL. *conidium* < Gr. *konídion* < *kónis*, dust + dim. suf. *-ídion* > L. *-idium*): a term applied to species that produce two morphological types of spores, in particular, the fusoid to oblong, biguttulate spore of *Phomopsis* (sphaeropsidaceous asexual fungi) and other conidial states of the Diaporthales. These fungi also generally produce **beta-conidia**. Alpha conidia often form first, are usually more common, and are germinable.

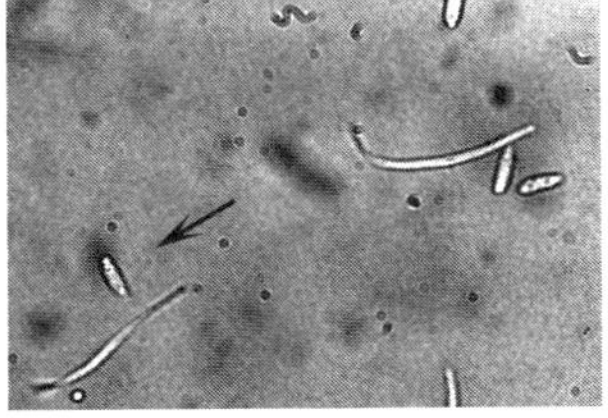

Alpha conidia from a pycnidium of *Phomopsis* sp., x 600 (*RTH*).

alternarioid (< gen. *Alternaria* < L. *alternare*, to alternate, do by turns + suf. *-ia*, which denotes quality + suf. *-oid* < Gr. *-oeídes*, similar to): a dictyosporous chlamydospore; i.e., with transverse and longitudinal septa, resembling the dictyospores of *Alternaria* (dematiaceous asexual fungi); present in some species of *Phoma* (sphaeropsidaceous asexual fungi), such as *Ph. glomerata* and *Ph. jolyana*.

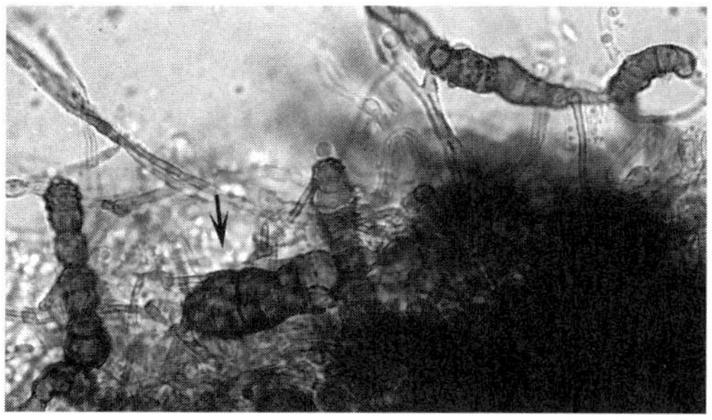

Alternarioid chlamydospores of *Phoma* sp., x 600 (*MU*).

alternate host (L. *alternatus*, alternate < *alternus* < *alter*, occurring or succeding by turns + suf. *-atus* > E. *-ate*, with the property of; *hospes*, a living organism affording subsistence or lodgment to a parasite): see **host**.

alutaceous (LL. *alutaceus*, *alutacius*, pertaining to soft leather < *aluta*, leather + *-aceus* > E. *-aceous*, of or pertaining to, with the nature of): **1.** *Color*. Having the color of soft, brown leather, light brown, pink-cinnamon color. **2.** *Texture*. Of tough consistency, but with a certain flexibility, like leather; also called **coriaceous**. Some basidiocarps, such as those of *Lentinus badius* (Aphyllophorales), and the thallus of *Cetraria richardsonii* (Lecanorales), are alutaceous.

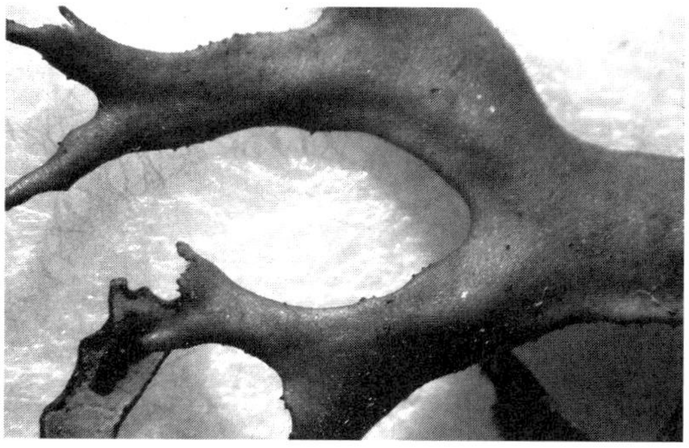

Alutaceous thallus of *Cetraria richardsonii*, x 2 (*MU*).

alveolate (L. *alveolatus*, hollowed, pitted < *alveolus*, pit, small hollow, dim. of *alveus*, hollow + suf. *-atus* > E. *-ate*, provided with or likeness): deeply pitted, like a honeycomb; e.g., the ascospores of *Gelasinospora* (Sordariales) have an alveolate wall, and the pileus of *Morchella esculenta* (Pezizales), the ascocarp of *Cyttaria hariotii* (Cyttariales), and the stipe of *Boletellus frostii* (Agaricales) are alveolate. Also called **faveolate**.

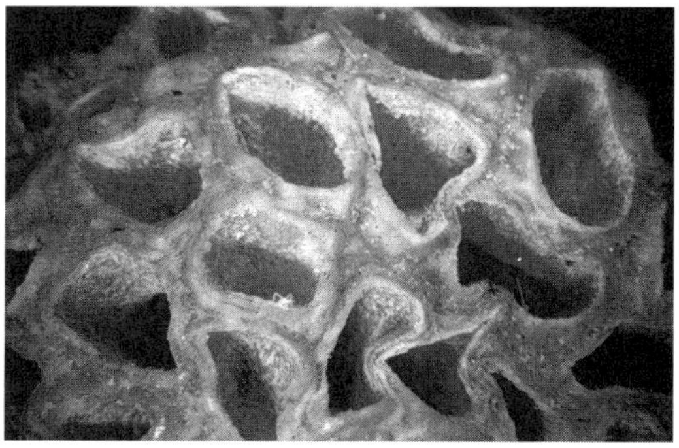

Alveolate ascocarp of *Cyttaria hariotii*, x 7.5 (*MU*).

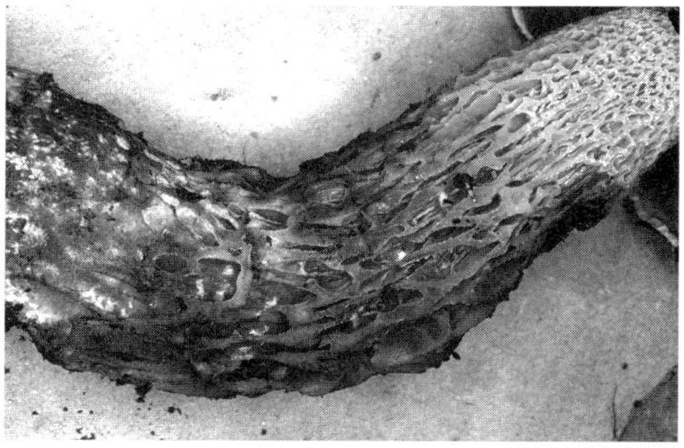

Alveolate stipe of the basidiocarp of *Boletellus frostii*, x 1 (*JC*).

amanitoid (< gen. *Amanita* < Gr. *amanós*, the name of an Asian mountain, or < Gr. *amánores*, pustules + L. suf. *-oide*, < Gr. *-oeídes*, similar to): *Agaricology*. A type of fruiting body (of 13 main types recognized) with the gills free or finely attached, and a ring and volva, represented by *Amanita* (Agaricales).

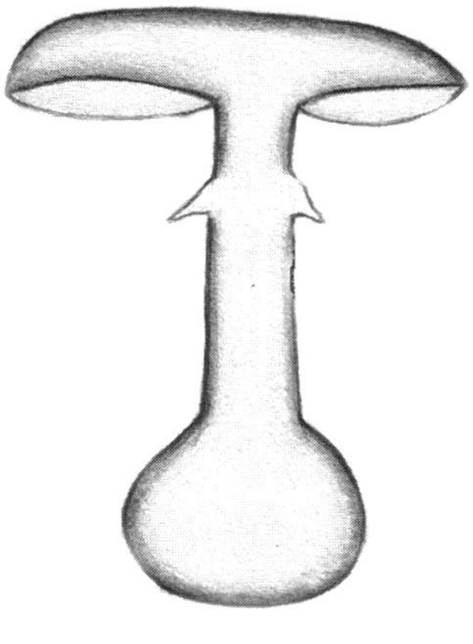

Amanitoid basidiocarp of *Amanita* sp. Diagram of a longitudinal section, x 0.5.

ambrosia (Gr. *ambrosía*, food of the Gods < *ámbrotos*, immortal): refers to certain fungi that some scolytic coleopterans (*Xyleborus*, *Corthylus*) cultivate in the galleries that they open in wood, and on which they feed.

amend (ME. *amenden* < OF. *amender* < L. *emendare* < *e*, *ex*, out + *menda*, fault): to change or alter, not necessarily to correct a fault or error. Cf. **emend.**

amensalism (Gr. *a-*, without + L. *mensa*, table + *-ismós* > L. *-ismus* > E. *-ism*, state, phase, tendency, action): the condition in which a species is favored by its ability to secrete metabolites that inhibit other fungi.

amerospore (Gr. *a-*, without + *méros*, part, portion + *sporá*, spore): a unicellular spore; a spore not divided into parts by septa; a Saccardoan term applied originally to the conidial fungi. E.g., the conidia of *Oedocephalum* (moniliaceous asexual fungi), among many others, are amerospores.

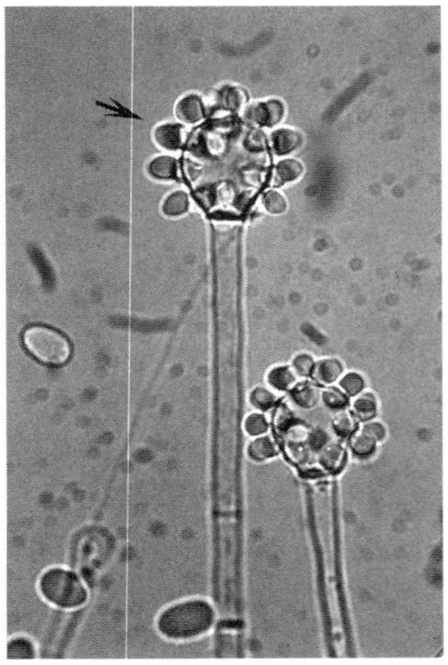

Amerospores of *Oedocephalum* sp., x 720 (*MU*).

amitosis, pl. **amitoses** (Gr. *a-*, without + *mítos*, filament + *-osis*, condition): simple, direct nuclear division by fragmentation; the nucleus constricts until it separates into two parts, often unequal. In this type of division the characteristic figures of mitosis, the formation of chromosomes, are not observed; it occurs in old cells and lower organisms.

amixis, pl. **amixes** (Gr. *a-*, without + *míxis*, mixture, intimate union): lack of mixis. A primitive cell state in which there is no sexual reproduction or amphimixis. Also called **apomixis**. Cf. **amphimixis**.

amoeboid (NL. *amoeba* < Gr. *amoibé*, change, alteration + L. suf. *-oide* < Gr. *-oeídes*, similar to): similar to an amoeba. The myxamoebae of the Dictyosteliales and the plasmodia and myxamoebae of the Myxomycetes are amoeboid, as are the zoospores of certain aquatic fungi that lack a cell wall and move by creeping, extending pseudopodia or prolongations into which the protoplasm flows, continually changing shape with the successive formation of the pseudopodia.

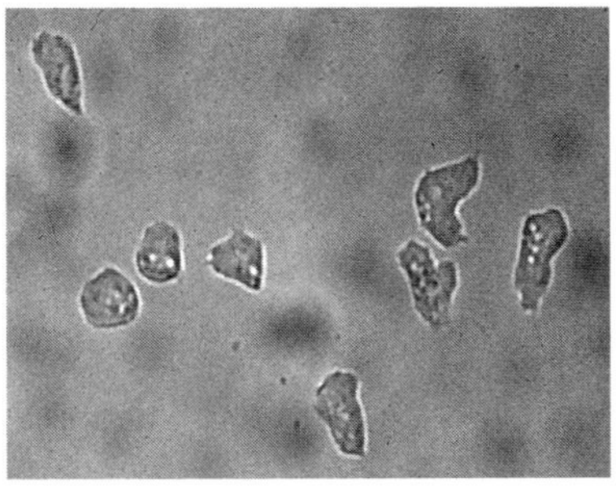

Amoeboid myxamoebae of *Dictyostelium discoideum*, x 840 (*RTH*).

amorphous (Gr. *amorphós* < *a*, without + *morphé*, shape, form + L. *-osus* > OF. *-ous*, *-eus* > E. *-ous*, having, possessing the qualities of): without a defined shape or structure.

amphigenous (Gr. *amphí*, from both sides + *génos*, origin + L. *-osus* > OF. *-ous*, *-eus* > E. *-ous*, having, possessing the qualities of): developing all around, growing on both sides or faces; e.g., a parasitic fungus that grows on both sides of a leaf (like *Puccinia graminis*, Uredinales), or a basidiocarp which forms a hymenium on all the surface of the branched aerial part, as occurs in *Ramaria* (Aphyllophorales).

amphigynous (Gr. *amphí*, both, on both sides, around both sides + *gyné*, woman; here, female sexual organ + L. *-osus* > OF. *-ous*, *-eus* > E. *-ous*, having, possessing the qualities of): *Peronosporales.* An antheridium which is penetrated by the oogonium in such a manner that the antheridium finally envelops the pedicel and base of the oogonium. Characteristic of some species of *Phytophthora*, such as *P. infestans* and *P. capsici* (Peronosporales). Cf. **paragynous.**

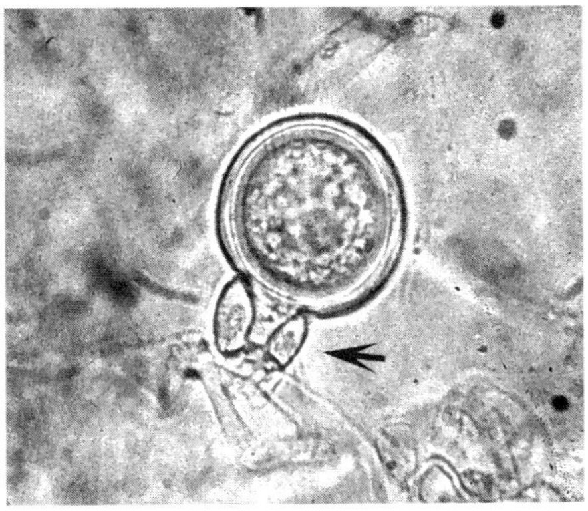

Amphigynous antheridium of *Phytophthora capsici*, x 200 (*JG*).

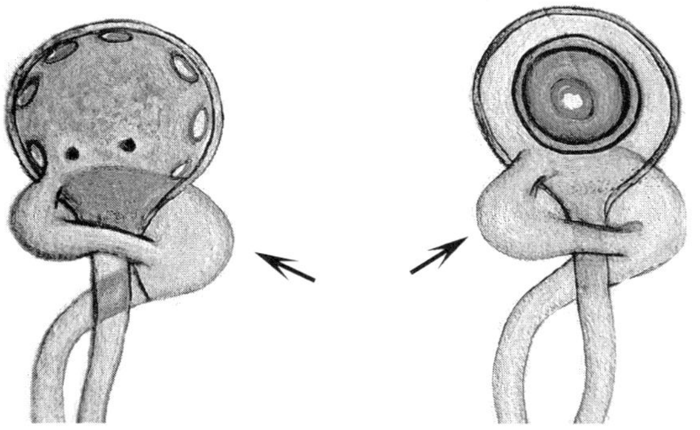

Amphigynous antheridium of *Phytophthora infestans*, x 300.

amphimixis, pl. **amphimixes** (Gr. *amphí*, both + *míxis*, mixture): the union of two germ plasms at the moment of fertilization, during which the two gametes fuse into a single cell, which is the zygote. In the simplest case, amphimixis is reduced to the union of two nuclei. Cf. **amixis** and **apomixis**. Also see **apogamy**, **parthenogenesis**, **automixis** (**parthenogamy** and **autogamy**), **pseudomixis** or **pseudogamy** (**adelphogamy** and **pedogamy**).

amphispore (Gr. *amphí*, both + *sporá*, spore): *Uredinales*. A special type of urediniospore, with a thickened wall, sometimes darkly pigmented, capable of passing through unfavorable ecological conditions in a dormant state; i.e., a urediniospore with the dormant or resistant function of a teliospore. Also called **mesospore**.

amphithallism (Gr. *amphí*, both, on both sides + *thallós*, thallus + *-ismós* > L. *-ismus* > E. *-ism*, state, phase, tendency, action): the phenomenon in which spores from the same sporophore give rise to both homothallic and heterothallic mycelia, as in some *Coprinus* species (Agaricales).

amphithecium, pl. **amphithecia** (NL. *amphithecium* < Gr. *amphí*, on both sides + *thekíon* < *thêke*, case, box; here, of the asci + dim. suf. *-íon* > L. *-ium*): *Lichens*. **1.** In the discolichens, it refers to the zone of the thallus that encloses the apothecium; it usually bears gonidia and is limited on the outside by the thalline cortex and thalline margin. **2.** In the pyrenocarpous lichens, it refers to the layer of hyaline or light-colored tissue that encloses the hymenium of the pseudothecium embedded in the thallus; e.g., as in *Dermatocarpon* (Verrucariales).

amplectant, **amplective** (L. *amplectens* < *amplectere*, to wind around, encircle + suf. *-ant* > E. *-ant*, one that performs, being; or + L. *-ivus* > E. *-ive*, quality or tendency, fitness): that which embraces or envelops, such as the volva of the basidiocarp of certain species of *Amanita* (Agaricales).

ampulla, pl. **ampullae** (NL. < L. *ampulla*, a small Roman glass vessel, oval in shape < Gr. *ámphora* + L. dim. suf. *-la*): a swollen conidiogenous cell at the apex of a conidiophore, on which the conidia are produced simultaneously, as occurs in *Ostracoderma* (=*Chromelosporium*), *Oedocephalum*, and other gen. of moniliaceous asexual fungi. See **botryoblastospore**.

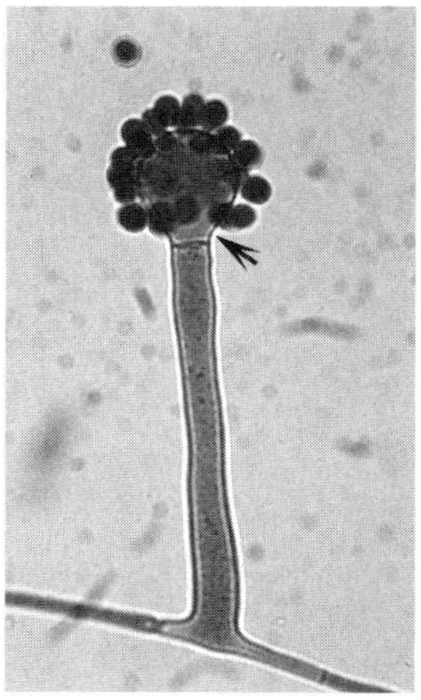

Ampulla of the conidiophore of *Oedocephalum* sp., x 480 (*MU*).

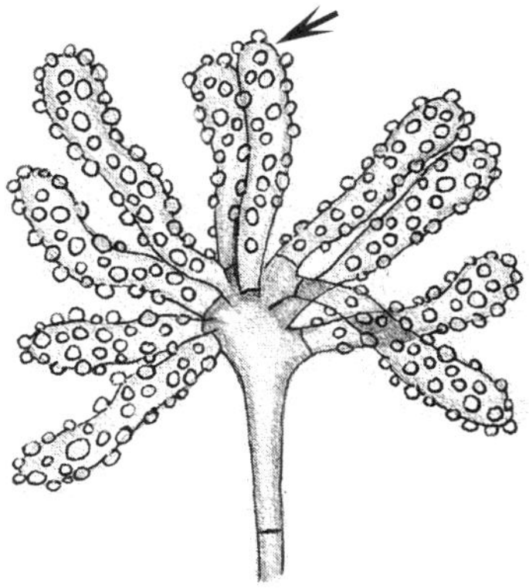

Ampullae of the conidiophore of *Ostracoderma* (=*Chromelosporium*) *ollare*, x 870.

ampulliform (L. *ampulliformis* < *ampulla*, bottle with swollen base + *-formis* < *forma*, shape): with the shape of a bottle, narrow above and swollen in the lower part; flask-shaped. It is applied especially to a conidiogenous cell called a **phialide**, like that of *Penicillium citrinum* and *Trichoderma viride* (moniliaceous asexual fungi). Also see **raduliform** and **rachiform**. In the Basidiomycetes, the cheilocystidia of *Galerina ampullaecystis* and *Agrocybe pediades*, among other Agaricales, are also ampulliform.

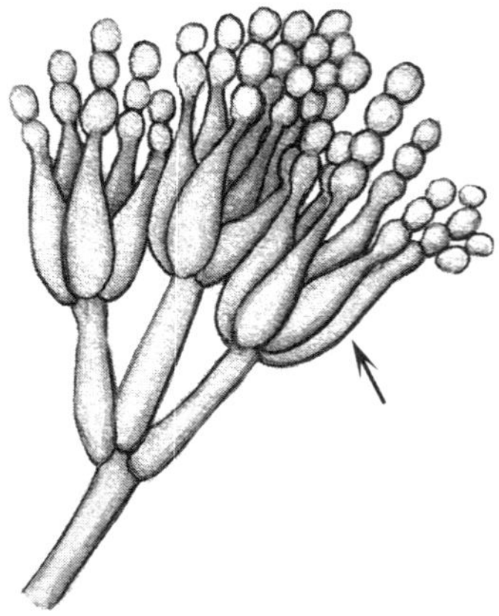

Ampulliform phialides of *Penicillium citrinum*, x 1 000.

amygdaloid (L. *amygdala*, almond < Gr. *amygdále*, almond + L. suf. *-oide* < Gr. *-oeídes*, similar to): similar to an almond, almond-shaped; e.g., like the ascospores of *Thielavia terricola* and *Chaetomium incomptum* (Sordariales), and the basidiospores of *Coprinus comatus* and *Panaeolina foenisecii* (Agaricales).

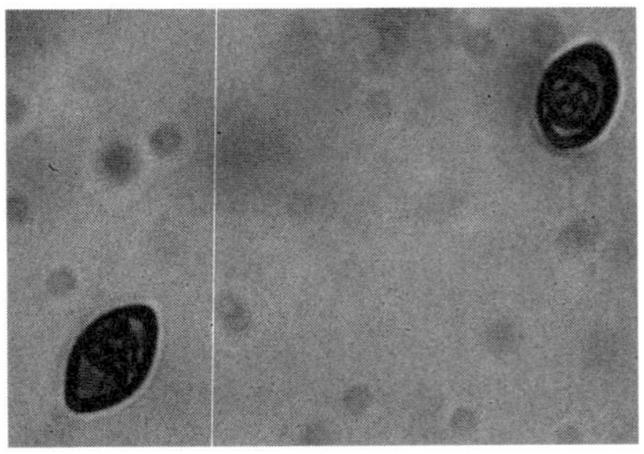

Amygdaloid ascospores of *Thielavia terricola*, x 650 (*MU*).

Amygdaloid basidiospores of *Coprinus comatus*, x 1 500.

amyloid (L. *amyloideus*, resembling starch < Gr. *ámylon*, starch + L. suf. *-oide* < Gr. *-oeídes*, similar to): turning blue on reacting with iodine, presumably due to containing amylaceous substances; e.g., the basidiospores of *Lactarius* and *Russula* (Agaricales) are amyloid. Cf. **anamyloid**, **dextrinoid**, and **pseudoamyloid**.

anabiosis, pl. **anabioses** (Gr. *aná*, again, another time + *bíosis* < *bíos*, life + *-osis*, condition or state): to revive; recovery of the active metabolism and growth from a condition of suspended life, similar to death; resuscitation of a dry or frozen state, such as cultures that are lyophilized or maintained on silica gel or in liquid nitrogen.

anaerobe (Gr. *a-*, without + *aér*, *aéros*, air + *bíos*, life): an organism capable of living in the absence of free oxygen; anaerobic organisms obtain their energy by the decomposition of diverse organic substances. An anaerobe is obligate when the presence of the smallest amount of oxygen makes life impossible; on the other hand, if it can resist it (as in some yeasts) it is called a facultative or discretional anaerobe. Numerous fungi can ferment sugars anaerobically, but since this capability is not essential for growth, they cannot be called anaerobes. There exist, however, some species of Chytridiomycetes that are strictly anaerobic, such as *Piromyces spiralis*, *P. minutus*, *P. mae* and *P. dumbonica* (Neocallismasticales), which live in the intestine of goat, deer, horse and elephant, respectively. Cf. **aerobe**.

anaholomorph (Gr. *aná*, again, another time + *hólos*, entire + *morphé*, shape): an asexually reproducing fungus that appears to lack a teleomorph. Anaholomorphs also are called **asexual fungi**, **conidial fungi**, **deuteromycetes**, **Fungi Imperfecti** or **mitosporic fungi**.

anamorph (Gr. *anamorphóo*, to transform): the asexual, conidial, or so-called imperfect state of a fungus, which produces its spores by mitosis, in contrast to the **teleomorph**, which is the sexual or so-called perfect state (ascogenous or basidiogenous state) and whose spores are produced by meiosis. The fungus in all its forms is called the **holomorph**.

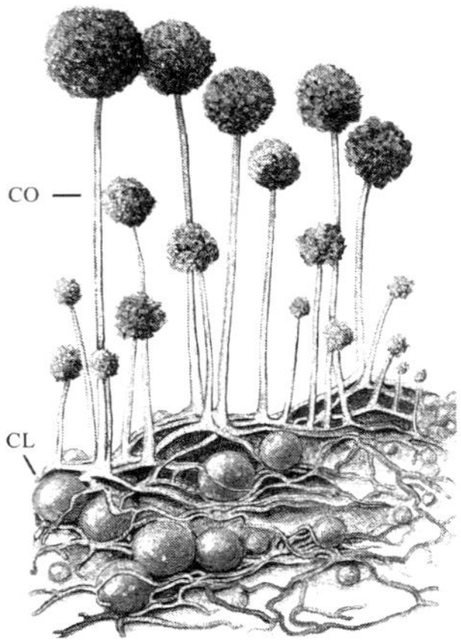

Anamorph (*Aspergillus rubrobrunneus*) of *Eurotium rubrum* (the teleomorph); the corresponding conidiophores (CO) and cleistothecia (CL) are shown, x 66.

anamorph-genus, pl. **genera** (Gr. *anamorphóo*, to transform; L. *genus*, race, kind): a term used to replace **form-genus**, which was applied to a gen. of fungi that reproduces asexually (deuteromycete).

anamorph-species (Gr. *anamorphóo*, to transform; L. *species*, species, kind): a term used to replace **form-species**, which was applied to a fungal species that reproduces asexually (deuteromycete).

anamyloid (Gr. *an*, without + *ámylon*, starch + L. suf. *-oide* < Gr. *-oeídes*, similar to): not amyloid. Applied to the color reaction of spores or cellular elements of fungi which take a yellow color or are almost colorless when treated with Melzer's Reagent (potassium iodide, chloral hydrate, water). Cf. **amyloid**, **dextrinoid**, and **pseudoamyloid**.

anaphysis, pl. **anaphyses** (Gr. *anáphysis*, action of being born again, new growth): a sterigma-like filament that arises from the apothecium of certain lichens, as in the gen. *Ephebe* (Lichinales).

anascosporic (Gr. *an-*, without + *askós*, sack, ascus + *sporá*, spore + suf. *-íkos* > L. *-icus* > E. *-ic*, belonging to, relating to): syn. of **anascosporogenous**. Cf. **ascosporic**.

anascosporogenous (Gr. *an-*, without + *askós*, sack, ascus + *sporá*, spore + *génos*, origin < *gennáo*, to engender, produce + L. *-osus* > OF. *-ous*, *-eus* > E. *-ous*, having, possessing the qualities of): not capable of producing ascospores. Cf. **ascosporogenous**.

anastomosis, pl. **anastomoses** (Gr. *anastómosis*, opening, fusion, connection): fusion of two cells or hyphae in contact that reabsorb their walls and fuse into one; this has great functional importance in fungi in which are recognized vegetative and sexual anastomoses. An example of the first is observed in the hyphal fusion of *Oedocephalum* (moniliaceous asexual fungi).

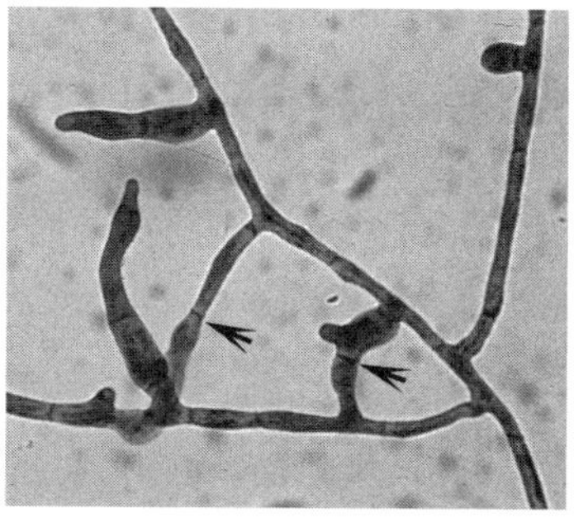

Anastomosis of somatic hyphae of *Oedocephalum* sp., x 200 (*MU*).

anastral (Gr. *an-*, without + *astér*, star + L. suf. *-alis* > E. *-al*, relating to or belonging to): without astral rays; applied to nuclear division in the myxomycetes and some fungi that lack these structures.

androgynous (L. *androgynus* < Gr. *andrós*, male + *gyné*, female, taken here as the masculine and feminine sex organs + L. *-osus* > OF. *-ous*, *-eus* > E. *-ous*, having, possessing the qualities of): applied to fungi with an antheridium and oogonium on the same hypha, as in *Pythium* (Peronosporales), *Achlya* and *Saprolegnia* (Saprolegniales), among other Oomycota.

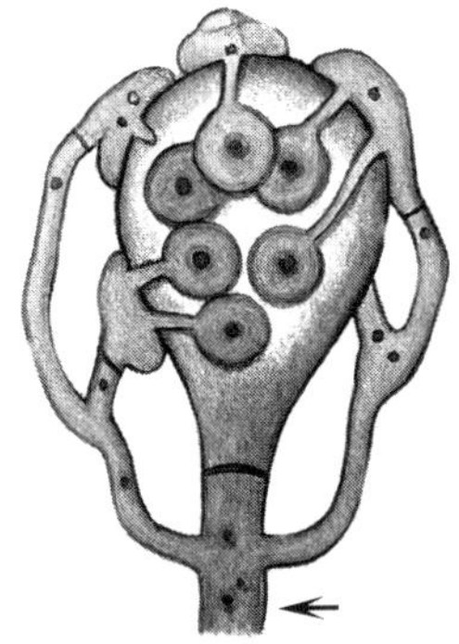

Androgynous hypha of *Saprolegnia parasitica*, with an oogonium and several antheridia, x 400.

andromorph (Gr. *andrós*, male + *morphé*, form): the morph for the spermatial state of a fungus.

androphore (Gr. *andrós*, male + *-phóros*, bearer): that which forms and bears the male sex organs, such as the hyphal branch that produces antheridia in certain ascomycetes (e.g., *Pyronema*, of the Pezizales).

anellarioid (< gen. *Anellaria* < L. *anellus*, *annellus*, ring + suf. *-aria*, which indicates possession or connection + suf. *-oide* < Gr. *-oeídes*, similar to): *Agaricology*. A type of fruiting body (13 main types are recognized) in which the gills adhere to the stipe, which is central, cartilaginous, and with a ring, but without a volva, as in *Anellaria* and some species of *Galerina*, *Psilocybe*, *Psathyrella* and *Coprinus* (Agaricales). Similar to the **omphalinoid** type, but differs in having a ring.

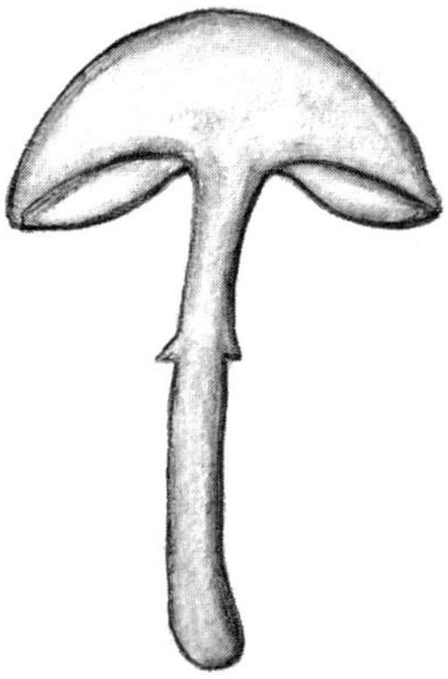

Anellarioid basidiocarp of *Anellaria* sp. Diagram of a longitudinal section, x 0.5.

anemochoric (Gr. *ánemos*, wind + *choréo*, change of place, to move away + suf. *-íkos* > L. *-icus* > E. *-ic*, belonging to, relating to): disseminated by the wind; depending on the wind for dispersal; the majority of fungi with dry spores (e.g., *Alternaria*, *Cladosporium*, dematiaceous asexual fungi, and *Aspergillus* and *Penicillium*, moniliaceous asexual fungi) are mainly wind-dispersed. Also called **anemophilous**. Cf. **hydrochoric**.

anemophilous (Gr. *ánemos*, wind + *phílos*, have an affinity for + L. *-osus* > OF. *-ous*, *-eus* > E. *-ous*, possessing the qualities of): syn. of **anemochoric**.

anemophily (Gr. *ánemos*, wind + *phílos*, have an affinity for + *-y*, E. suf. of concrete nouns): the condition of having spores or other propagules disseminated by wind.

angiocarpic, angiocarpous (L. *angio-*, vessel, container < Gr. *angeîon*, small vessel < *ángos*, vessel + dim. suf. *-eíon* + Gr. *karpós*, fruit + suf. *-íkos* > L. *-icus* > E. *-ic*, belonging to, relating to; or + L. *-osus* > OF. *-ous*, *-eus* > E. *-ous*, having, possessing the qualities of): a fruiting body that remains closed during the major part of its development; at maturity it opens in different ways in order to liberate the spores; some ascocarps and basidiocarps are angiocarpic. Also called angiocarpic development. Cf. **gymnocarpic** and **hemiangiocarpic**.

angiole, angiolum, pl. **angiola** (NL. *angiolum*, dim. of *angium* < Gr. *angeîon*, small vessel, closed receptacle + L. dim. suf. *-olum* > E. *-ole*): same as **peridiole**.

anguilliform (L. *anguilliformis* < *anguilla*, eel + dim. suf. *-illa* + *-formis* < *forma*, shape): in the shape of an eel; long, slender and curved; e.g., like the beta conidia of *Phomopsis* (sphaeropsidaceous asexual fungi), the conidia of *Anguillospora vermiformis* (moniliaceous asexual fungi), or the spores of certain lichens. Also called **vermiform**.

anguilluliform (L. *anguilluliformis* < *anguillula*, eel < *anguilla* + dim. suf. *-ula* + *-formis* < *forma*, shape): dim. of **anguilliform**.

Anguilliform conidia of *Anguillospora vermiformis*, x 530.

angular (L. *angulus*, angle, corner + suf. *-aris* > E. *-ar*, pertaining to): refers to a structure whose contour is not rounded but in the form of angles. These can be triangular, like the ascospores of *Microascus trigonosporus* (Microascales); quadrangular, like the ascospores of *Chaetomium quadrangulatum* (Sordariales); rhomboid, like the conidia of *Beltrania rhombica* (dematiaceous asexual fungi); hexagonal, like the basidiospores of *Psilocybe aerugineomaculans* (Agaricales); or polygonal, like the peridium of the sporangia of *Diderma radiatum* (Physarales) and the sporangiospores of *Mortierella ramanniana* var. *angulispora* (Mucorales).

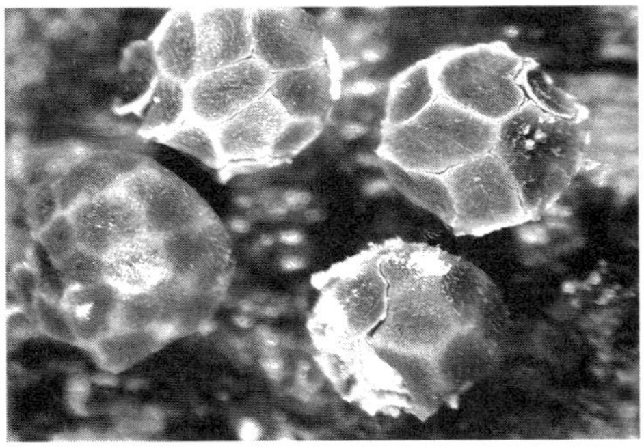

Angular (polygonal) peridium of the sporangia of *Diderma radiatum* growing on dead wood, x 20 (*MU*).

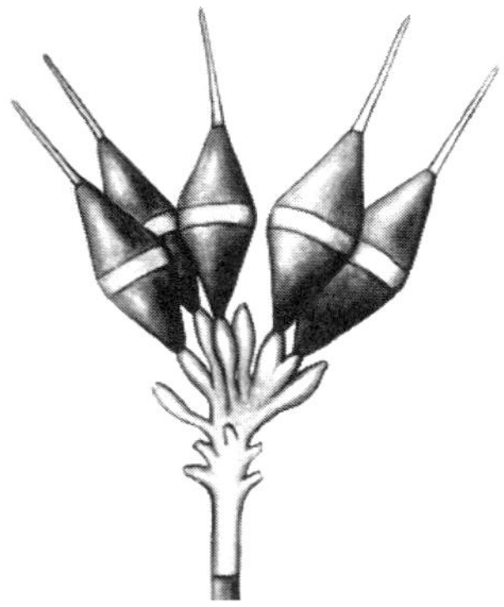

Angular (rhombic) conidia of *Beltrania rhombica*, x 700.

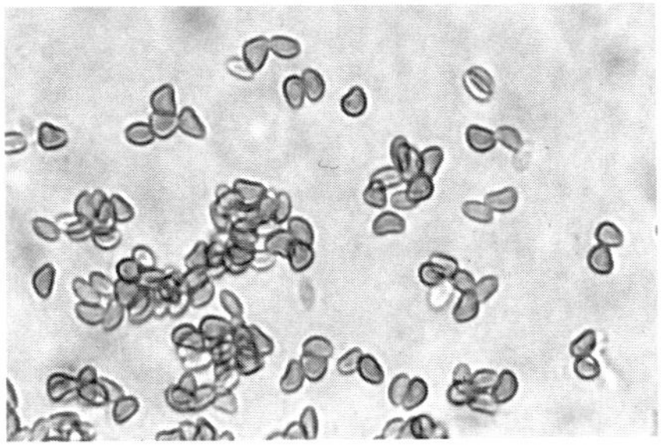

Angular (triangular) ascospores of *Microascus trigonosporus*, x 530 (*CB*).

angustate (L. *angustus*, narrow, small < *anguste*, within narrow bounds + suf. *-atus* > E. *-ate*, provided with or likeness): narrow.

anheliophilous (Gr. *an-*, without + *hélios*, sun + *phílos*, have an affinity for + L. *-osus* > OF. *-ous*, *-eus* > E. *-ous*, possessing the qualities of): see **heliophobic**.

anisodiametric (Gr. *ánisos*, unequal + *diámetros*, diameter < *diá*, through + *métron*, measure + suf. *-íkos* > L. *-icus* > E. *-ic*, belonging to, relating to): refers to a structure that is longer than it is wide, such as the hyphae that comprise the prosenchyma of various fungal organs. Cf. **isodiametric**.

anisogamete (Gr. *ánisos*, unequal + *gamétes*, husband): one of two types of gametes which differ phenotypically. See **anisogamy**.

anisogamy (Gr. *ánisos*, unequal + *gámos*, sexual union + *-y*, E. suf. of concrete nouns): fusion of two gametes similar in shape and structure but differing in size (**anisogametes**); the largest is regarded as female, the other as male; generally, both are flagellate, both are motile, and they are found in aquatic fungi of the order Blastocladiales, such as *Allomyces*. Cf. **heterogamy** and **isogamy**.

annellate (L. *anellus*, dim. of *anus*, ring + suf. *-atus* > E. *-ate*, provided with or likeness): possessing **annellations**, such as the phialides of *Exophiala jeanselmei* (dematiaceous asexual fungi).

Annellate phialides of *Exophiala jeanselmei*, a black yeast, x 2 500.

annellation (L. *anellus*, dim. of *anus*, ring + *-ationem*, action, state or condition, or result > E. suf. *-ation*): the ring left on the apex of an **annellide** by dehiscense of the conidium.

annellide (L. *anellus*, dim. of *anus*, ring): a type of conidiogenous cell (annellate phialide) which produces blastic conidia in basipetal succession (**annellospores**); with the production of each conidium the conidiogenous apex of the phialide, which in this type is not fixed, extends distally, and as each conidium detaches a scar is formed, similar to a ring or collar in the outer wall of the mouth of the phialide; the repetition of this process gives rise to the annellate phialide, as occurs, e.g., in *Scopulariopsis brevicaulis* and *Spilocaea pomi* (moniliaceous and melanconiaceous asexual fungi, respectively).

annellospore (L. *anellus*, dim. of *anus*, ring + Gr. *sporá*, spore): an asexual spore which originates from an annellate conidiogenous cell (annellate phialide or **annellide**), as in *Saccharomycodes ludwigii* (Saccharomycetales). *Scopulariopsis brevicaulis* and *Spilocaea pomi* (moniliaceous and melanconiaceous asexual fungi, respectively) are examples of conidial fungi with this type of conidium, considered a type of phialospore.

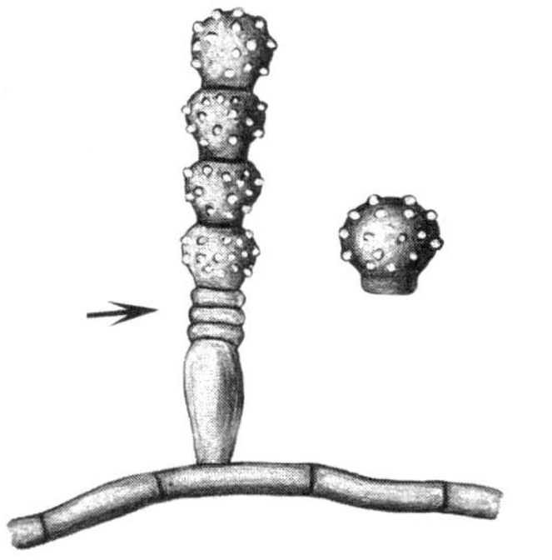

Annellide of *Scopulariopsis brevicaulis*, x 550.

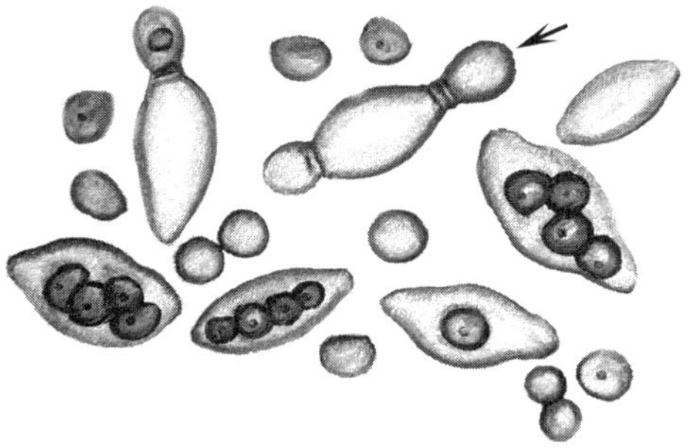

Annellospores of *Saccharomycodes ludwigii*, x 1 000.

annual (LL. *annualis* < *annuus*, yearly < *annus*, year + suf. *-alis* > E. *-al*, relating to or belonging to): refers to an organism that completes its life cycle in a single period that does not exceed one year. In the fungi it refers to the durability of the fruiting body, not the assimilative mycelium immersed in the substrate, and which in the majority of species is annual. Also see **biennial** and **perennial**.

annulate (L. *annulatus*, ringed < *annulus*, ring + suf. *-atus* > E. *-ate*, provided with or likeness): having a ring, like the stipe of the basidiocarp of many Agaricales (*Agaricus*, *Lepiota*, etc.). Cf. **exannulate**.

annulus, pl. **annuli** (L. *annulus*, ring): *Agaricales*. The fragments of the partial veil (which covers the hymenophore during development) which remains in part adhered to the upper portion of the foot or stipe of the basidiocarp, as happens in species of *Agaricus*, *Amanita*, *Chlorophyllum*, *Coprinus*, *Lepiota* and other gen.

Annulus of the basidiocarp of *Macrolepiota procera*, x 0.2.

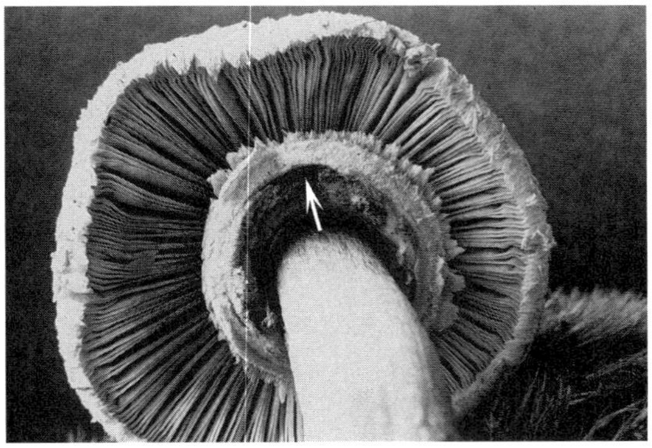

Annulus of a basidiocarp of *Chlorophyllum molybdites*, x 0.5 (*EPS*).

antagonism (Gr. *antagónisma*, to contend against, opposition to < *ant-*, against, opposite + *agón*, struggle + *-ismós* > L. *-ismus* > E. *-ism*, state, phase, tendency, action): *Ecology*. A relationship between two species of organisms in which each adversely affects the other, e.g., by partially or totally inhibiting its growth, including killing it. Examples of this occur in dual agar cultures in which *Agrobacterium azotophilum* inhibits the growth of *Rhizopus nigricans* (Mucorales), and *Dendrostilbella* (stilbellaceous asexual fungi) inhibits the growth of *Phymatotrichopsis omnivora* (moniliaceous asexual fungi). The types of parasitism present in many relationships between a parasite and a host are considered an antagonistic symbiosis. The opposing organisms are referred to as **antagonists**. Also see **commensalism**, **helotism**, **metabiosis**, **mutualism**, **parabiosis**, **parasymbiosis**, **symbiosis**, and **synergy.**

antagonistic symbiosis, pl. **symbioses**, or **parasitism** (Gr. *antagonistés* < *antí*, against + *agonistés* + suf. *-íkos* > L. *-icus* > E. *-ic*, belonging to, relating to; *symbíosis*, life in common; *parásitos*, parasite + *-ismós* > L. *-ismus* > E. *-ism*, state, phase, tendency, action): see **symbiosis**.

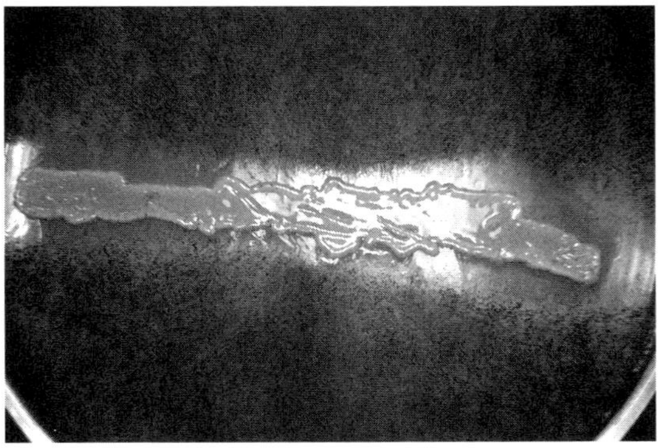

Antagonism *in vitro* of *Agrobacterium azotophilum* towards *Rhizopus nigricans*, x 1 (*TH*).

antheridiol (L. *antheridium* < NL. *anthera*, anther < L. < Gr. *antherós*, flowery + dim. suf. *-ídion* > L. *-idium*, with the suf. *-ol* < alcohol; a chemical compound containing hydroxyl): a sexual hormone (sterol) of *Achlya bisexualis* (Saprolegniales) which induces antheridial formation in male strains. Cf. **oogoniols.**

antheridium, pl. **antheridia** (L. *antheridium* < NL. *anthera*, anther < Gr. *antherós*, flowery + dim. suf. *-ídion* > L. *-idium*): the male gametangium, i.e., the organ in which are generated the antherozoids, as in *Monoblepharis polymorpha* (Monoblepharidales), or the gametic nuclei, as in the Saprolegniales, Peronosporales, and the ascomycetes.

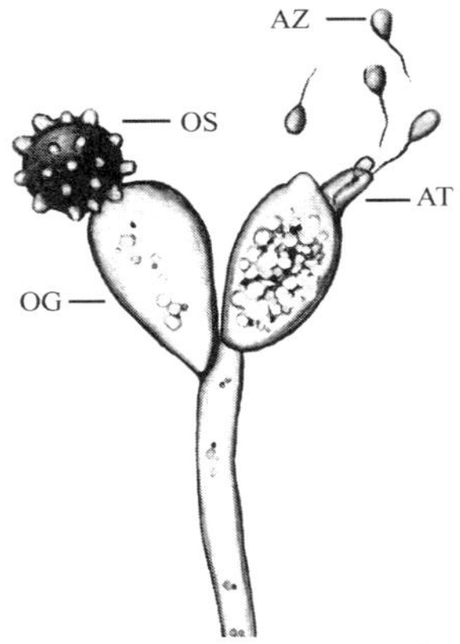

Antheridium (AT) and antherozoids (AZ) of *Monoblepharis polymorpha*. The oogonium (OG) and oospore (OS) are also shown, x 350.

antherozoid (NL. *anther*, anther < L. < Gr. *antherós*, flowery + *zõon*, animal + L. suf. *-oide* < Gr. *-oeídes*, similar to): a motile, uniflagellate gamete, especially of the Monoblepharidales. Syn. of **spermatozoid**.

anthracnose (Gr. *ánthrax*, coal, carbuncle + *nósos*, illness + L. *-osus*, full of, augmented, prone to > ME. *-ose*): a term initially utilized to designate a disease of plants caused by some species of melanconiaceous asexual fungi, such as *Colletotrichum*. Now the term is applied to necrotic diseases caused by gen. such as *Elsinoë* (Myriangiales), *Glomerella* (Phyllachorales), and *Sphaceloma* and *Marssonina* (=*Gloeosporium*) (melanconiaceous asexual fungi), among others.

anthracobiontic (Gr. *ánthrax*, coal, carbuncle + *bíos*, life + *óntos*, genit. of *ón*, a being + suf. *-íkos* > L. *-icus* > E. *-ic*, belonging to, relating to): refers to a fungus that obligately inhabits burned soil or wood and whose fructification is found in this type of habitat. See **phoenicoid fungus** and **pyrophilous.**

anthropophilic (Gr. *ánthropos*, man + *phílos*, have an affinity for + suf. *-íkos* > L. *-icus* > E. *-ic*, belonging to, relating to): *Med. Mycol.* A fungus that preferentially infects humans instead of other higher animals, as happens with the dermatophyte *Epidermophyton floccosum* (moniliaceous asexual fungi), the causal agent of athlete's foot. Cf. **geophilous** and **zoophilic**.

antibiosis, pl. **antibioses** (Gr. *antí*, against + *bíos*, life + *-osis*, condition, state): the inhibitory effect of one microorganism against the physiological processes of another. See **antibiotic**.

antibiotic (Gr. *antí*, against + *biotikós*, of or relating to life < *bíos*, life + *-tikós* > L. *-ticus* > E. *-tic*, relation, fitness, inclination or ability): a substance produced by a living organism which damages or kills another living organism, especially microorganisms; it refers, above all, to the substances produced by various species of molds (mainly species of *Penicillium* and *Aspergillus*, moniliaceous asexual fungi) and actinomycetes (*Actinomyces* and *Streptomyces*) that are harmful to other microorganisms. See **antibiosis**.

anticlinal (Gr. *antiklín*, to lean against each other < *antí*, against + *klínein*, to lean + L. suf. *-alis* > E. *-al*, relating to or belonging to): a membrane or cell wall of a structure that is perpendicular to the main axis. Cf. **periclinal**.

antigen (Gr. *antí*, contrary to, against + *génos*, origin < *gennáo*, to engender, produce): a foreign substance that on being introduced into the body of a vertebrate animal, stimulates the production of an antibody.

antiseptic (Gr. *antí*, against + L. *septicus*, causing decay < Gr. *sépsis*, putrefaction + *-tikós* > L. *-ticus* > E. *-tic*, relation, fitness, inclination or ability): antiputrefactant; a chemical substance that is utilized to destroy or restrict the growth of pathogenic or infectious microorganisms.

antrorse (NL. *antrorsus*, directed upward < L. *ante*, in front + *versum*, toward): directed forward or upward. Cf. **retrorse**.

aphanoplasmodium, pl. **aphanoplasmodia** (Gr. *aphanés*, invisible + L. *plasmodium* < Gr. *plásma*, formation, bland material with which a living being is formed + L. *-odium*, resembling < Gr. *ode*, like < *-oeídes*, similar to): *Myxomycetes*. A plasmodium that initially is constituted of a network of fine, transparent protoplasmic filaments that are not obviously differentiated into ectoplasm and endoplasm; the protoplasm is not very granulose, although the protoplasmic currents are rapid and reversibly rythmic. Characteristic of *Stemonitis* and related gen. (Stemonitales). Cf. **phaneroplasmodium** and **protoplasmodium**.

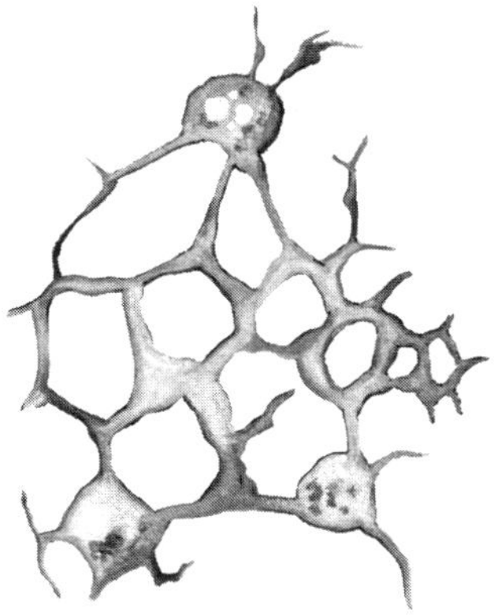

Aphanoplasmodium of *Stemonitis axifera*, x 100.

aphysoclastic (Gr. *a*, without + *physáo*, *physéo*, *physõ*, to throw + *klástos*, broken < *kláo*, to break + suf. *-íkos* > L. *-icus* > E. *-ic*, belonging to, relating to): applied to an ascus that liberates its spores without a previous rupture of the external wall layer (which is less elastic) and also without undergoing an extension of the inner wall layer. It is a term utilized in marine mycology; this type of ascus is present, e.g., in *Ceriosporopsis* and *Carbosphaerella* (Halosphaeriales), fungi saprobic on wood and other floating vegetable remains in the intertidal zone. Cf. **physoclastic**.

apical (L. *apicalis*, at the apex, tip < *apex*, *apicis*, apex > *apic-* + suf. *-alis* > E. *-al*, relating to or belonging to): located at the top or apex, such as the apical paraphyses of the ascocarp of *Nectria haematococca* (Hypocreales). Cf. **basal**.

apical body (L. *apicalis*, at the apex, tip < *apex*, *apicis* > *apic-* + suf. *-alis* > E. *-al*, relating to or belonging to; ME. < OE. *bodig*, body): see **Spitzenkörper**.

apiculate (NL. *apiculatus*, small or abruptly pointed < *apiculus*, dim. of L. *apic-*, *apex*, genit. *apicus*, apex, tip + suf. *-atus* > E. *-ate*, provided with or likeness):

ending abruptly in a small distinct point. E.g., the ascospores of *Hypomyces* (Hypocreales) and *Chaetomium* (Sordariales) are apiculate. See **apiculus** and **biapiculate**.

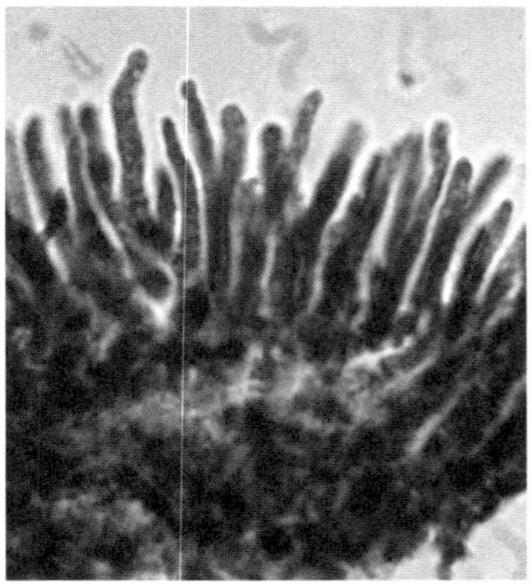

Apical paraphyses from the ascocarp of *Nectria haematococca*, x 1 500 (*RTH*).

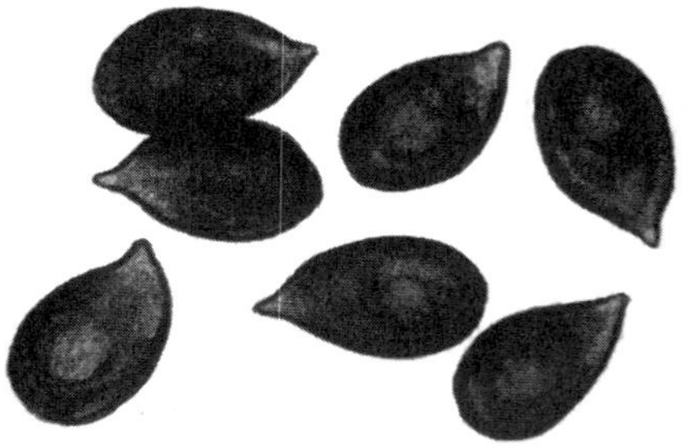

Apiculate ascospores (at one end) of *Chaetomium nigricolor*, x 1 500.

apiculus, pl. **apiculi** (NL. *apiculus* < L. *apic-*, *apex*, genit. *apicus*, apex, tip + dim. suf. *-ulus*): a short point or tip, but not stiff; generally refers to the short, sharp diverticulum found in the basal portion of basidiospores, by means of which they adhere to the sterigma of the basidium, and from which exudes a droplet at the moment of discharge of the basidiospore. Also see **hilar appendix**, which is a more correct term than apiculus to designate the diverticulum described, which is characteristic of all Agaricales.

apileate (Gr. *a-*, without + L. *pileatus*, provided with a pileus or cap + suf. *-atus* > E. *-ate*, provided with or likeness): applied to fungi whose basidiocarps lack a pileus; e.g., the resupinate species, as in *Poria*, *Stereum* and other thelephoraceous members of the order Aphyllophorales.

apiosporous (Gr. *ápion*, pear + *sporá*, spore + L. *-osus* > OF. *-ous*, *-eus* > E. *-ous*, having, possessing the qualities of): a pear-shaped, two- celled spore, with one cell markedly smaller than the other, e. g., the pyriform ascospores of *Apiosporina* (Pleosporales).

aplanetic (Gr. *a-*, without + *planétes*, wanderer + suf. *-íkos* > L. *-icus* > E. *-ic*, belonging to, relating to): immotile; refers to spores that are not motile. See **aplanospore**.

aplanogamete (Gr. *a-*, without + *plános*, wandering + *gamétes*, spouse): an immotile gamete, i.e., one lacking flagella. Cf. **planogamete**.

aplanospore (Gr. *aplanés*, immotile, fixed + *sporá*, spore): an asexual, immotile spore produced individually or in variable numbers inside a sporangium, as happens in members of the Mucorales. See **aplanetic**.

aplerotic (Gr. *a-*, without + *plerotikós* < *pleróo*, to fill + *-tikós* > L. *-ticus* > E. *-tic*, relation, fitness, inclination or ability): *Peronosporales*. Not filled up. Refers to species of the Pythiaceae in which the oospore does not fill the oogonium. E.g., *Pythium echinulatum* is aplerotic. Cf. **plerotic**.

Aplerotic oospore of *Pythium echinulatum*, x 430.

apobasidium, pl. **apobasidia** (Gr. *apó*, away from + L. *basidium*, small pedestal < Gr. *basídion* < *básis*, base + dim. suf. *-ídion* > L. *-idium*): *Gasteromycetes*. A type of basidium characterized by having its spores terminal and symmetrical or orthotropic on the axial prolongations of the sterigmata, such as occurs in *Sclerogaster* (Hymenogastrales).

apocyte (Gr. *apó*, away from, here it expresses the idea of something separated or hidden + *kýtos*, cavity, here cell): a multinucleate cell in which each of the nuclei is surrounded by a protoplasmic mass, thus constituting cellular units, but without separation by means of walls. The apocyte, which is an accidental transitory or secondary cellular condition in ascomycetes and basidiomycetes, is morphologically similar to a syncytium, but differs in that the former comes from an original single cell in which the nuclei have multiplied without establishing cellular separation of the protoplasts, whereas the syncytium originates from the fusion of two or more distinct cells.

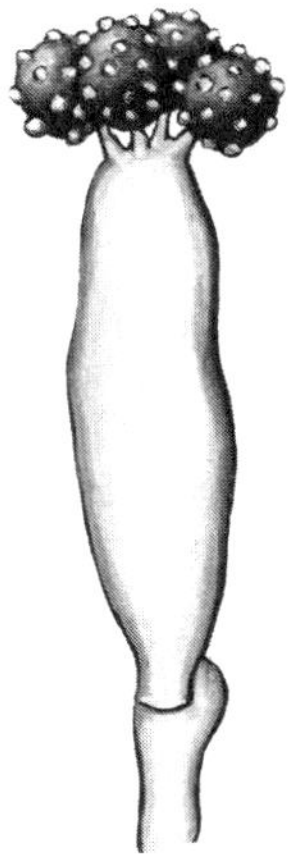

Apobasidium from the hymenium of *Sclerogaster* sp., x 1 270.

apodal, **apodial**, **apodous** (Gr. *ápous*, *ápodos* < *a-*, without + *poús*, *podós*, foot + L. suf. *-alis* > E. *-al*, relating to or belonging to; or + L. *-osus* > OF. *-ous*, *-eus* > E. *-ous*, having, possessing the qualities of): lacking a foot; without a pedicel or stalk; sessile. See **sessile**. Cf. **pedicellate** and **stipitate**.

apogamy (Gr. *apó*, away from + *gámos*, sexual union + *-y*, E. suf. of concrete nouns): away from sexuality, i.e., asexual development or apomixis of diploid cells. See **apomixis**. Cf. **amphimixis**.

apomictic (Gr. *apó*, away from + *miktós*, mixture + suf. *-íkos* > L. *-icus* > E. *-ic*, belonging to, relating to): related to the phenomenon of **apomixis**.

apomixis, pl. **apomixes** (Gr. *apó*, away from + *míxis*, mixture, union): asexual development of sexual cells without prior fertilization. Two types of apomixis are known: **parthenogenesis** and **apogamy**. See **amixis**. Cf. **amphimixis**.

apophysate (Gr. *apóphysis*, offspring + suf. *-atus* > E. *-ate*, provided with or likeness): having an **apophysis**.

apophysis, pl. **apophyses** (NL. *apophysis* < Gr. *apó*, away from + *phýsis*, growth < *phýein*, to bring forth, bear): **1**. Refers to a swollen protuberance that is found beneath the sporocarp of various fungi; e.g., some Protosteliales, such as *Nematostelium*, *Schizoplasmodium* and *Protostelium*. **2**. The subsporangial swelling present in the sporangia of certain aquatic fungi, such as *Phlyctochytrium* (Chytridiales). **3**. The widening apical part of the sporangiophore of some Mucorales, such as *Chlamydoabsidia*, *Gongronella* and *Rhizopus*. **4**. The ventral swelling of the endoperidium of *Geastrum* (Lycoperdales).

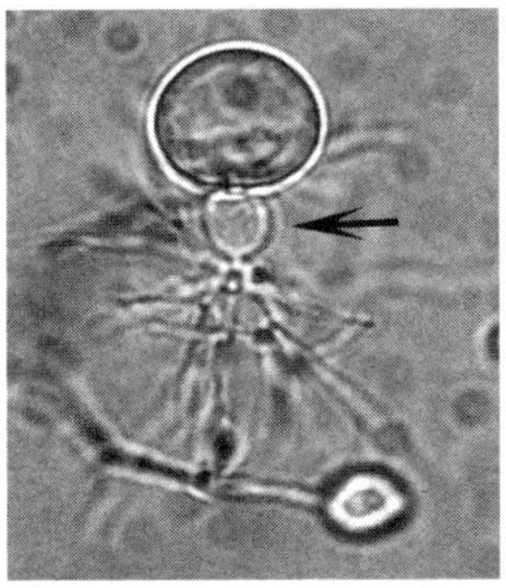

Apophysis of the zoosporangium of *Phlyctochytrium* sp., x 850 (*RTH*).

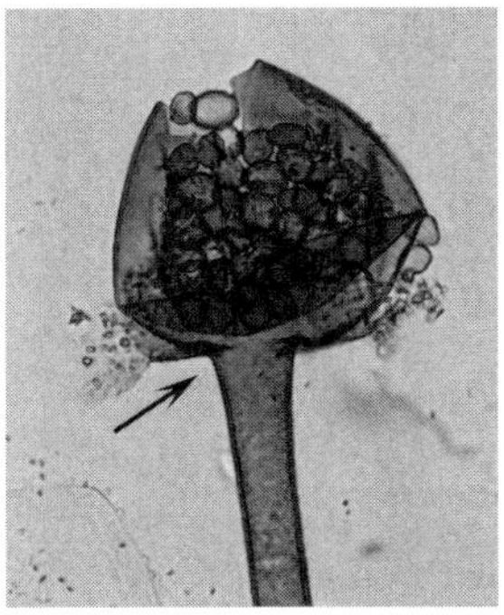

Apophysis of the sporangium of *Rhizopus arrhizus*, x 400 (*PL*).

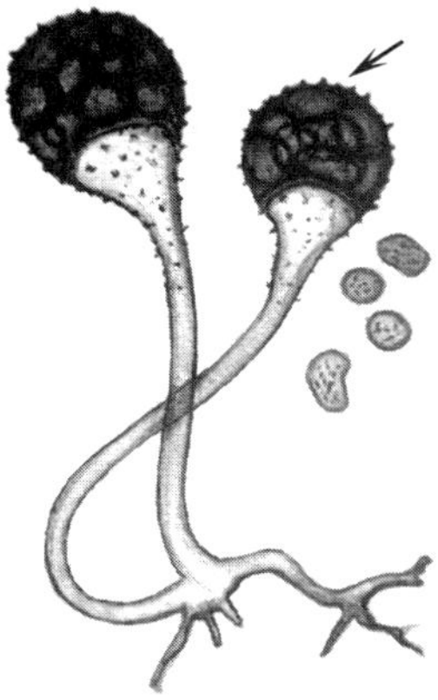

Apophysis of the sporangium of *Rhizopus oligosporus*, x 300.

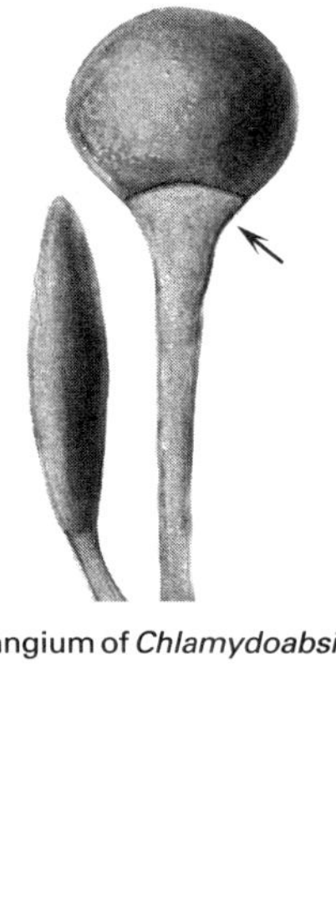

Apophysis of the sporangium of *Chlamydoabsidia pademii*, x 1 100.

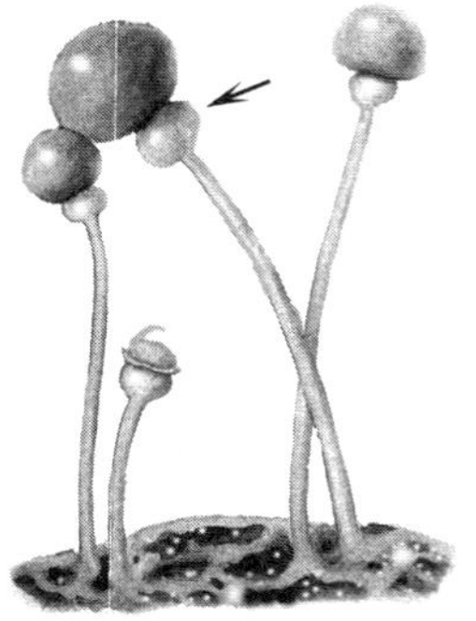

Apophysis of the sporangia of *Gongronella butleri*, x 460.

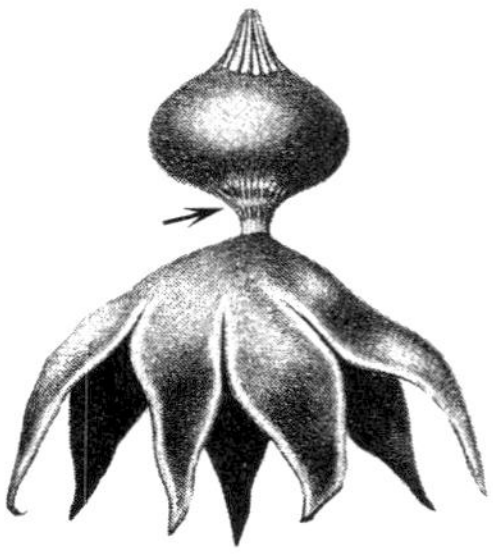

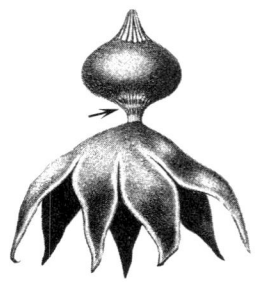

Apophysis of the fruiting body of *Geastrum pectinatum*, x 0.7.

apoplasmodial (Gr. *apó*, away from + *plasmodial* < L. *plasmodium* < Gr. *plásma*, bland material of which living beings are formed + L. *-odium*, resembling < Gr. *ode*, like < *-oeídes*, similar to + L. suf. *-alis* > E. *-al*, relating to or belonging to): not forming a true plasmodium, through lack of fusion of myxamoebae, as occurs in the pseudoplasmodium of Acrasiomycota.

apospory (Gr. *apó*, away from + *sporá*, spore + *-y*, E. suf. of concrete nouns): the incorporation of a diploid oogonial or antheridial nucleus and cytoplasm into a spore directly without the occurrence of meiosis.

apothecioid (NL. *apothecium* < Gr. *apothekíon*, dim. of *apotheke*, granery < *apó*, something that is separable + *thêke*, box, case; here, of the asci + L. suf. *-oide* < Gr. *-oeídes*, similar to): refers to the ascostroma of *Schizothyrium ptarmicae* (Dothideales) and *Myriangium* (Myriangiales), which on maturing acquires a shape similar to an apothecium.

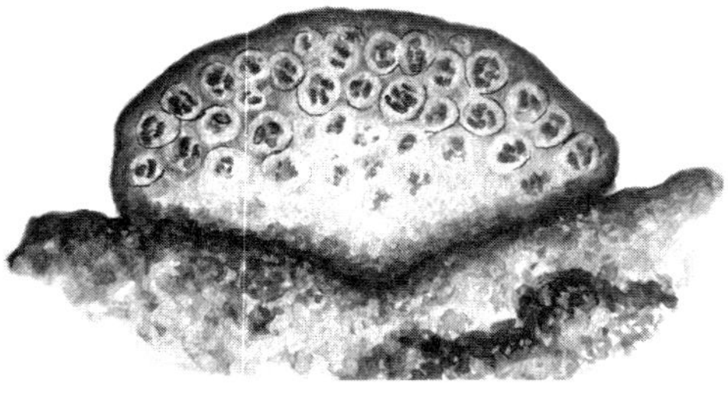

Apothecioid, multilocular ascostroma of *Myriangium* sp., seen in longitudinal section, x 50.

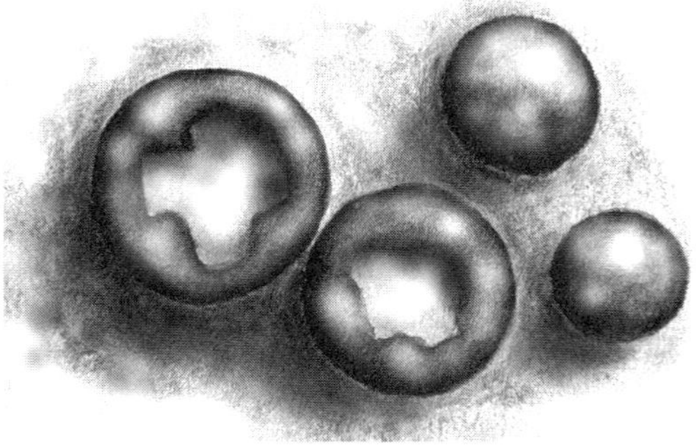

Apothecioid pseudothecia of *Schizothyrium ptarmicae*, two of them with the scutellum open, on a host leaf, x 80.

apothecium, pl. **apothecia** (NL. *apothecium* < Gr. *apothekíon* < *apotheke*, a storehouse < *apó*, away from, separate + *thêke*, case, box; here, of the asci + dim. suf. *-íon* > L. *-ium*): a type of ascocarp (fruiting body) which at maturity opens widely to expose the hymenium of asci; it commonly assumes the shape of a cup, disk, or saucer, but other shapes also occur. Characteristic ascocarp of the discomycetes, such as *Aleuria*, *Helvella*, *Peziza* (Pezizales) and *Monilinia* and *Sclerotinia* (Helotiales). Also called **discocarp** when saucer-shaped.

Apothecium of *Aleuria aurantia* in side view, x 3 (*CB*).

Apothecium of *Aleuria aurantia* in top view, x 3 (*CB*).

appendage (L. *appendage* < *appendere*, to append): a subordinate part projecting from some other structure, generally accessory. Distinctive structures present in very diverse fungal organs are called appendages. E.g., the appendages that form on the suspensors and which more or less cover the zygosporangia of many species of Mucorales, such as *Absidia glauca*, *Phycomyces nitens*, etc. Other examples of appendages include the hymenial elements of various Agaricales (such as digitate, echinulate, or diverticulate cystidia, the coprinoid appendages, etc.), the piliform appendages of the sporangiospores of *Choanephora cucurbitarum* (Mucorales) and of the trichospores of *Smittium* (Harpellales), and the membranaceous ones of the conidia of *Myrothecium verrucaria* (tuberculariaceous asexual fungi).

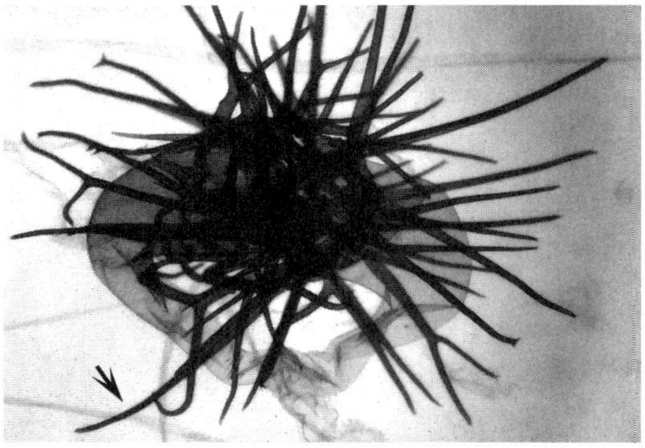

Appendages of the suspensors of *Phycomyces nitens*, x 100 (*MU*).

appendiculate (L. *appendiculatus*, furnished with appendages < *appendiculus*, small appendage < *appendix*, appendage + dim. suf. *-culus* + suf. *-atus* > E. *-ate*, provided with or likeness): having appendages or protruding parts, usually accessory or remnant. In fungi of the gen. *Hydnum* and *Sarcodon* (Aphyllophorales), each of the teeth of the sporophore, covered by hymenium, is called an appendage or **appendiculus**, as is the portion of the partial veil that remains hanging from the edge of the pileus after its rupture and separation, as happens in *Psathyrella*, *Boletellus ananas* (Agaricales) and *Polyporus tricholoma* (Aphyllophorales), whose pileus is appendiculate.

appendiculus, pl. **appendiculi** (L. *appendiculus*, small appendage < *appendix*, genit. *appendicis*, appendage + dim. suf. *-ulus*): a small appendage. See **appendiculate**.

apposed (MF. *aposer* < OF. < L. *appositus*, united, placed near, applied): placed in juxtaposition, side by side. See **parallel**.

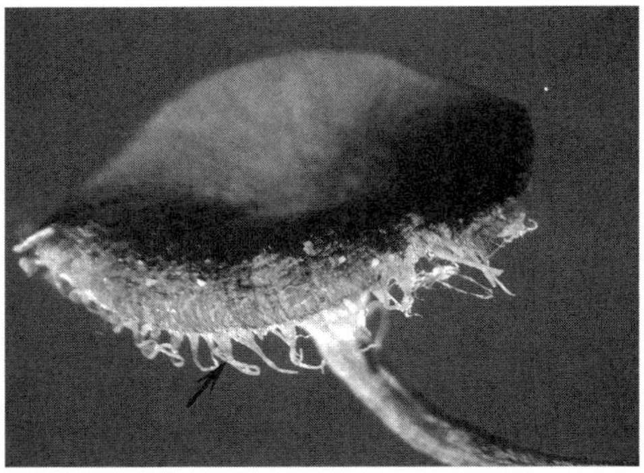

Appendiculate pileus of *Polyporus tricholoma*, x 2 (*MU*).

Appendiculi of the hymenophore of *Sarcodon* sp., x 30 (*MU*).

apposition (LL. *apposition*, genit. *appositionis*, the act of placing near < L. *appositus*, united, placed near, applied < *ap-*, in + *positus*, placed + *-io*, *-ionis* > E. suf. *-ion*, *-tion*, result of an action, state of): the growth in thickness of a cell wall by successive deposits of new molecules on the internal or external surface. Endogenous apposition is apposition in the strict sense by deposition of particles on the internal wall surface. Exogenous apposition involves centrifugal growth in thickness through deposition of particles on the exterior wall surface; various ornamentations which occur on the spore wall of many fungi form in this manner. Cf. **intususception**.

appressed (L. *appressus*, pressed closely against something < *apprimere*, to press): scales, fibers, hairs, etc., that are closely applied or flattened against the surface of the organ or part on which they are inserted. E.g., the surface of the pileus of *Agaricus placomyces* and *Macrolepiota procera* (Agaricales) is appressed squamulose in that the fibrils of the points of the hyphae agglutinate, forming applanate scales.

appressorium, pl. **appresoria** (NL. *appressorium* < *appress*, kept down + L. suf. *-orium*, a place of a thing): a flat swelling that forms on the end of a germ tube or vegetative hypha and which adheres to the surface of the host before penetrating it with an infection hypha that originates from the bottom of the swelling; characteristic of plant pathogenic fungi such as *Phytophthora* (Peronosporales) and *Colletotrichum* (melanconiaceous asexual fungi), and parasites of lower animals, such as *Entomophthora* (Entomophthorales).

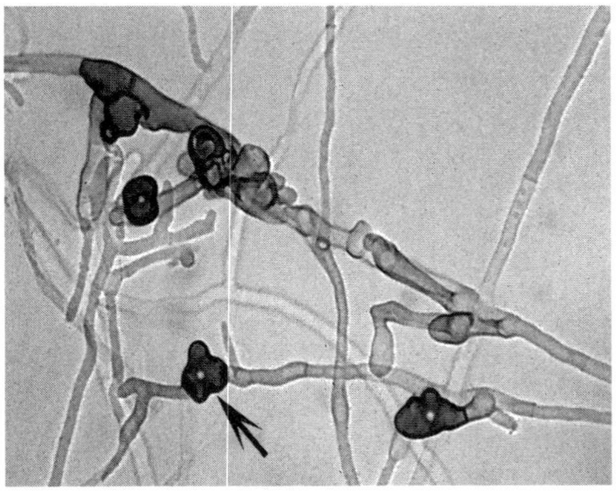

Appressoria of *Colletotrichum* sp., x 420 (*MU*).

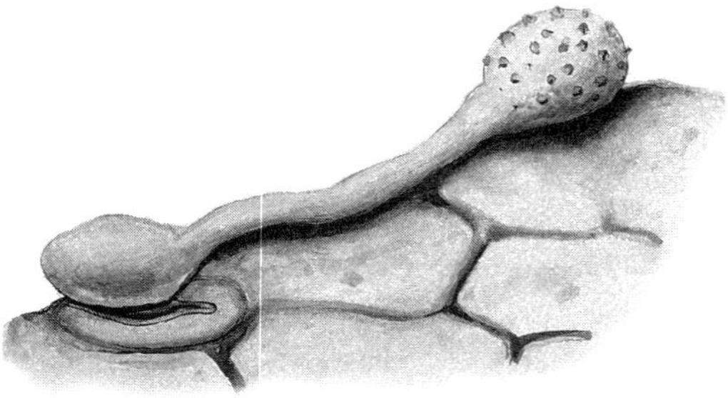

Appressorium of *Cercosporidium personatum*, developed from the tip of a germ tube that grew over a stroma on a host leaf, x 1 600.

arachnoid (NL. *arachnoides* < Gr. *arachnoeidés*, cobweb-like < *aráchne*, spider, cobweb + L. suf. *-oide* < Gr. *-oeídes*, similar to): similar to a spider's web, like the medullary layer of lichens when they are formed of loosely intertwined hyphae. The conidia of *Arachnophora fagicola* (moniliaceous asexual fungi) are arachnoid since their lateral protuberances have appendages that are incurved, spinose or claw-shaped and which confers on them a peculiar aspect, as if they were small arachnids, a characteristic implicit in the generic name. Also called **araneose** and **araneous**.

araneose, **araneous** (L. *araneosus*, pertaining to a spider's web < Gr. *aráne* < *aráchne*, spider + L. *-osus*, full of, augmented, prone to > ME. *-ose*; or + *-osus* > OF. *-ous*, *-eus* > E. *-ous*, having, possessing the qualities of): similar to a spider web. See **arachnoid**.

arboricolous (L. *arbor*, genit. *arboris*, tree + *-cola*, inhabitant < *colere*, to inhabit + L. *-osus* > OF. *-ous*, *-eus* > E. *-ous*, possessing the qualities of): living on trees, like many lichens.

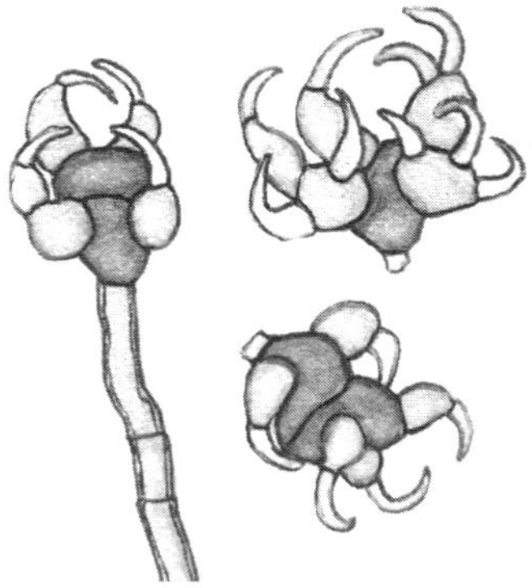

Arachnoid conidia of *Arachnophora fagicola*, x 1 000.

arbuscle, **arbuscule** (L. *arbusculum* < *arbor*, tree + dim. suf. *-culum* > E. *-ule*): hyphae with a shrubby or coralline aspect which mycorrhizal fungi form inside root cells of the associated plant. Arbuscules are one of the forms (vesicles constitute the other) produced by endotrophic mycorrhizae (endomycorrhizae). An example of an arbuscle is *Pythium ultimum* (Peronosporales) in cortical cells of onion root. Many Glomales, such as *Glomus* and *Gigaspora*, form vesicular-arbuscular mycorrhizae in many species of higher plants.

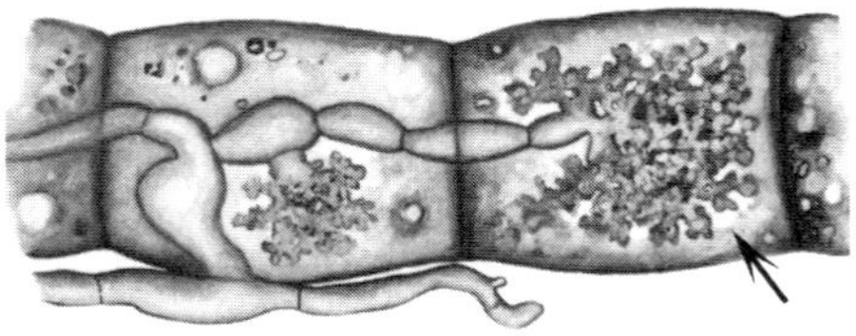

Arbuscles of *Glomus diaphanum* in root cells of red clover, x 370.

archecarpium, pl. **archecarpia** (Gr. *arché*, primitive, origin + *karpós*, fruit + dim. suf. *-íon* > L. *-ium*): the cells that form the ascus and pedicel in *Podosphaera* (Erysiphales).

archicarp (Gr. *arché*, primitive, origin + *karpós*, fruit): *Ascomycetes*. The ascogonium or other female sexual organ, which forms the first initial of the carpophore or ascocarp. The term is used indistinctly to designate the whole female organ, to the branch, the primary cell, to the group of initial cells, or to the coiled hypha that develops into the fruiting body or ascocarp.

Archiascomycetes (Gr. *arché*, origin, primitive + *askós*, sac + L. *-mycetes*, ending of class < Gr. *mýkes*, genit. *mýketos*, fungus): a class of fungi, formerly referred to in part as Hemiascomycetes, in which the asci are born free (i.e., not contained in any type of fruiting body or ascocarp), and ascogenous hyphae are lacking; the thallus is simple, and when mycelium is present it is poorly developed. The Archiascomycetes, together with the classes Plectomycetes, Pyrenomycetes, Discomycetes and Loculoascomycetes (along with some other ascomycetes, such as the orders Saccharomycetales, Erysiphales, Laboulbeniales and Spathulosporales) constitute the phylum Ascomycota of the kingdom Fungi.

Archimycetes (Gr. *arché*, primitive, origin + L. *-mycetes*, ending of class < Gr. *mýkes*, genit. *mýketos*, fungus): a class recognized by Gäumann and Dodge (1928) and Gäumann (1952) for certain unicellular, holocarpic chytrids which in the early stages of development lack cell walls; this characteristic made this class distinct from the rest of the chytrids, which were retained in the class Chytridiomycetes.

archontosome (Gr. *árchon*, genit. *árchontos*, a ruler + *sõma*, body): an electron-dense body occurring near nuclei at all stages from crozier formation to the development of young ascospores in *Xylaria polymorpha* (Xylariales).

arcuate (L. *arcuatus*, bent like a bow, curved < *arcus*, bow + suf. *-atus* > E. *-ate*, provided with or likeness): curved in an arc; e.g., the pileus of the basidiocarp of *Clitocybe infundibuliformis* (Agaricales) first develops upward and later bends downward, forming an arc in the lower part. The sporangia of *Comatricha typhoides* (Stemonitales) also are arcuate.

Arcuate sporangia of *Comatricha typhoides*, x 55 (*MU*).

ardella, pl. **ardellae** (L. *ardella*, little spot < Gr. *árda*, spot + L. dim. suf. *-ella*): a small apothecium in the form of a round spot, characteristic of lichens of the family Arthoniaceae (Arthoniales), such as *Arthothelium spilomatoides*. If the apothecia are elongate, they are **lirellae**.

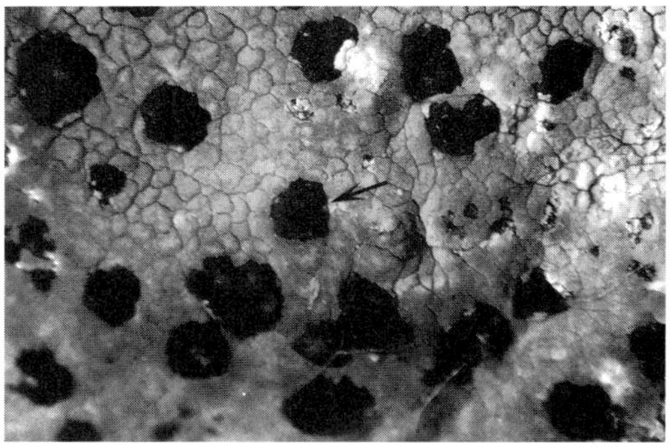

Ardellae (apothecia) of the thallus of *Arthothelium spilomatoides*, x 7 (*MU*).

arenarious (L. *arenarius*, pertaining to sand < *arena*, sand + *-arius*, belonging to): living in sand. Syn. of **arenicolous** and **psammophilous**.

arenicolous (L. *arena*, sand + *-cola*, inhabitant + *-osus* > OF. *-ous*, *-eus* > E. *-ous*, possessing the qualities of): an organism that lives among or on grains of sand, whether in a desert or beach. Also called **psammophilous**. In mycology it is applied most commonly to fungi that inhabit sand grains or their interstices of a marine beach, such as *Corollospora intermedia* and *Lulworthia lignoarenaria* (Halosphaeriales), among other species. See **endopsammon**.

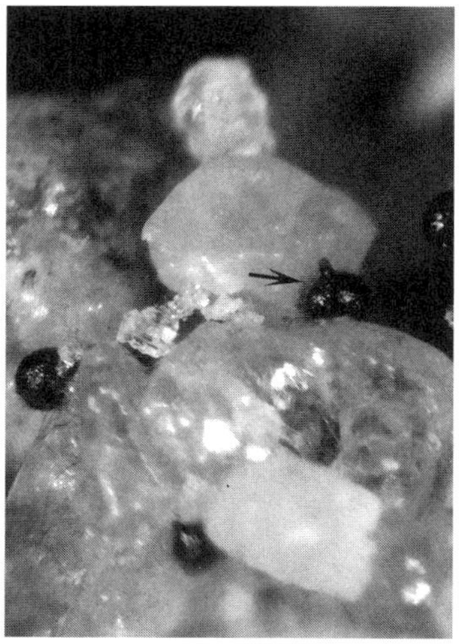

Arenicolous perithecia of *Corollospora intermedia* (arrow) adhered to sand grains, x 23 (*MU*).

areola, pl. **areolae** (L. *areola*, dim. of *area*, halo, open space): a space delimited on a surface, separated from other spaces by fissures or cracks. See **areolar**.

areolar, **areolate** (L. *areola*, dim. of *area*, halo, open space + suf. *-aris* > E. *-ar*, like, pertaining to; *areolatus*, having areoles < *areola* + suf. *-atus* > E. *-ate*, provided with or likeness): pertaining or relative to an **areola**; having a halo or ring of different color. E.g., in *Hypoxylon thouarsianum* and *H. michelianum* (Xylariales) halos are present around the ostioles of the perithecia embedded in the stroma. In the lichens, the thallus is said to be areolate when its surface is divided into small rounded or angular areas by means of cracks, as occurs in *Rhizocarpon geographicum* (Lecanorales) and *Thelomma santessonii* (Caliciales).

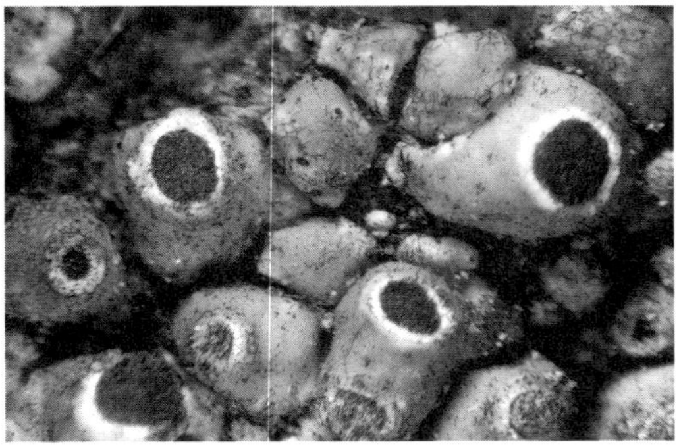

Areolate apothecia of the thallus of *Thelomma santessonii*, x 10 (*CB*).

arescent (L. *arescens*, genit. *arescentis* < *arescere*, to become dry, not deliquescent < *areere*, to be dry + *-escens*, genit. *-escentis*, that which turns, beginning to, slightly > E. *-escent*): dry, not moistened, like the gills of many agaricaceous fungi. Cf. **deliquescent**.

argentate, **argenteous** (L. *argenteus*, silvery < *argentum*, silver + L. suf. *-atus* > E. *-ate*, provided with or likeness; or + L. *-osus* > OF. *-ous*, *-eus* > E. *-ous*, having, possessing the qualities of): refers to organs or structures whose surface has a certain luster like silver, due to being covered by abundant hairs or soft and appressed scales, like the basidiocarps of *Geastrum argenteum* (Lycoperdales), *Amanita argentea* and *A. spreta* (Agaricales), and the peridiola of the fruiting body of *Cyathus intermedius* (Nidulariales), among others.

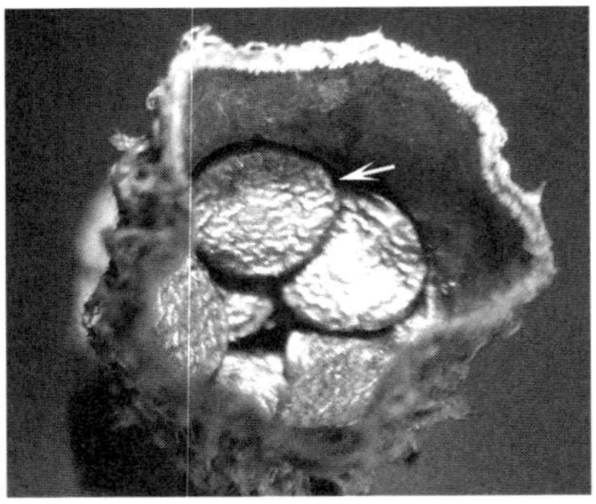

Argentate peridiola of the fruiting body of *Cyathus intermedius*, x 7 (*CB & MU*).

argentophilous (L. *argentum*, silver + Gr. *phílos*, have an affinity for + L. *-osus* > OF. *-ous*, *-eus* > E. *-ous*, possessing the qualities of): living on substrates containing silver or other heavy metals, and which not only resist them but capture the silver ions in their cells, such as *Cryptococcus albidus* var. *albidus* (cryptococcaceous asexual yeasts), an encapsulated yeast that has been isolated from the mud discharged by silver mines; these yeasts capture the silver and accumulate it in their capsules.

armilla, pl. **armillae** (L. *armilla*, bracelet, hoop): a ring situated on the upper part of the foot of certain agaricaceous fungi; e.g., in *Armillariella mellea*. See **annulus**.

armillarioid (< gen. *Armillaria* < L. *armilla*, *ring* + suf. *-aria*, which indicates possession or connection + L. suf. *-oide* < Gr. *-oeídes*, similar to): *Agaricology*. A fruiting body characterized by having gills adhering to the stipe, which is fibrous, fleshy, or chalky in consistency, and centrally attached to the pileus, with a ring, but lacking a volva. Typified by *Armillaria* and certain species of *Pholiota* (=*Flammula*) and *Gomphidius*, among other gen. of Agaricales.

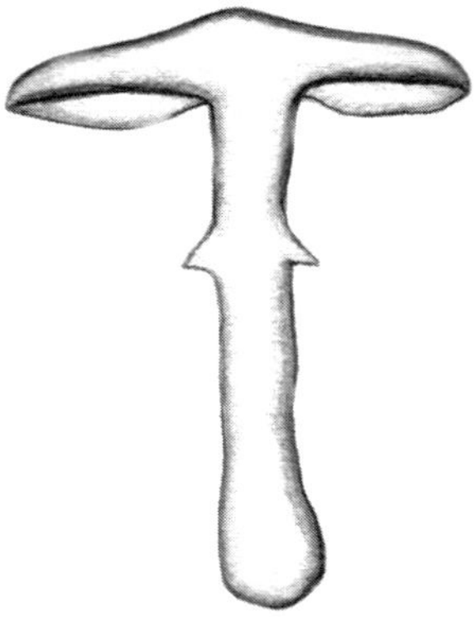

Armillarioid basidiocarp of *Armillaria* sp.
Diagram of a longitudinal section, x 0.5.

arrect (L. *arrectus*, set upright < *arrigere*, to erect): erect, pointing upward, very rigid; e.g., like the hairs of the basidiocarp of *Polyporus hirsutus* (Aphyllophorales).

arthric (Gr. *árthron*, joint, articulation + suf. *-íkos* > L. *-icus* > E. *-ic*, belonging to, relating to): conidia or spores that result from the fragmentation and disarticulation of fertile hyphae with determinate growth. In this type of thallic conidial development two types are distinguished: **holoarthric** and **enteroarthric**.

arthrocatenate (Gr. *árthron*, joint, articulation + L. *catenatus*, in a chain < *catena*, chain + suf. *-atus* > E. *-ate*, provided with or likeness): applied to thalloconidia formed in chains by the simultaneous or random fragmentation of a hypha.

arthrospore (Gr. *árthron*, articulation + *sporá*, spore): *Conidiogenesis*. A spore that results from the fragmentation of a hypha; in current terminology of conidial ontogeny, these spores are called holoarthric conidia and they are present, e.g., in some moniliaceous asexual fungi, such as *Trichosporon beigelii* and *Geotrichum candidum* (whose teleomorph is *Galactomyces geotrichum*, of the Saccharomycetales).

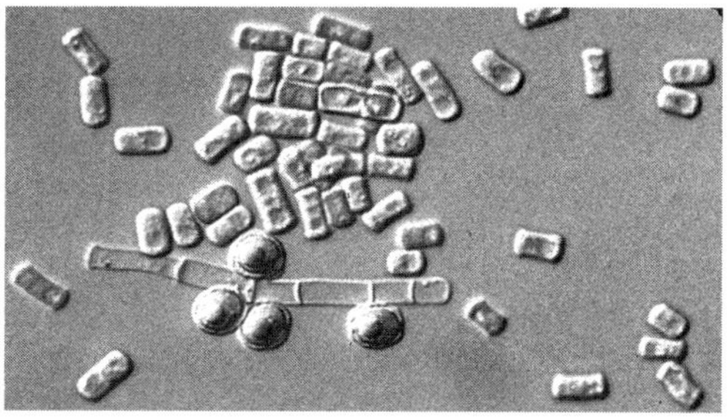

Arthrospores and ascospores of *Galactomyces* (=*Endomyces*) *geotrichum*, x 550 (*RTH*).

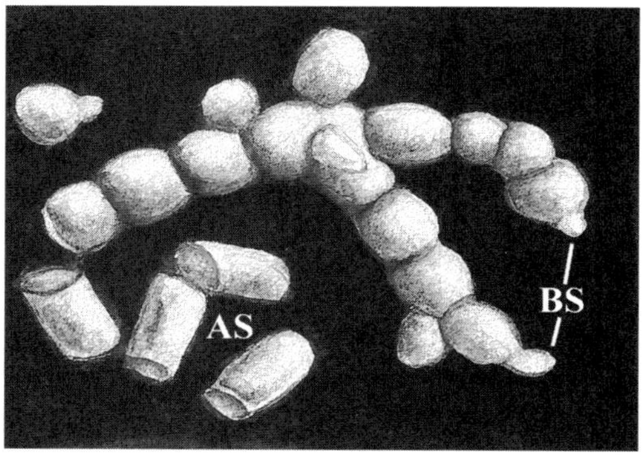

Arthrospores (AS) and blastospores (BS) of *Trichosporon beigelii*, x 1 450.

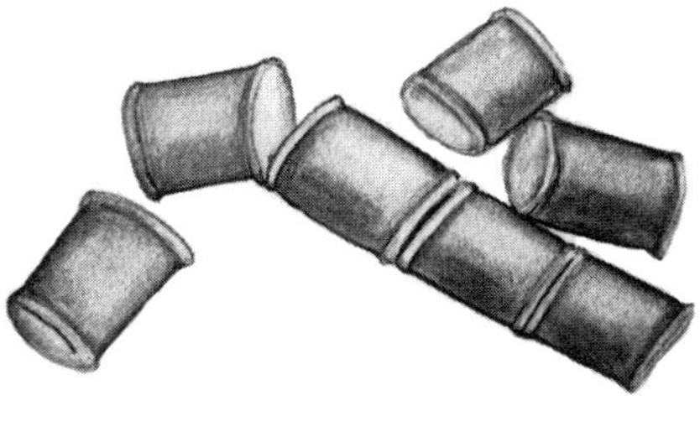

Arthrospores of *Geotrichum candidum*, x 1 170.

arthrosterigma, pl. **arthrosterigmata** (NL. *arthrosterigma* < Gr. *árthron*, joint, articulation + *stérigma*, support): the septate or multicellular filaments found in the interior of the pycnidia of some lichens which produce spermatia or pycnoconidia.

articulate (L. *articulatus*, divided into joints or nodes and internodes < *articulus*, dim. of *artus*, joint, node + suf. *-atus* > E. *-ate*, provided with or likeness): with articulations; having segments, like the hyphae that have nodes and internodes.

artist's conk (E. *artist*, one who professes and practices an imaginative art < L. *art-*, *ars-*, skill acquired by experience, study, or observations; E. *conk*, probably alteration of *conch* < L. *concha* < Gr. *kónche*, conch shell): the basidioma of *Ganoderma applanatum* (Aphyllophorales), has a smooth hymenial surface with small pores that is white when fresh, but which instantly turns brown when injured. This permanent browning reaction is utilized by artists, who use sharp-pointed instruments to make drawings on the surface. Cf. **conk**.

ascendent, **ascendant** (ME. *ascendent* < ML. *ascendent-*, that which goes up < *ascendens*, climbing up < *ascendere*, to climb up + *-entem* > E. *-ent* or *-ant*, one that performs, being): directed or curved upward. E.g., gills that are arranged vertically on the pileus, rather than horizontally, and which remain separated from the stipe, as is observed in *Agaricus* and other Agaricales. Conidiophores, pedicels, etc., are said to be ascendent when they first take a horizontal direction or nearly so, and then curve upward to an approximately vertical position; e.g., the sorocarp of *Polysphondylium* (Dictyosteliomycota) is ascendent, as during its development it raises up from its horizontal base (pseudoplasmodium).

asciferous (L. *ascifer* < *asci*, pl. of *ascus*, sack + *-ferous*, bearer < *ferre*, to bear, carry + L. *-osus* > OF. *-ous*, *-eus* > E. *-ous*, having, possessing the qualities of): *Ascomycetes*. Bearer of asci; applied to the form of the fungus that produces asci (in contrast to the conidiophorous form, which produces conidia), or to the hyphal layer of the fruiting body that gives rise to the asci. The asciferous layer supports the asci.

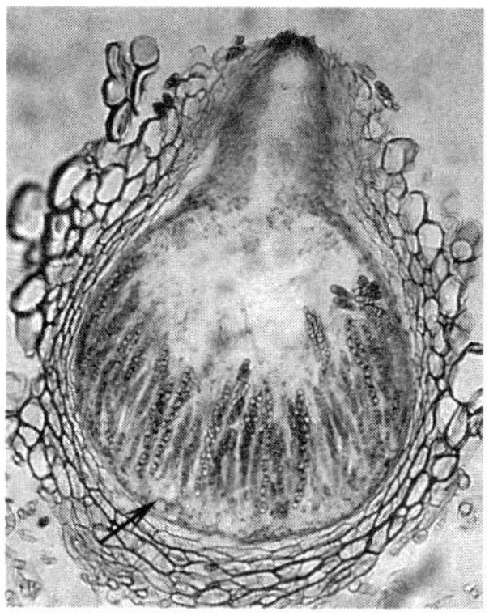

Asciferous hymenium of the perithecium of *Nectria haematococca*, seen in longitudinal section, x 140 (RTH).

ascigerous (L. *asci*, pl. of *ascus*, sack + E. suf. *-gerous* < L. *-ger*, to bear, carry + L. *-osus* > OF. *-ous*, *-eus* > E. *-ous*, having, possessing the qualities of): *Ascomycetes*. The structure or phase that produces the asci; the ascigerous state is the portion of the life cycle in which the asci are formed.

ascocarp (Gr. *askós*, sack + *karpós*, fruit): the sporiferous apparatus with asci and ascospores of the ascomycetes, whatever their form, e.g., the apothecia of *Hyaloscypha hyalina* (Helotiales). Also called **ascoma**.

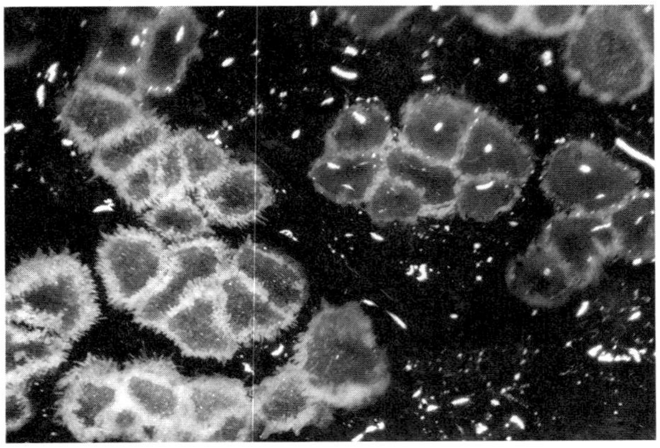

Ascocarps of *Hyaloscypha hyalina*, x 12 (*CB*).

ascoconidium, pl. **ascoconidia** (Gr. *askós*, sac + NL. *conidium* < Gr. *konídion* < *kónis*, dust + dim. suf. *-ídion* > L. *-idium*): a blastic, round or ovoid conidium produced by budding from an ascospore, as seen, e.g., in *Taphrina deformans* (Taphrinales), whose ascospores may bud before they are released from the asci and continue after the ascospores have been released to the substrate. In other ascomycetes, such as *Claviceps purpurea* (Hypocreales), ascospore germination occurs not by germ tube, but by production of conidia directly from conidiophores without an intervening mycelial stage. This process has been termed iterative germination or **microcyclic conidiation**.

ascogenous (Gr. *askós*, sack + *génos*, origin, birth + L. *-osus* > OF. *-ous*, *-eus* > E. *-ous*, having, possessing the qualities of): *Ascomycetes*. Pertaining to the production of asci. The specialized dikaryotic (ascogenous) cells or hyphae that form the asci; e.g., as in *Protomyces* (Taphrinales). Ascogenous hyphae generally terminate in the shape of a hook, or **crozier**; e.g., as in *Neurospora*, *Thielavia* (Sordariales) and *Nectria* (Hypocreales). Ascogenous hyphae are homologous to the basidiogenous hyphae of the basidiomycetes.

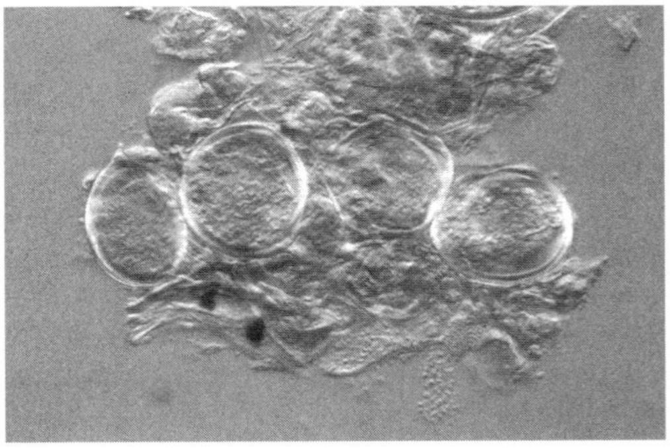

Ascogenous cells of *Protomyces* sp., x 1 000 (*RTH*).

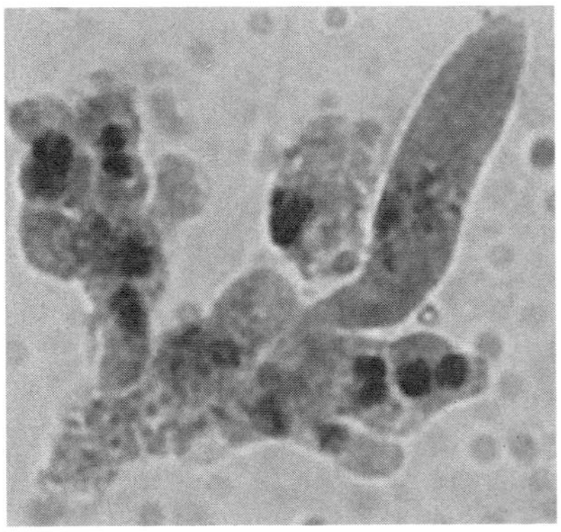

Ascogenous hyphae of *Nectria haematococca*, x 800 (*RTH*).

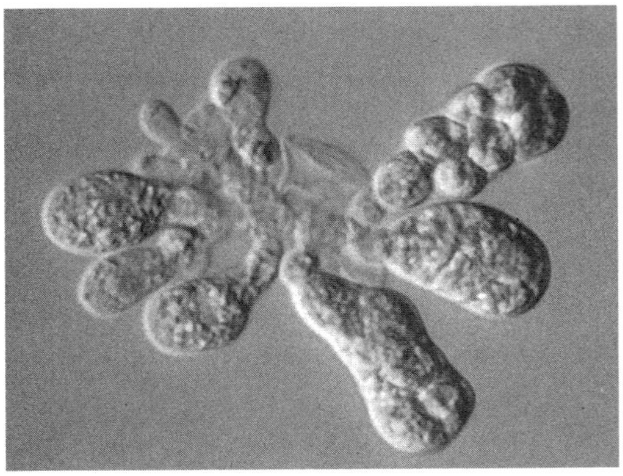

Ascogenous hyphae of *Thielavia terricola*, x 900 (*RTH*).

ascogonium, pl. **ascogonia**, **ascogone** (Gr. *askós*, sack + *gónos*, *góne*, the engendered, progeny, offspring + L. dim. suf. *-ium*): *Ascomycetes*. **1.** The cell or cells in an ascocarp from which the asci ultimately will be derived. **2.** The female gametangium, in those ascomycetes that form female and male (antheridium) gametangia. Ascogonia may be of various shapes, depending upon the species, but they are frequently coiled, e.g., in *Eurotium* (Eurotiales), or clavate, e.g., in *Pyronema* (Pezizales).

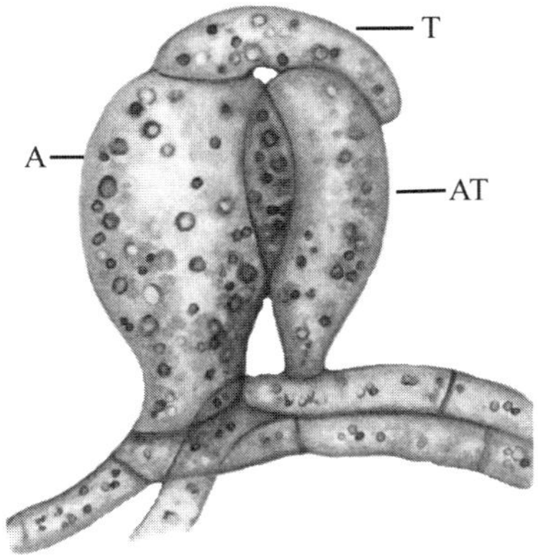

Ascogonium (A), antheridium (AT), and trichogyne (T) of *Pyronema omphalodes*, x 800.

ascohymenial (Gr. *askós*, sack + L. *hymenium* < Gr. *hyménion*, dim. of *hymén*, membrane + L. suf. *-alis* > E. *-al*, relating to or belonging to): *Ascocarp ontogeny*. A type of ascocarp development in which ascocarp formation is initiated by special, usually coiled, hyphal branches (the ascogonial initial), which are usually enveloped by investing hyphae that form the ascocarp wall. An antheridium may or may not be present. Characteristic of the Plectomycetes, Pyrenomycetes, and Discomycetes. Cf. **ascolocular**.

ascolocular, **ascoloculate** (Gr. *askós*, sack + L. *loculus*, dim. of *locus*, place, cavity, locule + L. suf. *-aris* > E. *-ar*, like, pertaining to; or + L. suf. *-atus* > E. *-ate*, provided with or likeness): *Ascocarp ontogeny*. A type of ascocarp development in which ascocarp formation is initiated by formation of a small stroma, in which ascogenous cells are later differentiated. The asci develop in cavities or locules, whose wall consists of stromal tissue. Characteristic of the Loculoascomycetes. Cf. **ascohymenial**.

ascoma, pl. **ascomata** (Gr. *askóma*, leather cushion < *askós*, sack + suf. *-oma*, which implies entirety): *Ascomycetes*. Any sporocarp or fruiting body with asci. Syn. of **ascocarp**.

ascomycetes (L. *ascomycetes* < Gr. *askós*, sack + L. *-mycetes*, ending of class < Gr. *mýkes*, genit. *mýketos*, fungus): a group of fungi characterized by the formation of asci. Once considered as a single taxonomic class of higher fungi, in the classification system followed in this dictionary, it is equivalent to the phylum Ascomycota (of the kingdom Fungi), which includes the classes Archiascomycetes, Plectomycetes, Pyrenomycetes, Discomycetes, Loculoascomycetes, as well as the ascosporogenous yeasts (Saccharomycetales), and other filamentous ascomycetes, such as the Erysiphales, Laboulbeniales and Spathulosporales.

ascomycote (L. *ascomycete* < Gr. *askós*, sac + *mýkes*, genit. *mýketos*, fungus): one of the Ascomycota, the largest group of fungi, considered as a phylum of kingdom Fungi in several recent books, e.g. *Five Kingdoms*, 2nd. ed., by Margulis and Schwartz, 1988, and *Introductory Mycology*, 4th. ed., by Alexopoulos *et al.*, 1996; the latter is the one followed in this dictionary.

ascophore (Gr. *askós*, sac + *-phóros*, bearer): *Ascomycetes*. A structure that produces or supports asci; in particular, the specialized, erect hypha in *Cephaloascus* and *Ascoidea* (Saccharomycetales) that forms asci at its apex.

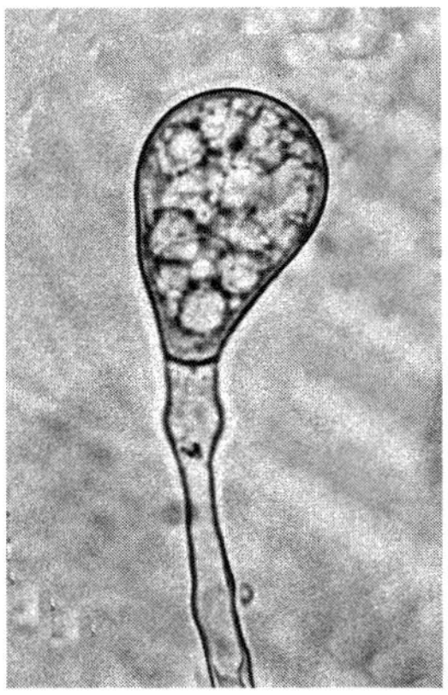

Ascophore of *Ascoidea* sp.; it corresponds to the hypha that supports a terminal, multisporate young ascus, x 570 (*RTH*).

ascospore (Gr. *askós*, sac + *sporá*, spore): *Ascomycetes*. A sexual haploid spore that is formed in the interior of an ascus, usually as the result of meiosis. Characteristic of the ascomycetous fungi; see the figures that represent some examples of asci with ascospores.

ascosporic (Gr. *askós*, sack, ascus + *sporá*, spore + suf. *-íkos* > L. *-icus* > E. *-ic*, belonging to, relating to): syn. of **ascosporogenous**. Cf. **anascosporic**.

ascosporogenous (Gr. *askós*, sack, ascus + *sporá*, spore + *génos*, origin < *gennáo*, to engender, produce + L. *-osus* > OF. *-ous*, *-eus* > E. *-ous*, having, possessing the qualities of): capable of producing ascospores. Cf. **anascosporogenous**.

ascostroma, pl. **ascostromata** (Gr. *askós*, sac + *strõma*, cushion): *Ascomycetes*. **1.** A stromatic ascocarp resulting from ascolocular ontogeny, with the asci produced in locules or cavities, the walls of which consist only of stromal tissue. No separable wall is formed around them. If a single cavity is present it is a **unilocular** (uniloculate) ascostroma, and if several locules are formed it is a **multilocular** (multiloculate) ascostroma. The ascostroma is characteristic of fungi in the class Loculoascomycetes, such as *Preussia* (=*Sporormiella*) (Pleosporales) and *Myriangium* (Myriangiales). **2.** Any stroma containing asci. Perithecial stroma is preferable for species with ascohymenial ontogeny.

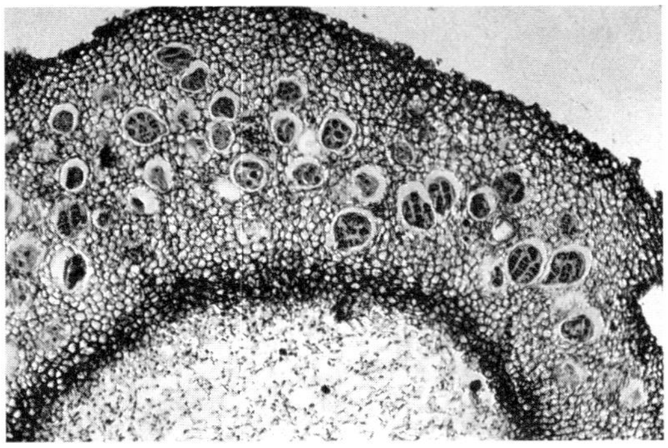

Ascostroma (multilocular) of *Myriangium durieae*, seen in longitudinal section, x 100 (*RTH*).

ascus, pl. **asci** (Gr. *askós*, sac): *Ascomycetes*. A specialized cell in which the ascospores are formed internally by a process called free-cell formation, usually following karyogamy and meiosis. The ascus is the definitive character of the ascomycetes; it generally contains a definite number of ascospores, typically eight. Asci are characteristic of such fungi as *Saccharomyces*, *Dipodascopsis* (Saccharomycetales), *Schizosaccharomyces* (Schizosaccharomycetales), *Emericella* (Eurotiales), *Sordaria* (Sordariales) and *Morchella* (Pezizales), among others. They are homologous to the basidia of the basidiomycetes, but they differ in that the latter form their spores exogenously, whereas the former form them endogenously.

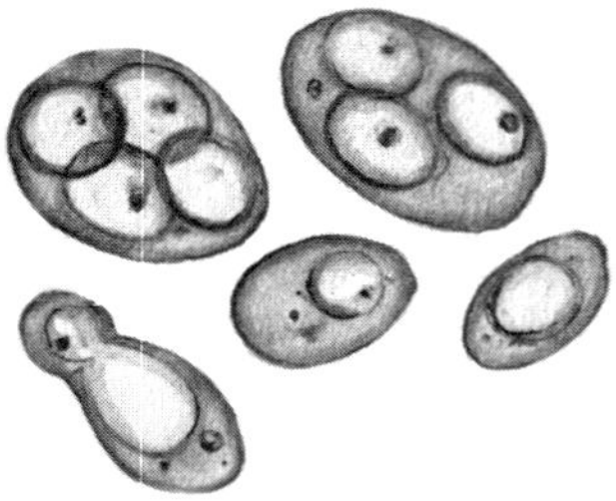

Asci (tri- and tetrasporate) of *Saccharomyces cerevisiae*, x 2 000.

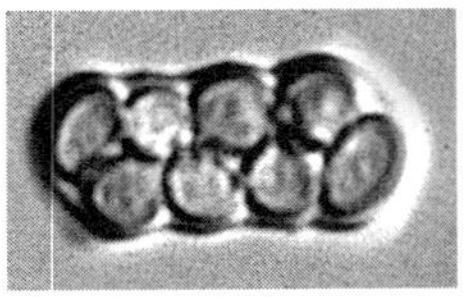

Ascus (octosporate) of *Schizosaccharomyces octosporus*, x 1 250 (*MU*).

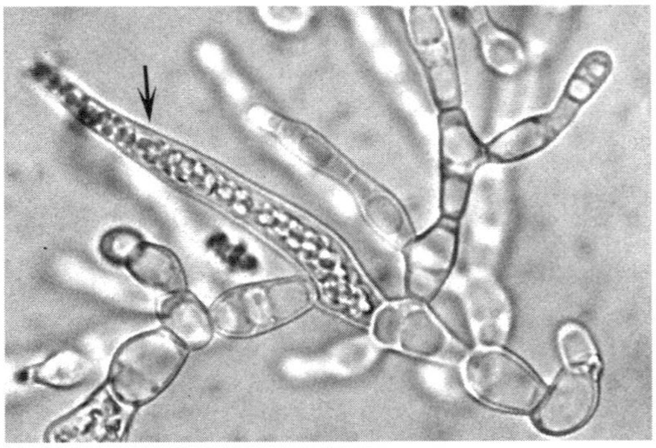

Ascus (multisporate) of *Dipodascopsis uninucleatus*, x 460 (*MU*).

Ascus (octosporate) of *Emericella nidulans*, x 3 500.

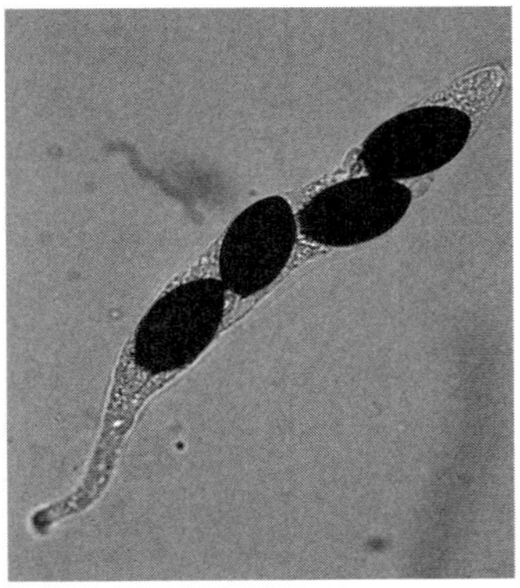

Ascus (tetrasporate) of *Podospora comata*, x 230 (*EAA*).

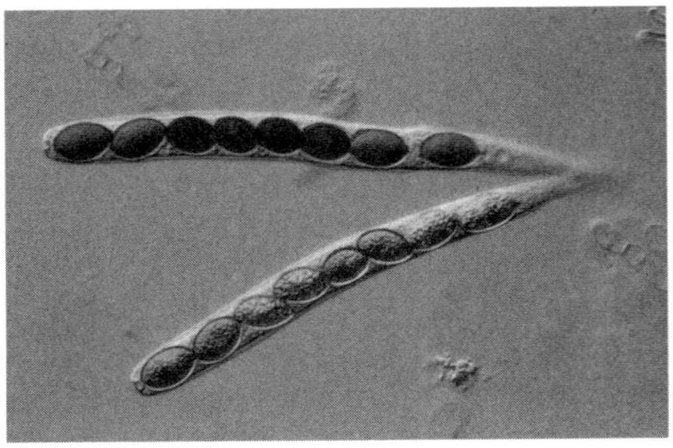

Ascus (octosporate) of *Sordaria fimicola*, x 340 (*RTH*).

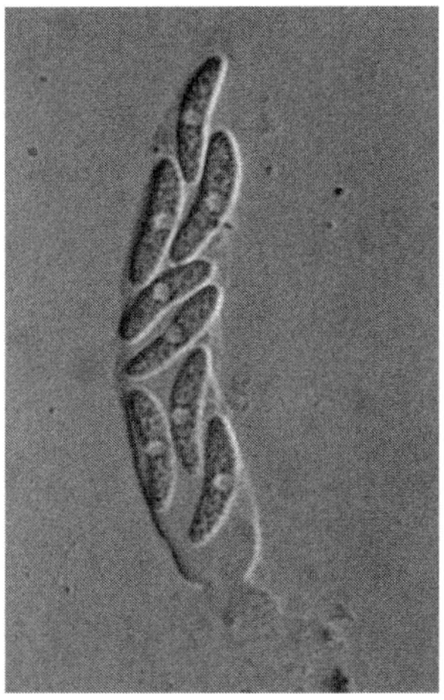

Ascus (octosporate) of *Glomerella* sp., x 1 120 (*RTH*).

ascus mother cell (Gr. *askós*, sac; ME. *mother*, *moder* < OE. *modor* < L. *mater*, genit. *matris* < Gr. *máter*, mother; L. *cella*, storeroom > ML. *cella* > OE. *cell*, cell, compartment): generally applied to the cells of yeasts that by budding give rise to one or several buds that constitute daughter cells. In the ascomycetes the diploid cell of the hook is called the ascus mother cell, which by meiosis forms the haploid nuclei of the future ascospores, as one sees, e.g., in *Nectria haematococca* (Hypocreales).

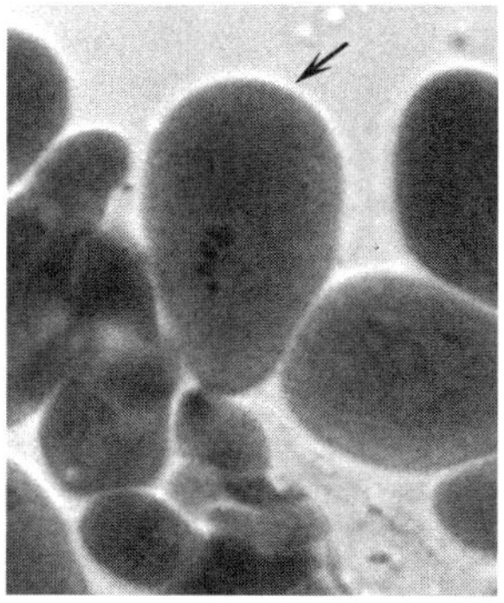

Ascus mother cells of *Thielavia terricola*, x 1 700 (*RTH*).

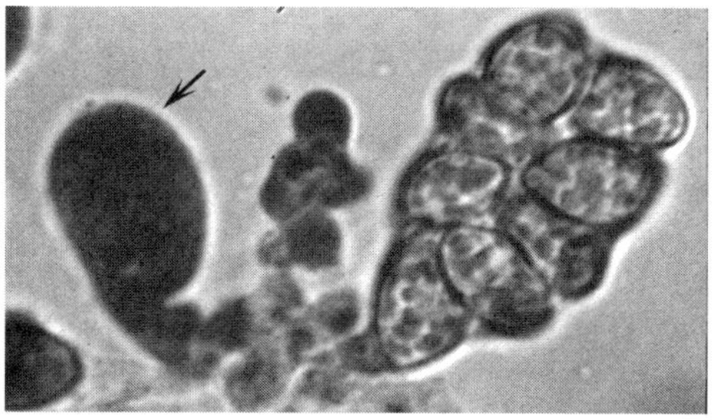

Ascus mother cells of *Thielavia terricola*, x 1 700 (*RTH*).

aseptate (L. *aseptatus* < Gr. *a-*, without + L. *septatus*, septate, divided + suf. *-atus* > E. *-ate*, provided with or likeness): without partitions (cross-walls, or septa). See **coenocytic**.

aseptic, **asepticous** (Gr. *a*, without + *septikós*, causing putrefaction < *sépsis*, decay + *-tikós* > L. *-ticus* > E. *-tic*, relation, fitness, inclination or ability; or + L. *-osus* > OF. *-ous*, *-eus* > E. *-ous*, possessing the qualities of): not septic; without contaminating. Cf. **septic**.

asexual (Gr. *a-*, without + L. *sexus*, sex + suf. *-alis* > E. *-al*, relating to or belonging to): a type of reproduction that does not involve karyogamy and meiosis; vegetative multiplication. Apomixis and amixis (if they lack amphimixis), are also asexual. Cf. **sexual**.

asexual fungus, pl. **fungi** (Gr. *a-*, without + L. *sexus*, sex + suf. *-alis* > E. *-al*, relating to or belonging to; L. *fungus*, fungus < Gr. *spóngos*, *sphóngos*, sponge): a fungus that lacks a means of sexual reproduction. Asexual fungi are also called **anaholomorphs**, **conidial fungi**, **deuteromycetes**, **Fungi Imperfecti** and **mitosporic fungi**.

asperate, **asperous**, **asperulous**, **asperulate** (L. *asperatus*, rough < *asperum*, dim. of *asper*, uneven or rough + suf. *-atus* > E. *-ate*, provided with or likeness; or *asper* + L. *-osus* > OF. *-ous*, *-eus* > E. *-ous*, full of, possessing the qualities of; or *asperulus*, dim. of *asper*, + *-ous*; or *asperulus* + L. suf. *-atus* > E. *-ate*, provided with or likeness): with a lightly roughened surface due to small spines or warts; generally applied to spores, conidiophore walls, etc., as seen under the microscope. E.g., the spores of *Hemitrichia stipitata* (Trichiales), the basidiospores of *Russula* spp. and *Lepiota asperula* (Agaricales), and of *Clavaria* (Aphyllophorales), and the spore sac of *Lycoperdon pulcherrimum* (Lycoperdales) are asperulate.

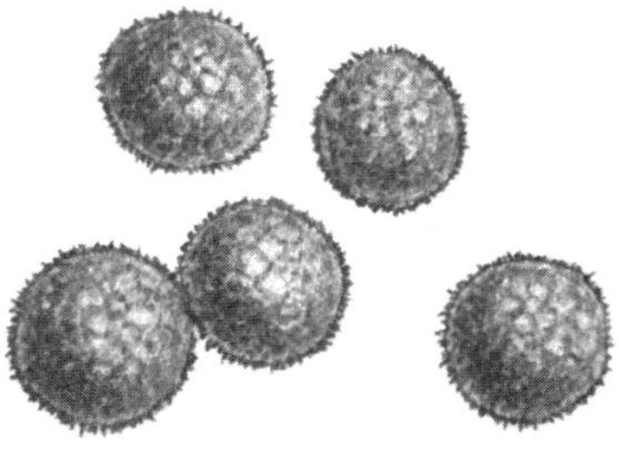

Asperate spores of *Hemitrichia stipitata*, x 1 800.

aspergillosis, pl. **aspergilloses** (< gen. *Aspergillus* < L. *aspergillo*, hyssop, sprinkler + Gr. *-osis*, state or condition): a wide range of human and animal diseases caused by several cosmopolitan spp. of *Aspergillus* (moniliaceous asexual fungi), particularly *A. fumigatus*,

A. flavus, and *A. niger*, which, as opportunistic invaders, provoke secondary infections in immunosuppressed patients (e.g., weakened by cancer and its chemotherapy). Symptoms of pulmonary aspergillosis can be acute (bronchopneumonia) or chronic (tuberculoid); other types are the so-called **fungus ball** (localized in preexisting lung cavities), the allergic aspergillosis, onychomycosis, otomycosis, and the systemic ones, affecting various organs of the body (heart, brain, nasal sinuses, etc.).

aspicilioid (< gen. *Aspicilia* < Gr. *aspís*, *aspídos*, a shield + L. suf. *-oide* < Gr. *oeídes*, similar to; < Gr. *aspís*, shield): *Lichens* (fam. Hymeneliaceae). Refers to lecanorine apothecia that are more or less immersed in the thallus, at least when young. Cf. **lecanorine** and **lecideine**.

asporogenous (L. *asporogenous* < Gr. *a-*, without + *sporá*, spore + *génos*, origin, birth + L. *-osus* > OF. *-ous*, *-eus* > E. *-ous*, having, possessing the qualities of): that which does not produce spores, i.e., sterile. Cf. **sporogenous**.

assimilative (ML. *assimilativus* < L. *assimilatus*, likened to, made like + *-ivus* > E. *-ive*, quality or tendency, fitness): to take in and incorporate, absorb, to bear a resemblance. See **vegetative**.

asterinoid (< gen. *Asterina* < L. *aster*, star + suf. *-ina*, which indicates possession or appearance + suf. *-oide* < Gr. *-oeídes*, similar to): a type of ectoparasitism in which the fungus develops a black, superficial mycelium in a radial arrangement on the living parts of the host plant, from which it obtains its nutrients by means of haustoria and later produces a uniloculate ascostroma, a radial structure. Asterinoid forms of mycelium are present in the Asterinaceae, Microthyriaceae, and related families of Loculoascomycetes that occur in humid climates.

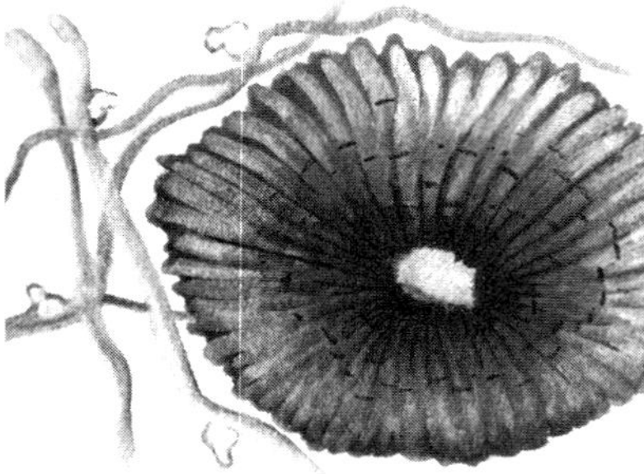

Asterinoid ascostroma of *Asterina veronicae*. Surface view of an ascostroma parasitic on a leaf of *Veronica*, x 500.

asterophysis, pl. **asterophyses** (Gr. *astér*, star + *phýsis*, a growth): a stellate, cystidioide hypha (with radiate branches), characteristic of the trama and hymenium of some gen. of Aphyllophorales, such as *Asterodon* (Hydnaceae) and *Asterostroma* (Thelephoraceae). Also called **asteroseta**.

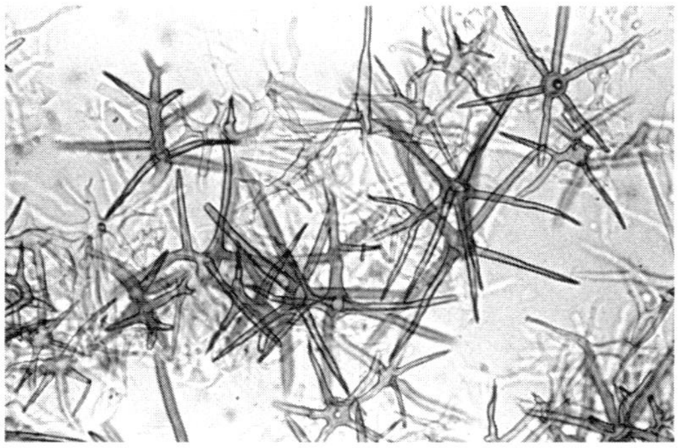

Asterophyses from the hymenophore of *Asterostroma* sp., x 130 (*MU*).

asteroseta, pl. **asterosetae** (Gr. *astér*, star + L. *seta*, bristle, stiff hair): a cellular element in the shape of a star, with a thick, pigmented wall, that is present in the cuticle of certain Aphyllophorales, as in *Asterostroma cervicolor*. Also called **asterophysis**.

asymmetric, **asymmetrical** (Gr. *asymmetría*, lack of symmetry; Gr. *a-*, without + *symmetría*, symmetry + suf. *-íkos* > L. *-icus* > E. *-ic*, belonging to, relating to; or + L. suf. *-alis* > E. *-al*, relating to or belonging to): an organ or structure that is not identical on both sides of a common axis, so that neither side can form an exact mirror image of the other. E.g., the ascospores of *Xylaria* and *Hypoxylon* (Xylariales) are asymmetric, since one side is convex and the other flat (see **inaequilateral**). Cf. **symmetric**.

Asymmetric ascospores of *Hypoxylon serpens*, x 1 100.

atomate (L. *atomatus*, sprinkled with minute particles < Gr. *átomos*, incapable of division + suf. *-atus* > E. *-ate*, provided with or likeness): refers to a surface that is covered with tiny, sharp, brilliant particles; e.g., the pileus and stipe of the basidiocarp of *Psathyrella atomata* (Agaricales) is atomate.

atopic (Gr. *atópos*, out of the way, uncommon < *a-*, not, without + *tópos*, place + suf. *-íkos* > L. *-icus* > E. *-ic*, belonging to, relating to): predisposed to some condition or material, as in certain allergic reactions; an organism that exhibits *atopy* (with *-y*, E. suf. of concrete nouns), i.e., a probably hereditary allergy characterized by symptoms (as asthma, hay fever, or hives) produced upon exposure to the exciting antigen, such as fungal spores, without inoculation. See **aeromycology**.

atramentarious (L. *atramentarius* < NL. *atramentum*, dark-colored ink < *ater*, black + L. *-arius*, belonging to): with the appearance of ink, like the deliquescent, black hymenophore of *Coprinus atramentarius* (Agaricales) and other related fungi.

attached (ME. *attachen* < MF. *attacher* < OF. estache, stake, to make fast): *Agarics*. Refers to lamellae that are fixed or united to the stipe of the fruiting body. Also called **adnate**. Cf. **decurrent**, **free** and **sinuate**.

attenuate (L. *attenuatus*, tapered < *attenuare*, to make thin < pref. *ad-*, toward + *tenuis*, thin + suf. *-atus* > E. *-ate*, provided with or likeness): **1.** Tapering gradually to a narrow end; e.g., as in certain phialides, perithecial necks and many other structures. **2.** Applied to pathogens that have lost their virulence, or that exhibit reduced or diminished pathogenecity.

aulaeate development (Gr. *aulaós*, fissure, furrow, to extend by narrowing + suf. *-atus* > E. *-ate*, provided with or likeness; MF. *développer* < OF. *desveloper* < *dés-* < L. *de-*, from, down, away + MF. *enveloper*, to enclose + L. *mentum* > OF. *-ment* > ME. *-ment*, action, process): *Gasteromycetes*. A type of fruiting body development characterized by a hymenium formed on the surface of furrows comprised of primordial tissue that grows downward and toward the center from the peridium, as in the Hymenogastrales and Lycoperdales. Cf. **forate** and **lacunar development**. See **aulaeothecium**.

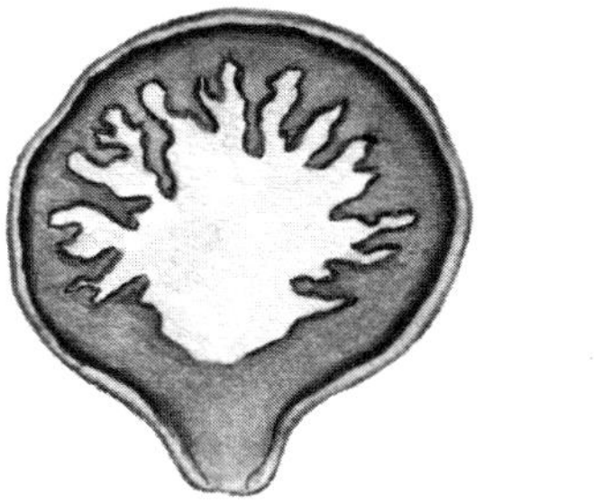

Aulaeate type of development of the hymenium in the fruiting body of Hymenogastrales (*Gastrosporium*) and Lycoperdales. Diagram of a longitudinal section, x 1.

aulaeothecium, pl. **aulaeothecia** (NL. *aulaeothecium* < Gr. *aulaós*, tube, furrow + *thekíon*, dim. of *thêke*, case, box; here, of the basidia + dim. suf. *-íon* > L. *-ium*): the downgrowing primordial tissue formed during **aulaeate development**.

auriculiform (L. *auricula*, dim. of *auris*, ear + suf. *-formis* < *forma*, shape): having the shape of an ear; e.g., as in the ascocarp of *Otidea onotica* (Pezizales) and the basidiocarp of *Auricularia* (Auriculariales).

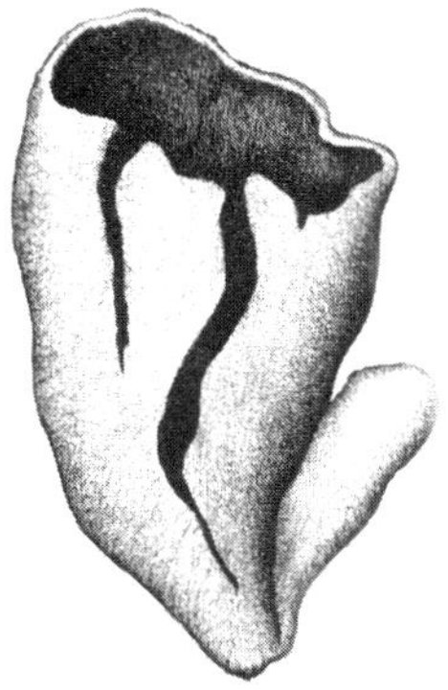

Auriculiform ascocarp of *Otidea concinna*, x 1.5.

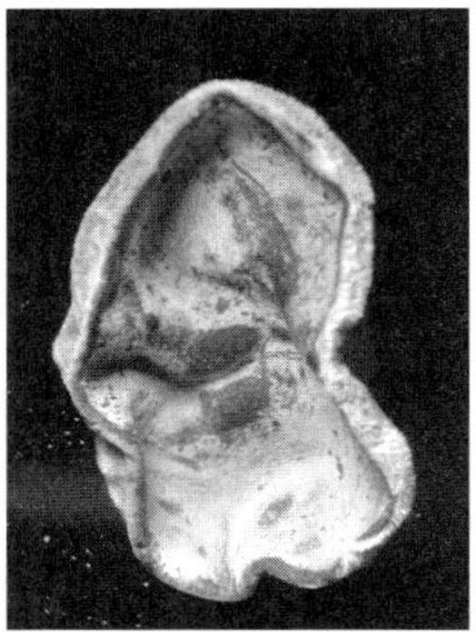

Auriculiform basidiocarp of *Auricularia polytricha*, x 1 (*MU*).

autapomorphy (Gr. *autós*, same + *apó*, away from + *morphé*, shape + *-y*, E. suf. of concrete nouns)**:** a term used in cladistics to refer to a unique, derived character state restricted to a single taxon.

auteuform (Gr. *autós*, same + *e* , true + L. *forma*, form, shape): *Uredinales*. Applied to the euform of the rust fungi when it is autoecious, i.e., the life cycle includes four of the five classes of spores (pycniospores, aeciospores, urediniospores, and teliospores) produced on the same host plant.

autochthonous (Gr. *autóchthon* < *autós*, same + *thón*, genit. *thonós*, land, country + L. *-osus* > OF. *-ous*, *-eus* > E. *-ous*, possessing the qualities of): indigenous; from the country, not introduced or naturalized; e.g., the "horse's tooth" (*Claviceps gigantea*, Hypocreales) of maize is native to Mexico. Cf. **allochthonous**.

autoecious (Gr. *autós*, same + *oîkos*, house, dwelling + L. *-osus* > OF. *-ous*, *-eus* > E. *-ous*, possessing the qualities of): *Uredinales*. Rust fungi that develop their entire life cycle on hosts belonging to the same gen., or the same species; e.g., *Phragmidium mucronatum* (rose rust). The capacity of a fungal parasite to complete its life cycle on a host of a single species is called **autoecism**. Cf. **heteroecious**.

autoecism (Gr. *autós*, same + *oîkos*, house, dwelling + -*ismós* > L. -*ismus* > E. -*ism*, state, phase, tendency, action): *Uredinales*. The capability to complete the entire life cycle on a single species of host plant. Species exhibiting autoecism are said to be **autoecious**, as in some rust fungi. Cf. **heteroecism**.

autogamous (Gr. *autós*, self + *gámos*, sexual union + L. -*osus* > OF. -*ous*, -*eus* > E. -*ous*, possessing the qualities of): having **autogamy**.

autogamy (Gr. *autós*, same + *gámos*, sexual union + -*y*, E. suf. of concrete nouns): sexual union of pairs of closely related nuclei in a cell of the female organ, without a fusion of cells having occurred. See **automixis** and **parthenogamy**.

autolysis, pl. **autolyses** (Gr. *autós*, self, the same + *lýsis*, dissolution): autodigestion, like that which occurs in the gills on the pileus of the gen. *Coprinus* (Agaricales) and which causes the removal of those parts of the gills that have already liberated their spores, thus permitting the pileus to continue opening, creating the spaces necessary for the spores to be discharged from the basidia. This autolysis or autodeliquescense gives rise to a black fluid mass that contributes to the dispersal of the spores. Also called **deliquescence**.

automixis, pl. **automixes** (Gr. *autós*, same, self + *míxis*, mixture, union): reproduction in which there is self-fertilization as a consequence of the fusion of two sexual cells of close parentage, or simply of two sexual nuclei. The first case is called **parthenogamy**, the second **autogamy**.

autotroph (Gr. *autós*, same, self + *trophós*, that which nourishes, which serves as food): an organism which, endowed with chlorophyll or another analogous pigment, is capable of synthesizing carbohydrates from carbon anhydride, so that it does not need to acquire them from exogenous sources; self nourishing. Cf. **heterotroph**.

autotropism (Gr. *autós*, same, self + *trópos*, a turn, change in manner < *tropé*, a turning < *trépo*, to revolve, turn towards + -*ismós* > L. -*ismus* > E. -*ism*, state or condition): an avoidance response between neighbouring hyphae at the colony margin of fungi.

auxanogram (Gr. *auxáno*, to grow, develop + *grámma*, describe): literally, a diagram of growth; it is applied to the culture of yeasts, bacteria and other microorganisms, principally on solid agar media, when there is deposited a determined quantity of certain substances with the object of seeing its effect on the growth of the microorganism seeded. If the latter is stimulated around the substance added, the auxanogram is positive; if, on the contrary, development is inhibited or not stimulated, the auxanogram is negative. Auxanograms are utilized to determine the types of carbon or nitrogen sources various species of yeasts can assimilate. Also called the auxanographic method. Cf. **zymogram**.

auxotroph (Gr. *aúxo*, to grow, develop + *trophós*, that which nourishes, serves as food): refers to an organism which, in order to grow, requires an external source of certain amino acids or vitamins, i.e., it is not capable of synthesizing said compounds, as occurs in specific isolates which by mutation differ from the wild isolate; e.g., there are strains of *Schizophyllum commune* (Aphyllophorales) that are auxotrophic for tryptophane. See **prototrophic**.

aversion (L. *aversus*, genit. *aversionis*, withdrawal, turned away, turned back + -*io*, -*ionis* > E. suf. -*ion*, result of an action, state of): inhibition of growth in the marginal zones of adjacent colonies of fungi or bacteria, due to the secretion of staling products, e.g., as occurs in *Aspergillus flavo-furcatis* (moniliaceous asexual fungi).

Aversion between colonies of *Aspergillus flavo-furcatis* on agar, x 0.8 (*MU*).

axenic (Gr. *a*-, without + *xénos*, stranger + suf. -*íkos* > L. -*icus* > E. -*ic*, belonging to, relating to): **1.** A culture that is free of contamination, i.e., which contains only one species and is free of any other living organism. **2.** An organism that is not capable of serving as a host to a parasite, i.e., that exhibits inhospitality or axenia, considered as passive resistance of the host toward the invading organism. Cf. **agnotobiotic**, **gnotobiotic** and **monoxenic**.

axoneme, **axonema**, pl. **axonemata** (Gr. *áxon*, *áxonos*, axis + *nêma*, thread): the central shaft, composed of microtubules, of a flagellum. In *Geotrichum candidumon*, vessel, closed receptacle): a structure equivalent to a zygosporangium but containing an **azygospore**, which develops parthenogenetically.

azonate (L. *a-*, without + *zona* < Gr. *zóne*, band, stripe + suf. *-atus* > E. *-ate*, provided with or likeness): without distinguishable zones. Cf. **zonate**.

azotodesmic (Gr. *azóton*, nitrogen + *desmós*, fixation, union + suf. *-íkos* > L. *-icus* > E. *-ic*, belonging to, relating to): an organism with the ability to fix elemental atmospheric nitrogen; some cyanophilic lichens (whose phycobionts are Cyanophyceae) are azotodesmic, as occurs, e.g., in *Collema* (Lecanorales).

azygosporangium (Gr. *a-*, without + *zygón*, yoke, pair + *sporá*, spore + *angeîon*, vessel, closed receptacle): a structure equivalent to a zygosporangium but containing an **azygospore**, which develops parthenogenetically, e.g., *Mucor bainieri* (Mucorales).

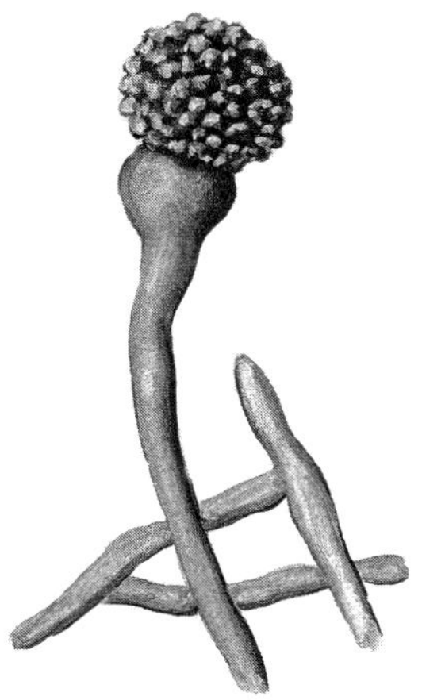

Azygosporangium of *Mucor bainieri*, x 200.

azygospore (Gr. *a-*, without + *zygón*, yoke, pair + *sporá*, spore): *Zygomycetes*. A spore morphologically equivalent to a zygospore but developed parthenogenetically, as occurs in many species of Mucorales and some species of Entomophthorales, including *Entomophthora muscae*. The structure that contains the azygospore is called an **azygosporangium**.

b

baccate (L. *baccatus* < *bacca*, berry + suf. *-atus* > E. *-ate*, provided with or likeness): similar to a berry, whether in shape or consistency; soft and juicy. Also refers to something adorned with berry-like excrescences, like the basidiocarp of *Amanita baccata* (Agaricales).

bacillar (L. *bacillus*, genit. *bacillaris*, dim. of *baculus*, walking stick, staff, support + suf. *-aris* > E. *-ar*, like, pertaining to): shaped like a little stick; rod-shaped. Applied to cylindrical structures whose length:width ratio is about 3:1. E.g., the conidia of *Myrothecium verrucaria* (tuberculariaceous asexual fungi) and *Polyscytalum foecundissimum* (moniliaceous asexual fungi) are bacillar. Also called **bacilliform**.

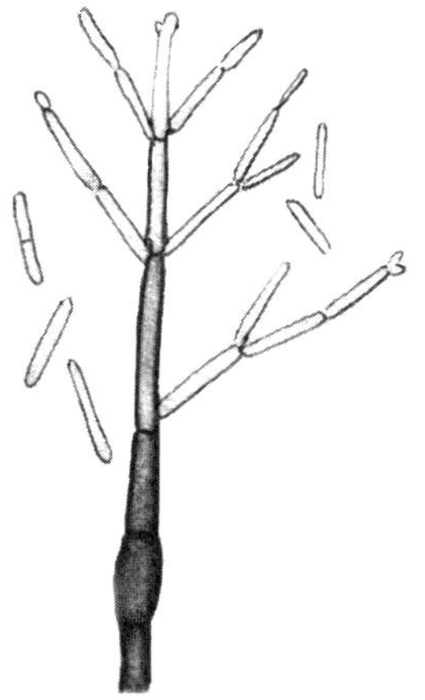

Bacillar conidia of *Polyscytalum foecundissimum*, x 500.

bacilliform (L. *bacillus*, dim. of *baculus*, walking stick, staff, support + *-formis* < *forma*, shape): rod-shaped. See **bacillar**.

bactericide (L. *bacterium* < Gr. *baktérion*, dim. of *báktron*, stick + MF. *-cide* < L. *-cida* < *caedere*, to cut, kill): capable of causing the death of bacteria. Some antibiotics of fungal origin are bactericides.

bacteriform, bactriform (L. *bacterium* < Gr. *baktérion*, dim. of *báktron*, stick + L. *-formis* < *forma*, shape): shaped like a bacterium. See **bacillar**.

bacterivorous (L. *bacterium* < Gr. *baktérion*, dim. of *báktron*, stick + L. *voro* < *vorare*, to eat, devour + *-osus* > OF. *-ous*, *-eus* > E. *-ous*, possessing the qualities of): an organism that feeds on bacteria, like the myxamoebae and plasmodia of the Myxomycota.

bacterized medium, pl. **media** (L. *bacterium* < Gr. *baktérion*, dim. of *báktron*, stick + suf. *-ized*; L. *medium*, medium, substrate on which an organism lives): a fluid or solid (agar) medium that contains living or dead bacteria as a source of food. Myxamoebae of the dictyostelids (Dictyosteliomycota) utilize a medium of this nature as source of nutrients.

baculate, **baculiform** (L. *baculatus*, stick-shaped < *baculus*, stick + suf. *-atus* > E. *-ate*, provided with or likeness; *baculus* + *-formis* < *forma*, shape): having the shape of a stick; e.g., like the ascospores of certain lichens. See **bacillar**.

ballistospore, **ballistosporic** (L. *ballista*, catapult < Gr. *ballistás* < *ballein*, to throw + *sporá*, spore; or + suf. *-íkos* > L. *-icus* > E. *-ic*, belonging to, relating to): a spore that is discharged forcibly from the basidium (sterigma) that supports it; this usually involves the explosion of droplets of liquid that form on the spore. The yeast gen. *Sporobolomyces* and *Itersonilia* (sporobolomycetaceous asexual yeasts), all of the the Hymenomycetes (Agaricales and Aphyllophorales), and the monotypic species *Schizoplasmodium cavostelioides* (a primitive member of the protostelid slime molds), all have ballistospores. Cf. **statismospore**.

barbate (L. *barbatus* < *barba*, beard + suf. *-atus* > E. *-ate*, provided with or likeness): furnished with hairs; bearded. E.g., like the thallus of the lichen *Usnea barbata* (Lecanorales).

barm (ME. *berme* < OE. *beorma*, akin to L. *fermentum*, yeast < *fervere*, to boil): the froth or foam rising on fermented malt liquor; brewer's yeast. It is used both as a leaven and to make liquors ferment.

barophilic (Gr. *barós*, weight + *phílos*, have an affinity for + suf. *-íkos* > L. *-icus* > E. *-ic*, belonging to, relating to): capable of living at great depths, like the marine fungi *Bathyascus* (Halosphaeriales) and *Oceanitis* (an ascomycete of uncertain affinities) which grow on submerged wood, enduring high hydrostatic pressures at depths of 1,720 and 3,975 m., respectively.

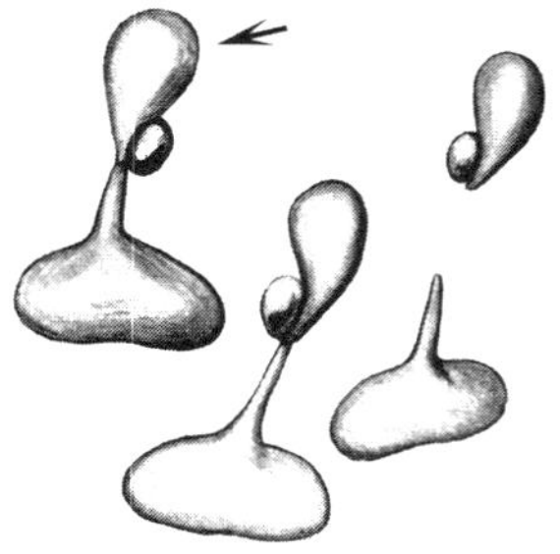

Ballistospores of *Itersonilia perplexans*, x 2 000.

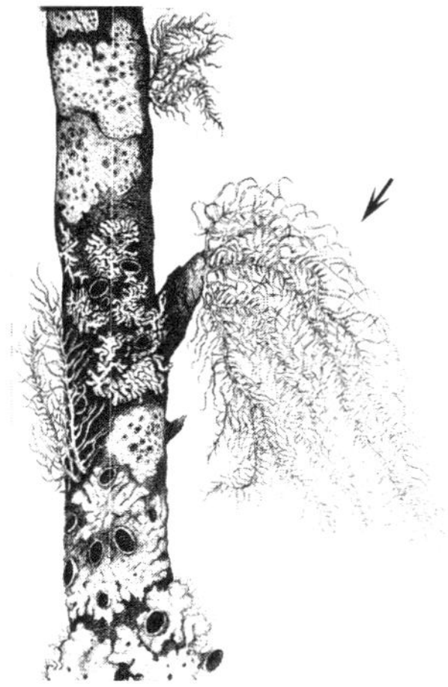

Barbate thallus of *Usnea* sp., x 0.3.

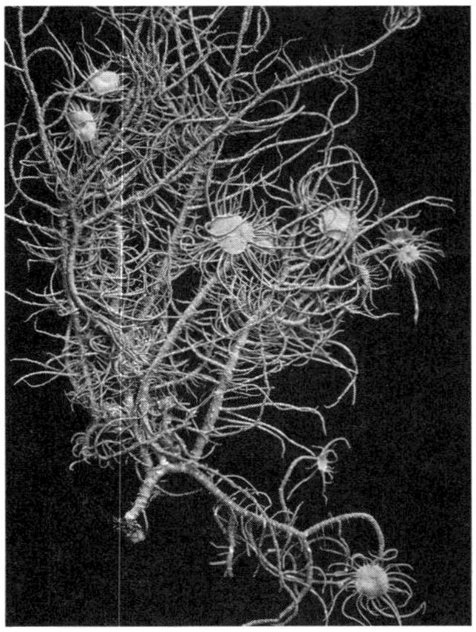

Barbate thallus of *Usnea* sp. with apothecia, x 0.5 (*MU*).

barrage (F. *barrage*, barrier): the phenomenon of mutual repulsion between certain haploid isolates of Hymenomycetes, which results in a zone between them that is barren of mycelium of the two paired isolates. This repulsion is manifested only in tetrapolar species when the two haploid isolates possess the 'a' factor in common, while one haplont possesses factor 'b' and the other 'b''.

basal (L. *basalis*, basal < *basis*, base + suf. *-alis* > E. *-al*, relating to or belonging to): near the base or relative to it; that which is found at the extreme bottom or near the point of attachment or adhesion. Cf. **apical**.

basal body (L. *basalis*, basal < *basis*, base + suf. *-alis* > E. *-al*, relating to or belonging to; ME. < OE. *bodig*, body): see **blepharoplast** and **kinetosome**.

basal cell (L. *basalis*, basal < *basis*, base + suf. *-alis* > E. *-al*, relating to or belonging to; L. *cella*, storeroom > ML. *cella* > OE. *cell*, cell, compartment): same as **foot cell**. The cell that supports the thallus of aquatic fungi of the order Blastocladiales (such as *Blastocladia* and *Blastocladiella*), which in the last gen. adopts the general form of a trunk, with branches bearing zoosporangia on the apical part, and rhizoids or attachment organs on the base.

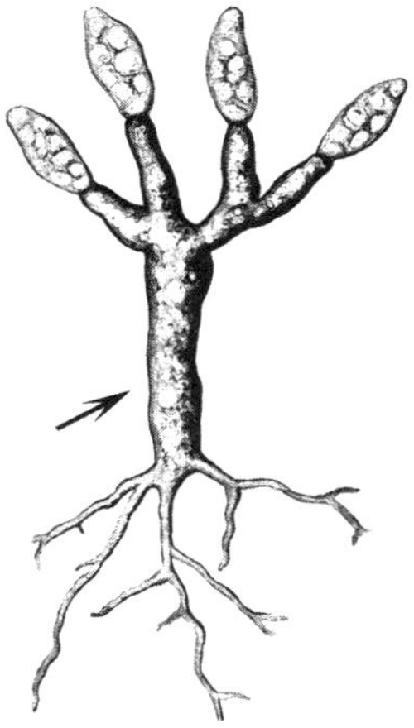

Basal cell of a thallus with zoosporangia of *Blastocladia ramosa*, x 50.

basal piece (L. *basalis*, basal < *basis*, base + suf. *-alis* > E. *-al*, relating to or belonging to; ME. *pece* < OF. < Ga. *pettia*, piece): see **sheath**.

basauxic (L. *basis*, base + Gr. *aúxo*, to grow, develop, increase + suf. *-îkos* > L. *-icus* > E. *-ic*, belonging to, relating to): *Conidiogenesis.* A type of development of a condiogenous cell in which there is elongation of a basal growth point of a fertile cell, after which a blastic conidium is formed at its apex. In the species of fungi that produce a basipetal succession of conidia, the conidiogenous cell can continue elongating from the basal point of growth after formation of the primary conidium, as happens in *Spegazzinia tessarthra* (moniliaceous asexual fungi).

basidial (L. *basidium* < Gr. *basídion*, dim. of *básis*, base + L. suf. *-alis* > E. *-al*, relating to or belonging to): pertaining or relative to basidia; possessing basidia.

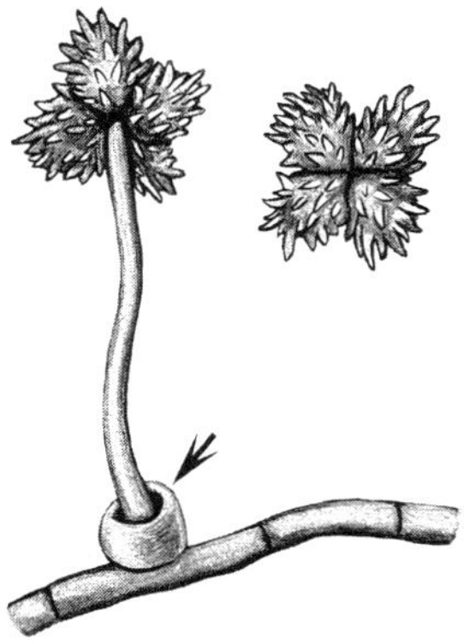

Basauxic conidiogenous cell of *Spegazzinia tessarthra*, x 1 300.

basidiocarp (Gr. *basídion*, basidium + *karpós*, fruit): *Basidiomycetes*. A sporophore or fruiting body on which basidia and basidiospores are formed. See the illustrations under the terms **adnate**, **adpressed**, **appendiculus**, **auriculiform**, **bulbose**, **calyptra**, **campanulate**, **clathrate**, **claviform**, **comate**, **conchate**, **connate**, **cortina**, **costate**, **crenate**, and **crispate**, for examples of basidiocarps showing various characteristics. One of the best known and attractive basidiocarps is that of *Amanita muscaria* (Agaricales). Also see **basidiome**.

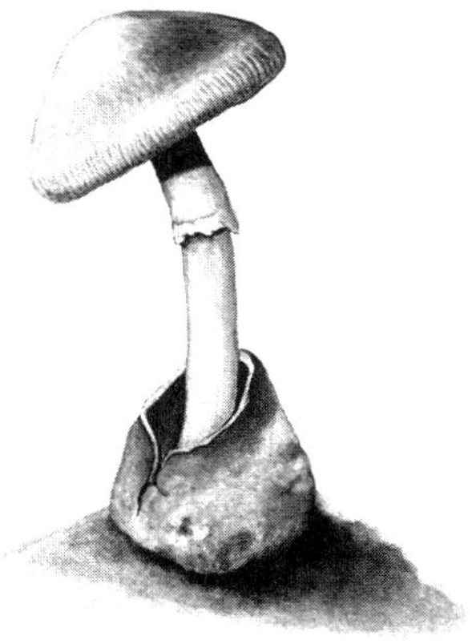

Basidiocarp of *Amanita caesarea*, x 0.5.

basidiole, **basidiolum**, pl. **basidiola** (NL. *basidiolum* < L. *basidium* < *basis*, base + dim. suf. *-idium* < Gr. *basídion* < *básis*, base + dim. suf. *-ídion* > L. *-idium*; the L. dim. suf. *-olum* > E. *-ole*): *Basidiomycetes*. A young binucleate basidium that is found in the hymenium of the Agaricales, e.g., in *Amanita*, *Boletus* and *Leptonia*. Basidiola are considered either as stages of development of basidia or as aborted basidia. Generally they are clavate or narrowly clavate and similar to basidia; the basidiola typically originate in the hymenium.

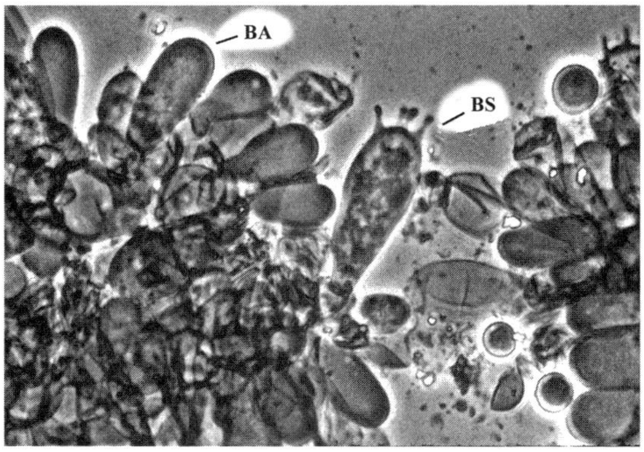

Basidiolum (BA) and **basidium** (BS) of the hymenium of *Amanita* sp., x 300 (*RMA*).

basidiome, **basidioma**, pl. **basidiomata** (Gr. *basídion*, dim. of *básis*, base + *-oma*, suf. implying the idea of entirety): *Basidiomycetes*. Any structure that produces basidia, whether it be a simple layer of basidia or a complex sporophore. E.g., in the Exobasidiales the basidia are formed superficially on the leaves of parasitized plants, without formation of a true fruiting body, whereas in the Agaricales they form on a fruiting body of varying complexity. Consequently, basidiome is more precise and is preferable to **basidiocarp**, which cannot be applied to all types of structures that bear basidia.

basidiomycetes (NL. *basidiomycetes* < L. *basidium* < Gr. *basídion*, dim. of *básis*, base + L. *-mycetes*, ending of class < Gr. *mýkes*, genit. *mýketos*, fungus): a group of fungi characterized by the formation of basidia. Once considered as a single taxonomic class of higher fungi, in the classification system adopted in this dictionary it is the equivalent of the phylum Basidiomycota, which includes the classes Hymenomycetes, Gasteromycetes, Ustilaginomycetes, and Urediniomycetes.

basidiomycote (L. *basidiomycete* < *basidium* + *mycete* < Gr. *basídion*, dim. of *básis*, base + *mýkes*, genit. *mýketos*, fungus): one of the Basidiomycota, a phylum of kingdom Fungi in the classification system adopted in some books, e.g. *Five Kingdoms*, 2nd. ed., by Margulis and Schwartz, 1988, and *Introductory Mycology*, 4th. ed., by Alexopoulos *et al.*, 1996; the latter is the one followed in this dictionary.

basidiospore (Gr. *basídion*, dim. of *básis*, base + *sporá*, spore): a type of sexual spore formed externally on a basidium, 1-2 or 4 in number, typical of the basidiomycetes. It is haploid and forms acrogenously on a sterigma and typically results from karyogamy and meiosis. When it germinates it produces a primary mycelium, which then produces secondary and tertiary mycelia, from which the fruiting bodies are formed.

basidium, pl. **basidia** (L. *basidium*, small pedestal; here, of the spores < *basis*, base + dim. suf. *-idium* < Gr. *basídion* < *básis*, base + dim. suf. *-ídion* > L. *-idium*): *Basidiomycetes*. A specialized cell on which basidiospores are formed externally on sterigmata, following karyogamy and meiosis. The basidium is the primary distinguishing characteristic of the basidiomycetes. The basidium is homologous to the ascus in regard to the mother cell and the cytological processes (karyogamy and meiosis) that occur in it, although they differ essentially in that the basidium forms exogenous spores, whereas the ascus forms them endogenously. See the figures of **abaxial** and **apobasidium**, where basidia with basidiospores are illustrated.

basipetal (L. *basis*, base + *petere*, to go forward, to grow toward + L. suf. *-alis* > E. *-al*, relating to or belonging to): *Conidiogenesis*. Developing or maturing from the apex toward the base; i.e., the degree of growth is greater with increased distance from the base. E.g., in a basipetal chain of conidia, the conidium at the base is the youngest and of most recent formation. *Aspergillus*, *Penicillium* and *Scopulariopsis* (moniliaceous asexual fungi) are examples of gen. of fungi that form conidia in this manner. *Albugo* (Peronosporales) also forms basipetal chains of sporangia. Cf. **acropetal**.

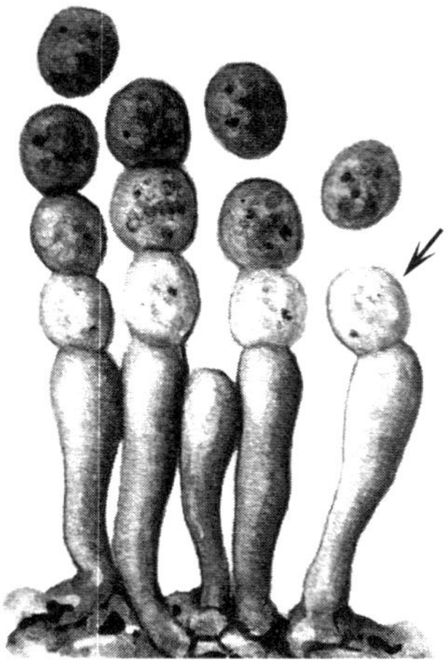

Basipetal succession of sporangia of *Albugo candida*, x 160.

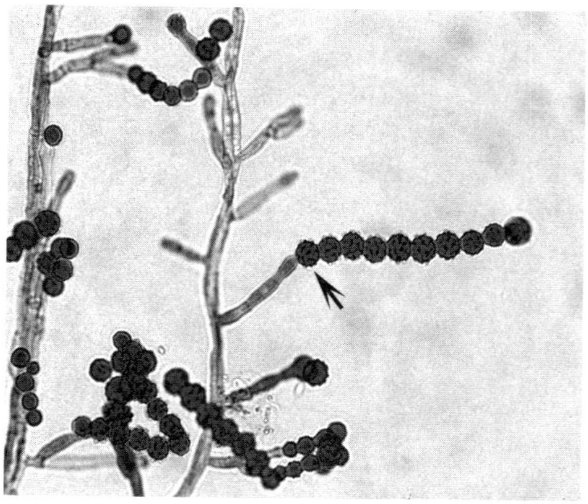

Basipetal succession of conidia in *Scopulariopsis brevicaulis*, x 460 (*CB*).

basophil, **basophile**, **basophilic** (Gr. *básis*, base + *phílos*, have an affinity for; or + Gr. suf. *-íkos* > L. *-icus* > E. *-ic*, belonging to, relating to): **1.** Having a preference for basic media or substrates. **2.** Also used for cellular organelles and bodies that stain with basic dyes (carmine, hematoxylin, thionine). Cf. **acidophilic**.

beard moss (ME. *berd* < OE. *beard*, beard; ME. *mos*, moss): species of *Alectoria*, *Bryoria*, *Ramalina* and *Usnea* (Lecanorales), characterized by long, pendant thalli.

beef-steak fungus, pl. **fungi** (E. *beef* < ME. < OF. *buef*, ox, beef < L. *bov-*, *bos*, head of cattle; E. *steak* < ME. *steke* < ON. *steik*, to roast on a stake, stik, stake; L. *fungus*, fungus): basidioma of the edible *Fistulina hepatica* (Aphyllophorales), the cause of brown rot of oak. Also called **liver fungus**.

benthic (Gr. *bénthos*, depth, generally marine + suf. *-íkos* > L. *-icus* > E. *-ic*, belonging to, relating to): forming part of the benthos, relative to the benthos, i.e., adapted to living on a solid substrate, as opposed to living in suspension (planctonic), whether in the littoral or abyssal benthos; an example of an abyssal benthic fungus is *Periconia abyssa* (dematiaceous asexual fungi), which lives on submerged wood at depths of 3,975-5,315 m.

beta-conidium, **ß-conidium**, pl. **conidia** (*beta*, *ß*, second letter of the Gr. alphabet; NL. *conidium* < Gr. *konídion* < *kónis*, dust + dim. suf. *-ídion* > L. *-idium*): a term applied to species that produce two morphological types of spores, in particular, the filiform spore of *Phomopsis* (sphaeropsidaceous asexual fungi) and other conidial states of the Diaporthales. These fungi also generally produce **alpha-conidia**. Beta conidia often form later than alpha conidia, are less common, and cannot be germinated.

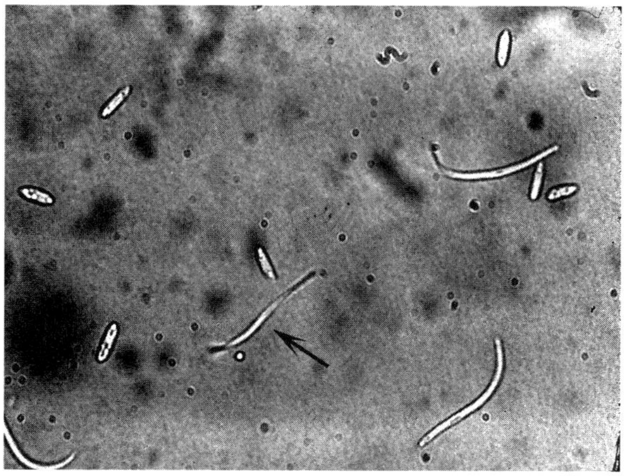

Beta conidia from a pycnidium of *Phomopsis* sp., x 800 (*RTH*).

biapiculate (L. *bis*, twice + *apiculatus*, furnished with an *apiculum*, dim. of *apex*, apex, end + suf. *-atus* > E. *-ate*, provided with or likeness): refers to a globose or elliptical structure with an apiculus at each end; i.e., more or less limoniform, like the ascospores of *Chaetomium cochliodes* (Sordariales), *Melanospora* (Melanosporales) and *Hypomyces lactifluorum* (Hypocreales). See **apiculate**.

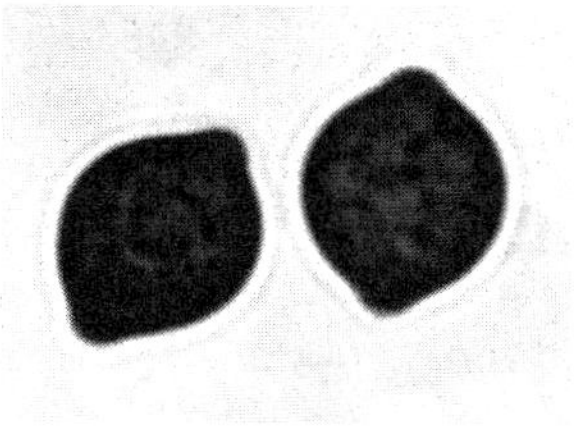

Biapiculate ascospores of *Melanospora* sp., x 1 200 (*RTH*).

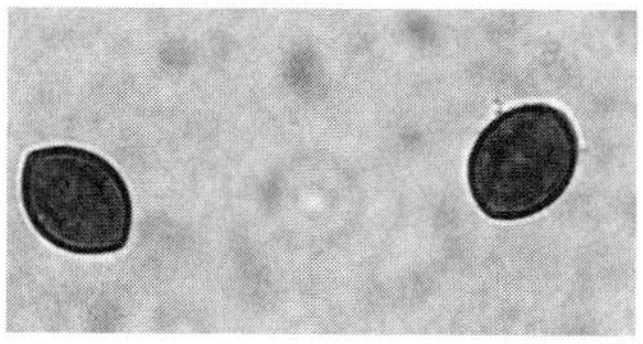

Biapiculate ascospores of *Chaetomium* sp., x 800 (*CB*).

bibulous (L. *bibulus*, fond of drinking, absorbent + *-osus* > OF. *-ous*, *-eus* > E. *-ous*, possessing the qualities of): capable of absorbing moisture, as in the surface of the pileus of some boletes (*Boletus*, *Suillus*) and agarics (*Hygrophorus*) of the Agaricales.

biconic, **biconical** (L. *bi-*, two + Gr. *konikós*, conic + suf. *-íkos* > L. *-icus* > E. *-ic*, belonging to, relating to; or + L. suf. *-alis* > E. *-al*, relating to or belonging to): having two cones opposed at their base, like the teliospores of *Puccinia graminis* (Uredinales).

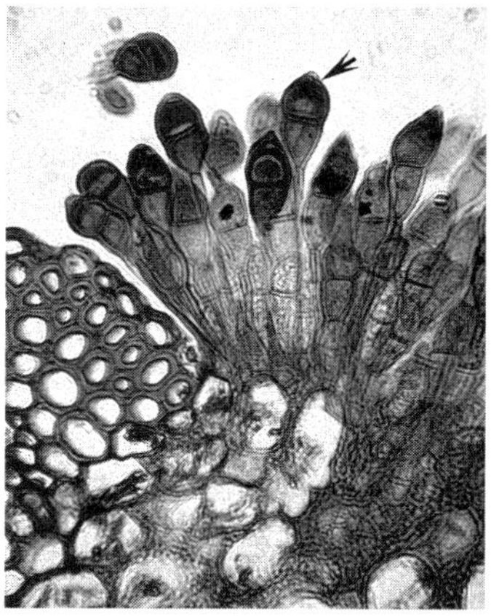

Biconic teliospores of *Puccinia graminis tritici*, seen in longitudinal section of a wheat leaf, x 250 (*GGA*).

biennial (L. *biennalis* < *bi-*, two + *annus*, year + suf. *-alis* > E. *-al*, relating to or belonging to; occurring every two years, continuing or lasting for two years): refers to an organism that lives more than one year without passing two; in the fungi it refers to the duration of the fruiting body, not to the vegetative mycelium immersed in the substrate. Cf. **annual** and **perennial**.

biforate (L. *biforatus*, *biforus*, having two doors or openings < *bis*, two + *forare*, to bore + suf. *-atus* > E. *-ate*, provided with or likeness): with two pores or openings, like the urediniospores of *Uromyces appendiculatus*, the bean rust (Uredinales), which have two pores or equatorial orifices.

bifurcate (ML. *bifurcatus*, *bifurcus*, with two points < *bifurcare*, to fork < *bi*, two + *furca*, fork + suf. *-atus* > E. *-ate*, provided with or likeness): forked or divided into two; dichotomous. E.g., the basidium with two sterigmata in *Dacrymyces deliquescens* (Dacrymycetales), or the branches of the sporangiophore of *Piptocephalis lepidula* (Zoopagales) are bifurcate in shape. See **bipartite**.

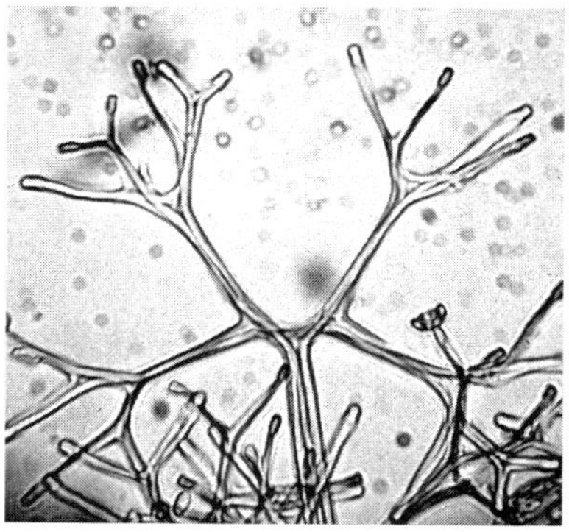

Bifurcate branches of the sporangiophore of *Piptocephalis lepidula*, x 900 (*EAA*).

bilateral (L. *bilateralis* < *bi*, two + *later-*, *latus*, side + suf. *-alis* > E. *-al*, relating to or belonging to): having two sides or arranged in two rows on both sides of an organ; e.g., as in the so-called bilateral symmetry of many spores (with two equal sides), or as in the bilateral gill trama of certain agarics (e.g., *Amanita muscaria*), which is characterized by having a central zone of columnar parallel cells (hyphae) that project downward from the pileus (the middle layer), and from which diverge obliquely two zones of cells, one on each side, called the lateral strata; this type of hymeniiferous trama is known as bilaterally divergent, and is found in Boletaceae (*Boletus*, *Paxillus*) and Gomphidiaceae (*Gomphidius*). See **divergent**.

binding hypha, pl. **hyphae** (ME. *binden* < OE. *bindan*, to fasten, secure, bind + E. suf. *-ing*, action or process; NL. *hypha*, hypha < Gr. *hyphé*, spider web; hypha): *Basidiomycetes*. Highly branched, thin- or thick-walled, cylindric flexuous hyphal cells which intertwine, creating a firm but often flexible tissue in sporophores; e.g., as in *Polyporus squamosus* and *Lentinus badius* (Aphyllophorales). Also called ligative hypha. See **generative hypha** and **skeletal hypha**.

binucleate (L. *binucleatus* < *bis*, two + *nucleus*, little nut, nucleus + suf. *-atus* > E. *-ate*, provided with or likeness): with two nuclei in each cell or spore. The dikaryophase of many fungi is the binucleate phase.

biocide (Gr. *bíos*, life + L. *caedere*, to kill): a substance which kills living organisms. Cf. **biostat**.

biocontrol or **biological control** (Gr. *bíos*, life + ME. *controllen* < MF. *controller* < L. *contra*, against + *rotulus*, little wheel; to check, test, or verify by evidence or experiments, to exercise restraining or directing influence over; or *bíos* + *lógos*, word or discourse + E. suf. *-ic* < L. *-icus* < Gr. *-íkos*, belonging to, relating to; or + L. suf. *-alis* > E. *-al*, relating to or belonging to): the use of one or more organisms to reduce the number of or eliminate another organism (pest) by interference with their ecology. Fungal pathogens, parasites and antagonists are being applied to control a range of agricultural pests, including arthropods (mainly insects and mites), nematodes, weeds and crop diseases. See **entomogenous**, **mycoinsecticide**, **mycoparasite**, and **nematophagous**.

bioconversion (Gr. *bíos*, life + ME. *conversion* < MF . < L. *conversion*, the act of converting or being converted < *con-*, with + *versus*, turned + *-io*, *-ionis* > E. suf. *-ion*, result of an action, state of): the conversion of organic materials, usually wastes, into one or more products of increased value (e.g., of lignocellulosic residues for ethanol production involving living organisms). Cf. **biodegradation**, **biodeterioration**, **bioremediation**, and **biotransformation**.

biodegradation (Gr. *bíos*, life + ME. *degradation*, the act or process of degrading < ME. *degraden* < MF. *degrader* < LL. *degradare* < L. *de-*, from, down, away + *gradus*, step, grade + E. suf. *-ation* < L. *-ationem*, action, state or condition, or result): the beneficial breakdown of materials, e.g., the removal of and/or utilization of wastes into more useful, innocuous or acceptable products by the action of living organisms. The use of fungal enzymes in solid or liquid state fermentations to improve digestibilities of lignocellulosic or other crop residues to ruminants is an example of biodegradation. Cf. **bioconversion**, **biodeterioration**, **bioremediation**, and **biotransformation**. See **tanninolytic**.

biodeterioration (Gr. *bíos*, life + ME. *deterioration*, the act or process of deteriorating < LL. *deterioratus* < *deteriorare* < L. *deterior*, worse + *-ationem*, action, state or condition, or result > E. suf. *-ation*): the undesirable change in the properties of a material caused by the activities of living organisms. Fungi play an important part in biodeterioration and their activities often damage grains, meat, all sorts of foods, animal feeding stuffs, wood, stone, building materials, fuel, electrical and optical equipment, leather, paper, textiles, and many other natural and man-made materials. Cf. **bioconversion**, **biodegradation**, **bioremediation**, and **biotransformation**.

biodiversity (Gr. *bíos*, life + E. *diversity*, the condition of being different < ME. *divers*, *diverse* < L. *diversus*, differing from one another, unlike, composed of distinct elements or qualities + E. suf. *-ty*, quality, condition): the variety and value of life on Earth from the genetic through the organismal to the ecological levels; the natural biotic resources of the globe.

biogenous (Gr. *bíos*, life + *génos*, origin < *gennáo*, to engender, produce + L. *-osus* > OF. *-ous*, *-eus* > E. *-ous*, possessing the qualities of): produced by living organisms; living on another living organism; parasitic.

biohazard (Gr. *bíos*, life + ME. *hazard* < MF. *hasard* < Ar. *z-zahr*, a source of danger): a biological agent or condition (such as an infectious organism or insecure and risky laboratory conditions) that constitutes a hazard to humans or their environment.

bioindication (Gr. *bíos*, life + ME. *indication*, something that serves to indicate < L. *indicatus*, ptp. of *indicare*, to point out or point to, to be a sign, symptom, or index of + *-ationem*, action, state or condition, or result > E. suf. *-ation*): the use of one or more organisms (bioindicators) which express particular symptoms or responses to indicate and monitor changes in some enviromental influence, e. g., lichens are used as bioindicators of air pollution; some other lichens and fungi are sensitive to acid rain, and still others are bioindicators of radioactivity in a given area since they absorb radioactive substances.

biomass (Gr. *bíos*, life + ME. *masse* < L. *massa* < Gr. *maza*, to knead, a quantity or aggregate of matter, mass): the amount of living material; the quantity (volume, weight, number of cells, etc.) of organisms or living material in a particular environment, e. g., fungi in soil, a fermented food or in any other substratum.

biont (Gr. *bíos*, life + *óntos*, genit. of *ón*, a being): a living organism; commonly used as a suf. to a word indicating the nature or position of the biont, e.g., **diplobiont**, **haplobiont**, **symbiont**.

bioprospecting (Gr. *bíos*, life + ME. *prospecting* < *prospect* < L. *prospectus*, view, prospect, to look forward, explore): the action of exploring or surveying natural ecosystems for economically valuable biotic products. For fungi, such products might include novel edible fungi, valuable enzymes for biotechnology companies or new biological control agents, among other examples.

bioremediation (Gr. *bíos*, life + *remediation*, the act or process of remedying < ME. *remedie* < L. *remedium* < *re*, again, against + *mederi*, to heal, relief + *-ationem*, action, state or condition, or result > E. suf. *-ation*): the use of microorganisms to remove, reduce or ameliorate pollution or potentially polluting materials from the environment, as, e. g., with naturally occurring microfungi that absorb heavy metals from the soil. *Cryptococcus albidus* var. *albidus* (cryptococcaceous asexual yeasts) is capable of capturing, and keeping in its capsule, the silver ions present in the soil near abandoned mines. This property may be exploited to remedy contaminated soils which inhibit the growth of vegetation. Cf. **bioconversion**, **biodegradation**, **biodeterioration**, and **biotransformation**.

biostat (Gr. *bíos*, life + *statós*, standing, placed): a substance which causes living organisms to stop growing. Cf. **biocide**.

biota (Gr. *bioté*, nature, the condition of life < *bíos*, life): all of the living beings of a locality, country, or of the Earth, consisting of the plants (flora), animals (fauna), microorganisms (microbiota) and fungi (mycobiota).

biotechnology (Gr. *bíos*, life + *technología*, systematic treatment of an art, applied science < *techné*, art + *lógos*, word, treatise + *-y*, E. suf. of concrete nouns): applied biological science by which products are obtained from raw materials with the aid of living organisms (as bioengineering or recombinant DNA technology, and all lines of industrial mycology). See **bioconversion**, **biodegradation**, **bioprospecting**, **bioremediation**, and **biotransformation**.

biotransformation (Gr. *bíos*, life + E. *transformation*, an act, process or instance of transforming or being transformed < ME. *transformen* < L. *transformare* < *trans*, across, beyond, through + *formare*, to form + *-ationem*, action, state or condition, or result > E. suf. *-ation*): also known as biological or microbial transformation, or more generally **bioconversion**; it involves the use of microorganisms to modify, by means of enzymatic activities, organic compounds to produce industrially, medically or environmentally important products. Among the most important reactions where a substrate is biotransformed are oxidations, reductions, hydrolysis, and condensation, among others, and a few of them are used industrially (e.g., transformation of antibiotics, steroids and sterols). *Rhizopus stolonifer* (Mucorales) has been used to produce 11a-hydroxyprogesterone from progesterone. Cf. **biodegradation**, **biodeterioration**, and **bioremediation**.

biotroph, **biotrophic** (Gr. *bíos*, life + *trophós*, that which nourishes, serves as food; or + Gr. suf. *-íkos* > L. *-icus* > E. *-ic*, belonging to, relating to): a parasite which can only grow and sporulate on living cells of the host; e.g., the piptocephalidaceous fungi that parasitize some mucors (e.g., *Piptocephalis* on *Mucor*), and the phytoparasitic fungi known as downy mildews and white rusts (Peronosporales), powdery mildews (Erysiphales) and rusts (Uredinales) are all biotrophic. Syn. of **obligate parasite**. Cf. **necrotroph** and **perthotroph**.

biotype (Gr. *bíos*, life + L. *typus*, type): a population of individuals of the same genotype that compose a *forma specialis* (see **form**). A group of biotypes with a similar virulence-avirulence pattern on a particular group of plants is called a **race**.

bipartite (L. *bi-*, two + *partire*, divide): divided into two, like the heterobasidium of *Dacrymyces* (Dacrymycetales). See **bifurcate**.

bipartition (L. *bipartitio*, divide into two < *bi-*, *bis-*, two + *partitus*, divided + *-io*, *-ionis* > E. suf. *-ion*, *-tion*, result of an action, state of): division into two equal parts, such as multiplication by cellular bipartition. See **fission** and **schizogenesis**.

bipolar (L. *bi-*, two + *polaris*, polar < *polus*, pole + suf. *-aris* > E. *-ar*, like, pertaining to): **1.** A type of sexuality in which the compatibility factors are of only two kinds (AB, ab), as occurs in the ascomycetes. Cf. **tetrapolar**. **2.** A type of spore germination in which a germ tube or bud is formed at each end or pole of the spore. E.g., bipolar germination occurs in the conidia of *Bipolaris* (dematiaceous asexual fungi); bipolar gemmation occurs in the yeast gen. *Kloeckera* (cryptococcaceous asexual yeasts) and *Nadsonia* and *Saccharomycodes* (Saccharomycetales), which produce a bud at each end of the mother cell. Cf. **monopolar** and **multipolar**.

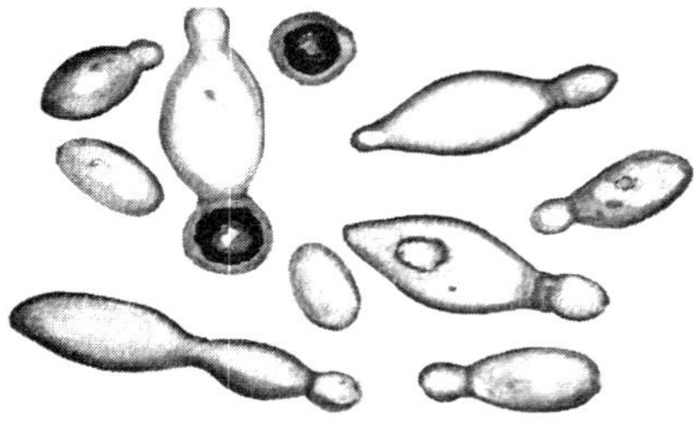

Bipolar budding of the vegetative cells of *Nadsonia fulvescens*, x 1 100.

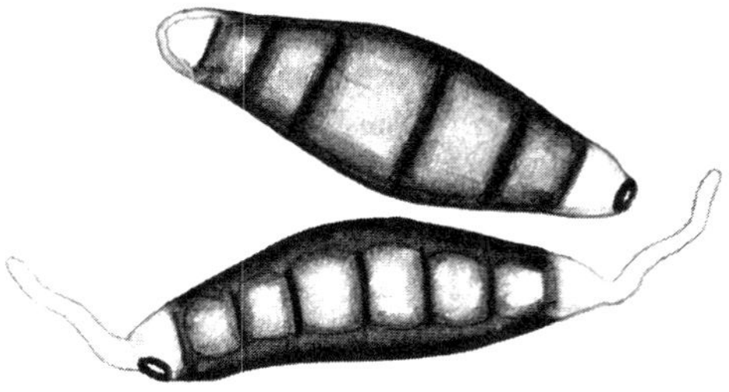

Bipolar germination of the conidia of *Drechslera* (*Bipolaris*) *iridis*, x 980.

bird's nest fungus, pl. **fungi** (ME. < OE. *bridd*, bird; ME. < OE. *nest*, nest; L. *fungus*, fungus): a member of the Nidulariales (such as *Cyathus* and *Nidularia*), with small, cupulate sporophores containing peridioles, thus resembling a bird's nest with eggs.

biscoctiform (ML. *biscoctus*, seamen's bread < L. *bis*, twice + *coctus* ptp. of *coquere*, to cook + *-formis* < *forma*, shape): in the shape of a biscuit; e.g., like the skin lesions produced by some dermatophilic fungi, and the colonies of the mold *Oidiodendron tenuissimum* (dematiaceous asexual fungi) when grown on certain culture media.

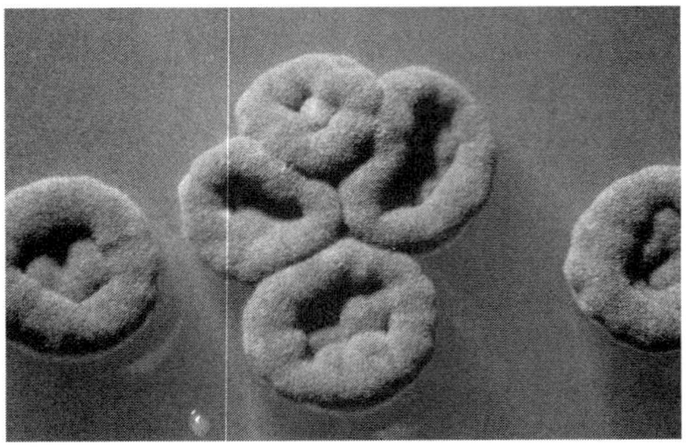

Biscoctiform colonies of *Oidiodendron tenuissimum* on agar, x 20 (*MU*).

biserial, **biseriate** (NL. *biseriatus* < *bi-*, two + *series*, series, row + suf. *-alis* > E. *-al*, relating to or belonging to; *bi-* + *series* + suf. *-atus* > E. *-ate*, provided with or likeness): arranged in two series or rows; e.g., in some species of *Aspergillus* (*A. candidus*, *A. flavus*, *A. niger*; moniliaceous asexual fungi) the conidial head is biseriate because the vesicle of the conidiophore bears a series or row of metulae on which is a series of phialides with conidia. Syn. of **distichous**. Cf. **uniseriate** and **monostichous**.

bisporic, **bisporous** (L. *bisporus* < *bi-*, two + Gr. *sporá*, spore + suf. *-ikos* > L. *-icus* > E. *-ic*, belonging to, relating to; or + L. *-osus* > OF. *-ous*, *-eus* > E. *-ous*, having, possessing the qualities of): with two spores, like the basidia of *Agaricus bisporus* (Agaricales).

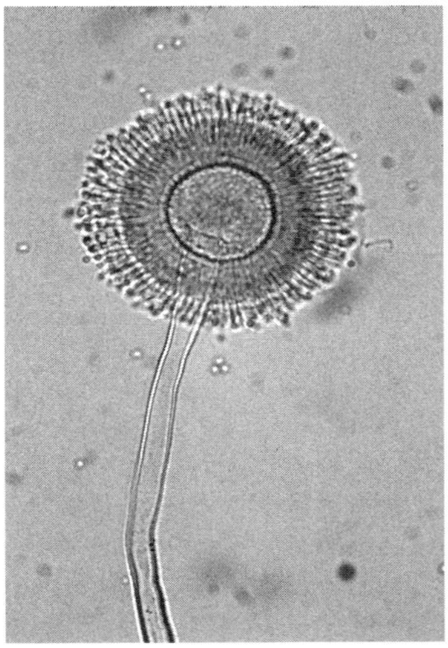

Biseriate conidial head of *Aspergillus candidus*, x 300 (*CB*).

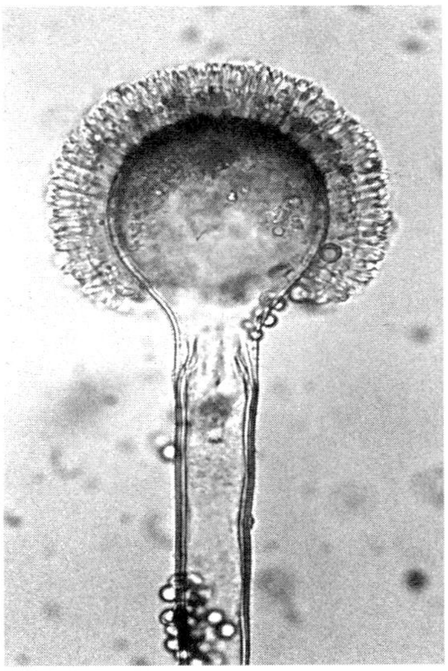

Biseriate conidial head of *Aspergillus flavus*, x 400 (*MU*).

bitunicate (L. *bitunicatus* < *bi-*, two + *tunica*, cloak, cover + suf. *-atus* > E. *-ate*, provided with or likeness): *Ascomycetes*. A type of ascus in which the wall has two functional layers, an internal wall layer (**endoascus** or endotunica) that is elastic and which extends beyond the external wall layer (**ectoascus** or ectotunica) at the time of spore liberation; bitunicate asci are characteristic of ascomycetes with ascoloculate ascomal ontogeny that comprise the Loculoascomycetes, e.g., *Preussia* (=*Sporormiella*) *australis* (Melanommatales) and *Leptosphaerulina crassiasca* (Dothideales). Cf. **prototunicate** and **unitunicate**.

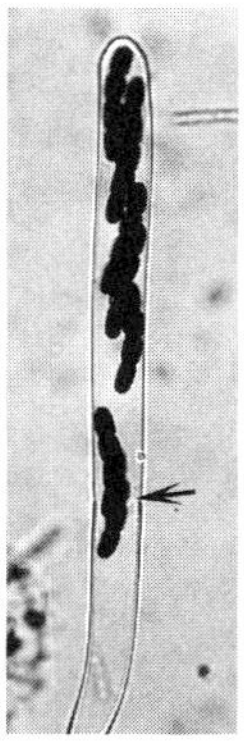

Bitunicate ascus of *Preussia* (=*Sporormiella*) *australis*, extended, x 300 (*RTH*).

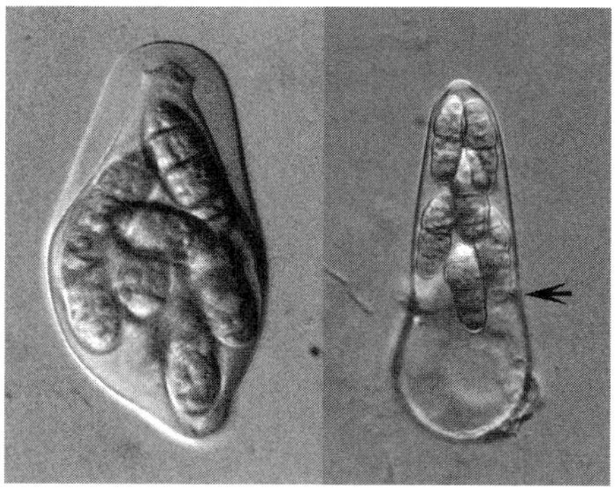

Bitunicate asci of *Leptosphaerulina crassiasca*, unextended and extended, x 400 (*RTH*).

bivalve (L. *bivalvis* < *bi-*, two + *valva*, valve, a leaf of a folding door): having two valves or convex faces, like the ascospores of *Eurotium chevalieri* and *E. rubrum* (Eurotiales).

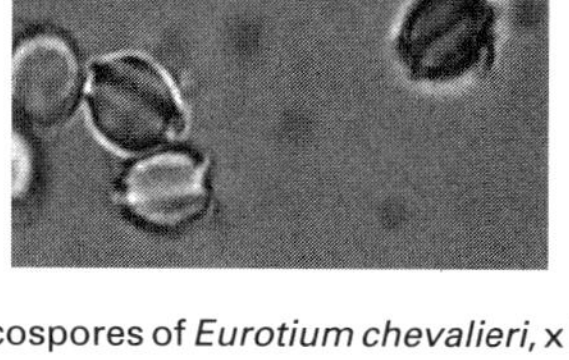

Bivalve ascospores of *Eurotium chevalieri*, x 2 000 (*PL*).

Bivalve ascospores of *Eurotium rubrum*, x 4 000.

biverticillate (L. *biverticillatus* < *bi-*, two + *verticillus*, verticil + suf. *-atus* > E. *-ate*, provided with or likeness): branched at two levels, i.e., with metulae and phialides, like the conidiophores of some species of *Penicillium* (*P. claviforme*, *P. islandicum*, *P. notatum*, of the moniliaceous asexual fungi) which Pitt considers in subgenus *Biverticillium*. See **monoverticillate**, **triverticillate**, **quadriverticillate** and **polyverticillate.**

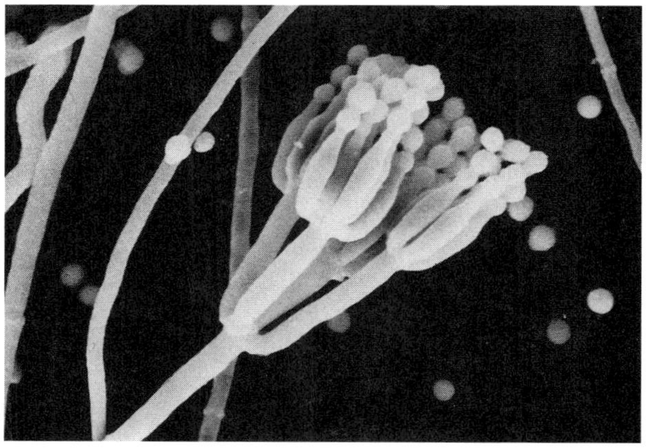

Biverticillate conidiophore of *Penicillium* sp., x 1 000 (*CM*).

black knot (ME. *black* < OE: *blaec*, black; ME. < OE. *cnotta*, knot, to press): the black ascostromata of *Apiosporina morbosa* (Pleosporales) formed on branches of plum and cherry trees.

Black knot - ascostromata of *Apiosporina morbosa* on a branch of a plum tree, x 2 (*RTH*).

black yeasts (ME. *black* < OE. *blaec*, black; ME. *yest* < OE. *gist*, yeast): yeast-like states of *Aureobasidium*, *Cladosporium* and *Moniliella*, among other dematiaceous asexual fungi, especially anamorphs of Herpotrichiellaceae, including *Exophiala*, *Ramichloridium* and *Rhinocladiella*, which form black colonies in culture.

blastesis, pl. **blasteses** (Gr. *blástesis*, germination, development): germination of yeast cells by the formation of a germ tube instead of a bud, which results in the formation of a membranous phase. Also applied to thallus development in the lichens.

blastic (Gr. *blastós*, sprout, bud, germ, shoot + suf. *-íkos* > L. *-icus* > E. *-ic*, belonging to, relating to): *Conidiogenesis*. A type of conidium development from a portion of a conidiogenous cell in which the conidium undergoes an obvious enlargement of the conidial primordium before it is delimited by a septum. If all the layers of the wall of the conidiogenous cell are involved in the synthesis of the conidium wall, the latter is called **holoblastic**, e.g., *Acremoniella verrucosa* (dematiaceous asexual fungi); if the outer layer of the wall of the conidiogenous cell is perforated during formation of the conidium (by evagination) and does not contribute to the synthesis of the condium wall, it is said to be **enteroblastic**, e.g., *Metarhizium anisopliae* (moniliaceous asexual fungi). Blastic conidia include **blastospores**, **botryoblastospores**, **sympodulospores**, **aleuriospores**, **phialospores**, **porospores**, and **meristem blastospores**. Blastic ontogeny is one of the two principal modes of conidial development among the conidial fungi, the other being the **thallic** mode.

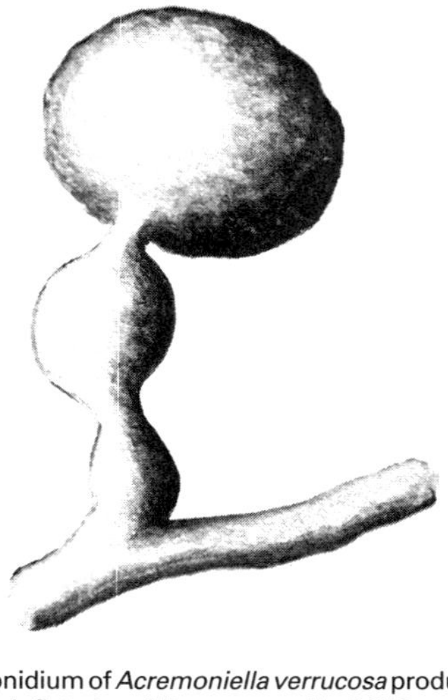

Holoblastic conidium of *Acremoniella verrucosa* produced from an inflated conidiogenous cell, x 2 350.

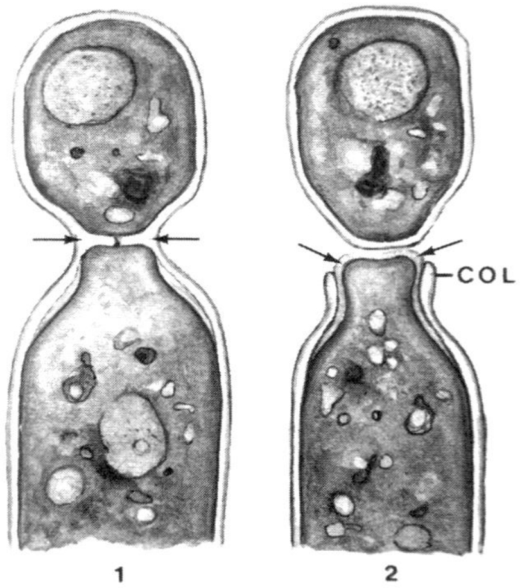

Enteroblastic conidia of *Metarhizium anisopliae*. Longitudinal sections through basal septa of differentiated conidia and apices of phialides, prior to initiation of new conidium (1, x 8 850) and at early stage of secondary conidial development (2, x 7 450). Arrows in 1 indicate continuity of inner phialide wall and outer conidial wall. The ring of torn wall material is the collarette (COL). Arrows in 2 indicate newly differentiated inner wall layer which will encompass next conidium. Drawing based on transmission electron micrographs published by Cole and Samson in their book *Patterns of Development in Conidial Fungi*, 1979.

blastidium, pl. **blastidia** (NL. *blastidium* < Gr. *blastídion* < *blastós*, sprout, bud, germ, shoot + dim. suf. *-ídion* > L. *-idium*): *Lichens*. A vegetative propagule, composed of mycobiont and phycobiont, which segments in a yeast-like manner, in the thallus of *Physcia opuntiella* (Lecanorales).

Blastomycetes (Gr. *blastós*, sprout, bud, germ, shoot + L. *-mycetes*, ending of class < Gr. *mýkes*, genit. *mýketos*, fungus): a class of conidial fungi that reproduces mainly by budding. It includes the imperfect or asporogenous yeasts of the orders Sporobolomycetales and Cryptococcales. Together with the Hyphomycetes and Coelomycetes, they were formerly considered, in some classification systems, as the Subdivision Deuteromycotina, with the general name of deuteromycetes. In the classification system followed in this dictionary, the asexual yeasts are not given a class rank and are classified as cryptococcaceous and sporobolomycetaceous asexual yeasts.

blastomycosis, pl. **blastomycoses** (< gen. *Blastomyces* < Gr. *blastós*, sprout, bud, germ, shoot + *mýkes*, fungus + *-osis*, state or condition): a primary pulmonary infection acquired by the inhalation of conidia of *Blastomyces dermatitidis* (moniliaceous asexual fungi), a dimorphic fungus inhabiting soil. Initially, the fungus causes a pulmonary blastomycosis, similar to tuberculosis, but it may disseminate to other parts of the body causing a cutaneous or a systemic infection.

blastospore (Gr. *blastós*, sprout, bud, germ, shoot + *sporá*, spore): *Conidiogenesis*. An asexual reproductive spore that originates by budding, as in the yeasts and some filamentous fungi. *Torulopsis*, *Kloeckera* and *Candida* (cryptococcaceous asexual yeasts) and *Monilia* and *Cladosporium* (moniliaceous and dematiaceous asexual fungi, respectively) are examples of gen. of fungi that reproduce asexually by means of blastospores. See **budding** and **gemmation**.

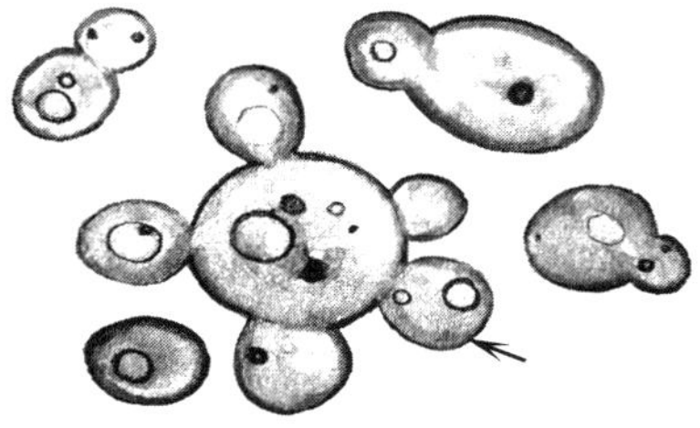

Blastospores of the somatic cells of *Torulopsis taboadae*, x 1 100.

blematogen, **blematogenous** (Gr. *blêma*, genit. *blêmatos*, cover + *génos*, origin, birth < *gennáo*, to engender, produce + L. *-osus* > OF. *-ous*, *-eus* > E. *-ous*, having, possessing the qualities of): see **protoblem**.

blepharoplast (NL. *blepharoplastus* < Gr. *blepharís*, cilium, eyelash + *plastós*, formed, organized particle or body): also called **basal body** and **kinetosome**.

bleu (**blue**) **cheese** (ME. < OF. *blou*, blue; F. *bleu*, blue; ME. *chese* < OE. *cese* < L. *caseus*, cheese): a type of cheese that is ripened and flavored by *Penicillium roquefortii* (moniliaceous asexual fungi), e. g., Roquefort, Stilton, Gorgonzola, Danish Blue and others.

blue stain (ME. < OF. *blou*, blue; ME. *steyne* < MF. *desteindre*, to discolor, and < ON. *steina*, to paint): a blue-grey coloration of wood caused by the growth of fungal hyphae, e. g., *Ceratocystis* (Microascales) and *Ophiostoma* (Ophiostomatales), in the tissues of the wood.

Blue stain of pine wood caused by *Ceratocystis* infection; cross section of a trunk, x 0.2 (*RTH*).

bolete (< gen. *Boletus* < Gr. *bolítes*, an old name for certain roots and edible mushrooms < *bõlos*, clod, gleba, lump): a fleshy, agaricoid fungus with a tubular hymenophore, in the family Boletaceae. The boletes include terrestrial, usually saprobic, occasionally fungicolous, or mycorrhizal fungi, mostly edible, a few poisonous, of wide geographical distribution.

boletinoid (< gen. *Boletinus* < gen. *Boletus* < Gr. *bolítes*, an old name for certain roots and edible mushrooms < *bõlos*, clod, gleba, lump + L. suf. *-oide* < Gr. *-oeídes*, similar to): having a hymenophore intermediate between pores and gills, as in *Boletinus* and *Gyrodon*, of the Boletaceae (Agaricales).

boletoid (< gen. *Boletus* < Gr. *bolítes*, an old name for certain roots and edible mushrooms < *bõlos*, clod, gleba, lump + L. suf. *-oide* < Gr. *-oeídes*, similar to): a mushroom with a capitate-stipitate fruiting body, but with a poroid hymenium, as in *Boletus*, *Boletellus* and *Suillus*, among other gen. of putrescent Agaricales.

bothrosome (Gr. *bóthros*, a hole, trench + *sõma*, body): an invaginated organelle at the cell surface which connects the plasma membrane to the network membranes in Labyrinthulomycota. See **sagenogen**.

botryoblastospore (Gr. *bótrys*, a cluster, bunch of grapes + *blastós*, sprout, bud, germ, shoot + *sporá*, spore): *Conidiogenesis*. The spore that results from a type of conidium formation in which numerous spores form simultaneously on the surface of an **ampulla**. E.g., *Botryosporium*, *Botrytis*, *Ostracoderma* (=*Chromelosporium*) and *Oedocephalum* (moniliaceous asexual fungi) are gen. that form botryoblastospores.

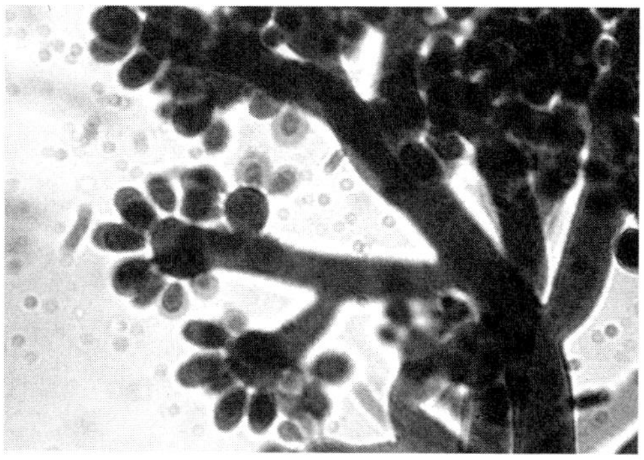

Botryoblastospores on conidiogenous cells (ampullae) of *Botrytis cinerea*, x 580 (*MU*).

botryose, **botryomorphic**, **botryomorphous** (Gr. *bótrys*, a cluster, bunch of grapes + L. *-osus*, full of, augmented, prone to > ME. *-ose*; or + Gr. *morphé*, shape + suf. *-íkos* > L. *-icus* > E. *-ic*, belonging to, relating to; or + L. *-osus* > OF. *-ous*, *-eus* > E. *-ous*, having, possessing the qualities of): arranged as a cluster of grapes; racemose, racimiform. E.g., like the arrangement of the conidia of *Botrytis cinerea* and *Oedocephalum* (moniliaceous asexual fungi) on the conidiophore. See **acinose**.

Botryose or botryomorphic arrangement of the conidia of *Botrytis cinerea*, x 240 (*CB*).

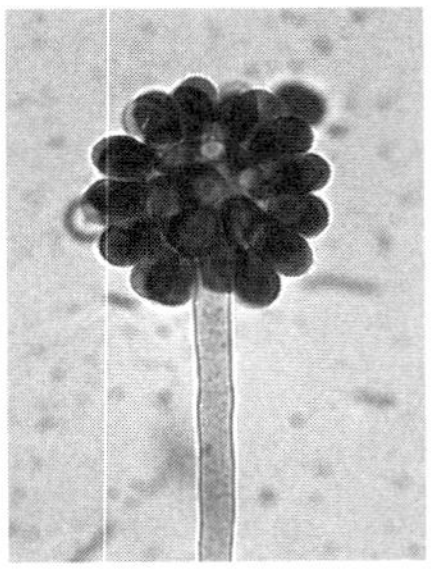

Botryose conidia on an ampulla of *Oedocephalum* sp., x 300 (*MU*).

botuliform (NL. *botuliformis* < *botulus*, sausage + *-formis* < *forma*, shape): sausage-shaped, i.e., cylindrical, plump, slightly curved, with both ends rounded. Syn. of **allantoid**.

bovistelloid (< gen. *Bovistella* < gen. *Bovista* < G. *Bovista*, common puffball + L. suf. *-oide* < Gr. *-oeídes*, similar to): *Basidiomycetes*. A type of fruiting body development (**forate**) in which a cavity is formed with a sterile base and with a capillitium composed of separate filaments, which taper toward their apices, as in *Bovistella* (Lycoperdales).

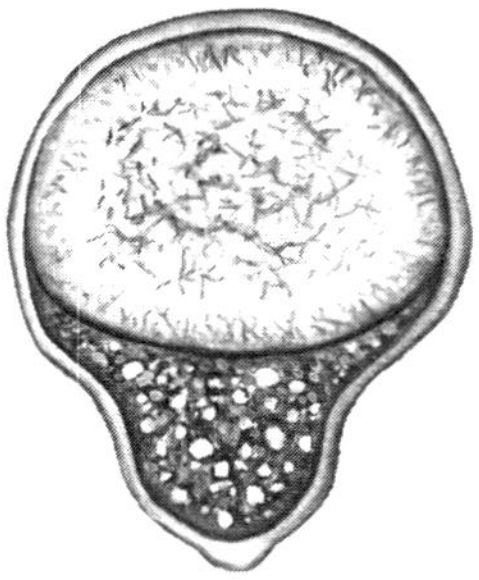

Bovistelloid development of the fruiting body of *Bovistella* sp., seen in longitudinal section, x 1.

bovistoid (< gen. *Bovista* < G. *Bovista*, common puffball + L. suf. *-oide* < Gr. *-oeídes*, similar to): *Basidiomycetes*. A type of fruiting body development resulting in a usually ovoid shape and lacking a sterile base, but with a capillitium of separate, branched filaments with tapered apices, as in *Bovista* (Lycoperdales).

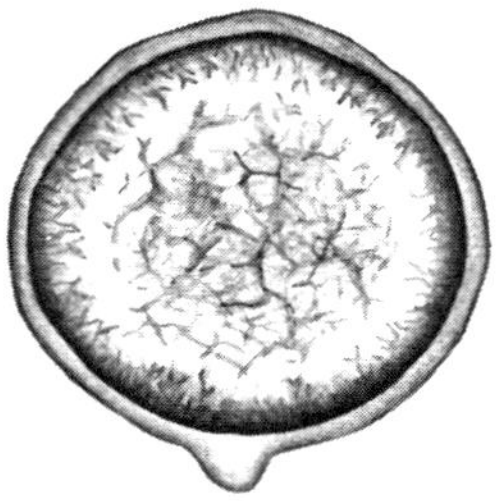

Bovistoid development of the fruiting body of *Bovista* sp., seen in longitudinal section, x 1.

brachiate (L. *brachiatus*, having arms < *brachium*, arm + suf. *-atus* > E. *-ate*, provided with or likeness): having arms.

brachybasidiole, brachybasidiolum, pl. **brachybasidiola** (Gr. *brachýs*, short + NL. *basidiolum* < L. *basidium* + dim. suf. *-olum* > E. *-ole*): *Basidiomycetes*. A hymenial element, especially in the gen. *Coprinus* (Agaricales), similar to a basidiolum in structure, but slightly more differentiated; it differs from a basidiolum mainly in being larger and more inflated. The brachybasidiola dominate to form a layer through which the basidia project; apparently the brachybasidiola act as spacing elements for the basidia. This type of hymenial structure is known as coprinoid. Also called **brachybasidium** and **brachycystidium**. Besides species of *Coprinus* (*C. atramentarius*, *C. comatus*, *C. micaceus*, etc.), brachybasidiola also are present in *Conocybe lactea* and *Bolbitius vitellinus*, among other Agaricales.

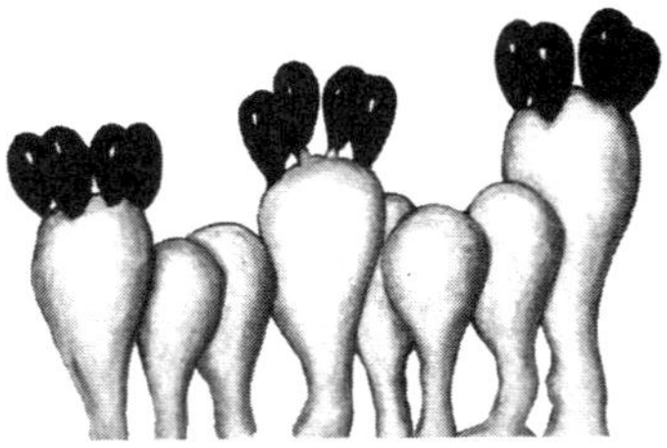

Brachybasidioles from the hymenium of *Coprinus comatus*, x 400.

brachybasidium, pl. **brachybasidia** (Gr. *brachýs*, short + L. *basidium* < Gr. *basídion* < *básis*, base + dim. suf. *-ídion* > L. *-idium*): see **brachybasidiole**.

brachycystidium, pl. **brachycystidia** (Gr. *brachýs*, short + *kystídion* < *kýstis*, bladder, vesicle, cell + dim. suf. *-ídion* > L. *-idium*): see **brachybasidiole**.

bromatium, pl. **bromatia** (NL. *bromatium* < Gr. *bromátion* < *bróma*, genit. *brómatos*, food + dim. suf. *-íon* > L. *-ium*): any of the swollen tips of hyphae that serve as food for ants that culture fungus gardens. Also called bromatia are the masses of yeasts with the appearance of cheese, that ants of the species *Cyphomyrmex rimsus* cultivate. Bromatia are also called **gongyli** (sing. **gongylus**) and **staphyla** (sing. **staphylum**). *Lepiota* (Agaricales) is an example of a fungus that forms bromatia.

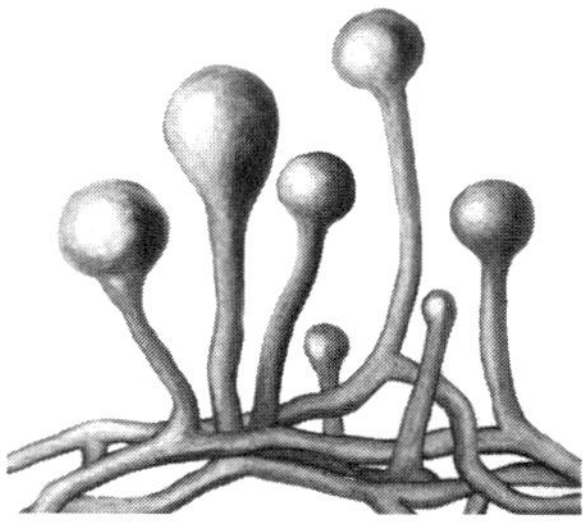

Bromatia of *Lepiota* sp., x 300.

broom cell (ME. *brome*, broom < OE. *brom*; L. *cella*, storeroom > ML. *cella* > OE. *cell*, cell, compartment): *Basidiomycetes.* Each of the cystidia that have on its apex various protuberances or simple or branched appendages, sharp-pointed and dark in color; e.g., like those in *Marasmius plicatilis* and *Mycena rulantiformis* (Agaricales).

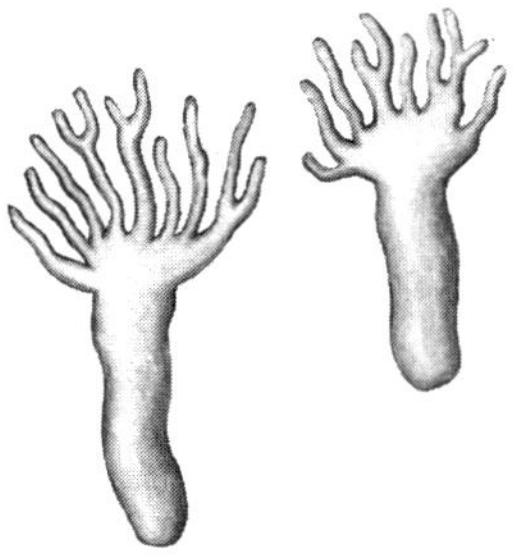

Broom cells from the hymenium of *Marasmius plicatilis*, x 1 120.

brush cell (ME. *brusshe* < MF. *broisse* < OF. *broce*, brush, a device composed of bristles; L. *cella*, storeroom > ML. *cella* > OE. *cell*, cell): a type of marginal hair found on the pileus of *Mycena* and *Marasmius* (Agaricales), clavate or sphaeropedunculate in shape, and bristled, echinulate or arborescent.

bryophilous (Gr. *brýon*, moss + *phílos*, have an affinity for + L. *-osus* > OF. *-ous*, *-eus* > E. *-ous*, possessing the qualities of): refers to an organism that develops in or on mosses. Syn. of **muscicole** or **muscicolous**. Various species of Myxomycetes, e.g., are bryophilous, among them *Craterium minutum* (Physarales).

budding (E. *budding*, to set or put forth buds < OE. *budda*, beetle, a small lateral or terminal protuberance; something not yet mature or at full development): see **gemmation**.

bulbil (L. *bulbilus*, dim. of *bulbus*, bulb): a small sclerotium, almost microscopic, formed of a few cells, such as the **papulospores** of *Papulaspora* (agonomycetaceous asexual fungi). Also called **microsclerotium**.

bulbillosis, pl. **bulbilloses** (L. *bulbilus*, dim. of *bulbus*, bulb + Gr. suf. *-osis*, condition): *Basidiomycetes.* A phenomenon observed in the gills of the agaricaceous fungi and which consists of their disintegration into rounded fragments, in the mode of bulbils, in whose cells meiosis occurs; i.e., it is the formation of bulbils in the basidia instead of basidiospores.

bulbous (L. *bulbosus*, full of bulbs, having bulbs, bulb-shaped < *bulbus*, bulb + *-osus* > OF. *-ous*, *-eus* > E. *-ous*, having): enlarged basally in the manner of a bulb; e.g., like the stipe of *Inocybe calospora* and *Lepiota* sp. (Agaricales).

Bulbous stipe of the basidiocarp of *Lepiota* sp., x 0.3 (*SA*).

bullate (L. *bullatus* < *bulla*, bubble + suf. *-ate*, provided with or likeness): having the surface covered with slight, irregular elevations, giving a blistered appearance. See **bulliform**.

Buller phenomenon (named after discoverer, Canadian mycologist A. H. R. Buller; LL. *phaenomenon* < Gr. *phainómenon*, appearance < *phaíno*, to show + *ménon*, disposition, force): the diploidization of a unisexual mycelium by a dikaryotic mycelium, in ascomycetes and basidiomycetes.

bulliform (L. *bulliformis* < *bulla*, bubble + *-formis* < *forma*, shape): with the shape of a bubble; bubble-like. E.g., the cystidia of certain species of *Leptonia* (Agaricales), *Peniophora* and *Corticium* (Aphyllophorales), have a tip that is bubble-like, evanescent, and which leave a kind of collar that consists of the basal portion of the original wall. Also called **bullate.**

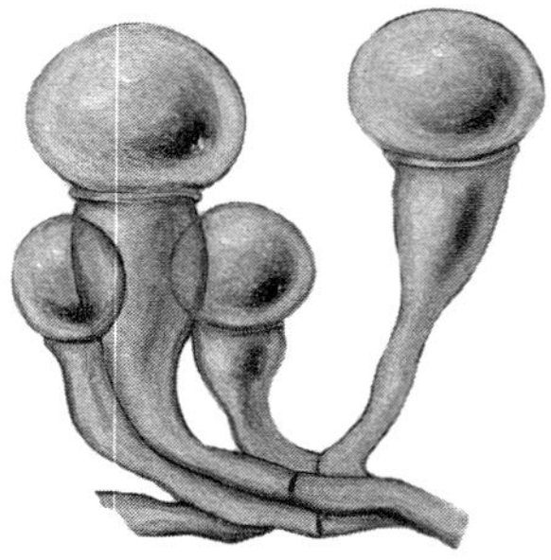

Bulliform cystidia from the hymenium of *Leptonia* sp., x 720.

bunt (unknown origin): a destructive disease of wheat (covered smut) caused by *Tilletia caries* and *T. laevis* (= *T. foetida*), of the Ustilaginales. Cf. **dwarf bunt.**

button (ME. *boton* < MF. *boton* < OF. *boter*, to thrust, project): *Basidiomycetes*. The youngest, unopened, immature stage in the development of the mushroom fruiting body.

butyraceous (L. *butyraceus*, buttery < *butyrum*, butter + suf. *-aceus* > E. *-aceous*, of or pertaining to, with the nature of): having the appearance or consistency of butter; e.g., like many of the cultures of yeasts on solid media, which appear moist and rather opaque or semibrilliant.

byssoid (Gr. *býssos*, tissue of fine linen of antiquity + L. suf. *-oide* < Gr. *-oeídes*, similar to): composed of very fine, delicate threads, like the mycelium of various fungi; e.g., the delicate pattern of radiating bundles of hyphae (=fibrils), characteristically found in the gleba of *Sclerogaster* (Hymenogastrales).

c

cactophilic (L. *cactus*, cactus < Gr. *káktos*, cardoon + *phílos*, have an affinity for + suf. *-íkos* > L. *-icus* > E. *-ic*, belonging to, relating to): developing preferentially on cactaceous plants, like the myxomycete *Badhamia gracilis* (Physarales), which forms its sporangia in the tissue of the cladodes of the prickly pear (*Opuntia ficus indica*), when this tissue begins to decompose and forms a kind of humid chamber where the plasmodia of this organism can grow. *Badhamia gracilis* and other species of Myxomycetes are considered cactophilic because they are often found associated with Cactaceae, although sporadically they can develop on other substrates.

caducous (L. *caducus*, perishable, falling early < *cadere*, to fall + *-osus* > OF. *-ous*, *-eus* > E. *-ous*, possessing the qualities of): sloughing off or falling away early; e.g., like the exoperidium of *Lycoperdon* (Lycoperdales) at maturity.

caeoma, pl. **caeomata** (NL. *caeoma* < Gr. *kaío*, to burn, be consumed by fire + *-óma*, a diseased state): *Uredinales*. A type of **aecium** that is surrounded by paraphyses instead of a peridium, or lacking both elements; as in the gen. *Caeoma* of the Uredinales.

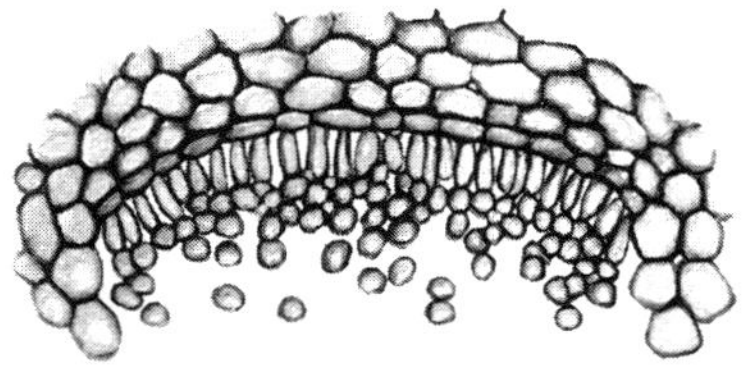

Caeoma of *Caeoma* sp. Drawing of a longitudinal section of this fungal structure formed on a host leaf, x 45.

caerulescent, **coerulescent** (L. *caeruleous*, dark blue, azure, bluish + *-escens*, genit. *-escentis*, that which turns, beginning to, slightly > E. *-escent*): turning bluish or greenish-blue; e.g., the basidiocarps of the hallucinogenic fungi of the gen. *Psilocybe* (Agaricales) stain bluish when they are rubbed or damaged. This appears to be an oxidation reaction.

caespitose, **caespitous**, **cespitose** (NL. *caespitosus*, tufted like sod-grass < *caespit-*, *caespes*, turf, grass + *-osus*, full of, augmented, prone to > ME. *-ose*; or + *-osus* > OF. *-ous*, *-eus* > E. *-ous*, having, possessing the qualities of): aggregated or clustered, but not fused; e.g., the basidiocarps of *Coprinus micaceus* (Agaricales) are caespitose.

Caespitose basidiocarps of *Coprinus micaceus*, x 0.3.

calcareous (L. *calcarius*, of the lime < *calx*, genit. *calcis*, lime, chalk, of calcium carbonate + *-arius*, belonging to): made up of or containing calcium carbonate; of a grayish-white color. The peridium, columella or capillitium of various species of Myxomycetes is calcareous in nature (*Mucilago*, *Badhamia*, *Fuligo*, and *Physarum*, among other Physarales).

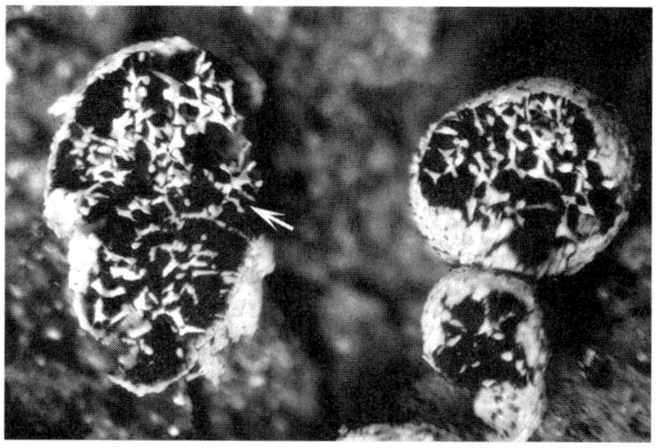

Calcareous capillitium of the sporangia of *Badhamia deanessii*, x 16 (*MU*).

Calcareous peridium of the sporangium of *Physarum* sp., x 15 (*MU*).

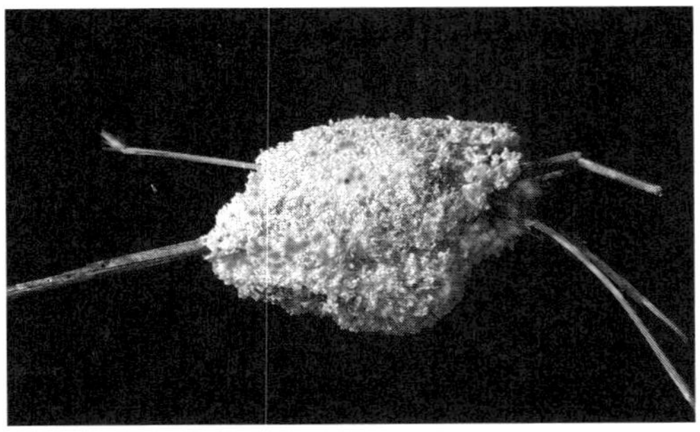

Calcareous peridium of the aethalium of *Mucilago crustacea*, x 2 (*MU*).

calcariferous (L. *calcar*, genit. *calcaris*, spur + *-ferous*, bearer < *ferre*, to bear, carry + L. *-osus* > OF. *-ous*, *-eus* > E. *-ous*, having, possessing the qualities of): having or bearing spines.

calcariform (L. *calcariformis* < *calcar*, genit. *calcaris*, spur + *-formis* < *forma*, shape): having the shape of a spur; e.g., like the appendages of the conidia of *Harpagomyces lomnicki* (moniliaceous asexual fungi).

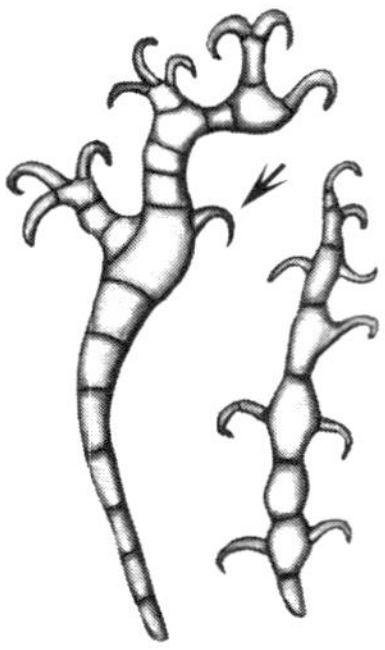

Calcariform appendages of the conidia of *Harpagomyces lomnicki*, x 2 160.

calceiform (L. *calceiformis* < *calceus*, shoe + *-formis* < *forma*, shape): with the shape of a shoe, like the conidia of *Calceispora hachijoensis* (moniliaceous asexual fungi).

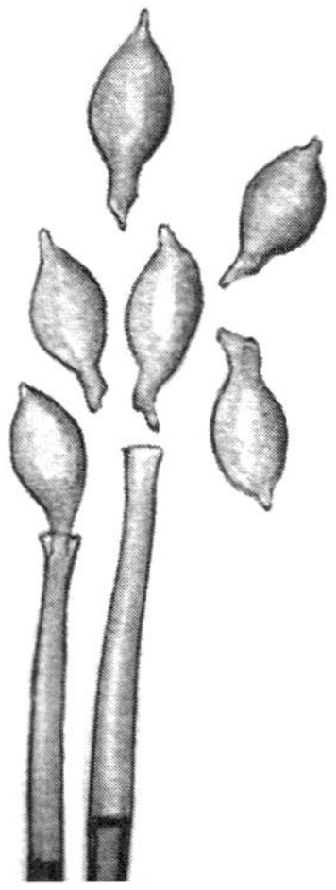

Calceiform conidia of *Calceispora hachijoensis*, x 350.

calciferous (L. *calx*, genit. *calcis*, lime, chalk, of calcium carbonate + *-ferre*, bearer < *fero*, to bear, carry + *-osus* > OF. *-ous*, *-eus* > E. *-ous*, having, possessing the qualities of): bearing calcium carbonate, like the capillitium of the Myxomycetes in the order Physarales.

calcivorous (L. *calx*, genit. *calcis*, lime, chalk, of calcium carbonate + *-voro* < *vorare*, to eat, devour + *-osus* > OF. *-ous*, *-eus* > E. *-ous*, possessing the qualities of): that which consumes calcium carbonate from the calcareous rocks on which it lives; e.g., the lichen *Lecidea calcivora* (Lecanorales) gradually dissolves this type of substrate.

caligate (L. *caliga*, military boot + suf. *-atus* > E. *-ate*, provided with or likeness): having a boot, wrapped in a veil. Syn. of **ocreate** and **peronate**.

callus, pl. **calli** (L. *callus*, hardened skin): *Agaricales*. The apical region of a basidiospore, which is thinner-walled, more or less convex and projecting, in which the germ pore is located; e.g., as in many species of *Galerina* and in *Psathyrella* (=*Lacrymaria*) *rigidipes*, among other Agaricales.

calo- (Gr. pref. *kalós-*, beautiful): used in generic names, such as *Calomyxa* (Trichiales), *Calonectria* (Hypocreales) and *Calocera* (Dacrymycetales).

calvescent (L. *calvescens*, genit. *calvenscentis*, becoming bald, prp. of *calvescere*, to become bald < *calvus*, hairless, smooth + *-escens*, genit. *-escentis*, that which turns, beginning to, slightly > E. *-escent*): becoming hairless, losing its hair.

calvous (L. *calvus*, bald + *-osus* > OF. *-ous*, *-eus* > E. *-ous*, having, possessing the qualities of): without hair or pubescence.

calyciform (L. *calyciformis*, calyx-like < *calyx*, genit. *calycis*, cup, goblet < Gr. *kályx*, husk, covering + L.

-formis < *forma*, shape): in the shape of a cup or goblet; e.g., the apothecia of many Discomycetes (*Cookeina*, *Peziza*, *Helvella* and other Pezizales), the basidiocarp of *Cotylidia aurantiaca* (Aphyllophorales), or the sporangia of some Myxomycetes (*Craterium* and other Physarales) are calyciform.

Calyciform apothecium of *Cookeina sulcipes*, x 1.

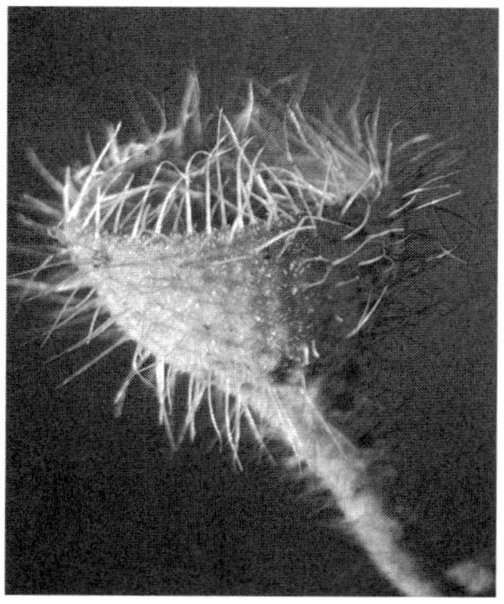

Calyciform apothecium of *Cookeina tricholoma*, x 2 (*MU*).

calyculus, pl. **calyculi** (L. *calyculus* < *calyx*, genit. *calycis*, goblet, cup + dim. suf. *-culus*): a structure shaped like a cup; e.g., the calyciform base of the sporangia of various common species of Myxomycetes, such as *Arcyria denudata* and *Hemitrichia clavata* (Trichiales) and related forms; the calyculus remains as a continuation of the pedicel after the upper portion of the sporangial peridium disappears.

calyptra, pl. **calyptrae** (NL. *calyptra* < Gr. *kalýptra*, cover): a structure or tissue that covers something; e.g., the perisporial cover that partially surrounds the basidiospore in certain agarics, such as *Galerina calyptrospora*; the rest of the universal veil that remains like a partial cover on the pileus of some Agaricales, such as *Amanita calyptratoides* and *A. fulva*; the tissue which covers the apex of the fruiting body of *Itajahya galericulata* (Phallales) and to which the gleba is attached. Cf. **utriculate**.

Calyculus of the sporangium of *Hemitrichia stipitata*, x 13 (*CB*).

Calyptra of the basidiocarp of *Amanita fulva*, x 0.3 (*EPS*).

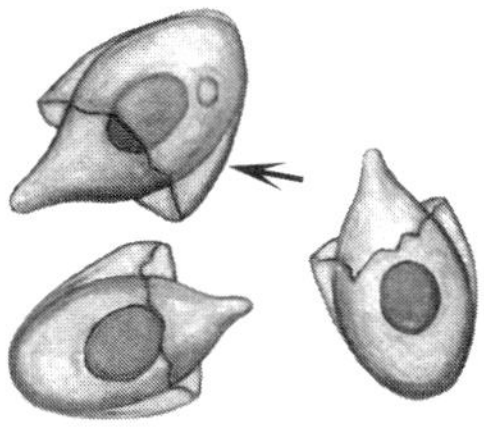

Calyptra of the basidiospores of *Galerina calyptrospora*, x 1 570.

calyptrate (NL. *calyptratus* < *calyptra* < Gr. *kalýptra*, cover + suf. *-atus* > E. *-ate*, provided with or likeness): furnished with a **calyptra**.

campanulate (NL. *campanulatus* < L. *campana*, bell + suf. *-atus* > E. *-ate*, provided with or likeness): shaped like a bell; e.g., like the pileus of certain basidiomycetes, such as *Panaeolus campanulatus*, *P. semiovatus* and *P. sphinctrinus* (Agaricales).

campylidium, pl. **campylidia** (L. *campylidium* < Gr. *kampylídion* < *kampýlos*, curved, bent + dim. suf. *-ídion* > L. *-idium*): a term used initially to designate a type of asexual fructification in certain lichens, such as *Aleurodiscus* (=*Cyphella*) *aeruginascens* (Aphyllophorales), but recognized later as a fungal parasite of this supposed lichen, since presently the latter is considered a corticiaceous fungus.

Campanulate pileus of the basidiocarps of *Panaeolus semiovatus*, x 1.

canaliculate (L. *canaliculatus*, channeled, grooved < *canaliculus*, dim. of *canalis*, canal, channel, conduit + suf. *-atus* > E. *-ate*, provided with or likeness): provided with one or several channels.

canaliculus, pl. **canaliculi** (L. *canaliculus* < *canalis*, canal, channel, conduit + dim. suf. *-culus*): a narrow passage like a small canal, such as those of the cell wall through which the plasmodesmata pass.

cancellate (L. *cancellatus*, made in the style of an iron-work screen or window lattice, latticed < *cancellus*, grating + suf. *-atus* > E. *-ate*, provided with or likeness): similar to a window lattice, or trellis; e.g., the peridium in sporangia of *Cribraria cancellata* (Liceales), which for being fleeting, leaves enlargements in the form of longitudinal ribs interconnected by delicate transverse bands, so that each mature sporangium has the appearance of a mesh or trellis.

Cancellate sporangia of *Cribraria cancellata*, x 12 (*MU*).

Cancellate sporangium of *Cribraria cancellata*, x 32 (*MU*).

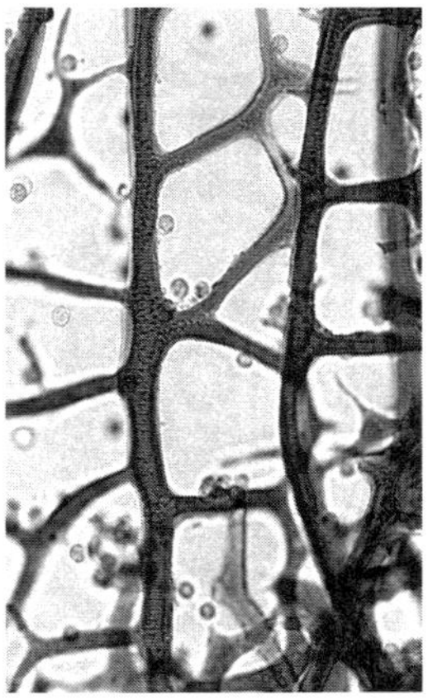

Cancellate peridial net of the sporangium of *Cribraria cancellata*, x 200 (*CB*).

candelabrum, pl. **candelabra** (L. *candelabrum*, candlestick < *candela*, candle + *labrum*, lip): a branched structure for holding several candles; e.g., like the conidiophore of *Candelabrella* (moniliaceous asexual fungi) and the dichotomous hyphae, called candlesticks or *favic candelabra*, characteristic of *Trichophyton schoenleinii* (moniliaceous asexual fungi), the causal agent of *tinea favosa* in man.

candidiasis (< gen. *Candida* < L. *candidus*, white + Gr. *-iasis*, suf. terminating names of diseases; it may also denote an action or process): a superficial infection of the skin, especially of the intertriginous areas (inframammary, periannal, etc.) or of the face, nails and mucose membranes of the mouth and the vagina. In debilitated hosts, yeasts causing this mycosis (*Candida albicans*, *C. tropicalis*, and *C. parapsilosis*, among other species of cryptococcaceous asexual yeasts) can become systemic. See **thrush**.

canescent (L. *canescens*, genit. *canescentis* < *canescere*, to become gray < *canus*, white, hoary, gray + *-escens*, genit. *-escentis*, that which turns, beginning to, slightly > E. *-escent*): covered with short, whitish or grayish hair; e.g., like the surface of the pileus of *Hygrophorus canescens* (Agaricales).

canker (ME. < ONF. *cancre* < L. *cancer*, crab, cancer): a sharply-limited area of necrosis of the cortical tissue of a plant, such as that caused by *Nectria galligena* (Hypocreales) in apple trees (apple canker) and by *Calosphaeria* (Calosphaeriales) in peach trees.

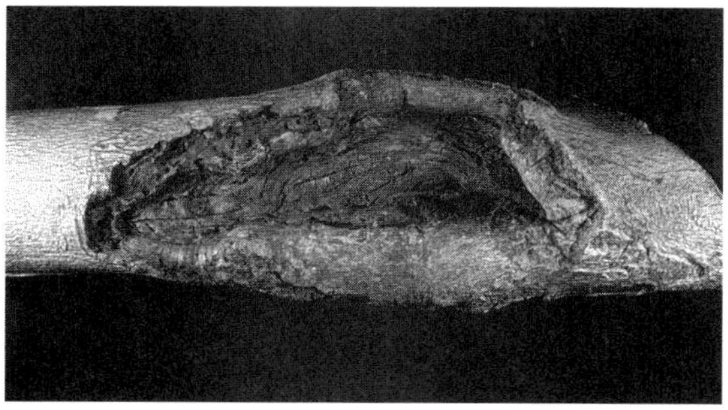

Canker on a branch of maple caused by *Nectria* infection, x 0.5 (*RTH*).

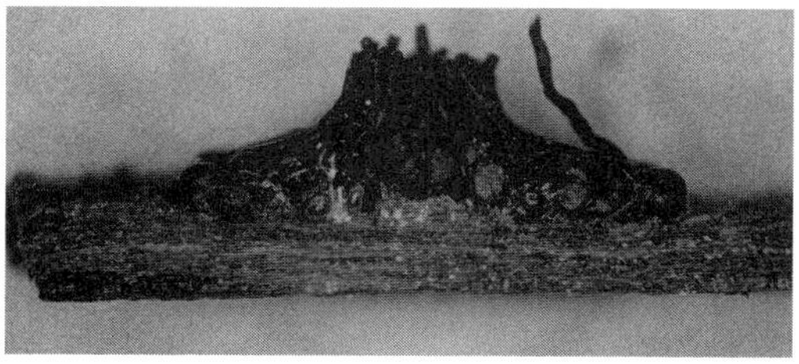

Canker on a trunk of peach caused by *Calosphaeria* sp. (*RTH*).

cantharelloid (< gen. *Cantharellus* < L. *cantharus* < Gr. *kántharos*, cup, pitcher + L. dim. suf. *-ellus* + L. suf. *-oide* < Gr. *-oeídes*, similar to): similar to the fruiting body of *Cantharellus* (Aphyllophorales), i.e., more or less turbinate or goblet-shaped, with decurrent lamellae.

capillary (L. *capillaris* < *capillus*, hair + suf. *-aris* > E. *-ar*, like, pertaining to + E. suf. *-y*, having the quality of): fine like a hair.

capillary spore (L. *capillaris* < *capillus*, hair + suf. *-aris* > E. *-ar*, like, pertaining to + E. suf. *-y*, having the quality of; Gr. *sporá*, spore): the only known reproductive structure of *Stylopage* (Zoopagales), a common gen. in dung where it is predaceous on amoebae. The capillary spore is almost identical to the **capilliconidia** of some entomophthoralean fungi. It has a haptor that develops at the tip when moisture is abundant, and by means of this structure the capillary spores adhere to mites and are dispersed in nature by the mites that ride along on dung beetles for their own dispersal. This hitchhiking mode of transport is known as **phoresy**.

capilliconidium, pl. **capilliconidia** (L. *capillus*, hair + NL. *conidium* < Gr. *konídion* < *kónis*, dust + dim. suf. *-ídion* > L. *-idium*): *Entomophthorales*. A secondary conidium formed at the end of a hairlike filament that results from the germination of a primary conidium, as happens in *Conidiobolus* and *Basidiobolus*. Cf. **capillary spore**.

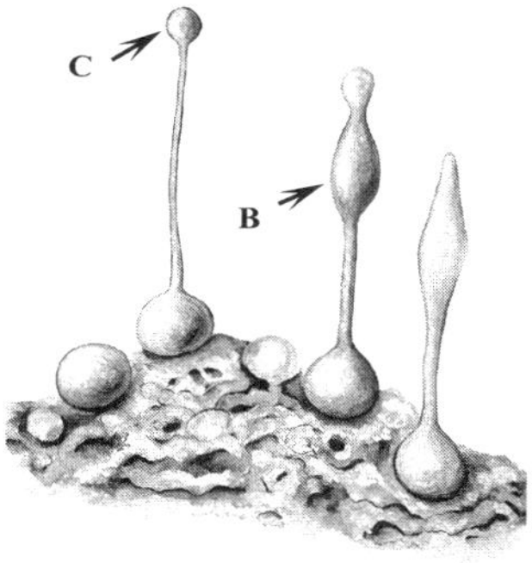

Capilliconidia of *Conidiobolus* sp. (C) and *Basisiobolus* sp. (B), x 1 200.

capilliform (L. *capilliformis* < *capillus*, hair + *-formis* < *forma*, shape): hair-like, i.e., long and slender.

capillitium, pl. **capillitia** (L. *capillitium*, long hair, in the collective sense < *capillus*, hair + *-itium*, adjectival ending, meaning provided with, having): **1.** *Myxomycetes*. Sterile filaments, free or anastomosed, found in the interior of many sporophores. In many species, such as *Hemitrichia stipitata* and *Arcyria cinerea* (Trichiales), the capillitium constitutes an elastic cottony or spongy mass that appears to have an important function in spore dispersal. Cf. **pseudocapillitium**. **2.** *Gasteromycetes*. In the gleba or fertile part of the basidiocarps, such as *Lycoperdon candidum* and *L. perlatum* (Lycoperdales) and *Scleroderma* spp. (Sclerodermatales), there also exist some filaments that constitute a capillitium, but in this case they are hyphal elements and not inert structures as in the Myxomycete capillitium. Such elements are coarse, thick-walled cells which mature late and may develop pores or slits in the thick secondary wall; they are usually branched. Cf. **paracapillitium**.

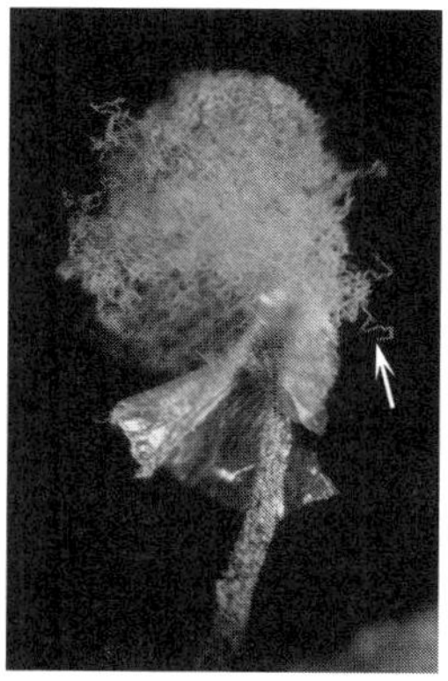

Capillitium of the sporangium of *Arcyria cinerea*, x 27 (*MU*).

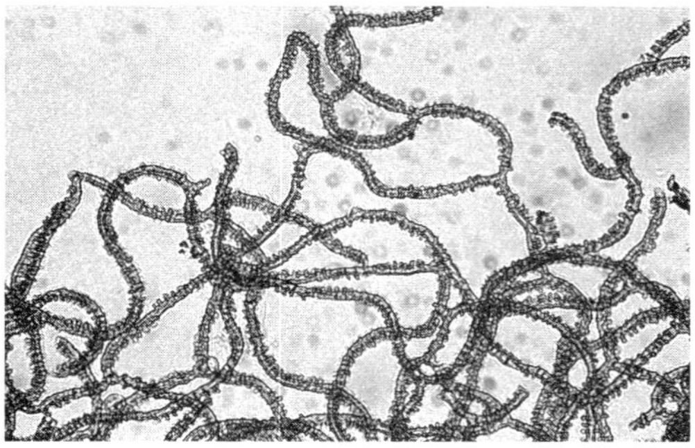

Capillitium from the sporangium of *Arcyria* sp., x 400 (*MU*).

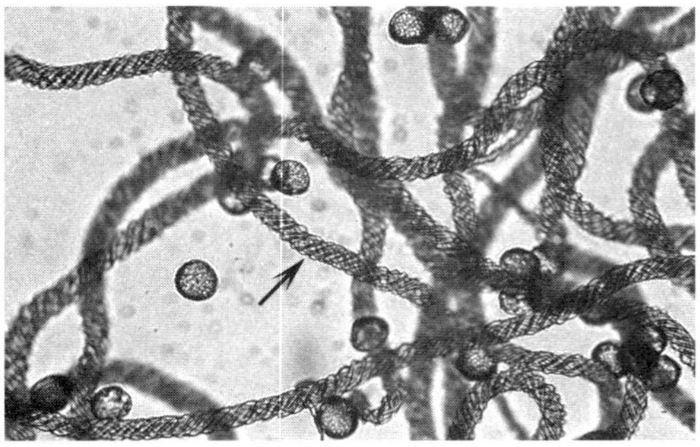

Capillitium and spores from the sporangium of *Hemitrichia stipitata*, x 600 (*MU*).

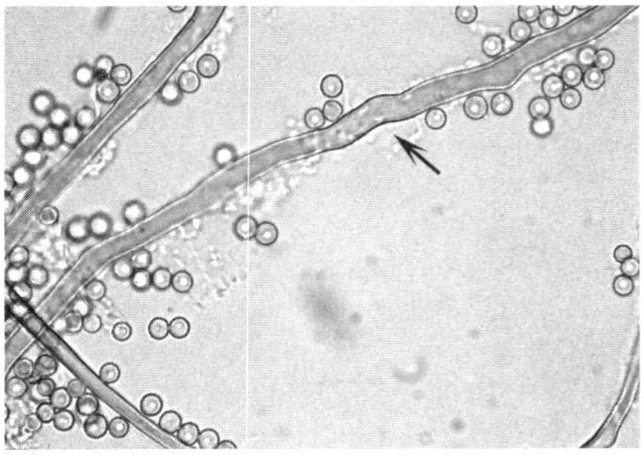

Capillitium and basidiospores from the basidiocarp of *Lycoperdon candidum*, x 600 (*ACV*).

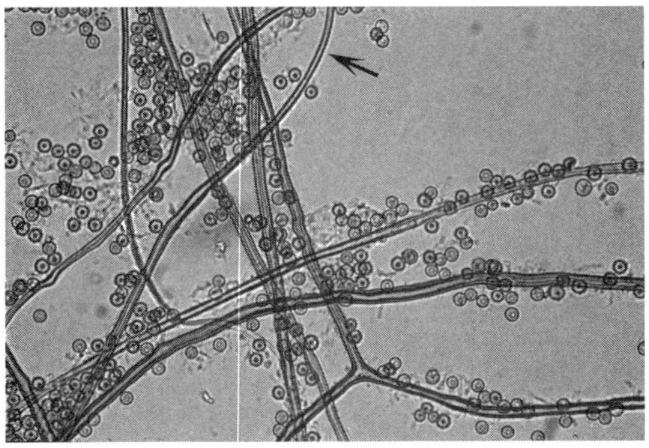

Capillitium and basidiospores from the basidiocarp of *Lycoperdon perlatum*, x 375 (*ACV*).

capitate (L. *capitatus*, having a head < *caput*, genit. *capitis*, head + suf. *-atus* > E. *-ate*, provided with or likeness): having an enlarged apex, like a small head; e.g., like some synnemata, stromata, cystidia, and spores. The synnemata of *Antromycopsis smithii* and the perithecial stromata of *Cordyceps melolonthae* var. *rickii* (Hypocreales) are capitate.

Capitate perithecial stromata of *Cordyceps melolonthae* var. *rickii* on a parasitized coleopteran larva, x 0.5 (*TH*).

Capitate synnema of *Antromycopsis smithii* on agar, x 10 (*CB & MU*).

capitellum, pl. **capitella** (L. *caput*, genit. *capitis*, head + dim. suf. *-ellum*): a little head.

capitulum, pl. **capitula** (L. *capitulum*, little head < *caput*, genit. *capitis*, head + dim. suf. *-ulum*): the apical, globose, pedunculate apothecium of lichens of the order Caliciales (*Calicium*). In some lichens, such as *Pertusaria* (Pertusariales), the soredia do not form on the surface of the thallus, but are arranged in sorediferous capitula, which occupy all of the excipulum of the apothecia, and which abort and do not produce asci. Also called **sphaeridium**.

capnodic, **capnoid** (Gr. *kapnós*, smoke, or *kapnódes*, smoky + suf. *-íkos* > L. *-icus* > E. *-ic*, belonging to, relating to; or + L. suf. *-oide* < Gr. *-oeídes*, similar to): smoke color, blackish. The dark mycelium of various fungi, known as sooty molds, that covers branches and leaves, and which develops on the thick, sweet liquid exuded by plant lice or mealy bugs or of vegetable origin; they are species of *Capnodium* (Capnodiales), which grow mainly on citrus, willows and poplars.

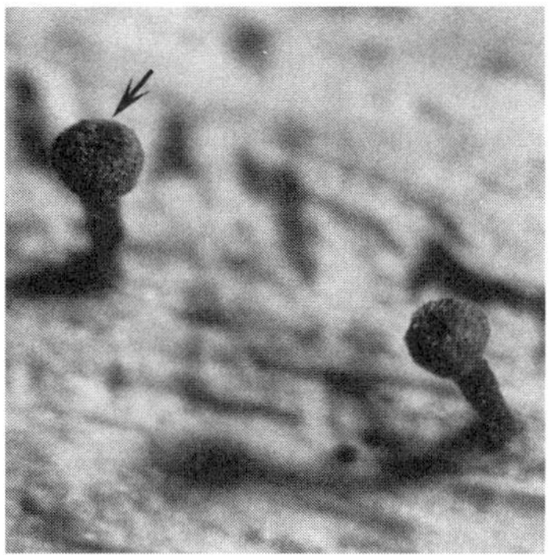

Capitula (apothecia) of the thallus of *Calicium farietinum*, x 13 (*MU*).

capsule (L. *capsula* < *capsa*, box + dim. suf. *-ula* > E. *-ule*): a hyaline, gelatinous covering that surrounds the exterior of the cell wall in certain yeasts (*Cryptococcus neoformans*, cryptococcaceous asexual yeasts), or of the ascospores of some ascomycetes (*Sordaria fimicola*, Sordariales; *Leptosphaerulina*, Pleosporales), in which instance it is often referred to as a sheath.

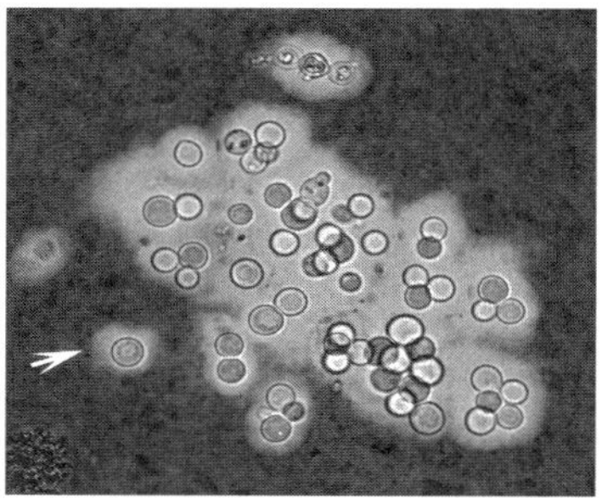

Capsule of the vegetative cells of *Cryptococcus albidus*, x 1 000 (*MU*).

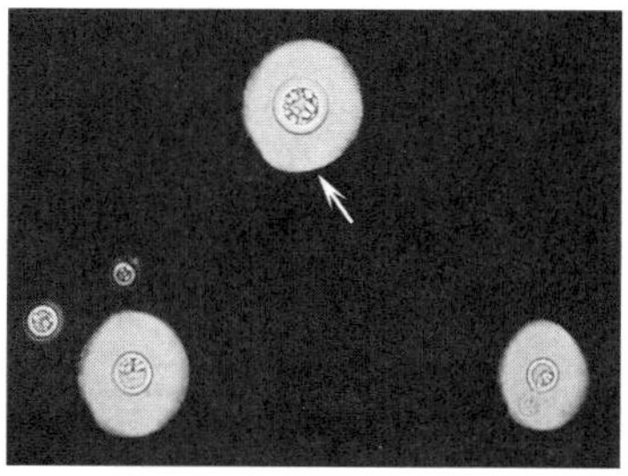

Capsule of the vegetative cells of *Cryptococcus neoformans*, x 630 (*RLM*).

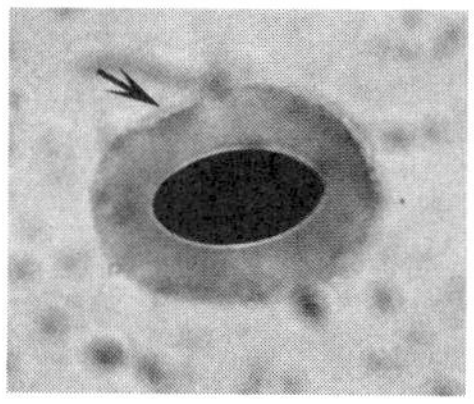

Capsule of the ascospores of *Sordaria fimicola*, x 550 (*RTH*).

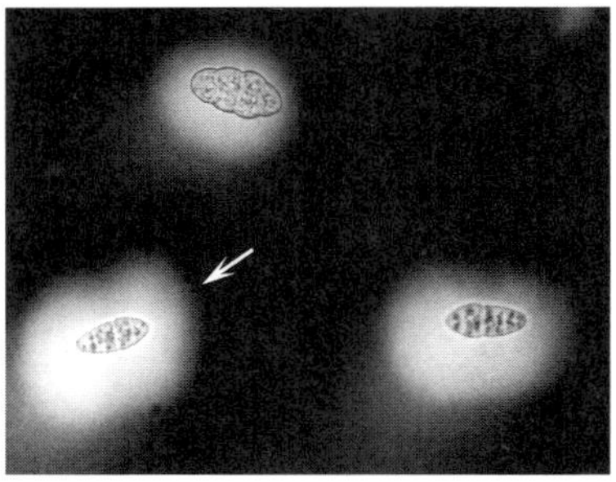

Capsule of the ascospores of *Leptosphaerulina crassiasca*, x 200 (*RTH*).

carbonaceous (L. *carbo*, genit. *carbonis*, carbon, charcoal + suf. *-aceus* > E. *-aceous*, of or pertaining to, with the nature of): similar to carbon, i.e., dark in color, almost black, and fragile or brittle; also used for substances composed predominantly of carbon. Many pyrenomycetes have carbonaceous perithecia or perithecial stromata, such as those of *Camillea*, *Hypoxylon*, *Phylacia* and *Xylaria* (Xylariales).

Carbonaceous stroma of *Camillea* sp., x 10 (*CB*).

Carbonaceous stromata of *Phylacia sagraeana*, x 14 (*CB*).

carbonicole, carbonicolous (L. *carbo*, genit. *carbonis*, coal + *-cola*, inhabitant; or + *-osus* > OF. *-ous*, *-eus* > E. *-ous*, possessing the qualities of): living on burned wood or soil, or on coal. See **anthracobiontic**, **phoenicoid fungus** and **pyrophilous**.

carinate (L. *carinatus*, keel-shaped < *carina*, keel + suf. *-atus* > E. *-ate*, provided with or likeness): equal to **cymbiform**, **navicular** and **scaphoid**.

carneous (L. *carneus*, fleshy, pulpy < *caro*, genit. *carnis*, meat + *-osus* > OF. *-ous*, *-eus* > E. *-ous*, having, possessing the qualities of): with the color of meat or flesh. See **carnose**.

carnose, **carnous** (L. *carnosus*, fleshy, pulpy < *caro*, genit. *carnis*, meat + *-osus*, full of, augmented, prone to > ME. *-ose*; or + *-osus* > OF. *-ous*, *-eus* > E. *-ous*, having, possessing the qualities of): fleshy, with the consistency of meat, like the majority of Agaricales, whose basidiocarps have a fleshy pileus. See **carneous**.

carpogenic (Gr. *karpós*, fruit + *génos*, origin + suf. *-íkos* > L. *-icus* > E. *-ic*, belonging to, relating to): the basal cell of the **carpogonium** in the thalli of Laboulbeniaceae, which eventually develops into the ascogonium.

carpogonium, pl. **carpogonia** (Gr. *karpós*, fruit + *gónos*, procreation, progeny, seed + L. dim. suf. *-ium*): the female sex organ of the thallus of Laboulbeniaceae, which is composed of three cells, **carpogenic**, **trichophore**, and the terminal one being the **trichogyne** (the latter also called **gynotrichous**). The carpogonium results from cell divisions of the developing ascocarp. The basal carpogenic cell of the carpogonium eventually produces the ascogonium.

Carpomycetes, **Carpomyceteae** (Gr. *karpós*, fruit + L. *-mycetes*, ending of class < Gr. *mýkes*, genit. *mýketos*, fungus): a name applied to the ascomycetes and basidiomycetes in general, which have more or less complex fruiting bodies (**carpophores**), in contrast to the fungi considered simpler and less advanced (phycomycetes).

carpophore (Gr. *karpós*, fruit + *-phóros*, bearer < *phéro*, to carry, sustain): the fruiting body of the higher fungi; a term used by French mycologists for fungi in which the conidiophores, ascophores and basidiophores tend to unite to form more complex fructifications that receive, respectively, the names conidiocarps, ascocarps, and basidiocarps. For that reason, these fungi are called **Carpomycetes**.

carpophoroid (Gr. *karpós*, fruit + *-phóros*, bearer < *phéro*, to carry, sustain + L. suf. *-oide*, < Gr. *-oeídes*, similar to): a false carpophore by reason of being sterile; found in some Agaricales, as occasionally happens in the basidiocarps of *Nyctalis parasitica* (Agaricales) that develop as mycoparasites on basidiocarps of *Russula* and *Lactarius* (Agaricales).

cartilagineus, **cartilaginous** (L. *cartilagineus*, gristly < *cartilago*, genit. *cartilaginis*, cartilage + *-osus* > OF. *-ous*, *-eus* > E. *-ous*, having, possessing the qualities of): refers to any structure with a consistency similar to that of animal cartilage, such as the peridium of the sporangium of *Leocarpus fragilis* (Physarales) and the stipe of the basidiocarp of *Clitocybe cartilaginea* (Agaricales).

Carpophoroids of *Nyctalis parasitica* on a basidiocarp of *Russula* sp., x 0.5.

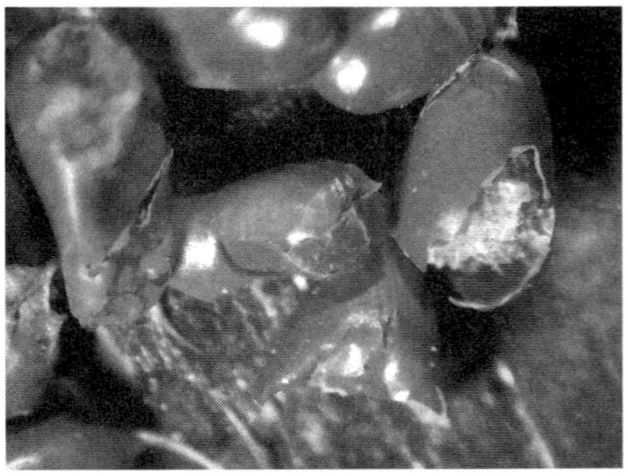

Cartilaginous peridium of the sporangia of *Leocarpus fragilis*, x 40 (*CB*).

caryallagic (L. *caryallagicus* < Gr. *káryon*, nucleus + *allagé*, permutation, commutation + suf. *-íkos* > L. *-icus* > E. *-ic*, belonging to, relating to): see **karyallagic**.

caryogamy (Gr. *káryon*, nucleus + *gámos*, sexual union + *-y*, E. suf. of concrete nouns): see **karyogamy**.

cata-species (Gr. *katá-*, pref. used in the formation of various terms in which it is desired to introduce the idea of descent or of an inferior, external, distant or opposed position; L. *species*, a shape, kind or sort, a particular kind): a term at times applied to the species of rusts (Uredinales) which in their life cycle have all of the spore stages except the spermogonium or pycnium.

catathecium, pl. **catathecia** (NL. *catathecium* < Gr. *katá*, downward, inverted + *thekíon*, dim. of *thêke*, case, box; here of the asci + dim. suf. *-íon* > L. *-ium*> L. *-ium*): syn. of **thyriothecium**.

catenate (L. *catenatus*, chained, arranged in chains < *catena*, chain + suf. *-atus* > E. *-ate*, provided with or likeness): arranged in chains. Cf. **catenulate**.

catenophysis, pl. **catenophyses** (L. *catena*, chain + Gr. *phýsis*, a being, growth): a persistent chain of thin-walled, utriculate cells formed by the vertical separation of the pseudoparenchyma in the centrum of some marine ascomycetes in the family Halosphaeriaceae, such as *Aniptodera* and *Lignincola* (Halosphaeriales). As the number of asci in the centrum of the maturing perithecium increases the catenophyses are pushed against the perithecial wall; at the time of ascospore discharge, the catenophyses are no longer evident.

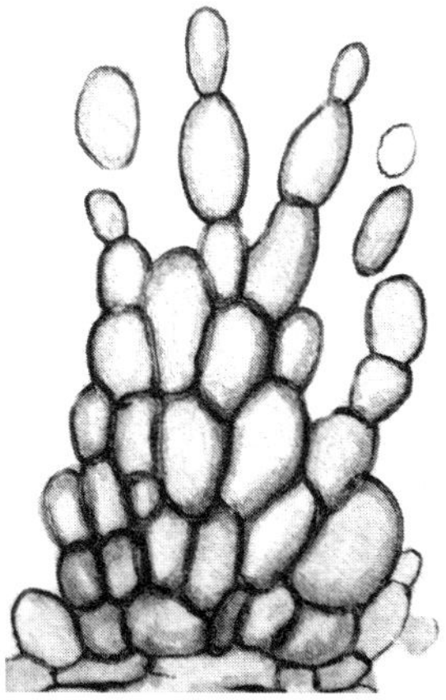

Catenophyses from the perithecium of *Aniptodera chesapeakensis*, x 325.

catenulate (L. *catenulatus*, arranged in little chains < *catenula*, dim. of *catena*, chain + suf. *-atus* > E. *-ate*, provided with or likeness): forming a chain, or having a chain-like structure; arranged end to end; e.g., like the sporangia of the aquatic fungus of the gen. *Catenaria* (Blastocladiales), or the conidia of some moniliaceous asexual fungi, such as *Aspergillus*, *Paecilomyces*, *Penicillium* and *Scopulariopsis*.

Catenulate conidia of *Aspergillus flavo-furcatis*, x 770 (*MU*).

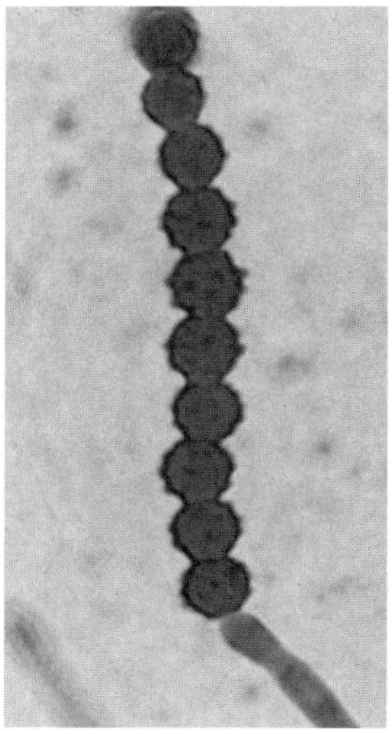

Catenulate conidia of *Scopulariopsis brevicaulis*, x 1 000 (*CB*).

caudate (L. *caudatus* < *cauda*, tail + suf. *-atus* > E. *-ate*, provided with or likeness): with a tail, like the pseudoplasmodium of *Dictyostelium* (Dictyosteliomycota), which on moving across a substrate leaves a tail or trail of mucilage behind it. Also applied to spores with a tail-like appendage.

caulescent (NL. *caulescens*, genit. *caulescentis* < *caulescere*, to sprout < *caulis*, stem + *-escens*, genit. *-escentis*, that which turns, beginning to, slightly > E. *-escent*): capable of forming a stem or pedicel.

caulicole, **caulicolous** (L. *caulis*, stem + *-cola*, dweller < *colere*, to inhabit; or + L. *-osus* > OF. *-ous*, *-eus* > E. *-ous*, possessing the qualities of): living or developing on the stem or branches of a host plant or plants that serve as support. Many parasitic and saprobic fungi, as well as various lichens, are caulicolous.

cauligenous, **caulogenous** (L. *caulis* < Gr. *kaulós*, stem + Gr. *génos*, origin + L. *-osus* > OF. *-ous*, *-eus* > E. *-ous*, possessing the qualities of): arising from or forming on the stems of plants.

caulocystidium, pl. **caulocystidia** (Gr. *kaulós*, stem, trunk; here, stipe + *kystídion* < *kýstis*, bladder, vesicle, cell + dim. suf. *-ídion* > L. *-idium*): *Agaricales*. A dermatocystidium that is found on the surface of the stipe of certain agarics; e.g., in *Pouzarella versatilis* and *Panaeolus sphinctrinus* (Agaricaceae), and Clavariaceae, such as *Pterula* spp. Cf. **pileocystidium**.

cavernose, **cavernous** (L. *cavernosus*, full of hollows < *caverna*, cavern, hollow + *-osus*, full of, augmented, prone to > ME. *-ose*; or + *-osus* > OF. *-ous*, *-eus* > E. *-ous*, having, possessing the qualities of): having holes or cavities, like the fruiting body of the hypogeous fungi or truffles (Pezizales). Cf. **farctate**, **fistular** and **solid**.

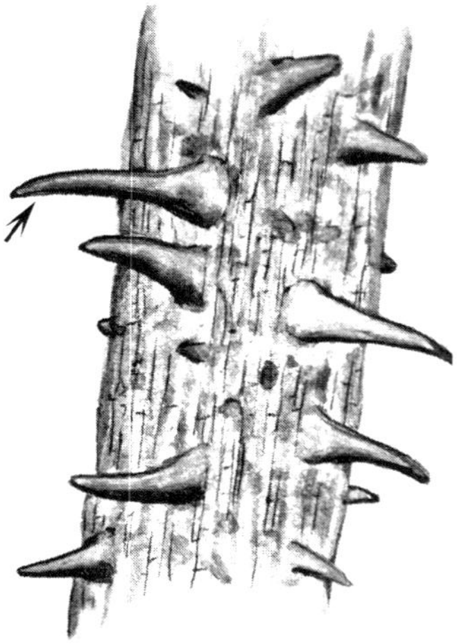

Caulocystidia of the filiform basidiocarp of *Pterula* sp., x 500.

cell wall (L. *cellula*, cell, living cell, dim. of *cella*, small room, compartment, cell of a honeycomb; ME. < OE. *weall* < L. *vallum*, rampart < *vallus*, stake, palisade, wall): the covering of fungal cells, whether individual or forming part of hyphae and tissues, that is located outside the plasma membrane; it is composed of a network of fibrils of variable composition, depending upon the group of fungi being considered, that are immersed in an amorphous matrix of glucans and proteins. In the Myxomycota only the spores have a cell wall and this contains cellulose. In the Chytridiomycetes the zoospores lack a wall; in the hyphae and spores the wall can contain chitin and chitosan in Zygomycota, cellulose and chitin in Oomycota, and chitin, protein and glucan in Ascomycota and Basidiomycota.

cellular dermis (NL. *cellularis* < *cellula*, cell, living cell, dim. of *cella*, small room, compartment, cell of a honeycomb + suf. *-aris* > E. *-ar*, like, pertaining to; Gr. *dérma*, skin, cortex): see **cystoderm**.

cellular slime mold (NL. *cellularis* < *cellula*, cell, living cell, dim. of *cella*, small room, compartment, cell of a honeycomb + suf. *-aris* > E. *-ar*, like, pertaining to; ME. < OE. *slim*, to smooth < L. *lima*, file; a soft, mucous or mucoid substance; ME. *mowlde* < *mowled*, ptp. of *moulen*, *molwen*, to grow moldy): one of the Acrasiomycota or Dictyosteliomycota, which form pseudoplamodia from the aggregation of myxamoebae that retain their individuality, i.e., they do not fuse to form a true plasmodium.

cellulin (L. *cellula*, cell, living cell, dim. of *cella*, small room, compartment, cell of a honeycomb + NL. *-in*, suf. used in chemistry to denote an activator or compound): a polysaccharide disposed in the form of granules or refringent disks, which transitorially plug the hyphae of the Oomycota of the order Leptomitales (*Leptomitus*, *Apodachlya*), mainly at the level of the constrictions. These plugs of cellulin constitute what are called **pseudosepta**.

cellulolytic (NL. *cellulose* < L. *cellula*, cell, living cell, dim. of *cella*, small room, compartment, cell of a honeycomb + suf. *-ose*, ending for carbohydrates + Gr. *lytikós*, able to loosen < *lýtos* dissolvable, broken + suf. *-íkos* > L. *-icus* > E. *-ic*, belonging to, relating to): having the capacity to degrade cellulose in substrates where it grows; many fungi are cellulolytic, e.g., the species of *Chaetomium* (Sordariales) and *Trichoderma* (moniliaceous asexual fungi).

centric (L. *centrum* < Gr. *kéntron*, center + Gr. suf. *-íkos* > L. *-icus* > E. *-ic*, belonging to, relating to): **1.** *Aquatic fungi*. A type of oospore in which the ooplast is in the center and the reserve substances of the periplasm (oil globules) are arranged regularly surrounding the ooplast, as in *Saprolegnia hypogyna* (Saprolegniales). **2.** *Agaricaceae*. A stipe that is united with the central part of the pileus of the basidiocarp. Cf. **eccentric**.

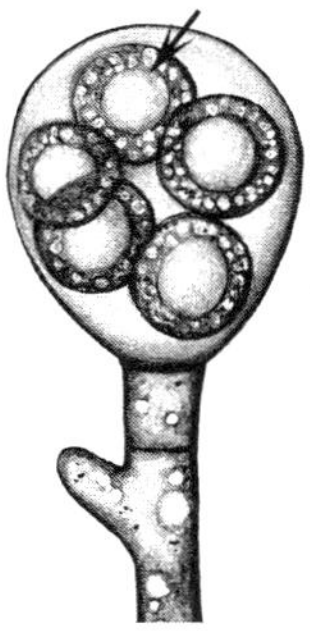

Centric oospores in the oogonium of *Saprolegnia hypogyna*, x 125.

centrifugal (NL. *centrifugus*, center-fleeing < L. *centrum*, center + *fugus* < *fugere*, to flee, go away from + L. suf. *-alis* > E. *-al*, relating to or belonging to): moving outward from the center toward the periphery; maturing or developing from the center outward. In lichenology it refers to the apothecia that preferentially occupy the borders of the thallus. Cf. **centripetal**.

centriole, **centriolum**, pl. **centriola** (L. *centriolum* < *centrum*, center + dim. suf. *-olum* > E. *-ole*): a cellular organelle that during mitosis forms the polar bodies of the achromatic spindle. In the zoosporic fungi it gives rise to the kinetosome or blepharoplast, in the base of the flagellum. Cf. **spindle pole body**.

centripetal (NL. *centripetus*, center-seeking < L. *centrum*, center + *petus* < *petere*, to seek + L. suf. *-alis* > E. *-al*, relating to or belonging to): moving inward from the periphery toward the center; maturing or developing from the outside inward. In lichenology it is applied to those apothecia that for the most part are grouped in the central portion of the thallus. E.g., the aggregation currents of the myxamoebae that form the pseudoplasmodium of the Dictyosteliomycota (*Dictyostelium*, *Polysphondylium*, etc.) are centripetal. Cf. **centrifugal**.

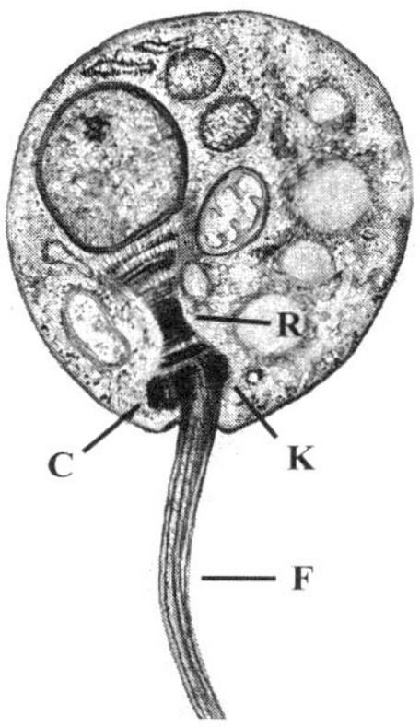

Centriolum (C) of the zoospore of *Rhizophlyctis rosea*. The kinetosome (K), rhizoplast (R), and flagellum (F) are also shown, x 14 000.

centrum, pl. **centra** (L. *centrum* < Gr. *kéntron*, center): the complex of fertile and sterile structures occupying the interior of an ascocarp. The type of centrum present in different ascomycetes is a fundamental criterion for modern classification of these fungi. The developing centrum can be seen in a young perithecium of *Sordaria fimicola* (Sordariales).

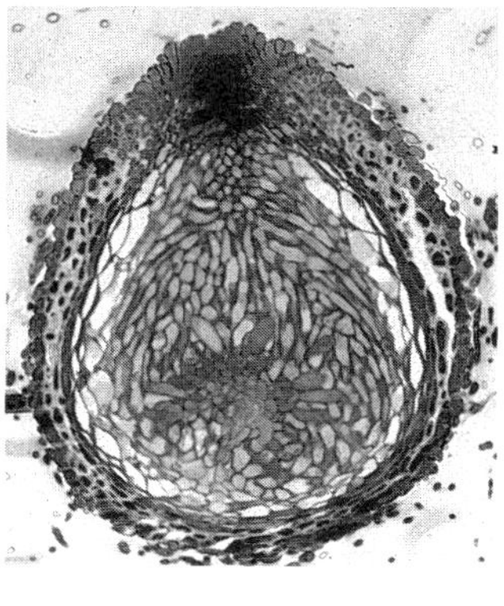

Centrum of the developing perithecium of *Sordaria fimicola*, x 130 (*RTH*).

cephalodium, pl. **cephalodia** (Gr. *kephalé*, head + L. *-odium*, resembling < Gr. *ode*, like < *-oeídes*, similar to): *Lichens*. A small globose body, irregular or verrucose, that contains an alga different from that present in the majority of the lichen thallus; i.e., the algae of the cephalodia are Cyanophyceae (they contribute to the nutrition of the lichen by fixing atmospheric nitrogen) and the algae of the rest of the thallus are Chlorophyceae; at times, the cephalodia function as propagules of vegetative multiplication. *Stereocaulon alpinum* (Lecanorales), e.g., has cephalodia.

Cephalodia of the thallus of *Stereocaulon saxatile*, x 12 (*MU*).

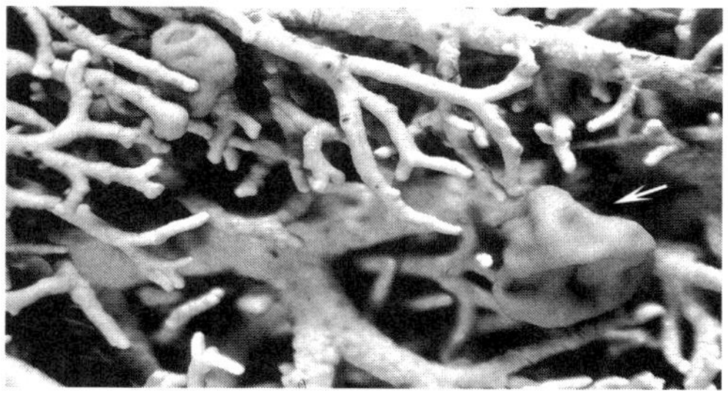

Cephalodia of the thallus of *Stereocaulon ramulosum* f. *elegans*, x 20 (*MU*).

cephalothecoid (< gen. *Cephalotheca* < Gr. *kephalé*, head + *thêke*, box, case + L. suf. *-oide* < Gr. *-oeídes*, similar to): having a peridium in which the cells split along preformed sutures, as in the cleistothecial ascocarps of the *Cephalotheca* (probably related to *Pseudeurotium*) of the Eurotiales, and a few unrelated gen.; the peridium is composed of radiating plates of cells that originate from various meristematic zones, as in some spp. of *Zopfiella* (Lasiosphaeriaceae of Sordariales), a marine saprobe.

ceraceous (L. *ceraceus*, waxy < *cera*, wax + *-aceus* > E. *-aceous*, of or pertaining to, with the nature of): having the appearance or consistency of wax, e.g., like the basidiocarps of some species of *Clavaria* (Aphyllophorales), which are white, delicate and soft.

cerebriform, **cerebroid**, **cerebrose** (L. *cerebriformis* < *cerebrum*, brain + *-formis* < *forma*, shape; or *cerebrum* + L. suf. *-oide* < Gr. *-oeídes*, similar to; or + L. *-osus*, full of, augmented, prone to > ME. *-ose*): with folds, bends and undulations which give a brain-like appearance, e.g., the thallus of *Psora cerebriformis* (Lecanorales). Syn. of **convolute**.

Cerebriform thallus of *Psora cerebriformis*, x 10 (*CB*).

cernuous (L. *cernuus*, turned toward the earth + L. *-osus* > OF. *-ous*, *-eus* > E. *-ous*, possessing the qualities of): hanging down; drooping, nodding.

chalaroplectenchyma, pl. **chalaroplectenchymata** (Gr. *chaláros*, loose, lax + *plektós*, intertwined + *énchyma*, stuffed): *Lichens*. A tissue composed of loosely interwoven hyphae, with distinctive holes present, found in the medulla of lichen thalli. The crystals of various lichen products often are found incrusting the walls of these medullary hyphae. Cf. **paraplectenchyma** and **prosoplectenchyma**.

character (ME. *caracter* < MF. *caractere* < L. *character*, mark, distinctive quality): a variable feature of a given taxon.

chartaceous (L. *chartaceus*, papery < *charta*, paper + *-aceus* > E. *-aceous*, of or pertaining to, with the nature of): with the consistency of paper or of parchment; e.g., like the basidiocarp of *Polyporus pergamenus* (Aphyllophorales).

cheilocystidium, pl. **cheilocystidia** (Gr. *cheílos*, lip, border, margin + *kystídion* < *kýstis*, bladder, vesicle; here, cell + dim. suf. *-ídion* > L. *-idium*): a marginal cystidium of diverse shapes according to the species, that arises from the edge of a gill, as in *Nolanea hirtipes* and *Marasmius* (Agaricales), in which case they are strangulate or tortuous.

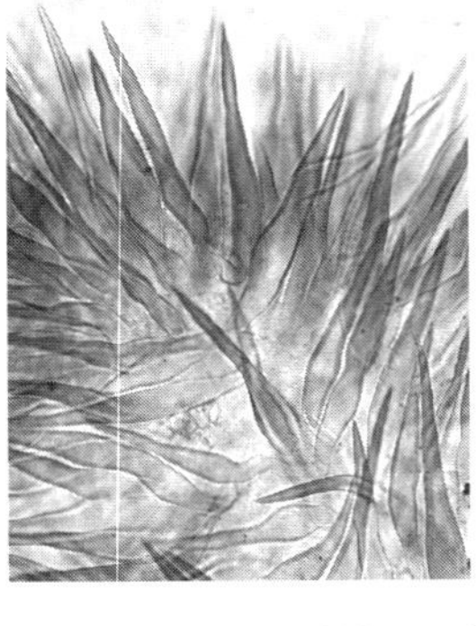

Cheilocystidia of the hymenophore of *Marasmius* sp., x 230 (*EPS*).

cheiroid, **chiroid** (Gr. *cheír*, *cheirós*, hand + L. suf. *-oide* < Gr. *-oeídes*, similar to): in the shape of a hand; e.g., like the spores of *Cheiromyces* (dematiaceous asexual fungi) or the haustoria of *Blumeria* (=*Erysiphe*) *graminis* (Erysiphales).

cheiropterocoprophilous or **chiropterocoprophilous** (Gr. *cheír*, *cheirós*, hand + *pterón*, wing + *kópros*, dung + *phílos*, have an affinity for + L. *-osus* > OF. *-ous*, *-eus* > E. *-ous*, possessing the qualities of): applied to organisms that live on bat's dung in caves, as *Histoplasma* (moniliaceous asexual fungi), *Guanomyces* (Sordariales), and *Coprinus* (Agaricales), among many others.

cheirospore, **cheiroid** or **chiroid spore**, **chirospore** (Gr. *cheír*, *cheirós*, hand + *sporá*, spore): a multicellular conidium whose appearance recalls that of a hand with the fingers more or less together, as is present in *Cheiromyces stellatus* and *Dictyosporium elegans* (dematiaceous asexual fungi).

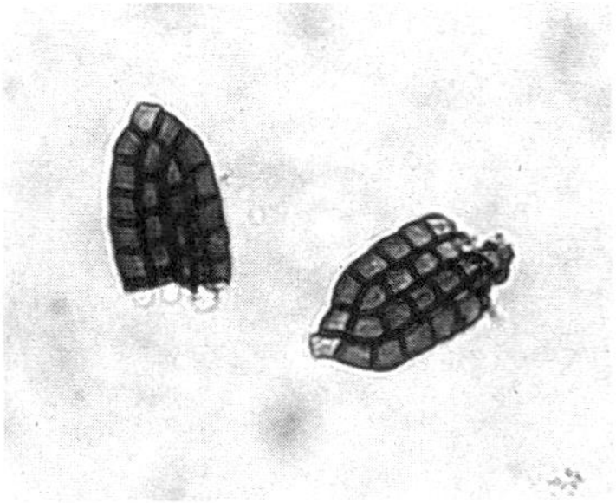

Cheirospores of *Dictyosporium elegans*, x 325 (*CC*).

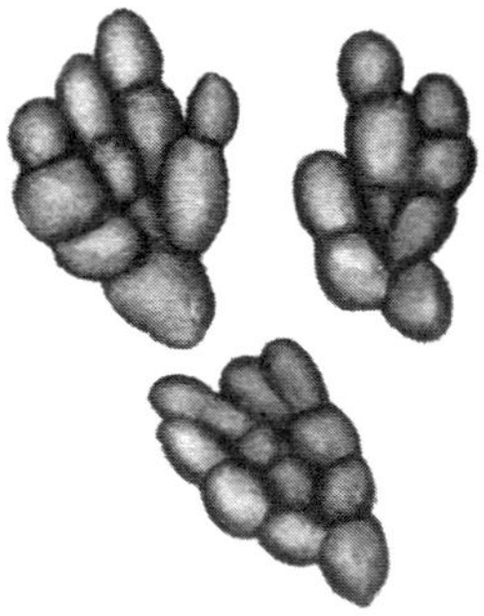

Cheirospores or **chirospores** of *Cheiromyces stellatus*, x 1 000.

chemotaxis, pl. **chemotaxes** (Gr. *chymós*, substance + *táxis*, orientation): a tactic phenomenon attributable to a chemical agent, as with the zoospores of *Saprolegnia* (Saprolegniales), which swim toward water extracts of dead insects and toward aqueous solutions of their decomposition products, such as proteins, phosphates, urea and organic acids, when they are adequately diluted, since a substance that attracts in weak solutions repels when concentrated.

Chemotactic reactions to this type of external stimulus can be related to the distribution and nutrition of the zoospores. Similar responses are present in the approach and fusion of the sexual organs, e.g., of antheridia and oogonia in the Oomycota, progametangia of Zygomycota and gametangia of ascomycetes, responses of obvious biological importance. See **chemotropism**.

chemotropism (Gr. *chymós*, substance + *trópos*, a turn, change in manner < *tropé*, a turning < *trépo*, to revolve, turn towards + *-ismós* > L. *-ismus* > E. *-ism*, state or condition): the movement or orientation of an organism, or part of an organism, due to the influence of a chemical agent. In essence, it is the same as chemotaxis, but the latter implies the free movement of the organisms in order to position themselves tactically (e.g., of zoospores). Chemotropism, like tropisms in general, is negative or positive, depending upon how the organism responds in orienting itself, whether toward the point from which the stimulus emanates, or away from it. An example of negative chemotropism is the response of hyphae and germ tubes of fungi which grow away from areas that have high concentrations of the products of their own metabolism, such as ammonia and potassium bicarbonate. It is known that, in general, depending upon the concentration, germ tubes and hyphae grow toward sources of salts, sugars and other nutritive substances (positive chemotropism), and away from acids, alkalis, and alcohols. See **chemotaxis**.

chiastobasidium, pl. **chiastobasidia** (Gr. *chiastós*, crossed, in the shape of the wings of a windmill + L. *basidium* < Gr. *basídion* < *básis*, base + dim. suf. *-ídion* > L. *-idium*): a type of basidium without septa (holobasidium), enlarged apically, in which the achromatic spindle is arranged perpendicularly to the major axis during the second meiotic division of the nucleus. This type of basidium is present in all of the agarics and boletes. Cf. **stichobasidium**.

chimeroid (L. *chimaera* < Gr. *chímaira*, she-goat, chimera, a she-monster in Greek mythology + L. suf. *-oide* < Gr. *-oeídes*, similar to): refers to lichen thalli, because they are compounded of apparently incongrous parts.

chionophilous (Gr. *chiónos*, snow + *phílos*, have an affinity for + L. *-osus* > OF. *-ous*, *-eus* > E. *-ous*, possessing the qualities of): also called **nivicolous**.

chitinoclastic (Gr. *chitón*, tunic, alluding to the exoskeleton of insects + *klástos*, broken + suf. *-íkos* > L. *-icus* > E. *-ic*, belonging to, relating to): having the ability to disintegrate chitin from the substrates on which it grows, as certain aquatic fungi in the order Chytridiales can do. Also called **chitinolytic**. *Abyssomyces hydrozoicus* is an abyssal marine ascomycete, of uncertain affinities, that lives as a parasite (or saprobe ?) on the chitinous exoskeleton of hydrozoarians. The hyphae of this fungus form canals on the surface of the hydrozoarian colonies, and even penetrate this substrate, which suggests that it is an important degrader of chitin in the deep sea.

chitinolytic (Gr. *chitón*, tunic, alluding to the dermatoskeleton of insects + *lytikós*, able to loosen < *lýtos*, dissolvable, broken + suf. *-íkos* > L. *-icus* > E. *-ic*, belonging to, relating to): see **chitinoclastic**.

chitosome (Gr. *chitón*, tunic, alluding to the dermatoskeleton of insects; chitin + *sõma*, body): a small spheroidal structure (40-70 nm in diam.) found in many fungi, containing chitin synthetase zymogen, involved in cell wall synthesis.

chlamydocyst (Gr. *chlamýs*, *chlamýdos*, cape, cloak + NL. *cystis* < Gr. *kýstis*, bag, bladder, vesicle): *Blastocladiales*. A long-lasting or latent zoosporangium, with two enveloping walls, found in a hypha.

chlamydospore (Gr. *chlamýs*, *chlamýdos*, cape, cloak + *sporá*, spore): a thick-walled, asexual, resistant spore; with respect to ontogeny, chlamydospores develop like holothallic conidia, but their function is survival rather than propagative. In the past this term was applied incorrectly to the teliospores of the heterobasidiomycetes, which are of sexual origin. Many fungi form chlamydospores of some sort, including, e.g., Mucorales (*Mucor racemosus*), *Fusarium oxysporum* (tuberculariaceous asexual fungi) and *Harposporium anguillulae* (moniliaceous asexual fungi), *Candida albicans* (cryptococcaceous asexual yeasts), and *Phoma eupyrena* (sphaeropsidaceous asexual fungi).

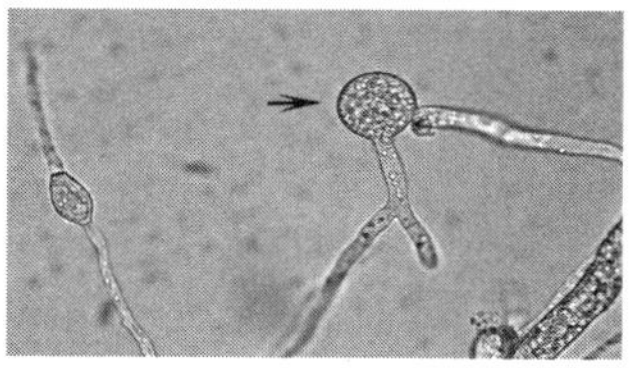

Chlamydospores of *Mucor racemosus*, x 680 (*MU*).

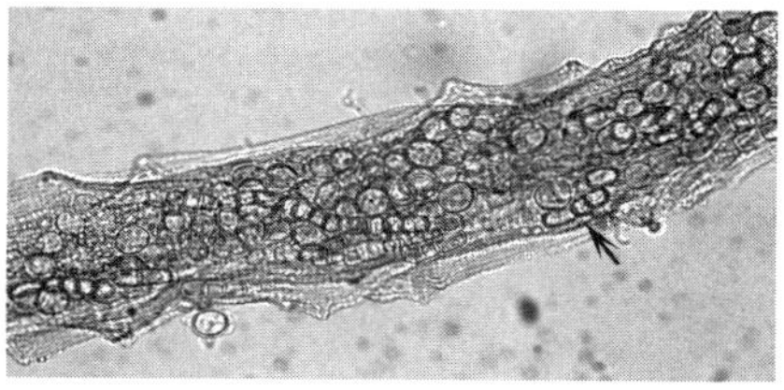

Chlamydospores of *Harposporium anguillulae* within a parasitized nematode, x 200 (*PL*).

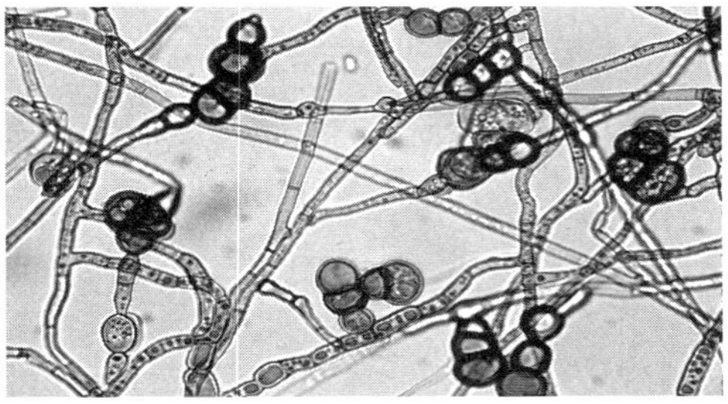

Chlamydospores of *Fusarium oxysporum*, x 250 (*MU*).

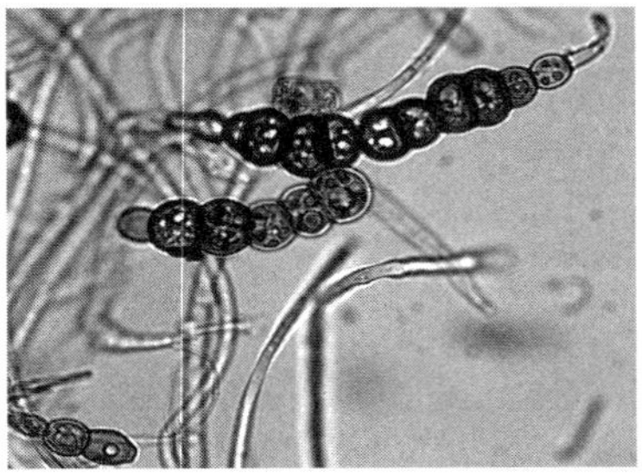

Chlamydospores of *Fusarium* sp., x 250 (*MU*).

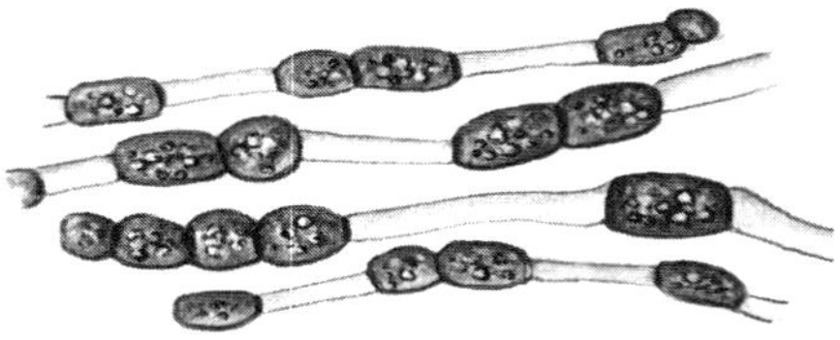

Chlamydospores of *Phoma eupyrena*, x 670.

chlorophycophilous (Gr. *chlorós*, green + *phýkos*, alga + *phílos*, have an affinity for + L. *-osus* > OF. *-ous*, *-eus* > E. *-ous*, possessing the qualities of): *Lichens*. Thalli in which the phycobiont belongs to the Chlorophyceae, as occurs in the majority of species. Cf. **cyanophycophilous**. See **diphycophilous**.

chondroid (Gr. *kóndros*, cartilage + L. suf. *-oide* < Gr. *-oeídes*, similar to): *Lichens*. The medullar layer of the thallus when the hyphae that constitute it are intimately united and have a cartilaginous consistency, e.g., as in *Usnea* (Lecanorales).

chondroid axis, pl. **axes** (Gr. *kóndros*, cartilage + L. suf. *-oide* < Gr. *-oeídes*, similar to; L. *axis*, axle, axis): *Lichens*. The cartilaginous axis occupying the central portion of the medulla in the thallus of *Usnea* (Lecanorales).

chromogenic, **chromogenous** (Gr. *chrõma*, color + *génos*, origin + suf. *-íkos* > L. *-icus* > E. *-ic*, belonging to, relating to; or + L. *-osus* > OF. *-ous*, *-eus* > E. *-ous*, having, possessing the qualities of): producing a colored material or substance.

chromomycosis, pl. **chromomycoses** (Gr. *chrõma*, color + *mýkes*, fungus + *-osis*, condition or state): a chronic, subcutaneous, granulomatous mycosis of man and higher animals caused by the implantation in the skin of any of several dematiaceous fungi of exogenous sources (wood, soil, and plant remains). In general, it is a localized infection of the skin and subepidermal tissues, especially of lower limbs (although it may occur elsewhere in the body), marked by verrucoid nodules and tumor-like masses. Among the principal causal agents of chromomycosis are *Phialophora verrucosa*, *Rhinocladiella* (=*Fonsecaea*) *pedrosoi* and *Cladosporium carrionii* (dematiaceous asexual fungi).

chromosome (Gr. *chrõma*, color + *sõma*, body): one of the linear or sometimes circular basophilic bodies of prokaryotic and eukaryotic organisms that contain protein and most or all of the DNA or RNA comprising the genes of the individual.

chrysocystidium, pl. **chrysocystidia** (Gr. *chrysós*, gold, golden-yellow + L. *cystidium* < Gr. *kystídion* < *kýstis*, bladder; here vesicle + dim. suf. *-ídion* > L. *-idium*): *Agaricales*. A clavate or fusiform-clavate pseudocystidium, with an inclusion of yellow color that accentuates on being treated with ammonia; a characteristic of *Hypholoma* (=*Naematoloma*), *Psilocybe* and *Stropharia*, among other gen. Cf. **lamprocystidium**, **leptocystidium** and **macrocystidium.**

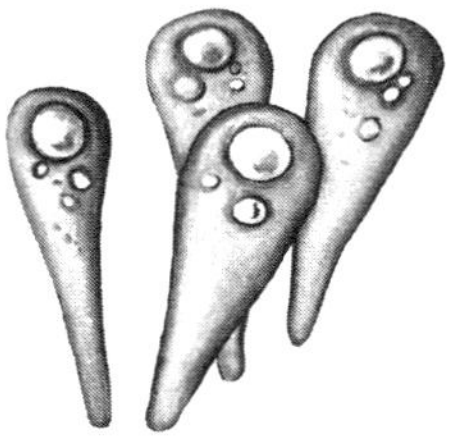

Chrysocystidia from the hymenium of *Stropharia* sp., x 500.

chytrid (Gr. *chytrídion*, dim. of *chytrís*, kettle): one of the Chytridiomycetes (a class) or Chytridiomycota (a phylum) of kingdom Fungi in some recent classification systems (e.g., Margulis and Schwartz, 1988, and Alexopoulos *et al.*, 1996).

Chytridiomycetes (Gr. *chytrídion*, dim. of *chytrís*, kettle + L. *-mycetes*, ending of class < Gr. *mýkes*, genit. *mýketos*, fungus): according to the classification system adopted in this dictionary, this is a group of aquatic organisms, considered as the phylum Chytridiomycota of the kingdom Fungi, with zoospores uniflagellate at the posterior end, a unicellular or filamentous thallus, and sexual reproduction by planogametic copulation. The class Chytridiomycetes includes the orders Spizellomycetales, Neocallismasticales, Chytridiales,

Blastocladiales and Monoblepharidales. Some authors, such as Margulis and Schwartz (1988), classify these organisms as a phylum in the kingdom Protoctista.

chitridiomycosis, pl. **chitrydiomycoses** (Gr. *chytrídion*, dim. of *chytrís*, kettle + *mýkes*, fungus + *-osis*, condition, state of something): a disease of animals caused by a chytrid; especially applied to the disease of epidermis of frogs caused by *Batrachochytrium dendrobatidis* (Chytridiales).

cingulate (L. *cingulatus* < *cingulum*, strap, belt + suf. *-atus* > E. *-ate*, provided with or likeness): having a belt or girdle-like zone, as if surrounded by a sash. With a border all around, like the ascospores of *Glomerella cingulata* (Phyllachorales), which have a central constriction.

circadian rhythm (L. *circa*, near to or around + *dies*, day + L. suf. *-anus* > E. *-an*, belonging to; Gr. *rhytmós*, rhythm, rhythmic movement): the periodic and cyclic modification, such as cell division, sporulation, maximum rate of photosynthesis, bioluminescence, or enzyme production, among other phenomena that are present in living cells and whose periodicity is ca. 24 hours.

circinate (L. *circinatus*, rounded off, rolled up < *circinare*, to make round + suf. *-atus* > E. *-ate*, provided with or likeness): coiled inward from the apex; e.g., like the sporangiophores of *Circinella umbellata* (Mucorales) and the hairs of *Chaetomium* sp. (Sordariales).

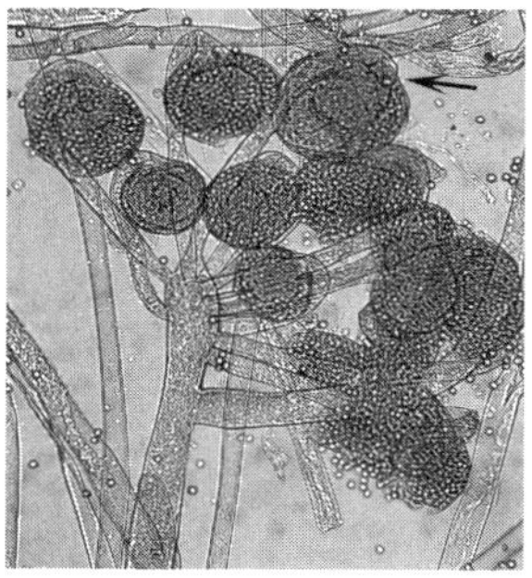

Circinate sporangiophores of *Circinella umbellata*, x 100 (*EAA*).

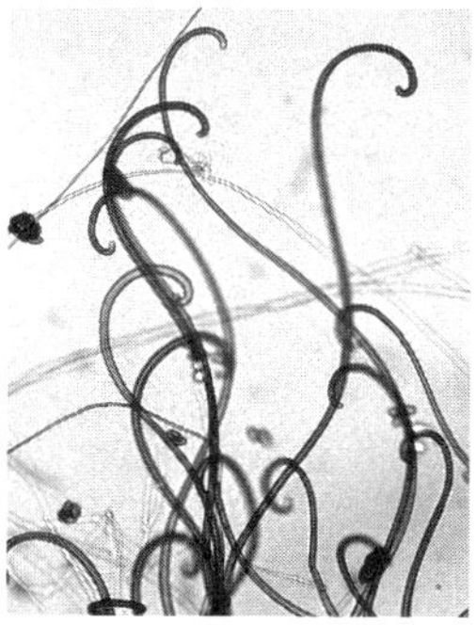

Circinate perithecial hairs of *Chaetomium* sp., x 200 (*MU*).

circumscissile, **circumscissile dehiscence** (NL. *circumcisilis* < *circumcisus*, cut all around < *circum*, around + *scissile*, capable of splitting < *scindere*, to cut + suf. *-ilis* > E. *-ile*, capable, of the character of; L. *dehiscentia* < *dehiscere*, to open itself, to part + suf. *-entia* > F. *-ence* < E. *-ence*, state, quality or action): opening or separating along a circular or equatorial line; e.g., the sporangia of the Myxomycete *Craterium minutum* (Physarales), the Zygomycetes *Choanephora cucurbitarum* and *Gilbertella persicaria* (Mucorales), and the basidiocarp of *Cyathus* (Nidulariales) have circumscissile dehiscence.

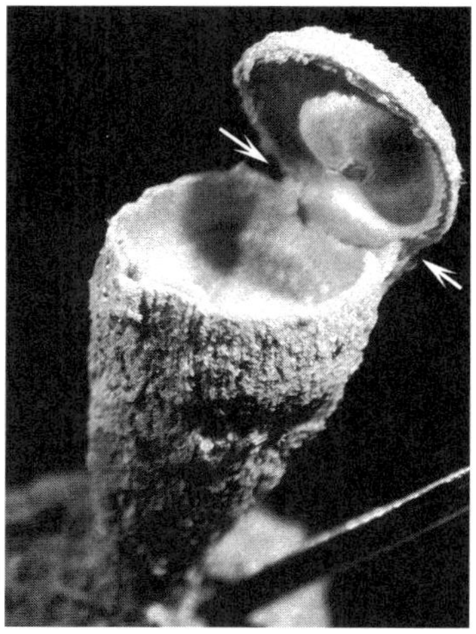

Circumscissile dehiscence of the fruiting body of *Cyathus* sp., x 6.5 (*MU*).

cirrate, **cirrhate** (L. *cirratus*, curly < *cirrus*, curl of hair + suf. *-atus* > E. *-ate*, provided with or likeness): arranged in coils. Also called **cirriferous**.

cirrhus, pl. **cirrhi**; **cirrus**, pl. **cirri** (L. *cirrus*, ringlet of hair): a slender column, often curled, of spores held together by mucilage that emerges from the ostiole of certain fungal fruiting bodies; e.g., the pycnidia of *Phoma* and *Phomopsis* (sphaeropsidaceous asexual fungi), and the perithecia of *Microascus* (Microascales) and *Chaetomium* (Sordariales).

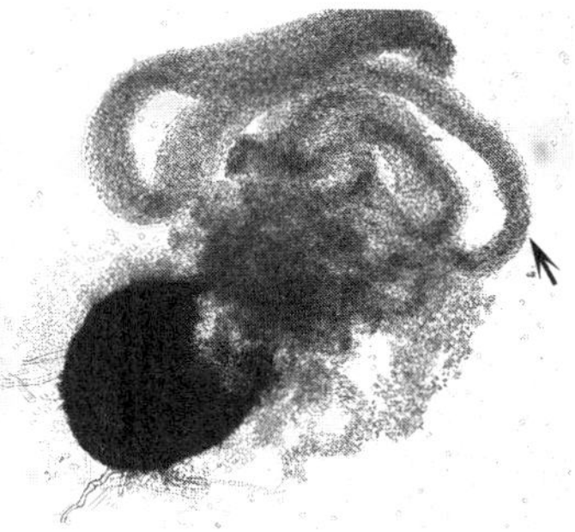

Cirrhus from the pycnidium of *Phoma sorghina*, x 150 (*PL*).

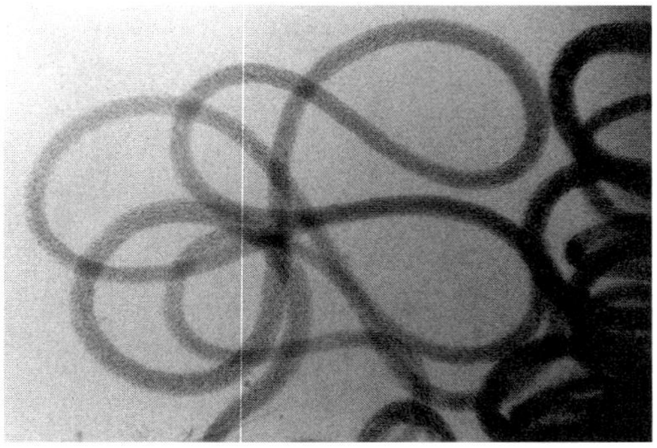

Cirrhus from the pycnidium of *Phoma* sp., x 220 (*RTH*).

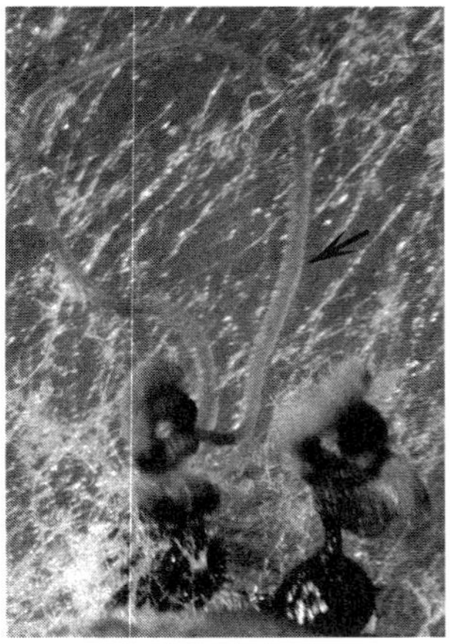

Cirrhus from the perithecium of *Microascus trigonosporus*, x 70 (*CB*).

cirriferous (L. *cirrus*, curl of hair + *-ferous*, bearer < *fero*, to bear, carry + *-osus* > OF. *-ous*, *-eus* > E. *-ous*, having, possessing the qualities of): syn. of **cirrate**.

citriform (L. *citrus*, citron, lemon + *-formis* < *forma*, shape): shaped like a lemon; e.g., like the sporangia of *Phytophthora* (Peronosporales).

cladistics (E. < Gr. *kládos*, branch + NL. *-istic*, pertaining to as agent): a method of systematics based on phylogenetic relationships, which aims to reconstruct the genealogical pathway of organisms by means of objective and repeatable analysis, and from this to propose a hypothesis of natural classification or phylogeny. The product of cladistic analysis is a branching diagram (**cladogram**) which shows the pattern of relationships among the organisms based on the characters used. See **classification**, **nomenclature**, **phylogeny**, **systematics**, and **taxonomy**.

cladogram (E. < Gr. *kládos*, branch + *grámma*, describe): see **cladistics.**

clamp connection (ME. *clamp* < MD. *clampe*, clamp; ME. *conneccioun*, *connexioun* < L. *connexion* < *connexus*, joined, connected + *-io*, *-ionis* > E. suf. *-ion*, *-tion*, result of an action, state of): *Basidiomycetes*. A slender hyphal branch that connects two adjacent cells at a septum. The passage created by the hyphal bridge allows one of the nuclei resulting from mitosis to move into the dorsal cell, bypassing the septum. Also called **fibula**.

classification (L. *classificare*, to classify; *classificatio*, genit. *classificationis*, the act of classifying or grouping together < *classis*, group + *-ationem*, action, state or condition, or result > E. suf. *-ation*): the systematic arrangement into groups or categories according to established criteria; **taxonomy**. The application of scientific names to the categories into which fungi may be placed and the relative order of those categories is governed by the International Code of Botanical Nomenclature. See **nomenclature**, **systematics**, and **taxonomy.**

clathrate, clathroid (L. *clathratus*, sieve-like, resembling a lattice < *clathrare*, to fit with bars < *clathra*, bar < Gr. *klãthra*, pl. of *kléithron*, bar + suf. *-atus* > E. *-ate*, provided with or likeness): *Gasteromycetes*. A latticed or interconnected receptaculum which supports the gleba; like a cage; e.g., like the fruiting bodies of the gen. *Clathrus* and *Laternea* (Phallales).

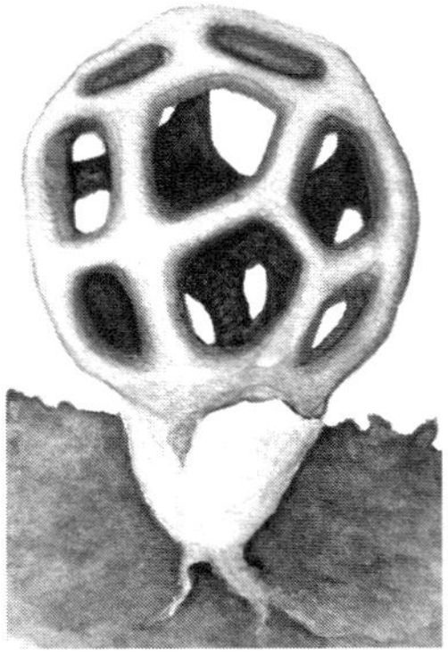

Clathrate fruiting body of *Clathrus ruber*, x 0.4.

Clathrate basidiocarp of *Laternea pusilla*, x 0.4.

clavate, **claviform** (NL. *clavatus*, club-shaped < L. *clava*, club + suf. *-atus* > E. *-ate*, provided with or likeness; L. *claviformis* < *clava* + suf. *-formis* < *forma*, shape): shaped like a club; e.g., like the conidial head of *Aspergillus clavatus* and *A. giganteus* (moniliaceous asexual fungi), the synnemata of *Hirsutella saussurei* (stilbellaceous asexual fungi), and the fruiting body of *Clavariadelphus truncatus* (Aphyllophorales).

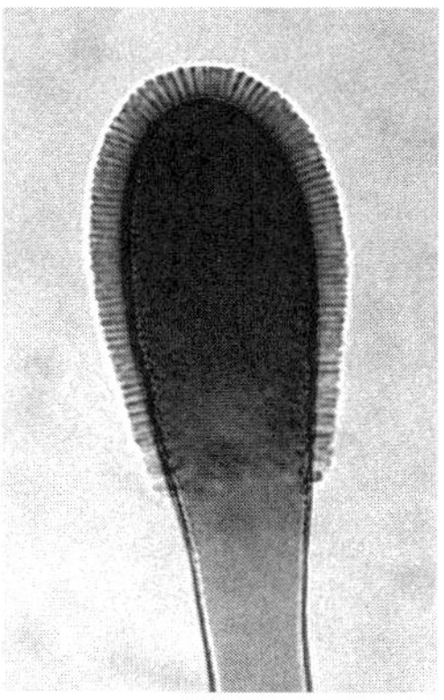

Clavate conidial head of *Aspergillus clavatus*, x 400 (*RTH*).

Clavate synnema of *Hirsutella saussurei*, x 220 (*MU*).

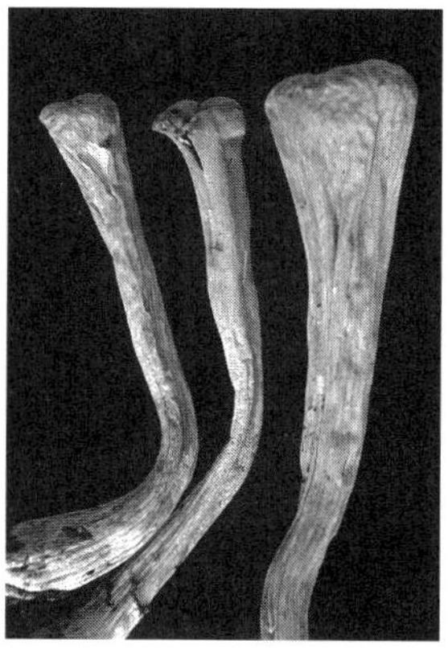

Clavate basidiocarps of *Clavariadelphus truncatus*, x 0.4 (*MU*).

Clavate basidiocarps of *Clavariadelphus truncatus*, x 0.3 (*MRO*).

clavule (L. *clavula*, small club < *clava*, club + dim. suf. *-ula* > E. *-ule*): a claviform fruiting body of fungi, such as *Clavaria pistillaris* (Aphyllophorales).

cleistocarp (Gr. *kleistós*, closed + *karpós*, fruit): a closed ascocarp, without a pore, whose ascospores are liberated only by rupture or degradation of the wall or peridium; e.g., as occurs in the truffles (*Tuber* and other gen. of Tuberaceae, Pezizales).

cleistohymenial (Gr. *kleistós*, closed + L. *hymenium* < Gr. *hyménion*, dim. of *hymén*, membrane + L. suf. *-alis* > E. *-al*, relating to or belonging to): one of the three types of apothecium development in the Discomycetes, characterized by being closed during most of the time. Sterile hyphae growing out of the stalk cell of the ascogonium or from adjacent mycelial hyphae completely surround the ascogonium in a loose weft, forming a ball with ascogenous hyphae in the center. In some species the ascocarp remains permanently closed, as in hypogeous forms; in others the cortical layer over the developing hymenium is ruptured sometime after its formation. This type of development has been well studied in *Inermisia* (=*Byssonectria*), of the Pezizales. Cf. **eugymnohymenial** and **paragymnohymenial**.

cleistothecioid (Gr. *kleistós*, closed + *thêke*, case, box + L. suf. *-oide* < Gr. *-oeídes*, similar to): **1**. Any enclosed (nonostiolate) ascoma. **2**. A closed ascoma, usually spherical, similar to a cleistothecium, but differing in having the asci arranged in a hymenium or fascicle; e.g., as in *Brasiliomyces malachrae* (Erysiphales). It represents a perithecium or an apothecium that lacks the typical opening.

cleistothecium, pl. **cleistothecia** (NL. *cleistothecium* < Gr. *kleistós*, closed + *thekíon* < *thêke*, case, box; here, for the asci + dim. suf. *-íon* > L. *-ium*): **1**. An enclosed ascocarp (fruiting body), lacking a pore and delimited by a wall that breaks irregularly or disintegrates at maturity to liberate the ascospores. **2**. *Plectomycetes*.

cleistothecium

A type of ascocarp with ascohymenial ontogeny and small, spherical asci that are irregularly distributed throughout the inside of the ascocarp, which is enclosed by a peridium or wall that lacks a pore. E.g., as in *Emericella variecolor*, *Eurotium chevalieri* and *Monascus purpureus* (Eurotiales).

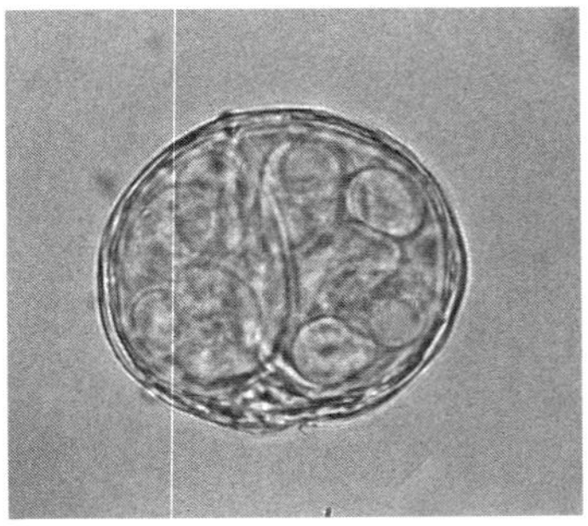

Cleistothecioid perithecium of *Brasiliomyces malachrae*, x 400 (*RTH*).

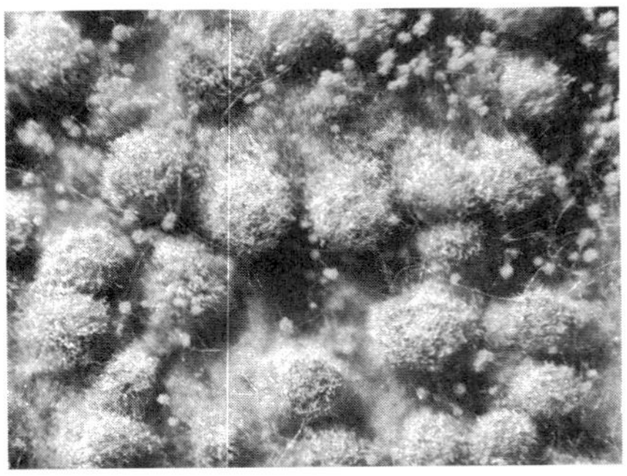

Cleistothecia of *Emericella variecolor* on agar, x 35 (*CB*).

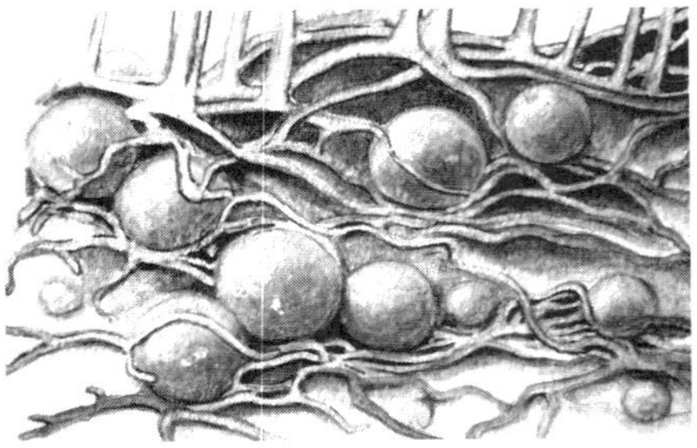

Cleistothecia of *Eurotium rubrum*, x 50.

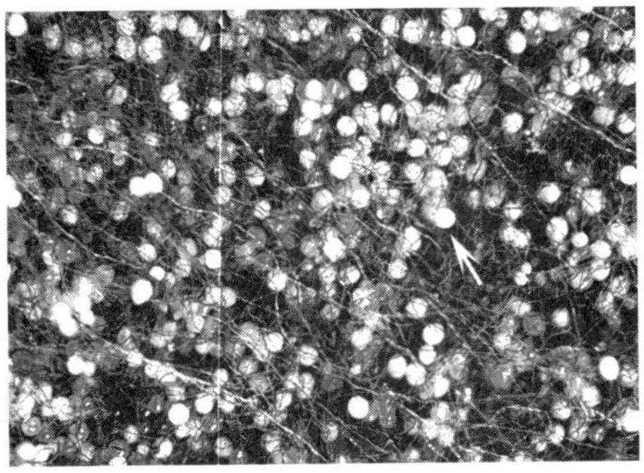

Cleistothecia of *Eurotium rubrum* growing on agar, x 8 (*MU*).

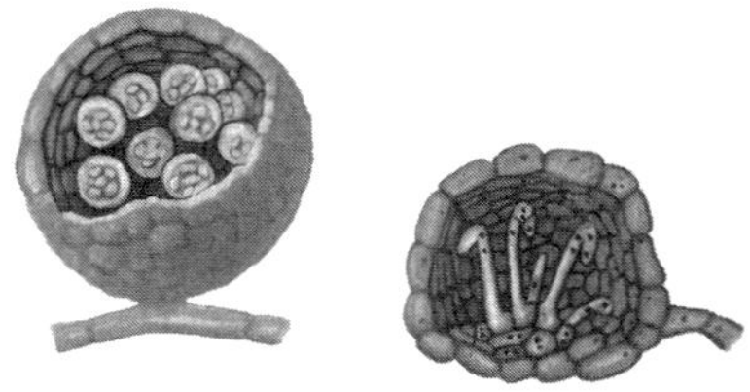

Cleistothecia of *Eurotium rubrum*. The young cleistothecium on the right is shown in cross section to reveal croziers inside. The one on the left shows globose asci inside, x 80.

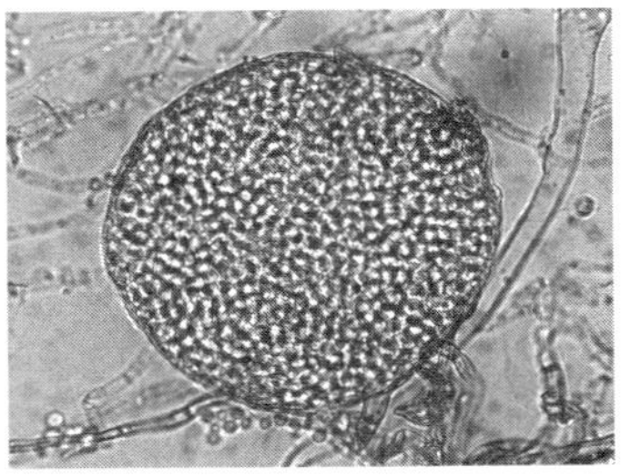

Cleistothecium of *Eurotium chevalieri*, x 120 (*PL*).

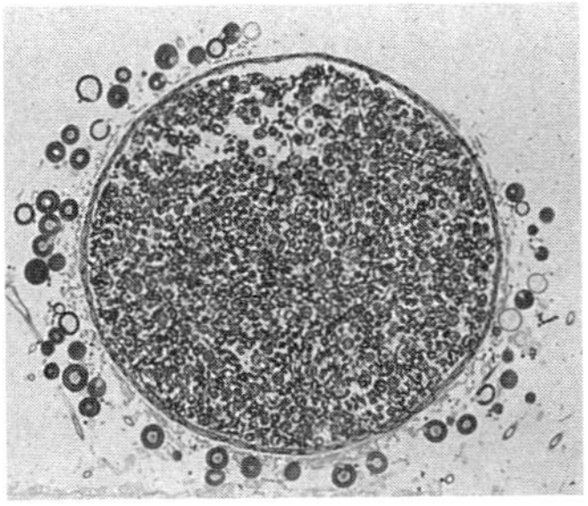

Cleistothecium of *Emericella nidulans* shown in cross section to reveal asci inside and Hülle cells outside the wall, x 130 (*RTH*).

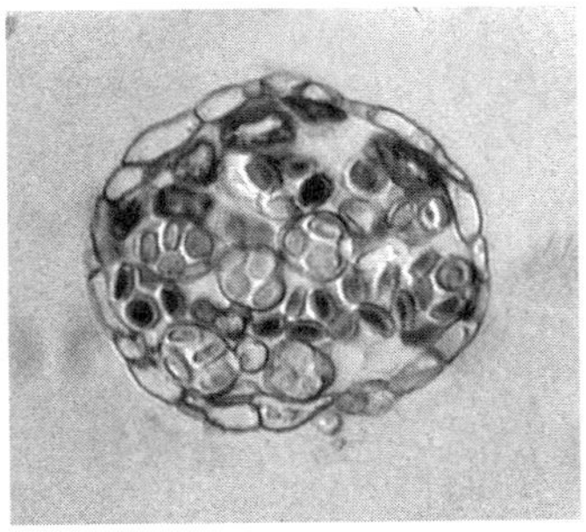

Cleistothecium of *Eurotium rubrum*, seen in cross section. The octosporate asci are scattered within, x 150 (*RTH*).

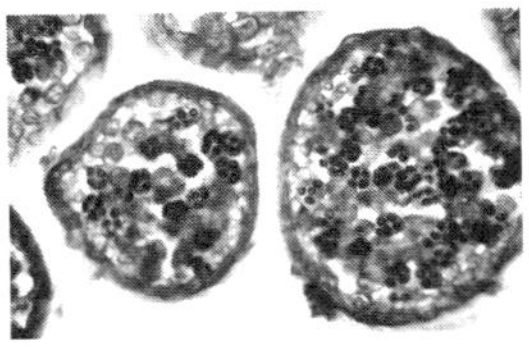

Cleistothecium of *Eurotium* sp. shown in cross section to reveal asci inside, x 95 (*RTH*).

clitocyboid (< gen. *Clitocybe* < Gr. *klitós*, inclined, slanting + *kýbe*, head + L. suf. *-oide* < Gr. *-oeídes*, similar to): *Agaricales*. A type of fruiting body (of 13 principal types recognized) with adherent, decurrent or subdecurrent gills on the stipe, the latter fibrous, fleshy or chalky in consistency, centrally attached to the pileus and lacking a ring and volva; e.g., like that of *Clitocybe* and some species of *Russula*, *Lactarius* and *Inocybe*, among other gen.

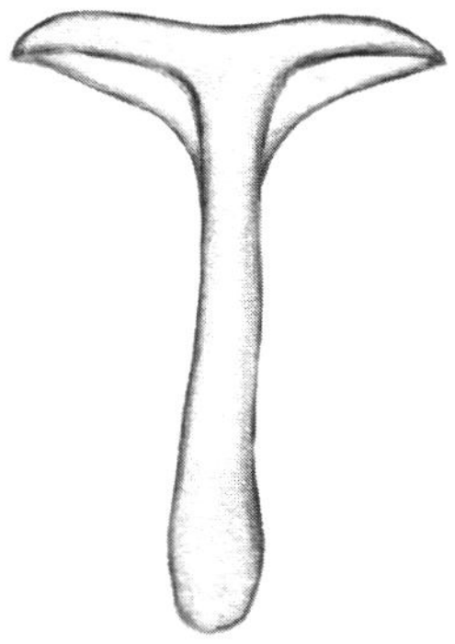

Clitocyboid basidiocarp of *Clitocybe* sp. Diagram of a longitudinal section, x 0.6.

clonal culture (E. *clone* < Gr. *klón*, offspring + L. suf. *-alis* > E. *-al*, relating to or belonging to; L. *cultus*, tilling): the cultivation of genetically identical individuals, all derived from an original individual, by asexual multiplication (division, parthenogenesis).

clone (Gr. *klón*, offspring): progeny or descendents produced from a single parental cell by asexual means.

closterospore (Gr. *klostér*, spindle + *sporá*, spore): a multinucleate, fusiform phragmospore (macroconidium); e.g., as in *Trichophyton rubrum* and *Microsporum canis* (moniliaceous asexual fungi).

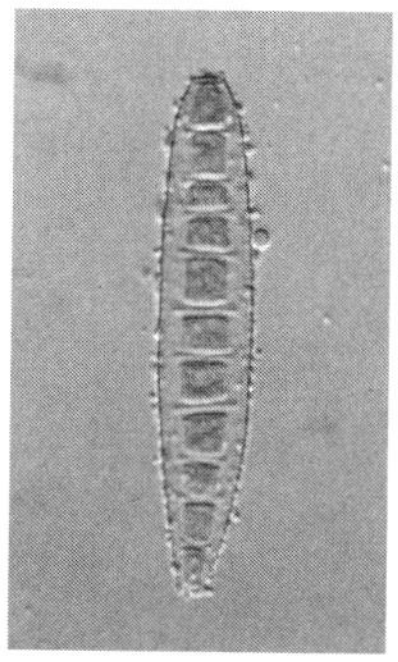

Closterospore (macroconidium) of *Microsporum cookei*, x 150 (*RTH*).

club fungus, pl. **fungi** (ME. *clubbe* < ON. *klubba*, club; L. *fungus*, fungus): the basidiomata of some members of the Clavariaceae, which are club-shaped, e. g., *Clavariadelphus*.

club root (ME. *clubbe* < ON. *klubba*, club; ME. < ON. *rot*, root): a disease of crucifers caused by *Plasmodiophora brassicae* (Plasmodiophorales), thus named because the hypertrophied roots are often club-shaped.

clypeate (L. *clypeatus*, armed with a shield < *clypeus*, shield + suf. *-atus* > E. *-ate*, provided with or likeness): possessing a **clypeus**. See **peltate** and **scutiform**.

clypeus, pl. **clypei** (NL. < L. *clypeus*, round shield): pseudostromatic tissue, often in the shape of a shield, generally dark brown or black in color, which surrounds the ostiole of conidiomata and perithecia in certain ascomycetes (e.g., *Phyllachora cibotii* and *Ophiodothella vaccinii*, Phyllachorales) and pycnidia (sphaeropsidaceous asexual fungi). In some species the clypeus also forms beneath the perithecium.

Clypeus above the perithecia of *Phyllachora cibotii*. Diagram of a longitudinal section of the perithecial stroma formed on a host leaf, x 70.

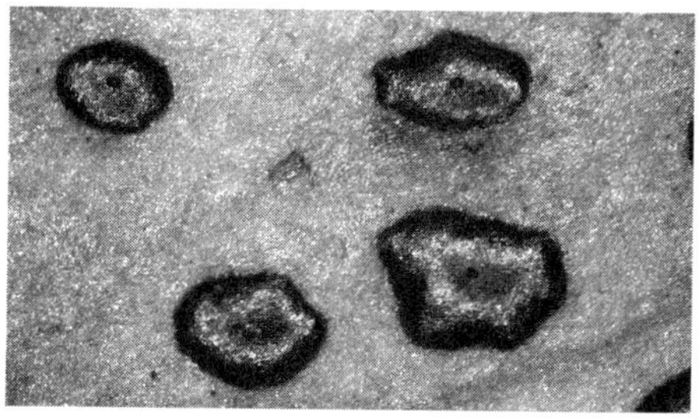

Clypeus around the ostiole of the conidiomata of *Ophiodothella vaccinii* on a parasitized leaf, x 40 (*RTH*).

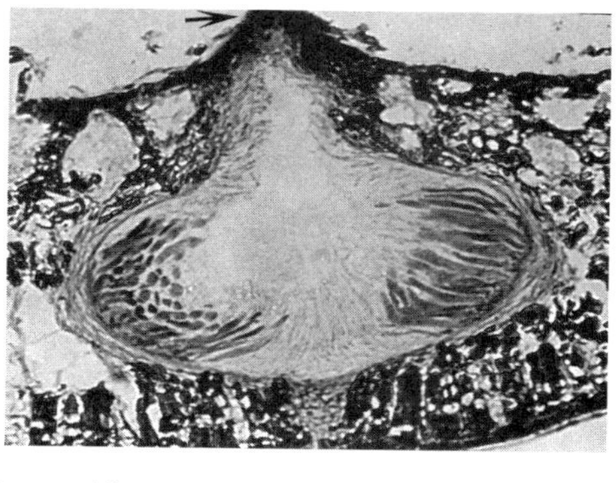

Clypeus of the perithecium of *Ophiodothella vaccinii*, seen in longitudinal section of the parasitized leaf, x 140 (*RTH*).

coalescent, **coalesced** (L. *coalescere*, to grow together, to unite + *-escens*, genit. *-escentis*, that which turns, beginning to, slightly > E. *-escent*); syn. of **concrescent**, **confluent**, and **connate**: to grow together; e.g., the sporangia of the myxomycete *Metatrichia vesparium* (Trichiales), which constitutes a pseudoaethalium, have the stipes united at the base. The gloeoid heads of the synnemata of *Antromycopsis smithii* (stilbellaceous asexual fungi) also are coalescent.

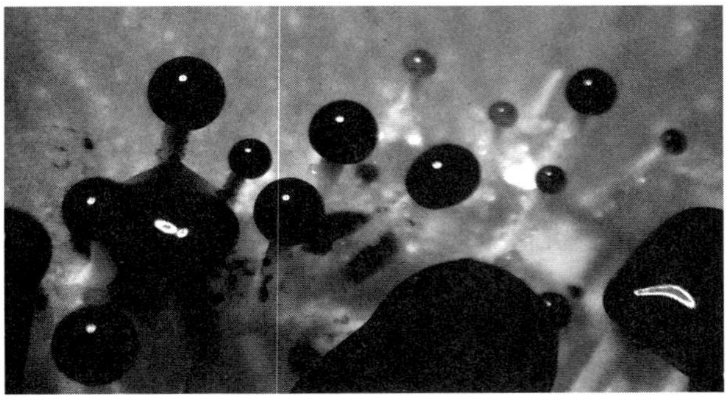

Coalescent gloeoid heads of the synnemata of *Antromycopsis smithii* on agar, x 5 (*CB & MU*).

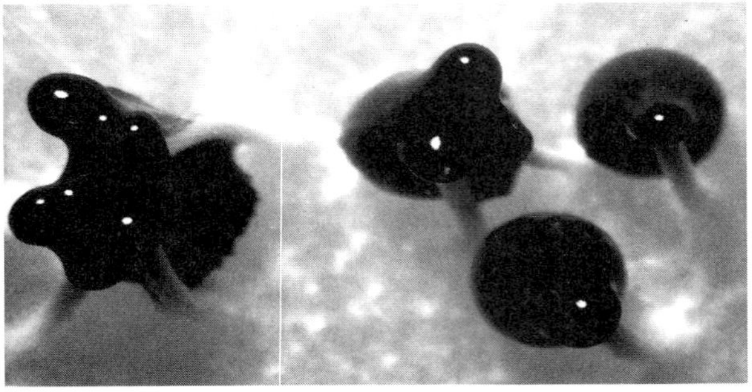

Coalescent gloeoid heads of the synnemata of *Antromycopsis smithii* on agar, x 9 (*CB*).

coccidioidomycosis, pl. **coccidioidomycoses** (< gen. *Coccidioides* < L. *coccidium* < Gr. *kokkíon*, small ball + L. suf. *-oides* < Gr. *-oeídes*, similar to + L. *mycosis* < Gr. *mýkes*, fungus + *-osis*, state or condition): an acute, self-limited respiratory infection or a chronic, progressive and generally fatal disease that can affect the skin, bones, joints, linfactic ganglions, suprarenal glands and central nervous system of man and some higher animals. The causal agent is the dimorphic fungus *Coccidioides immitis* (of uncertain taxonomic position, since it has been classified in Synchytriaceae, order Chytridiales, and among the moniliaceous asexual fungi) a saprobic inhabitant of soil in arid and semiarid areas of northern Mexico and southwestern USA. The infection is acquired by the inhalation of enteroarthric conidia.

cochleariform (L. *cochleariformis* < *cochleare*, spoon + *-formis* < *forma*, shape): having the shape of a spoon; e.g., the basidiocarp of *Dacryopinax elegans* and *D. spathularia* (Dacrymycetales) resembles a spoon or miniature spatula.

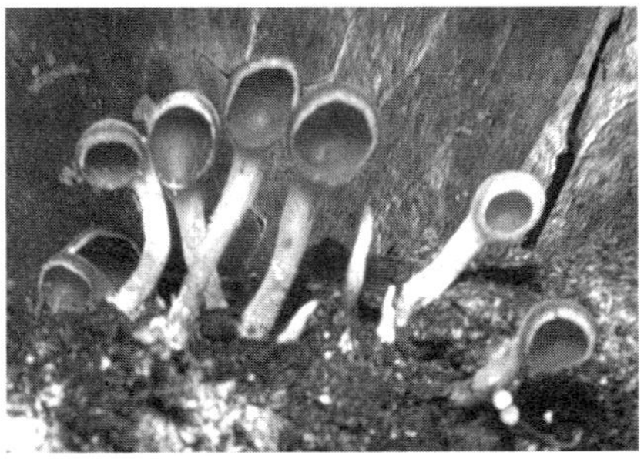

Cochleariform basidiocarps of *Dacryopinax elegans*, x 1.5 (*TH*).

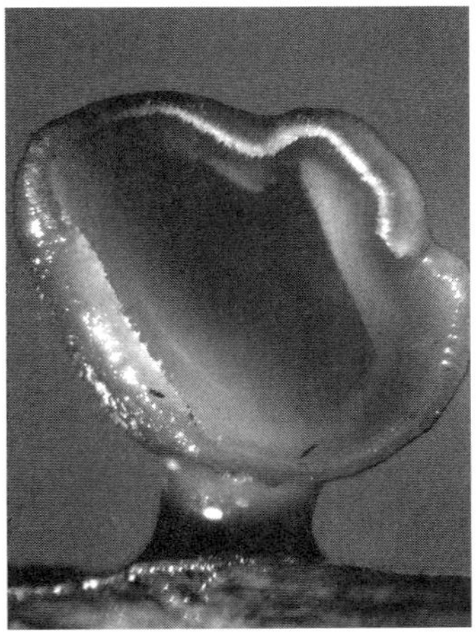

Cochleariform basidiocarp of *Dacryopinax elegans*, x 6 (*MU*).

cochleate (L. *cochleatus*, spiral < *cochlea*, snail, snail shell < Gr. *kochlías*, snail + L. suf. *-atus* > E. *-ate*, provided with or likeness): spirally coiled, like the shell of a snail; the perithecial hairs of *Chaetomium cochliodes* (Sordariales) are spirally coiled.

Coelomycetes (Gr. *koîlos*, hollow, cave, hole > L. *coelo*, hollow + L. *-mycetes*, ending of class < Gr. *mýkes*, genit. *mýketos*, fungus): a class of asexual fungi that produce their conidia on conidiophores contained in a fruiting body, either an acervulus or pycnidium. In some classifications the Coelomycetes (which includes the orders Melanconiales and Sphaeropsidales), together with the Blastomycetes and Hyphomycetes, constitute the subdivision Deuteromycotina, or deuteromycetes in a general sense. In the classification adopted in this dictionary, this group is considered as sphaeropsidaceous and melanconiaceous asexual fungi, with no class rank.

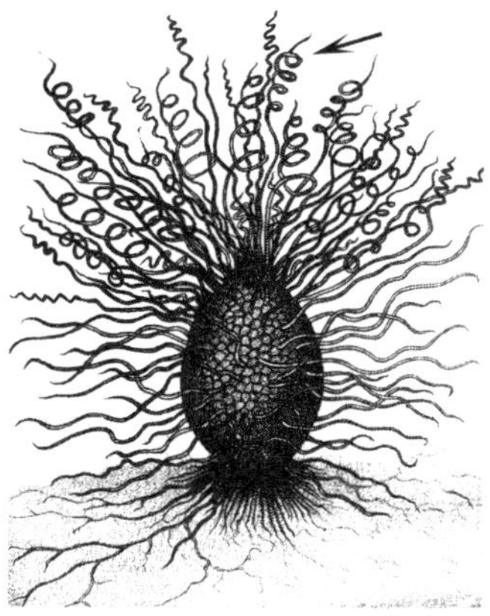

Cochleate perithecial hairs of *Chaetomium cochliodes*, x 50.

coelonema, pl. **coelonemata** (Gr. *koîlos*, hollow, hole + *nêma*, genit. *nêmatos*, filament): *Myxomycetes*. A hollow or tubular filament that constitutes a type of capillitium in the interior of the sporophores of many species, such as those of the gen. *Hemitrichia* (Trichiales); in some very evolved forms the coelonemoid capillitium has various ornamentations or reliefs (**elaters**) which contribute to spore dispersal. Cf. **stereonema**.

coenocentrum, pl. **coenocentra** (Gr. *koinós*, common + *kéntron*, center): refers to the multinucleate portion of the cytoplasm which surrounds the egg nucleus or oosphere during the oogonial development of certain leptomitaceous zoosporic fungi, such as *Apodachlya* and *Apodachlyella* (Leptomitales).

coenocytic (Gr. *koinós*, common + *kýtos*, cell, hollow + suf. *-íkos* > L. *-icus* > E. *-ic*, belonging to, relating to): aseptate; a thallus in which the nuclei are contained in a common cytoplasm, continual, without being separated by cross-walls or transverse septa. The majority of the aquatic fungi are coenocytic, such as *Saprolegnia* (Saprolegniales), as well as some terrestrial lower forms, such as *Rhizopus* (Mucorales). See **aseptate**.

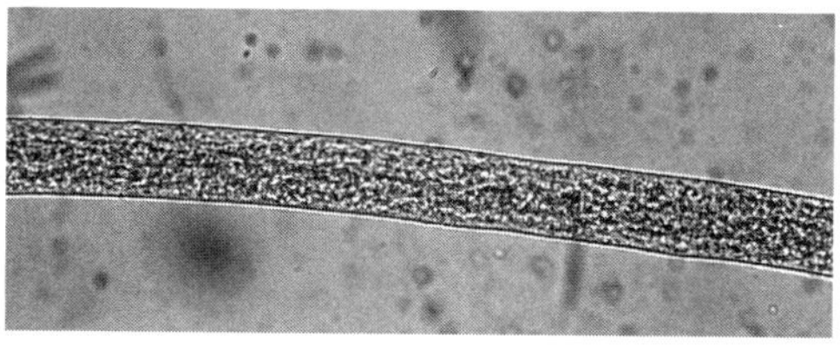

Coenocytic hypha of *Rhizopus nigricans*, x 330 (*MU*).

coetaneous (L. *co*, together + *aetas*, age + *-osus* > OF. *-ous*, *-eus* > E. *-ous*, having, possessing the qualities of): of the same age; existing for the same length of time.

collabent (L. *collabens*, genit. *collabentis*, to collapse, to fall to pieces, crumpling up + *-entem* > E. *-ent*, being): collapsing at maturity; e.g., as happens with many different types of sporiferous structures.

collar (L. *collare*, neckband): **1**. *Gasteromycetes*. The fringe of tissue that surrounds the pedicel of the fruiting body in some species of *Geastrum* (Lycoperdales), or the small, cup-like swelling (the "little cup" referred to by Coker) on the base of the basidiospores of *Rhizopogon* (Hymenogastrales) and Hysterangium (Phallales), which is formed on separating from the pedicels on the basidia. **2**. *Uredinales*. The more or less circular deposit in the host cell wall formed around the neck of the haustorium of *Melampsora* (Uredinales).

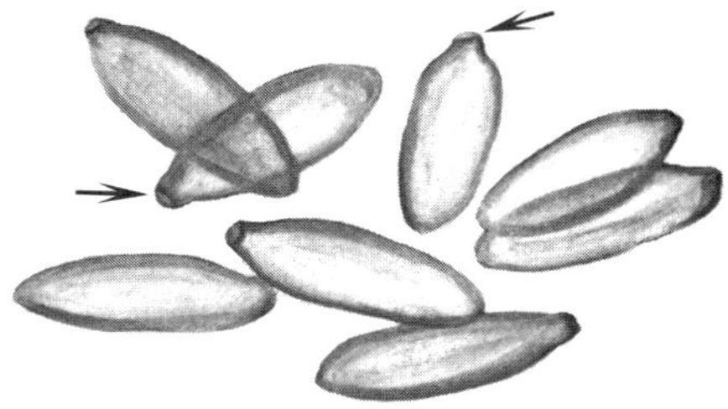

Collar of the orthotropic basidiospores of *Hysterangium* sp., x 1 400.

collarette (L. *collarette* < *collare*, neckband, collar + NL. dim. suf. *-etta*, *-ette*): **1**. *Conidial fungi*. A cup or goblet-shaped structure at the apex or subterminal portion of the phialide; characteristic of many gen., such as *Phialophora*, *Cyphellophora* and *Phialogeniculata* (dematiaceous asexual fungi). In the case of *Penicillium* (moniliaceous asexual fungi), the length of the collarette (more correctly named neck or **collulum**) is considered as an important taxonomic character for the distinction of species. **2**. *Mucorales*. The remains of the sporangial peridium that persist at the base of the columella; as one sees, e.g., in *Mucor*.

colliculose, **colliculous** (L. *colliculus*, little hill < *collis*, hill, mound + dim. suf. *-culus* + *-osus*, full of, augmented, prone to > ME. *-ose*; or + *-osus* > OF. *-ous*, *-eus* > E. *-ous*, having, possessing the qualities of): covered with small, rounded elevations; e.g., like the giant colony of *Candida colliculosa* (cryptococcaceous asexual yeasts).

collulum, pl. **collula** (L. *collulum* < *collum*, neck + dim. suf. *-ulum*): the part of the phialide of *Penicillium* (moniliaceous asexual fungi) located between the mouth and the base; i.e., the subapical region where the phialide tapers; the length of the phialide neck is considered an important taxonomic character in distinguishing species. See **collarette**.

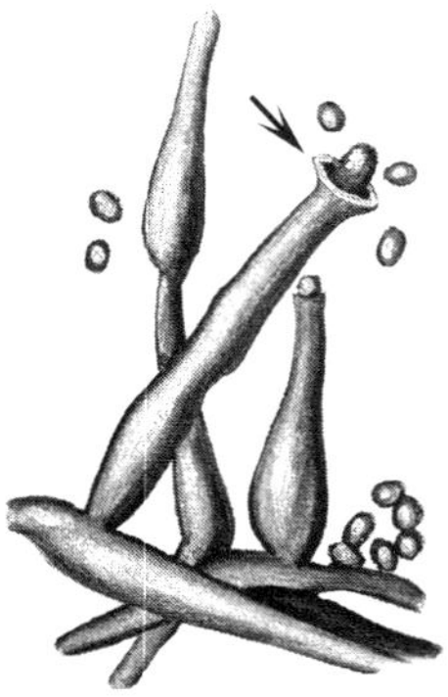

Collarette of the phialides of *Phialophora parasitica*, x 1 200.

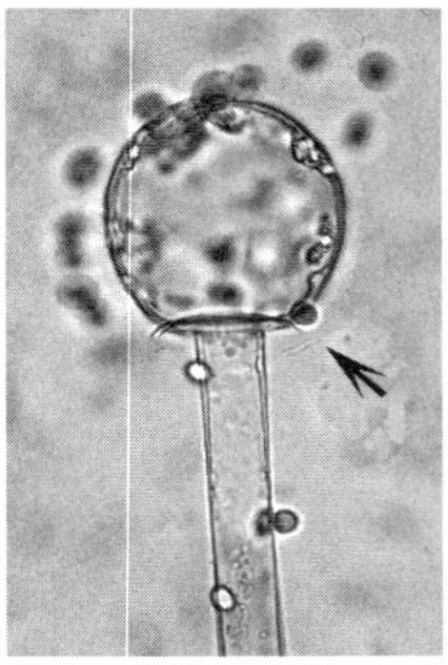

Collarette of the sporangium of *Mucor* sp., x 320 (*MU*).

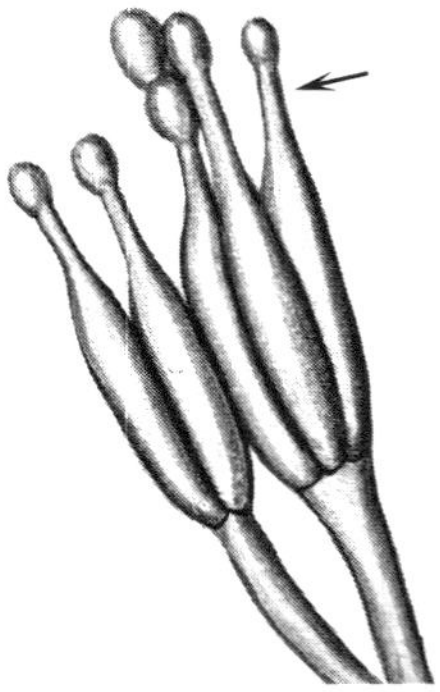

Collulum of the phialides of *Penicillium crustosum*, x 1 500.

collybioid (< gen. *Collybia* < Gr. *kóllybos*, a small coin + L. suf. *-oide* < Gr. *-oeídes*, similar to): *Agaricales*. A type of fruiting body (of 13 main types recognized) with gills adherent to the stipe, which is cartilaginous, central, and lacks a ring and volva; represented by *Collybia* and some species of *Marasmius*, *Psilocybe* and *Coprinus*, among other gen. of Agaricales. The collybioid type is similar to the mycenoid type but differs in its convex pileus, and initially incurved or inrolled margin.

colony (L. *colonia* < *colonus*, colonist < *colere*, to inhabit, cultivate): a group of individuals of the same species that live close together. In the fungi, the term generally refers to the mass of hyphae (mycelium) or individual cells that proliferate outward from a central point (centrifugally) to form an entity with the morphological characteristics typical of the species.

Collybioid basidiocarps of *Collybia* sp. Diagram of longitudinal sections, x 0.7.

columella, pl. **columellae** (L. *columella* < *columna*, column + dim. suf. *-ella*): a sterile, often columnar, supporting structure or tissue that is located inside a sporangium or other fructification; frequently it is an extension of the stalk that is surrounded by sporiferous tissues. The sporangia of various Myxomycetes, e.g., *Diachea* (Physarales) and *Stemonitis* (Stemonitales), and Mucorales, such as *Mucor*, *Rhizopus* and others, have a columella, as does the sporiferous apparatus of certain Gasteromycetes (*Lycoperdon*, of the Lycoperdales; *Scleroderma*, of the Sclerodermatales). See **dendritic** or **percurrent columella** and **stipe-columella**.

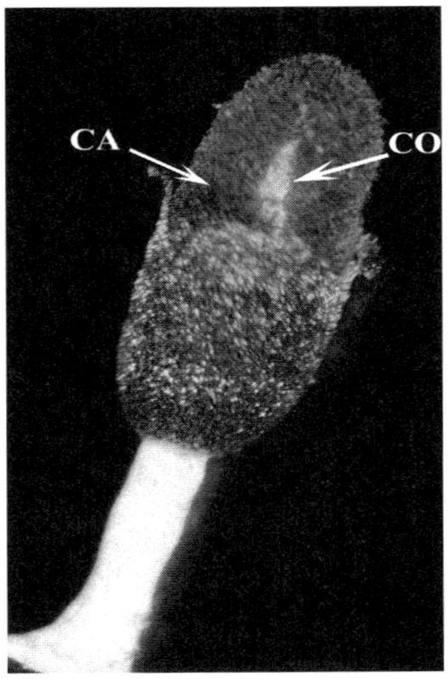

Columella (CO) and capillitium (CA) of the sporangium of *Diachea leucopodia*, x 45 (*MU*).

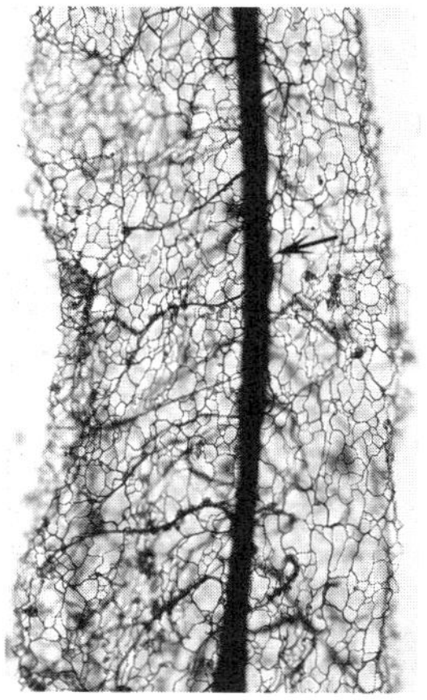

Columella of the sporangium of *Stemonitis* sp., x 35 (*MU*).

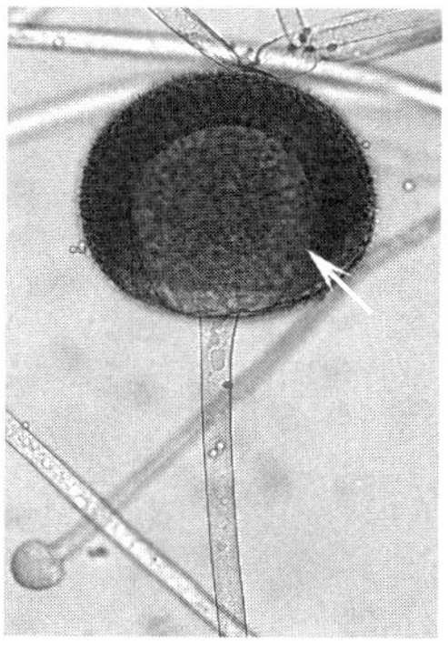

Columella of the sporangium of *Rhizopus arrhizus*, x 200 (*MU*).

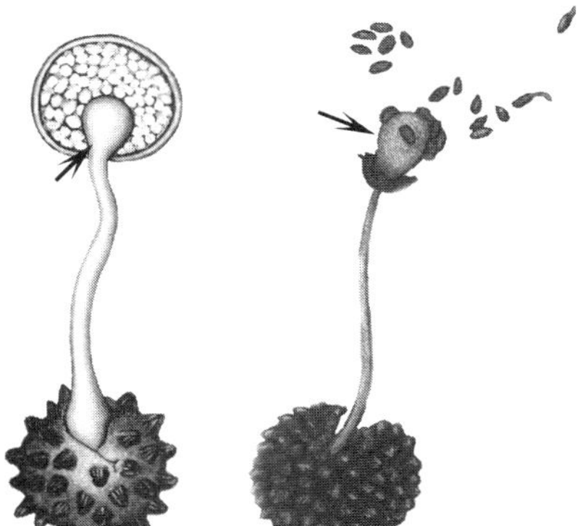

Columella of the germ sporangia of *Rhizopus nigricans*, x 170.

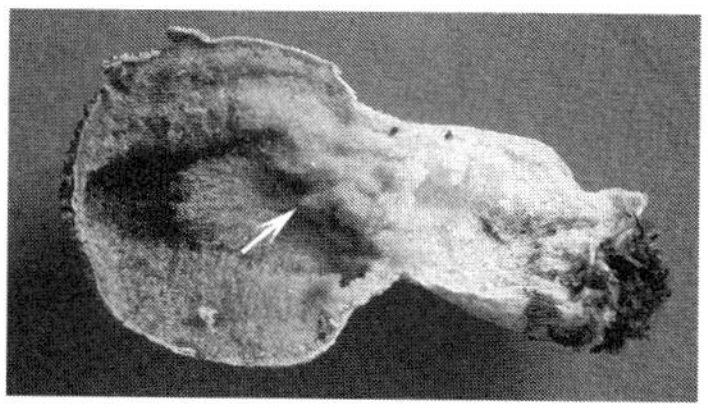

Columella of the basidiocarp of *Lycoperdon stellare*, seen in longitudinal section, x 0.6 (*MU*).

column (L. *columna*, column): *Gasteromycetes*. Each of the various pillars that support the sporiferous sack of the fruiting body of *Myriostoma* (Lycoperdales), separating it from the exoperidium.

Columns of the basidiocarp of *Myriostoma coliforme*, x 0.3.

Columns of the basidiocarp of *Myriostoma coliforme*, x 8 (*CB*).

comate (L. *comatus*, tufted < *coma*, tuft of hair, plume < Gr. *kóme* + L. suf. *-atus* > E. *-ate*, provided with or likeness): furnished with hairs grouped in woolly locks or tufts; e.g., like the pileus of *Coprinus comatus* and *C. echinosporus* (Agaricales).

Comate basidiocarp of *Coprinus lagopus*, x 2.

commensal (L. *commensalis* < *cum*, with + *mensa*, table + suf. *-alis* > E. *-al*, relating to or belonging to; one who shares a table with another): *Mycology*. A fungus that lives in, on or with an animal without injury to either organism. The fungus may be adhered to the outer surface (ectocommensal) or interior (endocommensal) of the animal. Commensalism (the phenomenon relative to the living together of the commensals) is present most clearly in the association that the Trichomycetes establish with various types of arthropods (insects, myriapods, crustaceans, etc.), an association in which the commensal fungus takes shelter and transportation in the host arthropod, feeding itself on the nutrients that pass through the anterior digestive tract of the animal or of nutrients that remain without being absorbed in the posterior digestive tract, but without causing any apparent damage to said host. On the other hand, it is not known with certainty if the host receives any benefit from the commensal, with respect to food, mating, change, development, maturation and reproduction. The species *Amoebidium parasiticum* (Amoebidiales) is an example of an ectocommensal Trichomycete on the antennae of *Daphnia*, a crustacean in the order Cladocera. See **parasitism**, **parasite**, **symbiont** and **symbiosis.**

commensalism (L. *commensalis* < *cum*, with + *mensa*, table + suf. *-alis* > E. *-al*, relating to or belonging to + Gr. *-ismós* > L. *-ismus* > E. *-ism*, state, phase, tendency, action): a phenomenon relative to the living together of commensals, in which an organism (the commensal) obtains certain benefits (shelter, transportation, food, etc.), and the other (the host) shows no apparent signs of benefit or harm. The living together that exists between various members of the class Trichomycetes (*Amoebidium*, *Harpella*, *Smittium*, etc.) and multiple species of mandibulate arthropods (especially insects, myriapods and crustaceans) qualifies it as commensalism. See **antagonism**, **helotism**, **metabiosis**, **mutualism**, **parabiosis**, **parasymbiosis**, **symbiosis** and **synergy.**

complanate (L. *complanatus*, level, flattened < *complanare*, to make level + suf. *-atus* > E. *-ate*, provided with or likeness): flattened on one plane, and smooth. The majority of the basidiospores (e.g., in *Coprinus comatus*, of the Agaricales) are slightly complanate in side view. See **compressed**.

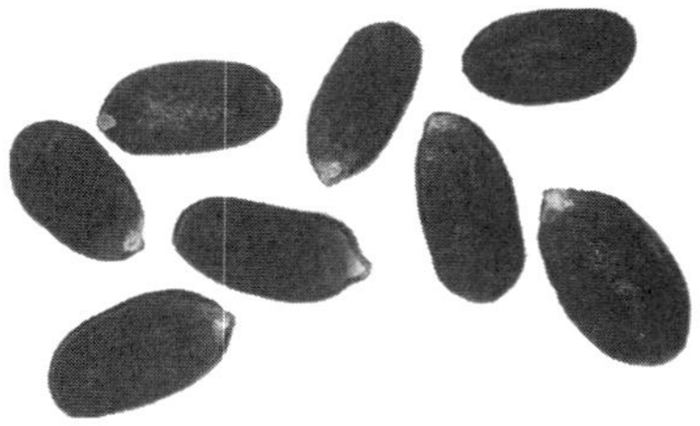

Complanate basidiospores of *Coprinus comatus*, x 1 500.

compost (E. *compost* < ME. < AF. < MF. < L. *compositus*, composite): a mixture of decaying organic materials, generally prepared by fermentation of straw and horse manure, in which the mycelium of the mushroom *Agaricus brunnescens* (Agaricales) is grown for industrial cultivation.

compressed (ME. < MF. *compresser* < LL. *compressore* < *comprimere*, squeeze together): flattened laterally. See **complanate**.

concentric (L. *con*, with, together + *centrum* < Gr. *kéntron*, center, central point of a circle + Gr. suf. -*íkos* > L. *-icus* > E. *-ic*, belonging to, relating to): refers to structures that have rings or zones arranged in parallel circles around a common center, as one can observe, e.g., in the zonate colonies formed by various molds, such as *Aspergillus* and *Penicillium* (moniliaceous asexual fungi), or in the perithecial stromata of *Daldinia concentrica* (Xylariales), which in vertical section show a series of concentric zones, with the perithecia embedded in the peripheral zone.

Concentric stromatic tissues of *Daldinia concentrica*, seen in longitudinal section, x 2 (*MU*).

concentric body (ML. *concentricus* < L. *con*, with, together + *centric* < *centrum*, center + Gr. suf. *-íkos* > L. *-icus* > E. *-ic*, belonging to, relating to; OE. *bodig*, body): a spherical concentric structure found in the hyphae of a great many lichenized fungi and in a few nonlichenized species. Concentric bodies are present in an area of cytoplasm devoid of other organelles, e.g., *Cercospora* (dematiaceous asexual fungi). In *Allomyces* (Blastocladiales) concentric bodies are found forming granules.

conceptacle (L. *conceptaculum*, a receiver or vessel, a place where something is conceived or engendered < *conceptus*, conceived + dim. suf. *-ulum*): any cavity or locule that opens to the exterior and in which are produced and lodged propagative cells; applied to pycnidia, ascigerous locules, structures that contain propagules in the lichens and other types of structures. The term is little used.

conchate (L. *conchatus*, shell-shaped < *concha*, shell < Gr. *kónche*, mussel, shell + L. suf. *-atus* > E. *-ate*, provided with or likeness); syn. of **conchiform** and **conchoid**: having the shape of a bivalve mollusk shell; e.g., like the basidiocarp of *Panus conchatus* (Aphyllophorales), or of *Pleurotus ostreatus* and *P. cornucopiae* (Agaricales), and the lichen thallus of *Dictyonema* (=*Cora*) *pavonia* (Aphyllophorales).

Conchate basidiocarps of *Pleurotus cornucopiae*, x 0.4 (*EPS*).

Conchate lichen thallus of *Dictyonema* (=*Cora*) *pavonia*, x 1.5 (*MU*).

conchiform (L. *concha*, shell < Gr. *kónche*, mussel, shell + L. *-formis* < *forma*, shape): with the shape of a mollusk shell. See **conchate**.

conchoid (L. *concha*, the shell of bivalve mollusks < Gr. *kónche*, mussel, shell + L. suf. *-oide* < Gr. *-oeídes*, similar to): similar to a mollusk shell. See **conchate**.

concinnous (L. *concinnus*, neatly arranged, pleasing + *-osus* > OF. *-ous*, *-eus* > E. *-ous*, having, possessing the qualities of): clean, pretty, neat, pure, natural, elegant.

concolor, **concolorous** (L. *con*, with + *color*, color, or + *-osus* > OF. *-ous*, *-eus* > E. *-ous*, having, possessing the qualities of): applied to structures that are the same color throughout; e.g., gills concolorous with the pileus.

concrescent (L. *concrescere*, to conglutinate, coagulate, unite < *con-*, with + *crescere*, to increase, grow + -*escens*, genit. *-escentis*, that which turns, beginning to, slightly > E. *-escent*): syn. of **confluent** and **connate**. See **coalescent**.

condidium, pl. **condidia** (L. *condere*, to hide + dim. suf. *-idium*): an old term for any fructification that bears conidia.

conferted (L. *confertus*, stuffed, packed): densely packed or crowded.

confervoid (< algal gen. *Conferva* + L. suf. *-oide* < Gr. *-oeídes*, similar to): composed of septate filaments with short cells, similar to those of the thallus of *Conferva*, a heterokontic filamentous alga.

confluent (L. *confluere*, to join + *-entem* > E. *-ent*, being): see **coalescent**.

congeneric (L. *con*, with + *genus*, genit. *generis*, race, lineage + Gr. suf. *-íkos* > L. *-icus* > E. *-ic*, belonging to, relating to): belonging to the same gen., a systematic unit in taxonomic classification, composed of species.

conglobate (L. *conglobatus*, gathered into a ball < *con-*, with + *globus*, globe, ball + suf. *-atus* > E. *-ate*, provided with or likeness): united into more or less globose, compact masses or groups; e.g., like the bases of the stipes of some polyporaceous basidiomycetes, such as *Polyporus frondosus* (Aphyllophorales), and the ascospores of various ascomycetes.

congregate (L. *congregatus* < *con*, with + *gregare*, to unite + suf. *-atus* > E. *-ate*, provided with or likeness): *Gasteromycetes*. Refers to a type of fruiting body, like that of *Broomeia congregata* (Sclerodermatales), composed of numerous sporiferous sacks (5-900) arranged in alveoli on the upper surface of a thick, fleshy stroma. At maturity the sporiferous sacks, which correspond to the endoperidium, are surrounded by the lower half of the exoperidium.

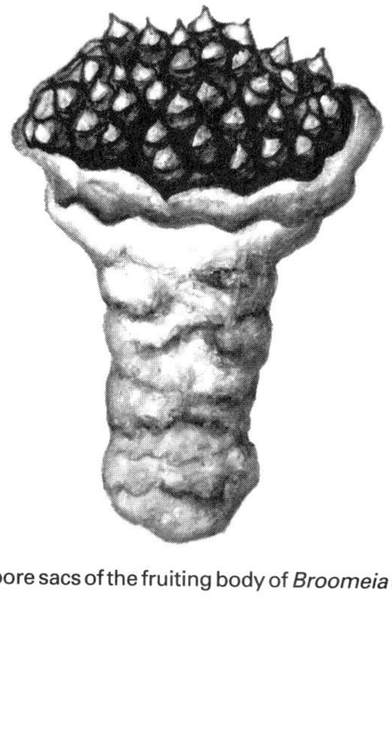

Congregate spore sacs of the fruiting body of *Broomeia congregata*, x 1.

conic, **conical** (L. *conicus*, conical < Gr. *konikós*, cone-shaped < *kónos*, cone + Gr. suf. *-íkos* > L. *-icus* > E. *-ic*, belonging to, relating to; or + L. suf. *-alis* > E. *-al*, relating to or belonging to): with the shape of, or resembling, a cone, i.e., a tapering structure whose height is greater than its width, with sharp-pointed apex; e.g., like the pileus of the basidiocarp of *Hygrocybe conica* (Agaricales).

conidial fungus, pl. **fungi** (NL. *conidium* < Gr. *kónis*, dust + dim. suf. *-ídion* > L. *-idium* + L. suf. *-alis* > E. *-al*, relating to or belonging to; L. *fungus* < Gr. *sphóngos*, fungus): a fungus that reproduces by conidia. Same as **anaholomorph**, **asexual fungi**, **deuteromycetes**, **Fungi Imperfecti** and **mitosporic fungi**.

conidial head (NL. *conidium* < Gr. *kónis*, dust + dim. suf. *ídion* + E. suf. *-al*, relating to or belonging to; ME. *hed*, head): the upper part of a conidiophore where the conidiogenous cells are produced; e.g., in *Aspergillus* (moniliaceous asexual fungi) the conidial head consists of a vesicle on whose surface are born phialides (the conidia-producing cells) or metulae and phialides, depending on the species. See **uniseriate** and **biseriate**.

Conidial head of the conidiophore of *Aspergillus nidulans*, x 550 (*CM*).

conidiogenesis, pl. **conidiogeneses** (NL. *conidium* < Gr. *kónis*, dust + dim. suf. *-ídion* > L. *-idium* + L. *genesis* < Gr. *génesis*, engenderment): the process by which conidia are formed by a conidiogenous cell on a conidiophore or an undifferentiated hypha. See **blastic** and **thallic**.

conidiogenous cell (NL. *conidium* < Gr. *kónis*, dust + dim. suf. *-ídion* > L. *-idium* + *génos*, origin < *gennáo*, to engender, produce + L. *-osus* > OF. *-ous*, *-eus* > E. *-ous*, having, possessing the qualities of; L. *cella*, storeroom > ML. *cella* > OE. *cell*, cell, compartment): in the conidial fungi, it refers to a fertile cell, specialized for the production of conidia. There exist various types of conidiogenous cells, depending upon the species, e.g., the ampulla of *Botrytis* and the phialide of *Aspergillus* (moniliaceous asexual fungi), among others.

conidiogenous locus, pl. **loci** (NL. *conidium* < Gr. *kónis*, dust + dim. suf. *-ídion* > L. *-idium* + *génos*, origin + L. *-osus* > OF. *-ous*, *-eus* > E. *-ous*, having, possessing the qualities of; L. *locus*, place, point): the particular site or point of a hypha or cell where conidia are generated, whether one or several; e.g., the conidiogenous locus of the phialides of *Chloridium chlamydosporis* (moniliaceous asexual fungi).

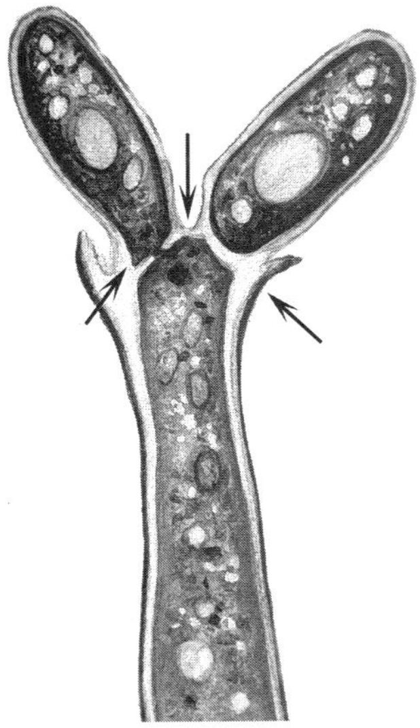

Conidiogenous loci of the conidiogenous cell of *Chloridium chlamydosporis*. Note the cross wall formation within the collarette. Drawing based on a transmission electron micrograph published by Beckett *et al.* in their *Atlas of Fungal Ultrastructure*, 1974, x 10 760.

conidiole, **conidiolum**, pl. **conidiola** (NL. *conidiolum* < *conidium* < Gr. *kónis*, dust + dim. suf. *-ídion* > L. *-idium* + L. dim. suf. *-olum* > E. *-ole*): syn. of **secondary conidium**; a small conidium formed from another conidium; e.g., as in *Entomophthora* (Entomophthorales).

conidioma, pl. **conidiomata** (NL. *conidium* < Gr. *kónis*, dust + dim. suf. *-ídion* > L. *-idium* + suf. *-óma*, which implies entirety): any structure formed by hyphae that produces conidia, whether a synnema, sporodochium, acervulus, or pycnidium. It was proposed especially for species in which the fructification is variable, as in *Pestalotiopsis* (melanconiaceous asexual fungi).

conidiophore (NL. *conidium* < Gr. *kónis*, dust + dim. suf. *-ídion* + > L. *-idium* + Gr. *-phóros*, bearer): a hypha, simple or branched, that is morphologically and/or physiologically differentiated from a somatic hypha to produce and bear conidia; the latter are generally on specialized **conidiogenous cells**, which can be arranged in many ways. In some species the terms conidiophore and conidiogenous cell are used interchangeably. Illustrations of conidiophores can be found in figures of the different types of conidia (e.g., **meristem arthrospore**, **blastospore**, **meristem blastospore** and **catenulate**), and of certain types of conidiophore branching (e.g., **biverticillate**, **monoverticillate**, **triverticillate** and **polyverticillate**), or of other structures related to conidiophores (e.g., **foot cell**).

conidiosporangium, pl. **conidiosporangia** (NL. *conidium* < Gr. *kónis*, dust + dim. suf. *-ídion* > L. *-idium* + NL. *sporangium* < Gr. *sporá*, spore + *angeîon*, vessel, receptacle): *Peronosporales*. A deciduous, oval or limoniform sporangium, which on maturing can either germinate directly by means of a germ tube to form a mycelium (as do the conidia of the higher fungi), or can produce biflagellate, reniform zoospores, thus behaving like a zoosporangium. This phenomenon is observed, e.g., in *Phytophthora infestans*, *Basidiophora entospora* and *Albugo candida*. Which mode of germination occurs is controlled by environmental conditions.

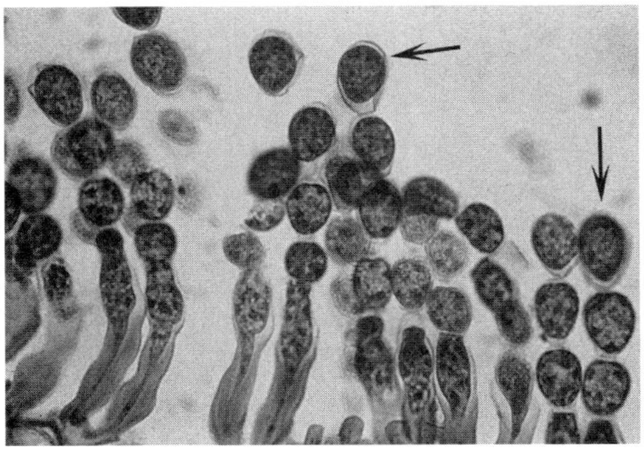

Conidiosporangia of *Albugo candida* in an erumpent pustule formed in the host leaf tissues, x 420 (*MU*).

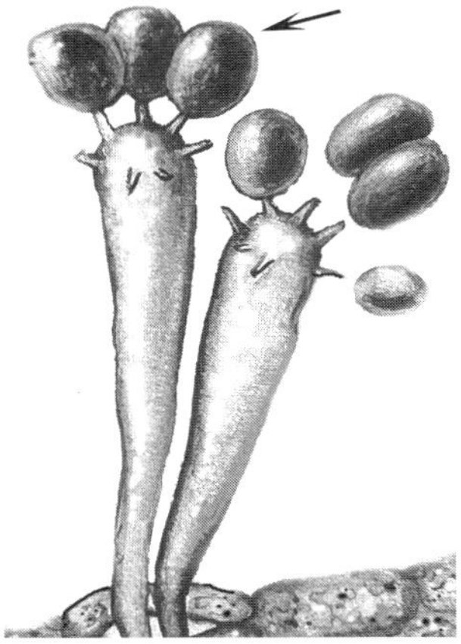

Conidiosporangia (conidium-like, deciduous zoosporangia) of *Basidiophora entospora*, borne on two zoosporangiophores that are growing out of a host stoma, x 110.

conidium, pl. **conidia** (NL. *conidium* < Gr. *kónis*, powder + dim. suf. *-ídion* > L. *-idium*): syn. of **conidiospore**: an asexual nucleate spore, immotile, generally formed at the apex or side of a specialized sporogenous (conidiogenous) cell. Conidia are of two basic types, **blastic** and **thallic.** Blastic conidia (the majority) are generated *de novo*, whereas thallic conidia form from preexisting hyphal cells. Conidia can be one, two, or multicellular, dry or mucous, but they are never produced within a structure with a rigid cell wall or peridium, as happens with sporangiospores. Conidia are the asexual spores of ascomycetes and basidiomycetes; typically they have been grouped together in the deuteromycetes.

conigenous (Gr. *kõnos*, pine cone + *génos*, origin + L. *-osus* > OF. *-ous*, *-eus* > E. *-ous*, possessing the qualities of): arising from or living on pine cones, like the basidiocarps of *Auriscalpium vulgare* (Aphyllophorales).

conjugation (L. *conjugatus*, united, ptp. of *conjugare*, to join, couple + *-ationem*, action, state or condition, or result > E. suf. *-ation*): see **gametangial copulation**.

conk (prob. < E. *conch* < L. *concha* < Gr. *kónche*, conch shell): the visible fruiting body (basidioma) of a wood-decay fungus, especially of a polypore. Cf. **artist's conk**.

connate (L. *connatus* < *connascor*, to be born at the same time with < *con-*, with + *natus*, born): syn. of **concrescent** and **confluent**. Applied to organs that form together and are united with one another, i.e., congenital adherence; e.g., like the stipes of the synnemata of *Antromycopsis smithii* (stilbellaceous asexual fungi), the stipes of the basidiocarps of *Lyophyllum connatum*, *L. decastes* and *Pleurotus ostreatus* (Agaricales), which grow together from the substrate. Connate is a variant of **caespitose**.

Connate basidiocarps of *Pleurotus ostreatus* var. *florida*, x 0.5 (*MU*).

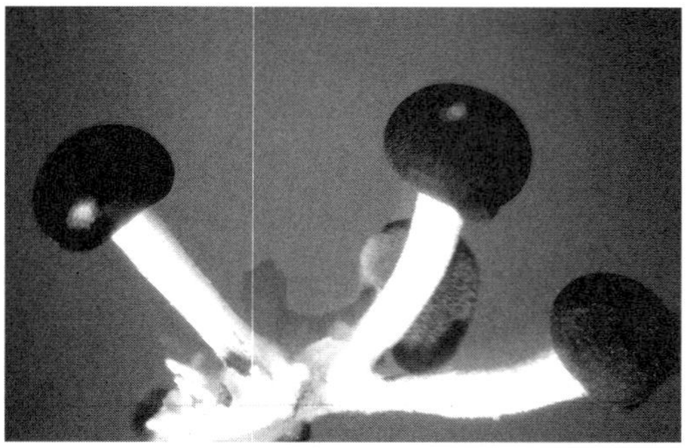

Connate synnemata of *Antromycopsis smithii* grown on pure culture, x 9 (*MU*).

connivent (L. *connivens*, genit. *conniventis*, winking at, overlooking < *connivere*, to approach, touch + *-entem* > E. *-ent*, being): refers to the border of the pileus in certain types of basidiocarps (such as *Coprinus*, of the order Agaricales) which through convergence touches the stipe but does not fuse with it. The minute fibrillae on the exoperidium of *Lycoperdon* and *Calvatia* (Lycoperdales) are also connivent, since their apices join together in a point, leaving a space below, giving the appearance of a cupule or vault to each group of fibrillae.

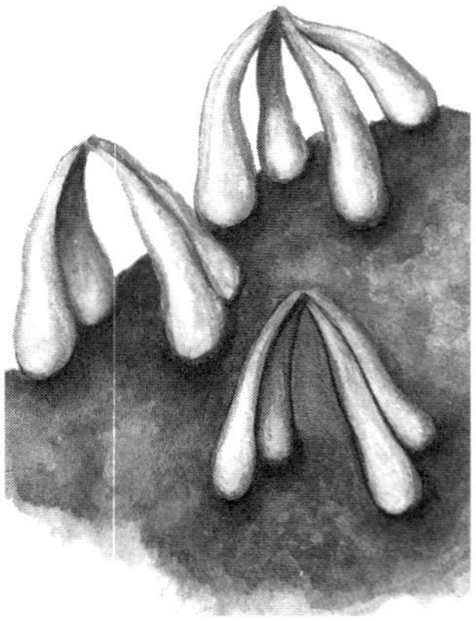

Connivent fibrils of the pyramidal scales formed on the exoperidium of the basidiocarp of *Lycoperdon* sp., x 10.

constricted (L. *constrictus* < *constringere*, to make narrow or draw together; to tie up): narrowed down, with a constriction; e.g., like the volva of *Amanita constricta* (Agaricales).

contaminate (L. *contaminatus*, ptp. of *contaminare*, to corrupt, infect + suf. *-atus* > E. *-ate*, provided with or likeness): to make impure or unfit by the introduction of undesirable elements, e.g., a not pure culture, seeds or soil with fungal spores, etc. Cf. **infect** and **infest.**

context (L. *contextus*, interwoven, united, connected < *contexere*, to intertwine): *Basidiomycetes*. The fibrous or fleshy tissue comprising the body of the pileus and stipe of the basidiocarp, not including the cortex and hymenium. The context of the pileus corresponds to the hymeniferous trama, which bears the hymenium. See **trama**.

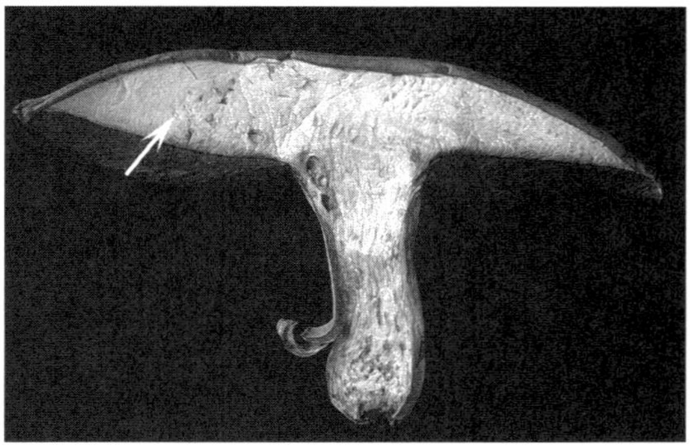

Context of the basidiocarp of *Boletus* sp., seen in longitudinal section, x 0.6 (*MU*).

contiguous (L. *contiguus*, near together, neighboring, adjoining < *contingere*, to have contact with + *-osus* > OF. *-ous*, *-eus* > E. *-ous*, possessing the qualities of): being in actual contact with; joining.

contingent (ME. < MF. < L. *contingent*, prp. of *contingere*, to have contact with, befall + *-entem* > E. *-ent*, being): touching.

continuous (L *continuus*, not interrupted + *-osus* > OF. *-ous*, *-eus* > E. *-ous*, possessing the qualities of): **1**. A unicellular spore, hypha, or capillitial filament that lacks septa. **2**. A stipe that merges imperceptibly with the pileus, peridium, etc., and which is composed of the same tissue. Cf. **septate**.

continuous culture (L. *continuus*, continuous + *-osus* > OF. *-ous*, *-eus* > E. *-ous*, possessing the qualities of; *cultus*, tilling): the cultivation of organisms or cells in a culture vessel that is constantly maintained through the continuous addition of new culture medium and the separation of spent medium from the medium that contains the organisms or cells.

contorted (L. *contortus*, twisted, crooked): twisted or crooked, e.g., like the sporangiolar branches of *Cokeromyces recurvatus* (Mucorales) and the perithecial hairs of *Chaetomium tortile* (Sordariales), with spirals in opposite directions.

convergent (L. *convergere*, to move toward one point or one another; to unite, proceeding from diverse points + *-entem* > E. *-ent*, being): *Agaricology*. A type of hymeniiferous trama in which the hyphae converge toward the center of the trama. Also called **inverted** because it appears V-shaped in vertical section. The hyphae are quite wide and long; the middle layer is evident only in young tissue, then disappears with age. It is present, e.g., in *Pluteus cervinus* (Agaricales). Cf. **divergent**.

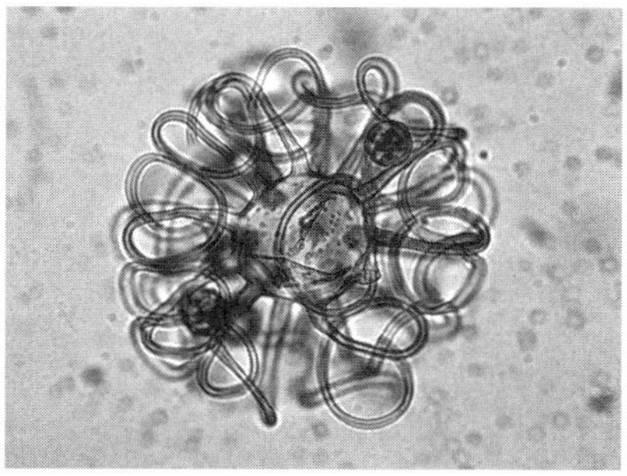

Contorted sporangiophores of *Cokeromyces recurvatus*, x 460 (*MU*).

Contorted perithecial hairs of *Chaetomium tortile*, x 1 000 (*MU*).

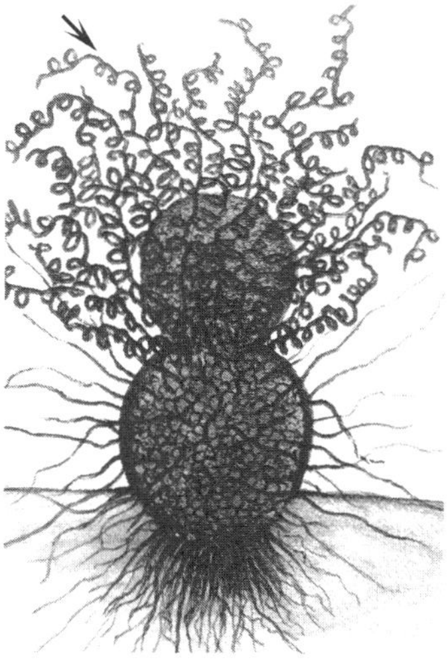

Contorted perithecial hairs of *Chaetomium tortile*, x 440.

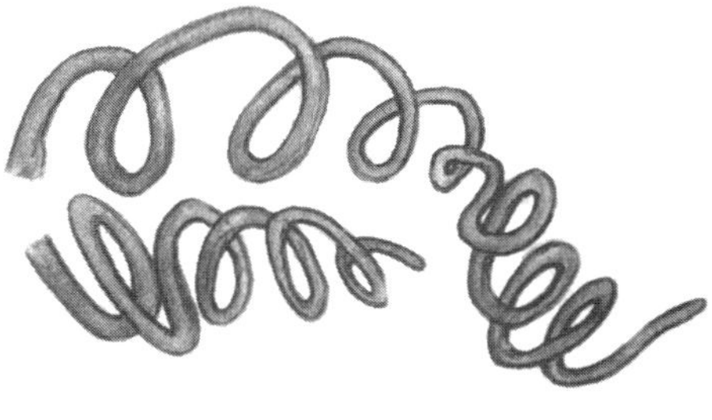

Contorted perithecial hairs of *Chaetomium tortile*, x 2 000.

convolute, **convoluted** (L. *convolutus*, rolled up, inrolled); syn. of **cerebriform**: with folds, bends and undulations which give a brain-like appearance; e.g., like the pileus of *Gyromitra esculenta* and *Helvella infula* (Pezizales) and the basidiocarp of *Tremella* (Tremellales).

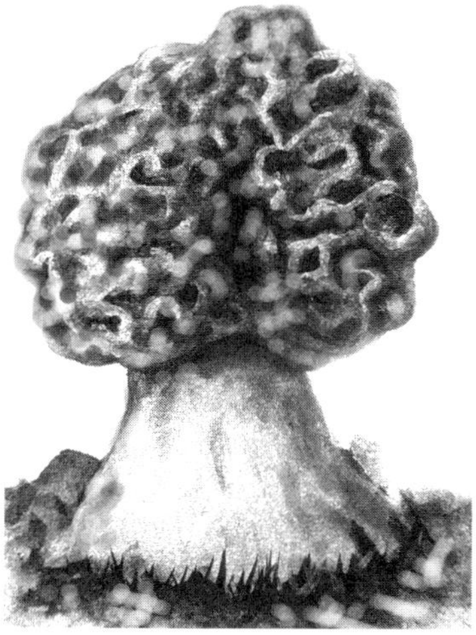

Convolute pileus of the ascocarp of *Gyromitra esculenta*, x 0.5.

Convolute basidiocarp of *Tremella* sp., x 1.6 (*TH*).

coprinoid (< gen. *Coprinus* < Gr. *kópros*, dung + L. suf. *-inus*, which indicates relationship + suf. *-oide* < Gr. *-oeídes*, similar to): a type of basidium found in *Coprinus* and other Agaricales. See **brachybasidium**.

coprogen (Gr. *kópros*, dung + *génes*, born, origin): a growth factor in dung required by *Pilobolus* spp. (Mucorales), which has been identified as an organic iron compound, similar to the hemine of the blood of vertebrates.

coprophilous (Gr. *kópros*, dung + *phílos*, have an affinity for + L. *-osus* > OF. *-ous*, *-eus* > E. *-ous*, possessing the qualities of): a fungus that develops preferentially on dung or manured soils; e.g., as occurs with *Pilaira* and *Pilobolus* (Mucorales), *Coprinus* and *Panaeolus* (Agaricales) and many other fungi.

Coprophilous sporangiophores of *Pilobolus kleinii*, developed on cow's dung maintained in a moist chamber, x 1.5 (*MU*).

coral fungus, pl. **fungi** (L. *corallium* < Gr. *korállion*, the calcareous or horny skeletal deposit produced by anthozoan or rarely hydrozoan polyps; coral; L. *fungus*, fungus): basidiomata of *Clavaria*, *Ramaria*, *Clavicorona*, and related gen. of Clavariaceae, which are coralloid, much branched, like coral in form.

coralloid (L. *corallium*, coral + suf. *-oide*, < Gr. *-oeídes*, similar to): having the appearance of coral, because of the manner of branching; e.g., as in the basidiocarps of *Calocera viscosa*, *Clavicorona pyxidata*, *Clavaria vermicularis*, and other clavariaceous species (Aphyllophorales), or like the coralloid subhymenium of *Amanita sublutea* (Agaricales). In some Gasteromycetes, it also refers to a type of fruiting body development, known as **coralloid** or **forate development**.

Coralloid basidiocarps of *Calocera viscosa*, x 2.

corbicula, pl. **corbiculae** (L. *corbicula*, little basket): *Uredinales*. A peripheral, paraphysis-like structure present in the telia of some rusts; it consists of cells with a thick, brown wall, which adhere to the internal wall of a host epidermal cell, and which fuse laterally with others to form a basket-like structure that protects, and assists in the dissemination of, the teliospores.

cordiform (L. *cordis*, heart + *-formis* < *forma*, shape): heart-shaped; e.g., like the vesicle of the sporangiophore of *Syncephalis cordata* (Zoopagales) and the basidiospores of *Psathyrella hirsutosquamulosa* (Agaricales).

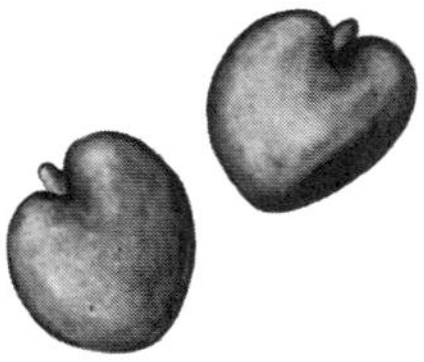

Cordiform basidiospores of *Psathyrella hirsutosquamulosa*, x 1 500.

coremiform (NL. *coremiformis* < *coremium* < Gr. *kórema*, broom + L. *-formis* < *forma*, shape): having the shape of a coremium (synnema), as in *Penicillium claviforme* (moniliaceous asexual fungi).

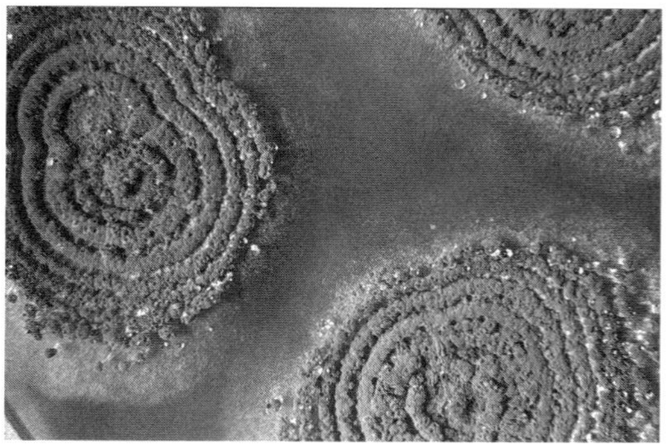

Coremiform texture of the colony of *Penicillium claviforme* on agar, x 1 (*TH*).

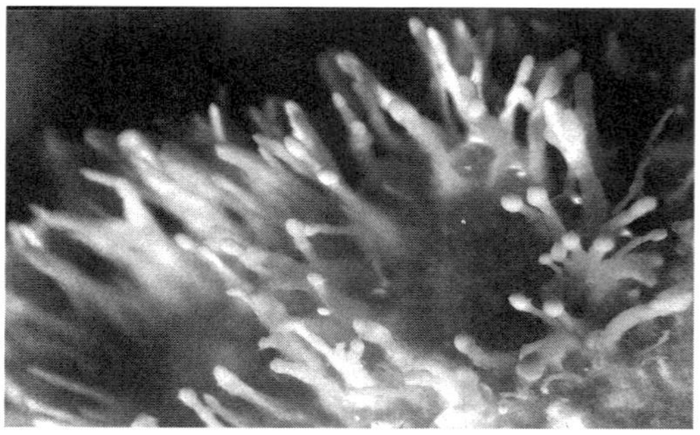

Coremiform mycelium of the colony of *Penicillium claviforme* on agar, x 2 (*MU*).

coremium, pl. **coremia** (NL. *coremium* < Gr. *kórema*, broom + L. dim. suf. *-ium*): an erect, loosely arranged bundle of conidiophores with a broom-like apex, e.g., *Penicillium claviforme*. A type of **synnema**.

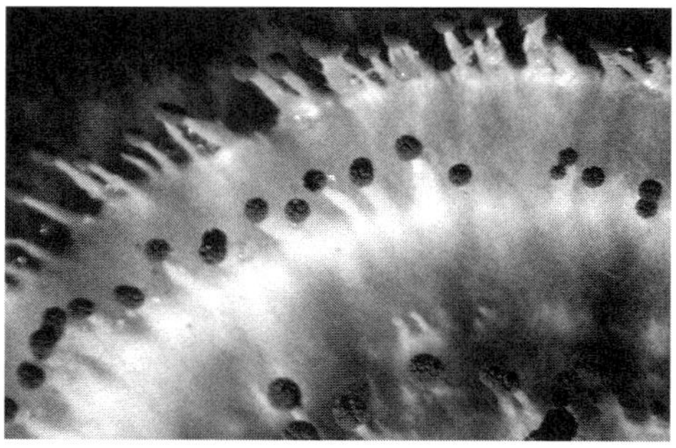

Coremia of *Penicillium claviforme* on agar, x 2 (*MU*).

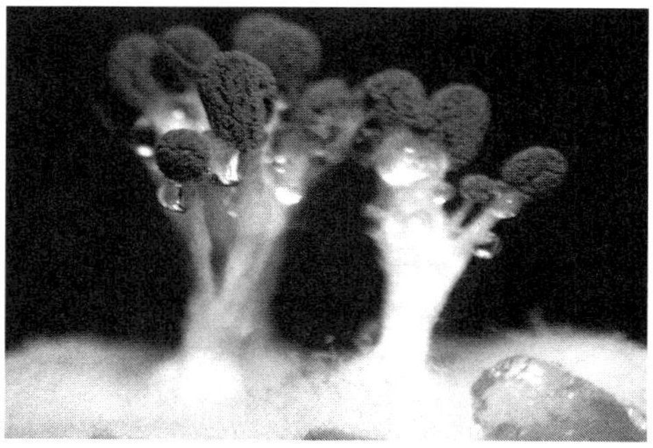

Coremia of *Penicillium claviforme* on agar, x 8 (*MU & CB*).

coriaceous (LL. *coriaceus*, leather-like < *corium*, leather + *-aceus* > E. *-aceous*, of or pertaining to, with the nature of): having the consistency of leather, tough but with a certain flexibility; e.g., like that of the fruiting body of some species of Aphyllophorales, especially in the families Corticiaceae and Stereaceae. Also called **leathery**.

corky (L. *cortex*, genit. *corticis*, cork + E. suf. *-y*, having the quality of): syn. of **suberose**.

corneous (L. *corneus*, horny < *cornu*, horn + L. *-osus* > OF. *-ous*, *-eus* > E. *-ous*, having, possessing the qualities of): horn-like; of a hard consistency, like horn.

corniculate (L. *corniculatus*, horned < *corniculum*, dim. of *cornu*, horn + suf. *-atus* > E. *-ate*, provided with or likeness): having a small horn-like structure; curved like a horn; e.g., like the cystidia of *Pluteus cervinus* (Agaricales), which have apical appendages in the shape of little horns.

corniculiform (L. *corniculiformis* < *corniculum*, little horn + *-formis* < *forma*, shape): having the shape of a little horn; e.g., like the sclerotia of *Claviceps purpurea* (Hypocreales).

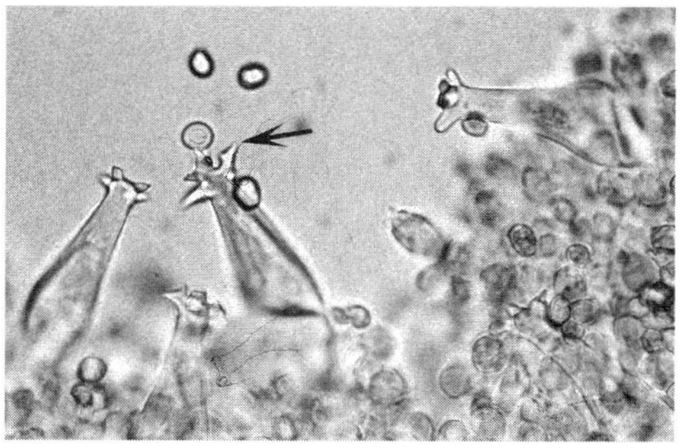

Corniculate cystidia of the hymenium of *Pluteus cervinus*, x 300 (*MU*).

corniform (L. *cornu*, horn + *-formis* < *forma*, shape): having the shape of a horn; e.g., like the appendages of *Phycomyces nitens* (Mucorales).

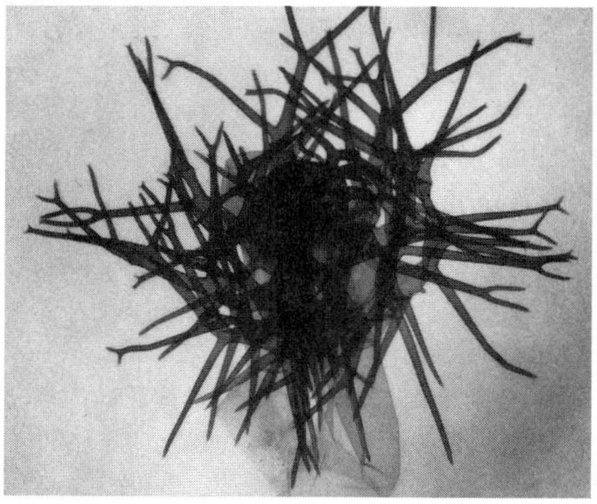

Corniform appendages of the suspensors of *Phycomyces nitens*, x 90 (*MU*).

cornute (L. *cornutus*, having horns, having the shape of a horn): *Uredinales*. A type of aecium, horn-like and sharp-pointed, with the peridium composed of characteristically imbricate cells, which break by longitudinal fissures; e.g., like that of *Roestelia*.

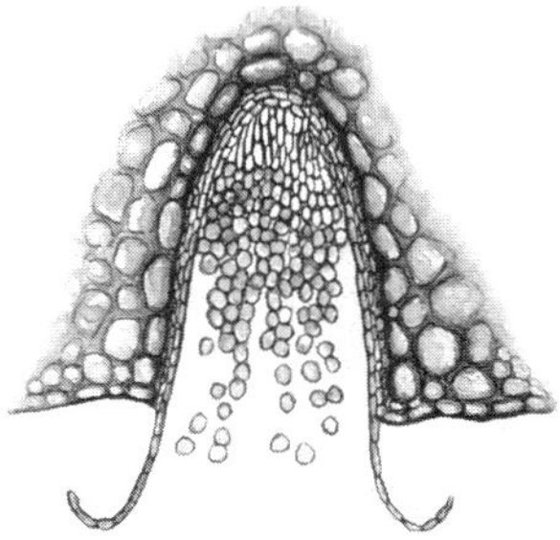

Cornute aecium of *Roestelia* sp. on a host leaf, seen in longitudinal section, x 110.

coronate (L. *coronatus*, crowned, ptp. of *coronare*, to crown < *corona*, crown + suf. *-atus* > E. *-ate*, provided with or likeness): **1**. Provided with a crown;

e.g., like the teliospore of *Puccinia coronata* (Uredinales), or the ascocarp of *Phyllactinia corylea* (Erysiphales), which has a crown composed of short, branched, gleoid apical appendages. **2**. Splitting to form a crown-like structure; e.g., like the exoperidium of *Sphaerobolus* (Nidulariales). Syn. of **crowned**.

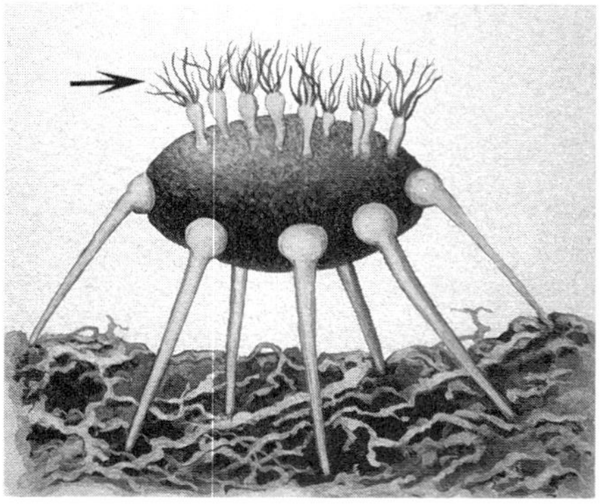

Coronate ascocarp of *Phyllactinia corylea* on the surface of a host leaf, x 190.

corrugate (L. *corrugatus*, ptp. of *corrugare* < *com-*, with + *ruga*, wrinkle + suf. *-atus* > E. *-ate*, provided with or likeness): wrinkled or folded, with alternating ridges and grooves.

cortex (L. *cortex*, bark, rind): the outer layer or layers of a sporocarp or somatic structure, including the peridium; e.g., in Nidulariales, the main wall of the peridiole.

corticicolous, **corticolous** (L. *cortex*, genit. *corticis*, cortex + *-cola*, inhabitant + L. *-osus* > OF. *-ous*, *-eus* > E. *-ous*, possessing the qualities of): living on the bark of trees and shrubs, as in many lichens, e.g., *Caloplaca* (Teloschistales). Cf. **lignicolous**.

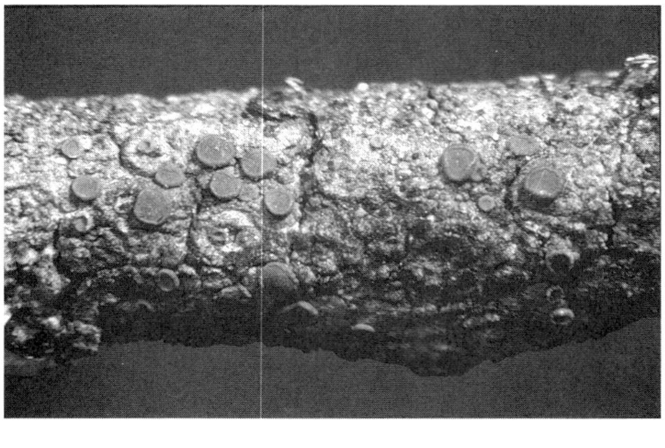

Corticicolous thallus of *Caloplaca* sp., x 7 (*MU*).

cortina, pl. **cortinae** (NL. < LL. *cortina*, veil, curtain): *Agaricology*. A thin, cloth-like tissue that hangs from the edge of the pileus of the basidiocarp in some agarics (e.g., the gen. *Cortinarius)*. The cortina comes from the partial veil that initially unites the edge of the pileus with the foot of the basidiocarp, but which because of its delicate consistency breaks and eventually disappears, leaving at most a peripheral filamentous remnant on the pileus and another on the stipe. Cf. **pellicular veil**.

Cortina of the basidiocarp of *Cortinarius* sp., x 0.8 (*TH*).

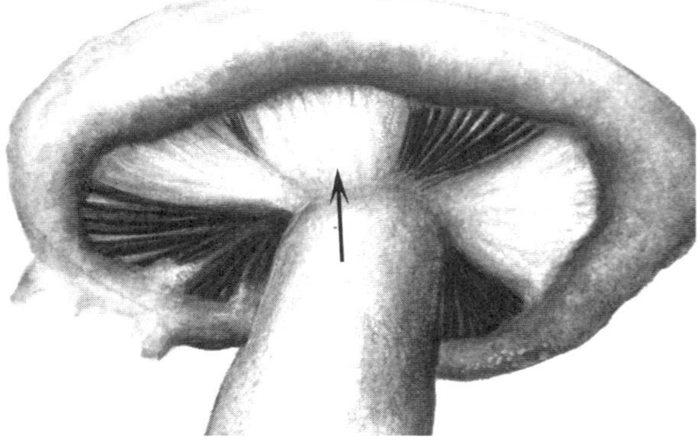

Cortina of the basidiocarp of *Cortinarius* sp., x 1.5.

cortinate (NL. < LL. *cortina*, veil, curtain + suf. *-atus* > E. *-atus*, provided with or likeness): having a cobweb-like, fibrillose partial veil; typically found in the Cortinariaceae (Agaricales).

corymbiferous, **corymbose** (L. *corymbus* < Gr. *kórymbos*, cluster of flowers + *-ferous*, bearer < *ferre*, to bear, carry + L. *-osus* > OF. *-ous*, *-eus* > E. *-ous*, having, possessing the qualities of; or + *-osus*, full of, augmented, prone to > ME. *-ose*): borne in a corymb, i.e., with the sporogenous branches arising at different points along the vertical axis but all terminating at the same level at the apex; e.g., as in the sporangiophore of *Absidia corymbifera* (Mucorales).

coscinocystidium, pl. **cosinocystidia** (Gr. *kóskinion*, sieve + *kystídion* < *kýstis*, bladder, vesicle, cell + dim.

suf. *-ídion* > L. *-idium*): a special type of gloeocystidium which has a cribose or porous wall and a sponge-like interior. The coscinocystidia are the terminal cells of a **coscinoid**. They are present, e.g., in *Gloeocantharellus* (Aphyllophorales); absent in the Agaricaceae.

coscinoid (Gr. *kóskinion*, sieve + L. suf. *-oide* < Gr. *-oeídes*, similar to): a specialized conductive hypha, brown in color, with a cribose surface and spongy contents; characteristic of *Gloeocantharellus* (=*Linderomyces*) *lateritius* (Aphyllophorales). See **coscinocystidium**.

costate (L. *costatus*, ribbed < *costa*, rib, flank + suf. *-atus* > E. *-ate*, provided with or likeness): having ribs or rib-like borders, **ribbed**; e.g., like the basidiocarp of *Cantharellus cibarius* (Aphyllophorales).

Costate basidiocarp of *Cantharellus cibarius*, x 0.4.

cothurnate (L. *cothurnatus* < *cothurnus*, shoe + suf. *-atus* > E. *-ate*, provided with or likeness): *Agaricales*. Having a tissue which sheathes the base of the stipe and usually has a thickened roll of tissue (like a rolled sock) at the top, as in *Amanita pantherina* var. *velatipes*.

cottony (E. *cotton* < ME. *coton* < OF. < OIt. *cotone* < Ar. *qutun*, *qutn*, cotton + E. suf. *-y*, having the quality of): with long, white, soft hairs, like cotton. The mycelium of many fungi is cottony; e.g., such as that of *Penicillium camembertii* (moniliaceous asexual fungi).

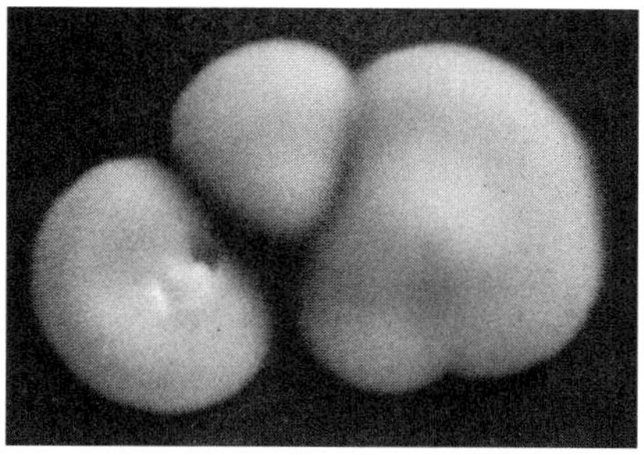

Cottony colonies of *Penicillium camembertii* on agar, x 1 (*MU*).

cotyliform (Gr. *kotýle*, concavity, a cup-shaped hollow > L. *cotyla*, *cotula* + L. suf. *-formis* < *forma*, shape): syn. of **cotylimorphous**. Having the shape of a cavity, concave; e.g., like the apothecia of *Peziza* (Pezizales), which have the shape of a saucer or cup, with the edge erect or turned upward.

cotylimorphous (Gr. *kotýle*, concavity, a cup-shaped hollow + *morphé*, shape + L. *-osus* > OF. *-ous*, *-eus* > E. *-ous*, having, possessing the qualities of): see **cotyliform**.

crateriform (L. *crateriformis* < *crater*, vessel, cup + *-formis* < *forma*, shape): bowl-shaped, hemisphaerical and concave; e.g., like the sporangia of *Craterium minutum* (Physarales).

Crateriform sporangium of *Craterium minutum*, x 13 (*MU*).

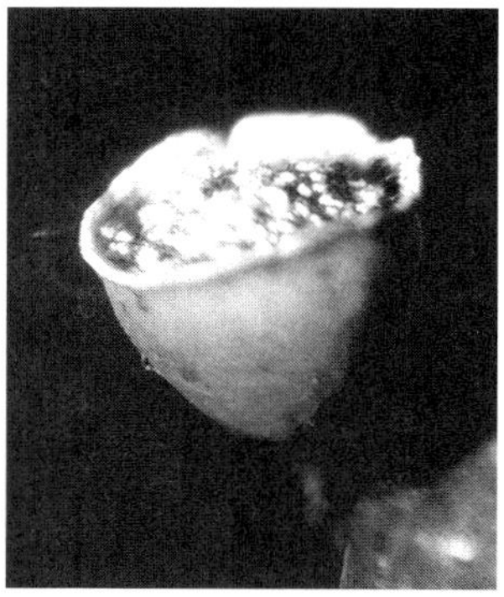

Crateriform sporangium of *Craterium minutum*, x 13 (*MU*).

crenate (L. *crenatus* < *crena*, notch + suf. *-atus* > E. *-ate*, provided with or likeness): with the margin edged with notches or scallops; e.g., like the giant colony of some yeast strains when grown on solid media (*Saccharomyces cerevisiae* of the Saccharomycetales), or like the margin of the gills of certain agarics (e.g., *Cortinarius alboviolaceus*). Also called **festooned**.

crenulate (L. *crenulatus*, minutely notched < *crenula*, dim. of *crena*, notch or low cut + suf. *-atus* > E. *-ate*, provided with or likeness): like crenate, but with smaller notches; e.g., like the margin of the pileus of *Mycena pura* (Agaricales).

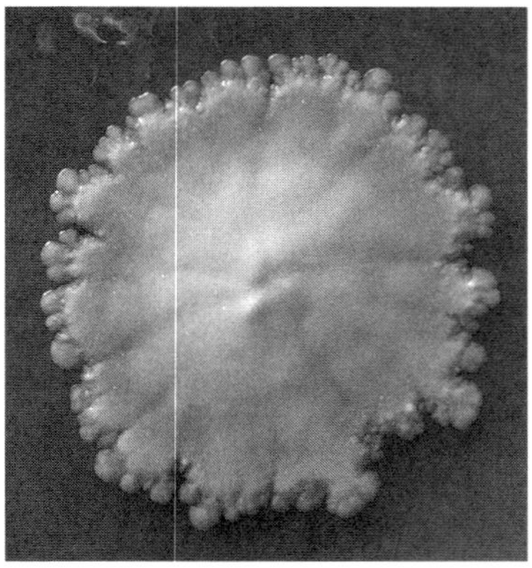

Crenate border of the giant colony of *Saccharomyces cerevisiae* on agar, x 2 (*MU*).

crescentic, **crescentiform** (L. *crescere*, to increase, grow + *-escens*, genit. *-escentis*, beginning to, slighty > E. *-escent*, + Gr. suf. *-íkos* > L. *-icus* > E. *-ic*, belonging to, relating to; or L. suf. *-formis* < *forma*, shape): crescent-shaped; half-moon-shaped. Syn. of **lunate**.

cretaceous (L. *cretaceus*, chalky < *creta*, chalk, gypsum + *-aceus* > E. *-aceous*, of or pertaining to, with the nature of): with the color or consistency of chalk, like the medulla of many crustaceous lichens, which is white, compact, pimply, with few hyphae.

cribrate, **cribrose** (L. *cribrum*, sieve + L. suf. *-atus* > E. *-ate*, provided with or likeness; or + L. *-osus*, full of, augmented, prone to > ME. *-ose*): perforated; sometimes used to describe the perforated and reticulate appearance of some boletes, a condition that results from the decurrence of the walls of the pileus tubes.

crinate, **crinite** (L. *crinatus*, hairy, long-haired < *crinus*, hair + suf. *-atus* > E. *-ate*, provided with or likeness): having long hair or hair-like structures; e.g., like the fruiting body of *Panus crinitus* (Aphyllophorales).

crispate, **crisped** (L. *crispatus* < *crispus*, curly, wavy + suf. *-atus* > E. *-ate*, provided with or likeness): finely curled or wavy; e.g., like the gills of fungi of the gen. *Trogia* (Agaricales).

Crispate lamellae of the basidiocarp of *Trogia* sp. Diagram of a longitudinal section, x 0.6.

crista, pl. **cristae** (NL. < L. *crista*, crest): any of the inwardly projecting folds of the inner membrane of a mitochondrion; cristae, which can be tubular, pouch-like or shelf-like according to the group of fungi, are the site of ATP production during aerobic metabolism. See **mitochondrion**.

cristate (L. *cristatus* < *crista*, crest, relief + suf. *-atus* > E. *-ate*, provided with or likeness): having one or more crests.

crowned (ME. *croune*, *crune* < AF. *croune* < L. *corona*, wreath): see **coronate**.

crozier (variant of G. *crosier* < ME. *croser*, crosier bearer < MF. *crossier* < *crosse*, crosier): a structure with a hooked end. See **uncinulum**. The term is commonly applied to the initial stage in ascus formation of most ascomycetes.

cruciate (L. *cruciatus*, crossed < *crux*, genit. *crucis*, cross + suf. *-atus* > E. *-ate*, provided with or likeness): having the shape of a cross, like the nuclear division spindle of the Plasmodiophoromycetes. Syn. of **cruciform**.

cruciform (L. *cruciformis* < *crux*, genit. *crucis*, cross + *-formis* < *forma*, shape): having the shape of a cross; e.g., like the conidia of *Valdensia heterodoxa* (moniliaceous asexual fungi), or the spindle in the nuclear division (cruciform division) of *Plasmodiophora* and other Plasmodiophorales, such as *Sorosphaera*, which adopt this figure because the chromosomes form a ring that surrounds perpendicularly the elongated nucleolus, which appears as a cross when viewed from one side. This type of nuclear division is also called **promitosis**.

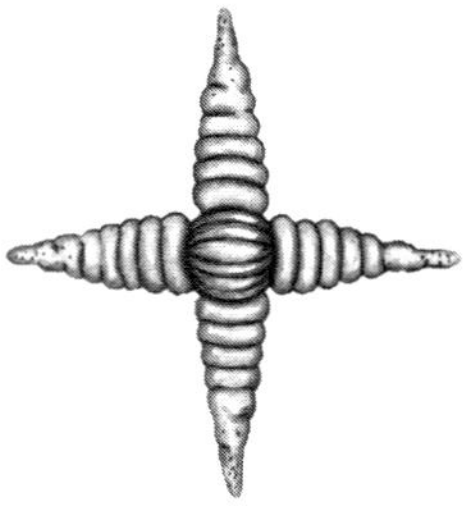

Cruciform conidium of *Valdensia heterodoxa*, x 650.

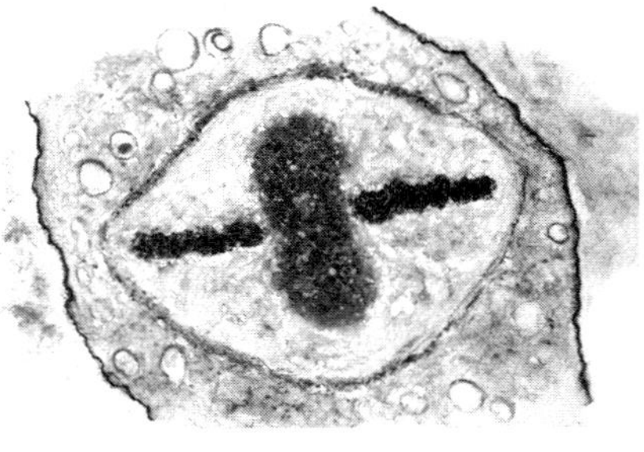

Cruciform nuclear division in the endobiotic plasmodium of *Sorosphaera veronicae*, x 12 500.

cruciform division (L. *cruciformis* < *crux*, genit. *crucis*, cross + *-formis* < *forma*, shape; L. *divisio*, genit. *divisionis*, division < Gr. *di-* < *dís*, two, double + L. *visus*, seen, viewed + *-io*, *-ionis* > E. suf. *-ion*, result of an action, state of): see **cruciform** and **promitosis**.

crustaceous, **crustose** (L. *crustaceus*, having a shell or rind < *crusta*, crust, cortex + *-aceus* > E. *-aceous*, of or pertaining to, with the nature of; or + L. *-osus*, full of, augmented, prone to > ME. *-ose*): *Lichens*. A thallus that has the appearance of a powdery, warted, areolate or granulose crust, closely adhered to the substrate and lacking a lower cortex and rhizines; e.g., as occurs in *Ochrolechia parella* (Pertusariales), *Caloplaca elegans* (Teloschistales), *Haematomma subpuniceum* and *Lecanora frustulosa* (Lecanorales).

Crustaceous thallus of *Caloplaca saxicola*, x 5 (*CB*).

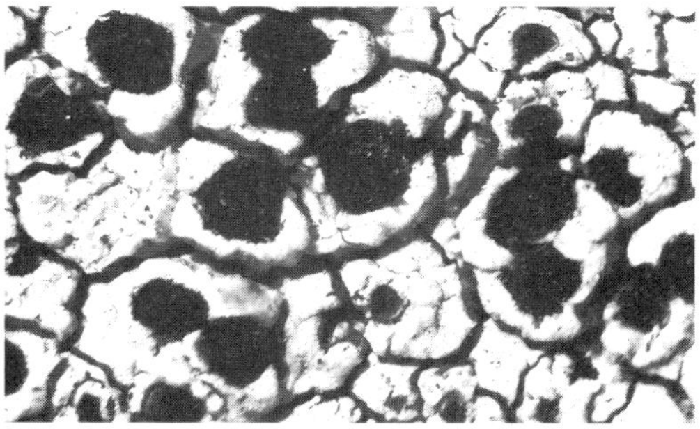

Crustaceous thallus of *Haematomma subpuniceum* on rock; the dark bodies are apothecia, x 5 (*MU*).

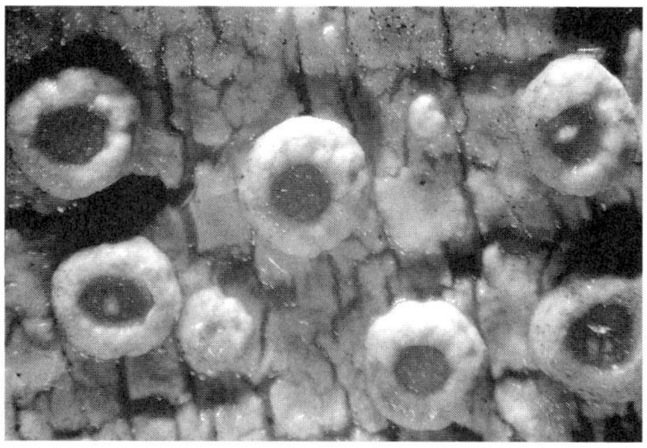

Crustaceous thallus of *Ochrolechia parella* on bark; the apothecia are evident, x 7 (*MU*).

crypta, pl. **cryptae** (L. *crypta* < Gr. *kryptós*, hidden, covered): a sleeve-like formation around the roots of certain trees, formed by some Agaricales, which serves as shelter for various epibiotic, symbiotic or parasitic insects, or with a relationship to the tree that involves a combination of these modes of life.

cryptobiosis, pl. **cryptobioses** (Gr. *kryptós*, hidden + *bíosis* < *bíos*, life + *-osis*, condition or state): the condition of suspended life, similar to death, that generally occurs through starvation, dessication or freezing, a condition that can revert through **anabiosis**.

cryptococcosis, pl. **cryptococcoses** (< gen. *Cryptococcus* < Gr. *kryptós*, hidden + *kókkos* < *kokkíon*, small ball + *-osis*, state or condition): a subacute or chronic infection of the central nervous system of humans, with meningitis and cerebral tumors, although other organs can be affected (skin, bones, lungs). The infection is acquired by the inhalation of dust containing yeast cells of the causal agent, *Cryptococcus neoformans* (cryptococcaceous asexual yeasts), an ubiquitous soil inhabitant, in particular that enriched with bird droppings (chickens, doves, and pigeons).

cryptoendolithic (Gr. *kryptós*, hidden + *éndon*, inside + *líthos*, stone + suf. *-íkos* > L. *-icus* > E. *-ic*, belonging to, relating to): an organism, especially a lichen, surviving at low temperatures through modification of the thallus so that it can exist inside of a rock between the rock crystals. Cf. **endolithic**.

cteinomycete (Gr. *kteíno*, to kill + *mýkes*, genit. *mýketos*, fungus): a parasitic fungus that produces the immediate death of the cells that it penetrates to obtain food.

cuboid (L. *cubus* < Gr. *kýbos*, cube + L. suf. *-oide* < Gr. *-oeídes*, similar to): like a cube; the brown rot of wood caused by *Poria* and related gen. of Aphyllophorales results in cuboid wood fragments.

cucullar, **cucullate** (L. *cuculla*, hood, cowl, cloak + suf. *-aris* > E. *-ar*, like, pertaining to; L. *cucullatus*, hooded < *cuculla* + suf. *-atus* > E. *-ate*, provided with or likeness): having the shape of a cowl or cloak, hooded; e.g., like the pileus of *Phaeocollybia christinae* (Agaricales). Also called **cuculliform**.

cuculliform (NL. *cuculliformis* < *cuculla*, cowl + *-formis* < *forma*, shape): see **cucullate**.

culmicole, **culmicolous** (L. *culmus*, grass stem + *cola*, dweller < *colere*, to inhabit + *-osus* > OF. *-ous*, *-eus* > E. *-ous*, possessing the qualities of): living and developing on the stems of cereals and other grasses; e.g., such as the fungi commonly called rusts (Uredinales). Also called **culmigenous**.

culmigenous (L. *culmus*, grass stem + *-genous*, to

produce, engender + *-osus* > OF. *-ous*, *-eus* > E. *-ous*, possessing the qualities of): see **culmicole**.

Cucullar pileus of the basidiocarps of *Phaeocollybia christinae*, x 1.

cultrate, **cultriform** (L. *cultratus* < *culter*, knife + suf. *-atus* > E. *-ate*, provided with or likeness; L. *cultriformis* < *culter* + *-formis* < *forma* shape): knife-shaped.

culture (ME. < MF. < L. *cultura*, tilling < *cultus*, culture, cultivation): **1**. a population of organisms, especially microorganisms, maintained in the laboratory, where they are grown on or in a culture medium and transferred to other media by inoculation. **2**. The act of growing organisms in the laboratory on appropriate media. See **agnotobiotic**, **axenic**, **isolate**, **monoxenic**, and **strain.**

cumulate (L. *cumulatus*, ptp. of *cumulare* < *cummulus*, mass + suf. *-atus* > E. *-ate*, provided with or likeness): massed together; heaped up; combined into one.

cuneate, **cuneiform** (L. *cuneate* < *cuneus*, wedge + suf. *-atus* > E. *-ate*, provided with or likeness; *cuneus* + *-formis* < *forma*, shape): having the shape of a wedge; wedge-shaped in longitudinal section, when used for laminar structures. For example, the basidiospores of *Lepiota cuneatispora* (Agaricales) are cuneiform.

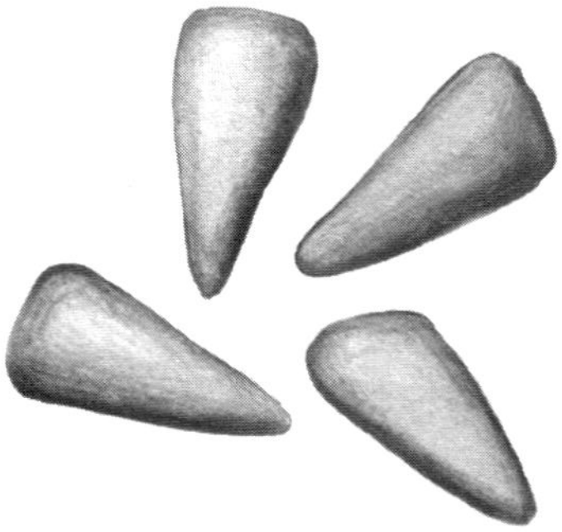

Cuneiform basidiospores of *Lepiota cuneatispora*, x 2 000.

cuniculate (L. *cuniculatus* < *cuniculus*, den, burrow + suf. *-atus* > E. *-ate*, provided with or likeness): traversed on one end by a long passage.

cup fungus, pl. **fungi** (ME. *cuppe* < OE. < LL. *cuppa*, cup, goblet < L. *cupa*, tub; L. *fungus*, fungus): a cup-shaped discomycete ascoma, especially of Pezizales and Helotiales.

cupulate, **cupular** (L. *cupulatus*, tub-shaped < *cupula*, dim. of *cupa*, tub, cask + suf. *-atus* > E. *-ate*, provided with or likeness; or + L. suf. *-aris* > E. *-ar*, like, pertaining to): having the shape of a tub (cup); e.g., like the aecia of *Aecidium* and *Puccinia graminis* (Uredinales), with a peridium in the shape of a cup, cylindric, with a recurved edge. The ascocarps of *Cookeina aurantiaca* and *C. tricholoma* (Pezizales) also are cupulate, as are the podetia of the lichens *Cladonia chlorophaea* and *C. pyxidata* (Lecanorales).

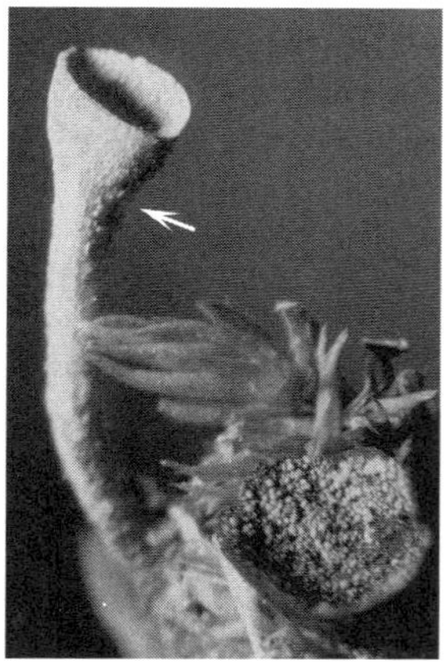

Cupulate podetium of the thallus of *Cladonia chlorophaea*, x 9 (*MU*).

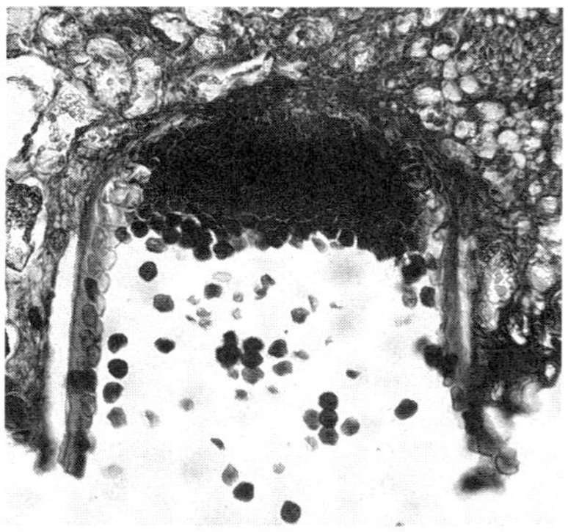

Cupulate aecium of *Puccinia graminis tritici*, seen in longitudinal section of a host leaf, x 175 (*MU*).

cuspidate (NL. *cuspidatus* < L. *cuspis*, genit. *cuspidis*, point + suf. *-atus* > E. *-ate*, provided with or likeness): ending in a sharp point; e.g., like the pileus of *Phaeocollybia* or the cystidia of *Panaeolus cyanescens* (Agaricales).

cyanobiont (Gr. *kýanos*, a greenish blue color + *bíos*, life + *óntos*, genit. of *ón*, a being): the photosynthetic cyanobacterial symbiont in a lichen. See **photobiont**.

cyanophilic, **cyanophilous** (Gr. *kýanos*, substance with a dark blue color + *phílos*, have an affinity for < *philein*, to love, prefer + suf. *-íkos* > L. *-icus* > E. *-ic*, belonging to, relating to; or + L. *-osus* > OF. *-ous*, *-eus* > E. *-ous*, possessing the qualities of): refers to fungal structures, especially cell or spore walls, which stain dark blue in 1% cotton blue in lactic acid (Amman's solution, commonly known as cotton blue or aniline blue). For example, the hyphae of *Crinipellis piceae*, the basidia of *Lyophyllum multiceps* (Agaricales) and the basidiospores of *Gomphus* (Aphyllophorales) are all cyanophilic. In some *Gomphus* spores the ornamentation stains distinctly and contrasts with the interior which stains lightly.

cyanophilous (Gr. *kýanos*, a greenish blue color + *phílos*, have an affinity for + L. *-osus* > OF. *-ous*, *-eus* > E. *-ous*, possessing the qualities of): readily absorbing a blue stain such as cotton blue or gentian violet.

cyanophycophilous (Gr. *kýanos*, substance with a dark blue color + *phýkos*, alga + *phílos*, have an affinity for < *philein*, to love, prefer, appreciate + L. *-osus* > OF. *-ous*, *-eus* > E. *-ous*, possessing the qualities of): *Lichens.* Refers to thalli that have Cyanophyceae as phycobionts, as occurs in the gen. *Collema* (Lecanorales). Cf. **chlorophycophilous**. See **diphycophilous**.

cyathiform (L. *cyathus*, cup < Gr. *kyáthos*, ladle + L. suf. *-formis* < *forma*, shape): having the shape of a cup, e.g., like the receptacles of the Gasteromycetes of the gen. *Cyathus* (Nidulariales) and the podetia of some lichens of the gen. *Cladonia*, such as *C. pyxidata* (Lecanorales).

Cyathiform podetium of the thallus of *Cladonia chlorophaea*, x 25 (*CB*).

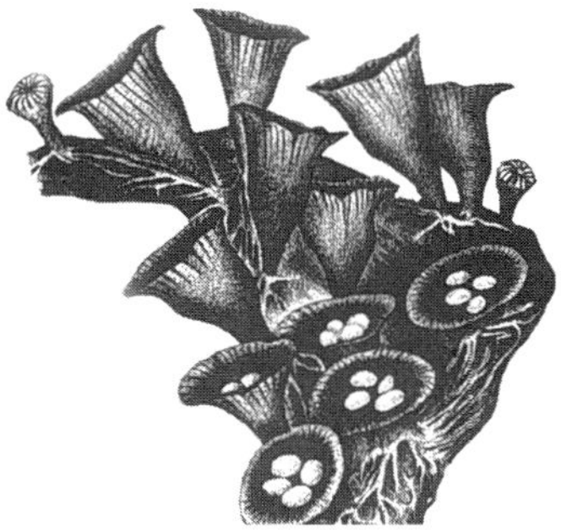

Cyathiform basidiocarp of *Cyathus striatus*, x 1.

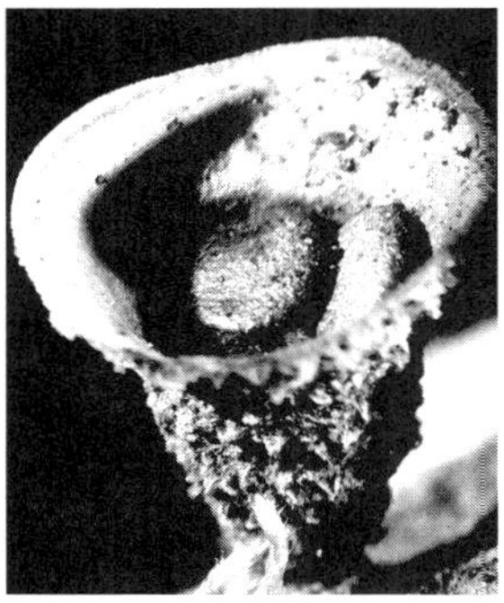

Cyathiform basidiocarp of Cyathus sp., x 5 (*CB*)

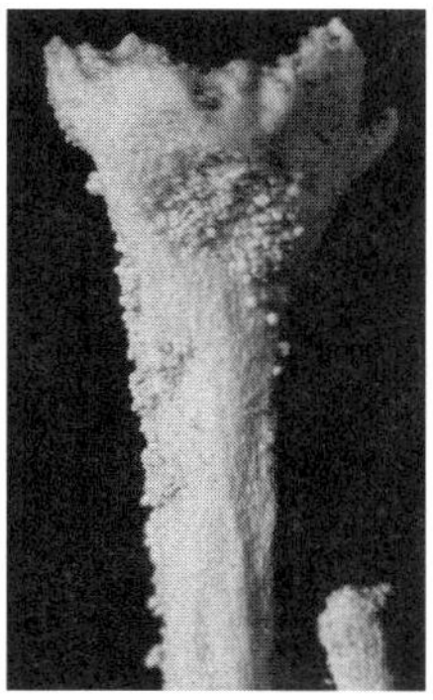

Cyathiform podetium of *Cladonia chlorophaea*, x 25 (*CB*).

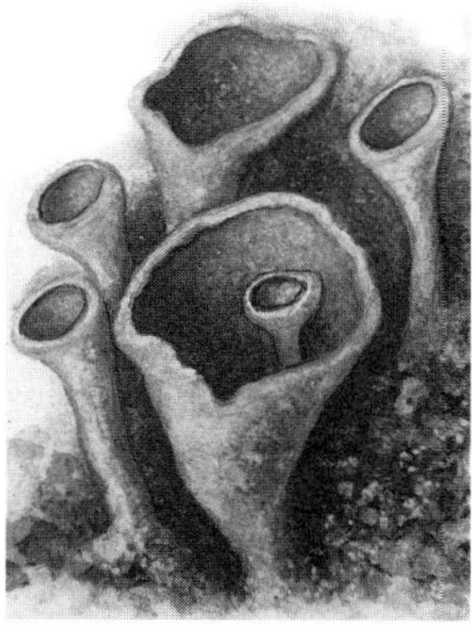

Cyathiform podetium of *Cladonia chlorophaea*, x 25.

cyclomoid (< gen. *Cyclomyces* < Gr. *kýklos*, circle + L. suf. *-oide* < Gr. *-oeídes*, similar to): *Basidiomycetes*. With the hymenium formed on structures similar to gills or plates that are arranged concentrically, and which represent transformed pores, not true gills, as in the gen. *Cyclomyces* and *Coltricia*, of the Aphyllophorales.

cyclosis, pl. **cycloses** (Gr. *kýklosis*, action of enclosing, wrapping up): the cytoplasmatic circulatory current characteristic of eukaryotic cells; this internal motility of the cells depends on the complex of actinomiosin fibers, which are distinct from those present in animal muscle.

cygneous (L. *cycnus* < Gr. *kýknos*, swan + L. *-osus* > OF. *-ous*, *-eus* > E. *-ous*, having): curved like the neck of a swan.

cylindric, **cylindrical**, **cylindroidal** (NL. *cylindricus* < L. *cylindrus* < Gr. *kylindrikós* < *kýlindros*, roller < *kylíndein*, to roll + suf. *-íkos* > L. *-icus* > E. *-ic*, belonging to, relating to; or + L. suf. *-alis* > E. *-al*, relating to or belonging to; or + L. suf. *-oide* < Gr. *-oeídes*, similar to): having the shape of a cylinder, with a length:width ratio of between 2:1 and 3:1. For example, the sporangiola of *Syncephalis depressa* (Zoopagales) and the conidia of *Cylindrocarpon olidum*, *Geosmithia* (moniliaceous asexual fungi) and *Blumeriella haddonii* (Helotiales) are cylindrical.

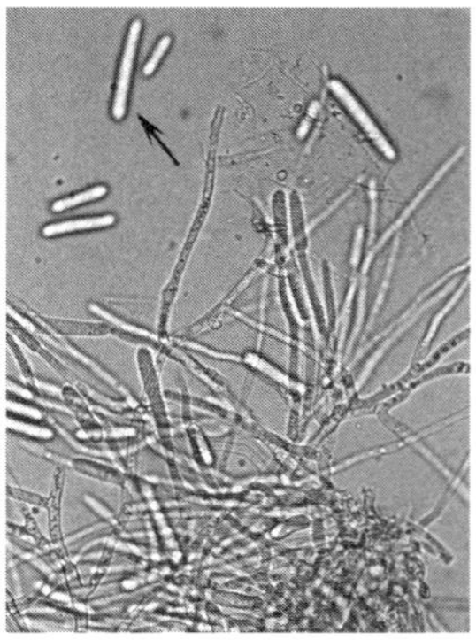

Cylindric conidia of *Cylindrocarpon olidum*, x 250 (*JAS*).

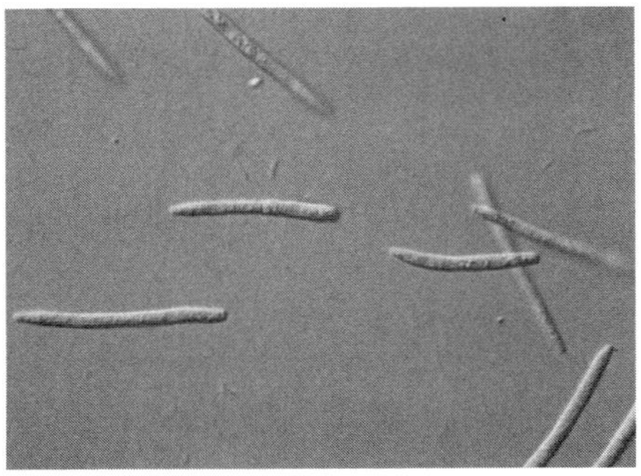

Cylindric conidia of *Blumeriella haddonii*, x 1 300 (*RTH*).

cymbiform (NL. *cymbiformis* < *cymba*, boat + *-formis* < *forma*, shape): boat-shaped; equal to **carinate**, **navicular** and **scaphoid**.

cyme (L. *cyma*, green cabbage < Gr. *kýma*, swelling, bud, yoke): racime; a bouquet of sporangial hyphae in which each principal axis ends in a sporangium; each of the secondary and tertiary axes also can end in a sporangium, as is seen in some Oomycota sporangiophores.

cyphella, pl. **cyphellae** (L. *cyphella*, the concavity of the ear): *Lichens*. The small cavities or hollow depressions, delimited by a cortical layer, that are only found on the lower surface of the majority of lichens of the family Lobariaceae (*Sticta*), of the Peltigerales. See **cyphelloid** and **pseudocyphella**.

Cyphellae in the lower surface of the thallus of *Sticta weigelii*, x 5 (*MU*).

cyphelloid (L. *cyphella*, the concavity of the ear + suf. *-oide* < Gr. *-oeídes*, similar to): *Lichens*. A concavity or pit in the upper surface of the thallus of some lichens. Cf. **cyphella** and **pseudocyphella**.

cyst (NL. *cystis* < Gr. *kýstis*, bag, bladder; here, cell, vesicle): a spore protected by a resistant cell wall; it is especially applied to the structure that forms when a zoospore encysts, as happens, e.g., in the aquatic fungi of the orders Chytridiales and Saprolegniales (such as *Phlyctochytrium* and *Saprolegnia*, respectively), or to the structure resulting from the encystment of an amoeba, as in *Amoebidium parasiticum* (Amoebidiales). Also used as a suf., e.g., **heterocyst**, **macrocyst**, **microcyst**, etc.

cystidiole (NL. *cystidiolum* < *cystidium* + dim. suf. *-olum* < Gr. *kystídion*, dim. of *kýstis*, bag, bladder, vesicle + dim. suf. *-ídion* > L. *-idium*): a primordial cystidium, of superficial origin, that originates at the same level as the basidia in the hymenium of some Agaricales, (e.g., *Conocybe tenera*); cystidioles differ slightly in size and shape from the basidia and brachybasidia.

Cysts of *Amoebidium parasiticum*, x 730.

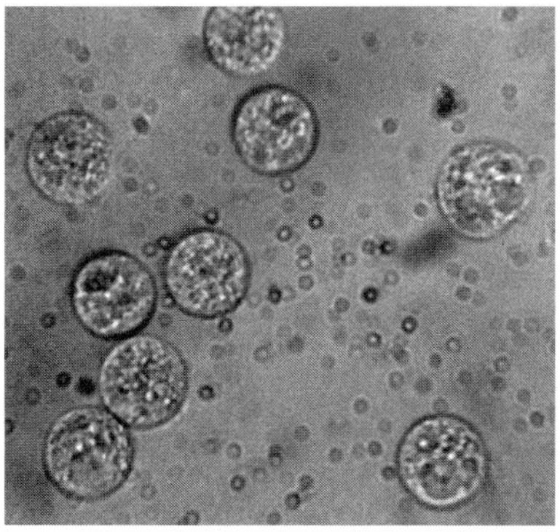

Cysts of *Saprolegnia parasitica*, x 600 (*MU*).

cystidium, pl. **cystidia** (NL. *cystidium* < Gr. *kystídion* < *kýstis*, bag, bladder, vesicle, cell + dim. suf. *-ídion* > L. *-idium*): *Agaricales*. A large, inflated, sterile cell that grows among the basidia and paraphyses; cystidia are usually larger in size, so that they project from the hymenial layer. Their function is not well understood, although in some cases they appear to serve as spacers between the basidia. See **brachycystidium**, **caulocystidium**, **cheilocystidium**, **chrysocystidium**, **coscinocystidium**, **gloeocystidium**, **lamprocystidium**, and **pleurocystidium.**

cystoderm (NL. *cystis* < Gr. *kýstis*, vesicle, cell + *derma* < Gr. *dérma*, skin): *Agaricales*. A type of pellicle (**pellis**, in particular of the **suprapellis**), composed of a single layer of subglobose or globose, vesiculose cells (**sphaerocysts**) that cover the pileus of certain agarics; e.g., as in *Cystoderma amianthinium*, *Coprinus stercorarius* and *C. micaceus*. To this cuticle or pilear cortex (**pileipellis**) the term **cellular dermis** is also applied.

cystogenous (NL. *cystis* < Gr. *kýstis*, vesicle, cell, cyst + *génos*, origin < *gennáo*, to engender, produce + L. *-osus* > OF. *-ous*, *-eus* > E. *-ous*, having, possessing the qualities of): *Plasmodiophorales*. Refers to the plasmodium of the sexual phase which is transformed into immotile spores or cysts, which can remain separate or grouped into sori; some species form a definite number of cysts, e.g., four in *Tetramyxa* and eight in *Octomyxa*, whereas others (*Polymyxa* and *Spongospora*) have numerous cysts. See **sporangiogenous**.

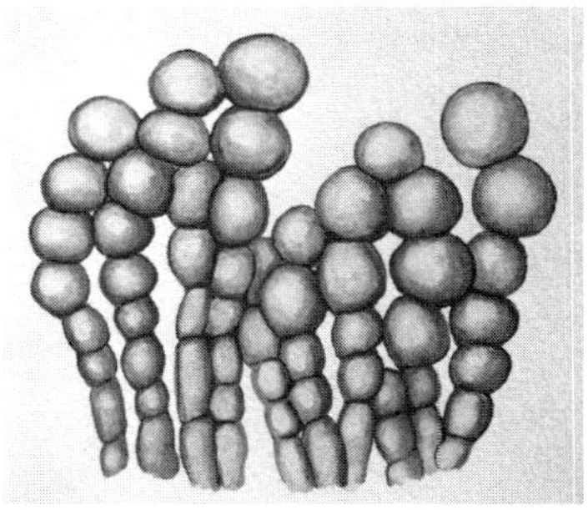

Cystoderm of the pileus of *Cystoderma amianthinum*, x 600.

cystosorus, pl. **cytosori** (NL. *cystis* < Gr. *kýstis*, bag, bladder, vesicle + NL. *sorus* < Gr. *sorós*, heap, pile): a more or less compact aggregate of latent spores or sporangiocysts, which are derived from a single protoplast; as occur, e.g., in *Woronina* and *Spongospora* (Plasmodiophorales).

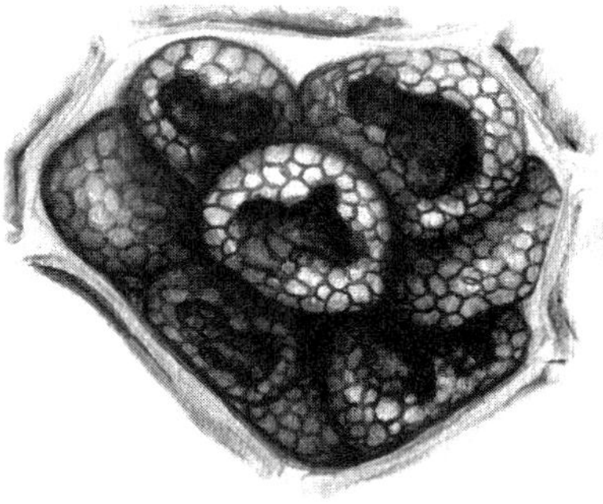

Cystosorus of *Spongospora subterranea* within a cell of potato tuber, x 600.

cystospore (NL. *cystis* < Gr. *kýstis*, bag, bladder, vesicle + *sporá*, spore): *Trichomycetes*. The spore that results from the encystment of an amoeba; as happens, e.g., in *Amoebidium* (Amoebidiales). Also used for the encysted zoospore of certain Chytridiales.

d

dacryoid (Gr. *dácryon*, tear + L. suf. *-oide* < Gr. *-oeídes*, similar to): tear-shaped, more or less rounded at one end and prolonged and sharp-pointed at the other; e.g., like the conidia of *Spiniger meineckellus* (moniliaceous asexual fungi) and the basidiospores of *Clitocybe gibba* (Agaricales). Also called **lacrimoid**.

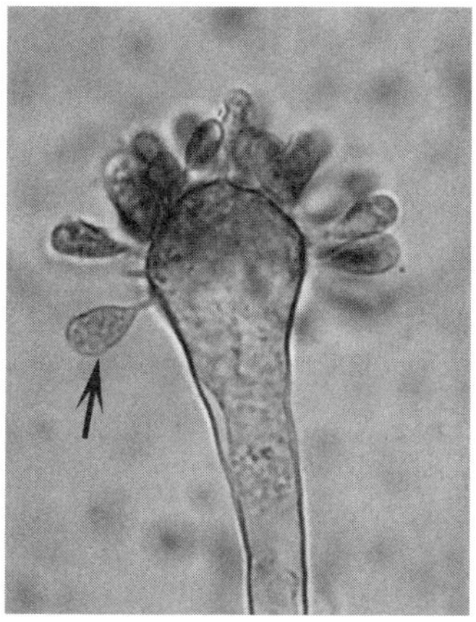

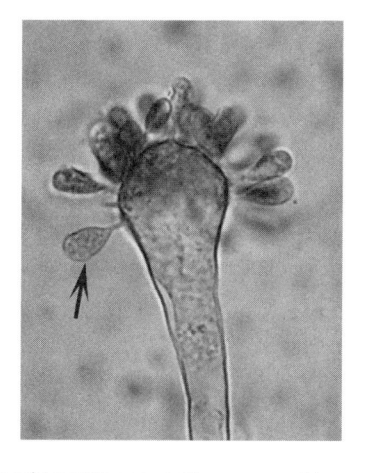

Dacryoid conidia on a conidiogenous cell (ampulla) of *Spiniger meineckellus*, x 500 (*RTH*).

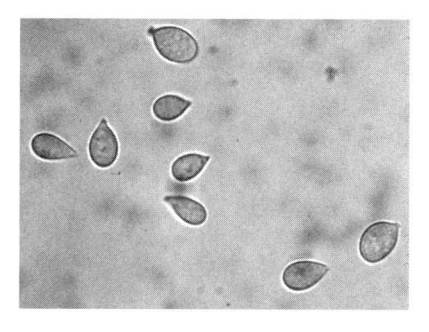

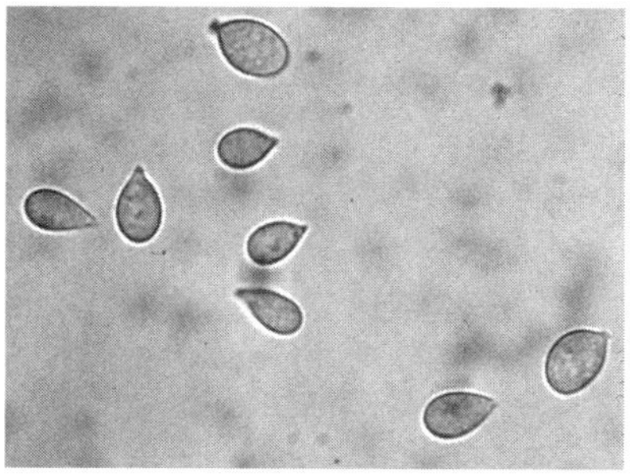

Dacryoid conidia of *Spiniger meineckellus*, x 590 (*RTH*).

dactyloid (Gr. *dáktylos*, finger + L. suf. *-oide* < Gr. *-oeídes*, similar to): divided into finger-like structures; e.g., like the haustoria of *Blumeria* (=*Erysiphe*) *graminis* (Erysiphales). Also called **digitate**.

daedaleoid, **daedaloid** (< gen. *Daedalea* < Gr. *daidáleos*, labyrinth + L. suf. *-oide* < Gr. *-oeídes*, similar to): with the tube mouths enlarged and sinuous; labyrinthiform, like the hymenophore of *Daedalea* (Aphyllophorales).

damping-off (E. *damp*, moist < MD. < MHG., *damph*, vapor; OE. *off*, removal): a disease condition of seedlings or cuttings caused by fungi (e.g. *Pythium*, of the Peronosporales), and characterized by collapse and death of the plants.

de novo (L. *de novo*, anew): *Conidiogenesis*. Refers to conidia that are formed from newly synthesized protoplasm (**blastic**), rather than from already existing cells. Cf. **arthric.**

dead man's fingers (ME. *deed* < OE. *dead*, dead; ME. < OE. *man*, man, a human being; an adult male human; ME. < OE. *finger*, finger; a digit of the forelimb, especially other than the thumb): the black stromata of *Xylaria polymorpha* (Xylariales).

deciduous (L. *deciduus*, tending to fall off < *decidere*, to fall off + *-osus* > OF. *-ous*, *-eus* > E. *-ous*, possessing the qualities of): falling off; being easily detached from the structure that bears it; e.g., like the sporangia of the higher Peronosporales (*Peronospora*, *Sclerospora*). Cf. **persistent**.

deciduous wart (L. *deciduus*, tending to fall off < *decidere*, to fall off + *-osus* > OF. *-ous*, *-eus* > E. *-ous*, possessing the qualities of; ME. < OE. *wearte*, wart): a wart that can be readily detached from the exoperidium, usually leaving a visible scar; e.g., in *Lycoperdon* (Lycoperdales).

declinate (L. *declinatus*, bent downward, turned aside < *declinare*, to incline, bend + suf. *-atus* > E. *-ate*, possessing the properties of or likeness): bent or curved downward. See **decurved**.

decumbent (L. *decumbens*, genit. *decumbentis*, reclining < *decumbere*, to lie down + *-entem* > E. *-ent*, being): lying on the substrate, with the margin tending to turn upward; e.g., like the fruiting body of the corticiaceous and stereaceous fungi of the order Aphyllophorales.

decurrent (L. *decurrens*, genit. *decurrentis*, running down < *decurrere*, to run down + *-entem* > E. *-ent*, being): used for gills that adhere to and extend along the stipe toward its base, as in the basidiocarp of *Clitocybe dealbata* and *C. senilis* (Agaricales), and *Lentinus badius* (Aphyllophorales).

Decurrent lamellae of the basidiocarp of *Clitocybe* sp., x 0.7 (*EPS*).

Decurrent lamellae of the basidiocarp of *Lentinus badius*, x 1.7 (*MU*).

decurvate, **decurved** (L. *decurvatus*, curved or bent downward < *de-*, from, down, away + *curvus*, curve + suf. *-atus* > E. *-ate*, provided with or likeness): bent or curved downward; e.g., the margin of the pileus of *Xerocomus chrysenteron*, *Boletus edulis* and other species of the gen. *Boletus* (Agaricales) is decurved, since its edge is curved downward, more or less parallel to the stipe or pointed slightly toward it.

decussate (ML. *decussatus* < *decussare*, to cross in the form of an X < L. *decussis*, # ten + suf. *-atus* > E. *-ate*, provided with or likeness): *Lichens*. Having a surface that is crossed by dark lines that intersect to form an X, as in some lichen thalli.

dehiscence (NL. *dehiscentia* < L. *dehiscere*, to open itself, to part + suf. *-entia* > F. *-ence* < E. *-ence*, state, quality or action): the spontaneous opening of an organ by splitting, on reaching maturity, to release the contents. In the fungi, many types of sporangia, gametangia and other sporophores have some type of characteristic dehiscence which permits the liberation of the spores, gametes, etc., e.g.; the splitting of the peridium in mature fruiting bodies of many gasteromycetes in order to expose the gleba and facilitate discharge of the spores. Also see **circumscissile**.

Decurvate pileus of the basidiocarp of *Xerocomus chrysenteron*, x 0.4.

dehiscent, **dehiscing** (L. *dehiscens*, genit. *dehiscentis*, parting, dividing, yawning < *dehiscere*, to divide, split + *-entem* > E. *-ent*, being): opening at maturity, by a pore or a slit. Cf. **indehiscent**.

delimited peristome (F. *délimiter* < L. *delimitare*, to limit; NL. *peristoma* < Gr. pref. *perí-*, around + *stóma*, mouth): a zone around the peristome, usually small and sometimes slightly raised, which is distinct from the exoperidium and often different in color.

delineate (L. *delineatus* < *delineare*, to delineate < *de-*, from, down, away + *linea*, line + suf. *-atus* > E. *-ate*, provided with or likeness): outlined; a figure sketched with lines.

deliquescent, **deliquescence** (L. *deliquescens*, genit. *deliquescentis*, the process of becoming fluid when mature < *deliquescere*, to become liquid + suf. *-escens*, genit. *-escentis*, that which turns, beginning to, slightly > E. *-escent*; or + L. suf. *-entia* > F. *-ence* < E. *-ence*, state, quality or action): becoming a liquid mass through autodigestion or autolysis at maturity; e.g., dissolving into an "inky" fluid, as in the basidiocarps of the gen. *Coprinus* (Agaricales). See **autolysis**.

Deliquescent basidiocarps of *Coprinus comatus*, x 2 (*TH*).

Deliquescent basidiocarps of *Coprinus comatus*, x 0.7.

delitescent (L. *delitescens*, genit. *delitescentis*, hiding, prp. of *delitescere*, to hide, to conceal oneself + *-escens*, genit. *-escentis*, that which turns, beginning to, slightly > E. *-escent*): hidden.

deltoid (L. *delta* < Gr. *delta*, Δ, fourth letter of the Gr. alphabet + L. suf. *-oide* < Gr. *-oeídes*, similar to): triangular, like the letter; such as the conidia of *Anthopsis deltoidea* (moniliaceous asexual fungi).

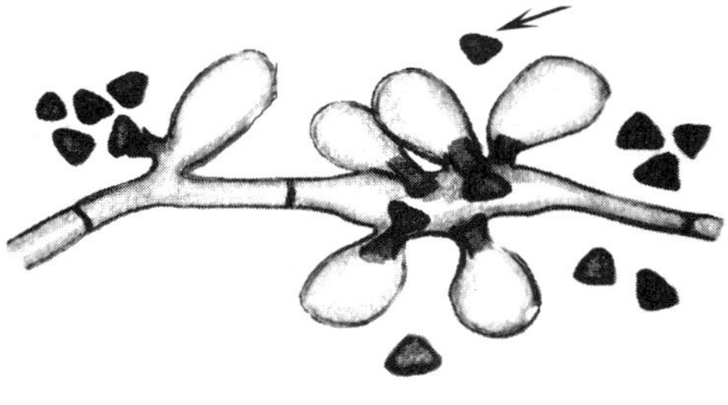

Deltoid conidia of *Anthopsis deltoidea*, x 1 500.

dematioid (< gen. *Dematium* < Gr. *demátion*, small sheaf + L. suf. *-oide* < Gr. *-oeídes*, similar to): similar to *Dematium* (a dematiaceous asexual fungus which corresponds to a *nomen confusum*), i.e., dark in color, almost black, filamentous, with a layer or felt of dark hyphae, as in the hyphae of the sooty molds (Dothideales). See **fuliginous**.

demicyclic (L. *demi*, half + Gr. *kyklikós*, circular, cyclic < *kýklos*, circle + suf. *-íkos* > L. *-icus* > E. *-ic*, belonging to, relating to): *Uredinales*. A rust in whose life cycle there is no uredinial state (state II), but with the spermogonial, aecial and telial states (states 0, I, III and IV, respectively) present, as occurs in *Gymnosporangium juniperi-virginianae*, the cedar-apple rust; also termed **hemicyclic**. Cf. **macrocyclic** and **microcyclic**.

dendritic (L. *dendriticus* < Gr. *dendrítes*, *dendrítis*, pertaining to trees + Gr. suf. *-íkos* > L. *-icus* > E. *-ic*, belonging to, relating to): see **dendroid**.

dendritic columella, pl. **columellae** (L. *dendriticus* < Gr. *dendrítes*, *dendrítis*, pertaining to trees, treelike + Gr. suf. *-íkos* > L. *-icus* > E. *-ic*, belonging to, relating to; L. *columella*, dim. of *columna*, column): *Gasteromycetes*. Sterile tissue with repeated branches into the gleba, usually from a solid basal tissue. Also called **percurrent columella** and **stipe-columella**.

dendroid (Gr. *déndron*, tree + L. suf. *-oide* < Gr. *-oeídes*, similar to): with tree-like branches; e.g., like the sporangiophores of *Peronospora destructor* (Peronosporales), the plasmodium of *Physarum polycephalum* (Physarales), and the conidiophores of *Botrytis cinerea* and *Metarhizium brunneum* (moniliaceous asexual fungi). Also called **dendritic** and **dendromorphous.**

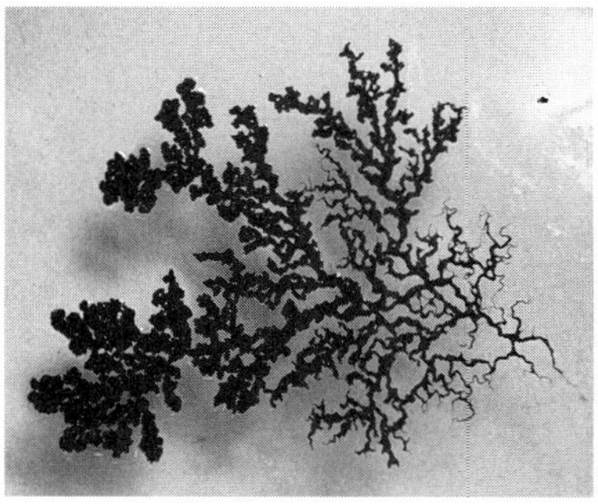

Dendroid phaneroplasmodium of *Physarum polycephalum* on agar, x 1 (*MU*).

Dendroid conidiophore of *Botrytis cinerea*, x 50 (*MU*).

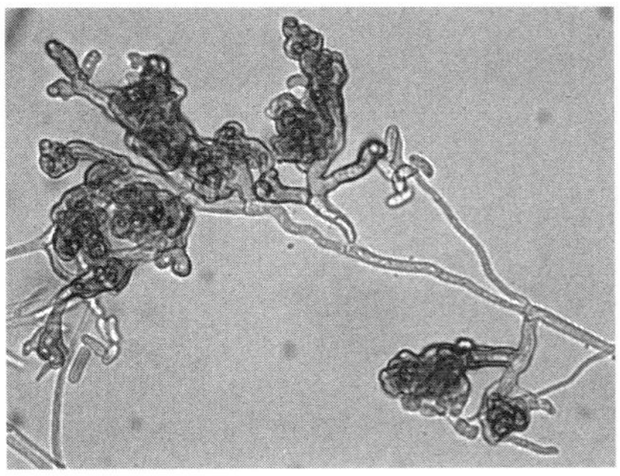

Dendroid conidiophore of *Metarhizium anisopliae*, x 270 (*MU*).

dendromorphous (Gr. *déndron*, tree + *morphé*, shape + L. *-osus* > OF. *-ous*, *-eus* > E. *-ous*, having, possessing the qualities of): see **dendroid**.

dendrophysis, pl. **dendrophyses** (Gr. *déndron*, tree + *phýsis*, action of budding or being born, a growth): *Aphyllophorales*. A type of paraphysis, cystidium or hyphoid present in the hymenium; it can be simple or branched, more or less compact, of equal or unequal length, and with all or part of the surface provided with spiny ornamentations. Dendrophyses are found, e.g., in *Aleurodiscus*, *Echinochaete* (=*Dendrochaete*) *russiceps* and *E. vallata* (Aphyllophorales). They are one of the two types of sterile formations belonging to the gen. *Aleurodiscus* (the other is the **pseudophysis**).

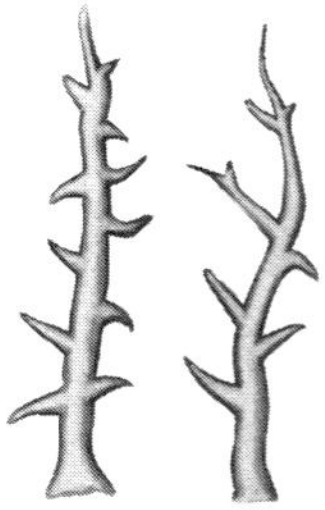

Dendrophyses from the hymenium of *Echinochaete* (=*Dendrochaete*) *russiceps*, x 650.

dense body vesicle (L. *densus*, dense, thick; ME. < OE. *bodig*, body; L. *vesicula*, dim. of *vesica*, bladder): a vesicle delimited by a membrane and associated with the metabolism of phosphoglucans, which are found in some cells of Oomycota and other heteroconts. They can appear electron transparent, with an opaque zone that is central or eccentrically located, at times with tight laminar formations between the electron transparent and opaque zones. At the time of oospore formation, the vesicles of the dense body coalesce to give rise to the ooplast, an inclusion bordered by a membrane, which first originates in the oosphere and later in the oospore.

dentate (L. *dentatus*, toothed < *dens*, tooth + suf. *-atus* > E. *-ate*, provided with or likeness): **1**. Applied to solid structures with tooth-like protuberances, e.g., like the basidiocarps of *Auriscalpium vulgare*, *Hydnum imbricatum*, *Hericium erinaceum* and *Sarcodon* (Aphyllophorales), or to the lamellate structures that have them on the border, resembling the teeth of a saw, but less pointed, as is seen in the gills of the basidiocarps of *Leptonia serrulata* (Agaricales). **2**. Used for the margin of certain colonies of fungi grown on solid media.

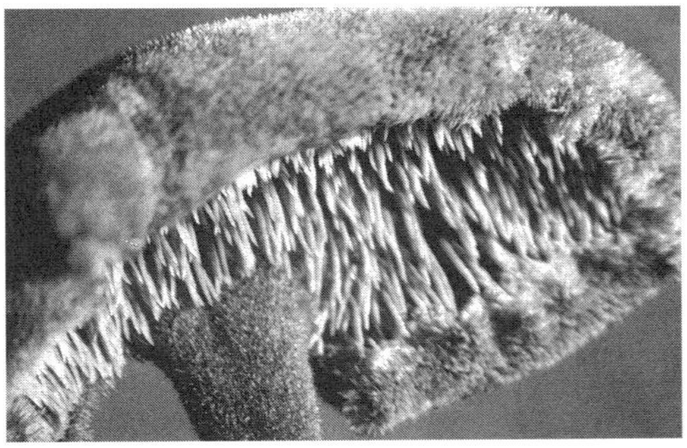

Dentate hymenophore of *Auriscalpium vulgare*, x 4 (*MU*).

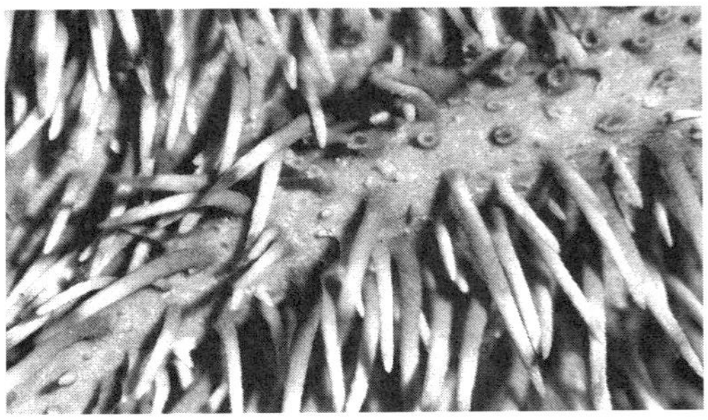

Dentate hymenophore of *Sarcodon* sp., x 6 (*MU*).

denticle (L. *denticulus*, little tooth < *dens*, tooth): *Conidiogenesis*. A small, tapered, tooth-like projection, on which is borne a spore. Denticles are characteristic of the conidiogenous cells known as **ampullae**, which form botryoblastospores, as in *Botrytis cinerea*, *Oedocephalum glomerulosum*, *Spiniger meineckellus* and other moniliaceous asexual fungi, but they occur in other species as well. See **rachiform** and **raduliform.**

denticulate (L. *denticulatus* < *denticula*, dim. of *dentis*, tooth + suf. *-atus* > E. *-ate*, provided with or likeness): having small teeth.

denudate (L. *denudatus*, bare, naked < *de-*, from, down, away + *nudus*, bare + suf. *-atus* > E. *-ate*, provided with or likeness): applied to structures that lack hairs, scales, spines, etc.; e.g., a denudate pileus.

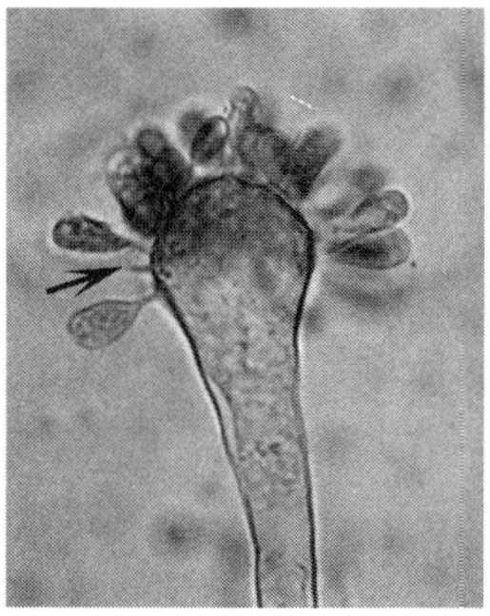

Denticles of the conidiogenous cell (ampulla) subtending conidia of *Spiniger meineckellus*, x 560 (*RTH*).

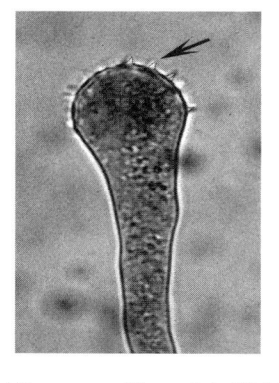

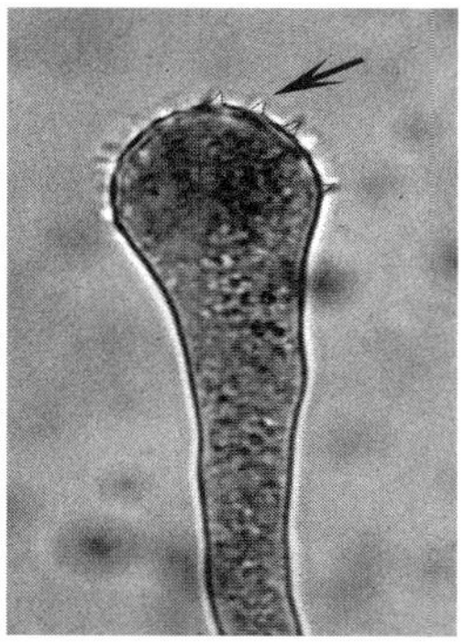

Denticles of the conidiogenous cell (ampulla) of *Spiniger meineckellus*, after releasing the conidia, x 650 (*RTH*).

depauperate (ME. *depauperat* < ML. *depauperatus*, ptp. of *depauperare* < L. *de-*, from, down, away + *pauperare*, to impoverish + suf. *-atus* > E. *-ate*, provided with or likeness): poorly developed.

depressed (ME. *depressen* < AF. < OF. *depresser* < L. *depressus*, pressed down): flattened; greater in width than in height. *Mycology*. A pileus that has a central depression or concavity, which does not become infundibuliform or omphalloid; it can be greatly or slightly depressed, as is seen in *Lacccaria laccata* and *Russula lepida* (Agaricales), respectively.

Depressed pileus of the basidiocarp of *Russula* sp., x 0.5 (*EPS*).

dermatobasidium, pl. **dermatobasidia** (Gr. *dérma*, genit. *dérmatos*, skin, leather + *basídion* < *básis*, base + dim. suf. *-ídion* > L. *-idium*): *Basidiomycetes*. A cell with the shape of a basidium which is present in the cuticular layer of the basidiocarp, not in the hymenium. In the pileus of various species of *Russula* and in the stipe of other species of Agaricales (such as *Leccinum scabrum*) dermatobasidia are frequent.

dermatocystidium, pl. **dermatocystidia**, **dermatocyst** (Gr. *dérma*, genit. *dérmatos*, skin + *kystídion* < *kýstis*, bladder; here, vessel or cell + dim. suf. *-ídion* > L. *-idium*): *Basidiomycetes*. A cystidium that is formed on the cuticular layer of a basidiocarp, such as occurs in *Pluteus*, *Panaeolus*, *Coprinus* and other gen. of Agaricales. Two types of dermatocystidia are known: **pileocystidia** (on the surface of the pileus) and **caulocystidia** (on the surface of the stipe).

dermatomycete (Gr. *dérmato*, skin + *mýkes*, genit. *mýketos*, fungus): see **dermatophyte**.

dermatomycosis, pl. **dermatomycoses** (Gr. *dérmato*, skin + *mýkes*, fungus + suf. *-osis*, condition or state of something; here, the state of the skin subject to the action of the parasitic fungi): a broad gamut of cutaneous infections of man and higher animals, caused by very diverse species of molds and yeasts from the soil; at times they are present as the result of a secondary invasion from a systemic mycosis. See **mycosis**.

dermatophyte (Gr. *dérmato*, skin + *phytón*, plant): a member of a group of moniliaceous asexual fungi that develop on the skin, hair and nails of man and other higher animals; these keratinophilic fungi are classified in three gen.: *Microsporum*, *Trichophyton* and *Epidermophyton*, and they cause a series of diseases called *dermatophytoses*. If one considers that the fungi do not belong to the plant kingdom, dermatomycete is a more accurate term, although the word dermatophyte is of very general use. See **tinea**.

dermatophytid (*dermatophyte* < Gr. *dérmato*, skin + *phytón*, plant + suf. *-id*): a pustular allergic eruption or rash (id-reaction) of the skin at a distance from a primary infection by a dermatophyte.

desquamate (L. *desquamatus* < *desquamare* < *de-*, from, down, away + *squama*, scale + suf. *-atus* > E. *-ate*, likeness): lacking scales. Same as **alepidote**. Cf. **squamulose**.

determinate (L. *determinatus* < *determinare*, to terminate, end + suf. *-atus* > E. *-ate*, with the property of): used for a sporogenous structure that does not continue developing after the spore has detached from its point of origin; thus, there are determinate sporangiophores and conidiophores that produce sporangia and conidia only once, such as the sporangiophores of *Plasmopara viticola*, *Peronospora destructor* and other species of the family Peronosporaceae (Peronosporales). Cf. **indeterminate**.

detersile (L. *detersilis* < *detersus*, cleaned, removed < *detergere*, to wipe away + suf. *-ilis* > E. *-ile*, capable, of the character of): refers to something that detaches, such as the pubescence of the pileus of certain agarics, which when mature are almost glabrous or bare due to the loss of their pubescence.

deuteromycetes (Gr. *déuteros*, second + L. *-mycetes*, ending of class < Gr. *mýkes*, genit. *mýketos*, fungus): conidial fungi that lack a means of sexual reproduction (although they undergo the phenomenon of **parasexuality**); they represent the asexual states of ascomycetes and basidiomycetes. Formerly considered as a single taxonomic class, and as **Fungi Imperfecti**, in some classifications they were frequently separated into three classes, Blastomycetes, Hyphomycetes and Coelomycetes, in the subdivision Deuteromycotina. In this dictionary the deuteromycetes are considered as asexual ascomycetes along with other asexual fungi, some of them with basidiomycetous teleomorphs, but not as a recognized and valid taxonomic group of class rank. Some recent authors, however, no longer recognize a separate, formal taxonomic status for these fungi, since they are in fact the anamorphs of teleomorphic fungi, preferring instead to classify them in appropriate teleomorphic orders. Taking into consideration the fact that deuteromycetes are an important, widely spread and large group of fungi, which for a long time have been classified as different families, orders, and classes, mainly following Saccardoan criteria of practical value, in this dictionary the orders Sporobolomycetales and Cryptococcales of the Blastomycetes are respectively referred to as sporobolomycetaceous and cryptococcaceous asexual yeasts. The order Moniliales of the Hyphomycetes, with its families Moniliaceae, Dematiaceae, Stilbellaceae and Tuberculariaceae, are here-in considered as agonomycetaceous, moniliaceous, dematiaceous, stilbellaceous, and tuberculariaceous asexual fungi, respectively. The orders Melanconiales and Sphaeropsidales of former classifications are referred to as melanconiaceous, and sphaeropsidaceous and pycnothyriaceous asexual fungi, respectively. Deuteromycetes are also called **anaholomorphs**, **asexual fungi**, **conidial fungi** and **mitosporic fungi**.

deuteromycote (L. *deuteromycete* < Gr. *déuteros*, second + *mýkes*, genit. *mýketos*, fungus): one of the deuteromycetous, asexual, conidial or mitosporic fungi.

dextrinoid (L. *dextra*, right, referring to the deflection of polarized light by crystals + suf. *-oide* < Gr. *-oeídes*, similar to): applied to the color reaction of spores or cell walls that stain yellowish-brown or reddish-brown in Melzer's reagent (potassium iodide, chloral hydrate and water). This color reaction is also called **pseudoamyloid** or false amyloid. For example, the basidiospores of *Macrolepiota procera* and the hyphae of *Mycena* (Agaricales) are dextrinoid or pseudoamyloid. Cf. **amyloid** and **anamyloid**.

diageotropism (Gr. *diá*, through + *geo* < *gê*, earth, soil + *trópos*, a turn, change in manner < *tropé*, a turning < *trépo*, to revolve, turn towards + *-ismós* > L. *-ismus* > E. *-ism*, state, phase, tendency, action): a transverse geotropism in which the organ affected by the geotropic stimulus remains in perpendicular position with respect to the stimulus, i.e., horizontal. The pilei of many basidiomycetes, such as Agaricales and Aphyllophorales are diageotropic, since on developing they expand in a direction parallel to the surface of the soil or substrate, whereas the stipe of the fruiting body is negatively geotropic. This phenomenon is also called **plagiotropism** and the organ or organism affected is plagiotropic. Cf. **orthogeotropism**.

diaphanous (ML. *diaphanous* < Gr. *diaphanés*, transparent < *diá*, through + *phaíno*, to make appear, to show < *phanós*, light, lantern + L. *-osus* > OF. *-ous*, *-eus* > E. *-ous*, possessing the qualities of): transparent or almost so, like many hyaline fungal structures.

diaphragm (LL. *diaphragma* < Gr. *diáphragma*, separation, partition): *Gasteromycetes*. A membrane or homogeneous wall, composed of hyphae, that separates the gleba from the sterile base or subgleba in the fruiting body of *Calvatia*, *Lycoperdon*, *Vascellum* and other Lycoperdales.

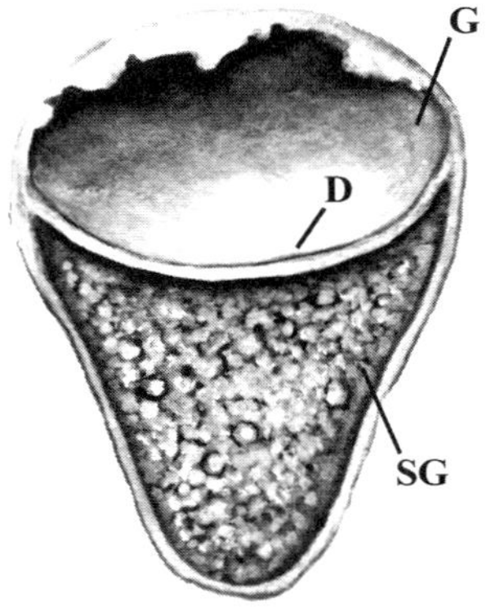

Diaphragm (D) of the fruiting body of *Vascellum pratense*, separating the gleba (G) from the subgleba (SG), seen in longitudinal section, x 1.

diaspore (Gr. *diasporá*, dissemination, dispersal): a propagule produced by a lichen thallus, such as an isidium or a soredium. In general, it is used for any spore, propagule or other portion of a fungus that constitutes a dispersal phase with the capacity to produce a new organism.

diastole (LL. *diastole* < Gr. *diastolé*, expansion, dilation): dilation movement of a contractile vacuole, such as occurs in certain plasmodia and zoospores. Cf. **systole**.

diatrypoid (< gen. *Diatrype* < Gr. *diatrypáo*, to perforate + L. suf. *-oide* < Gr. *-oeídes*, similar to): *Ascomycetes*. A type of stroma composed of both fungal and host tissues (**pseudostroma**), with a poorly developed ectostroma that often bears the conidial state, and a well developed entostroma in which the perithecia are embedded. The perithecial necks extend straight up and project through the erumpent upper part (disc) of the stroma, as occurs, e.g., in Pyrenomycetes of the gen. *Diatrype* (Xylariales), common on the dry branches of many trees. Cf. **eutypoid** and **valsoid**.

Diatrypoid and erumpent perithecial stromata of *Diatrype disciformis*, x 25 (*CB*).

dichophysis, pl. **dichophyses** (Gr. *dícho* < *dícha*, in two + *phýsis*, a growth): *Basidiomycetes*. A sterile cellular or hyphal element with dichotomous branching, thick, pigmented wall and narrow lumen that is present in the hymenium of the thelephoraceous Aphyllophorales, such as *Vararia* sp. and *Vararia* (=*Asterostromella*) *investiens*; frequently dichophyses have the appearance of antlers.

dichotomous (LL. *dichotomos* < Gr. *dichótomos*, cut into two < *dícha*, in two parts + *tómos*, a cut + L. *-osus* > OF. *-ous*, *-eus* > E. *-ous*, having, possessing the qualities of): a type of branching in which an axis divides into two equal parts, so that it produces a fork of approximately equal branches, at least at first, as is seen, e.g., in the sporangiophore of *Dicranophora fulva* (Mucorales), *Piptocephalis virginiana* (Zoopagales) and in the perithecial appendages of *Microsphaera* (Erysiphales). Cf. **polychotomous**.

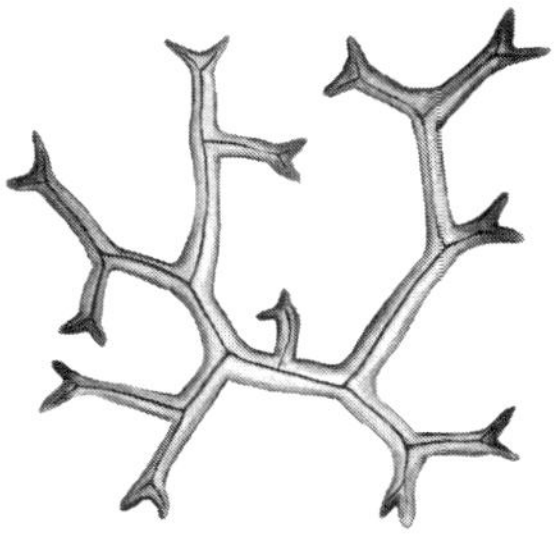

Dichophyses from the hymenium of *Vararia* sp., x 600.

Dichotomous branching of the sporangiophore of *Dicranophora fulva*, x 165.

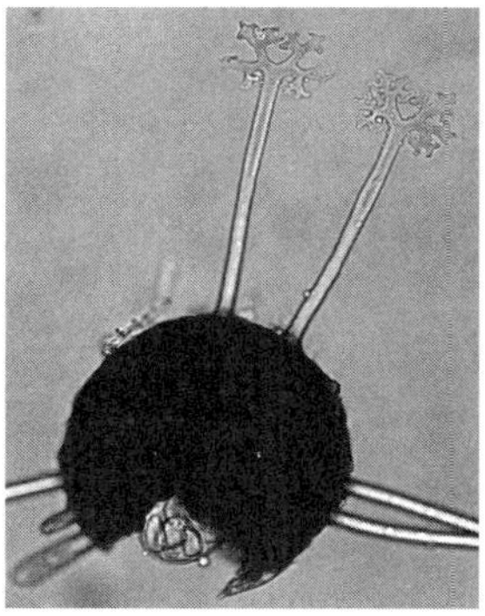

Dichotomous perithecial appendages of *Microsphaera* sp., x 98 (*MU*).

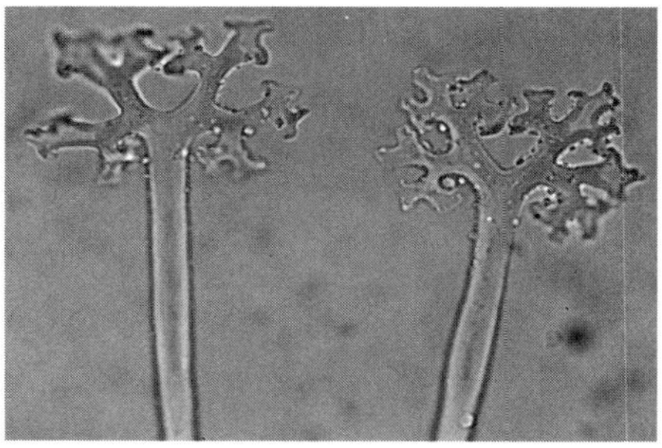

Dichotomous perithecial appendages of *Microsphaera* sp., x 260 (*MU*).

diclinous (Gr. *dís*, two + *klíne*, bed; here, thallus > L. *clinare*, to lean + L. *-osus* > OF. *-ous*, *-eus* > E. *-ous*, having, possessing the qualities of): refers to fungi that have an antheridium on one hypha or branch and the oogonium on another; as happens in various aquatic species, e.g., *Achlya bisexualis* (Saprolegniales). Cf. **monoclinous**.

dictydine (< gen. *Dictydium* < Gr. *díktyon*, net + L. dim. suf. *-idium* + L. suf. *-inus* > E. *-ine*, of or pertaining to): microscopic structures, generally dark in color, separate, aggregated or arranged in rows, that comprise the so-called dictydine or plasmodial granules found on the surface of the peridium or network, and often also on the spores of the cribrariaceous Myxomycetes, such as *Cribraria* (=*Dictydium*) (Liceales).

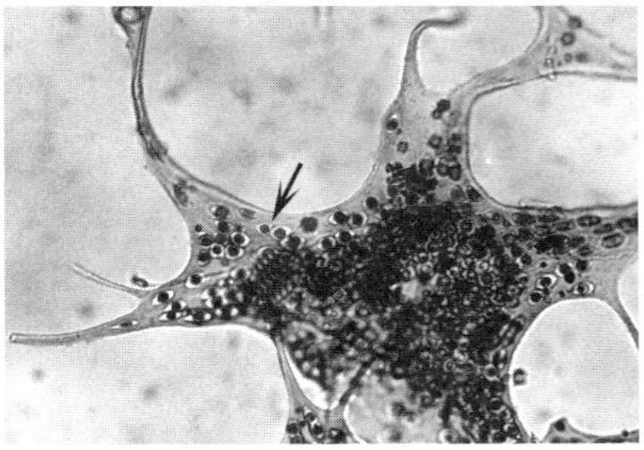

Dictydine (or plasmodic) granules on the peridial net of the sporangium of *Cribraria purpurea*, x 300 (*MU*).

dictyochlamydospore (Gr. *díktyon*, net + *chlamýs*, *chlamýdos*, cloak, mantle + *sporá*, spore): a dictyosporous chlamydospore, i.e., with transverse and vertical septa, such as are formed in *Verticillium chlamydosporum* (moniliaceous asexual fungi) and *Phoma glomerata* (sphaeropsidaceous asexual fungi). Also called alternarioid chlamydospore, due to its resemblance to the conidia of *Alternaria* (dematiaceous asexual fungi). A distinctive characteristic of this multicellular chlamydospore is that the outer wall layer is separable from the component cell walls, which can readily separate from one another.

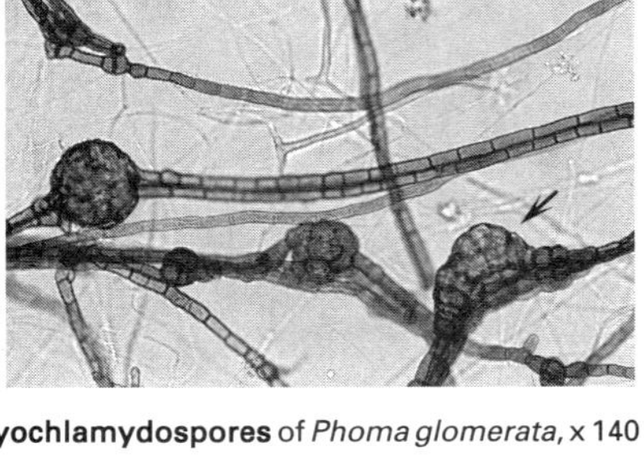

Dictyochlamydospores of *Phoma glomerata*, x 140 (*CB*).

dictyosome (Gr. *díktyon*, net + *sõma*, body): each of the bodies that form part of the **Golgi apparatus**, more or less globose and comprised of a series of cisternae or cavities. They are complexes of lipids and proteins and are found in young cells or in the elaboration of cytoplasm or secretions. During the apical growth of a hypha the precursor materials for the synthesis of the cell wall are transferred from the endoplasmic reticulum (ER) in small vesicles to the dictyosomes, where the vesicles coalesce to form a cisterna. From the cisternae of the dictyosomes the materials and membranes migrate as secretory vesicles toward the hyphal apex, where they fuse with the plasma membrane and liberate their contents in the formation of the new cell wall.

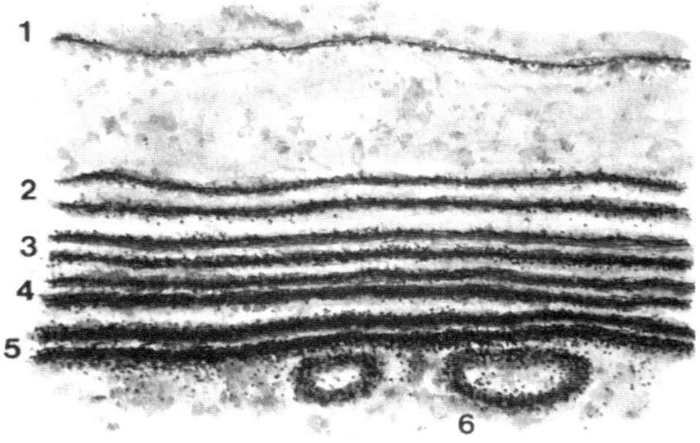

Dictyosomes of the Golgi apparatus of a hyphal cell of *Pythium ultimum*. Cross section of a Golgi body showing the way in which the structure of the membranes changes from that of the nuclear envelope (1), which is comparable to the endoplasmic reticulum, through the cisternae of the Golgi body (2-5), and finally to that of the Golgi vesicle (6). Drawing based on a transmission electron micrograph published by Beckett *et al.* in their *Atlas of Fungal Ultrastructure*, 1974, x 150 000.

dictyosporangium, pl. **dictyosporangia** (Gr. *díktyon*, net + *sporá*, spore + *angeion*, vessel, receptacle): *Saprolegniales.* A term applied to the zoosporangium of *Dictyuchus*, so named because the cell walls of the encysted zoospores, when these have been liberated, remain inside in the form of a net, since they are polyhedric.

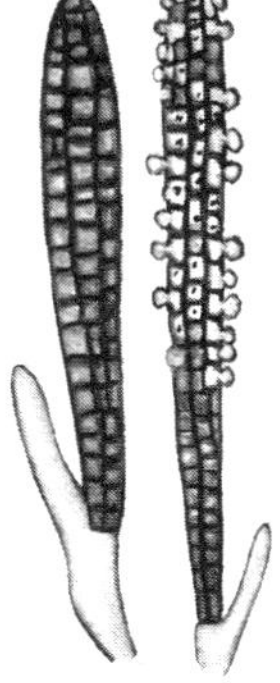

Dictyosporangia of *Dictyuchus monosporus*, x 200.

dictyospore (Gr. *díktyon*, net + *sporá*, spore): a spore that has both transverse and longitudinal septa, giving it a reticulate appearance; e.g., as in *Alternaria alternata* (dematiaceous asexual fungi). See **muriform**.

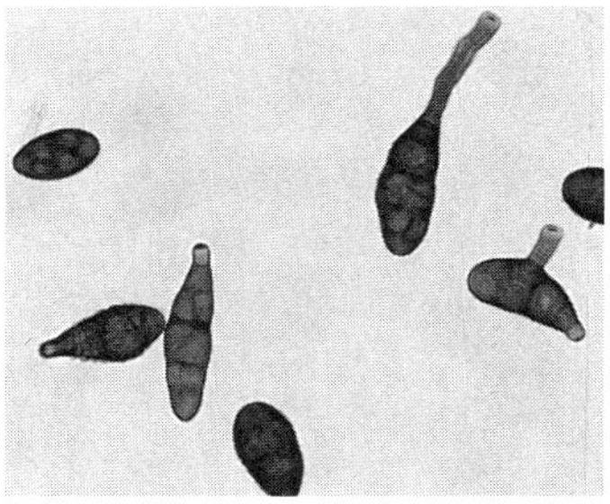

Dictyospores of *Alternaria alternata*, x 390 (*MU*).

dictyosporous (Gr. *díktyon*, net + *sporá*, spore + L. *-osus* > OF. *-ous*, *-eus* > E. *-ous*, having, possessing the qualities of): refers to fungi having spores with both transverse and vertical septa; such spores are **dictyospores**, e.g., the conidia of *Alternaria*, *Stemphylium* and other gen. of dematiaceous asexual fungi, and the ascospores of *Leptosphaerulina crassiasca* (Dothideales).

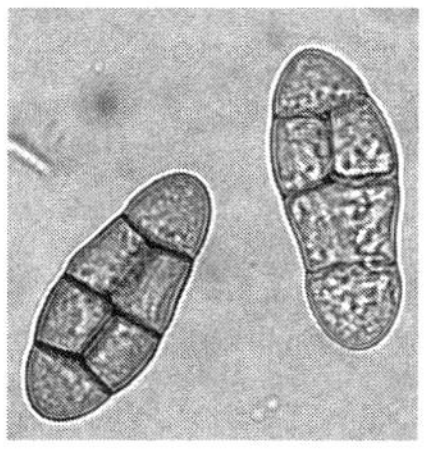

Dictyosporous ascospores of *Pyrenophora* sp., x 410 (*CB*).

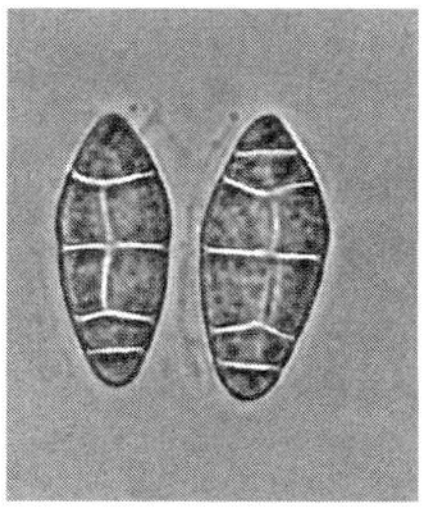

Dictyosporous ascospores of *Leptosphaerulina crassiasca*, x 660 (*RTH*).

didymospore (NL. *didymospora* < Gr. *dídymos*, twin + *sporá*, spore): a two-celled spore (divided by a septum); a Saccardoan term originally applied mainly to conidial fungi; as, e.g., in the conidia of *Arthrobotrys oligospora*, *Trichothecium roseum* (moniliaceous asexual fungi), *Lasiodiplodia theobromae* (sphaeropsidaceous asexual fungi), and the ascospores of *Venturia liriodendri* (Pleosporales).

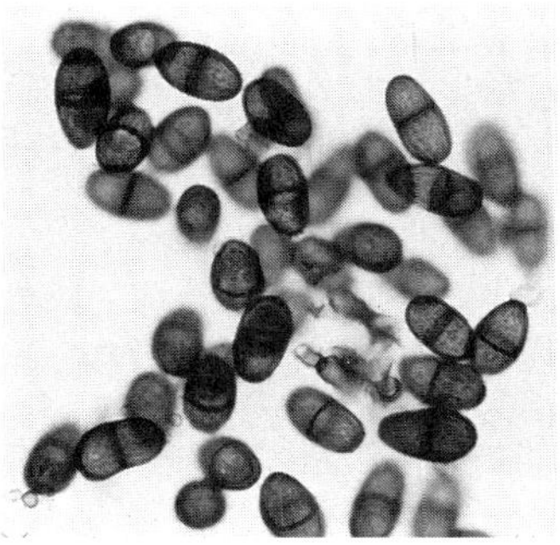

Didymospores (pycnidiospores) of *Lasiodiplodia theobromae*, x 320 (*CB*).

didymous (Gr. *dídymos*, twin, double + L. *-osus* > OF. *-ous*, *-eus* > E. *-ous*, having, possessing the qualities of): applied to structures formed in pairs; double.

diffluent (L. *diffluens*, genit. *diffluentis*, flowing apart < *diffluere*, to flow apart + *-entem* > E. *-ent*, being): something that dissolves easily or degrades in the presence of water; e.g., like the peridium of the sporangia of *Mucor hiemalis*, *M. pusillus* and other species of *Mucor* and other gen. of Mucorales.

diffracted (L. *diffractus*, broken up < *diffringere*, to break up): refers to the surface of a fruiting body that is fragmented into areolas; e.g., like the exoperidium of *Lycoperdon caelatum* (Lycoperdales).

diffuse (L. *diffusus*, spread): scattered, spread out thinly, without a defined margin or limit; e.g., like the thallus of some crustaceous lichens, such as *Caloplaca cerina* (Teloschistales).

digitate, **digitiform** (L. *digitatus*, having fingers < L. *digit*, finger + suf. *-atus* > E. *-ate*, provided with or likeness; *digit* + *-formis* < *forma*, shape): provided with finger-like protuberances, such as the haustoria of *Blumeria* (=*Erysiphe*) *graminis* (Erysiphales), or like the cystidia of *Mycena epipterygia* and *Agrocybe arvalis* (Agaricales). Also called **dactyloid**.

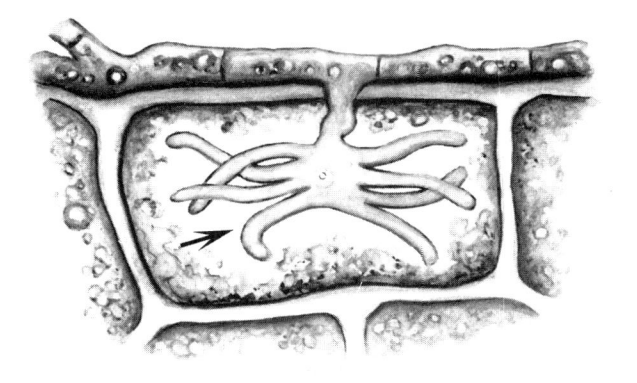

Digitate haustorium of *Blumeria graminis*, x 2 000.

dikaryon, **dicaryon** (Gr. *dís*, two + *káryon*, nucleus): see **dikaryotic**.

dikaryotic, **dicaryotic**, **dikaryontic**, **dicaryontic** (Gr. *dís*, two + *káryon*, nucleus + suf. *-íkos* > L. *-icus* > E. *-ic*, belonging to, relating to): containing a dikaryon (n + n), i.e., two nuclei capable of complementing each other genetically; some examples of dikaryotic structures are, e.g., the ascogenous hyphae of the Plectomycetes (such as *Eurotium rubrum* of the Eurotiales) and Pyrenomycetes (such as *Neurospora crassa* of the Sordariales), the young teliospores and infective mycelium of *Ustilago maydis* (Ustilaginales), or the secondary mycelium and the young basidia of *Coprinus comatus* (Agaricales). Cf. **monokaryotic**.

dilacerate (ME. < OF. < L. < Gr. *dís*, twice, twofold + L. *laceratus*, ptp. of *lacerare*, to tear + suf. *-atus* > E. *-ate*, provided with or likeness): torn apart.

dimerous (Gr. *dís*, two + *méros*, part + L. *-osus* > OF. *-ous*, *-eus* > E. *-ous*, having, possessing the qualities of): composed of two parts; generally applied to basidia that have a constriction between the probasidium (lower part) and the metabasidium (upper part), as in *Brachybasidium* (Exobasidiales).

dimidiate (L. *dimidiatus*, halved, divided in half < *dimidiare*, to divide + suf. *-atus* > E. *-ate*, provided with or likeness): **1**. *Basidiomycetes*. A semicircular basidiocarp, with only one side developed; e.g., as one sees in *Ganoderma applanatum*, *Heterobasidion annosum*, *Lenzites saepiaria* and other Aphyllophorales. **2**. *Ascomycetes*. A shield-shaped ascocarp that lacks a bottom wall.

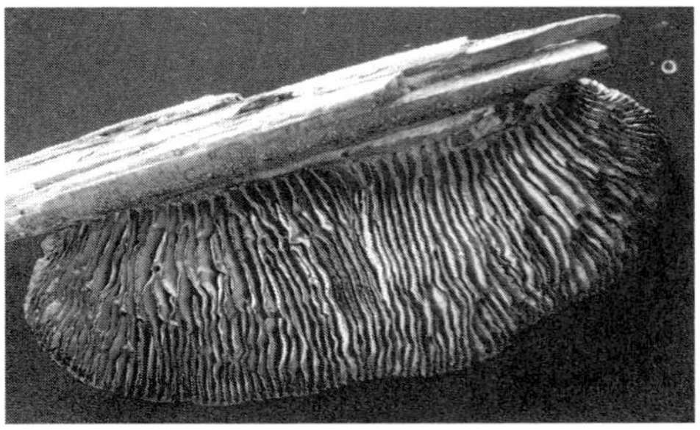

Dimidiate basidiocarp of *Lenzites saepiaria*, x 0.5 (*MU*).

dimitic (Gr. *dís*, two + *mítos*, filament; here, hypha + suf. *-íkos* > L. *-icus* > E. *-ic*, belonging to, relating to): *Basidiomycetes*. A fructification with two types of hyphae, generative or fertile (producing basidia) hyphae and either skeletal or binding hyphae; e.g.,like those of *Lentinus tigrinus* and *Ramaria stricta* (Aphyllophorales). Cf. **monomitic** and **trimitic**.

dimitic system (Gr. *dís*, two + *mítos*, filament; here, hypha + suf. *-íkos* > L. *-icus* > E. *-ic*, belonging to, relating to; LL. *systema* < Gr. *sýstema*, composed of several parts): *Gasteromycetes*. Used to describe the development of true capillitium and paracapillitium within the same gleba.

dimorphic (Gr. *dís*, two + *morphé*, form + suf. *-íkos* > L. *-icus* > E. *-ic*, belonging to, relating to): **1**. Refers to an aquatic fungus that has two morphological types of zoospores (primary or pyriform and secondary or reniform) in the life cycle, as occurs, e.g., in *Saprolegnia* (Saprolegniales). Cf. **monomorphic**. **2**. A fungus that is capable of growing in mycelial or yeast-like form, depending upon growth conditions. Usually applied to pathogenic fungi that form mycelium in the host and a yeast-like colony in culture, as is seen in the plant pathogens *Taphrina* (Taphrinales) and *Ustilago* (Ustilaginales). In the animal pathogen *Histoplasma* (moniliaceous asexual fungi) the yeast state is formed in the host and the mycelium develops in culture and in its substrate, where it grows as a saprobe.

dioecious (Gr. *dís*, two + *oikía*, house; here, thallus + L. *-osus* > OF. *-ous*, *-eus* > E. *-ous*, having, possessing the qualities of): an organism in which one thallus produces either male or female sexual organs, but not both; i.e., with the sexes segregated on different individuals. Also called **unisexual**. Cf. **monoecious** and **hermaphrodite**.

diorchidioid (Gr. *dís*, two + *órchis*, testicle + L. suf. *-oide* < Gr. *-oeídes*, similar to): a bicellular spore with the septum vertical or nearly so; e.g., as the teliospores of the rust fungus *Diorchidium spinulosum* (Uredinales).

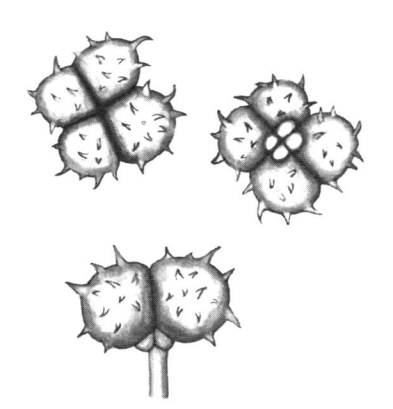

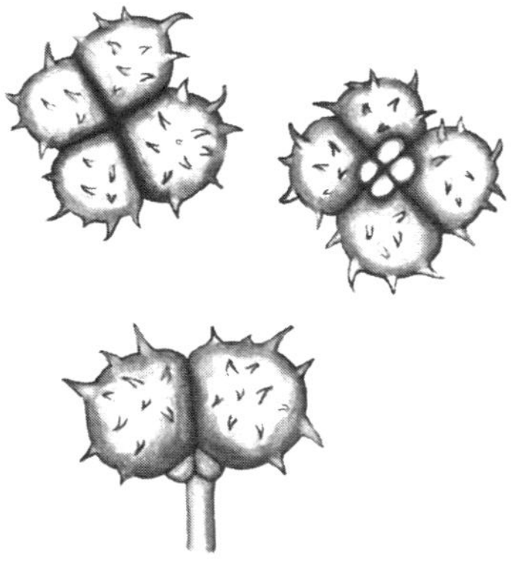

Diorchidioid teliospores of *Diorchidium spinulosum*, x 250.

diphycophilous (Gr. *dís*, two + *phýkos*, alga + *phílos*, have an affinity for + L. *-osus* > OF. *-ous*, *-eus* > E. *-ous*, having, possessing the qualities of): *Lichens*. A lichen that has two types of phycobionts, one green alga (Chlorophyceae) and one blue-green alga (Cyanophyceae). In *Peltigera aphthosa* (Peltigerales), e.g., the associated green alga (*Coccomyxa*) is found throughout the thallus, whereas the blue-green alga (*Nostoc*), is confined to the cephalodia. Cf. **chlorophycophilous** and **cyanophycophilous**.

diplanetic (Gr. *dís*, two + *planétes*, roving, wandering + suf. *-íkos* > L. *-icus* > E. *-ic*, belonging to, relating to): *Oomycota*. A species in whose life cycle there are two types of zoospores (dimorphic), primary and secondary, and two periods of swimming, one for each type of zoospore; e.g., as in some species of *Saprolegnia* (Saprolegniales). Some monomorphic species, i. e., with only one type of zoospores, can also be diplanetic. Cf. **monoplanetic** and **polyplanetic**.

diplobiont, **diplobiontic** (Gr. *diplóos*, double + *bíos*, life + *óntos*, genit. of *ón*, a being; or + Gr. suf. *-íkos* > L. *-icus* > E. *-ic*, belonging to, relating to): an organism in whose life cycle both the haploid and diploid phases occur as free-living individuals. Cf. **haplobiont**. Distinguish from **diplont** and **haplont**.

diploconidium, pl. **diploconidia** (Gr. *diplóos*, double + *kónis*, dust + dim. suf. *-ídion* > L. *-idium*): *Basidiomycetes*. A binucleate conidium that forms in the fruiting body of the jelly fungi (Tremellales). Cf. **haploconidium**.

diploid (Gr. *diplóos*, double + NL. *-oid* < Gr. *-oeídes*, similar to): having a double set of chromosomes (2N). Cf. **haploid**.

diplont (Gr. *diplóos*, double + *ón*, genit. *óntos*, a being): an organism whose existence is spent all, or mostly, in the diploid phase (diplophase), as happens in the Oomycota and in some yeasts, such as *Saccharomycodes* (Saccharomycetales). Cf. **haplont**. Distinguish from **haplobiont** and **diplobiont**.

diplostromatic (Gr. *diplóos*, double + *strõma*, cushion + suf. *-íkos* > L. *-icus* > E. *-ic*, belonging to, relating to): *Ascomycetes*. Having both ectostroma and endostroma; e.g., as happens in certain Xylariales and Diatrypales. Cf. **haplostromatic**. See **placodium**.

disc (Gr. *dískos*, disc): **1**. The curved or flattened portion of an apothecium that bears the hymenium. **2**. The exposed, erumpent portion of a diatrypoid stroma. Also see **delimited peristome**.

discocarp (Gr. *dískos*, disc + *karpós*, fruit): see **apothecium**.

discoid (Gr. *dískos*, disc + L. suf. *-oide* < Gr. *-oeídes*, similar to): similar to a disc, flat and circular, like the sporangia of *Diderma haemisphaericum* (Physarales), and the apothecia of many discolichens, e.g., *Lecanora* (Lecanorales). See **lecanorine**.

discolichens (L. *discus* < Gr. *dískos*, disc + L. *lichen* < Gr. *leichén*, lichen): a subgroup of ascolichens that is characterized by an apothecial fruiting body with a structure analogous to that of the Discomycetes. The great majority of the species of lichens are discolichens, with discoid apothecia.

Discoid head of the sporangia of *Diderma haemisphaericum*, x 11 (*MU*).

Discomycetes (Gr. *dískos*, disc, tray + L. *-mycetes*, ending of class < Gr. *mýkes*, genit. *mýketos*, fungus): a class of ascomycetes that are characterized by an apothecial ascocarp in which the asci are formed in a hymenium that is exposed at maturity. In some classifications they are regarded as a subclass (Discomycetidae) of the Euascomycetes. The discomycetes, as considered in this dictionary, include the orders Medeolariales, Rhytismatales, Ostropales, Cyttariales, Helotiales, Neolectales, Gyalectales, Lecanorales, Lichinales, Peltigerales, Pertusariales, Teloschistales, Caliciales and Pezizales. Together with the ascosporogenous yeasts, the plectomycetous, pyrenomycetous and loculoascomycetous forms, they constitute the phylum Ascomycota of kingdom Fungi.

discothecium, pl. **discothecia** (NL. *discothecium* < Gr. *dískos*, disc + *thekíon*, dim. of *thêke*, case, box; here, of the asci): a type of ascostroma (an apothecial pseudothecium), similar to an apothecium but bearing bitunicate asci; it differs from a hysterothecium in that its covering tissue disintegrates at maturity, as is seen in the Loculoascomycetes of the family Patellariaceae (of the Patellariales), such as *Rhytidhysteron rufulum*.

discrete (L. *discretus*, separate, discontinuous): a conidiogenous cell that is clearly distinct from the conidiophore. Cf. **integrated**.

disjunctor (L. *disjunctor* < *disjungere*, to separate): a connective, consisting of either a cell or cell wall material, that is found between the spores of certain species that form their spores in chains, and which serve as a link, often ephemeral. For example, the conidial chains of *Amblyosporium botrytis* (moniliaceous asexual fungi) and the species of *Oidiodendron* (dematiaceous asexual fungi) have disjunctor cells between adjacent conidia.

dispirous (Gr. *dís*, two + *speîra*, spiral + L. *-osus* > OF. *-ous*, *-eus* > E. *-ous*, having, possessing the qualities of): with two spirals; e.g., like the sporangiolar branches of *Dispira* (Dimargaritales), which are markedly coiled or recurved.

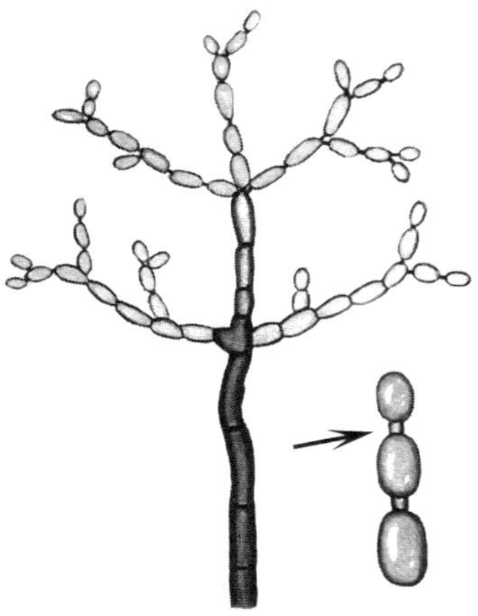

Disjunctors between the conidia in a conidiophore of *Oidiodendron* sp., x 1 600.

Dispirous branches of the sporangiophore of *Dispira simplex*, x 850.

dissepiment (L. *dissepimentum*, a partition, a dividing; that which separates two contiguous things < *dissaepiere*, to divide + *-men* > E. *-ment*, result, object, agent of an action): *Hymenomycetes.* The plectenchyma tissue found between the pores of the poroid Hymenomycetes, such as *Boletus* (Agaricales) or *Polyporus* (Aphyllophorales).

distichous (Gr. *dístichos*, arranged in two rows or lines < *dís*, two + *stichós*, row, line + L. *-osus* > OF. *-ous*, *-eus* > E. *-ous*, having, possessing the qualities of): occurring in two rows; e.g., like the ascospores in the ascus of *Cudoniella* (=*Helotium*) *herbarum* (Helotiales), *Ascobolus michaudii* and *Saccobolus truncatus* (Pezizales), which are arranged partially or totally in two rows of four each. Also called **biserial**. Cf. **monostichous** and **uniseriate**.

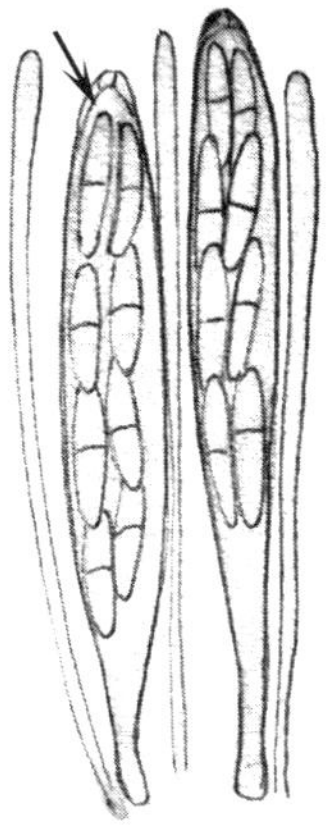

Distichous ascospores within asci of *Cudoniella* (=*Helotium*) *herbarum*, x 650.

distorted (Gr. *dýs-*, pref. that expresses the idea of difficulty, inferiority + L. *torquere*, twist, roll up): twisted or rolled up, so that the normal shape is lost; deformed.

distoseptate (L. *disto*, to be distant, separated from + *septatus*, provided with septa, partitions < *septum*, wall + suf. *-atus* > E. *-ate*, provided with or likeness): *Conidial fungi.* A type of multicellular conidium in which each cell has a thick secondary wall (**distoseptum**) that is readily separable from the outer wall of the conidium, so that when the outer wall is broken the individual thick-walled cells separate; this occurs, e.g., in *Drechslera avenae* and *Exserohilum* (=*Helminthosporium*) *rostratum* (dematiaceous asexual fungi). Cf. **euseptate**.

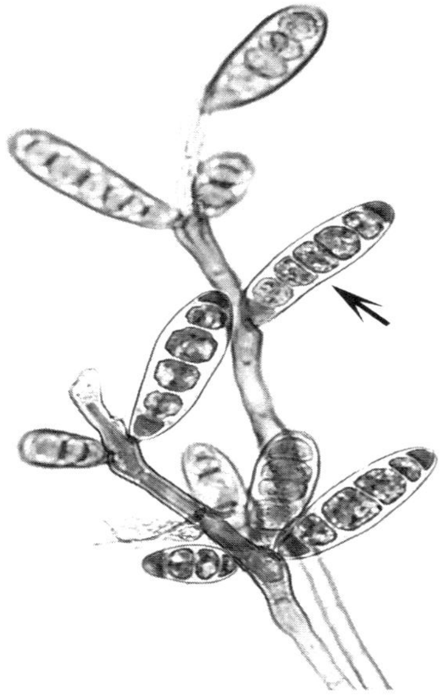

Distoseptate conidia on a conidiophore of *Drechslera avenae*, x 540 (*CB*).

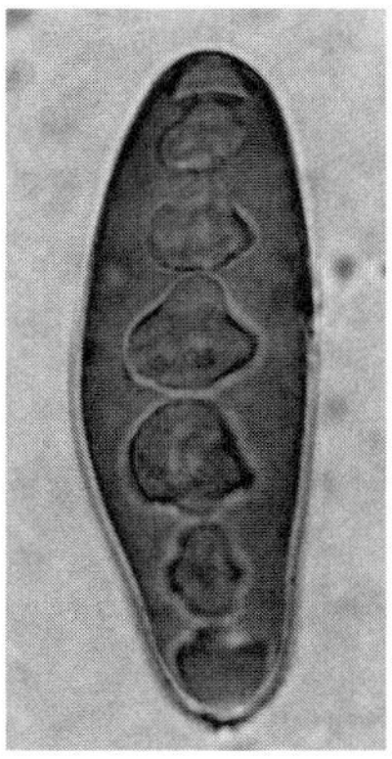

Distoseptate conidium of *Drechslera avenae*, x 1 340 (*CB*).

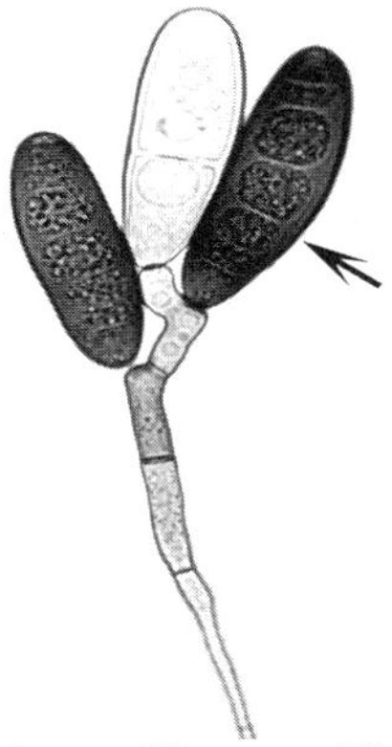

Distoseptate conidia on a conidiophore of *Helminthosporium* sp., x 630 (*CB*).

distoseptum, pl. **distosepta** (L. *disto*, to be distant, separated from + *septum*, septum, partition): see **distoseptate**.

distrophy (Gr. *dýs-*, pref. that expresses the idea of difficulty, inferiority + *trophós*, what nourishes or serves as food + *-y*, E. suf. of concrete nouns): a phenomenon that is caused by poor or inappropriate nutrition of an organism, resulting in structural modifications or abnormalities. Cf. **hypertrophy**.

divaricate (L. *divaricatus* < *divaricare*, to spread apart + suf. *-atus* > E. *-ate*, provided with or likeness): extremely divergent; branching at a wide angle; e.g., like the conidiophore of *Penicillium oxalicum*, *P. citrinum*, *P. melinii*, *P. waksmanii* and other species of the subgenus *Furcatum* (moniliaceous asexual fungi) whose branchlets form very open angles with the main axis.

divergent (L. *divergens*, genit. *divergentis*, widespreading, bending apart < *di*, *dis*, twice + *vergere*, to bend + *-entem* > E. *-ent*, being): *Agaricology*. A type of hymeniiferous trama in which there is a central zone of hyphae (the middle layer) from which diverge obliquely two rows of cells (the lateral layers). It is present, e.g., in *Amanita muscaria* and *Hygrophorus bakerensis* (Agaricales). Cf. **convergent**.

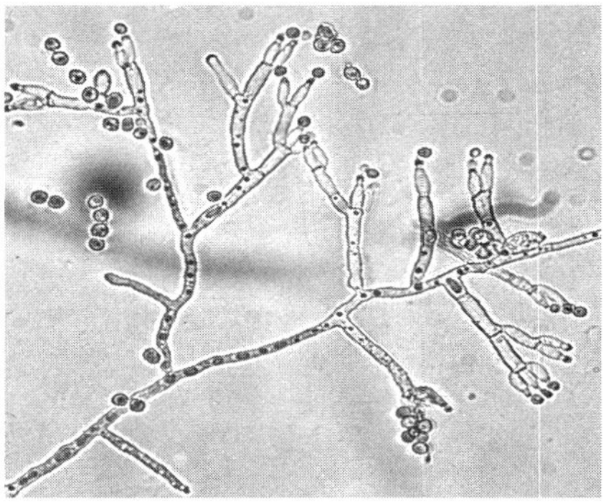

Divaricate conidiophore of *Penicillium waksmanii*, x 250 (*PL*).

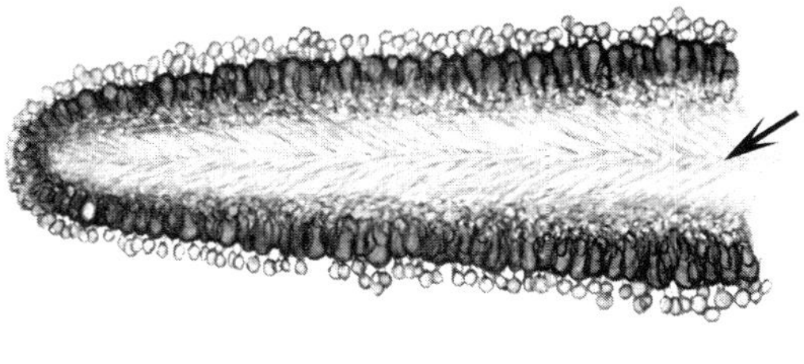

Divergent bilateral hymeniiferous trama of *Amanita muscaria*, seen in longitudinal section of a lamella, x 160.

diverticulum, pl. **diverticula** (L. *diverticulum*, byway, tributary < *de-*, from, down from, away, off + *vertare*, to turn + dim. suf. *-culum*): a blind, tubular appendage that branches off of a larger cavity; e.g., as occurs in the mycelium of *Pythium* (Peronosporales) or in the four sterigmata of the typical basidium, which are formed from four apical diverticula.

dolabriform (L. *dolabriformis* < *dolabra*, adze + *-formis* < *forma*, shape): having the shape of an adze.

doliiform (L. *doliiformis* < *dolium*, keg, barrel + *-formis* < *forma*, shape): having the shape of a barrel; e.g., like the arthrospores of *Geotrichum candidum* (moniliaceous asexual fungi), the dictyospores of *Pithomyces chartarum* (dematiaceous asexual fungi), or the catenulate conidia of *Ceratocystis paradoxa* (Microascales).

dolipore septum, pl. **septa** (L. *dolium*, keg, barrel, earthenware jar + ME. *poore* < LL. *porus* < Gr. *póros*, passage; L. *septum*, enclosure): a septum or partition with a central pore surrounded by a barrel-shaped swelling of the septal wall and covered on both sides by perforate membranes called septal pore caps or parenthesomes; common in many basidiomycetes, e.g., in *Heterobasidion annosum* (Aphyllophorales).

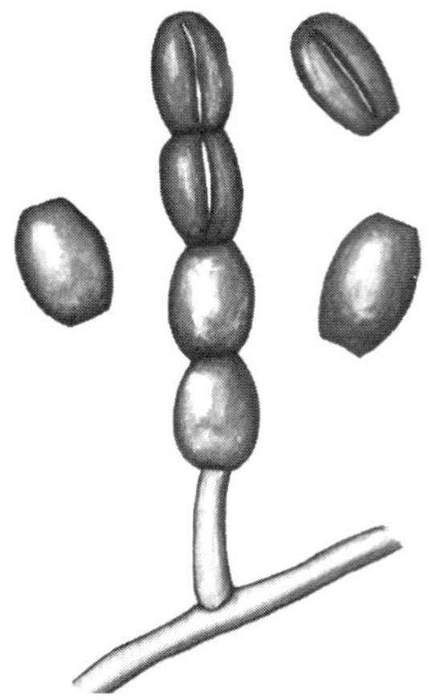

Doliiform conidia of *Ceratocystis paradoxa*, x 600.

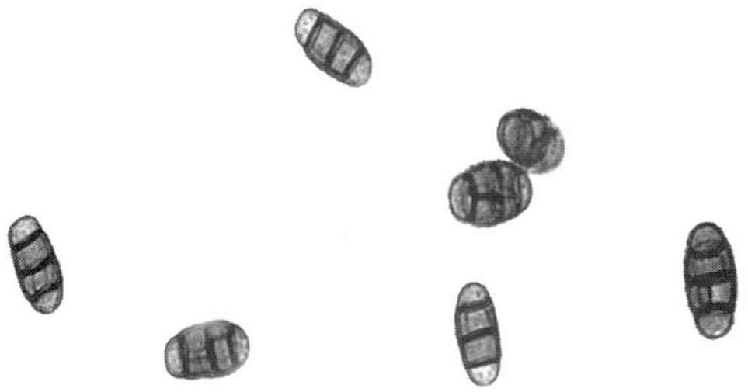

Doliiform conidia of *Pithomyces chartarum*, x 500 (*CB*).

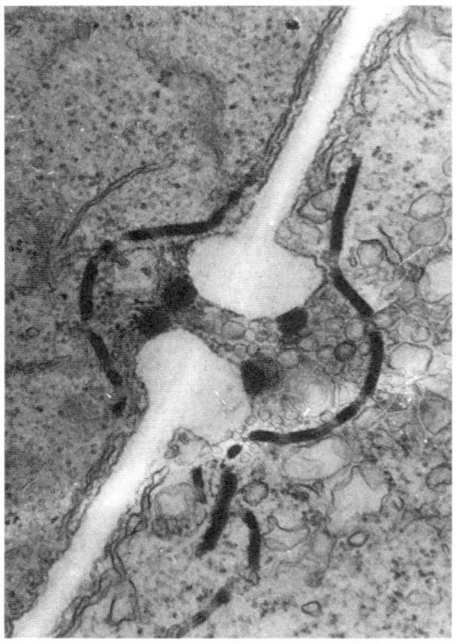

Dolipore septum of a hypha of *Heterobasidion annosum*, seen in longitudinal section with the transmission electron microscope, x 64 500 (*RTH*).

dome (L. *domus*, dome, vault, cupola): the apical part of a bitunicate (fissitunicate) ascus, where the wall is thickened and forms a vault in which is formed a digitiform projection (the **nasse**) which extends from the endotunica. This can be seen clearly, e.g., in the ascus of *Leptosphaerulina* (Dothideales).

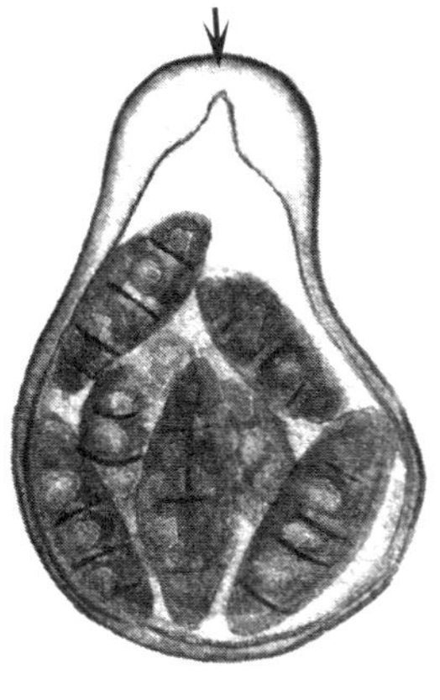

Dome of the bitunicate ascus of *Leptosphaerulina crassiasca*, x 600.

dormant (L. *dormire*, to sleep < *dormiens*, genit. *dormientis*, sleeping, resting + suf. *-ant* > E. *-ant*, one that performs, being): a biological state characterized by lack of growth and minimal metabolic activity; this may result from adverse environmental conditions or be part of the normal life cycle of an organism.

dormancy (L. *dormire*, to sleep < *dormiens*, genit. *dormientis*, sleeping, resting + E. suf. *-cy*, state, condition): the act of being **dormant**.

druse (G. *Druse*, mineral crystals): a stellate or rounded group of incomplete crystals (generally composed of calcium oxalate) which form in the thallus of some lichens.

dryophilous (Gr. *drýs*, *drýos*, tree, especially oak, + *phílos*, have an affinity for + L. *-osus* > OF. *-ous*, *-eus* > E. *-ous*, possessing the qualities of): living on, beneath or near oak trees; e.g., like the basidiocarps of *Collybia dryophila* (Agaricales).

dulcious (L. *dulcis*, sweet + L. *-osus* > OF. *-ous*, *-eus* > E. *-ous*, having, possessing the qualities of): with a sweet flavor or taste.

duplex (L. *duo*, two + *plex*, fold; twofold): refers to a context made up of two layers, with the layer adjacent to the lamellae or tubes being harder than the one over it.

duvet (F. *duvet*, down): a soft, thick layer of hyphae like brushed-up cloth, formed by some dermatophyte colonies when grown on agar.

dwarf bunt (ME. *dwerg*, *dwerf*, dwarf; bunt is of unknown origin): a wheat disease (**smut**) caused by *Tilletia controversa* (Ustilaginales). Cf. **bunt**.

dysgonic (Gr. *dýs-*, apart, bad, difficult + *gónos*, generation, offspring + suf. *-íkos* > L. *-icus* > E. *-ic*, belonging to, relating to): applied to those strains of dermatophytes which grow more slowly in culture, often with less aerial mycelium, than normal (**eugonic**) strains.

e

earth ball (ME. *erthe* < OE. *eorthe*, earth; ME. *bal*, *balle* < OF. < G. *ballax*, spherical structure): common name for the members of the Sclerodermatales, such as *Scleroderma*, and Hymenogastrales, such as *Sclerogaster*, which form globose fructifications in the soil.

earth star (ME. *erthe* < OE. *eorthe*, earth; ME. *sterre* < OE. *steorra*, star): a basidioma of *Geastrum* (Lycoperdales), in which the exoperidium splits into several more or less triangular segments that fold back in a star-like manner.

earth tongue (ME. *erthe* < OE. *eorthe*, earth; ME. *tunge*, tongue): an ascoma of *Geoglossum* (Helotiales).

eccentric, **excentric** (ML. *eccentricus* < Gr. *ékkentros*, not in the center + Gr. suf. *-íkos* > L. *-icus* > E. *-ic*, belonging to, relating to): **1**. *Oomycota*. A type of oospore in which the ooplast is off-center and in contact with the inner wall of the oogonium, because of which it is not completely surrounded by the oil globules that constitute the reserve substances; this is seen, e.g., in *Saprolegnia asterophora* (Saprolegniales). **2**. *Basidiomycetes*. A stipe is said to be eccentric when it is not attached to the center of the pileus of the basidiocarp, as occurs in some Agaricaceae and Polyporaceae, e.g., *Amauroderma* sp. and *Ganoderma curtisii*, of the Aphyllophorales. Cf. **centric**.

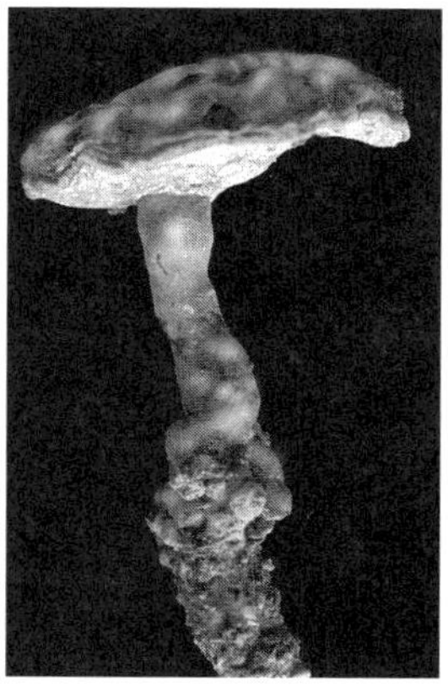

Eccentric stipe of the basidiocarp of *Amauroderma* sp., x 0.5 (*MU*).

eccrinid (< gen. *Eccrinidus* < Gr. *ekkríno*, to separate, expel + L. suf. *-idus*, state or action): one of the Eccrinales, Trichomycetes that live as endocommensals in the gut of arthropods.

echinate (L. *echinatus* < *echinus* < Gr. *echínos*, porcupine, sea urchin + L. suf. *-atus* > E. *-ate*, provided with or likeness): covered with spines or bristles that recall the porcupine. For example, the basidiocarps of *Lycoperdon echinulatum* (Lycoperdales) and the ascospores of *Tuber melanosporum* (Pezizales) are echinate.

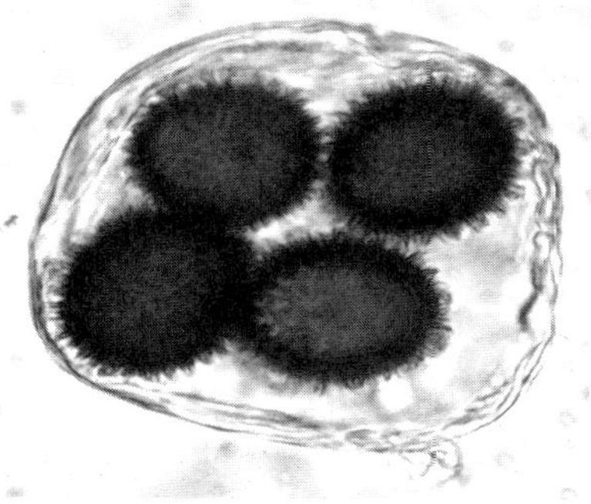

Echinate ascospores within an ascus of *Tuber melanosporum*, x 1 200 (*EAA*).

echinid, **echinidium**, pl. **echinidia** (L. *echinus* < Gr. *echínos*, porcupine, sea urchin + L. dim. suf. *-idium*): a type of marginal hair characteristic of the covering of the pileus of *Mycena* and *Marasmius* (Agaricales), not of the hymenium, although at times it is present in portions next to the hymenium due to an invasion; varied in shape from clavate to sphaeropedunculate, fusiform or lageniform, and present as spines, diverticula or bristles which give a peculiar appearance.

echinulate (NL. *echinulatus* < *echinus* < Gr. *echínos*, porcupine, sea urchin + L. suf. *-atus* > E. *-ate*, provided with or likeness): like echinate, but with the spines or bristles smaller and less rigid. Applied mainly when it refers to microscopic structures or to microorganisms. For example, the sporangial peridium of *Actinomucor elegans* and the sporangiola of *Cunninghamella echinulata* (Mucorales), the conidia of *Penicillium echinulatum* (moniliaceous asexual fungi), and the urediniospores of *Puccinia graminis tritici* (Uredinales) are echinulate.

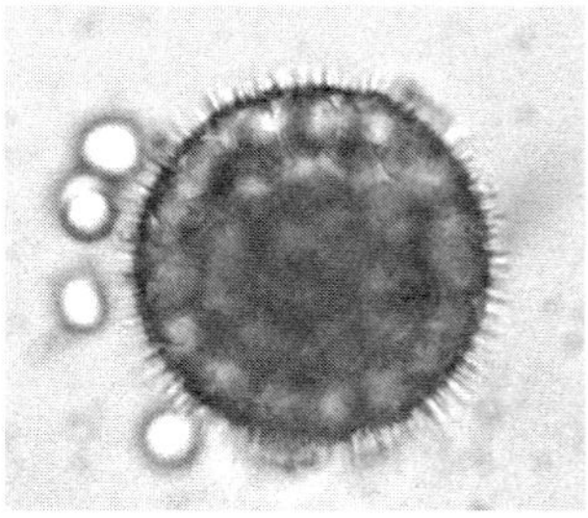

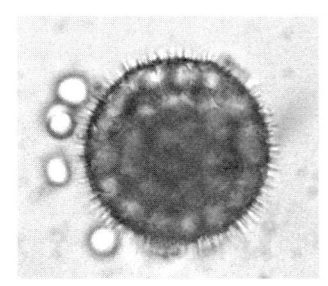

Echinulate sporangium of *Actinomucor elegans*, x 850 (*MU*).

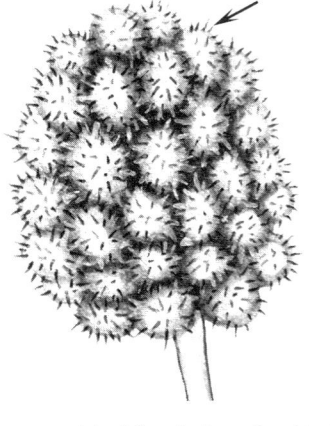

Echinulate sporangiola of *Cunninghamella echinulata*, x 900.

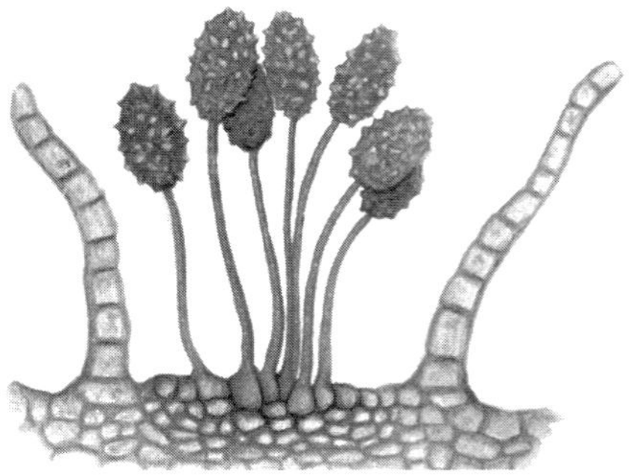

Echinulate urediniospores of *Puccinia graminis tritici*, x 400.

eclosion (F. *éclosion* < *éclore*, to hatch, prob. < VL. *exclaudere*, alteration of L. *excludere*, to hatch out, exclude): an explosive series of movements which result in the release of a germinating inner spore from a rigid exosporium, as in the ascospores of *Hypoxylon fragiforme* (Xylariales).

ecotype (Gr. *oîkos*, house, dwelling + *týpos*, type): a part of an ecospecies whose morphological, physiological and other characteristics appear to be genetically determined but influenced by environmental selection and isolation. Some consider an ecotype as a taxonomic subspecies, while others do not think that it has major taxonomic significance.

ectal (Gr. *ektós*, outside + L. suf. *-alis* > E. *-al*, relating to or belonging to): external, the outermost, superficial. Cf. **ental**.

ectangial, **ectoangial** (Gr. *ektós*, external + *angeîon*, vessel, receptacle + L. suf. *-alis* > E. *-al*, relating to or belonging to): forming or developing outside of a conceptacle, like the zoospores in the vesicle of *Pythium* (Peronosporales). Cf. **entangial**.

ectoascus, pl. **ectoasci** (L. *ecto* < Gr. *ektós*, outside + L. *ascus* < Gr. *askós*, sack, wine-skin): the outer layer of the ascus wall which on breaking permits the extension of the **endoascus**. It is present in fungi of the class Loculoascomycetes, such as *Preussia* (=*Sporormiella*) *australis* (Pleosporales). See **bitunicate**.

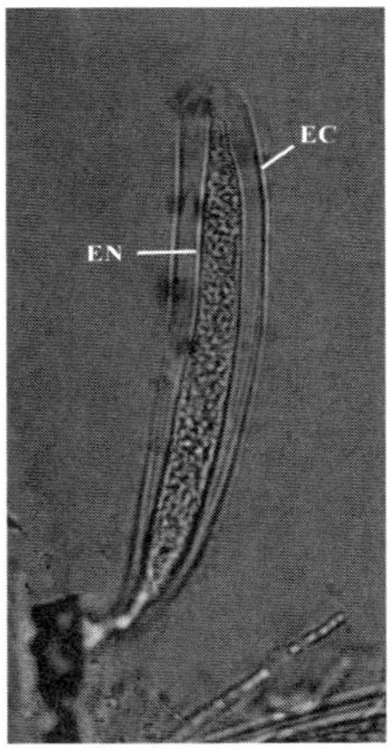

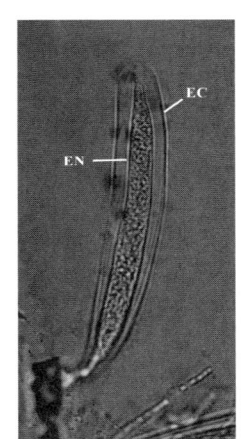

Ectoascus (EC) and **endoascus** (EN) of a young bitunicate ascus of *Preussia* (=*Sporormiella*) *australis*, x 600 (*RTH*).

ectocommensal (Gr. *ektós*, outside + L. *commensalis* < *cum*, with + *mensa*, table + suf. *-alis* > E. *-al*, relating to or belonging to): see **commensal**.

ectomycorrhiza, pl. **ectomycorrhizae** (Gr. *ektós*, outside + *mýkes*, fungus + *rhíza*, root): a mycorrhiza in which the hyphae of the associated fungus only grow intercellularly, never within the cells of the associated plant; e.g., the symbiosis between *Pisolithus tinctorius* (Sclerodermatales) and the roots of pine seedlings. Cf. **endomycorrhiza**.

ectoparasite (Gr. *ektós*, outside + *parásitos* < *pará*, at the side of + *sîtos*, bread, food): a parasitic organism which for the most part develops on the exterior of another organism (the host), from which it is nourished by means of haustoria that penetrate the cells or tissues of the host. For example, the

Laboulbeniales (such as the species of *Rickia*, common on coleopterous Passalidae, and of *Laboulbenia*, found on ants) are ectoparasites of arthropods; also the powdery mildews (Erysiphales), which attack various vascular plants, are ectoparasites. Cf. **endoparasite.** See **epibiotic**, **endobiotic** and **interbiotic.**

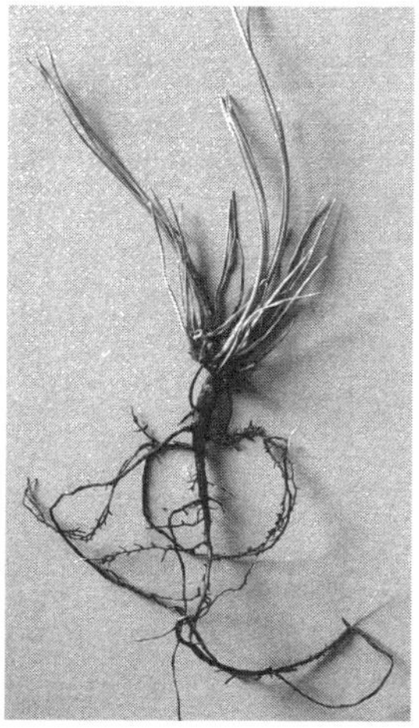

Ectomycorrhiza of pine roots formed by *Pisolithus tinctorius*, x 0.6 (*MU*).

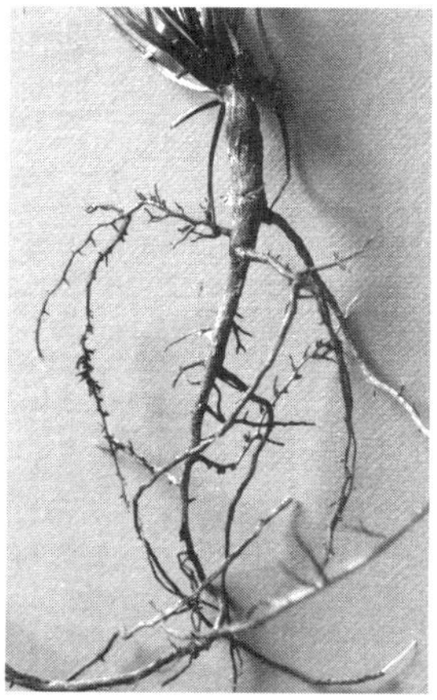

Ectomycorrhiza of pine roots formed by *Pisolithus tinctorius*, x 1 (*MU*).

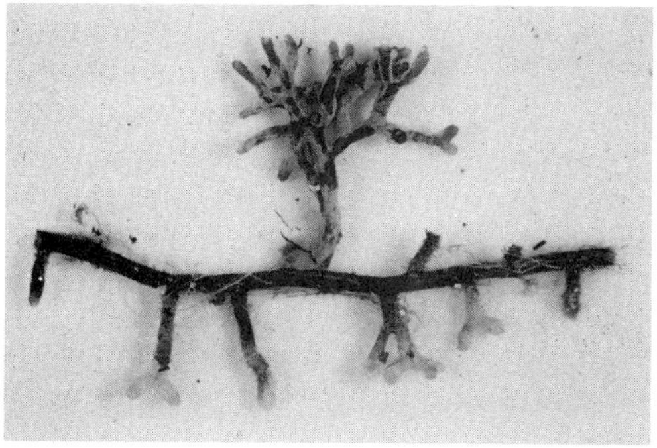

Ectomycorrhiza of pine root formed by *Pisolithus tinctorius*, x 3 (*DM*).

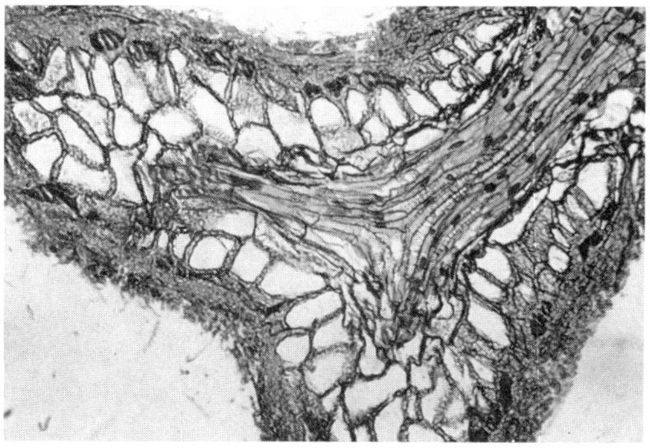

Ectomycorrhiza of pine root formed by *Pisolithus tinctorius*, seen in longitudinal section, x 300 (*DM*).

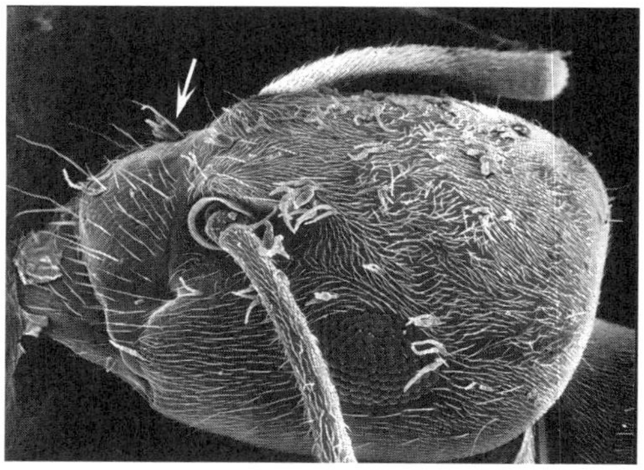

Ectoparasite - thalli of *Laboulbenia formicarum* on ant's head, x 65 (*GVD*).

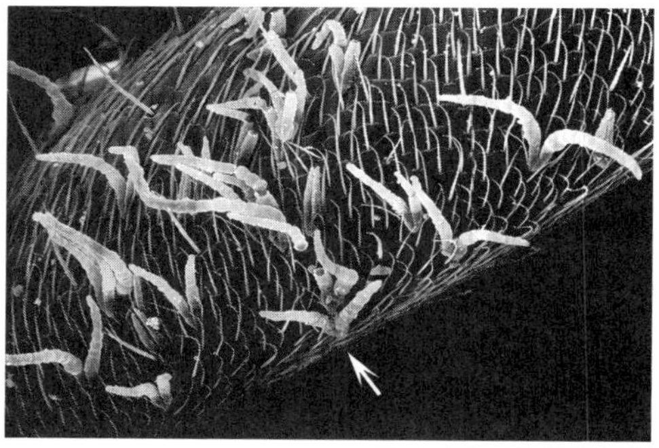

Ectoparasite - thalli of *Laboulbenia formicarum* on ant's head, x 130 (*GVD*).

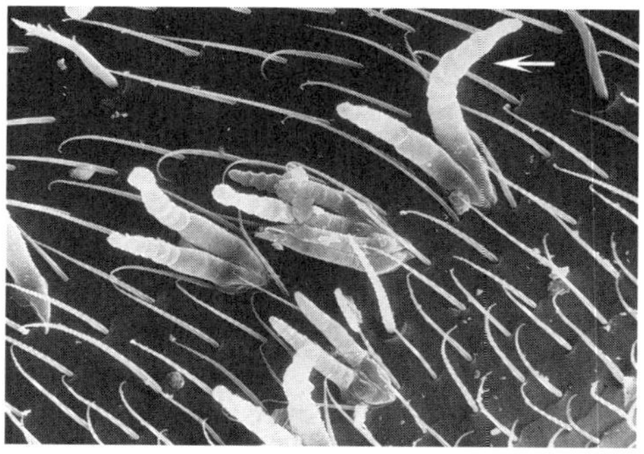

Ectoparasite - thalli of *Laboulbenia formicarum*, x 220 (*RTH*).

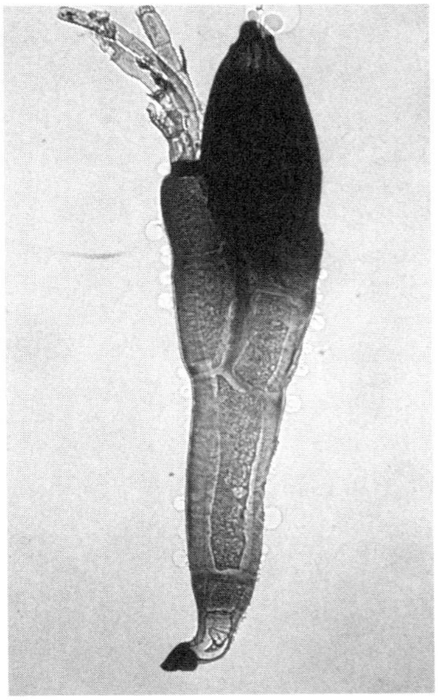

Ectoparasite - thallus of *Laboulbenia elongata*, x 180 (*RTH*).

ectoplacodium, pl. **ectoplacodia** (Gr. *ektós*, outside + NL. *placodium* < Gr. *plakós*, plaque + L. *-odium*, resembling < Gr. *ode*, like < *-oeídes*, similar to): see **placodium**.

ectostroma, pl. **ectostromata** (Gr. *ektós*, outside + *strõma*, bed): in the stromatic pyrenomycetes, such as species of *Eutypa* (Xylariales) and *Cryphonectria* (Diaporthales), it represents the stroma situated in the periderm of the host plant, on top of the **endostroma** (internal stroma in which the perithecia form); generally, the ectostroma produces conidia. In other stromatic pyrenomycetes, such as the species of *Xylaria* (Xylariales), the ectostroma (generally dark colored) is penetrated by the ostiolar necks of the perithecia, well separated from the endostroma which is much less pigmented.

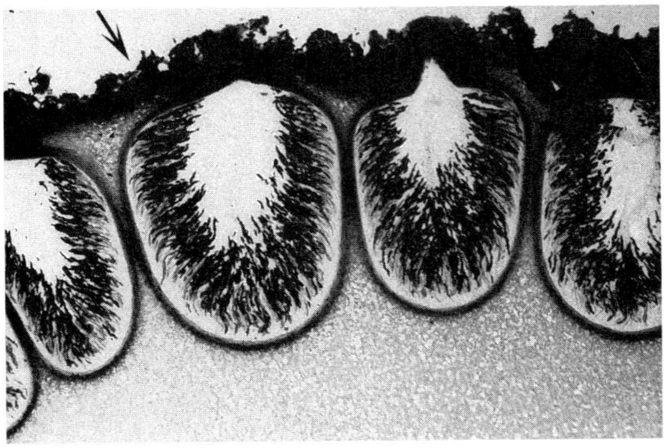

Ectostroma and endostroma of the perithecial stroma of *Xylaria* sp., seen in cross section, x 80 (*MU*).

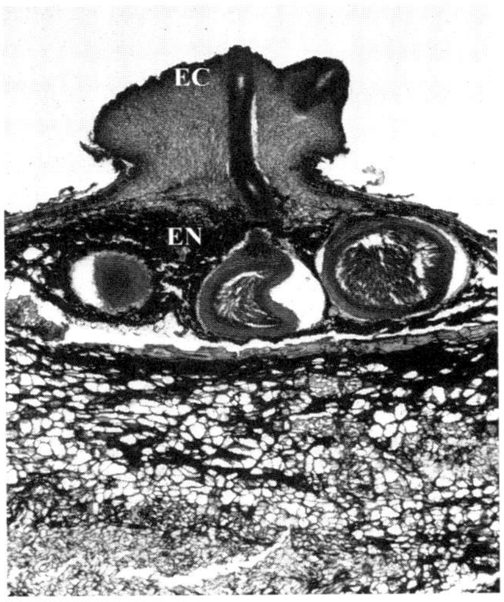

Ectostroma (EC) and **endostroma** (EN) of *Cryphonectria parasitica*; the perithecia are embedded in the endostroma and their necks protrude through the ectostroma, x 25 (*MU*).

ectothecal (Gr. *ektós*, outside + *theca* < Gr. *thêke*, case + L. suf. *-alis* > E. *-al*, relating to or belonging to): *Ascomycetes*. Having the hymenium exposed.

ectothrix (Gr. *ektós*, outside + *thríx*, hair): one of the two modes of invasion by fungi of the hair of man and animals, in which the parasite forms a layer of arthrospores in a mosaic outside the axis of the hair, as happens, e.g., in the ectothritic species *Microsporum audouinii* and *M. canis* (moniliaceous asexual fungi). Cf. **endothrix**.

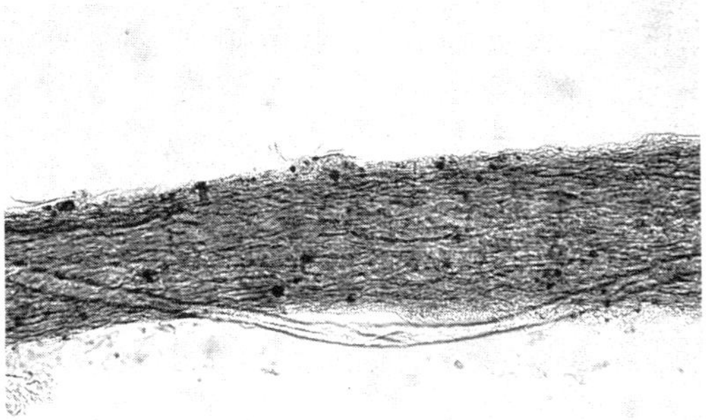

Ectothrix invasion of a human hair by *Microsporum canis*, x 260 (*MU*).

ectotroph (Gr. *ektós*, outside + *trépho*, to feed, nourish): **1**. *Mycorrhizae*. A mycorrhizal fungus that does not penetrate the cells of the host plant. The fungus remains permanently outside the root of the host, or at most, insinuates itself between the cells; also applied to the mycorrhiza formed by a fungus with this behavior. The name ectotroph is due to the fungus obtaining from the external medium the nutrients that it supplies to the host. **2**. *Lichens*. A cephalodium that forms on the lichen thallus. Cf. **endotroph**.

ectozoic (Gr. *ektós*, outside + *zõon*, animal + suf. *-íkos* > L. *-icus* > E. *-ic*, belonging to, relating to): fungi that develop on the exterior of living animals as commensals, symbionts or parasites, e.g., as happens in the Laboulbeniales. Cf. **endozoic**.

edaphosphere (Gr. *édaphos*, soil + *sphaîra*, sphere): the area of the soil outside any effect of plant roots, beyond the rhizosphere. The microbial composition of the edaphosphere differs qualitatively and quantitatively from that of the rhizosphere; the latter is generally richer in fungal species, including mycorrhizae. See **rhizosphere**.

effete (L. *effetus* < *ex-*, out of + *fetus*, fruitful): no longer fertile, overmature, exhausted, as when a fruiting body becomes empty.

efflorescent (L. *efflorescens*, genit. *efflorenscentis*, flourishing, blooming, prp. of *efflorescere*, to flourish, bloom + *-escens*, genit. *-escentis*, that which turns, beginning to, slightly > E. *-escent*): bursting out of.

effuse (L. *effusus*, spread out, extended < *effundere*, poured out): spread out or scattered, especially toward one side, like the thallus of some lichens (e.g., *Chiodecton sanguineum* (Arthoniales) and some species of *Lecanora*, Lecanorales), which do not have a well-defined outline. The term **effuse-reflex** is commonly used to refer to the pileus of certain basidiomycetes which have a margin that is extended and bent backwards, as happens in *Corticium* (Aphyllophorales).

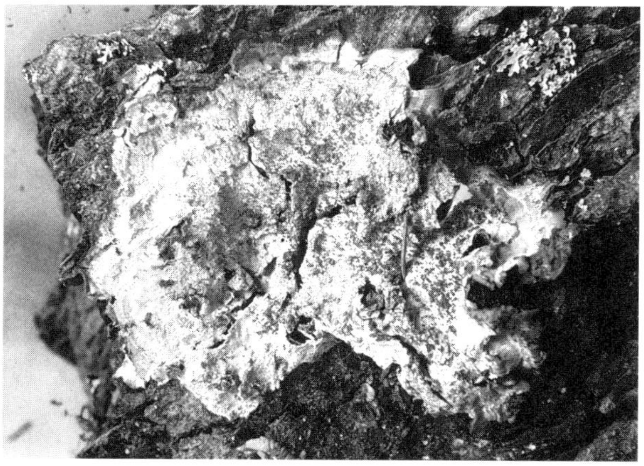

Effuse thallus of *Chiodecton sanguineum*, x 1 (*MU*).

egg (ME. *egge* < ON. *egg* < L. *ovum* < Gr. *oón*): **1**. The female oomycota gamete (**oosphere**). **2**. The young basidioma (primordium) of phallaceous fungi, and of *Amanita* and other agarics, before the volva is broken.

eguttulate (L. *e*, lacking + *guttulatus* < *guttula*, dim. of *gutta*, drop + suf. *-atus* > E. *-ate*, provided with or likeness): lacking oil or lipid droplets, like the conidia of *Sphaeropsis malorum* (sphaeropsidaceous asexual fungi) and *Marssonina juglandis* (melanconiaceous asexual fungi), among many such species. Cf. **guttulate**.

elastic (NL. *elaster*, spring, elasticity < Gr. *elastikós* < *elatés*, *elatér*, a driver, dispersant + suf. *-tikós* > L. *-ticus* > E. *-tic*, relation, fitness, inclination or ability): the ability of a body that can recover more or less completely its shape and size after the action that altered it ceases; e.g., as in the context of the basidiocarp of *Marasmius* (Agaricales). Syn. of **resilient**.

elater (Gr. *elatér*, a driver, dispersant): hygroscopic capillitial filaments that contribute to spore dispersal; e.g., as occur in Myxomycetes of the order Trichiales (such as *Hemitrichia serpula* and *Arcyria denudata*) or the gleba of *Battarraea* (Tulostomatales). Elaters are often ornamented or spirally thickened.

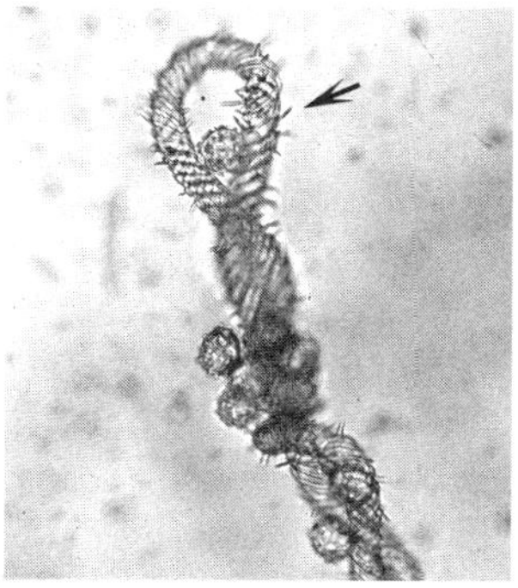

Elaters of the capillitium of *Hemitrichia serpula*, x 610 (*MU*).

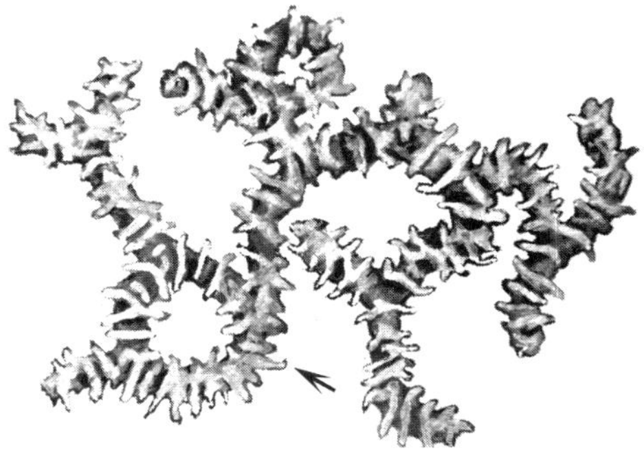

Elaters of the capillitium of *Arcyria denudata*, x 650.

electrotaxis, pl. **electrotaxes** (E. *electr-*, *electro-* < NL. *electricus*, electric, electricity; produced from amber by friction < L. *electrum*, amber, electrum < Gr. *élektron* + Gr. *táxis*, arrangement, disposition): a tactic movement in response to a electrically charged field, e.g., the zoospores of *Phytophthora palmivora* (Peronosporales). Cf. **gravitaxis**.

eleohypha, pl. **eleohyphae** (Gr. *elaía*, olive, or *alaío*, oil + *hyphé*, hypha, tissue, spider's web): *Lichens*. Enlarged hyphae, whose cells are filled with oil, that occur in the endolithic calcicolous lichens.

ellipsoid, **ellipsoidal**, **elliptic** (L. *ellipsis*, ellipse < Gr. *éllepsis*, a closed curve + L. suf. *-oide* < Gr. *-oeídes*, similar to; or + L. suf. *-alis* > E. *-al*, relating to or belonging to; or + Gr. suf. *-íkos* > L. *-icus* > E. *-ic*, belonging to, relating to): refers to a solid body that forms an ellipse in the longitudinal plane and a circle in cross section. Many fungal spores are ellipsodal or elliptic.

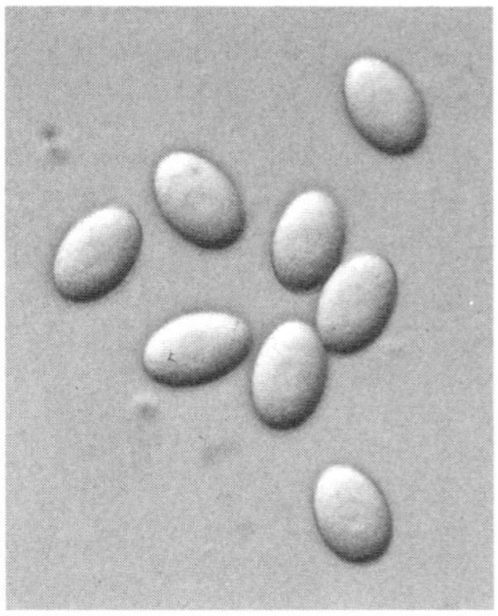

Ellipsoidal ascospores of *Pyronema domesticum*, x 600 (*RTH*).

emarginate (L. *emarginatus* < *e*, without + *marginatus*, margined < *margo*, margin + suf. *-atus* > E. *-ate*, provided with or likeness): lacking a margin; with a poorly defined notch or groove. Generally used for gills that have a notch near their attachment to the stipe, as happens in *Tricholoma flavovirens* (Agaricales).

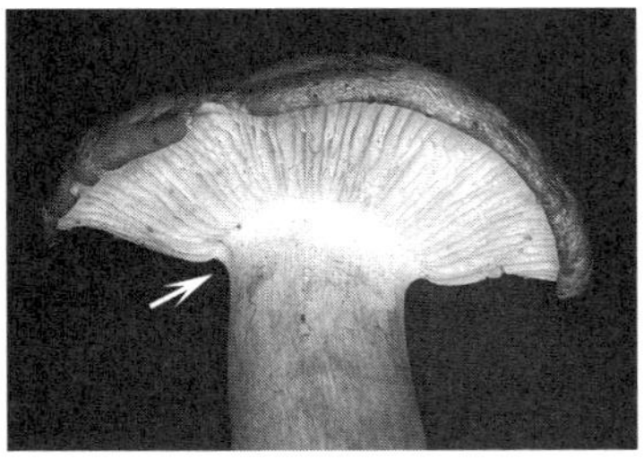

Emarginate lamellae of the basidiocarp of *Tricholoma* sp., x 0.6 (*JC*).

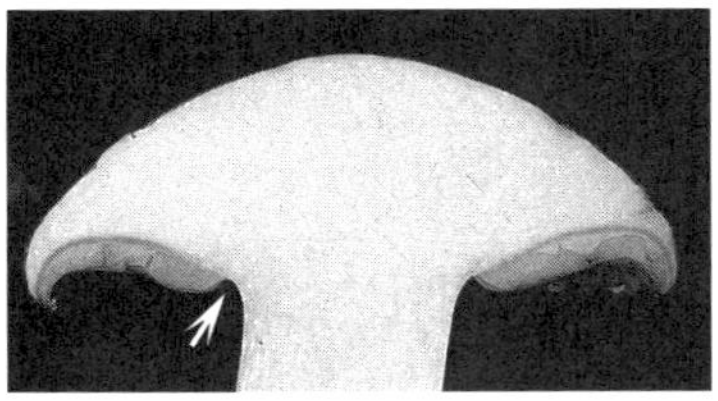

Emarginate lamellae of the basidiocarp of *Tricholoma* sp., seen in longitudinal section, x 0.6 (*JC*).

embedded (*em-*, var. of *en* < ME. < OF. < L. *in*, in + *bedded*, ptp. of *bed* < ME. < OE. *bedd*, bed): immersed, completely enclosed or surrounded; e.g., perithecia (such as the perithecia in the stroma of *Claviceps purpurea* and *Epichloë typhina*, of the Hypocreales), as well as pycnidia of the sphaeropsidaceous asexual fungi, can be partially or totally embedded in the tissues of the host organism or substrate on which they grow, or in a fungal stroma.

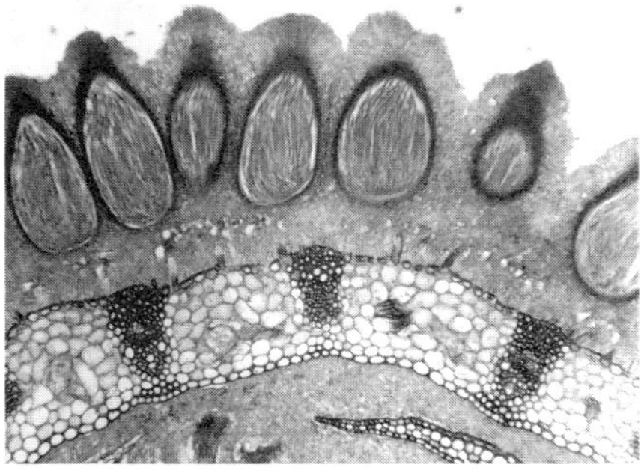

Embedded perithecia in a stroma of *Epichloë typhina*, seen in longitudinal section, x 60 (*RTH*).

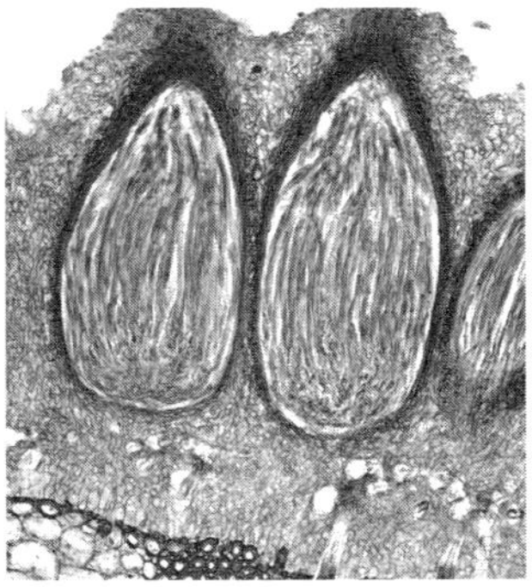

Embedded perithecia in a stroma of *Epichloë typhina*, seen in longitudinal section, x 125 (*RTH*).

emend (ME. *emenden* < L. *emendare* < *e*, *ex*, out + *menda*, fault): to correct an error. Cf. **amend**.

emplacement (F. *emplacer*, to place + L. *-men* > E. *-ment*, result, object, agent of an action): a cushion or pad of closely packed hyphae beneath the sporocarp, usually incorporating bits of substrate; as occurs in the Nidulariales.

endemic (Gr. *endemía* < *en*, in + *démos*, people + suf. *-íkos* > L. *-icus* > E. *-ic*, belonging to, relating to): native to a certain region; found exclusively in one place. The pyrenomycete *Claviceps gigantea* (Hypocreales), e.g., is endemic to the region near the volcanos Popocatépetl and Iztaccíhuatl in Mexico.

endoascus, pl. **endoasci** (L. *endo* < Gr. *éndon*, inside + L. *ascus* < Gr. *askós*, sack, wine-skin): the internal wall layer of a bitunicate ascus, present in

Loculoascomycetes, e.g., *Preussia* (=*Sporormiella*) *australis* (Pleosporales); the endoascus expands before liberating the ascospores when the **ectoascus** breaks. See **bitunicate**.

endobiotic (Gr. *éndon*, inside + *biotikós*, vital < *bíos*, life + suf. + *-tikós* > L. *-ticus* > E. *-tic*, relation, fitness, inclination or ability): a parasite or saprobic organism that develops its somatic as well as its reproductive structures in the interior of the host or substrate. *Plasmodiophora brassicae* (Plasmodiophorales) forms its spores within the cells of cabbage root, *Synchytrium endobioticum* (Chytridiales) does the same in the cells of potato tubers; *Rozella allomycis* (Spizellomycetales) lives in the cells of the aquatic fungus *Allomyces javanicus* (Blastocladiales), and *Nucleophaga amoebae* (Spizellomycetales) forms its sporangium within the nucleus of *Amoeba verrucosa*. Equivalent to **intramatrical**. Cf. **epibiotic** and **interbiotic**.

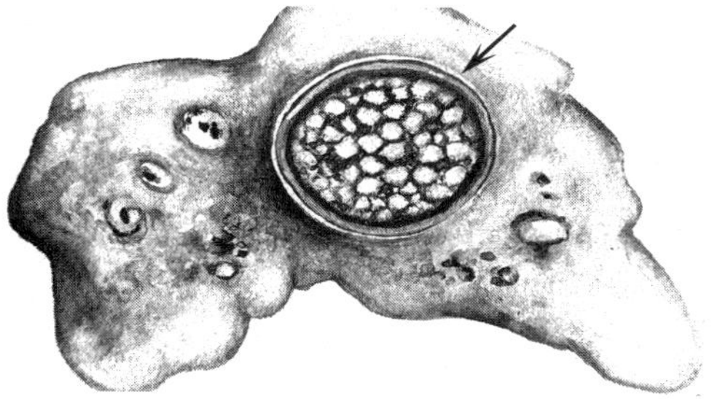

Endobiotic sporangium of *Nucleophaga amoebae*, within the nucleus of the host (*Amoeba verrucosa*), x 500.

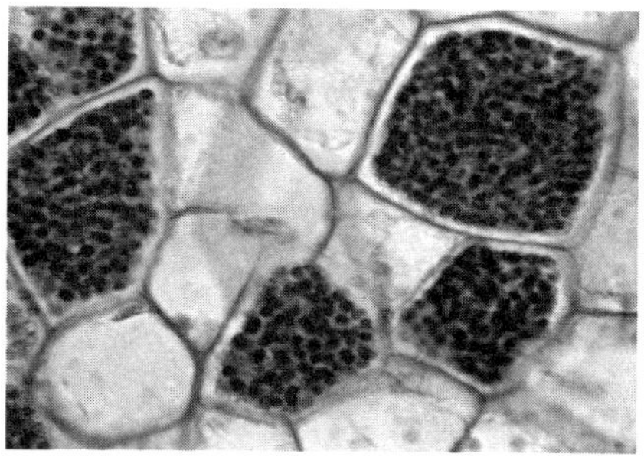

Endobiotic spores of *Plasmodiophora brassicae* in cells of cabbage root, seen in cross section, x 700 (*MU*).

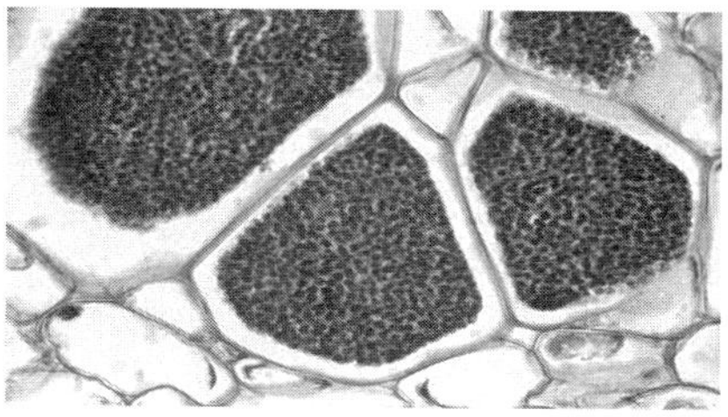

Endobiotic spores of *Plasmodiophora brassicae* in cells of cabbage root, seen in cross section, x 700 (*MU*).

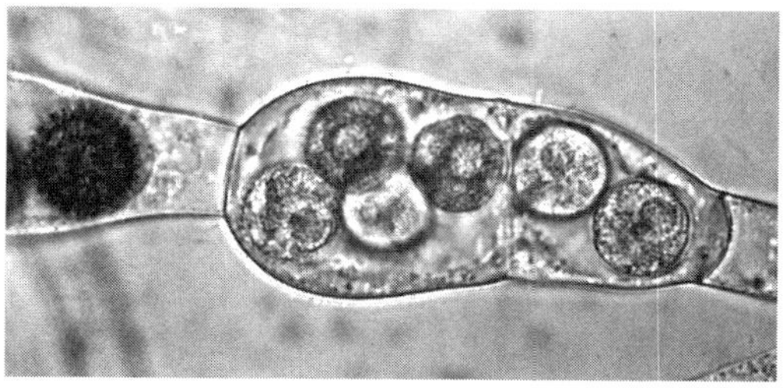

Endobiotic young thalli of *Rozella allomycis* in a hyphal cell of *Allomyces javanicus*, x 410 (*RTH*).

endocarpic (Gr. *éndon*, inside + *karpós*, fruit + suf. *-íkos* > L. *-icus* > E. *-ic*, belonging to, relating to): *Lichens.* Applied to species that have the perithecia inserted in the thallus, as is seen in *Endocarpon* and *Dermatocarpon* (Verrucariales).

endocommensal (Gr. *éndon*, inside + L. *commensalis* < *cum*, with + *mensa*, table + suf. *-alis* > E. *-al*, relating to or belonging to): see **commensal.**

endoconidium, pl. **endoconidia** (Gr. *éndon*, inside + *kónis*, dust + dim. suf. *-ídion* > L. *-idium*): a conidium produced in the interior of a conidiogenous cell, as occurs in *Chalara basicola* (dematiaceous asexual fungi), which has phialides with chains of endoconidia that emerge already mature from the mouth of the phialide.

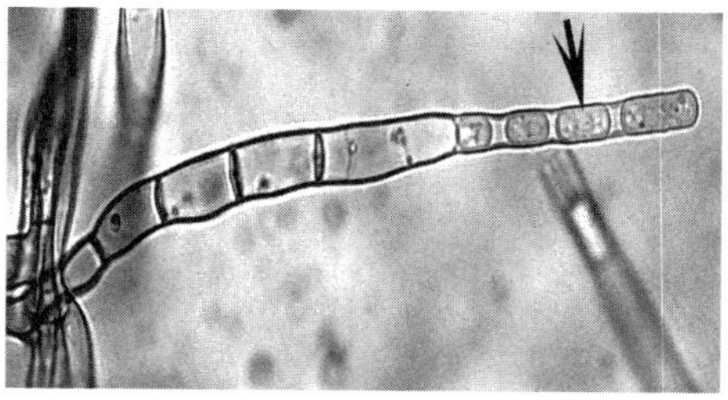

Endoconidia within a phialide of *Chalara basicola*, x 275 (*MU*).

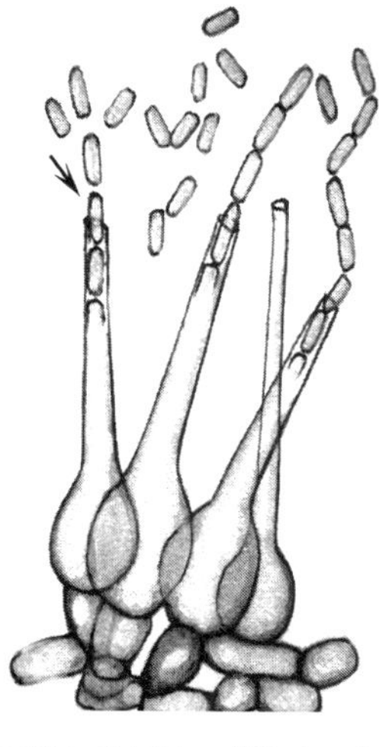

Endoconidia within phialides of *Chalara fusidioides*, x 190.

endocyanosis, pl. **endocyanoses** (Gr. *éndon*, inside + *kýanos*, substance with a dark blue color + suf. *-osis*, state or condition): the condition of having cyanobacteria inside the cells of another organism, as happens in the gen. *Geosiphon* (an uncertainly classified member of Zygomycota), the only known example of endocyanosis in fungi. Sometimes considered lichenized, as it includes cyanobacteria in vesicles.

endocyclic rust fungi (Gr. *éndon*, within + *kyklikós*, circular, cyclic < *kýklos*, circle + suf. *-íkos* > L. *-icus* > E. *-ic*, belonging to, relating to; ME. < OE. *rust*, red ; L. *fungi*, pl. of *fungus*, fungus): microcyclic species of rust fungi (Uredinales) that lack stages I and II (spermogonia and aecia) and the primary host, and whose teliospores are morphologically identical to aeciospores. However, upon germination the teliospores give rise to basidia and basidiospores rather than germ tubes.

endocystidium, pl. **endocystidia** (Gr. *éndon*, inside + *kystídion* < *kýstis*, bladder, vesicle, cell + dim. suf. *-ídion* > L. *-idium*): *Agaricales*. A chrysocystidium that is embedded in the trama of the pileus, stipe or hymenophore of many Agaricales. For example, in some species of *Pholiota*, *Stropharia* and *Hypholoma*, endocystidia are present in the trama of the hymenophore.

endoexogenous (Gr. *éndon*, inside + *éxo*, outside + *génos*, origin < *gennáo*, to engender, produce + L. *-osus* > OF. *-ous*, *-eus* > E. *-ous*, possessing the qualities of): a mode of growth and development of the aquatic fungi of the order Chytridiales, which is characterized first by a flow of cytoplasm and food and a direction of growth from the zoospore toward the extramatrical rhizoidal system, but later this flow of protoplasm and food reverses and is directed from the rhizoids toward the developing zoospore.

endogenous (Gr. *éndon*, inside + *génos*, origin < *gennáo*, to engender, produce + L. *-osus* > OF. *-ous*, *-eus* > E. *-ous*, possessing the qualities of): any structure produced or developed in the interior of another structure; e.g., the spores produced in sporangia, pycnidia, asci, etc., are endogenous. Cf. **exogenous**.

endohypha, pl. **endohyphae** (Gr. *éndon*, inside + *hyphé*, web, hypha): the intraconidiophoric hypha of *Brachysporiella* (=*Monotosporella*) *sphaerocephala* (moniliaceous asexual fungi), which originates from the internal wall of the conidiophores (which is electron transparent), and which on growing make the conidiophores appear superficially like annellides or annellidic phialides.

endokapylic (Gr. *éndon*, inside + *kapyleío*, to falsify, adulterate + suf. *-íkos* > L. *-icus* > E. *-ic*, belonging to, relating to): a thallus of a lichenicolous fungus that does not form a morphologically distinct lichenized structure. Cf. **epikapylic**.

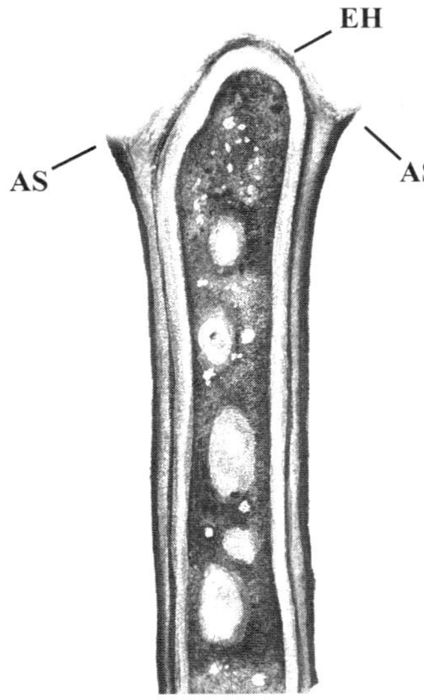

Endohypha (EH) in a conidiophore of *Brachysporiella* (=*Monotosporella*) *sphaerocephala*. Longitudinal section of part of a conidiophore showing proliferation as a result of the penetration of an endohypha through the remaining layer of septum, which had delimited the former conidium. Note the annular scar (AS). Drawing based on a transmission electron micrograph published by Beckett *et al.* in their *Atlas of Fungal Ultrastructure*, 1974, x 1 900.

endolithic (Gr. *éndon*, inside + *líthos*, stone + suf. *-íkos* > L. *-icus* > E. *-ic*, belonging to, relating to): *Lichens*. A crustaceous lichen that develops almost in the interior of rocks, such as *Caloplaca elegans* (Teloschistales) and other related species.

endomycorrhiza, pl. **endomycorrhizae** (Gr. *éndon*, inside + *mýkes*, fungus + *rhíza*, root): a mycorrhiza in which the hyphae of the associated fungus penetrate the cells of the associated plant; also called vesicular-arbuscular mycorrhiza; e.g., the symbiosis between certain species of *Endogone* (Endogonales) and the roots of onion plants. Cf. **ectomycorrhiza**.

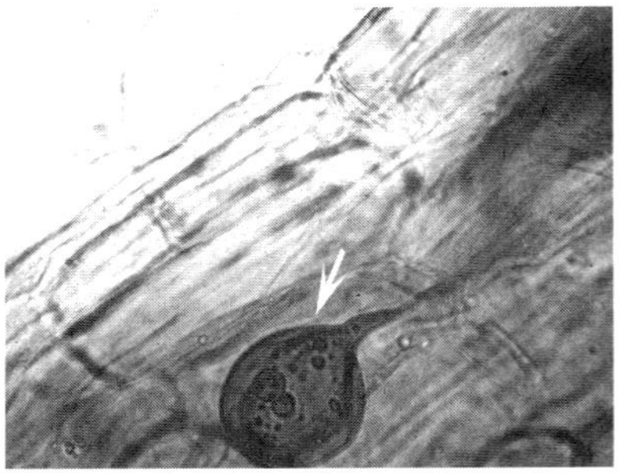

Endomycorrhiza - intracellular vesicle of *Endogone* sp. in onion root, x 245 (*SP*).

endoparasite (Gr. *éndon*, inside + *parásitos* < *pará*, at the side of, near + *sîtos*, bread, food): a parasitic organism that develops in the interior of a cell (intracellular parasite) or of the tissues (intercellular parasite) of another organism (the host) on which it feeds. Some primitive aquatic fungi (Chytridiales) are endoparasites of algae, of plants and of other fungi, e.g., *Synchytrium endobioticum* is endoparasitic in potato tubers. Among the Saccharomycetales of the family Metschnikowiaceae, *Coccidiascus legeri* represents a rare and interesting endoparasite that lives in the intestinal cells of the fly *Drosophila funebris*. Cf. **ectoparasite**.

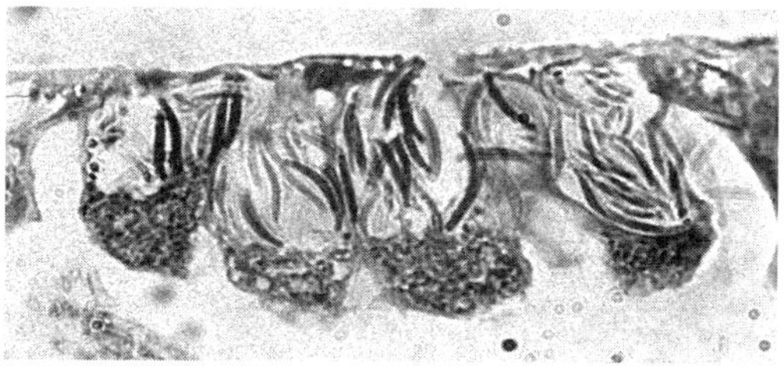

Endoparasite - asci of *Coccidiascus legeri* within intestinal cells of the fly *Drosophila funebris*, x 400 (*RTH*).

endoperidium, pl. **endoperidia** (Gr. *éndon*, inside + *péridion*, a small leather purse): the inner wall of a multi-walled peridium, as in the Gasteromycetes of the gen. *Geastrum* (Lycoperdales); e.g., *G. saccatum* and *G. triplex*; the external layer is the **exoperidium**.

Endoperidium (EN) and exoperidium (EX) of the basidiocarp of *Geastrum saccatum*, x 0.5 (*TH*).

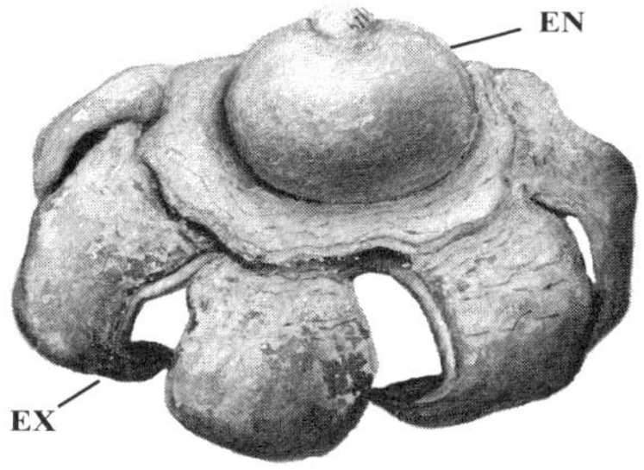

Endoperidium (EN) and exoperidium (EX) of the basidiocarp of *Geastrum saccatum*, x 1.

endophyte (Gr. *éndon*, inside + *phytón*, plant): **1**. A fungus that lives inside plant tissues without showing external signs of its presence, like certain fungal parasites of pasture grasses that are toxic to animals that feed on them. **2**. The internal, systemic mycelium of clavicipitaceous fungi, such as *Epichloë* (Hypocreales).

endoplacodium, pl. **endoplacodia** (Gr. *éndon*, inside + NL. *placodium* < Gr. *plakós*, plaque + L. *-odium*, resembling < Gr. *ode*, like < *-oeídes*, similar to): see **placodium**.

endoplasmic reticulum, pl. **reticula** (Gr. *éndon*, inside + *plásma*, genit. *plásmatos*, bland material with which is formed a living being + suf. *-íkos* > L. *-icus* > E. *-ic*, belonging to, relating to; L. *reticulum*, little net < *rete*, net + dim. suf. *-culum*): an extensive network of double membranes, connected to the nuclear envelope, that is arranged in flat, perforated, laminar or tubular chambers called cisternae, that are diffused in the cytoplasm. Ribosomes can be attached to the cytoplasmic side of the membranes (rough reticulum) or they can be lacking (smooth reticulum). Usually abbreviated as ER.

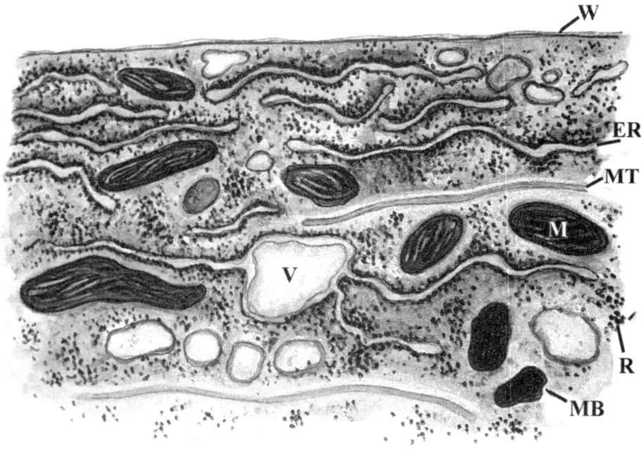

Endoplasmic reticulum (ER) of a young hypha of *Armillaria mellea*. Ribosomes (R) of the cisternae of the ER, microbodies (MB), mitochondria (M), microtubules (MT), vesicles (V), and cell wall (W) are also seen. Drawing composed from various transmission electron micrographs published by Beckett *et al.* in their *Atlas of Fungal Ultrastructure*, 1974, x 22 200.

endopsammon (Gr. *éndon*, inside + *psámmos*, sand): the microhabitat among the grains of sand in marine beaches, where arenicolous and psammophilous organisms, including some fungi, such as *Corollospora* and *Carbosphaerella* (Halosphaeriales), and *Nia* (Melanogastrales), are found. See **arenicolous** and **psammophilous**.

endospore, **endosporium**, pl. **endosporia** (Gr. *éndon*, inside + *sporá*, spore + L. dim. suf. *-ium*): **1**. The innermost layer, adjacent to the cytoplasmic membrane, of the five layers that form the wall of the most complex basidiospore. The endospore can vary from very thick (in which case it can be subdivided into internal and outer portions) up to apparently or truly absent (e.g., the spores of *Tubaria*, Agaricales, are thin-walled and lack an endospore). Over the endospore are four other layers: **mesospore**, **epispore**, **exospore** and **perispore**. **2**. Also applied to the asexual spores formed by internal budding of the hyphae of some species of yeasts with true mycelium, such as *Trichosporon capitatum* and *T. fermentans* (cryptococcaceous asexual yeasts). In these yeasts 1-15 and 2-3 endospores are formed, respectively, in intercalary or terminal hyphal segments, either from the septa or in the tip of the intrahyphal tubular growths. In *Aureobasidium pullulans* (dematiaceous asexual fungi) one can also observe endospores or endoconidia in the hyphae.

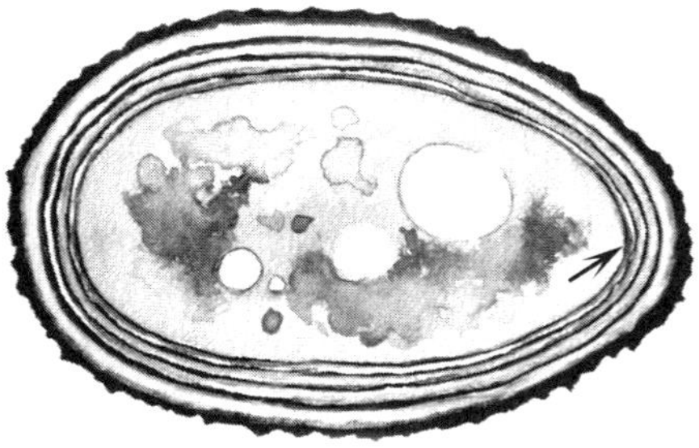

Endospore, mesosporium, epispore, exospore, and perispore of the cell wall of a basidiospore of *Tubaria* sp., shown in this diagram of a longitudinal section, x 5 500.

endostroma, pl. **endostromata** (Gr. *éndon*, inside + *strõma*, bed): *Pyrenomycetes*. The portion of the perithecial stroma situated beneath the **ectostroma**; generally it is sunken in the tissues of the host plant and surrounds the perithecia. Stromatic species, such as *Diatrype*, *Eutypa* and *Xylaria* of the Xylariales, have an endostroma.

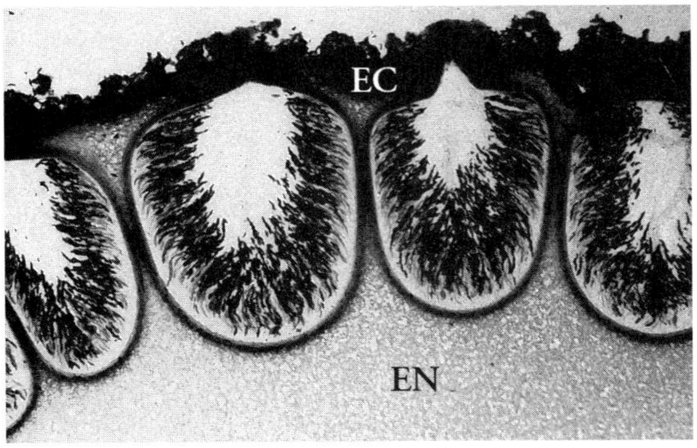

Endostroma (EN) and ectostroma (EC) of the perithecial stroma of *Xylaria* sp., seen in cross section; the perithecia are seen in longitudinal section, x 80 (*MU*).

endothrix (Gr. *éndon*, inside + *thríx*, hair): one of the two modes of invasion by fungi of the hair of man and animals, in which the parasite forms parallel chains of arthrospores inside the axis of the hairs, as is seen, e.g., in the endothritic species *Trichophyton tonsurans* and *T. violaceum* (moniliaceous asexual fungi). Cf. **ectothrix**.

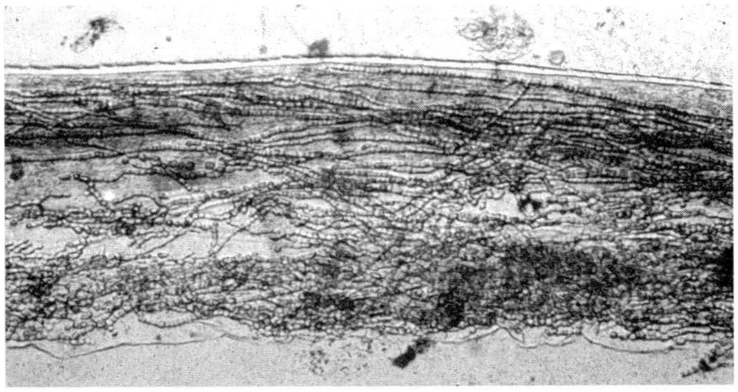

Endothrix invasion of a human hair by *Trichophyton tonsurans*, x 400 (*RLM*).

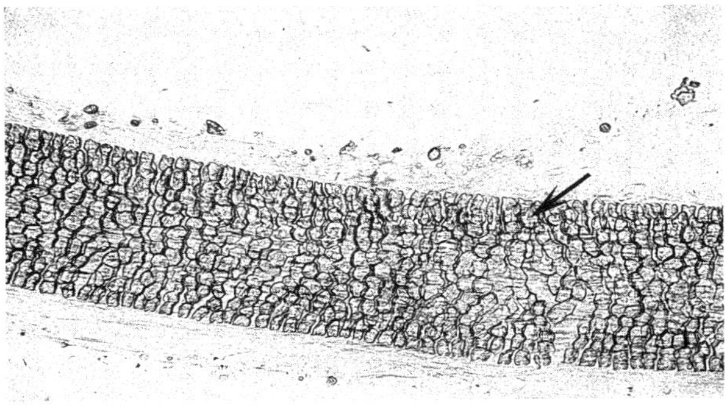

Endothrix type of infection caused by *Trichophyton mentagrophytes* on hair of white rat; the arthrospores of the fungus are seen inside the hair shaft, x 115 (*MU*).

endotroph (Gr. *éndon*, inside + *trépho*, to feed): **1**. *Mycorrhizae*. A mycorrhizal fungus that penetrates the cells of the host plant, and of the mycorrhiza that results from the association of both. In this case, the exchange of nutrients between the two occurs inside the cells of the host. Opposite this term is **ectotroph**. **2**. *Lichens*. A cephalodium that originates in the interior of the thallus.

endozoic (Gr. *éndon*, inside + *zõon*, animal + suf. *-íkos* > L. *-icus* > E. *-ic*, belonging to, relating to): a fungus that develops as a commensal, symbiont or parasite in the interior of living animals, i.e., it is endozoic, like the majority of species of Trichomycetes, e.g., *Smittium* (Harpellales). Cf. **ectozoic**.

ensate, **ensiform** (L. *ensis*, sword + suf. *-atus* > E. *-ate*, provided with or likeness; *ensis* + *-formis* < *forma*, shape): having the shape of a sword, i.e., long, flat, with parallel, sharp edges and ending in a point; e.g., like some hymenial cystidia of certain Aphyllophorales. Also called **gladiate**.

ental (Gr. *entós*, inside + L. suf. *-alis* > E. *-al*, relating to or belonging to): of, or in, the interior. Cf. **ectal**.

entangial, **entoangial** (Gr. *entós*, inside + *angeîon*, vessel, receptacle + L. suf. *-alis* > E. *-al*, relating to or belonging to): formed or developed inside a conceptacle, like the sporangiospores of the Chytridiomycota and Zygomycota. Cf. **ectangial**.

enteroarthric (Gr. *énteron*, intestine + *árthron*, articulation + suf. *-íkos* > L. *-icus* > E. *-ic*, belonging to, relating to): *Conidiogenesis*. A type of conidial development in which the outer wall of the conidiogenous cell generally separates from the wall of the conidium which forms in its interior; e.g., as occurs in *Sporendonema* (moniliaceous asexual fungi). Cf. **holoarthric**.

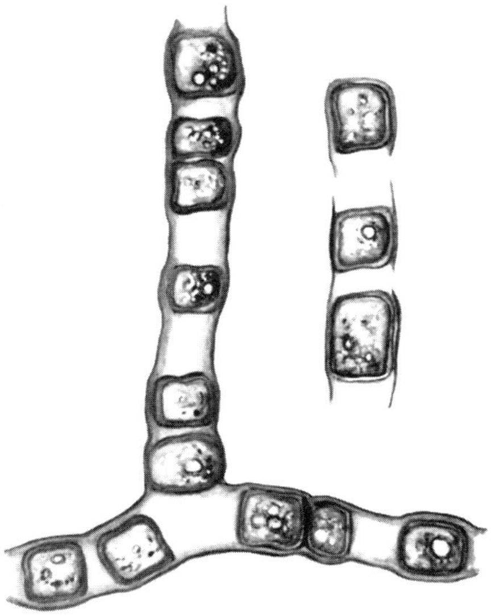

Enteroarthric conidia in hyphae of *Sporendonema purpurascens*, x 1 100.

enteroblastic (Gr. *énteron*, intestine + *blastós*, sprout, bud, germ, shoot + suf. *-íkos* > L. *-icus* > E. *-ic*, belonging to, relating to): see **blastic**.

enterophilous (Gr. *énteron*, intestine + *phílos*, have an affinity for + L. *-osus* > OF. *-ous*, *-eus* > E. *-ous*, possessing the qualities of): an organism that lives in the intestine of animals; e.g., like the Trichomycetes of the gen. *Enterobryus* (Eccrinales), which live as endocommensals, adhering to the chitinous lining of the intestine of coleopterans, diplopods and decapods.

enterothallic (Gr. *énteron*, intestine + *thallós*, shoot, thallus + suf. *-íkos* > L. *-icus* > E. *-ic*, belonging to, relating to): see **thallic**.

entheogen (Gr. *en*, in + *theós*, God + *génos*, engenderment): "God generated within". A term used for hallucinogens (plants, fungi) with strong religious connotations, as part of a theory sustaining that, very early in the evolution of humans, the weird, unearthly effects of hallucinogenic substances (such as psilocybin of *Psilocybe*, Agaricales) introduced man to believe in the supernatural, giving rise to prototypical ideas of the religious experience.

entire (ME. *entere* < MF. *entier* < L. *integrum* < *integer*, whole): having absolute marginal integrity, i.e., without teeth, notches or indentations that make it uneven; applied to the margin of a colony on culture media, to the edge of gills on an agaric pileus, etc.

entomogenous (Gr. *éntomos*, cut, divided into segments, insect + *génos*, origin < *gennáo*, to engender + L. *-osus* > OF. *-ous*, *-eus* > E. *-ous*, possessing the qualities of): growing on or obtaining nourishment, especially as pathogens, from insects (entomophagous), such as the fungi of the gen. *Entomophthora* (Entomophthorales), *Verticillium* (moniliaceous asexual fungi), *Hirsutella* (stilbellaceous asexual fungi), and *Cordyceps* (Hypocreales).

Entomogenous synnemata of *Hirsutella saussurei* arising from a parasitized wasp, x 1.5 (*MU*).

Entomogenous synnemata of *Hirsutella saussurei* arising from a parasitized wasp, x 5 (*MU & CB*).

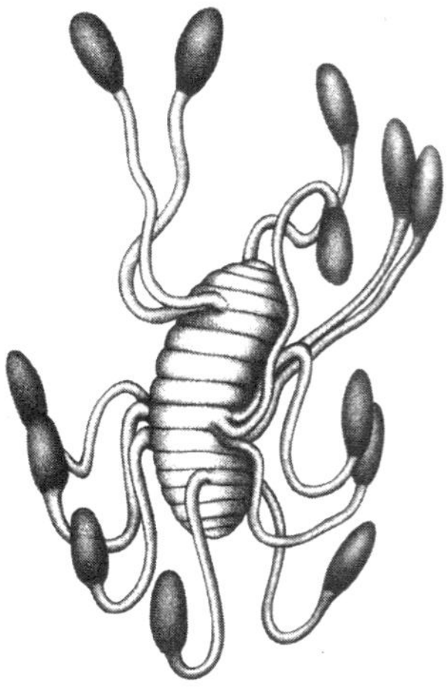

Entomogenous perithecial stromata of *Cordyceps melolonthae* var. *rickii* on a parasitized coleopteran larva, x 1.

A white-fly nymph parasitized by *Verticillium lecanii*, on the under surface of a bean leaf. The fungus is **entomogenous** and is used for the biological control of this insect, x 20 (*MU*).

entomophilous (Gr. *éntomos*, cut, divided into segments, insect + *phílos*, have an affinity for + L. *-osus* > OF. *-ous*, *-eus* > E. *-ous*, possessing the qualities of): applied to fungi which have their spores distributed by insects.

epapillate (L. *e*, without, lacking + *papilla*, papilla, nipple + suf. *-atus* > E. *-ate*, provided with or likeness): without a papilla; e.g., in the gen. *Didymella* (Dothideales) there are marine species with an epapillate ascocarp (such as *D. fucicola*, a perthotroph that develops in the middle veins and air vesicles of the thallus of *Fucus*), and species with the ascocarp epapillate or with a short papilla (like *D. gloiopeltidis*, a parasite of the red alga *Gloiopeltis furcata*).

epibasidium, pl. **epibasidia** (Gr. *epí*, upon, superior + *basídion* < *básis*, base + dim. suf. *-ídion* > L. *-idium*): see **heterobasidium**.

epibiotic (Gr. *epí*, upon + *biotikós*, vital < *bíos*, life + *-tikós* > L. *-ticus* > E. *-tic*, relation, fitness, inclination or ability): an organism that develops its reproductive organs on the surface of the host or substrate, although part or all of the thallus or somatic structures are in the interior; many aquatic fungi of the classes Chytridiomycetes (e.g., *Phlyctochytrium* and *Chytriomyces*, of the Chytridiales) and Hyphochytriales, are epibiotic. Cf. **endobiotic** and **interbiotic**.

epiblastesis, pl. **epiblasteses** (Gr. *epí*, upon + *blástesis*, germination, development): growth of lichens from gonidia that develop in the parent lichen thallus.

epicutis, pl. **epicutes** (Gr. *epí*, upon + L. *cutis*, skin): a layer external to the cuticle of the pileus, when the cuticle or cuticular layer is composed of two or more layers.

epidemic (F. *épidémique* < LL. *epidemia* < Gr. *epidemía*, among the people < *epí*, upon + *démos*, people + suf. *-íkos* > L. *-icus* > E. *-ic*, belonging to, relating to): a disease that suddenly and extensively affects many people. When an infectious illness causes true epidemics or plagues in cultivated or wild plants, it is called an **epiphytotic**; if the epidemic affects animals it is called an **epizootic**.

epigeal, **epigean**, **epigeic** (L. *epigeus* < Gr. *epí*, upon + *geo*, earth + L. suf. *-alis* > E. *-al*, relating to or belonging to; or + L. suf. *-anus* > E. *-an*, belonging to; or + Gr. suf. *-íkos* > L. *-icus* > E. *-ic*, belonging to, relating to): growing or developing on the surface of the soil. Cf. **hypogeal**.

epigynous (Gr. *epí*, upon + *gyné*, female; here, female sex organ + L. *-osus* > OF. *-ous*, *-eus* > E. *-ous*, having, possessing the qualities of): *Aquatic fungi*. The formation of the antheridium on the oogonium, as in *Monoblepharis polymorpha* of the Monoblepharidales, or the male gametangium on the female gametangium, as in *Allomyces javanicus* (Blastocladiales). Cf. **hypogynous**. See **exogynous**.

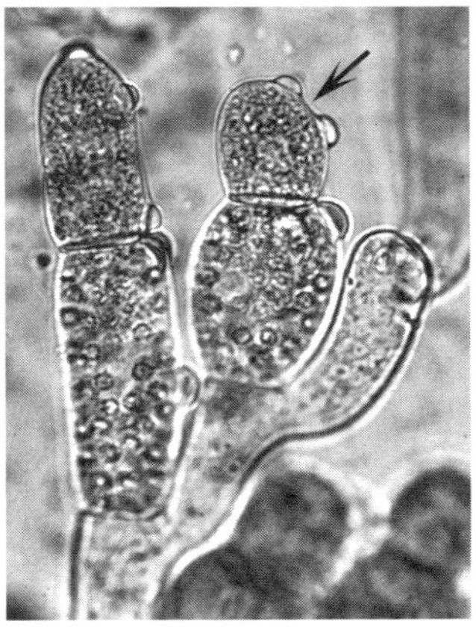

Epigynous male gametangium in a thallus of *Allomyces javanicus*, x 575 (*RTH*).

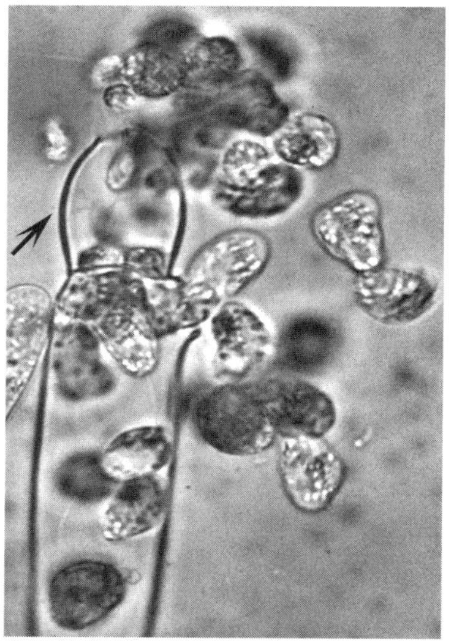

Epigynous male gametangium in a thallus of *Allomyces javanicus*. Male and female gametes are being released through pores in the cell wall of the gametangia, x 575 (*RTH*).

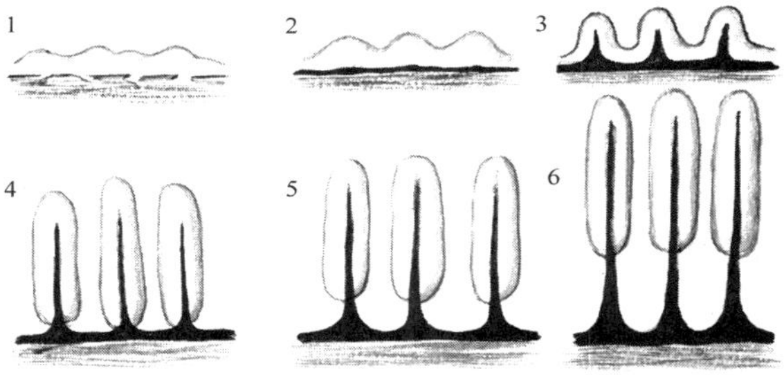

Epihypothallic type of sporangial development in the species of Stemonitales (Myxomycetes). 1. Emerging plasmodium. 2. Formation of subplasmodial hypothallus. 3. Differentiation into primordia ; stalk initial secreted internally as protoplasm mounds up. 4. Protoplasm segregates into discrete primordia, each secreting a stalk. 5. Continuous deposition of material at stalk apex causes stalk elongation and elevates prespore mass of protoplasm. 6. Full-size sporangia prior to capillitial formation ; interior stalk continuous with subplasmodial hypothallus. Peridium, if present, secondary. Diagram based on figures and information published by I. K. Ross in *Mycologia 65* :477-485, 1973, x 10.

epihypothallic (Gr. *epí*, on + L. *hypothallus* < Gr. *hypó*, beneath + *thallós*, offspring, thallus + suf. *-íkos* > L. *-icus* > E. *-ic*, belonging to, relating to): refers to a type of sporocarp development present in Myxomycetes of the order Stemonitales (also known as stemonitoid development), and characterized by the formation of a hypothallus on the lower surface of the plasmodium, which separates it from the substrate. Certain equidistant points on the hypothallus are converted into centers of deposition of materials for the formation of pedicels, each one of which elongates upward, carrying a mass of prespore protoplasm, which continues secreting more pedicelar material at the apex. This differs from subhypothallic development, of the orders Liceales, Trichiales, Echinosteliales and Physarales, in which there is no continuity of the hypothallus with the pedicel and the sporocarp peridium; the hypothallus forms beneath the plasmodium, and any connection between the pedicel and the peridium of the mature sporangium is secondary. It is called epihypothallic development because the sporangium is formed on the hypothallus. Cf. **subhypothallic**.

epikapylic (Gr. *epí*, upon + *kapyleío*, to falsify, adulterate + suf. *-íkos* > L. *-icus* > E. *-ic*, belonging to, relating to): a thallus of a lichenicolous fungus that develops a morphologically distinct lichenized structure. Cf. **endokapylic**.

epilithic (Gr. *epí*, upon + *líthos*, stone + suf. *-íkos* > L. *-icus* > E. *-ic*, belonging to, relating to): growing on rock. Also called **petricolous**, **rupestral** and **saxicolous**.

epiphloeodic, **epiphloedal** (Gr. *epí*, upon + *phloiós*, bark > L. *phloeo*, bark + Gr. *-oeídes*, like + suf. *-íkos* > L. *-icus* > E. *-ic*, belonging to, relating to; or + L. suf. *-alis* > E. *-al*, relating to or belonging to): *Lichens*. A crustaceous lichen that develops on the surface of the bark of trees; e.g., such as *Opegrapha* of the Arthoniales. Cf. **hypophloeodic**.

epiphragm (Gr. *epíphragma*, plug, lid): a round and membranous apical lid that covers the upper part of the peridium of the receptacle or carpophore of the nidulariaceous Gasteromycetes, and which tears at maturity to permit liberation of the peridioles, which contain the basidiospores; e.g., as happens in *Cyathus olla* and *Crucibulum vulgare* (Nidulariales).

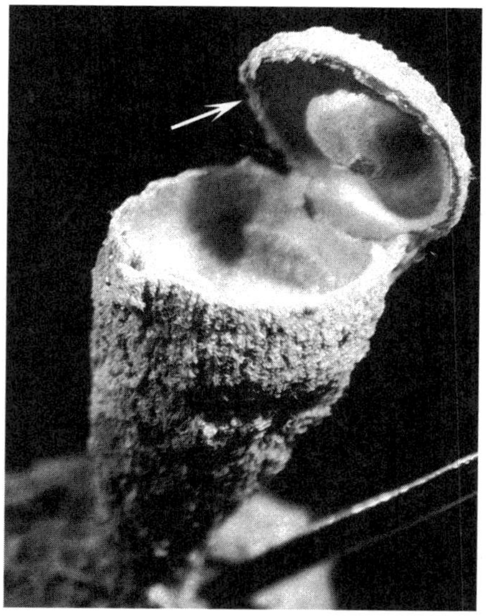

Epiphragm of the fruiting body of *Cyathus* sp., x 10 (*MU*).

Epiphragm of the fruiting body of *Cyathus* sp., x 10 (*MU*).

Epiphragm of the fruiting body of *Cyathus* sp., x 12 (*MU*).

epiphyllous (L. *epiphyllus* < Gr. *epí*, upon + *phýllon*, leaf + L. *-osus* > OF. *-ous*, *-eus* > E. *-ous*, possessing the qualities of): living on, borne on, or developing on, the leaves of plants; e.g., the lichens of the gen. *Strigula* (Chaetothyriales), which live on tropical plants, and *Porina epiphylla* (Trichotheliales) on fern leaves. Also epiphyllous are the apothecia of certain lichens of the gen. *Cladonia* (Lecanorales), which are produced on the leaflets of the primary thallus, instead of being on podetia.

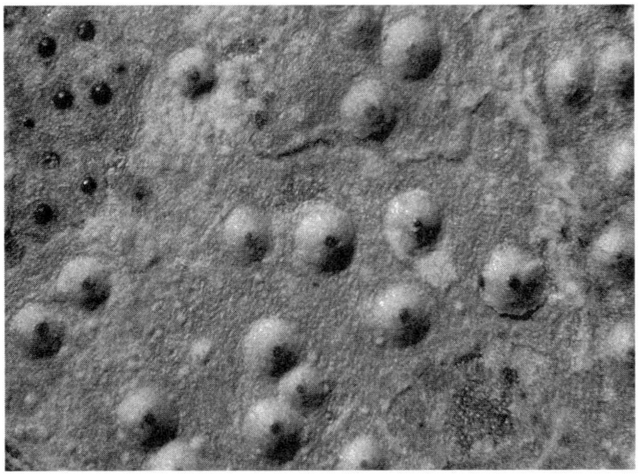

Epiphyllous thallus of *Porina epiphylla* on a fern leaf; the lichen pseudothecia are evident, x 25 (*MU*).

epiphyte (Gr. *epí*, upon + *phytón*, plant): an organism that lives on a plant without obtaining nutrients from it, since the plant only provides support; many lichens are epiphytes.

epiphytotic (Gr. *epí*, upon + *phytón*, plant + *-tikós* > L. *-ticus* > E. *-tic*, relation, fitness, inclination or ability): an infectious disease that suddenly and destructively affects wild or cultivated plants in a determinate locality. Cf. **epidemic** and **epizootic**.

epiplasm (Gr. *epí*, upon + *plásma*, formation, a bland material with which a living being is formed): the portion of the cytoplasm in an ascus that is not incorporated into the ascospores when they are delimited; it may contribute to their nutrition during their development, or condense to form ornamentations on the epispore. Epiplasm can be seen, e.g., in the asci of many Sordariales, such as *Neurospora crassa* and *Sordaria fimicola*, and Rhytismatales, such as *Lophodermella sulcigena*.

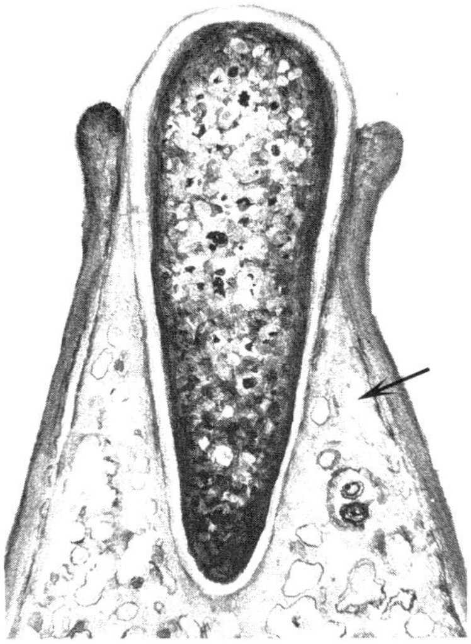

Epiplasm of the ascus apex of *Lophodermella sulcigena*, evident at the time of ascospore release. Drawing based on a transmission electron micrograph published by Beckett *et al.* in their *Atlas of Fungal Ultrastructure*, 1974, x 9 500.

epispore, **episporium**, pl. **episporia** (Gr. *epí*, upon + *sporá*, spore + L. dim. suf. *-ium*): the wall layer of the basidiospore just beneath the **perispore** and the **exospore**. It is the thickest layer of the spore wall, which gives the spores their shape and rigidity, as well as their color and ornamentation. In the basidiomycetes the epispore, which is a continuation of the outer wall of the sterigma and basidium, is bounded on the inside by the **mesospore** and **endospore**, the two innermost wall layers of the most structurally complex basidiospores.

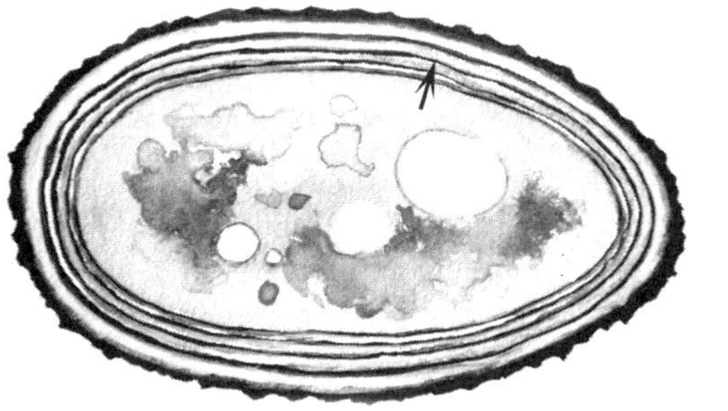

Epispore of the cell wall of the basidiospore of *Tubaria* sp.; the endospore, mesosporium, exospore, and perispore also are shown in this diagram of a longitudinal section, x 5 500.

epithecium, pl. **epithecia** (NL. *epithecium* < Gr. *epí*, upon + *thekíon*, dim. of *thêke*, case, box; here, of the asci): *Discomycetes*. A layer of tissue above the surface of the hymenium of an apothecium, formed by union of the tips of the paraphyses which project from among the asci; as is seen, e.g., in *Peziza* (Pezizales) and in the apothecia of many lichens.

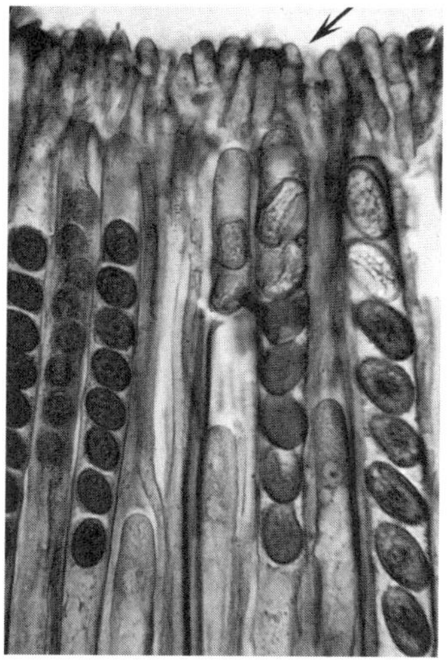

Epithecium of the apothecium of *Peziza* sp., seen in longitudinal section, x 200 (*RTH*).

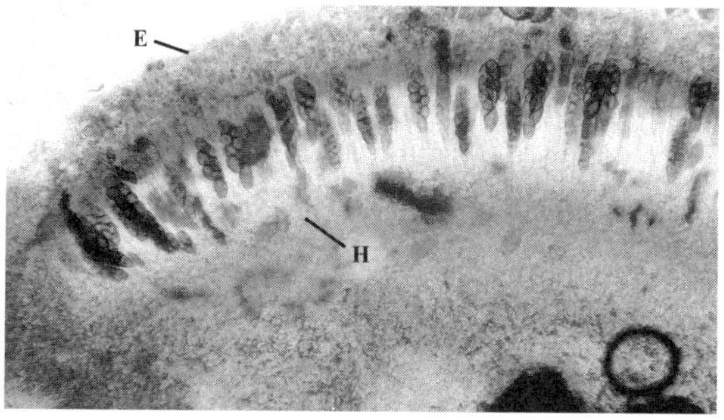

Epithecium (E) and hypothecium (H) of a lecanorine apothecium of a lichen, seen in longitudinal section, x 50 (*RV*).

epithelium, pl. **epithelia** (NL. *epithelium* < Gr. *epí*, upon + *thelé*, nipple + L. dim. suf. *-ium*): *Agarics*. A trichodermis composed of small globose or subglobose elements arranged in one or several layers.

epixylous (Gr. *epí*, upon + *xýlon*, wood + L. *-osus* > OF. *-ous*, *-eus* > E. *-ous*, possessing the qualities of): syn. of **lignicolous**; living or developing on wood; both terms refer only to the presence of the fungus on wood, and not to its nutrition. In the case of fungi that degrade wood, i.e., that take nourishment from it, it is more accurate to use the term **xylophagous**. Many Aphyllophorales are epixylous.

epizoic (Gr. *epí*, upon + *zõon*, animal + suf. *-íkos* > L. *-icus* > E. *-ic*, belonging to, relating to): fungi that develop as commensals, saprobes or parasites on living animals, as is seen, e.g., in *Amoebidium parasiticum* (Trichomycetes, Amoebidiales), which lives on the antennae of crustaceans of the gen. *Daphnia*. Cf. **endozoic**.

epizootic (Gr. *epí*, upon + *zõon*, animal + *-tikós* > L. *-ticus* > E. *-tic*, relation, fitness, inclination or ability): a disease that suddenly and extensively affects many animals. Cf. **epidemic** and **epiphytotic**.

equilateral (LL. *aequilateralis*, of equal sides < *aequus*, equal + *later-*, *latus*, side + suf. *-alis* > E. *-al*, relating to or belonging to): *Of spores*. A spore that is bilaterally symmetrical, i.e., that has two identical sides. Examples of equilateral shapes are globose, dacryoid, ovate, pyriform, amygdaliform, lenticular, cylindrical and bacilliform spores, among others. Cf. **inaequilateral**.

erasure phenomenon (L. *erasus*, ptp. of *eradicare* < *e-*, *ex-*, out + *radere*, to scratch, scrape, delete; LL. *phaenomenon* < Gr. *phainómenon*, appearance < *phaínein*, to show): refers to the myxoamoebae of *Dictyostelium* (Dictyosteliomycota) which lose their capacity to recapitulate when their developing cultures are disaggregated and placed on a culture medium.

ergot (F. *ergot*, ergot, spur): a term applied to the disease of cereals caused by *Claviceps* spp. (Hypocreales), to the fungus in general, and to the sclerotia of the fungus. The intoxication induced in humans and animals by these sclerotia is known as ergotism.

ericoid mycorrhiza, pl. **mycorrhizae** : see **monotropoid mycorrhiza**.

erose, **eroded** (L. *erosus*, eroded < *erodere*, to eat away < *ex*, contracted to *e*, out, beyond + *rodere*, to gnaw + *-osus*, full of, augmented, prone to > ME. *-ose*): having unequal borders, with small teeth or sinuosities, as if nibbled or gnawed; e.g., like the border of the giant colonies of the yeasts

Kluyveromyces marxianus (Saccharomycetales) and *Candida lambica* (cryptococcaceous asexual yeasts), when grown on certain media.

Ergot - sclerotia of *Claviceps lionelium-paspali*, x 4 (*MU & CB*).

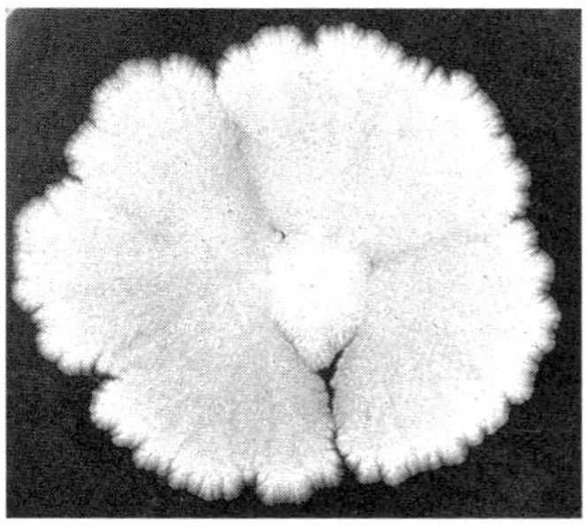

Erose border of a giant colony of *Candida lambica* on agar, x 1 (*MU*).

erotactin (Gr. *éros*, love + *taktós*, a handling, touch + NL. *-in*, suf. used in chemistry to denote an activator or compound): a sperm attractant. See **sirenin**.

erratic (ME. < MF. *erratique* < L. *erraticus* < *erratus*, ptp. of *errare*, to miss, wonder, err + Gr. *-tikós* > L. *-ticus* > E. *-tic*, relation, fitness, inclination or ability): having no fixed course; refers to lichen thalli which are not fixed to the substratum and often blow around, e.g., *Chondropsis semiviridis* and *Aspicilia* (=*Sphaerothallia*) *esculenta* ("manna"), of the Lecanorales.

erubescent (L. *erubescens*, genit. *erubescentis*, becoming red, prp. of *erubescere*, to become red, blush + *-escens*, genit. *-escentis*, that which turns, beginning to, slightly > E. *-escent*): turning reddish. Erubescence is present in some fungi when they are cut or bruised, such as *Amanita rubescens* (Agaricales); in plant pathology it is equivalent to eritrosis.

erumpent (L. *erumpens*, genit. *erumpentis*, breaking out, bursting forth < *erumpere*, to break forth + *-entem* > E. *-ent*, being): syn. of **perrumpent**; an internal sporiferous structure that breaks through the overlying host tissue at maturity to liberate the spores; this happens, e.g., with the acervuli of melanconiaceous asexual fungi (such as *Colletotrichum*), with the perithecial stromata of *Endothia parasitica* (Diaporthales), *Diatrype disciformis* (Xylariales), the ascotroma |*Botryosphaeria ribis* (Dothideales), and with the aecia and uredinia of the Uredinales.

Erumpent perithecial stromata of *Diatrype disciformis*, x 15 (*CB*).

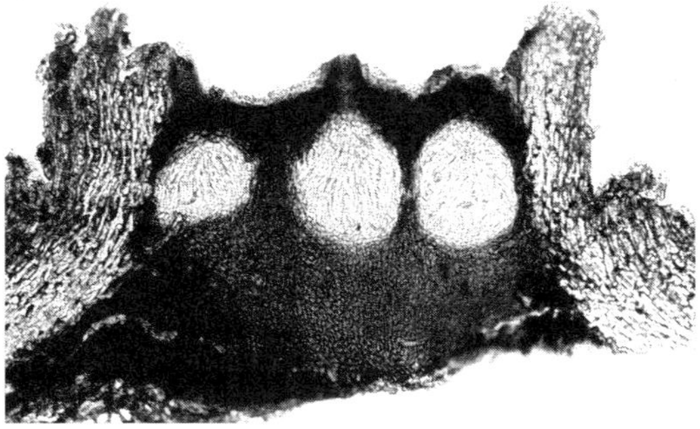

Erumpent ascostroma of *Botryosphaeria ribis*, x 40 (*RTH*).

esculent (L. *esculentus*, edible, good to eat < *esca*, food < *edere*, to eat + suf. *-ulentus*, that abounds in): edible, without causing harmful effects; such as *Morchella esculenta* (Pezizales).

ethnomycology (Gr. *éthnos*, people, race + *mýkes*, fungus + *lógos*, study, treatise + *-y*, E. suf. of concrete nouns): a special branch of science that links mycology with ethnology, and which studies the various aspects related to ethnic groups and the uses they make of fungi, i.e., edible, medicinal, ceremonial, etc.

etiology (ML. *aetiologia*, statement of causes < Gr. *aitíologia* < *aitía*, cause + *lógos*, word, discourse, treatise + *-y*, E. suf. of concrete nouns): cause or origin of a disease or abnormal condition; the science of the causes of disease. Also called **aetiology**.

Euascomycetes (Gr. *eû*, true + *askós*, sack + L. *-mycetes*, ending of class < Gr. *mýkes*, genit. *mýketos*, fungus): in some classifications systems, the Euascomycetes were regarded as a class of fungi with well developed, septate mycelium and unitunicate asci produced in some type of fruiting body or ascocarp (cleistothecium, perithecium or apothecium) formed by the ascohymenial pattern of ascocarp development. It included the subclasses Plectomycetidae, Pyrenomycetidae and Discomycetidae, which together with the Hemiascomycetes, Laboulbeniomycetes and Loculoascomycetes, used to compraise the subdivision Ascomycotina. In the classification followed in this dictionary, the subclasses are recognized as classes (Plectomycetes, Pyrenomycetes and Discomycetes), the Hemiascomycetes now correspond to the Archiascomycetes and Saccharomycetes; the Laboulbeniomycetes are only regarded as the order Laboulbeniales, and the Loculoascomycetes remain as such.

eucarpic (Gr. *eû*, true + *karpós*, fruit + suf. *-íkos* > L. *-icus* > E. *-ic*, belonging to, relating to): an organism in which only one part of the thallus is transformed into reproductive organs, and the rest of the thallus itself continues to carry out its somatic functions; the great majority of the fungi are eucarpic, and are said to reproduce by **eucarpy**. In general they are considered more evolved than the **holocarpic** species, which reproduce by **holocarpy**.

eucarpy (Gr. *eû*, true + *karpós*, fruit + *-y*, E. suf. of concrete nouns): the condition of being **eucarpic**.

eucaryotic, **eucaryontic** (Gr. *eû*, true + *káryon*, nut, nucleus + *-tikós* > L. *-ticus* > E. *-tic*, relation, fitness, inclination or ability): see **eukaryotic**.

euform (Gr. *eû*, true + L. *forma*, form): see **auteuform**.

eugonic (Gr. *eû*, true + *gónos*, generation, offspring + suf. *-íkos* > L. *-icus* > E. *-ic*, belonging to, relating to): a dermatophyte strain which grows faster in culture, often with more aerial mycelium, than an abnormal (**dysgonic**) strain.

eugymnohymenial (Gr. *eû*, true + *gymnós*, naked + L. *hymenium* < Gr. *hyménion*, dim. of *hymén*, membrane + L. suf. *-alis* > E. *-al*, relating to or belonging to): a type of development in some Discomycetes in which the developing ascocarp is completely open from the beginning; the ascogonium is never overarched by the investing hyphae that continue to grow upward as a palisade layer and develop into paraphyses. Some species develop an excipulum, but many lack an excipulum entirely. Cf. **cleistohymenial** and **paragymnohymenial**.

eukaryotic, **eukaryontic** (Gr. *eû*, true + *káryon*, nut, nucleus + *-tikós* > L. *-ticus* > E. *-tic*, relation, fitness, inclination or ability): an organism whose cells possess true nuclei (surrounded by a perforated double membrane), mitochondria, endoplasmic reticulum, dictyosomes, vacuoles and ribosomes, among other characteristics, i.e., with a level of cellular organization more complex than that of a prokaryotic organism. Only the organisms belonging to the kingdom Monera are prokaryotic; those of the kingdoms Protista, Stramenopila, Fungi, Plantae and Animalia are eukaryotic. Cf. **prokaryotic**.

Eumycota (Gr. *eû*, true + L. *-mycota*, ending of division or phylum < Gr. *mýkes*, genit. *mýketos*, fungus): the division (phylum) of the kingdom Fungi that contains the true fungi. In some classifications the Eumycota, together with the Myxomycota, constitute the kingdom Fungi. In the classification adopted in this dictionary, the Eumycota has been replaced by the various phyla that constitute the kingdom Fungi (Chytridiomycota, Zygomycota, Ascomycota and Basidiomycota). The other phyla, which are now considered separated from the Fungi belong to the kingdom Stramenopila (Oomycota, Hyphochytriomycota and Labyrinthulomycota) and to the kingdom Protista (Plasmodiophoromycota, Dictyosteliomycota, Acrasiomycota and Myxomycota).

eurychoric (Gr. *eurýs*, broad, wide + *chóra*, country + suf. *-íkos* > L. *-icus* > E. *-ic*, belonging to, relating to): a cosmopolitan species, i.e., one that is widely distributed throughout the world. Cf. **stenochora**.

euryoecious (Gr. *eurýs*, broad, wide area + *oîkos*, house + L. *-osus* > OF. *-ous*, *-eus* > E. *-ous*, possessing the qualities of): applied to an organism found in the most diverse ecological conditions, such as the common molds of the gen. *Aspergillus*, *Penicillium* and *Cladosporium*, among other moniliaceous and dematiaceous asexual fungi. Cf. **stenoecious**.

euryxenous (Gr. *eurýs*, broad, wide area + *xénos*, stranger, host + L. *-osus* > OF. *-ous*, *-eus* > E. *-ous*, possessing the qualities of): an organism capable of parasitizing a more or less large number of hosts; among the fungi, e.g., the moniliaceous asexual fungus *Phymatotrichopsis omnivora* has the potential of parasitizing more than 2 000 species of dicotyledoneous plants. Cf. **stenoxenous**.

euseptate (Gr. *eû*, true + L. *septatus*, having partitions, septa < *septum*, wall + suf. *-atus* > E. *-ate*, provided with or likeness): a type of multicellular conidium in which the cells that compose it are separated by true septa, which consist of diaphragms that merge peripherally with the lateral wall of the conidium. Cf. **distoseptate**.

eustroma, pl. **eustromata** (Gr. *eû*, true + *strõma*, bed): a stroma that consists only of fungal tissue, as occurs in the majority of species of stromatic fungi. Cf. **pseudostroma**.

euthecium, pl. **euthecia** (NL. *euthecium* < Gr. *eû*, true + *thekíon*, dim. of *thêke*, box, case; here, of the asci): an ascoma (**cleistothecium**, **perithecium**, **apothecium**) of an euascomycete. Cf. **pseudothecium**.

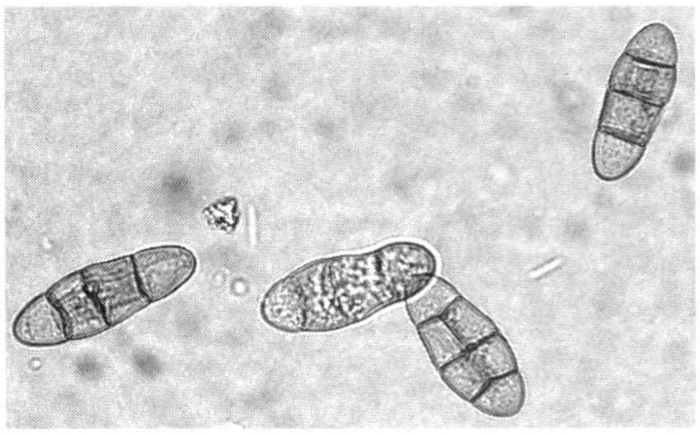

Euseptate ascospores of *Pyrenophora* sp., x 270 (*CB*).

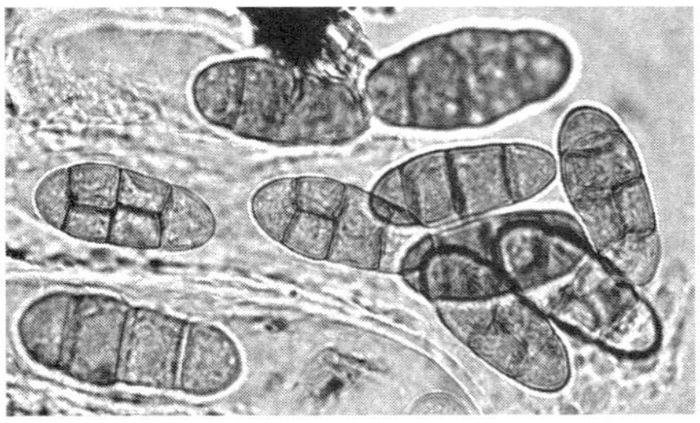

Euseptate ascospores of *Pyrenophora* sp., x 270 (*CB*).

eutypoid (< gen. *Eutypa* < Gr. *eû*, true + *typé*, *týpos*, mark, scar + L. suf. *-oide* < Gr. *-oeídes*, similar to): a fungus or sporiferous apparatus that has the perithecia embedded in an effuse endostroma; the perithecial necks project from the ectostroma without converging toward a central point, as happens in Pyrenomycetes of the gen. *Eutypa* (Xylariales). This type of stroma generally is formed on decorticated branches of many trees. Cf. **diatrypoid** and **valsoid**.

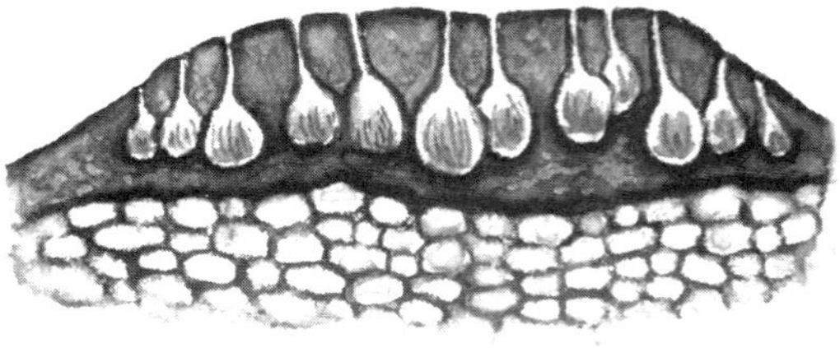

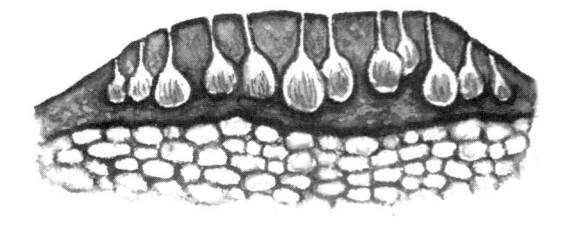

Eutypoid perithecial stroma of *Eutypa armeniacae* on host tissue, drawn as seen in longitudinal section, x 5.

evanescent (L. *evanescens*, genit. *evanescentis*, withering away, loosing strength, shrivelling < *evanescere*, to wither, shrink + *-escens*, genit. *-escentis*, that which turns, beginning to, slightly > E. *-escent*): ephemeral or of short duration; e.g., like the asci of the fungi of the gen. *Chaetomium* (Sordariales), or the thin, fragile exoperidium of *Gautieria* (Gautieriales).

eversion (L. *evertion* < *evertere* < *ex-*, out + *vertere*, to turn + *-io*, *-ionis* > E. *-ion*, state, result of): the act of turning inside out, like the membrane found at the base of the sporiferous sack of *Sphaerobolus* (Nidulariales), which is concave but suddenly becomes convex at the moment of discharging the sack into the air.

everted (L. *evertus* < *evertere*, turn inside out, reverse): with the interior thrown outward.

ex (Gr. *ex*, out of, from): in citations, e.g., *Clathrus* Micheli *ex* Persoon, from; first validly published by the second author, but based on specimens or materials of the first author.

ex situ (L. *ex situ* < *ex*, out of, from; *situ*, site, place): an organism taken from its natural habitat, as with living cultures isolated from nature and maintained in culture collections. The term is also applied to non-viable material held in reference collections. Cf. **in situ**.

exannulate (L. *e*, *ex*, without + *annulus*, ring + suf. *-atus* > E. *-ate*, provided with or likeness): lacking a ring; like the stipe of *Russula brunneoviolacea* (Agaricales). Cf. **annulate**.

excipulum, pl. **excipula** (L. *excipulum*, receptacle): generally it is a layer of hyphae on which rests the asci or basidia of the hymenium (hymenophore); in the apothecium of the Discomycetes the excipulum comprises most of the tissue of the apothecium; e.g., in *Ascobolus* and *Peziza* (Pezizales). In the lichens the excipulum is sometimes referred to as the **hypothecium**, other times as the **parathecium** or **amphithecium**, but always to the parts that surround the **thecium** (hymenium or hymenial layer); the thalline excipulum of the discolichens (such as *Physcia*, of the Lecanorales) is the basal zone of the apothecium on which rests the hypothecium. The so-called marginal excipulum or proper excipulum corrresponds to the parathecium. The term excipulum also is applied to the hyphal tissue of the pycnidial wall of the sphaeropsidaceous asexual fungi.

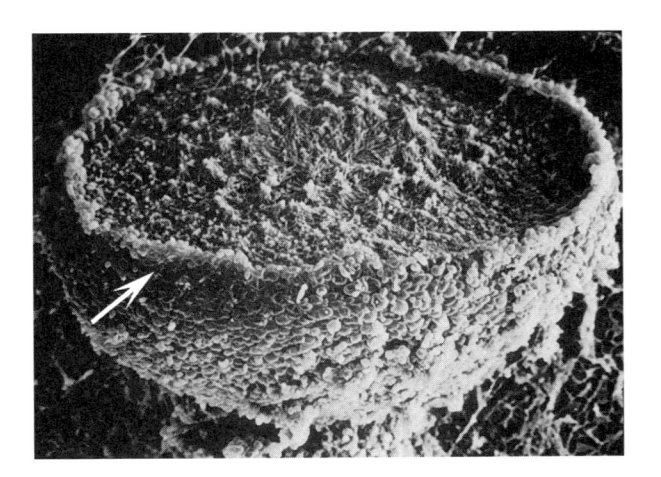

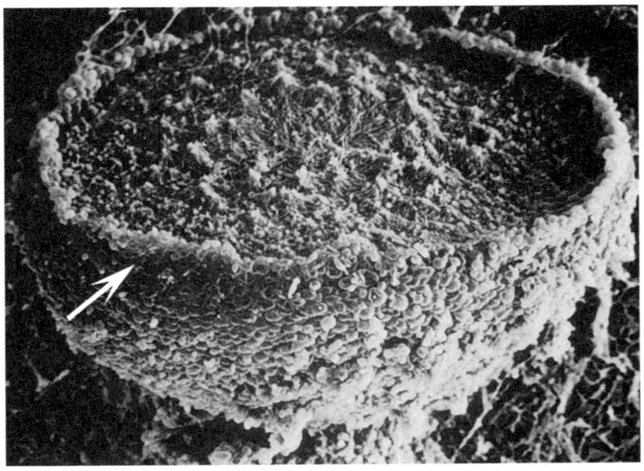

Excipulum of the apothecium of *Ascobolus furfuraceus*, x 25 (*RTH*).

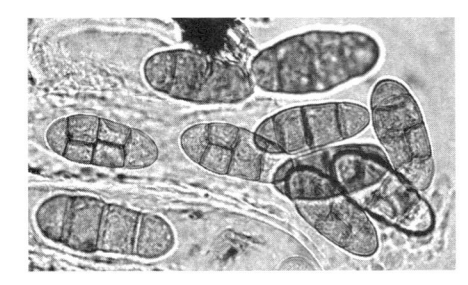

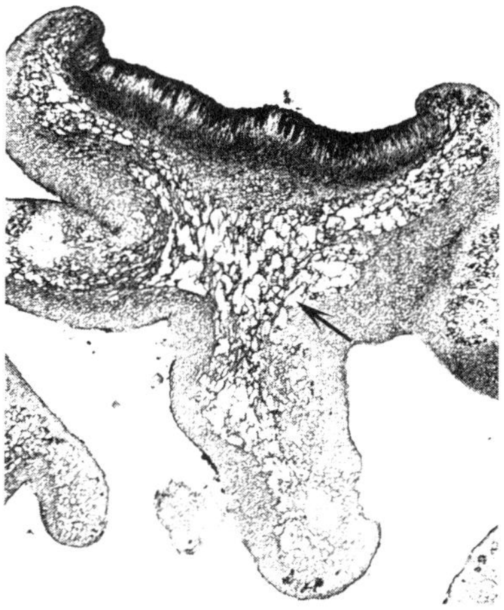

Excipulum of the apothecium of *Physcia* sp., seen in longitudinal section, x 65 (*RTH*).

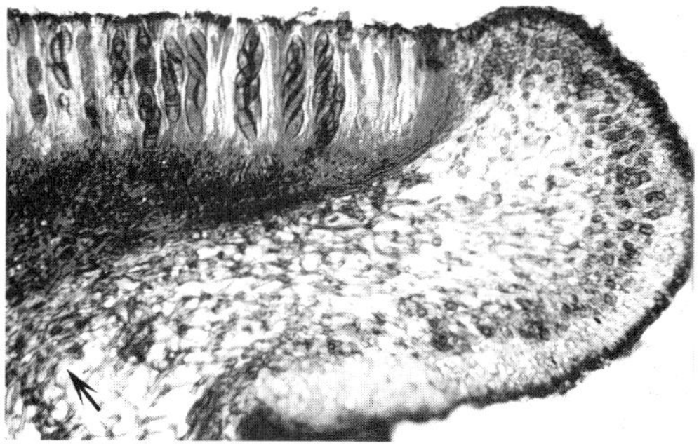

Excipulum of the apothecium of *Physcia* sp., seen in longitudinal section, x 130 (*RTH*).

exigynous (NL. *exigynus* < L. *ex*, out of, from + Gr. *gyné*, woman; here, female sex organ + L. *-osus* > OF. *-ous*, *-eus* > E. *-ous*, having, possessing the qualities of): *Aquatic fungi*. Having the antheridial stalk arising directly from the oogonial cell above the basal septum, as in *Monoblepharis polymorpha* (Monoblepharidales). See **epigynous** and **hypogynous**. Cf. **exogynous**.

exit tube (L. *exitus*, ptp. of *exire*, to go out or away; L. *tubus*, tube): an extension of the sporangium, produced prior to or during sporangial discharge, through which the zoospores are released outside the host or substrate, as in *Olpidium* (Spizellomycetales).

exogenous (Gr. *éxo*, outside, to the exterior + *génos*, origin < *gennáo*, to generate, produce + L. *-osus* > OF. *-ous*, *-eus* > E. *-ous*, having, possessing the qualities of): any structure produced or developed on the exterior of another structure; e.g., the spores of the Myxomycetes of the order Ceratiomyxales are exogenous, as are the majority of the conidia of the asexual fungi. Cf. **endogenous**.

exogynous (NL. *exogynus* < Gr. *éxo*, outside, to the exterior + *gyné*, woman; here, female sex organ + L. *-osus* > OF. *-ous*, *-eus* > E. *-ous*, having, possessing the qualities of): *Aquatic fungi*. The zygote of *Monoblepharis* (Monoblepharidales) which migrates from the interior of the oogonium and becomes situated on top of it, remaining at rest until maturity, when it is transformed into an oospore. Cf. **exigynous**.

Exogynous oospore of *Monoblepharis polymorpha*, x 500.

exolete (L. *exoletus* < *exolescere*, to stop growing, degenerate, fall into disuse): applied to pycnidia, perithecia and other sporophores that on maturing remain empty and unused once they have liberated their spores.

exomycology (Gr. *éxo*, outside + *mýkes*, fungus + *lógos*, study, treatise + *-y*, E. suf. of concrete nouns): mycology of outer space.

exon (E. *ex*pressed sequence + *-on*, in contact with or supported by): a term used in molecular biology to refer to a sequence of nucleotides in DNA or RNA that is expressed as all or part of the polypeptide chain of a protein. Cf. **intron**.

exoperidium, pl. **exoperidia** (Gr. *éxo*, outside, to the exterior + *péridion*, a small leather purse): the outer wall of a multi-walled peridium of a basidiocarp, as in *Geastrum* and *Disciseda* (Lycoperdales), and *Astraeus* (Sclerodermatales); the internal layer is the **endoperidium**.

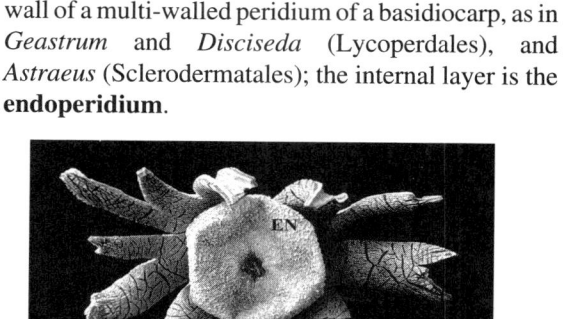

Exoperidium (EX) of *Astraeus hygrometricus* after the stellate dehiscence; the endoperidium (EN) contains the basidiospores, x 1 (*MU*).

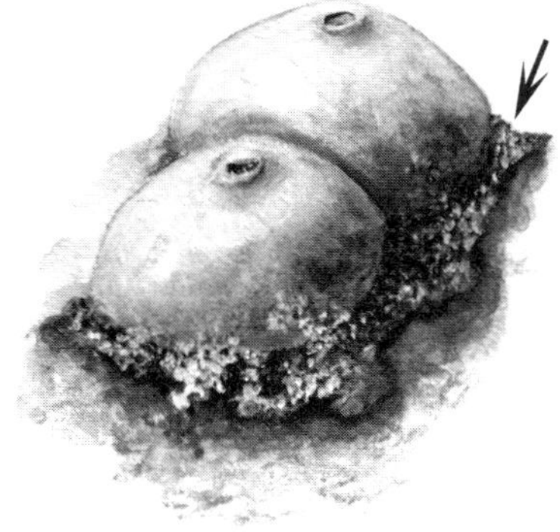

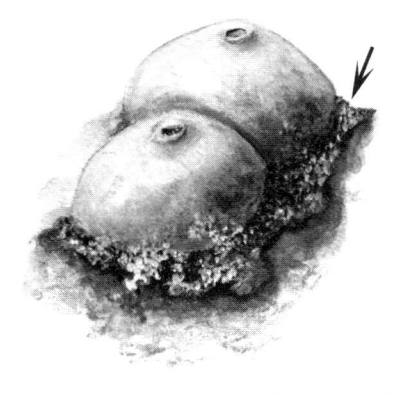

Exoperidium of the fruiting body of *Disciseda candida*, x 1.

exospore, **exosporium**, pl. **exosporia** (Gr. *éxo*, outside, to the exterior + *sporá*, spore + L. dim. suf. *-ium*): the second layer, beneath the **perispore**, in the wall of the most complex basidiospore, which is formed of five layers. It becomes the outer layer of the wall after the perispore disappears and usually forms the spore ornamentation. The other three layers, **epispore**, **mesospore** and **endospore** are beneath the exospore, which generally lacks pigment but frequently can be distinguished chemically from the other adjacent layers.

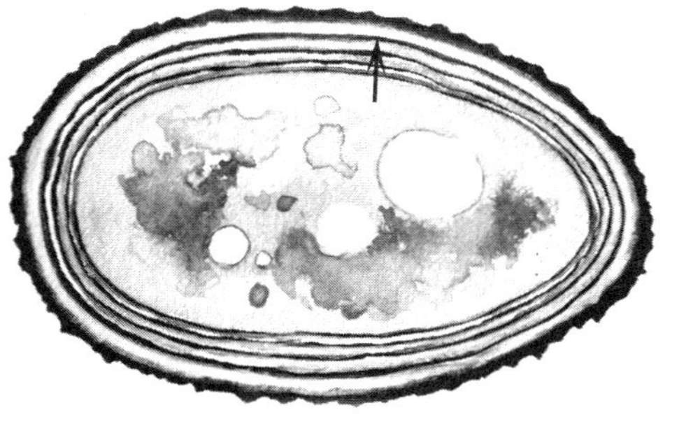

Exospore of the cell wall of the basidiospore of *Tubaria* sp. The endospore, mesosporium, epispore, and perispore also are shown in this diagram of a longitudinal section, x 5 500.

exotic, **exotical** (Gr. *exotikós*, from outside < *éxo*, out, outside + *-tikós* > L. *-ticus* > E. *-tic*, relation, fitness, inclination or ability; or + L. suf. *-alis* > E. *-al*, relating to or belonging to): see **allochthonous**. Cf. **autochthonous**.

expallant (L. *ex*, out of, from + ME. *pallen*, short for *appallen*, to become pale + suf. *-ant* > E. *-ant*, one that performs, being): becoming pale on drying, as in the pileus of *Laccaria* (Agaricales).

expersate (L. *ex*, out of, from + *sparsus*, spread out < *spargere*, to scatter + suf. *-atus* > E. *-ate*, provided with or likeness): refers to oospores of Saprolegniaceae that have one large refractive body surrounded by a homogeneous cytoplasm.

explosive (E. relating to, characterized by, or operated by explosion; tending to explode < L. *explosion*, act of driving off by clapping + *-ivus* > E. *-ive*, quality or tendency, fitness): a type of ascus dehiscence that allows the rapid, violent discharge of ascospores, as in *Sordaria* (Sordariales).

exserted (L. *exsertus*, thrown outward, uncovered): refers to asci that extend out above the epithecium of the ascocarp, since their tips project in order to discharge the ascospores, e.g., as happens in *Ascobolus michaudii*, *A. furfuraceus* and *Saccobolus truncatus* (Pezizales).

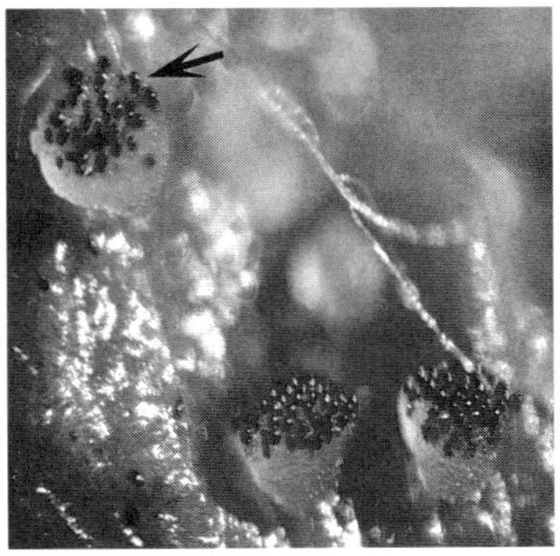

Exserted asci of the apothecium of *Saccobolus truncatus*, x 37 (*EAA*).

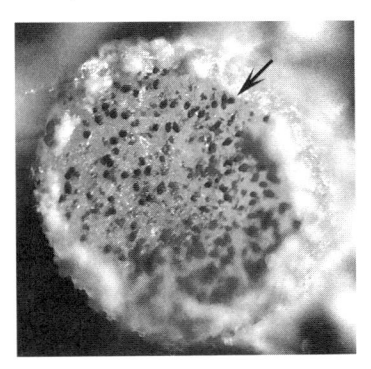

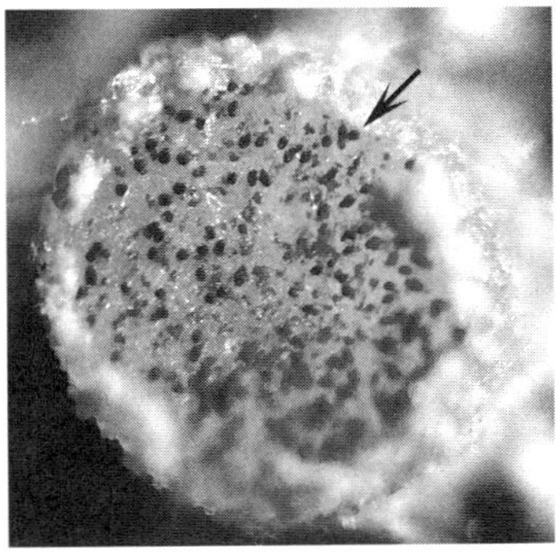

Exserted asci of the apothecium of *Ascobolus michaudii*, x 20 (*EAA*).

Exserted asci of *Ascobolous* sp. protruding above surface of apothecium, x 100.

exsiccatus, pl. **exsiccata**, **exsiccatae**, **exsiccati** (L. *exsiccare*, to dessicate + *-atus*, provided with): applied to an herbarium specimen properly dried and labelled. Exsiccata refers to a collection of plant specimens, and exsiccati is utilized for a collection of fungi.

extramatrical (L. *extra*, beyond, on the outside + *matrix*, matrix, womb; often used in the sense of a place where anything is generated + L. suf. *-alis* > E. *-al*, relating to or belonging to): outside the matrix or substrate, such as the extramatrical filaments of an epibiotic fungus, which grows on the surface of the substrate. Cf. **intramatrical**.

extrorse (L. *extrorsus* < *extra*, outside + *versus*, toward): arranged toward the outside, the exterior, or the edge; e.g., like the pileus margin in *Coprinus* (Agaricales). Cf. **introrse**.

exudate (L. *exudare* < *ex-*, out + *sudare*, to sweat + suf. *-atus* > E. *-ate*, provided with or likeness): any liquid or fluid that, through exudation, oozes out of the hyphae of many molds, accummulating in droplets of diverse shades according to the species and conditions of the medium on which it is growing. The characteristics of the exudate are taken into account in species determination of isolates of *Penicillium*, *Aspergillus* and other gen. of moniliaceous asexual fungi.

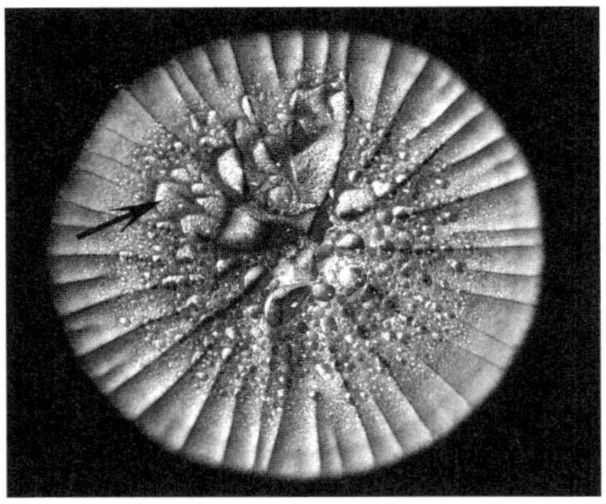

Exudate of a colony of *Aspergillus candidus* growing on agar, x 1 (*MU*).

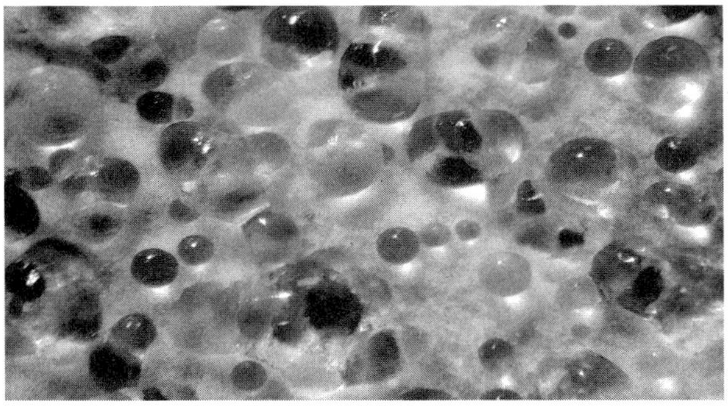

Exudate of a colony of *Penicillium aurantiogriseum* on agar, x 7.5 (*MU*).

f

fabiform (L. *faba*, broad bean + *-formis* < *forma*, shape): with the shape of a broadbean, e.g., the ascospore of *Petriella* sp. (Microascales). See **reniform**.

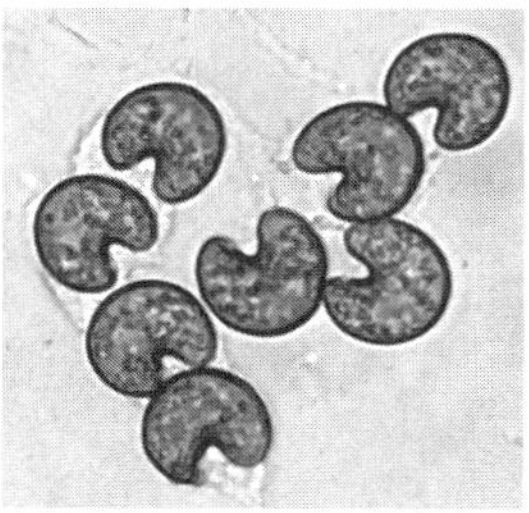

Fabiform ascospores of *Petriella* sp., x 1 450 (*CB*).

facultative (NL. *facultativus* < *facultas*, capacity, ability + *-ivus* > E. *-ive*, quality or tendency, fitness): the ability to live under more than one set of environmental conditions; e.g., fungi that can live saprobically or as parasites (facultative parasites), or microorganisms that can live in a medium without air (oxygen), commonly being aerobic (facultative aerobes), or vice versa, if they tolerate the presence of air, when they normally live without it (facultative anaerobes).

facultative parasite (L. *facultas*, capacity, ability + *-ivus* > E. *-ive*, quality or tendency, fitness; Gr. *parásitos* < *pará*, close, at the side of + *sîtos*, food): an organism that has the capacity to invade and live on another living organism, or to live on dead organic material, depending upon the circumstances; in the fungi, this capacity is restricted to a smaller number of species than the opposite condition (**facultative saprobe**). Also see **obligate parasite**.

facultative saprobe (NL. *facultativus* < L. *facultas*, ability + *-ivus* > E. *-ive*, quality or tendency, fitness; Gr. *saprós*, rotten + *bíos*, life): an organism that has the capacity to live on dead organic material or of infecting another living organism, according to the circumstances; the majority of the fungi have this ability. See **facultative parasite** and **obligate parasite**.

fairy ring (ME. *fairie*, fairyland, fairy people < OF. *faerie* < *feie*, *fee*, fairy; ME. < OE. *hring*, ring): a ring of fungus fruiting bodies, generally of basidiomycetes, produced at the periphery of a body of mycelium which has grown centrifugally from an initial growth point, especially the mushroom *Marasmius oreades* (Agaricales). Some 60 species have been recorded as frequent formers of rings in grass and grassland, and not uncommon in woods.

Fairy ring of mushrooms on a lawn, x 0.02 (*BL*).

falcate (L. *falcatus* < *falx*, genit. *falcis*, sickle + suf. *-atus* > E. *-ate*, provided with or likeness): with a flat, curved shape, like a sickle; e.g., like the conidia of *Harposporium* (moniliaceous asexual fungi) and the macroconidia of *Fusarium oxysporum* (tuberculariaceous asexual fungi). Also called **falciform** and **lunate**.

falciform (L. *falciformis* < *falx*, genit. *falcis*, sickle + *-formis* < *forma*, shape): see **falcate** and **lunate**.

falciphore (L. *falx*, genit. *falcis*, sickle + *-phóros*, bearer): a hypha bearing a sickle, i. e., a sickle-shaped conidiogenous cell, characteristic of the gen. *Zygosporium* (moniliaceous asexual fungi).

farctate (L. *farctum*, *farctus*, stuffed + suf. *-atus* > E. *-ate*, provided with or likeness): turgent, solid; used for the stipe of a basidiocarp that is not hollow, but filled with more or less compact hyphae. Also called **solid**. Cf. **cavernose** and **fistulose**.

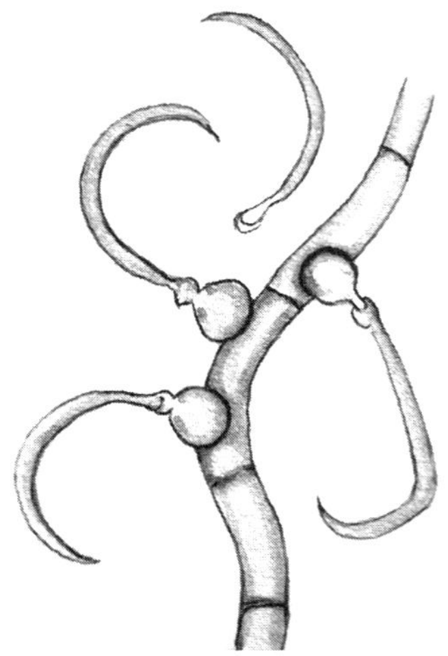

Falcate conidia of *Harposporium*, borne on ampulliform conidiogenous cells, x 1 170.

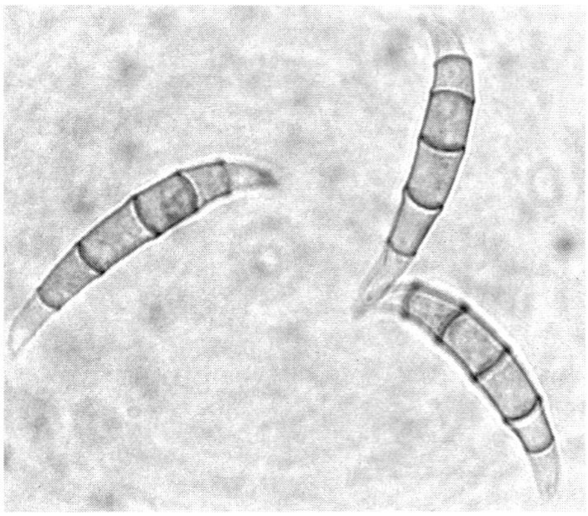

Falcate conidia of *Fusarium*, x 840 (*CB*).

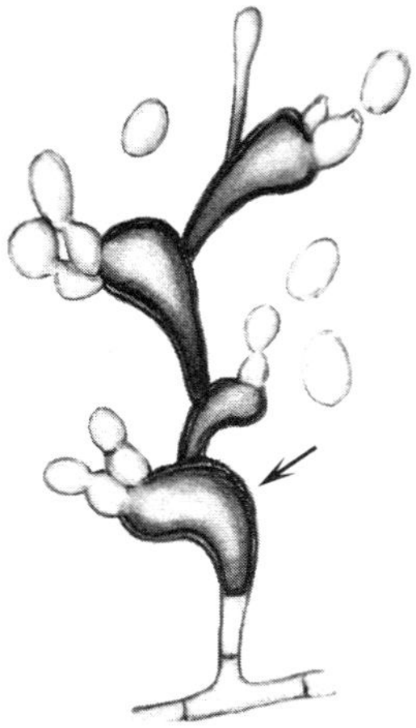

Falciphores of the conidiophore of *Zygosporium masonii*, x 790.

farinaceous, **farinose** (L. *farinaceus*, similar to meal in appearance or color, mealy < L. *farina*, flour, meal + suf. *-aceus* > E. *-aceous*, of or pertaining to, with the nature of; LL. *farinosus*, mealy < L. *farina* + *-osus*, full of, augmented, prone to > ME. *-ose*): covered with a mealy powder; the surface of the colonies of some fungi on agar have a farinaceous texture; e.g., certain dermatophytes (such as *Microsporum* and *Trichophyton*) and other moniliaceous asexual fungi (such as *Geotrichum*).

fasciate, **fasciated** (L. *fasciatus*, bundled < *fascia*, bundle, band, ribbon, sash + suf. *-atus* > E. *-ate*, provided with or likeness): **1**. Concrescent, growing together, as is seen in some fungi whose stipes or pilei grow united. **2**. Having wide, band or ribbon-like structures, like the thallus of certain fruticose lichens whose branches are flat (e.g., *Ramalina leptocarpa* and *Cetraria islandica* of the Lecanorales).

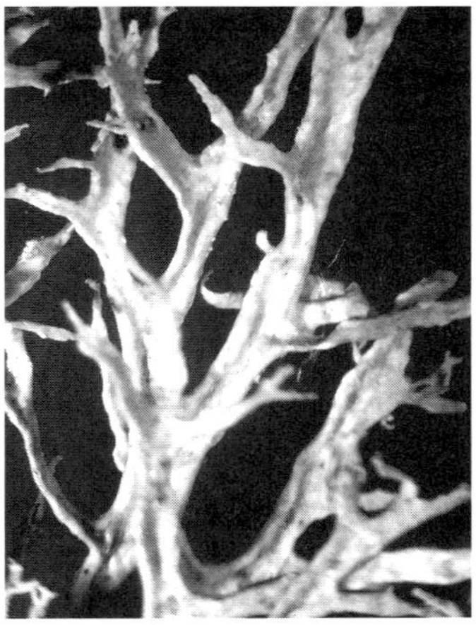

Fasciate branches of the thallus of *Ramalina homalea*, x 2 (*MU*).

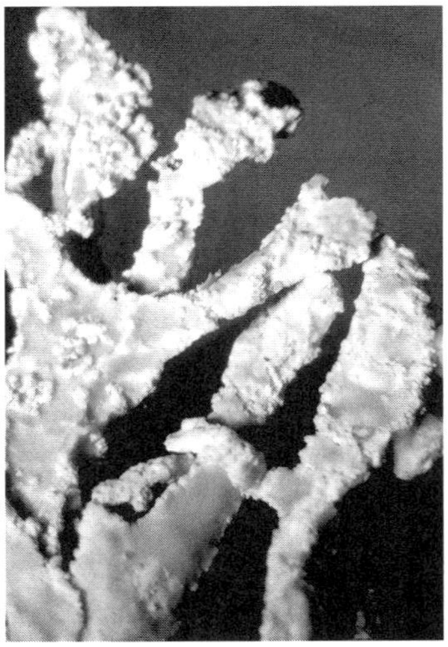

Fasciate branches of the thallus of *Ramalina homalea*, x 2 (*MU*).

fascicle (L. *fasciculus*, dim. of *fascis*, bundle, raceme, faggot): a small cluster, as of hyphae, asci, conidiophores, etc.

fasciculate (NL. *fasciculatus*, arranged in small bundles < *fasciculus*, dim. of *fascis*, bundle, raceme, faggot + suf. *-atus* > E. *-ate*, provided with or likeness): grouped in bundles; e.g., like the sporangia of *Stemonitis* (Stemonitales), the hyphae of *Oedocephalum*, *Acremonium*, some species of *Penicillium* (moniliaceous asexual fungi), and *Stachybotrys* (dematiaceous asexual fungi); the conidiophores of *Cercospora* (also dematiaceous) are fasciculate too.

Fasciculate sporangia of *Stemonitis axifera*, x 4.5 (*MU*).

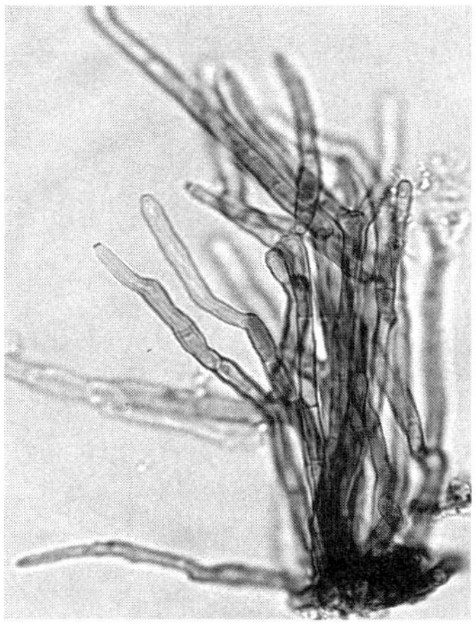

Fasciculate conidiophores of *Cercospora coffeicola*, x 750 (*CB*).

fastigiate, **fastigiated** (NL. *fastigiatus*, that which ends in a point < L. *fastigium*, narrowing toward the top, having upright usually clustered branches + suf. *-atus* > E. *-ate*, provided with or likeness): **1**. Adhering structures that taper to a point; e.g., hyphae, conidiophores, etc., whose branches, pedicels, etc., end in a point. **2**. *Lichens*. A cortical zone whose hyphae are perpendicular to the main axis of the thallus. Cf. **fibrous**.

faveolate (L. *faveolatus*, honey-combed < *faveolus*, dim. of *favus*, honeycomb + suf. *-atus* > E. *-ate*, provided with or likeness): syn. of **alveolate**; with honeycomb-like cavities or cells; e.g., like the hymenophore of the basidiomycete *Hexagonia tenuis* (Aphyllophorales).

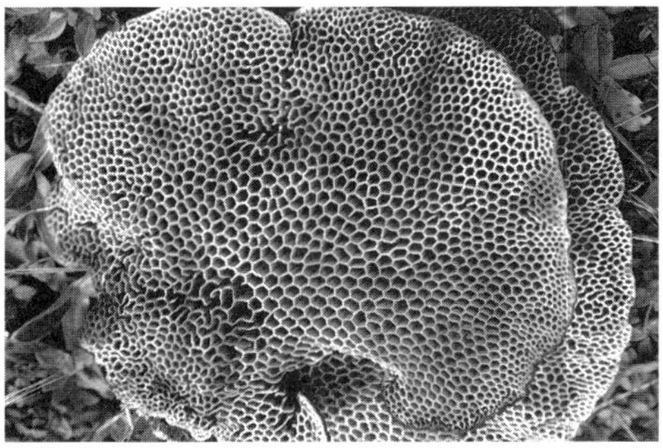

Faveolate hymenophore of *Hexagonia tenuis*, x 0.7 (*CB*).

Faveolate hymenophore of *Hexagonia papyracea*, x 10 (*MU*).

favus, pl. **favi** (L. *favus*, honeycomb): syn. of *tinea favosa*; *Med. Mycol*. A clinical entity characterized by the formation of a dense mass of mycelium and arthrospores which originate in the hair follicules and which take on the shape of a rhomb or inverted cone; this type of tinea in man and animals is caused by *Trichophyton shoenleinii*, *T. violaceum* and *Microsporum gypseum* (moniliaceous asexual fungi). See **tinea**.

fenestrate (L. *fenestratus*, that which has windows < *fenestra*, window + suf. *-atus* > E. *-ate*, provided with or likeness): provided with symmetrically arranged perforations or gaps; e.g., like the indusium of *Dictyophora* (Phallales).

fermentation (L. *fermentatio*, genit. *fermentationis* < *fermentare*, to undergo fermentation < *fermentum*, yeast, enzyme + *-ationem*, action, state or condition, or result > E. suf. *-ation*): a chemical change with effervescence (if the fermented substrate is liquid); an enzymatically controlled anaerobic breakdown of an energy-rich compound (such as a carbohydrate to carbon dioxide and alcohol or to an organic acid).

Broadly, an enzymatically controlled transformation of an organic compound, as seen in the elaboration of beer, bread, and other fermented products made with the intervention of yeasts, as well as of antibiotics, acids, steroids, and a wide assemblage of chemicals obtained by means of molds and other microorganisms used in the food-processing industry and the pharmaceutical/medical industry. See **zymosis**.

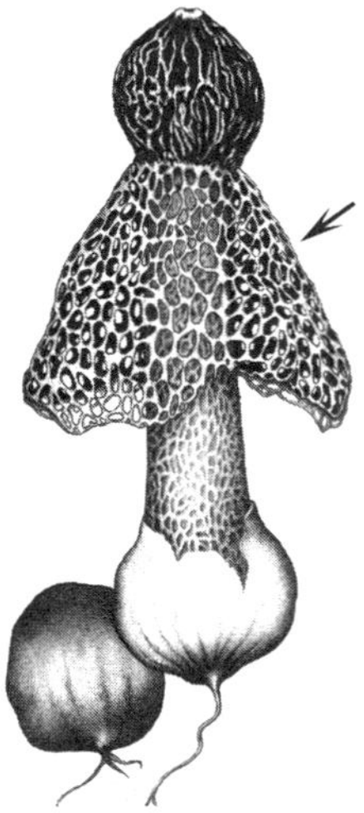

Fenestrate indusium of the fruiting body of *Dictyophora duplicata*, x 0.3.

fertilization (L. *fertilis*, fertile < *ferre*, to bear + *-ationem*, action, state or condition, or result > E. suf. *-ation*): the action and effect of fertilizating; the fusion of gametes.

fertilization tube (L. *fertilis*, fertile + *-ationem*, action, state or condition, or result > E. suf. *-ation*; *tubus*, tube): a tube that originates from the male gametangium (antheridium) and penetrates the female (oogonium) and through which the male gametic nucleus migrates to unite with the female to produce the zygote nucleus, as occurs, e.g., in *Saprolegnia* (Saprolegniales).

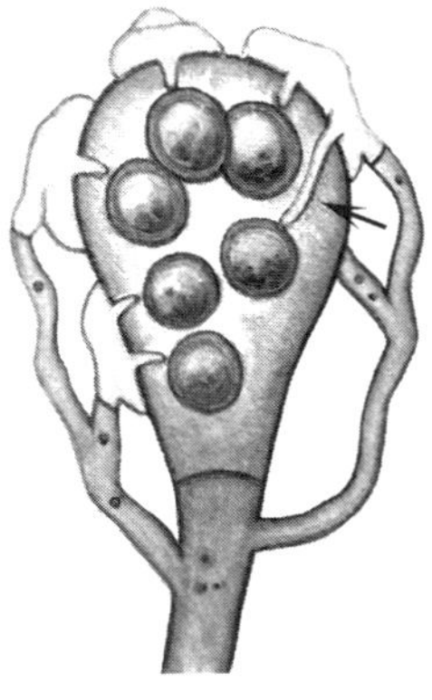

Fertilization tubes of the antheridia of *Saprolegnia parasitica*, x 260.

festooned (L. *festa*, party, wavy adornment at the edge of something): syn. of **crenate**; i.e., edged with scallops or indentations; e.g., like the gill margin of *Cortinarius alboviolaceus*, or like the margin of the pileus of *Laccaria laccata* (Agaricales).

fibercle (F. *fibre* < L. *fibra*, fiber, filament, thread + *-cle*, ending of *tubercle* < L. *tuberculum*, dim. of *tuber*, lump, prominence): any of the more or less protuberant scars left on the branches of some lichen thalli after the fibrils have broken; e.g., as in *Usnea nashii* (Lecanorales).

fibril (NL. *fibrilla*, dim. of *fibra*, fiber, filament, thread): a very thin fiber; **1.** *Lichens*. A short, simple branchlet, containing both the mycobiont and phycobiont, that arises perpendicularly from the main branch of the fruticose thallus; e.g., as in *Usnea* (Lecanorales), and which can serve as a vegetative propagule. **2.** *Gasteromycetes*. Exoperidial ornamentation in the form of a thin, hair-like structure; typical of *Lycoperdon* and *Calvatia* (Lycoperdales).

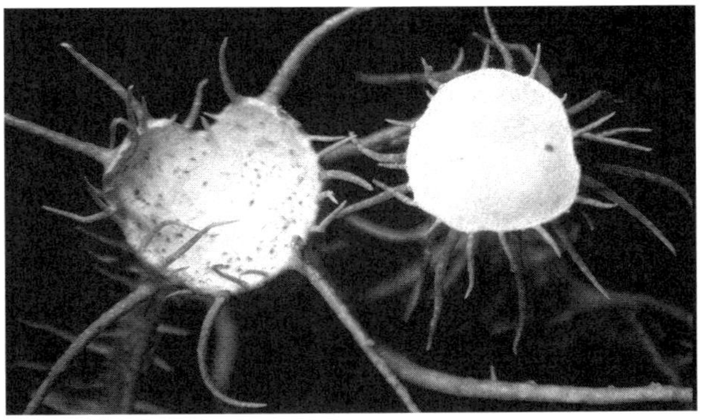

Fibrils of the branches and of the apothecia of *Usnea* sp., x 3 (*CB & MU*).

fibrillose (L. *fibrillosus*, having fibers < *fibrilla*, dim. of *fibra*, fiber, filament + *-osus*, full of, augmented, prone to > ME. *-ose*): having thin fibrils, with the appearance of very fine silk threads; e.g., like the pileus of *Amanita phalloides* (Agaricales).

fibrillose-truncate wart (L. *fibrillosus*, having fibers < *fibrilla*, dim. of *fibra*, fiber, filament + *-osus*, full of, augmented, prone to > ME. *-ose*; *truncatus* < *truncare*, to lop off + suf. *-atus* > E. *-ate*, provided with or likeness; ME. < OE. *wearte* < L. *verruca*, wart): a flat-topped wart adorned with fibrils.

fibrosin bodies (L. *fibra*, fiber, filament, thread; OE. *bodig*, body): structures of unknown exact nature present in the cytoplasm of the conidia of some powdery mildew fungi (Erysiphales).

fibrous (L. *fibrosus*, composed of threads < *fibra*, fiber, filament, thread + L. *-osus* > OF. *-ous*, *-eus* > E. *-ous*, having, possessing the qualities of): having fibers or being thin like a fiber. In the lichens, it refers to the cortical zone which is constituted of hyphae parallel to the thalline axis; in this last case it is the opposite of **fastigiate**.

fibula, pl. **fibulae** (L. *fibula*, clasp, buckle): also called clamp connection or hook; a bridge-like hyphal connection between two adjacent cells through which nuclei migrate to maintain the dikaryon; characteristic of the secondary or dikaryotic mycelium of many basidiomycetes (such as *Coprinus comatus*, of the Agaricales).

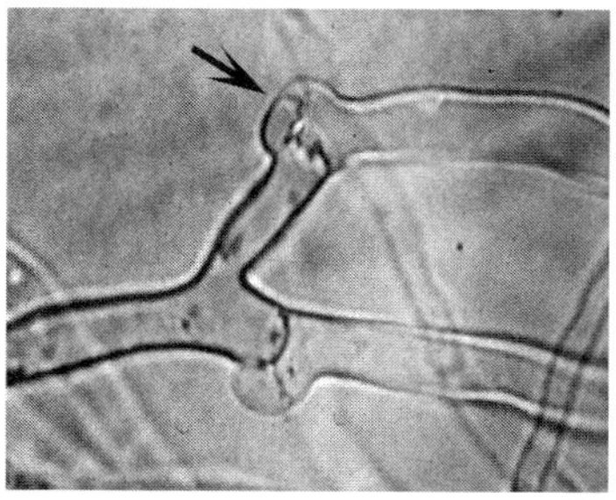

Fibulae of the secondary mycelium of a basidiomycete, x 500 (*DM*).

filament (ML. *filamentum*, filament < *filare*, to spin < *filum*, linen thread + *-amentum*, strap): a fine thread; a term loosely used to refer to a hypha or any other fungal element that is thread-like in appearance.

filamentous (ML. *filamentosus*, composed of filaments < *filare*, to spin < *filum*, linen thread + *-amentum*, strap + L. *-osus* > OF. *-ous*, *-eus* > E. *-ous*, having, possessing the qualities of): composed of, or similar to, a filament.

filiform (L. *filiformis* < *filum*, linen thread + suf. *-formis* < *forma*, shape): with the shape of a thread, thin and slender, like a linen fiber; e.g., like the ascospores of *Ophiodothella* (Phyllachorales), the appendages of the ascospores of *Cercophora palmicola* (Sordariales), and the ß-conidia of *Phomopsis* (sphaeropsidaceous asexual fungi).

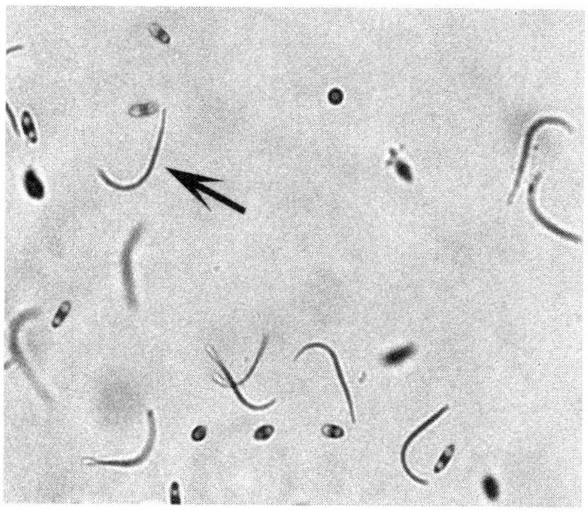

Filiform beta-conidia from a pycnidium of *Phomopsis* sp., x 450 (*CB*).

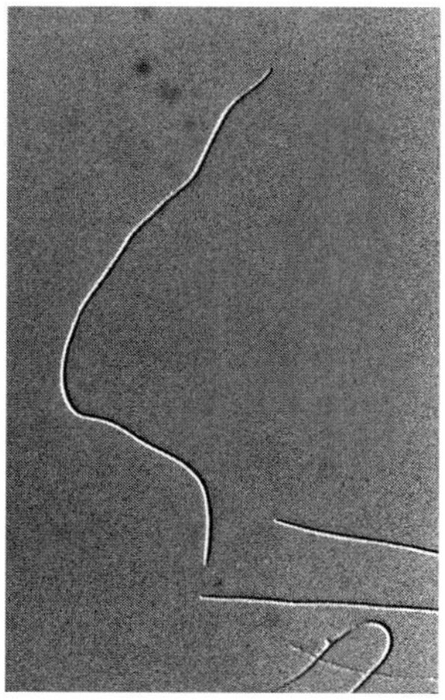

Filiform ascospore of *Epichloë typhina*, x 550 (*RTH*).

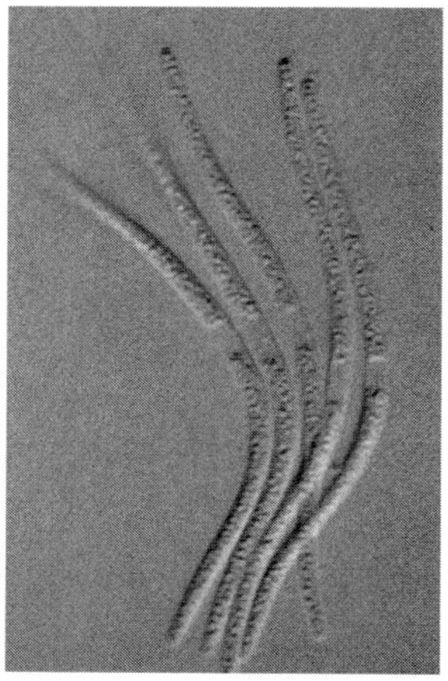

Filiform ascospores of *Ophiodothella vaccinii*, x 650 (*RTH*).

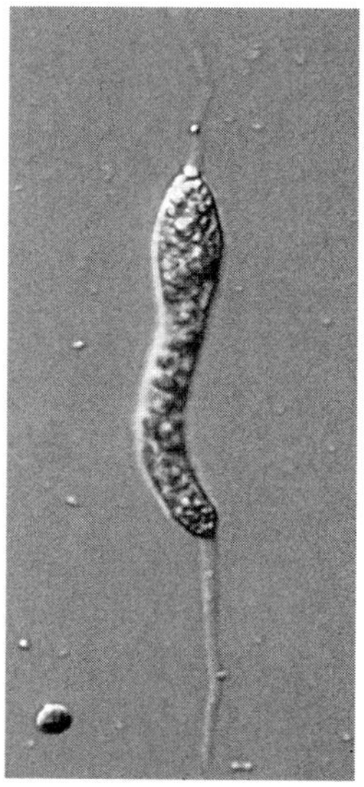

Filiform appendages of the young ascospore of *Cercophora palmicola*, x 770 (*RTH*).

filoplasmodium, pl. **filoplasmodia** (L. *filum*, linen thread + Gr. *plásma*, the bland material of living beings + L. *-odium*, resembling < Gr. *ode*, like < *-oeídes*, similar to): a reticulate, filamentous pseudoplasmodium which consists of a colony of individual fusiform cells that move and grow within the filaments or tunnels of mucilage that they themselves produce. Characteristic of the Labyrinthulales, such as *Labyrinthula marina*. Also known as viscous network.

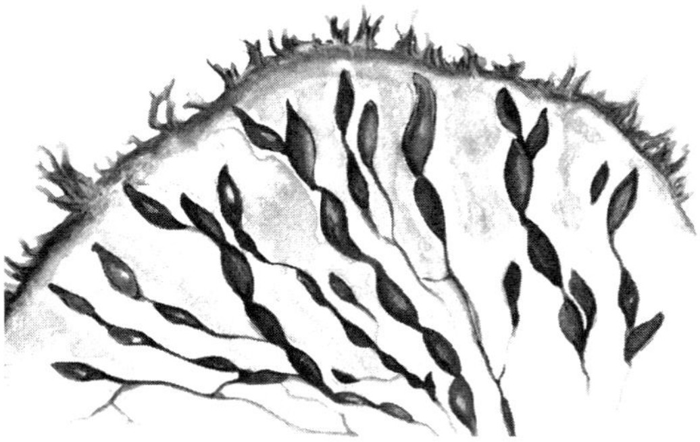

Filoplasmodium of *Labyrinthula marina*, x 370.

filopod, **filopodium**, pl. **filopodia** (L. *filum*, linen thread + NL. *podium* < Gr. *pódion*, dim. of *poús*, *podós*, foot): a pseudopodium or filiform extension of a plasmodium or of a protoplast; the myxamoebae of *Dictyostelium* (Dictyosteliomycota) and the reticulate plasmodia of the Labyrinthulales form filopodia. Cf. **lobopod**.

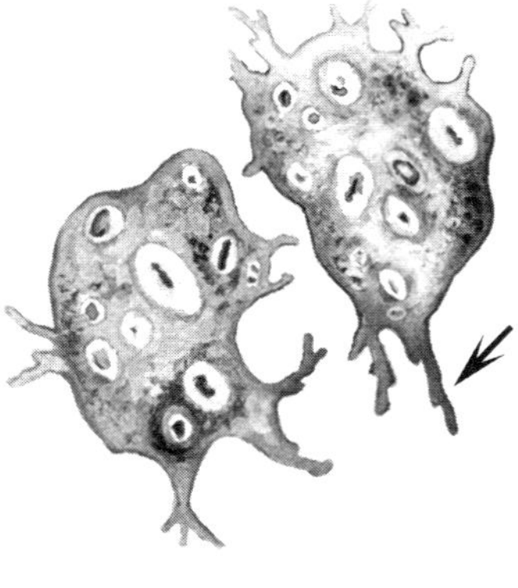

Filopods of the myxamoebae of *Dictyostelium discoideum*, x 500.

fimbriate (L. *fimbriatus*, fibrous, fringed, bordered with hairs < *fimbria*, fiber, fringe, extremity + suf. *-atus* > E. *-ate*, provided with or likeness): **1**. Provided with a fringe; e.g., like the edge of the colony of some yeasts (such as *Candida tropicalis* and *C. krusei*, of the cryptococcaceous asexual yeasts), which have slender lobes or very fine segments, with a pointed apex. Also, the ostioles of the perithecia of *Ceratocystis fimbriata* (Microascales) and *Melanospora zamiae* (Melanosporales) are fimbriate. **2**. Delicately toothed; often used in reference to a stoma. Also called **laciniate**.

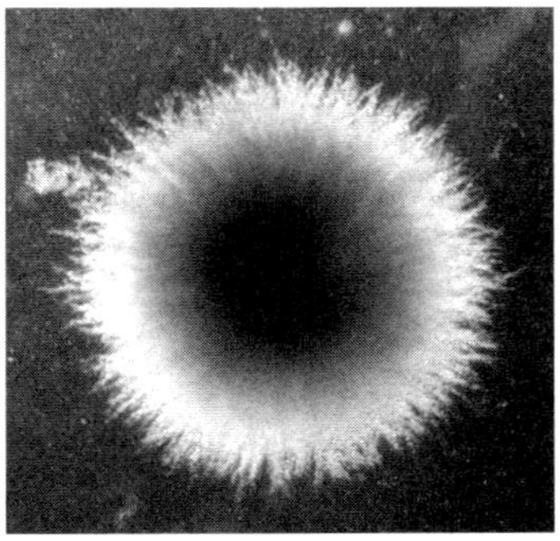

Fimbriate border of a colony of *Candida krusei* on agar, x 25 (*TH*).

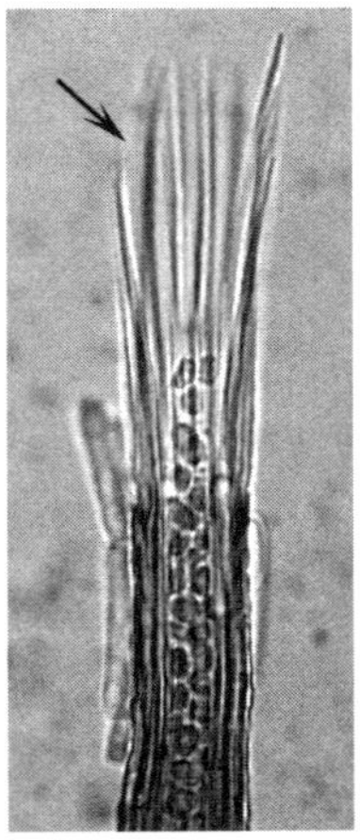

Fimbriate ostiole of the ascocarp of *Ceratocystis fimbriata*, x 300 (*RTH*).

Fimbriate ostiole of the perithecium of *Melanospora zamiae*, x 100 (*RTH*).

fimbrillate (NL. *fimbrillatus*, bordered with minute fibers or hairs < *fimbrilla*, dim. of *fimbria*, fiber, fringe + suf. *-atus* > E. *-ate*, provided with or likeness): minutely **fimbriate**.

fimicole, **fimicolous** (L. *fimus*, *fimo*, dung + *-cola*, inhabitant < *colere*, to inhabit; or + *-osus* > OF. *-ous*, *-eus* > E. *-ous*, possessing the qualities of): living or developing on dung; e.g., like many Mucorales (*Mucor*, *Pilobolus*, etc.) and other fungi that belong to diverse groups.

fission (L. *fissio*, genit. *fissionis*, a cleaving, partition, scission + *-io*, *-ionis* > E. suf. *-ion*, result of an action, state of): see **bipartition** and **schizogenesis**.

fissitunicate (L. *fissio*, genit. *fissionis*, a cleaving, partition, scission + *tunicatus*, covered by a tunic < *tunica*, cover, cloak + suf. *-atus* > E. *-ate*, provided with or likeness): a type of bitunicate ascus (Luttrellian term) in which the elastic internal layer(s) of the wall (endotunica) are separate from the rigid outer wall layer(s) (exotunica, ectotunica), through which they extend prior to ascospore discharge; e.g., *Leptosphaerulina crassiasca* (Pleosporales). The fissitunicate ascus corresponds to the **physoclastic ascus** of the literature on marine ascomycetes. The non-fissitunicate ascus (**aphysoclastic ascus** of marine mycology) lacks the separation and extension of the ascus wall layers before spore liberation. Inasmuch as electron microscopy has shown that all ascus walls are multilaminate, it is preferable to utilize **physoclastic** and **fissitunicate** instead of **bitunicate**, and non-fissitunicate and **aphysoclastic** instead of **unitunicate** for these types of ascus dehiscence. Cf. **rostrate dehiscence** and **semifissitunicate**.

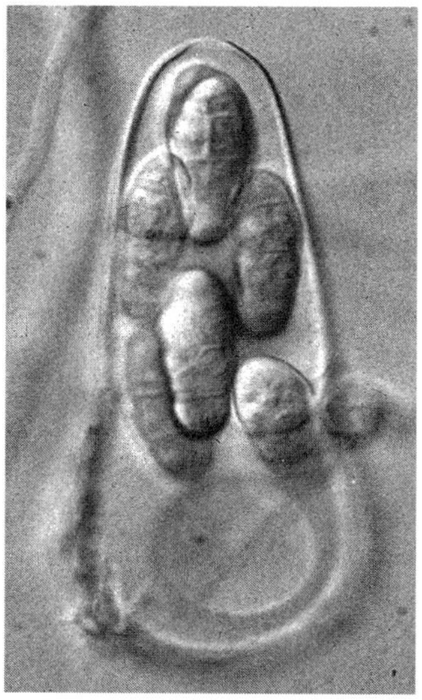

Fissitunicate ascus of *Leptosphaerulina crassiasca*, x 600 (*RTH*).

fistular, **fistulose** (L. *fistularis*, like a pipe < *fistula*, a hollow reed-stalk, pipe, tube + suf. *-aris* > E. *-ar*, like, pertaining to; L. *fistulosus*, full of pipes < *fistula* + *-osus*, full of, augmented, prone to > ME. *-ose*): something that has one or several tubes, i.e., with a tubular structure, hollow; e.g., like the stipe of *Amanita caesarea* (Agaricales). Cf. **cavernose**, **farctate** and **solid**.

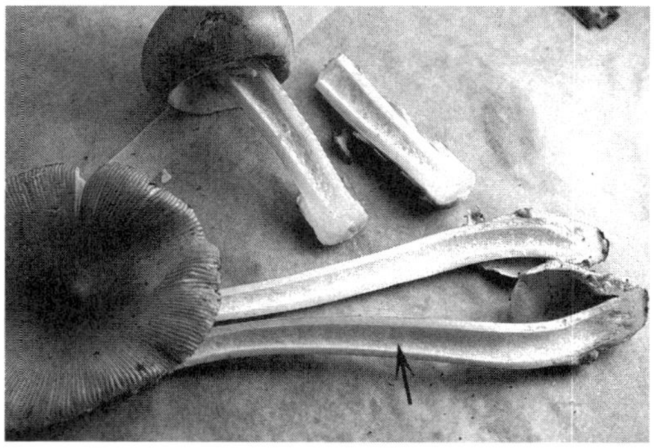

Fistulose stipe of the basidiocarps of *Amanita caesarea*, x 0.4 (*RMA*).

flabellate, **flabelliform** (L. *flabellatus*, furnished with a fan-like structure < *flabellum*, fan + suf. *-atus* > E. *-ate*, provided with or likeness; *flabellum* + *-formis* < *forma*, shape): having the shape of a fan; e.g., like the fruiting body of *Sparassis crispa* (Aphyllophorales).

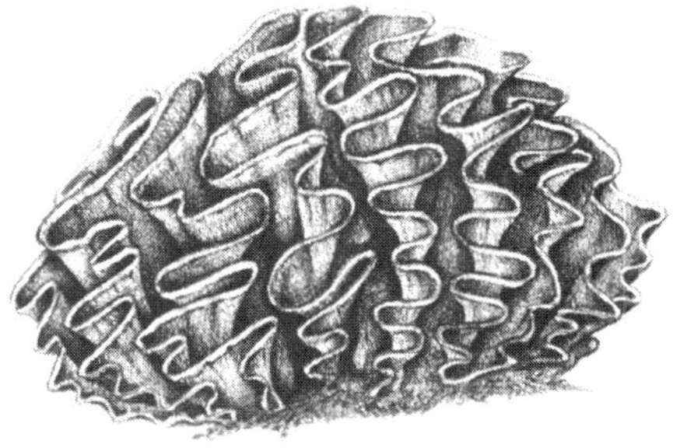

Flabellate fruiting body of *Sparassis crispa*, x 0.5.

flaccid (L. *flaccidus*, *flaccus*, loose, limp, flimsy, flabby): limp, soft, lacking strength or turgescence, unable to remain erect and hanging due its own weight.

flagellate (L. *flagellatus*, whipped < *flagellum*, whip + suf. *-atus* > E. *-ate*, provided with or likeness): furnished with one or more flagella.

flagellum, pl. **flagella** (L. *flagellum*, whip): a mobile, flexible protoplasmic filament, in the form of a little whip, with or without mastigonemes, which serves to propel a motile cell or zoospore. Some authors apply the term flagellum to propelling filaments of prokaryotic cells (without a 9 + 2 construction of microtubules, composed of a single protein called flagelline), whereas they use **undulipodium** for the motile cells of eukaryotic organisms (with a 9 + 2 construction of microtubules, composed of more than 100 proteins). The amoeboid cells of some species of Myxomycota (such as the myxozoospores of *Physarum* of the Physarales) and the zoospores of the Chytridiomycetes (such as *Allomyces*, of the Blastocladiales), Hyphochytriomycota (such as *Hyphochytrium*, of the Hyphochytriales) and the Oomycota (such as *Saprolegnia*, of the Saprolegniales) also have flagella.

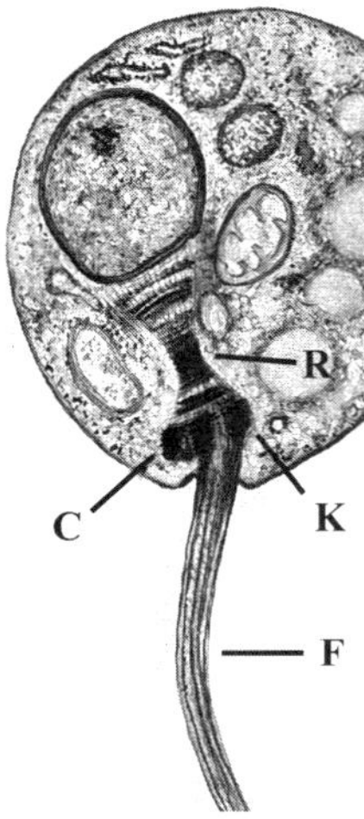

Flagellum (F) of the zoospore of *Rhizophlyctis rosea*. The centriolum (C), kinetosome, (K), and rhizoplast (R) are also shown, x 16 000.

flexible (L. *flexibilis*, flexible): not rigid or firm, easily bent; e.g., like the basidiocarp of *Collybia* (Agaricales).

flexuose, **flexuous** (L. *flexuosus*, twisted, tortuous < *flexus*, bent, turned, curved + *-osus*, full of, augmented, prone to > ME. *-ose*; or + *-osus* > OF. *-ous*, *-eus* > E. *-ous*, possessing the qualities of): winding or alternately bent in opposite directions, in a zigzag; e.g., like the pedicle of the sporangium of *Physarum polycephalum* (Physarales).

flimmergeissel (G. *flimmer*, to vibrate, tremble + *Geissel*, flagellum): a flagellum bearing two rows of tripartite tubular hairs. Cf. **straminipilous**.

flimmer hairs or **flimmers** (G. *flimmer*, to vibrate, tremble; OE. *haer*, hair, a slender threadlike outgrowth of the epidermis of an animal): same as **mastigonemes**.

floccose (L. *floccosus*, full of flocks of wool < *floccus*, flock or tuft of wool + *-osus*, full of, augmented, prone to > ME. *-ose*): tomentose, loosely cottony or woolly, or more densely agglomerated into small bundles like flannel. The colonies of many fungi, which form in natural conditions as well as on agar culture media, have a floccose appearance, as occurs, e.g., in *Monilia sitophila* and *Microsporum gypseum* (moniliaceous asexual fungi), among others. The basidiocarp of *Coprinus niveus* (Agaricales) also is noticeably floccose.

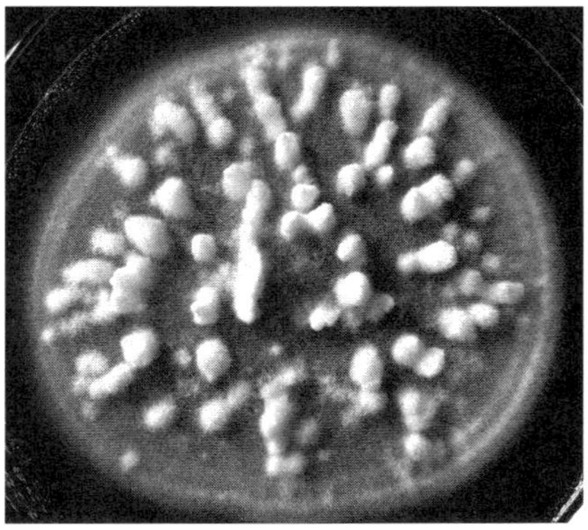

Floccose texture of a colony of *Aspergillus* sp. on agar, x 0.7 (*MU*).

flocculose, **flocculent** (LL. *flocculosus*, full of diminutive flocks of wool < *flocculus*, dim. of *floccus*, flock of wool + *-osus*, full of, augmented, prone to > ME. *-ose*; or + L. *-entem* > E. *-ent*, being): slightly **floccose**. In cultures of yeasts or bacteria developed in liquid media, it refers to small groups or bundles of microorganisms distributed throughout its mass or accummulated in the bottom.

flock (ME. < OE. *floc*, short for *floccule*, flock, group): a group of hyphae that resembles a small tassel of cotton and which together form the floccose mycelium of many fungi.

floricole, **floricolous** (L. *flos*, *florus*, flower + *-cola*, inhabitant < *colere*, to inhabit; or + L. *-osus* > OF. *-ous*, *-eus* > E. *-ous*, possessing the qualities of): an organism that lives in flowers or inflorescences ecologically compatible to it; e.g., like certain yeasts which live in flower nectaries.

flower of tan (ME. *flour*, flower, best of anything < OF. *flor* < L. *flor*, *flos*, flower; ME. *tannen* < MF. *tanner* < ML. *tannare* < *tanum*, *tannum*, tanbark, tannin): the aethalium of the Myxomycete *Fuligo septica* (Physarales), which is sometimes found growing on the tannin-containing tubs used to convert hide into leather.

fluorescent (NL. *fluor*, mineral belonging to a group used as fluxes and including fluorite < L. *fluere*, to flow + *-escens*, genit. *-escentis*, that which turns, beginning to, slightly > E. *-escent*): having fluorescence, i.e., the property of emitting light in the presence of certain radiations, because of which the structure or compound appears different in color than with reflected or transmitted light. For example, the aflatoxins of *Aspergillus flavus* (moniliaceous asexual fungi) and the mycotoxins of other fungi have a characteristic fluorescence in a chromatogram under ultraviolet light.

flush (MF. *flus*, *fluz* < L. *fluxus*, flow, flux): the sudden and usually abundant development of mycelium or a periodic surge of fruiting body emergence, especially in mushroom cultures.

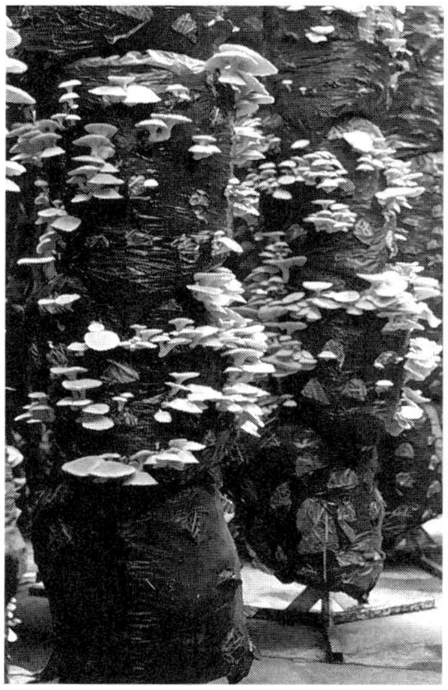

Flush of basidiocarps in a cultivated mushrooms factory; *Pleurotus ostreatus* growing on wheat straw, x 0.05 (*CB*).

fly agaric (ME. *flie* < OE. *fleoge*, to fly; a fly, a winged insect; *agaric*, one of the Agaricales): the basidioma of *Amanita muscaria* (also called **fly mushroom**), which has insecticidal properties.

fly fungus, pl. **fungi** (ME. *flie* < OE. *fleoge*, to fly; fly, a winged insect; L. *fungus*, fungus): *Entomophthora muscae* (Entomophthorales), the entomopathogenic fungus which attacks house flies.

fly mushroom (ME. *flie* < OE. *fleoge*, to fly; fly, a winged insect; ME. *musseroum* < MF. *mousseron* < LL. *mussirion*, mushroom): see **fly agaric**.

fly-speck fungus, pl. **fungi** (ME. *flie* < OE. *fleoge*, to fly; fly, a winged insect; ME. *specke* < OE. *specca*, speck; L. *fungus*, fungus): a fungus whose ascomata appear as small dark spots on the leaves and/or fruits of the host plant; e.g., the fly-speck disease of apple is caused by *Schizothyrium pomi* (Dothideales).

foetid (L. *foetidus* < *foetere*, to smell bad): stinking; with a nauseating odor, like some Phallaceae.

foliicole, **foliicolous** (L. *folium*, leaf + *-cola*, inhabitant < *colere*, to inhabit; or + *-osus* > OF. *-ous*, *-eus* > E. *-ous*, possessing the qualities of): an organism that lives or develops on leaves; there are many kinds of foliicolous fungi, e.g., species of the gen. *Erysiphe* and other Erysiphales, as well as some lichens, e.g., *Strigula complanatum* (Chaetothyriales) and *Porina epiphylla* (Trichotheliales).

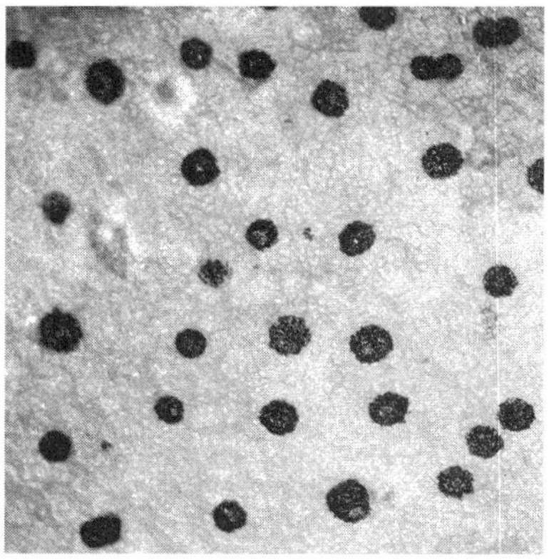

Fly speck of apple caused by *Schizothyrium pomi* infection; the dark spots are ascomata, x 5 (*RTH*).

Foliicole lichen thalli of *Strigula complanata* on a leaf, x 1 (*TH*).

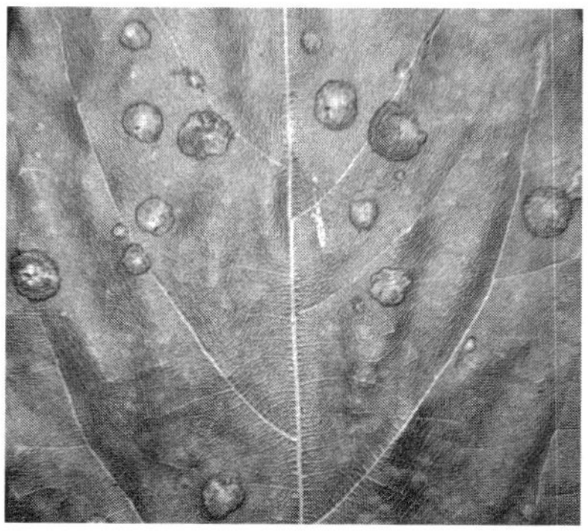

Foliicole lichen thalli of *Porina epiphylla* on a fern leaf, x 1.2 (*MU*).

foliole, **foliolum**, pl. **foliola** (L. *foliolum* < *folium*, leaf + dim. suf. *-olum* > E. *-ole*): each one of the small excrescences similar to leaves that form on the surface of a foliose lichen, as *in Parmelia imbricatula* (Lecanorales).

Folioles of the thallus of *Parmelia* sp., x 1.5 (*MU*).

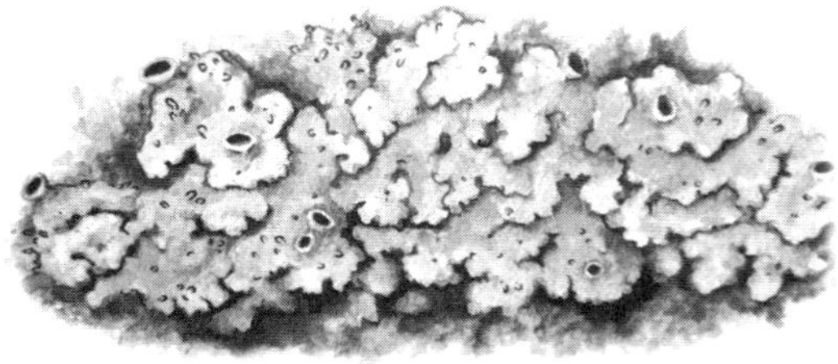

Folioles of the thallus of *Parmelia imbricatula*, x 1.

foliose (L. *foliosus*, leafy < *folium*, leaf + *-osus*, full of, augmented, prone to > ME. *-ose*): *Lichens*. A type of thallus constituted of numerous small leaflike lobes that extend in a more or less circular pattern from a center of growth; e.g., *Parmelia imbricatula* and *Anaptychia palmulata* (Lecanorales). Many other lichens have this type of thallus.

Foliose thallus of *Anaptychia palmulata* on bark, x 9 (*MU*).

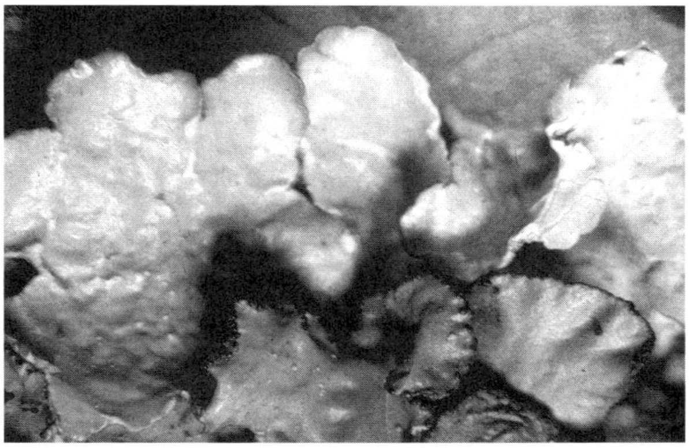

Foliose thallus of *Parmelia subrudecta*, x 2 (*MU*).

follicular (L. *folliculus*, dim. of *follis*, leather purse + suf. *-aris* > E. *-ar*, like, pertaining to): relative to a follicle; having follicles or similar to a follicle or small bag.

fomite (L. *fomitis*, genit. of *fomes*, tinder): **1**. Tinder, of combustible consistency, woody, like the basidiocarps of *Fomes* (Aphyllophorales), whose combustible fruiting bodies have been utilized to produce fire by rubbing them with other hard objects. **2**. An inanimate object that transmits the infective stages of parasites or pathogens.

foot cell (ME. < OE., *fot*, foot; L. *cella*, storeroom > ML. *cella* > OE. *cell*, cell, compartment): also called **basal cell**; this is the hyphal cell that this supports a sporogenous cell or thallus; e.g., a basal cell supports the conidiophore in *Aspergillus* (moniliaceous asexual fungi), the macroconidium of *Fusarium* (tuberculariaceous asexual fungi), and the thallus of aquatic fungi of the order Blastocladiales (such as *Blastocladia* and *Blastocladiella*), which in the last gen. adopts the general form of a trunk, with branches bearing zoosporangia on the apical part, and rhizoids or attachment organs on the base.

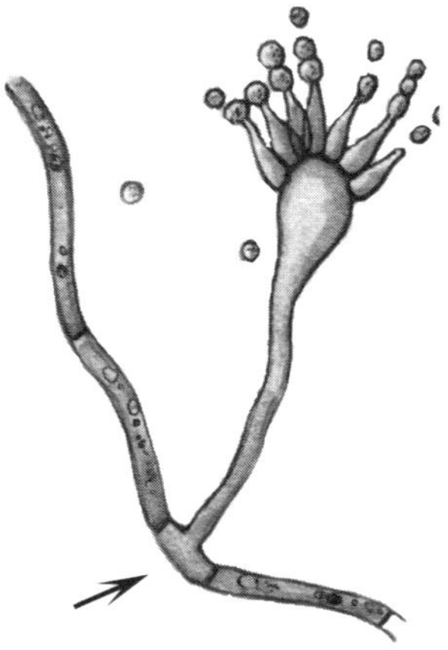

Foot cell of the conidiophore of *Aspergillus japonicus*, x 190.

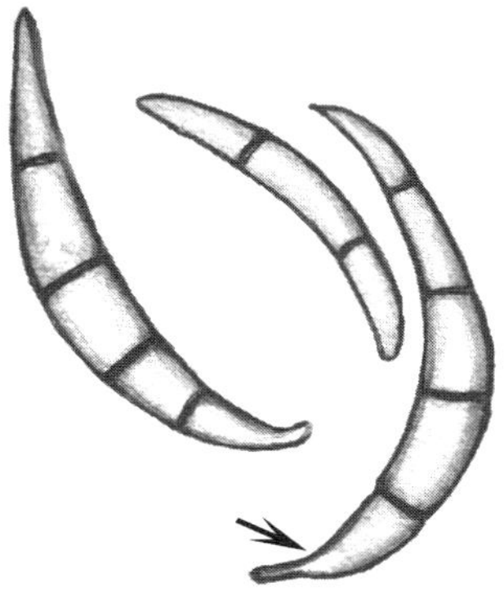

Foot cell of the falcate conidia of *Fusarium solani*, x 1 800.

foramen (L. *foramen*, orifice < *forare*, to perforate): a small orifice or opening.

forate development (L. *forare*, to bore + suf. *-atus* > E. *-ate*, provided with or likeness; MF. *développer* < OF. *desveloper* < *dés-* < L. *de-*, from, down, away + MF. *enveloper*, to enclose + L. *mentum* > OF. *-ment* > ME. *-ment*, action, process): *Gasteromycetes*. A type of fruiting body development in which the differentiation of the plectobasidia is carried out among the primordial tissues, within tiny locules that are formed near the peridium and mature centripetally. The development occurs from the periphery of the gleba, near the peridium, or in a center of the sterile basal region, and continues toward the interior of the fructification. Characteristic of the orders Gautieriales and Lycoperdales (such as *Lycoperdon*). Also called **coralloid**. Cf. **aulaeate** and **lacunar development**.

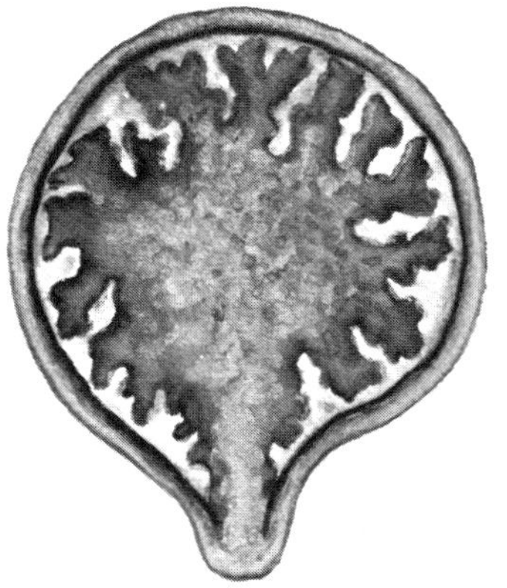

Forate development of the hymenium in the fruiting body of Gautieriales and Lycoperdales, seen in longitudinal section, x 1.2.

forcipate, **forcipiform** (L. *forceps*, tong, forceps, nipper + suf. *-atus* > E. *-ate*, provided with or likeness; L. *forceps* + *-formis* < *forma*, shape): tong-shaped, like the gametangia of *Phycomyces nitens* (Mucorales).

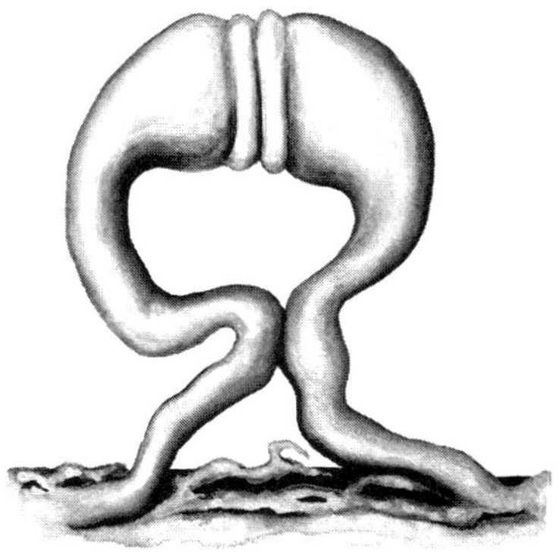

Forcipiform gametangia of *Phycomyces nitens*, x 45.

form (L. *forma*, form, shape): *Botany*. A somewhat imprecise systematic category, generally considered inferior to variety and subvariety. The special form (*forma specialis*, abb. f. sp.) is utilized when treating of parasitic fungi that, without having morphological differences, are physiologically adapted to particular hosts; it is recommended that a special form bear the name of the respective host in the genitive: *Puccinia graminis* f. sp. *tritici* (i.e., from *Triticum aestivum*), the wheat rust; or *Fusarium oxysporum* f. sp. *lycopersici* (i.e., from *Lycopersicum esculentum*), a causal agent of tomato wilt.

form-genus, pl. **genera** (ME. *forme* < OF. < L. *forma*, form; L. *genus*, race, kind): see **anamorph-genus**.

form-species (ME. *forme* < OF. < L. *forma*, form; L. *species*, species, kind): see **anamorph-species**.

fornicate, **forniculate** (NL. *fornicatus*, arched-over < *fornix*, vault, arch, or NL. *forniculatus* < *forniculus*, dim. of *fornix*, + suf. *-atus* > E. *-ate*, provided with or likeness): arched, vaulted; refers to the fruiting body of some geastraceous fungi, such as *Geastrum fornicatum* (Lycoperdales), which on maturing has fibrous and fleshy layers of the peridium arched on the cupulate mycelial layer, forming a kind of vault.

Fornicate fruiting body of *Geastrum fornicatum*, x 0.8.

fornicate ray (NL. *forniculatus* < *forniculus*, dim. of *fornix*, *fornicis*, vault, arch + suf. *-atus* > E. *-ate*, provided with or likeness; ME. *raie*, *raye* < OF. *rai* < L. *radius*, radius): *Gasteromycetes*. An arched ray formed when the mesoperidium separates from the exoperidium, adhering only at the margin; the ray arches to support the elevated endoperidium, e.g., in *Geastrum* (Lycoperdales).

forophyte (Gr. *-phóros*, bearer + *phytón*, plant): a host or supporting plant of a particular epiphytic organism, e.g., trees that support lichens on their bark.

fossil fungus, pl. **fungi** (L. *fossilis*, dug up < *fossus*, ptp. of *fodere*, to dig; L. *fungus*, fungus): a remnant, impression, or trace of a fungus of past geologic ages that has been preserved in the earth's crust. The fossil record of fungi extends back to the early Phanerozoic and into the Proterozoic. The generic names of fossil fungi are generally based on extant names modified by a suf., usually *-ites*, e.g., *Boletellites*, *Papulosporonites*, etc.

foveate (L. *foveatus*, pitted < *fovea*, hole + suf. *-atus* > E. *-ate*, provided with or likeness): having holes or perforations; pitted.

foveolate (L. *foveolatus*, pitted < *foveola*, dim. of *fovea*, hole, concavity + suf. *-atus* > E. *-ate*, provided with or likeness): applied to surfaces that have ornamentations in the shape of small holes, as is observed in the ascospore of *Gelasinospora* (Sordariales), or like the basidiospores of *Boletellus betula* (Agaricales).

Foveolate ascospore of *Gelasinospora* sp., x 960 (*MU*).

fragmentation (L. *fragmentatio*, genit. *fragmentationis*, to reduce to fragments < *fragmentum*, a piece, a part broken off, detached + *-ationem*, action, state or condition, or result > E. suf. *-ation*): in the fungi it refers to the separation of a thallus into segments, each of which has the capacity to grow and form a new individual; it is a type of asexual reproduction or vegetative propagation, in natural as well as artificial conditions; artificial fragmentation is commonly utilized in the laboratory to transfer cultures of fungi onto various media.

free (ME. < OE. *freo*, free, not united with or attached to): *Agarics*. Applied to lamellae or tubes, not joined to the stipe of the fruiting body. Cf. **adnate** or **attached**, **decurrent**, and **sinuate**.

free cell formation (ME. < OE. *freo*, free; NL. *cellula*, cell, living cell < L. dim. *cella*, small room; L. *forma*, form + *-ationem*, action, state or condition, or result > E. suf. *-ation*; giving form or shape to something or of taking form): *Ascomycetes*. In the ascomycetes with ascohymenial type of ascocarp development, e.g., *Ascodesmis*, of the Pezizales, it refers to the process by which ascospores are formed in the asci. Following meiosis the 8 nuclei, each with some surrounding cytoplasm, are delimited by portions of the ascus vesicle that enclose them, followed by the deposition of wall material to form separate ascospores in the mature ascus.

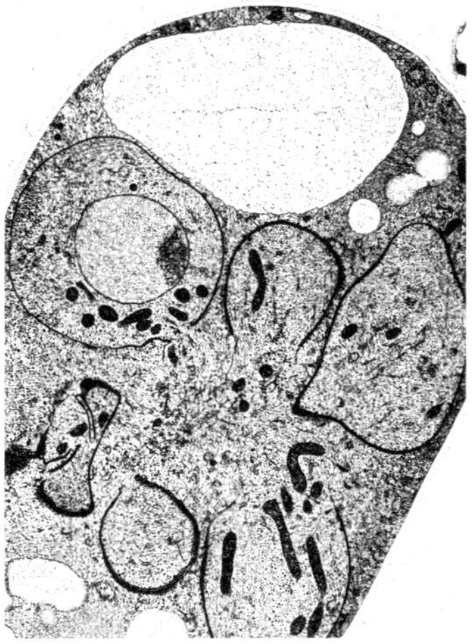

Free-cell formation in ascus of *Ascodesmis* sp., seen in longitudinal section, x 2 350 (*CM*).

friable (L. *friabilis*, breakable < *friare*, to break up, crumble + *-abilis* > E. *-able*, tendency toward, able to be): breaking up or crumbling easily, like the volva of certain species of *Amanita* (Agaricales), which is fleeting, for which reason it rarely leaves any trace on the mature stipe.

fruticole, **fruticolous** (L. *frutex*, shrub + *-cola*, inhabitant + L. *-osus* > OF. *-ous*, *-eus* > E. *-ous*, possessing the qualities of): living on trees or shrubs.

fructicole, **fructicolous** (L. *fructus*, fruit + *-cola*, inhabitant < *colere*, to inhabit + *-osus* > OF. *-ous*, *-eus* > E. *-ous*, possessing the qualities of): an organism that lives or develops in or on fruits, such as *Ciboria carunculoides* and *Monilinia fructicola* (Helotiales), which parasitize mulberry and peach fruits, respectively.

fructification (LL. *fructificatio*, genit. *fructificationis*, the bearing of fruit < L. *fructificare*, to bear fruit < *fructus*, fruit + *-ationem*, action, state or condition, or result > E. suf. *-ation*): **1**. The action of forming or producing fruit. **2**. The formation of the sporiferous apparatus or fruiting body of fungi. The fructification is any fungal structure that produces or bears spores, either sexual or asexual. Syn. of **fruiting body**.

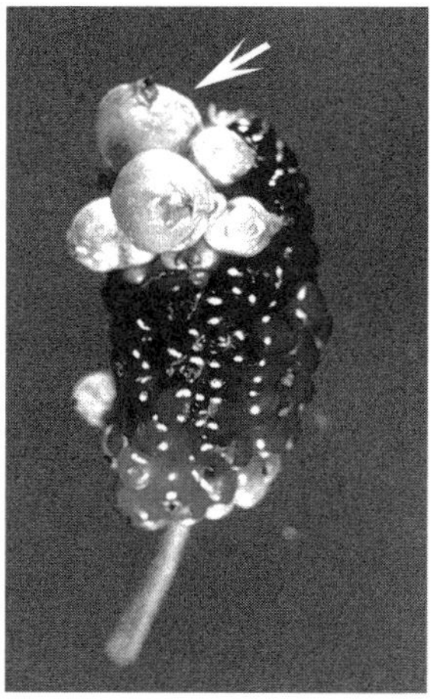

Fructicolous stromatized ovaries on mulberry fruit infected by *Ciboria carunculoides*, x 2 (*RTH*).

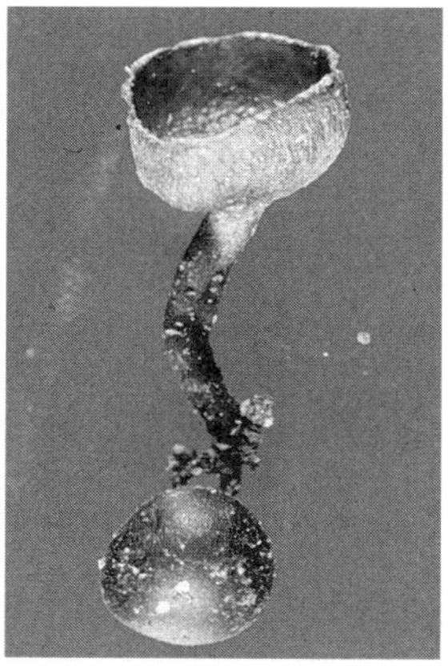

Fructicolous stromatized ovary of mulberry fruit with an apothecium of *Ciboria carunculoides*, x 2.5 (*RTH*).

fruiting body (L. *fructus*, fruit; ME. < OE. *bodig*, body): a general term for spore-bearing organs in both macro- and micromycetes. The more precise terms **apothecium**, **perithecium**, **ascostroma**, **conidioma**, **basidioma**, etc., are applied to specific types of fruiting bodies.

frustulose (L. *frustulosus*, made out of bits < *frustum*, little piece or fragment + *-osus*, full of, augmented, prone to > ME. *-ose*): with the surface cracked or fissured in a more or less regular pattern, like the polygonal pyramids on the surface of a pileus, spore, etc. For example, the surface of the fruiting body of *Boletus frustulosus* (Agaricales), *Stereum frustulatum* (Aphyllophorales), and the thallus of *Arthothelium spilomatoides* (Arthoniales) are frustulose.

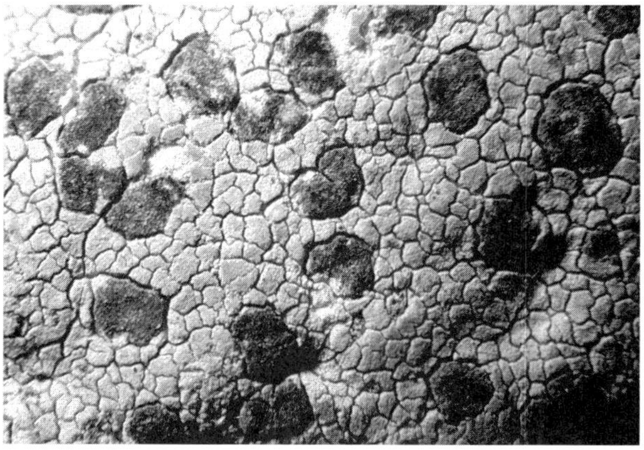

Frustulose thallus of *Arthothelium spilomatoides*, x 8.5 (*MU*).

fruticose (L. *fruticosus*, shrubby < *frutex*, shrub + *-osus*, full of, augmented, prone to > ME. *-ose*): *Lichens*. A thallus that is attached to the substrate by a single point and which develops branches and becomes shrub-like; many Lecanorales have this type of thallus, which can be upright, as in *Cladonia rangiferina*, pendant, as in *Usnea florida*, or intermediate, as in *Teloschistes* (Teloschistales).

fruticulose, **fruticulous** (L. *fruticulosus*, shrubby < *fruticulus*, dim. of *frutex*, shrub + *-osus*, full of, augmented, prone to > ME. *-ose*; or + *-osus* > OF. *-ous*, *-eus* > E. *-ous*, having, possessing the qualities of): *Lichens*. With a minutely shrub-like thallus, e.g., *Cladonia rangiferina* (Lecanorales).

Fruticulose thallus of *Pseudevernia* sp., x 1.3 (*MU*).

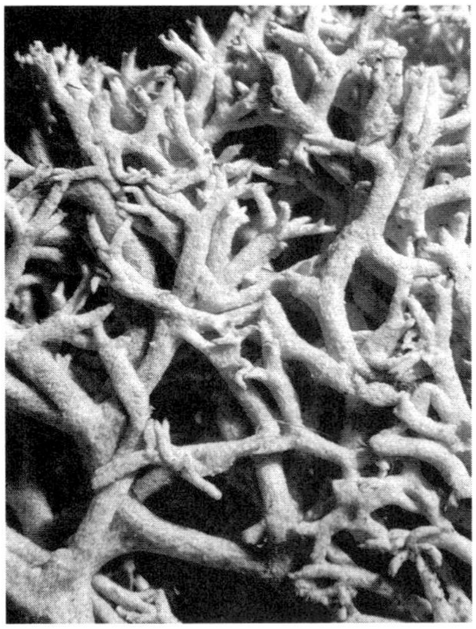

Fruticulose thallus of *Cladonia stellaris*, x 5 (*MU*).

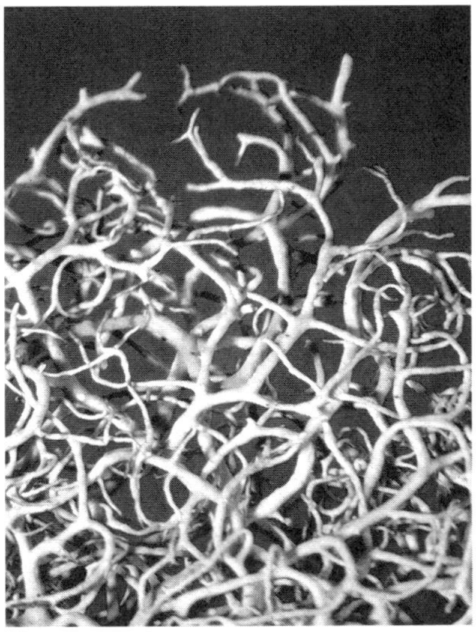

Fruticulose thallus of *Alectoria ochroleuca*, x 4 (*MU*).

Fruticulose thallus of *Cladonia rangiferina*, x 6 (*MU*).

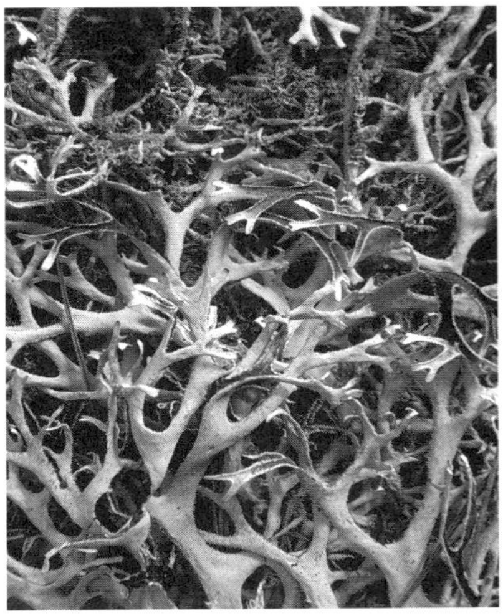

Fruticulose thallus of *Pseudevernia* sp., x 1.3 (*MU*).

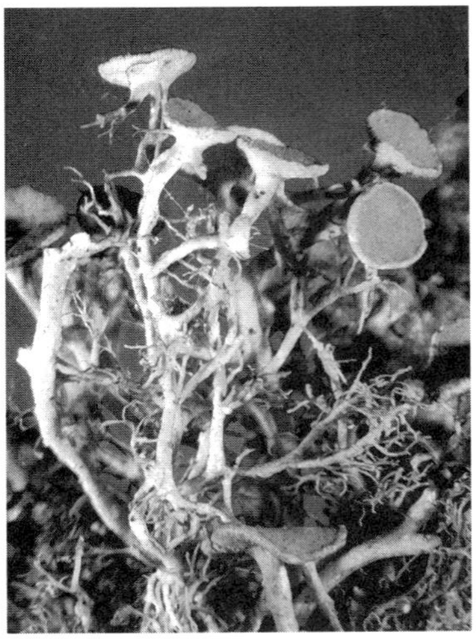

Fruticulose thallus of *Teloschistes exilis*, with branches ending in apothecia, x 6 (*MU*).

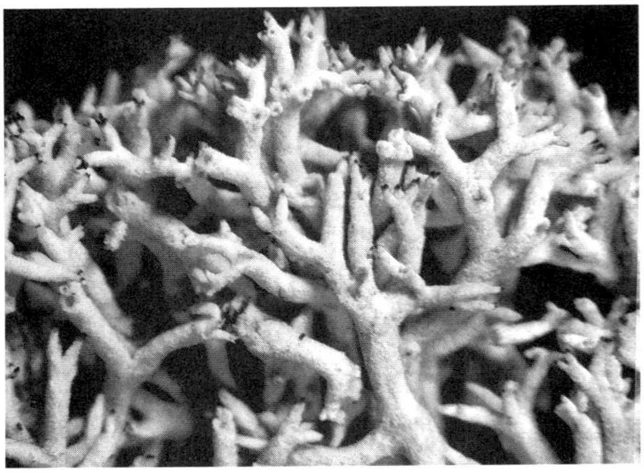

Fruticulose thallus of *Cladina stellaris*, x 7 (*MU*).

fugaceous (L. *fugax*, genit. *fugacis*, unstable, fleeting + suf. *-aceus* > E. *-aceous*, of or pertaining to, with the nature of): a structure of very short duration, like the ring on the basidiocarp of *Coprinus comatus* (Agaricales), which disappears very soon after becoming evident during development (deciduous, ephemeral).

fulcrum, pl. **fulcra** (L. *fulcrum*, a prop < *fulcire*, to support + the adjectival ending *-um*): **1**. A term used for the sporophore of lichens. **2**. A supporting structure, such as the perithecial appendages of *Erysiphe*, *Blumeria*, *Phyllactinia* and other Erysiphales.

fuligineous, **fuliginous** (L. *fuliginosus*, full of soot < *fuligo*, genit. *fuliginis*, soot + *-osus* > OF. *-ous*, *-eus* > E. *-ous*, having, possessing the qualities of): blackish, like soot smudge. In phytopathology it is applied to plants with a dunnish or black covering on leaves, branches and fruits, due to the superficial growth of certain Loculoascomycetes of the gen. *Capnodium* (Capnodiales) and *Dothidea* (Dothideales), among others, known commonly by the names of smut or sooty molds, this last from the gen. *Fumago* (a *nomen confusum*, considered as the asexual or conidial state of *Capnodium*).

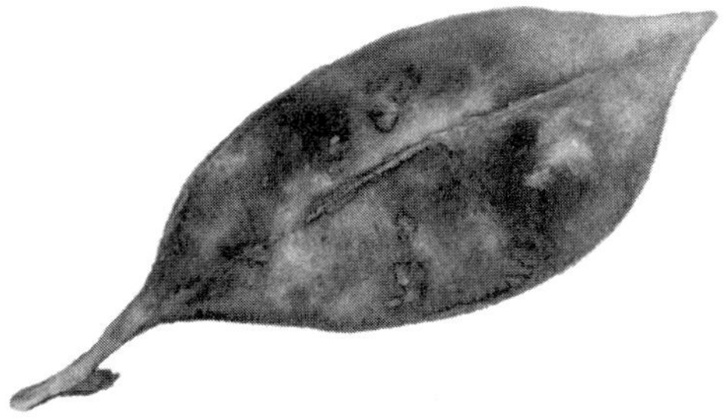

Fuligineous parasitic mycelial growth of *Capnodium walteri* on the upper surface a leaf of *Arbutus menziesii*, x 1.

fumaceous, **fumacious**, **fumaginous** (L. *fumus*, smoke + *-aceus* > E. *-aceous*, of or pertaining to, with the nature of; F. *fumagine* < gen. *Fumago* (a conidial fungus) < *fumus* + *-ago*, resemblance + *-osus* > OF. *-ous*, *-eus* > E. *-ous*, having, possessing the qualities of): smoke-colored; dark gray.

fumagoid (L. *fumago*, smoke-like < *fumus*, smoke + *-ago*, resemblance + *-oide* < Gr. *-oeídes*, similar to): *Med. Mycol.* Refers to a type of rounded cell, isolated or in groups, with thick walls pigmented gray or dark brown, which are developed by *Phialophora* and other gen. of dematiaceous fungi (causing chromomycosis in man and higher animals) in infected tissues.

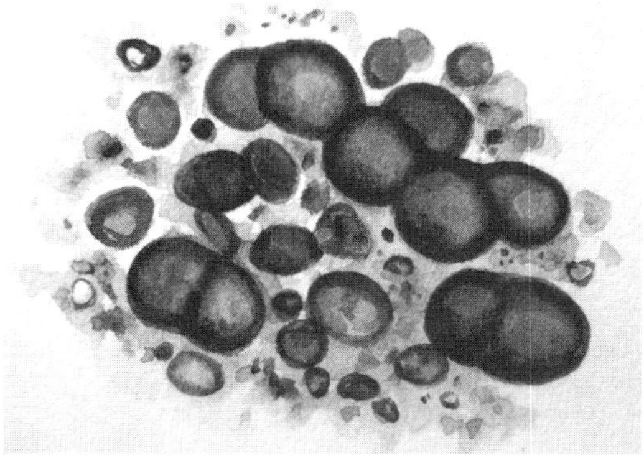

Fumagoid budding cells of *Phialophora pedrosoi* infecting human tissues, x 1 500.

fungal (L. *fungus*, fungus + suf. *-alis* > E. *-al*, relating to or belonging to): pertaining to or belonging to fungi, e.g., fungal infection.

Fungi Imperfecti (L. *fungi*, pl. of *fungus*, fungus; ME. *imperfit* < MF. *imparfait* < L. *imperfectus* < *in-*, not + *perfectus*, perfect): imperfect or conidial fungi. See **anaholomorph**, **asexual fungus**, **deuteromycetes** and **mitosporic fungus**.

fungicide (L. *fungus*, fungus + *caedere*, to kill): any substance capable of destroying fungi, such as the chemicals captan, maneb, thiabendazole and others. Cf. **fungistatic**.

fungicole, **fungicolous** (L. *fungus*, fungus + *-cola*, inhabitant < *colere*, to inhabit + *-osus* > OF. *-ous*, *-eus* > E. *-ous*, possessing the qualities of): that which lives or develops on fungi; there exist many organisms that live on fungi (bacteria, nematodes, insects), including other fungi, such as those of the gen. *Hypomyces* (Hypocreales), whose perithecial stromata form on basidiocarps of *Russula* and *Lactarius* (Agaricales), or like the carpophores of *Nyctalis parasitica*, which grow on the basidiocarps of *Russula* (both gen. in the Agaricales).

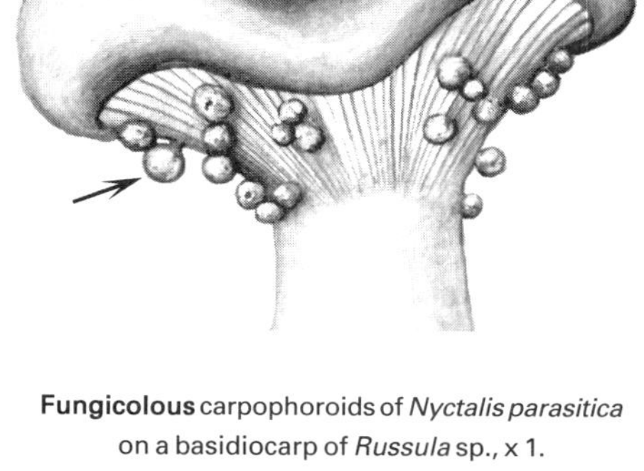

Fungicolous carpophoroids of *Nyctalis parasitica* on a basidiocarp of *Russula* sp., x 1.

fungiferous (L. *fungiferous* < *fungus*, fungus + *-ferous*, bearer < *ferre*, to bear, carry + *-osus* > OF. *-ous*, *-eus* > E. *-ous*, having, possessing the qualities of): that which has fungi, such as the fungiferous layer of mycorrhizae.

fungistatic (L. *fungus*, fungus + Gr. *statós*, standing, placed + Gr. suf. *-íkos* > L. *-icus* > E. *-ic*, belonging to, relating to): syn. of **mycostatic**. Cf. **fungicide**.

fungistic (L. *fungus*, fungus + NL. *-istic*, pertaining to as agent): relative to or belonging to the fungi. A term of recent introduction that is used to refer to the taxonomic study of fungi in a definite geographic region, which is done with the object of inventorying the systematic entities, giving the area of each and information relative to their habitat, abundance or scarcity, time of fruiting and other data. Previously the term mycofloristic was utilized (or the floristic study of fungi), but considering that fungi are not plants, and even more, that they lack flowers, the use of fungistic seems more appropriate.

fungivore, **fungivorous** (L. *fungus*, fungus + *vorare*, to devour, eat + *-osus* > OF. *-ous*, *-eus* > E. *-ous*, possessing the qualities of): that which eats fungi. Equivalent to **mycetophagous**.

fungoid (L. *fungus*, fungus + *-oide* < Gr. suf. *-oeídes*, similar to): similar to a fungus in texture or morphology. Also called **fungous**.

fungous (L. *fungus*, fungus + L. *-osus* > OF. *-ous*, *-eus* > E. *-ous*, possessing the qualities of): belonging to or relative to fungi; generally refers to the nature of the flesh of the sporiferous apparatus of many fungi. Also called **fungoid**.

fungus, pl. **fungi** (L. *fungus*, *sfungus*, fungus < Gr. *spóngos*, *sphóngos*, sponge): **1**. An old term used by Pliny to designate all the fungi that developed on the outside of a covering called a volva. **2**. In modern systems of classification of living beings (such as the five kingdoms of Whittaker, 1969, and of Margulis, 1983), *Fungi* is the name of the kingdom that includes the higher fungi, leaving excluded the Myxomycota and all of the zoosporic fungi. These classifications, among others, broke the tradition of a three-kingdom system of classification of all living organisms as prokaryotes, animals, or plants (including fungi). The addition of the kingdoms Fungi and Protista attempted to place organisms in kingdoms that more nearly reflected their presumed evolutionary relationships. This was an important beginning in the attempt to establish monophyletic groups (groups that contain an ancestor and all its descendants) and to develop a hierarchical classification to reflect the relationships of these groups, i.e., a phylogenetic classification. In an attempt to recognize monophyletic groups, the organisms once classified as fungi now are considered in three different groups, the monophyletic kingdoms *Fungi* and *Stramenopila*, and four protist phyla. This is the classification system adopted in this dictionary (Alexopoulos *et al*., 1996), which recognizes the fact that the organisms that have been called "fungi" are not all closely related. Although these organisms do not share a common evolutionary history, they do form a closely knit group on the basis of their morphology, nutritional modes, and ecology. With advances in ultrastructural, biochemical, and especially molecular biology, the organisms studied by mycologists are now established as polyphyletic (i.e., with different phylogenies) and have to be referred to at least three different kingdoms: kingdom *Fungi* includes the phyla Chytridiomycota, Zygomycota, Ascomycota and Basidiomycota; kingdom *Stramenopila* the phyla Oomycota, Hyphochytriomycota and Labyrinthulomycota, and kingdom *Protista* the phyla Dictyosteliomycota, Acrasiomycota, Myxomycota and Plasmodiophoromycota. **3**. The organisms included in the category of fungi are so diverse that it is difficult to give a concise differential diagnosis. The following definition applies to all of the phyla mentioned in **2**: all are heterotrophic (never photosynthetic) and absorbent. The thallus varies from an amoeboid myxamoeba or plasmodium lacking a cell wall, to a unicellular or filamentous thallus (**hypha**) delimited by a rigid cell wall. In the species that have a filamentous thallus (**mycelium**), the latter can be septate or not, but those with septa are functionally coenocytic, since the septa are perforate. The thalli can be within or on a host or substrate. The wall of the thallus is typically chitinous, but occasionally it is cellulosic (Oomycota); only rarely do chitin and cellulose occur together. According to the group, they can have other polysaccharides (mannan, glucan, etc.) in the wall. The fungi are generally immotile, although they have cytoplasmic flow within the mycelium; in some groups motile states (zoospores) are produced. The nucleus is eukaryotic. Thalli can be uni- or multinucleate, with cells homo- or heterokaryotic, haploid, dikaryotic, or diploid, the last generally of short duration. The life cycle can be simple or complex, and reproduction asexual or sexual. The fungi are cosmopolitan and they live in almost all types of habitats, as saprobes, parasites, or symbionts.

fungus ball (L. *fungus*, fungus; ME. *bal*, ball, a round or roundish body or mass): a roundish mass of hyphae of *Aspergillus* (especially *A. fumigatus*, of the moniliaceous asexual fungi) and cellular debris formed in a preexisting cavity of the lung or a bronchus; a clinical type of **aspergillosis.**

funicle, **funiculus**, pl. **funiculi** (L. *funiculus*, little cord < *funis*, cord + dim. suf. *-culus*): *Gasteromycetes* (order Nidulariales). In fungi such as *Cyathus*, it refers to the slender cord (composed of a sheath, middle piece, and cord), that connects the peridioles to the internal peridium (endoperidium) of the receptacle or basidiocarp. In certain lichens, such as those of the gen. *Umbilicaria* (Lecanorales), it corresponds to the **umbilical cord**.

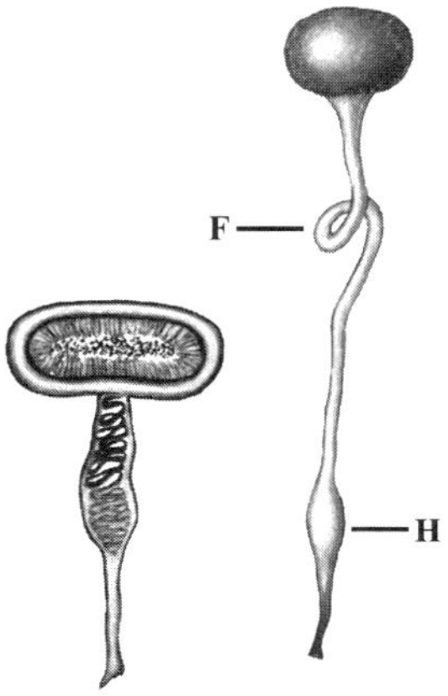

Funiculus (F) and hapteron (H) of the peridiola of *Cyathus striatus*. The peridiolum on the left is shown in longitudinal section and with the funiculus unexpanded, x 8.

Funiculus of the thallus of *Umbilicaria* sp., x 0.5.

funicular, **funiculose** (L. *funiculus*, little cord, dim. of *funis*, cord + L. suf. *-aris* > E. *-ar*, like, pertaining to; or + L. *-osus*, full of, augmented, prone to > ME. *-ose*): being aggregated into little cords or bundles, like those that give rise to conidiophores, as on the mycelium of *Acremonium polichromum* and some species of *Penicillium (P. funiculosum*, moniliaceous asexual fungi); also applied to structures provided with a funiculus, such as the peridioles of the nidulariaceous basidiomycetes (such as *Cyathus striatus* and *Crucibulum laeve*), which have a pedicel that unites them to the internal peridium of the receptacle.

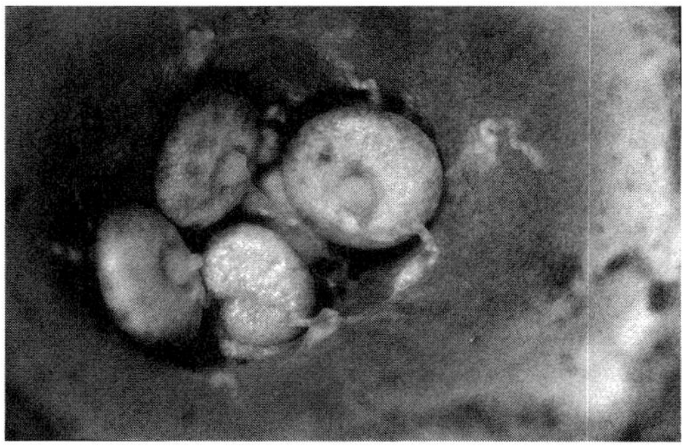

Funiculose peridiola within the fruiting body of *Crucibulum laeve*, x 8 (*CB*).

furcate (L. *furcatus*, forked < *furca*, fork + suf. *-atus* > E. *-ate*, provided with or likeness): forked or bifurcate, like the cystidia of *Paxillus involutus* and *Russula* species (Agaricales), or like the basidia of fungi of the gen. *Dacrymyces* (Dacrymycetales). Also called **lituate**.

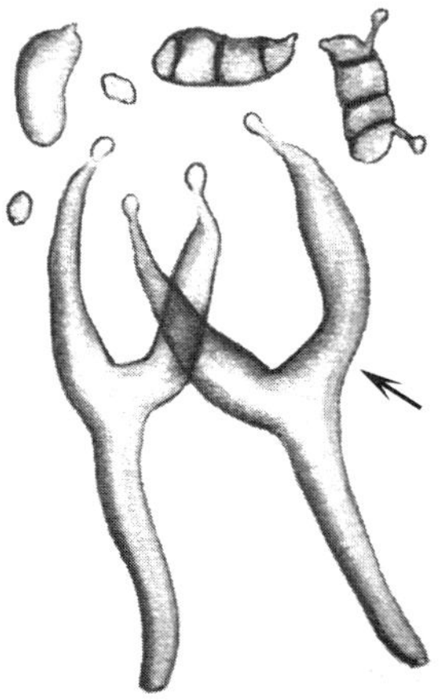

Furcate basidia of *Dacrymyces ellisii*, x 450.

furfuraceous (LL. *furfuraceus*, flaky < L. *furfur*, bran, dandruff + suf. *-aceus* > E. *-aceous*, of or pertaining to, with the nature of): covered with little scales similar to those of bran or dandruff, like the basidiocarps of *Tubaria furfuracea* and *Lepiota clypeolaria* (Agaricales).

fusiform (L. *fusiformis*, spindle-shaped < *fusus*, spindle + *-formis* < *forma*, shape): like a spindle, tapered at the ends, like the conidia of *Paecilomyces fumosoroseus* and *Trichophyton ajelloi* (moniliaceous asexual fungi), the conidia of *Fusarium* (tuberculariaceous asexual fungi), the ascospores of *Melanospora zamiae* (Melanosporales) and *Leptosphaeria* (Pleosporales), or like the basidiospores of *Pleurotus ostreatus* (Agaricales). It is called **fusoid** when it is somewhat fusiform.

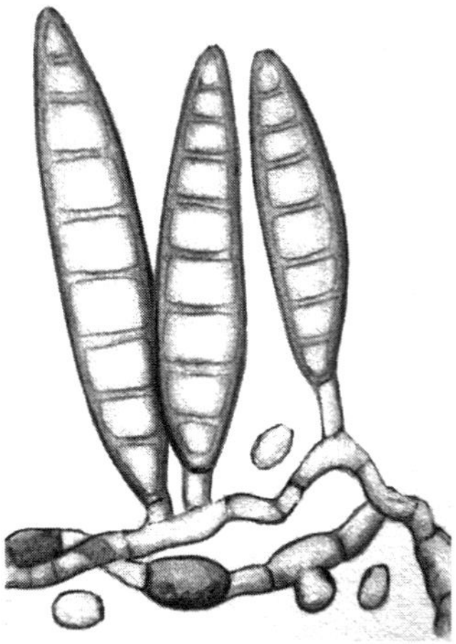

Fusiform macroconidia of *Trichophyton ajelloi*, x 1 000.

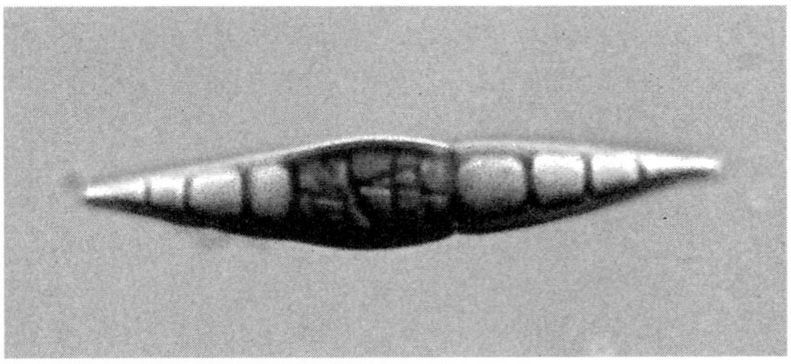

Fusiform ascospore of *Leptosphaeria* sp., x 1 700 (*RTH*).

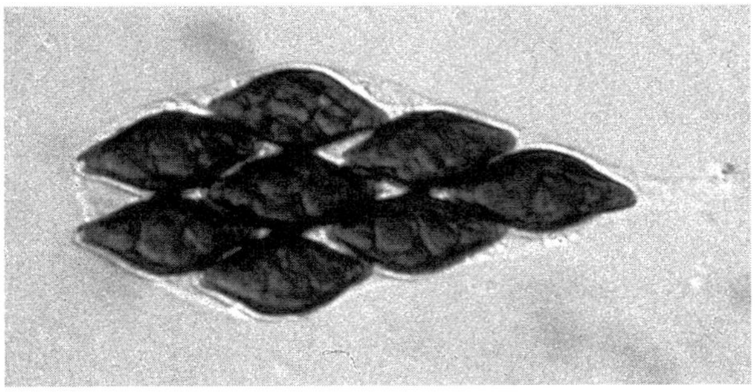

Fusiform ascospores of *Melanospora zamiae*, x 1 250 (*RTH*).

fusion biotrophism (L. *fusion-*, *fusio* < *fusus*, a union by or as if by melting; a merging of diverse elements into a unified whole + *-io*, *-ionis* > E. suf. *-ion*, result of an action, state of; Gr. *bíos*, life + *trophós*, that which nourishes, serves as food + *-ismós* > L. *-ismus* > E. *-ism*, state, phase, tendency, action): a type of interaction displayed by species of *Melanospora* (Melanosporales), whose hyphae may cause the erosion of the host hyphal walls followed by the fusion with fungal host protoplasts, a nutritional interaction that saps the host nutrients and slows its growth.

fusoid (L. *fusus*, spindle + *-oide* < Gr. *-oeídes*, similar to): see **fusiform**.

g

galeate (L. *galeatus*, covered with a helmet < *galea*, helmet + suf. *-atus* > E. *-ate*, provided with or likeness): provided with a helmet-shaped structure.

galeiform (L. *galea*, helmet + *-formis* < *forma*, shape): shaped like a helmet or hat, e.g., the ascospores of *Saccharomycopsis fibuligera* (Saccharomycetales) and *Ceratocystis fimbriata* (Microascales).

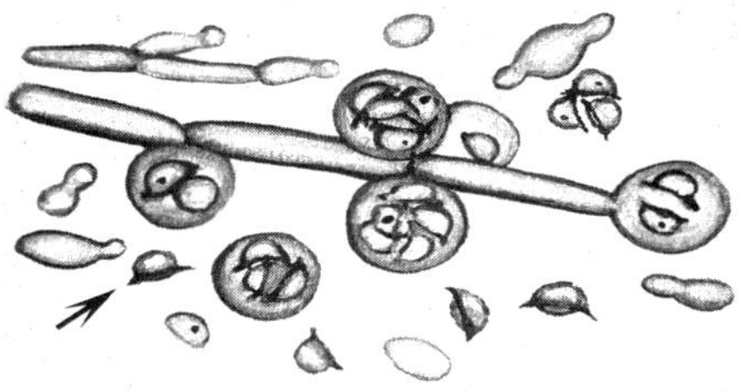

Galeiform or **galeate** ascospores of *Saccharomycopsis fibuligera*, x 450.

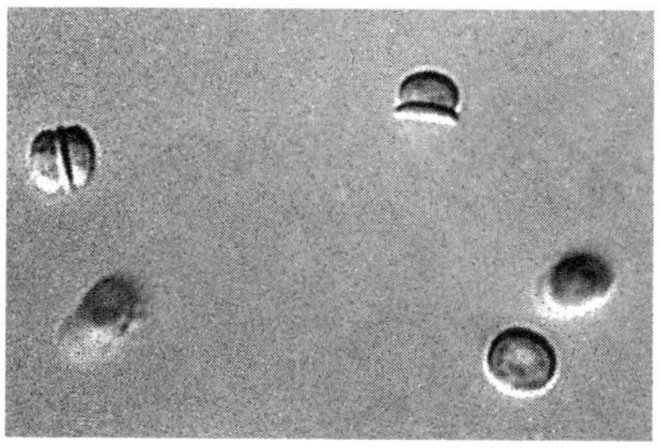

Galeiform ascospores of *Ceratocystis fimbriata*, x 1 000 (*RTH*).

galeriform (L. *galerus* or *galerum*, cap, helmet-like covering for the head < *galea*, helmet + *-formis* < *forma*, shape): having the shape of a cap or helmet, like the pileus of the basidiocarp of *Galerina* (Agaricales).

galvanotropism (E. *galvanic* < F. *galvanisme* < It. *galvanismo* < Luigi Galvani; galvanic current, a direct current of electricity, especially when produced by chemical action + Gr. *trópos*, a turn, change in manner < *tropé*, a turning < *trépo*, to revolve, turn towards + *-ismós* > L. *-ismus* > E. *-ism*, state, phase, tendency, action): a reaction to an electrical field.

gametangial contact (Gr. *gamétes*, husband + *angeîon*, vessel, recipient + L. suf. *-alis* > E. *-al*, relating to or belonging to; L. *contactus*, ptp. of *contingere*, to have contact with): a type of sexual reproduction in which the two gametangia come in contact but do not fuse, at least not totally (both gametangia more or less retain their individuality); the male nucleus migrates through a pore or fertilization tube and penetrates the female gametangium; as happens, e.g., in aquatic fungi of the order Saprolegniales (*Achlya americana*, *Saprolegnia parasitica*, *Dictyuchus monosporus*).

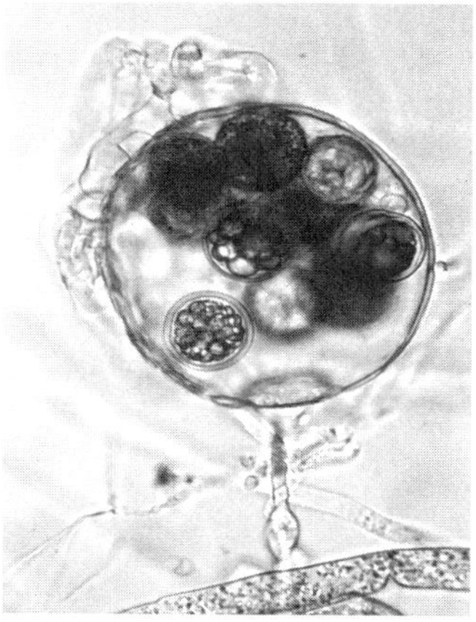

Gametangial contact in *Achlya americana*, x 490 (*RTH*).

gametangial copulation (NL. *gametangium* < Gr. *gamétes*, husband + *angeîon*, vessel, receptacle + L. suf. *-alis* > E. *-al*, relating to or belonging to; L. *copulatio*, genit. *copulationis*, union, link < *copula*, band or link + *-ationem*, action, state or condition, or result > E. suf. *-ation*): a type of sexual reproduction in which the two gametangia or their protoplasts fuse (**plasmogamy**) and give rise to a zygote (**karyogamy**) which develops into a latent spore (the zygospore); as occurs, e.g., in *Rhizopus nigricans* and *Mucor miehei* (Mucorales) and *Schizosaccharomyces octosporus* (Schizosaccharomycetales). This type of sexual union also is known as **conjugation**.

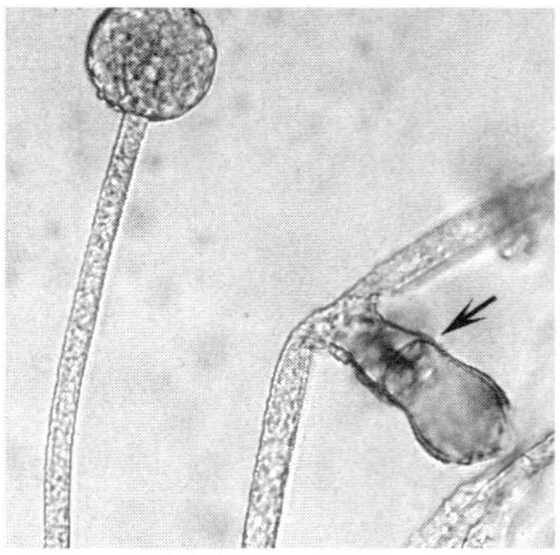

Gametangial copulation in *Mucor miehei*, x 840 (*MU*).

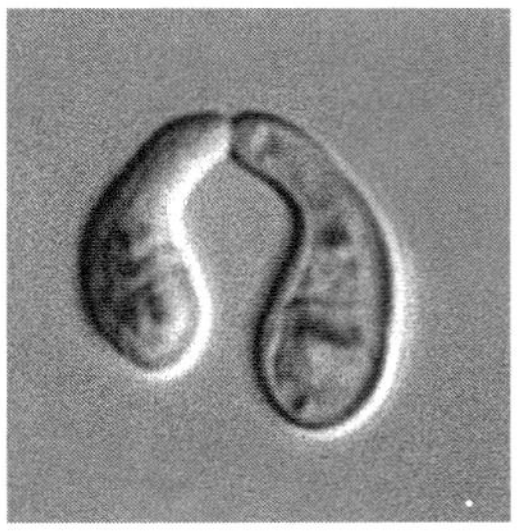

Gametangial copulation of two yeast cells of *Schizosaccharomyces octosporus*, x 2 000 (*RTH*).

gametangium, pl. **gametangia** (NL. *gametangium* < Gr. *gamétes*, husband + *angeîon*, vessel, receptacle): a sexual organ that contains gametes, which can be motile cells (planogametes) or gametic nuclei, depending upon the species.

gametangy (NL. *gametangium* < Gr. *gamétes*, husband + *angeîon*, vessel, receptacle + *-y*, E. suf. of concrete nouns): a sexual process in which two compatible gametangia copulate, which can be equal or not in size and shape; two types of gametangy are **gametangial contact**, which occurs in the Oomycota (e.g., Saprolegniales and Peronosporales), and **gametangial copulation**, which is observed in the Zygomycota (e.g., Mucorales). This process also is called gametangiogamy.

gamete (Gr. *gamétes*, husband): a sexual cell, or a sexual nucleus, which fuses with another complementary one in sexual reproduction to give rise to the zygote. See **aplanogamete** and **planogamete**.

gametogenesis, pl. **gametogeneses** (Gr. *gamétes*, husband, gamete + *génesis*, engenderment): the production of gametes.

gametothallus, pl. **gametothalli** (Gr. *gamétes*, husband + *thallós*, sprout, thallus): a thallus that produces gametes; present in aquatic fungi in whose life cycle occurs an alternation of generations, as in *Allomyces javanicus* (Blastocladiales), which has gametothalli that alternate with sporothalli, these latter producers of asexual spores. See **sporothallus**.

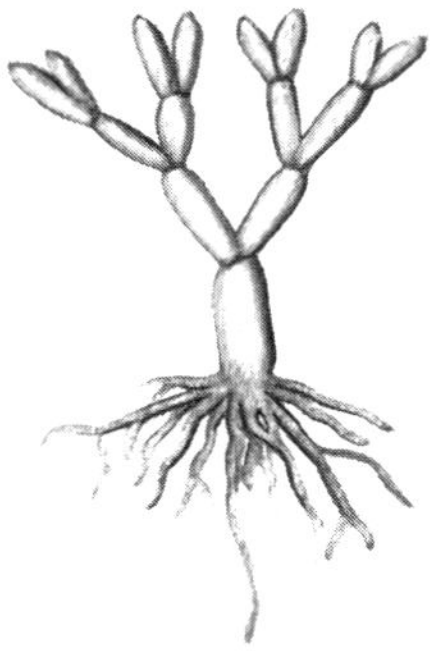

Gametothallus of *Allomyces javanicus*, x 480.

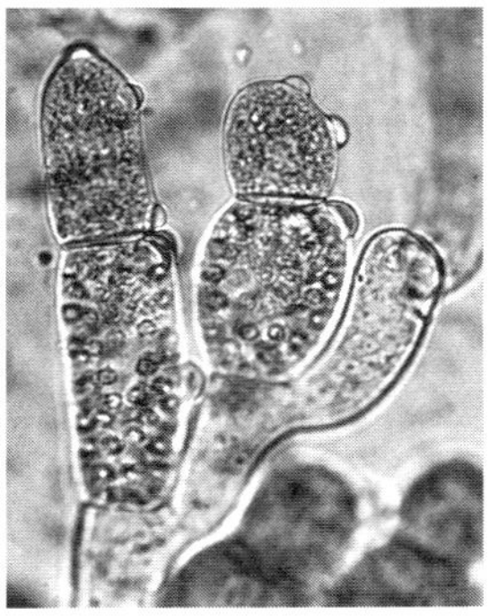

Gametothallus of *Allomyces javanicus*, x 840 (*RTH*).

gamma particle (*gamma*, γ, third letter of Gr. alphabet; L. *particula* < *part*, piece + dim. suf. *-cula*): each of the small globose structures, delimited by membranes, that contain a dense body, of approximately 0.5 m diam., and which appear to be involved in the formation of the wall of the cyst or encysted zoospore; characteristic of the zoospores of *Blastocladiella emersonii* (Blastocladiales).

gamont (Gr. *gámos*, sexual union + *óntos*, genit. of *ón*, a being): an individual producer of gametes, especially in a holocarpic form, or a separate portion that produces gametes.

gangliform (Gr. *gánglion*, swelling + L. *-formis* < *forma*, form): having swellings or knots; knotted, as the hyphae of some fungi.

gangliospore (Gr. *gánglion*, swelling + *sporá*, spore): a term used to designate an aleuriospore, considering that by its origin it is a holoblastic conidium. See **aleuriospore**.

gasteroid (Gr. *gastér*, stomach, belly + L. suf. *-oide* < Gr. *-oeídes*, similar to): with the hymenium in cavities, i.e., enclosed, not arranged on gills, or without any hymenium, as in the Gasteromycetes. Same as **gastroid.** Cf. **agaricoid**.

gasteromorphic (Gr. *gastér*, stomach, belly + *morphé*, form + suf. *-íkos* > L. *-icus* > E. *-ic*, belonging to, relating to): having more or less, or completely, the habit, ontogeny, shape and characteristics of gasteromycete spores.

Gasteromycetes (Gr. *gastér*, abdomen + L. *-mycetes*, ending of class < Gr. *mýkes*, genit. *mýketos*, fungus): a group of basidiomycetes with the spores contained in the interior of the sporiferous apparatus, closed during the major part of its development or indehiscent, like the Lycoperdaceae. Corresponds to the class Gasteromycetes, of the phylum Basidiomycota, and contains the orders Lycoperdales, Tulostomatales, Sclerodermatales, Phallales, Nidulariales, and Hymenogastrales.

gasterospore (Gr. *gastér*, abdomen + *sporá*, spore): a globose spore, with a thick wall (chlamydospore); probably formed by apomixis in the interior of the tissues or tubes of a fruiting body, not on the exterior, as happens in *Ganoderma* (Aphyllophorales).

gastroid (Gr. *gastrós*, stomach, abdomen + L. suf. *-oide* < Gr. *-oeídes*, similar to): a general adj. applied to all basidiomycetes which have angiocarpic development. Same as **gasteroid**. Cf. **agaricoid**.

gel tissue (E. *gelatin* < F. *gélatine*, edible jelly, gelatin < ML. *gelatina* < L. *gelatus*, frozen < *gelare*, to freeze; ME. *tissu*, a rich fabric < OF. ptp. of *tistre*, to weave < L. *texere*): *Leotiales* and *Tremellales*. A mixture of gel and hyphae found in the fruiting bodies, that may originate either by direct secretion or by disintegration of hyphae. Cf. **gliatope**.

gelatinous (E. *gelatin* < F. *gélatine*, edible jelly, gelatin < ML. *gelatina* < L. *gelatus*, frozen < *gelare*, to freeze + *-osus* > OF. *-ous*, *-eus* > E. *-ous*, having, possessing the qualities of): jelly-like; having the appearance and consistency of gelatin, a glutinous substance obtained from animal tissues by prolonged boiling. In the fungi it is applied to tissues whose hyphae become glutinous and partially dissolve under conditions of high atmospheric humidity; when mounted in water and viewed under the microscope, they appear more transparent, wider and looser than in unhydrated tissues. The fungi of the order Tremellales, such as *Tremella lutescens*, are commonly called jelly fungi or trembling fungi. The gleba of the Gasteromycetes of the order Phallales also is gelatinous, as well as the hygroscopic gel that some lichens, such as *Collema* and *Leptogium* (Lecanorales), have in the hymenium.

gelatinous peridium, pl. **peridia** (E. *gelatin* < F. *gélatine*, edible jelly, gelatin < ML. *gelatina* < L. *gelatus*, frozen < *gelare*, to freeze + L. *-osus* > OF. *-ous*, *-eus* > E. *-ous*, having, possessing the qualities of; NL. < Gr. *péridion* < *péra*, purse + dim. suf. *-ídion* > L. *-idium*): a peridium composed of a hyaline, pectinous matrix with embedded hyaline hyphae or hyphal fragments, infrequent; e.g. in *Calostoma* (Tulostomatales).

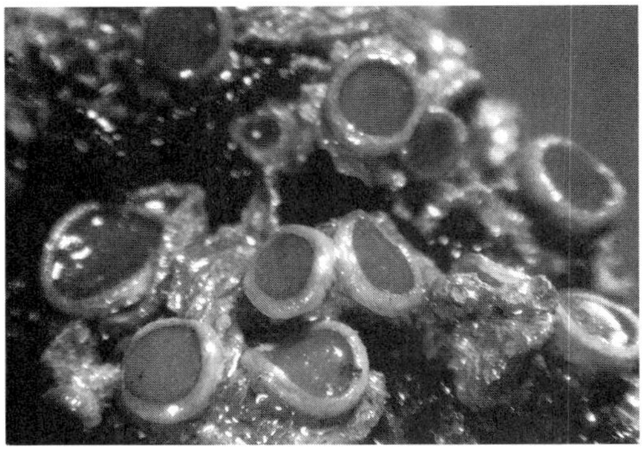

Gelatinous thallus with apothecia of *Collema collocarpum*, x 4 (*MU*).

Gelatinous thallus with apothecia of *Leptogium corticola*, x 10 (*MU*).

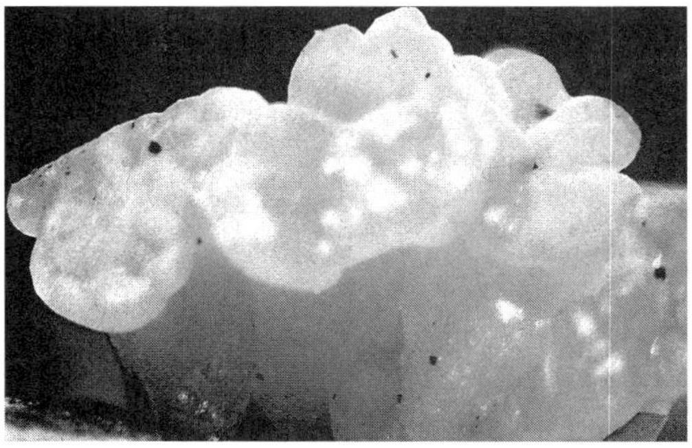

Gelatinous basidiocarp of *Tremella lutescens*, x 2 (*MU*).

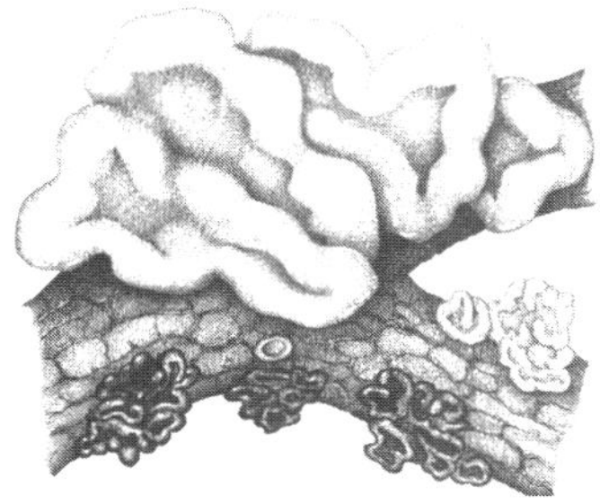

Gelatinous basidiocarp of *Tremella mesenterica*, x 1.5.

gelatinous tissue (E. *gelatin* < F. *gélatine*, edible jelly, gelatin < ML. *gelatina* < L. *gelatus*, frozen < *gelare*, to freeze + L. *-osus* > OF. *-ous*, *-eus* > E. *-ous*, having, possessing the qualities of; ME. *tissew* < MF. < OF. *tissu* < *tistre*, to weave < L. *texere*, to weave): a translucent, pectinous matrix with or without embedded hyphae; common in Phallales, e.g., *Kobayasia*.

gemifer (L. *gemmifer* < *gemma*, bud, shoot + *-fer* < *ferre*, to bear): bearing a bud; sometimes in *Omphalina flavida* (Agaricales), in a small and incompletely developed basidiocarp, the pileus functions as a bud.

gemma, pl. **gemmae** (L. *gemma*, bud < Gr. *yémo*, to be full): **1**. A bud or propagule that is produced by budding, as in the yeasts. **2**. In the aquatic saprolegniaceous fungi (such as *Saprolegnia*), the structures called gemmae are formed like the chlamydospores of other fungi, from intercalary cells in the mycelium; in this case they also function as thick-walled propagules of asexual reproduction.

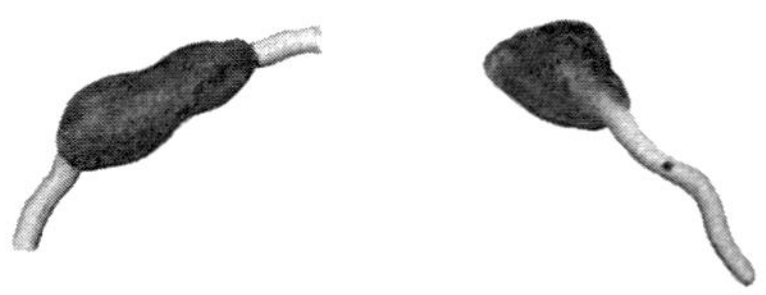

Gemmae of the mycelium of *Saprolegnia parasitica*, x 280.

gemmation (L. *gemmare*, to sprout, bud < *gemma*, bud < Gr. *yémo*, to be full + L. *-ationem*, action, state or condition, or result > E. suf. *-ation*): same as **budding**. A type of cellular, asexual reproduction characterized by the formation of small superficial protuberances or evaginations in the generative cells; the protuberances enlarge, become individual, and on separating from the mother cell, constitute other independent individuals. Budding is typical of the yeasts (e.g., in *Hansenula anomala*, *Nadsonia* sp. and *Saccharomyces cerevisiae* of the Saccharomycetales), but also present in many molds. The spores or cells produced by budding are known as **blastospores**. See **bipolar**, **monopolar** and **multipolar**, for budding with two, one, or many buds.

gene (Gr. *gen*, short for *pangen* < *pan*, all, every, completely + *-gen*, *-genes*, born, or *génos*, origin, birth, a race, kind, descent): an element of the germ plasm having a specific function in inheritance that is determined by a specific sequence of purine and pyrimidine bases in DNA or sometimes in RNA and that serves to control the transmission of a hereditary character by specifying the structure of a particular protein, such as an enzyme, or by controlling the function of other genetic material.

generative hypha, pl. **hyphae** (L. *generatum* < *generare*, to engender + *-ivus* > E. *-ive*, quality or tendency, fitness; Gr. *hyphé*, tissue, spider web; hypha): *Basidiomycetes*. A slender (1.5-10 µm diam.), branched, cylindrical, undifferentiated hypha, with a thin wall, regularly septate, and capable of generating fertile cells, the basidia of the hymenium and related tissues; e.g., as in *Polyporus versicolor* (Aphyllophorales). It stains intensely in cotton blue. See **binding hypha** and **skeletal hypha**.

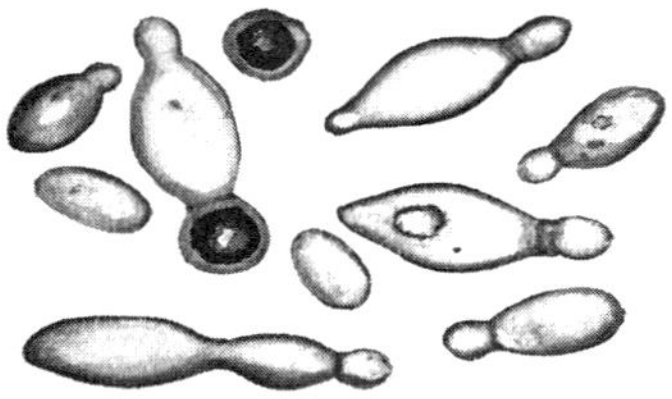

Gemmation of the vegetative cells and asci of *Nadsonia fulvescens*, x 1 000.

genestasis, pl. **genestases** (Gr. *gennáo*, to engender + *stásis*, fixity, immobility): see **genistat**.

geniculate (L. *geniculatus*, having a knot or protuberance like a knee or elbow, knotted < *geniculum*, dim. of *genu*, knee + suf. *-atus* > E. *-ate*, provided with or likeness): having kneelike joints or bends; applied to the part of a hypha or of a conidiophore that forms bends due to changes in direction resulting from sympodial growth, such as one sees in the *Drechslera* state of *Cochliobolus bicolor*, and in *Helminthosporium*, *Alternaria* and *Cercospora*, among many other dematiaceous asexual fungi.

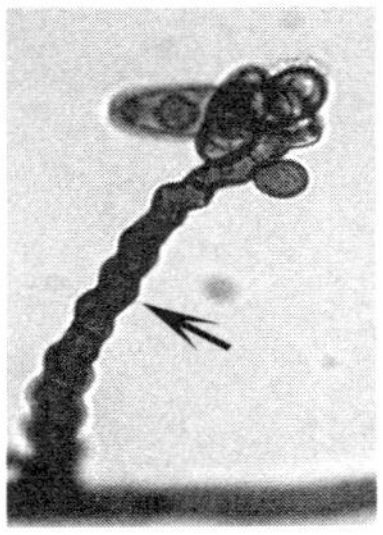

Geniculate conidiophore of *Drechslera* sp., x 380 (*JAS*).

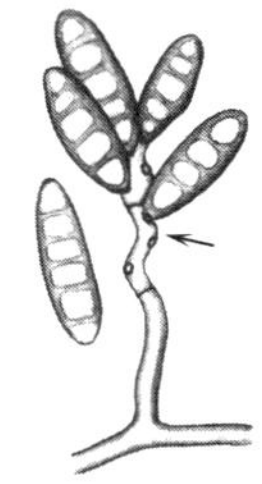

Geniculate conidiophore of the *Drechslera* state of *Cochliobolus bicolor*, x 630.

genistat (Gr. *gennáo*, to engender + *statós*, standing, placed): a substance that prevents or reduces sporulation in fungi, but without affecting the growth of somatic or assimilative hyphae; i.e., which has **genestasis**. Among some species one can observe this phenomenon, e.g., *Phymatotrichopsis omnivora* inhibits the sporulation of *Trichoderma viride* (moniliaceous asexual fungi), and the formation of microsclerotia by *Papulaspora* (agonomycetaceous asexual fungi), when they develop in dual cultures on agar.

geophilic, **geophilous** (Gr. *geo* < *gê*, soil, earth + *phílos*, have an affinity for + suf. *-íkos* > L. *-icus* > E. *-ic*, belonging to, relating to; or + L. *-osus* > OF. *-ous*, *-eus* > E. *-ous*, possessing the qualities of): **1.** *Med. Mycol.* Refers to a fungus that naturally lives in the soil, where probably it grows as a saprobe in remnants of keratin; e.g., *Microsporum cookei* (moniliaceous asexual fungi) which, although it is related to species of keratinophilic dermatophytes, is not pathogenic to man and higher animals, or is very rare and accidental. Cf. **anthropophilic** and **zoophilic**. **2.** Also used for fungi that produce their fruiting bodies underground from assimilative mycelium that also develops underground, such as is observed in the truffles (Pezizales).

geotropism (Gr. *geo* < *gê*, earth, soil + *trópos*, a turn, change in manner < *tropé*, a turning < *trépo*, to revolve, turn towards + *-ismós* > L. *-ismus* > E. *-ism*, state, phase, tendency, action): a tropic phenomenon in which the stimulating factor is gravity, either positive or negative, depending on whether the organism moves toward it or away from it. The sporangiophores of *Phycomyces* and of *Mucor* (Mucorales) are negatively geotropic, although the mycelium appears indifferent to gravity. The fructifications of many ascomycetes and basidiomycetes have a negatively geotropic stalk, but the tramal plates of the fertile region are positively geotropic; e.g., the teeth of the Hydnaceae and the pores of the Polyporaceae (Aphyllophorales). In some species, such as *Polyporus rufescens*, it is notable that on occasion abnormal fruiting bodies develop, larger and more or less spherical, due to the loss of the capacity to respond to the morphogenic stimulus of gravity; these abnormal fructifications lack a differentiated stipe and pileus and the hymenial tubes are irregularly arranged over the whole surface. Normal fruiting bodies have stipes with negative geotropism, pilei with diageotropism, and hymenial tubes with positive geotropism. Also called **gravitropism**.

germ pore (L. *germen*, bud, offshoot, sprout; *porus*, pore < Gr. *póros*, passage): a hole or opening in the wall of a spore (conidium, ascospore, basidiospore, etc.) through which it germinates, producing a germ tube that procedes to develop the mycelium if conditions are favorable. Generally the germ pore of basidiospores is apical and oblique, as in *Coprinus comatus* and other species of the gen., or apical and central, as in *Panaeolina foenisecii* (Agaricales). In the ascospores of *Sordaria fimicola* and *Podospora* sp. (Sordariales) the germ pore is apical. In some spores there is a germ slit in place of a pore, as happens in the ascospores of *Hypoxylon* (Xylariales). Cf. **germ slit**.

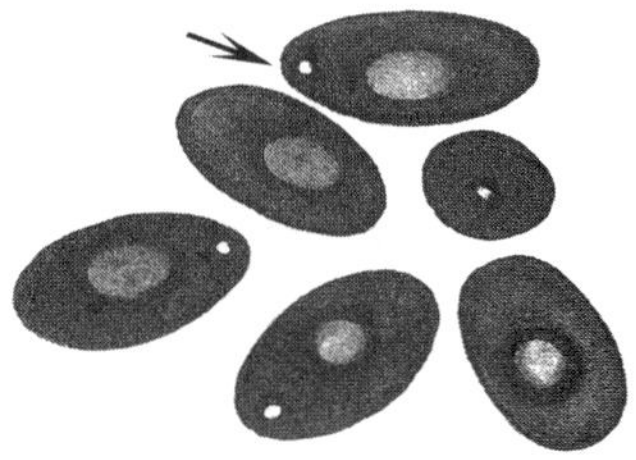

Germ pore of the ascospores of *Podospora* sp., x 360.

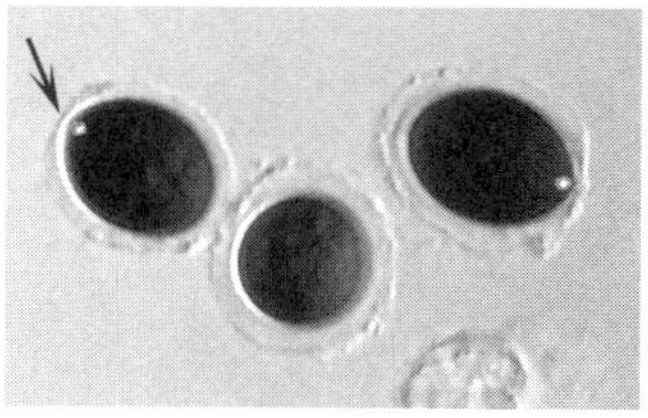

Germ pore of the ascospores of *Sordaria fimicola*, x 570 (*RTH*).

germ slit (L. *germen*, bud, offshoot, sprout; ME. *slitte* < *slitten*, to split along a line): a longitudinal fissure in a spore wall through which the germ tube emerges upon germination of the spore, e.g., *Hypoxylon serpens* (Xylariales). Cf. **germ pore**.

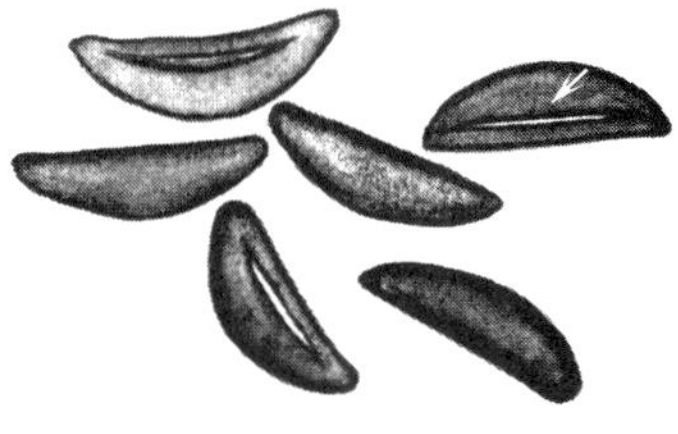

Germ slit of the ascospores of *Hypoxylon serpens*, x 570.

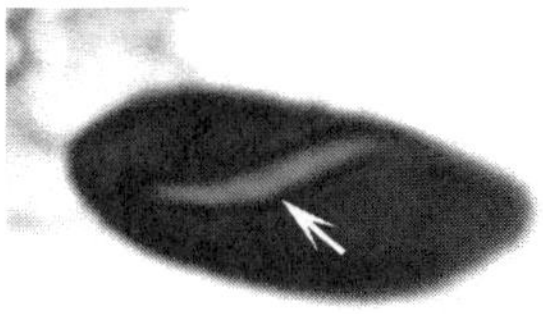

Germ slit of the ascospore of *Hypoxylon* sp., x 1 000 (*CB*).

germ sporangium, pl. **sporangia** (L. *germen*, bud, offshoot, sprout; Gr. *sporá*, spore + *angeîon*, vessel, receptacle): a sporangium that forms at the end of a germ tube or hypha which originates by the germination of an oospore (as in *Phytophthora infestans*, Peronosporales) or of a zygosporangium (as in *Rhizopus nigricans*, Mucorales).

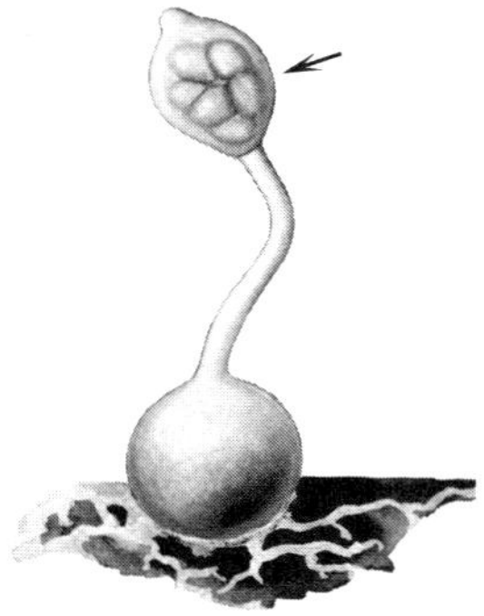

Germ sporangium from a germinating oospore of *Phytophthora infestans*, x 450.

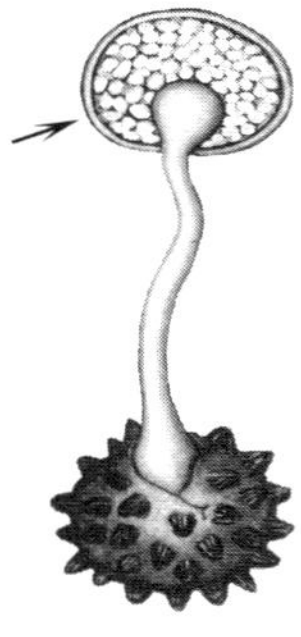

Germ sporangium from a germinating zygosporangium of *Rhizopus nigricans*, x 180.

germ tube (L. *germen*, bud, offshoot, sprout; *tubus*, tube): a short hypha that sprouts from the germ pore or slit of a spore during germination, and which on continuing its development under favorable conditions forms a hypha of larger size, including a mycelium. In many plant pathogenic fungi, e.g., the germ tube forms an **appressorium**, from which is originated the infective hypha that penetrates the tissues of the host. The germ tube represents the initiation of a new assimilative phase of the fungus; it can be clearly seen in the germinating conidia of *Conidiobolus thromboides* (Entomophthorales), and in the germinating ascospores of *Leptosphaerulina crassiasca* (Dothideales).

germicide (L. *germen*, bud, offshoot, sprout + *-cide* < *caedere*, to kill): a substance that kills microorganisms, although the term is utilized particularly for bacteria. See **fungicide**.

germling (L. *germen*, bud, offshoot, sprout + ME. *-ling*, young, small): a bud or newly developed propagule capable of growing into an adult organism.

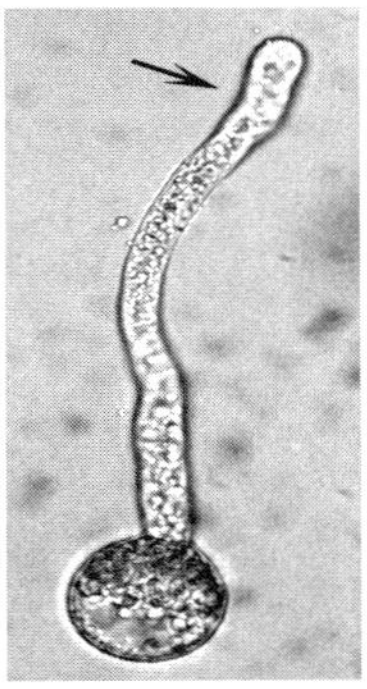

Germ tube of a germinating conidium of *Conidiobolus thromboides*, x 480 (*MU*).

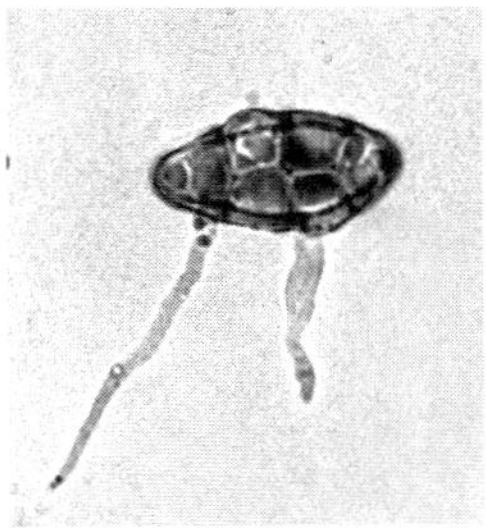

Germ tubes of a germinating ascospore of *Leptosphaerulina crassiasca*, x 560 (*RTH*).

gibbose, **gibbous** (L. *gibbosus*, hunched, humped < *gibbus*, bent, hunched < *gibber*, hunch, hump + *-osus*, full of, augmented, prone to > ME. *-ose*; or + *-osus* > OF. *-ous*, *-eus* > E. *-ous*, having, possessing the qualities of): with one or several assymetrical convexities, such as the sporangium of *Physarum polycephalum* (Physarales). Also used for a structure that has the upper part convex and the lower flat, or which has a boss or swelling, such as the pileus of *Psilocybe mazatecorum* (Agaricales).

Gibbose head of the sporangium of *Physarum polycephalum*, x 15 (*MU*).

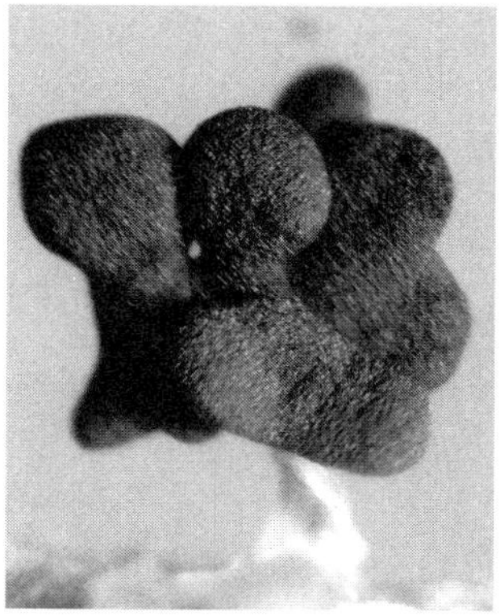

Gibbose head of the sporangium of *Physarum polycephalum*, x 17 (*MU*).

gill (ME. *gile*, *gille* < Scand., gill): *Agaricales*. One of the radiating plates forming the hymenophore on the undersurface of the cap of a mushroom. See **lamella**, which is a more international term than gill. **Gill fungi** is the common name for agarics.

gill fungus, pl. **fungi** (ME. *gile*, *gille* < Scand., gill; L. *fungus*, fungus): one of the Agaricales.

gill trama, pl. **tramae** (ME. *gile*, *gille* < Scand., gill; L. *trama*, trama, warp, texture): see **trama**.

glabrescent (L. *glabrescens*, genit. *glabrescentis* < *glabrescere*, to become hairless < *glabrare*, to make smooth + *-escens*, genit. *-escentis*, that which turns, beginning to, slightly > E. *-escent*): becoming glabrous, such as certain areas of the basidiocarp of *Polyporus hirsutus* (Aphyllophorales).

glabrous (L. *glabrare*, to make smooth, deprive of hair and bristles + L. *-osus* > OF. *-ous*, *-eus* > E. *-ous*, possessing the qualities of): lacking hair. Cf. **pubescent**.

gladiate (L. *gladiatus* < *gladius*, sword + suf. *-atus* > E. *-ate*, provided with or likeness): syn. of **ensiform**.

glaireous (E. *glair* or *glaire* < ME. *gleyre*, egg white < MF. *glaire*, modification of VL. *claria* < L. *clarus*, clear + L. *-osus* > OF. *-ous*, *-eus* > E. *-ous*, having, possessing the qualities of): viscous or slimy, suggestive of an egg white, as in some plasmodia of Myxomycetes.

gleba, pl. **glebae** (NL. < L. *gleba*, lump, mass): the central, internal portion of the fruiting body of the Gasteromycetes, composed of a fertile hymeniiferous part, which produces the basidiospores, and a sterile part, consisting of a trama of pseudotissue. The central zone of the basidiocarp of *Scleroderma* (Sclerodermatales) corresponds to the gleba. In the Nidulariales the gleba is disposed in a glebal chamber or **peridiole**, and in the Phallales (such as *Phallus*) it is present as a viscous mass in the receptacle, localized in the apex of the stipe of the basidiocarp.

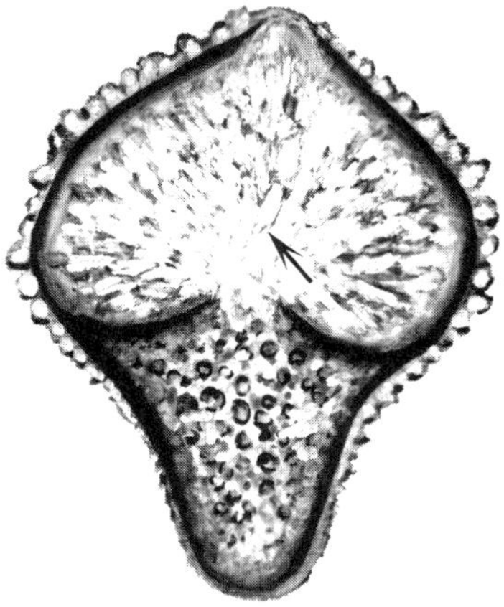

Gleba of the fruiting body of *Lycoperdon perlatum*, seen in longitudinal section, x 1.

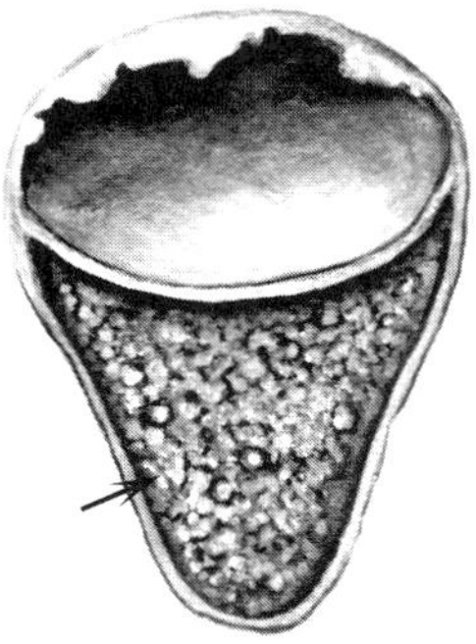

Gleba of the fruiting body of *Vascellum pratense*, seen in longitudinal section, x 1.

glebal chamber (L. *gleba*, lump, mass + L. suf. *-alis* > E. *-al*, relating to or belonging to; ME. *chambre* < OF. < L. *camara*, room, vault): a cavity within the gleba that is usually lined with hymenial elements, either empty or gel-filled, e.g., in *Hymenogaster* (Hymenogastrales) and *Melanogaster* (Melanogastrales)

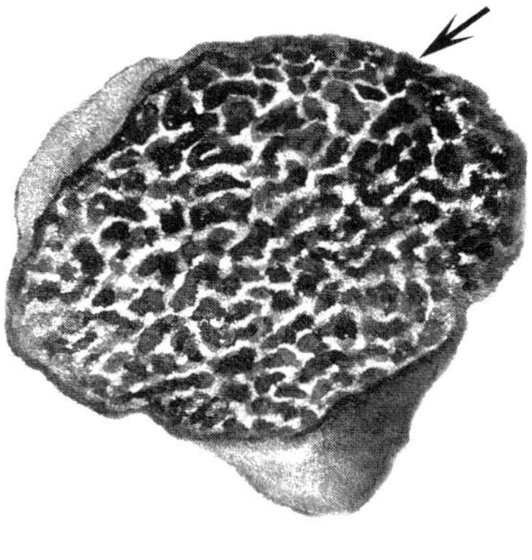

Glebal chambers of a basidiocarp of *Hymenogaster sulcatus*, seen in longitudinal section, x 4.5.

glebifer (L. *gleba*, lump, mass + *-fer* < *ferre*, to bear): *Gasteromycetes*. A special hymenial structure borne in the arch of the receptaculum, called a **lantern**. The glebifer is held in place by either a piece of tissue or by **trabeculae** which anchor it to the arch, as in *Laternea* (Phallales).

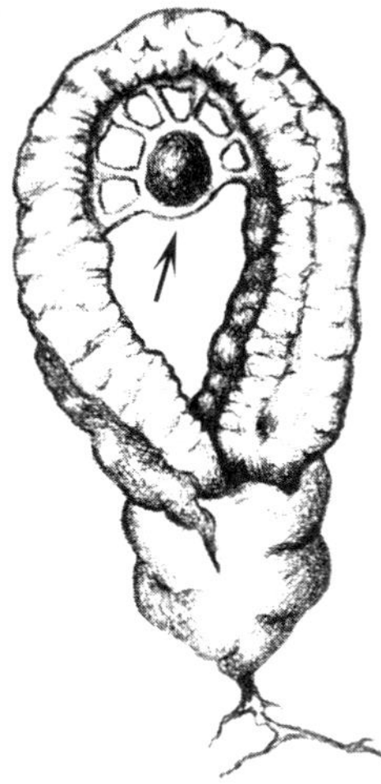

Glebifer of the fruiting body of *Laternea pusilla*, x 1.

glebose, **glebous** (L. *glebosus*, full of lumps < *gleba*, lump + *-osus*, full of, augmented, prone to > ME. *-ose*; or + *-osus* > OF. *-ous*, *-eus* > E. *-ous*, having, possessing the qualities of): full of lumps; lumpy.

gleocystidium, pl. **gleocystidia**, **gloeocystidium**, pl. **gloeocystidia** (Gr. *gloiós*, gum and other glutinous substances + *kystídion* < *kýstis*, bladder + dim. suf. *-ídion* > L. *-idium*): a cystidium with oily contents present in some species of *Russula*, such as *R. polyphylla* (Agaricales) and other fungi. With cresil blue, the gloeocystidia stain intensely blue, except for the wall which aquires a pale violet tone. The gloeocystidia are versiform cystidia, of which four types are known: **pseudocystidia** or **macrocystidia**, **chrysocystidia**, **phaeocystidia** and **coscinocystidia**.

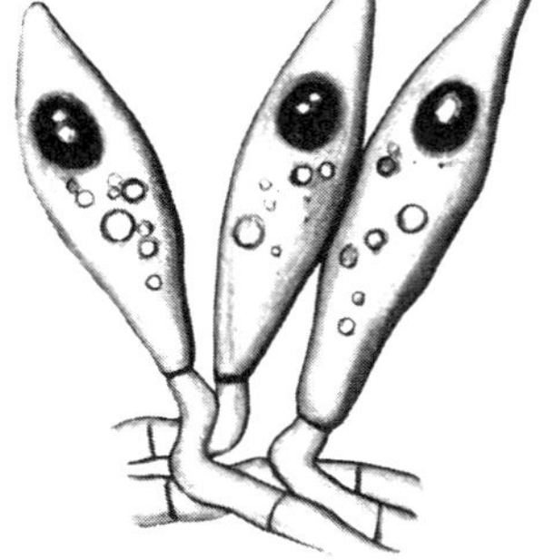

Gleocystidia from the hymenium of *Russula polyphylla*, x 740.

gleoid (Gr. *gloiós*, gum, glue and other sticky substances + *-oeídes*, similar to): any glutinous substance; applied to gelatinous, viscid or adhesive structures, such as hyphae, spores, etc.

gleoid head (Gr. *gloiós*, gum, glue and other sticky substances + *-oeídes*, similar to; E. *head* < ME. *heved* < OE. *heafod* < IE. *kauput*, *kaupet*, head): the mucilaginous, globose mass containing embedded spores that forms at the apex of conidiogenous cells, such as phialides. The conidia accummulate around the mouth of the phialides as they are formed. The mucilage of the gleoid heads, which keeps the spores together, eventually dissolves in atmospheric moisture, permitting the dissemination of the spores. Gleoid heads may be of different colors, depending upon the species. Such heads occur, e.g., on the phialides *of Acremonium* (= *Cephalosporium*), of the moniliaceous asexual fungi, and on the synnemata of *Antromycopsis smithii* and *Didymostilbe* sp. and *Stilbella flavida* (stilbellaceous asexual fungi). See **slime spores**.

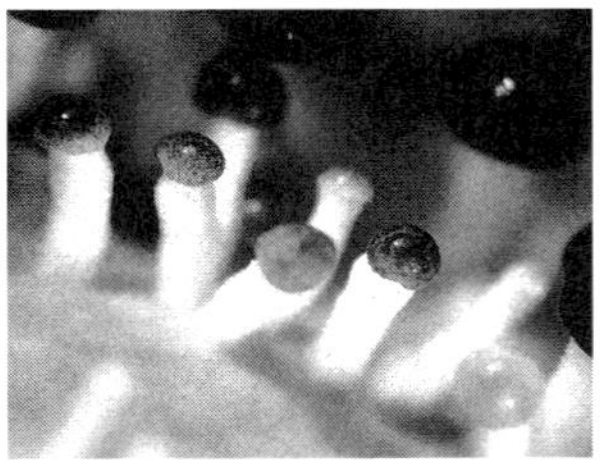

Gleoid head of the synnemata of *Antromycopsis smithii* on agar, x 7 (*RV*).

Gleoid head of the synnemata of *Antromycopsis smithii* on agar, x 7 (*CB & MU*).

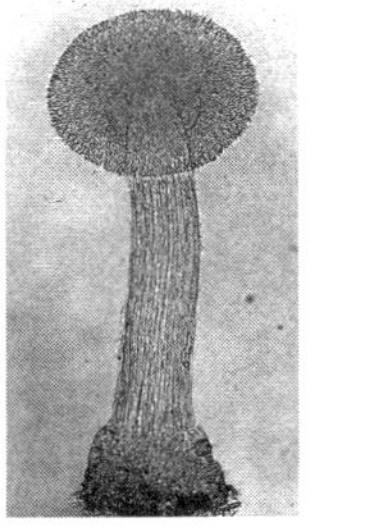

Gleoid head of a synnema of *Stilbella flavida*, with embedded conidia, x 48 (*RTH*).

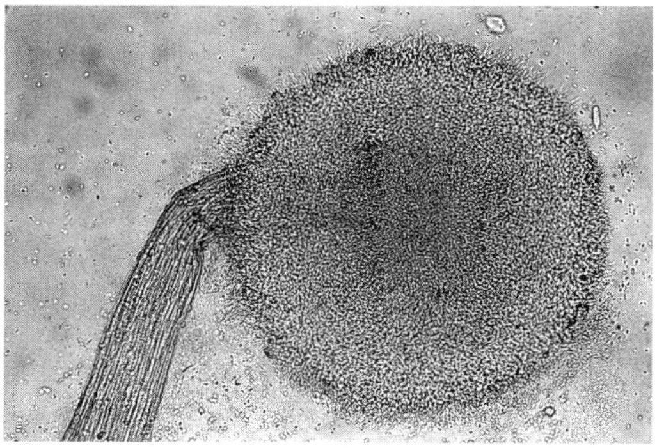

Gleoid head of a synnema of *Stilbella flavida*, with embedded conidia, x 105 (*RTH*).

Gleoid head of the synnemata of *Didymostilbe* sp. on agar, x 1.5 (*CB*).

gleolichen, **gloeolichen** (Gr. *gloiós*, gum, mucilage and other glutinous substances + *leichén* > L. *lichen*, lichen): a homomerous lichen whose gonidia or algal cells belong to *Chroococcus*, *Gleocapsa* and other Chroococcales, which have a mucilaginous capsule.

gleoplerous hypha, pl. **hyphae** (Gr. *gloiós*, gum and other glutinous substances + *plerés*, to fill + L. *-osus* > OF. *-ous*, *-eus* > E. *-ous*, having, possessing the qualities of; Gr. *hyphé*, tissue, spider web; hypha): refers to hyphae with long cells containing numerous oil droplets in the cytoplasm.

gleospore, **gloeospore** (Gr. *gloiós*, gum and other glutinous substances + *sporá*, spore): a viscous or mucilaginous, moist spore, like that of *Gliocladium roseum* (moniliaceous asexual fungi) and *Marssonina* (=*Gloeosporium*), of the melanconiaceous asexual fungi, adapted for dissemination by water drops, simple contact, insects and other similar means. Cf. **xerospore**.

gliatope (NL. *glia* < Gr. *glía*, glue; a gelatinous substance that absorbs water to form a viscous solution + *tópos*, place): a site of abundant gel production in gelatinous fungi. Cf. **gel tissue**.

globose, **globoid**, **globular** (L. *globosus*, round as a ball < *globus*, ball + *-osus*, full of, augmented, prone to > ME. *-ose*; or + L. *-oide* < Gr. *-oeídes*, similar; L. *globulus*, dim. of *globus* + L. suf. *-aris* > E. *-ar*, like, pertaining to): having the shape of a globe or globule; sphaerical or almost spherical. Applied to structures (spores, sporocarps, etc.) whose length:width ratio is between 1:1.0 and 1:1.05.

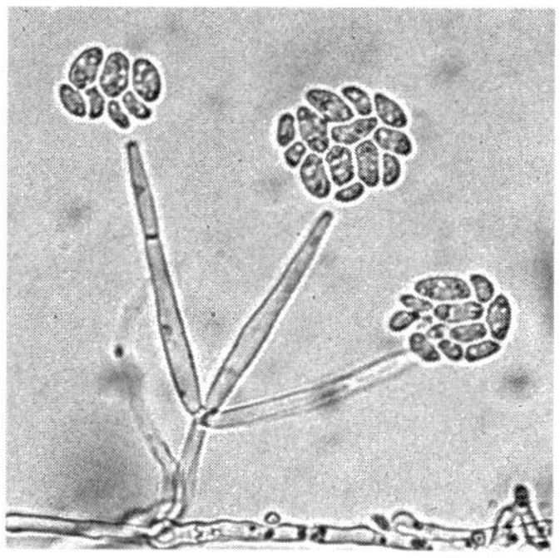

Gleospores of the phialides of *Gliocladium roseum*, x 620 (*JAS*).

glochid, **glochidium**, pl. **glochidia** (NL. *glochidium* < Gr. *glochídion* < *glochís*, arrowhead or trident + dim. suf. *-ídion* > L. *-idium*): a hair or bristle with small barbs.

glomerulus, pl. **glomeruli** (NL. *glomerulus*, dim. of *glomer-*, *glomus*, ball of thread or yarn): a dense mass formed by an aggregate of cells or massed cells, as is observed on the surface of certain lichens. Also used for the haustoria of some species of *Septobasidium* (Septobasidiales), and for the mass of hyphae formed by mycorrhizogenous fungi in the interior of the cells of the associated plant, in which case it is also known as **peloton**.

glossoid cell (Gr. *glõssa*, tongue + L. suf. *-oide* < Gr. *-oeídes*, similar to; NL. *cellula*, cell, living cell < L. dim. *cella*, small room): an elongate (tongue-shaped) cell in *Haptoglossa* (Myzocytiopsidales), a member of the Oomycota. See **trichocyst**.

gluten (L. *gluo*, genit. *glutinis*, glue, adhesive): a viscous substance, very sticky (glutinous), which is derived from the dissolution of the gelatinous hyphae that form part of certain fungal tissues, such as those of the cuticle of the pileus and stipe, or like those of the universal veil, such as are observed in fungi of the gen. *Suillus* (Agaricales).

glutinose, **glutinous** (MF. *glutineux* < L. *glutinosus*, sticky < *gluten*, glue + *-osus*, full of, augmented, prone to > ME. *-ose*; or + *-osus* > OF. *-ous*, *-eus* > E. *-ous*, having, possessing the qualities of): with an adhesive surface; having the consistency of eggwhite or liquid glue; e.g., as in the basidiocarp of *Gomphidius glutinosus* (Agaricales). Equivalent to **viscid**.

gnotobiotic (Gr. *gnótos*, well known + *biotikós*, pertaining to life < *bíos*, life + *-tikós* > L. *-ticus* > E. *-tic*, relation, fitness, inclination or ability): a term applied to axenic cultures, when it is certain that it contains a single species, free of contamination; pure. Cf. **agnotobiotic**, **axenic** and **monoxenic**.

Golgi apparatus, **Golgi body** (named after discoverer Camilo Golgi; L. *apparatus*, equipment < *apparare*, to prepare; ME. < OE. *bodig*, body): a group of flat cisternae or packets delimited by double membranes, often with the surface dilated or vesiculose, from which are generated secretory vesicles that are involved in the synthesis of the plasma membrane and of the cell wall in the subapical region of a growing hypha. These stacked cisternae also are called **dictyosomes**.

gongylidius, pl. **gongylidia** (dim. of Gr. *gongýlos*, round): a bulbous or globular structure that develops in fungi cultivated by termites.

gongylus, pl. **gongyli** (Gr. *gongýlos*, round, alluding to the spherical shape of the majority of spores): **1**. A globular propagative body that is formed in the thallus of certain lichens; an example is the the **soredium**. **2**. The swollen cells that are formed in the apices of hyphae and which serve as food for the ants that cultivate fungal gardens. See **bromatium** and **staphylum**.

gonidium, pl. **gonidia** (NL. < Gr. *gonídion* < *gónos*, offspring, seed + dim. suf. *-ídion* > L. *-idium*): a lichenic alga (preferably a Chlorophyceae) to which is attributed a role in the sexual reproduction of the lichen thalli, a process that only the mycobiont undergoes; notwithstanding, there are hymenial gonidia that are found in the hymenium of the reproductive bodies of the fungus and which are disseminated at the same time as the spores of the latter, assuring the formation of a new lichen, since these spores germinate together with the appropriate algal cells, in order to establish symbiosis with them. See **gonimium**.

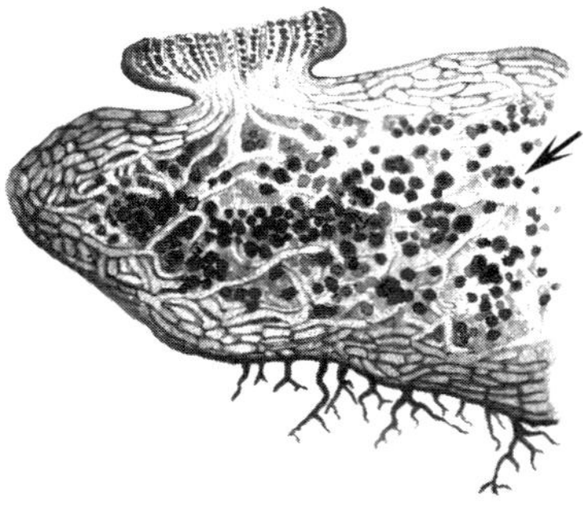

Gonidia of the thallus of *Parmelia imbricatula*, seen in longitudinal section, x 10.

gonimium, pl. **gonimia** (Gr. *gónimos*, fertile, able to produce + L. dim. suf. *-ium*): *Lichens.* An algal cell (gonidium) with a thin cell wall and blue color, belonging to the Cyanophyceae, that is found in certain lichen thalli. See **gonidium**.

goniocyst (NL. *goniocyst* < *gonium* < Gr. *gónos*, offspring + *kýstis*, bladder, vesicle, cell): a vegetative propagule of the thallus of some tropical foliicolous lichens, composed of a cell of the photobiont and its progeny enveloped by the hyphae of the mycobiont; it differs from a soredium because it is formed in a special organ called the **goniocystangium**.

goniocystangium, pl. **goniocystangia** (NL. *goniocyst* < *gónos*, offspring + *kýstis*, bladder, vesicle, cell + *angeîon*, vessel, receptacle): a special organ of some foliicolous lichens where **goniocysts** are produced.

goniospore (Gr. *gonía*, angle + *sporá*, spore): an angular spore; e.g., like the basidiospores of *Inocybe* (Agaricales).

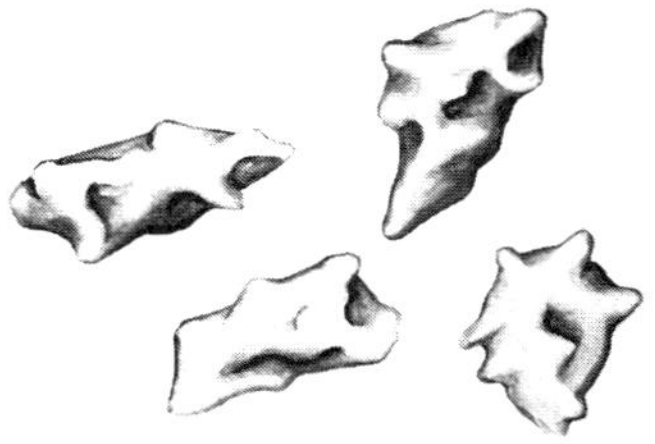

Goniospores of *Inocybe* sp., x 1 800.

gonoplasm (Gr. *gónos*, procreation, progeny + *plásma*, the bland material with which is formed an organ or living being): *Peronosporales*. The central mass of protoplasm of the antheridium which passes through the fertilization tube into the oogonium, where it fuses with the oosphere.

gonotaxis, pl. **gonotaxes**, **gonotropism** (Gr. *gónos*, procreation, as a pref. implying the idea of sexuality or engenderment + *táxis*, disposition; or *gónos* + *trópos*, a turn, change in manner < *tropé*, a turning < *trépo*, to revolve, turn towards + *-ismós* > L. *-ismus* > E. *-ism*, state, phase, tendency, action): the movement of antherozoids toward the female sexual organ, which contains an oosphere; this is supposedly due to the unilateral effect of a stimulating factor, as in *Monoblepharis polymorpha* (Monoblepharidales). Because the movement is toward the center from which the excitation emmanates, it is considered a positive tropism.

gonotocont, **gonotokont** (Gr. *gónos*, that which aludes to the offspring + *tókos*, childbirth): a diploid cell or organ in which, by reduction division or meiosis, the offspring are formed.

gossamer (ME. *gossomer* < *gos*, goose + *somer*,

summer; a film of cobwebs floating in air in calm clear weather; something light, delicate, tenuous): the fine, floating mycelial nets produced by fungi on culture media lacking added carbon.

grain (Fr. *grain* < L. *granum*, grain, seed, small kernel, small particle): each of the minute granules formed by several pathogenic species of actinobacteria (e.g., *Nocardia brasiliensis* and *Streptomyces madurae*) and fungi (e.g., *Madurella mycetomi* and *M. grisea*, of the dematiaceous asexual fungi) that cause the mycetoma of humans and higher animals. Such granules are contained in the pus of the lesions; they vary from microscopic in size to more than 2 mm in diameter. The size, color, shape and texture of the granules, as well as the dimensions of the fungus hyphae within the granule, vary greatly with the species of fungus; the characteristics of the grains are used in the differential diagnosis of the causative agents. For example, in *M. mycetomi* the grains have a homogeneous brown pigmentation of the matrix.

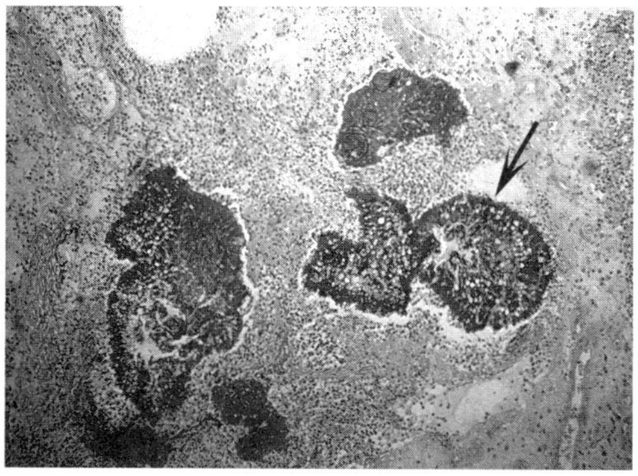

Grains of *Madurella mycetomi*, seen in a cross section of infected host tissue, x 30 (*RLM*).

graminicole, **graminicolous** (L. *gramen*, grass + *-cola*, inhabitant + L. *-osus* > OF. *-ous*, *-eus* > E. *-ous*, possessing the qualities of): growing on grasses, like the perithecial stromata of *Balansia henningsiana* and *Claviceps purpurea* (Hypocreales).

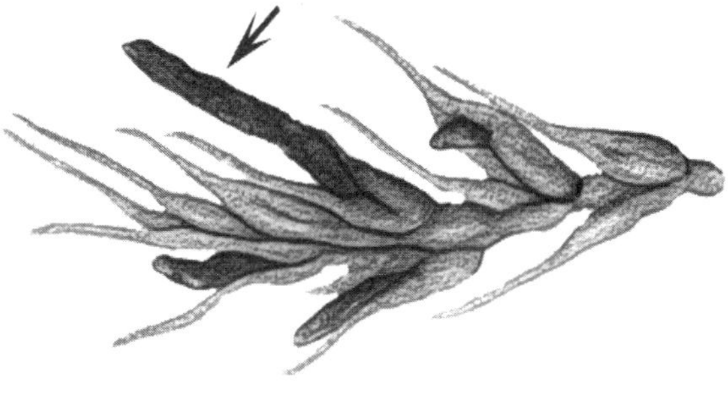

Graminicolous sclerotia of *Claviceps purpurea* (ergot), developed on a rye spike, x 1.

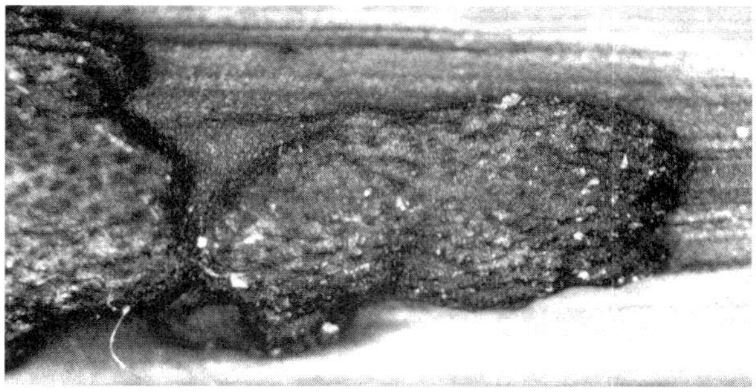

Graminicolous perithecial stroma of *Balansia henningsiana*, x 47 (*RTH*).

granular, **granulate**, **granulose** (L. *granulus*, dim. of *granum*, grain, seed of wheat, barley and other grasses + L. suf. *-aris* > E. *-ar*, like, pertaining to; or + L. suf. *-atus* > E. *-ate*, provided with or likeness; or + L. *-osus*, full of, augmented, prone to > ME. *-ose*): having or composed of grains.

gravitaxis, pl. **gravitaxes** (MF. < L. *gravis*, heavy + Gr. *táxis*, arrangement, disposition): a tactic movement in response to the gravitational attraction or force of the mass of the Earth, e.g., the zoospores of *Phytophthora palmivora* (Peronosporales). Cf. **electrotaxis**.

gravitropism (L. *gravis*, heavy + Gr. *trópos*, a turn, change in manner < *tropé*, a turning < *trépo*, to revolve, turn towards + *-ismós* > L. *-ismus* > E. *-ism*, state, phase, tendency, action): a reaction to gravity, e.g., the sporangiophores of Mucorales and stipes, gills and tubes of basidiomycetes. Syn. of **geotropism**.

gregarious (L. *gregarius*, grouped, belonging to a herd or flock < *grex*, genit. *gregis*, flock, group + *agora*, assembly + *-arius*, belonging to): an organism that lives in close proximity to others in a small area; not very scattered. The sporangia of a majority of the Myxomycetes, such as *Perichaena* sp. (Trichiales), *Diachea leucopodia*, *Leocarpus fragilis* and *Physarum cinereum* (Physarales), are gregarious, as are many Agaricales and some Gasteromycetes, such as *Lycoperdon pyriforme* (Lycoperdales).

Gregarious sporangia of *Leocarpus fragilis*, x 6 (*MU & CB*).

Gregarious sporangia of *Perichaena* sp., x 25 (*MU*).

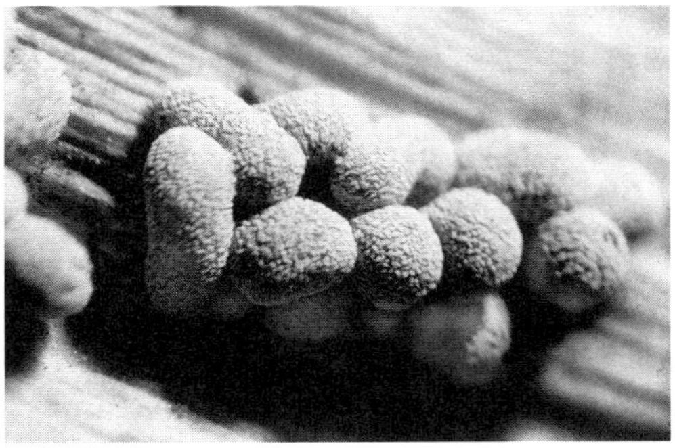

Gregarious sporangia of *Physarum cinereum* on a plant leaf, x 9 (*MU*).

Gregarious fruiting bodies of *Lycoperdon pyriforme*, x 0.4 (*RV*).

grex (L. *grex*, flock, group): a term for the **pseudoplasmodium** of the Dictyosteliomycota, which forms by the aggregation of myxamoebae, therefore also called aggregation plasmodium.

grumose, **grumous** (L. *grumus*, lump + *-osus*, full of, augmented, prone to > ME. *-ose*; or + *-osus* > OF. *-ous*, *-eus* > E. *-ous*, having, possessing the qualities of): piled up; composed of heaped grains or granules.

gummose (L. *gummosus*, gummy < *gummi*, gum + *-osus*, full of, augmented, prone to > ME. *-ose*): gummy, mucilaginous, like the exudate of various sources that is produced by some plants infected by fungi, bacteria, viruses, etc.

gusset (ME. < OF. *gousset* < *gousse*, pod, husk): each of the segments resulting from the splitting of the fruiting body of *Montagnea* (Hymenogastrales). The gussets are black, hardened, recurved plates on which the spores are borne; they represent regions of the tramal plate that has thickened at the point of contact with the peridium or with another tramal plate.

Gussets of the fruiting body of *Montagnea arenaria*, x 1.

guttulate (L. *guttulatus*, containing drops or drop-like masses < *guttula*, dim. of *gutta*, drop + suf. *-atus* > E. *-ate*, provided with or likeness): with droplets or globules of oil. There are uni-, bi-, tri- or multiguttulate structures. E. g., the ascospores of *Podospora comata* (Sordariales) are uniguttulate; the conidia of *Phoma glomerata* (sphaeropsidaceous asexual fungi) and the ascospores of some *Xylaria* spp. (Xylariales) are biguttulate, since they contain an oil droplet in each end. Among the basidiomycetes, the basidiospores of *Bolbitius vitellinus* (Agaricales) are uniguttulate, and among the conidial fungi each of the two central, pigmented cells of the conidia of *Pestalotia* (melanconiaceous asexual fungi) are occasionally uniguttulate.

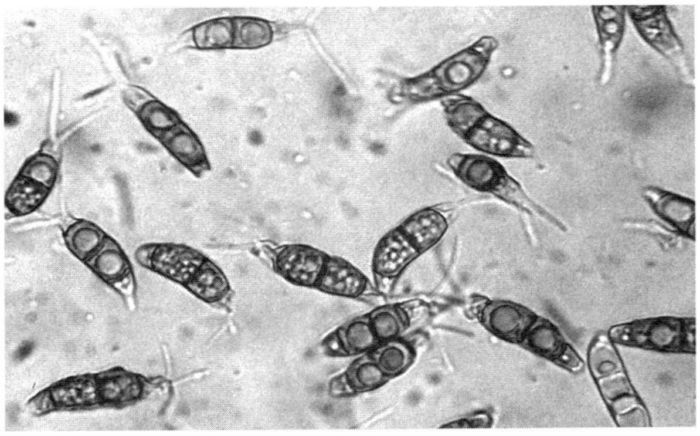

Guttulate conidia of *Pestalotia* sp., x 830 (*MU*).

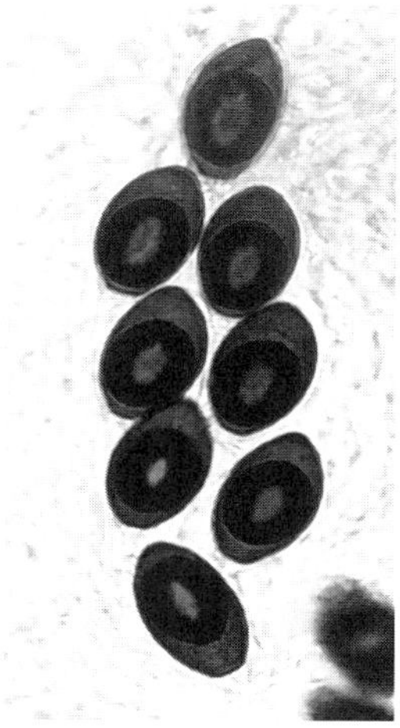

Guttulate (biguttulate) ascospores of *Xylaria* sp., x 500 (*RTH*).

gymnocarpic, **gymnocarpous** (Gr. *gymnós*, naked + *karpós*, fruit + suf. *-íkos* > L. *-icus* > E. *-ic*, belonging to, relating to; or + L. *-osus* > OF. *-ous*, *-eus* > E. *-ous*, having, possessing the qualities of): a fruiting body that is open from the early stages of its development, exposing the fertile layer that produces the spores; some ascomata (apothecia), such as those of *Leotia* and *Geoglossum* (Helotiales), and basidiocarps, such as those of *Ramaria* and *Clavaria* (Aphyllophorales), are gymnocarpic. Cf. **angiocarpic** and **hemiangiocarpic**.

gymnothecium, pl. **gymnothecia** (NL. *gymnothecium* < Gr. *gymnós*, naked + *thekíon*, dim. of *thêke*, case, box; here, of the asci): an ascoma composed of small clusters of asci surrounded by a loose network of hyphae, with or without characteristic appendages of various types; typically found in gen. of the families Arthrodermataceae, Myxotrichaceae, and Gymnoascaceae (Onygenales), which includes, among others, *Arthroderma*, *Myxotrichum*, *Eidamella*, *Gymnoascus* and *Pseudogymnoascus*.

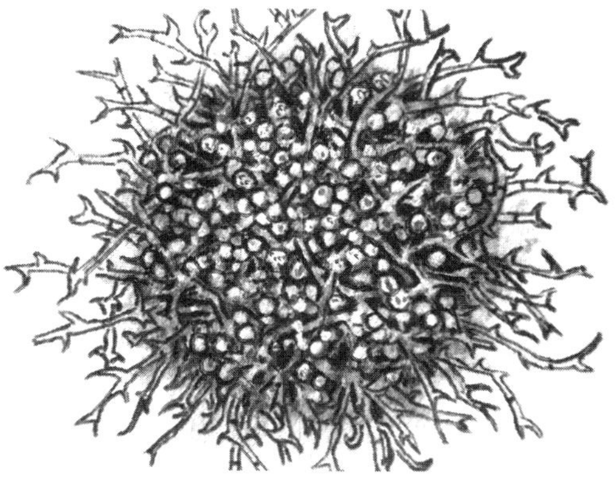

Gymnothecium of *Gymnoascus reessii*, x 430.

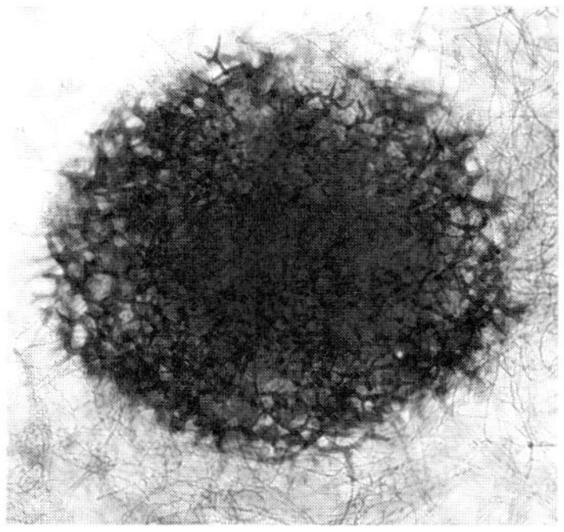

Gymnothecium of *Pseudogymnoascus roseus*, x 260 (*CB*).

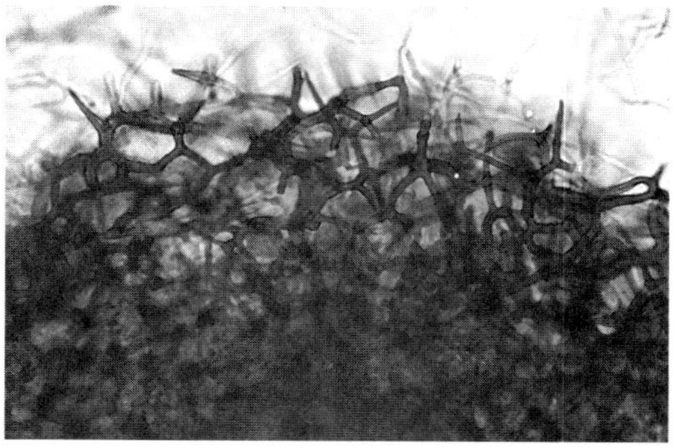

Gymnothecium of *Pseudogymnoascus roseus*; a close up of the peripheral area, x 760 (*CB*).

gynophore (Gr. *gyné*, female; here, female sexual organ + *-phóros*, bearer): that which bears or is borne by the female sexual organ; accomplished in a feminine way. Applied to the fructification of *Pyronema* (Pezizales) originating from the multinucleate female structure that develops numerous helical ascogonia.

gynotrichous (Gr. *gyné*, female, female sexual organ + *thríx*, *trichós*, hair + L. *-osus* > OF. *-ous*, *-eus* > E. *-ous*, having, possessing the qualities of): see **trichogyne**.

gyrodisc (L. *gyrus* < Gr. *gýros*, rotation + *dískos*, disc): *Lichens*. An apothecium that has concentric circles on its upper surface; e.g., as in *Umbilicaria cylindrica* (Lecanorales).

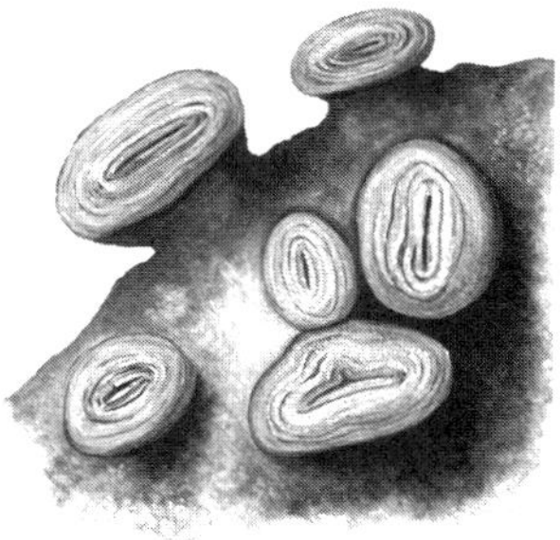

Gyrodiscs of the thallus of *Umbilicaria cylindrica*, x 7.5.

gyroma, pl. **gyromata** (Gr. *gyrós*, round, circular + *-oma*, suf. which implies entirety): an apothecium with concentric circular ridges, as in the gen. *Gyrophora* (syn. of *Umbilicaria*, lichen of the order Lecanorales). Appears to be equivalent to **gyrodisc**.

gyrose, gyrate (Gr. *gyrós*, round, circular + L. suf. *-osus*, full of, augmented, prone to > ME. *-ose*; or + L. suf. *-atus* > E. *-ate*, provided with or likeness): curved toward the rear and then forward; doubled and undulate, cerebriform or convolute; e.g., like the sporangium of *Physarum polycephalum* (Physarales), the pileus of *Gyromitra* (Pezizales) and the stipe of the fruiting body of *Tulostoma* (Tulostomatales).

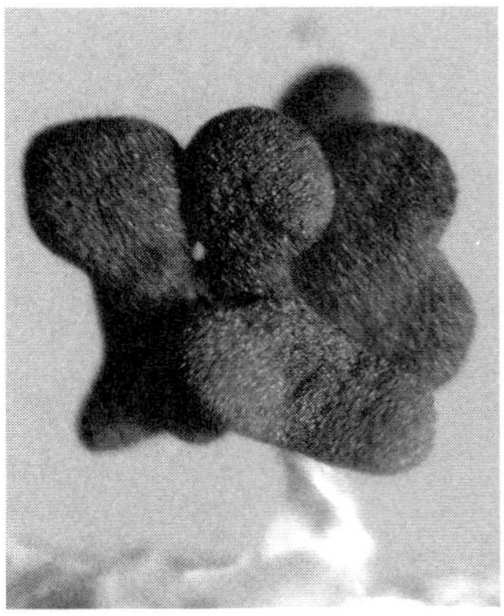

Gyrose sporangial head of *Physarum polycephalum*, x 17 (*CB*).

Gyrose stipe of the fruiting body of *Tulostoma* sp., x 6.5 (*MU & CB*).

h

habit, **habitus** (L. *habitus*, condition, exterior appearance, attire, nature): the general external appearance, or manner of growth, of a fungus.

habitat (L. *habitat* < *habitare*, to inhabit, to dwell, reside): the natural place where an organism grows.

haerangium, pl. **haerangia** (L. *haerere*, to be adhered + Gr. *angeîon*, vessel, receptacle): a sporulating organ of certain ascomycetes, such as *Ceratostomella* (Xylariales), in which the eight ascospores that are developed from the **octophore** remain covered by a membrane and are surrounded by a ring of hairs, the **tenacle** around the ostiole of the perithecium.

halonate, **haloniferous** (L. *halos*, circle, halo + suf. *-atus* > E. *-ate*, provided with or likeness; or + L. *-ferous*, bearer < *ferre*, to bear, carry + *-osus* > OF. *-ous*, *-eus* > E. *-ous*, having, possessing the qualities of): used for certain lichen spores that appear to be surrounded by a halo, due to the presence of a mucilaginous, hyaline epispore.

halophilic, **halophilous** (Gr. *halós*, salt, especially when alluding to salty soils or water + *phílos*, have an affinity for + suf. *-íkos* > L. *-icus* > E. *-ic*, belonging to, relating to; or + L. *-osus* > OF. *-ous*, *-eus* > E. *-ous*, possessing the qualities of): that which prefers salty media (soil or water). The marine and salt fungi are halophilic, e.g., *Catenochytridium carolinianum*, a zoosporic fungus (of the Chytridiales), parasitic on the marine alga *Cladophora japonica*, as well as some species of yeasts, e.g., *Candida marina* (cryptococcaceous asexual yeasts). Cf. **halophobic**.

halophobic (Gr. *halós*, salt + *phóbos*, that which fears or avoids < *phobéo*, to fear + suf. *-íkos* > L. *-icus* > E. *-ic*, belonging to, relating to): that which does not tolerate the presence of salt solutions, even in small proportions; the majority of the lichens are halophobic. Cf. **halophilic**.

hallucinogenic fungus, pl. **fungi** (L. *hallucinatus*, ptp. of *hallucinari*, *allucinari*, to prate, dream < Gr. *alýein*, to be distressed, to wander + *génos*, engenderment + suf. *-íkos* > L. *-icus* > E. *-ic*, belonging to, relating to; L. *fungus*, fungus): basidiomata of *Psilocybe* (Agaricales), mainly *P. mexicana* and *P. caerulescens*, which are eaten by several Mexican ethnic groups during religious ceremonies to induce visions or imaginary perceptions that are interpreted as magical. See **entheogen**. *Psilocybin* and *psilocin* are the active substances found in these fungi. Also called **magic mushrooms**.

hamate, **hamose**, **hamous** (L. *hamus*, fishhook + L. suf. *-atus* > E. *-ate*, provided with or likeness; or + L. *-osus*, full of, augmented, prone to > ME. *-ose*; or + *-osus* > OF. *-ous*, *-eus* > E. *-ous*, having, possessing the qualities of): similar to a fishhook, uncinate, hooked; e.g., like the perithecial appendages of *Uncinula macrospora* (Erysiphales). Also called **hamulose**.

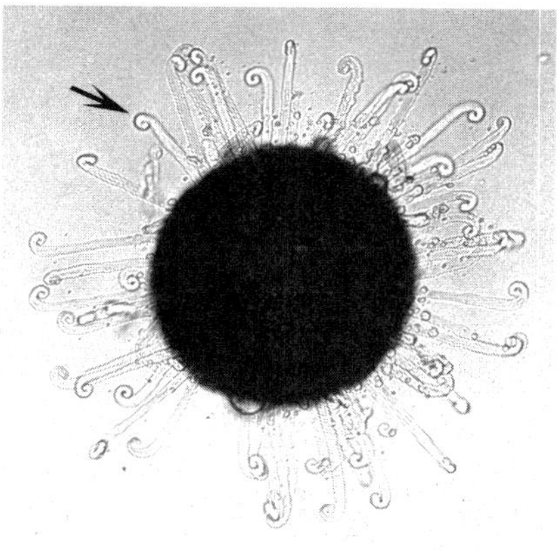

Hamose perithecial appendages of *Uncinula macrospora*, x 200 (*RTH*).

hamathecium, pl. **hamathecia** (NL. *hamathecium* < Gr. *háma*, all together, at the same time + *thekíon*, dim. of *thêke*, box, case; here, of the asci): a general term that is applied to the tissues that separate the asci in the ascomycetes, and whose nature depends on the ontogeny of the ascoma; these tissues include the paraphyses, paraphysoids, pseudoparaphyses, periphyses, and periphysoids, which occur in distinct taxonomic groups of these fungi. The different types of hamathecium are: 1) **interascal pseudoparenchyma**, 2) **paraphyses**, 3) **paraphysoids**, 4) **pseudoparaphyses** 5) **periphysoids**, and 6) **periphyses**; or the hamathecium can be absent. See these terms.

hamulate, **hamulose** (L. *hamulus*, small fishhook, dim. of *hamus*, hook + L. suf. *-atus* > E. *-ate*, provided with or likeness; *hamulosus* < *hamulus* + *-osus*, full of, augmented, prone to > ME. *-ose*): see **hamate**.

hanging collar (ME. < OE. *hangande*, hanging < ME. *hangen* < OE. *hangian*, to hang; ME. *coler* < AF. < OF. *colier* < L. *collare*, neckband, collar): *Gasteromycetes*. A ridge of tissue surrounding the stalk and attached to the bottom of the endoperidium, typical of some species of *Geastrum* (Lycoperdales).

haplobiont, **haplobiontic** (Gr. *haplóos*, simple + *bíos*, life + *óntos*, genit. of *ón*, a being; or + Gr. *-tikós* > L. *-ticus* > E. *-tic*, relation, fitness, inclination or ability): an organism whose life cycle has a single type of cell or thallus, haploid in nature, such as species of conidial fungi (deuteromycetes) which reproduce only by means of asexual mechanisms, or at most a **parasexual** mechanism (never truly sexual, i.e., without meiosis). Cf. **diplobiont**. Do not confuse with **haplont**.

haploconidium, pl. **haploconidia** (Gr. *haplóos*, simple + *kónis*, dust + dim. suf. *-ídion* > L. *-idium*): a conidium with a haploid nucleus that is formed on the mycelium of the Tremellales. See **diploconidium**.

haplodiplobiont, **haplodiplobiontic** (Gr. *haplóos*, simple + *diplóos*, double + *bíos*, life + *óntos*, genit. of *ón*, a being; or + Gr. *-tikós* > L. *-ticus* > E. *-tic*, relation, fitness, inclination or ability): an organism that has alternation of generations and develops its life cycle in two phases, one haploid and the other diploid; this occurs, e.g., in the Chytridiomycete *Allomyces* (Blastocladiales) and in the yeast *Saccharomyces cerevisiae* (Saccharomycetales). Since it has two types of cells (N and 2N), it is a **diplobiontic** organism (literally, two modes of life).

haploid (Gr. *haplóos*, simple, single, alluding to the chromosome number): an organism, or phase in the life cycle (such as the gametes), in which the cells have only one set (N) of chromosomes; normally this is one-half the **diploid** (2N) number.

haplont (Gr. *haplóos*, simple, single + *óntos*, genit. of *ón*, a being): an organism whose existence occurs all, or mostly, in the haploid phase; this occurs, e.g., in yeasts of the gen. *Schizosaccharomyces* (Schizosaccharomycetales). Cf. **diplont**. Do not confuse with **diplobiont**.

haplostromatic (Gr. *haplóos*, simple, single + *strõma*, mattress + *-tikós* > L. *-ticus* > E. *-tic*, relation, fitness, inclination or ability): composed of only one type of stroma; e.g., in certain Xylariales the placodium forms only from the ectostroma. Cf. **diplostromatic**. See **placodium**.

hapteron (Gr. *hápto*, to fix, fasten): **1**. *Nidulariales*. A ball-like mass of adhesive hyphae that form an organ of attachment, located at the base of the funicular cord attached to the peridioles (as in *Cyathus striatus*). **2**. *Entomophthorales*. An adhesive appendage on the apex of the conidium, as in *Basidiobolus* and *Neozygites*. **3**. *Lichens*. Filaments that serve to fasten the thallus to the substrate in some fruticose species, such as *Cladonia*, *Usnea* and *Ramalina* (Lecanorales). They differ from the rhizines and from lichen hairs (which also are structures of attachment) by their place of origin. See **funicle**.

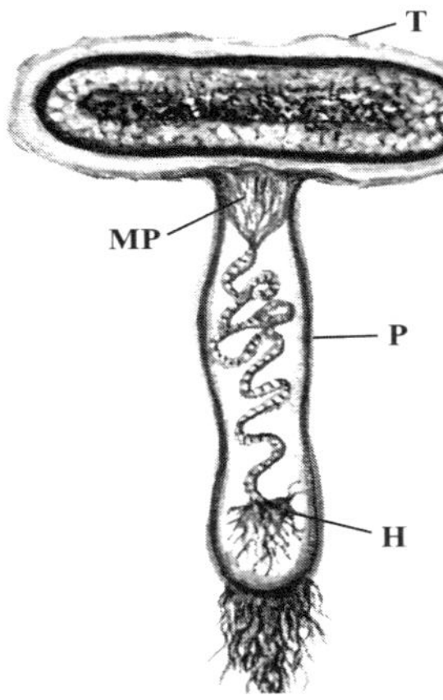

Hapteron (H) of a peridiolum of *Cyathus striatus*. The middle piece (MP), purse (P) and tunica (T) also are shown in the diagram of a longitudinal section, x 20.

haptomorphosis, pl. **haptomorphoses** (Gr. *hápto*, to grasp, touch, attach + *mórphosis*, the action of taking shape, of giving form): a structure whose shape is determined by a contact stimulus, as, e.g., in the formation of appressoria and of various sexual organs.

haptonema, pl. **haptonemata** (Gr. *hápto*, to grasp, touch, attach + *nêma*, filament): in the flagellated forms, it refers to a filamentous appendage (usually coiled) consisting of the plasma membrane, a sheath of endoplasmic reticulum, and a core of microtubules anchored near the kinetosome.

hastate (L. *hastatus*, armed with a spear < *hasta*, spear + suf. *-atus* > E. *-ate*, provided with or likeness): a flat, sharp-pointed, structure shaped like a lance or arrowhead. The **Stachel**, found in the encysted zoospore of *Plasmodiophora brassicae* (Plasmodiophorales), has a hastate apex. The synnemata of *Doratomyces stemonitis* (stilbellaceous asexual fungi) are also hastate.

Hastate synnemata of *Doratomyces stemonitis* on agar, x 8 (*MU*).

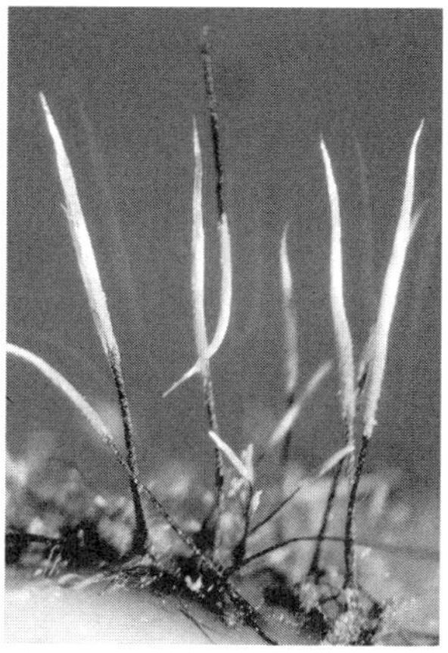

Hastate synnemata of *Doratomyces stemonitis* on agar, x 33 (*CB*).

haustorial cap (NL. *haustorium* < L. *haustor*, a drawer of water + dim. suf. *-ium* + L. suf. *-alis* > E. *-al*, relating to or belonging to; ME. *cappe* < LL. *cappa*, head covering, cloak, cap): an electron-dense, cap-like mass at the end of a lobe of the haustorium of *Exobasidium camelliae* (Exobasidiales).

haustorial mother cell (NL. *haustorium* < L. *haustor*, a drawer of water + dim. suf. *-ium* + L. suf. *-alis* > E. *-al*, relating to or belonging to; ME. *moder*, mother; L. *cella*, cell of a honeycomb, storeroom, chamber, cell): the cell from which the haustorium of plant pathogenic fungi originates, as in *Melampsora* (Uredinales). This cell comes in contact with the host cell wall and a layer of adhesive material is deposited between them. The infective hypha originates from the inner layer of the haustorial mother cell and effects the invagination of the parasitized host cell. During the development of the haustorium a **haustorial neck** is formed, in which three distinctive zones are diferentiated: 1) a zone that originates from the wall of the haustorial mother cell, which it resembles; 2) a ring or **collar**, which is stained dark; and 3) the lower haustorial neck which is surrounded by the **sheath**, which is the structure that separates the cytoplasm of the haustorium from that of the host cell, and which is the zone of the host-parasite interface.

haustorium, pl. **haustoria** (L. *haustor*, a drawer of water + dim. suf. *-ium*): an absorbent organ that originates from the hypha of a parasitic fungus and which penetrates into a host cell; the host protoplast is not breached by the haustorium, only invaginated. Haustoria are principally formed by obligately parasitic fungi, such as *Melampsora lini* (Uredinales), but also by facultative parasites and mycorrhizae. The shape of the haustoria varies depending upon the species, bulbous to pyriform or digitate; e.g., in *Blumeria* (=*Erysiphe*) *graminis*, of the Erysiphales, they are digitate or dactyloid.

Haustorium of *Blumeria graminis* within a host cell, x 1 250.

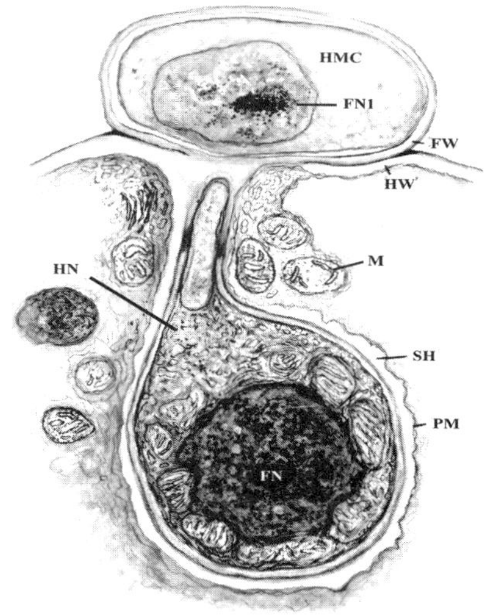

Haustorium of *Melampsora lini*. Longitudinal non-median section through the **haustorial mother cell** (HMC), at the point where it penetrates the cell wall of its host (flax, *Linum usitatissimum*). The fungus wall (FW) is in contact with the host cell wall (HW). The haustorium and **haustorial neck** (HN) are surrounded by the invaginated host plasma membrane (PM). One fungal nucleus (FN) is in the haustorium, and the other fungal nucleus (FN1) appears to be partly in the haustorial neck and partly in the haustorial mother cell. The host cell mitochondria (M) and sheath (SH) also are shown. Drawing based on a transmission electron micrograph published by Beckett *et al.* in their *Atlas of Fungal Ultrastructure*, 1974, x 9 800.

helicospore (Gr. *hélix*, *hélikos*, spiral + *sporá*, spore): a cylindrical spore coiled into a spiral or helix; helicospores are commonly multicellular, as in *Helicosporium linderi*, *Ceratophorum helicosporum* (moniliaceous asexual fungi) and *Helicoma* sp. (dematiaceous asexual fungi).

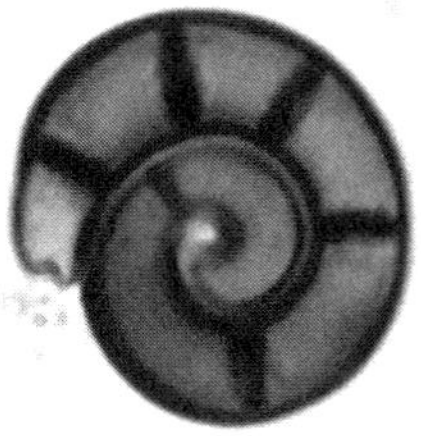

Helicospore of *Helicoma* sp., x 1 300 (*CC*).

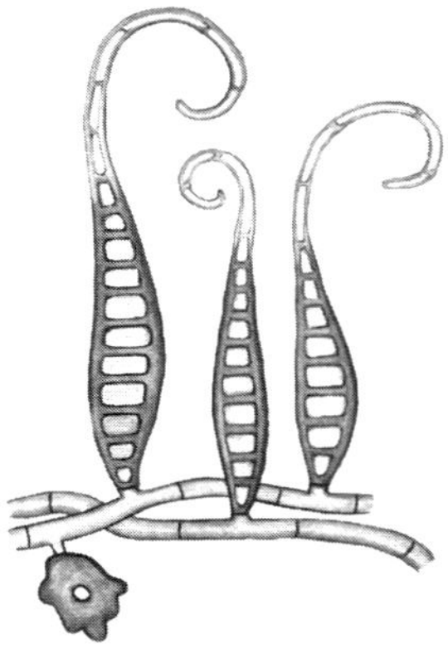

Helicospores of *Ceratophorum helicosporum*, x 260.

heliophilous (Gr. *hélios*, sun + *phílos*, have an affinity for + L. *-osus* > OF. *-ous*, *-eus* > E. *-ous*, possessing the qualities of): that which prefers sunny habitats, or which requires them in order to develop; e.g., such as species of *Panaeolus* (Agaricales). Cf. **heliophobic** and **anheliophilic**.

heliophobic, **heliophobous** (Gr. *hélios*, sun + *phóbos*, fearing or avoiding < *phobéo*, to fear + suf. *-íkos* > L. *-icus* > E. *-ic*, belonging to, relating to; or + L. *-osus* > OF. *-ous*, *-eus* > E. *-ous*, possessing the qualities of): that which avoids light and requires shade; e.g., such as species of *Mycena* (Agaricales). Also called **anheliophilous**. Cf. **heliophilous**.

heliotropism (Gr. *hélios*, sun + *trópos*, a turn, change in manner < *tropé*, a turning < *trépo*, to revolve, turn towards + *-ismós* > L. *-ismus* > E. *-ism*, state, phase, tendency, action): a tropism exhibited by heliotropic organs and organisms, which turn or move on being stimulated by sunlight. Positive heliotropism occurs, e.g., in the curvature of the stipes of the majority of coprinaceous (Agaricales) fruiting bodies, the sporangiophores of *Mucor*, *Phycomyces* and *Pilobolus* (Mucorales), the asci of the Ascobolaceae (Pezizales) and the perithecial necks of the Sordariaceae (Sordariales).

helotism (Gr. *Helos*, a city in Peloponesia < *heilótis*, *hilota* or *ilota*, a Greek slave + *-ismós* > L. *-ismus* > E. *-ism*, state, phase, tendency, action): a type of association between two living beings, in which one of the members obtains the major benefit; the association that exists between a fungus and an alga to form a lichen has at times been considered helotism, as well as mutualistic symbiosis. Cf. **antagonism**, **commensalism**, **metabiosis**, **mutualism**, **parabiosis**, **parasymbiosis**, **symbiosis** and **synergy**.

hemiangiocarpic (Gr. *hémi*, half + *angeîon*, vessel, receptacle + *karpós*, fruit + suf. *-íkos* > L. *-icus* > E. *-ic*, belonging to, relating to): a type of ontogeny in ascocarps (Pezizales) and basidiocarps (Agaricales) in which development is initially angiocarpic, but later becomes gymnocarpic. Cf. **angiocarpic** and **gymnocarpic**.

Hemiascomycetes (Gr. *hémi*, half + *askós*, wine bag, sack + L. *-mycetes*, ending of class < Gr. *mýkes*, genit. *mýketos*, fungus): in some classifications, a class of fungi in which the asci are formed free (i.e., not contained in any type of fruiting body or ascocarp), and ascogenous hyphae are lacking; the thallus is simple, and when mycelium is present, it is poorly developed. In the classification system adopted in this dictionary, the Hemiascomycetes correspond in part to the class Archiascomycetes (which includes the orders Taphrinales, Schizosaccharomycetales); together with the classes Plectomycetes, Discomycetes and Loculoascomycetes, plus some other filamentous ascomycetes (such as the Eurotiales, Laboulbeniales and Spathulosporales) the constitute the phylum Ascomycota of the kingdom Fungi.

hemibiotroph, **hemibiotrophic** (Gr. *hémi*, half + *bíos*, life + *trophós*, something that nourishes, serves as food; or + suf. *-íkos* > L. *-icus* > E. *-ic*, belonging to, relating to): applied to the entomogenous fungi, such as the Clavicipitaceae (Hypocreales), that have a parasitic phase composed of yeast-like cells that inhabit the hemocele of the host insect, and a well delimited saprobiotic phase corresponding to the mycelial colonization of the entire insect body after it has died due to the infection. The insect is transformed into a hard, sclerotiform structure, in which the fungus can survive for long periods and produce synnemata and perithecial stromata when conditions are favorable and while food reserves last. This situation is observed, e.g., in species of *Cordyceps* that infect larvae of Scarabaeidae coleopterans.

hemicyclic (Gr. *hémi*, half + *kyklikós*, circular, cyclic < *kiklós*, circle + suf. *-íkos* > L. *-icus* > E. *-ic*, belonging to, relating to): *Uredinales*. Species of rust fungi in whose life cycle one of the dikaryotic spore stages is lacking; e.g., *Gymnosporangium juniperi-virginianae*, the cedar-apple rust, lacks urediniospores. These rusts also are called **demicyclic**. Cf. **macrocyclic** and **microcyclic**.

herbicolous (ME: *herbe* < OF. < L. *herba*, herb + L. *-cola*, inhabitant + *-osus* > OF. *-ous*, *-eus* > E. *-ous*, possessing the qualities of): living on herbs.

hermaphrodite (Gr. *hermaphróditos*, son of Hermes and Aphrodite, with the attributes of both sexes): a species or thallus having both male and female sexual organs, which may or may not be compatible; also called **monoecious**. For example, the gametothallus of *Allomyces javanicus* (Blastocladiales) is hermaphroditic (or monoecious), since it has female gametangia, and on these, male gametangia; *Achlya americana* and *Saprolegnia parasitica* (Saprolegniales), among others, has oogonia and antheridia on the same thallus. Cf. **dioecious**.

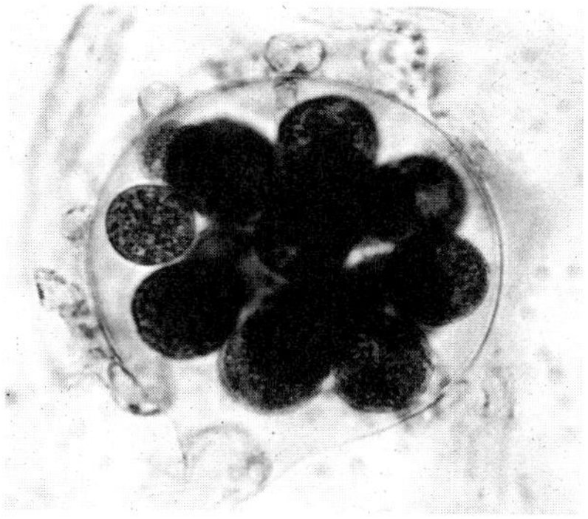

Hermaphrodite thallus of *Saprolegnia parasitica*, with both oogonium and antheridium, x 660 (*MU*).

Heterobasidiomycetes (Gr. *héteros*, another, distinct + *basídion*, dim. of *básis*, base + L. *-mycetes*, ending of class < Gr. *mýkes*, genit. *mýketos*, fungus): in some classifications, a class of fungi that includes those basidiomycetes with heterobasidia and basidiospores that typically germinate by forming secondary spores instead of mycelium. In the classification adopted in this dictionary, the heterobasidiomycetes are included in the classes Ustilaginomycetes (with the order Ustilaginales) and Urediniomycetes (with the order Uredinales) of the phylum Basidiomycota. Other orders that include forms with heterobasidia, but which are not considered within a recognized taxonomic class, are the Tremellales, Auriculariales, Dacrymycetales, Ceratobasidiales, Tulasnellales, Sporidiales, Septobasidiales and Exobasidiales.

heterobasidium, pl. **heterobasidia** (Gr. *héteros*, different + *basídion* < *básis*, base + dim. suf. *-ídion* > L. *-idium*): a type of basidium composed of two different regions, a septate basal portion, or **hypobasidium**, and the upper portion, or **epibasidium**, on which the spores are borne. Because the basidium is divided, it is called a **phragmobasidium**. Such basidia are found, e.g., in *Puccinia* (Uredinales), *Ceratobasidium* (Ceratobasidiales), and *Ustilago* (Ustilaginales). Some authors also consider the tuning-fork basidium of the Dacrymycetales a type of heterobasidium. Cf. **holobasidium** and **homobasidium**.

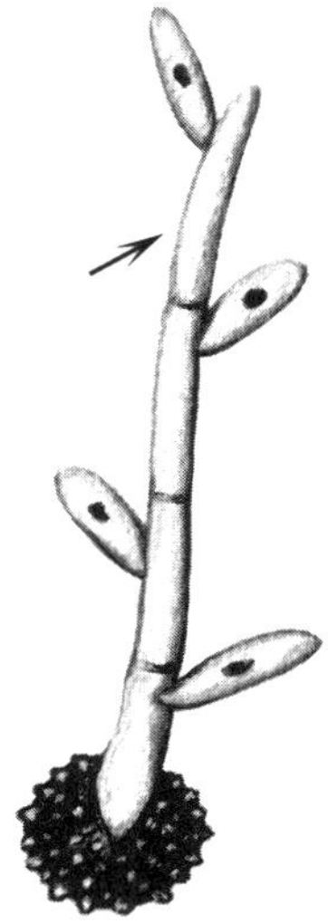

Heterobasidium of the germinating teliospore of *Ustilago maydis*, x 1 360.

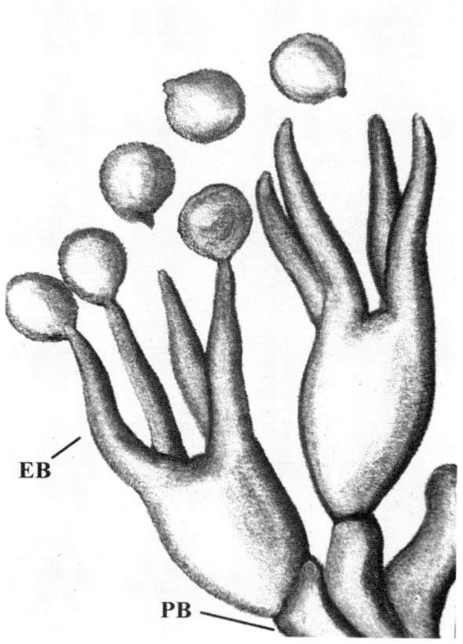

Heterobasidia of *Ceratobasidium plumbeum*; each epibasidium (EB) derives from a probasidium (PB) and produces a basidiospore at its apex, x 1 540.

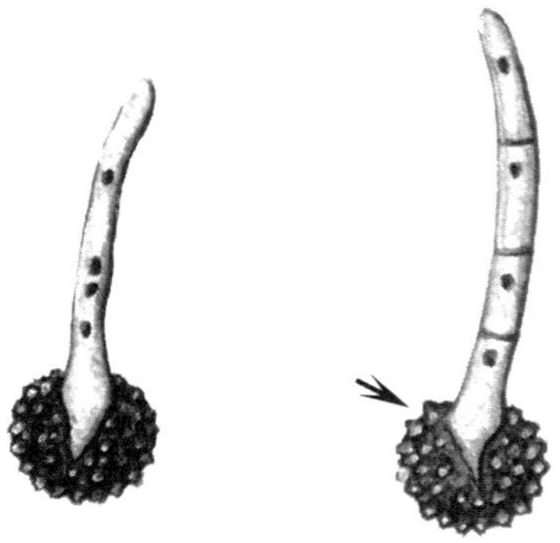

Heterobasidia (the germinating teliospores) of *Ustilago maydis*, x 1 250.

heterocyst (Gr. *héteros*, another, different + *kýstis*, bladder; here, cell): *Lichens*. A special cell that originates from a vegetative cell, in some families of Cyanophyceae, by a thickening of the wall and disappearance of the assimilating pigment; heterocysts, like hormogonia, are involved in the multiplication of the Cyanophyceae that constitute the phycobiont of some lichens, but in addition it is believed that they function in the fixation of atmospheric nitrogen by the lichen thallus.

heteroecious (Gr. *héteros*, another, distinct + *oîkos*, house, dwelling + L. *-osus* > OF. *-ous*, *-eus* > E. *-ous*, possessing the qualities of): a parasitic fungus that develops one portion of its life cycle on one host and the other portion on a different species of host; e.g., the wheat rust, *Puccinia graminis* (Uredinales), whose urediniospores and teliospores develop on cereals, and the pycniospores (or spermatia) and aeciospores on barberry; the phenomenon of requiring two species to complete the life cycle of the parasite is called **heteroecism**. **Heteroxenous** is a syn. of heteroecious. Cf. **autoecious**.

heteroecism (Gr. *héteros*, another, one distinct + *oîkos*, house or inn + *-ismós* > L. *-ismus* > E. *-ism*, state, phase, tendency, action): the development of different stages of a parasitic fungus on two different host plants; such fungi are said to be **heteroecious**. Many rust fungi (Uredinales) are heteroecious. Cf. **autoecism**.

heterogametangium, pl. **heterogametangia** (Gr. *héteros*, another, different + *gamétes*, husband + *angeîon*, vessel, receptacle): male and female gametangia that are morphologically different, whether in size or shape; e.g., the gametangia of *Zygorhynchus vuilleminii* (Mucorales), differ in size, whereas the gametangia, differentiated into antheridia and ascogonia, of the ascocarpic ascomycetes, often differ in both size and shape. Cf. **isogametangium**.

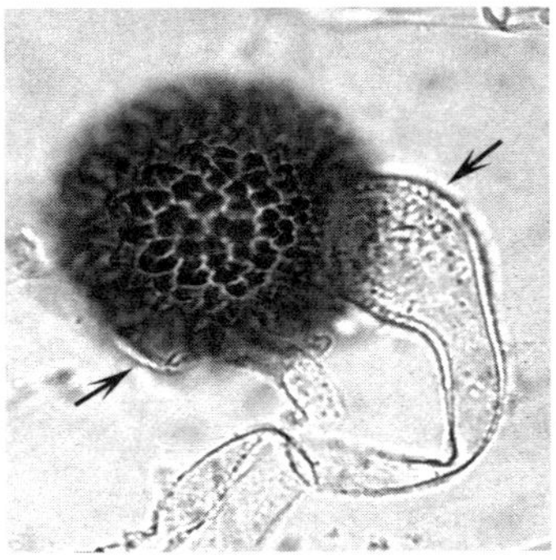

Heterogametangia of *Zygorhynchus moelleri*, x 360 (*MU*).

heterogamete (Gr. *héteros*, another, different + *gamétes*, husband): male and female gametes that are morphologically different; e.g., in the aquatic fungi of the order Monoblepharidales (such as *Monoblepharis polymorpha*), which has immotile female gametes (oospheres contained in oogonia) and motile male gametes (antherozoids produced in antheridia). Cf. **isogamete**. This type of sexual reproduction is called **heterogamy**.

heterogamy (Gr. *héteros*, another, different + *gámos*, sexual union + *-y*, E. suf. of concrete nouns): the union of two dissimilar gametes. Cf. **anisogamy** and **isogamy**. See **heterogamete**.

heterokaryontic, **heterocaryontic**, **heterokaryotic**, **heterocaryotic** (Gr. *héteros*, another, different + *káryon*, nucleus + *-tikós* > L. *-ticus* > E. *-tic*, relation, fitness, inclination or ability): an individual that exhibits **heterocaryosis**. Cf. **homocaryontic**.

heterokaryosis, pl. **heterokaryoses**, **heterocaryosis**, pl. **heterocaryoses** (Gr. *héteros*, another, different + *káryon*, nucleus + *-osis*, state or condition): the condition in which genetically different nuclei are associated in the same protoplast or in the same mycelium. Such organisms are said to be **heterocaryontic**. Cf. **homocaryosis**.

heterokont, **heterocont**, **heterokontic**, **heterocontic** (Gr. *héteros*, another, different + *kontós*, oar; here, flagellum; or + suf. *-îkos* > L. *-icus* > E. *-ic*, belonging to, relating to): with flagella of different length. This occurs in many algae and some fungi, whose biflagellate zoospores have one flagellum longer than the other; e.g., the Plasmodiophoromycetes (*Plasmodiophora*, *Spongospora*, *Woronina*) have zoospores with two unequal flagella, both of the whiplash type (amastigonemate), located in the anterior end of the cells. Also considered as heterocont are the zoospores of the Oomycota (such as *Achlya*, *Saprolegnia*, *Plasmopara*, etc.), which have one whiplash type flagellum and the other mastigonemate. Cf. **isokont**.

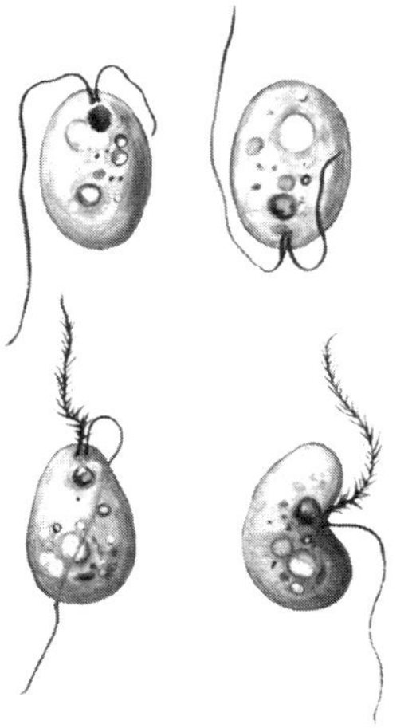

Heterokont zoospores of *Plasmodiophora brassicae* (above) and of *Saprolegnia parasitica* (below); of the latter, a primary zoospore is shown on the left, and a secondary zoospore is shown on the right, x 4 600.

heteromerous (Gr. *héteros*, another, different + *méros*, part + L. *-osus* > OF. *-ous*, *-eus* > E. *-ous*, having, possessing the qualities of): **1**. *Agaricales (Russulaceae)*. A type of tramal context in the gills of the basidiocarps, which is composed of clusters of spherical cells (**sphaerocysts**) intermixed with hyphae (as in the gen. *Russula*), or of tissue containing normal hyphae and **lactiferous** hyphae (latex conductors), as in the gen. *Lactarius*. **2**. *Lichens*. A thallus in which the **mycobiont** and the **phycobiont** are arranged in definite layers or strata, whether with horizontal stratification or with radial stratification. The majority of the crustaceous, squamulose, and foliose lichens are heteromerous, with the layers of hyphae and algal cells well delimited; as, e.g., in *Parmelia* (Lecanorales). Cf. **homomerous**.

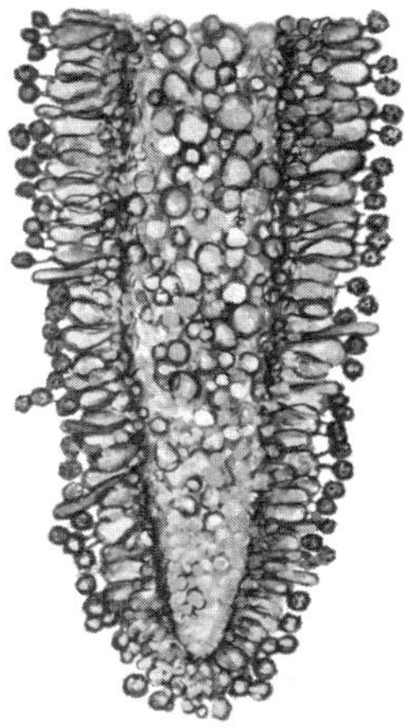

Heteromerous hymeniiferous trama of a lamella of *Russula* sp., seen in longitudinal section, x 180.

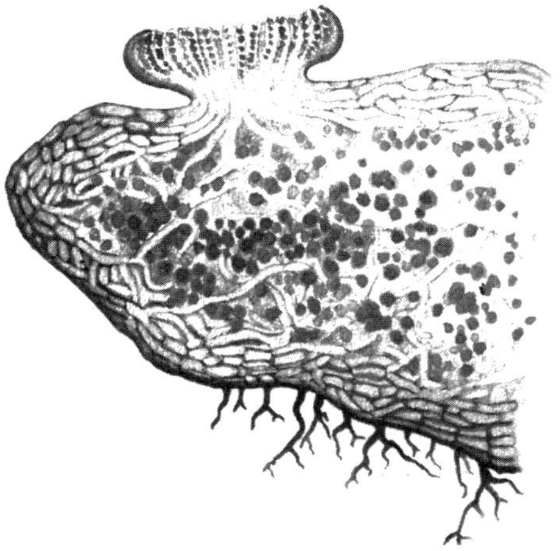

Heteromerous thallus of the foliose lichen *Parmelia imbricatula*, seen in longitudinal section, x 12.5.

heteromorphic, **heteromorphous** (Gr. *héteros*, another, distinct + *morphé*, shape + suf. *-íkos* > L. *-icus* > E. *-ic*, belonging to, relating to; or + L. *-osus* > OF. *-ous*, *-eus* > E. *-ous*, having, possessing the qualities of): something that varies from the structure considered normal (of different size, with different elements, etc.). In general, it is applied to organs or structures different in shape and size, e.g., thalli, gametangia, etc. In agaricology, it refers to the margins of the gills or of the borders of the tubes which are sterile or predominantly sterile, due to the presence of a type of cystidium or marginal hair, which does not occur on the sides of the gills or in the interior of the tubes. Cf. **homomorphic**.

heterosporic, **heterosporous** (Gr. *héteros*, another, distinct + *sporá*, spore + suf. *-íkos* > L. *-icus* > E. *-ic*, belonging to, relating to; or + L. *-osus* > OF. *-ous*, *-eus* > E. *-ous*, having, possessing the qualities of): producing asexual spores of more than one type (heterospory); e.g., *Fonsecaea* (dematiaceous asexual fungi), a pathogen of man, produces three types of conidia (phialospores, catenulate blastospores and blastospores on a rachis), which are placed in the gen. *Phialophora*, *Cladosporium* and *Rhinocladiella*, respectively. These conidia are formed in the same colony, or even on the same conidiophore. The term also is applied to fungi, like *Inocybe* (Agaricales), that have a polymorphism of the basidiospores. Also called heterosporial. Cf. **homosporic** and **isosporic**.

heterothallic (Gr. *héteros*, different + *thallós*, thallus + suf. *-íkos* > L. *-icus* > E. *-ic*, belonging to, relating to): this term has been defined in two different ways: **1**. Having the male and female organs segregated in distinct thalli, so that two thalli are required for sexual reproduction to occur; i.e., **dioecious**. **2**. Species in which both male and female organs are borne on the same thallus (**hermaphroditic**, **monoecious**), but they are self-sterile (self-incompatible), because of which it is necessary to have two compatible thalli (cross fertilization) in order to achieve sexual reproduction. Cf. **homothallic**.

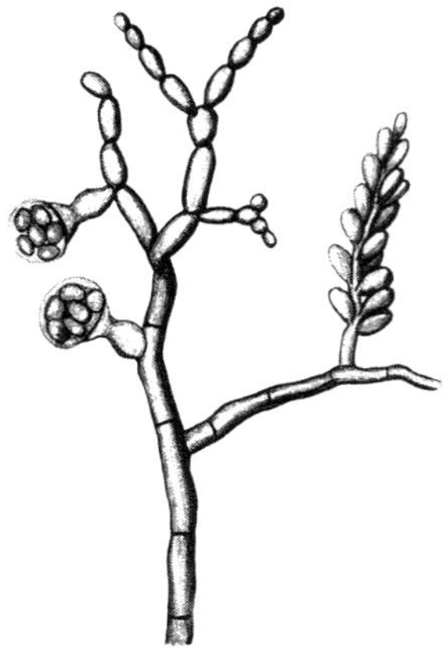

Heterosporic conidiophore of *Rhinocladiella* (=*Fonsecaea*) *pedrosoi*, x 600.

heterothallism (Gr. *héteros*, different + *thallós*, thallus + *-ismós* > L. *-ismus* > E. *-ism*, state, phase, tendency, action): the condition of being **heterothallic**. Cf. **homothallism**.

heterotroph, **heterotrophic** (Gr. *héteros*, another, different + *trophós*, that which nourishes, serves as food; or + suf. *-íkos* > L. *-icus* > E. *-ic*, belonging to, relating to): an organism which must feed on organic material elaborated by other organisms, because it is incapable of synthesizing carbohydrates from inorganic elements. All fungi are heterotrophs, because of which they live as saprobes or as parasites. Cf. **autotroph**.

heterotropic (Gr. *héteros*, another, distinct + *trópos*, rotation, turn + suf. *-íkos* > L. *-icus* > E. *-ic*, belonging to, relating to): *Basidiomycetes*. A type of spore attachment in which the basidiospore is borne obliquely on the sterigma, as in the Agaricales. This arrangement facilitates spore discharge. Cf. **orthotropic**.

heteroxenous (Gr. *héteros*, another, distinct + *xénos*, host + L. *-osus* > OF. *-ous*, *-eus* > E. *-ous*, having, possessing the qualities of): with another host. Syn. of **heteroecious**.

heterozygous (Gr. *héteros*, different + *zygón*, marriage tie, pair + L. *-osus* > OF. *-ous*, *-eus* > E. *-ous*, having, possessing the qualities of): having heterokaryosis resulting from the fusion of gametes.

hilar appendix, pl. **appendixes** or **appendices** (L. *hilum*, a trifle, navel; used for the scar on a bean seed; here, the scar on the wall of a spore that indicates its point of attachment to the sporophore, sterigma, etc. + suf. *-aris* > E. *-ar*, like, pertaining to; *appendix*, an appendage): *Basidiomycetes*. A minute appendage that protrudes from a basidiospore adjacent to the hilum of the spore. This term is preferable to **apiculus**, as the appendix is basal and not apical. It can be observed clearly in basidiospores of *Amanita abrupta* (Agaricales). The area present above the hilar appendix, in which the ornamentation of the wall is reduced or absent is known as **suprahilar depression**, **suprahilar disc**, **suprahilar plage** or **suprahilar spot**; it is particularly found in the basidiospores of *Lactarius*, *Russula*, *Phaeocollybia* and *Galerina* (Agaricales).

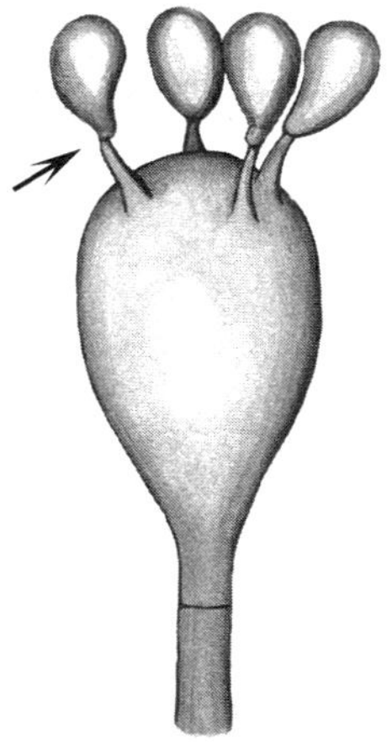

Heterotropic basidiospores on a basidium of *Amanita* sp., x 1 100.

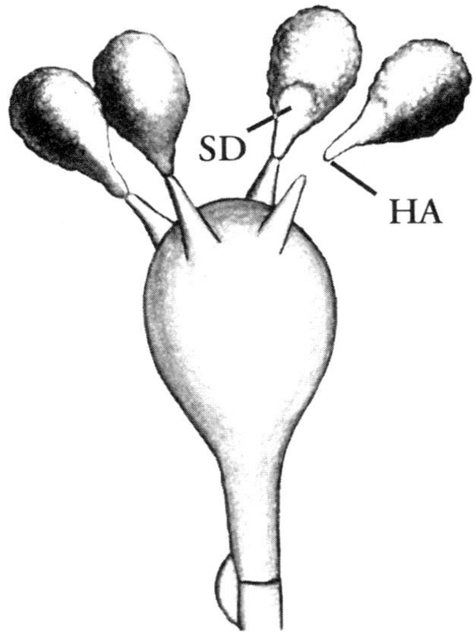

Hilar appendix (HA) and suprahilar depression (SD) of the basidiospores of *Galerina phillipsi*, x 1 530.

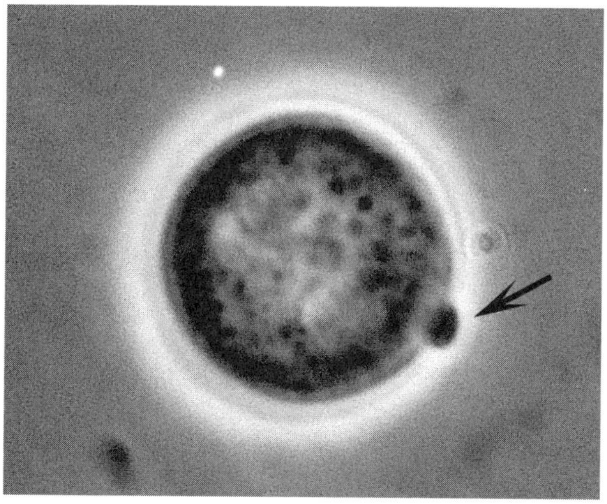

Hilar appendix of the basidiospore of *Amanita abrupta*, x 4 600 (*RMA*).

hilum, pl. **hila** (L. *hilum*, mask, navel, seed scar of bean): a scar, area or mark on spores (conidia, basidiospores, etc.) which indicates the point of attachment to the conidiophore, sterigma, etc. The hilum is a distinctive structure, like a dark-colored ring, e.g., in conidia of *Alternaria*, *Drechslera* and *Helminthosporium* (dematiaceous asexual fungi).

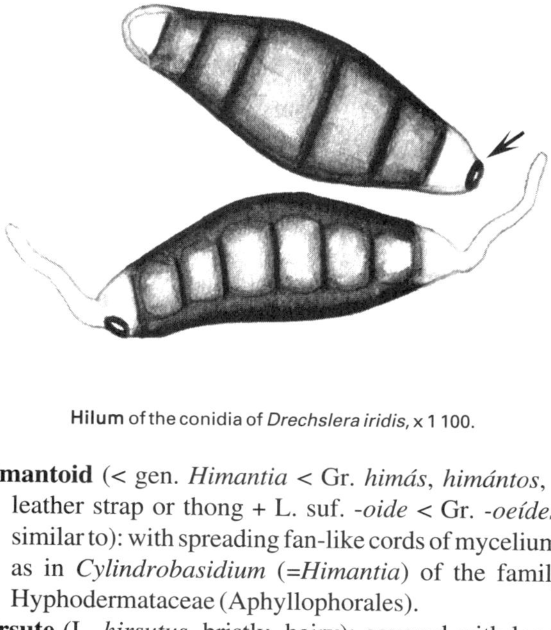

Hilum of the conidia of *Drechslera iridis*, x 1 100.

himantoid (< gen. *Himantia* < Gr. *himás*, *himántos*, a leather strap or thong + L. suf. *-oide* < Gr. *-oeídes*, similar to): with spreading fan-like cords of mycelium, as in *Cylindrobasidium* (=*Himantia*) of the family Hyphodermataceae (Aphyllophorales).

hirsute (L. *hirsutus*, bristly, hairy): covered with long, rigid hairs, rough to the touch. For example, the base of the stipe of some species of *Peziza* (Pezizales) and of *Marasmius alliaceus* (Agaricales), as well as the pileus of *Lentinus velutinus* (Aphyllophorales), are hirsute. Also called **hispid** and **strigose**.

Hirsute pileus of the basidiocarp of *Lentinus velutinus*, x 0.8 (*MU*).

Hirsute pileus (a close up) of the basidiocarp of *Lentinus velutinus*, x 12 (*MU*).

hispid (L. *hispidus*, coarsely bristly, hairy): having large, very stiff hairs, almost spiny, very rough to the touch; e.g., like the pileus of *Pluteus hispidus* (Agaricales). Syn. of **hirsute** and **strigose**.

histogenous (Gr. *hístos*, tissue + *génos*, origin < *gennáo*, to engender, produce; literally, tissue producing + L. *-osus* > OF. *-ous*, *-eus* > E. *-ous*, having, possessing the qualities of): **1**. Conversion of the masses of *Ustilago maydis* (Ustilaginales) tissue in the host into basidiospores. **2**. Openings or cavities that are formed as a result of the development of tissues.

histoplasmosis, pl. **histoplasmoses** (< gen. *Histoplasma* < Gr. *hístos*, tissue + *plásma*, the primordial matter of living beings + *-osis*, state or condition): a primary lung infection acquired by inhalation of conidia and hyphal fragments of the dimorphic fungus *Histoplasma capsulatum* (moniliaceous asexual fungi), abundant in some soils, especially those in caves contaminated with bird and bat dung. Histoplasmosis may be benign or severe, even mortal, and several organs can be infected (linfactic ganglions, spleen, liver, lungs, kidneys, central nervous system, and oral and intestinal mucose membranes).

hoary (ME. *hor*, hoariness < *hor*, frost + E. suf. *-y*, having the quality of): with a thick covering of grayish or whitish, silky hairs; **canescent**, e.g., the pubescent pileus of *Coprinus lagopus* (Agaricales).

holdfast (E. *holdfast* < ME. *holden* < OE. *haldan*, to hold + prob. ME. *fest*, ptp. of *festen*, to fasten): *Trichomycetes*. The basal structure of the thallus, mucilaginous and adhesive, which can be discoid, branched or pectinate, and which serves to attach the thallus to the cuticle of the host, with which it has an obligate commensal relationship (the hosts of Trichomycetes are mandibulate arthropods of aquatic or terrestrial habitat). The holdfast is indispensable for the growth and sporulation of the thallus; e.g., in *Eccrinidus flexilis* (Eccrinales), which lives as endocommensal of myriapods, the holdfast is discoid.

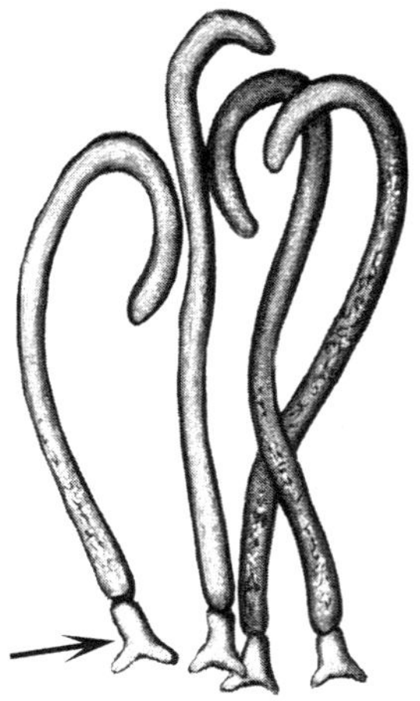

Holdfast of the thalli of *Eccrinidus flexilis*, x 20.

holoarthric (Gr. *hólos*, entire + *árthron*, articulation, joint + suf. *-íkos* > L. *-icus* > E. *-ic*, belonging to, relating to): *Conidiogenesis*. A type of arthric conidium development in which all of the wall layers of the conidiogenous cell participate in the formation of the conidium wall; e.g., as in the arthrospores of *Geotrichum* (moniliaceous asexual fungi). Cf. **enteroarthric**.

Holoarthric conidia (arthrospores) of *Geotrichum candidum*, x 300.

Holobasidiomycetes (Gr. *hólos*, entire + *basídion*, dim. of *básis*, base + L. *-mycetes*, ending of class < Gr. *mýkes*, genit. *mýketos*, fungus): same as **Homobasidiomycetes**. In some classifications, a class of fungi with holobasidia (homobasidia) and basidiospores that germinate by forming a mycelium instead of secondary spores. In the classification followed in this dictionary, these fungi are not recognized as a taxonomic class, but they would correspond to the basidiomycetes of the orders Agaricales and Aphyllophorales of the class Hymenomycetes, as well as the orders of Gasteromycetes (Lycoperdales, Tulostomatales, Sclerodermatales, Phallales, Nidulariales and Hymenogastrales). Together with the heterobasidiomycetous classes (Ustilaginomycetes and Urediniomycetes), along with the orders Tremellales, Auriculariales, Dacrymycetales, Ceratobasidiales and Tulasnellales (which in this system are included in the class Hymenomycetes), they constitute the phylum Basidiomycota of the kingdom Fungi.

holobasidium, pl. **holobasidia** (Gr. *hólos*, entire + *basídion* < *básis*, base + dim. suf. *-ídion* > L. *-idium*): same as **homobasidium**. A single-celled basidium, typically club-shaped; characteristic of the Holobasidiomycetes. Cf. **heterobasidium**.

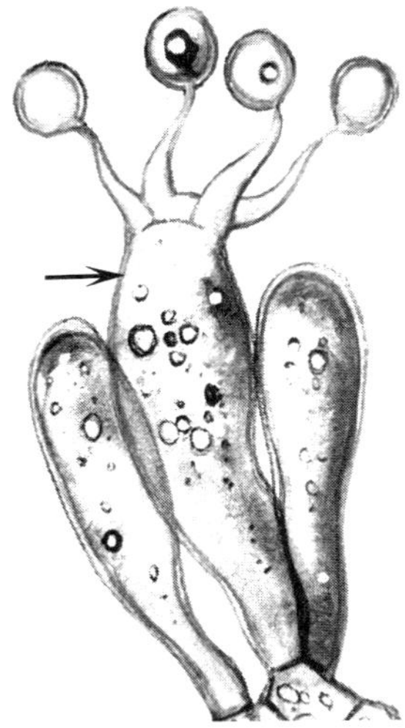

Holobasidium from the hymenium of *Lactarius fuliginosus*, x 1 400.

holoblastic (Gr. *hólos*, entire + *blastós*, sprout, bud, germ, shoot + suf. *-íkos* > L. *-icus* > E. *-ic*, belonging to, relating to): see **blastic**.

holocarpic (Gr. *hólos*, entire + *karpós*, fruit + suf. *-íkos* > L. *-icus* > E. *-ic*, belonging to, relating to): an organism whose entire thallus is converted into one or more reproductive organs, so that the somatic and reproductive phases do not coexist; this condition is called **holocarpy**. The aquatic zoosporic fungi of the families Olpidiaceae (Spizellomycetales) and Synchytriaceae (Chytridiales), e.g., are holocarpic. Cf. **eucarpic**.

holocarpy (Gr. *hólos*, entire + *karpós*, fruit + *-y*, E. suf. of concrete nouns): the condition of being **holocarpic**.

hologamy (Gr. *hólos*, entire + *gamía* < *gámos*, sexual union + *-y*, E. suf. of concrete nouns): a type of fertilization consisting of the union of two entire individuals, i.e., both thalli are converted into a single gametangium; e.g., as is observed in the fusion of two mature individuals of *Polyphagus euglenae* (Chytridiales).

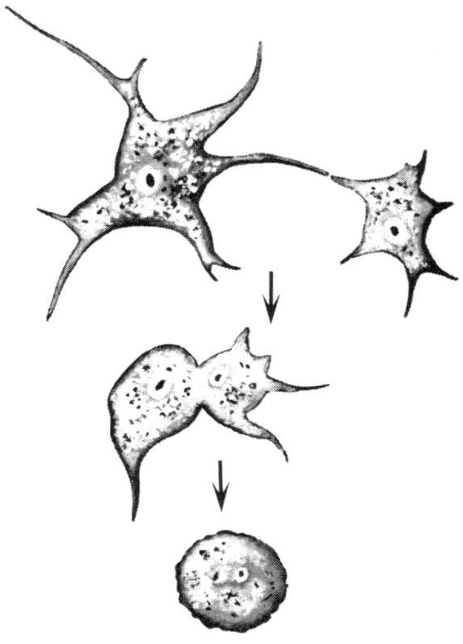

Hologamy in *Polyphagus euglenae*, x 620.

holomorph (Gr. *hólos*, entire + *morphé*, shape): the fungus as a total organism, i.e., in all its forms and phases. The holomorph includes the **teleomorph** (the sexual form, characterized by the presence of an ascoma, or of a basidioma, according to the species being considered) and the **anamorph** (the asexual phase, represented by some type of conidioma), with its corresponding binomials.

holosporic, **holosporous** (Gr. *hólos*, entire + *sporá*, spore + suf. *-íkos* > L. *-icus* > E. *-ic*, belonging to, relating to; or + L. *-osus* > OF. *-ous*, *-eus* > E. *-ous*, having, possessing the qualities of): a type of conidial maturation in which the conidium reaches its final shape and size before the cells that compose it are delimited and it matures as a whole. The holothallic conidia of *Microsporum* and *Trichophyton* (moniliaceous asexual fungi) are examples of holosporic conidia.

holothallic (Gr. *hólos*, entire + *thallós*, shoot, thallus + suf. *-íkos* > L. *-icus* > E. *-ic*, belonging to, relating to): see **thallic**.

holozoic (Gr. *hólos*, entire + *zoikós* < *zõon*, animal + suf. *-íkos* > L. *-icus* > E. *-ic*, belonging to, relating to): an organism that ingests solid food particles by means of phagocytosis, as do the amoebae and other protozoans; e.g., the myxamoebae of *Cavostelium*, *Protostelium*, *Ceratiomyxella* and many other Protosteliales are holozoic. Also called **phagotroph**.

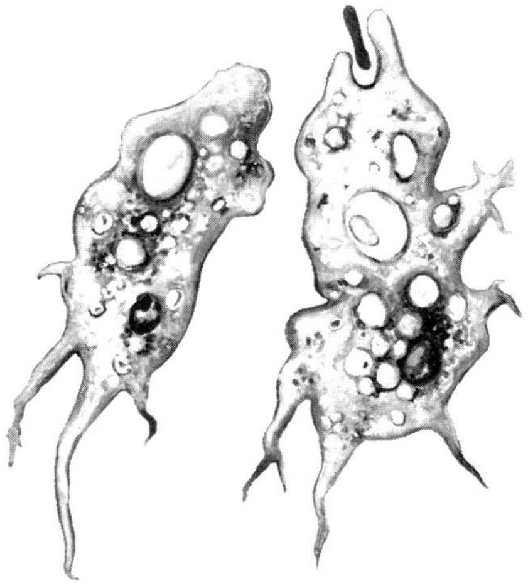

Holozoic myxamoebae of *Cavostelium bisporum*, x 1 280.

Homobasidiomycetes (Gr. *homós*, similar, equal + *basídion*, dim. of *básis*, base + L. *-mycetes*, ending of class < Gr. *mýkes*, genit. *mýketos*, fungus): see **Holobasidiomycetes**.

homobasidium, pl. **homobasidia** (Gr. *homós*, similar, equal + *basídion* < *básis*, base + dim. suf. *-ídion* > L. *-idium*): see **holobasidium**.

homobium, pl. **homobia** (NL. *homobium* < Gr. *homós*, similar, equal + *bíos*, life + L. dim. suf. *-ium*): an autotrophic, symbiotic association of a fungus and an alga, as in the lichens.

homogeneous development (ML. *homogeneus* < Gr. *homogenés*, of the same kind + L. *-osus* > OF. *-ous*, *-eus* > E. *-ous*, having, possessing the qualities of; F. *développement* < MF. *développer* < OF. *desveloper* + L. *mentum* > OF. *-ment* > ME. *-ment*, action, process): *Gasteromycetes*. Differentiation of the basidia in primordial tissue evenly throughout the gleba; typically found in species of *Bovista* (Lycoperdales).

homokaryontic, **homocaryontic**, **homokaryotic**, **homocaryotic** (Gr. *homós*, similar, same + *káryon*, nut, nucleus + *-tikós* > L. *-ticus* > E. *-tic*, relation, fitness, inclination or ability): an individual or cell that has genetically similar nuclei. Cf. **heterokaryontic**.

homokaryosis, **homocaryosis**, pl. **homokaryoses**, **homocaryoses** (Gr. *homós*, similar, same + *káryon*, nut, nucleus + *-osis*, condition): the condition in which genetically identical nuclei occur in the same protoplast or in the same mycelium. Cf. **heterokaryosis**.

homomerous (Gr. *homós*, similar, same + *méros*, part + L. *-osus* > OF. *-ous*, *-eus* > E. *-ous*, having, possessing the qualities of): **1**. *Agaricales*. A type of context or trama in the gills of the basidiocarp which is composed only of hyphae, e.g., in *Coprinus*. **2**. *Lichens*. A type of thallus in which the cells of the phycobiont (*Nostoc*) are distributed in a homogeneous manner in the interior of the thallus, intermixed with the hyphae of the mycobiont, without forming a definite layer; the homomerous thallus is characteristic of the gelatinous cyanophycophilous lichens, such as *Leptogium* and *Collema* (Lecanorales), although not exclusively so. Cf. **heteromerous**.

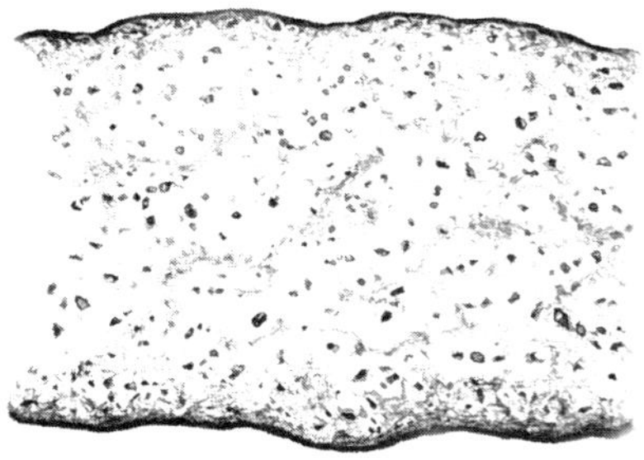

Homomerous thallus of *Collema tenax*, seen in longitudinal section, x 470.

homomorphic, **homomorphous** (Gr. *homós*, similar, equal + *morphé*, shape + suf. *-íkos* > L. *-icus* > E. *-ic*, belonging to, relating to; or + L. *-osus* > OF. *-ous*, *-eus* > E. *-ous*, having, possessing the qualities of; i.e., of similar shape): *Agaricales*. It refers to margins of the gills or the borders of the tubes which have the same tissue composition as the hymenium. Cf. **heteromorphic**.

homosporic, **homosporous** (Gr. *homós*, similar, equal + *sporá*, spore + suf. *-íkos* > L. *-icus* > E. *-ic*, belonging to, relating to; or + L. *-osus* > OF. *-ous*, *-eus* > E. *-ous*, having, possessing the qualities of): having only one kind of spore, as in the majority of the fungi. Also called **isosporic** or **isosporous**. Cf. **heterosporic**.

homothallic (Gr. *homós*, similar, same + *thallós*, shoot, thallus + suf. *-íkos* > L. *-icus* > E. *-ic*, belonging to, relating to): a fungus in which sexual reproduction occurs in a single thallus, which is therefore self compatible; this phenomenon is known as **homothallism**. Cf. **heterothallic.**

homothallism (Gr. *homós*, similar, same + *thallós*, shoot, thallus + *-ismós* > L. *-ismus* > E. *-ism*, state, phase, tendency, action): the condition of being **homothallic**.

honey-dew (ME. *hony* < OE. *hunig*, honey, the sweet viscid material elaborated out of the nectar of flowers in the honey sac of various bees; ME. *dew* < OE. *deaw*, dew, moisture condensed in minute droplets): a secretion attractive to insects, associated with *Sphacelia* (tuberculariaceous asexual fungi), the anamorph of *Claviceps* (Hypocreales). A secretion by aphids living on the leaves of citrous plants, where some fungi may grow, e.g., the **sooty molds**.

hormesis, pl. **hormeses** (Gr. *hormáo*, to excite, stimulate + suf. *-esis*, tendency): the excitation or stimulus produced in an organism by means of a toxic substance employed in innocuous concentrations.

hormocyst (Gr. *hórmos*, collar + *kýstis*, bladder, cell): *Lichens*. A propagule or **diaspore** that is composed of hyphae of the fungus and a few algal cells and is produced in a special organ called a **hormocystangium**; they occur in certain species of gelatinous lichens in the family Collemataceae, e.g., *Lempholemma cladodes* and *L. vesiculiferum* (Lichinales).

hormocystangium, pl. **hormocystangia** (Gr. *hórmos*, collar + *kýstis*, bladder, cell + *angeîon*, vessel, receptacle): see **hormocyst**.

hormogonium, pl. **hormogonia** (NL. *hormogonium* < Gr. *hórmos*, collar, pearl bearing ornamentation of the neck + NL. *gonium* < Gr. *gónos*, offspring + L. dim. suf. *-ium*): *Lichens*. A fragment, generally motile, of the trichome of a Cyanophyceae, composed of a variable number of cells, which can fluctuate between two and several hundred and which is naturally separated from the thallus by virtue of a process of budding, or proceeds directly from the germination of a hypnocyst. The hormogonia are destined for the propagation of the species, as they are capable of growing and forming a new filament or thallus; the Cyanophyceae or cyanobacteria that constitute the phycobionts or bacteriobionts of some lichens can multiply by means of hormogonia.

host (L. *hospes*, one who receives a stranger as guest): *Biology*; *Mycology*. A plant, animal or fungus from which a parasite obtains its nourishment, either internally or externally. The hosts or victims of the parasitic fungi are, in order of frequency: angiosperms, gymnosperms, pteridophytes, insects and other arthropods, mollusks, other fungi, algae, bryophytes, nematodes, protozoans, and man and higher animals. An **alternate host** is a plant in which the parasite (generally a uredinial micromycete) carries out the complementary phases of its life cycle; an **intermediate host** is one that serves as a link for the infection by a fungus of two plants of different species; a **primary host** is the plant on which the parasite develops its sexual state, and almost always it corresponds to the plant of major economic importance.

Hülle cell (G. *Hülle*, covering, envelopment; L. *cella*, storeroom > ML. *cella* > OE. *cell*, cell, compartment): a cell with a thick wall and narrow lumen, terminal or intercalary, that surrounds the peridium of the cleistothecia of certain plectomycetous ascomycetes, e.g., *Emericella nidulans* (Eurotiales) and in the mycelium of *Aspergillus puniceus* (moniliaceous asexual fungi).

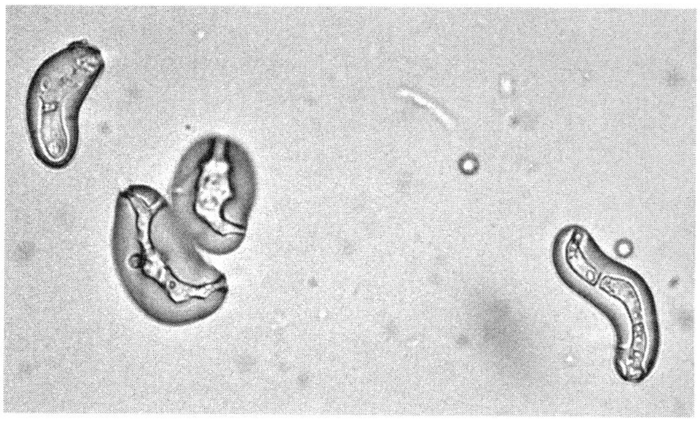

Hülle cells of *Aspergillus puniceus*, x 400 (*MU*).

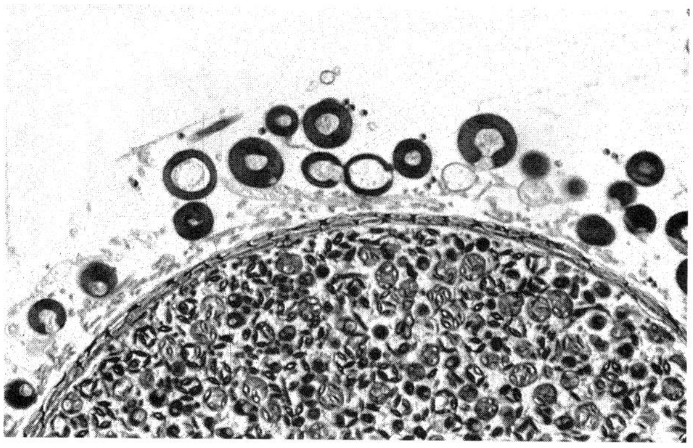

Hülle cells of *Emericella nidulans* on the outside of the cleistothecial wall, seen in cross section, x 530 (*RTH*).

humicole, **humicolous** (L. *humus*, earth, soil + *-cola*, inhabitant < *colere*, to inhabit; or + *-osus* > OF. *-ous*, *-eus* > E. *-ous*, possessing the qualities of): living or developing in humus on or in the soil; e.g., as with the gen. *Humicola* (dematiaceous asexual fungi).

hyaline (L. *hyalinus* < Gr. *hyálinos* < *hýalos*, crystal, glass + L. suf. *-inus* > E. *-ine*, of or pertaining to): transparent and colorless; glassy, like crystal. Initially, the hyphae, spores and other types of structures are hyaline in most fungi, and in many they remain so at maturity, e.g., in *Brasiliomyces malachrae* (Erysiphales).

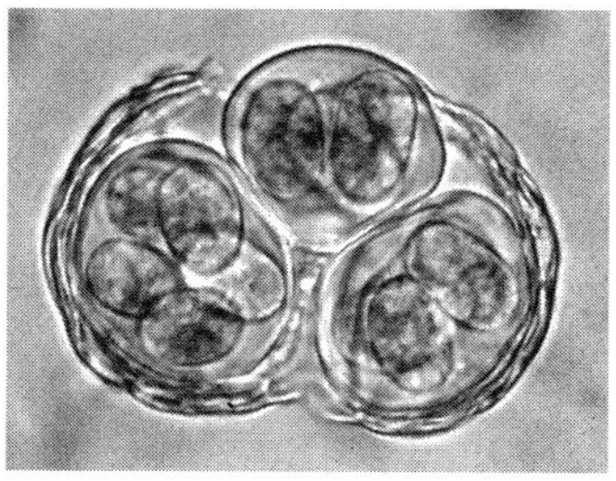

Hyaline ascocarp of *Brasiliomyces malachrae*, x 505 (*RTH*).

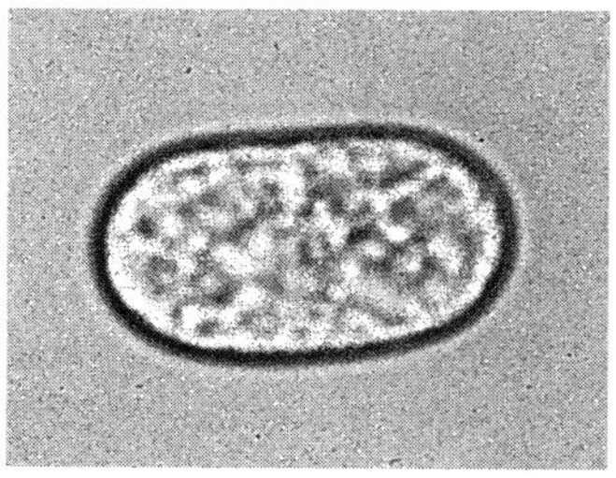

Hyaline ascospore of *Brasiliomyces malachrae*, x 1 345 (*RTH*).

hyalospore (L. *hyalinus* < Gr. *hyálinos* < *hýalos*, crystal, glass + *sporá*, spore): a hyaline asexual reproductive spore; a Saccardoan term that is applied principally to conidial fungi. The conidia of *Botrytis*, *Beauveria* and many other gen. of moniliaceous asexual fungi, e.g., are hyaline at maturity. Cf. **phaeospore**.

hydathode (Gr. *hydatódes*, aqueous, full of water): a secretory organ that secretes very dilute aqueous solutions, a little less than pure water, like the trichomes of certain agarics.

hydnoid (< gen. *Hydnum* < Gr. *hýdnon*, an old name applied to a certain type of edible fungi + L. suf. *-oide* < Gr. *-oeídes*, similar to): a type of fruiting body with a dentate hymenophore, i.e., arranged on teeth or prongs, stipitate or not (sessile); e.g., as in *Hydnum*, *Sarcodon* and *Hericium*, among other Hydnaceae (Aphyllophorales).

hydrocarbonoclastic (E. *hydrocarbon* < L. *hydro* < Gr. *hýdro-* < *hýdor*, water + L. *carbo*, genit. *carbonis*, coal, any substance composed exclusively of carbon and hydrogen + Gr. *klástos*, broken + suf. *-íkos* > L. *-icus* > E. *-ic*, belonging to, relating to): capable of decomposing hydrocarbons, such as the yeasts *Trichosporon* (cryptococcaceous asexual yeasts) and *Pichia ohmeri* (Saccharomycetales), isolated from oil and petroleum spills in contaminated marine environments.

hydrochoric (Gr. *hýdro-* < *hýdor*, water + suf. *-coro* < *choréo*, to change place, move away from + suf. *-íkos* > L. *-icus* > E. *-ic*, belonging to, relating to): applied to organisms that are disseminated by water. Many fungi have appendaged spores that facilitate their floating in water (e.g., the marine fungus *Orbimyces spectabilis*, of the dematiaceous asexual fungi); other species depend on the kinetic energy of raindrops for dispersal of their spores (e.g., the Nidulariaceae, such as *Cyathus* and *Nidularia*, of the Nidulariales); still others form swimming spores in which water is an essential factor in dispersal. Cf. **anemochoric.**

hydrophilous (Gr. *hýdro* < *hýdor*, water + *phílos*, have an affinity for + L. *-osus* > OF. *-ous*, *-eus* > E. *-ous*, possessing the qualities of): from an aquatic habitat; being disseminated by water, although the term **hydrochoric** is more appropriate to define this latter concept. The zoosporic fungi are hydrophilous. Cf. **hydrophobic**.

hydrophobic (L. *hydrophobus* < Gr. *hýdro* < *hýdor*, water + *phóbos*, fright, terror + suf. *-íkos* > L. *-icus* > E. *-ic*, belonging to, relating to): avoiding or repelling water, like the dry conidia of many fungi (e.g., *Penicillium* and *Aspergillus*, of the moniliaceous asexual fungi) which are not easily wettable, and the thallus of numerous crustose and foliose lichens that live in ecological niches (such as hot deserts) where the water is only available in the form of water vapor or in liquid form in the substrate (e.g., *Ramalina maciformis*, of the Lecanorales), from which they absorb it to become physiologically active, for which reason they are termed **aerohygrophilic** lichens. Cf. **hydrophilous**.

hydrotropism (Gr. *hýdro* < *hýdor*, water + *trópos*, a turn, change in manner < *tropé*, a turning < *trépo*, to revolve, turn towards + *-ismós* > L. *-ismus* > E. *-ism*, state, phase, tendency, action): a chemotrophic phenomenon in which water is the stimulating agent. The sporangiophores of the Mucorales, e.g., adopt a position perpendicular to the substrate due to negative hydrotropism. In air saturated with water, the sporangiophores of *Phycomyces* are oriented irregularly and at times are slightly curved, moving away from moist paper when this is placed vertically near them.

hygrophanous (Gr. *hygrós*, humid + *phaíno*, to show < *phanós*, light, lantern + L. *-osus* > OF. *-ous*, *-eus* > E. *-ous*, possessing the qualities of): becoming translucent when embedded in water; in certain Agaricales (e.g., *Psathyrella*, *Oudemansiella* and *Hygrophorus*) the pileus and other fungal organs acquire a translucent appearance due to imbibition.

hygroscopic, **hygroscopical** (Gr. *hygrós*, wet, humid + *skópein* < *skopiáo*, to observe, see + suf. *-íkos* > L. *-icus* > E. *-ic*, belonging to, relating to; or + *-ic* + E. suf. *-al* < L. *-alis*, relating to or belonging to): that which absorbs water readily (from the atmosphere), thus becoming soft and expanded, in order to facilitate spore dissemination; e.g., as in the capillitium of some Myxomycetes (*Trichia*, *Arcyria* and other Trichiales) or like the basidiocarp of *Astraeus hygrometricus* (Sclerodermatales), in which the exoperidium recurves to enclose the endoperidium when dry, but opens when moistened to expose and elevate the inner spore case.

Hygrophanous pileus of the basidiocarps of *Hygrophorus psittacinus*, x 1.5.

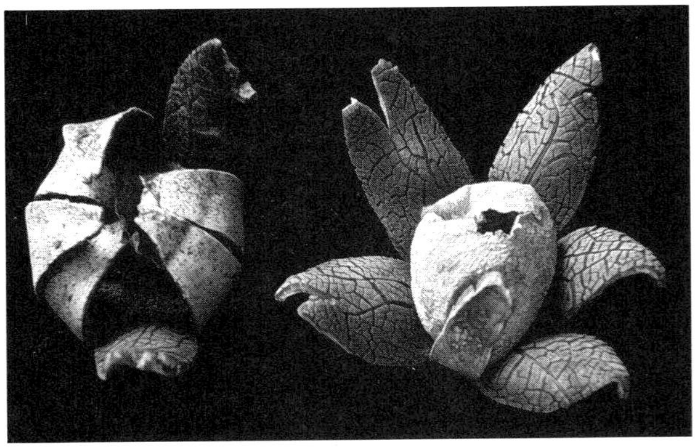

Hygroscopic fruiting bodies of *Astraeus hygrometricus*, x 1 (*MU*).

hymenial veil (L. *hymenium* < Gr. *hyménion*, dim. of *hymén*, membrane + L. suf. *-alis* > E. *-al*, relating to or belonging to; L. *velum*, veil): see **veil**.

hymeniform (L. *hymenium* < Gr. *hyménion*, dim. of *hymén*, membrane + *-formis* < *forma*, shape): refers to a layer of narrowly clavate cells that appear similar to basidioles, and which make up the hymenium.

hymeniiferous (L. *hymeniifer*, that which bears the hymenium < *hymenium* + *-ferous*, bearer < *ferre*, to bear, carry + *-osus* > OF. *-ous*, *-eus* > E. *-ous*, having, possessing the qualities of): see **hymenophore**.

hymeniiferous trama, pl. **tramae** (L. *hymeniifer*, that which bears the hymenium < *hymenium* + *-ferous*, bearer < *ferre*, to bear, carry + L. *-osus* > OF. *-ous*, *-eus* > E. *-ous*, having, possessing the qualities of; *trama*, trama, warp, texture): see **context** and **trama**.

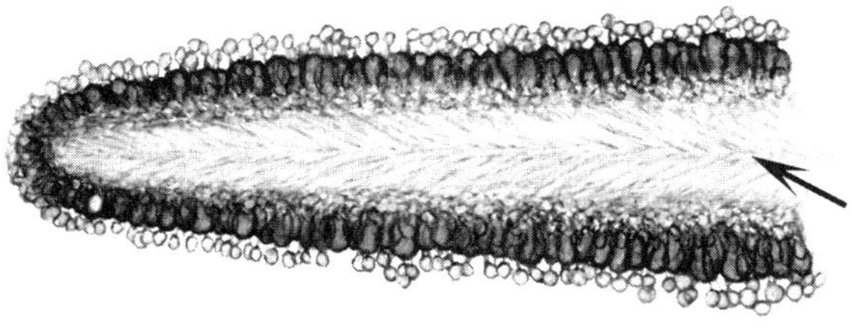

Hymeniiferous trama of a lamella of *Amanita muscaria*, seen in longitudinal section, x 160.

hymenium, pl. **hymenia** (L. *hymenium* < Gr. *hyménion* < *hymén*, membrane + dim. suf. *-íon* > L. *-ium*): a layer or stratum of highly diverse organization but always constituted of hyphae specialized for the production of spores (ascogenous, or producers of asci, in the ascomycetes, or basidiogenous, for the production of basidiospores, in the basidiomycetes), arranged in the shape of a palisade and frequently intermixed with sterile elements (paraphyses in the ascomycetes, and cystidia in the basidiomycetes). In the basidiomycetes the hymenium covers determinate parts of the sporophores which are called **hymenophores**, which can be gills (in Agaricales, such as *Coprinus*), tubes (in Agaricales, such as *Boletus*), or teeth or spines (in Aphyllophorales, such as *Hydnum*).

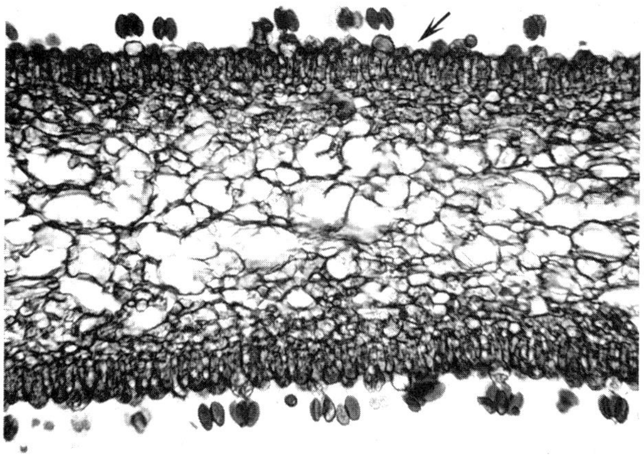

Hymenium of a lamella of *Coprinus* sp., seen in longitudinal section, x 200 (*MU*).

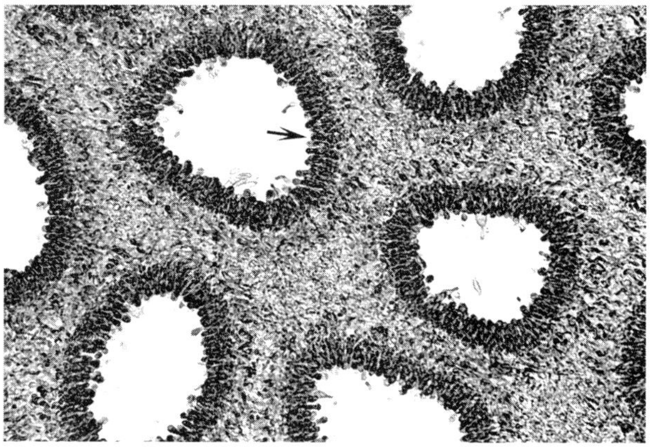

Hymenium of the tubes of *Boletus* sp., seen in transverse section, x 265 (*MU*).

hymenolichen (L. *hymenolichen* < Gr. *hyménion*, dim. of *hymén*, membrane + L. *lichen* < Gr. *leichén*, lichen): a member of a subgroup of basidiolichens in which a hymenomycete is associated with an alga. For example, the species *Dictyonema* (=*Cora*) *pavonia* (association of a thelephoraceous fungus with algal cells of *Croococcus*) and *Dictyonema* (association of the same Thelephoraceae with algal cells of *Scytonema*) are hymenolichens.

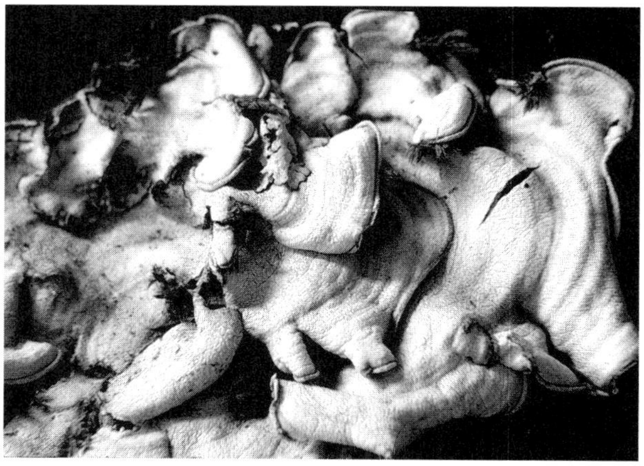

Hymenolichen thallus of *Dictyonema* (=*Cora*) *pavonia*, x 2 (*MU*).

Hymenomycetes (L. *hymenium* < Gr. *hyménion*, dim. of *hymén*, membrane + L. *-mycetes*, ending of class < Gr. *mýkes*, genit. *mýketos*, fungus): in the classification system followed in this dictionary, it is a class of basidiomycetes with the basidia originating in a hymenium which remains exposed at maturity. The hymenium can form freely on the surface of the host or in some type of fruiting body or basidiocarp. The Hymenomycetes includes the orders Agaricales, Aphyllophorales, Tremellales, Auriculariales, Dacrymycetales, Ceratobasidiales, Tulasnellales,

Sporidiales, Septobasidiales and Exobasidiales, as well as the orders of Gasteromycetes (Lycoperdales, Tulostomatales, Sclerodermatales, Phallales, Nidulariales, Hymenogastrales, Gautieriales and Melanogastrales). Together with the Urediniomycetes (Uredinales) and Ustilaginomycetes (Ustilaginales) they constitute the phylum Basidiomycota of the kingdom Fungi.

hymenophore (L. *hymenium* < Gr. *hyménion*, dim. of *hymén*, membrane + Gr. suf. *-phóros*, carrier < *phéro*, to carry, sustain): part of the sporiferous apparatus, composed of sterile tissue, on which is supported the hymenium, such as the trama of the gills of the Agaricaceae (e.g., *Coprinus*) and that of the tubes of the Boletaceae. Also called **hymeniiferous**. The term hymenophore is at times applied incorrectly to refer to the whole basidiocarp.

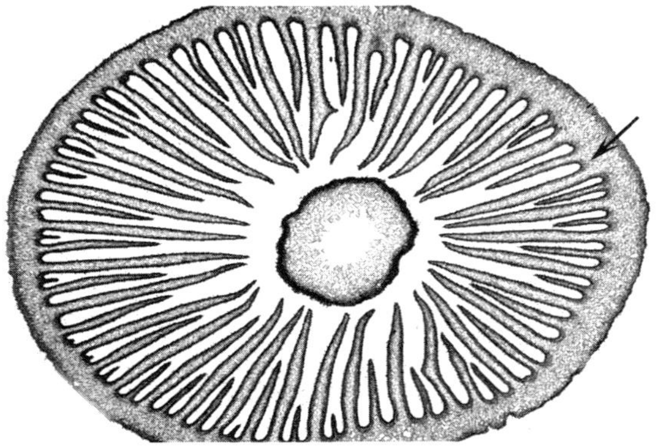

Hymenophore (lamellae) of the pileus of *Coprinus* sp., seen in transverse section of the pileus, x 9 (*MU*).

hymenopode, **hymenopodium**, pl. **hymenopodia** (Gr. *hyménion*, dim. of *hymén*, membrane + NL. *podium* < Gr. *pódion* < *poús*, *podós*, foot, support + dim. suf. *-íon* > L. *-ium*): the layer of tissue found between the trama and the subhymenium in the pezizaceous ascomycetes and in the agaricaceous basidiomycetes; it is composed of thin, filiform elements, with connectives densely intertwined and little differentiated, as in the lateral layer in forms with a bilateral trama. In some species, the hymenopodium can be gelatinized, as in *Pholiota decorata* (Agaricales). Syn. of **hypothecium** in the ascomycetes.

hyperparasite (Gr. *hypér*, above, on, or in a metaphoric sense, beyond the normal, exaggerated + *parásitos* < *pará*, at the side of + *sîtos*, bread, food): a parasite that lives at the expense of another parasite, on it or in it. For example, the species of *Septobasidium* (Septobasidiales) are hyperparasites, as they develop on coccids parasitic on vascular plants. *Verticillium lecanii* (moniliaceous asexual fungi) develops on *Hemileia vastatrix* (Uredinales), the rust of coffee. *Phycomelaina laminariae* (Phyllachorales) is a fungal parasite of marine Phaeophyceae of the species *Laminaria agardhii*, *L. digitata* and others, and its ascocarps are at the same time hyperparasitized by thalli of Oomycota of the gen. *Petersenia* (Myzocytiopsidales) and *Atkinsiella* (Salilagenidiales), whose presence impedes ascus formation in *Ph. laminariae*.

hyperplasia, pl. **hyperplasiae** (Gr. *hypér*, on, excess + *plásis*, action of forming): an abnormal increase in cell multiplication, which can cause an increase in the size of organs affected due to the large number of cells produced. This condition can be induced by certain plant parasitic fungi, such as *Plasmodiophora brassicae*, the causal agent of club root of cabbage (Plasmodiophorales), *Synchytrium endobioticum*, the causal agent of black wart of potato (Chytridiales), and *Taphrina deformans* (Taphrinales). Hyperplasia usually is accompanied by **hypertrophy**, or exaggerated cell growth.

Hyperplasia and **hypertrophy** of peach leaf tissues infected with *Taphrina deformans*, x 1 (*RTH*).

hypersaprobe (Gr. *hypér*, above, on, or in a metaphoric sense, beyond the normal, exaggerated + *saprós*, rotten + *bíos*, life): a saprobe that develops only in substrates that have been invaded by other saprobes, as is observed, e.g., in *Lasiosphaeria* (Sordariales) and in *Nectria* (Hypocreales). Saprobic fungi that grow on other dead saprobic fungi, like *Massarina* (Dothideales), whose ascocarps are at times associated with the sphaeropsidaceous asexual fungi *Coniothyrium*, *Ceratophoma* and *Microsphaeropsis* (anamorphs of Leptosphaeriaceae), among other saprobic fungi, and *Halosphaeria hamata*

(Halosphaeriales), which develops its ascocarps like a hypersaprobe, inside the old or dead, decomposing fruiting bodies of *Leptosphaeria oreamaris* (Pleosporales), a saprobic marine fungus.

hypersensitivity (Gr. *hypér*, above, beyond, excessive + E. *sensitivity* < ML. *sensitivus* < L. *sensus*, sense + E. suf. *-ty*, quality, condition) : a resistant reaction in plants to the attack of biotrophic parasites that involves a rapid death of a limited number of host cells, which becomes visible as flecklike lesions. Flecking can effectively prevent the establishment of a biotrophic parasite, such as a rust fungus, on a particular plant.

hypertonic (Gr. *hypér*, above normal, excess + *tonikós* < *tónos*, tension, tone + suf. *-íkos* > L. *-icus* > E. *-ic*, belonging to, relating to): having an osmotic pressure higher than that of the organism cultured. Cf. **hypotonic**.

hypertrophy (Gr. *hypér*, on, above, excess + *trophía*, to enlarge + *-y*, E. suf. of concrete nouns): an abnormal increase in the size of cells, resulting in the swelling of the affected tissue or organ and the formation of a tumor or canker; hypertrophy usually is accompanied by **hyperplasia** (abnormal cell multiplication); certain plant parasitic fungi (Plasmodiophoromycetes, Chytridiomycetes and Taphrinales) are capable of causing this pathological process. Cf. **distrophy**.

hypha, pl. **hyphae** (NL. *hypha* < Gr. *hyphé*, tissue, spider web; hypha): a tubular filament that represents the structural entity (thallus) of the majority of the fungi. Hyphae can be somatic or fertile and from their differentiation are derived a great diversity of structures related to the assimilative and reproductive functions, including, e.g., rhizoids, stolons, haustoria, appresoria and all of a broad gamut of asexual and sexual sporophores, such as sporangiophores, conidiophores, ascocarps and basidiocarps, all of them manifesting diverse forms depending upon the species. The two general forms of hyphae present in fungi are the **coenocytic** (in many of the so-called lower fungi) and the **septate** (in the Zygomycetes and Trichomycetes of the Zygomycota, and in the asexual fungi, Ascomycota, and Basidiomycota). See **binding**, **generative**, **inflated**, **lactiferous**, **receptive** and **skeletal hypha**.

hyphal (NL. *hypha* < Gr. *hyphé*, tissue, spider web, hypha + L. suf. *-alis* > E. *-al*, relating to or belonging to): belonging to or pertaining to hyphae.

hyphal body (NL. *hypha* < Gr. *hyphé*, tissue, spider web, hypha + L. suf. *-alis* > E. *-al*, relating to or belonging to; OE. *bodig*, body): *Zygomycetes.* Each of the cells resulting from the fragmention of the mycelium of the Entomophthorales (such as *Entomophthora muscae* and *Conidiobolus thromboides*). Hyphal bodies reproduce asexually by means of conidia, or they may fuse in pairs to form zygosporangia.

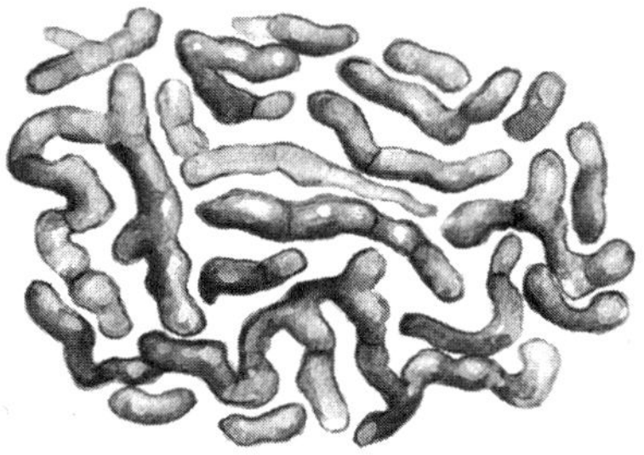

Hyphal bodies of *Conidiobolus thromboides*, x 69.

hyphal net (NL. *hypha* < Gr. *hyphé*, tissue, spider web, hypha + L. suf. *-alis* > E. *-al*, relating to or belonging to; ME. *net* < OE. *nett*): an organ of attachment of some squamulose lichens (placodioid), such as *Psora decipiens* (Lecanorales), consisting of a delicate network of hyphae that penetrate the substrate. Also known as the rhizoidal cord.

hyphal peg (NL. *hypha* < Gr. *hyphé*, tissue, spider web, hypha + L. suf. *-alis* > E. *-al*, relating to or belonging to; ME. *pegge*, peg): **1**. A fascicle of hyphae that projects from the hymenium, consisting of two or more parallel or intertwined hyphae, incrusted or gelatinized; present in some Aphyllophorales, such as *Lentinus* and *Megasporoporia mexicana*. **2**. A hyphal projection or extension that fuses or anastomoses with another hypha.

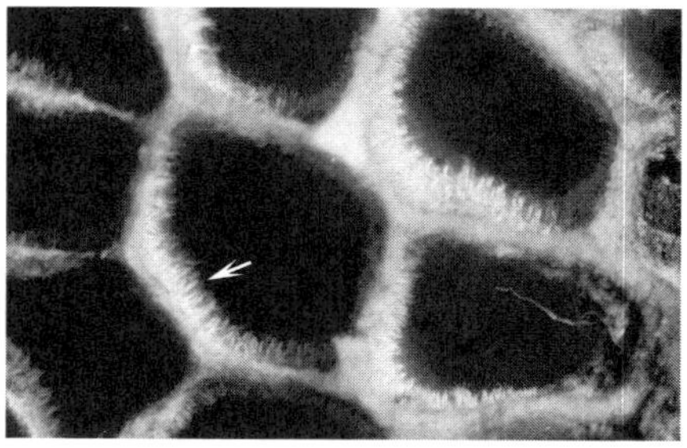

Hyphal pegs of the porous hymenophore of *Megasporoporia mexicana*, x 12 (*CB*).

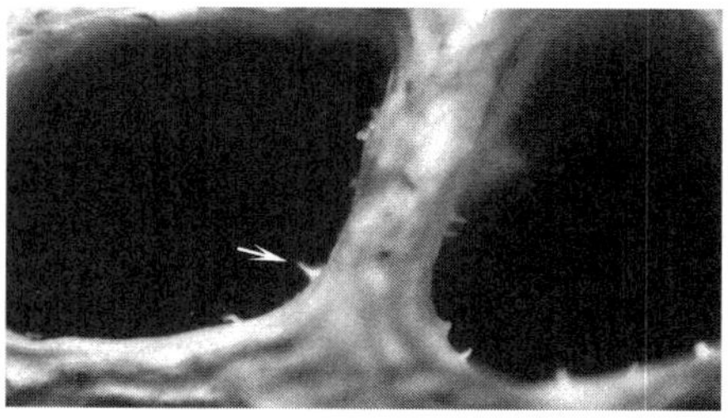

Hyphal pegs of the porous hymenophore of *Megasporoporia mexicana*, x 21 (*CB*).

hyphal string (NL. *hypha* < Gr. *hyphé*, tissue, spider web, hypha + L. suf. *-alis* > E. *-al*, relating to or belonging to; ME. *string* < OE. *streng*, string): a loose cord of soft hyphae that extends into the central part of the stipe cavity, from the apical portion to the base, in the fruiting body of *Coprinus comatus* and *C. sterquilinus* (Agaricales). As the stipe continues growing, the hyphal string stretches, becomes thinner, and breaks.

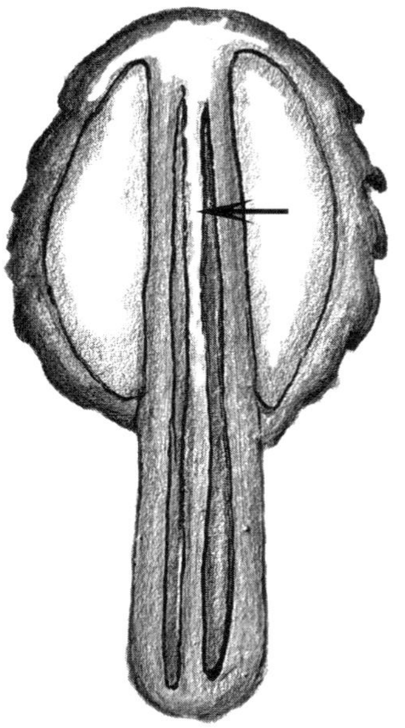

Hyphal string of the basidiocarp of *Coprinus comatus*, seen in longitudinal section, x 1.5.

hyphidium, pl. **hyphidia** (NL. *hyphidium*, small hypha < *hypha*, tissue, spider web, hypha + dim. suf. *-idium*): syn. of **hyphoid**. **1**. *Aphyllophorales*. Characteristic hyphae found in the **catahymenium** of some basidiomycetes (such as *Aleurodiscus*), that are thin or thick-walled, generally lacking cytoplasmic contents, and frequently branched in a distinctive and intricate manner, producing complex forms. The hyphidia, which include a variety of cystidia, develop in early stages and envelop the basidia in distinct levels, because of which they have to open a passage among themselves in order to reach the surface of the catahymenium. Various types of hyphidia are recognized: **acanthophyses**, **asterophyses** and **dendrophyses**. **2**. Another term for **spermatium**.

hyphochytrid (Gr. *hyphé*, tissue, spider web, hypha + *chytrídion*, dim. of *chytrís*, kettle): one of the Hyphochytriomycota, a phylum of the kingdom Stramenopila. This phylum, with the single order Hyphochytriales, is considered by Margulis and Schwartz (1988) within the kingdom Protoctista, and in the kingdom Stramenopila by Alexopoulos *et al.* (1996).

Hyphochytriomycetes (Gr. *hyphé*, tissue, spider web, hypha + *chytrídion*, dim. of *chytrís*, pot + L. *-mycetes*, ending of class < Gr. *mýkes*, genit. *mýketos*, fungus): in some classifications, a class of fungi with anteriorly uniflagellate zoospores (mastigonemate flagellum); the general structure is similar to that of the Chytridiomycetes. There is a single order, Hyphochytriales. In the classification adopted in this dictionary (Alexopoulos *et al.*, 1996), this group is regarded as the phylum Hyphochytriomycota, which along with the phyla Oomycota and Labyrinthulomycota, form part of the kingdom Stramenopila.

hyphoid (Gr. *hyphé*, tissue, spider web, hypha + L. suf. *-oide* < Gr. *-oeídes*, similar to): **1**. Similar to a hypha; hypha-like. **2**. *Uredinales*. A type of aecium, characterized by having colorless, branched hyphae, that form a mold-like colony on the epidermis of the host plant, with each hypha forming a single colorless aeciospore at its apex; e.g., as in *Dasyspora*.

Hyphomycetes (Gr. *hyphé*, tissue, spider web, hypha + L. *-mycetes*, ending of class < Gr. *mýkes*, genit. *mýketos*, fungus): in some classifications, it corresponds to a class of conidial fungi characterized by the formation of asexual spores (conidia) on conidiophores that are not contained in a fruiting body; i.e., without acervuli or pycnidia, although some form sporodochia and synnemata. The Hyphomycetes (with the single order Moniliales), together with the Blastomycetes and Coelomycetes, used to comprise the subdivision Deuteromycotina, or deuteromycetes in a general sense. However, in the classification adopted in this dictionary all of these forms are regarded as asexual ascomycetes and other asexual fungi, with no class rank, and the Moniliales have been changed to agonomycetaceous, moniliaceous, dematiaceous, tuberculariaceous, and stilbellaceous asexual fungi, after the names of the families Agonomycetaceae, Moniliaceae, Dematiaceae, Tuberculariaceae and Stilbellaceae of the order Moniliales. The same has been done with the Blastomycetes and Coelomycetes, herein not recognized as taxonomic classes, but treated as cryptococcaceous and sporobolomycetaceous asexual yeasts, and sphaeropsidaceous and melanconiaceous asexual fungi, respectively.

hyphopodium, pl. **hyphopodia** (Gr. *hyphé*, tissue, spider web, hypha + NL. *podium* < Gr. *pódion* < *poús*, *podós*, foot + dim. suf. *-íon* > L. *-ium*): a short, lateral hyphal branch consisting of one or two cells, characteristic of the obligately parasitic fungi belonging to the order Meliolales (such as *Meliola caesariae* and *Asteridiella homalii-angustifolii*). Two types of hyphopodia are formed: the capitate hyphopodium is flattened, sometimes lobed, with a rounded end cell, which adheres to the host surface; it gives rise to a tube which penetrates the cuticle to reach an epidermal cell, in which it develops a haustorium that absorbs nutrients. It thus functions as an appressorium. The mucronate hyphopodium is a tapered, erect cell that has been shown to be conidiogenous. See **stigmatocyst** and **stigmatopod**.

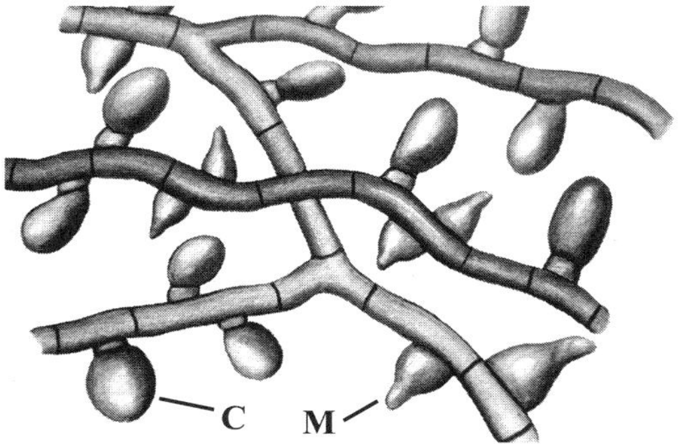

Hyphopodia of the capitate (C) and mucronate (M) types formed by *Meliola caesariae*, x 350.

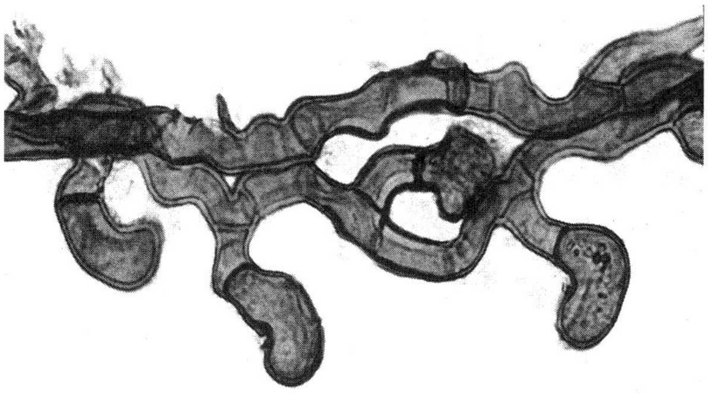

Hyphopodia of *Meliola* sp., x 480 (*RTH*).

hypnocyst (Gr. *hýpnos*, sleep + *kýstis*, vessel, cell): a group of cells, resembling *Alternaria* (dematiaceous asexual fungi) spores, capable of dormancy.

hypnospore (Gr. *hýpnos*, sleep + *sporá*, spore): a thick-walled, resistant spore (such as chlamydospores, latent spores, and teliospores), that does not germinate immediately, but must undergo a period of rest before developing. Also called **memnospore**. Cf. **xenospore**.

hypobasidium, pl. **hypobasidia** (Gr. *hypó*, below + *basídion* < *básis*, base + dim. suf. *-ídion* > L. *-idium*): see **heterobasidium**.

hypocreaceous (< gen. *Hypocrea* < Gr. *hypó*, beneath + *kréas*, flesh + L. suf. *-aceus* > E. *-aceous*, of or pertaining to, having the nature of): fleshy and brightly colored (orange, red), like the perithecial stromata and perithecia of *Hypocrea* (Hypocreales).

hypochnoid (< gen. *Hypochnus* < Gr. *hypó*, beneath + *chnóos*, wool + L. suf. *-oide* < Gr. *-oeídes*, similar to): similar to *Hypochnus*, a syn. of *Tomentella* (Aphyllophorales), which has an effused, resupinate, and dry basidioma, composed of rather loosely intertwined hyphae, like a tomentum.

hypoderm (Gr. *hypó*, below + *dérma*, skin): see **subcutis**.

hypogeal, **hypogean**, **hypogeic**, **hypogeous** (Gr. *hypó*, beneath + *geo*, soil, earth + L. suf. *-alis* > E. *-al*, relating to or belonging to; or L. suf. *-anus* > E. *-an*, belonging to; or + suf. *-íkos* > L. *-icus* > E. *-ic*, belonging to, relating to; or + L. *-osus* > OF. *-ous*, *-eus* > E. *-ous*, possessing the qualities of): growing or developing below the surface of the ground, as occurs in the fungi called truffles (Pezizales). Cf. **epigeal**.

hypogynous (Gr. *hypó*, beneath + *gyné*, woman; here, female sex organ + L. *-osus* > OF. *-ous*, *-eus* > E. *-ous*, having, possessing the qualities of): an arrangement in which the male gametangium is located beneath the female one; e.g., as in *Allomyces arbuscula* (Blastocladiales). Cf. **epigynous**.

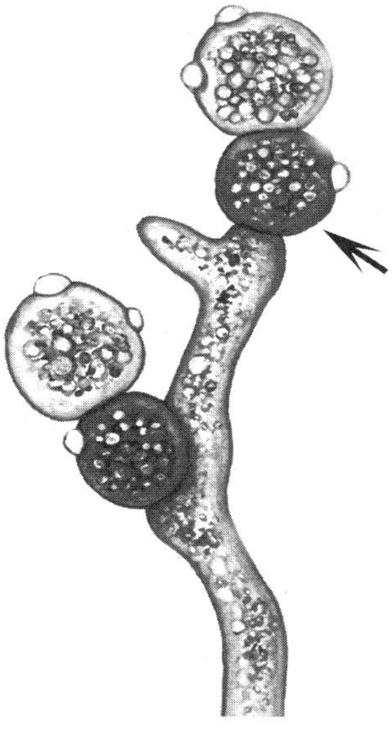

Hypogynous male gametangia of *Allomyces arbuscula*, x 500.

hypoparasite (Gr. *hypó*, beneath + *pará*, at the side, above + *sîtos*, food): a concealed parasite; a pathogen dispersed with another pathogen, such as a mycovirus in *Olpidium viciae* (the chytrid vector, of the Spizellomycetales) that affects plants of *Vicia unijuga*.

hypophloeodic, **hypophloeodal** (Gr. *hypó*, beneath + *phloiós* > L. *phloeo*, cortex + Gr. *-oeídes*, like + suf. *-íkos* > L. *-icus* > E. *-ic*, belonging to, relating to; or + L. suf. *-alis* > E. *-al*, relating to or belonging to): *Lichens*. Crustaceous species that develop almost immersed in the bark of trees, such as various species of *Thelenella* (=*Microglaena*) and other gen. of the Thelenellaceae, an ascomycetous family of uncertain affinities, previously in the Pyrenulales. Cf. **epiphloeodic**.

hypophoretic (Gr. *hypó*, beneath + *-phóros*, bearer < *phéro*, to bear + *-tikós* > L. *-ticus* > E. *-tic*, relation, fitness, inclination or ability): refers to fungi which are carried by the mites that ride along on insects, like *Pyxidiophora* (Laboulbeniales). See **phoresy**.

hypophyllous (Gr. *hypó*, below + *phýllon*, leaf + L. *-osus* > OF. *-ous*, *-eus* > E. *-ous*, having, possessing the qualities of): occurring beneath a leaf or on its lower surface; such as parasitic fungi whose fructifications appear in the underside of leaves (e.g., the aecia of *Puccinia graminis tritici*, of the Uredinales, which develop on the underside of the leaves of barberry).

hypostroma, pl. **hypostromata** (Gr. *hypó*, underneath + *strõma*, bed, mattress, cushion): the pseudoparenchymatous base on which are seated the ascocarps of some marine Dothideales, such as *Manglicola guatemalensis*, which inhabits the bark of the dead roots of mangrove (*Rhizophora*). The thallus of *Spathulospora* (Spathulosporales) also develops a hypostroma which, like a foot formed by infective cells that merge to make room for the assimilative cells, is introduced into the cells of the algal host.

hypothallus, pl. **hypothalli** (Gr. *hypó*, beneath + *thallós*, shoot, thallus): **1**. *Myxomycetes*. A thin, transparent, filmlike membrane, commonly with calcareous deposits, formed by the plasmodium on the substrate, at the base of the fructifications. In some species the hypothallus is not very conspicuous, but in others it may be red, as in *Metatrichia vesparium* (Trichiales), yellow, as in *Leocarpus fragilis* and *Physarum polycephalum* (Physarales), or yellowish-green, as in *Stemonitis nigrescens* (Stemonitales). **2**. *Lichens*. In the crustaceous species, it corresponds to the lower layer of the thallus, blackish in color, which produces the rhizines, such as occurs in *Parmelia imbricatula* (Lecanorales). **3**. *Clavicipitaceae* (Hypocreales). The thin layer of fungal tissue formed on the host surface and upon which the perithecial stroma develops, as in *Balansia*.

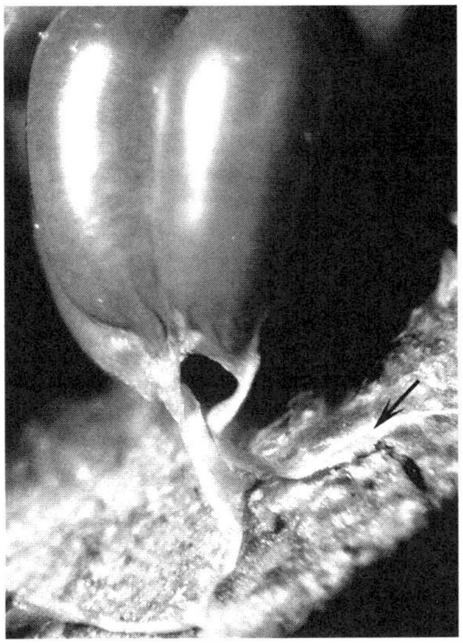

Hypothallus of *Leocarpus fragilis*, x 20 (*CB*).

Hypothallus of *Physarum polycephalum*, x 26 (*MU*).

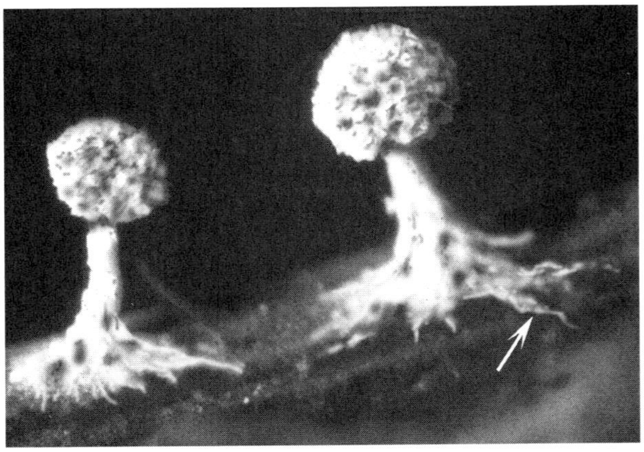

Hypothallus of the sporangia of *Physarum leucopus*, on the upper surface of a grass leaf, x 12 (*MU*).

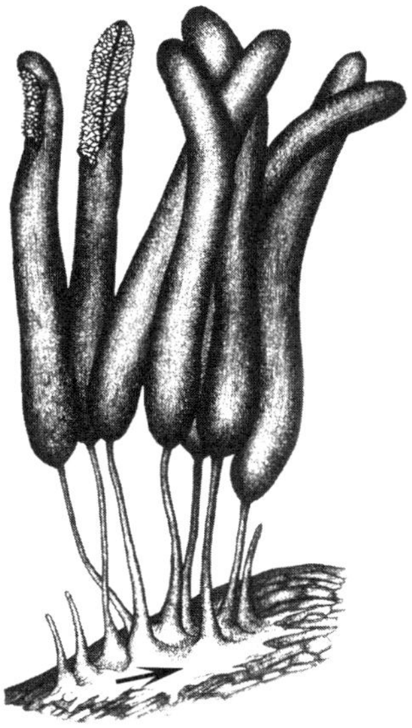

Hypothallus of *Stemonitis nigrescens*, x 15.

hypothecium, pl. **hypothecia** (NL. *hypothecium* < Gr. *hypó*, beneath + *thekíon*, dim. of *thêke*, case, box; here, of the asci): **1**. *Discomycetes*. The thin upper layer of the apothecial tissues on which the asci rest; as in the Pezizales and related fungi. **2**. *Discolichens*. The subhymenial layer which extends laterally, forming the wall of the apothecium (parathecium).

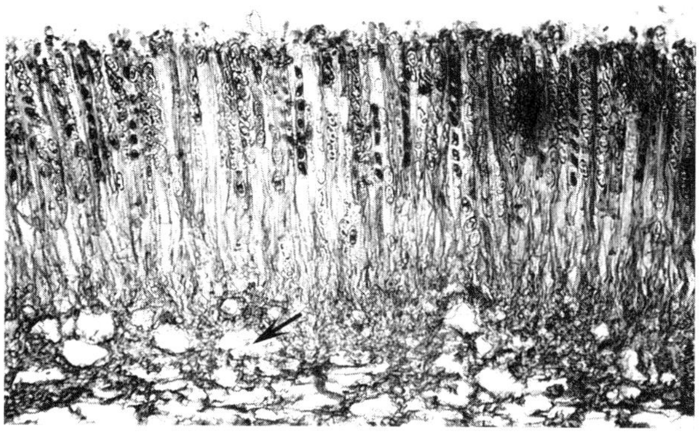

Hypothecium of the apothecium of *Peziza* sp., seen in longitudinal section, x 80 (*RTH*).

hypotonic (Gr. *hypó*, beneath, below normal + *tonikós* < *tónos*, tension, tone + suf. *-íkos* > L. *-icus* > E. *-ic*, belonging to, relating to): having an osmotic pressure lower than that of the organism cultured. Cf. **hypertonic**.

hysterothecium, pl. **hysterothecia** (NL. *hysterothecium* < Gr. *hýsteros*, uterus, womb, cavity + *thekíon*, dim. of *thêke*, case, box; here, of the asci): an elongated, boat-shaped ascocarp, which is closed at first and later opens at maturity by means of a longitudinal fissure along a line of dehiscence, exposing the hymenium of bitunicate asci. The hysterothecium is characteristic of the Loculoascomycetes of the family Hysteriaceae, order Dothideales, such as *Glonium*, *Hysterium* and *Hysterographium*, and of certain Loculoascolichenes (Arthoniales and Pyrenulales).

Hysterothecium of an unidentified loculoascomycete, erumpent through the bark of a twig, x 30 (*CB*).

i

ianthinosporous (Gr. *iánthinos*, purplish-blue + *sporá*, spore + L. *-osus* > OF. *-ous*, *-eus* > E. *-ous*, having, possessing the qualities of): applied to fungi with spores that are purplish-brown when seen in mass in a spore print, as from the basidiocarps of some species of Agaricaceae, Strophariaceae and Coprinaceae (Agaricales).

icon, pl. **icones** (L. < Gr. *eikón*, image): printed figures, plates and illustrations, generally in color. An example of the use of this term in mycological works is the *Icones fungorum hucusque cognitorum* by A. C. J. Corda, published in six parts from 1837-1854.

iconography (ML. *iconographia* < Gr. *eikonographía*, sketch, description < *eikón*, image + *graphía* < *grápho*, describe + *-y*, E. suf. of concrete nouns): pictorial material relating to or illustrating a subject.

idiomorph (Gr. *ídios*, one's own + *morphé*, form): in the ascomycetes and basidiomycetes that have been studied the genes at the mating-type loci are completely dissimilar and cannot be regarded as homologous alleles. The term idiomorph has been coined to describe such alternative forms of a locus that lack significant sequence homology but occupy the same site in the genome.

igniarious (L. *igniarius*, belonging to or relative to fire < *igneus*, fire, fiery + *-arius*, belonging to): tinder; very dry, easily flammable material, such as pieces of wood, thistle, or dry fungi; of the latter, *Fomes igniarius* and other woody Aphyllophorales, wood decomposers, as well as some Lycoperdaceae, are examples of this type of tinder or punk.

imbricate, **imbricated** (L. *imbricatus*, with the shape of a roof tile < *imbricare*, to cover with roof tile + suf. *-atus* > E. *-ate*, provided with or likeness): superimposed, like the tiles in a tiled roof or the scales of fish which overlap one another. For example, the colonies of many molds, such as *Phoma destructiva* (sphaeropsidaceous asexual fungi), the leaflets that constitute the foliose thallus of lichens such as *Parmelia* (Lecanorales), or the basidiocarps of *Hydnum imbricatum*, *Polyporus versicolor* and *Trametes cervina* (Aphyllophorales) are imbricate.

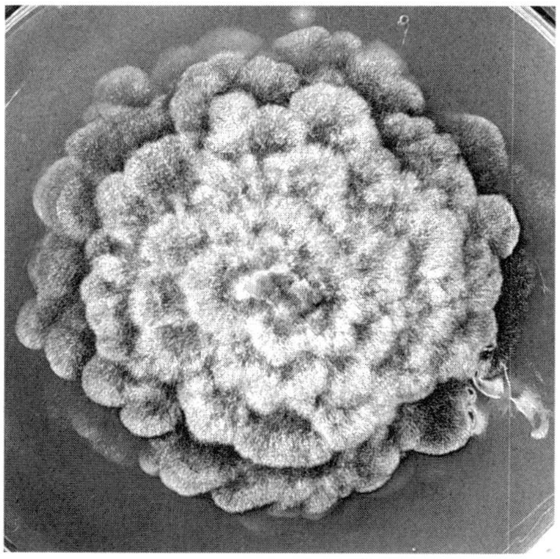

Imbricate colony of *Phoma destructiva* on agar, x 0.8 (*MU*).

Imbricate basidiocarps of *Polyporus versicolor*, x 0.7.

immersed (L. *immersus*, immersed, embedded): beneath the surface; embedded in the surrounding parts. For example, perithecia that are within the substrate or stromatic tissue, such as those of *Xylaria*, *Daldinia* and other Xylariales, are immersed.

in situ (L. *in situ* < *in*, in, inside + *situ*, site, place): an organism living in its natural habitat. Cf. **ex situ**.

in vitro (NL. *in vitro*, literally in glass): outside the living body, or out of its natural habitat, and in an artificial environment, as with microbial cultures grown in glass containers in the laboratory. Cf. **in vivo**. See **ex situ** and **in situ**.

in vivo (NL. *in vivo*, literally in the living): in the living body of a plant or animal, as with parasitic microorganisms maintained in contact with the living cells of their hosts for research purposes. Cf. **in vitro**. See **in situ** and **ex situ**.

inaequihymenial (L. *inaequi*, unequal + *hymenium*, hymenium, membrane + L. suf. *-alis* > E. *-al*, relating to or belonging to): *Agaricales*. A type of hymenial development in which the basidia develop in zones across the gill, instead of all at once. Cf. **aequihymenial**. See **inaequihymeniiferous**.

inaequihymeniiferous (L. *inaequi*, unequal < *in-*, not + *aequus*, equal + *hymenium*, hymenium, membrane + *-ferous*, bearer < *ferre*, to bear, carry + *-osus* > OF. *-ous*, *-eus* > E. *-ous*, having, possessing the qualities of): *Agaricales*. Having inaequihymenial development, in which the basidia develop in zones of different age; e.g., as happens in the gen. *Coprinus* (Agaricales). Cf. **aequihymeniiferous** and **aequihymenial**.

inaequilateral (L. *inaequi*, unequal < *in-*, not + *aequus*, equal + *later*, *latus*, side + suf. *-alis* > E. *-al*, relating to or belonging to): *Of spores*. A spore that is bilaterally asymmetrical, i.e., with one side different from the other. For example, the ascospores of *Xylaria* (Xylariales) have one side convex and the other flat (see **asymmetrical**). Cf. **equilateral**.

incised (L. *incisus*, cut): refers to the margin of a pileus that on maturing is divided into irregular, deep segments, as in *Oudemansiella canarii* and *Inocybe fastigiata* (Agaricales), or also to the lobes of the thallus of a foliose lichen, such as *Parmelia* (Lecanorales).

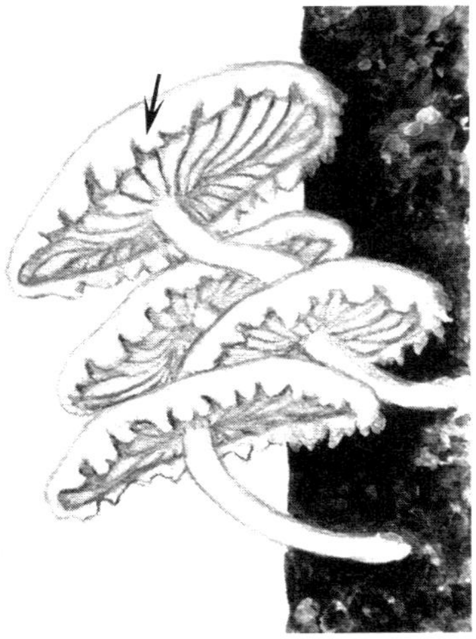

Incised margin of the pileus of *Oudemansiella canarii*, x 0.9.

incrassate (L. *in*, in, toward + *crassus*, thick, gross + suf. *-atus* > E. *-ate*, provided with or likeness): thickened.

incrusted (L. *incrustare*, to incrust, cover with a rind): hardened by interposition of mineral materials among the organic particles. For example, the wall or peridium of the sporangium of some species of *Mucor*, such as *M. mucedo* and *M. racemosus* (Mucorales), is incrusted with aciculiform crystals of calcium oxalate. Also, the sporangial peridium of certain Myxomycetes, such as those of *Badhamia gracilis* and *Physarum compressum* (Physarales) is incrusted with calcium carbonate, which forms a calcareous crust on the surface. In the Agaricales there exist various types of hyphae whose wall is incrusted with diverse materials that give rise to distinctive ornamentations (spirals, warts, crusts, needles, etc.).

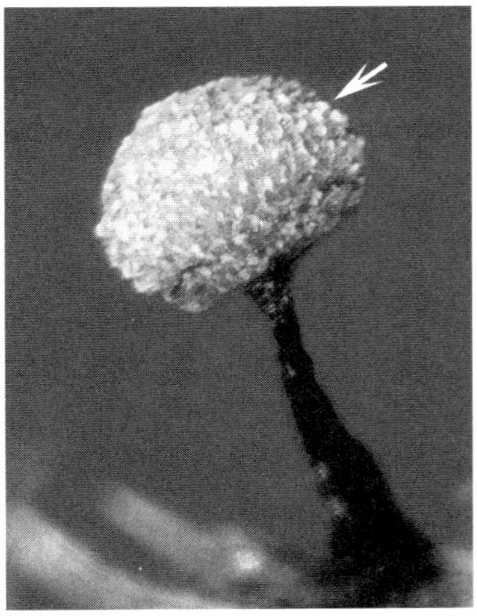

Incrusted peridium of the sporangium of *Badhamia gracilis*, x 50 (*MU*).

incurvate, **incurved** (L. *incurvatus*, incurved < *incurvare*, to bend in < *in-*, in + *curvare*, to curve < *curvus*, curved + suf. *-atus* > E. *-ate*, provided with or likeness): Syn. of **inflexed**; curved inward, i.e., with the concavity on the inner side; e.g. like the border of the pileus of *Entoloma lividum*, *Lactarius indigo*, *L. vellereus* and *Russula delica* (Agaricales). Cf. **recurvate** and **reflexed**.

indehiscence (NL. *in*, not + *dehiscentia*, bursting open < L. *dehiscere*, to open, gape + suf. *-entia* > F. **-ence** < E. *-ence*, state, quality or action): the condition of not rupturing or splitting at maturity; often refers to the peridium. See **dehiscence**.

indehiscent (L. *indehiscens*, genit. *indehiscentis* < pref. *in-*, not + *dehiscere*, to open + *-entem* > E. *-ent*, being): not opening naturally. Cf. **dehiscent**.

indeterminate (LL. *indeterminatus*, not determined, indefinite < *in-*, not + *determinatus*, limited, defined < *determinare*, to limit, determine + suf. *-atus* > E. *-ate*, provided with or likeness): **1**. The continued development of sporangia or conidia due to elongation of the producing hypha; i.e., no terminal sporangia or conidia are formed, allowing indefinite growth. For example, in the Peronosporales of the family Pythiaceae (such as *Pythium* and *Phytophthora*) the sporangia form on somatic hyphae or in sporangiophores of indeterminate growth. **2**. *Lichens*. It refers to certain crustaceous thalli whose margin is difficult to define precisely due to being immersed in the substrate where it grows; e.g., as in some species of *Lecanora*, of the Lecanorales. Cf. **determinate**.

Incurvate pileus of the basidiocarp of *Entoloma lividum*, x 0.8.

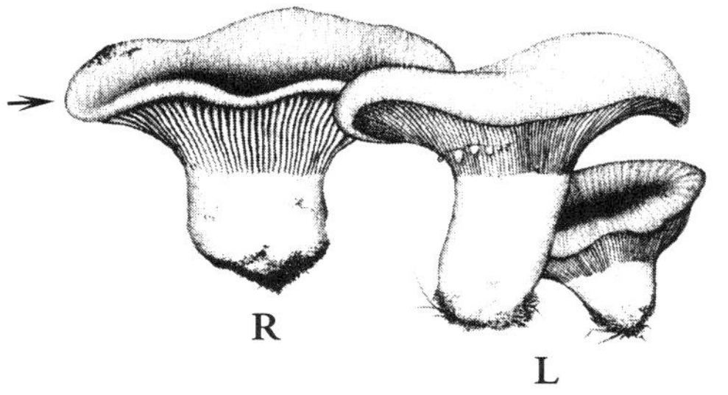

Incurvate pileus of the basidiocarps of *Russula brevipes* (R) and *Lactarius piperatus* (L), x 0.3.

indigenous (L. *indigena*, native, not foreign + *-osus* > OF. *-ous*, *-eus* > E. *-ous*, possessing the qualities of): belonging to the place or country where it is found; not having come from outside. Syn. of **autochthonous**.

indumentum, pl. **indumenta** (L. *indumentum*, indument < *indumentus*, a garment): the ensemble of hairs, scales, etc. that cover the surface of some organs or structures, such as the basidiocarp of *Polyporus versicolor* and other Aphyllophorales.

indurate, **indurated trama**, pl. **tramae** (L. *induratus*, hardened < *indurare* < *in-*, in, toward + *durare*, to harden < *durus*, hard + suf. *-atus* > E. *-ate*, provided with or likeness; *trama*, warp of a loom): a trama which hardens with age, such as the tramal plates in mature *Montagnea* (Hymenogastrales) sporocarps.

indusium, pl. **indusia** (L. *indusium*, tunic, shirt): a white, net-like veil which flares out in a pendant fashion from the base of the receptaculum of the expanded fruiting body of *Dictyophora indusiata*, a Gasteromycete of the order Phallales.

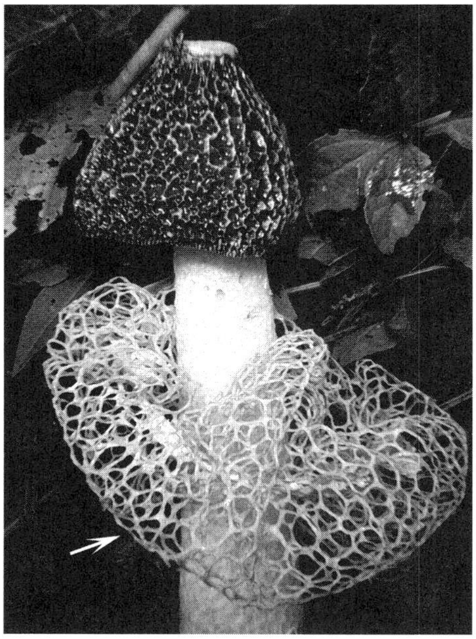

Indusium of the fruiting body of *Dictyophora indusiata*, x 0.8 (*TH*).

inerm, **inermous** (L. *inermis*, unarmed, without arms; or + L. *-osus* > OF. *-ous*, *-eus* > E. *-ous*, having, possessing the qualities of): without spines or thorns.

infarctate (L. *in*, in, into + *farctum*, *farctus*, stuffed + suf. *-atus* > E. *-ate*, provided with or likeness): see **farctate**.

infect (ME. *infecten* < L. *infectus*, ptp. of *inficere* < *in*, in + *facere*, to make, do): *Pathology*. To enter, invade, and establish a pathogenic relationship with a host organism: to communicate a pathogen or a disease to; to make an attack on an organism; to make infection of an organism take place. *Infected*: an organism attacked by a pathogen. *Infection*: the act or result of infecting. *Infectious*: capable of causing infection; communicable by infection; able to be handed on by touch (contagious) or by inoculum; spreading or capable of spreading rapidly to others. *Infective*: a pathogen able to make an attack on a living organism; a vector, medium, etc., capable of effecting the transmission of a pathogen. Cf. **contaminate**.

infest (MF. *infester* < L. *infestare* < *infestus*, hostile): to spread or swarm in or over in a troublesome manner, as with insects, rodents and other animals that attack crops. Sometimes used for fungi in soil or other substrata in the sense of **contaminate**.

inflated (L. *inflatus*, inflated, ptp. of *inflare*, to inflate, blow into, puff out < *in-*, in toward + *flare*, to blow + suf. *-atus* > E. *-ate*, provided with or likeness): swollen like a vesicle, like some basidiomycete cystidia.

inflated hypha, pl. **hyphae** (L. *inflatus*, inflated, ptp. of *inflare*, to inflate, blow into, puff out < *in-*, in, toward + *flare*, to blow + suf. *-atus* > E. *-ate*, provided with or likeness; Gr. *hyphé*, tissue, spider web, hypha): *Basidiomycetes*. A hypha that has subapical cells that enlarge and undergo the apparent rapid growth that characterizes the majority of the basidiocarps of the Agaricales and Gasteromycetes.

inflexed (L. *inflexus*, curved inward): syn. of **incurvate**: curved inward with respect to the pedicle, branch, etc. on which it is borne; e.g., the margins of the pileus of the basidiocarp of some species of *Clitocybe* and *Coprinus* (Agaricales) are inflexed. Cf. **reflexed**.

infundibuliform (L. *infundibulum*, funnel + suf. *-formis* < *forma*, shape): with the shape of a funnel, like the fruiting body of *Clitocybe infundibuliformis* (Agaricales), *Cantharellus cibarius*, *Lentinus badius* and *Polyporus tuberaster* (Aphyllophorales).

Infundibuliform basidiocarps of *Clitocybe infundibuliformis*, x 0.5 (*EPS*).

Infundibuliform basidiocarp of *Lentinus badius*, x 2 (*CB*).

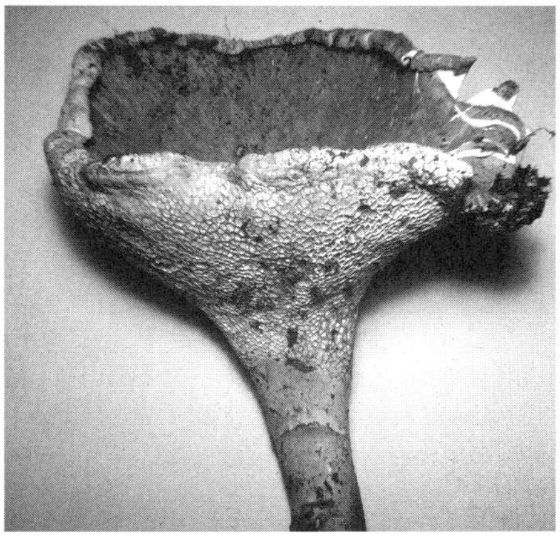

Infundibuliform basidiocarp of *Polyporus tuberaster*, x 1 (*TH*).

ingroup (L. *in*, in, into + F. *groupe* < It. *gruppo*, group, a number of individuals assembled together or having some unifying relationship): a term used in systematics to refer to a group of the taxa that generally are considered to be monophyletic; usually the group under study. Cf. **outgroup**.

ink-cap fungus, pl. **fungi** (ME. *enke*, ink < OF. < LL. *encaustum* < L. *encaustus*, burned in; ME. *cappe* < OE. *caeppe* < LL. *cappa*, head covering, cap; L. *fungus*, fungus): the deliquescent or autodigesting basidiocarp of *Coprinus* spp. (Agaricales). See **autolysis** and **deliquescence**.

innate (L. *innatus*, natural, inborn): applied to scales, fibrils, etc. that form part of the tissue, i.e., they are immersed in the tissue and not on its surface.

inoculate (L. *inoculatus*, implanting, ingrafting, ptp. of *inoculare*, to graft, implant, sow < *in*, in, into + *oculus* eye, with the suf. *-atus* > E. *-ate*, to form the verb): to introduce or put a microorganism (usually pathogenic), or a substance containing one, into an organism (the host) or a substratum. *Inoculation*: the act or processs of inoculating; especially the introduction of a pathogen into a living organism. See **inoculum**.

inoculum, pl. **inocula** (L. *inoculare*, to graft, implant, sow < *in*, in, into + *oculus*, eye, with the adjectival ending *-um*): fungal material that is transferred to a substrate or culture medium in order to propagate the fungus. In phytopathology and medical mycology it refers to the portion of the pathogen that is transferred to a host, either in an artificial or natural way, in order to cause infection in the host.

inoperculate (L. *in-*, not + *operculatus* < *operculum*, cover, lid + suf. *-atus* > E. *-ate*, provided with or likeness): lacking an operculum; e.g., as in asci and sporangia whose dehiscence is not by means of an operculum. In the Discomycetes, the asci of the gen. classified in the orders Rhytismatales (such as

Rhytisma), Leotiales (such as *Vibrissea*) and Helotiales (e.g., *Monilinia*) are inoperculate. Cf. **operculate**.

inserted (L. *insertus* < *inserere*, to insert, attach): refers to a structure that originates directly from the substrate, without being adhered by rhizoids or fibrils of any kind, such as the stipe of the basidiocarp of *Hygrophorus chrysodon* (Agaricales).

integrated (L. *integratus* < *integrare*, to form, coordinate, or blend into a functioning or unified whole + suf. *-atus* > E. *-ate*, provided with or likeness): *Conidiogenesis.* A conidiogenous cell that is incorporated into the main axis or branches of the conidiophore; it generally resembles the conidiophore. Cf. **discrete**.

interbiotic (L. *interbioticus* < *inter-*, between + *bioticus* < Gr. *biotikós*, relative to or pertaining to *biósis* < *bíos*, life + *-tikós* > L. *-ticus* > E. *-tic*, relation, fitness, inclination or ability): a parasitic organism whose situation with respect to the host is intermediate between endobiotic and epibiotic; if the organism is saprobic it refers to its position in the substrate. Thus, there are some aquatic fungi, such as *Chytridium olla*, *Chytriomyces hyalinus* and *Rhizophydium sphaerotheca* (Chytridiales), that are interbiotic. The former forms its sporangium on the outside of an algal thallus and the rhizoids within; the second forms a rhizoidal system within insect molts or pieces of chitin, and their sporangia on the external part of these substrates; the third forms similar structures on pine pollen grains. Also, the Laboulbeniales are interbiotic, since their thalli or receptacles are on the surface of the exoskeleton of the host insects, but the haustoria are located in the interior of the tissues of said hosts. Cf. **endobiotic** and **epibiotic**.

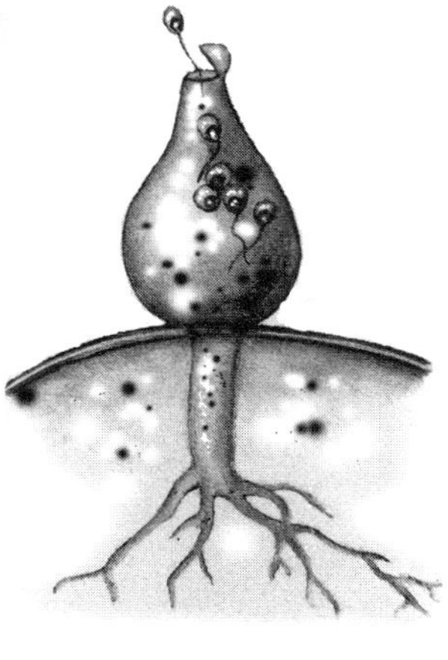

Interbiotic thallus of *Chytridium olla* on an algal host; the epibiotic part of the thallus corresponds to the sporangium, and the interbiotic part to the rhizoids, x 475.

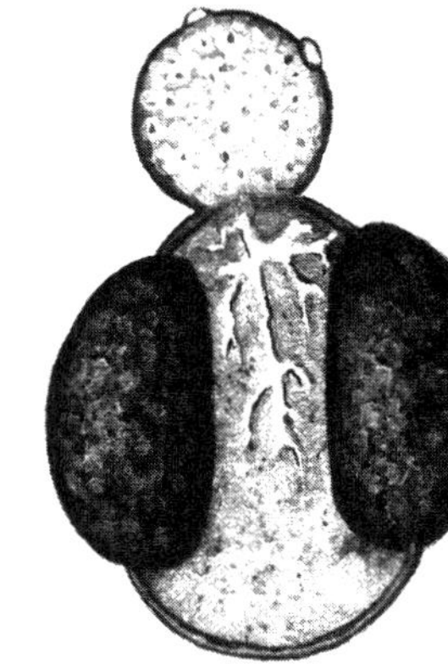

Interbiotic thallus of *Rhizophydium sphaerotheca* growing on a grain of pine pollen, x 300.

intercalary (L. *intercalaris*, of or for insertion < *intercalare*, to insert + *-aris* > E. *-ar*, pertaining to + E. suf. *-y*, having the quality of): positioned between the apex and base, i.e., neither terminal nor basal. It refers to resistant or reproductive structures, such as chlamydospores, sporangia, etc., when they are confined on both ends by vegetative cells. Also applied to growth that is not apical, but which is localized in the base, or better, the middle part of the organ or structure treated. The asci of *Saccharomycopsis synnaedendra* and occasionally of *Pichia membranaefaciens* (Saccharomycetales), e.g., are intercalary.

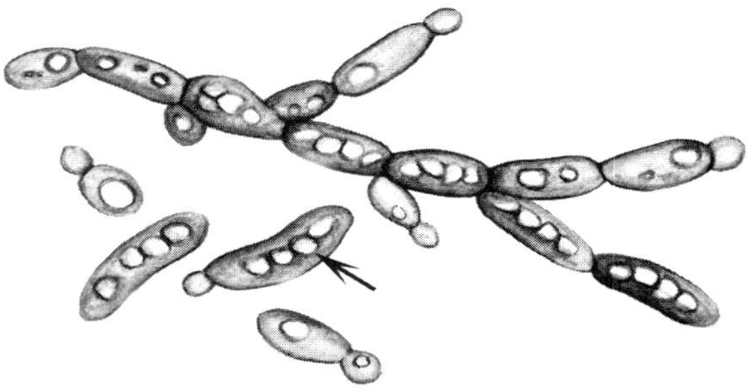

Intercalary asci of *Pichia membranaefaciens*, x 920.

intermediate host (ML. *intermediatus* < L. *intermedius* < *inter-*, between + *medius*, mid, middle + suf. *-atus* > E. *-ate*, with the property of; L. *hospes*, one who receives a stranger as guest): see **host**.

internode (L. *internodium*, internode): the portion encompassed between two nodes or consecutive joints, as occurs between the radulae of the percurrent conidiophore of *Arthrobotrys oligospora* (moniliaceous asexual fungi) and *Gonatobotryum apiculatum* (dematiaceous asexual fungi).

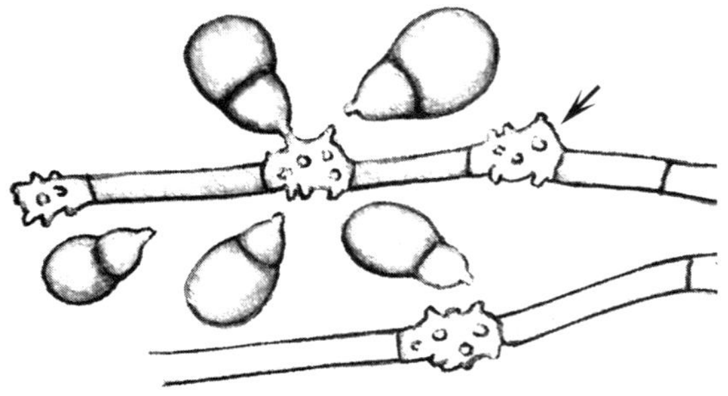

Internodes of the conidiophore of *Arthrobotrys oligospora*, x 1 000.

interspace (L. *interspatum* < *inter-*, between + *spatium*, space): the space that is found between two gills of the sporophore of agarics; laminar interspace is the more precise term for this.

interthecial (L. *inter-*, between + *thecicus* < *theca* < Gr. *thêke*, case; here, ascus + L. suf. *-alis* > E. *-al*, relating to or belonging to): that which is found between the asci, or between the ascigerous locules, such as the plectenchyma of the stroma of the Myriangiales.

intrahyphal (L. *intra*, in, inside + NL. *hypha* < Gr. *hyphé*, tissue, spider web, hypha + L. suf. *-alis* > E. *-al*, relating to or belonging to): the development of one hypha inside another, as in the somatic hyphae of *Myrioconium* (moniliaceous asexual fungi).

intramatrical (L. *intra*, within + *matrix*, the matrix, a place where anything is generated, or the host plant + E. suf. *-ic* < L. *-icus* < Gr. *-íkos*, belonging to, relating to + L. suf. *-alis* > E. *-al*, relating to or belonging to): an organism that lives in the interior of the host, within the substrate. Equivalent to **endobiotic**. Cf. **extramatrical**.

intraparietal (L. *intra*, within + *parietalis* < *paries*, genit. *parietis*, wall + suf. *-alis* > E. *-al*, relating to or belonging to): occurring in the wall or membrane; e.g., such as the segments that occur in the internal portion (of the hyphal wall) in the form of spirals, rings or irregular heaps, resulting in a certain type of internal ornamentation, as in *Panaeolus sphinctrinus* (Agaricales).

intrasporangial (L. *intra*, in, inside + NL. *sporangium*, belonging or relative to the sporangium < Gr. *sporá*, spore + *angeîon*, receptacle + L. suf. *-alis* > E. *-al*, relating to or belonging to): see **proliferation**.

intricate (L. *intricatus*, ptp. of *intricare*, to entangle + suf. *-atus* > E. *-ate*, provided with or likeness): entangled, but not coalesced; having many complexly interrelating parts or elements. In lichenology, it is applied to the type of thallus cortex that is composed of hyphae that are twisted and entangled with one another.

intron (E. *inter*vening, to come between + *-on*, in contact with or supported by): a term used in molecular biology to refer to a polynucleotide sequence in a nucleic acid that does not code information for protein synthesis and is removed before translation of messenger RNA. Cf. **exon**.

introrse (L. *introrsus*, toward the inside): next to the central axis, toward the inside. Cf. **extrorse**.

intumescence (L. *intumescens*, intumescent < *intumescere*, to swell + suf. *-escens*, genit. *-escentis*, that which turns, beginning to, slightly > E. *-escence*): a swelling due to the flow of watery liquids into a tissue.

intussusception (L. *intus*, within + *susceptio*, genit. *susceptionis*, a taking in hand, to receive < *susceptus*, undertaken, ptp. of *suscipiere*, to support, to take up + *-io*, *-ionis* > E. suf. *-ion*, result of an action, state of): applied to growth in thickness of a cell wall by intercalation, between the molecules or the mycelia, of other elements of the same nature. The growth occurs, therefore, without internal or external superficial deposits of molecules or of layers; the new wall materials are laid down within the original thickness of the wall. The subapical growth of fungal hyphae occurs by intussusception. Cf. **apposition**.

invaginated, **invagination** (ML. *invaginatus*, sheathed, enclosed < L. *invaginare*, to sheath < *in*, in + *vagina*, covering, sheath + suf. *-atus* > E. *-ate*, provided with or likeness; or + *-ationem*, action, state or condition, or result > E. suf. *-ation*): **1**. Inserted or contained in a covering or sheath; e.g., like the stylet (originally called **Stachel** in German), which is included in a tube (the **Rohr** in German) that in turn is formed in the interior of the encysted zoospore of *Plasmodiophora brassicae* (Plasmodiophorales). **2**. *Gasteromycetes*. An inward growth from the peridium, as in forate development, to form multiple cavities in which the clusters of basidia are found. See **forate**.

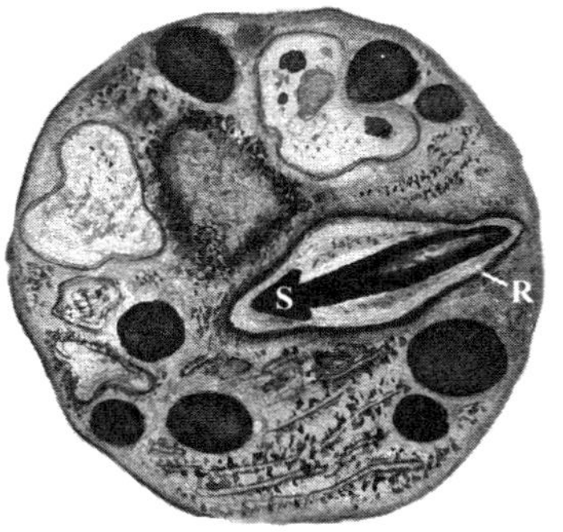

Invaginated Stachel (S) in the Rohr (R) of an encysted zoospore of *Plasmodiophora brassicae*. Drawing based on a transmission electron micrograph published by Beckett *et al.* in their *Atlas of Fungal Ultrastructure*, 1974, x 20 500.

inverse (L. *inversus*, opposite in order, nature or effect): see **convergent**.

involucellum, pl. **involucella**, **involucrellum**, pl. **involucrella** (L. *involucellum* < *involucrum*, a wraper, envelope, cover + dim. suf. *-ellum*): refers to the upper region of the ascocarp of some marine fungi, such as *Stigmidium* (=*Pharcidia*), of the Dothideales, whose peridium is especially thick around the ostiole, and is composed of thick-walled cells, subglobose or somewhat flattened in the central part of the pseudothecium. Some loculoascolichens, such as *Verrucaria* and *Staurothele* (Verrucariales), have an involucrellum that extends outward from the ostiole, forming a carbonaceous, shield-like layer.

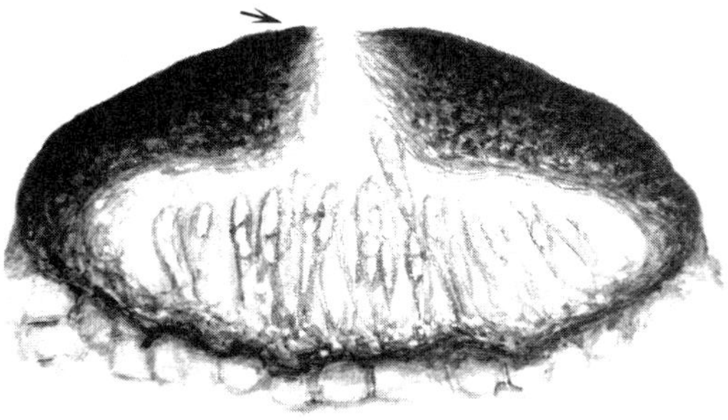

Involucellum of the peridium of a pseudothecium of *Stigmidium* (=*Pharcidia*) *laminariicola* parasitizing the alga *Laminaria digitata*. Drawing of a longitudinal section, x 520.

involute (L. *involutus* < *involvere*, to roll up): *Agaricales*. A pileus in which the borders are more or less rolled toward the upper face of the pileus; involute is like **decurvate**, but rolled, and it can be exemplified by the pileus of the basidiocarp of some species of *Marasmius* and *Coprinus* (such as *C. lagopus*, of the Agaricales). Cf. **revolute**.

Involute margin of the pileus of *Coprinus lagopus*, x 1.

involute ray (L. *involutus* < *involvere*, to roll up; ME. *raie*, *raye* < OF. *rai* < L. *radius*, radius): *Gasteromycetes*. A ray which in profile has the outer peridium concave and inrolled at the edge, as in the fruiting body of *Astraeus* (Sclerodermatales).

Involute exoperidial **rays** of the fruiting body of *Astraeus hygrometricus*, x 2 (*MU*).

iridescent (L. *iris*, genit. *iridis*, rainbow + *-escens*, genit. *-escentis*, that which turns, beginning to, slightly > E. *-escent*): showing a range of colors, similar to those of the rainbow; e.g., like the peridium of the sporangia of the Myxomycetes of the gen. *Diachea* (Physarales) and *Lamproderma* (Stemonitales), which have the property of reflecting various chromatic tones of light.

irpiciform (< gen. *Irpex* < L. *irpex*, dentate + L. suf. *-formis* < *forma*, shape): with flattened teeth, similar to those in the hymenophore of *Irpex* (Aphyllophorales).

isarioid (< gen. *Isaria* < Gr. *ísis*, coral + L. suf. *-aria*, alike + L. suf. *-oide* < Gr. *-oeídes*, similar to): similar to the synnemata of *Isaria* (stilbellaceous asexual fungi), in forming a cylinder of hyphae.

isidiiferous (NL. *isidium* < Gr. *ísis*, coral + L. dim. suf. *-idium* + *-ferous*, bearer < *ferre*, to bear, carry + *-osus* > OF. *-ous*, *-eus* > E. *-ous*, having, possessing the qualities of): a lichen thallus bearing **isidia**.

isidium, pl. **isidia** (NL. *isidium* < Gr. *ísis*, coral + L. dim. suf. *-idium*): *Lichens*. A microscopic outgrowth formed on the upper surface of the thalli of some species that serves to vegetatively propagate the species; e.g., *Pseudevernia consocians*, *Parmotrema* and *Parmelia imbricatula* (Lecanorales). Isidia differ from soredia by being covered by the cortical layer of the thallus, as well as in shape, which can be cylindrical, claviform, scale-shaped or coralloid. Cf. **soredium**.

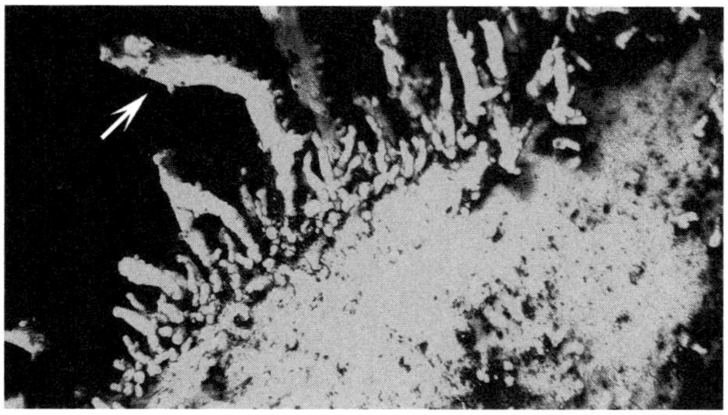

Isidia of the thallus of *Pseudevernia* sp., x 60 (*MU*).

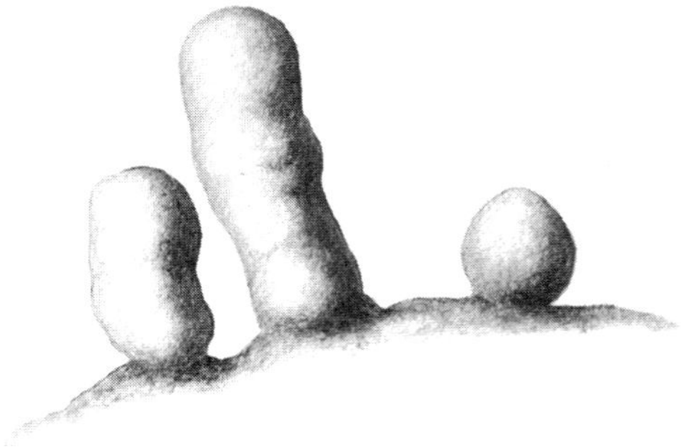

Isidia of the thallus of *Parmelia imbricatula*, x 175.

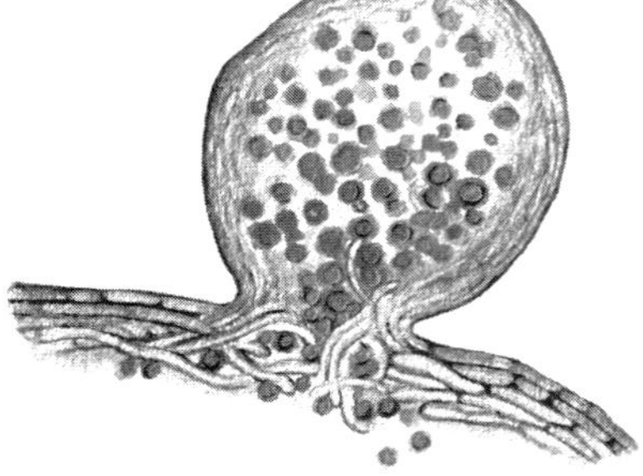

Isidium of *Parmelia imbricatula*, seen in longitudinal section, showing the gonidia inside, x 170.

isodiametric (Gr. *ísos*, equal + *diámetron*, diameter < *diá*, through + *métron*, measure + suf. *-îkos* > L. *-icus* > E. *-ic*, belonging to, relating to): refers to a structure whose length is equal to its width; e.g., like the cells that compose the pseudoparenchyma of various fungal organs. Cf. **anisodiametric**.

isogametangium, pl. **isogametangia** (Gr. *ísos*, equal + *gamétes*, husband + *angeîon*, vessel, receptacle): a gametangium or organ producing gametes that cannot be distinguished morphologically from another of the opposite sex; e.g., as is observed in *Rhizopus* and other Mucorales, such as *Rhizomucor miehei*. Cf. **heterogametangium**.

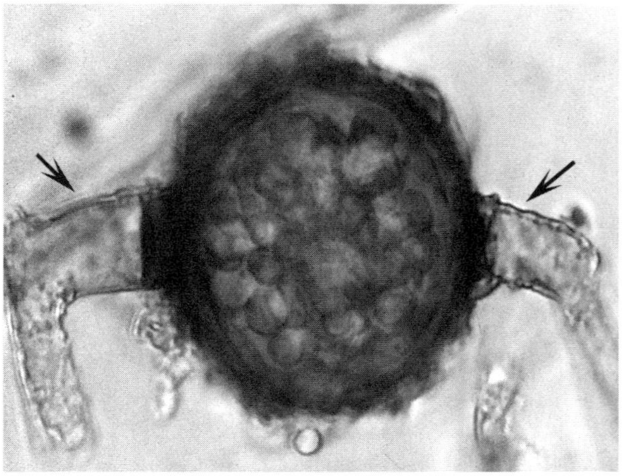

Isogametangia of *Rhizomucor miehei*, x 350 (*MU*).

isogamete (Gr. *ísos*, equal + *gamétes*, husband): a gamete (sexual cell or nucleus) that cannot be distinguished morphologically from another belonging to the opposite sex; generally isogametes are flagellate and motile (isoplanogametes) and they occur in aquatic fungi; e.g., as in *Olpidium* (Spizellomycetales). Cf. **heterogamete**.

isogamy (Gr. *ísos*, equal + *gámos*, wedding, sexual union + *-y*, E. suf. of concrete nouns): fertilization of two gametes that are similar in shape, size and structure (**isogametes**). The gametes are indistinguishable from one another, for which reason they are referred to as + and - isogametes; they commonly are flagellate and motile and are formed in aquatic fungi, such as *Olpidium* (Spizellomycetales) and *Synchytrium* (Chytridiales). Cf. **anisogamy** and **heterogamy**.

isokont, **isocont**, **isokontous**, **isocontous** (Gr. *ísos*, equal + *kontós*, oar; here, flagellum; or + L. *-osus* > OF. *-ous*, *-eus* > E. *-ous*, having, possessing the qualities of): with flagella of equal length; e.g., as in the zoospores of *Ectrogella bacillariacearum* (Myzocytiopsidales) and of some other species of aquatic zoosporic fungi. One must take note that although in this case the two flagella are of equal length or more or less equal in length, only one of them is mastigonemate. Cf. **heterokont**.

isolate (E. *isolated*, separated, set apart, alone < F. *isolé* < It. *isolato* < *isola* < L. *insula*): a strain or population of an organism that is maintained in pure culture, either active or preserved, under laboratory conditions, for research, teaching or industrial production. Cf. **strain**.

isomorphic (Gr. *ísos*, equal + *morphé*, shape + suf. *-îkos* > L. *-icus* > E. *-ic*, belonging to, relating to): applied to organs or structures similar in shape; e.g., thalli, gametangia, etc. Cf. **heteromorphic**.

isosporic, **isosporous** (Gr. *ísos*, equal + *sporá*, spore + suf. *-íkos* > L. *-icus* > E. *-ic*, belonging to, relating to; or + L. *-osus* > OF. *-ous*, *-eus* > E. *-ous*, having, possessing the qualities of): see **homosporic**.

isotomic (Gr. *ísos*, equal + *tómikos*, of or for cutting < *tómos*, a cut, slice + suf. *-íkos* > L. *-icus* > E. *-ic*, belonging to, relating to): refers to the length of the branches of a dichotomy, when it is almost the same, even in the oldest parts of the thallus, as occurs in the type of isotomic dichotomous branching of the lichen *Cladonia evansii* (Lecanorales).

isthmospore (L. *isthmus* < Gr. *isthmós*, narrowed at the middle + *sporá*, spore): an asexual spore constituted of four cells with a thick wall, separated by cells with a thin wall, as in *Isthmospora spinosa* (dematiaceous asexual fungi), the asexual state of *Trichothyrium asterophorum* (Dothideales). The ascospores of *Vialaea* (of uncertain taxonomic position) also are isthmospores; their end cells are connected by a narrow cell.

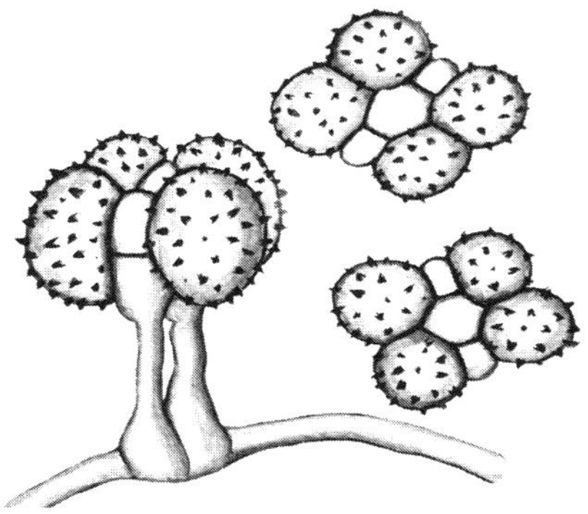

Isthmospores of *Isthmospora spinosa*, x 740.

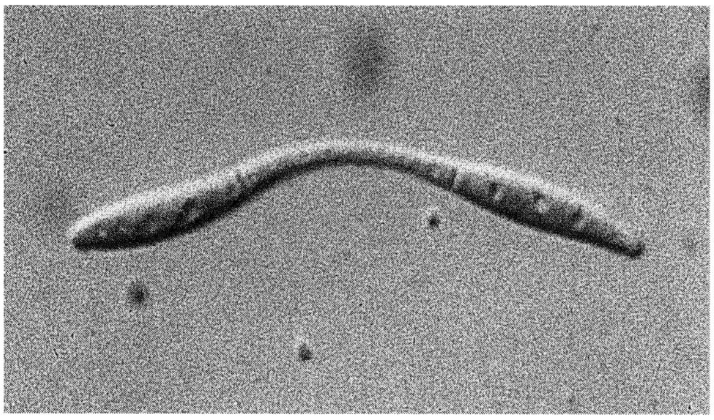

Isthmospore of *Vialaea* sp., x 1 000 (*RTH*).

isthmus, pl. **isthmi** (L. *isthmus* < Gr. *isthmós*, narrowed at the middle): *Lichens*. With the middle portion narrowed and the ends swollen, as in the perforate septum of a polarilocular or polocellate ascospore, such as occurs in *Caloplaca citrina* and in *Teloschistes chrysophthalmus*, of the order Teloschistales.

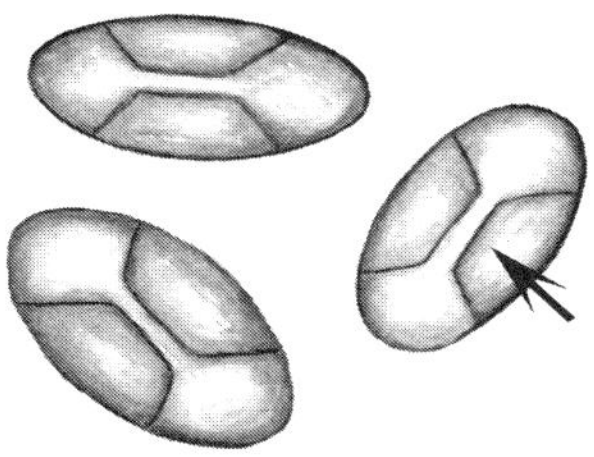

Isthmus of the ascospores of *Teloschistes chrysophthalmus*, x 900.

ixocutis (Gr. *ixós*, viscosity + L. *cutis*, skin): see **ixotrichoderm**.

ixotrichoderm (Gr. *ixós*, viscosity + *thríx*, *trichós*, hair + *dérma*, skin): a pileus covering composed of the apices of more or less erect hyphae, undulate and intertwined, which come to gelatinize in a certain degree and make the surface of the pileus turn viscous, sticky or mucilaginous; e.g., as in *Suillus* and *Hygrophorus* (Agaricales). Also called **ixocutis**.

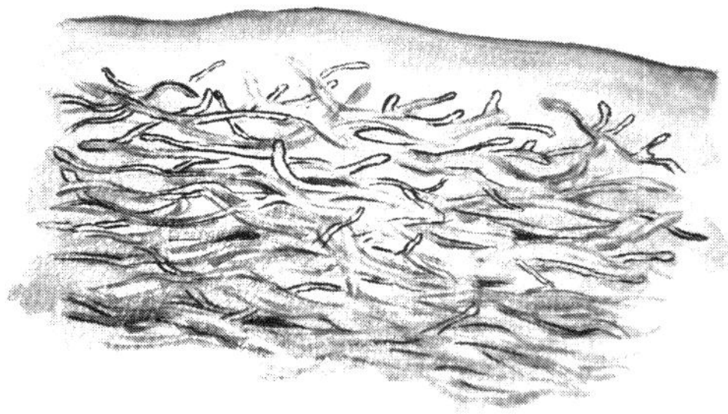

Ixotrichoderm of the pileus of *Hygrophorus* sp., x 55.

j

jelly fungus, pl. **fungi** (ME. *gelly* < MF. *gelee* < fem. of *gelé*, ptp. of *geler*, to freeze, congeal < L. *gelare*; L. *fungus*, fungus): a common name for the Tremellales, also called the trembling fungi, due to their soft, elastic consistency, similar to gelatin or jelly.

Jod (G. *iodine* < Gr. *iódes*, violetlike < *íon*, a violet + L. suf. *-oide* < Gr. *-oeídes*, similar to) : often used in descriptions of Dyscomycetes to indicate a positive (J+) or negative (J-) reaction of the asci to treatment with and iodine solution; in a positive (amyloid) reaction, a portion of the ascus apex takes a blue stain. In some recent works the English I+ or I- is used instead.

k

karyallagic (L. *caryallagicus* < Gr. *káryon*, nucleus + *allagé*, permutation, commutation + suf. *-íkos* > L. *-icus* > E. *-ic*, belonging to, relating to): refers to reproduction that occurs by means of exchange of nuclear material, i.e., true or sexual reproduction. Cf. **akaryallagic**.

karyogamy (Gr. *káryon*, nucleus + *gámos*, sexual union + *-y*, E. suf. of concrete nouns): fusion of two nuclei to form the zygote nucleus; it corresponds to the second phase of sexual reproduction, after plasmogamy but before meiosis.

kefir grains (Russian *kefir*, a beverage of fermented cow's milk; ME. < MF. *grain*, cereal grain < L. *granum*, grain): granular masses (microbiogloeae) composed of lactic acid bacteria and yeasts in symbiosis, embedded in kefiran (a polysaccharide of bacterial origin), used as a starter culture or inoculum for fermenting milk in order to get *kefir*, a slightly alcoholic, lactic acid and effervescent beverage that is consumed in Russia and other European countries. Cf. **tea fungus** and **tibi grains**. See **microbiogloea**.

keratolytic (Gr. *kératos*, horn, of a horny nature + *lytikós*, able to loosen < *lýtos*, dissolvable, broken + suf. *-íkos* > L. *-icus* > E. *-ic*, belonging to, relating to): a fungus that has the capacity to decompose keratin and assimilate it in order to grow, as do, e.g., the dermatophytes of the gen. *Epidermophyton*, *Microsporum* and *Trichophyton* (moniliaceous asexual fungi), which infect man and higher animals, causing various types of tinea on hair, skin and nails.

kerion (Gr. *kérion*, a honeycomb): an inflammatory form of ringworm of the scalp; *tinea kerion*. See **favus** and **tinea**.

kinetid (Gr. *kinetós*, motile): basal apparatus; flagellar (undulipodial) apparatus. Kinetids always consist of at least one kinetosome, but may have pairs or occasionally more than two. Details of the kinetid are essential for taxonomic and evolutionary studies of motile protoctists.

kinetosome (Gr. *kinetós*, motile + *sõma*, body): the base of the zoospore flagellum, which consists of a cylinder with nine triplets of microtubules; also called **blepharoplast** or **basal body.** Kinetosomes generally originate from the centrioles and are connected to the nuclear membrane by means of a **rhizoplast**. They occur, e.g., in *Rhizophlyctis* of the Chytridiales, or *Blastocladia*, of the Blastocladiales.

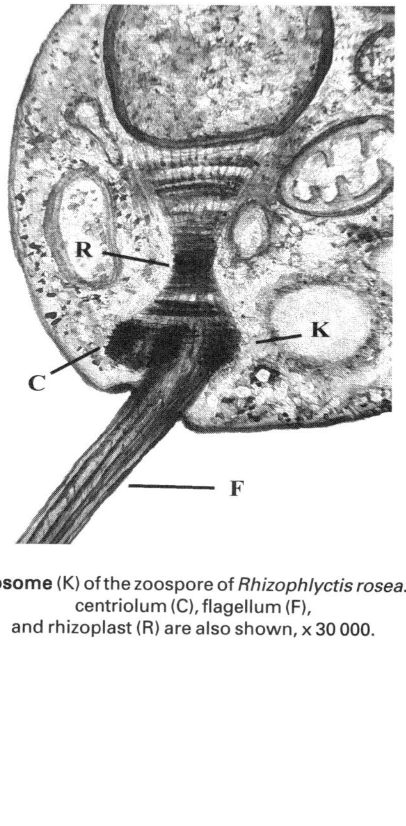

Kinetosome (K) of the zoospore of *Rhizophlyctis rosea*. The centriolum (C), flagellum (F), and rhizoplast (R) are also shown, x 30 000.

l

labiate (NL. *labiatum*, lipped < L. *labium*, lip + suf. *-atus* > E. *-ate*, provided with or likeness): provided with two lips; e.g., like the hysterothecia of the Loculoascomycetes of the family Hysteriaceae, order Dothideales (such as *Hysterium* and *Glonium*) and of certain Loculoascolichens (Arthoniales and Pyrenulales), which have a longitudinal line of dehiscence bordered by a lip on each side.

Laboulbeniomycetes (named in honor of *A. Laboulbène*, French entomologist of the 19th century + L. *-mycetes*, ending of class < Gr. *mýkes*, genit. *mýketos*, fungus): in some classifications, a class of perithecial fungi that live as ectoparasites of insects, arachnids and myriapods. The thallus or receptacle is simple, of few cells, and it adheres to the host by a foot from which the haustorium develops. Conidia are not known. In this dictionary this group is not given a class rank, but is treated as the order Laboulbeniales. The Spathulosporales, which are parasitic on marine Rhodophyceae, and formerly considered in the class Laboulbeniomycetes, are herein regarded, as with the Laboulbeniales, as separate orders within the phylum Ascomycota of the kingdom Fungi.

labyrinthiform (L. *labyrinthiformis* < *labyrinthos*, tortuous passage + *-formis* < *forma*, shape): long irregular, tortuous cavities, repeatedly branched and interconnected; e.g., like those of the hymenophore of the basidiocarp of *Daedalea confragosa* (Aphyllophorales), or the irregular glebal chambers of *Gautieria* (Gautieriales).

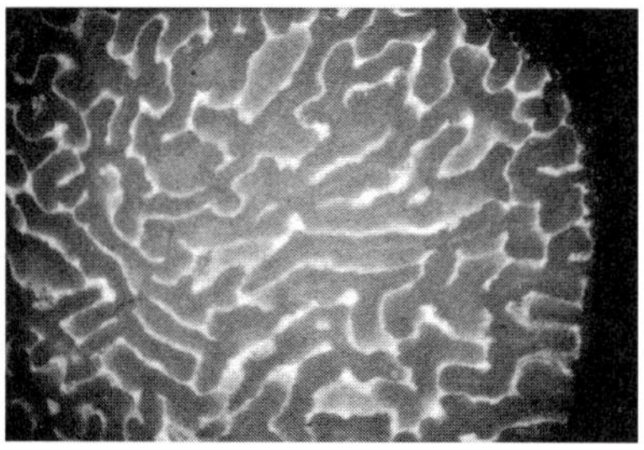

Labyrinthiform hymenophore of the ascocarp of *Tuber texense*, seen in transverse section, x 2 (*RTH*).

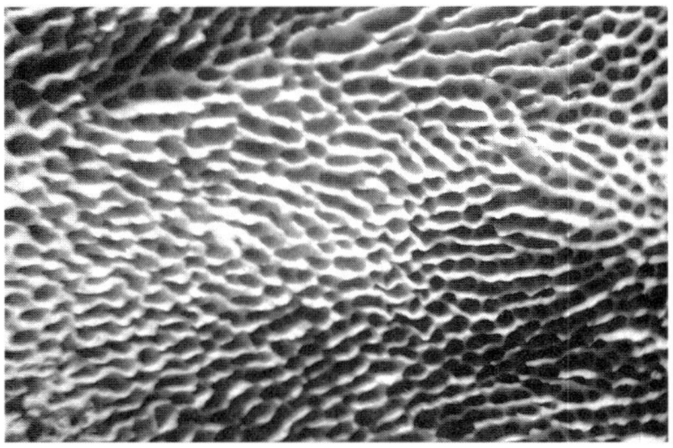

Labyrinthiform hymenophore of the basidiocarp of *Daedalea elegans*, x 2 (*MU*).

laccate (L. *laccatus*, varnished < Hind. *lakh*, reddish resin < Sanskrit *laksha*, the lac insect + suf. *-atus* > E. *-ate*, provided with or likeness): polished, burnished, as if varnished; e.g., like the basidiocarp upper surface of *Laccaria laccata* (Agaricales) and of *Ganoderma lucidum* (Aphyllophorales).

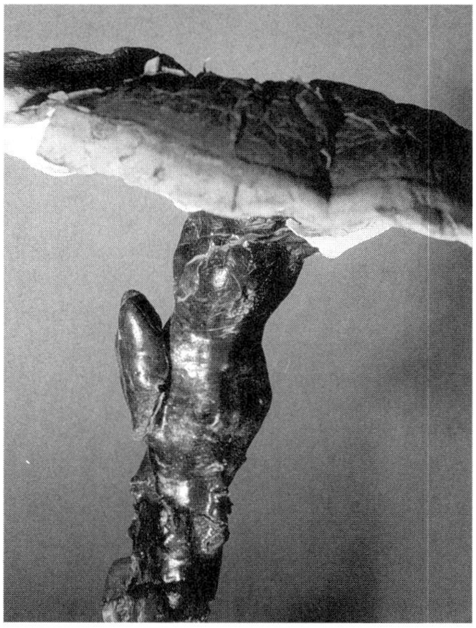

Laccate surface of the basidiocarp of *Ganoderma lucidum*, x 0.5 (*MU*).

lacerate, **lacerated** (L. *laceratus*, torn to pieces, mangled < *lacerus*, mangled, torn + suf. *-atus* > E. *-ate*, provided with or likeness): refers to the ring, pileus, scales, etc., that appear as if they had been torn up; e.g., like the pileus scales of *Lepiota clypeolaria* (Agaricales) and the pore walls of the hymenophore of *Irpex* (Aphyllophorales).

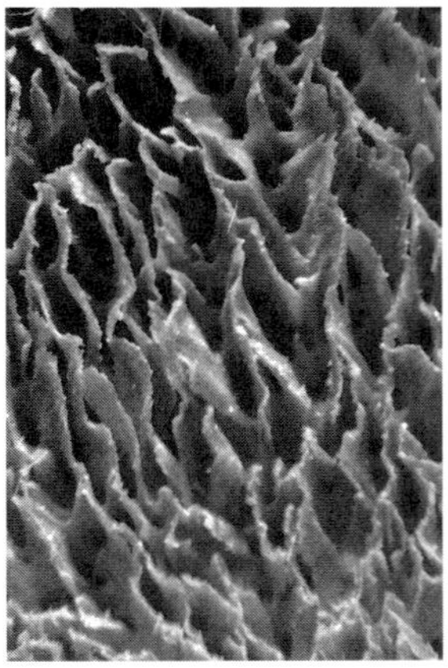

Lacerate pore walls of the hymenophore of *Irpex* sp., x 5 (*MU*).

lacerate peristome (L. *laceratus*, torn to pieces, mangled < *lacerus*, mangled, torn + suf. *-atus* > E. *-ate*, provided with or likeness; Gr. *perí*, around + *stóma*, mouth): a peristome with an irregularly torn edge around the mouth (**stoma**); e.g., as in *Geastrum* (Lycoperdales).

lacinia, pl. **laciniae** (L. *lacinia*, fringe): a delicate branch of a foliose lichen thallus, with the anatomy typical of a foliose lichen.

laciniate, **laciniated** (L. *laciniatus* < *lacinia*, fringe + suf. *-atus* > E. *-ate*, provided with or likeness): with the edge segmented or split into narrow lobes, generally deep, with a sharp tip. Applied to foliar or laminar structures, the border of some fruiting bodies (e.g., the apothecia of *Hyaloscypha hyalina*, of the Helotiales), and to the colony margin of some molds and yeasts when they are growing on solid media; also called **fimbriate**.

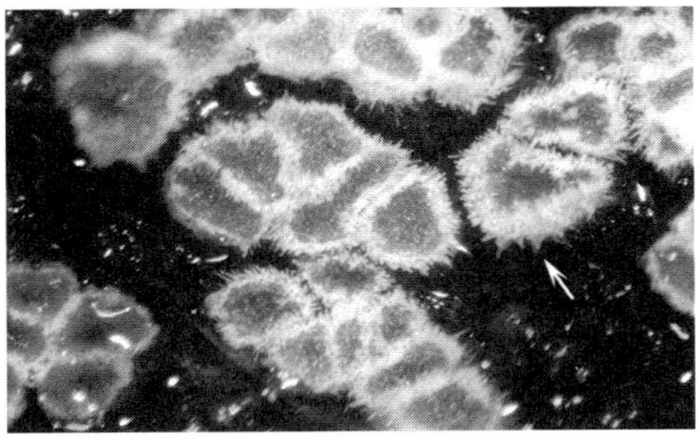

Laciniate border of the apothecia of *Hyaloscypha hyalina*, x 40 (*CB*).

lacrimiferous (L. *lacrima*, tear + *-ferous*, bearer < *ferre*, to bear, carry + *-osus* > OF. *-ous*, *-eus* > E. *-ous*, having, possessing the qualities of): having small droplets similar to tears; e.g., like the gills of the basidiocarps of *Psathyrella* (=*Lacrymaria*) *lacrimoabunda* (Agaricales).

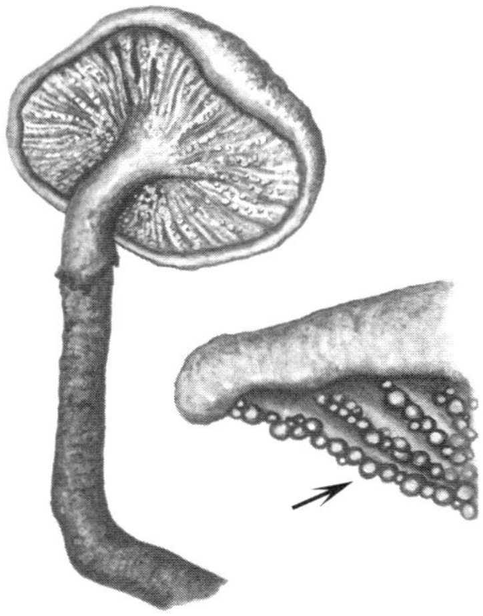

Lacrimiferous lamellae of the basidiocarp of *Psathyrella* (=*Lacrymaria*) *lacrimoabunda*, x 2.

lacrimoid (L. *lacrima*, tear drop + suf. *-oide* < Gr. *-oeídes*, similar to): shaped like a dropping tear. Also called **dacryoid**.

lactarioid (< gen. *Lactarius* < L. *lactarius*, lacteal < *lacteus*, of milk + L. suf. *-oide* < Gr. *-oeídes*, similar to): with characters similar to those of *Lactarius* (Agaricales), i.e., with fragile gills, sphaerocysts in the trama, lactiferous ducts and amyloid spores, among others.

lactescent (L. *lactescens*, genit. *lactescentis*, ptp. of *lactescere*, to produce milk < *lact-*, *lac*, milk + *-escens*, genit. *-escentis*, that which turns, beginning to, slightly > E. *-escent*): milky; a structure that secretes latex on being cut, such as the basidiocarps of *Lactarius* (Agaricales).

lactiferous hypha, pl. **hyphae** (L. *lacti-*, *lac*, milk + *-ferous*, bearer < *ferre*, to bear, carry + *-osus* > OF. *-ous*, *-eus* > E. *-ous*, having, possessing the qualities of; Gr. *hyphé*, tissue, spider web; hypha): *Basidiomycetes*. A hypha that contains latex or milky juice, as in *Lactarius* (Agaricales).

lacunar, **lacunose**, **lacunous** (L. *lacuna*, pit, hole, gap + L. suf. *-aris* > E. *-ar*, like, pertaining to; L. *lacunosus*, full of pits or holes < *lacuna* + *-osus*, full of, augmented, prone to > ME. *-ose*; or + *-osus* > OF. *-ous*, *-eus* > E. *-ous*, having, possessing the qualities of): having gaps or cavities, like the stipe of the apothecium of *Helvella lacunosa* and *H. crispa*

(Pezizales), or the basidiospores of *Strobilomyces floccopus* (Agaricales); in the latter the projecting striations in the wall are quite large and constitute interconnected borders, forming a network with large holes or pits between the borders. The basidiospores of *Calostoma* (Tulostomatales) also are lacunose, with pitted walls.

Lacunose stipe of the ascocarp of *Helvella crispa*, x 3 (*CB*).

lacunar development (L. *lacunosus*, having lacunae < *lacuna*, pit, hole, gap + suf. *-aris* > E. *-ar*, like, pertaining to; MF. *développer* < OF. *desveloper* < *dés-* < L. *de-*, from, down, away + MF. *enveloper*, to enclose + L. *mentum* > OF. *-ment* > ME. *-ment*, action, process): *Gasteromycetes*. A type of fruiting body development in which the gleba is differentiated into a series of chambers that are more or less regular in shape and lined by a well developed hymenium; characteristic of the Gautieriales (such as *Gautieria*), the Melanogastrales (such as *Melanogaster*), the Nidulariales (such as *Nidularia*), and the Sclerodermatales (such as *Pisolithus*).Cf. **aulaeate** and **forate development**.

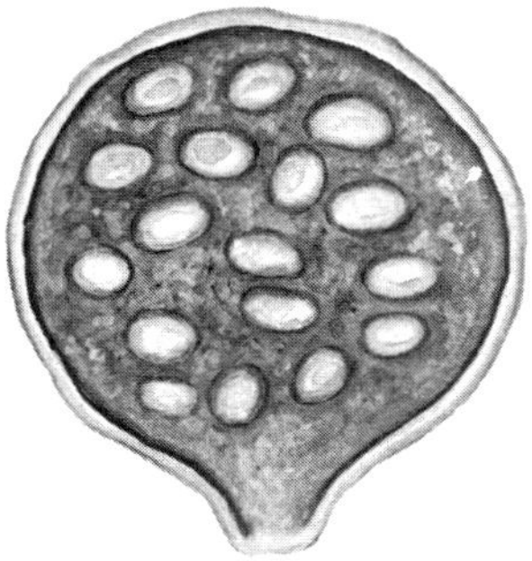

Lacunar development of the hymenium of the fruiting body of Melanogastrales and Nidulariales. Diagram of a longitudinal section, x 1.

laevigate, **levigate** (L. *levigatus* < *levigare*, to polish, make smooth + suf. *-atus* > E. *-ate*, provided with or likeness): having a smooth, polished surface; e.g., like the upper part of the pileus of *Hydnum laevigatum* (Aphyllophorales).

lageniform (L. *lageniformis* < *lagena*, bottle, jar + *-formis* < *forma*, shape): applied to an organ or structure enlarged at the base and with the upper part narrower, like the young sporangia of *Arcyria* (Trichiales), the zoosporangium of *Lagenidium* (Lagenidiales), and the ascocarps of *Microascus* (Microascales) and *Lagenulopsis* (Coryneliales). Also called **sicyoid**, **utriculose** and **ventricose-rostrate**.

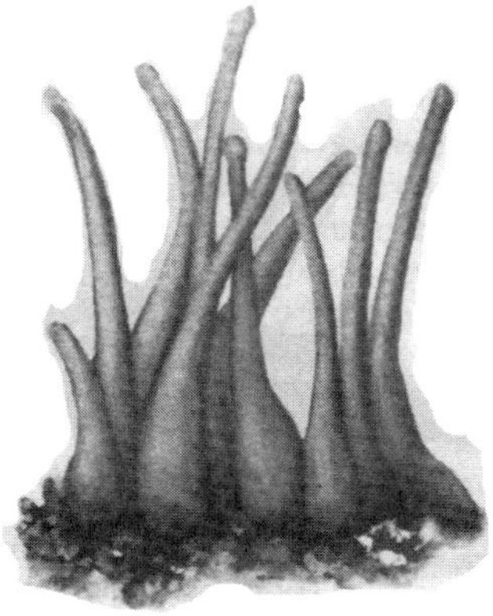

Lageniform ascocarps of *Lagenulopsis bispora*, borne on a stroma formed on a leaf of the host plant (*Podocarpus milanjianus*), x 28.

Lageniform young sporangia of *Arcyria* sp., x 6 (*CB & MU*).

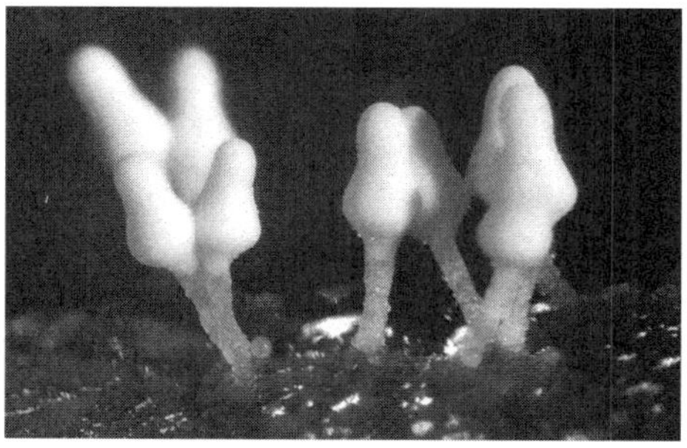

Lageniform young sporangia of *Arcyria* sp., x 9 (*CB & MU*).

lamella, pl. **lamellae** (L. *lamella*, plate, sheet): *Agaricales*. Each one of the narrow, vertical, radial plate-like structures (gills) found on the lower side of the pileus of mushrooms, and on which are borne the hymenium of basidia and basidiospores. Gill characters are important in delimiting taxonomic categories, as in *Macrolepiota procera*, *Lactarius vellereus* and *Mycena pura*.

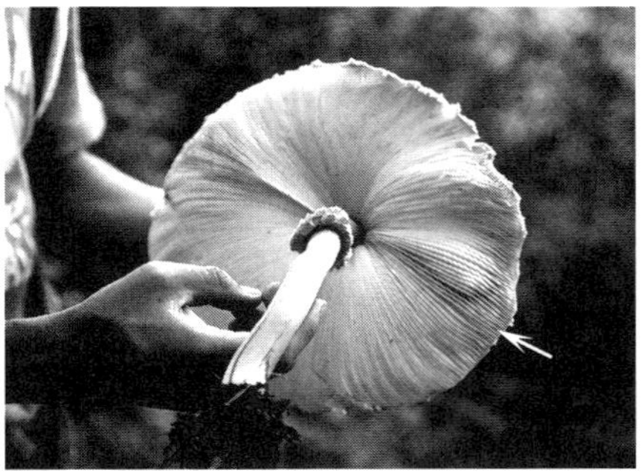

Lamellae of the basidiocarp of *Macrolepiota procera*, x 0.1 (*EPS*).

Lamellae of the basidiocarp of *Mycena pura*, x 4 (*MU*).

lamellate (L. *lamella*, little sheet, plate + suf. *-atus* > E. *-ate*, provided with or likeness): composed of or having small plates or thin sheets, i.e., gills; e.g., like the pileus of the Agaricales.

lamellula, pl. **lamellulae** (L. *lamellula*, little plate, sheet): a small gill that does not extend all the way from the pileus margin to the stipe; e.g., *Amanita caesarea*, *Oudemansiella canarii*, *Pleurotus ostreatus* and *Xeromphalina tenuipes* (among other Agaricales) have lamellulae.

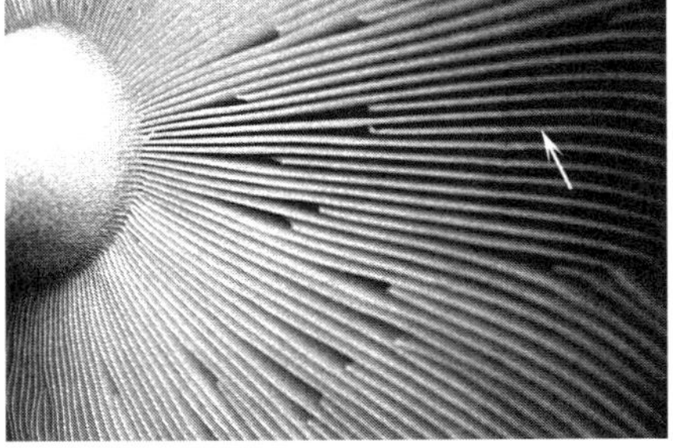

Lamellulae of the basidiocarp of *Amanita* sp., x 1 (RMA).

Lamellulae of the basidiocarp of *Pleurotus ostreatus* var. *florida*, x 1 (*MU*).

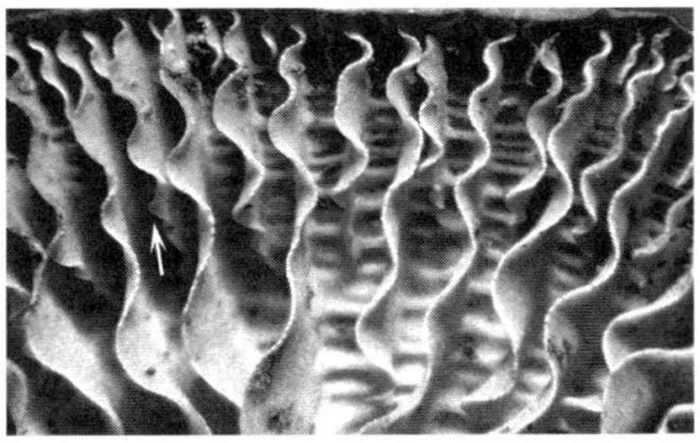

Lamellulae of the basidiocarp of *Xeromphalina tenuipes*, x 4 (*MU*).

lamina, pl. **laminae** (L. *lamina*, a thin plate or scale, blade): the main part of the thallus in foliose lichens; the epithecium along with the hymenium and the subhymenium in an apothecium. *Laminal* or *laminar* (*lamina* + E. suf. *-al*, relating to or belonging to; or + L. suf. *-aris* > E. *-ar*, like, pertaining to): on laminae, arranged in, consisting of, or resembling laminae.

lamprocystidium, pl. **lamprocystidia** (Gr. *lamprós*, brilliant + *kystídion* < *kýstis*, bladder; here, cell + dim. suf. *-ídion* > L. *-idium*): a type of cystidium, close to the leptocystidium, with a thick, refringent, frequently spiny wall, with crystals in the upper portion (in some species), and with a hamate, corniform or proliferating apex; it is characteristic of the gen. *Inocybe*, but also found in other gen. of Agaricales, such as *Galerina*, *Hohenbuehelia*, *Psathyrella* (=*Drosophila*) and *Pluteus*. Cf. **chrysocystidium** and **macrocystidium**.

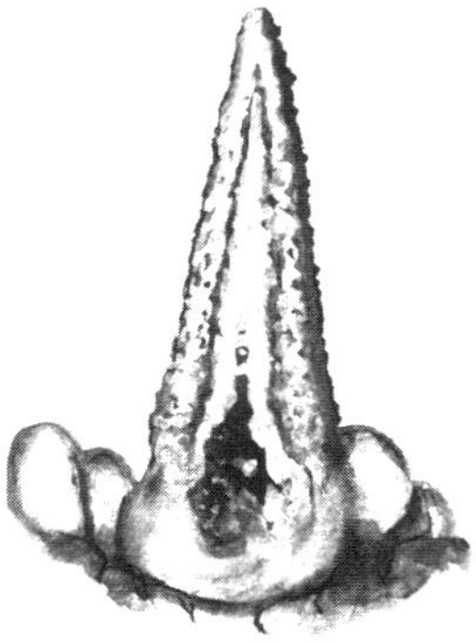

Lamprocystidium from the hymenium of *Hohenbuehelia niger*, x 1 520.

lanceolate (L. *lanceolatus* < *lanceola*, dim. of *lancea*, lance + suf. *-atus* > E. *-ate*, provided with or likeness): applied to laminar structures that are shaped like a lance, narrowly elliptical and pointed at both ends, like certain spores, phialides, cystidia, etc.; e.g., the cystidia of *Mycena corticaticeps* and *Marasmius* (Agaricales).

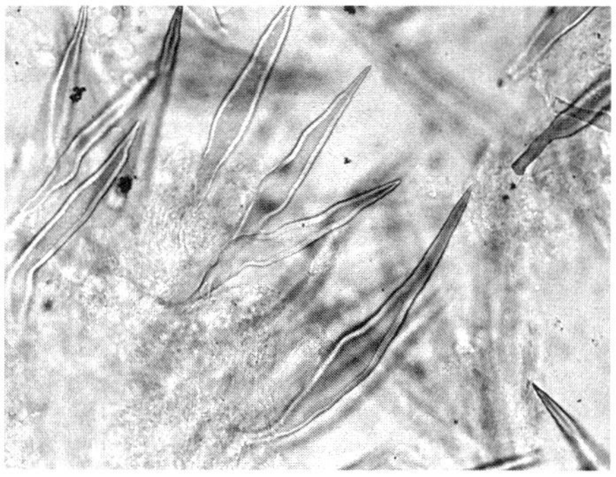

Lanceolate pleurocystidia of the hymenium of *Marasmius* sp., x 250 (*EPS*).

lanose (L. *lanosus*, woolly, with soft hairs < *lana*, wool + *-osus*, full of, augmented, prone to > ME. *-ose*): covered with hairs similar to the fibers of wool; woolly; e.g., like the colonies of *Linderina pennispora* (Kickxellales), *Fusarium* (tuberculariaceous asexual fungi) and many other fungi.

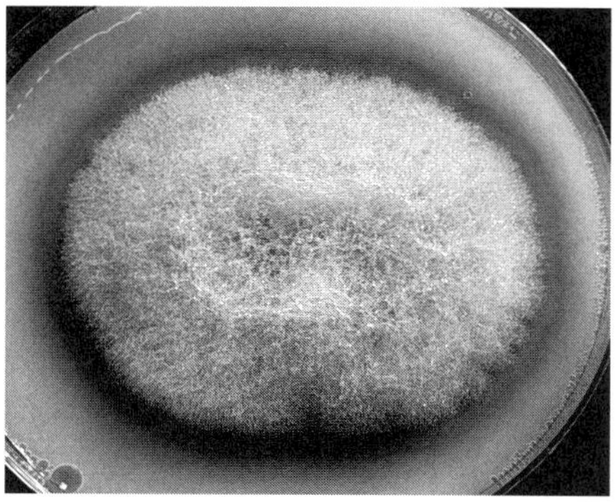

Lanose colony of *Linderina pennispora* on agar, x 0.5 (*MU*).

lantern (ME. *lanterne* < L. *lanterna* < Gr. *lamptér*, lamp, light): *Gasteromycetes*. An ovoid, deep-orange **glebifer** suspended from the columnar receptacle and held in place by **trabeculae**, often numerous, and which at maturity is coated with a blackish-green basidial layer, as in *Laternea* (Phallales).

lanuginose, **lanuginous** (L. *lanuginosus*, downy, woolly < *lanugo*, woolly substance, down + *-osus*, full of, augmented, prone to > ME. *-ose*; or + *-osus* > OF. *-ous*, *-eus* > E. *-ous*, having, possessing the qualities of): covered with soft downy hair, like the basidiocarp of *Inocybe lanuginosa* (Agaricales).

Lantern of the fruiting body of *Laternea pusilla*, x 0.4.

larviform (L. *larva*, larva, spectre, shadow + *-formis* < *forma*, shape): having the appearance of an insect larva; e.g., such as the conidia of *Polyschema larviformis* (dematiaceous asexual fungi).

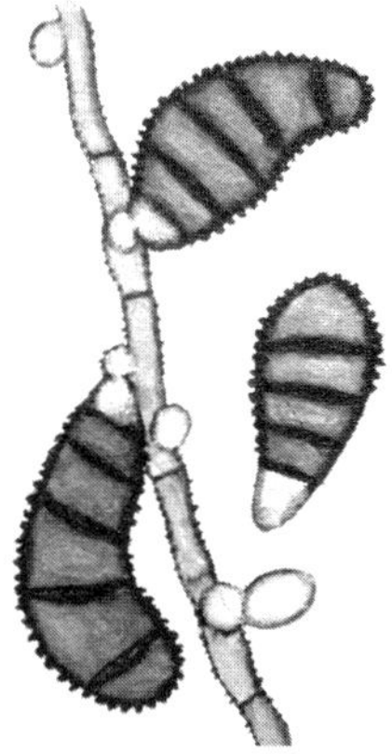

Larviform conidia of *Polyschema larviformis*, x 480.

latent spore or **resting spore** (L. *latens*, genit. *latentis* < *latere*, to lurk, lie hid, hidden + *-entem* > E. *-ent*, being; E. *resting*, dormant < ME. *resten* < OE. *restan*, to rest; NL. *spora* < Gr. *sporá*, spore): *Aquatic fungi*. The latent spore or sporangium, which results from the transformation of the zygote, and which remains dormant for a time before germination; as happens, e.g., in *Rozella* (Spizellomycetales), *Allomyces* (Blastocladiales) and *Chytriomyces* (Chytridiales). The supposed zygospore of the Trichomycetes is also called a resting spore. The resting spore is also sometimes called a resistant spore.

lateral (L. *lateralis*, of the side < *later*, genit. *latus*, side + suf. *-alis* > E. *-al*, relating to or belonging to): belonging to the side or situated on it, flank of something. In the fungi it is applied, e.g., to the disposition of the stipe on the fructification, which is lateral in *Pleurotus* (Agaricales) and other basidiomycetes.

Lateral stipe of the basidiocarp of *Pleurotus ostreatus*, x 0.6 (*MU*).

Lateral stipe of the basidiocarp of *Pleurotus ostreatus*, x 0.5 (*MU*).

latex, pl. **latices** or **latexes** (NL. *latic-*, milk < L. *latex*, fluid, juice): a fluid, generally milky, commonly white in color, at times yellow or other colors, which flows or exudes from various parts of certain fungi when they are cut or damaged, as happens in the gen. *Lactarius* (Agaricales), whose species have latex of distinctive colors and properties.

Latex exuding from the cut lamellae of *Lactarius volemus*, x 0.5 (*VBM*).

leaf curl (ME. *leef* < OE. *leaf*, leaf < L. *liber*, bast, book; ME. *curlen* < *crul*, curly, prob. < MD., to form into coils or ringlets): a disease symptom of peach and almond leaves infected by *Taphrina deformans*, and of cherry (*T. minor*), of the Taphrinales.

leathery (OE. < ME. *lether*, leather + E. suf. *-y*, partaking of the nature of): syn. of **coriaceous**. With the consistency of leather; tough and flexible.

lecanorine (< gen. *Lecanora* < Gr. *lekáne*, dim. of *lekánion*, plate, kettle + L. suf. *-ora* < Gr. *hóra*, *horaîos*, beautiful, a crustaceous lichen of the order Lecanorales + L. suf. *-inus* > E. *-ine*, of or pertaining to): *Lichens*. Rounded apothecia that are bordered by a protruding thalline margin, like those of *Lecanora frustulosa*. Cf. **lecideine**. See **discoid**.

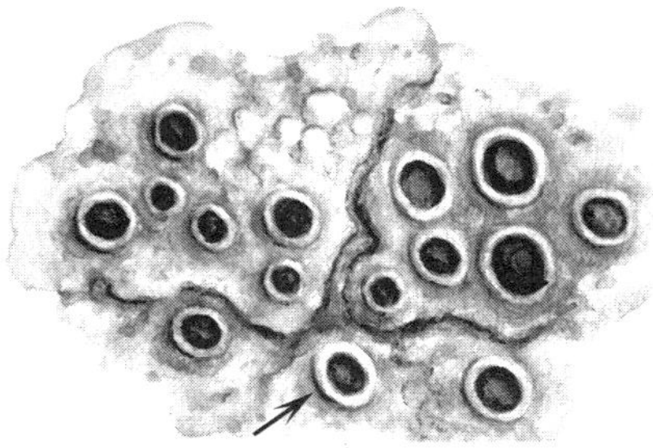

Lecanorine apothecia of the thallus of *Lecanora subfusca*, x 12.

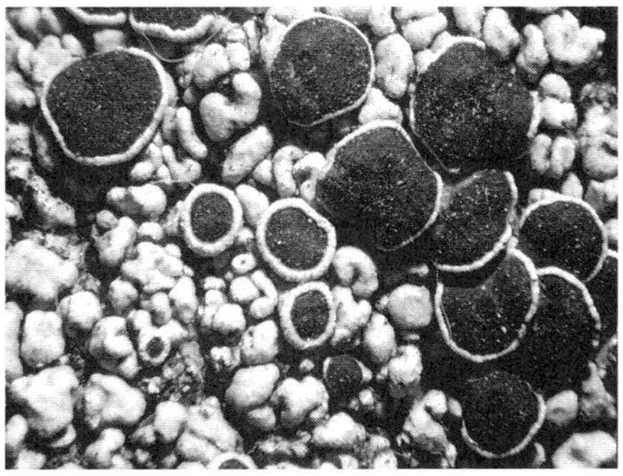

Lecanorine apothecia of the thallus of *Lecanora frustulosa* x 12 (*MU*).

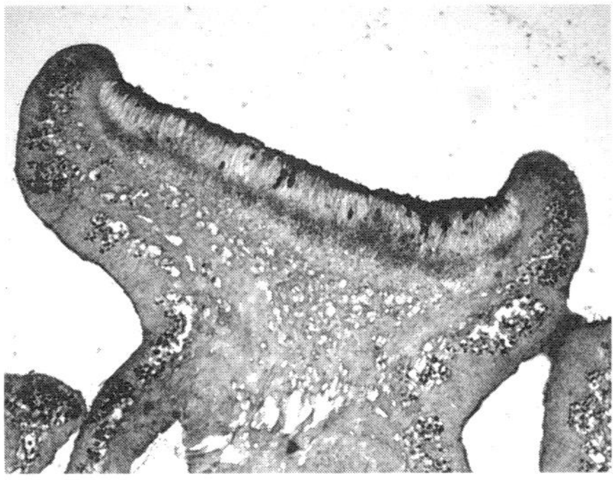

Lecanorine apothecium of the thallus of a discolichen, seen in longitudinal section, x 57 (*MU*).

lecideine (< gen. *Lecidea* < Gr. *lékos*, plate, shield + suf. *ídes*, alikeness + L. suf. *-inus* > E. *-ine*, of or pertaining to): *Lichens*. Orbicular apothecia that lack a thalline margin; commonly the margin is composed of the excipulum itself, such as those of *Lecidea albocaerulescens*, a crustaceous lichen of the order Lecanorales. Cf. **lecanorine**.

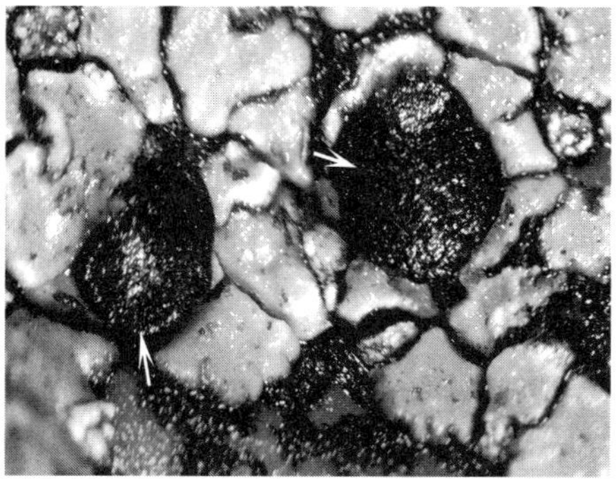

Lecideine apothecia of the thallus of *Lecidea atrobrunea*, x 30 (*MU*).

lecythiform (Gr. *lékythos*, oil jar + L. *-formis* < *forma*, shape): with the shape of a bottle having a narrow base, ellipsoid body, a narrow neck, and a flanged mouth with a ball-like stopper; applied to the cystidia (lamprocystidia) of certain basidiomycetes (such as the gen. *Agrocybe* and *Conocybe*, Agaricales) that are markedly ventricose in the basal part and attenuated in the upper part, where they form a neck that terminates in a rounded or enlarged button in the shape of a small pear. Also, the sporangium of *Saksenaea vasiformis* (Mucorales) is lecythiform.

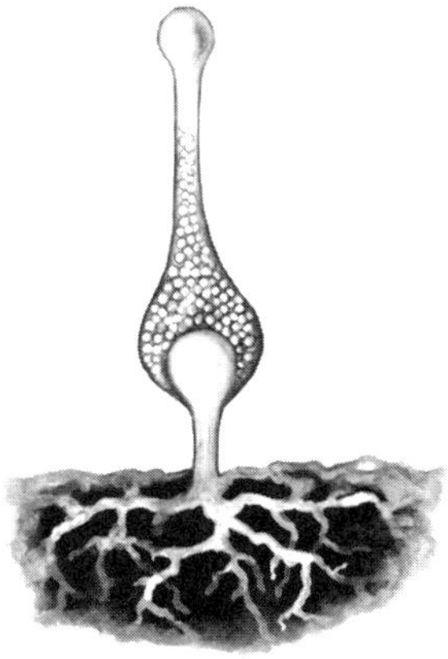

Lecythiform sporangium of *Saksenaea vasiformis*, x 470.

leiodisc (Gr. *leîos*, smooth + *dískos*, disc): *Lichens*. An apothecium with the appearance of a smooth, shiny disk, as is observed in *Umbilicaria rigida* (Lecanorales).

leiospore (Gr. *leîos*, smooth + *sporá*, spore): a spore that is characterized by having a smooth wall; fungi that have this type of spore are known as leiosporic.

lentic (L. *lentus*, sluggish + suf. *-íkos* > L. *-icus* > E. *-ic*, belonging to, relating to): of, relating to, or living in still waters (as lakes, ponds or swamps). Cf. **lotic.**

lenticular (L. *lenticularis*, of or pertaining to a lentil < *lenticula*, a lentil, dim. of *lens*, genit. *lentis*, disc + suf. *-aris* > E. *-ar*, like, pertaining to): having the shape of a disk or watch glass, like a double convex lens, wider than ellipsoid but not as long. Also called **lentiform**. For example, the basidiospores of *Psilocybe atrorufa* and *Coprinus comatus* (Agaricales) are lenticular or lentiform.

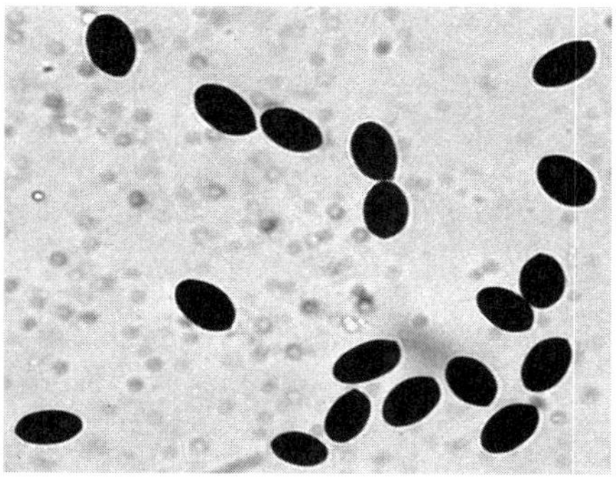

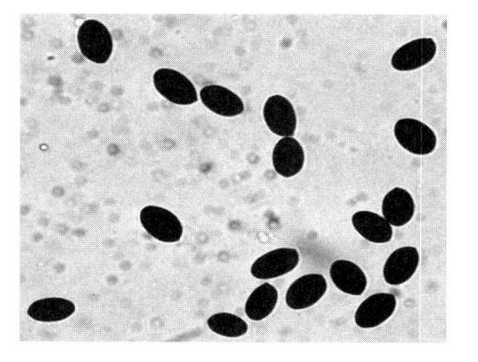

Lenticular basidiospores of *Coprinus comatus*, x 650 (*MU*).

lentiform (L. *lens*, genit. *lentis*, disk + *-formis* < *forma*, shape): see **lenticular**.

lentiginose, **lentiginous** (L. *lentiginosus*, freckled < *lentigo*, mole, freckle + *-osus*, full of, augmented, prone to > ME. *-ose*; or + *-osus* > OF. *-ous*, *-eus* > E. *-ous*, having, possessing the qualities of): having small spots, similar to small freckles; e.g., like the thallus of the lichen *Melaspilea lentiginosula* (of the Melaspileaceae, an ascomycetous family of uncertain afinities, previously included in the Hysteriales), which has minute apothecia on its surface.

lepidote (NL. *lepidotus* < Gr. *lepidotós*, covered with scales): scaly or covered with scale-like hairs, like the pileus of *Lepiota* (Agaricales). Same as **squamulose**. Cf. **alepidote**.

lepiotoid (< gen. *Lepiota* < Gr. *lepís*, scale + L. suf. *-otus*, *-ota*, that indicates possession or alikeness, + L. suf. *-oide* < Gr. *-oeídes*, similar to): *Agaricales*. A type of fruiting body (13 principal types are considered) with free or finely adhered gills, and with an annulus but without a volva, as represented by *Lepiota* and some species of *Agaricus* and *Coprinus* (Agaricales).

leprose (L. *leprosus*, scurfy, scaly < *lepras* < Gr. *lépra*, leprosy < *leprós*, rough + L. *-osus*, full of, augmented, prone to > ME. *-ose*): *Lichens*. A type of thallus whose surface has a large number of soredia, or when all of the thallus disintegrates and gives place to soredia, as occurs in *Lepraria* (Lichenes Imperfecti, Deuterolichenes or imperfect lichens).

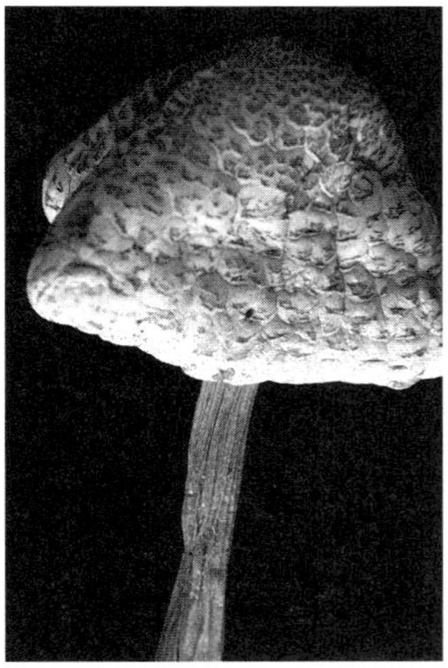

Lepidote pileus of the basidiocarp of *Lepiota* sp., x 0.6 (*MU*).

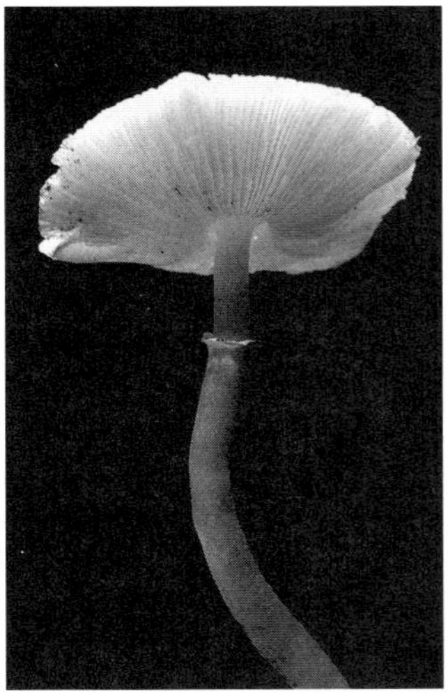

Lepiotoid fruiting body of *Lepiota* sp., x 0.7 (*MU*).

Leprose thallus of *Lepraria membranacea*, x 10 (*MU*).

leptocystidium, pl. **leptocystidia** (Gr. *leptós*, minute, fine, delicate + *kystídion* < *kýstis*, bladder; here, vesicle, cell + dim. suf. *-ídion* > L. *-idium*): a cellular element, similar to a **cystidiole**, that originates from the lower part of the subhymenium or from the trama itself, at a level lower than the basidia. When it has a thickened, pigmented wall it is called a **seta**, **setula**, or **setuloid cystidium**. Leptocystidia are found in various species of *Amanita* (such as *A. herrerae*), *Psathyrella* (=*Drosophila*) and *Leptonia* (Agaricales). Cf. **chrysocystidium**, **lamprocystidium** and **macrocystidium**.

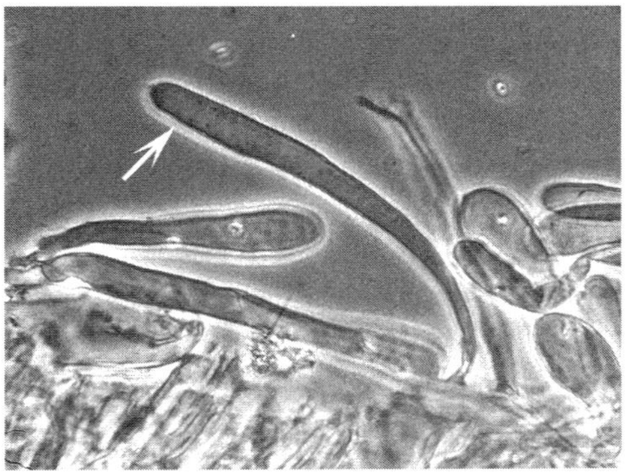

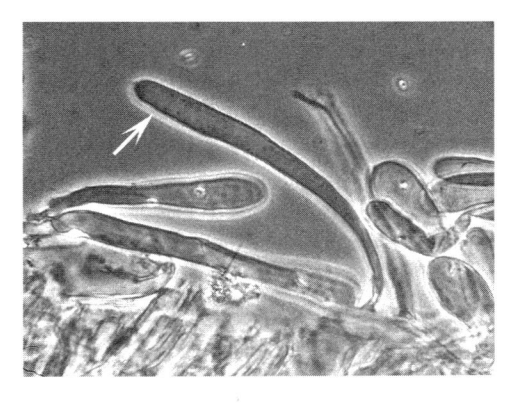

Leptocystidia of the hymenium of *Amanita herrerae*, x 360 (*RMA*).

leptoderm, **leptodermatous**, **leptodermous** (Gr. *leptós*, delicate, fine + *dérma*, skin, cortex; or + *dérma*, genit. *dérmatos* + L. *-osus* > OF. *-ous*, *-eus* > E. *-ous*, having, possessing the qualities of): a hypha with the external wall thinner than the lumen. Cf. **mesoderm** and **pachyderm**.

lesion (ME. < MF. < L. *laesion*, *laesio*, genit. *laesionis* < *laesus*, ptp. of *laedere*, to injure, to wound + *-io*, *-ionis* > E. suf. *-ion*, result of an action, state of): injury, wound, an abnormal change in structure of an organ or part due to injury or disease, especially one that is a well-marked or circumscribed but limited diseased-area.

leucosporous (Gr. *leukós*, white + *sporá*, spore + L. *-osus* > OF. *-ous*, *-eus* > E. *-ous*, having, possessing the qualities of): having white spores, like the basidiocarps of *Amanita* (Agaricales).

lichen (L. *lichen* < Gr. *leichén*, lichen): **1**. A dual organism, constituted by the symbiotic association of fungi and algae; a type of parasitism on the part of the fungus on the alga. The majority of lichens belong to the ascomycetes (Ascolichens); a minority are basidiomycetes (Basidiolichens) and a few are deuteromycetes (Deuterolichens or Lichenes Imperfecti). The algae that participate in the lichen association belong, according to the species of lichen, to the Chlorophyceae and the Cyanophyceae. **2**. A cutaneous disease caused by dermatophytes (formerly called ringworm), which due to its scaly appearance resembles the thallus of certain lichens.

lichenicolous (L. *lichen* < Gr. *leichén*, lichen + L. suf. *-cola*, inhabitant < *colere*, to inhabit + *-osus* > OF. *-ous*, *-eus* > E. *-ous*, possessing the qualities of): that which lives on lichens, as a parasite, parasymbiont, or saprobe. More than 1,000 species of lichenicolous fungi are known, belonging to very diverse groups; there even exist lichenicolous lichens.

licheniferous (L. *lichen* < Gr. *leichén*, lichen + *-ferous*, bearer < *ferre*, to bear, carry + L. *-osus* > OF. *-ous*, *-eus* > E. *-ous*, full of, possessing the qualities of): bearing lichens, like certain beetles of New Guinea, including *Gymnopholus lichenifera*, on whose back grow foliose and crustose lichen thalli, which also have been found on the shell of the giant tortoise (*Geochelone elephantopus*) in the Galapagos Islands.

lichenoid (L. *lichen*, < Gr. *leichén*, lichen + L. suf. *-oide* < Gr. *-oeídes*, similar to): a type of symbiotic association present in some primitive marine lichens, in which the algal components are capable of living free, i.e., independent of the fungal component. The algae involved in the lichenoid association are usually microscopic and the fungi belong to such gen. as *Chadefaudia* and *Stigmidium* (=*Pharcidia*), of the Dothideales. *S. laminariicola* is associated with epiphytic phaeophyceous algae, of the species *Ectocarpus fasciculatus*, on pedicels of *Laminaria digitata*. *Chadefaudia corallinarum* is associated with epiphytic crustose rhodophyceous algae of the gen. *Dermatolithon* and *Epilithon*. The fact that this fungus can live with different phycobionts indicates that the symbiosis is not very narrow. This association is only found in submerged conditions and the fungus grows on macroalgae (Chlorophyceae, Phaeophyceae and Rhodophyceae) and leaves of marine grasses, whose tissues are never penetrated by hyphae of the fungus, although the ascocarps are often found partially covered by the calcareous thallus of the phycobiont.

lichenometry (L. *lichen* < Gr. *leichén*, lichen + E. *metry* < ME. *-metrie* < L. *metria* < Gr. *métrein*, to measure) : the use of lichens as a measure of minimum elapsed time of exposure of a substrate, since lichens first developed on the substrate in question; important in dating geomorphological or archeological events of relatively recent occurrences, such as deposition of glacial moraines or erection of Easter Island stone heads. This method relies on accurate data of growth rates of long-lived species of lichens under specific conditions.

life history (ME. *lif* < OE. *lif*, life, the quality that distinguishes a vital and functional being from a dead body; L. *historia* < Gr. *istoría*, inquiry, history, information < *hístor*, *ístor*, knowing, learned + *-y*, E. suf. of concrete nouns): in fungi, the stages or series of stages (often characterized by different spore states) between one spore form and the development of the same spore form again. Also called *life cycle* (Gr. *kýklos*, circle).

ligneous (L. *ligneus*, wooden < *lignum*, wood + L. *-osus* > OF. *-ous*, *-eus* > E. *-ous*, having, possessing the qualities of): having the consistency or the nature of wood, like the fruiting body of *Fomes*, *Ganoderma* and other Aphyllophorales.

lignicole, **lignicolous** (L. *lignicola* < *lignum*, wood + *-cola*, inhabitant < *colere*, to inhabit; or + *-osus* > OF. *-ous*, *-eus* > E. *-ous*, possessing the qualities of): living on wood, but not necessarily deriving nourishment from it, as would a **xylophagous** organism. Many Myxomycetes, basidiomycetes and lichens are lignicolous, such as *Trichia* (Trichiales), *Poria* (Aphyllophorales), *Evernia* and certain species of *Lecanora* (Lecanorales).

Lignicolous sporangia of *Trichia decipiens*, x 15 (*MU*).

ligulate, **liguliform** (L. *ligula* < *lingua*, tongue + suf. *-atus* > E. *-ate*, provided with or likeness; *ligula* + *-formis* < *forma*, shape): tongue-shaped; flat and wide like a belt; e.g., like the small, long and narrow thalline branches called **lorulae**, sing. **lorula**, of some lichens (e.g., certain species of *Evernia*, *Letharia* and *Ramalina*, of the Lecanorales). Also called **linguiform** and **loriform**.

limb (L. *limbus*, band, fringe, border): a border or margin that is distinguished in a structure.

limbate (LL. *limbatus* < L. *limbus*, hem, belt, fringe + suf. *-atus* > E. *-ate*, provided with or likeness): with the border fringed or edged with another color; e.g., like the pileus of *Psilocybe coprophila* and the volva of *Amanita plumbea* (Agaricales).

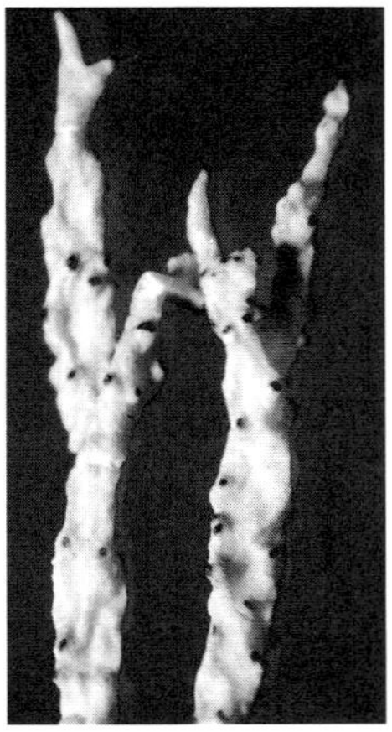

Ligulate branches of the thallus of *Letharia vulpina* var. *californica*, x 16 (*MU*).

Limbate margin of the pileus of *Psilocybe pelliculosa*, x 0.9.

limoniform (LL. *limoniformis* < ML. *limon*, lemon + *formis* < *forma*, shape): lemon-shaped; i.e., broadly ellipsoidal with apiculate ends, like the fruit of lemon (*Citrus*). Also called **citriform**. For example, the sporangia of *Phytophthora infestans* (Peronosporales), the ascospores of *Melanospora* (Melanosporales) and the basidiospores of *Phaeocollybia* sp. (Agaricales) are limoniform or citriform.

Limoniform sporangia of *Phytophthora infestans*, x 114.

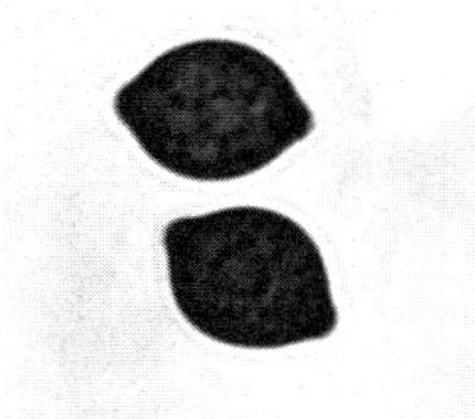

Limoniform ascospores of *Melanospora* sp., x 800 (*RTH*).

lineate (L. *lineatus*, lined < *lineare*, to mark with lines < *linea*, line + suf. *-atus* > E. *-ate*, provided with or likeness): having lines, usually lengthwise and parallel, distinguishable by their color or texture; striped; e.g., as in the conidia of *Stachybotrys cylindrospora* (dematiaceous asexual fungi).

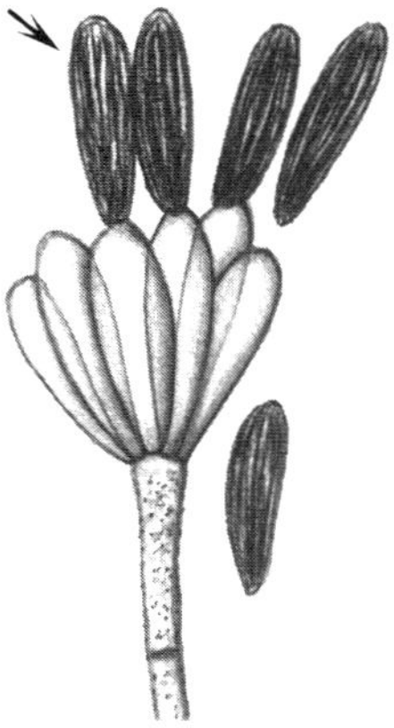

Lineate wall of the conidia of *Stachybotrys cylindrospora*, x 800.

linguiform (L. *linguiformis* < *lingua*, tongue + *-formis* < *forma*, shape): having the shape of a small tongue, such as the apical projection of the perithecia of *Rickia passalina* (Laboulbeniales). Also called **ligulate** and **loriform**.

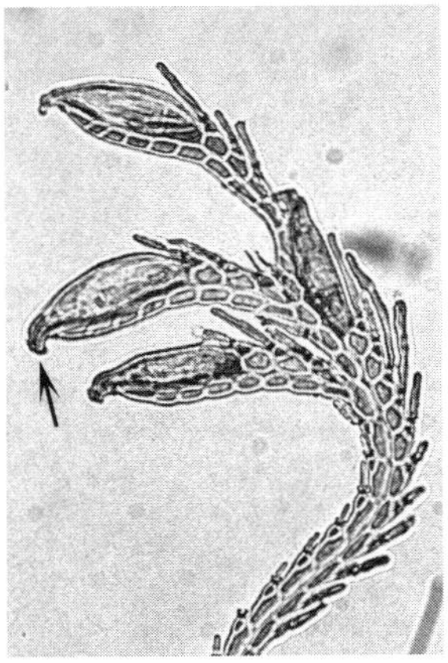

Linguiform process at the apex of the perithecia of *Rickia passalina*, x 700 (*MU*).

lipsanenchyma, pl. **lipsanenchymata** (Gr. *leipsánon*, residue, relic + *énchyma*, tissue): the primordial tissue of a basidiocarp that is found between the stipe and the pileus covering the hymenium, but which does not correspond to the universal veil.

lirella, pl. **lirellae** (NL. *lirella*, dim. of L. *lira*, furrow): the lirellae, also called hysterothecia or lirelline apothecia, are linear ascocarps, straight or curved and flexuous, simple or branched and sunken or somewhat erumpent, with a longitudinal slit and a well-defined proper exciple, unitunicate asci and simple paraphyses or bitunicate asci and pseudoparaphyses. The lirellae are characteristic of crustaceous lichens of the orders Ostropales (such as *Graphis scripta*) and Arthoniales (such as *Chiodecton*).

Lirellae of the thallus of *Graphis scripta*, x 14 (*MU*).

lithotroph (Gr. *líthos*, stone + *trépho*, to feed): capable of utilizing rocks as nourishment, as some lichens.

lituate (L. *lituus*, augural baton, battle trumpet, bugle, hunting horn + suf. *-atus* > E. *-ate*, provided with or likeness): forked, with the points arching slightly outward, like the basidium of *Dacrymyces* (Tremellales). Also called **furcate**.

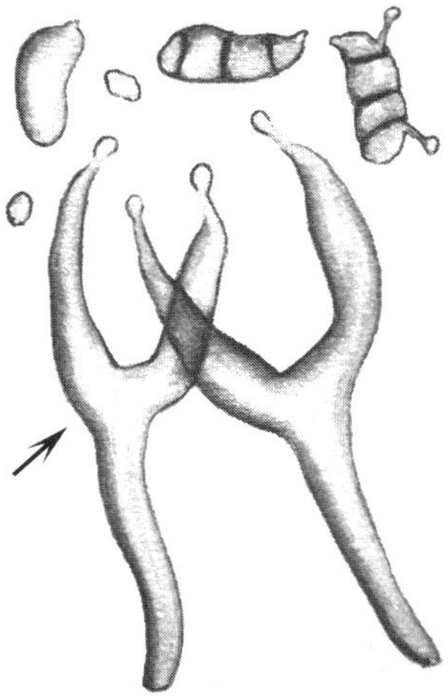

Lituate basidia from the hymenium of *Dacrymyces ellisii*, x 490.

liver fungus (ME. < OE. *lifer*, liver): the fruiting body of *Fistulina hepatica* (Aphyllophorales). Also called **beef-steak fungus**.

lobate (NL. *lobatus* < LL. *lobus* < Gr. *lobós*, lobe + L. suf. *-atus* > E. *-ate*, provided with or likeness): divided into lobes, resembling a lobe, i.e., in portions not very deep and more or less rounded; applied to laminar organs as well as larger structures. When the lobes are smaller (lobule) they are said to be **lobulate**. For example, the basidiospores of *Inocybe grammata* (Agaricales) are lobate and those of *I. oblectabilis* are lobulate. Also applied to the border of macrocolonies of yeasts in culture, like that of certain isolates of *Saccharomyces cerevisiae* (Saccharomycetales).

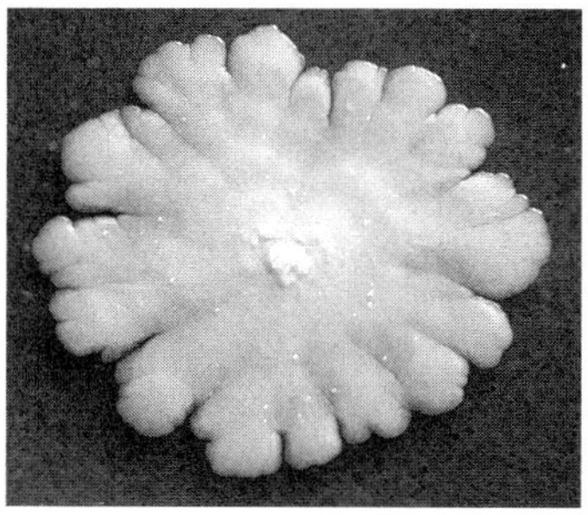

Lobate border of a giant colony of *Saccharomyces cerevisiae* on agar, x 2 (*MU*).

lobomycosis, pl. **lobomycoses** (< gen. *Loboa*, after the Brazilian physician Jorge Lobo, + NL. *mycosis* < Gr. *mýkes*, fungus + *-osis*, state or condition): a chronic subcutaneous, localized infection of man characterized by nodular and verrucose, tumor-like lesions in arms, legs, face and ears, which contain the yeast-like cells of the causal agent, *Loboa loboi*, of uncertain taxonomic position. Besides man, lobomycosis has been reported only from dolphins.

lobopod, **lobopodium**, pl. **lobopodia** (Gr. *lobós*, lobe + NL. *podium* < Gr. *pódion* < *poús*, *podós*, foot + dim. suf. *-ídion* > L. *-idium*): a pseudopodium or thick and rounded protoplasmic elongation of a protoplast or of a plasmodium; e.g., the myxamoebae of *Acrasis rosea* (Acrasiomycota) move and feed (phagocyte) by means of lobopodia. See **filopod**.

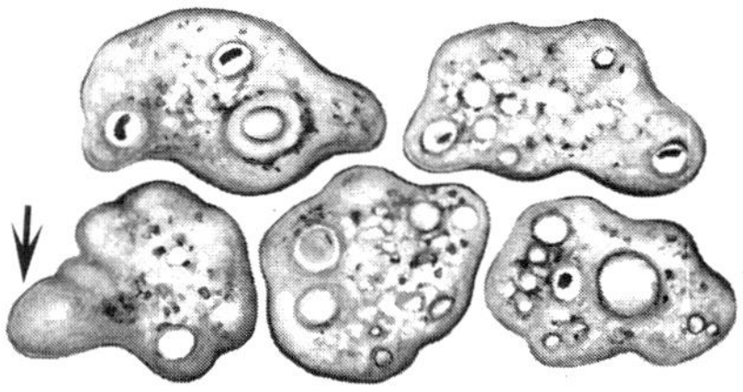

Lobopods of the myxamoebae of *Acrasis rosea*, x 1 000.

lobulate (NL. *lobulatus*, divided into small lobes < *lobulus*, dim. of LL. *lobus*, lobe + suf. *-atus* > E. *-ate*, provided with or likeness): see **lobate**.

loculate (L. *loculus*, dim. of *locus*, point, place, cavity + suf. *-atus* > E. *-ate*, provided with or likeness): containing locules or chambers.

locule, loculus, pl. **loculi** (L. *loculus* < *locus*, point, place + dim. suf. *-ulus* > E. *-ule*): an unwalled cavity or chamber within a stroma where asci are formed; a stroma can be uniloculate or multiloculate and the locules can be uniascal or multiascal, depending upon the species. Locules are characteristic of the Loculoascomycetes. In *Myriangium* (a gen. in the Myriangiales that parasitizes scale insects), e.g., the locules are uniascal, whereas in *Dothidea* (Dothideales) the locules are multiascal. The term is also applied to the cavity of the pycnidium of the sphaeropsidaceous asexual fungi, in which are produced the conidiophores and conidia.

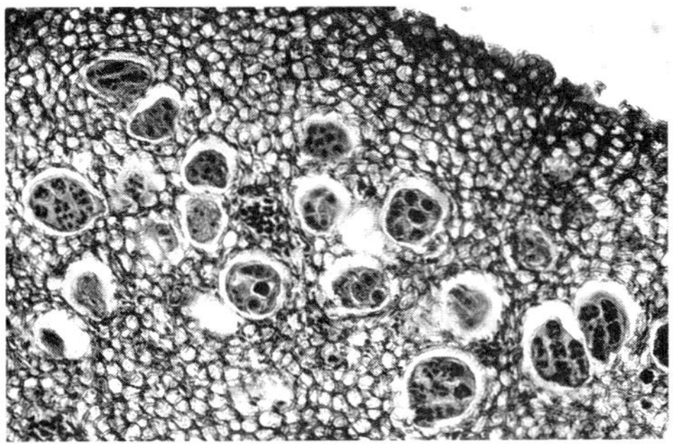

Locules of an apothecioid ascostroma of *Myriangium duriaei*, seen in longitudinal section; each locule contains one ascus, x 140 (*RTH*).

Loculoascomycetes (L. *loculus*, dim. of *locus*, point, place, cavity + L. *-mycetes*, ending of class < Gr. *mýkes*, genit. *mýketos*, fungus): a class of fungi with bitunicate asci produced in a unilocular or multilocular ascostroma that has an ascolocular pattern of development. The Loculoascomycetes (which embraces the orders Coryneliales, Dothideales, Myriangiales, Arthoniales, Asterinales, Capnodiales, Chaetothyriales, Patellariales, Pleosporales, Melanommatales and Verrucariales) together with the Archiascomycetes, Saccharomycetes, Plectomycetes, Pyrenomycetes, Discomycetes, and other filamentous ascomycetes (such as the Erysiphales, Laboulbeniales and Spathulosporales) constitute the phylum Ascomycota of the kingdom Fungi.

locus, pl. **loci** (L. *locus*, point, place): **1**. *Genetics*. A point or place on a chromosome in which is situated a gene. **2**. *Conidiogenesis*. The particular site on a hypha or a cell, i.e., the conidiogenous locus, which represents the fertile point where the conidia are generated. For example, in a phialide the conidiogenous locus is found in the subapical part, almost in the mouth through which the conidia emerge.

lomasome (Gr. *lõma*, fringe, border, edge + *sõma*, body): a polymorphic structure, but more commonly composed of membranous tubules or vesicles, that is found between the plasma membrane and the cell wall, although at times the vesicles can be embedded in the cell wall. It has been considered that lomasomes are implicated in the synthesis of the cell wall, although it also has been suggested that they represent excess plasma membrane in relation to that needed to line the cell wall, in which case their functions would be

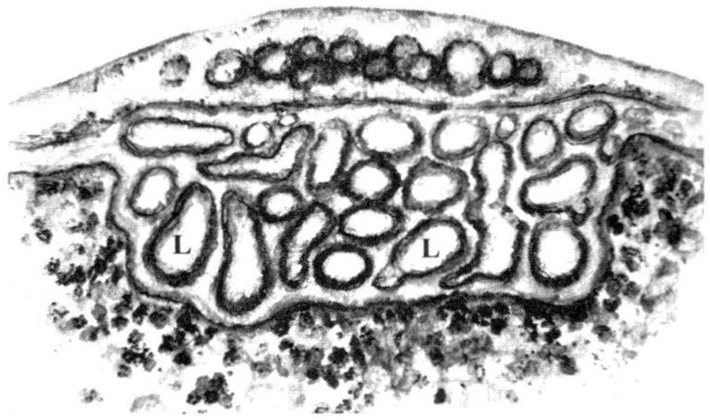

Lomasome (L) of a hyphal cell of *Saprolegnia ferax* seen in longitudinal section. Drawing based on a transmission electron micrograph published by Beckett *et al.* in their *Atlas of Fungal Ultrastructure*, 1974, x 61 000.

longitudinal (L. *longitudinalis*, belonging to or relative to length < *longitudo*, *longitudin-*, length + suf. *-alis* > E. *-al*, relating to or belonging to): formed or positioned parallel to the long axis of a structure, such as the longitudinal dehiscence of the hysterothecium of *Hysterium* (Dothideales) and the longitudinal septa of many types of spores. When united with the transverse septa they form dictyospores (e.g., in *Alternaria*, of the dematiaceous asexual fungi). A longitudinal section is one made parallel to the major dimension of the cell, organ, etc. Cf. **transversal**.

Longitudinal dehiscence of the hysterothecium of an unidentified loculoascomycete, x 25 (*MU*).

loriform (L. *loriformis* < *lorum*, leash, bridle + *-formis* < *forma*, shape): see **ligulate** and **linguiform**.

lotic (L. *lotus*, washing + suf. *-íkos* > L. *-icus* > E. *-ic*, belonging to, relating to): of, relating to, or living in running waters. Cf. **lentic**.

lubricous (L. *lubricus*, smooth and slippery < *lubricare*, to make slippery + *-osus* > OF. *-ous*, *-eus* > E. *-ous*, having, possessing the qualities of): a surface, usually shiny, that feels waxy or oily to the touch, slippery; e.g., the ascocarp of *Leotia lubrica* (Helotiales) and the basidiocarp of *Mycena galericulata* (Agaricales). See **oleaginous**.

lumen, pl. **lumina** or **lumens** (L. *lumen*, light, window): the cavity enclosed by the cell walls of a cell or a hypha; also used for the interior space of the tubes of the Polyporaceae, such as *Megasporoporia mexicana* (Aphyllophorales).

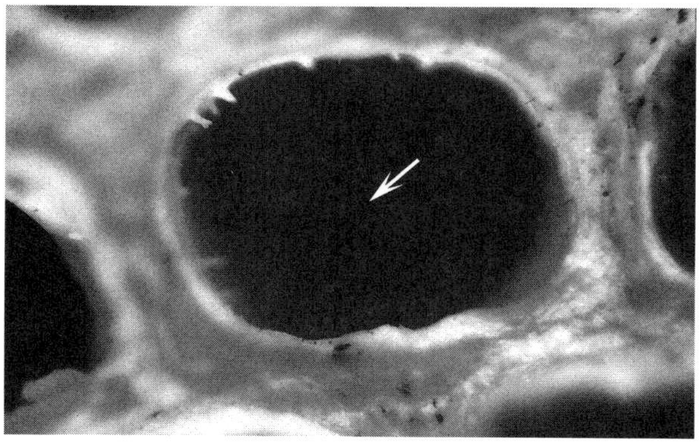

Lumen of the tubes in the porous hymenophore of *Megasporoporia mexicana*, x 23 (*MU*).

luminescent (L. *luminescens*, genit. *luminescentis* < *lumen*, light + *-escens*, genit. *-escentis*, that which turns, beginning to, slightly > E. *-escent*): giving off light without being incandescent, i.e., which has the phenomenon of luminescense, which is the production of light of weak intensity, due to slow oxidations, mediated by special enzymes, that produce light but do not elevate the temperature (therefore also called chemoluminescense). The mycelium of some fungi, such as that of *Armillaria mellea*, and the sporiferous apparatus of other Agaricales, such as that of *Omphalotus olearius* or that of *Mycena lux-coeli*, are luminescent.

lunate (L. *luna*, moon + suf. *-atus* > E. *-ate*, provided with or likeness): having the shape of a half moon, or lunar crescent; e.g., like the macroconidia of *Fusarium* (tuberculariaceous asexual fungi) in side view. Also called **lunulate**, **falcate** and **falciform**.

lunulate (L. *lunula*, dim. of *luna*, moon + suf. *-atus* > E. *-ate*, provided with or likeness): see **lunate**.

lurid (L. *luridus*, pale yellow): pale yellow in color, like the basidiocarp of *Boletus luridus* (Agaricales).

lycoperdoid (< gen. *Lycoperdon* < Gr. *lýkos*, wolf + *pérdo*, to break wind + L. suf. *-oide* < Gr. *-oeídes*, similar to): refers to a globose or subglobose fruiting body having a gleba which is powdery at maturity and usually supported by a sterile tissue; characteristic of *Lycoperdon*, *Bovista*, *Calvatia* and other gen. of Lycoperdales.

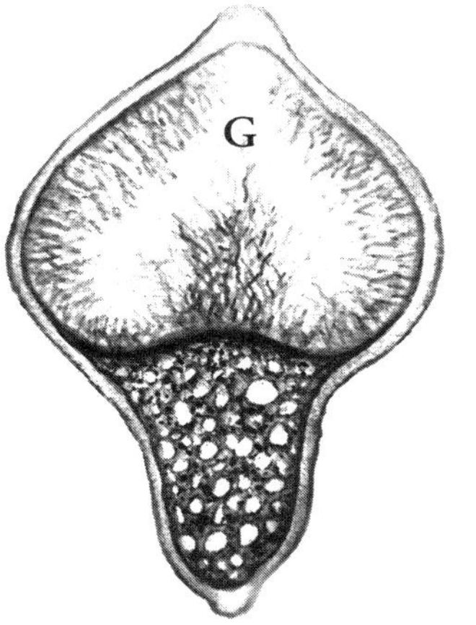

Lycoperdoid type of development. Diagram of a longitudinal section of a fruiting body of *Lycoperdon* sp. showing the gleba (G) or fertile part and a supporting stalk, x 1.5.

lymabiont (Gr. *lýma*, filth, impurity + *bíos*, life + *óntos*, genit. of *ón*, a being): an organism only found in sewage.

lymaphile (Gr. *lýma*, filth, impurity + *phílos*, have an affinity for): an organism commonly found in sewage.

lymaphobe (Gr. *lýma*, filth, impurity + *phóbos*, that which fears or avoids < *phobéo*, to fear): an organism never found in sewage.

lymaxene (Gr. *lýma*, filth, impurity + *xénos*, stranger): an organism rarely found in sewage.

lyrate (L. *lyratus* < *lyra*, lyre or lute + suf. *-atus* > E. *-ate*, provided with or likeness): having the shape of a lyre, i.e., whose contour recalls the musical instrument, with the narrowest part at the apex.

lysigenous (Gr. *lýsis*, dissolution, disintegration + *génos*, origin + L. *-osus* > OF. *-ous*, *-eus* > E. *-ous*, possessing the qualities of): forming through the dissolution of cells, as happens during the formation of the ostiolar canal of many ascomycetes, e.g., *Haloguignardia* (Phyllachorales), and of the cavities (locules) that originate during the morphogenesis of loculoascomycete ascocarps. Cf. **schizogenous**.

m

macroconidium, pl. **macroconidia** (Gr. *makrós*, large + *kónis*, dust + dim. suf. *-ídion* > L. *-idium*): a conidium or asexually reproduced spore that is distinguished from the microconidium by its larger size, as well as by being multicellular; macroconidia are present in some asexual fungi, e.g., in various species of *Fusarium* (such as *F. moniliforme*, a tuberculariaceous form), dermatophytes like *Epidermophyton floccosum* and *Trichophyton mentagrophytes* (moniliaceous asexual fungi), and in *Neurospora* (Sordariales). Cf. **microconidium**.

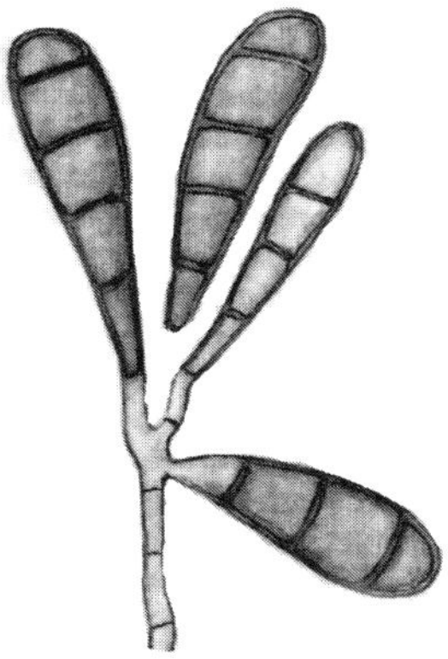

Macroconidia of *Epidermophyton floccosum*, x 1 000.

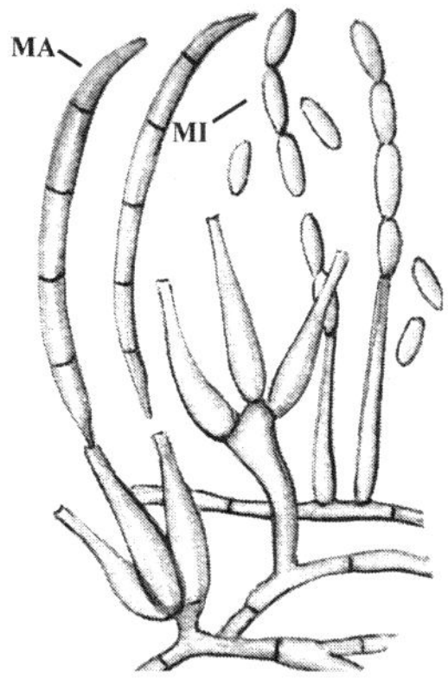

Macroconidia (MA) and microconidia (MI) produced from phialides of *Fusarium moniliforme*, x 790.

macrocyclic (Gr. *makrós*, large + *kyklikós*, circular, cyclic < *kiklós*, circle + suf. *-íkos* > L. *-icus* > E. *-ic*, belonging to, relating to): *Uredinales*. Species of rust fungi with a long life cycle, which contains five phases or stages of development, each with a characteristic type of spore (spermatia, aeciospores, urediniospores, teliospores and basidiospores); as, e.g., *in Puccinia graminis tritici*, the rust of wheat. The macrocyclic rusts can be autoecious or heteroecious, depending upon whether they complete their life cycle on a single host or on two different hosts, respectively. Cf. **microcyclic** and **hemicyclic**.

macrocyst (Gr. *makrós*, large + *kýstis*, bladder, cell): **1**. *Dictyosteliomycota*. A cell with a thick cellulosic wall, which originates from the fusion of a pair of haploid myxamoebae, and in which karyogamy and meiosis occur to form again the haploid myxamoebae of the vegetative stage. Thus the macrocysts represent the sexual state of these organisms, such as *Dictyostelium discoideum*. **2**. An old term for the ascogonium of *Pyronema* (Pezizales). Cf. **microcyst** and **paracyst**.

macrocystidium, pl. **macrocystidia** (Gr. *makrós*, large + *kystídion* < *kýstis*, bladder; here, vesicle, cell + dim. suf. *-ídion* > L. *-idium*): *Agaricales*. A type of cystidium that originates from the deep zone of the trama; it is very long, fusiform or claviform, often with an apical appendage of variable shape, and with a peduncle that generally is connected with a lactiferous hypha in the trama; it is characteristic of *Macrocystidia*, *Lactarius* and *Russula* (Agaricales). The macrocystidium is considered a very evolved form of pseudocystidium, which contains refringent structures by way of needles or masses, and at times is muricated with crystals that are soluble in ammoniac. Cf. **chrysocystidium**, **lamprocystidium** and **leptocystidium**.

macromycete (L. *macromycete* < Gr. *makrós*, large + *mýkes*, genit. *mýketos*, fungus): a fungus with macroscopic reproductive bodies or sporiferous apparatus; they represent the higher fungi, including certain ascomycetes (principally Xylariales and Pezizales) and the major part of the basidiomycetes (Tremellales, Agaricales, Aphyllophorales and the orders of the Gasteromycetes). Cf. **micromycete**.

Macrocystidia of the hymenium of *Macrocystidia* sp., x 96 (*MU*).

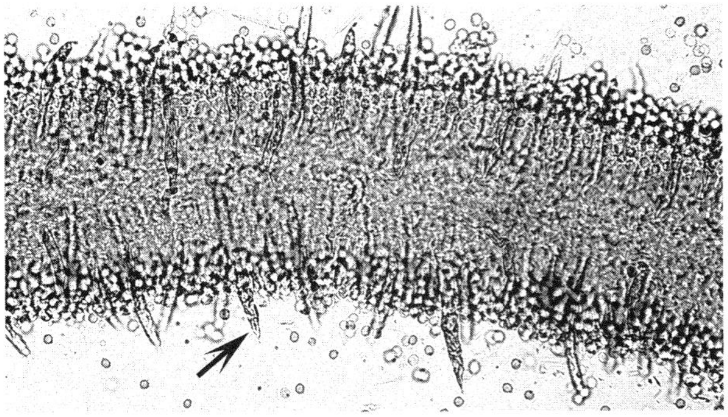

Macrocystidia of the hymenium of *Russula* sp., x 110 (*MU*).

macronematous (Gr. *makrós*, large + *nêma*, genit. *nêmatos*, filament + L. *-osus* > OF. *-ous*, *-eus* > E. *-ous*, having, abounding in, possessing the qualities of): refers to a specialized conidiophore, that is distinguished morphologically from the assimilative hyphae in the rest of the mycelium. Cf. **micronematous**. These terms, as well as **mononematous** and **synnematous**, are used frequently to describe species of dematiaceous asexual fungi .

macroscopic (Gr. *makrós*, large + *skópein*, to see + suf. *-íkos* > L. *-icus* > E. *-ic*, belonging to, relating to): something large enough to be seen with the naked eye, without the help of a microscope. Cf. **microscopic**.

maculate (L. *maculatus* < *macula*, spot + suf. *-atus* > E. *-ate*, provided with or likeness): with spots, spotted. In lichenology it refers to species in which the apothecia on the upper surface of the thallus appear as small spots, as, e.g., in *Arthonia radiata* (Arthoniales) and *Melaspilea maculosa* (of the Melaspileaceae, an ascomycetous family of uncertain affinities, previously included in the Hysteriales).

Maculate thallus of *Arthonia radiata*, x 1.7.

maculicole, **maculicolous** (L. *macula*, spot + *-cola*, inhabitant < *colere*, to inhabit + *-osus* > OF. *-ous*, *-eus* > E. *-ous*, possessing the qualities of): occurring in spots; many plant pathogenic fungi sporulate in leaf spots caused by the infection of the host plant. *Fusarium stoveri* (tuberculariaceous asexual fungi) has been found in leaf spots of plantain, caused by the disease called sigatoka, whose etyological agent is *Cercospora musicola* (dematiaceous asexual fungi).

magic mushroom (ME. *magik* < MF. *magique* < L. *magice* < Gr. *magiké*, fem. of *magikós*, Magian, magical < *mágos*, magus, sorcerer + suf. *-íkos* > L. *-icus* > E. *-ic*, belonging to, relating to; ME. *musseroum* < MF. *mousseron* < LL. *mussirion*, mushroom): syn. of **hallucinogenic fungus**.

malacoid (Gr. *malakós*, soft + L. suf. *-oide* < Gr. *-oeídes*, similar to): soft or mucilaginous in texture; e.g., like the zoogleae or microbiogleae called tibicos in Mexico (tibi grains or sugary kefir grains, which are macrocolonies of yeasts and bacteria embedded in dextrans) which are utilized popularly to manufacture the fermented beverages known as tepache and vinegar of tibicos. The disease known as leucoencephalomalacia, which is caused by some toxins of *Fusarium moniliforme* (tuberculariaceous asexual fungi), receives its name from the white, mucilaginous appearance of the lesions in the equine encephalic mass in animals that have suffered this type of mycotoxicosis due to the ingestion of contaminated feed.

mammiform (L. *mammiformis*, breast-shaped < *mamma*, breast + *-formis* < *forma*, shape): breast-shaped, i.e., conical or rounded with a protruding apex; e.g., like the ostiolate perithecia of *Hypoxylon mammatum* (Xylariales), the apothecia of the thallus of *Thelomma santessonii* (Caliciales), and the pileus of *Psilocybe caerulescens* and *P. sanctorum* (Agaricales).

manglicolous (Sp. *mangle*, mangrove + L. *-cola*, inhabitant + L. *-osus* > OF. *-ous*, *-eus* > E. *-ous*, possessing the qualities of): living on mangrove trees (spp. of gen. *Rhizophora*), like *Manglicola* *guatemalensis* (Dothideales), a saprobic marine fungus that forms its ascocarps in the bark of dead roots of *Rhizophora mangle*. *Mycosphaerella pneumatophorae* (Dothideales) also is manglicolous, since it is parasitic (or saprobic?) in the bark of the living pneumatophores of the mangroves *Avicennia africana* and *A. germinans*.

Mammiform ostiole of the perithecium of *Hypoxylon mammatum*, x 12 (*EPS*).

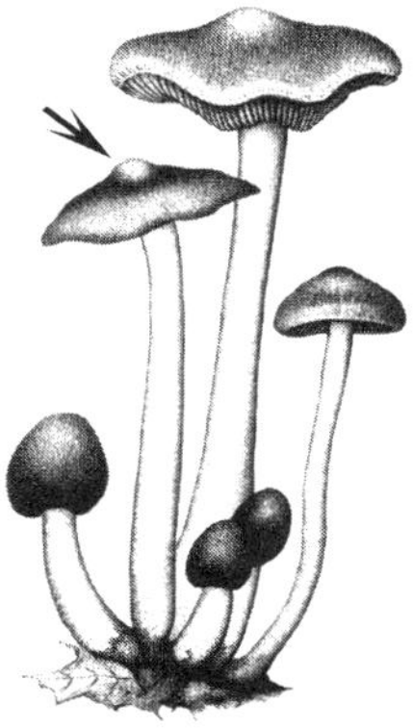

Mammiform pileus of the mature basidiocarps of *Psilocybe caerulescens*, x 0.9.

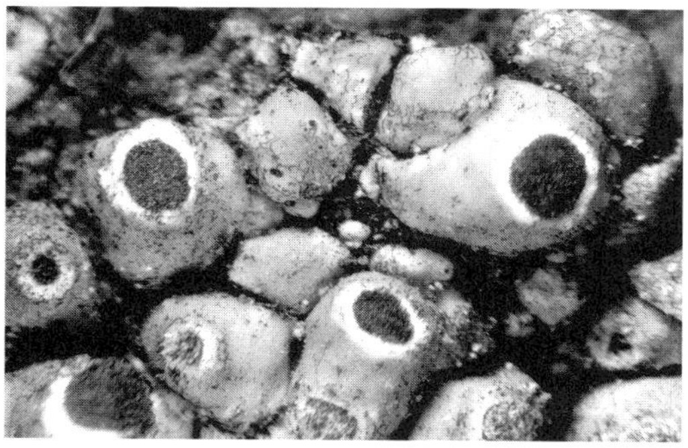

Mammiform apothecia of the thallus of *Thelomma santessonii*, x 10 (*CB*).

manocyst (Gr. *manós*, thin + *kýstis*, bladder, cell): *Peronosporales*. An oogonial papilla that extends into the antheridium during gametangial contact in species of the gen. *Phytophthora*.

mantle (ME. *mantel* < L. *mantellum*, cloak, cover, envelope): a compact layer of hyphae enveloping short feeder roots of ectomycorrhizal plants, connected to the Hartig net on the inside, and to the extramatrical hyphae on the outside; acts as a nutrient sink.

marcescent (L. *marcescens*, genit. *marcescentis*, becoming dry, withering < *marcens*, withering, feeble + *-escens*, genit. *-escentis*, that which turns, beginning to, slightly > E. *-escent*): refers to structures that become dry and remain attached to the organs or parts that gave rise to them, or to those that, being dehydrated, can revive and recover their normal appearance when they are remoistened, such as occurs with the basidiocarps of *Marasmius* (Agaricales).

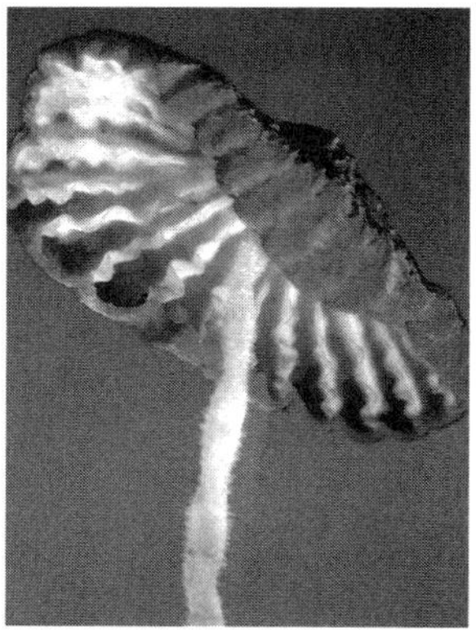

Marcescent basidiocarp of *Marasmius siccus*, x 3 (*MU & CB*).

marginal cilium, pl. **cilia** (ML. *marginalis*, situated on the edge or margin < L. *margin-*, *margo*, edge, margin, border + suf. *-alis* > E. *-al*, relating to or belonging to; *cilium*, the edge of the eyelid where the eyelashes are implanted; generally applied to the eyelash proper or a very fine hair): *Lichens*. The fine prolongations, composed of hyphae, present in the laminar or foliose thallus of certain lichen species, such as *Parmelia perlata* (Lecanorales).

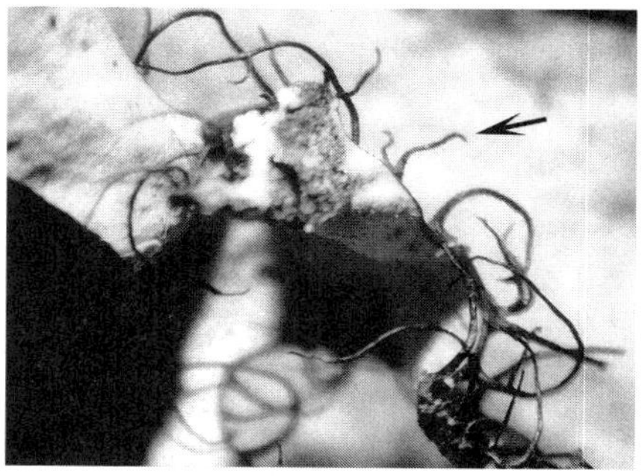

Marginal cilia of the thallus of a foliose lichen, x 3 (*MU*).

marginal veil (ML. *marginalis*, situated on the edge or margin < L. *margin-*, *margo*, edge, margin, border + suf. *-alis* > E. *-al*, relating to or belonging to; *velum*, veil): a secondary proliferation of hyphae on the margin of the pileus, which is destined to protect the hymenium during development, as occurs in agarics and boletes. See **partial veil**.

marginate (L. *marginatus*, bordered < *marginare*, to provide with a border, edge < *margin-*, *margo*, edge, margin, border + suf. *-atus* > E. *-ate*, provided with or likeness): having a distinct border or edge; e.g., like the reddish border of the gills of *Mycena rubromarginata* (Agaricales), or like the basal bulb of the stipe of *Leucocoprinus bulbiger* (Agaricales), which has a border similar to a dropper or duct, the volva of *Amanita abrupta* (Agaricales), which is separated abruptly from the base of the stipe, and the basidiocarp of *Fomes marginatum* (Aphyllophorales).

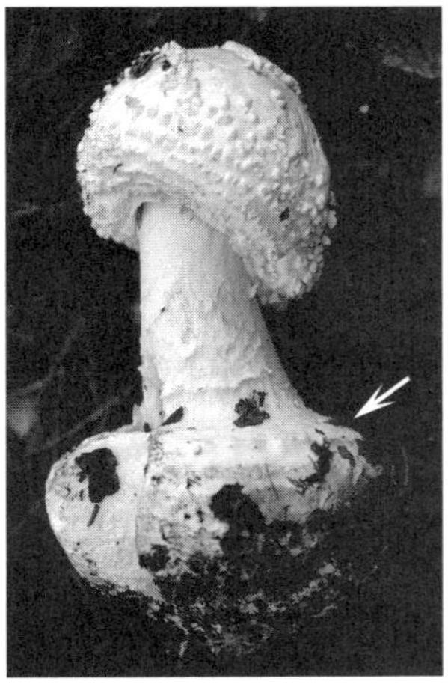

Marginate volva of the basidiocarp of *Amanita abrupta*, x 0.8 (*RMA*).

Marginate basidiocarp of *Fomes* (=*Ungulina*) *marginata* var. *pinicola*, x 0.5 (*MU*).

mastigonemate (Gr. *mástix*, *mástigos*, whip + *nêma*, filament + suf. *-atus* > E. *-ate*, provided with or likeness): furnished with **mastigonemes**.

mastigoneme (Gr. *mástix*, *mástigos*, whip + *nêma*, filament): each one of the small and numerous secondary hairlike fibrils of a tinsel flagellum, that are implanted on the axoneme (main axis); the mastigonemate flagella, which are also called pantonemate, are present in zoospores of the Hyphochytriomycota, such as *Hyphochytrium* and *Rhizidiomyces*, and the Oomycota, such as *Saprolegnia*. Also called **flimmers** or **flimmer hairs**. See **axoneme** and **pantonema**.

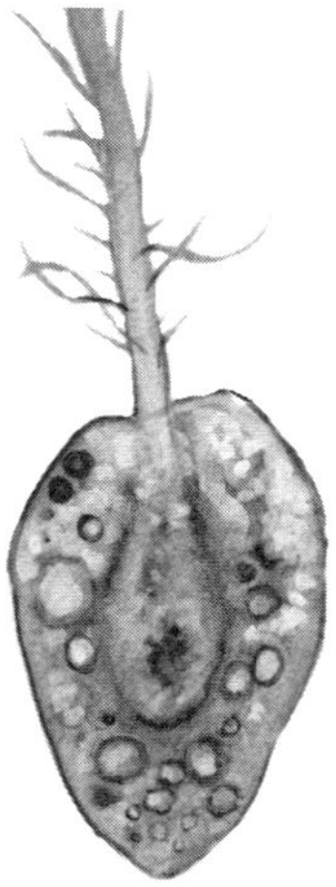

Mastigoneme of the flagellum of a zoospore of *Rhizidiomyces apophysatus*, x 4 800.

matrix, pl. **matrices** or **matrixes** (L. *matrix*, womb, a place where anything is generated < *matr-*, *mater*, mother): 1. The substratum in or on which an organism is living. 2. Mucilaginous material in which spores (conidia, ascospores) are embedded, and which influences dissemination, survival, germination, etc. See **extramatrical** and **intramatrical**.

mazaedium, pl. **mazaedia** (L. *mazaedium* < Gr. *máza*, kneaded flour + L. *-odium*, resembling < Gr. *ode*, like < *-oeídes*, similar to): refers to the mucilaginous and powdery mass that appears in the ascocarp of certain ascomycetes, such as *Onygena corvina* (Onygenales), composed of the tips of the paraphyses and their secretions, and which normally contain large quantities of ascospores. The mazaedium is also a type of fructification of the lichens of the order Caliciales.

mediostratum, pl. **mediostrata** (NL. *mediostratum* < L. *medius*, middle + *stratum*, layer): the central portion of the hymenophorous trama of the gills (in agarics) or of the wall of the tubes (in Boletaceae and Polyporaceae of the order Agaricales), distinct from the lateral layers. In *Boletus*, e.g., the mediostratum is found in the center of a bilateral trama.

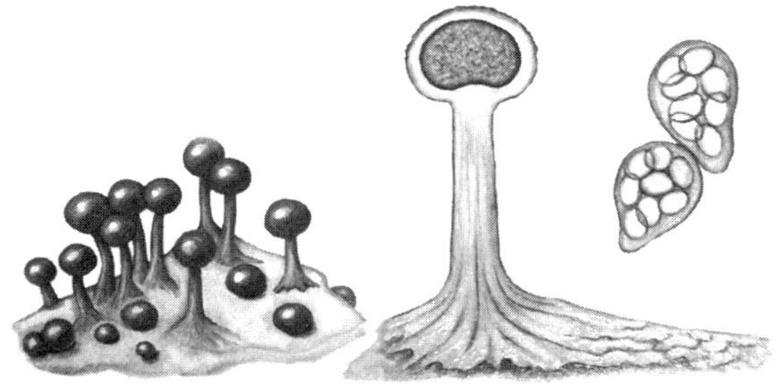

Mazaedia of *Onygena corvina*.
Two octosporate asci are also shown, x 3.5.

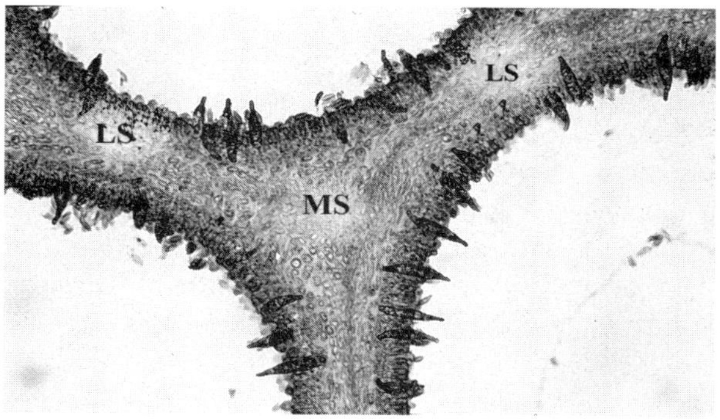

Mediostratum (MS) and lateral strata (LS) in the wall of the tubes of *Boletus* sp., seen in transverse section, x 85 (*MU*).

medipellis (L. *medius*, middle + *pellis*, skin): see **pellicule**.

medium, pl. **media** (L. *medium*, neuter of *medius*, middle; in this case, a surrounding or enveloping substance; a culture medium): a nutrient system or substratum for the artificial cultivation of fungi and other microorganisms. A wide range of liquid and solid media is used in mycology for stimulating growth, sporulation and preservation of fungi.

medulla, pl. **medullae** (L. *medulla*, marrow, pith, center): **1.** *Agaricales*. The internal tissues of a thallus or of an organ, soft or spongy in nature, but different from that of the outermost or peripheral zone; in *Amanita casearea* (Agaricales), e.g., the medulla of the stipe of the basidiocarp has hyphae with inflated, claviform cells. **2.** *Lichens*. In species with a heteromerous thallus, the medulla is the central layer, beneath the algal layer, and it is usually composed of looser hyphae than the external layers.

meiosis, pl. **meioses** (Gr. *meíosis*, diminution, alluding to the number of chromosomes): a series of two successive nuclear divisions in which the number of chromosomes is reduced by one-half, from the diploid number to the haploid. It represents the last phase of sexual reproduction, after **plasmogamy** and **karyogamy**. Cf. **mitosis**.

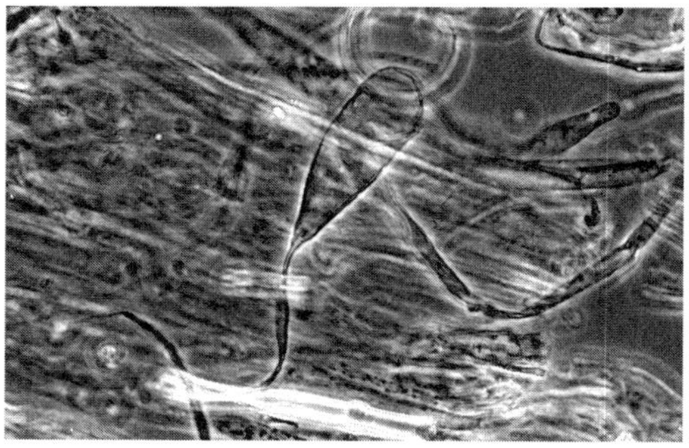

Medulla of the stipe of *Amanita caesarea*, containing inflated cells, x 540 (*RMA*).

meiosporangium, pl. **meiosporangia** (Gr. *meíosis*, diminution + *sporá*, spore + *angeîon*, vessel, receptacle): a sporangium in which meiosis occurs. Used for thick-walled sporangia (meiosporangia) formed on sporothalli in the aquatic fungus *Allomyces* (Blastocladiales) in which the diploid nuclei undergo meiosis and haploid zoospores (meiospores) are produced; these initiate a new haploid generation consisting of a gametothallus with gametangia, since in this fungus there is an alternation of generations. The sporothallus also forms a second type of sporangium, the latent zoosporangia. Cf. **mitosporangium**.

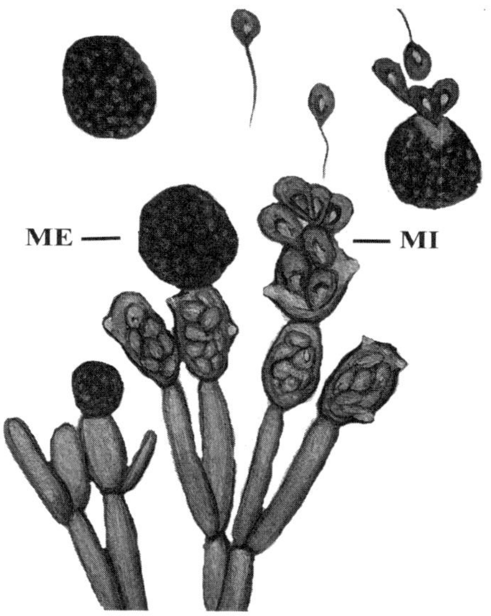

Meiosporangia (ME) of a sporothallus of *Allomyces javanicus*. The mitosporangia (MI) also are shown, x 600.

melanosporous (Gr. *mélas*, *mélanos*, black + *sporá*, spore + L. *-osus* > OF. *-ous*, *-eus* > E. *-ous*, having, possessing the qualities of): having black spores; e.g., the basidiocarps of *Coprinus* (Agaricales) form black or dark gray spores and spore prints.

membranaceous (L. *membranaceus* < *membrana*, skin, parchment, membrane + *-aceus* > E. *-aceous*, of or pertaining to, with the nature of): similar to a membrane, i.e., a thin pliable sheet or layer; e.g., like the hypothallus of *Physarum polycephalum* (Physarales) and the polar appendage of the conidia of *Myrothecium verrucaria* (tuberculariaceous asexual fungi). Cf. **membranous**.

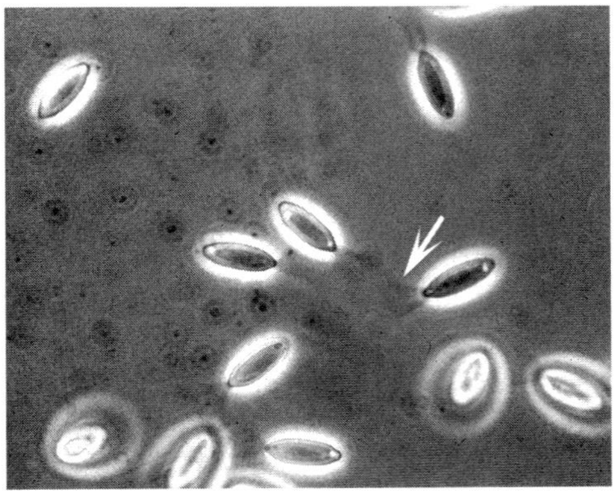

Membranaceous appendage of the conidia of *Myrothecium verrucaria*, x 830 (*MU*).

membranous (L. *membrana*, skin, parchment, membrane + L. *-osus* > OF. *-ous*, *-eus* > E. *-ous*, having, abounding in, possessing the qualities of): having one or more membranes; e.g., like the fruiting body of *Vascellum* (Lycoperdales), which has a membrane between the gleba and the subgleba. Cf. **membranaceous**.

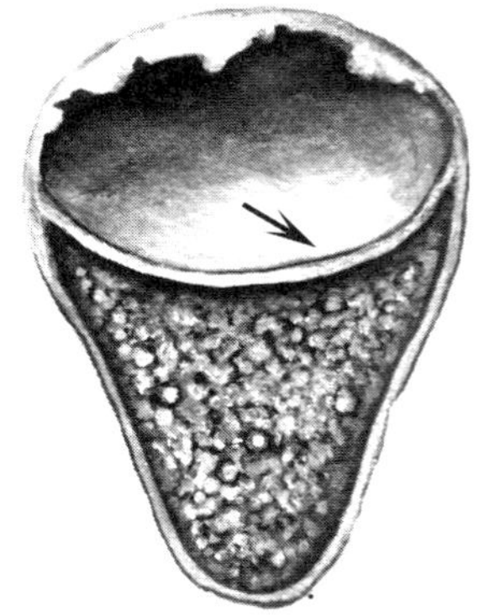

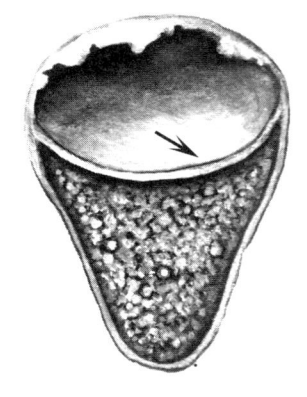

Membranous diaphragm of the fruiting body of *Vascellum pratense*, x 1.

memnonious (L. *memnonius*, black, brownish-black, dull gray + *-osus* > OF. *-ous*, *-eus* > E. *-ous*, having, possessing the qualities of): dark brown in color, almost black, like the conidial chains of *Memnoniella* (dematiaceous asexual fungi), a gen. segregated from *Stachybotrys*.

memnospore (Gr. *mémnon* < *méno*, to remain, persist + *sporá*, spore): a term applied to a latent spore that is surrounded by a thick, resistant wall, such as chlamydospores (asexual) and teliospores (sexual), which can remain at rest for a long time (often more than one year); such spores do not participate in the dispersal of the species as do other types of spores (conidia, sporangiospores, etc.), generally thin-walled, that disseminate the fungus during the same season. Also called **hypnospore**. Cf. **xenospore**.

merismatoid, **merismoid** (Gr. *mérisma*, genit. *merísmatos*, part < *merís*, portion, and *merismós*, division + L. suf. *-oide* < Gr. *-oeídes*, similar to): refers to a pileate fructification that is formed from many small pilei, or which is divided into slender lobes. Sometimes the first term is applied to the Polyporaceae and the second to the Agaricaceae.

meristem arthrospore (Gr. *merísto*, to divide + suf. *-emo*, that made or created; *árthron*, joint, articulation + *sporá*, spore): *Conidiogenesis*. An asexual spore that originates from the meristematic growth of the apical region of the conidiophore, and its concurrent basipetal conversion into spores; e.g., as in *Oidium*, the moniliaceous asexual state of *Blumeria* (=*Erysiphe*) *graminis* (Erysiphales).

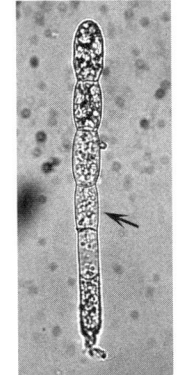

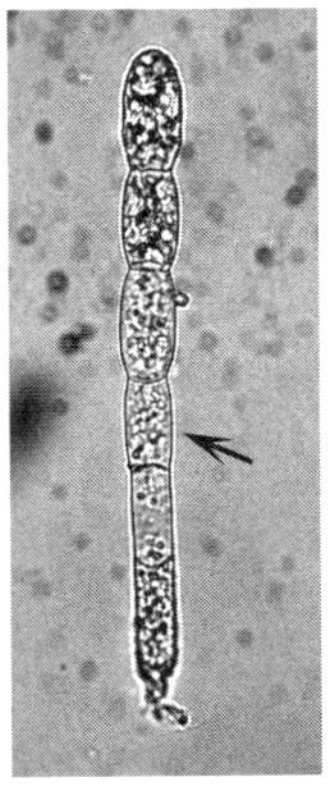

Meristem arthrospores of the conidiophores of *Oidium*, the asexual state of *Erysiphe* sp., x 670 (*MU*).

Meristem arthrospores of the conidiophores of *Oidium*, the asexual state of *Blumeria* (=*Erysiphe*) *graminis*, x 590 (*WKW*).

meristem blastospore (Gr. *merísto*, to divide + suf. *-emo*, that made or created; Gr. *blastós*, sprout, bud, germ, shoot + *sporá*, spore): *Conidiogenesis*. An asexual reproductive spore that originates by budding of a conidiophore that extends from a basal meristematic zone (**basauxic**), as, e.g., in *Arthrinium phaeospermum* (dematiaceous asexual fungi).

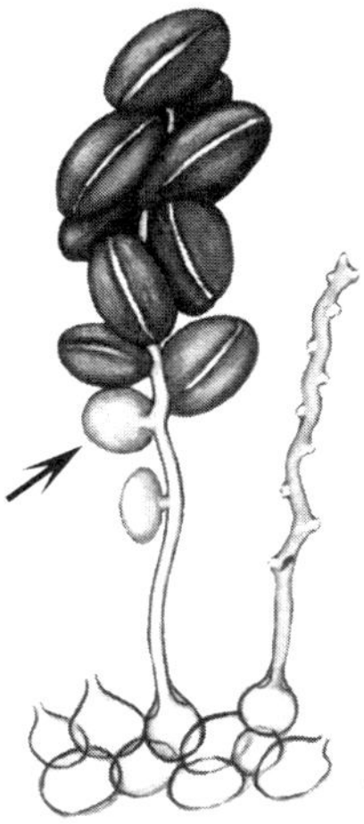

Meristem blastospores of a conidiophore of *Arthrinium phaeospermum*, x 1 000.

meristem spore (Gr. *merísto*, to divide + suf. *-emo*, that made or created; Gr. *sporá*, spore): *Conidiogenesis*. A spore that detaches after having matured in basipetal succession from the tip of a conidiophore, from a phialide or from a hypha, in a manner that said tip is considered as an open growing point.

meristogenous (Gr. *merísto*, to divide + *génos*, origin < *gennáo*, to engender, reproduce + L. *-osus* > OF. *-ous*, *-eus* > E. *-ous*, having, possessing the qualities of): a type of development resulting from repeated divisions of hyphae (meristogenesis); generally applied to the development of a pycnidium. In simple meristogenous development the pycnidium forms by growth and division of a single hypha, sometimes including branches of the same hypha, to form a stroma in which a cavity is later produced; e.g., as in *Phoma herbarum* (sphaeropsidaceous asexual fungi). In compound meristogenous development, which is present in *Phoma pirina*, several hyphae aggregate and unite to form a sphere of tissue in which a cavity is subsequently differentiated. Cf. **symphogenous**.

meront (Gr. *méros*, part + suf. *ón*, genit. *óntos*, a being): a part of something; it refers especially to a daughter myxamoeba that is separated from the periphery of the mother myxamoeba, a phenomenon that occurs in succession in the myxamoebae and plasmodia of the Myxomycetes.

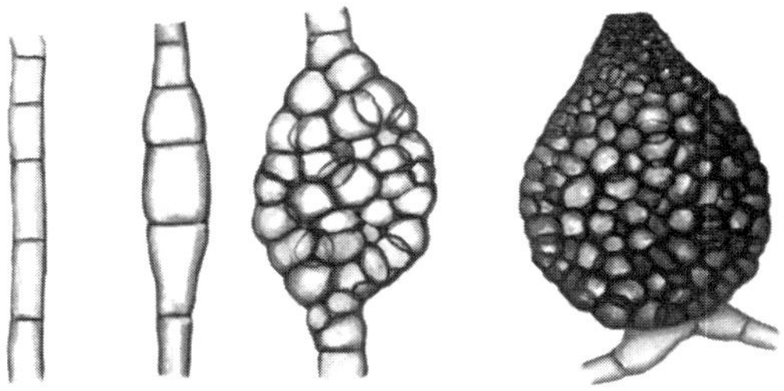

Meristogenous (simple) development of the pycnidium of *Phoma herbarum*, x 145.

merosporangiophore (Gr. *méros*, part, portion + *sporá*, spore + *angeîon*, receptacle, vessel + *-phóros*, bearer): a specialized hypha that produces and supports one or several merosporangia; it is characteristic of *Syncephalastrum racemosum* (Mucorales) and equivalent to the sporangiophore of other Mucorales.

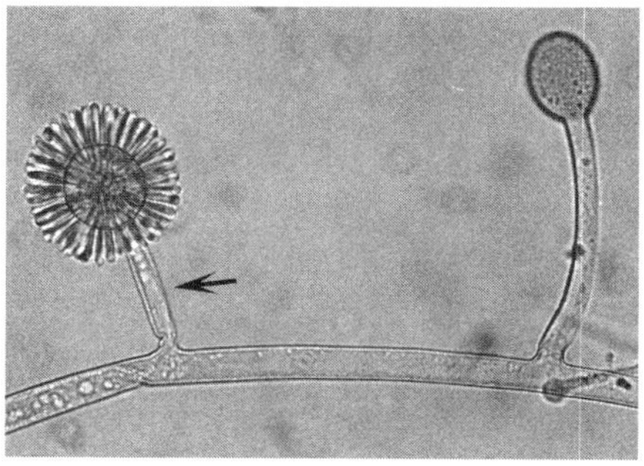

Merosporangiophores of *Syncephalastrum racemosum*, x 140 (*MU*).

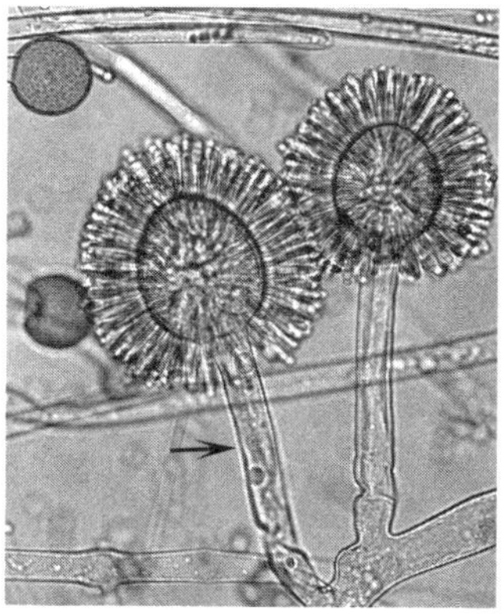

Merosporangiophores of *Syncephalastrum racemosum*, x 200 (*MU*).

merosporangium, pl. **merosporangia** (Gr. *méros*, part, portion + *sporá*, spore + *angeîon*, vessel, receptacle): a cylindrical sporangiolum, with the spores in a row, that is produced on the swollen end of a sporangiophore, as is observed, e.g., in *Syncephalastrum racemosum* (Mucorales).

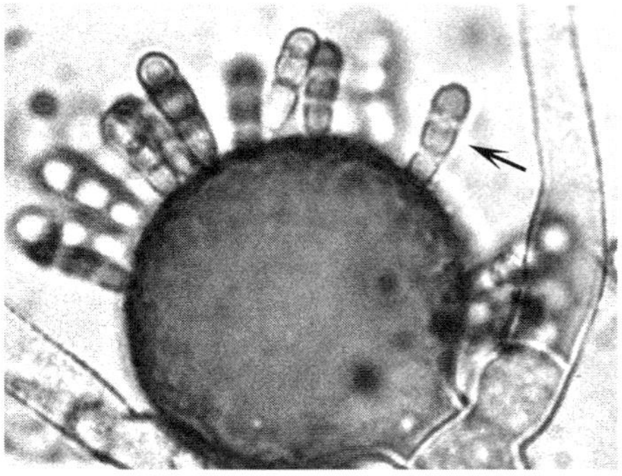

Merosporangia of *Syncephalastrum racemosum*, x 680 (*MU*).

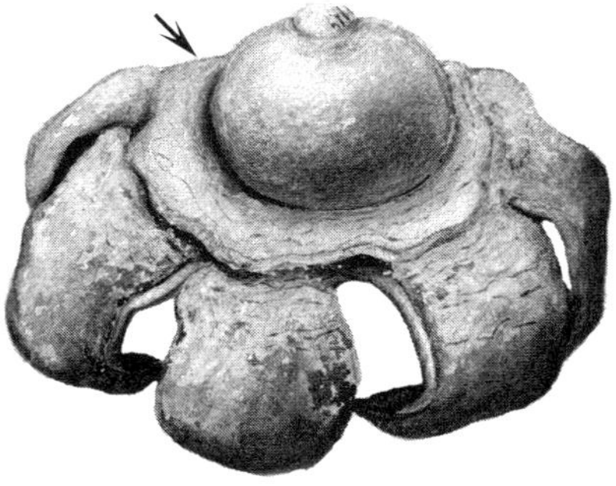

Mesoperidium of the fruiting body of *Geastrum saccatum*, x 1.

mesendogenous (Gr. *mésos*, middle, intermediate + *éndon*, within + *génos*, origin < *gennáo*, to engender, produce + L. *-osus* > OF. *-ous*, *-eus* > E. *-ous*, having, possessing the qualities of): *Conidiogenesis*. A mode of conidium formation in which the secondary cell wall of the conidium is formed inside the primary wall of the phialide or conidiogenous cell; this happens, e.g., in *Penicillium* and *Aspergillus* (moniliaceous asexual fungi).

mesoderm, **mesodermous**, **mesodermatous** (Gr. *mésos*, middle, intermediate + *dérma*, skin, cortex; or + L. *-osus* > OF. *-ous*, *-eus* > E. *-ous*, having, possessing the qualities of; or + *-osus* > OF. *-ous*, *-eus* > E. *-ous*): a hypha in which the external wall and the lumen have almost the same thickness. Cf. **leptoderm** and **pachyderm**.

mesoperidium, pl. **mesoperidia** (Gr. *mésos*, middle, intermediate + *péridion*, small leather purse): the middle layer of a peridium composed of three layers: exoperidium, mesoperidium and endoperidium, as in many Gasteromycetes; e.g., *Geastrum triplex* and *Disciseda candida* (Lycoperdales).

mesophile, **mesophilic**, **mesophilous** (Gr. *mésos*, middle, intermediate + *phílos*, have an affinity for; or + suf. *-íkos* > L. *-icus* > E. *-ic*, belonging to, relating to; or + L. *-osus* > OF. *-ous*, *-eus* > E. *-ous*, possessing the qualities of): an organism that develops at average temperatures; the majority of fungi are mesophilic, since they grow at temperatures between 10-40°C, with an optimum around 25-35°C. Cf. **psychrophile** and **thermophile**.

mesospore (Gr. *mésos*, middle, intermediate + *sporá*, spore): **1**. *Uredinales*. A unicellular teliospore that is found among normal bicellular teliospores, as sometimes occurs in *Puccinia*. The term also is applied to the intermediate cell of tricellular spores, and to urediniospores of the rusts that germinate only after a latent period. **2**. The latter type of mesospore is also called **amphispore**, for having a double function, that of germinating quickly and that of germinating after a latent period. **3**. In the wall of the most complex basidiospores, which is composed of five layers, the **mesospore** or **mesosporium**, pl. **mesosporia**, is a delicate and scarcely distinguishable structure; it is found beneath the three outer layers (**perispore**, **exospore** and **epispore**) and above the **endospore**.

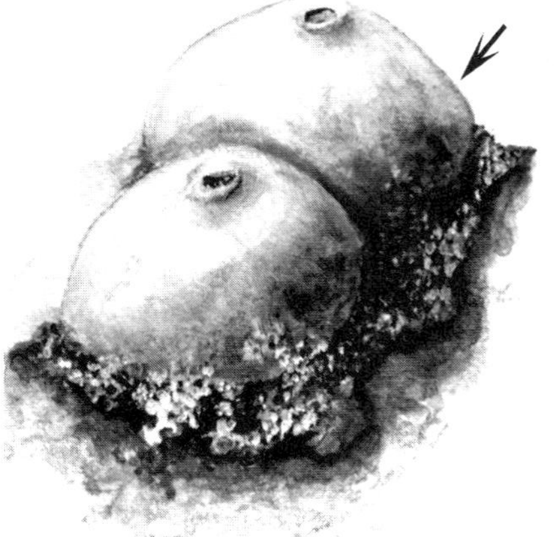

Mesoperidium of the fruiting bodies of *Disciseda candida*, x 1.5.

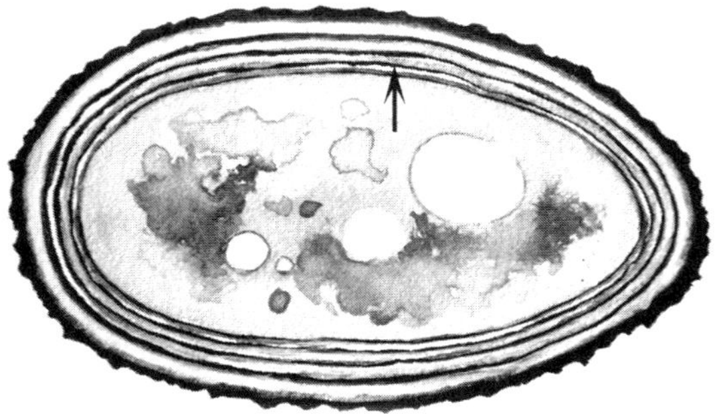

Mesospore of the basidiospore wall of *Tubaria* sp. In this diagram of a transverse section, the endospore, epispore, exospore and perispore also are shown, x 8 000.

metabasidium, pl. **metabasidia** (Gr. *metá*, afterwards, beyond + *basídion* < *básis*, base + dim. suf. *-ídion* > L. *-idium*): a basidium in the phase of development in which the diploid zygote nucleus divides by meiosis and gives rise to the four haploid nuclei that migrate through the sterigmata to each basidiospore, such as occurs in *Amanita* and *Coprinus* (Agaricales).

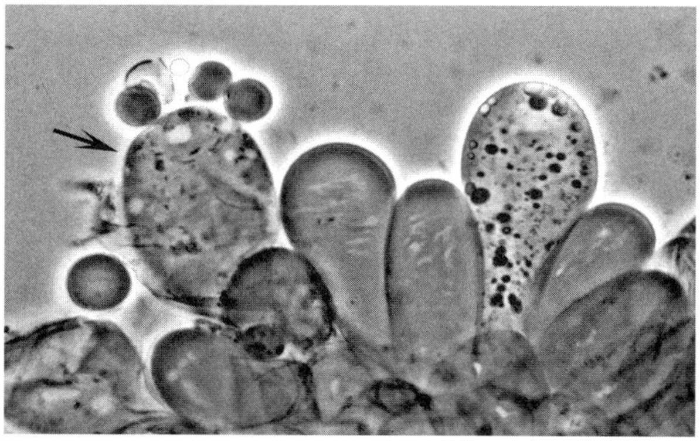

Metabasidium and basidioles of the hymenium of *Amanita* sp., x 700 (*RMA*).

metabiosis, pl. **metabioses** (Gr. *metá*, afterwards, beyond + *bíosis* < *bíos*, life + suf. *-osis*, state, condition): the interdependence or association of two organisms that act one after the other, i.e., one of them prepares a substrate which later can accommodate or support growth of the second; this can occur with the former as a symbiont or it can develop only when the first has disappeared due to incompatibility with the now altered medium. An example of metabiosis is seen in the transformation of starch into sugar by *Aspergillus oryzae* (moniliaceous asexual fungi), which makes the resulting sugar medium available to the yeasts that produce the alcoholic fermentation of the sugar. Cf. **antagonism**, **commensalism**, **helotism**, **mutualism**, **parabiosis**, **parasymbiosis**, **symbiosis** and **synergy**.

metachromatic (Gr. *metá*, afterwards, beyond + *chromatikós*, pertaining to color < *chrõma*, color + *-tikós* > L. *-ticus* > E. *-tic*, relation, fitness, inclination or ability): refers to a structure, such as a cell wall or inner spore wall, which stains violet to reddish with cresyl blue.

metamorphosis, pl. **metamorphoses** (Gr. *metamórphosis* < *metá*, farther on, later + *morphé*, shape, form + *-osis*, state, condition): all of the morphological and structural modifications undergone by an organism or an organ during the course of its development. For example, the myxozoospores of the Myxomycetes can originate either by direct germination of the spores or by metamorphosis of the myxamoebae, which can be transformed into myxozoospores, depending on environmental conditions.

metula, pl. **metulae** (L. *metula*, dim. of *meta*, pyramid, tip of something): a small branch of a conidiophore that gives rise to phialides or conidiogenous cells, as is observed in the gen. *Penicillium* and *Gliocladium* (moniliaceous asexual fungi), among others. The metula also has been called a **prophialide**, since it is found beneath the phialide. The term "primary sterigma" was formerly used for metula, but this usage is no longer considered appropriate. Cf. **sterigma**.

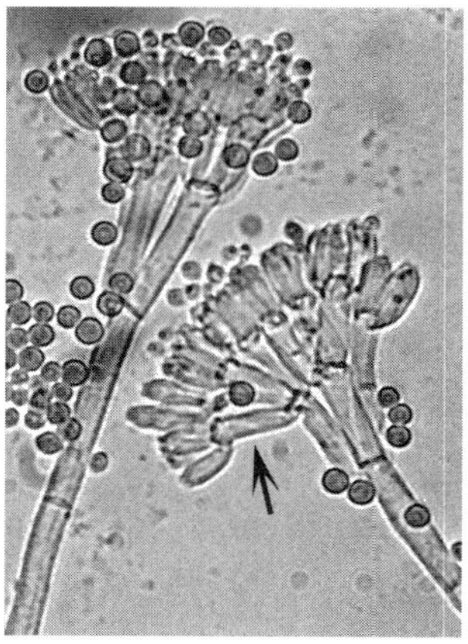

Metulae of the conidiophore of *Penicillium lanoso-coeruleum*, x 500 (*PL*).

metuloid (L. *metula*, dim. of *meta*, pyramid, tip of something + suf. *-oide* < Gr. *-oeídes*, similar to): *Basidiomycetes*. A cystidium incrusted with calcium carbonate, which is deeply inserted in the trama; it has a thickened wall and is generally hyaline or lightly straw-colored. It is present, e.g., in *Peniophora*, *Lentinus rudis* and some species of *Hydnum* (Aphyllophorales), as well as in certain Agaricales, such as *Panaeolus cyanescens*.

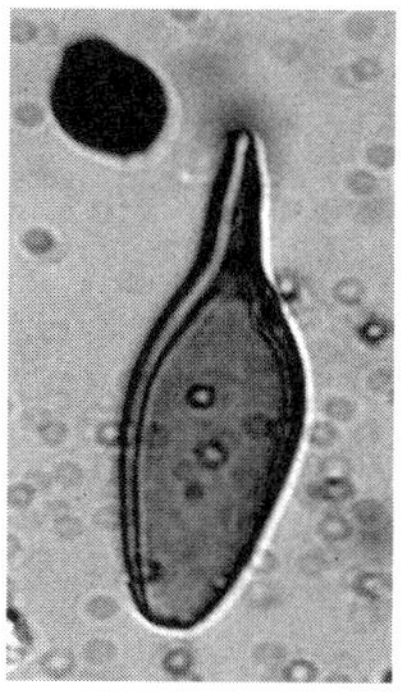

Metuloid cystidium from the hymenium of *Panaeolus cyanescens*, x 890 (*MU*).

micaceous (L. *micaceus*, resembling mica < *mica*, glittering sand + *-aceus* > E. *-aceous*, of or pertaining to, with the nature of): covered with shiny particles; e.g., like the pileus of *Coprinus micaceus* (Agaricales).

microbe (Gr. *mikrós*, small, minute + *bíos*, life): a **microorganism**.

microbial (Gr. *mikrós*, small, minute + *bíos*, life + L. suf. *-alis* > E. *-al*, relating to or belonging to): relating to, or pertaining to microbes.

microbiogloea, pl. **microbiogloeae** (Gr. *mikrós*, small, minute + *bíos*, life + *gloiós*, glutinous, viscous): a group of bacteria and yeasts that comprise a macrocolony with a gelatinous consistency, yellowish-white in color, like the so-called tibi grains which are utilized as inoculum for the lactic acid-alcohol-acetic acid fermentation of sugary liquids that form tepache and vinegar of tibi in Mexico. In the U. S. the tibi grains are given the names *American bees*, *California bees*, *ginger beer plant* and *sugary kefir grains*. Another example of microbiogloeae is the so-called *kefir grains* in the former USSR, and *búlgaros* in Mexico, which are used to ferment milk for the elaboration of *kefir*. They are composed of lactic bacteria (*Lactobacillus* spp., *Streptococcus spp.*) and yeasts (*Saccharomyces cerevisiae* and *Kluyveromyces marxianus*, of the Saccharomycetales, *Candida kefir*, of the cryptococcaceous asexual yeasts, and others), embedded in a matrix of polysaccharides produced by the lactic bacteria; they are brilliant white in color and their consistency, although somewhat gelatinous, is quite elastic. See **kefir grains**, **tea fungus** and **tibi grains**.

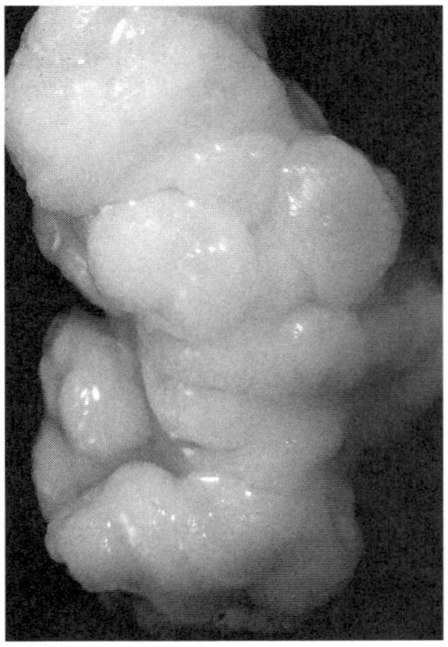

Microbiogloea - a búlgaro grain (so-called in Mexico) or kefir grain, constituting a symbiotic association of lactic bacteria and yeasts, that is used to ferment milk, x 3 (*MU*).

microbiology (Gr. *mikrós*, small, minute + *bíos*, life + *lógos*, treatise + *-y*, E. suf. of concrete nouns): a branch of biological science dealing especially with microorganisms.

microbiotic (Gr. *mikrós*, small, minute + *biotikós*, vital < *bíos*, life + *-tikós* > L. *-ticus* > E. *-tic*, relation, fitness, inclination or ability): applied to the crusts of soil lichens, bryophytes, and free-living cyanobacteria present in certain types of ecosystems, such as deserts, the *Stereocaulon* (Lecanorales) heaths of the tundra, and the old-growth Douglas fir forests, where lichens and cyanobacterial photobionts are important in providing significant amounts of fixed nitrogen to the ecosystem, exceeding nitrogen input from other sources.

microconidium, pl. **microconidia** (Gr. *mikrós*, small, minute + *kónis*, dust + dim. suf. *-ídion* > L. *-idium*): a small conidium, generally unicellular. Microconidia in the conidial fungi function as asexual spores; e.g., in *Fusarium moniliforme* (tuberculariaceous asexual fungi) and in the dermatophytes, such as *Trichophyton* (moniliaceuos asexual fungi). In certain ascomycetes they can also function as spermatia in sexual reproduction, e.g., in *Neurospora* (Sordariales). Cf. **macroconidium.**

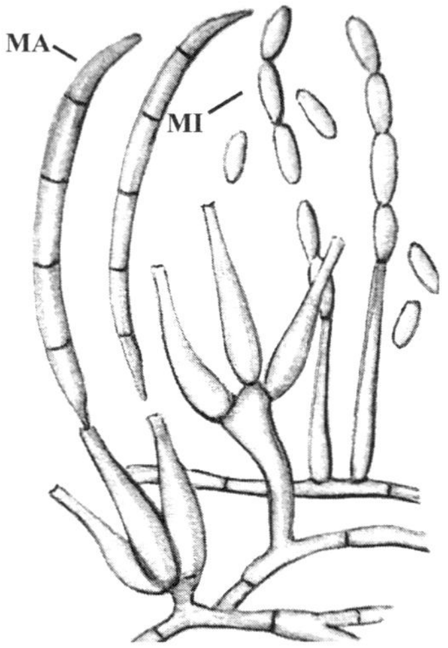

Microconidia (MI) and macroconidia (MA) formed in phialides of *Fusarium moniliforme*, x 940.

microcycle conidiation (Gr. *mikrós*, small, minute + LL. *cyclus* < Gr. *kýklos*, circle, cycle; NL. *conidium* < Gr. *kónis*, dust + dim. suf. *-ídion* > L. *-idium* + Gr. *-ationem*, action, state or condition, or result > E. suf. *-ation*): the production of conidia directly by a spore without the intervention of hyphal growth.

microcyclic (Gr. *mikrós*, small, minute + *kyklikós*, circular, cyclic < *kýklos* + suf. *-íkos* > L. *-icus* > E. *-ic*, belonging to, relating to): refers to species of rusts

(Uredinales) in whose life cycles the aeciospores and urediniospores (and sometimes spermogonia) are lacking; the teliospores are the only dikaryotic spores that are developed; an example is *Puccinia delicatula*, the salvia rust. The microcyclic rusts are **autoecious**, i.e., they develop all of their life cycle on a single host. The plant pathogenic smut fungi (Ustilaginales) also are microcyclic. Cf. **macrocyclic**.

microcyst (Gr. *mikrós*, small, minute + *kýstis*, bladder; here, cell): an encysted protoplast, protected by a wall resistant to conditions of unfavorable moisture; the term is generally applied to the encysted myxamoeba of the Acrasiomycota, such as *Copromyxa arborescens* (Acrasiales) and Myxomycota, although at times also to the encysted zoospore of aquatic fungi (Chytridiomycetes). Cf. **macrocyst** and **paracyst**.

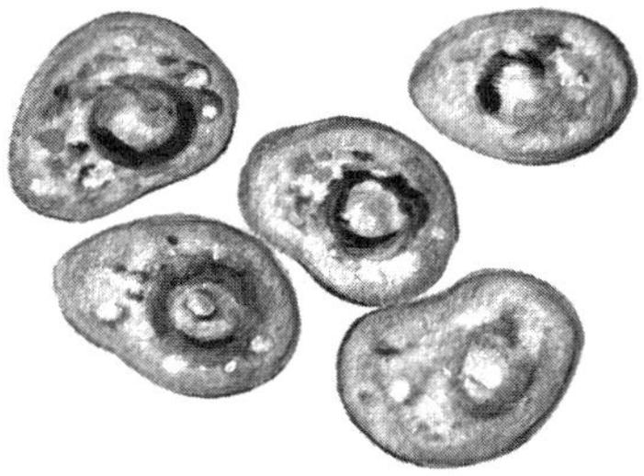

Microcysts (encysted myxamoebae) of *Copromyxa arborescens*, x 1 300.

microendospore (Gr. *mikrós*, small, minute + *éndon*, inside + *sporá*, spore): a minute cytoplasmic particle in a hypha or spore that is capable of developing into a structure typical of that from which it arose; found in *Graphium ulmi* (stilbellaceous asexual fungi).

micrometer (Gr. *mikrós*, small, minute + ME. *meter* < OE. *meter* < L. *metrum* < Gr. *métron*, measure, meter): abb. µm; a unit of measurement equal to one thousandth of a millimeter or one millionth of a meter. It is equivalent to the *micron*, abb. µ, that was used before. Cf. **nanometer**.

micromycete (L. *micromycetes* < Gr. *mikrós*, small, minute + *mýkes*, genit. *mýketos*, fungus): a fungus of very small size, usually with microscopic sporiferous structures; micromycetes occur in all major groups of fungi. Cf. **macromycete**.

micronematous (Gr. *mikrós*, small, minute + *nêma*, genit. *nêmatos*, filament + L. *-osus* > OF. *-ous*, *-eus* > E. *-ous*, having, abounding in, possessing the qualities of): a conidiophore that is not distinguished morphologically from the assimilative hyphae in the rest of the mycelium; e.g., in the gen. *Scytalidium* (dematiaceous asexual fungi), the conidiophores with arthrospores are very similar to the hyphae that do not generate spores. Cf. **macronematous**.

microorganism (Gr. *mikrós*, small, minute + ME. *organism* < *organ*, partly < OE. *organa* < L. *organum* < Gr. *órganon*, tool, instrument, organ, partly < OF. *organe* < L. *organum*, akin to Gr. *érgon*, work + E. suf. *-ism* < L. *-ismus* < Gr. *-ismós*, state, phase, tendency, action): a microbe; an organism of microscopic or ultramicroscopic size, which either cannot be seen with the unaided eye or requires microscopic examination and/or growth in pure culture for its identification. Microorganisms include all unicellular prokaryotes and eukaryotes, and also some multicellular eukaryotes, i.e., microscopic algae, bacteria, fungi (including yeasts), protozoa and viruses. The term microorganism is sometimes used only for prokaryotes (bacteria and viruses).

micropycnidium, pl. **micropycnidia** (Gr. *mikrós*, small, minute + NL. *pycnidium* < Gr. *pyknós*, dense, tight, concentrated + dim. suf. *-ídion* > L. *-idium*): a small pycnidium that develops from a dictyochlamydospore or alternarioid chlamydospore, as happens in *Phoma destructiva* and *Ph. glomerata* (sphaeropsidaceous asexual fungi).

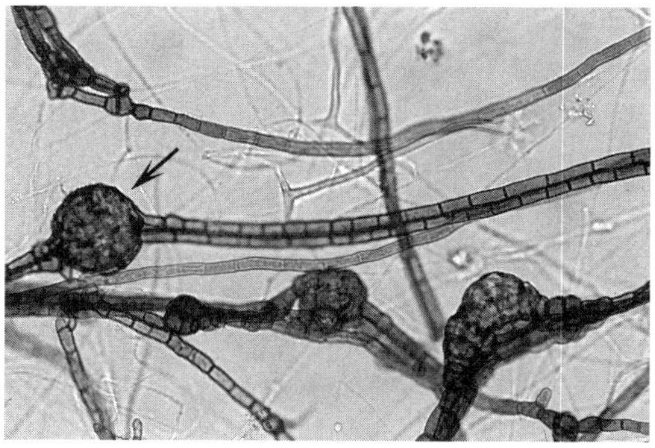

Micropycnidia and alternarioid chlamydospores of *Phoma destructiva*, x 200 (*CB*).

microsclerotium, pl. **microsclerotia** (Gr. *mikrós*, small, minute + L. *sclerotium* < Gr. *sklerotés*, hardness < *sklerós*, hard + L. dim. suf. *-ium*): a small group of dark-colored cells with more or less thickened walls, each one of which is viable and can form a germ tube or hypha. Microsclerotia can form under natural conditions as well as on culture media, and they are characteristic of many asexual fungi, such as *Verticillium* (moniliaceous) and *Rhizoctonia* (agonomycetaceous). The **papulospores** or **bulbils** of *Papulaspora* (agonomycetaceous asexual fungi) also are called microsclerotia.

microscopic (Gr. *mikrós*, small, minute + *skópein*, to see + suf. *-íkos* > L. *-icus* > E. *-ic*, belonging to, relating to): small enough in size that a microscope is needed to be seen with clarity. Cf. **macroscopic**.

microtubule (Gr. *mikrós*, small, minute + L. *tubulus* < *tubus*, tube + dim. suf. *-ulus* > E. *-ule*): a proteinaceous tubule, apparently rigid, 20-25 nm in diam., that is found in the cytoplasm, in achromatic spindles and in the axonemes of flagella. In hyphae, the microtubules are arranged parallel to the major axis; they are commonly found associated with the nuclei and the mitochondria, for which reason it has been considered there is an interaction between the organelle and the microtubules that is related to motility in the organelle, which is independent of the cytoplasmic currents. Microtubules also can constitute a cytoskeleton in various fungal cells, e.g., in zoospores and basidia.

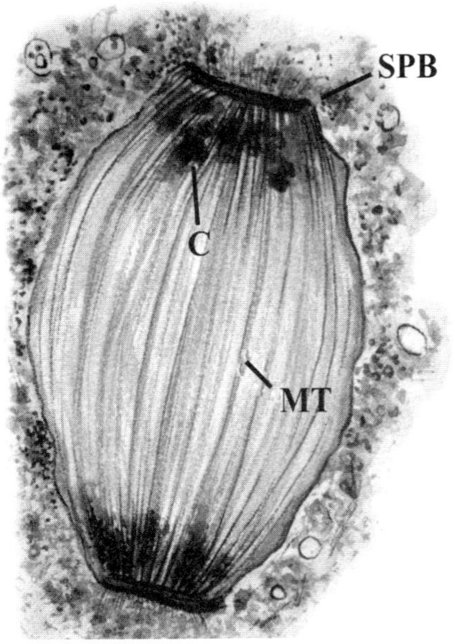

Microtubules (MT) of the nuclear division spindle of *Ascobolus stercorarius.* Longitudinal section through a nucleus at anaphase; note the relationship between the spindle microtubules, the spindle pole bodies (SPB), and the chromosomes (C). Drawing based on a transmission electron micrograph published by Beckett *et al.* in their *Atlas of Fungal Ultrastructure*, 1974, x 7 620.

middle piece (ME. < OE. *middel*, central, middle; ME. *pece* < OF. < Ga. *pettia*, piece): **1**. The central part that connects the two ends of a bipolar achromatic spindle; it is a band of electron opaque material, similar in shape and position to the structure called the **plaque bridge**, which interconnects the two globular daughter halves that result from the replication of the spindle polar body during mitosis in yeast cells, such as *Saccharomyces* (Saccharomycetales). The middle piece is typical of the basidiomycetes. **2**. A stout, short tissue which anchors the purse to the sheath in the peridioles of some species of Nidulariales, such as *Cyathus*.

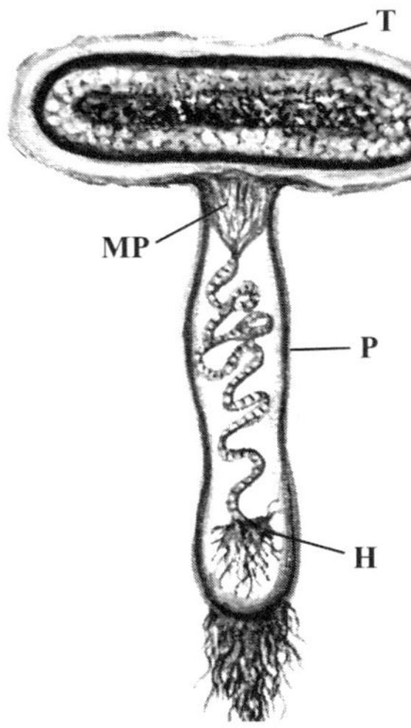

Middle piece (MP) of a peridiolum of *Cyathus striatus.* In this diagram of a longitudinal section, the hapteron (H), purse (P), and tunica (T) also are shown, x 20.

mildew (ME. < OE. *meledeaw*, honeydew): a superficial, usually whitish growth produced on organic matter or living plants by fungi. A *powdery mildew* (E. *powder* < ME. *poudre* < L. *pulver*, genit. *pulveris*, dust, powder + *-y*, abounding in, full of) caused by one of the Erysiphaceae (e.g. *Uncinula necator* on grape vine, and *Erysiphe cichoracearum* on cucurbits); a *downy mildew* (ME. *doun* < ON. *dunn*, a covering of soft fluffy feathers) by one of the Peronosporaceae (e.g. *Peronospora tabaci* on tobacco); a *dark* or *black mildew* (ME. *derk* < OE. *deorc*, hide, devoid or partially devoid of light; ME. *blak* < OE. *blaec*, black) by one of the Meliolales (e.g., *Meliola palmicola* on palms).

Powdery **mildew** (conidial state of *Blumeria graminis*) on the surface of a wheat leaf; note the erect conidiophore with an acropetal conidial chain, x 900 (*WKW*).

mischoblastiomorph (Gr. *míschos*, pedicel, peduncle + *blastós*, sprout, bud, germ, shoot + *morphé*, form): applied to a type of ascospore present in some lichens, such as *Rinodina sophodes* (Lecanorales), characterized by its thick wall and two Y-shaped cellular compartments or locules that are opposed by its base, resembling gymnasium weights.

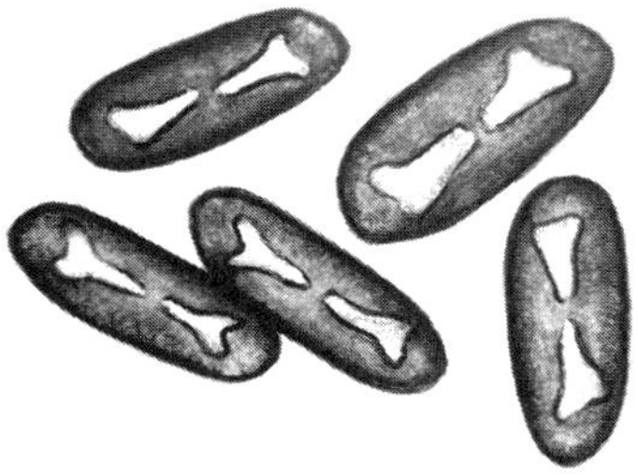

Mischoblastiomorphous ascospores of *Rinodina sophodes*, x 890.

mitochondrion, pl. **mitochondria** (NL. *mitochondrium* < Gr. *mítos*, filament + *kondríon*, dim. of *kóndros*, grain): an intracellular organelle, generally granular in form, in which is located part of the enzymatic system necessary for respiration and energy production. It is delimited by a double membrane, the external one smooth and the internal one convoluted and forming digital or laminar invaginations called **cristae**.

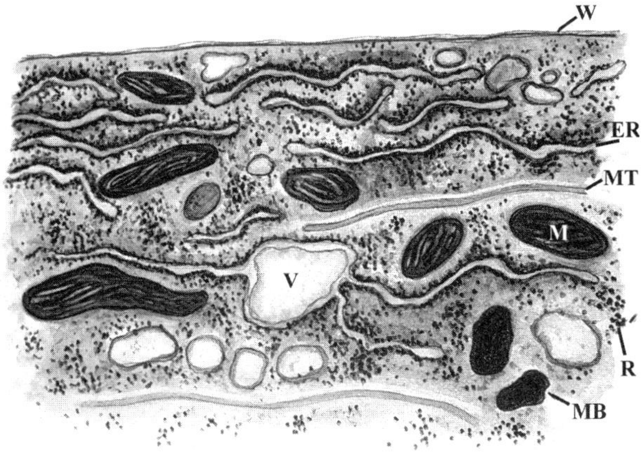

Mitochondria (M) of a young hypha of *Armillaria mellea*. Endoplasmic reticulum (ER), ribosomes (R) of the cisternae of the ER, microbodies (MB), microtubules (MT), vesicles (V), and cell wall (W) are also seen. Drawing composed from various transmission electron micrographs published by Beckett *et al.* in their *Atlas of Fungal Ultrastructure*, 1974, x 15 800.

mitosis, pl. **mitoses** (Gr. *mítos*, filament + *-osis*, condition, state): normal division of the nucleus in which the chromosomes are differentiated, then divide and separate, resulting in two daughter nuclei with the same chromosome number as the parent nucleus. Cf. **meiosis**.

mitosporangium, pl. **mitosporangia** (Gr. *mítos*, filament + *sporá*, spore + *angeîon*, vessel, receptacle): a sporangium in which the nuclei divide by mitosis. In the aquatic fungus *Allomyces* (Blastocladiales) the thin-walled mitosporangia (zoosporangia) are formed on the sporothallus, thus the nuclei are diploid. Mitosis results in additional diploid nuclei that form the zoospores (mitospores), which repeat the diploid generation when they encyst and germinate, giving rise to another sporothallus. A second type of sporangium, the meiosporangium, also is formed on the sporothallus. Cf. **meiosporangium**.

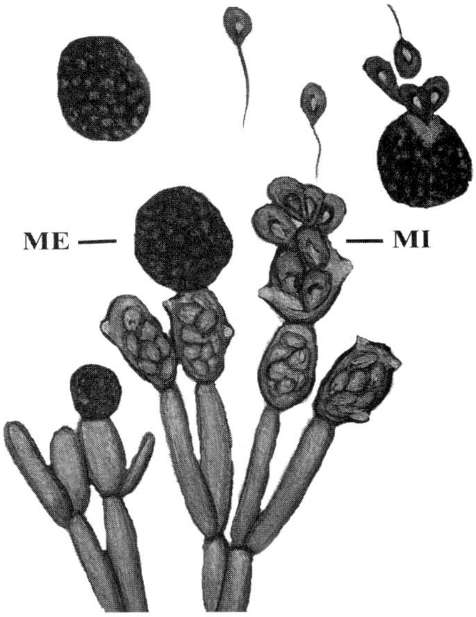

Mitosporangia (MI) and mieosporangia (ME) of the sporothallus of *Allomyces macrogynus*, x 630.

mitosporic fungi (Gr. *mítos*, filament + *sporá*, spore + suf. *-íkos* > L. *-icus* > E. *-ic*, belonging to, relating to; L. *fungi*, fungi, pl. of *fungus*, fungus): an artificial assemblage of fungi, constituiting the second largest group of fungi (ca. 2 600 gen. + 1 500 syn., 15 000 spp.); also referred to as **deuteromycetes**, **Fungi Imperfecti**, Deuteromycotina, **asexual fungi**, and **conidial fungi**. Mitosporic fungi are so-called because they reproduce asexually and/or parasexually and may produce spores by mitosis or presumed mitosis (they have not been correlated with any meiotic states). Many have been correlated with teleomorphs in the ascomycetes and basidiomycetes, so mitosporic fungi can be termed anamorph or anamorphic states of those groups. See **anaholomorph**.

mitriform (L. *mitra*, miter, mitre, head-gear + *-formis* < *forma*, shape): shaped like a miter; e.g., like the ascocarps of the gen. *Mitrula* (Helotiales).

Mitriform pileus of the apothecium of *Mitrula* sp., x 1.

molar (L. *mola*, millstone + suf. *-aris* > E. *-ar*, like, pertaining to): a structure composed of intertwined mycelium with particles of soil (like a pseudosclerotium) that *Termitomyces* (Agaricales) forms in termite nests and which gives rise to the basidiocarps that emerge from the nest. Molars vary in shape and size, depending upon the species, but generally they are more or less globose, with the surface rough and convoluted, similar to the teeth or stones that are used to grind cereal grains.

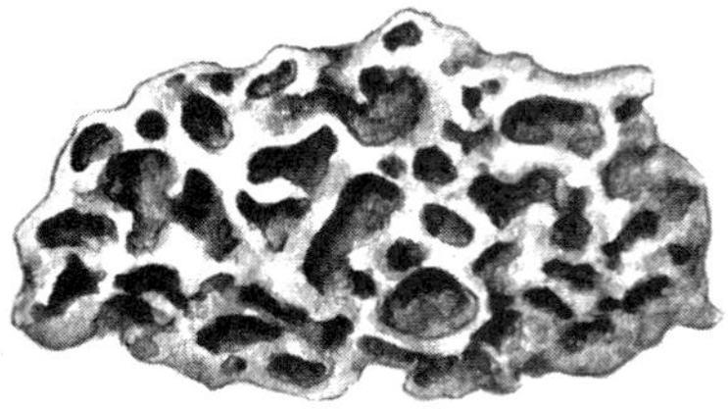

Molar of *Termitomyces microcarpus*, x 1.

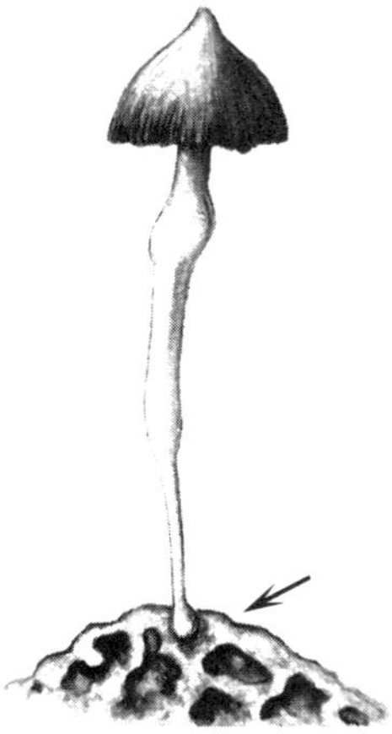

Molar of *Termitomyces striatus* giving rise to a basidiocarp, which has a basal disc of insertion of the pseudorhiza, x 1.

mold, **mould** (ME. *mowlde* < *mowled*, ptp. of *moulen*, *malwen*, to grow moldy): applied to any micromycete, principally of the Mucoraceae (Zygomycetes), mainly the gen. *Mucor* and *Rhizopus*, and the asexual, conidial fungi (agonomycetaceous, moniliaceous, dematiaceous, tuberculariaceous and stilbellaceous hyphomycetes), in particular the moniliaceous gen. *Penicillium*, *Aspergillus*, *Monilia* and *Botrytis*, which develop on organic materials in decomposition, covering them with a more or less dense turf, commonly with a downy or velvety appearance due to the mass of mycelium and spores. The following are examples of common names of molds: *blue mold of citrus fruits*: *Penicillium italicum*; *apple blue mold*: *P. expansum*; *white mold of cheese*: *P. candidum* (=*P. caseicolum*); *common mold*: *Mucor mucedo* which, accompanied by other Mucorales and certain moniliaceous asexual fungi, cause the decay of many kinds of fruit; *snow* mold: *Sclerotinia borealis* (Helotiales), sexual state of *Monilia*; *gray mold*: *Botrytis cinerea* of grapes and other plants (*Rosa*, *Dahlia*, etc.); *black mold*: *Aspergillus niger*; *black bread mold*: *Rhizopus nigricans*; *green mold*: *Penicillium digitatum* of citrus fruits, and *P. glaucum* of cheese; *pink bread mold*: *Monilia sitophila*; *spider web mold*: *Corticium koleroga* (Aphyllophorales) of coffee, which also is known as spider catcher of coffee.

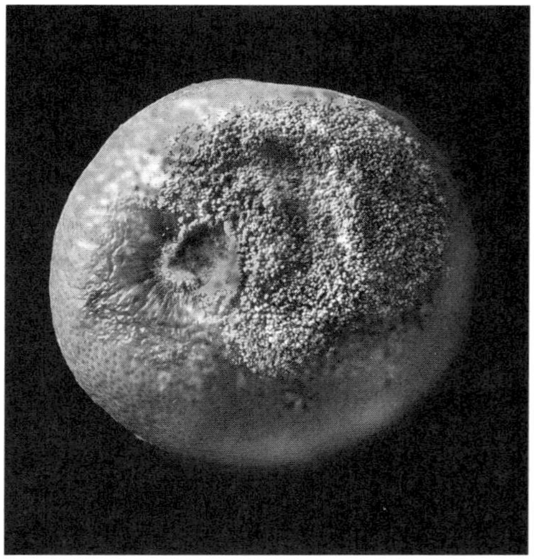

Blue citrus **mold** (*Penicillium italicum*) growing on an orange, x 0.7 (*MU*).

monandrous (Gr. *mónos*, only + *andrós*, male; here, male sexual organ + L. *-osus* > OF. *-ous*, *-eus* > E. *-ous*, having, possessing the qualities of): *Oomycota*. Applied to oospores that develop when fertilized by a single functional antheridium; e.g., this occurs in some gen., such as *Ectrogella* (Myzocytiopsidales) and *Pythiella* (Lagenaceae, *familia incertae sedis*). Cf. **polyandrous**.

monaxial (Gr. *mónos*, only, sole + L. *axialis*, axial < *axis*, axis, axle + suf. *-alis* > E. *-al*, relating to or belonging to): see **uniaxial**.

Monera (Gr. *monéres*, single): a special group, within the kingdom Protista proposed by Haeckel in 1866, that includes the bacteria. In 1956, in his system of classification of living beings into four kingdoms, Copeland recognized the bacteria and Cyanophyceae or cyanobacteria (prokaryotes) in the kingdon Monera, which was still sustained in the classification systems with five kingdoms, such as those of Whittaker (1959) and of Margulis and Schwartz (1988).

moniliform (L. *moniliformis* < gen. *Monilia* < L. *monile*, genit. *monilis*, chain + *-formis* < *forma*, shape): in a chain, composed of a series of segments more or less rounded and linked together; e.g., like the conidiophore of *Monilia* (moniliaceous asexual fungi), composed of branched chains of blastic conidia, or like the microconidia of *Fusarium moniliforme* (tuberculariaceous asexual fungi).

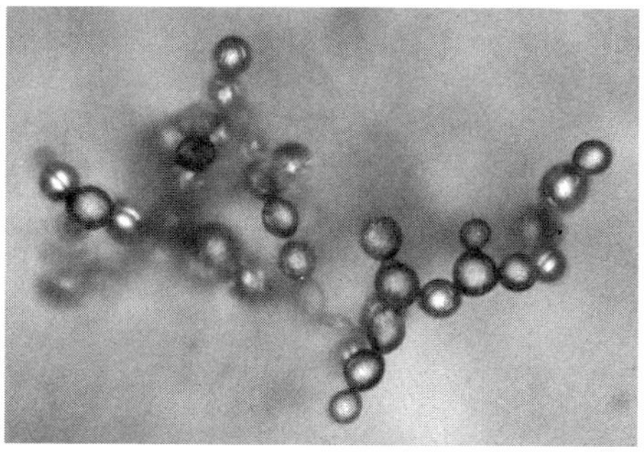

Moniliform conidial chains of *Monilia sitophila*, x 600 (*MU*).

monoblastic (Gr. *mónos*, one + *blastós*, sprout, bud, germ, shoot + suf. *-íkos* > L. *-icus* > E. *-ic*, belonging to, relating to): a type of holoblastic conidiogenous cell that produces a conidium at only one point or site, as is seen in *Acrogenospora*, *Humicola* and other dematiaceous asexual fungi. Cf. **polyblastic**.

monocarpic (Gr. *mónos*, single + *karpós*, fruit + suf. *-íkos* > L. *-icus* > E. *-ic*, belonging to, relating to): refers to infections by *Exobasidium* (Exobasidiales) that are circumscribed and annual. Cf. **polycarpic**.

monocentric (Gr. *mónos*, single + *kéntron*, center + suf. *-íkos* > L. *-icus* > E. *-ic*, belonging to, relating to): a thallus that has a single center of growth, in which is formed a reproductive organ (sporangium or resting spore), as is observed in *Chytriomyces* (Chytridiales) and other aquatic fungi. Cf. **polycentric**.

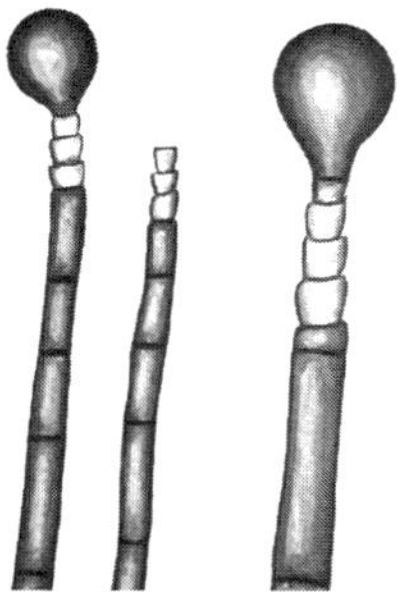

Monoblastic conidiogenous cells of *Acrogenospora sphaerocephala*. These cells also are called **monopodia**, x 320.

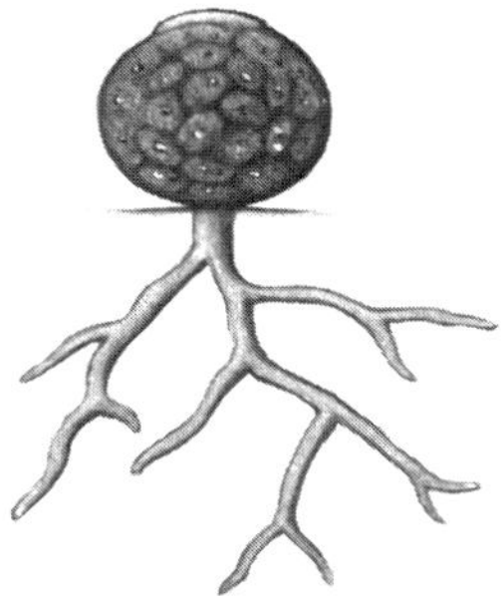

Monocentric thallus of *Chytriomyces hyalinus*, x 525.

monoclinous (Gr. *mónos*, single + *klíne*, bed; here, chamber, thallus > L. *clinare*, to lean + L. *-osus* > OF. *-ous*, *-eus* > E. *-ous*, having, possessing the qualities of): a fungus that has the antheridium on the same small branch or pedicel that bears the oogonium, as occurs in many aquatic species, e.g., *Monoblepharis polymorpha* (Monoblepharidales). In a strict sense, monoclinous is a syn. of **hermaphrodite**. Cf. **diclinous**.

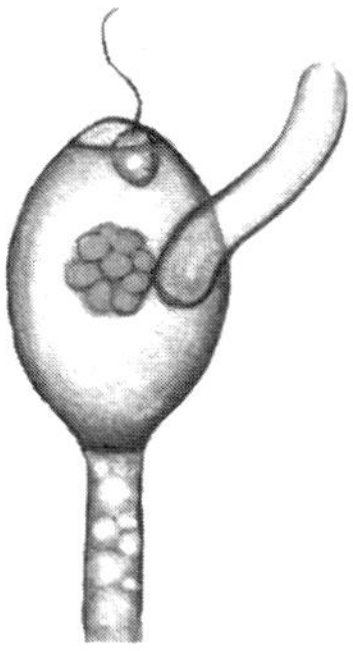

Monoclinous thallus of *Monoblepharis polymorpha*, x 440.

monoecious (Gr. *mónos*, single + *oikía*, house + L. *-osus* > OF. *-ous*, *-eus* > E. *-ous*, having, possessing the qualities of): literally, with a single house, alluding to fungi that form male and female organs on the same mycelium or thallus, because of which they are also called bisexual or hermaphroditic. Cf. **dioecious**. See **hermaphrodite**.

monokaryotic, **monocaryotic**, **monokaryontic**, **monocaryontic** (Gr. *mónos*, alone + *káryon*, nut, nucleus + *-tikós* > L. *-ticus* > E. *-tic*, relation, fitness, inclination or ability): an organism whose cells contain a single nucleus each, such as the primary mycelium of the Agaricales. Cf. **dikaryotic**.

monomitic (Gr. *mónos*, single + *mítos*, filament + suf. *-íkos* > L. *-icus* > E. *-ic*, belonging to, relating to; here, hypha): a fructification with a single type of hyphae, the generative hyphae, which have a thin wall, septa, frequent clamp connections, and are branched and fertile, since they have the capacity of producing basidia. Examples of monomitic fructifications are *Clavariadelphus* (Aphyllophorales) and all of the fleshy agarics. Cf. **dimitic** and **trimitic**.

monomorphic, **monomorphous** (Gr. *mónos*, single + *morphé*, shape + suf. *-íkos* > L. *-icus* > E. *-ic*, belonging to, relating to; or + L. *-osus* > OF. *-ous*, *-eus* > E. *-ous*, having, possessing the qualities of): *Oomycota.* Species of aquatic fungi in whose life cycle there is only a single type of zoospore (called primary), as happens, e.g., in the gen. *Pythiopsis* (Saprolegniales). If a species produces only secondary zoospores, it also is monomorphic. Cf. **dimorphic**.

mononematous (Gr. *mónos*, single + *nêma*, genit. *nêmatos*, thread, filament + L. *-osus* > OF. *-ous*, *-eus* > E. *-ous*, having, abounding in, possessing the qualities of): a conidiophore composed of a single hypha or filament, as in the majority of the gen. of hyphomycetous asexual fungi. Cf. **synnematous**.

monophyletic (Gr. *mónos*, single + *phyletikós* < *phyletés*, one of the same tribe, race, stock < *phylé*, tribe + *-tikós* > L. *-ticus* > E. *-tic*, relation, fitness): derived from a single common ancestral form; of a single line of descent. Cf. **paraphyletic** and **polyphyletic**.

monopileate (Gr. *mónos*, one + L. *pileatus*, capped < *pileus*, pileus, cap, hat + suf. *-atus* > E. *-ate*, provided with or likeness): a type of fruiting body development in certain Gasteromycetes, such as *Podaxis* (Hymenogastrales), in which a single stipe is formed that supports the gleba, arranged as in the embryonic button of an agaric, with laminar hymeniiferous plates. The term monopileate is also applied to the fruiting body of *Dictyophora* and *Mutinus* (among other Phallales), whose gleba forms a cylinder of hymenial tissue that is differentiated around a central sterile stipe. Cf. **multipileate**.

monoplanetic (Gr. *mónos*, single + *planétes*, wandering + suf. *-íkos* > L. *-icus* > E. *-ic*, belonging to, relating to): *Oomycota.* Species in whose life cycle there is a single type of zoospore (called primary) with a single period of swimming; as happens, e.g., in the gen. *Pythiopsis* (Saprolegniales). However, this is not necessarily so; *Aphanomyces* (Saprolegniales) has monomorphic, monoplanetic, secondary zoospores. Cf. **diplanetic** and **polyplanetic**.

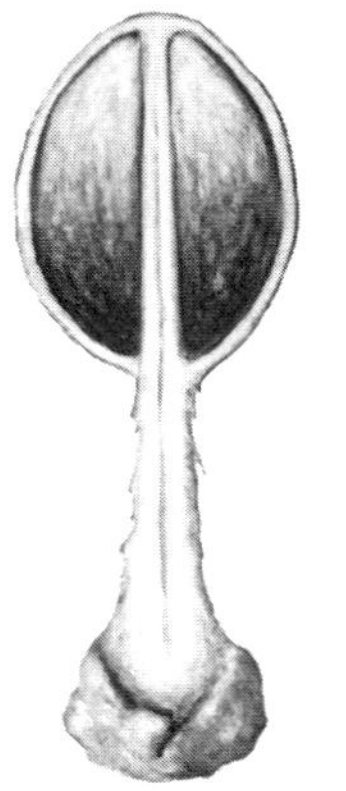

Monopileate basidiocarp of *Podaxis pistillaris*, seen in longitudinal section, x 1.5.

monopodial (Gr. *mónos*, single + NL. *podium* < Gr. *pódion*, dim. of *poús*, *podós*, foot; here, support + L. suf. *-alis* > E. *-al*, relating to or belonging to): adj. of **monopodium**.

monopodium, pl. **monopodia** (NL. *monopodium* < Gr. *mónos*, single + NL. *podium* < Gr. *pódion*, dim. of *poús*, *podós*, foot; here, support): a type of branching in which a hypha continues growing at the apex without change in direction, while lateral branches of the same type are produced below in acropetal succession. Among the fungi, there are many types of sporangiophores and conidiophores that have monopodial branching, e.g., the conidiophore of *Acrogenospora sphaerocephala* (=*Acremoniella atra*=*Monopodium*), a dematiaceous asexual fungus. Cf. **sympodium**.

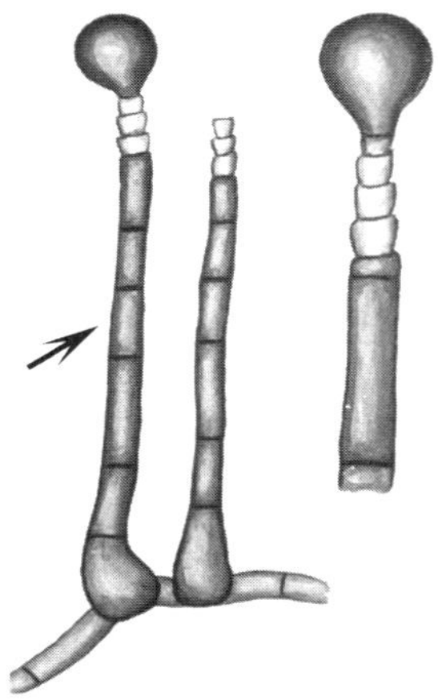

Monopodium - conidiophore of *Acrogenospora sphaerocephala*, x 300.

monopolar (Gr. *mónos*, single + *polaris*, polar < *polus*, pole + suf. *-aris* > E. *-ar*, like, pertaining to): **1**. A type of germination in which a spore forms only one germ tube; characteristic of the majority of fungi. **2**. A type of budding in yeasts in which a single bud forms at one end of the mother cell, as in *Malassezia* (=*Pityrosporum*), of the cryptococcaceous asexual yeasts. Cf. **bipolar** and **multipolar**.

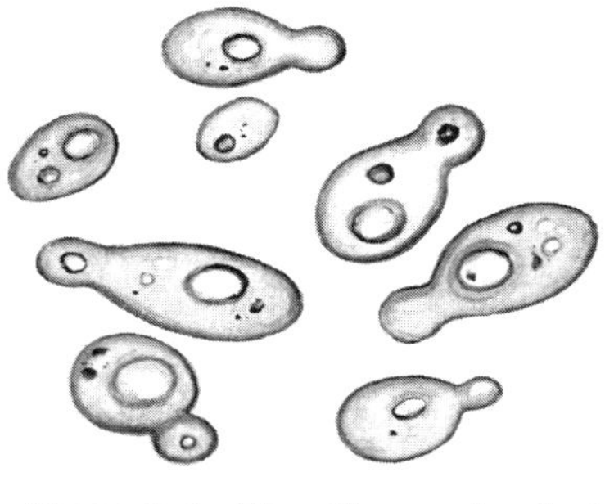

Monopolar budding of the somatic cells of *Malassezia* (=*Pityrosporum*) sp., x 1 900.

monostic, **monostichate**, **monostichous** (Gr. *mónos*, single + *stíchos*, row, line + suf. *-íkos* > L. *-icus* > E. *-ic*, belonging to, relating to; or + suf. *-atus* > E. *-ate*, provided with or likeness; or + L. *-osus* > OF. *-ous*, *-eus* > E. *-ous*, having, possessing the qualities of): in a single row, like the ascospores within the asci of *Sordaria* (Sordariales). Syn. of **uniserial**. Cf. **distichous**.

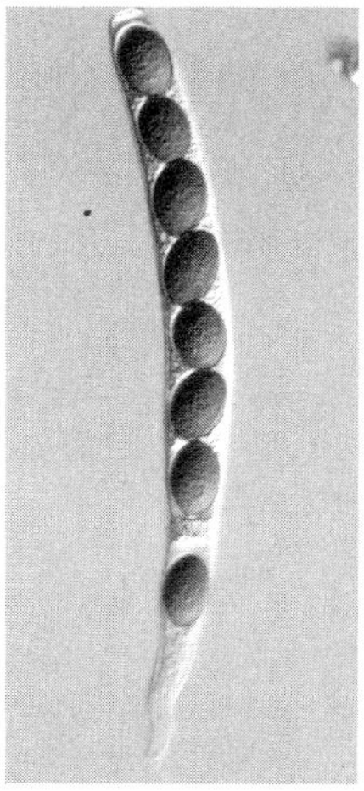

Monostichous ascus of *Sordaria fimicola*, x 270 (*RTH*).

monotretic (Gr. *mónos*, single + *tretós*, perforate + suf. *-íkos* > L. *-icus* > E. *-ic*, belonging to, relating to): *Conidiogenesis*. A type of enteroblastic conidiogenous cell that produces conidia through a single canal or perforation in the outer wall; the protruding internal wall of the conidiogenous cell becomes the outer wall of the young conidium. Such conidia are generally solitary, but sometimes form in acropetal chains. For example, the conidiogenous cells of *Corynespora cassiicola* (dematiaceous asexual fungi) are monotretic. Cf. **polytretic**.

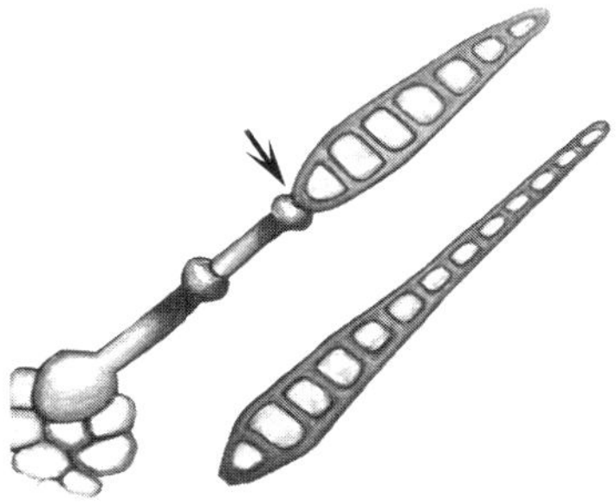

Monotretic conidiogenous cell of *Corynespora cassiicola*, x 310.

monotrophic (Gr. *mónos*, single + *trophós*, that which nourishes, serves as food + suf. *-íkos* > L. *-icus* > E. *-ic*, belonging to, relating to): an organism that is only able to feed on a single substrate, such as a nutritive medium (thiobacteria) and many autoecious parasitic rusts (Uredinales). Cf. **polytrophic**. See **omnivorous**.

monotropoid mycorrhiza, pl. **mycorrhizae** (< plant gen. *Monotropa* + L. suf. *-oide* < Gr. *oeídes*, similar to; Gr. *mýkes*, fungus + *rhíza*, root): another term for the so-called **ericoid mycorrhizae**, which are named for their mycorrhizal plant associates, such as *Vaccinium* (blueberries), and several angiosperm parasites, such as *Monotropa*, all members of the heath order, Ericales. The parasitic plants actually are connected to their hosts by a bridge of ericoid ectendomycorrhizal fungi, which include species of inoperculate *Thelebolus* (=*Pezizella*) and hypogeous *Elaphomyces* (Pezizales).

monotypic (Gr. *mónos*, single + L. *typicus* < Gr. *typikós*, conforming to a type + suf. *-íkos* > L. *-icus* > E. *-ic*, belonging to, relating to): having a single representative; i.e., a monotypic gen. has a single species, a monotypic family a single genus, etc.

monoverticillate (Gr. *mónos*, single + NL. *verticillatus* < *verticillus*, verticil + suf. *-atus* > E. *-ate*, provided with or likeness): generally applied to a conidiophore that strictly or predominantly has a single point of branching between the pedicel and the chain of conidia, i.e., that bears a single verticil of phialides, which are formed directly from the pedicel; very rarely the phialides are formed on metulae in some species. Some examples of monoverticillate conidiophores are those of *Penicillium citreonigrum*, *P. glabrum* and *P. restrictum* (moniliaceous asexual fungi), which certain authors consider within the subgenus *Aspergilloides*. Cf. **biverticillate**, **triverticillate**, **quadriverticillate** and **polyverticillate**.

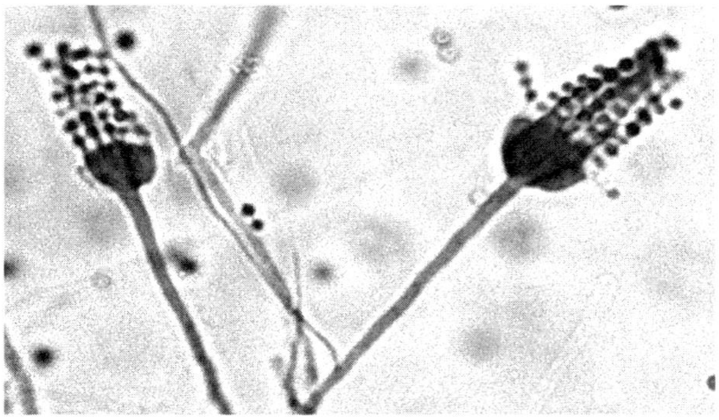

Monoverticillate conidiophores of *Penicillium citreonigrum*, x 585 (*CB*).

monoxenic, **monoxenous** (Gr. *mónos*, one, single + *xénos*, stranger + suf. *-íkos* > L. *-icus* > E. *-ic*, belonging to, relating to; or + L. *-osus* > OF. *-ous*, *-eus* > E. *-ous*, having, possessing the qualities of): **1.** The laboratory growth of two species of organisms together; it can involve, e.g., a ciliate plus another organism, such as a bacterium, alga, yeast, or another ciliate. The second organism can be an undesired contaminant, or it can be present to serve as food for the first one. Cf. **agnotobiotic**, **axenic** and **gnotobiotic**. **2.** With a single host, completing the life cycle on a single host; e.g., like certain plant parasitic fungi (the rusts, classified in the order Uredinales). In this sense it is syn. of **autoecious**. Cf. **heteroxenous** or **heteroecious**.

morbose (L. *morbosus*, sickly < *morbus*, sickness + *-osus*, full of, augmented, prone to > ME. *-ose*): illness; causing or concerned with illness.

morel (F. *morille* < OHG. *morhila*, morel): the edible ascoma of *Morchella*, especially *M. esculenta* (Pezizales).

moriform (L. *moriformis* < *morus*, mulberry + *-formis* < *forma*, shape): shaped like a mulberry, like the clusters of spores present in some fungi, e.g., *Oedocephalum* (moniliaceous asexual fungi).

morphogenesis, pl. **morphogeneses** (Gr. *morphé*, shape + *génesis*, origin): the ensemble of phenomena relative to the creation of the shape and structure of an organism, as well as to the evolution of morphological characters. The study of morphogenesis of the ascocarp, for example, has served to better define the affinities or the differences among the groups of ascomycetes.

mother (ME. *mother*, *moder* < OE. *modor* < L. *mater*, genit. *matris* < Gr. *máter*, mother): in mycology, and in biology in general, it is often employed with the meaning of cause, root, or origin, source of something, e.g., the ascus mother cell, mother of vinegar, etc.

mother cell (ME. *mother*, *moder* < OE. *modor* < L. *mater*, genit. *matris* < Gr. *máter*, mother; L. *cella*, storeroom > ML. *cella* > OE. *cell*, cell, compartment): generally applied to the cells of yeasts that by budding give rise to one or several buds that constitute daughter cells. In the ascomycetes the diploid cell of the hook is called the mother cell of the ascus, which by meiosis forms the haploid nuclei of the future ascospores, as one sees, e.g., in *Nectria haematococca* (Hypocreales).

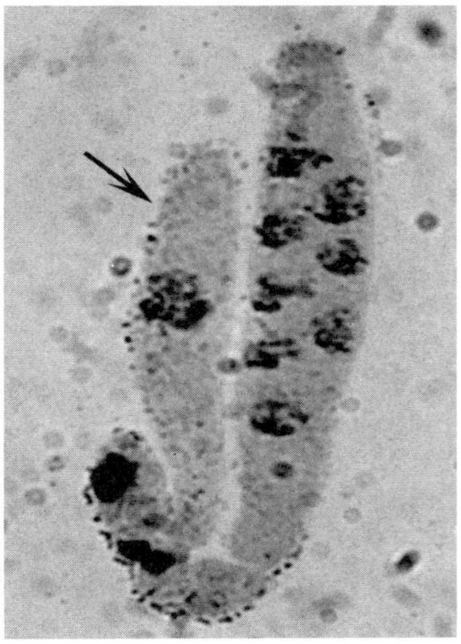

Mother cell of ascus, with a single diploid nucleus, and young ascus, with eight haploid nuclei, of *Nectria haematococca*, x 1 840 (*RTH*).

mucedinoid, **mucedinous** (L. *mucedus*, *mucidus*, moldy < *muceere*, to mold < *mucus*, mucus + *-oide* < Gr. *-oeídes*, similar to; or + L. *-osus* > OF. *-ous*, *-eus* > E. *-ous*, having, abounding in, possessing the qualities of): applied to fungi with a loose, filamentous, white, cottony mycelium, common in many molds, e.g., *Mucor mucedo* (Mucorales), *Botrytis* and *Verticillium* (moniliaceous asexual fungi).

mucid (L. *mucedus*, *mucidus*, moldy, musty < *mucus*, mucus): moldy; mucilaginous or mucous when wet.

mucilaginous (LL. *mucilaginosus* < *mucilagin*, mucilage + L. *-osus* > OF. *-ous*, *-eus* > E. *-ous*, having, possessing the qualities of): sticky, viscous or mucous when wet, due to the chemical composition (music acid, pentoses and hexoses); in water mucilage becomes viscous or swells to form a gelatinous pseudosolution. These materials are present in many fungi, e.g., in members of the phylum Myxomycota, which are known as the mucilaginous or slime molds. Colonies of *Aureobasidium pullulans* (dematiaceous asexual fungi) also are mucilaginous or slimy.

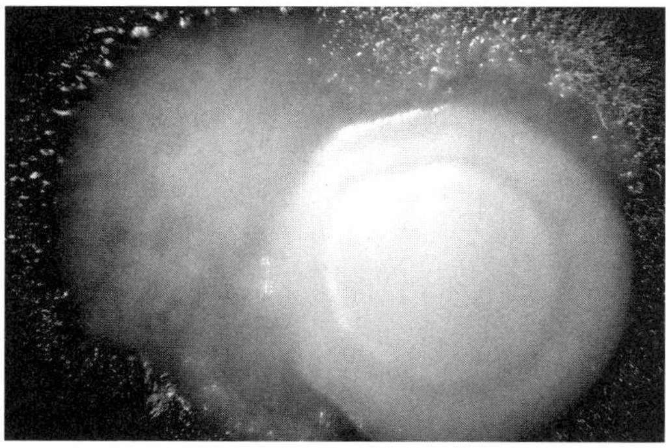

Mucilaginous colony of *Aureobasidium pullulans* on agar, x 3 (*MU*).

mucoid, **mucous** (L. *mucus*, mucus + *-oide* < Gr. *-oeídes*, similar to; or + L. *-osus* > OF. *-ous*, *-eus* > E. *-ous*, having, abounding in, possessing the qualities of): having the appearance and consistency of mucilage.

mucormycosis, pl. **mucormycoses** (< gen. *Mucor* < L. *mucor*, mold < *muceere*, to be moldly + Gr. *-osis*, state or condition): an opportunistic infection of the rhinocerebral tissues, from which it may disseminate to the brain and meninges, therefore causing the death of the patient. Other types of mucormycosis vary from the subcutaneous (localized) to the gastrointestinal (systemic). The most common causal agents are the widely distributed *Mucor circinelloides*, *M. pusillus*, *M. racemosus*, *Rhizopus oryzae*, *Rh. arrhizus*, and *Absidia corymbifera* (Mucorales). These types of infections are most common in immunosuppressed patients (due to chemotherapy with corticosteroids and antimicrobial substances).

mucronate (L. *mucronatus* < *mucro*, genit. *mucronis*, point + suf. *-atus* > E. *-ate*, provided with or likeness): ending in an abrupt point, often sharp; as in the hyphopodia of *Meliola caseariae* (Meliolales), the pileus of the basidiocarps of *Phaeocollybia*, *Weraroa* and *Galeropsis*, or the cheilocystidia of *Mycena haematopus* and *Panaeolus cyanescens* (Agaricales).

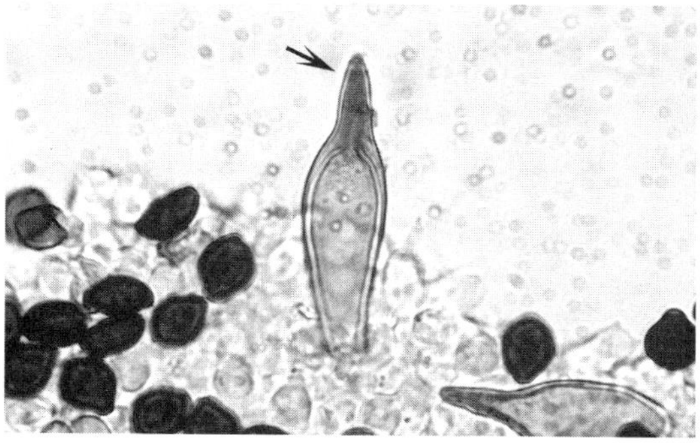

Mucronate cystidium, and basidiospores, of the hymenium of *Panaeolus cyanescens*, x 700 (*EAA*).

multiaxial (L. *multus*, many + *axialis*, axial, relative to the axis < *axis*, axis, axle + L. suf. *-alis* > E. *-al*, relating to or belonging to): refers to a structure that has various axes, more or less radially branched and approximate, so that they come together into a general compact mass. The fruiting body of some ascomycetes, such as *Wynnea* (Pezizales), and of some basidiomycetes, such as *Clavicorona pyxidata*, *Ramaria flava* and *R. stricta* (Aphyllophorales) are multiaxial since their stipes are divided into many branches or axes. Cf. **uniaxial**.

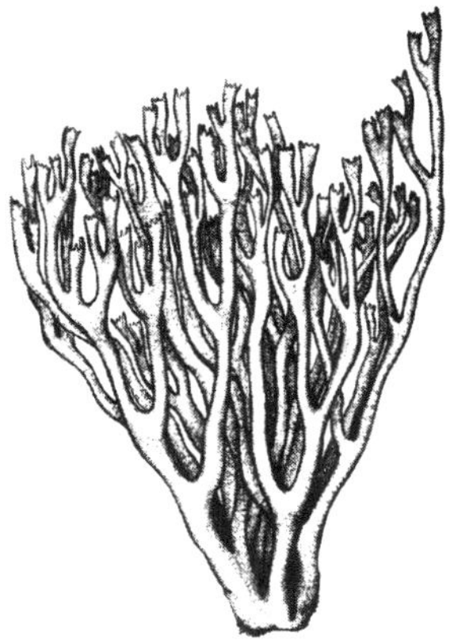

Multiaxial basidiocarp of *Clavicorona pyxidata*, x 0.5.

multilateral (L. *multilateralis* < *multi* < *multus*, many + *later*, *latus*, side, flank + suf. *-alis* > E. *-al*, relating to or belonging to): see **multipolar**.

multilocular, **multiloculate** (L. *multilocularis* < *multi* < *multus*, many + *loculus*, dim. of *locus*, cavity, locule + suf. *-aris* > E. *-ar*, like, pertaining to; or + suf. *-atus* > E. *-ate*, provided with or likeness): having several or many cavities; e.g., the ascostroma of *Elsinoë* and *Myriangium* (Myriangiales) has numerous ascigerous locules. Also called **plurilocular**. Cf. **unilocular**.

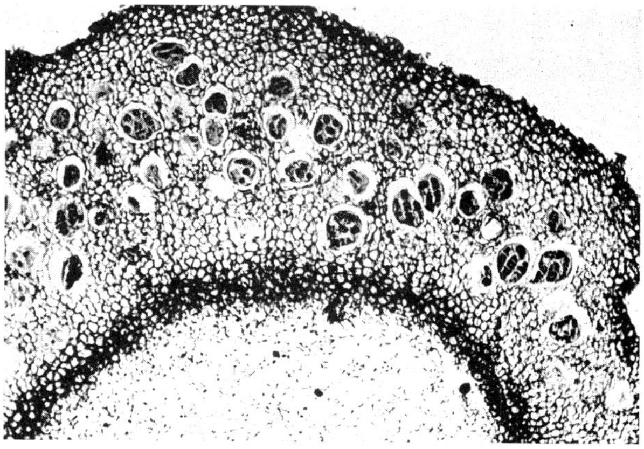

Multilocular ascostroma of *Myriangium duriaei*, seen in longitudinal section, x 80 (*RTH*).

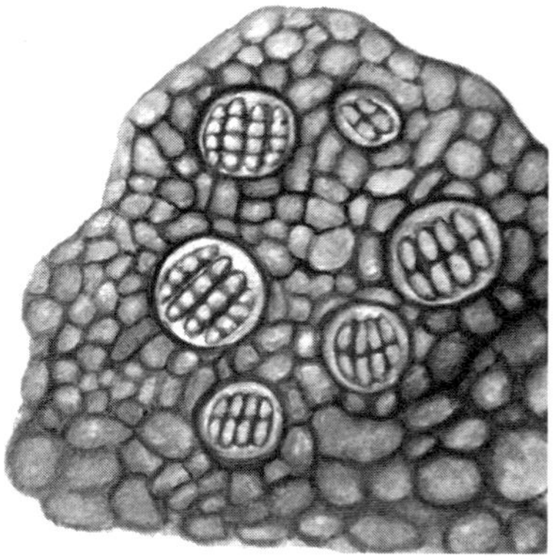

Multilocular ascostroma of *Elsinoë ampelina*, x 400.

multipileate (L. *multus*, many + *pileatus*, capped < *pileus*, pileus, cap, hat, helmet, pileus + suf. *-atus* > E. *-ate*, provided with or likeness): a type of fruiting body development characterized by the formation of three or more individual secondary growths composed of sterile tissue, that originate from a central stipe. Each of the secondary stipes or pedicels later generates its own hymenial tissues, whether a slender layer or as a concentrated glebal mass. Characteristic of *Clathrus* and *Aseroë*, among other Phallales of the family Clathraceae. Cf. **monopileate**.

Multipileate type of fruiting body development, characteristic of *Clathrus*. Diagram of longitudinal section, x 1.

multipileate development (L. *multus*, many, multiple + *pileatus*, capped < *pileus*, cap, hat, helmet, pileus + suf. *-atus* > E. *-ate*, provided with or likeness; F. *développement* < MF. *développer* < OF. *desveloper* + L. *mentum* > OF. *-ment* > ME. *-ment*, action, process): *Agaricales*. Development in which many individual stalks with small pilei, each with its own hymenial tissue, form from a single stalk.

multipolar (L. *multus*, many, multiple + *polaris*, polar < *polus*, pole + suf. *-aris* > E. *-ar*, like, pertaining to): syn. of **multilateral**. **1**. The formation of multiple buds on the surface of the mother cell in certain yeasts, such as those of the gen. *Candida*, *Torulopsis* (cryptococcaceous asexual yeasts) and *Hansenula* (Saccharomycetales). **2**. The formation of several germ tubes during germination by ascospores, as in *Leptosphaerulina* (Dothideales), or by conidia, as in *Alternaria* (dematiaceous asexual fungi). Cf. **bipolar** and **monopolar**.

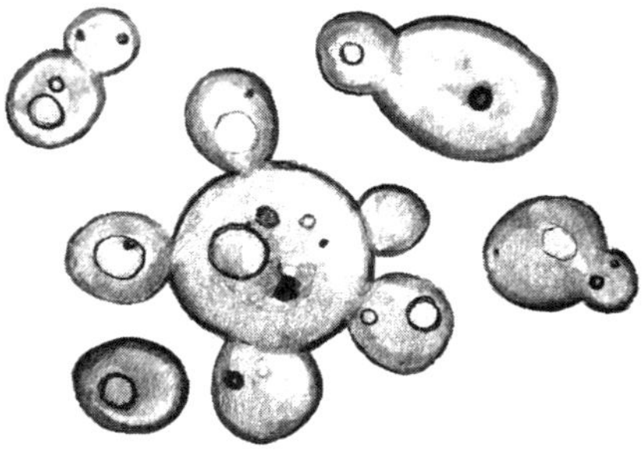

Multipolar budding of the somatic cells of *Torulopsis taboadae*, x 1 500.

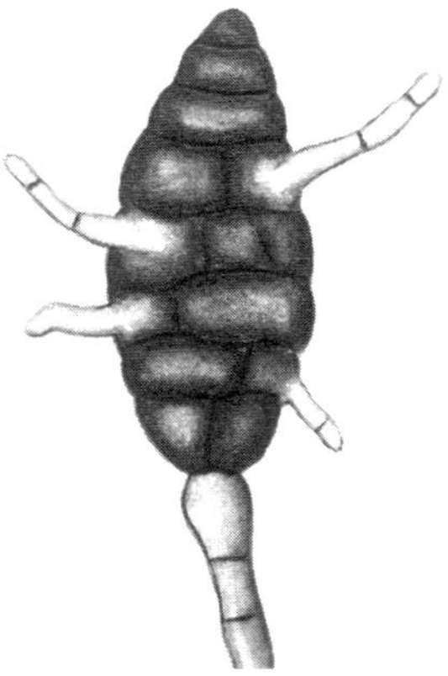

Multipolar germination of a conidium of *Alternaria* sp., x 1 200.

muricate, **muricated** (L. *muricatus*, sharp-pointed < *murex*, genit. *muricis*, sharp-pointed stone + suf. *-atus* > E. *-ate*, provided with or likeness): covered with short, hard, sharp points; e.g., like the cystidia of *Inocybe* and *Melanoleuca melaleuca* (Agaricales), which are incrusted with crystals of oxalate.

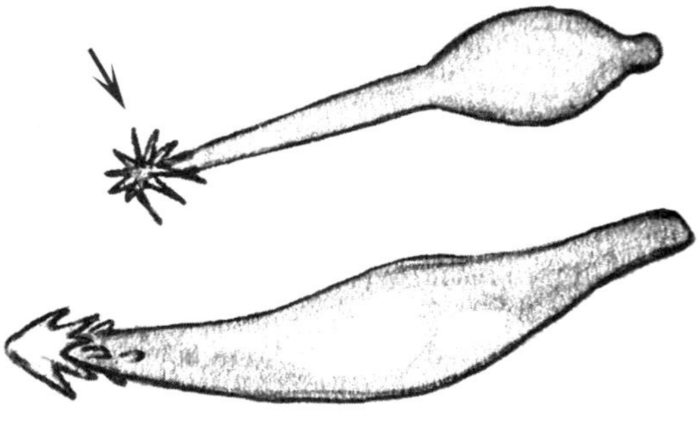

Muricate cystidia from the hymenium of *Melanoleuca* sp., x 1 360.

muriform (L. *muriformis* < *murus*, wall + *-formis* < *forma*, shape): with both transverse and longitudinal septa, like the partitions of a brick wall. For example, the conidia of *Epicoccum purpurascens* (tuberculariaceous asexual fungi) are muriform. See **dictyospore**.

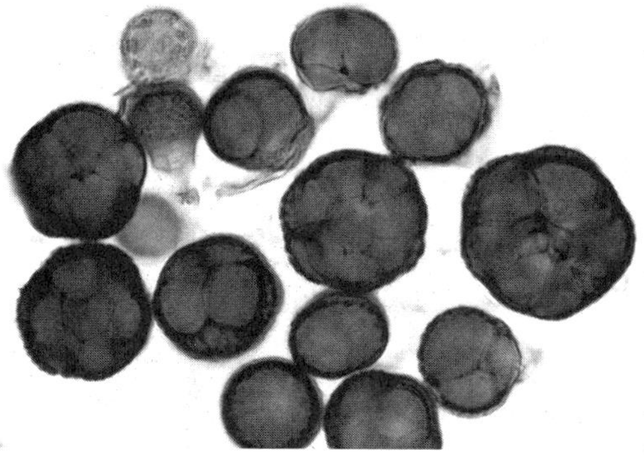

Muriform conidia of *Epicoccum purpurascens*, x 800 (*CB*).

murogenous (L. *murus*, wall + Gr. *génos*, descent, generation + L. *-osus* > OF. *-ous*, *-eus* > E. *-ous*, having, possessing the qualities of): *Conidiogenesis*. A type of blastic conidium development in which the tip of a hypha enlarges and is then delimited by a septum. The width of the attachment is equal to that of the hypha.

muscicole, **muscicolous** (L. *muscus*, moss + *-cola*, inhabitant; or + L. *-osus* > OF. *-ous*, *-eus* > E. *-ous*, possessing the qualities of): living upon, or among, mosses. See **bryophilous**.

mushroom (ME. *musseroum* < MF. *mousseron* < LL. *mussirion*, mushroom): *Basidiomycetes*. **1**. An agaric (or a bolete) fleshy fruiting body that consists typically of a stem bearing a cap, especially one that is edible. **2**. Any agaric; a macrofungus with a distinctive fruiting body (either epigeous or hypogeous), large enough to be seen with the naked eye and to be picked by hand. Cf. **toadstool**.

mushroom stones (ME. *musseroum* < MF. *mousseron* < LL. *mussirion*, mushroom; ME. < OE. *stan*, stone): mushroom-like stone sculptures, with different effigies (including animal and anthropomorphic), that have been associated with ancient Mayan religious cults involving hallucinogenic fungi. See **mycolatry**.

musiform (< gen. *Musa*, the banana + L. suf. *-formis* < *forma*, form): banana-shaped, like the basidiospores of *Exobasidium* (Exobasidiales).

muticate, **muticous** (L. *muticus*, cut, mutilated, without hair or beard + L. suf. *-atus* > E. *-ate*, provided with or likeness; or + L. *-osus* > OF. *-ous*, *-eus* > E. *-ous*, having, possessing the qualities of): applied to a structure lacking a point; blunt, like the conidia of *Drechslera*, among other dematiaceous asexual fungi.

Mushroom stones of the Mayan Culture. From left to right: a stone sculpture with a feline effigy (Early Preclassic, Kaminaljuyú, Guatemala); an antropomorphic stone mushroom (Middle Preclassic, Guatemala), and a stone sculpture with the effigy of an animal (Early and Middle Preclassic, Kaminaljuyú, Guatemala), x 0.2.

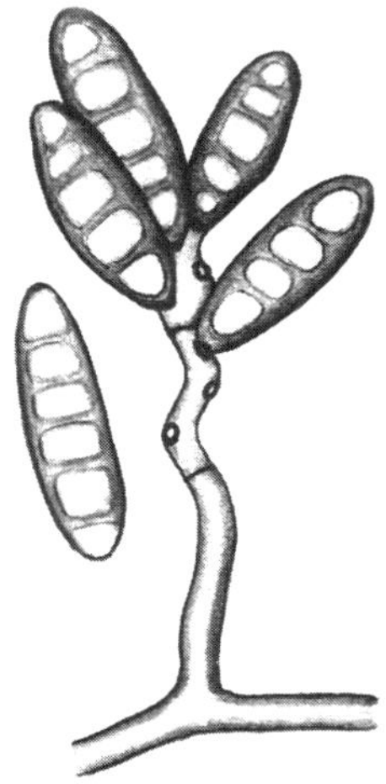

Muticous conidia on a conidiophore of the *Drechslera* state of *Cochliobolus bicolor*, x 700.

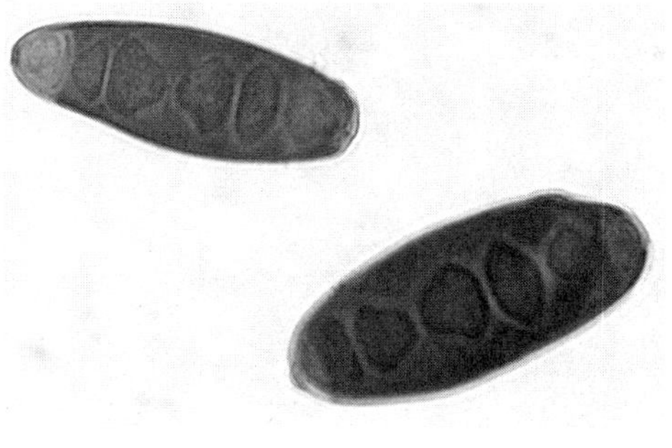

Muticous conidia of *Drechslera* sp., x 1 360 (*CB*).

mutualism (ME. < MF. *mutuel* < L. *mutuus*, mutual, reciprocal + Gr. *-ismós* > L. *-ismus* > E. *-ism*, state, phase, tendency, action): a type of association between two living beings that gives mutual benefits; it represents a type of mutualistic symbiosis, such as exists in the lichens. Cf. **antagonism**, **commensalism**, **helotism**, **metabiosis**, **parabiosis**, **parasymbiosis**, **symbiosis** and **synergy**.

mutualistic symbiosis, pl. **symbioses** (ME. < MF. *mutuel* < L. *mutuus*, mutual, reciprocal + NL. *-istic*, pertaining to as agent; Gr. *symbíosis*, life in common): see **symbiosis**.

myc-, **mycet-**, **-mycete**, **myceto-** (L. < Gr. *mýkes*, genit. *mýketos*, fungus): fungus, fungal; combining prefixes and suffixes used in the formation of terms, such as mycangium, mycetism, ascomycetes, basidiomycetes, mycetophagous, etc.

mycangium, pl. **mycangia** (Gr. *mýkes*, fungus + *angeîon*, vessel, receptacle): the small pouches, located in various parts of the exoskeleton of certain insects (ambrosia beetles and wood wasps), that contain the fungal spores (or yeast-like cells) with which the particular insect has established a mutualistic relationship. In the wood boring beetles (*Xyleborus*), the symbiotic fungi contained in the mycangia (and which, depending upon the species of beetle, can be located in the base of each elytron or in the oral cavity) include *Graphium* spp. (stilbellaceous asexual fungi) and *Acremonium* (=*Cephalosporium*) spp. (moniliaceous asexual fungi). The fungus *Stereum* (Aphyllophorales) is disseminated by wood wasps.

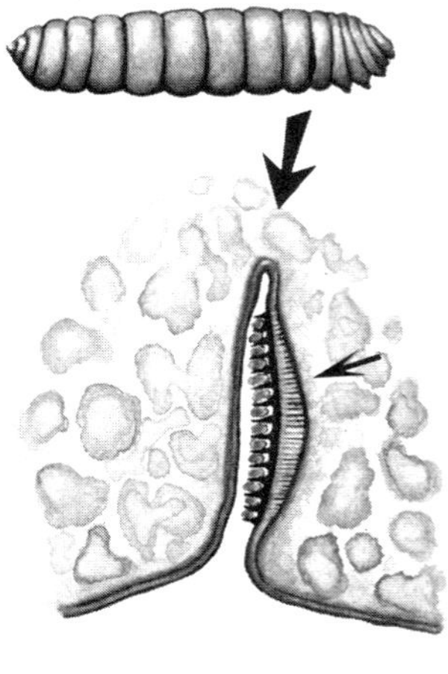

Mycangium of *Stereum* sp. between the two first abdominal segments of a larva of wood wasp, x 28.

mycelial (NL. *mycelium* < Gr. *mýkes*, fungus + L. suf. *-elis*, pertaining to + L. suf. *-alis* > E. *-al*, relating to or belonging to): pertaining to or belonging to the mycelium.

mycelial strand (NL. *mycelium* < Gr. *mýkes*, fungus + L. suf. *-elis*, pertaining to + L. dim. suf. *-ium* + L. suf. *-alis* > E. *-al*, relating to or belonging to; ME. *strond*, fibers or filaments): a compact group of parallel hyphae that form a bundle or packet, with apical growth analogous to that of vascular plant roots; mycelial strands are present in the majority of the basidiomycetes, as well as in other groups. When these mycelial strands have a more complex structure, with a hard cortex, in which the hyphae lose their identity, they are known as a **rhizomorphs**.

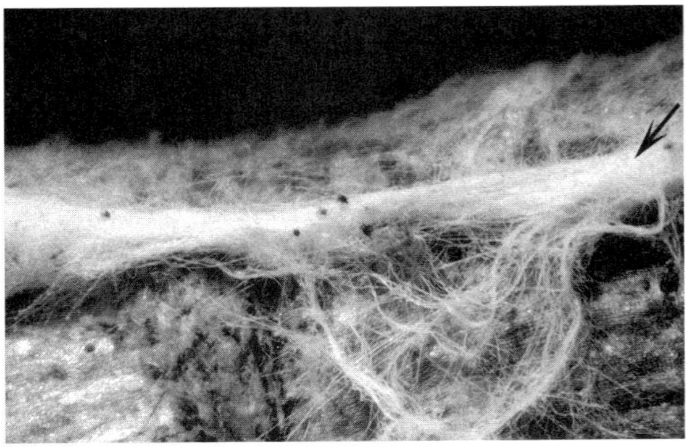

Mycelial strand of a basidiomycete, x 2 (*SA*).

mycelioid (NL. *mycelium* < Gr. *mýkes*, fungus + L. suf. *-elis*, pertaining to + L. suf. *-oide* < Gr. *-oeídes*, similar to): having the characteristics of a mycelium, similar to a mycelium; e.g., like the perithecial appendages of *Blumeria* (=*Erysiphe*) *graminis* (Erysiphales).

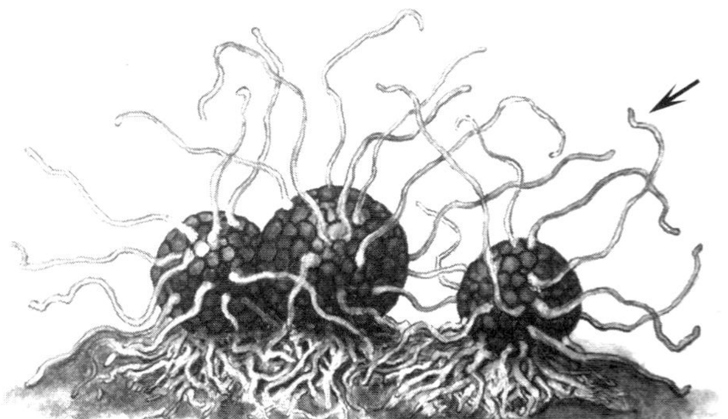

Mycelioid perithecial appendages of *Erysiphe* sp., x 190.

mycelium, pl. **mycelia** (NL. *mycelium* < Gr. *mýkes*, fungus + L. suf. *-elis*, pertaining to + L. dim. suf. *-ium*): the entire mass of hyphae that constitutes the vegetative body or thallus of a fungus.

mycenoid (< gen. *Mycena* < Gr. *mýkes*, fungus + *kenós*, hollow + L. suf. *-oide* < Gr. *-oeídes*, similar to):

Agaricales. A type of fruiting body (13 principal types are recognized) with the gills adhered to the stipe, which is cartilaginous, central and lacks a ring and volva, like that of *Mycena* and some species of *Marasmius*, *Nolanea*, *Psilocybe* and *Panaeolus*, among other gen. The mycenoid type is similar to the collybioid type but differs in its conical to campanulate pileus with a margin initially decurved.

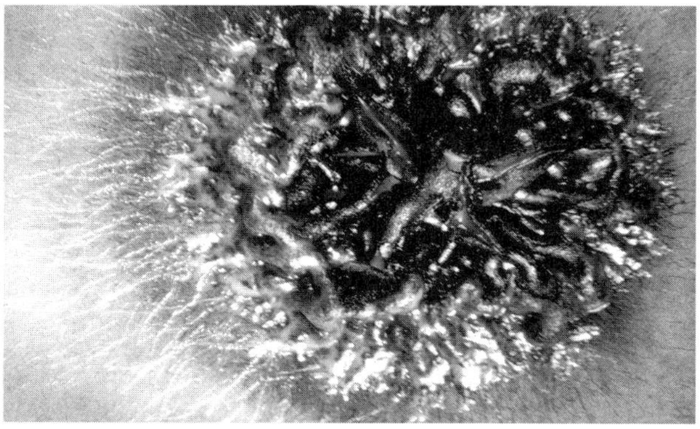

Mycelium of a colony of *Aureobasidium pullulans* on agar, x 3 (*MU*).

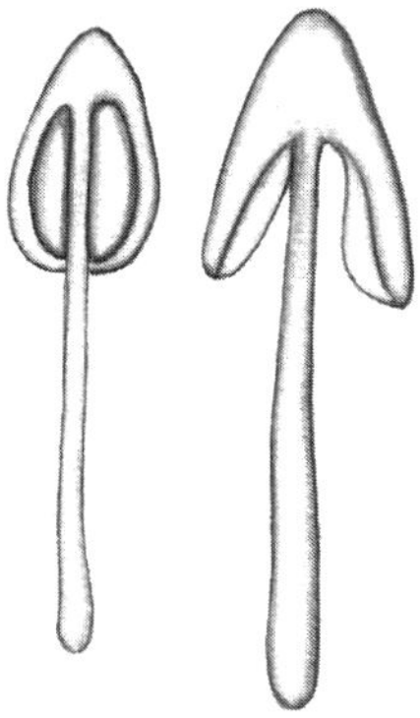

Mycenoid type of basidiocarp, represented by *Mycena* sp. Diagram of longitudinal sections, x 1.

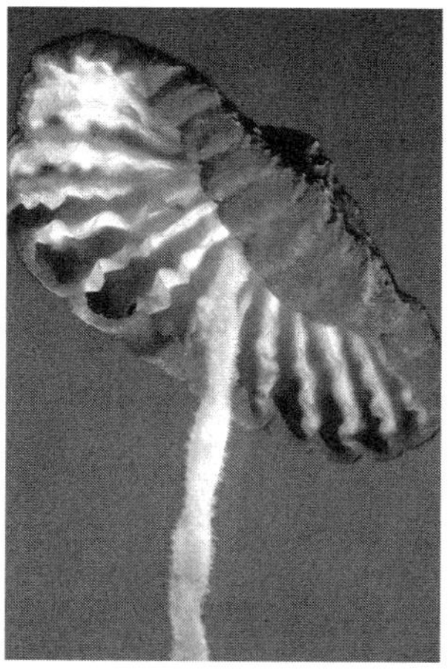

Mycenoid basidiocarp of *Marasmius siccus*, x 3 (*MU*).

-mycetes (Gr. *mýkes*, genit. *mýketos*, fungus > L. genit. *mycetem* > *mycetes*): ending of taxonomic classes, e.g., Zygomycetes, Archiascomycetes, etc.

-mycetidae (Gr. *mýkes*, genit. *mýketos*, fungus + L. *-idae*, fem. pl. adjectival suf.): ending of taxonomic subclasses, i.e., Pyrenomycetidae, Discomycetidae, etc.

mycetism, mycetismus, pl. **mycetismi** (L. *mycetismus* < Gr. *mýkes*, genit. *mýketos*, fungus + *-ismós* > L. *-ismus* > E. *-ism*, state, phase, tendency, action): intoxication provoked by the ingestion of certain fungi. Recognized mycetisms include phalloidian, paraphalloidian, muscarinic, gastrointestinal, inconstant or conditional, and cerebral, which are induced respectively by *Amanita phalloides*, *Cortinarius orellanus*, *Amanita muscaria*, *Russula emetica*, *Coprinus atramentarius* and *Psilocybe caerulescens* (Agaricales), to cite only some of the species involved. Cf. **mycotoxicosis**.

mycetization (Gr. *mýkes*, genit. *mýketos*, fungus + *-ationem*, action, state or condition, or result > E. suf. *-ation*): the action and effect of becoming fungal; used for the infection of the roots of plants with hyphae of mycorrhizogenous fungi, which invade to form mycorrhizae.

myceto- (Gr. *mýkes*, genit. *mýketos*, fungus): pertaining to fungi. Equivalent to **myco-**.

mycetocyte (Gr. *mýkes*, genit. *mýketos*, fungus + *kýtos*, cavity, cell): refers to cells that contain symbiotic yeasts, such as are found in the blind sacks of the digestive tracts of larvae of *Lasioderma*, in which are found *Torulopsis ernobii*, *T. buchnerii* and *Symbiotaphrina buchnerii* (cryptococcaceous asexual yeasts).

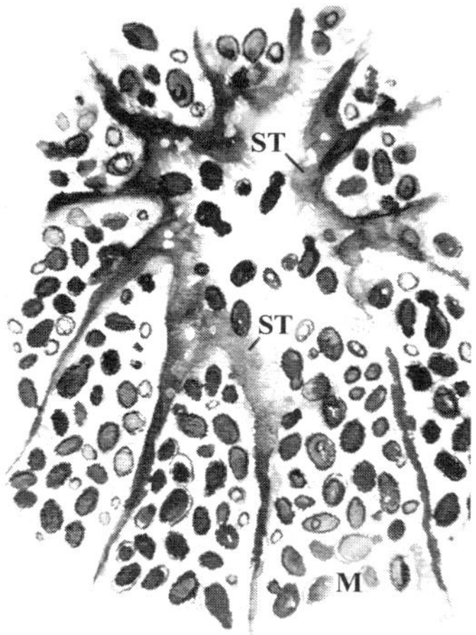

Mycetocytes (endosymbiotic yeasts) of *Symbiotaphrina buchnerii*. Cross section through a blind sac of *Lasioderma* larva. Note broad mycetocytes (M) and narrower sterile (ST) columnar cells, the latter with denser protoplasm. In the apical parts of the mycetocytes the cell wall is lysed and its yeasts cells are expelled into the lumen. Drawing based on a transmission electron micrograph by Jurzitza, 1977, x 1 250.

mycetology (Gr. *mýkes*, genit. *mýketos*, fungus + *lógos*, study, treatise + *-y*, E. suf. of concrete nouns): see **mycology**.

mycetoma (Gr. *mýkes*, genit. *mýketos*, fungus + *-óma*, suf. indicating a pathological state, usually in reference to a tumor): literally, a fungal tumor; a type of lesion caused by actinomycetes, which are really actinobacteria (actinomycetoma), or by molds (eumycotic mycetoma), in the subcutaneous tissues of man and the higher animals; the organisms generally involved are conidial fungi (e.g., *Madurella* and *Phialophora*, of the dematiaceous asexual fungi) and ascomycetes (such as *Pseudallescheria*, of the Microascales).

mycetophagy (Gr. *mýkes*, genit. *mýketos*, fungus + *phágos*, to eat + *-y*, E. suf. of concrete nouns): see **mycophagy**.

mycetotheca (Gr. *mýkes*, genit. *mýketos*, fungus + *thêke*, case, reservoir): see **mycotheca**.

Mycetozoa (Gr. *mýkes*, genit. *mýketos*, fungus + *zõon*, animal): *Zoology*. An order of the class Sarcodina, phylum Sarcomastigophora, subkingdom Protozoa, kingdom Animalia. In mycology, equivalent to the phylum Myxomycota of the kingdom Fungi.

mycetozoan (Gr. *mýkes*, genit. *mýketos*, fungus + *zoárion*, animalculum + L. suf. *-anus* > E. *-an*, belonging to): a name given to a myxomycete (class Myxomycetes, phylum Myxomycota, kingdom Fungi) by biologists who classify them with the protists (subphylum Mycetozoa, phylum Gymnomyxa, kingdom Protista). The mycetozoans combine, within the same life cycle, characteristics of animals (amoeboid trophic state, heterotrophic, holozoic) and of fungi (sporiferous apparatus).

mycin (Gr. *mýkes*, fungus + E. suf. *-in* < L. *-in*, of or belonging to; neutral chemical compound, enzyme or antibiotic): the recommended ending for names coined for antibiotics produced by actinomycetes (actinomycin, streptomycin, etc.).

myco- (Gr. *mýkes*, fungus): a prefix employed in the formation of compound terms, such as mycology (the study of fungi), mycobiota (mycological biota), mycobacteriosis (a complex disease entity, formed by the concomitant parasitism, although not necessarily symbiotic, of fungi and bacteria), mycorrhizogenous (a fungus capable of forming mycorrhizae), and mycotic (belonging or relative to mycosis), among others. **Myceto-** is technically the more correct, but less popular, term.

mycobacteriosis, pl. **mycobacterioses** (Gr. *mýkes*, fungus + L. *bacteri-* < *bacterium*, bacterium + Gr. *-osis*, condition, state): an infection caused by microorganisms of the gen. *Mycobacterium* (Actinomycetales) or by the concomitant parasitism, not necessarily symbiotic, of bacteria and fungi.

mycobiont (Gr. *mýkes*, fungus + *bíos*, life + *óntos*, genit. of *ón*, a being): the fungal component of a lichen; together with the algal component (phycobiont) it constitutes the lichen symbiosis.

mycobiota (Gr. *mýkes*, fungus + NL. *biota*, the living beings of a particular region or period < *biotós*, life, manner of living < *bíos*, life): fungal life; the ensemble of fungi indigenous to a place or habitat. To avoid the implication that the fungi are considered as plants, it is preferable to use the term mycobiota (or mycetobiota) rather than **mycoflora** (or mycetoflora).

mycocecidium, pl. **mycocecidia** (Gr. *mýkes*, fungus + *kekís*, *kekîdos*, gall + L. dim. suf. *-ium*): a plant gall originated by the parasitism of a fungus, such as the dark warts or tumors produced in potato by *Synchytrium endobioticum* (Chytridiales), the galls formed by *Protomyces* (Taphrinales) on the leaves of wild carrot, or the galls of cedar caused by the rusts *Gymnosporangium clavipes* and *G. juniperi-virginianae* (Uredinales).

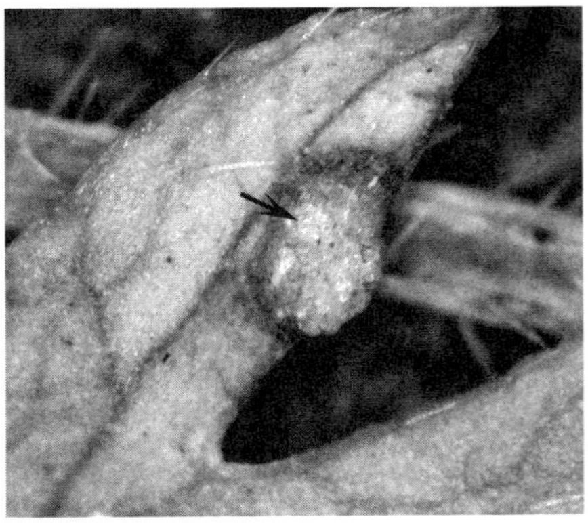

Mycocecidium formed by *Protomyces* sp. on a leaf of wild carrot, x 30 (*RTH*).

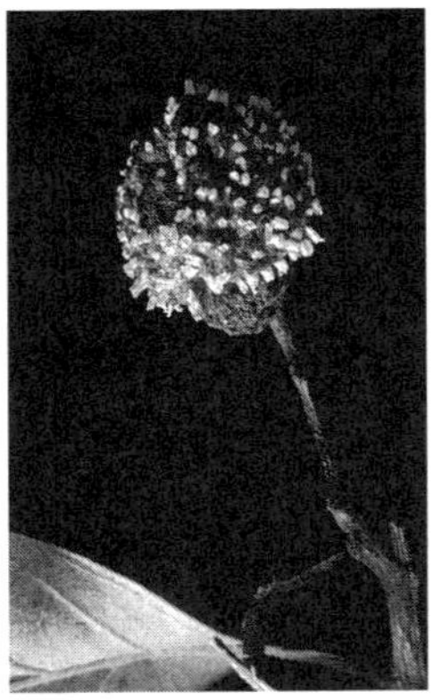

Mycocecidium (a close up) formed by *Gymnosporangium clavipes* on a fruit of hawthorn (*Crataegus*). Note the aecial peridia protruding from the infected fruit, x 1 (*MU*).

Mycocecidium (a close up) formed by *Gymnosporangium clavipes* on a fruit of hawthorn (*Crataegus*). Note the aecial peridia protruding from the infected fruit, x 17 (*MU*).

mycoclena, pl. **mycoclenae** (NL. *mycochlaena* < Gr. *mýkes*, fungus + *chlaína*, blanket, cover): the fungal investment of a root in an ectotrophic mycorrhiza.

mycoderm (Gr. *mýkes*, fungus + *dérma*, skin, membrane): **1**. The membranous scum or film that is formed on the surface of liquids under fermentation, a process in which yeasts participate (the old gen. *Mycoderma* is now a syn. of the gen. *Saccharomyces*, of the Saccharomycetales). **2**. A fungal growth on the skin, which forms a crust, as in some dermatophytoses. **3**. The mother of vinegar, which is a thick scum composed of a cellulosic matrix in which are embedded the bacteria (*Acetobacter aceti* subsp. *xylinum*) that generate the matrix, together with some yeasts (*Saccharomycodes*, of the Saccharomycetales, *Candida*, of the cryptococcaceous asexual yeasts, and others). See **tea fungus**.

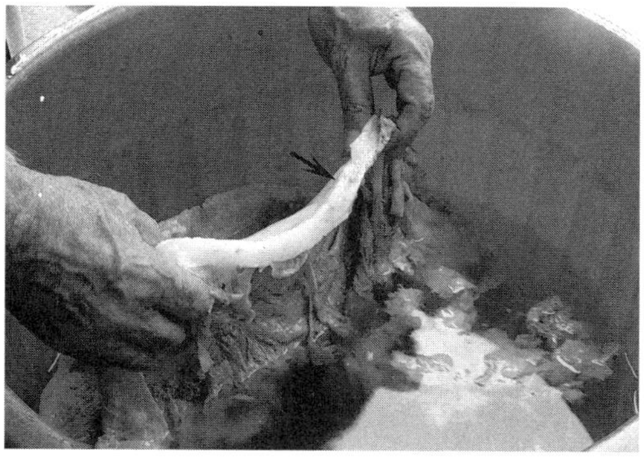

Mycoderm - thick scum (mother of vinegar) formed by the growth of bacteria and yeasts living symbiotically in a fermentable sugary liquid, x 0.1 (*MU*).

mycodomatium, pl. **mycodomatia** (Gr. *mýkes*, fungus + *dóma*, genit. genit. *dómatos*, house, roof + L. dim. suf. *-ium*): a dwelling in which are lodged fungi; a term used for certain nodules on the roots of *Alnus* and some legumes, and apparently related to mycorrhizae.

mycoecology (Gr. *mýkes*, fungus + *oîkos*, house, dwelling + *lógos*, study, treatise + *-y*, E. suf. of concrete nouns): ecology of fungi, the science that studies the interrelationships of fungi and their environment.

mycogenous (Gr. *mýkes*, fungus + *génos*, origin < *gennáo*, to engender, produce + L. *-osus* > OF. *-ous*, *-eus* > E. *-ous*, possessing the qualities of): of fungal origin, or which lives on fungi; e.g., like the basidiocarps of the species of *Nyctalis* (Agaricales) that develop on the basidiocarps of ***Russula*** (Agaricales), or like the perithecial stromata of *Cordyceps capitata* (Hypocreales), which develop on the ascocarps of *Elaphomyces* (Pezizales). See **mycoparasite**.

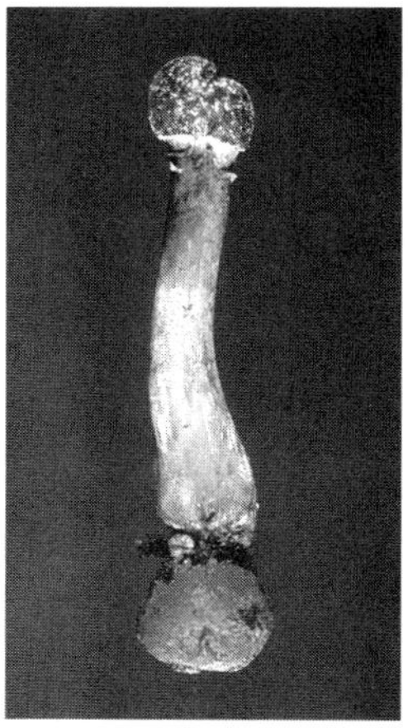

Mycogenous perithecial stroma of *Cordyceps capitata* arising from an ascocarp of *Elaphomyces* sp., x 0.8 (RTH).

Mycogenous perithecial stroma of *Cordyceps capitata* arising from an ascocarp of *Elaphomyces granulatus*. A non-infected ascocarp of *E. granulatus* is also shown, x 0.9 (*MU*).

mycogeography (Gr. *mýkes*, fungus + *geo*, soil < *gê*, earth + *graphía*, description < *grápho*, drawing + *-y*, E. suf. of concrete nouns): the branch of mycology that studies the geographical distribution and organization of species and populations of fungi.

mycography (Gr. *mýkes*, fungus + *graphía*, description < *grápho*, drawing + *-y*, E. suf. of concrete nouns): a mycological work based principally on illustrations.

mycoherbicide (Gr. *mýkes*, fungus + L. *herba*, grass + E. *-cide* < L. *caedere*, to kill): applied to a pathogenic fungus that is utilized in the biological control of certain plants considered as undesirable weeds, from an anthropocentric point of view. For example, various spp. of fungi belonging to the gen. *Cercospora*, *Alternaria* (dematiaceous asexual fungi) and *Cylindrocarpon* (moniliaceous asexual fungi) are utilized in the biological control of water hyacinth (*Eichornia crassipes*).

mycoinsecticide (Gr. *mýkes*, fungus + L. *insectum* < neuter of *insectus*, ptp. of *insecare*, to cut into < *in*, in + *secare*, to cut; insect + E. *-cide* < L. *caedere*, to kill): a fungus used to control insects. E.g., *Verticillium lecanii* (moniliaceous asexual fungi) is used for the biological control of white fly (*Trialeurodes vaporariorum*), a pest of many vegetable crops.

mycolatry (Gr. *mýkes*, fungus + *latreía* < *látreyma*, religious cult, worship < *látris*, servant, minister + *-y*, E. suf. of concrete nouns): the worship of fungi, especially hallucinogenic ones, an ancient cult of several ethnic groups of Mexico and Central America. See **mushroom stones**.

mycolith (Gr. *mýkes*, fungus + *líthos*, stone): a mass of sand grains bound together by mycelial filaments of *Lithomyces nidulans* (syn. of *Melanospora*, of the Melanosporales), that are found in the vineyards of Palestine.

mycology (NL. *mycologia* < Gr. *mýkes*, fungus + *lógos*, study, treatise + *-y*, E. suf. of concrete nouns): the branch of science that treats of the study of fungi. Actually, the word **mycology** is an improperly coined term. Etymologically, the correct word is **mycetology**, inasmuch as the combining form of *mýkes* is **myceto-**, in accordance with the principles of Greek grammar.

mycolysis, pl. **mycolyses** (Gr. *mýkes*, fungus + *lýsis*, dissolution, disintegration): the lysis of fungal cells, particularly due to the action of a virus.

mycoparasite (Gr. *mýkes*, fungus + *parásitos* < *pará*, together + *sîtos*, bread, food): a fungus parasitic on another fungus; there exist biotrophic mycoparasites, which cause little damage, or no apparent damage in the host, such as *Piptocephalis* and *Syncephalis* (Zoopagales), which parasitize *Mucor* and *Cokeromyces* (Mucorales), and necrotrophic mycoparasites, which kill part or all of the host tissues, or at least render them sterile, as occurs with *Cordyceps capitata* (Hypocreales) on *Elaphomyces* (Pezizales), or the species of *Nyctalis* (Agaricales), which live on other Agaricales, such as *Russula*. See **mycogenous**.

mycophagist (Gr. *mýkes*, fungus + *phágesis*, to eat): one who eats fungi, who takes advantage of them as food; especially a gourmet with a predilection for fungi. Cf. **mycophobist**.

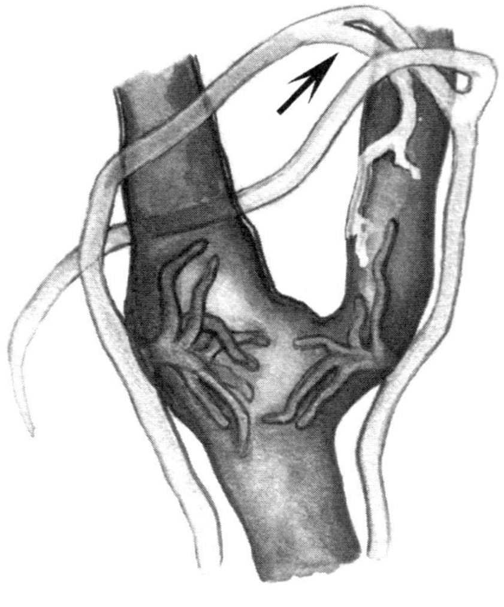

Mycoparasitic hyphae and haustoria of *Piptocephalis virginiana* in a hypha of *Mucor* sp., x 1 500.

mycophagy (Gr. *mýkes*, fungus + *phágos*, to eat + *-y*, E. suf. of concrete nouns): **1**. To eat fungi, to take advantage of fungi as food; applied to wild animals or ethnic groups that consume fungi. Also called mycetophagy. **2**. The phenomenon in which the cells of a mycorrhizal or mycorrhizogenous fungus are consumed by the host root, degenerating the symbiosis into parasitism. Cf. **mycophobia**.

mycophilic (Gr. *mýkes*, fungus + *phílos*, have an affinity for + suf. *-íkos* > L. *-icus* > E. *-ic*, belonging to, relating to): fond of fungi in general, or of mushrooms in particular, not only as food.

mycophobia (Gr. *mýkes*, fungus + *phóbos*, to repel): aversion to fungi. Cf. **mycophagy**.

mycophobist (Gr. *mýkes*, fungus + *phóbos*, to repel): one who has an aversion to fungi, who rejects them as food. Cf. **mycophagist**.

mycophycobiosis, pl. **mycophycobioses** (Gr. *mýkes*, fungus + Gr. *phýkos*, alga + *bíosis*, the act of living < *bíos*, life + suf. *-osis*, state, condition): the obligate symbiotic association between a systemic marine fungus and a macroalga, whose predominant habit is that of the fungus, as is present, e.g., in the association of the ascomycete *Mycosphaerella ascophylli* (Dothideales) and the brown alga *Ascophyllum nodosum* (Phaeophyceae), and between *Blodgettia bornetii* (asexual or mitosporic fungi) and the green alga *Cladophora fuliginosa* (Chlorophyceae).

mycoprotein (Gr. *mýkes*, fungus + F. *protéine* < Gr. *prõteios*, primary < *prõtos*, first): commercially processed fungal mycelium, e.g. "Quorn", the trade name under which mycoprotein manufactured from non-pathogenic strains of *Fusarium graminearum* (tuberculariaceous asexual fungi) is marketed.

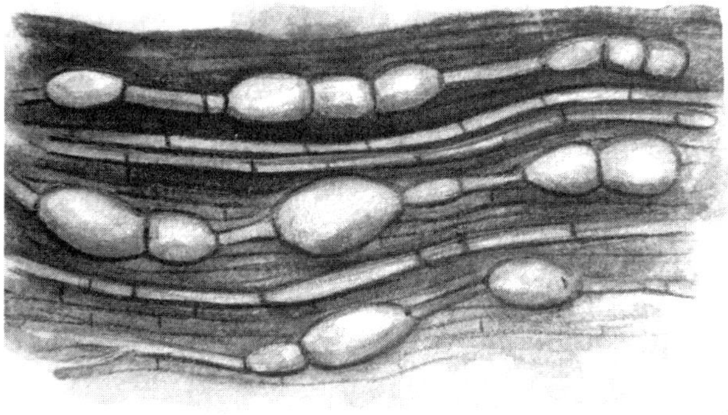

Mycophycobiosis formed by *Blodgettia bornetii* and *Cladophora fuliginosa*; the hyphae and chlamydospores of the fungus are embedded between the outer and inner layers of the algal cell wall, x 430.

mycorrhiza, pl. **mycorrhizae** (Gr. *mýkes*, fungus + *rhíza*, root): the symbiotic association between the hyphae of certain fungi and the roots of vascular plants, e.g., *Boletus* (Agaricales) + oak. See **endomycorrhiza** and **ectomycorrhiza**.

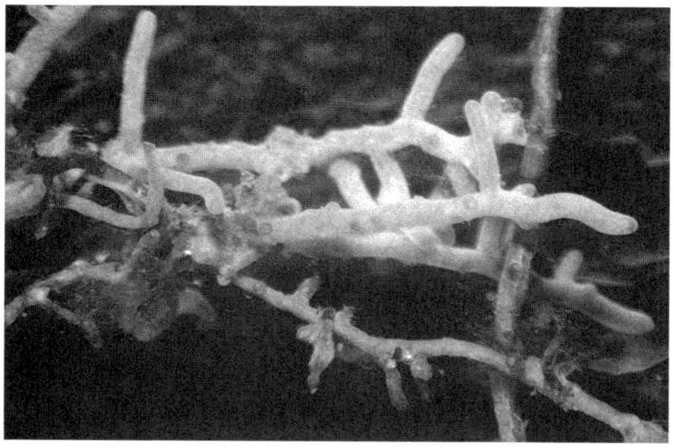

Mycorrhizae resulting from the symbiosis between oak roots and hyphae of *Boletus* sp., x 4 (*MU*).

mycorrhizal (Gr. *mýkes*, fungus + *rhíza*, root + L. suf. *-alis* > E. *-al*, relating to or belonging to): having a mycorhiza in the roots; i.e., refers to a plant whose roots have been invaded by a **mycorrhizogenous** fungus.

mycorrhizogenous (Gr. *mýkes*, fungus + *rhíza*, root + *génos*, origin < *gennáo*, to engender, produce + L. *-osus* > OF. *-ous*, *-eus* > E. *-ous*, possessing the qualities of): a fungus that is capable of forming a mycorrhiza, which participates in its formation.

mycosclerid (Gr. *mýkes*, fungus + *sklerós*, hard): an irregular, thick-walled skeletal hypha, occasionally present in the peridial layers or around the ostiole of the fruiting body of *Vascellum* (Lycoperdales).

mycosis, pl. **mycoses** (Gr. *mýkes*, fungus + suf. *-osis*, condition, state of something; i.e., the state of the organism subject to the harmful effects of the parasitic fungi): refers to a broad range of infections, of humans and higher animals, caused by diverse species of fungi. The names of mycoses may derive from the generic names of the causal agents (e.g., aspergillosis, histoplasmosis and blastomycosis derive from *Aspergillus*, *Histoplasma* and *Blastomyces*, respectively), from the name of the affected organ (for example otomycosis, the mycosis of the ear canal), and from the group of fungi causing the infection (e.g., dermatophytoses, which are caused by the dermatophytes). Mycoses are usually categorized based on the degree of anatomical depth of the infection, i.e., as superficial, cutaneous, subcutaneous or systemic.

mycosporin (Gr. *mýkes*, fungus + *sporá*, spore + L. suf. *-inus* > E. *-ine*, of or pertaining to): a fungal secondary metabolite that absorbs UV light at near 310 nm. Mycosporines are produced by several fungal species at the time of sporulation. The production of these compounds is light dependent and always associated with fungal sporulation. Although substantial information on the distribution of these metabolites is available its function remains to be determined. Sporulation factor, UV protective and self-inhibitor of germination are some of the functions attributed to them. E.g., in *Colletotrichum graminicola* and *C. lindemuthianum* (melanconiaceous asexual fungi) this inhibitor has mycosporine-alanine in the mucilage and also produces a second inhibitor, a volatile acetate. In the rust fungi, the inhibitor present in many species is cis-3,4dimethoxycinnamate. The role of self-inhibitors in imposing dormancy is frequently apparent when high concentrations of spores fail to germinate, while the frequency of germination increases if the spore concentration is decreased. See **self-inhibitor**.

mycostatic (Gr. *mýkes*, fungus + *statós*, standing, placed + suf. *-íkos* > L. *-icus* > E. *-ic*, belonging to, relating to): a compound that stops or inhibits the growth of fungi. Also called **fungistatic**, a hybrid term made of a Latin component (*fungus*) and a Greek component (*statós*). These hybrid terms are common in biological lexicology, but their coining is not recommended. It is preferable to use terms formed by components from the same linguistic nature. Cf. **fungicide**.

mycostatin (Gr. *mýkes*, fungus + *statós*, standing, placed + NL. *-in*, suf. used in chemistry to denote an activator or compound): the trade name for nystatin, an antibiotic from the actinomycete *Streptomyces noursei*, widely used against *Candida albicans* (cryptococcaceous asexual yeasts) infections of man.

-mycota (Gr. *mýkes*, genit. *mýketos*, fungus + NL. *-ota*, a suf. meaning having, used as a terminal ending for a group of organisms): ending of taxonomic phyla, or division in plant taxonomy, e.g., Chytridiomycota, Ascomycota, etc.

mycothallus, pl. **mycothalli** (Gr. *mýkes*, fungus + *thallós*, thallus): the symbiotic association of a fungus with a hepatic (liverwort) or fern gametophyte.

mycotheca (Gr. *mýkes*, fungus + *thêke*, case): a collection of fungi prepared, labelled, and mounted on sheets organized in a book or folder for distribution.

mycotic (Gr. *mýkes*, fungus + *-tikós* > L. *-ticus* > E. *-tic*, relation, fitness, inclination or ability): caused by fungi; belonging to or relative to mycoses.

-mycotina (Gr. *mýkes*, genit. *mýketos*, fungus + L. *-ina*, a suf. denoting likeness): ending of taxonomic subphyla, or subdivisions in plant taxonomy, e.g., Deuteromycotina, Basidiomycotina, etc.

mycotoxicosis, pl. **mycotoxicoses** (Gr. *mýkes*, fungus + *toxikós*, poison + suf. *-osis*, condition, state): poisoning induced by the ingestion of foods contaminated by toxins (**mycotoxins**) produced by various species of molds, principally of the gen. *Aspergillus*, *Penicillium* (moniliaceous asexual fungi) and *Fusarium* (tuberculariaceous asexual fungi), which respectively produce, among many others, aflatoxin and ochratoxin, citrinin and patulin, and zearalenone and nivalenol. Cf. **mycetism**.

mycotoxin (Gr. *mýkes*, fungus + E. *toxin* < *toxic* < LL. *toxicus* < L. *toxicum* < Gr. *toxikón*, poison, especially of a dart or arrow + NL. *-in*, suf. used in chemistry to denote an activator or compound): a toxin produced by a fungus, especially one affecting humans or animals, in which it may cause a **mycotoxicosis**.

mycotroph (Gr. *mýkes*, fungus + *trépho*, to nourish, feed): **1**. Refers to plants that form mycorrhizae and need the mycorrhizal fungus in order to obtain its nutrition. There are obligate mycotrophs, which require the presence of the mycorrhizal fungus, and facultative mycotrophs, which can live without it under certain circumstances. **2**. A fungus that obtains its food from another fungus.

mycovirus, pl. **mycoviruses** (Gr. *mýkes*, fungus + L. *virus*, toxin): a virus that infects fungi.

myriosporous (Gr. *myríos*, innumerable + *sporá*, spore + L. *-osus* > OF. *-ous*, *-eus* > E. *-ous*, having, possessing the qualities of): having a large number of spores; e.g., the myriosporous sporangium of *Mucor hiemalis* (Mucorales). Also called **pleiosporous** or **pliosporous**. Cf. **oligosporous**.

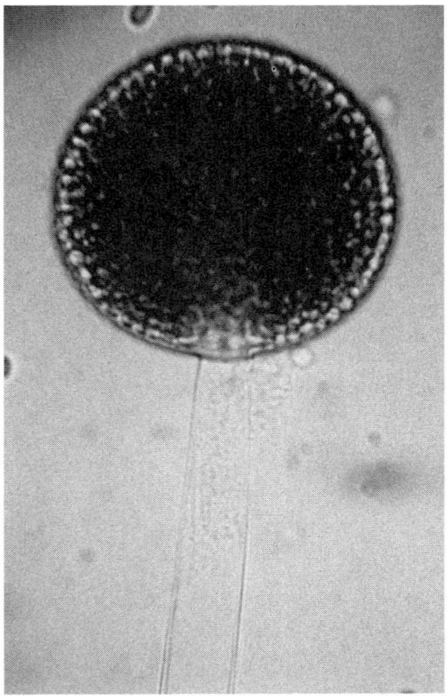

Myriosporous sporangium of *Mucor hiemalis*, x 500 (*CB*).

myriostomes (Gr. *myríos*, innumerable + NL. *stome* < Gr. *stóma*, mouth): having various pores or stomas, which result from the dehiscence of the endoperidium in *Myriostoma* (Lycoperdales), and which appear as small lacerate ostioles in the mature fructification. See **stoma**.

Myriostomes of the fruiting body of *Myriostoma coliforme*, x 2.

myrmecophilous (Gr. *mýrmex*, *mýrmekos*, ant + *phílos*, have an affinity for + L. *-osus* > OF. *-ous*, *-eus* > E. *-ous*, possessing the qualities of): refers to the fungi that afford shelter or food to certain species of ants. For example, *Rozites gongylophora* (Agaricales) is a fungus that lives symbiotically with various species of ants of the gen. *Atta*, and *Lepiota* sp. (Agaricales) is the symbiotic fungus of the ant *Cyphomyrmex costatus*.

mytiliform (Gr. *mýtilos*, marine mussel, mytilus + L. *-formis* < *-forma*, shape): having the shape of a mussel shell; e.g., like the ascocarp (hysterothecium) of certain Loculoascomycetes of the family Mytilinidiaceae (such as *Lophium* and *Mytilinidion* of the Melanommatales).

Mytiliform hysterothecia of *Mytilinidion decipiens*, x 58.

myxamoeba, pl. **myxamoebae** (Gr. *mýxa*, mucosity, viscosity + *amoibé*, change, transformation): *Myxomycetes*. The amoeba that results from germination of a spore. A myxamoeba can be transformed into a **myxozoospore** by formation of flagella, and vice versa; together with the plasmodia the myxamoebae comprise the trophic phases of these organisms, e.g., *Physarum polycephalum* (Physarales). Celular slime molds, such as *Dictyostelium discoideum* (Dictyosteliomycota) and *Acrasis rosea* (Acrasiomycota), among many others, also have myxamoebae.

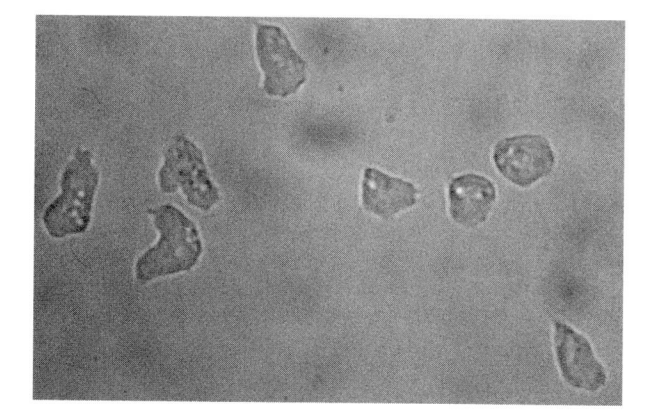

Myxamoebae of *Dictyostelium discoideum*, x 900 (*RTH*).

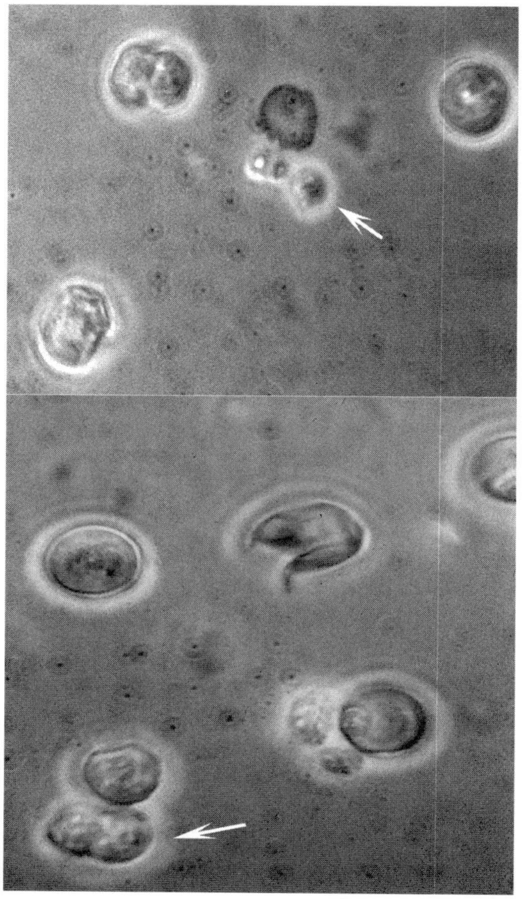

Myxamoebae of *Physarum polycephalum* being released from the spores, x 1 350 (*MU*).

myxoflagellate (Gr. *mýxa*, mucosity, viscosity + L. *flagellum*, whip, flagellum + suf. *-atus* > E. *-ate*, provided with or likeness): the amoeboid, flagellate cell (myxozoospore) of the Myxomycetes; a term used especially by protozoologists. Also called **myxomonad**.

myxolichen (Gr. *mýxa*, mucosity, viscosity + *leichén*, lichen): the association of Myxomycete plasmodia and green algal cells (*Chlorella*), artificially produced in the laboratory.

myxomonad (Gr. *mýxa*, mucosity, viscosity + L. *monas*, genit. *monadis*, unit, entity < Gr. *monás*, *monádos* < *mónos*, only, alone): another term for **myxozoospore**; rarely used in botany and mycology.

Myxomycetes (Gr. *mýxa*, mucosity, viscosity + L. *-mycetes*, ending of class < Gr. *mýkes*, genit. *mýketos*, fungus): a class of viscous or mucilaginous fungi, also called slime molds, whose myxamoebae or myxozoospores coalesce to form plasmodia that fruit by producing sporangia and other types of sporiferous apparatus. In the classification system adopted in this dictionary, the Myxomycetes (which includes the orders Ceratiomyxales, Echinosteliales, Liceales, Trichiales, Physarales, and Stemonitales) are the only class of the phylum Myxomycota of the kingdom Protista (Alexopoulos *et al.*, 1996). In other classification systems the Myxomycetes are included in the kingdom Protoctista (Margulis and Schwartz, 1988) or in the kingdom Fungi (Herrera and Ulloa, 1990).

myxomyceticolous (L. *myxomycete* < Gr. *mýxa*, mucosity, viscosity + *mýkes*, genit. *mýketos*, fungus + L. suf. *-cola*, inhabitant + L. *-osus* > OF. *-ous*, *-eus* > E. *-ous*, possessing the qualities of): living on Myxomycetes, like some spp. of *Nectria* (Hypocreales), whose perithecia, mycelia and conidiophores develop on sporangia and aethalia of Myxomycetes; e.g., *N. myxomyceticola*, which is found on sporangia of *Arcyria cinerea*, *A. nutans* (Trichiales), *Stemonitis fusca*, and *S. nigrescens* (Stemonitales), and on aethalia of *Fuligo septica* (Physarales).

Myxomycota (Gr. *mýxa*, mucosity, viscosity + L. *-mycota*, ending of division or phylum < Gr. *mýkes*, genit. *mýketos*, fungus): Myxomycetes and other mucous or mucilaginous fungi, which mycologists classify in the division (phylum) Myxomycota, either within the kingdom Fungi or as part of the kingdom Protista. Some authors (such as Margulis and Schwartz, 1988; Whittaker, 1959) consider the Myxomycota as part of the kingdom Protoctista.

myxozoospore (Gr. *mýxa*, mucosity, viscosity + *zõon*, animal, motile cell + *sporá*, spore): also called **myxomonad**, mainly by protozoologists. *Myxomycetes*. The zoospore that originates by germination of a spore or by metamorphosis of the myxamoeba. The myxozoospores, which also can transform themselves into myxamoebae, together with the plasmodia, comprise the trophic phases of these organisms, e.g., in *Echinostelium minutum* (Echinosteliales).

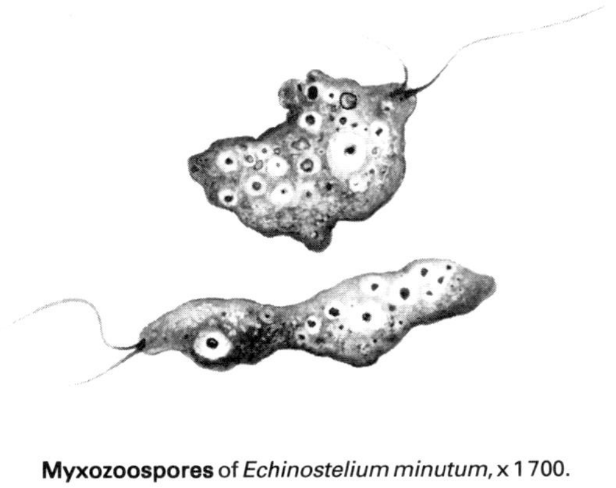

Myxozoospores of *Echinostelium minutum*, x 1 700.

n

nacreous (MF. < OIt. *naccara*, nacre < Ar. *naggarah* + L. *-osus* > OF. *-ous*, *-eus* > E. *-ous*, having, possessing the qualities of): with a pearly color or luster, like mother-of-pearl.

naked (ME. *nakede* < OE. *nacod*, naked): a term applied to the cells that lack a wall (e.g., the myxamoebae of slime molds and the zoospores of aquatic fungi); also refers to the surface of the pileus or stipe of higher fungi that is devoid of fibrils, scales, spines, or other covering, and to a stoma which does not have a differentiated peristome, e.g., in *Tulostoma* (Tulostomatales). Cf. **ornamented**.

nanometer (Gr. *nános*, *nánnos*, dwarf + ME. *meter* < OE. *meter* < L. *metrum* < Gr. *métron*, measure, meter): one thousandth of a micrometer, or one billionth of a meter, which is abb. nm. Cf. **micrometer**.

napiform (L. *napiformis* < *napus*, turnip + *-formis* < *forma*, shape): similar in shape to a turnip, i.e., swollen or bulbose at the top and tapering abruptly toward the base; e.g., like the volva of the basidiocarp of *Amanita perpasta* and the pileocystidia of *Leptonia fulva* (Agaricales).

Napiform volva of the basidiocarp of *Amanita perpasta*, x 0.7.

nasse, **nassace** (F. *nasse*, fish-trap, cylindrical in shape; basket): a dactyloid projection of the internal part of a bitunicate ascus, inward from the endoascus or endotunica; the apical internal peak of the bitunicate ascus. It is present in the Loculoascomycetes, e.g., *Leptosphaerulina* (Dothideales). Also called *ocular chamber* and *nasse apicale*.

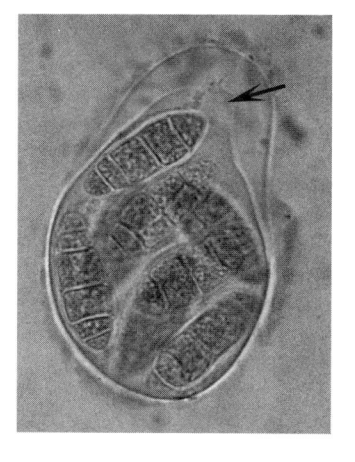

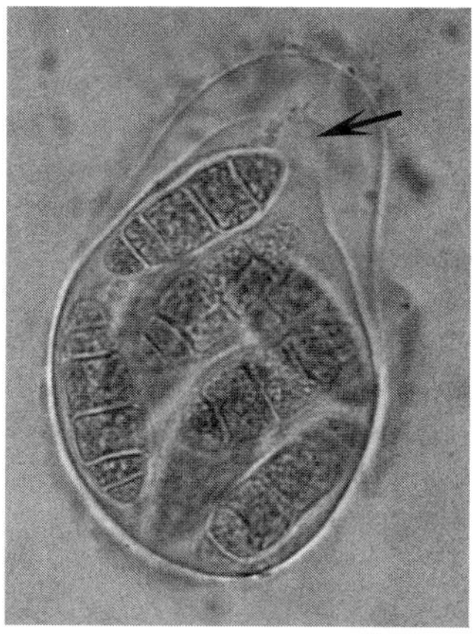

Nassace of the bitunicate ascus of *Leptosphaerulina arachidicola*, x 450 (*MU*).

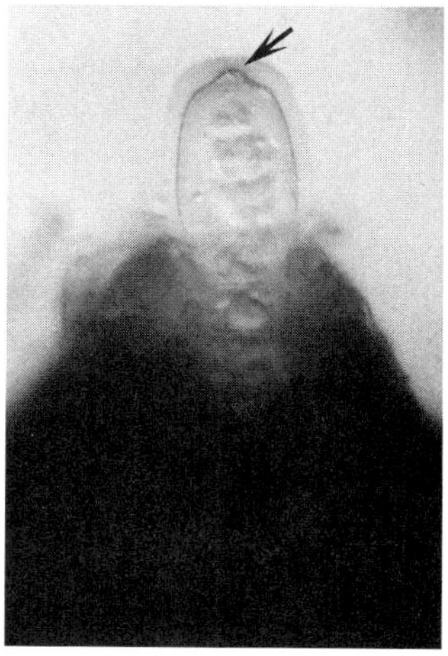

Nassace of the bitunicate ascus of *Leptosphaerulina crassiasca*, x 435 (*RTH*).

naucorioid (< gen. *Naucoria* < Gr. *naus*, a ship + L. *corium*, leather, skin + L. suf. *-oide* < Gr. *-oeídes*, similar to): *Agaricales*. A type of fruiting body (13 principal types are recognized) with gills adnate or emarginate, stipe of fleshy consistency, fibrous or gypseous, centrally implanted in the pileus, and lacking a ring and volva, as in *Naucoria* and some species of *Laccaria*, *Russula* and *Cortinarius*, among other gen. of Agaricales.

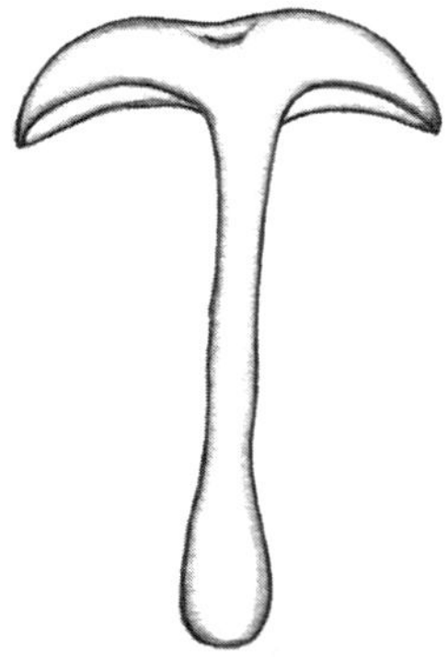

Naucorioid type of basidiocarp, represented by *Naucoria* sp., x 2.

navicular, **naviculate** (L. *navicularis*, relative to a little boat < *navicula*, dim. of *navis*, boat + suf. *-aris* > E. *-ar*, like, pertaining to; or + suf. *-atus* > E. *-ate*, provided with or likeness): having the shape of a small boat; e.g., like the macroconidia of *Fusarium* (tuberculariaceous asexual fungi), the pseudothecia (hysterothecia) of the Loculoascomycetes of the orders Arthoniales and Patellariales, or like the basidiospores of *Mycena lactea* and *Marasmius rotula* (Agaricales). Also called **carinate**, **cymbiform** and **scaphoid**.

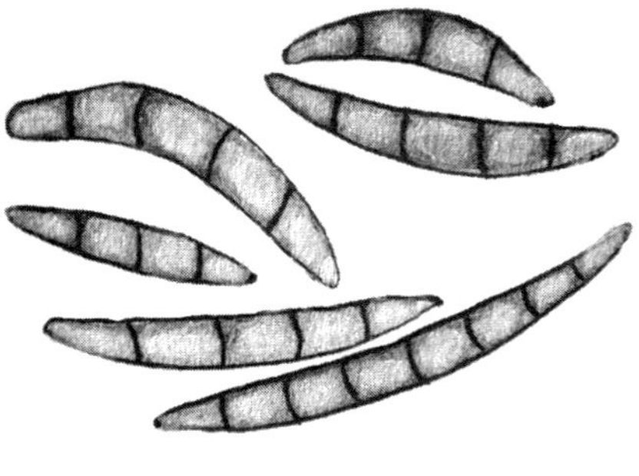

Naviculate macroconidia of *Fusarium* sp., x 870.

necrophoral (Gr. *nekrós*, dead + *-phóros*, bearer < *phéro*, to carry, sustain + L. suf. *-alis* > E. *-al*, relating to or belonging to): *Lichens*. The layers of dead cells, without distinct lumens, that can function as resevoirs of water; they are located near the cortex of the thallus and can be above or below the algal or gonidial layer.

necrosis, pl. **necroses** (Gr. *nékrosis*, death < *nekrós*, dead + *-osis*, condition or state): partial or total death of cells or tissues.

necrotic (Gr. *nekrós*, dead + *-tikós* > L. *-ticus* > E. *-tic*, relation, inclination): pertaining to or relative to **necrosis**.

necrotroph, **necrotrophic** (Gr. *nekrós*, dead + *trophós*, that which nourishes, serves as food; or + suf. *-íkos* > L. *-icus* > E. *-ic*, belonging to, relating to): a parasite that feeds on the dead parts of the host. In the fungi, the majority of the plant parasitic species are necrotrophs. Cf. **biotroph** and **perthotroph**.

nematophagous (Gr. *nêma*, genit. *nêmatos*, thread + *phágos*, eater < *phágomai*, to eat + L. *-osus* > OF. *-ous*, *-eus* > E. *-ous*, possessing the qualities of): nematode-feeding, e.g., the predacious fungi that trap, kill, and feed on soil nematodes, such as *Arthrobotrys* and *Dactylella* (moniliaceous asexual fungi). Syn. of **vermivorous**.

nemoral (L. *nemoralis*, of a wood or grove, sylvan < *nemus*, genit. *nemoris*, grove, forest + suf. *-alis* > E. *-al*, relating to or belonging to): inhabiting woods or forests.

net (ME. *net*, *netten* < OE. *nett*, net): applied to the group of thick longitudinal ribs interconnected by delicate transverse bands, like a mesh or grille-work, present in the mature sporangia of the Myxomycete *Cribraria* (=*Dictydium*), of the Liceales, when the peridium, which is fugaceous, detaches. The spaces or interstices of the reticular mesh are almost rectangular, although they can be somewhat irregular or polygonal. Attached to the network are the dictydine granules, along with spores on and around it.

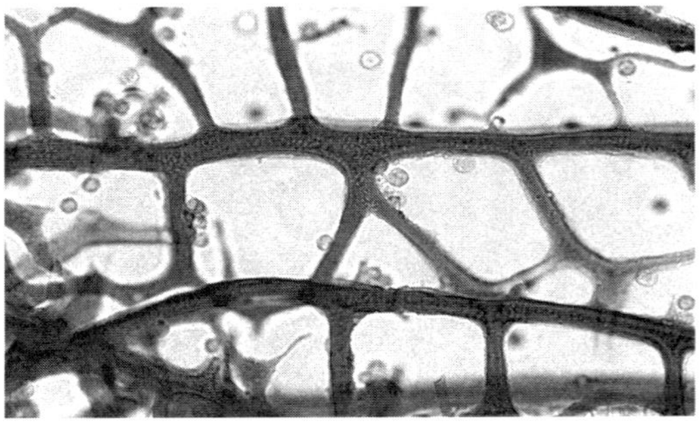

Net of the sporangial peridium of *Cribraria cancellata*, x 250 (*CB*).

nidose, **nidorose** (L. *nidorosus*, steaming, reeking as with a bad odor < *nidor*, a penetrating and not very agreeable odor + *-osus*, full of, augmented, prone to > ME. *-ose*): with a disagreeable odor, similar to rotted meat, given off by fungi of the gen. *Phallus* and *Mutinus* (Phallales).

nidulant, **nidulate** (L. *nidulatus*, nestled < *nidulus*, dim. of *nidus*, nest + suf. *-ant* > E. *-ant*, one that performs, being; or + suf. *-atus* > E. *-ate*, provided with or likeness): placed or inserted, or situated freely, in a cavity or structure similar to a nest; e.g., like the peridiola of the fruiting body of *Crucibulum vulgare* (Nidulariales).

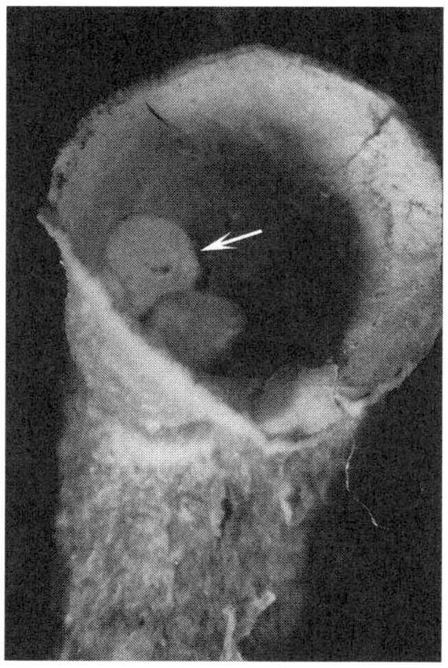

Nidulant peridiola of the fruiting body of *Crucibulum vulgare*, x 5 (*MU*).

nigricant (L. *nigricans*, black, swarthy, with the suf. *-ant* > E. *-ant*, one that performs, being; L. *nigricantis*, being black): becoming or turning black; e.g., like the mature sporangia of *Rhizopus nigricans* (Mucorales).

nimbospore (L. *nimbus*, dense cloud + Gr. *sporá*, spore): a spore with a gelatinous wall, apparently of many layers, such as that of *Histoplasma capsulatum* (moniliaceous asexual fungi).

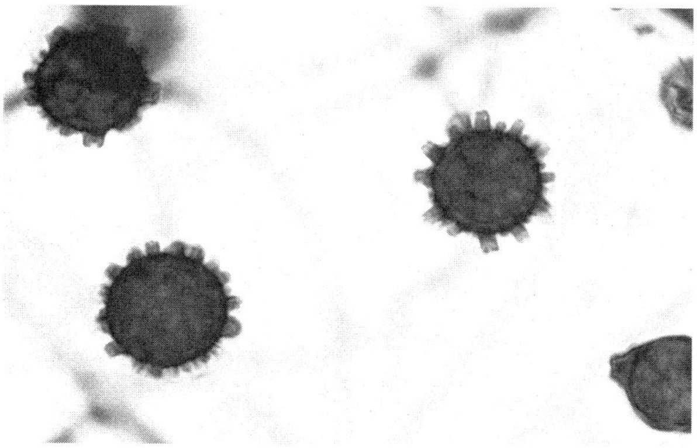

Nimbospores (macroconidia) of *Histoplasma capsulatum*, x 1 500 (*CB*).

nitid, **nitidous** (L. *nitidus*, bright, lustrous < *nitere*, to shine + L. *-osus* > OF. *-ous*, *-eus* > E. *-ous*, having, possessing the qualities of): smooth, clear or dark but shiny or polished, like the black, glistening sporodochium of *Schizoxylon* (=*Agyriella*), of the Ostropales, and the thallus of *Pyrenula nitida* (Pyrenulales).

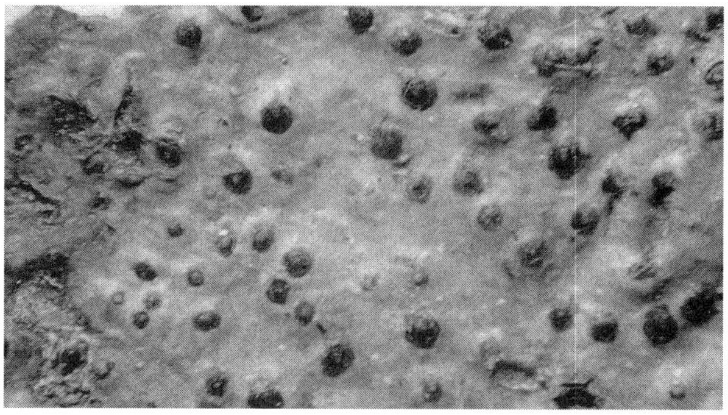

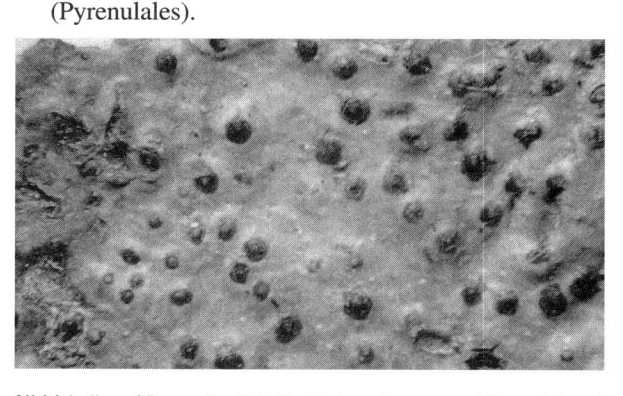

Nitid thallus of *Pyrenula nitida*; the dark spots represent the ostioles of the perithecia, x 4 (*MU*).

nival (L. *nivalis*, relative to snow < L. *nix*, genit. *nivis*, snow + suf. *-alis* > E. *-al*, relating to or belonging to): living in or near snow, or being white in color like snow. For example, *Fusarium poae* and *F. sporotrichioides* (tuberculariaceous asexual fungi) develop very well in soil and vegetable remains covered by snow, and *F. nivale* forms a white mycelium.

niveous (L. *niveous*, snowy < *nix*, genit. *nivis*, snow + *-osus* > OF. *-ous*, *-eus* > E. *-ous*, possessing the qualities of): snow white, pure, without spots, like the basidiocarp of *Coprinus niveus* (Agaricales).

nivicolous (L. *nix*, genit. *nivis*, snow + *-cola*, inhabitant + *-osus* > OF. *-ous*, *-eus* > E. *-ous*, possessing the qualities of): living in snow, like some species of Myxomycetes that are associated with *Pinus silvestris* vegetation, on detritus near the snow; e.g., *Trichia sordida* var. *sordidoides* (Trichiales) and *Comatricha alpina* (Stemonitales). Also called **chionophilous**.

noctilucent (L. *nox*, genit. *noctis*, night + *lucentis*, shining, bright, visible < *lucere*, to shine < Gr. *leukós*, white, bright + L. *-entem* > E. *-ent*, being): emitting light and shining during the night; e.g., like the basidiocarps of *Mycena lux-coeli* (Agaricales), which have the phenomenon of luminescence, i.e., the production of light of weak intensity without being incandescent.

node cell (L. *nodus*, knot; L. *cella*, storeroom > ML. *cella* > OE. *cell*, cell, compartment): *Meliolales*. A stigmatocyst that is on a hypha, in order to differentiate it from that which forms in a hyphopodium, a case in which the stigmatocyst constitutes the terminal cell provided with a haustorium.

noduliferous (L. *nodulus*, dim. of *nodus*, knot + *-ferous*, bearer < *ferre*, to bear, carry + *-osus* > OF. *-ous*, *-eus* > E. *-ous*, full of, having, possessing the qualities of): see **nodulose**.

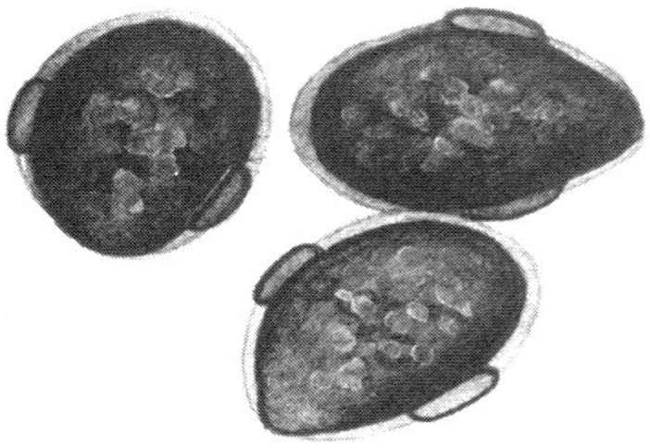

Noduliferous ascospores of *Sordaria nodulifera*, x 1 400.

nodulose (L. *nodulosus*, knotty, nobby < *nodulus*, dim. of *nodus*, knot + *-osus*, full of, augmented, prone to > ME. *-ose*): generally applied to spores whose wall has blunt excrescences with a broad base, similar to warts; e.g., like the aleuriospores of *Mycogone perniciosa* and the macroconidia of *Histoplasma capsulatum* (moniliaceous asexual fungi), and the ascospores of *Sordaria nodulifera* (Sordariales).

nomen (L. *nomen*, name): name; refers to scientific names; i. e., a genus, species, binomial, etc. *Nomen ambiguum* (L. *ambiguus*, ambiguous < *ambigere*, to wander about): a doubtful or uncertain name that can be understood in two or more possible senses or ways. *Nomen confusum* (L. *confusus*, ptp. of *confundere*, to disturb, make indistinct or mix indiscriminately): a taxon based on two or more different elements. *Nomen conservandum* (L. *conservatus*, ptp. of *conservare*, to conserve, preserve): a name authorized for use by a decision of an International Botanical Congress. *Nomen dubium* (L. *dubius*, doubtful, questionable < *dubare*, to vacillate): a doubtful name, of uncertain sense. *Nomen illegitimum* (ME. suf. *-il* < MF. < L. *in*, not + L. *legitimus*, legitimate, being exactly as purposed): a validly published name but contravening particular Articles in the Botanical Code. *Nomen invalidum* (L. *invalidus*, weak < *in*, not + *validus*, strong; not valid): one not validly published. *Nomen novum* (L. *novus*, new): a new name; a replacement. *Nomen nudum* (L. *nudus*, naked, devoid): one for a taxon having no description. *Nomen provisorium* (L. *provisus*, ptp. of *providere*, to see ahead): a name serving for the time being; proposed provisionally. *Nomen rejiciendum* (L. *rejectus*, ptp. of *reicere*, to reject, refuse to accept): one rejected by a Botanical Congress.

nomenclature (L. *nomenclatura*, nomina): a group of voices or technical names relating to a faculty or science. *Binary nomenclature* (LL. *binarus* < L. *bini*, compounded or consisting of two things or parts): a binary name for each species formed by two terms, one generic and another specific. See **systematics**.

non-persistent attachment (L. *non*, not + *persistens*, genit. *persistentis*, continuing, persevering, persisting < *persistere*, to continue + *-entem* > E. *-ent*, being; ME. *attachement*, seizure < AF. *atacher*, to seize < OF. *atachier*, to fasten + L. *-entem* > E. *-ent*, being): occurs in fruiting bodies which break the attachment to the ground and are free at maturity to be blown about by the wind, e.g., in *Bovista* (Lycoperdales).

notate (L. *notatus*, designated, pointed out < *notare*, to mark + suf. *-atus* > E. *-ate*, provided with or likeness): with the surface provided with lines or marks notable for their color or some other characteristic, like the colony of *Penicillium notatum* (moniliaceous asexual fungi) when it develops on a solid medium, which appears zonate, with concentric zones of conidia alternating with non-sporulating areas of mycelium.

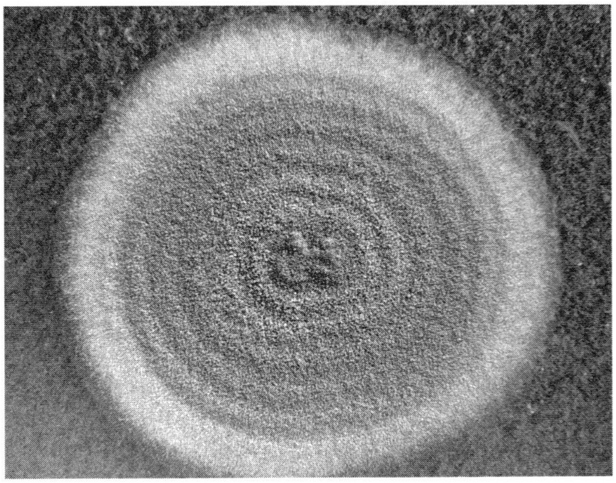

Notate colony of *Penicillium notatum* on agar, x 2 (*MU*).

nuclear cap (E. *nuclear* < *nucleus* < L. *nuculeus*, kernel < *nucula*, little nut < *nux*, nut + suf. *-aris* > E. *-ar*, like, pertaining to; LL. *cappa*, hooded cloak, cap): a structure composed of the ribosomes of the cell which aggregate near the nucleus and surrounded by a membrane that appears to be an extension of the nuclear envelope; characteristic of the zoospores of the Blastocladiales (*Allomyces*, *Blastocladia*, *Blastocladiella*).

nucleus, pl. **nuclei** (E. *nucleus* < L. *nuculeus*, kernel, dim. of *nux*, genit. *nucis*, nut): in biology, an organelle present in the protoplasm of most plant and animal cells (eukaryotic), except in certain forms of low organization (prokaryotic), and regarded as an essential factor in the constructive metabolism, growth, and reproduction and in the hereditary transmission of characters. It typically consists of a rounded or oval mass of protoplasm inclosed in a delicate perforated membrane (the nuclear membrane); it contains one or more nucleoli and the chromatin which during nuclear division gives rise to the chromosomes. The nuclei of most fungi are quite small, and are better seen after being fixed and stained with dyes such as Giemsa, iron-hematoxylin, acetoorcein, and acetocarmin, as well as fluorescent stains including 4',6'-diamidino-2-phenylindole and mitramycin. Nuclear division in the fungi is basically intranuclear, that is, the bulk of the nuclear envelope remains intact until late telophase when it breaks in the interzonal regional and then re-forms around the daughter nuclei. The ultrastructural details of mitosis and meiosis in fungi are of value as phylogenetic characters.

Nuclear cap of the male gamete of *Allomyces macrogynus*, x 1 350.

nummiform (L. *nummus*, money, coin + *-formis* < *forma*, shape): shaped like a coin, like the argentate peridioles of *Cyathus intermedius* (Nidulariales) and the perithecial stromata of *Nummularia* (Xylariales).

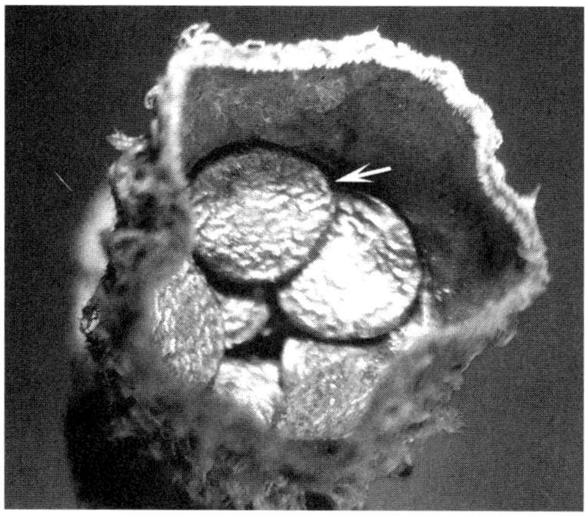

Nummiform peridiola in a fruiting body of *Cyathus intermedius*, x 7 (*CB & MU*).

nutriocyte (L. *nutrire*, to nourish, feed + Gr. *kýtos*, a hollow vessel, now often taken to mean a cell): the inflated portion of the ascogonium of *Ascosphaera* (Ascosphaerales), which receives the contents of the trichogyne hypha after plasmogamy has occurred; eventually, the nutriocyte develops into a **sporocyst**.

O

ob- (L. pref. with various meanings: before, toward, in front, against, on, above): used to give words an intensive sense or opposite or inverted meaning, as in **obclavate**, **obconic**, etc.

obclavate, **obclaviform** (L. *ob-*, inverted + NL. *clavatus*, club-shaped < L. *clava*, club + suf. *-atus* > E. *-ate*, provided with or likeness; L. *obclaviformis* < *ob-* + *claviformis* < *clava* + *-formis* < *forma*, shape): having the shape of a club, but with the broad part at the base; e.g., like the conidia of *Triramulispora obclavata* and the synnemata of *Beauveria cretacea* and *Meria coniospora* (moniliaceous asexual fungi), and the cheilocystidia of *Mycena sanguinolenta* (Agaricales). Cf. **clavate**.

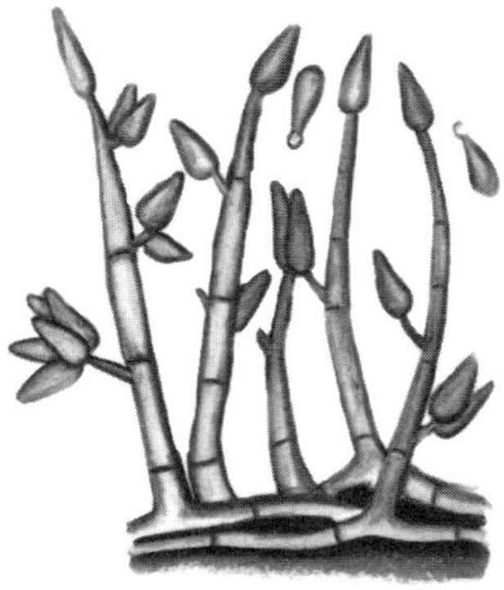

Obclavate conidia of *Meria coniospora*, x 660.

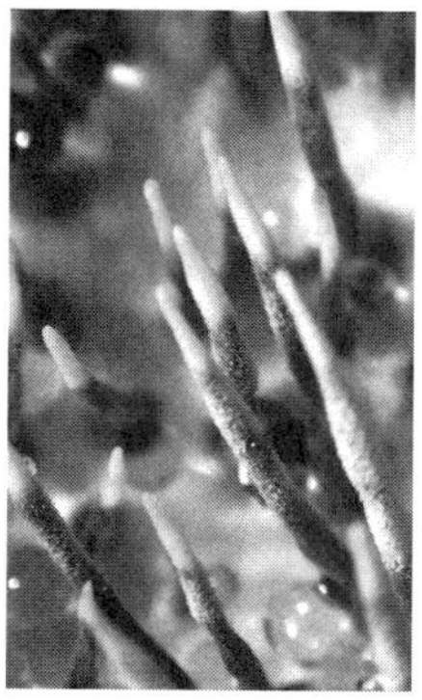

Obclavate synnemata of *Beauveria cretacea* on agar, x 5 (*MU*).

obconic (L. *ob-*, inverted + Gr. *konikós*, conic < *kónos*, cone, pine cone + suf. *-íkos* > L. *-icus* > E. *-ic*, belonging to, relating to): inverted cone-shaped, with the widest part at the top; e.g., like the conidia of *Acaulopage tetraceros* (Zoopagales). Similar to **turbinate**.

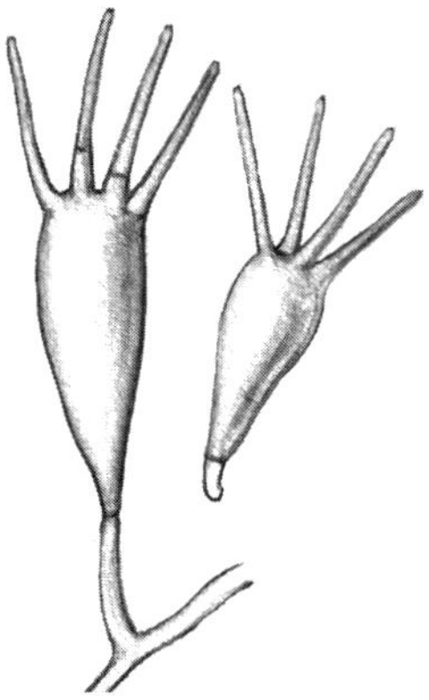

Obconic spores of *Acaulopage tetraceros*, x 800.

oblate (L. *oblatus* < *ob-*, intensive sense + *latus*, wide): literally, very wide; wider than long. An oblate spore is wide laterally and flattened at the poles; similar to the shape of an orange. Cf. **prolate**.

obligate parasite (L. *obligatus*, obligate, obligative; Gr. *parásitos* < *pará*, near, at the side of + *sîtos*, food): an organism that can be a parasite on another living organism. Among the fungi, the minority of the species are obligate parasites, including the gen. in the family Peronosporaceae (*Peronospora*, *Albugo*, etc.), those of the order Erysiphales (*Blumeria*, *Erysiphe*, *Podosphaera*, etc.), and the rusts of the order Uredinales (*Puccinia*, *Uromyces*, etc.). Although a few species of obligately parasitic fungi have been induced to form distinct phases of their life cycle in culture on synthetic laboratory media, they are still considered obligate parasites because they behave so in nature. Cf. **facultative parasite** and **obligate saprobe**.

obligate saprobe (L. *obligatus*, obligate, obligative; Gr. *saprós*, rotten + *bíos*, life): an organism that can only live on dead organic matter, for which reason it is incapable of infecting a living being, i.e., it cannnot live as a parasite. Cf. **facultative saprobe** and **obligate parasite**.

oblong (L. *oblongus* < *ob-*, intensive sense + *longus*, long): literally, very long; longer than wide. An oblong spore is elongate, with the sides almost parallel and the corners rounded, somewhat flattened at the ends; e.g., like the sporangiospores of *Pilobolus crystallinus* (Mucorales), the conidia of some species of *Colletotrichum* (melanconiaceous asexual fungi), the ascospores of *Peziza ammophila* (Pezizales) and the basidiospores of *Tubaria furfuracea* (Agaricales). An oblong-ellipsoid spore has the long sides parallel and the ends almost hemispheric; e.g., like the basidiospores of *Schizophyllum commune* (Aphyllophorales).

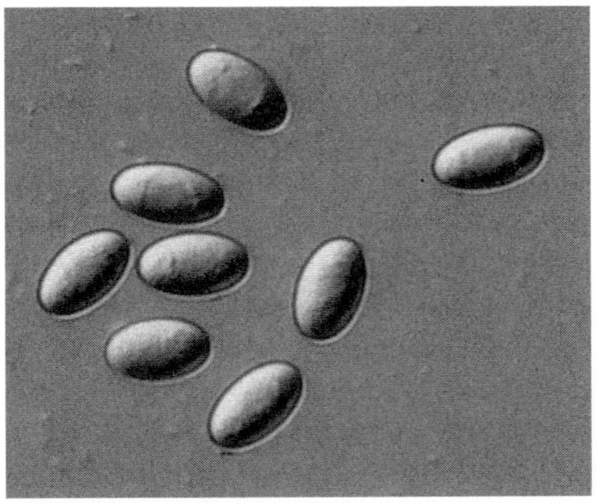

Oblong ascospores of *Peziza ammophila*, x 730 (*RTH*).

obovate (L. *ob-*, inverted + *ovatus* < *ovum*, egg + suf. *-atus* > E. *-ate*, provided with or likeness): inversely egg-shaped; oval with a broad apex and narrow base, like an egg standing on end. Cf. **ovate**.

obovoid (L. *ob-*, inverted + NL. *ovoides*, egg-shaped < *ovum*, egg + *-oide* < Gr. *-oeídes*, like): inversely egg-shaped; ovoid, but with the widest part at the apex; e.g., like the sporangia of *Leocarpus fragilis* (Physarales), the conidia of *Brachysporium obovatum* and *Cordana musae* (dematiaceous asexual fungi), and the urediniospores of *Uromyces fabae* (Urediniales). Same as **obovate**. Cf. **ovoid**.

obpyriform (L. *ob-*, inverted + *pyriformis* < *pyrum*, pear + *-formis* < *forma*, shape): inversely pear-shaped, i.e., with the widest part at the point of attachment; e.g., like the primordium of the sorocarp of *Dictyostelium discoideum* (Dictyosteliomycota) and the majority of the pycnidia (sphaeropsidaceous asexual fungi) and perithecia (Pyrenomycetes). Cf. **pyriform**.

Obovoid sporangia of *Leocarpus fragilis*, x 50 (*MU & CB*).

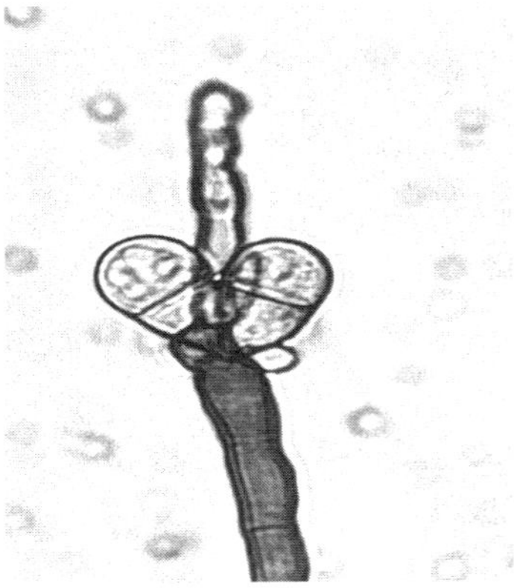

Obovoid conidia of *Cordana musae*, x 740 (*MU*).

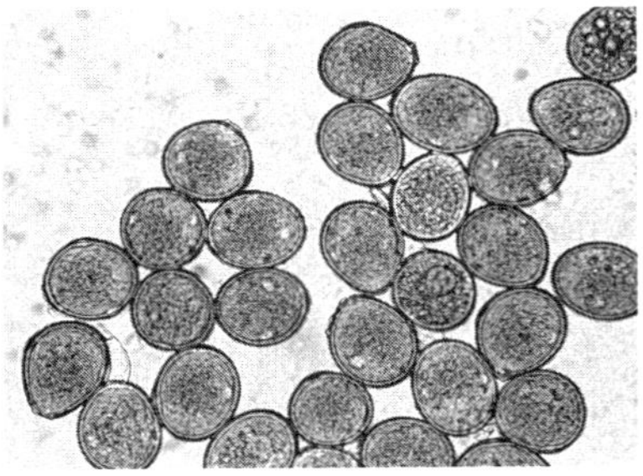

Obovoid urediniospores of *Uromyces fabae*, x 270 (*MU*).

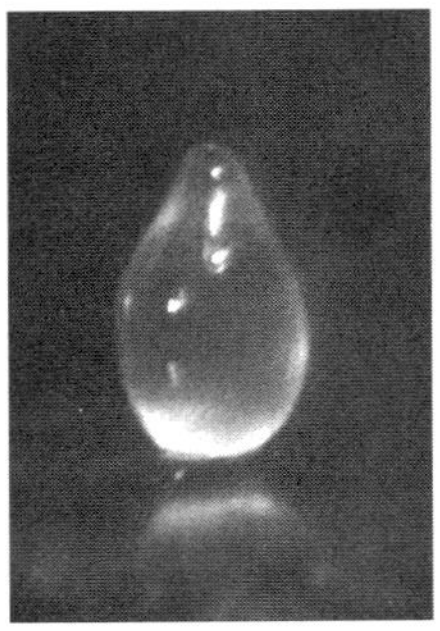

Obpyriform primordium of a sorocarp of *Dictyostelium discoideum*, x 90 (*RTH*).

obsubulate (L. *ob-*, inverted + *subula*, awl, chisel + suf. *-atus* > E. *-ate*, provided with or likeness): inversely subulate; very narrow and sharp-pointed at the base and widening a little toward the apex; e.g., like the conidia of *Flagellospora curvula* (moniliaceous asexual fungi). Cf. **subulate**.

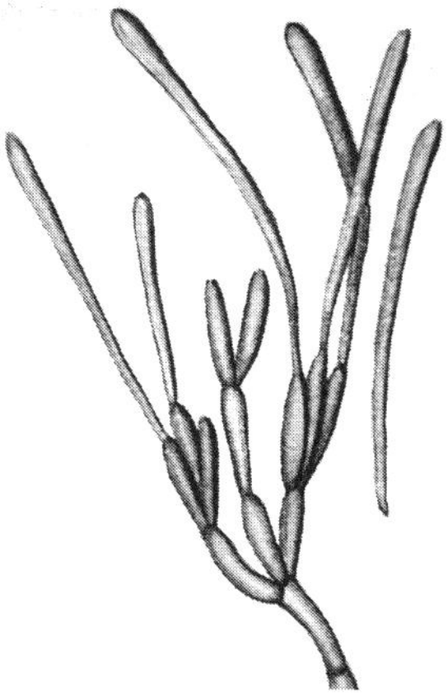

Obsubulate conidia of *Flagellospora curvula*, x 400.

obtuse (L. *obtusus*, blunt, without an edge): **1**. A solid, blunt structure that is rounded at the end. **2**. A structure whose edges form at the apex an obtuse angle (more than a right angle); applied to pilei, cystidia, spores, etc. For example, the cylindrical cystidia of *Leptonia trichomata* (Agaricales) are obtuse.

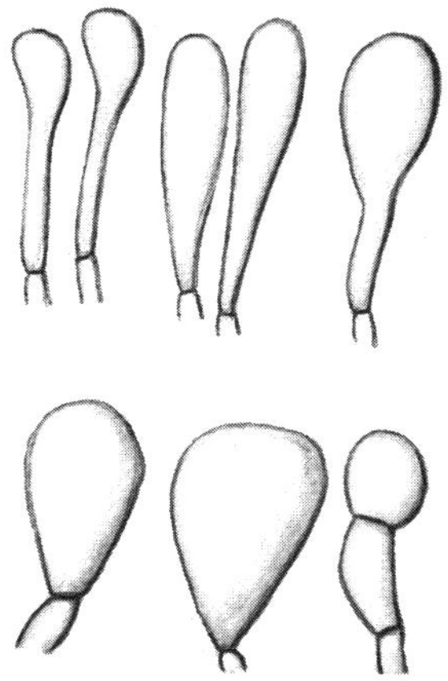

Obtuse apex of the cystidia from the hymenium of *Entoloma sericatum* (cylindro-clavate), *Leptonia decolorans* (clavate), *Leptonia lividocyanula* (sphaeropedunculate), and *Leptonia fulva* (napiform, turbinate and vesiculate), x 500.

ocellate, **ocellated** (L. *ocellatus*, spotted as with little eyes < *ocellus*, little eye + suf. *-atus* > E. *-ate*, provided with or likeness): with spots, patches or points, that resemble small eyes; e.g., like the thallus of the lichen *Schismatomma ocellatum* (Arthoniales), whose appearance and name are due to the minute apothecia that are scattered on its surface.

ochrosporous (Gr. *ochrós*, pale yellow + *sporá*, spore + L. *-osus* > OF. *-ous*, *-eus* > E. *-ous*, having, possessing the qualities of): refers to a fungus that gives a yellow, ochre, or ochreous-brown spore print; e.g., like the basidiocarp of *Armillaria* (=*Armillariella*) (Agaricales).

ocreate (L. *ocreatus*, sheathed < *ocrea*, sheath + suf. *-atus* > E. *-ate*, provided with or likeness): having a sheath, such as the volva that envelopes the base of the stipe of the basidiocarp of some agarics, such as *Amanita caesarea* and *A. pantherina* var. *velatipes* (Agaricales). This term appears to be a syn. of **caligate** and **peronate**.

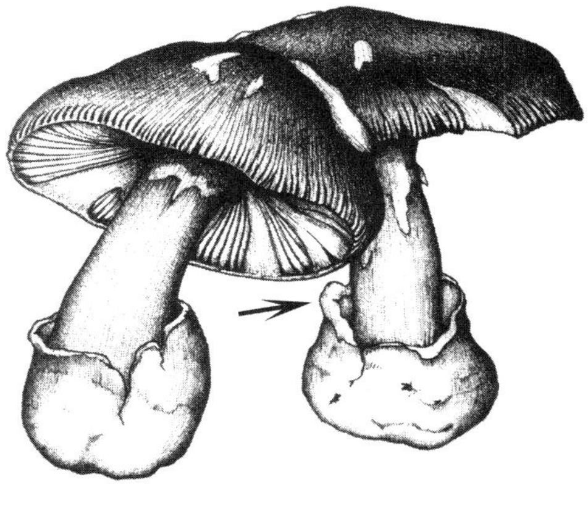

Ocreate volva of the basidiocarp of *Amanita caesarea*, x 0.5.

octophore (Gr. *októ*, eight + *-phóros*, bearer): see **haerangium**.

oidiophore (Gr. *oidión*, small egg or small ovule < *oón*, egg, female sexual cell + *-phóros*, bearer): a specialized hypha that produces and bears oidia, as is observed, e.g., in *Coprinus lagopus* (Agaricales).

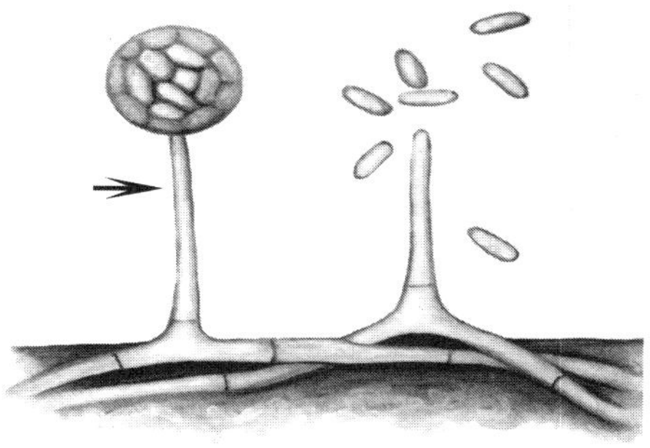

Oidiophore with **oidia** of *Coprinus lagopus*, x 480.

oidiospore (Gr. *oidión*, small egg or small ovule < *oón*, egg + suf. dim. *-ídion* > L. *-idium* + Gr. *sporá*, spore): see **oidium**.

oidium, pl. **oidia** (Gr. *oidión*, small egg or small ovule < *oón*, egg + dim. suf. *-ídion* > L. *-idium*): a small hyphal cell, free, with a thin cell wall, that is derived from the fragmentation of somatic hyphae, or of oidiophores, and behaves like an asexual spore in some Agaricales, such as *Coprinus lagopus*; at times the oidia function as spermatia and bring about the dikaryotization of the somatic hyphae with which they fuse (oidization). The term oidium gave rise to the gen. *Oidium*, the asexual or conidial phase of *Erysiphe* (Erysiphales). Also called **oidiospores**; also called **arthrospores**, as in *Geotrichum* (moniliaceous asexual fungi).

oily (ME. *olie*, *oile* < OF. < L. *oleum*, *olivum*, olive oil < Gr. *oleíum* < *elaíwa*, olive + E. suf. *-y*, having the quality of): resembling oil. Same as **oleaginous**.

oleaginous (L. *oleaginosus*, of the olive < *olea*, *oleagin*, olive + *-osus* > OF. *-ous*, *-eus* > E. *-ous*, having, possessing the qualities of): oily. See **lubricous**.

oleiferous hypha, pl. **hyphae** (L. *oleifer*, oil + *-ferous*, bearer < *ferre*, to bear, carry + *-osus* > OF. *-ous*, *-eus* > E. *-ous*, full of, having, possessing the qualities of; Gr. *hyphé*, tissue, spider web; hypha): **1**. *Lichens*. A hypha that contains guttulate (with oil), torulose cells, as in the submerged hyphae of endolithic species. **2**. *Basidiomycetes*. A hypha that contains resinous substances, such as *Russula emetica* and *Amanita vaginata* (Agaricales).

oleocystidium, pl. **oleocystidia** (L. *oleum*, oil + Gr. *kýstis*, vessel; here, cell + dim. suf. *-ídion* > L. *-idium*): a cystidium with an oily or resinous exudate, present in the hymenium of *Resinomycena* and other Agaricales.

oligosporous (Gr. *olígos*, few + *sporá*, spore + L. *-osus* > OF. *-ous*, *-eus* > E. *-ous*, having, possessing the qualities of): having few spores; e.g., like the sporangia of *Rhizopus oligosporus* (Mucorales). Cf. **myriosporous**.

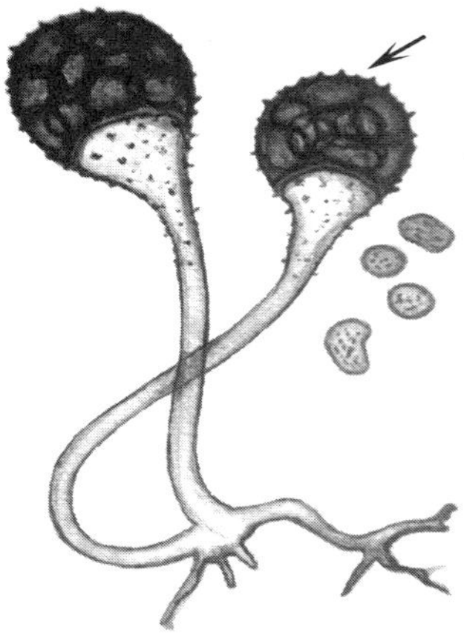

Oligosporous sporangia of *Rhizopus oligosporus*, x 400.

omnivorous (L. *omnivorous*, eating everything < *omnis*, all + *vorare*, to eat + *-osus* > OF. *-ous*, *-eus* > E. *-ous*, possessing the qualities of): capable of developing on all kinds of substrates or nutritive media. Rigorously speaking, this concept is false, although there are **plurivorous** or **polytrophic** fungi, i.e., which can utilize diverse substances, or thrive on the most diverse substrates, whether as saprobes or parasites, like the common molds of the gen. *Aspergillus* and *Penicillium* (moniliaceous asexual fungi).

omphalinoid (< gen. *Omphalina* < Gr. *omphalós*, navel + L. suf. *-ina*, likeness + L. suf. *-oide* < Gr. *-oeídes*, similar to): *Agaricales*. A type of fruiting body (13 principal types are recognized) with gills adherent, decurrent or subdecurrent on the stipe, which is cartilaginous, central and lacks a ring and volva, like that of *Omphalina* and certain species of *Marasmius* and *Psilocybe*. It is similar to the mycenoid and collybioid types but differs in its broadly convex to plane pileus, at times umbilicate, and by its decurrent gills. It is similar to the anellarioid type but differs in lacking a ring. See **omphaloid**.

omphalodisc (Gr. *omphalós*, navel + *dískos*, disc): *Lichens*. An apothecium that has a central prominence composed of sterile hyphae, which gives the appearance of a navel, such as occurs in *Umbilicaria* (=*Omphalodiscus*) *virginis* and *U. crustulosa* (Lecanorales).

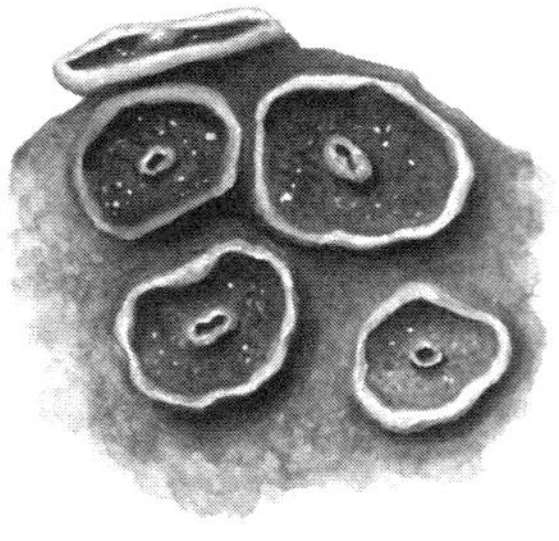

Omphalodiscs (apothecia) of the thallus of *Umbilicaria virginis*, x 10.

omphaloid (< gen. *Omphalia* < Gr. *omphalós*, navel + L. suf. *-oide* < Gr. *-oeídes*, similar to): with the characteristics of the agaricaceous gen. *Omphalina* (=*Omphalia*). See **omphalinoid**.

one-spored basidium, pl. **basidia** (ME. *oon* < OE. *an*, one + E. *spored* < NL. *spora* < Gr. *sporá*, spore; L. *basidium*, small base < Gr. *basídion* < *básis*, base + dim. suf. *-ídion* > L. *-idium*): a basidium which forms a single sterigma on which a single spore develops.

ontogeny (Gr. *óntos*, genit. of *ón*, a being + *géneia*, act of being born < *génos*, origin, birth, engenderment + *-y*, E. suf. of concrete nouns): the development or course of development of an individual organism. Cf. **phylogeny**.

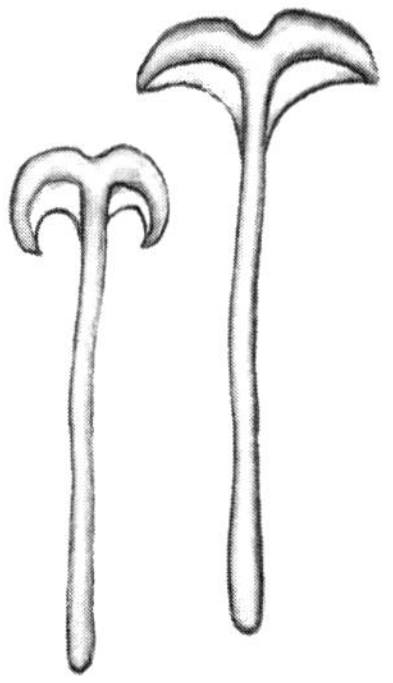

Omphaloid type of basidiocarp, represented by *Omphalina* (=*Omphalia*) sp.; diagram of longitudinal sections, x 1.

oogamy (Gr. *oón*, egg, female sex cell + *gámos*, sexual union + *-y*, E. suf. of concrete nouns): fertilization of an immobile female gamete by a motile male gamete; the female (oosphere or ovule) is larger and originates in the female organ called an oogonium. The male gametes are smaller (spermatozoids or antherozoids) and are produced in the sexual organs called antheridia. Among the fungi, this type of sexual reproduction is exclusive of the aquatics belonging to the order Monoblepharidales, such as *Monoblepharis polymorpha*.

oogenesis, pl. **oogeneses** (Gr. *oón*, egg, female sex cell + *géneia*, *génesis*, generating): development of the oogonium after fertilization, such as occurs in the Oomycota.

oogon, **oogone**, **oogonium**, pl. **oogonia** (Gr. *oón*, egg, female sex cell + NL. *gonium* < Gr. *gónos*, the engendered, offspring): the female gametangium in the aquatic and amphibious fungi of the phylum Oomycota, which can contain one oosphere, as in *Monoblepharis* (Monoblepharidales), or several oospheres, as in *Achlya* and *Saprolegnia* (Saprolegniales).

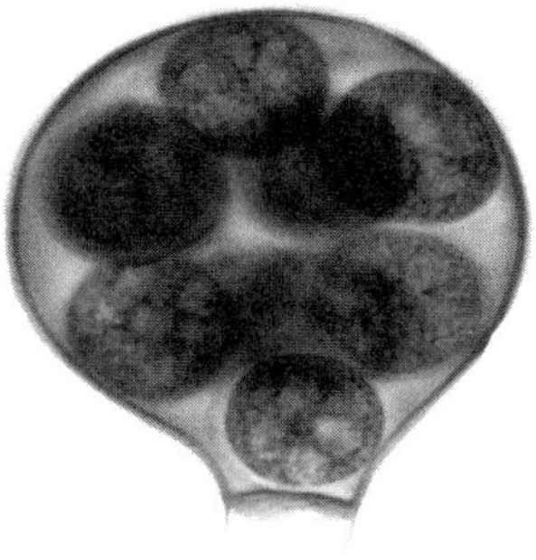

Oogonium with oospheres of *Saprolegnia* sp., x 440 (*MU*).

oogoniols (Gr. *oón*, egg, female sex cell + NL. *gonium* < Gr. *gónos*, the engendered, offspring, with the suf. *-ol* < alcohol; a chemical compound containing hydroxyl): *Achlya* (Saprolegniales) hormones (sterols) which induce oogonial formation. Cf. **antheridiol**.

Oomycetes (Gr. *oón*, egg, female sex cell; here, oosphere + L. *-mycetes*, ending of class < Gr. *mýkes*, genit. *mýketos*, fungus): in some classification systems, a class of fungi with biflagellate zoospores, coenocytic mycelium, and sexual reproduction by oogamy. In the classification adopted in this dictionary (mainly that of Alexopoulos *et al.*, 1996), these organisms are regarded as the phylum Oomycota of the kingdom Stramenopila, with the orders Saprolegniales, Salilagenidiales, Lagenidiales, Leptomitales, Myzocytiopsidales, Rhipidiales and Peronosporales. In some other classification schemes, like that of Margulis and Schwartz (1988), the Oomycetes are considered as a phylum within the kingdom Protoctista.

oomycote (L. *oomycete* < Gr. *oón*, egg, female sex cell + *mýkes*, genit. *mýketos*, fungus): one of the Oomycota, a phylum of kingdom Protoctista in the classification system of Margulis and Schwartz, 1988, and a phylum of the kingdom Stramenopila in the scheme of Alexopoulos *et al.*, 1996. The latter scheme is the one followed in this dictionary.

ooplasm (Gr. *oón*, egg, female sex cell + *plásma*, bland material with which is formed an organ or living being): *Oomycota*. The protoplasm in the center of the oogonium that first gives rise to the oosphere and afterwards to the oospore. Cf. **periplasm**.

ooplast (Gr. *oón*, egg, female sex cell + *plastós*, formed, organized particle or body): *Oomycota*. A cellular inclusion, surrounded by a membrane, in the oospore of the Saprolegniales; the ooplast comes from the ooplasm and forms the oosphere, which is surrounded by the periplasm of the oogonium.

oosphere (Gr. *oón*, egg, female sex cell + *sphaîra*, sphere): *Oomycota*. A gamete (or gametic nucleus), without a cell wall, immotile, contained in the oogonium, as is observed, e.g., in the aquatic fungi of the gen. *Monoblepharis* (Monoblepharidales) and *Saprolegnia* (Saprolegniales).

oospore (Gr. *oón*, egg, female sex cell + *sporá*, spore): a spore, with a thick, resistant wall, that results from the development of an oosphere, whether by fertilization or by parthenogenesis; the oospore is the resting spore of the aquatic fungi of the orders Monoblepharidales (Chytridiomycota), Saprolegniales, Lagenidiales, Leptomitales, Rhipidiales, and Peronosporales (Oomycota).

operculate (L. *operculatus*, covered < *operculum*, lid + suf. *-atus* > E. *-ate*, provided with or likeness): having an **operculum**, as the asci of *Saccobolus* (Pezizales). Cf. **inoperculate**.

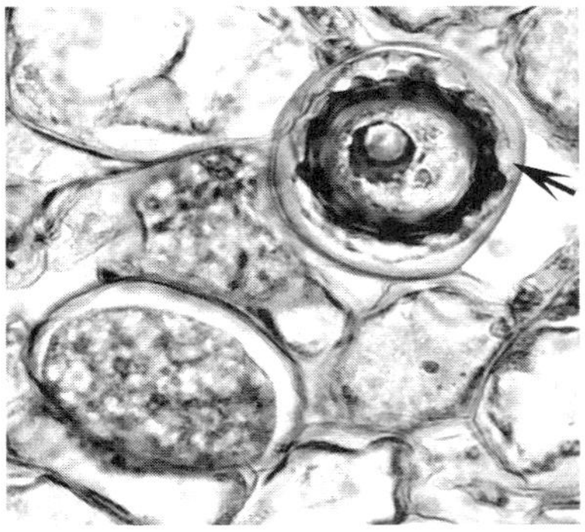

Oospore of *Albugo candida* within a cabbage root cell, x 675 (*MU*).

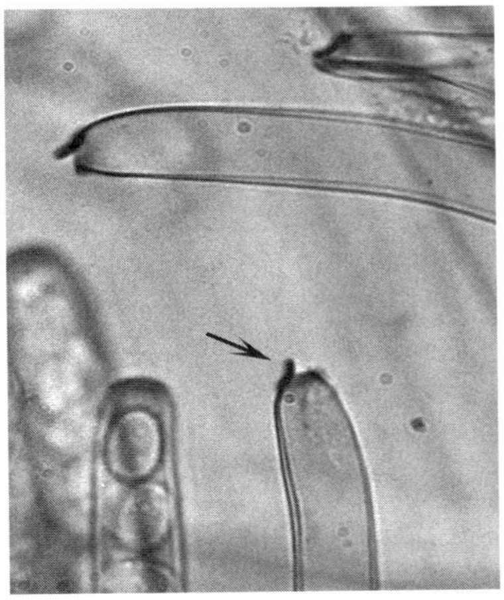

Operculum of the asci in the hymenium of *Peziza ammophila*, x 500 (*RTH*).

operculum, pl. **opercula** (L. *operculum*, lid, cover): the lid-like apex of a sporangium or ascus, which opens on a hinge by a circumscissile dehiscence; there are operculate sporangia in the Chytridiales, such as *Chytridium* and *Nowakowskiella*, and operculate asci in the Pezizales, such as *Peziza* and *Saccobolus*.

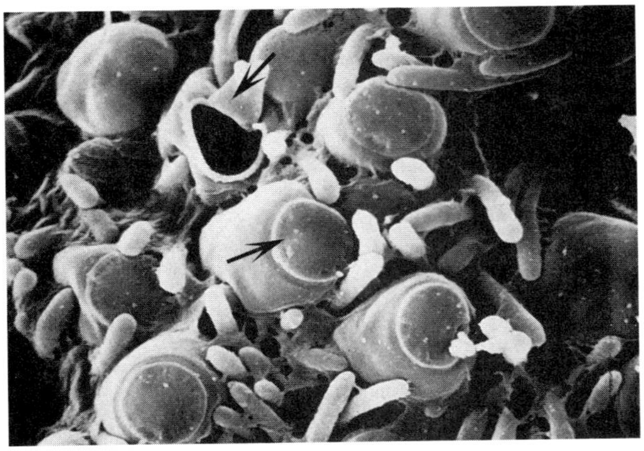

Operculate asci of the hymenium of *Saccobolus* sp.; note the **operculum** at the apex of each ascus, x 1 000 (*KO*).

opisthokont, opisthocont, opisthokontic, opisthocontic, opisthokontous, opisthocontous (Gr. *ópisthe*, *ópisthen*, behind + *kontós*, oar; here, flagellum; or + suf. *-íkos* > L. *-icus* > E. *-ic*, belonging to, relating to; or + L. *-osus* > OF. *-ous*, *-eus* > E. *-ous*, having, possessing the qualities of): a motile cell that has an impelling flagellum inserted in the posterior part, as occurs with the zoospores and planogametes of the fungi of the class Chytridiomycetes, e.g., *Chytriomyces*, *Phlyctochytrium* and other Chytridiales. Cf. **acrocont** and **pleurocont**.

opportunistic (ME. < MF. *opportun* < L. *opportunus*, suitable or convenient for a particular occurrence + NL. *-istic*, pertaining to as agent): applied to fungi that normally are saprobic and frequently common but on occasion capable of causing disease, especially of humans and other animals. E.g., *Candida albicans*, *C. tropicalis* and other species of *Candida* (cryptococcaceous asexual yeasts) are opportunistic yeasts that can cause candidiasis in human and animal hosts rendered susceptible by some predisposing factor(s). See **facultative saprobe**.

orbicular (L. *orbicularis* < *orbiculus*, dim. of *orbis*, circle + suf. *-aris* > E. *-ar*, like, pertaining to): circular, round; e.g., the pileus of the basidiocarp of *Stropharia semiglobata* (Agaricales) is orbicular.

orchil (ME. *orchil*, pigment): a general term for lichens that produce pigments, like the orcein of *Roccella tinctoria* (Arthoniales).

orculiform (L. *orculiformis* < *orcula*, dim. of *orca*, cask, earthen jar + *-formis* < *forma*, shape): having the shape of a cask, like the apothecium of *Geopora* (=*Sepultaria*) *arenicola* (Pezizales). Equal to **urceolate**. Similar to **polarilocular** and **polocellate**.

Orculiform apothecia of *Geopora* (=*Sepultaria*) *arenicola*, x 2 (*MU*).

orismology (Gr. *horismós*, limitation, definition < *horízein*, to delimit < *hóros*, limit, term + *lógos*, discourse, study + *-y*, E. suf. of concrete nouns): the science of defining terms, especially technical terms in a particular subject.

ornamented (ME. < OF. *ornement* < *ornamentum*, adornment, a useful accessory): applied to spores, hyphae, organs, etc., having the surface marked or sculptured with spines, ridges, warts, striations, reticulations, wrinkles, fibrils, scales, etc.; not smooth. Cf. **naked**.

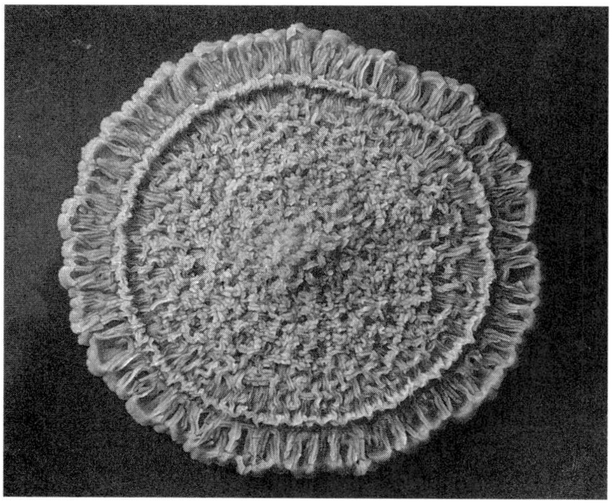

Ornamented giant colony (with a wrinkled surface and a fringed border) of *Rhodotorula rubra* on agar, x 2 (*MU*).

ornithocoprophilous (Gr. *órnis*, genit. *órnithos*, bird + *kópros*, dung + *phílos*, have an affinity for + L. *-osus* > OF. *-ous*, *-eus* > E. *-ous*, possessing the qualities of): a coprophile of bird droppings; e.g., like certain lichens that utilize such substrates to obtain nitrogen, or like the human pathogenic yeast, *Cryptococcus neoformans* (cryptococcaceous asexual yeasts), which inhabits poultry and pigeon droppings.

orphan anamorph (LL. *orphanus*, orphan < Gr. *orphanós*, deprived of one or usually both parents; Gr. *anamorphóo*, to transform): an asexual form that has obvious connections to sexual fungi that have similar conidial states.

orthogeotropism (Gr. *orthós*, straight + *geo* < *gê*, earth + *trópos*, a turn, change in manner < *tropé*, a turning < *trépo*, to revolve, turn towards + *-ismós* > L. *-ismus* > E. *-ism*, state, phase, tendency, action): vertical geotropism concerning orthogeotropic organisms and organs, i.e., orthotropic with respect to the stimulus of gravity, like the majority of pedicels and stipes (negative orthogeotrops), and rhizoids, rhizines and pseudorhizae (positive orthotrops). Cf. **diageotropism**.

orthotropic (Gr. *orthós*, straight + *trópos*, gyration + suf. *-íkos* > L. *-icus* > E. *-ic*, belonging to, relating to): growing vertically from a point of origin; e.g., like the gametangial suspensors of *Phycomyces* (Mucorales). Also refers to the disposition of basidiospores on the basidium of Gasteromycetes; see **orthotropic spore**. Cf. **heterotropic**.

orthotropic spore (Gr. *orthós*, straight + *trópos*, gyration + suf. *-íkos* > L. *-icus* > E. *-ic*, belonging to, relating to; NL. *spora* < Gr. *sporá*, spore): a spore which is borne erect and centered on the sterigma. Discharge is passive and a spore collar is often visible at the point of detatchment from the sterigma. Common in the Gasteromycetes, such as *Podaxis* and *Sclerogaster* (Hymenogastrales).

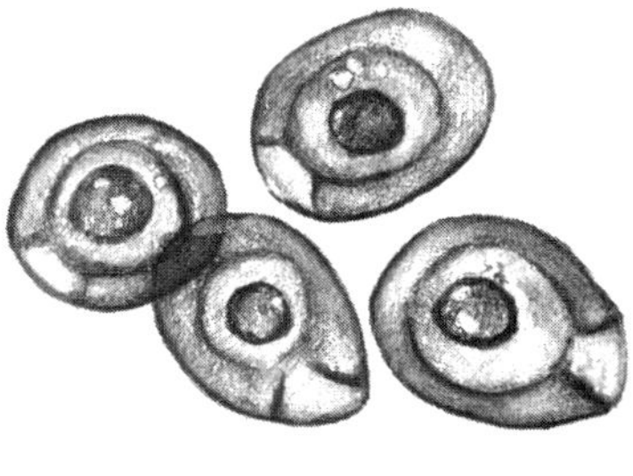

Orthotropic basidiospores of *Podaxis* sp., with an apical germ pore, thick wall, and central lipid body, x 1 500.

oscule (L. *osculum* < *os*, mouth, opening + dim. suf. *-culum* > E. *-ule*): generally applied to the germ pore of urediniospores of the rusts (Uredinales). The number and location of the pores or oscules varies according to the species. In *Uromyces appendiculatus* (bean rust), e.g., the urediniospores have two equatorial pores.

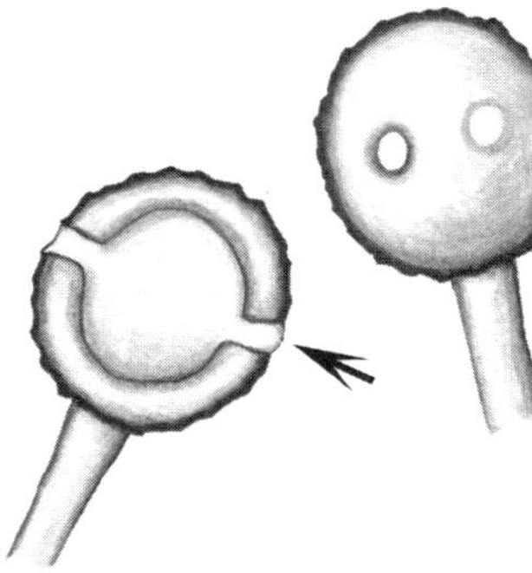

Oscules of the urediniospores of *Uromyces appendiculatus*, x 820.

osmiophilic (NL. *osmium* < Gr. *osmé*, smell, a hard brittle blue-gray or blue-black polyvalent metallic element + Gr. *phílos*, have an affinity for + suf. *-íkos* > L. *-icus* > E. *-ic*, belonging to, relating to): having the tendency to stain black with osmium tetroxide (a biological fixative), such as the lipids in the cells of Hyphochytriales, when prepared for observation with the electron microscope.

osmophile, **osmophilic**, **osmophilous** (Gr. *osmós*, to push, action of going through, a thrusting + *phílos*, have an affinity for; or + suf. *-íkos* > L. *-icus* > E. *-ic*, belonging to, relating to; or + L. *-osus* > OF. *-ous*, *-eus* > E. *-ous*, possessing the qualities of): applied to fungi that can live on substrates with high osmotic pressures. For example, the yeasts of the gen. *Zygosaccharomyces* (Saccharomycetales) have a marked tolerance to elevated concentrations of sugar in the medium. Some species of *Aspergillus*, such as *A. halophilicus* and *A. restrictus*, and of *Penicillium*, such as *P. citrinum*, (moniliaceous asexual fungi), require high concentrations of sugar or salt in the substrate.

osmosis, pl. **osmoses** (Gr. *osmós*, to push, action of going through, a thrusting + *-osis*, state of): a phenomenon concerning the diffusion that occurs through a semipermeable membrane that separates two solutions with different concentration of solute; osmosis tends to equalize the concentrations on both sides of the membrane. The fungi are typically osmotrophs, since they are nourished by osmosis.

osmotroph, **osmotrophic** (Gr. *osmós*, to push, action of going through + *trophós*, that which nourishes, serves as food; or + suf. *-íkos* > L. *-icus* > E. *-ic*, belonging to, relating to): acquiring nourishment by osmosis or absorption. The fungi are typically osmotrophs, although certain structures lacking a cell wall, such as the zoospores of aquatic species, and the myxamoebae and plasmodia of the Myxomycota, can be nourished by phagocytosis, in addition to absorption by osmosis.

ossiform (L. *oss*, *os*, bone + *form*, form): see **phalangoid**.

ostiolar canal (L. *ostiolum*, dim. of *ostium*, door + suf. *-aris* > E. *-ar*, like, pertaining to; L. *canalis*, canal, channel, conduit): a tubular anatomical passage or canal located within the **ostiolar neck** of perithecia. This canal is usually lined with periphyses, which are short unbranched hyphae that apparently serve to direct the asci toward the tip of the ostiole prior to ascospore discharge; e.g., in *Cercophora palmicola* (Sordariales).

ostiolar disc (L. *ostiolum*, dim. of *ostium*, door + suf. *-aris* > E. *-ar*, like, pertaining to; L. *discus*, disk): see **placodium**.

ostiolar neck (L. *ostiolum*, dim. of *ostium*, door + suf. *-aris* > E. *-ar*, like, pertaining to; ME. *nekke*, neck): an apical extension of some perithecia that contains the **ostiolar canal** and terminates in the ostiole; e.g., in *Gnomonia comari* (Diaporthales).

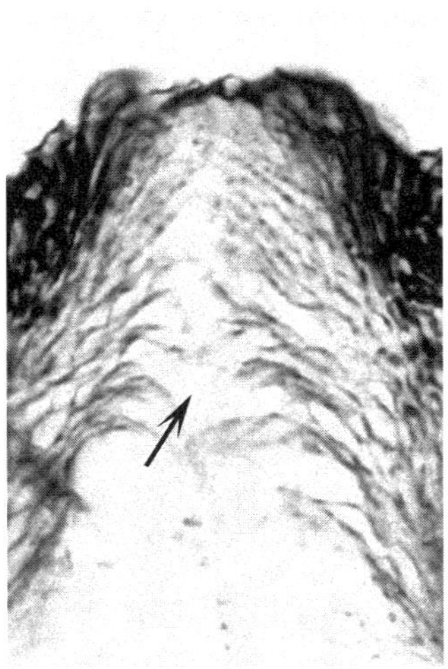

Ostiolar canal of the perithecium of *Cercophora palmicola*, x 375 (*RTH*).

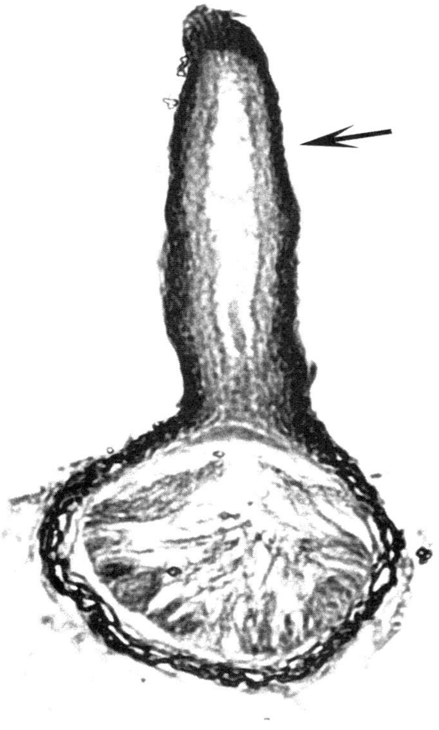

Ostiolar neck of the perithecium of *Gnomonia comari*, x 190 (*RTH*).

ostiole (L. *ostiolum* < *ostium*, door + dim. suf. *-olum* > E. *-ole*): the pore or opening, usually at the top, of diverse reproductive structures, such as the pycnidia of *Phoma* and other sphaeropsidaceous asexual fungi, or the perithecia of *Sordaria fimicola* (Sordariales), among many other Pyrenomycetes (Sordariales, Melanosporales, Diaporthales, Hypocreales). See **stoma**.

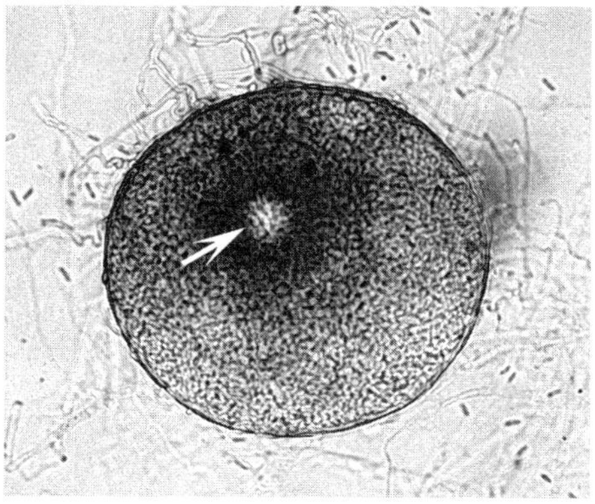

Ostiole of the pycnidium of *Phoma* sp., x 280 (*MU*).

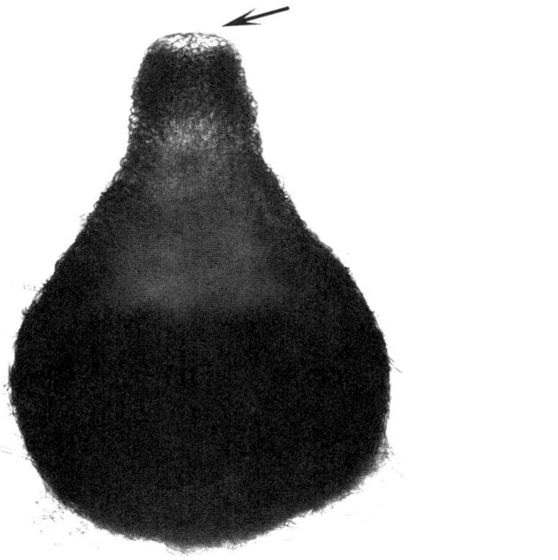

Ostiole of the perithecium of *Sordaria fimicola*, x 125 (*RTH*).

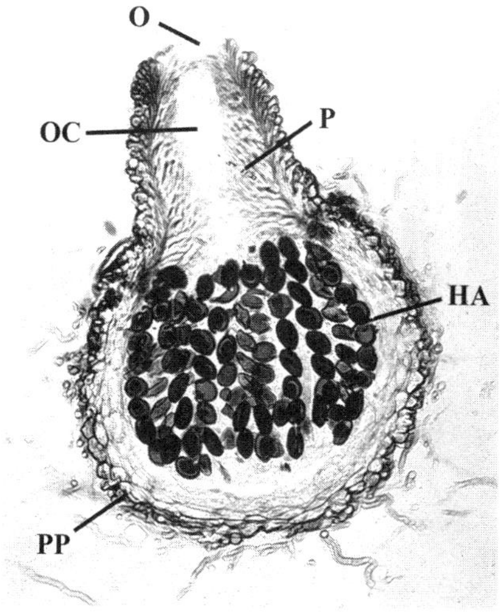

Ostiole (O) of the perithecium of *Sordaria fimicola*, seen in longitudinal section. The ostiolar canal (OC), periphyses (P), pseudoparenchymatous peridium (PP), and a hymenium with octosporate asci (HA) also are shown, x 125 (*RTH*).

outgroup (ME. < OE. *ut*, out, in a direction away from the inside or center + F. *groupe* < It. *gruppo*, group, a number of individuals assembled together or having some unifying relationship): a term used in systematics to refer to one or more taxa considered to be outside the monophyletic group (ingroup) of interest. Cf. **ingroup**.

oval (NL. *ovalis*, egg-shaped < *ovum*, egg + suf. *-alis* > E. *-al*, relating to or belonging to): shaped like an egg. See **ovate**.

ovariicolous (NL. *ovariicola* < *ovarium*, ovary + *-cola*, inhabitant + L. *-osus* > OF. *-ous*, *-eus* > E. *-ous*, possessing the qualities of): developing in the ovary of higher plants; e.g., such as *Claviceps* (Hypocreales), which infects ovaries of rye and other grains, in which are formed the sclerotia of the fungus.

ovate (L. *ovatus* < *ovum*, egg + suf. *-atus* > E. *-ate*, provided with or likeness): egg-shaped; a laminar structure with the shape of an egg, i.e., with the widest part at the base. In referring to a solid body, it is better to employ **ovoid**. Also called oval, which is perprolate, according to the polar axis, and peroblate, according to the equatorial diameter. For example, the basidiospores of *Panaeolus sphinctrinus* (Agaricales) are ovoid in face view. Cf. **obovate**.

ovoid (NL. *ovoides* < L. *ovum*, egg, ovule + *-oide* < Gr. *-oeídes*, similar to): a solid structure with the shape of an egg. Cf. **ovate**.

oyster mushroom (ME. *oistre* < MF. < L. *ostrea* < Gr. *óstreon*, shell; bivalve mollusk shell; ME. *musseroum* < MF. *mousseron* < LL. *mussirion*, mushroom): the basidioma of the edible *Pleurotus ostreatus* (Agaricales), which is industrially cultivated in many parts of the world.

P

pachyderm, **pachydermate**, **pachydermatous**, **pachydermous** (Gr. *pachýs*, thick + *dérma*, skin + L. suf. *-atus* > E. *-ate*, provided with or likeness; or + L. *-osus* > OF. *-ous*, *-eus* > E. *-ous*, having, possessing the qualities of): a hypha in which the external wall is thicker than the lumen. Cf. **leptoderm** and **mesoderm**.

paleomycology (Gr. *pálai*, *palaiós*, ancient + *mýkes*, fungus + *lógos*, study + *-y*, E. suf. of concrete nouns): the study of fossil fungi.

palisade (F. *palissade*, fence < OPr. *palissada* < *palissa*, paling < *pal* < L. *palus*, stake): a row of parallel structures, elongate in shape, arranged next to one another; e.g., such as the conidiophores of *Entomophthora muscae* (Entomophthorales).

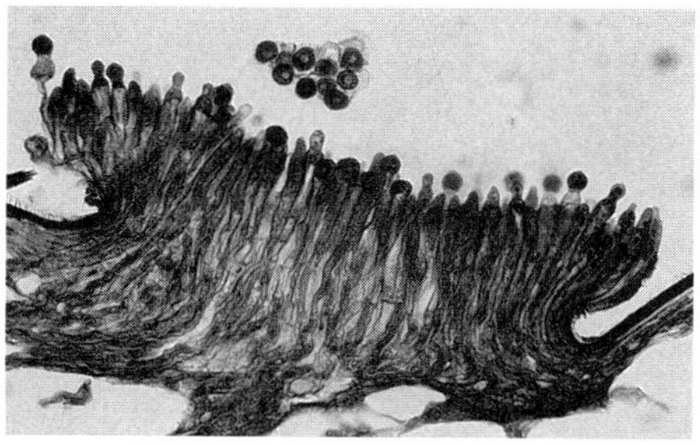

Palisade of conidiophores of *Entomophthora muscae* protruding from the infected tissues of a house fly, seen in longitudinal section, x 55 (*MU*).

palisade cell (F. *palissade*, fence < OPr. *palissada* < *palissa*, paling < *pal* < L. *palus*, stake; L. *cella*, storeroom > ML. *cella* > OE. *cell*, cell, compartment): *Lichens*. Terminal cell of the hyphae of the fastigiate cortex that are perpendicular or anticlinal to the plane of the thallus.

palisade plectenchyma, pl. **plectenchymata** (F. *palissade*, fence < OPr. *palissada* < *palissa*, paling < *pal* < L. *palus*, stake; Gr. *plektós*, braided, intertwined + *énchyma*, stuffed): *Lichens*. A tissue in the cortex composed of hyphae that are parallel and perpendicular to the surface of the thallus.

paludal, **paludicoline**, **paludicolous**, **paludine** (L. *palus*, genit. *paludis*, swamp + L. suf. *-alis* > E. *-al*, relating to or belonging to; or + L. suf. *-cola*, inhabitant < *colere*, to inhabit + L. suf. *-inus* > E. *-ine*, of or pertaining to; or + L. *-osus* > OF. *-ous*, *-eus* > E. *-ous*, possessing the qualities of): applied to organisms that live in low, flooded terrain, such as lagoons, bogs and swamps; e.g., like many zoosporic fungi (Chytridiomycota, Hyphochytriomycota and Oomycota) that inhabit these environments.

palustrine (L. *paluster*, genit. *palustris*, marshy < *palus*, *paludis*, swamp, marsh + L. suf. *-inus* > E. *-ine*, of or pertaining to): marshy, swampy.

pandemia (Gr. *pandemía* < *pándemos*, belonging to all the people): an epidemic or epiphytotic that extends to many countries, including continents, producing massive death of the organisms affected.

pandurate, **pandurated**, **panduriform** (L. *panduratus*, fiddle-shaped < *pandura*, a musical instrument with three strings + suf. *-atus* > E. *-ate*, provided with or likeness; L. *panduriformis* < *pandura* + *-formis* < *forma*, shape): oblong, but narrow in the middle or near the base; similar in shape to a guitar or violin. A rare shape, found in the columella of the sporangium of *Circinella muscae* (Mucorales).

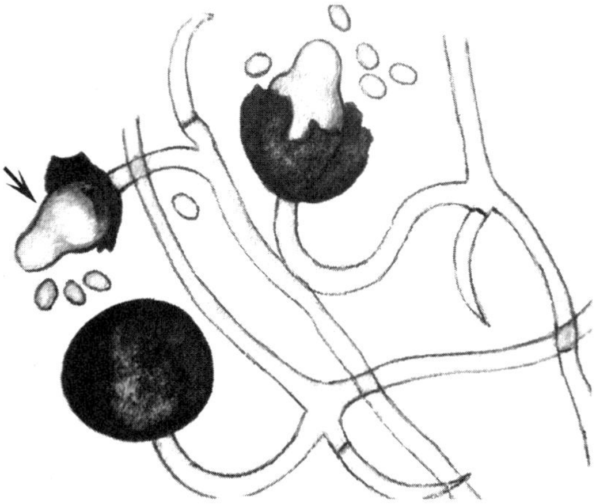

Pandurate columella of the sporangia of *Circinella muscae*, x 360.

pannose, **panniform** (L. *pannosus*, full of rags < *pannus*, a piece of cloth, rag + *-osus*, full of, augmented, prone to > ME. *-ose*; or + L. *-formis* < *forma*, form): having the appearance of felt or woollen cloth, like the fruiting body of *Lentinus* (=*Panus*), of the Aphyllophorales.

pantonema, pl. **pantonemata** (Gr. *pantós*, genit. of *pán*, everthing + *nêma*, genit. *nêmatos*, filament): a term applied to the tinsel or mastigonemate type of flagellum that is present in the zoospores of the Hyphochytriomycota and Oomycota. See **mastigoneme**.

papilionaceous (L. *papilio*, genit. *papilionis*, butterfly + *-aceus* > E. *-aceous*, of or pertaining to, with the nature of): speckled or variegated, with marks of different colors, like the gills of some species of *Panaeolus* (Agaricales).

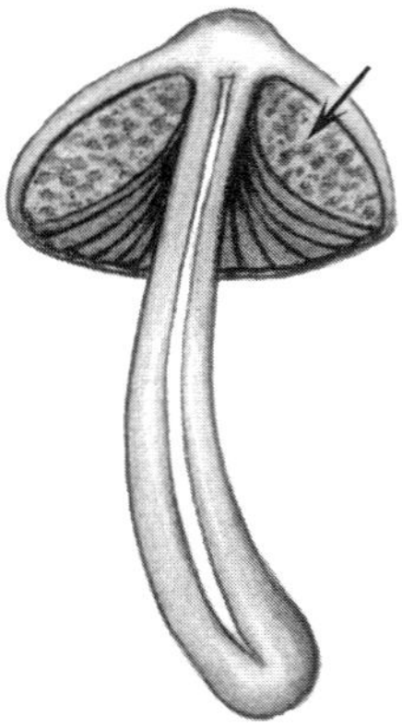

Papilionaceous lamellae of the basidiocarp of *Panaeolus campanulatus*, x 0.6.

papilla, pl. **papillae** (L. *papilla*, nipple): a small rounded or conic elevation, generally translucent, of the wall of sporangia and gametangia, which on breaking serves as the point of exit of zoospores and planogametes. For example, the sporangia and gametangia of *Allomyces* (Blastocladiales) have one or more papillae, and are said to be **papillate** or **papillose**. The columella of the sporangium of *Absidia cylindrospora* (Mucorales) has a papilla at its tip.

papillate, **papillose** (L. *papillatus* <*papilla*, nipple + suf. *-atus* > E. *-ate*, provided with or likeness; or + L. *-osus*, full of, augmented, prone to > ME. *-ose*): **1**. *Agaricales.* A pileus with a well delineated but not very elongated umbo, so that it appears mammiform; e.g., as in *Nolanea papillata* and *Psilocybe caerulescens*. **2**. *Aquatic fungi.* Small, thin-walled evaginations (papillae), that at maturity open and permit the exit of zoospores or planogametes, as in the zoosporangium of *Phlyctochytrium* (Chytridiales) and the gametangium of *Allomyces* (Blastocladiales).

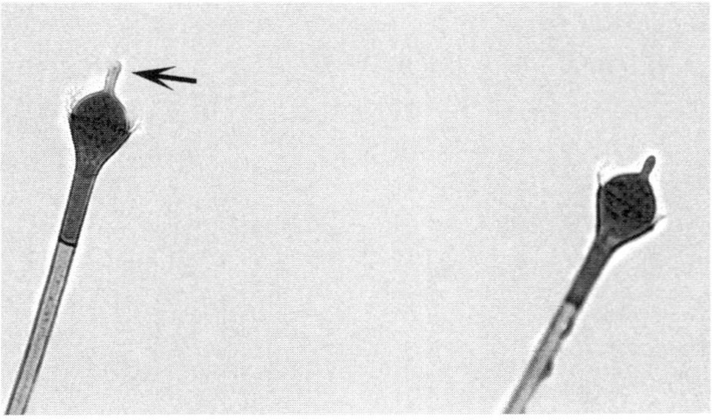

Papilla at the tip of the columella in the sporangia of *Absidia cylindrospora*, x 260 (*CB*).

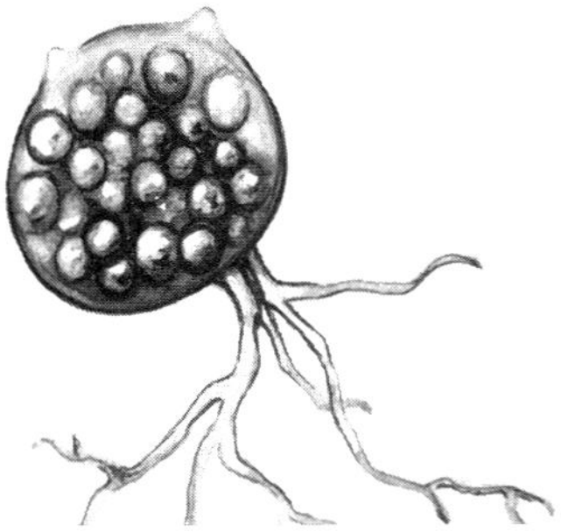

Papillate zoosporangium of *Phlyctochytrium* sp., x 1 000.

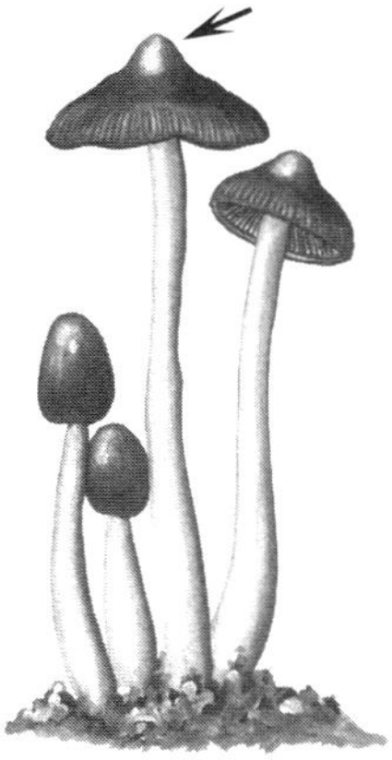

Papillate pileus of the basidiocarps of *Psilocybe caerulescens*, x 0.6.

papilliform (L. *papilliformis* <*papilla*, nipple + *-formis* <*forma*, shape): having the shape of a nipple.

papulose (L. *papulosus* <*papula*, papule, protuberance + *-osus*, full of, augmented, prone to > ME. *-ose*): covered with papules or pustules, like the surface of the sporangium of *Enerthenema papillatum* (Stemonitales) and the pileus of *Cortinarius* (=*Phlegmacium*) *papulosus* (Agaricales).

papulospore (L. *papula*, papule, protuberance + Gr. *sporá*, spore): a group of enlarged cells that function as a spore; they form a compact group, with each cell germinating independently, as in *Papulaspora* (agonomycetaceous asexual fungi). Also called **bulbil** and **microsclerotium**.

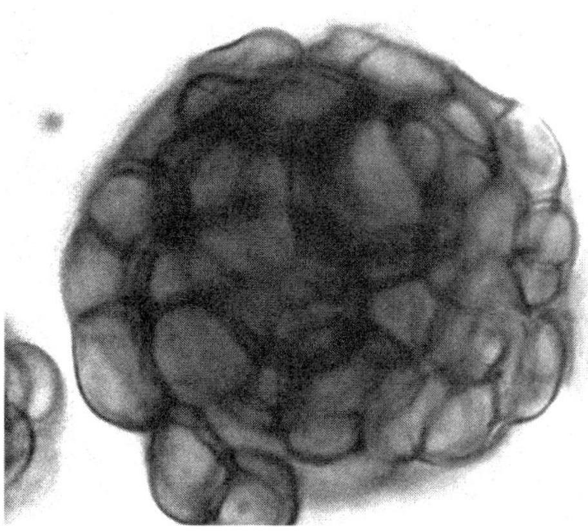

Papulospores of *Papulaspora* sp., x 1 500 (*MU*).

papyraceous (L. *papyraceus*, papery < *papyrus* < Gr. *papýros*, paper made from the papyrus plant + L. suf. *-aceus* > E. *-aceous*, of or pertaining to, with the nature of): papery, parchment-like, with the consistency and thinness of paper; e.g., like the endoperidium of *Bovista* (Lycoperdales).

par-, **para-** (Gr. pref. *pará*, with various meanings): at the side, above, beyond, toward, against, almost; used in the formation of biological terms, such as **parabiont** and **parabiosis**.

parabiont (Gr. *pará*, at the side, above + *bionte*, living being < *bión*, to live + *óntos*, genit. of *ón*, a being): refers to an organism that lives in symbiosis with two other organisms, as happens in certain lichen thalli, which consist of the association (**parabiosis**) of two distinct fungi and the same alga, or of one fungus and two different algae. For example, *Peltigera aphthosa* (Peltigerales) has the algae *Rivularia* (in the cephalodia) and *Nostoc* or *Dactylococcus* (in the other parts of the thallus). Cf. **parasymbiont**.

parabiosis, pl. **parabioses** (Gr. *pará*, at the side, above, beyond + *bíosis* < *bíos*, life): *Lichens*. The phenomenon in which a single lichen thallus is composed of two distinct fungi in symbiosis with the same alga, or of one fungus and two different algae. In *Sticta amplissima* (Lecanorales), e.g., two different algae (*Protococcus* and *Nostoc*) coexist with the fungus to constitute the lichen thallus. The organisms involved in this association are called **parabionts**. Cf. **antagonism**, **commensalism**, **helotism**, **metabiosis**, **mutualism**, **parasymbiosis** and **symbiosis**.

paracapillitium, pl. **paracapillitia** (Gr. *pará*, together, at the side of + L. *capillitium,* long hair, in the collective sense < *capillus*, hair + *-itium*, adjectival ending, meaning provided with, having): *Gasteromycetes*. A group of thin-walled, colorless, septate skeletal hyphae in the gleba that functions as a capillitium, but which originates from generative hyphae with a thin, or sometimes thickened, wall. It differs from the capillitium in not being cyanophilic; it is present, e.g., in *Vascellum* (Lycoperdales) and *Astraeus* (Sclerodermatales). Cf. **capillitium**.

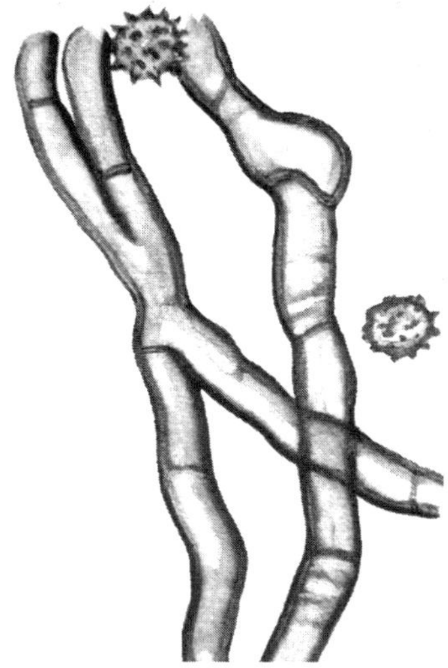

Paracapillitium from a fruiting body of *Astraeus* sp., x 1 000.

paracoccidiodomycosis, pl. **paracoccidiodomycoses** (< gen. *Paracoccidioides* < Gr. *pará*, at the side of + gen. *Coccidioides* < L. *coccidium* < Gr. *kokkíon*, small ball + L. suf. *-oides* < Gr. *-oeídes*, similar to + NL. *mycosis* < Gr. *mýkes*, fungus + *-osis*, state or condition): a primary pulmonary infection, acquired, as with other systemic mycoses, by the inhalation of conidia. The visible signs are mucocutaneous lesions, especially in the mouth, but the linfatic dissemination of this dimorphic organism (*Paracoccidioides brasiliensis*, of the moniliaceous asexual fungi) causes damage to skin, gastrointestinal tract, lungs, liver, and other organs. Also called South American blastomycosis because of its main occurrence in South America; it is rare in Central America, Mexico, and the Caribbean.

paracyst (Gr. *pará*, at the side of + *kýstis*, vesicle, cell): an old term for the antheridial cell which is found next to the ascogonium or macrocyst in the gen. *Pyronema* (Pezizales). Cf. **macrocyst** and **microcyst**.

paragymnohymenial (Gr. *pará*, at the side, above + *gymnós*, naked + L. *hymenium* < Gr. *hyménion*, dim. of *hymén*, membrane + L. suf. *-alis* > E. *-al*, relating to or belonging to): a type of ascocarp development in some Discomycetes, similar to the eugymnohymenial type; however, in this case the hymenium is partially protected in a manner that is analogous to cleistohymenial development. The sterile hyphae arise near the ascogonium, but do not envelop it completely, never forming a closed sheath. As the ascocarp expands, it remains open, with its side walls surrounding the developing asci among the paraphyses. Cf. **cleistohymenial** and **eugymnohymenial**.

paragynous (Gr. *pará*, at the side of, together + *gyné*, woman; here, the female sexual organ + L. *-osus* > OF. *-ous*, *-eus* > E. *-ous*, having, possessing the qualities of): *Peronosporales*. An antheridium that is formed at the side of the oogonium, as is observed in *Pythium* and *Plasmopara viticola* (Oomycota). Cf. **amphigynous**.

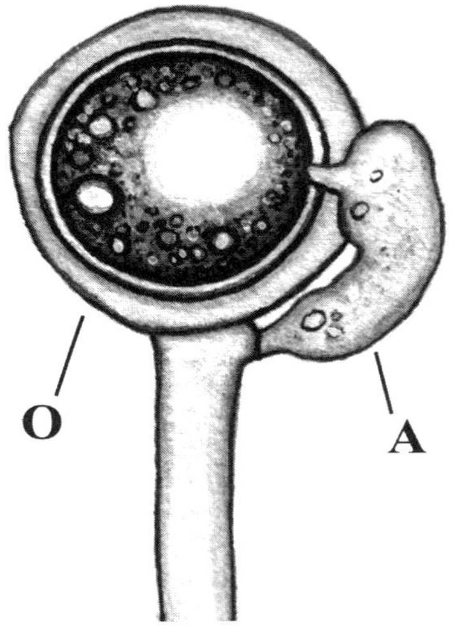

Paragynous antheridium (A), at the side of an oogonium (O) of *Pythium ultimum*, x 1 440.

parallel (L. *parallelus* < Gr. *parállelos*, side by side): **1.** *Agarics*. A type of hymeniiferous trama in which the hyphal elements are arranged next to one another, side by side, parallel to the long axis of the gills of the basidiocarp, as is seen in *Hygrocybe acutoconica* (Agaricales). **2.** *Mucorales*. In some gen., such as *Pilobolus*, the suspensors of the zygosporangium are parallel, unilateral and close, i.e., they are not opposite as in the majority of the gen. of the order. Also called **apposed**.

paramorph (Gr. *pará*, at the side, toward + *morphé*, form): a term proposed for a fossil fungus that, even though it lacks the characteristics necessary for its proper classification, is believed to have affinities with a particular non-fossil group.

paraphyletic (Gr. *pará*, at the side of + *phyletikós* < *phyletés*, one of the same tribe, race, stock < *phylé*, tribe + *-tikós* > L. *-ticus* > E. *-tic*, relation, fitness): a term used in cladistics to refer to a taxon or group of taxa in which some descendants of a taxon are not included. Cf. **monophyletic** and **polyphyletic**.

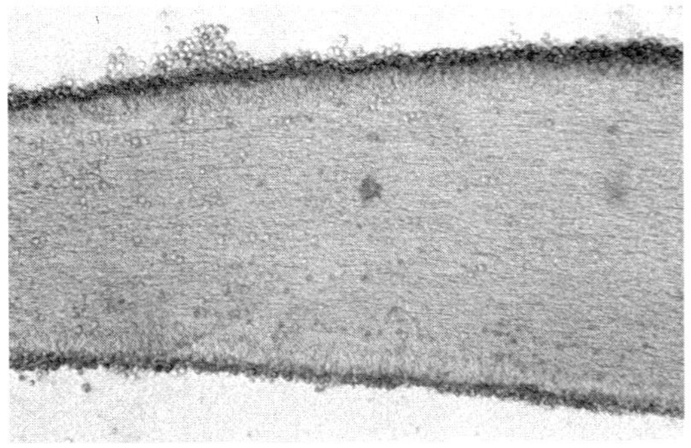

Parallel hymeniiferous trama in a gill of a basidiocarp of *Hygrocybe acutoconica*, x 150 (*RV*).

paraphysis, pl. **paraphyses** (Gr. *paráphysis*, offspring < *pará*, at the side of + *phýsis*, action of budding, of being born): a sterile, basally attached, hypha-like structure present in the hymenium of many ascomycetes and some pycnidial fungi; in *Geoglossum* and *Trichoglossum* (Helotiales), e.g., the paraphyses are long and slender, and are intermixed with the asci in the hymenium. The term paraphysis also has been applied, probably in the wrong sense, to the brachybasidia in the agarics that form a particular layer of the hymenium through which the basidia protrude. See **brachybasidium**.

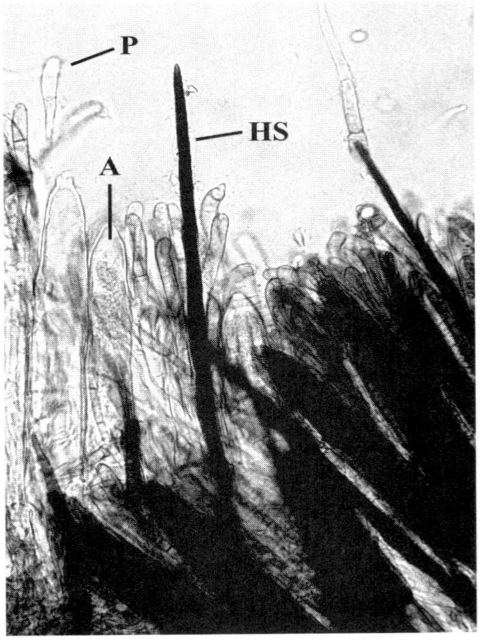

Paraphyses (P) of the hymenium of *Trichoglossum* sp. The asci (A) and hymenial setae (HS) also are shown, x 250 (*RTH*).

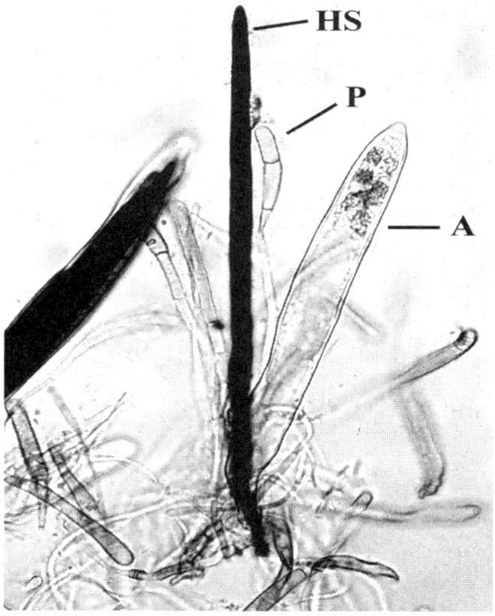

Paraphyses (P) of the hymenium of *Trichoglossum* sp. The asci (A) and hymenial setae (HS) also are shown, x 300 (*RTH*).

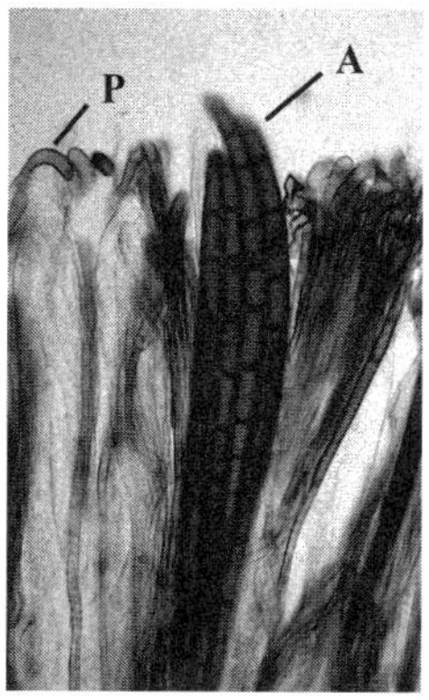

Paraphyses (P) of the hymenium of *Trichoglossum velutipes*. An ascus (A) with septate ascospores is also evident, x 200 (*VBM*).

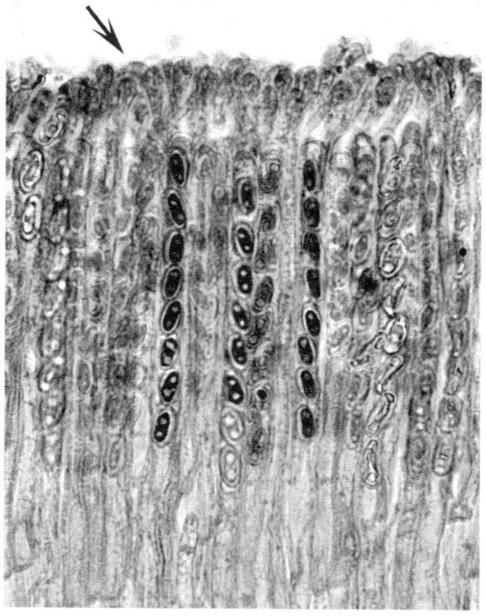

Paraphyses of the hymenium of *Peziza* sp. forming the epithecium over the layer of asci, x 100 (*MU*).

paraphysoid (Gr. *paráphysis*, offspring + L. suf. *-oide* Gr. < *-oeídes*, similar to): a sterile, thread-like hyphal structure, similar to a paraphysis, but often branched and forming a kind of network (paraphysoid network); found principally in the pseudothecia of some pyrenolichens with a crustaceous thallus. Also called paraphysoids are the slender hyphae that are extended and insinuated among the asci and stick out above them, in the manner of slender paraphyses, such as occurs in *Bagnisiella* (Dothideales) and *Myriangium* (Myriangiales).

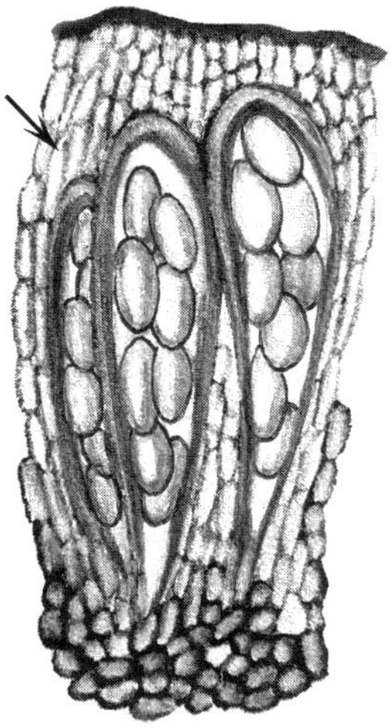

Paraphysoids of a pseudothecium of *Bagnisiella mirabilis*, embedded in the tissues of a dead twig, seen in longitudinal section, x 520.

paraplectenchyma, pl. **paraplectenchymata** (Gr. *pará*, together, at the side of + *plektós*, intertwined + *énchyma*, stuffed): a type of tissue found in the cortex of many macrolichens, in which the hyphae are oriented in all directions, giving a cellular appearance. It corresponds to the parenchyma of other fungi. Cf. **chalaroplectenchyma** and **prosoplectenchyma**.

parasexuality (Gr. *pará*, at the side of, together + L. *sexualis*, belonging to or relative to sex + E. suf. *-ty*, quality, condition): a process in which plasmogamy, karyogamy, and meiosis occur in sequence, but not in sexually differentiated organs nor at a specified time in the life cycle of the organism; in conidial fungi with heterokaryotic individuals, parasexuality may be important as a source of genetic variation that permits them to adapt to changing environmental conditions, as would occur with sexual reproduction. Cf. **perittogamy**.

parasite (Gr. *parásitos* < *pará*, together + *sîtos*, food; that which is fed near another): an organism that derives its nourishment from another living organism, whether plant, animal or fungus; if it invades and causes disease, it is considered a **pathogen**.

parasitism (Gr. *parásitos* < *pará*, together + *sîtos*, food + *-ismós* > L. *-ismus* > E. *-ism*, state, phase, tendency, action): the action of being parasitic. See **antagonistic symbiosis**, under **symbiosis**.

parasol mushroom (F. < OIt. *parasole* < *parare*, to shield + *sole*, sun; ME. *musseroum* < MF. *mousseron* < LL. *mussirion*, mushroom): the basidioma of the edible *Macrolepiota* (=*Lepiota*) *procera* (Agaricales).

parasymbiont (Gr. *pará*, at the side of, beyond, above + *symbióo*, life in common of two or more organisms + *óntos*, genit. of *ón*, a being): refers to an organism involved in the phenomenon of **parasymbiosis**, such as occurs in certain lichens, in which the same alga coexists with the lichenogenic fungus and another invading fungus, a parasite of the lichen (although not directly harmful to the algal cells). Cf. **parabiont**.

parasymbiosis, pl. **parasymbioses** (Gr. *pará*, beyond, above + *symbíosis*, life in common of two or more organisms): the state of coexistence of a lichen alga with two fungi, one the lichen mycobiont and the other a fungal parasite of the lichen, but which is not harmful to the gonidia. The organisms involved are termed **parasymbionts**. Cf. **antagonism**, **commensalism**, **helotism**, **metabiosis**, **mutualism**, **parabiosis** and **symbiosis**.

parathecium, pl. **parathecia** (NL. *parathecium* < Gr. *pará*, at the side of, beyond + *thekíon*, dim. of *thêke*, case, box; here, of the asci): *Lichens*. The peripheral extension of the hypothecium or subhymenial layer upward along the margin of the hymenium in some discolichens, e.g., *Ochrolechia parella* (Pertusariales). Also called marginal excipulum or proper exciple. See **excipulum**.

Parathecium of the apothecia of a thallus of *Ochrolechia parella*, x 12 (*MU*).

parenthesome (LL. *parenthesis* < Gr. *parénthesis*, putting in beside + Gr. *sõma*, body): a curved double membrane (shaped like a parenthesis) that is found on each side of a dolipore septum; it can be perforate, imperforate or vesiculose. Also called **septal pore cap**. It is present in most basidiomycetes, as, e.g., in *Corticium solani* (Aphyllophorales). See **dolipore**.

parietal (L. *parietalis* < *paries*, genit. *parietis*, wall + L. suf. *-alis* > E. *-al*, relating to or belonging to): belonging to or relative to the wall; formed in the wall. In mycology it is used to refer to something found in or on the walls of cells, asci, ascocarps, etc. For example, some perithecia have parietal asci, since they are attached to the internal wall of the perithecia.

parmuliform (L. *parmula*, dim. of *parma*, shield + *-formis* < *forma*, shape): shield-shaped; a small round shield with the margins turned upward; e.g., like the foliaceous thallus of the lichen *Parmelia* (Lecanorales).

part spore (L. *pars*, piece, portion; NL. *spora* < Gr. *sporá*, spore): each of the unicellular spores that results from the fragmentation of a bicellular or multicellular ascospore of *Hypocrea* and other Hypocreales, as well as of the ascospores of *Cordyceps militaris* and other Clavicipitaceae, such as *Balansia henningsiana* (Hypocreales), as well as *Preussia* (=*Sporormiella*) *australis* (Pleosporales).

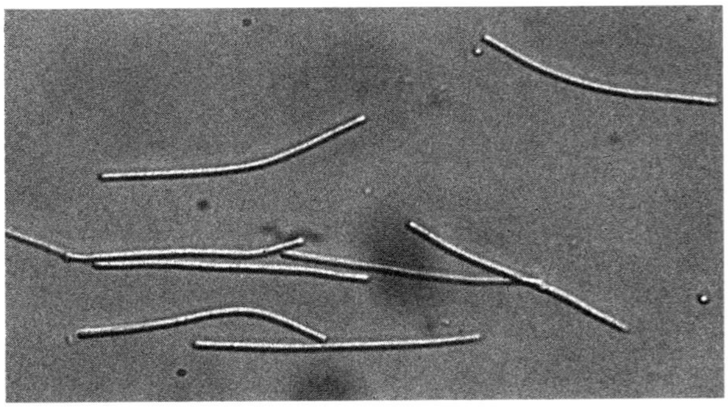

Part spores derived from the fragmentation of the ascospores of *Balansia henningsiana*, x 570 (*RTH*).

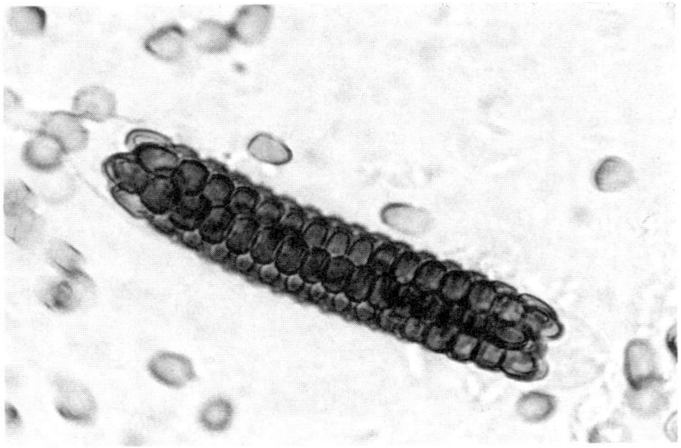

Part spores derived from the fragmentation of the septate ascospores of *Preussia* (=*Sporormia*) *fimetaria*, x 530 (*CB*).

Ascospore of *Preussia* (=*Sporormiella*) *australis*; its fragmentation originates **part spores**, x 1 300 (*RTH*).

parthenogamy (Gr. *párthenos*, virgin + *gámos*, sexual union + *-y*, E. suf. of concrete nouns): a type of copulation characterized by the union of two female sexual cells, especially of an archicarp; parthenogamy is a special case of **automixis**.

parthenogenesis, pl. **parthenogeneses** (Gr. *párthenos*, virgin + *génesis*, origin, production): generation of an individual from a female gamete that has not been fertilized by a male gamete (apomictic development), as can happen, e.g., in *Monoblepharis* (Monoblepharidales), and in many Mucorales, which can develop zygospores parthenogenetically without gametangial copulation having occurred. See **apogamy**, **apomixis** and **azygospore**. Cf. **amphimixis**.

parthenospore (Gr. *párthenos*, virgin + *sporá*, spore): *Entomophthorales*. An azygospore which originates as a bud from the hyphal bodies and which on maturing acquires the appearance and structure of a zygospore, but with parthenogenetic development; e.g., as occurs in *Entomophthora muscae*. An aboospore (an oospore with apomictic development) also is a parthenospore. See **aboospore** and **azygospore**.

partial veil (L. *partialis* < *pars*, genit. *partis*, part + suf. *-alis* > E. *-al*, relating to or belonging to; L. *velum*, veil): see **veil**.

passage (E. *pass* < ME. *passen* < OF. *passer* < VL. *passare* < L. *passus*, step, move, proceed; the action or process of passing from one place, condition, or stage to another): the experimental infection of a host with a parasite which is subsequently reisolated; a method used to increase the virulence of a parasite or to confirm the pathogenicity of a particular fungus.

patella, pl. **patellae** (L. *patella*, dim. of *patina*, plate, frying pan): a sessile, orbicular apothecium with a border that is distinguished from the thallus; present in some lichens.

patelliform, **pateniform** (L. *patelliformis* < *patella*, dim. of *patina*, plate, frying pan + *-formis* < *forma*, shape): in the shape of a round plate, with a well delimited border, like the slightly concave disciform apothecia of some lichens, e.g., *Scutellinia* (=*Patella*) *albida* (Pezizales) and *Arthonia patellulata* (Arthoniales).

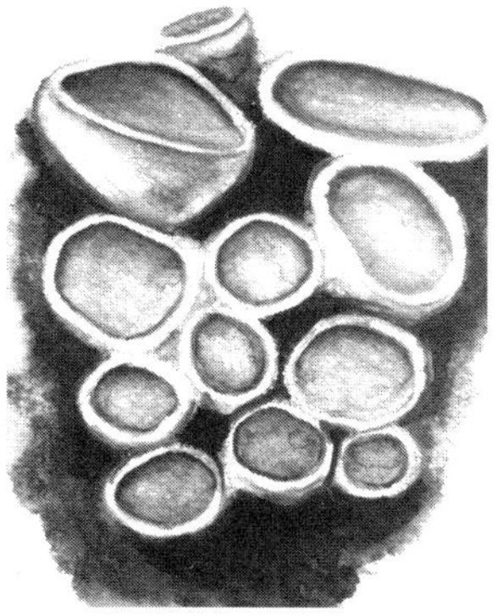

Patelliform apothecia of *Scutellinia* (=*Patella*) *albida*, x 2.

pathogen (Gr. *páthos*, disease + *génos*, origin < *gennáo*, to engender, produce): an organism capable of causing a disease; not necessarily as a parasite. Cf. **parasite**.

p-body (< *p*, for pigment, or for protein, and OE. *bodig*, body): a distinctive organelle that is associated with the rough endoplasmic reticulum which surrounds the platelike cristae of the mitochondria of the myxamoebae of *Acrasis rosea* (Acrasiomycota). The function of the p-body is unknown but has been suggested to be a pigment granule, or a proteinaceous organelle, since much of the p-body can be removed with proteases.

pectinate (L. *pectinatus*, hair-dresser < *pecten*, genit. *pectinis*, comb + suf. *-atus* > E. *-ate*, provided with or likeness): having a comb-shaped structure or being comb-like; a hypha that ends in a comb-shaped structure, somewhat swollen at the apex and with short processes on one side, as occurs in the dermatophytes, e.g., *Microsporum audouinii* (moniliaceous asexual fungi).

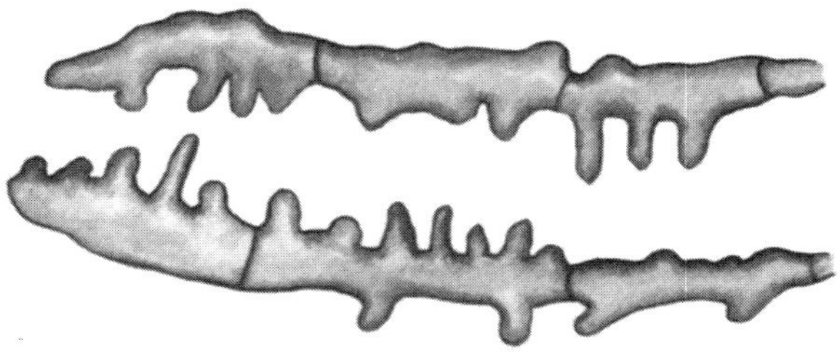

Pectinate hyphae of *Microsporum audouinii*, x 1 150.

pedicel (NL. *pedicellus*, dim. of L. *pediculus*, dim. of *pes*, genit. *pedis*, foot; here, support): a slender stalk or support of spores, sporangia, cystidia, asci, etc. In general, **pedicle** and **peduncle** are equivalent terms.

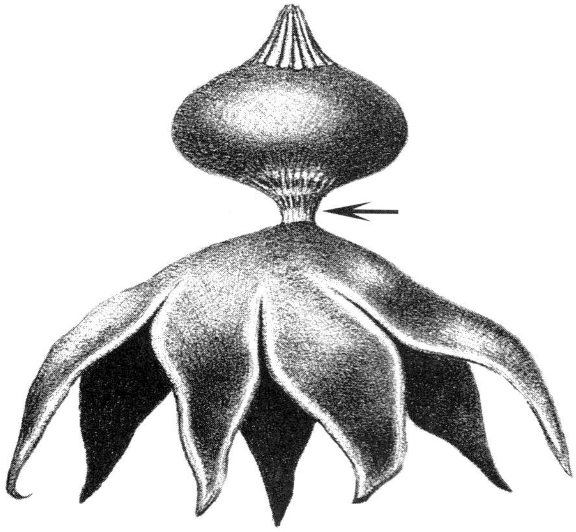

Pedicel supporting the spore case of a fruiting body of *Geastrum pectinatum*, x 2.

pedicellate, **pediculate**, **pedunculate** (L. *pedicellus*, dim. of *pediculus*, dim. of *pes*, genit. *pedis*, foot; here, support + suf. *-atus* > E. *-ate*, provided with or likeness): having a pedicel, or peduncle. For example, the sporangia of *Physarum polycephalum* (Physarales) and *Lamproderma columbinum* (Stemonitales), the spores of *Gigaspora* (Glomales), the asci of *Preussia* (=*Sporormiella*) *australis* (Pleosporales), and the teliospores of the species of *Puccinia*, *Gymnosporangium* and *Phragmidium* (Uredinales) are pedicellate. In some species of basidiomycetes a portion of the pedicel remains attached to the basidiospore after discharge; such species are described as pedicellate, e.g., in *Bovista*, *Lycoperdon* and *Geastrum* (Lycoperdales). Cf. **apodal**, **sessile** and **stipitate**.

Pedicellate sporangia of *Lamproderma columbinun*, x 8 (*MU*).

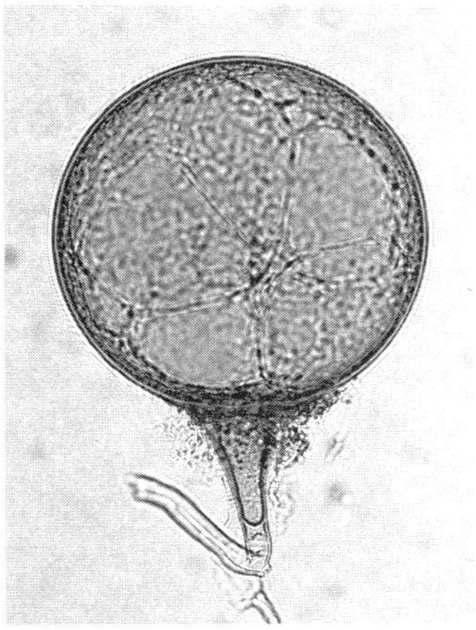

Pedicellate spore of *Gigaspora* sp., x 400 (*CB*).

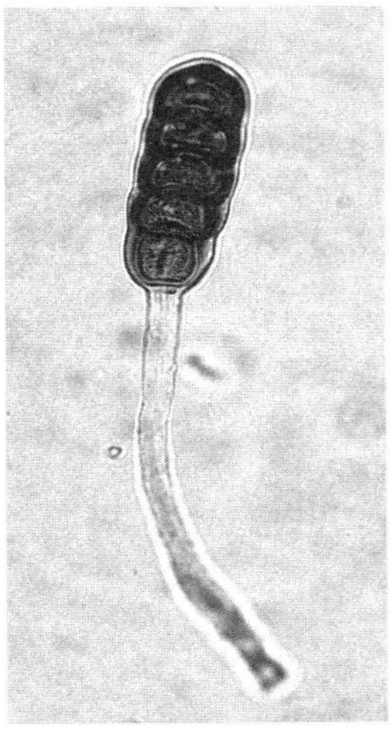

Pedicellate teliospore of *Phragmidium potentillae*, x 1 600 (*CB*).

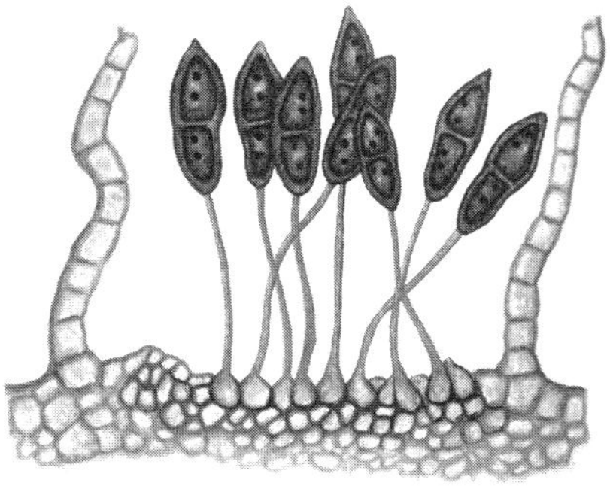

Pedicellate teliospores of a telium of *Puccinia graminis tritici*, x 480.

pedogamy (Gr. *país*, *paidós*, child, infant + *gámos*, sexual union + *-y*, E. suf. of concrete nouns): a type of pseudomictic sexuality, which consists of the union of an adult mother cell with a recently formed daughter cell; e.g., as happens in yeasts of the gen.

Debaryomyces and *Nadsonia* (Saccharomycetales), in which copulation occurs between a mother cell and its bud, which is still attached. Cf. **adelphogamy**.

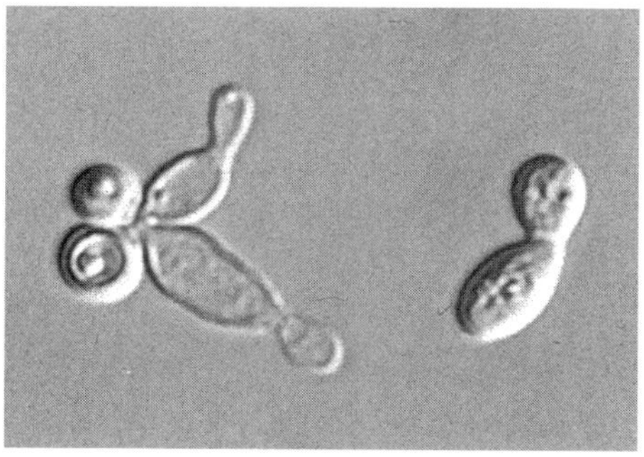

Pedogamy in *Nadsonia fulvescens*, between a mother cell and its attached bud, resulting in the formation of an ascus containing an ascospore, x 1 000 (*RTH*).

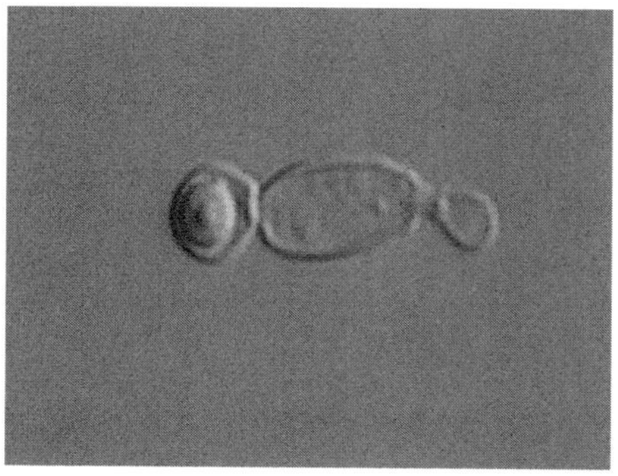

Pedogamy in *Nadsonia fulvescens*, between a mother cell and its attached bud, resulting in the formation of an ascus containing an ascospore, x 1 100 (*RTH*).

pedogenesis, pl. **pedogeneses** (Gr. *país*, *paidós*, child, infant + *génesis*, origin, production): reproduction in young or immature organisms.

pedogenesis, pl. **pedogeneses** (Gr. *pédon*, soil, earth + *génesis*, engenderment): the formation and development of soil by the intervention of lichen secondary metabolites (so-called lichen acids), which include fatty acids, lactones, polyketides such as dibenzofurans, depsides, and anthraquinones, and metabolites produced by the mevalonate and shikimate pathways.

pelagic (Gr. *pélagos*, sea, ocean + suf. *-îkos* > L. *-icus* > E. *-ic*, belonging to, relating to): a term applied to organisms that inhabit the sea, beyond the neritic zone, or away from the coast or shore. Among the fungi, there are zoosporic species (Chytridiomycota, Hyphochytriomycota and Oomycota) and conidial species (hyphomycetous and blastomycetous asexual fungi), as well as certain basidiosporogenous yeasts, that are pelagic.

peloton (F. *peloton*, ball): syn. of **glomerulus**; a dense mass, composed of a group of hyphae, formed by mycorrhizogenous fungi in the interior of the root cells of the associated plant, e.g., in *Allium ursinum*. The pelotons are one of the forms (the others correspond to the arbuscles, or to the vesicles) that the endotrophic mycorrhizae have.

pelta, pl. **peltae** (L. *pelta* < Gr. *pélte*, small shield): an apothecium without a differentiated border, flat and not very prominent, in the shape of a small shield, like that of certain peltigerous lichens (e.g., *Peltigera* and *Solorina*) of the order Peltigerales.

peltate (L. *peltatus*, armed with a **pelta** < *pelta*, shield < Gr. *pélte*, small shield + L. suf. *-atus* > E. *-ate*, provided with or likeness): applied to a flat, rounded structure, with the stalk or support inserted in the center of the lower side; e.g., like the sporangium of *Physarum flavicomum* (Physarales). Also called **clypeate** and **scutiform**.

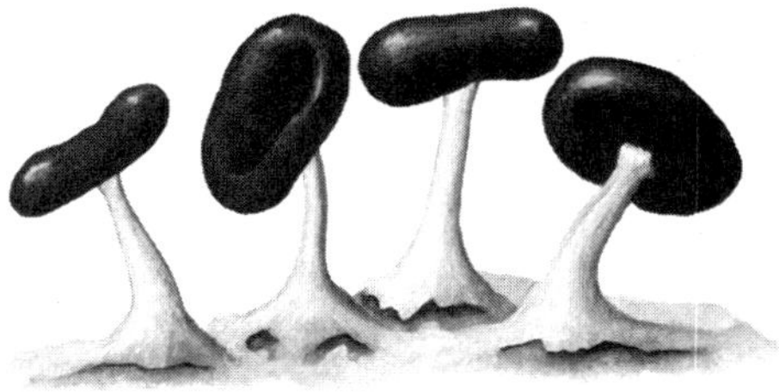

Peltate sporangia of *Physarum flavicomum*, x 10.

pellet (ME. *pelote* < MF. < VL. *pilota*, dim. of L. *pila*, ball): a spheroidal colony developed in a liquid culture, particularly a shaken culture.

pellicle, **pellicula**, pl. **pelliculae**, **pellis** (L. *pellicula*, dim. of *pellis*, skin): **1**. A thin, delicate, external membrane, similar to cuticle or skin, that can break off of the pileus of the basidiocarp of certain agarics. **2**. The thin membrane that forms on the surface of some cultures of bacteria or yeasts when they develop in liquid nutritive media.

pellicular veil (L. *pellicula*, dim. of *pellis*, skin + suf. *-aris* > E. *-ar*, like, pertaining to; L. *velum*, veil, curtain, cloth): a very thin partial veil that is present on a sporophore lacking a stipe. Cf. **cortina**.

pelliculate, **pelliculose** (L. *pellicula*, dim. of *pellis*, skin + suf. *-atus* > E. *-ate*, provided with or likeness; or + L. *-osus*, full of, augmented, prone to > ME. *-ose*): pellicle-like; having the shape of a pellicle or crust, like the hymenial layer of the thelephoraceous basidiomycetes.

pellicule, **pellis**, **cortex** (MF. *pellicule* < L. *pellicula* < *pellis*, skin + dim. suf. *-cula* > E. *-cule*; *cortex*, cortex): *Agaricales*. The cortical or superficial sterile layer of the basidiocarp. The term pellicule is applied to the cortex, which in a viscid pileus can detach easily. The skin of the stipe is called the **stipitipellis** (or cuticle of the stipe), whereas the skin of the pileus is called **pileipellis** (or cuticle of the pileus). The skin or cortex of the basidiocarp is composed of layers of cells, whose number varies in accordance with the species of the fungus. If it is composed of a single layer it is called **suprapellis**; if it is of two layers, the upper one is the suprapellis and the lower one the **subpellis**; when an intermediate layer exists, it is called the **medipellis**; see these terms.

pellucid (L. *pellucidus*, transparent): transparent or translucent. Refers to a pileus which is translucent, so that the gills appear as lines or striae (striate-pellucid) from above, such as occurs in various basidiocarps of Agaricales (*Mycena*, *Coprinus*, *Oudemansiella*).

pendulous (L. *pendulus*, hanging down + *-osus* > OF. *-ous*, *-eus* > E. *-ous*, having, possessing the qualities of): hanging down, like the hymenophore teeth of *Hydnum erinaceus* (Aphyllophorales), which develop with positive geotropism, and the thalli of some lichens, such as *Usnea* (Lecanorales).

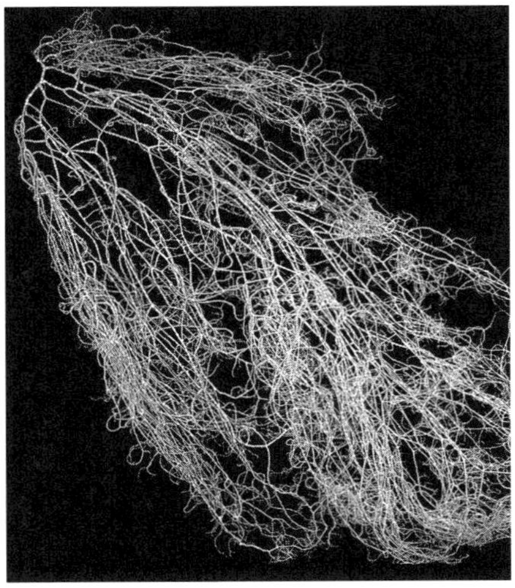

Pendulous lichen thallus of *Usnea ceratina*, x 0.2 (*MU*).

penicillate, **penicilliform** (L. *penicillatus* < *penicillus*, brush + suf. *-atus* > E. *-ate*, provided with or likeness; NL. *penicilliformis* < *penicillus*, brush + suf. *-formis* < *forma*, shape): in the shape of a brush, e.g., like the conidiophores of the gen. *Trichoderma*, *Penicillium*, and *Gliocladium* (moniliaceous asexual fungi).

penicillus, pl. **penicilli** (L. *penicillus*, brush): an asexual conidial head in the shape of a brush, consisting of a conidiophore or pedicel that supports a cluster of condiogenous cells (phialides), and which also can have other elements (such as ramae and metulae), depending upon the species; characteristic of the gen. *Penicillium* and of other related gen. of moniliaceous asexual fungi (*Paecilomyces*, *Gliocladium*).

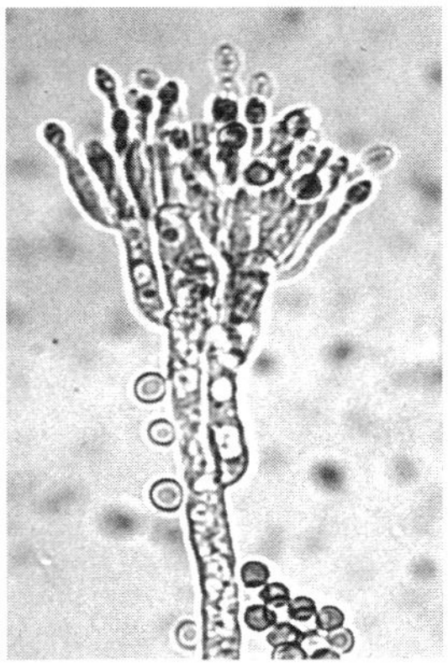

Penicillate conidiophore of *Penicillium* sp., x 460 (*RTH*).

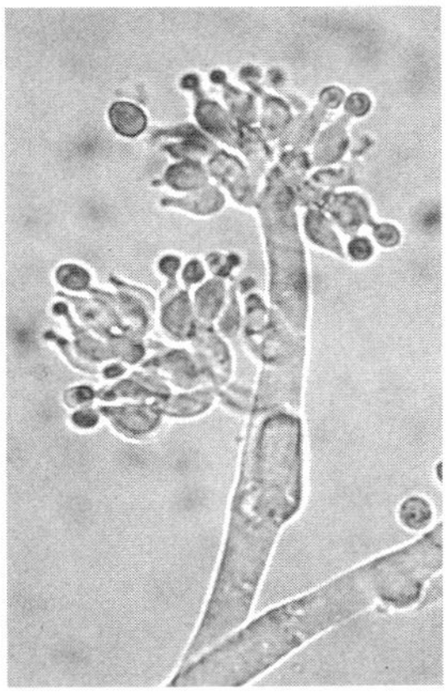

Penicillate conidiophore of *Trichoderma viride*, x 400 (*RTH*).

pennate, **penniform** (L. *pennatus* < *penna*, feather + suf. *-atus* > E. *-ate*, provided with or likeness; L. *-formis* < *forma*, shape): possessing feather-like structures or shaped like a feather; e.g., like the merosporangia of *Linderina pennispora* (Kickxellales), referring to the arrangement of the pseudophialides on the sporocladium.

Pennate merosporangia formed on a sporocladium of *Linderina pennispora*, x 500.

percurrent (L. *percurrere*, to run through, to pass over + *-entem* > E. *-ent*, being): to grow or extend through; refers to the growth of a conidiophore (percurrent proliferation) or germ tube (percurrent germination) through a preexisting pore; e.g., as occurs in the conidiophore of *Corynespora* or in the phialides of some species of *Phialophora* (both dematiaceous asexual fungi). See **proliferation**.

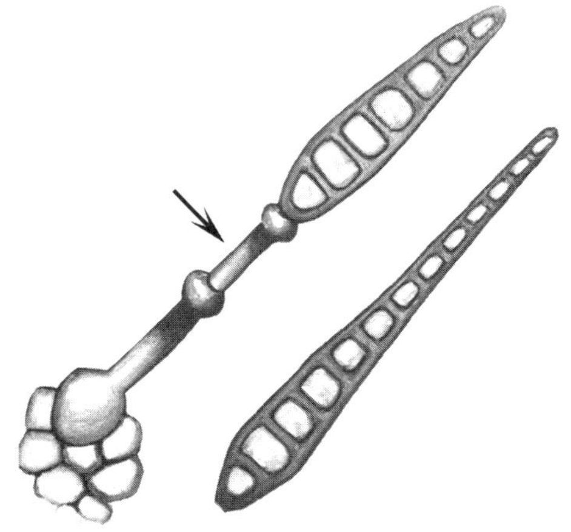

Percurrent conidiogenous cell of *Corynespora cassiicola*, x 530.

percurrent columella, pl. **columellae** (L. *percurrere*, to run through, to pass over + *-entem* > E. *-ent*, being; L. *columella*, dim. of *columna*, column): a sterile column of tissue which extends from the base of the gleba through it to the peridium at its apex, as in *Podaxis* (Hymenogastrales). Also called **dendritic columella** and **stipe-columella**.

perennial (L. *perennis*, persistent, durable, eternal + L. suf. *-alis* > E. *-al*, relating to or belonging to): refers to a fungus that lives three or more years; in reality, it refers to the durability of the fruiting body, not of the assimilative mycelium in the substrate. If it lives two years, or more than one, it is biennial; if it lives one year or less it is annual. Many basidiomycetes in the Aphyllophorales have perennial sporophores because they are woody and not easily decayed. Cf. **annual** and **biennial**.

perforate at maturity (L. *perforatus*, pierced < *perforare*, to bore through + suf. *-atus* > E. *-ate*, provided with or likeness; *maturitas*, ripeness + E. suf. *-ty*, quality, condition): *Gasteromycetes*. Refers to a thin peridium which splits or develops holes and thereby exposes the mature gleba, e.g., in *Phallogaster* (Phallales).

periclinal (Gr. *perí*, around + *klíno*, to incline + L. suf. *-alis* > E. *-al*, relating to or belonging to): refers to a membrane or cell wall of a structure that is parallel to the surface of that structure. Cf. **anticlinal**.

peridermioid (Gr. *perí*, around + *dérma*, skin + L. suf. *-oide* < Gr. *-oeídes*, similar to): *Uredinales*. An aecium of the type characterized by the gen. *Peridermium*. See **peridermium**.

peridermium, pl. **peridermia** (Gr. *perí*, around + *dérma*, skin + L. dim. suf. *-ium*): *Uredinales*. A type of aecium with a peridium that is cylindrical or shaped like a tongue or blister, and which forms on the outside of the host; such aecia are placed in the gen. *Peridermium*, such as *P. ipomoeae*, formed on the leaves of sweet potato.

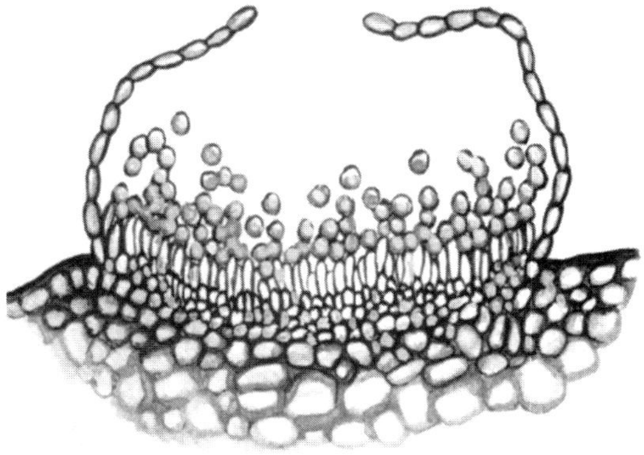

Peridermium (aecium) of *Peridermium* sp. on a host leaf, seen in longitudinal section, x 75.

peridiole, **peridiolum**, pl. **peridiola** (L. *peridiolum*, small purse < *peridium* < Gr. *péridion*, leather purse, knapsack + dim. suf. *-olum* > E. *-ole*): *Gasteromycetes*. The small oval to rounded bodies ("eggs"), with a thick wall, or tunica, that contain the glebal elements, i.e., basidia and basidiospores, in the orders Nidulariales (such as *Cyathus*, *Nidularia* and *Crucibulum*) and Sclerodermatales (such as *Pisolithus tinctorius*). Also called **angiole**.

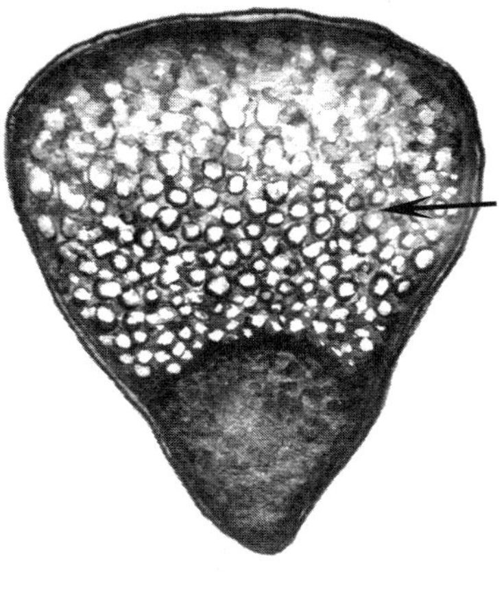

Peridioles of the fruiting body of *Pisolithus tinctorius*, seen in longitudinal section, x 1.5.

Peridioles in the gleba of a basidiocarp of *Pisolithus tinctorius*, seen in longitudinal section, x 1 (*RV*).

Peridioles of the fruiting body of *Cyathus* sp., x 6 (*MU*).

Peridioles of the fruiting body of *Crucibulum laeve*, x 6 (*CB*).

peridium (pl. **peridia**) **types**: *reticuloperidium* (L. *reticulum*, mesh + NL. *peridium* < Gr. *péridion*, small leather purse) - a mesh-like peridium composed of much-branched, often anastomosed, thick-walled hyphae; e.g., as occurs in *Auxarthron* (Onygenales); *incompositoperidium* (L. pref. *in-*, not, without + *compositus*, put together, joined) - a peridium composed of an unorganized mass of thick-walled, contorted cells; e.g., as found in *Shanorella* (Onygenales); *telaperidium* (L. *tela*, web) - a cobweb-like peridium of thin-walled, undifferentiated hyphae that resembles vegetative hyphae; e.g., as occurs in *Byssoascus* (Onygenales); *cleistoperidium* (Gr. *kleistós*, closed, enclosed) - a thin membranous peridium composed of angular cells; e.g., as occurs in *Aphanoascus* (Onygenales).

peridium, pl. **peridia** (NL. *peridium* < Gr. *péridion*, small leather purse): a wall that envelopes a fruiting body (microscopic or macroscopic); the peridium can be acellular or be composed of plectenchyma, depending upon the type of fungus. For example, the peridium is acellular in the sporangia of the Myxomycetes and Zygomycetes, but the cleistothecia, perithecia, pseudothecia and some basidiocarps possess a peridium composed of plectenchyma.

periphysis, pl. **periphyses** (Gr. *perí*, around + *phýsis*, a being, growth, natural form of something): short, sterile hyphae that line the internal wall of the ostiolar canal of a perithecium; e.g., as is seen in *Sordaria fimicola* (Sordariales) and *Meliola floridensis* (Meliolales).

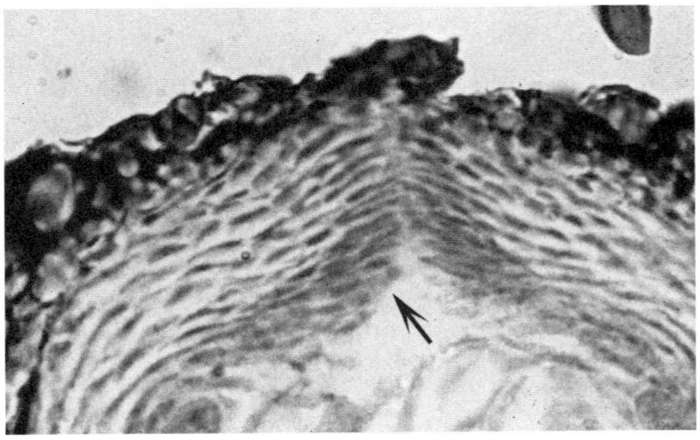

Periphyses lining the ostiolar canal of a perithecium of *Meliola floridensis*, x 800 (*RTH*).

periphysoid (Gr. *perí*, around, action of budding + *phýsis*, a being, growth + L. suf. *-oide* < Gr. *-oeídes*, similar to): lateral periphyses which are present in the pseudothecia of some pyrenolichens with a crustaceous thallus.

periplasm (Gr. *perí*, around + *plásma*, bland material with which is formed a living being): the peripheral plasma of a gametangium or of a sporangium, which is not utilized in the formation of the gametes or spores; in the oogonium of the Oomycota two zones are distinguished in the protoplasm, a dense central zone, the ooplasm (which first generates the oosphere and afterwards the oospore), and another, peripheral zone, more fluid, which is the periplasm, as is observed, e.g., in *Apodachlya* (Leptomitales) and *Rhipidium* (Rhipidiales). Cf. **ooplasm**.

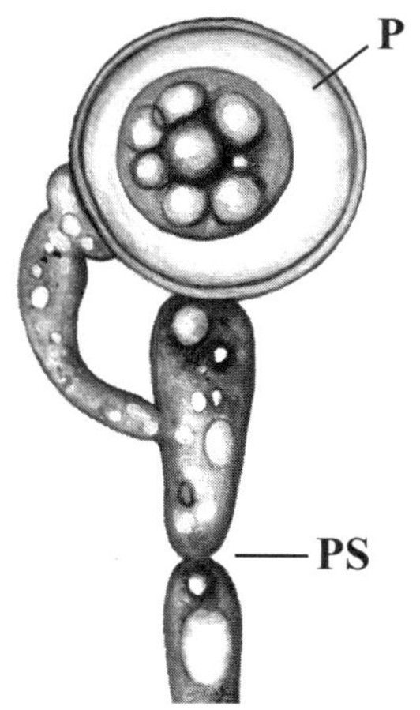

Periplasm (P) of the oogonium of *Apodachlya* sp. The pseudosepta (PS) of the supporting hypha are also shown, x 890.

perispore, perisporium, pl. **perisporia** (NL. *perisporium* < Gr. *perí*, around + *sporá*, spore + L. dim. suf. *-ium*): in the wall of the most complex basidiospore, which consists of five layers or strata, e.g., *Tubaria* (Agaricales), the perispore is the outermost layer, which is not pigmented and generally envelops the spore like a sack, which can disappear in an early stage of development (as in *Coprinus narcoticus* and related species of Agaricales). Beneath the perispore are found the four other layers: **exospore**, **epispore**, **mesospore** and **endospore**.

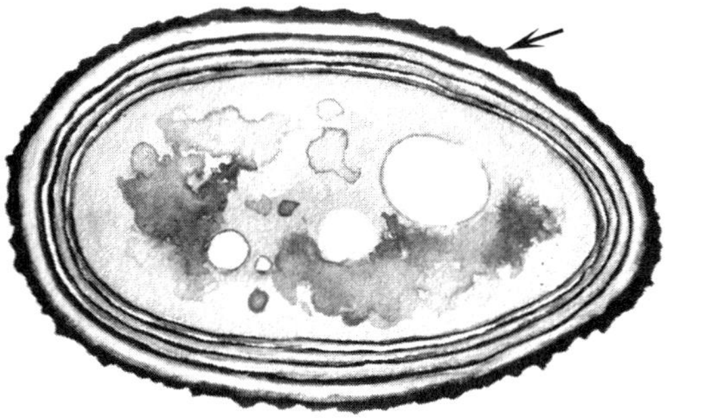

Perispore of the basidiospore wall of *Tubaria* sp. In this diagram of a transverse section, the endospore, mesosporium, epispore, and exospore also are shown, x 6 000.

peristome (Gr. *perí*, around + *stóma*, mouth): *Gasteromycetes*. A circular area, ornamented in various ways, around the opening or stoma in the peridium of the fruiting body of *Geastrum (G. saccatum, G. triplex)* and related gen. (Lycoperdales). See **stoma**.

perithecial stroma, pl. **stromata** (Gr. *perithekíon*, dim. of *perithêke*, helmet < *perí*, around + *thêke*, case + L. suf. *-alis* > E. *-al*, relating to or belonging to; Gr. *strõma*, bed, cushion): a stroma bearing perithecia, of variable morphology; e.g., in *Hypocrea schweinitzii* (Hypocreales) it is fleshy. Among the Xylariales, some examples are *Camillea*, which has carbonous stromata; *Diatrype*, with woody stromata, and *Xylaria*, with stromata that can be carbonous, woody or fleshy, according to the species. Many ascomycetes, belonging to various orders, have perithecial stromata.

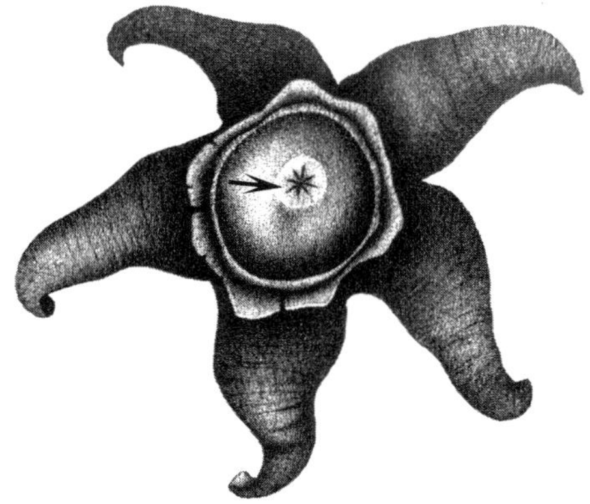

Peristome of the fruiting body of *Geastrum triplex*, x 1.

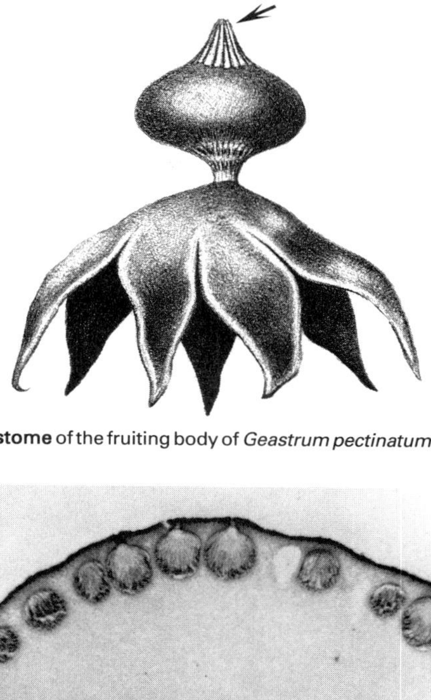

Peristome of the fruiting body of *Geastrum pectinatum*, x 1.

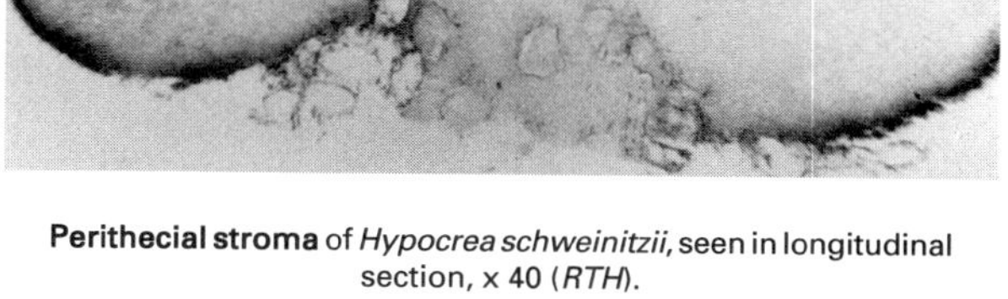

Perithecial stroma of *Hypocrea schweinitzii*, seen in longitudinal section, x 40 (*RTH*).

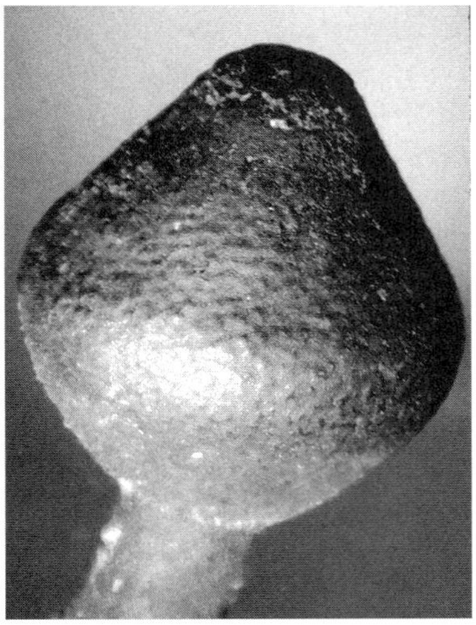

Perithecial stroma of *Camillea* sp., x 5 (*CB*).

Perithecial stroma of *Xylaria* sp., seen in longitudinal section, x 3 (*RTH*).

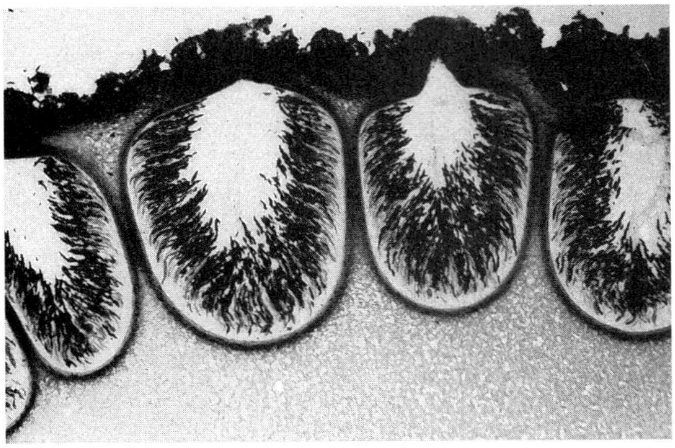

Perithecial stroma of *Xylaria* sp., seen in cross section, x 78 (*RTH*).

Perithecial stromata of *Diatrype disciformis* bursting out of the bark of a branch, x 20 (*MU*).

perithecioid (NL. *perithecium* < Gr. *perithekíon*, dim. of *perithêke* < *perí*, around + *thekíon*, dim. of *thêke*, case, box; here, of the asci + L. suf. *-oide* < Gr. *-oeídes*, similar to): similar to a perithecium, like the unilocular ascostroma of *Pyrenophora* (Pleosporales), which is a pseudothecium or pseudoperithecium, for its similarity to a perithecium.

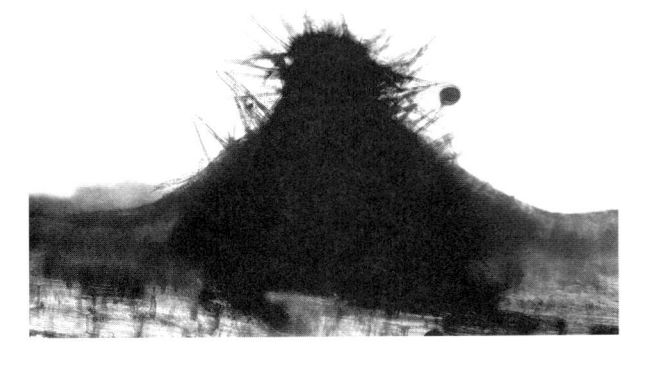

Perithecioid ascostroma of *Pyrenophora* sp., x 80 (*CB*).

perithecium, pl. **perithecia** (NL. *perithecium* < Gr. *perithekíon*, dim. of *perithêke*, helmet, cap < *perí*, around + *thêke*, case, box; here, of the asci): *Ascomycetes*. A type of ascocarp formed by ascohymenial development in which the asci are arranged in a fascicle or hymenium and are surrounded by a distinct wall. Typically the perithecium is obpyriform and possesses an ostiole through which the ascospores are freed, but it may also be spherical and remain closed (nonostiolate), when it can be confused with a cleistothecium. Perithecia are characteristic of the Pyrenomycetes and of the pyrenolichens. Some examples are *Guanomyces*, *Sordaria fimicola* (Sordariales), and *Melanospora zamiae* (Melanosporales).

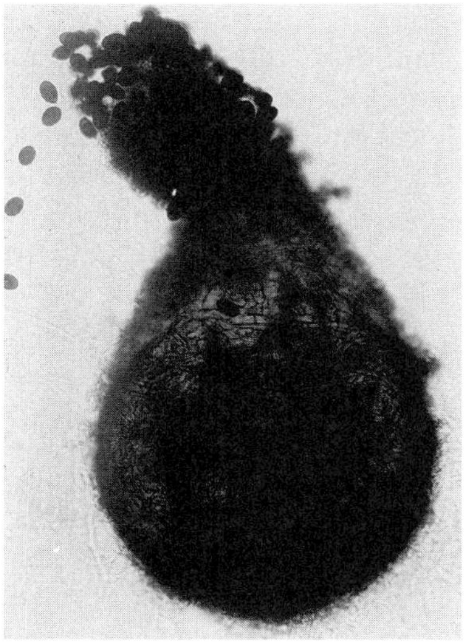

Perithecium of *Sordaria fimicola* with ascospores coming out of the ostiole, x 125 (*MU*).

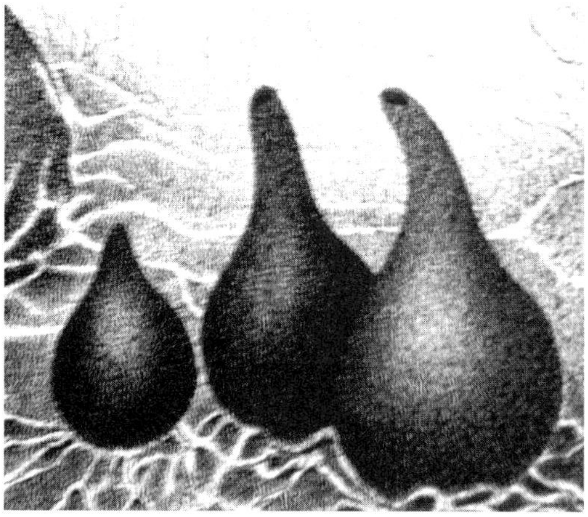

Perithecia of *Sordaria fimicola*, bending their neck towards the light, x 90.

perittogamy (Gr. *perittós*, unequal + *gámos*, sexual union + *-y*, E. suf. of concrete nouns): diffuse sexuality, with a flexible genetic rhythm and a not very precise behavior for undertaking the sexual act, since neither sexual organs nor gametes are differentiated, although there is somatogamy and karyogamy. The basidiomycetes are perittogamous, since in them fertilization is effected in an imprecise manner with regard to a determinate point and moment. Cf. **parasexuality**.

peroblate (L. pref. *per-*, which emphasizes the meaning of the following word + *oblatus* < *ob-*, intensive sense + *latus*, wide): generally applied to spores oval or ovate in shape, whose length:width ratio is less than 0.5.

peronate (L. *peronatus*, rough-booted < *pero*, genit. *peronis*, sandal + suf. *-atus* > E. *-ate*, provided with or likeness): *Agaricales*. Agarics in which the lower part of the stipe is covered by the volva or by the universal veil, as in *Amanita pantherina*. Syn. of **caligate** and **ocreate**.

perprolate (L. pref. *per*, which emphasizes the meaning of the following word + *prolatus*, bringing forward, elongated in the direction of the major axis): applied to structures markedly elongated (e.g., spores), whose length:width ratio is greater than 2.0.

perrumpent (L. *perrumpere*, to emerge by breaking + *-entem* > E. *-ent*, being): syn. of **erumpent**.

persistent (L. *persistens*, genit. *persistentis*, continuing, persevering, persisting < *persistere*, to continue, to last, stand firm + *-entem* > E. *-ent*, being): persisting, lasting. Applied to structures that remain after their function has ceased; e.g., sporangia that remain in place after sporulation (such as those of the saprolegniaceous fungi, *Saprolegnia*, *Achlya*, etc.), or the perennial fruiting bodies that live three or more years, such as those of many woody Aphyllophorales that do not rot easily. Cf. **deciduous**.

persistent attachment (L. *persistens*, genit. *persistentis*, continuing, persevering, persisting < *persistere*, to continue, to last, stand firm + *-entem* > E. *-ent*, being; ME. *attachement*, seizure < *atachen*, < AF. *atacher*, to seize < OF. *atachier*, to fasten + L. *-entem* > E. *-ent*, being): refers to fruiting bodies which remain attached to the substrate after they are mature, e.g., *Bovistella* (Lycoperdales). See **non-persistent attachment**.

perthotroph, **perthotrophic** (Gr. *pértho*, to seize the booty after a defeat + *trophós*, that which nourishes, serves as food; or + suf. *-íkos* > L. *-icus* > E. *-ic*, belonging to, relating to): a parasite that kills the cells of the host, on which it then lives saprobically. Certain plant pathogenic fungi, such as the species of *Sclerotinia* (Helotiales), are perthotrophs. *Colletotrichum lindemuthianum* (melanconiaceous asexual fungi) is a fungus that during its life cycle changes its nutritional relationship with the host plant, since it has the double character of biotroph/perthotroph. The perthotrophs are **necrotrophs**. Cf. **biotroph**.

pertusate (L. *pertusus*, perforate, punctured < *pertundere*, to pierce + suf. *-atus* > E. *-ate*, provided with or likeness): perforate, with holes, like the thallus of the lichen *Pertusaria* (Pertusariales), whose numerous small, black apothecia make it appear perforate.

pervious (L. *pervius*, open, passible + *-osus* > OF. *-ous*, *-eus* > E. *-ous*, possessing the qualities of): permeable, permitting passage; e.g., like the cyphellae and scyphi of the lichens that are open or perforate at the base and continuous with the thallus tissue on which they are formed.

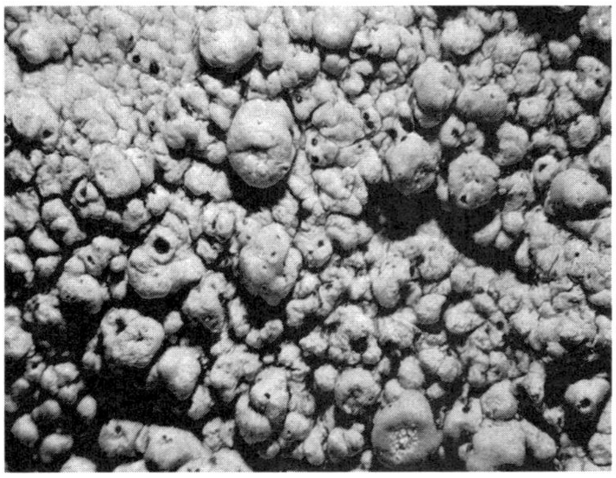

Pertusate thallus of *Pertusaria wulfenioides*, x 5 (*MU*).

petaloid (Gr. *pétalon*, leaf + L. suf. *-oide* < Gr. *-oeídes*, similar to): similar to a petal, petal-like; e.g., like the pileus of the basidiocarp of *Panellus mitis* (Agaricales) or the peridial lobes of the mature, dehisced sporangia of *Diderma radiatum* (Physarales).

Petaloid peridium of the dehisced sporangia of *Diderma radiatum*, x 13 (*MU*).

petricolous (L. *petra*, rock, stone + *-cola*, inhabitant + *-osus* > OF. *-ous*, *-eus* > E. *-ous*, possessing the qualities of): growing on rock; e.g., like certain lichens. Also called **epilithic**, **rupestral** and **saxicolous**. Various species of *Parmelia* (Lecanorales) and *Caloplaca* (Teloschistales), e.g., inhabit distinct types of rocks.

petrous (MF. *petreux* < L. *petrosus*, rocky < *petra*, rock, stone + *-osus* > OF. *-ous*, *-eus* > E. *-ous*, possessing the qualities of): very hard, similar to stone in consistency; the grains formed by the fungi causing mycetoma in man are petrous.

phaeocystidium, pl. **phaeocystidia** (L. *phaeo* < Gr. *phaiós*, blackish, dark + Gr. *kystídion* < *kýstis*, bladder, cell + dim. suf. *-ídion* > L. *-idium*): a lightly pseudoamyloid gloeocystidium, with brown contents, like that of *Fayodia deusta* (Agaricales).

phaeospore (L. *phaeo* < Gr. *phaiós*, blackish, dark + Gr. *sporá*, spore): an asexual reproductive spore of dark color; a Saccardoan term applied principally to conidial fungi, in particular those of the dematiaceous asexual fungi. Cf. **hyalospore**.

phagocytic (Gr. *phágos* < *phágomai*, to eat + *kýtos*, cavity, cell + suf. *-íkos* > L. *-icus* > E. *-ic*, belonging to, relating to): ingesting solid food particles by means of **phagocytosis**. Syn. of **phagotroph** or **phagotrophic** and **holozoic**.

phagocytosis, pl. **phagocytoses** (Gr. *phágos* < *phágomai*, to eat + *kýtos*, cavity, cell + suf. *-osis*, condition): the ingestion of solid food particles by means of pseudopods or protoplasmic extensions, as occurs in amoebae and other protozoans, as well as the myxamoebae and plasmodia of the Myxomycetes, which represent the trophic phases of these latter organisms.

phagotroph, **phagotrophic** (Gr. *phágos* < *phágomai*, to eat + *trophós*, what nourishes, serves as food; or + suf. *-íkos* > L. *-icus* > E. *-ic*, belonging to, relating to): an organism feeding by means of **phagocytosis.**

phalangoid (Gr. *phálange*, phalanx + L. suf. *-oide* < Gr. *-oeídes*, similar to): shaped like the phalanges or bones of the fingers and toes; e.g., like the cells of the peridial hyphae which are present in ascomata (gymnothecia) of *Arthroderma incurvata* and *Nannizzia gypsea* (Onygenales). Also called **ossiform**.

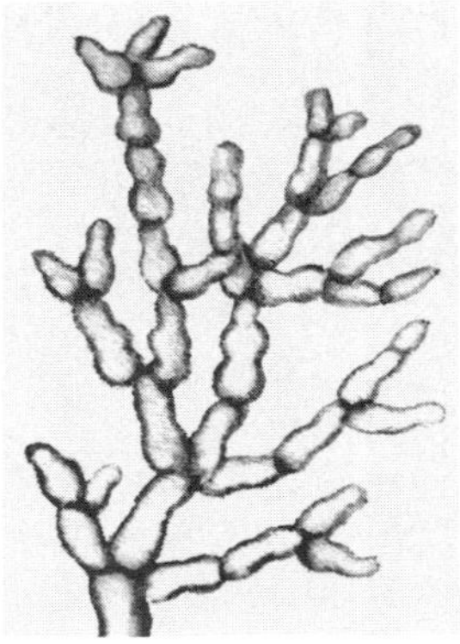

Phalangoid cells of the peridial appendages of a gymnothecium of *Arthroderma* (=*Nannizzia*) *incurvata*, x 900.

phalloid (Gr. *phallós*, phallus + L. suf. *-oide* < Gr. *-oeídes*, similar to): with the shape of a phallus, like the fruiting body of the gen. *Phallus* and *Mutinus* (Phallales).

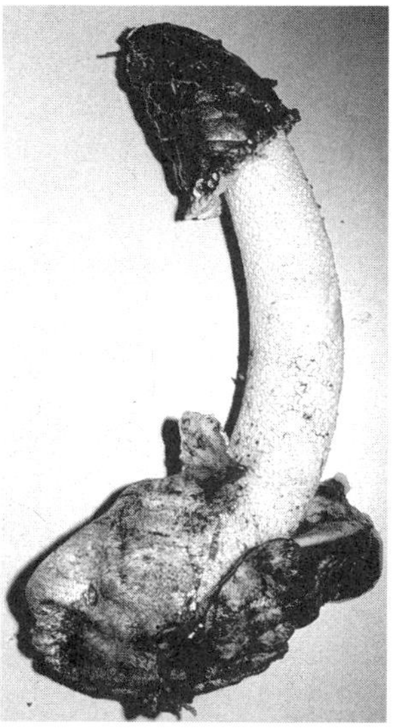

Phalloid fruiting body of *Phallus impudicus*, x 0.5 (*TH*).

phaneroplasmodium, pl. **phaneroplasmodia** (Gr. *pháneros*, visible, conspicuous + L. *plasmodium* < Gr. *plásma*, bland material with which a living being takes shape + L. *-odium*, resembling < Gr. *ode*, like < *-oeídes*, similar to): a plasmodium that is macroscopic at maturity, conspicuous, frequently reticulate, with a front in the shape of a well differentiated fan and typically with reversible currents of rapid, rhythmic, endoplasm within a system of veins consisting of ectoplasm. On fruiting, a phaneroplasmodium can form several or numerous sporangia; it is characteristic of Myxomycetes of the order Physarales, like, e.g., *Physarum polycephalum*. Cf. **aphanoplasmodium**, **protoplasmodium** and **pseudoplasmodium**.

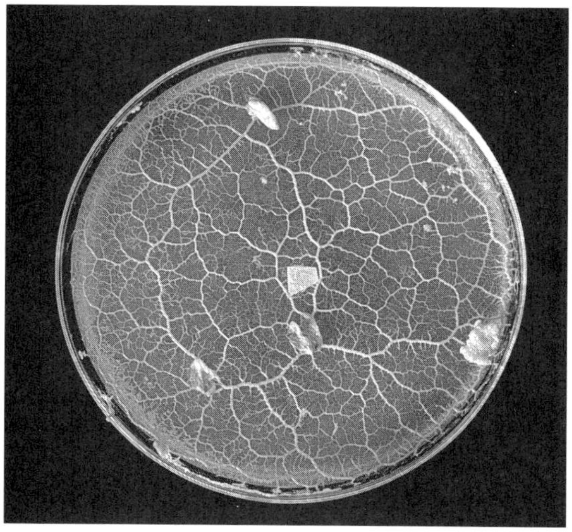

Phaneroplasmodium of *Physarum polycephalum* on agar, x 0.6 (*MU*).

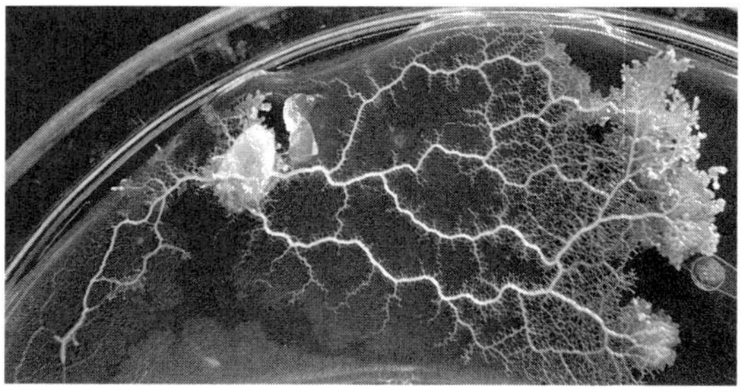

Phaneroplasmodium of *Physarum polycephalum* on agar, x 1.5 (*MU*).

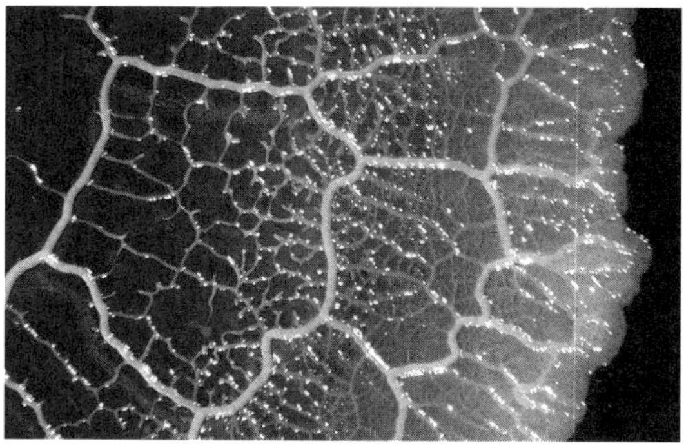

Phaneroplasmodium of *Physarum polycephalum* on agar, x 5 (*MU*).

phellophagy (Gr. *phellós*, cork + *phágia*, eating < *phágos*, to eat + *-y*, E. suf. of concrete nouns): ability to grow on and derive nourishment from cork cells.

phialide (Gr. *phialís*, genit. *phialídis*, dim. of *phiále*, glass, jar): a type of conidiogenous cell, bottle shaped, that produces blastic conidia (phialoconidia or phialospores) in basipetal succession. Phialides are the most common conidiogenous cells among the conidial fungi, such as *Aspergillus*, *Dendrodochium*, *Paecilomyces* (moniliaceous asexual fungi) *Chalara* (dematiaceous asexual fungi) and many other gen.

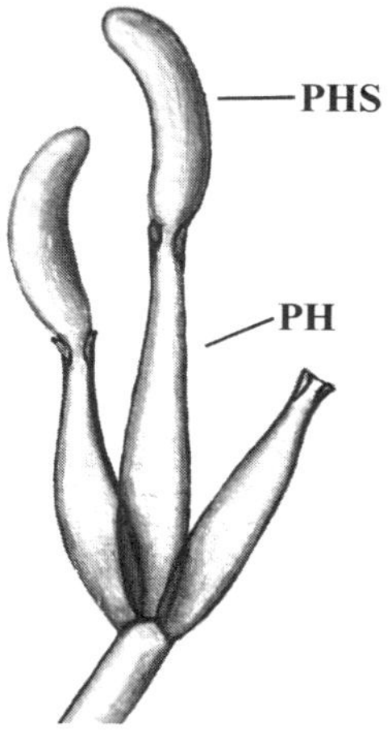

Phialides (PH) and phialospores (PHS) of the *Dendrodochium* state of *Nectria magnusiana*, x 1 340.

phialospore (Gr. *phialís*, dim. of *phiále*, small glass, bottle + *sporá*, spore): *Conidiogenesis*. An asexual reproductive spore formed by abstriction from the tip of a phialide; applied mainly to conidial and spermatial states of fungi. *Aspergillus*, *Penicillium* (moniliaceous asexual fungi) and *Chalara* (dematiaceous asexual fungi) are examples of gen. of fungi that produce phialospores. A terminal phialospore is the spore of a monosporial phialide, i.e., with its production phialide activity terminates.

phoenicoid fungus, pl. **fungi** (E. *phoenix* < ME. *fenix* < OE. < L. *phoenix* < Gr. *phoínix*, purple, crimson; a legendary bird of Egypt that lived 500 years, burned itself to ashes on a pyre, and rose alive from the ashes to live another period + L. suf. *-oide* < Gr. *-oeídes*, similar to; L. *fungus*, fungus): a fungus growing among the ashes of former fires. See **anthracobiontic**, **carbonicole** and **pyrophilous**.

phoresy (NL. *phorus* < Gr. *-phóros*, carrying < *phoreín*, to carry + *-y*, E. suf. of concrete nouns): a phenomenon occurring in certain habitats, such as dung, which are characterized by the presence of fungi, beetles, and mites, where the latter are fairly immobile and hitchhike on beetles and flies. Ascospores of *Pyxidiophora* (Laboulbeniales) are carried by the mites that ride along on insects and, thus, are said to be **hypophoretic**.

photobiont (Gr. *phós*, genit. *photós*, light + *bionte*, living being < *bíos*, life + *ón*, genit. *óntos*, a being): a photosynthetic symbiont of a lichen, which according to the species can be a member of the Chlorophyceae (phycobiont) or a cyanobacterium (bactobiont or cyanobiont).

photogen (Gr. *phós*, genit. *photós*, light + *génos*, producer, generator): a substance that generates light in fungi and other luminescent organisms; also known as luciferin, which, in the presence of the other photogenic component, the enzyme luciferase, is oxidized, producing light. It is present, e.g., in the fruiting bodies of *Lentinus stypticus luminescens* (Aphyllophorales), *Pleurotus japonicus*, *P. candescens* and other bioluminescent Agaricales. The mycelium of *L. stypticus luminescens* and *Clitocybe illudens*, as well as the basidiocarps, also is luminescent. In *Pleurotus japonicus* only the gills have this property. In *Armillaria mellea* (Agaricales) the mycelium and the rhizomorphs are luminescent, but not the fruiting bodies. The photogen is absolutely dependent on the presence of oxygen and moisture in order to emit light.

photomorph (Gr. *phós*, genit. *photós*, light + *morphé*, form): an organism whose form is determined by the nature of its photosynthesis; e.g., the very different thalli formed by some lichens, depending upon whether the photobiont is a green alga or a cyanobacterium.

photophile, **photophilic**, **photophilous** (Gr. *phós*, genit. *photós*, light + *phílos*, have an affinity for; or + suf. *-íkos* > L. *-icus* > E. *-ic*, belonging to, relating to; or + L. *-osus* > OF. *-ous*, *-eus* > E. *-ous*, possessing the qualities of): that which has a preference for a well illuminated habitat. Cf. **photophobic**.

photophobic, **photophobous** (Gr. *phós*, genit. *photós*, light + *phóbos*, fright, aversion + suf. *-íkos* > L. *-icus* > E. *-ic*, belonging to, relating to; or + L. *-osus* > OF. *-ous*, *-eus* > E. *-ous*, having, possessing the qualities of): having an aversion to light. Cf. **photophile**. See **umbraticous**.

photosporogenic (Gr. *phós*, genit. *photós*, light + *spóros*, spore + *génos*, birth + suf. *-íkos* > L. *-icus* > E. *-ic*, belonging to, relating to): requiring light for sporogenesis.

phototaxis, pl. **phototaxes** (Gr. *phós*, genit. *photós*, light + *táxis*, movement): a tactic phenomemon determined by the action of light radiation, such as the movement of the zoospores of some aquatic fungi, which is influenced by light, e.g., *Rhizophlyctis vorax* (Chytridiales), which attacks *Sphaerella lacustris*, a motile, photosynthetic organism that inhabits illuminated areas.

phototropic (Gr. *phós*, genit. *photós*, light + *trópos*, a turn, change in manner < *tropé*, a turning < *trépo*, to revolve, turn towards + suf. *-íkos* > L. *-icus* > E. *-ic*, belonging to, relating to): an organism exhibiting **phototropism**, e.g., *Pilobolus kleinii* (Mucorales) and *Beauveria cretacea* (moniliaceous asexual fungi).

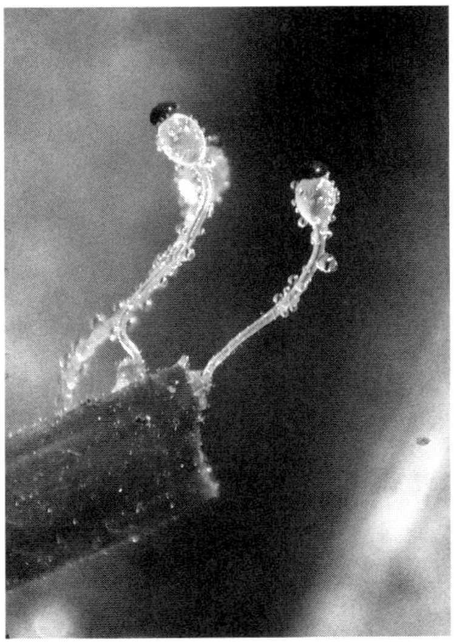

Phototropic (positively) sporangiophores of *Pilobolus kleinii*, growing on cow's dung, x 8 (*MU*).

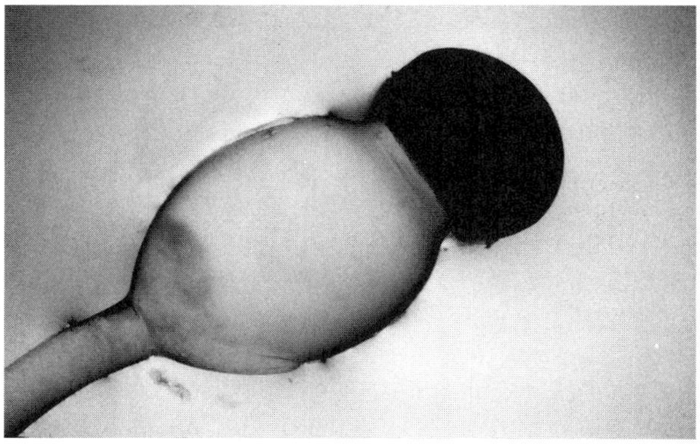

Phototropic sporangiophores of *Pilobolus kleinii*, x 500 (*MU*).

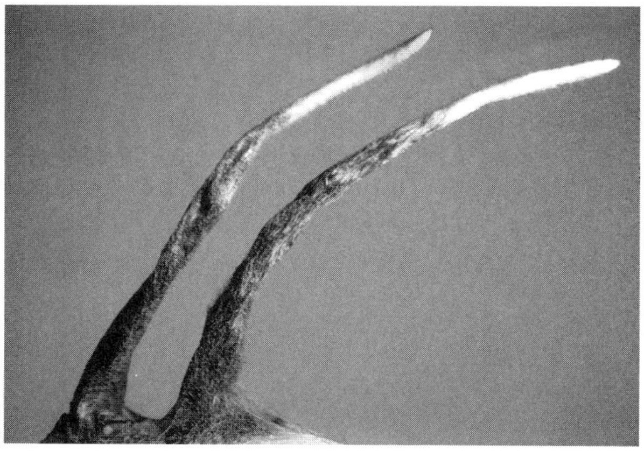

Phototropic synnemata of *Beauveria cretacea* on agar, x 7 (*CB*).

Phototropic synnema of *Beauveria cretacea* on agar, x 8 (*MU*).

phototropism (Gr. *phós*, genit. *photós*, light + *trópos*, a turn, change in manner < *tropé*, a turning < *trépo*, to revolve, turn towards + *-ismós* > L. *-ismus* > E. *-ism*, state, phase, tendency, action): a tropic phenomenon in which the stimulating factor is light. For example, the sporangiophores of *Pilobolus* (Mucorales) show positive phototropism because they direct themselves toward the light when the latter is unilateral; this is due to the fact that the side of the sporangiophore that receives more light grows less rapidly than the opposite side, and as a consequence the sporangiophore curves toward the light. The conidiophores of *Aspergillus giganteus* and the synnemata of *Beauveria cretacea* (moniliaceous asexual fungi) also are positively phototropic. The opposite phenomenon is negative phototropism.

phragmobasidium, pl. **phragmobasidia** (Gr. *phragmós*, fence + L. *basidium* < Gr. *basídion* < *básis*, base + dim. suf. *-ídion* > L. *-idium*): *Hymenomycetes*. A basidium that is initially unicellular and which, after the divisions of the diploid nucleus, forms partitions that divide it into four compartments or cells; the partitions can be vertical (as in *Tremella*, of the Tremellales) or transverse (as in *Auricularia*, of the Auriculariales). See **heterobasidium**.

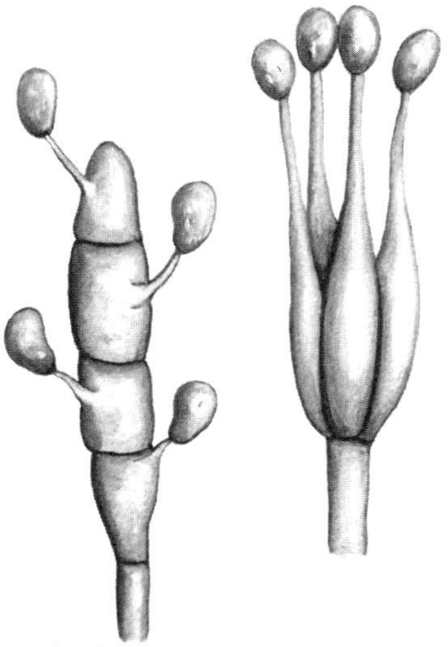

Phragmobasidia of *Tremella* (vertical septa) and *Auricularia* (transverse septa), x 1 670.

phragmospore (Gr. *phragmós*, fence, palisade + *sporá*, spore): an asexual reproductive spore, multicellular, partitioned (divided by means of several transverse septa); a Saccardoan term principally applied to conidial fungi. *Bipolaris maydis*, *Curvularia inaequalis*, *Drechslera iridis* and *Helminthosporium avenae* (dematiaceous asexual fungi) are examples of fungi with phragmospores.

phycobiont (Gr. *phýkos*, alga + *bíos*, life + *óntos*, genit. of *ón*, a being): the algal component of a lichen. In the majority of lichen species the phycobionts are Chlorophyceae; in some cases the algae incorporated into the lichen association are Cyanophyceae or cyanobacteria.

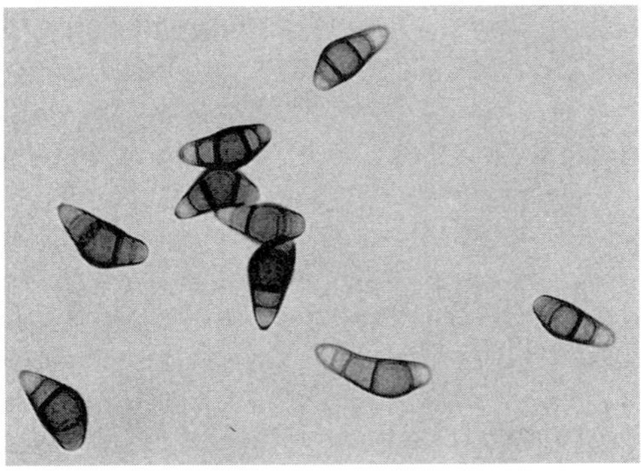

Phragmospores of *Curvularia inaequalis*, x 600 (*RTH*).

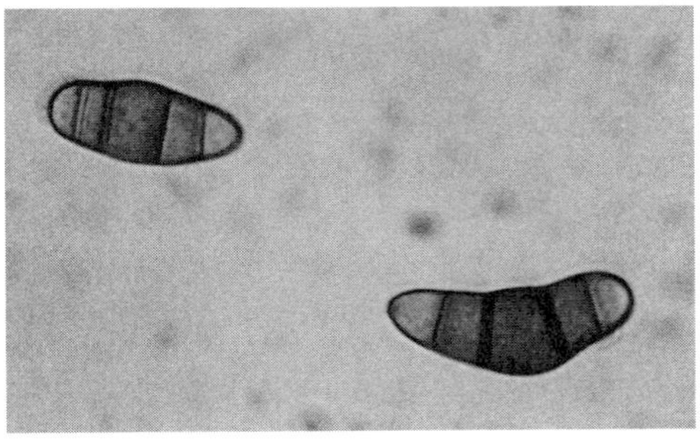

Phragmospores of *Curvularia* sp., x 900 (*CB*).

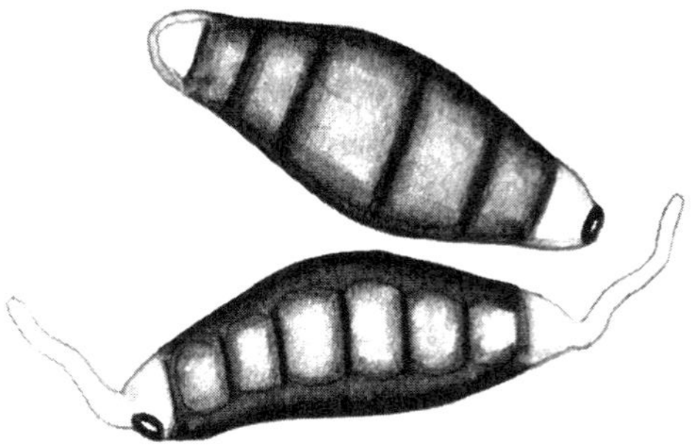

Phragmospores of *Drechslera* (=*Bipolaris*) *iridis*, one with a germ tube at each end, x 1 670.

phycocecidium, pl. **phycocecidia** (Gr. *phýkos*, alga + dim. of *kekís*, *kekîdos*, gall + L. dim. suf. *-ium*): a gall produced by cyanophyceous algae in some chlorophycophilous lichens; some authors interpret the cephalodia as a parasitism in which the phycocecidia are formed.

phycomycetes (Gr. *phýkos*, alga + L. *-mycetes*, ending of class < Gr. *mýkes*, genit. *mýketos*, fungus): formerly considered as a single taxonomic class of lower fungi, in some classifications it is equivalent to the subdivision Phycomycotina (of the division Eumycota), which comprises five classes (Chytridiomycetes, Hyphochytridiomycetes, Oomycetes, Zygomycetes and Trichomycetes). In some recent classification systems the subdivision Phycomycotina has disappeared as such, and some of the classes that it contained have come to be distinct subdivisions, such as the Mastigomycotina and Zygomycotina, or distinct phyla, such as the Chytridiomycota and Zygomycota (of the kingdom Fungi), and Hyphochytriomycota and Oomycota (of the kingdom Stramenopila); the Trichomycetes have been included, along with the Zygomycetes, in the phylum Zygomycota. This latter classification (Alexopoulos *et al.*, 1996), which considers phyla, is the one followed in this dictionary.

phycomycosis, pl. **phycomycoses** (Gr. *phýkos*, alga + NL. *mycosis* < Gr. *mýkes*, fungus + *-osis*, state or condition): a subcutaneous infection of humans, mostly children, characterized by the appearance of large, hard, and painless nodules or swellings in the neck, arms and chest. The causal agent is *Basidiobolus haptosporus* (Entomophthorales), an ubiquitous, saprobic microorganism that inhabits soil and plant remains, as well as the gut and dung of reptiles and amphibians.

phycosymbiodeme (Gr. *phýkos*, alga + *symbíosis*, life in common + *dému*, to build, construct): joined lichen thalli with a single mycobiont but different photobionts.

phylogeny (NL. *phylum* < Gr. *phýlon*, tribe, race + Gr. *géneia*, act of being born < *génos*, origin, birth, engenderment + *-y*, E. suf. of concrete nouns): the history of a kind of organism; the evolution through time of a genetically related group of organisms as distinguished from the development of the individual organism (**ontogeny**). Considering the possibility of transformation of some living beings into others, the phylogeny studies the genesis of the phyla of the organic world and the probable derivation of some organisms from others, in order to construct a genealogic tree of all living beings that exist or have existed on Earth. Views on phylogeny of fungi have been based traditionally on comparative morphology, cytology, hyphal wall chemistry, ultrastructure, and to a lesser degree fossils. The present use of cladistic and molecular approaches suggest that fungi may have originated from protozoan ancestors before the Kingdoms Animalia and Plantae split. According to molecular evidence, fungi appear to be closer to Animalia than Plantae.

phyllidium, pl. **phyllidia** (NL. *phyllidium* < Gr. *phyllídion* < *phýllon*, leaf + dim. suf. *-ídion* > L. *-idium*): applied to the small laminar or squamiform portion of foliose lichens that separates from the thallus by abstriction and functions as a vegetative propagule composed of the mycobiont and phycobiont; e.g., as in *Collema flaccidum* (Lecanorales) and *Peltigera praetextata* (Peltigerales).

phyllocladium, pl. **phyllocladia** (Gr. *phýllon*, leaf + *kládos*, branch + L. dim. suf. *-ium*): a small branch or expansion of a fruticose lichen thallus, very specialized and assimilative, since it contains the phycobiont; present in the species of *Stereocaulon* (Lecanorales) and can be granulose, verrucose, coralloid, scaly or squamulose, digitate, peltate, or foliose. For example, they are flat and squamulose in *S. saxatile*, and cylindrical and coralloid in *S. dactylophyllum*.

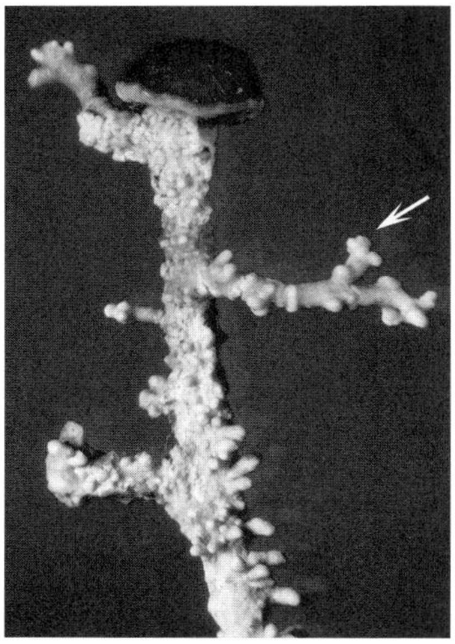

Phyllocladia of a thallic branch of *Stereocaulon ramulosum*, x 9 (*CB*).

phylloplane (Gr. *phýllon*, leaf + L. *planus*, flat, level surface): the surface of a leaf, with a special microclimate, in which exists a characteristic mycobiota, composed of species of saprobic molds and yeasts. The term **phyllosphere**, frequently is used in the same sense as phylloplane. The species of *Sporobolomyces*, *Tilletiopsis* and *Bullera* (sporobolomycetaceous asexual yeasts), as well as *Rhodotorula* and *Cryptococcus* (cryptococcaceous asexual yeasts), are common in the phyllosphere.

phyllosphere (Gr. *phýllon*, leaf + *sphaîra*, sphere): see **phylloplane**.

physoclastic (Gr. *physáo*, *physéo*, *physõ*, to throw, cast + *klástos*, broken < *kláo*, to break + suf. *-íkos* > L. *-icus* > E. *-ic*, belonging to, relating to): a type of ascus that liberates its spores by rupture of the external portion, which is less extensible than the internal wall of the ascus. It is applied to certain marine fungi, such as *Capronia* (=*Herpotrichiella*), of the Chaetothyriales, and *Didymella* (Dothideales). Cf. **aphysoclastic**.

piedra (Sp. *piedra* < L. *petra*, stone, rock): a superficial stone-like concretion formed by parasitic fungi on human and/or animal hairs. *White piedra* (ME. < OE. *hwit*, white) is caused by *Trichosporon cutaneum* (cryptococcaceous asexual yeasts) which infects human hairs forming whitish masses of arthrospores in the hair shaft. *Black piedra* (ME. *blak* < OE. *blaec*, black) is caused by *Piedraia hortae* (Myrangiales), the teleomorph of *Trichosporon*. The black concretions formed by *Piedraia* on the hairs of humans and gorillas are stromata containing asci and ascospores.

pileate, **pileated** (L. *pileatus* < *pileus* < Gr. *pílos*, hat, helmet, cap; here, pileus + L. suf. *-atus* > E. *-ate*, provided with or likeness): provided with a pileus.

pileate development (L. *pileatus* < *pileus* < Gr. *pílos*, hat, helmet, cap; here, pileus + L. suf. *-atus* > E. *-ate*, provided with or likeness; MF. *développer* < OF. *desveloper* < *dés-* < L. *de-*, from, down, away + MF. *enveloper*, to enclose + L. *mentum* > OF. *-ment* > ME. *-ment*, action, process): a type of development resulting in a fruiting body with a single stalk and pileus, with hymenial tissue often in the form of gill-like tramal plates.

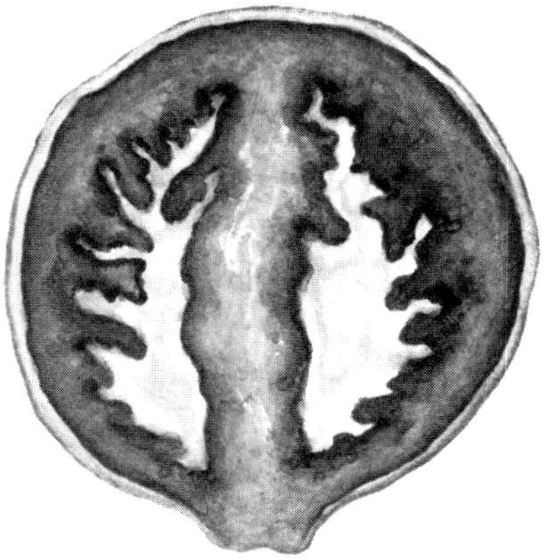

Pileate type of development of the fruiting body of *Podaxis pistillaris*, x 2.

pileiform (L. *pileus* < Gr. *pílos*, hat, helmet, cap; here, pileus + L. suf. *-formis* < *forma*, shape): having the shape of a pileus or small hat; characteristic of the mushrooms (Agaricales), or the zygospore of *Harpella melusinae* and *Stipella vigilans* (Harpellales).

pileipellis (L. *pileus* < Gr. *pílos*, hat, helmet, cap; here, pileus + L. *pellis*, skin): *Agaricales*. The outer layer of the pileus, usually composed of anticlinal or decumbent hyphae. See **pellicule**.

pileocystidium, pl. **pileocystidia**, **pilocystidium**, pl. **pilocystidia** (L. *pileus* < Gr. *pílos*, hat, helmet, cap; here, pileus + *kystídion*, dim. of *kýstis*, bladder, vesicle, cell): *Agaricales*. A dermatocystidium that protrudes from the sterile surface of the pileus; e.g., as happens in *Leptonia lividocyanula*, *Pouzarella versatilis* and some species of *Coprinus*, among other Agaricales. Cf. **caulocystidium**.

Pileiform zygospore of *Stipella vigilans*, x 800.

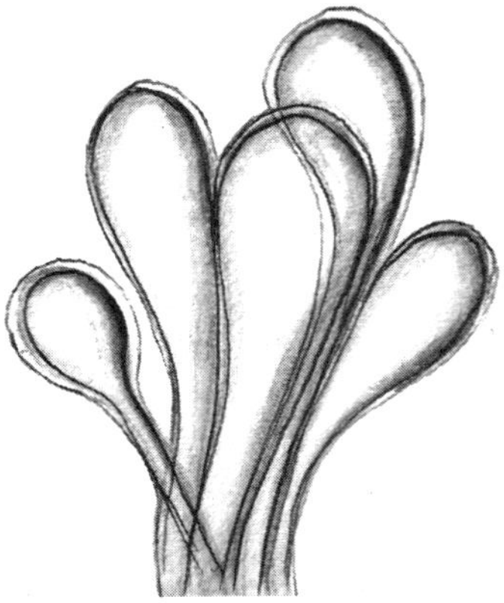

Pileocystidia from the hymenium of *Leptonia lividocyanula*, x 1 000.

pileolus, pl. **pileoli** (L. *pileolus*, dim. of *pileus* < Gr. *pílos*, hat, helmet, cap; here, pileus): a small pileus.

pileus, pl. **pilei** (L. *pileus* < Gr. *pílos*, hat, helmet, cap; here, pileus): the expanded upper part of certain types of ascocarps and basidiocarps, on which is formed the fertile hymenium, which produces the ascospores or basidiospores, according to the type of fruiting body.

pilidium, pl. **pilidia** (L. *pilidium* < Gr. *pilídion* < *pílos*, hat, helmet, cap + dim. suf. *-ídion* > L. *-idium*): an orbicular, hemisphaerical shield of certain lichens, whose external part becomes powdery, such as occurs in the gen. *Calicium* (Caliciales).

piliform (L. *piliformis*, hair-shaped < *pilus*, hair + *formis* < *forma*, shape): shaped like a hair; e.g., like the appendages of the spores of *Choanephora cucurbitarum* and *Gilbertella persicaria* (Mucorales), and of the ascospores of *Nematospora coryli* (Saccharomycetales).

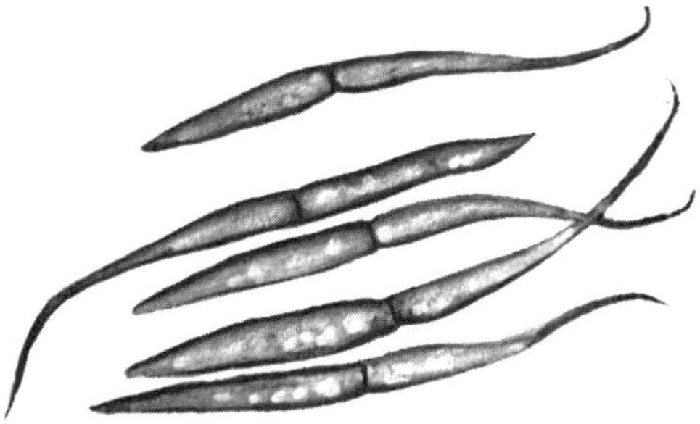

Piliform appendages of the ascospores of *Nematospora coryli*, x 1 400.

pilose (L. *pilosus*, hairy < *pilus*, hair + *-osus*, full of, augmented, prone to > ME. *-ose*): covered with long, soft filaments, similar to hairs, such as the basidiocarp of *Oudemansiella pilosa* (Agaricales).

pinocytosis, pl. **pinocytoses** (Gr. *pinó*, to drink + *cytosis* < Gr. *kýtos*, cavity, here, cell + *-osis*, condition or state): an intracellular process, present in eukaryotic cells, that utilizes microfibrils to "drink" small droplets of liquid from the surrounding medium. See **phagocytosis**.

pionnote (Gr. *píon*, fat, thick juice + *notís*, moisture): an effuse, gelatinous or viscous sporodochium. The term applied to the masses of spores with an oily or mucilaginous appearance that are produced by fungi such as *Fusarium* and *Myrothecium* (tuberculariaceous asexual fungi), and *Pestalotiopsis* (melanconiaceous asexual fungi).

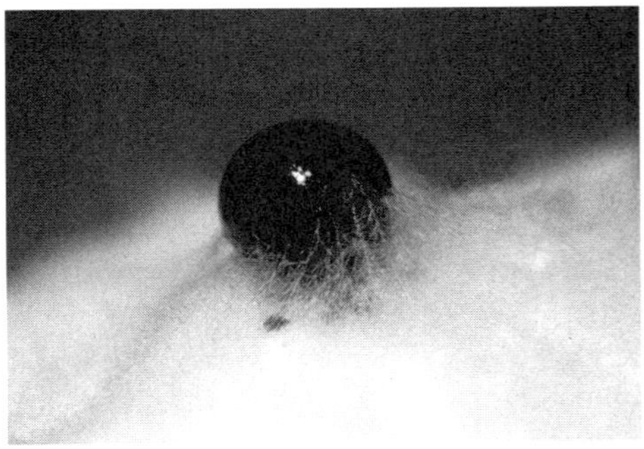

Pionnote of *Pestalotiopsis* sp. on agar, x 20 (*CB*).

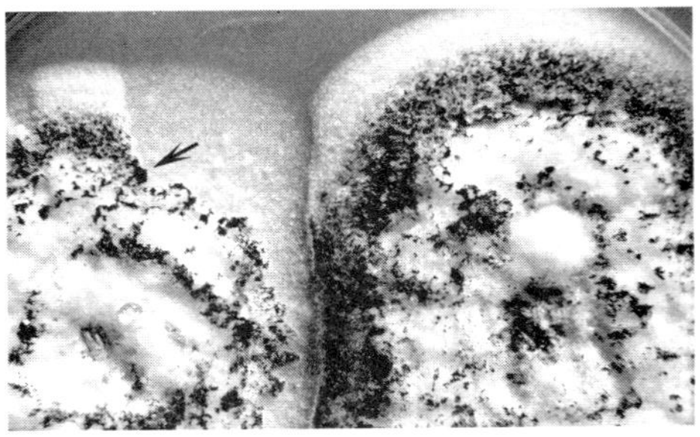

Pionnotes of *Myrothecium verrucaria* on agar, x 1 (*MU*).

Pionnotes of *Myrothecium verrucaria* on agar, x 5 (*MU*).

pit (ME. *pit* < OE. *pytt* < L. *puteus*, well, shaft, pit): a shallow hole or depression in the surface of a structure; e.g., as in the capillitial elements with round holes or the surface depressions of the peridium, such as the endoperidium of *Lycoperdon perlatum* (Lycoperdales).

pit connection (ME. *pit* < OE. *pytt* < L. *puteus*, well, shaft, pit; ME. *conneccioun*, *connexioun* < MF. < L. *connexion*, *connectus*, ptp. of *connectere*, to connect + *-io*, *-ionis* > E. suf. *-ion*, result of an action, state of): the very small canal that traverses the cell wall of various fungi and through which pass the plasmodesmata, the fine cytoplasmic filaments that connect one cell with another. See **plasmodesma**.

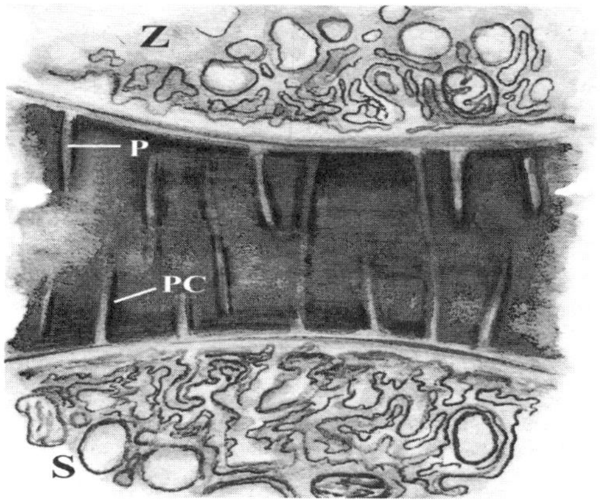

Pit connections (PC) of the septum between the zygospore (Z) and suspensor (S) of *Rhizopus sexualis*, seen in longitudinal section. The plasmodesmata (P) pass from one cell to the other through the pitt connections. Drawing based on a transmission electron micrograph obtained by Beckett *et al.*, and published in their *Atlas of Fungal Ultrastructure*, 1974, x 23 000.

pitted (E. *pitted* < OE. *pytted*, ptp. of *pytt* < L. *puteus*, well, shaft, pit): having small and shallow holes (visible with a lens) in the surface of the lichen thallus of *Letharia* (Lecanorales), which to the naked eye appear as dark points, and of the ascospore wall of *Gelasinospora* (Sordariales).

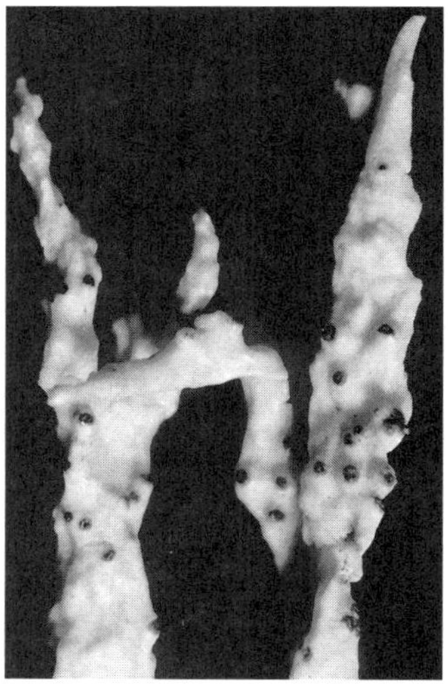

Pitted branches of the thallus of *Letharia vulpina* f. *californica*, x 12 (*MU*).

placodiomorph (Gr. *plakós*, plaque, tablet + *dís*, twice + *morphé*, shape): *Lichens*. A type of thallus that is crustaceous in the center and lobate-radiate on the edges, making it appear foliaceous; e.g., as is present in *Buellia oidalea* (Lecanorales).

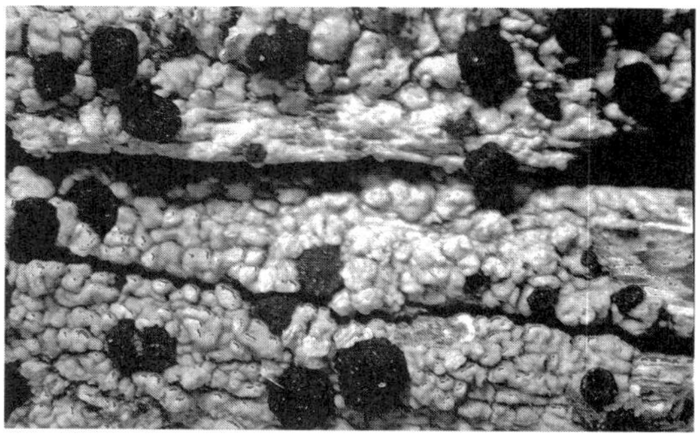

Placodiomorph - thallus of *Buellia oidalea*; the dark spots are apothecia, x 8 (*MU*).

placodium, pl. **placodia** (NL. *placodium* < Gr. *plakós*, plaque, tablet + L. *-odium*, resembling < Gr. *ode*, like < *-oeídes*, similar to): the sclerotial cortical stroma that surrounds the perithecial necks in xylariaceous ascomycetes such as *Diatrype*. Also called **ostiolar disc**. If derived from the endostroma it is called **endoplacodium**, and **ectoplacodium** if it forms from the ectostroma; if it is derived from both types of stroma it is called **diplostromatic**, and if it is only of one type it is called **haplostromatic**.

plagiotropism (Gr. *plagiós*, oblique, slanting, transversal + *trópos*, a turn, change in manner < *tropé*, a turning < *trépo*, to revolve, turn towards + *-ismós* > L. *-ismus* > E. *-ism*, state, phase, tendency, action): syn. of **diageotropism**.

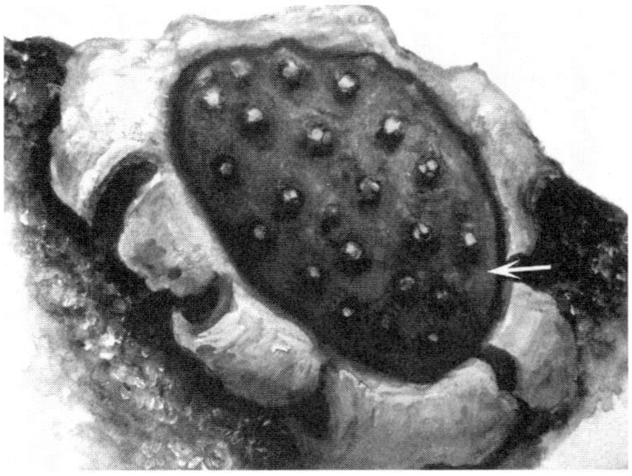

Placodium of the perithecial stroma of *Diatrype virescens*, x 30.

plaited, **plicate** (ME. *pleit* < VL. *plictus* < L. *plicare*, to fold, crease; L. *plicatus* < *plicare* < *plica*, crease, fold + L. suf. *-atus* > E. *-ate*, provided with or likeness): folded like a fan; e.g., like the pileus of the basidiocarp of *Coprinus impatiens* and *C. plicatilis* (Agaricales), which has creases between the lines that furrow the surface. The colonies of some *Penicillium* species, e.g., *P. puberulum* (moniliaceous asexual fungi), and *Aureobasidium pullulans* (dematiaceous asexual fungi) are plicate. Also called **sulcate**.

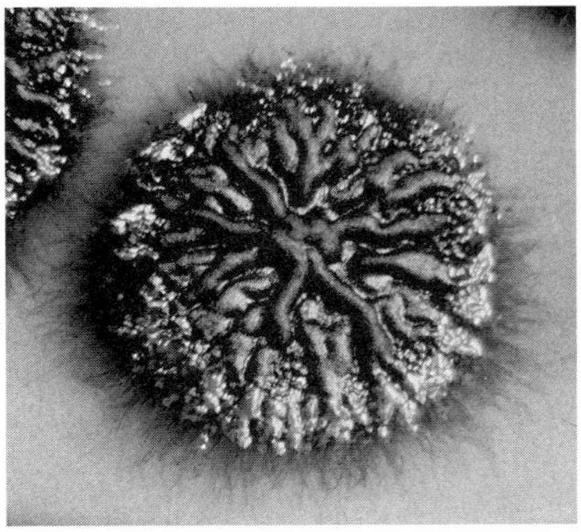

Plaited colony of *Aureobasidium pullulans* on agar, x 2 (*MU*).

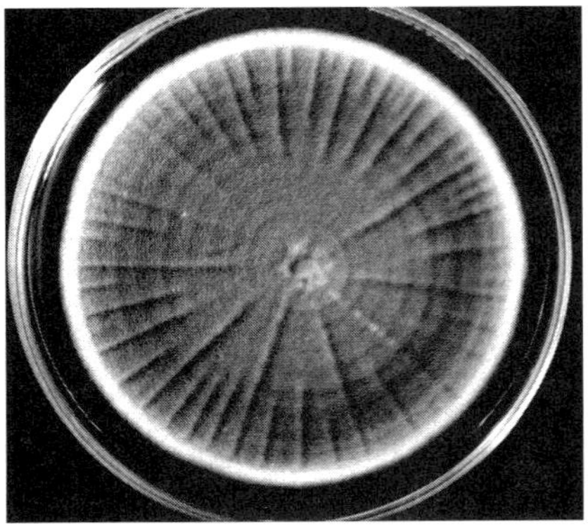

Plaited colony of *Penicillium chrysogenum* on agar, x 0.5 (*MU*).

planocyte (Gr. *plános*, wanderer + *kýtos*, cavity; here, cell): a motile (flagellate) cell destined for multiplication or reproduction, whether as a zoospore, planogamete or planozygote.

planogamete (Gr. *plános*, wanderer + *gamétes*, husband, sexual cell): also called **zoogamete**; a flagellate, motile gamete, characteristic of aquatic fungi of the class Chytridiomycetes. Cf. **aplanogamete**.

planospore (Gr. *plános*, wanderer + *sporá*, spore): a motile spore; zoospore.

planozygote (Gr. *plános*, wanderer + *zygotós* < *zygós*, marriage, pair): a motile zygote. A flagellated cell and consequently active, which results from the fusion of two flagellate gametes, as happens in some Chytridiales (*Synchytrium*) and Blastocladiales (*Allomyces*).

plaque bridge (F. *plaquer*, to plate < MD. *placken*, to patch; ME. *brigge* < OE. *brycg*, bridge): the part that interconnects the two globular daughter halves that result from the replication of the spindle polar body during mitosis in yeast cells, such as *Saccharomyces* (Saccharomycetales). See **middle piece**.

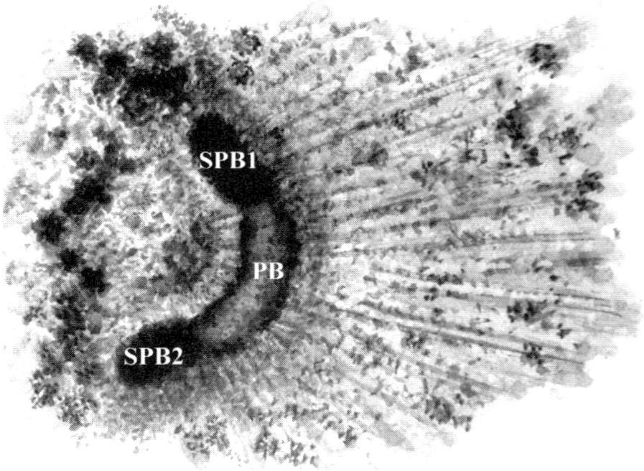

Plaque-bridge (PB) of a meiotic prophase nucleus of *Saccharomyces cerevisiae*. The replicated spindle pole bodies (SPB1 and SPB2) lying side by side are connected by the plaque bridge. Drawing based on a transmission electron micrograph published by Beckett *et al.* in their *Atlas of Fungal Ultrastructure*, 1974, x 140 000.

plasmodesma, pl. **plasmodesmata** (Gr. *plásma*, bland material that forms a living being + *desmós*, filament): an extremely slender plasmatic filament that passes through a fine perforation in the cell wall (**pit connection**), so that it establishes a continuity between adjacent cells, as has been observed, e.g., in *Rhizopus sexualis* (Mucorales); several plasmodesmata are formed through the wall that separates the suspensor from the zygospore during development. These plasmodesmata are often associated with large accumulations of endoplasmic reticulum on both sides of the small canals that traverse the wall of the gametangial septum, for which reason it is considered that soluble nutrients and other substances pass through them from the suspensors toward the developing zygospore, utilizing the membranes of the endoplasmic reticulum as a transport system.

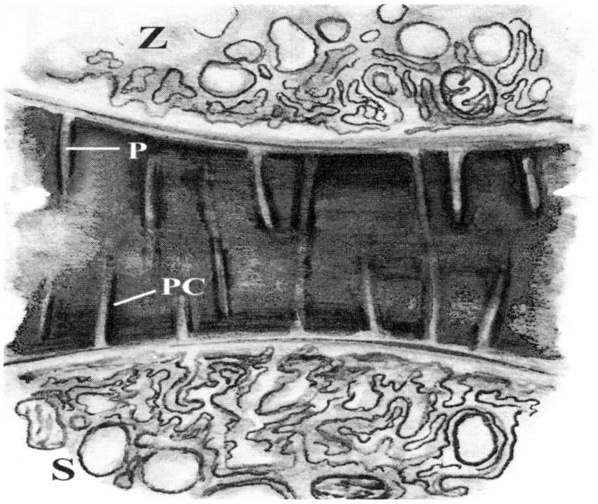

Plasmodesmata (P) of the septum between the zygospore (Z) and suspensor (S) of *Rhizopus sexualis*, seen in longitudinal section. Plasmodesmata pass from one cell to the other through the pitt connections (PC). Drawing based on a transmission electron micrograph obtained by Beckett *et al.*, and published in their *Atlas of Fungal Ultrastructure*, 1974, x 23 000.

plasmodiocarp (NL. *plasmodium* < Gr. *plásma*, bland material with which is formed a living being + L. *-odium*, resembling < Gr. *ode*, like < *-oeídes*, similar to + Gr. *karpós*, fruit): *Myxomycetes*. A type of sporangium that preserves the reticulate or veined shape of the plasmodium from which it forms, as is observed in *Hemitrichia serpula* (Trichiales).

Plasmodiocarp of *Hemitrichia serpula*, x 14 (*MU*).

Plasmodiocarp of *Hemitrichia serpula* on dead wood, x 14 (*MU*).

plasmodiophagous (NL. *plasmodium* < Gr. *plásma*, bland material with which is formed a living being + L. *-odium*, resembling < Gr. *ode*, like < *-oeídes*, similar to + Gr. *phágos*, one that eats + L. *-osus* > OF. *-ous*, *-eus* > E. *-ous*, possessing the qualities of): plasmodium-eating organisms, such as the insects that feed on Myxomycete plasmodia. Cf. **sporophagous**.

Plasmodiophoromycetes (NL. *plasmodium* < Gr. *plásma*, bland material with which is formed a living being + L. *-odium*, resembling < Gr. *ode*, like < *-oeídes*, similar to + Gr. *-phóros*, bearer + L. *-mycetes*, ending of class < Gr. *mýkes*, genit. *mýketos*, fungus): a class of viscous or mucilaginous fungi, endoparasitic on plants, with two types of endobiotic plasmodia, one related to the asexual reproductive phase (sporangiogenous plasmodium) and the other with sexual reproduction (cystogenous plasmodium). In some classifications, the Plasmodiophoromycetes were formerly included, together with the Protosteliomycetes, Acrasiomycetes and Myxomycetes, in the division Myxomycota of the kingdom Fungi. In the classification scheme that is followed in this dictionary (Alexopoulos *et al.*, 1996), these organisms are all regarded as different phyla of the kingdom Protista, i.e., Plasmodiophoromycota, Dictyosteliomycota, Acrasiomycota, and Myxomycota. Others (e.g., Margulis and Schwartz, 1988) include the class Plasmodiophoromycetes in the kingdom Protoctista.

plasmodium, pl. **plasmodia** (NL. *plasmodium* < Gr. *plásma*, bland material with which is formed a living being + L. *-odium*, resembling < Gr. *ode*, like < *-oeídes*, similar to): a multinucleate protoplasmic mass, without a cell wall, with amoeboid movement and phagocytic nutrition; it is the somatic and trophic phase of the Myxomycetes, of the Plasmodiophoromycetes and of some Protosteliales.

See **aphanoplasmodium**, **phaneroplasmodium** and **protoplasmodium**.

plasmogamy (Gr. *plásma*, bland material with which is formed a living being + *gámos*, sexual union + *-y*, E. suf. of concrete nouns): the first phase of sexual reproduction in which two protoplasts fuse and there is juxtaposition of the nuclei, forming a dikaryon; this phase precedes karyogamy and meiosis.

plasmoptysis, pl. **plasmoptyses** (Gr. *plásma*, bland material with which is formed a living being + *ptýsis*, to spit, throw out): *Mycorrhizae*. The process in endomycorrhizae in which the tips of the hyphae (which are inside the host cells of the plant) break and emit a globose mass of cytoplasm called a **ptyosome**, which is then digested and assimilated by the host cell.

plasmotomy (Gr. *plásma*, bland material with which is formed a living being + *tomía* < *tomé*, action of cutting + *-y*, E. suf. of concrete nouns): *Protosteliales*, *Acrasiomycota*. The process of cell division present in the myxamoebae, which consists of a thinning of the central part of the protoplast until it breaks the fine protoplasmic bridge that for a brief time connects the two cells, so that from one, two myxamoebae originate. During the process, the nucleus and the cytoplasmic organelles divide by amitosis.

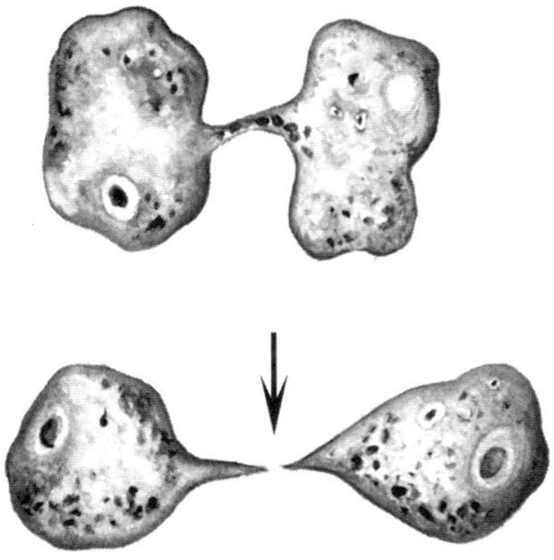

Plasmotomy of the myxamoebae of *Cavostelium bisporum*, x 1 800.

plectenchyma, pl. **plectenchymata** (Gr. *plektós*, braided, intertwined + *énchyma*, stuffed): a general term used to refer to all types of fungal tissues; the two most common types are **prosenchyma** and **pseudoparenchyma**.

plectobasidium, pl. **plectobasidia** (Gr. *plektós*, intertwined + *basídion*, dim of. *básis*, base): *Gasteromycetes*. A type of basidium characterized by being extremely variable in shape and by its irregular origin from hyphal tissues of the gleba. Plectobasidia are not oriented to discharge the spores toward an air space from an organized hymenium (they are statismosporic basidia), but they form layers or fertile areas called pseudohymenia, which can surround small locules in the gleba, be concentrated within peridioles, or form layers on the phallaceous receptacle. Plectobasidia are present in Hymenogastrales, such as *Podaxis*.

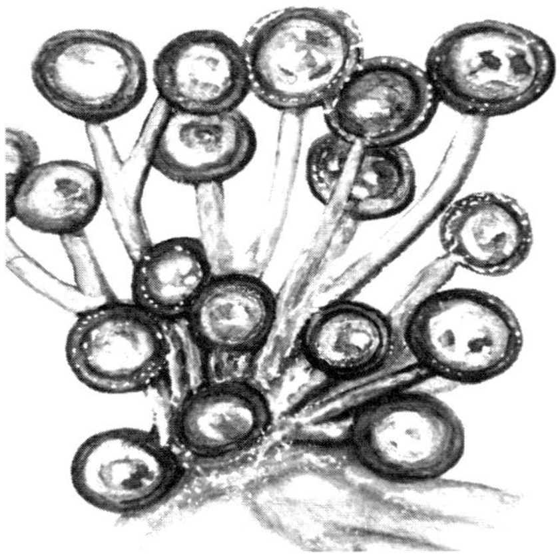

Plectobasidia, with orthotropic basidiospores, from the hymenium of *Podaxis pistillaris*, x 850.

Plectomycetes (Gr. *plektós*, braided, intertwined + L. *-mycetes*, ending of class < Gr. *mýkes*, genit. *mýketos*, fungus): a class of ascomycetes with small, evanescent asci, produced at different levels within a cleistothecium. The Plectomycetes (which includes the orders Eurotiales, Ascosphaerales and Onygenales) together with the classes Archiascomycetes, Pyrenomycetes, Discomycetes and Loculoascomycetes, plus the order Saccharomycetales (ascosporogenous yeasts), constitute the phylum Ascomycota of the kingdom Fungi.

plectonematogenous (Gr. *plektós*, braided, intertwined + *nêma*, genit. *nêmatos*, thread, filament + *génos*, origin < *gennáo*, to engender, produce + L. *-osus* > OF. *-ous*, *-eus* > E. *-ous*, having, possessing the qualities of): *Conidial fungi*. Fungi in which the conidiophores and conidiogenous cells originate from cords or funiculae of intertwined hyphae (but not of a single hypha), but without forming a synnema; e.g., as is seen in *Acremonium* (=*Cephalosporium*), *Oedocephalum* (moniliaceous asexual fungi) and *Stachybotrys* (dematiaceous asexual fungi), which form fasciculate hyphae.

pleiosporous, **pliosporous** (Gr. *pleíon*, more numerous + *sporá*, spore + L. *-osus* > OF. *-ous*, *-eus* > E. *-ous*, having, possessing the qualities of): syn. of **myriosporous**.

pleomorphic, **pleiomorphic** (Gr. *pléo* < *pleíon*, more numerous + *morphé*, shape + suf. *-íkos* > L. *-icus* > E. *-ic*, belonging to, relating to): having more than one form in the life cycle; usually with reference to having more than one spore state. Also called **polymorphic** or **versiform**. This last term is mainly applied to fungal structures that change shape with age, such as the basidiospores of *Conocybe lactea* and the cheilocystidia of *Bolbitius vitellinus* (Agaricales).

plerotic (Gr. *plerotikós* < *plerés*, to fill + *-tikós* > L. *-ticus* > E. *-tic*, relation, fitness, inclination or ability): to be full; e.g., a plerotic sporangium is one which is full of spores; in some species of Pythiaceae, such as *Pythium monospermum* (Peronosporales), a plerotic oospore is one that fills the oogonium. Cf. **aplerotic**.

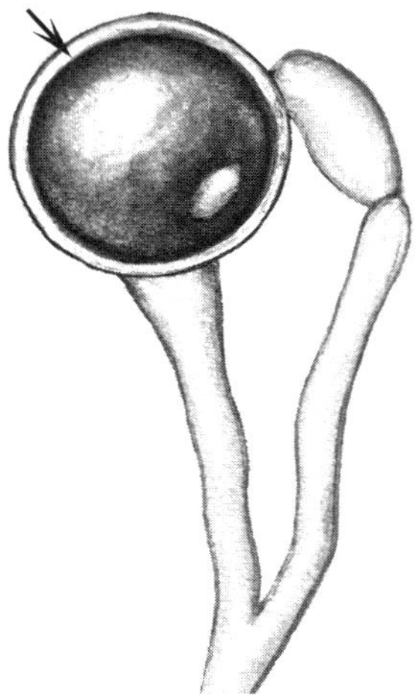

Plerotic oospore in the oogonium of *Pythium monospermum*, x 1 000.

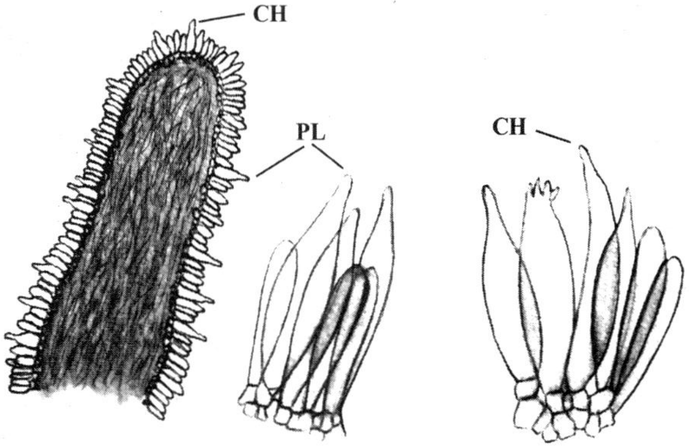

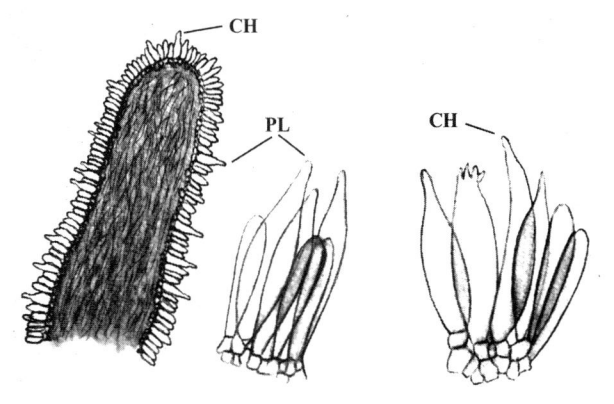

Pleurocystidia (PL) and cheilocystidia (CH) on the hymenium of *Pouzarella versatilis*, x 600. Longitudinal section of a lamella, x 160.

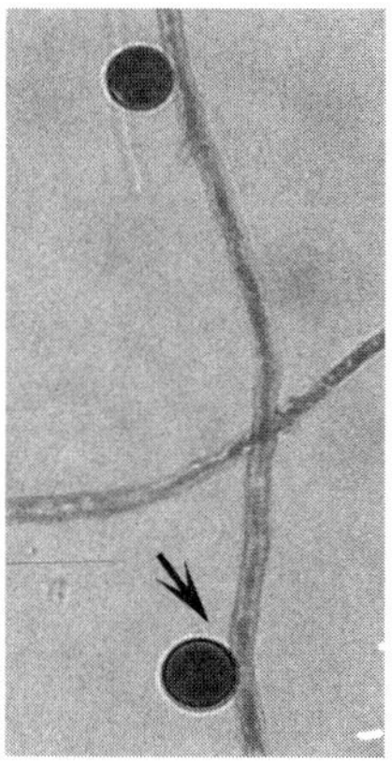

Pleurogenic conidia of *Trichosporiella cerebriformis*, x 1 000 (*CB*).

pleuroacrogenous (Gr. *pleurá*, side, flank + *ákros*, apex + *génos*, origin, birth + L. *-osus* > OF. *-ous*, *-eus* > E. *-ous*, having, possessing the qualities of): see **acropleurogenous**.

pleurobasidium, pl. **pleurobasidia** (Gr. *pleurá*, side, flank + *basídion*, dim. of *básis*, base): a basidium that is relatively wide at the base, with prolongations that extend it, such as occurs in the gen. *Pleurobasidium* (Aphyllophorales).

pleurocystidium, pl. **pleurocystidia** (Gr. *pleurá*, side, flank + *kystídion*, dim. of *kýstis*, bladder, vessel, cell): a cystidium that emerges from the flank of the gill, projecting from the level of the basidia, as occurs in certain agarics, e.g., *Marasmius ferrugineus* and *Pouzarella versatilis* (Agaricales).

pleurogenous, **pleurogenic** (Gr. *pleurá*, side, flank + *génos*, origin < *gennáo*, to engender, produce + L. *-osus* > OF. *-ous*, *-eus* > E. *-ous*, having, possessing the qualities of; or + Gr. suf. *-íkos* > L. *-icus* > E. *-ic*, belonging to, relating to): formed or borne on the sides, as the conidia on the conidiophores of *Trichosporiella cerebriformis* (moniliaceous asexual fungi). See **acropleurogenous**.

pleurokont, **pleurocont**, **pleurokontic**, **pleurocontic**, **pleurokontous**, **pleurocontous** (Gr. *pleurá*, side + *kontós*, oar; here, flagellum; or + suf. *-íkos* > L. *-icus* > E. *-ic*, belonging to, relating to; or + L. *-osus* > OF. *-ous*, *-eus* > E. *-ous*, having, possessing the qualities of): refers to a motile cell that has the flagella or swimming organs on one side, as is observed in the secondary (reniform) zoospores of the Oomycota, which have both flagella (one whiplash and the other tinsel) inserted in the lateral cavity, e.g., *Achlya* and *Saprolegnia* (Saprolegniales). Cf. **acrokont** and **opisthokont**.

pleurosporous (Gr. *pleurá*, side, flank + *sporá*, spore + L. *-osus* > OF. *-ous*, *-eus* > E. *-ous*, having, possessing the qualities of): having the spores inserted laterally, like the basidium of the Uredinales.

pleurotoid (< gen. *Pleurotus* < Gr. *pleurá*, *pleurón*, side, flank + L. suf. *-otus* < Gr. *otós*, ear + L. suf. *-oide*, < Gr. *-oeídes*, similar to): *Agaricales*. A type of fruiting body (13 principal types are recognized) without a stipe, or with the stipe eccentric or lateral, and different types of gills according to the gen. that have this type of fruiting body: *Pleurotus*, *Paxillus*,

Crepidotus (Agaricales), *Panus* and *Lentinus* (Aphyllophorales). This type of fruiting body or habit is considered artificial because it includes gen. from several different families.

Pleurotoid basidiocarp of *Pleurotus* sp., x 0.5.

plurilocular (L. *plurilocularis* < *plus*, genit. *pluris*, more, larger number + *loculus*, dim. of *locus*, cavity, locule + suf. *-aris* > E. *-ar*, like, pertaining to): with several locules or cavities; e.g., like the stroma of the Myriangiales and Dothideales, or the multicellular ascospores of many fungi and lichens. Also called **multilocular**. Cf. **unilocular**.

pluriostiolate (L. *plus*, genit. *pluris*, more, greater number + *ostiolatus*, provided with an ostiole < *ostiolum*, dim. of *ostium*, door + L. suf. *-atus* > E. *-ate*, provided with or likeness): having several ostioles, like the eustromatic conidioma of *Dichomera* (sphaeropsidaceous asexual fungi), which is subepidermal, erumpent, dark and multilocular, and which develops in branches and leaves of various plants. There also are ascomycetes with pluriostiolate fructifications, e.g., *Camillea* (Xylariales).

Pluriostiolate perithecia of *Camillea* sp., x 5 (*CB*).

plurivorous (L. *plus*, genit. *pluris*, several + *vorare*, to devour + *-osus* > OF. *-ous*, *-eus* > E. *-ous*, possessing the qualities of): capable of living on diverse substrates, either as saprobe or parasite; e.g., like the common molds of the gen. *Aspergillus* and *Penicillium* (moniliaceous asexual fungi). Same as **polytrophic**. Cf. **omnivorous**.

pluteotoid (< gen. *Pluteus* < L. *pluteus*, a shed, to eject spores from a natural receptacle + L. suf. *-oide* < Gr. *-oeídes*, similar to): *Agaricales*. A type of fruiting body (13 principal types are recognized) with free gills but without a ring and a volva, like that of *Pluteus* and certain species of *Hygrophorus*, *Lepiota* and *Bolbitius*, among other gen. of Agaricales.

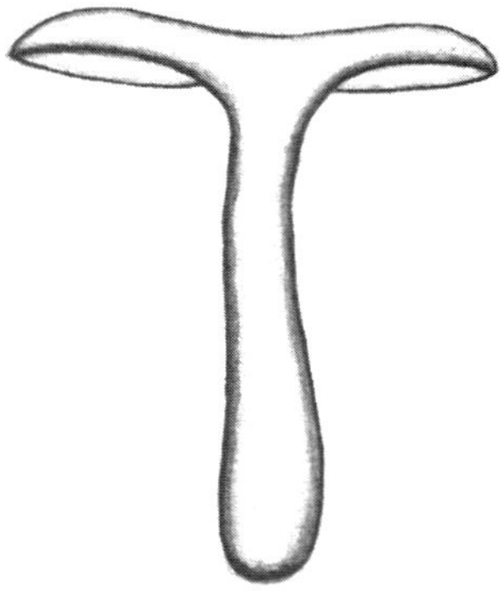

Pluteotoid basidiocarp of *Pluteus* sp. Diagram of longitudinal section, x 1.

podetium, pl. **podetia** (NL. *podetium* < Gr. *poús*, *podós*, foot + L. dim. suf. *-ium*): *Lichens*. An erect, stem-like, simple or branched, hollow structure, on which apothecia are formed; podetia constitute the secondary thallus, derived from the crustaceous primary thallus, in lichens of the family Cladoniaceae, such as *Cladonia pyxidata*, *C. cristatella* and *C. didyma* (Lecanorales). Cf. **pseudopodetium**.

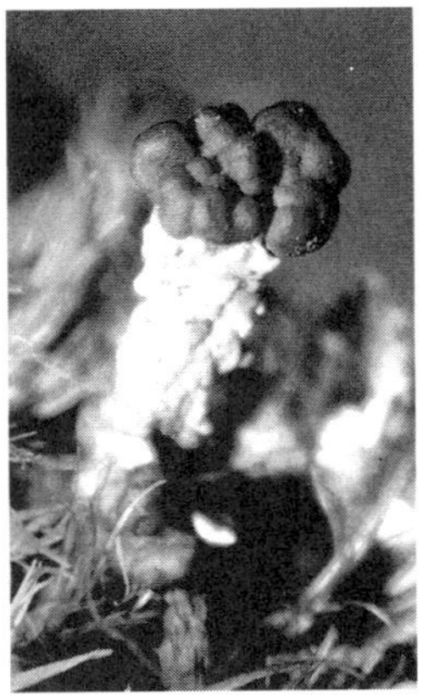

Podetium with an apical apothecium of *Cladonia didyma*, x 5 (*MU & CB*).

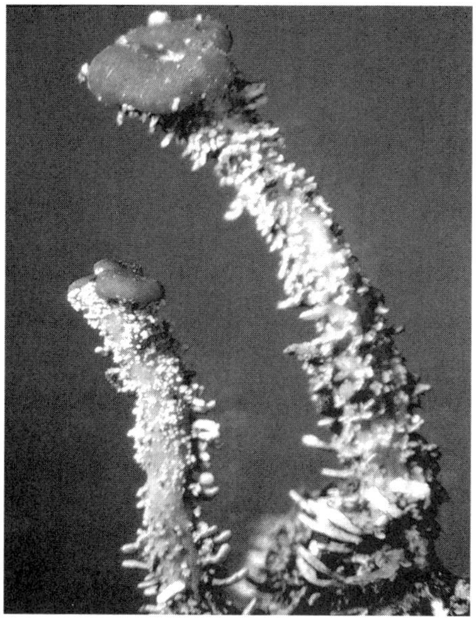

Podetia with an apical apothecium of *Cladonia didyma*, x 8 (*CB*).

poikilohydric (Gr. *poikílos*, changing, variable + pref. *hýdro-* < *hýdor*, water + suf. *-íkos* > L. *-icus* > E. *-ic*, belonging to, relating to): capable of functioning or surviving under changing conditions of water content, like the majority of lichens, a physiological property not present in the majority of the other nutritional groups of fungi. Lichen thalli can accommodate repeated periods of wetness and drought, and their tolerance to drought is much greater than in the majority of other fungi studied.

polarilocular (L. *polaris* < *polus*, pole + suf. *-aris* > E. *-ar*, pertaining to + *loculus*, dim. of *locus*, cavity, locule + suf. *-aris* > E. *-ar*, like, pertaining to): *Lichens*. A spore that has two end cavities or locules united by a narrow canal or isthmus. Also called **polocellate**, a term that is considered more correct than polarilocular, as it refers to the exact structure of this type of spore. *Caloplaca citrina* and *Teloschistes chrysophthalmus* (Teloschistales) are examples of lichens with this type of spore. Similar to **orculiform** or **urceolate**. See **isthmus**.

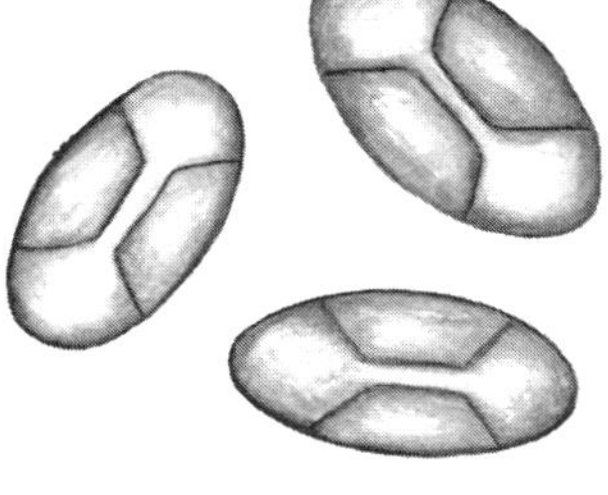

Polarilocular ascospores of *Teloschistes chrysophthalmus*, x 975.

poleophilous (Gr. *pólis*, city + *phílos*, have an affinity for + L. *-osus* > OF. *-ous*, *-eus* > E. *-ous*, possessing the qualities of): adapted to live in cities, like the lichens of urban areas (e.g., *Lecanora conizaeoides*, Lecanorales).

polocellate (L. *polocelatus* < *polus*, pole, the end of an axis + *celatus*, concealed < *celare*, to conceal + suf. *-atus* > E. *-ate*, provided with or likeness): equal to **polarilocular**. Similar to **orculiform** and **urceolate**.

polyandrous (Gr. *polýs*, many, numerous + *andrós*, male; here, the male sexual organ + L. *-osus* > OF. *-ous*, *-eus* > E. *-ous*, having, possessing the qualities of): *Oomycota*. Applied to oospores that develop when fertilized by several functional antheridia, as occurs in *Achlya* (Saprolegniales). Cf. **monandrous**.

polyascous (Gr. *polýs*, many + *askós*, pouch, sack + L. *-osus* > OF. *-ous*, *-eus* > E. *-ous*, having, possessing the qualities of): having many asci; applied especially to species that have the asci in a hymenium, not separated by sterile bands, as occurs in the Pezizales, and to the locules of some Loculoascomycetes, such as *Dothidea* (Dothideales).

polyblastic (Gr. *polýs*, many + *blastós*, sprout, bud, germ, shoot + suf. *-íkos* > L. *-icus* > E. *-ic*, belonging to, relating to): a type of holoblastic conidiogenous cell that produces buds (the conidium primordia) at various points or sites, as in *Gonatobotryum* (dematiaceous asexual fungi). Cf. **monoblastic**.

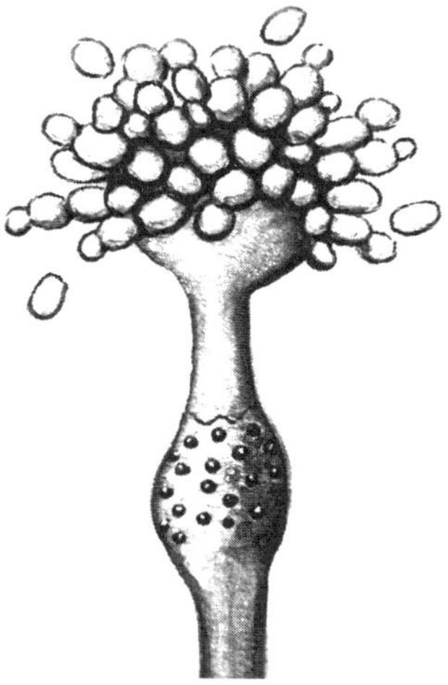

Polyblastic conidiogenous cells of a conidiophore of *Gonatobotryum apiculatum*, x 600.

polycarpic (Gr. *polýs*, many + *karpós*, fruit + suf. *-íkos* > L. *-icus* > E. *-ic*, belonging to, relating to): refers to infections caused by *Exobasidium* (Exobasidiales) when they are systemic (or circumscribed) and perennial. Cf. **monocarpic**.

polycentric (Gr. *polýs*, many, numerous + *kéntron*, center + suf. *-íkos* > L. *-icus* > E. *-ic*, belonging to, relating to): a thallus that has several interconnected centers of growth, in which are formed the reproductive organs (sporangia or resting spores), as happens in aquatic fungi of the family Megachytriaceae (Chytridiales), e.g., *Megachytrium westonii*. Cf. **monocentric**.

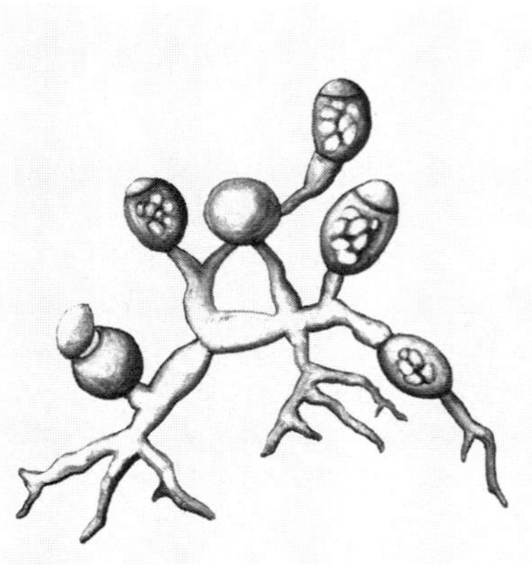

Polycentric thallus of *Megachytrium westonii*, x 800.

polycephalic, **polycephalous** (Gr. *polýs*, many, numerous + *kephalé*, head + suf. *-íkos* > L. *-icus* > E. *-ic*, belonging to, relating to; or + L. *-osus* > OF. *-ous*, *-eus* > E. *-ous*, having, possessing the qualities of): having many heads; e.g., like the sporangia of the Myxomycete *Physarum polycephalum* (Physarales).

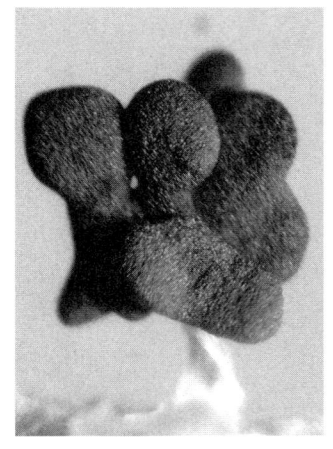

Polycephalous sporangial head of *Physarum polycephalum*, x 20 (*CB*).

polycystoderm (Gr. *polýs*, much, many + *kýstis*, vesicle, cell + *dérma*, skin): a group of spherocysts (of globose and various other shapes) that comprise the connivent fibrils of the exoperidium of *Calvatia cretacea* (Lycoperdales).

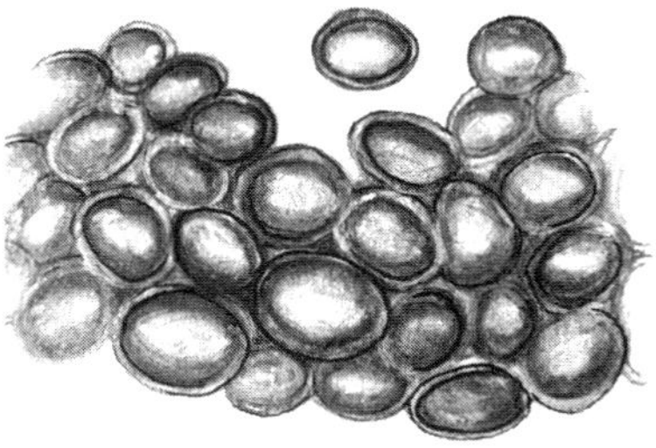

Polycystoderm from the peridium of the fruiting body of *Calvatia cretacea*, x 500.

polychotomous (Gr. *polýschistos*, split, divided into many parts + *témeo*, to cut, divide + L. *-osus* > OF. *-ous*, *-eus* > E. *-ous*, having, possessing the qualities of): with the apex divided simultaneously into more than two branches, like the conidiophore of *Penicillium atrovenetum* (moniliaceous asexual fungi), whose branches give rise to three, five or more metulae at the distal end. Cf. **dichotomous**.

polygonal, **polygonous** (Gr. *polýs*, many + *gónia*, angle + L. suf. *-alis* > E. *-al*, relating to or belonging to; or + L. *-osus* > OF. *-ous*, *-eus* > E. *-ous*, having, possessing the qualities of): having many angles, like the sporangiospores of *Mortierella ramanniana* var. *angulispora* (Mucorales).

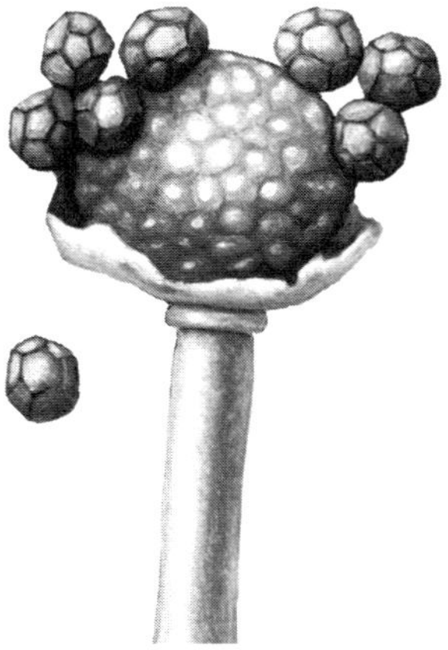

Polygonal sporangiospores of *Mortierella ramanniana* var. *angulispora*, x 2 900.

polymorphic, **polymorphous** (Gr. *polýs*, many + *morphé*, shape + suf. *-íkos* > L. *-icus* > E. *-ic*, belonging to, relating to; or + L. *-osus* > OF. *-ous*, *-eus* > E. *-ous*, having, possessing the qualities of): having more than one form or shape. See **pleomorphic**.

polyphagous (Gr. *polyphagós*, eating a lot, voraceous < *polýs*, many + *phágos*, great eater + L. *-osus* > OF. *-ous*, *-eus* > E. *-ous*, possessing the qualities of): applied to organisms that can utilize various types of food, like the aquatic fungi (Chytridiales) that live on different hosts and whose mycelium occupies various host cells.

polyphialide (Gr. *polýs*, many + *phialís*, genit. *phialídis*, dim. of *phiále*, glass, jar): a bottle-shaped conidiogenous cell that produces phialoconidia (phialospores) through several openings that are formed in synchronous or irregularly sympodial succession in the cell wall, as is observed in *Fusarium moniliforme* var. *subglutinans* (tuberculariaceous asexual fungi) and in *Helicoma inflatum* (dematiaceous asexual fungi).

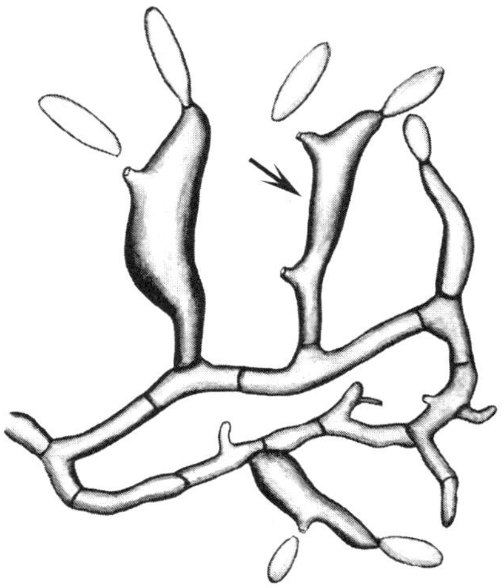

Polyphialides of *Fusarium moniliforme* var. *subglutinans*, x 1 000.

polyphyletic (Gr. *polýs*, many + *phyletikós* < *phyletés*, one of the same tribe, race, stock < *phylé*, tribe + *-tikós* > L. *-ticus* > E. *-tic*, relation, fitness): derived from more than one ancestral line; not sharing a common ancestry, i.e., including taxa from another lineage. Cf. **monophyletic** and **paraphyletic**.

polyphyllous (Gr. *polýs*, many + *phýllon*, leaf + L. *-osus* > OF. *-ous*, *-eus* > E. *-ous*, having, possessing the qualities of): with many leaves; multifoliate, like the thallus of certain lichens that have many foliose lobules, e.g., *Parmelia* (Lecanorales) and *Peltigera* (Peltigerales).

polyplanetic (Gr. *polýs*, many + *planétes*, wanderer + suf. *-íkos* > L. *-icus* > E. *-ic*, belonging to, relating to): *Oomycota*. A species in whose life cycle the zoospores have several swimming periods, alternating with an encystment phase, although the zoospores are of the same type, called secondary (reniform and biflagellate heteroconts), as in the gen. *Dictyuchus* (Saprolegniales). Cf. **monoplanetic** and **diplanetic**.

polypore (L. *polyporus* < Gr. *polýs*, many, numerous + *póros*, pore, passage): one of the Polyporaceae, the basidiomycetous, cosmopolitan gen. with tubulate hymenophore, stipitate, annual or perennial, dimitic basidiomata; lignicolous or terrestrial.

polyporoid (< gen. *Polyporus* < L. *polyporus* < Gr. *polýs*, many + *póros*, pore, passage + L. suf. *-oide* < Gr. *-oeídes*, similar to): *Aphyllophorales*. Refers to a fruiting body, stipitate or not (sessile), with the hymenophore lining the interior of tubes which have pores at the ends, as in *Polyporus*, *Fomes* and other related gen.

polytretic (Gr. *polýs*, many + *tretós*, perforate + suf. *-íkos* > L. *-icus* > E. *-ic*, belonging to, relating to): *Conidiogenesis*. A type of enteroblastic conidiogenous cell that produces conidia through several pores or perforations in the external wall, by protrusion of the internal wall layer through the pores; such conidia may be solitary or in chains, according to the species; e.g., as in the gen. *Helminthosporium*, *Alternaria* and *Curvularia*, among other dematiaceous asexual fungi. This characteristic is observed clearly in *Ulocladium atrum*. Cf. **monotretic**.

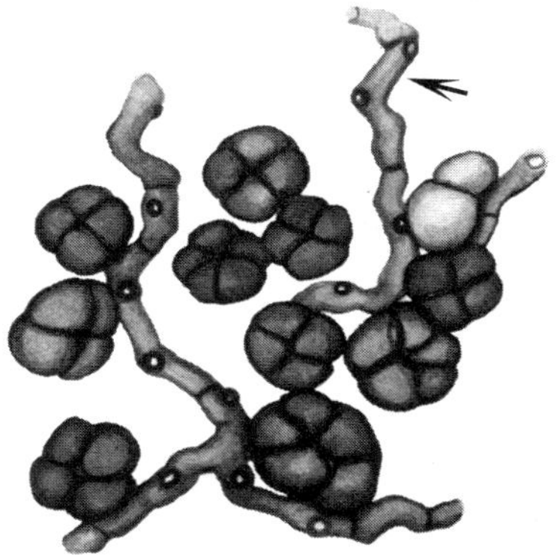

Polytretic conidiogenous cells of *Ulocladium atrum*, x 625.

polytrophic (Gr. *polýs*, many + *trophós*, that which nourishes, serves as food + suf. *-íkos* > L. *-icus* > E. *-ic*, belonging to, relating to): same as **plurivorous**. Cf. **monotrophic** and **omnivorous**.

polyverticillate (Gr. *polýs*, many + NL. *verticillatus* < L. *verticillus*, verticil + suf. *-atus* > E. *-ate*, provided with or likeness): *Conidial fungi*. Refers to a complex conidiophore that has branching at more than three levels, as is observed in several species of *Penicillium* (moniliaceous asexual fungi), such as *P. arenicola*, which can have four (**quadriverticillate**, five or even more points of branching in an irregular pattern; these types of conidiophores have phialides, metulae, ramae and ramillae. Cf. **monoverticillate**, **biverticillate**, **triverticillate** and **quadriverticillate**.

pore (L. *porus* < Gr. *póros*, pore, passage): a small hole or opening in a cell wall. Pores are common in fungi; many spores germinate through a pore (germ pore) in the spore wall; tretic conidia are formed through a pore in the wall of the conidiogenous cell (see **monotretic** and **polytretic**). The mouths of the tubes that contain the hymenium in Aphyllophorales (such as *Pycnoporus*) and in Boletaceae (such as *Boletus*) also are called pores. The germ pores of the ascospores of *Podospora commata* and *Gelasinospora seminuda* (Sordariales), and the pores of the hymenophore of *Poria incrassata*, *Meripilus tropicalis* and *Laetiporus sulphureus* (Aphyllophorales), are some examples.

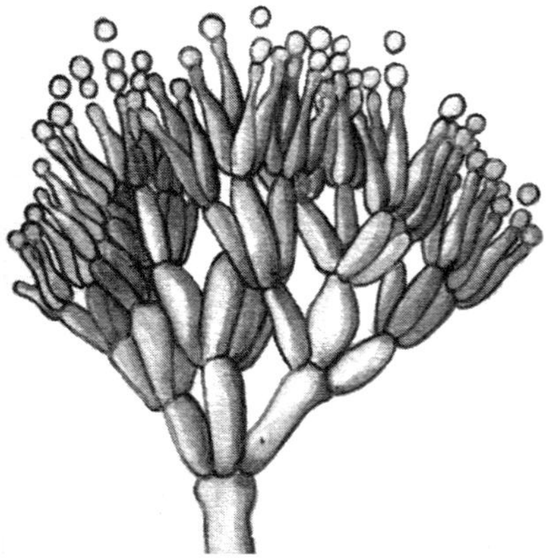

Polyverticillate conidiophore of *Penicillium arenicola*, x 1 000.

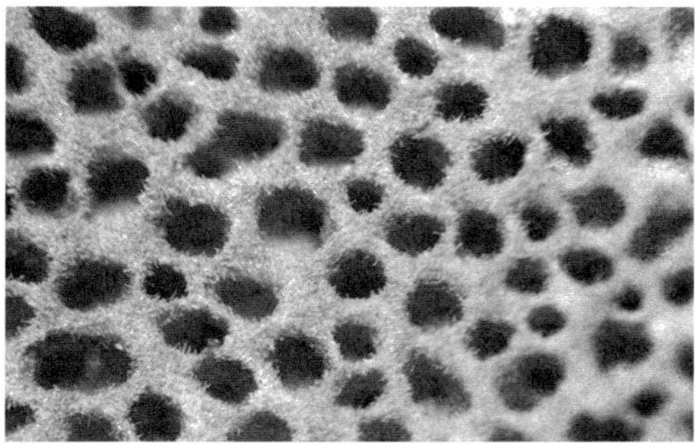

Pores constituting the hymenophore of *Laetiporus sulphureus*, x 15 (*MU*).

porogenous (Gr. *póros*, way, pore + *génos*, origin < *gennáo*, to engender, produce + L. *-osus* > OF. *-ous*, *-eus* > E. *-ous*, having, possessing the qualities of): *Conidiogenesis*. A type of blastic conidium development in which the inner wall of the conidiogenous cell pushes out through a pore in the outer wall to form the conidium; common in many dematiaceous asexual fungi, such as *Alternaria*, *Helminthosporium* and *Ulocladium*. See **monotretic** and **polytretic**.

porospore (Gr. *póros*, orifice, pore + *sporá*, spore): *Conidiogenesis*. An asexual reproductive spore that originates through a pore in the wall of the conidiogenous cell or of another porospore; a term applied to conidial fungi. *Alternaria*, *Bipolaris*, *Curvularia*, *Drechslera*, *Helminthosporium* and *Ulocladium* (dematiaceous asexual fungi) are examples of gen. of fungi that produce porospores.

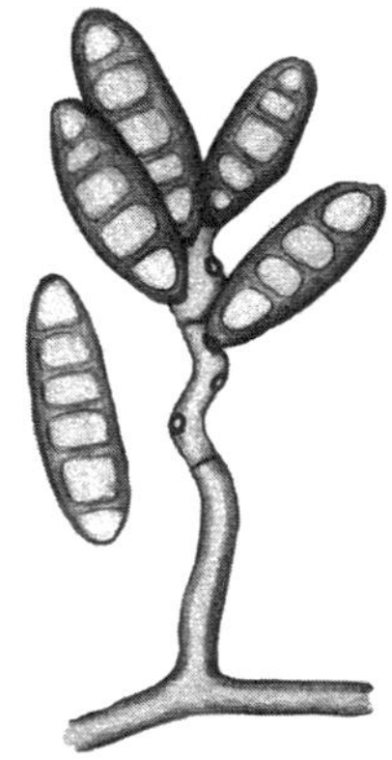

Porospores of *Drechslera* (asexual state of *Cochliobolus bicolor*), x 700.

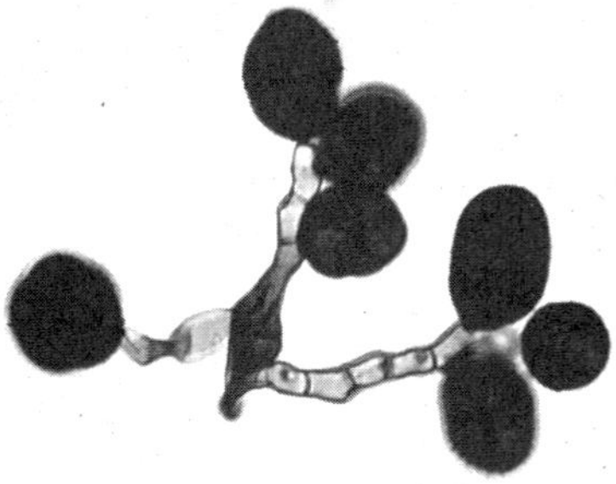

Porospores of *Ulocladium atrum*, x 650 (*MU*).

porous (L. *porus*, pore < Gr. *póros*, way, pore + L. *-osus* > OF. *-ous*, *-eus* > E. *-ous*, having, possessing the qualities of): provided with pores, spongy, like the hymenophore of the polypores, e.g., *Megasporoporia*, *Polyporus* and *Poria* (Aphyllophorales), and boletes, such as *Boletus* and *Strobilomyces* (Agaricales).

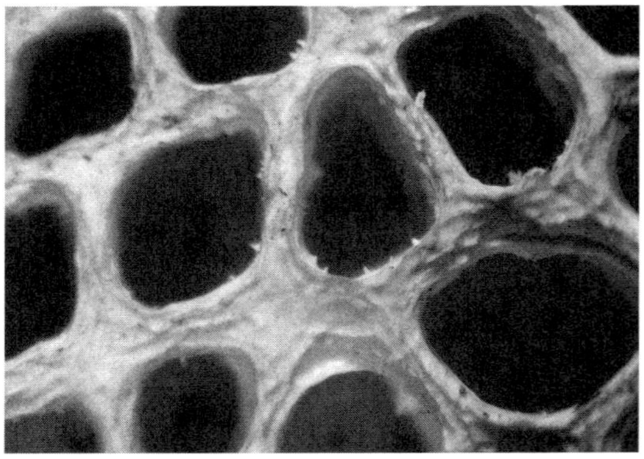

Porous hymenophore of *Megasporoporia mexicana*, x 15 (*MU*).

powdery (E. *powder* < ME. *poudre* < L. *pulver*, powder + E. suf. *-y*, partaking of the nature of, abounding in, full of, having): covered with fine powder, which in the fungi corresponds to spores, such as the powdery mildews (species of Erysiphales that live as obligate plant parasites). The mycelium of other fungi also has this characteristic, e.g., that of the colonies of *Microsporum nanum* and *Beauveria bassiana* (moniliaceous asexual fungi), which is due to the abundant production of dry conidia.

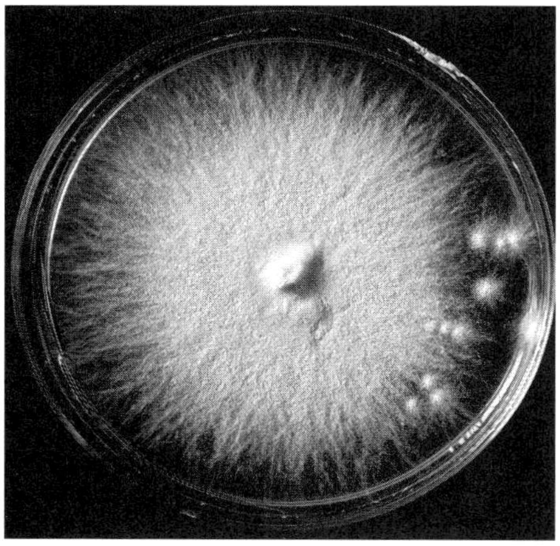

Powdery colony of *Microsporum nanum* on agar, x 6 (*MU*).

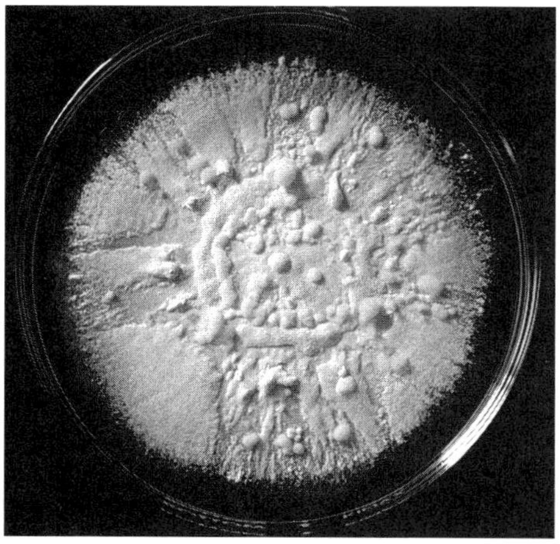

Powdery colony of *Beauveria bassiana* on agar, x 6 (*MU*).

pozol (Azt. *pozolli*, foamy): a fermented maize dough of Mayan origin used for the preparation of a staple, non alcoholic beverage that is consumed by several ethnic groups of Southern Mexico. The microbiota involved in the fermentation of the dough is a complex one, composed of lactic acid bacteria (*Lactobacillus*), yeasts (*Candida* and *Trichosporon*, of the cryptococcaceous asexual yeasts), and molds (*Mucor*, of the Mucorales, *Cladosporium*, of the dematiaceous asexual fungi, and *Geotrichum*, of the moniliaceous asexual fungi, among others).

praemorse (L. *praemorsus* < *praemordere*, to bite something by the tip): more or less irregularly truncate; in general it is applied to the base of the stipe when it seems as if it had been broken off, as occurs in various species of *Lactarius* (Agaricales).

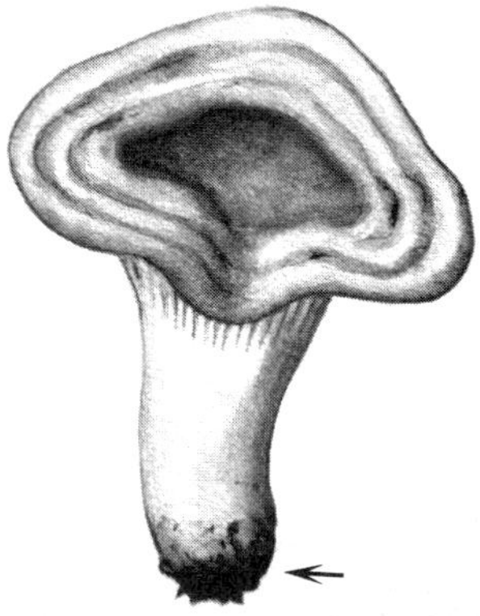

Praemorse stipe of the basidiocarp of *Lactarius zonarius*, x 0.7.

predaceous or **predacious fungus**, pl. **fungi** (L. *praedari*, to prey upon < *praeda*, prey + *-aceus* > E. *-aceous*, of or pertaininig to, with the nature of; L. *fungus*, fungus): living by preying on amoebae, nematodes, rotifers, or other small terrestrial or aquatic animals. The most important are Zoopagales (*Zoopage*, *Cochlonema*, etc.), moniliaceous asexual fungi (*Arthrobotrys*, *Dactylella*, *Harposporium*, *Rotiferophtora*, *Verticillium*, etc.), and Saprolegniales (*Sommerstorffia*).

prespore, **prespore cell** (L. *prae*, in advance, priority + Gr. *sporá*, spore; L. *cellula* < dim. of *cella*, a cell of a honeycomb; cell): an amoeboid, uninucleate protoplast that gives rise to a sporocarp; characteristic of the gen. *Protosporangium* (Protosteliales). Also called prespore cells are the posterior cells of the pseudoplasmodium of species of Dictyosteliomycota that are destined to become the spores of the sorocarp; they have distinctive vacuoles in which spore wall polysaccharides are contained. Cf. **prestalk cell**.

prestalk cell (L. *prae*, in advance, priority + ME. *stalke*, stem; L. *cellula* < dim. of *cella*, a cell of a honeycomb; cell): each of the anterior cells of the pseudoplasmodium of species of Dictyosteliomycota that are destined to become the stalk of the sorocarp. Cf. **prespore cell**.

primary (L. *primarius*, principal, first in order or grade + E. suf. *-y*, having the quality of): the first or main thing. **Primary mycelium**: haploid mycelium of the basidiomycetes, which results from germination of the basidiospores. See **secondary** and **tertiary mycelium**. **Primary thallus**: the first squamules of the thallus of *Cladonia* (Lecanorales) from which originate the podetia (with apothecia) that constitute the **secondary thallus**. **Primary universal veil**: see **protoblem**.

primary appendage (L. *primarius*, principal, first in order or grade; *appendage* < *appendere*, to append): *Laboulbeniales*. An outgrowth that develops from the upper ascospore cell. In some species of *Laboulbenia*, e.g., the primary appendage or its lateral branches become antheridial branches that produce phialides and spermatia; in others, these appendages may be sterile. Cf. **secondary appendage**.

primary host (L. *primarius*, principal, first in order or grade + E. suf. *-y*, having the quality of; *hospes*, one who receives a stranger as guest): see **host**.

primary marine fungus, pl. **fungi** (L. *primarius*, principal, first in order or grade + E. suf. *-y*, having the quality of; L. *marinus*, of the sea < *mare*, sea + L. suf. *-inus* > E. *-ine*, of or pertaining to; L. *fungus*, fungus < Gr. *spóngos*, *sphóngos*, sponge): a species derived from a marine ancestor and which has remained in the marine environment, like *Spathulospora* (Spathulosporales). Cf. **secondary marine fungus**.

primary mycelium, pl. **mycelia** (L. *primarius*, first + E. suf. *-y*, having the quality of; NL. *mycelium* < Gr. *mýkes*, fungus + L. *-elis*, pertaining to + dim. suf. *-ium*): see **primary**.

primary septum, pl. **septa** (L. *primarius*, principal, first + E. suf. *-y*, having the quality of; *septum*, barrier, partition): see **septum**.

primary thallus, pl. **thalli** (L. *primarius*, principal, first in order or grade + E. suf. *-y*, having the quality of; *thallus*, shoot, thallus): see **primary**.

primordial cuticle, **primordial covering** (L. *primordium*, beginning, origin + L. suf. *-alis* > E. *-al*, relating to or belonging to; L. *cuticula*, dim. of *cutis*, skin; ME. *coveren* < OF. *covrir* < L. *cooperire*, to close, cover): see **protoblem.**

primordial veil (L. *primordium*, principle, origin + L. suf. *-alis* > E. *-al*, relating to or belonging to; *velum*, veil): see **protoblem**.

primordium, pl. **primordia** (L. *primordium*, principle, beginning, origin < *primus*, first + *ordior*, to begin + dim. suf. *-ium*): the initial stage of development of any structure; e.g., the primordia of cultivated mushrooms (buttons), *Agaricus brunnescens* (Agaricales). The primordia of the sorocarp of *Dictyostelium mucoroides* (Dictyosteliomycota), of the ascocarps of *Peziza quelepidotia* (Pezizales), and of the basidiocarps of *Coprinus* and *Pleurotus ostreatus* (Agaricales) are some examples illustrated.

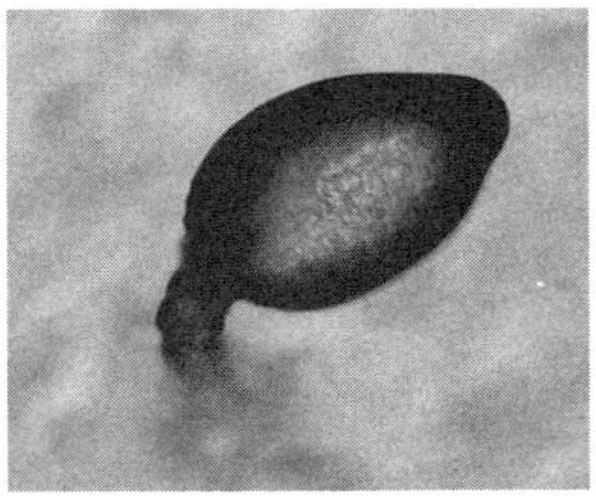

Primordium of a sorocarp of *Dictyostelium mucoroides* on agar, x 100 (*MU*).

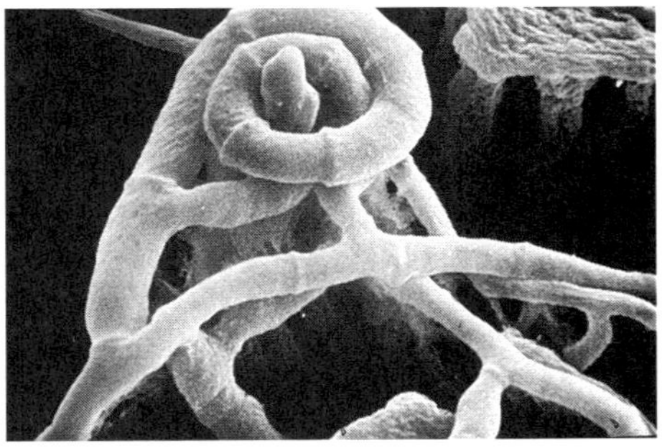

Primordia-ascocarp initials of *Peziza quelepidotia*, seen with the scanning electron microscope. The antheridium is coiled around the ascogonium, x 1 000 (*KO*).

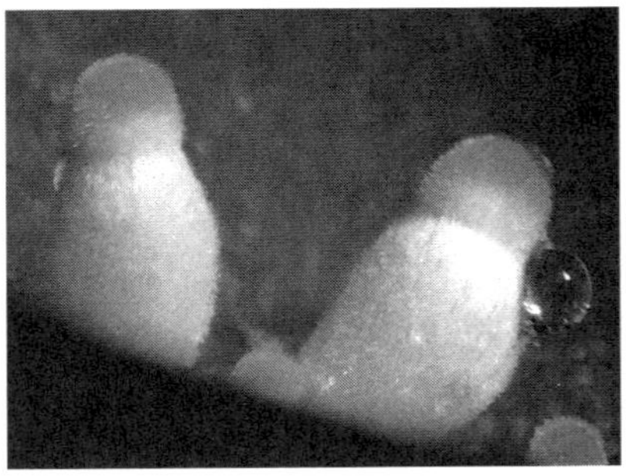

Primordia of two basidiocarps of *Coprinus* sp., x 6 (*MU*).

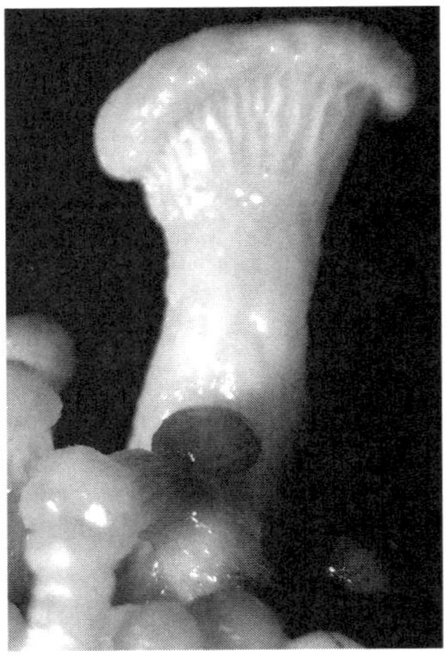

Primordia of several basidiocarps of *Pleurotus ostreatus*, x 5 (*MU*).

pristine (L. *pristinus*, primitive < *priscus*, of or belonging to former times + suf. *-inus* > E. *-ine*, of or pertaining to): primitive, first, old, original.

probasidium, pl. **probasidia** (Gr. *pró*, before, previous + *basídion*, dim. of *básis*, base): *Basidiomycetes*. It refers to the young basidium in a primordial state, from karyogamy to the initiation of the formation of the sterigmata and spores; e.g., as occurs in *Septobasidium* (Septobasidiales), *Coprinus* and *Amanita* (Agaricales). Cf. **protobasidium**.

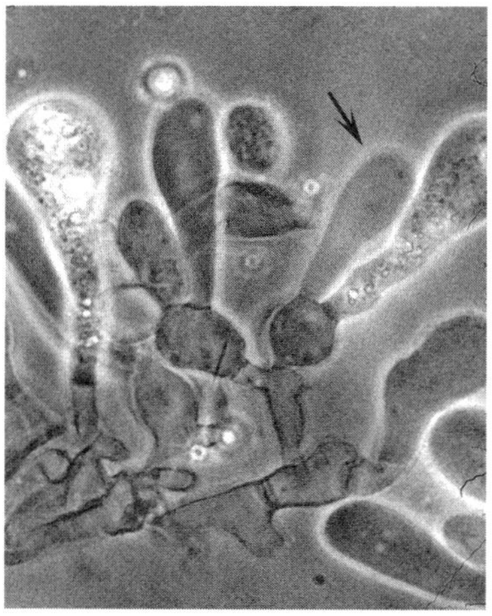

Probasidia of the hymenium of *Amanita* sp., x 560 (*RMA*).

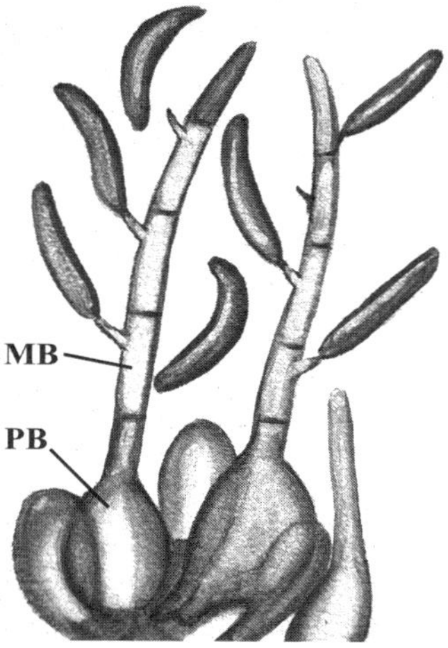

Probasidia (PB) and metabasidia (MB) with basidiospores of *Septobasidium burtii*, developed in symbiosis with the scale-insect *Aspidiotus osborni*, which parasitizes oak branches, x 490.

procumbent (L. *procumbens*, genit. *procumbentis*, prostrate + *-entem* > E. *-ent*, being): prostrate, lying on the substrate.

progametangium, pl. **progametangia** (Gr. *pró*, before, previous + *gamétes*, husband + *angeîon*, vessel, receptacle): *Mucorales*. A small, lateral hyphal branch that first differentiates into a gametangium and afterwards into a suspensor of the zygosporangium, e.g., *Rhizopus nigricans*.

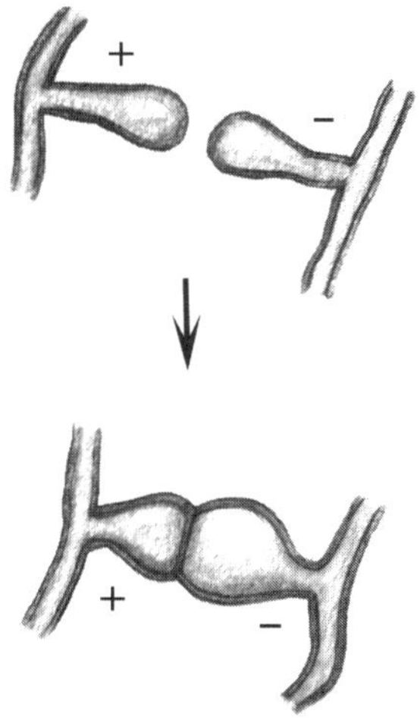

Progametangia of *Rhizopus nigricans*; the + and - zygophores grow toward one another and when they make contact their tips swell to form progametangia, x 100.

progamones (Gr. *pró*, before + *gámos*, marriage, sexual union): a group of sex hormones of Zygomycetes.

prokaryotic, **procaryotic**, **prokaryontic**, **procaryontic** (Gr. *pró*, primitive, before, previous + *káryon*, nucleus + *-tikós* > L. *-ticus* > E. *-tic*, relation, fitness, inclination or ability): an organism whose cells do not possess true nuclei (i.e., the chromosomes are not surrounded by a membrane) nor the organelles of the eukaryotic cells (mitochondria, endoplasmic reticulum, dictyosomes, vacuoles and ribosomes, among other characteristics), because of which they have a type of organization less complex than that of a eukaryotic organism. Only the organisms of the kingdom Monera are prokaryotic, those of the kingdoms Protoctista (or Protista for some), Fungi, Plantae and Animalia are **eukaryotic**. According to the classification adopted in this dictionary, eukaryotic would also include the organisms considered as the kingdom Stramenopila (which includes the phyla Oomycota, Hyphochytriomycota and Labyrinthulomycota) that has been separated from the kingdom Fungi.

prolate (L. *prolatus*, bringing forward < *pro-*, forward + *latus*, wide): elongated along the polar axis; applied to spores of elongate or ellipsoid shape, whose length:width ratio is between 6:1 and 3:1; e.g., like the ascospores of *Ascobolus scatigenus* (Pezizales). Cf. **oblate**.

proliferation (F. *prolifération* < L. *prolifer*, *prolificus*, having the capacity to engender, proliferate < *prole*, progeny + *-fer*, *-ferous*, bearing, producing + *-ationem*, action, state or condition, or result > E. suf. *-ation*): the action of proliferating; growth through the development of new parts. Intrasporangial proliferation is the formation of a new sporangium inside an old sporangium, following zoospore release, from the septum that delimits the sporangium from the supporting hypha, in the saprolegniaceous Oomycota (such as *Saprolegnia parasitica*). A similar proliferation occurs in the asci of *Ascoidea* (Saccharomycetales). There also are proliferating conidiogenous cells, which are called **percurrent**.

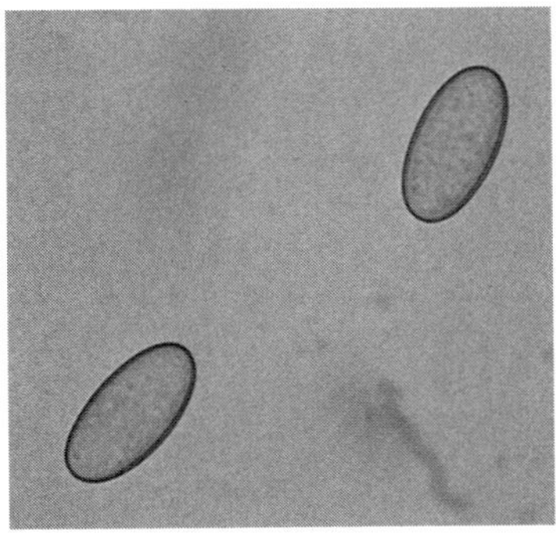

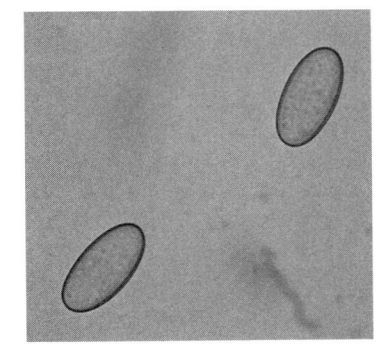

Prolate ascospores of *Ascobolus scatigenus*, x 1 160 (*EAA*).

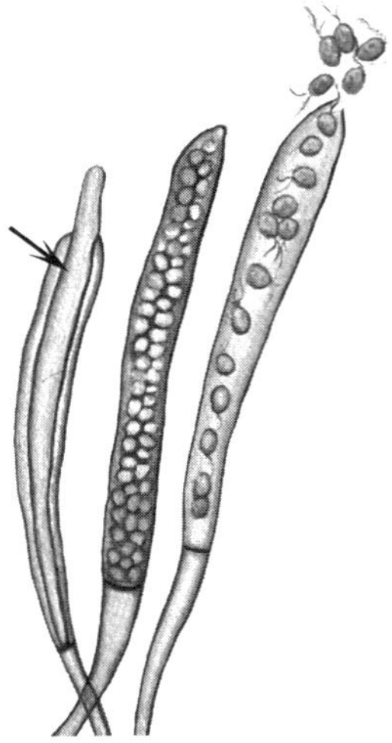

Intrasporangial **proliferation** of *Saprolegnia*, x 150.

promitosis, pl. **promitoses** (Gr. *pró*, before, previous + *mítos*, filament + *-osis*, condition or state): see **cruciform division**.

promycelium, pl. **promycelia** (Gr. *pró*, before, previous + NL. *mycelium* < Gr. *mýkes*, fungus + L. *-elis*, pertaining to + dim. suf. *-ium*): the portion of the basidium that results from the germination of the teliospore, in which occur karyogamy and meiosis, prior to the formation of the basidiospores, as happens in the rusts (e.g., *Puccinia graminis tritici*, of the Uredinales), the smuts (e.g., *Ustilago maydis*, of the Ustilaginales), and other basidiomycetes (e.g., *Aessosporon salmonicolor* and *Rhodosporidium sphaerocarpum*, of the Sporidiales).

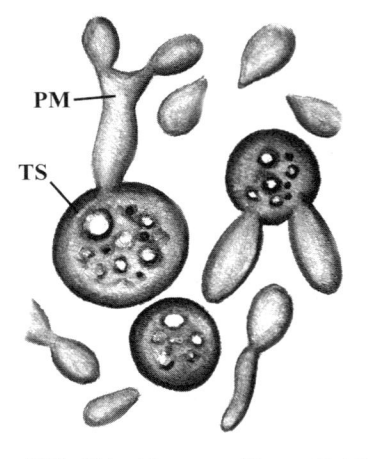

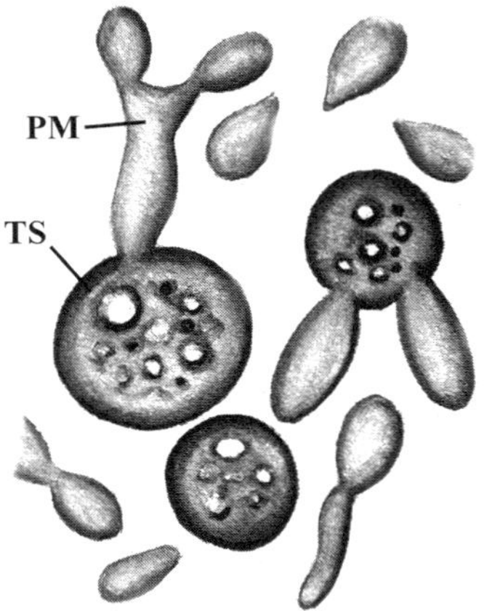

Promycelium (PM) with basidiospores of *Aessosporon salmonicolor*, resulting from the germination of teliospores (TS), x 1 600.

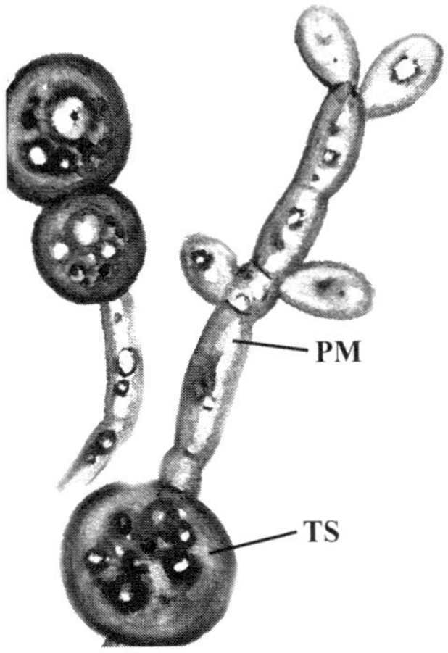

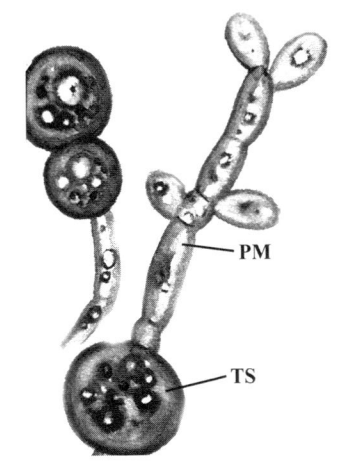

Promycelium (PM), also called metabasidium, with basidiospores of *Rhodosporidium sphaerocarpum*, resulting from the germination of teliospores (TS), x 1 360.

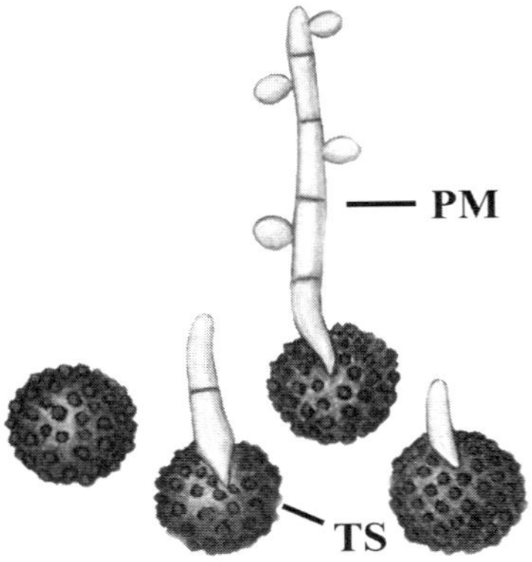

Promycelium (PM) growing from the germinating teliospores (TS) of *Ustilago maydis*, x 1 150.

propagule, **propagulum**, pl. **propagula** (NL. *propagulum* < *propages*, slip, shoot + dim. suf. *-ulum* > E. *-ule*): in general, any structure that serves to propagate or vegetatively multiply an organism; in the fungi the propagules can be spores, bulbils, fragments of mycelium, isidia, and soredia, among others.

prophialide (Gr. *pró*, before, previous + *phialís*, genit. *phialídis*, dim. of *phiále*, glass, jar): see **metula**.

prophylactic (Gr. *prophylaktikós* < *pró*, forward + *phylaktikós*, vigilant, cautious, guarding < *phýlax*, genit. *phylactós*, guard + suf. *-íkos* > L. *-icus* > E. *-ic*, belonging to, relating to): pertaining to or relative to **prophylaxis**.

prophylaxis, pl. **prophylaxes** (Gr. *prophýlaxis*, caution, foresight < *pró*, forward + *phýlax*, guard + *-sis*, act of): measures or treatments designed to prevent diseases.

prosenchyma, pl. **prosenchymata** (Gr. *prós*, near, close to + *énchyma*, stuffed): a type of plectenchyma in which the hyphae that compose it are arranged parallel to one another, intertwining or anastomosing, but retaining their individuality. For example, in the pedicle of the perithecial stroma of *Claviceps purpurea* (Hypocreales) one can observe clearly prosenchyma tissue. Cf. **pseudoparenchyma**.

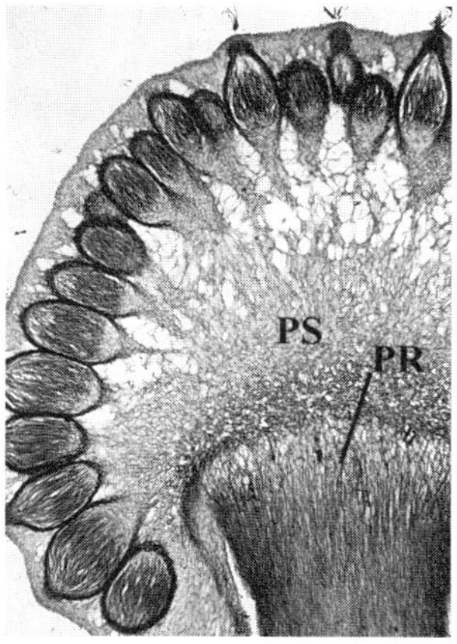

Prosenchyma (PR) and pseudoparenchyma (PS) of the perithecial stroma of *Claviceps purpurea*, x 25 (*RTH*).

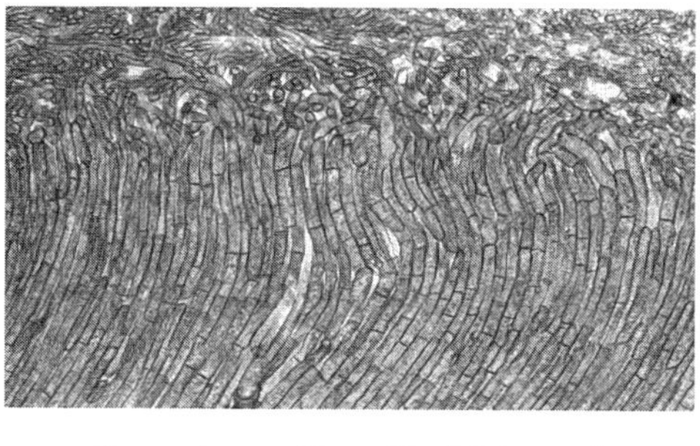

Prosenchyma of the perithecial stroma of *Claviceps purpurea*, x 130 (*CB*).

prosoplectenchyma, pl. **prosoplectenchymata** (Gr. *prós*, near together + *plektós*, intertwined + *énchyma*, stuffed): *Lichens*. A type of tissue found in the thallus cortex of many macrolichens in which the hyphae are oriented in a particular direction; it corresponds to the prosenchyma of other fungi. Cf. **chalaroplectenchyma** and **paraplectenchyma**.

prosorus, pl. **prosori** (Gr. *pró*, before, previous + *sorós*, pile, heap): a cell, derived from an infective zoospore, that eventually germinates and divides to give rise to a sorus or group of sporangia, as happens in *Synchytrium endobioticum* (Chytridiales), an endoparasite of potato tubers.

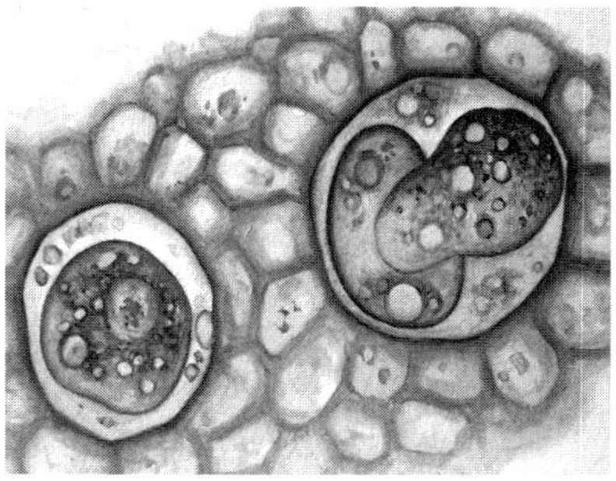

Prosori (one germinating) of *Synchytrium endobioticum* parasitizing potato tuber cells, seen in transverse section, x 510.

prosporangium, pl. **prosporangia**, **presporangium**, pl. **presporangia** (Gr. *pró* > L. *prae* > E. *pre-*, before, previous + *sporá*, spore + *angeîon*, vessel, receptacle): *Aquatic fungi*. A structure similar to a sporangium, from which is produced a vesicle (the sporangium proper) in which the zoospores develop and from which they are liberated, as is observed in *Chytriomyces* (Chytridiales). The prosporangium is present in diverse gen. of aquatic fungi whose resting sporangia on germinating give rise to a vesicle, as happens in *Rhizidium* and *Rhizophlyctis* (Chytridiales).

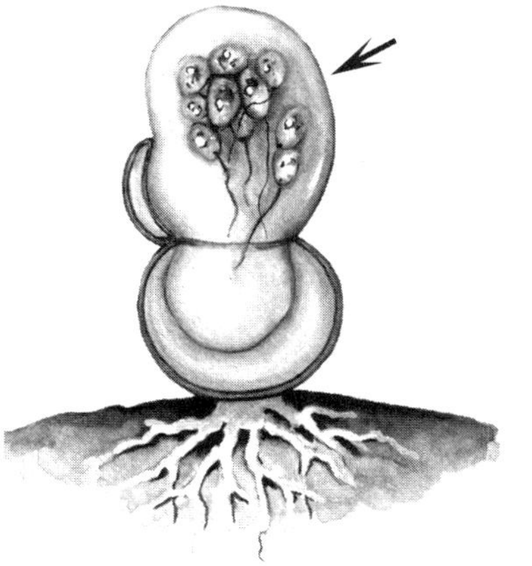

Prosporangium of *Chytriomyces hyalinus*, x 900.

prothallus, pl. **prothalli** (Gr. *pró*, before, previous + *thallós*, thallus > L. *thallus*): *Lichens*. The zone free of algae (containing only the mycobiont), white, brown or black in color, found between the areoles and the growing margins of the thallus of crustaceous lichens.

Protista (L. *protista*, pl. of *protistum* < Gr. *prótiston*, the very first): a kingdom proposed by Haeckel, in 1866, to include the majority of the primitive unicellular organisms (ambiguous organisms that do not fit clearly in either the plant or animal kingdoms), among them the protozoans and various groups of lower algae and fungi. In the classification system adopted in this dictionary (Alexopoulos *et al*., 1996), the phyla Plasmodiophoromycota, Dictiosteliomycota, Acrasiomycota, and Myxomycota, which were formerly considered within the kingdom Fungi, now have been included in the kingdom Protista. Haeckel also distinguished a special group within the Protista, called the **Monera**, to incorporate the bacteria.

protobasidium, pl. **protobasidia** (Gr. *prõtos*, first + *basídion*, dim. of *básis*, base): *Fungi with heterobasidia*. The young basidium or primordial cell that later divides into four cells by the formation of transverse septa, each of which forms a basidiospore on a sterigma, as occurs in the Uredinales and Auriculariales; in the Tremellales the protobasidium divides longitudinally and each of the four cells terminates in a long, tubular sterigma. Cf. **probasidium**.

protoblem (Gr. *prõtos*, first + *blêma*, cover): a layer of pachydermous hyphae, not very dense, that covers the general or universal veil (first covering) of the primordium of the sporiferous apparatus of some basidiomycetes, as in the gen. *Amanita* (Agaricales). Also called **primary universal veil**, **primordial cuticle**, **primordial veil**, and **teleblem** or **teleoblem**.

Protoctista (Gr. *prõtos*, first + *ktistýs*, *ktisís*, creation, thing created, creature): a kingdom proposed by Copeland in 1956, basing it on the term introduced by Hogg in 1861, to include the eukaryotic organisms, including the protozoans, algae, and fungi in their entirety. In later systems of classification, such as that of Whittaker (1959) and of Margulis and Schwartz (1988), the kingdom Protoctista is one of five kingdoms, and includes algae, protozoans, mucilaginous molds, aquatic and amphibious fungi with flagellate forms, and many other parasitic aquatic organisms. According to these authors, groups of organisms customarily classified as fungi, such as Protosteliomycetes, Acrasiomycetes, Myxomycetes, Plasmodiophoromycetes, Chytridiomycetes, Hyphochytriomycetes and Oomycetes, would be in the kingdom Protoctista. The molds, the mushrooms and other macroscopic fungi, and the lichens, are separated in the kingdom Fungi. In the classification system adopted in this dictionary (Alexopoulos *et al*., 1996), the Oomycetes and Hyphochytriomycetes are treated as phyla Oomycota and Hyphochytriomycota, respectively, and are included, along with the Labyrinthulomycota, in the kingdom Stramenopila.

protocyst (Gr. *prõtos*, first, primitive + *kýstis*, bladder, vesicle): the bulbous part of a sterile hypha that forms part of the immature gleba of the basidiocarp of *Nia vibrissa*, a saprobic marine Gasteromycete (Melanogastrales); the protocysts, sometimes also called "primary basidia", are replaced by chance by basidia, as the fructification continues maturing.

protoperithecium, pl. **protoperithecia** (NL. *protoperithecium* < Gr. *prõtos*, first + *perithekíon*, dim. of *perithêke*, helmet, cap; also *perí*, around + *thêke*, case, box; here, of the asci): a perithecial primordium that develops into a perithecium after fertilization (spermatization) occurs, as happens in *Neurospora sitophila* (Sordariales).

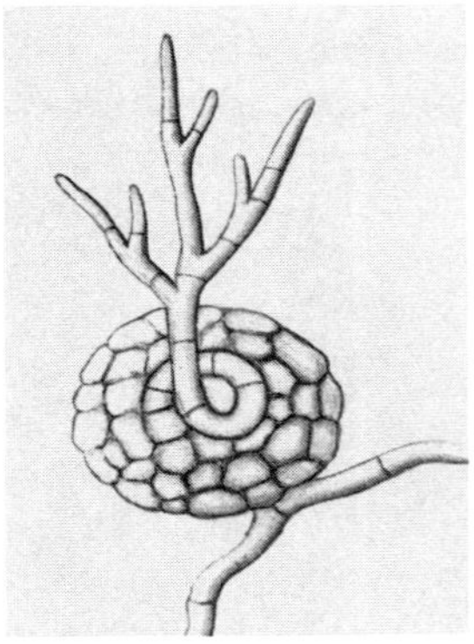

Protoperithecium of *Neurospora sitophila*, x 375.

protoplasm (Gr. *prõtos*, first + *plásma*, bland material that forms a living being): a living substance, semifluid, of the cells, that constitutes the physical and chemical basis of life.

protoplasmodium, pl. **protoplasmodia** (Gr. *prõtos*, first + L. *plasmodium* < Gr. *plásma*, bland material with which is formed a living being + L. *-odium*, resembling < Gr. *ode*, like < *-oeídes*, similar to): a microscopic plasmodium, without filaments or differentiated fan-shaped regions, which has slow and irregular protoplasmic currents, and which gives rise to a single small fruiting body; the protoplasmodium is typical of the Echinosteliales (such as *Echinostelium minutum*) and of *Licea parasitica* (Liceales), although it also occurs in other Myxomycetes. Cf. **aphanoplasmodium** and **phaneroplasmodium**.

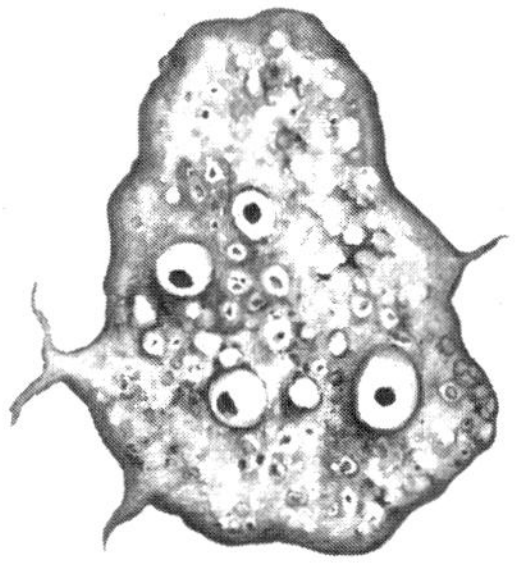

Protoplasmodium (with four nuclei) of *Echinostelium minutum*, x 1 360.

protopseudothecium, pl. **protopseudothecia** (NL. *protopseudothecium* < Gr. *prõtos*, first, primitive + *pseûdos*, false + *thekíon*, dim. of *thêke*, case, box; here, of the asci): a small stroma that will develop into a pseudothecium (ascostroma) following fertilization.

protosexuality (Gr. *prõtos*, first + L. *sexualis*, belonging to or pertaining to sex + E. suf. *-ty*, quality, condition): a type of reproduction in certain yeasts, in which haploid cells are formed from diploid cells in culture, without having produced ascospores; also, the chlamydospores that these species of yeasts form can function as asci, as in the species of *Metschnikowia* (Saccharomycetales),

protospore (Gr. *prõtos*, first + *sporá*, spore): *Mucorales.* The multinucleate mass of cytoplasm that is separated in developing sporangia due to the primary cleavage planes, and which with the formation of the subsequent planes of cleavage gives rise to multiple portions of uninucleate cytoplasm, which are transformed into spores. This process is present in *Phycomyces* and other Mucorales, and in the spherule (sporangium) of *Coccidioides* (of uncertain taxonomic position, classified in the family Synchytriaceae, of the order Chytridiales, as well as in the moniliaceous asexual fungi). The uninucleate portion of the protoplasm that eventually is converted into the sporangium is also called a protospore.

Protosteliomycetes (Gr. *prõtos*, first + *stéle*, column + L. *-mycetes*, ending of class < Gr. *mýkes*, genit. *mýketos*, fungus): a class of mucilaginous or viscous fungi, whose myxamoebae do not aggregate to form pseudoplasmodia, but sporulate directly or form plasmodia that produce fructifications called sporocarps or sporophores. The Protosteliomycetes (with the single order Protosteliales) together with the Acrasiomycetes, Myxomycetes and Plasmodiophoromycetes were formerly included in the division Myxomycota of the kingdom Fungi. Some authors (e.g., Margulis and Schwartz, 1988) classify the Protosteliomycetes as a phylum of the kingdom Protoctista. In the classification system followed in this dictionary (Alexopoulos *et al.*, 1996), this class is not included, so, in the cases in which these kinds of organisms are considered for explaining some of the concepts, such organisms are referred to as Protosteliales. Some other authors treat the Protosteliomycetes (and the order Protosteliales) as part of the kingdom Protista.

protothecium, pl. **protothecia** (NL. *protothecium* < Gr. *prõtos*, first + *thekíon*, dim. of *thêke*, case, box; here, of the asci): an incompletely differentiated ascoma containing neither asci nor ascospores.

prototroph (Gr. *prõtos*, first; here, taken in the sense of a chemical element + *trépho*, to nourish): **1**. An organism capable of developing on a medium composed only of inorganic substances, or at least competent for the assimilation of free nitrogen when it can dispose of hydrocarbon compounds. The aerobic fungi are prototrophs with respect to oxygen, since they can assimilate it directly as a simple chemical element, without it having to be in combination with other elements. **2**. Mutant strains that do not differ from the wild strain in nutritional requirements relative to particular vitamins or amino acids, i.e., like the wild strain, they are capable of synthesizing these compounds. Cf. **auxotroph**.

prototunicate (Gr. *prõtos*, first + *tunica*, covering, blanket + L. suf. *-atus* > E. *-ate*, provided with or likeness): *Ascomycetes*. An ascus with a thin delicate wall that releases its spores by deliquescing. Cf. **bitunicate** and **unitunicate**.

pruinate, **pruinose** (L. *pruinatus*, frosted < *pruina*, frost, snow + suf. *-atus* > E. *-ate*, provided with or likeness; L. *pruinosus*, frosted < *pruina* + *-osus*, full of, augmented, prone to > ME. *-ose*): having a very tenuous waxy covering, similar to frost or flour, that gives it a characteristic appearance; this covering, formed of granules or little crusts, somewhat masks the true color of the respective structure. The basidiocarps of *Coprinus comatus* and *C. micaceus* (Agaricales) are pruinose.

psammophilous (Gr. *psámmos*, sand + *phílos*, have an affinity for + L. *-osus* > OF. *-ous*, *-eus* > E. *-ous*, possessing the qualities of): syn. of **arenicolous**.

pseudoaethalium, pl. **pseudoaethalia** (Gr. *pseûdos*, false + *aíthalos*, soot + L. dim. suf. *-ium*): a mass of sporangia, with very compacted stipes, that simulates an aethalium, but is distinguished from the latter because individual sporangia are apparent at maturity, as is seen in *Metatrichia vesparium* (Trichiales).

pseudoamyloid (Gr. *pseûdos*, false + *ámylon*, starch + L. suf. *-oide* < Gr. *-oeídes*, similar to): falsely amyloid. Applied to the color reaction that spores or cell walls have when treated with Melzer's reagent (potassium iodide, chloral hydrate, and water), which finally take a brown or purple-brown color (the amyloid reaction turns them pale gray in color, at times so dark they are almost black). The pseudoamyloid reaction also is called **dextrinoid**. Cf. **amyloid** and **anamyloid**.

Pruinose surface of the basidiocarps of *Coprinus micaceus*, x 1.5.

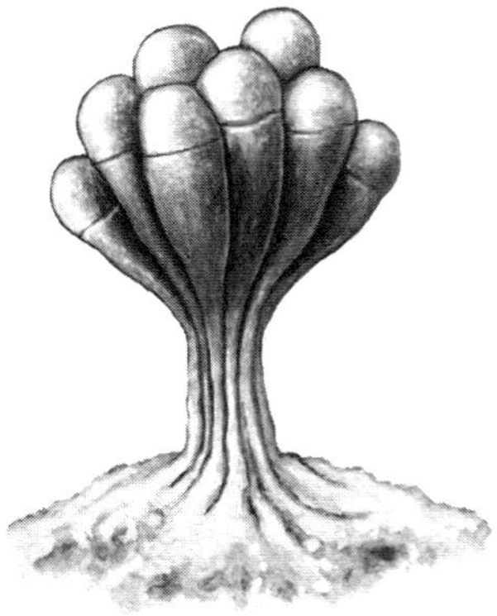

Pseudoaethalium of *Metatrichia vesparium*, x 20.

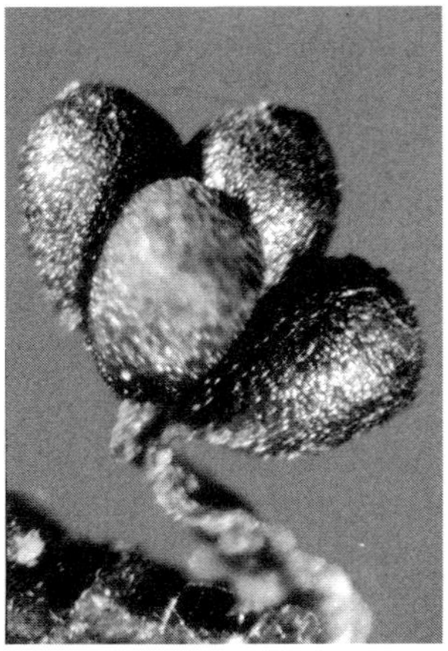

Pseudoaethalium of *Metatrichia vesparium*, x 25 (*MU*).

pseudocapillitium, pl. **pseudocapillitia** (Gr. *pseûdos*, false + L. *capillitium*, long hair, in the collective sense < *capillus*, hair + *-itium*, adjectival ending, meaning provided with, having): *Myxomycetes*. Filaments that are irregular in shape, diameter and ornamentation, perforated plates or other structures, similar to true capillitium (whose elements tend to be more uniform), that are found among the spores inside the fructifications of some species, such as *Lycogala* and *Dictydiaethalium* (Liceales). Cf. **capillitium**.

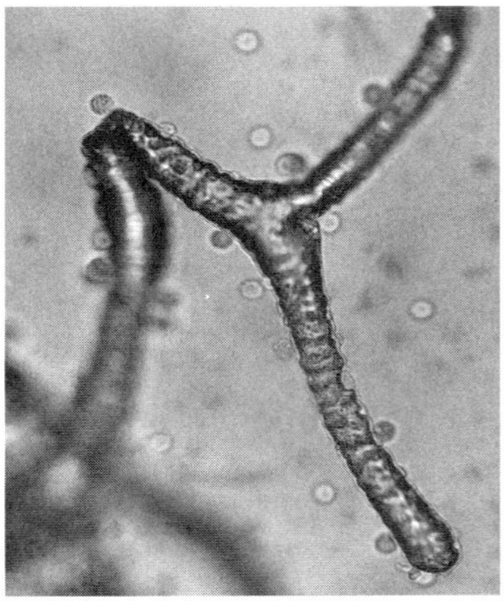

Pseudocapillitium from an aethalium of *Lycogala epidendrum*, x 270 (*MU*).

pseudoclypeus, pl. **pseudoclypei** (Gr. *pseûdos*, false + L. *clypeus*, round shield): a clypeus in which there is included host tissue, as in *Heliascus* (Dothideales), a marine fungus whose black, carbonaceous, lenticular stromata possess locules covered by a pseudoclypeus that is composed of a black tissue that includes host cells, in this case, of the dead roots of mangrove (*Rhizophora mangle*).

pseudocolumella, pl. **pseudocolumellae** (Gr. *pseûdos*, false + L. *columella*, small column < *columna*, column + dim. suf. *-ella*): **1**. *Myxomycetes*. The calcareous nodules of the capillitium which in mass form a kind of small column in the center of the sporangium, as occurs in some species of *Physarum* (Physarales), such as *Ph. nicaraguense* and *Ph. mutabile*. **2**. *Gasteromycetes*. A stalk composed of sterile tissue, often chambered, that is surrounded by the gleba; e.g., in *Geastrum*, *Lycoperdon* and *Radiigera* (Lycoperdales).

pseudocyphella, pl. **pseudocyphellae** (Gr. *pseûdos*, false + L. *cyphella*, the concavities of the ears): *Lichens*. Each of the small open cavities in the lower part of the thallus, similar to the cyphellae, found in the family Stictaceae of the order Lecanorales. Cf. **cyphella** and **cyphelloid**.

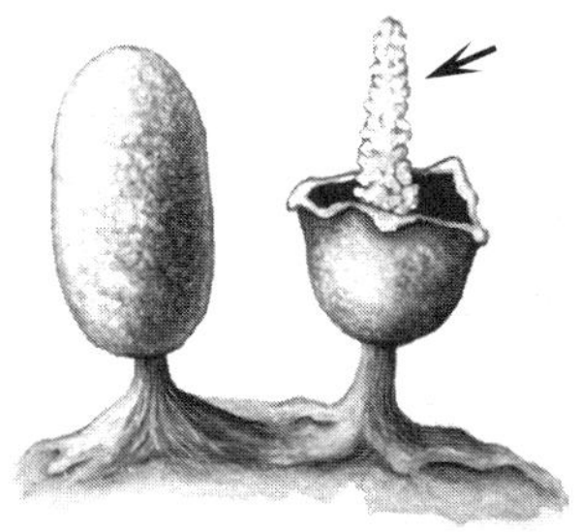

Pseudocolumella of the sporangia of *Physarum mutabile*, x 30.

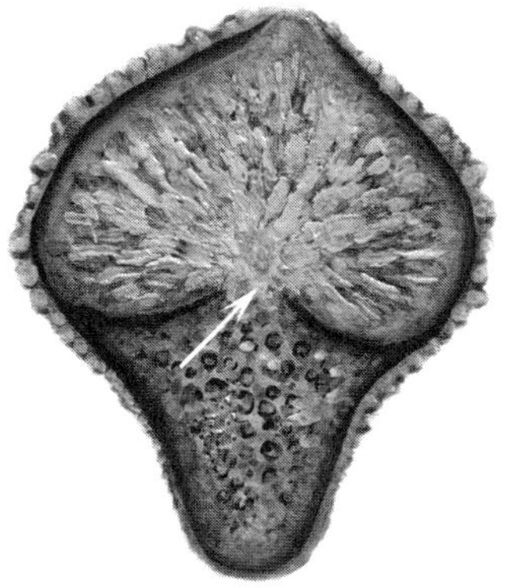

Pseudocolumella of the fruiting body of *Lycoperdon perlatum*, x 1.

pseudocystidium, pl. **pseudocystidia** (Gr. *pseûdos*, false + *kystídion*, dim. of *kýstis*, bladder; here, vesicle, cell): a prolongation, similar to a cystidium, of conducting elements (gloeocystidia, chrysocystidia, phaeocystidia, and macrocystidia) that reach the hymenium, the epicutis, or the upper layer of the stipe, in the basidiocarps of *Russula* and *Lactarius* (Agaricales).

pseudoflagellum, pl. **pseudoflagella** (Gr. *pseûdos*, false + L. *flagellum*, whip): hyaloplasmatic structures which frequently form at the flagellated tip of the myxozoospores of certain Protosteliales (such as *Cavostelium*); these fleeting prolongations, which are not flagellate, continue forming and disappearing in succession toward the posterior end of the myxozoospores. They appear to be filose pseudopods or extremely long and slender filopods.

pseudogamy (Gr. *pseûdos*, false + *gámos*, sexual union + *-y*, E. suf. of concrete nouns): see **amphimixis** and **pseudomixis**.

pseudohymenium, pl. **pseudohymenia** (Gr. *pseûdos*, false + *hyménion* < *hymén*, membrane + dim. suf. *-íon* > L. *-ium*): *Gasteromycetes*. A hymenial layer of basidia, generally very convolute, that has not evolved for forcible spore discharge and has no air space to allow for forcible discharge; e.g., as in the orders Phallales and Lycoperdales. In the order Sclerodermatales, the basidia are produced either as isolated cells distributed throughout the gleba, or as nodules or islands of fertile cells. Pseudohymenia differ from the true hymenia that are present in the hymenomycetous basidiomycetes (Agaricales, Aphyllophorales, Auriculariales, Dacrymycetales, Ceratobasidiales, Tulasnellales, and Tremellales), in which there is forcible spore discharge into an air space that allows for spore dissemination.

pseudomixis, pl. **pseudomixes** (Gr. *pseûdos*, false + *míxis*, mixture, intimate union): substitution of the true sexual union of reproductive cells by a process of pseudosexual copulation between two cells not specifically differentiated as sexual. **Adelphogamy** and **pedogamy** are two types of pseudomictic sexuality. Pseudomixis also is known as **pseudogamy**.

pseudomycelium, pl. **pseudomycelia** (Gr. *pseûdos*, false + NL. *mycelium* < Gr. *mýkes*, fungus + L. *-elis*, petaining to + dim. suf. *-ium*): a grouping of loosely united cells arranged in short chains, as occurs in various species of yeasts, ascosporogenous (such as *Saccharomyces*, of the Saccharomycetales) as well as anascosporogenous (e.g. *Candida*, of the cryptococcaceous asexual yeasts). The chain of cells of the pseudomycelium is originated by budding and is distinguished from true mycelium, which forms by the continuous extension of a hypha that usually is not constricted at the septa that delimit the cells.

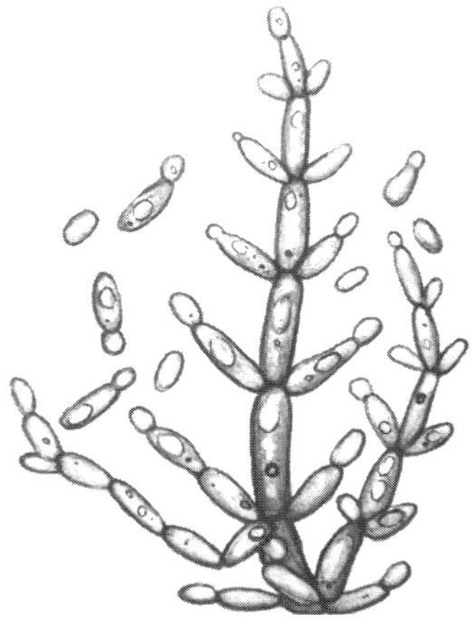

Pseudomycelium of *Saccharomyces kluyveri*, x 800.

pseudoparaphysis, pl. **pseudoparaphyses** (Gr. *pseûdos*, false + *paráphysis*, sprout < *pará*, at the side of + *phýsis*, action of budding, to be born, a growth): sterile hyphae that form in the top of the centrum of an ascostroma and grow downward to the bottom of the centrum, where they may intermingle with the ascogenous system. The asci grow up among the pseudoparaphyses, which may or may not persist until maturity of the asci. Pseudoparaphyses are present, e.g., in the orders Pleosporales, Microthyriales and Hysteriales, of the Loculoascomycetes, such as *Pleospora*, *Myocopron* (=*Ellisiodothis*) *smilacis* and *Hysterium*, respectively.

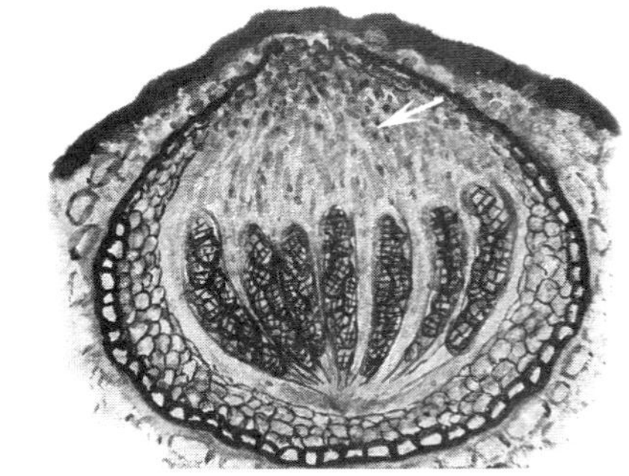

Pseudoparaphyses of a pseudothecium of *Pleospora herbarum*, seen in longitudinal section, developing underneath onion epidermis, x 100.

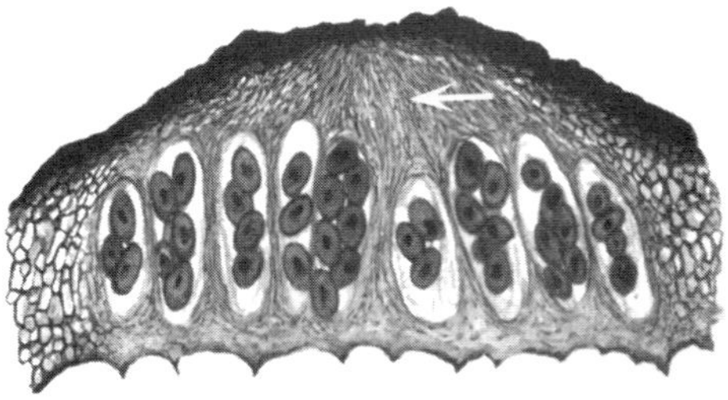

Pseudoparaphyses of the pseudothecium of *Muyocopron* (=*Ellisiodothis*) *smilacis*, seen in longitudinal section, x 230.

pseudoparenchyma, pl. **pseudoparenchymata** (Gr. *pseûdos*, false + *parénchyma*, a tissue of higher plants < *para*, beside, near + *énchyma*, stuffed): a type of plectenchyma composed of closely united angular or isodiametric cells; the hyphae involved in the formation of a pseudoparenchyma lose their individuality. Pseudoparenchyma is present in sclerotia (e.g., *Claviceps purpurea*, of the Hypocreales), stromata, fruiting bodies, and other fungal structures. In the agarics pseudoparenchyma refers to the still undifferentiated tissue from the primordium of the basidiocarp. Cf. **prosenchyma**.

pseudoperithecium, pl. **pseudoperithecia** (NL. *pseudoperithecium* < Gr. *pseûdos*, false + *perithekíon*, dim. of *perithêke*, cap, helmet; also from *perí*, around + *thêke*, case, box; here, of the asci + dim. suf. *-íon* > L. *-ium*): see **pseudothecium**.

pseudophialide (Gr. *pseûdos*, false + *phialís*, genit. *phialídis*, dim. of *phiále*, glass, jar): a small ovoid to obpyriform cell that is found in the concavity of a sporocladium, and from which is formed a unispored sporangiolum, as occurs in *Kickxella*, *Coemansia* and other Kickxellales.

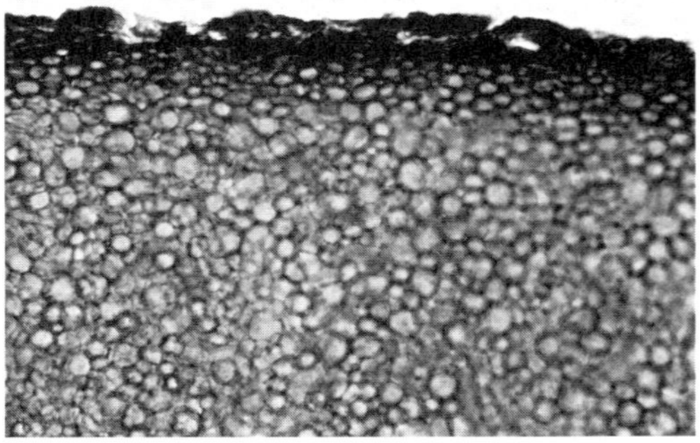

Pseudoparenchyma of the sclerotium of *Claviceps purpurea*, seen in transverse section, x 670 (*MU*).

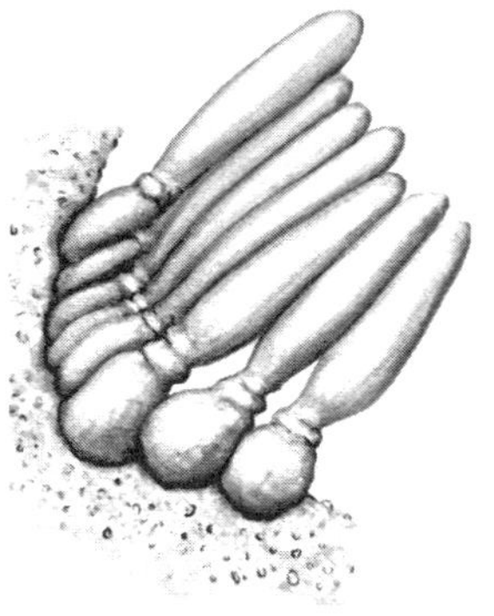

Pseudophialides, each with a single-spored merosporangium, of *Kickxella alabastrina*, x 1 750.

pseudophysis, pl. **pseudophyses** (Gr. *pseûdos*, false + *phýsis*, action of budding or being born; a growth): a type of paraphysis or sterile structure, which is characterized by being thin-walled and smooth, with moniliform constrictions at the apex, formed in the hymenium of the gen. *Aleurodiscus* (Aphyllophorales), along with another structure, the dendrophysis. In the agarics, the pseudophyses correspond to the brachycystidia or brachybasidia. See **dendrophysis**.

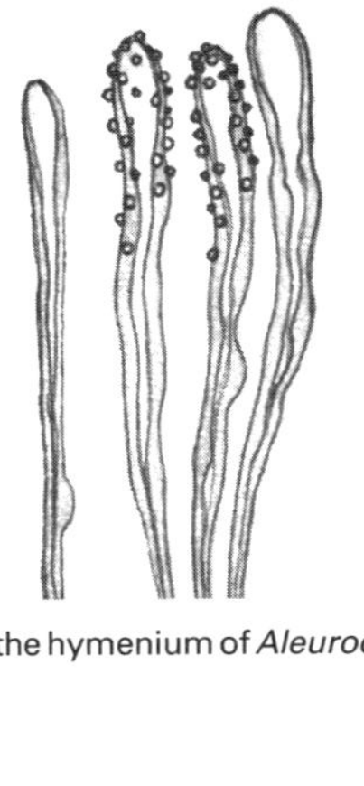

Pseudophyses from the hymenium of *Aleurodiscus scutellatus*, x 375.

pseudopionnote (Gr. *pseûdos*, false + *pión*, fat, thick juice + *notís*, moisture): a continuous and effuse mass of numerous small sporodochia, that form a more or less continuous mucilaginous layer of conidia, as in *Fusarium solani* (tuberculariaceous asexual fungi).

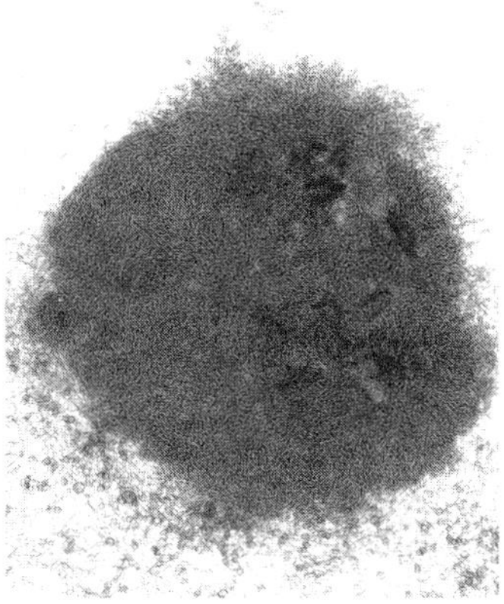

Pseudopionnote of *Fusarium* sp., x 100 (*CB*).

pseudoplasmodium, pl. **pseudoplasmodia** (Gr. *pseûdos*, false + NL. *plasmodium* < Gr. *plásma*, bland material with which is formed a living being + L. *-odium*, resembling < Gr. *ode*, like < *-oeídes*, similar to): also called aggregation plasmodium or **grex**; it is formed by the union of myxamoebae that, however, maintain their individuality. It is characteristic of the acraseous or cellular slime molds of the phylum Dictyosteliomycota, such as *Dictyostelium* and *Polysphondylium*. Cf. **aphanoplasmodium**, **phaneroplasmodium** and **protoplasmodium**.

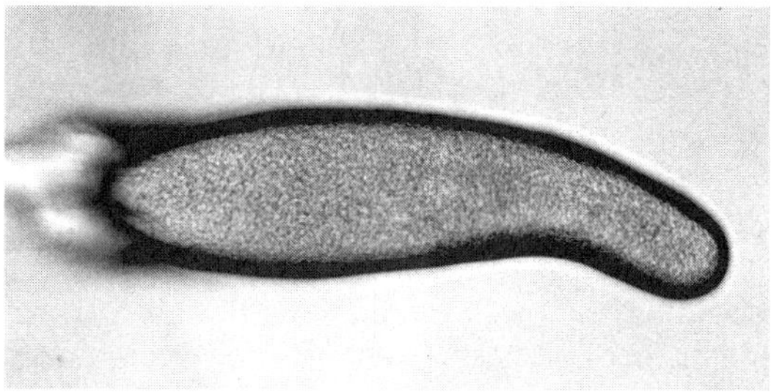

Pseudoplasmodium of *Dictyostelium discoideum* on agar, x 40 (*RTH*).

pseudopod, **pseudopodium**, pl. **pseudopodia** (Gr. *pseûdos*, false + NL. *podium* < Gr. *pódion*, dim. of *poús*, *podós*, foot): a protoplasmic prolongation, which varies continually in shape, of a myxamoeba or a plasmodium, and which serves as an organ of locomotion and for the capture of food, e.g., *Cavostelium* (Protosteliales). Cf. **filopod** and **lobopod**.

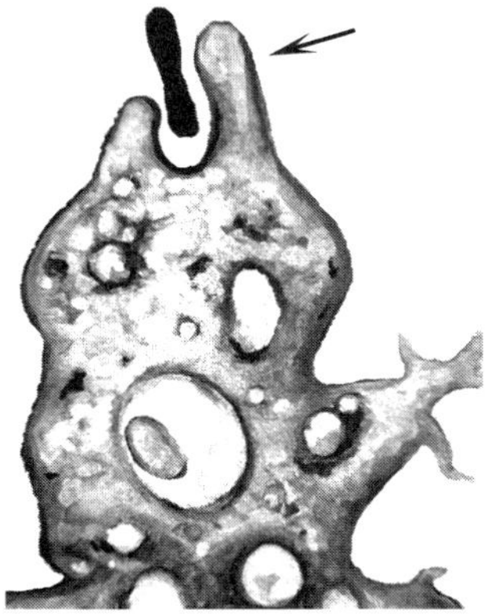

Pseudopods of a myxamoeba of *Cavostelium bisporum*, about to ingest a bacterial cell, x 3 950.

pseudopodetium, pl. **pseudopodetia** (Gr. *pseûdos*, false + NL. *podetium* < Gr. *podotés*, footed < *poús*, *podós*, foot + L. dim. suf. *-ium*): *Lichens.* Pseudopodetia are erect, podetium-like, solid branches of vegetative origin; when ascogonia are present they arise on pseudopodetia, and the latter do not originate on a preformed granular or squamulose thallus, e.g., in *Stereocaulon* (Lecanorales). Cf. **podetium**.

pseudorhiza, pl. **pseudorhizae** (Gr. *pseûdos*, false + *rhíza*, root): a root-like extension at the base of a stipe, which functions as an organ of union between the basidiocarp and the mycelium in the soil, as well as an organ of conduction and storage. It is present, e.g., in *Oudemansiella radicata* (Agaricales) and *Fistulina radicata* (Aphyllophorales).

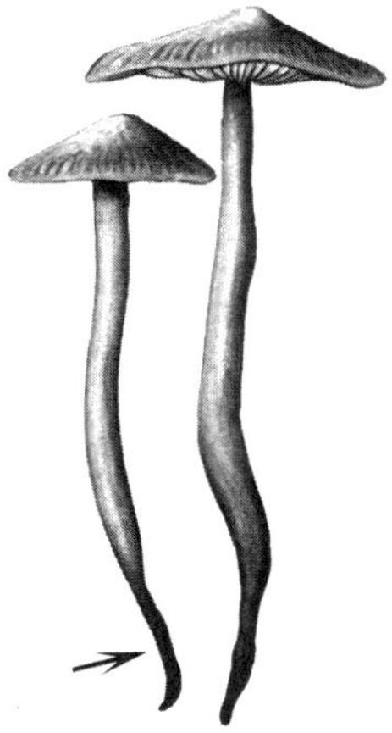

Pseudorhiza of the basidiocarps of *Oudemansiella radicata*, x 1.

pseudosclerotium, pl. **pseudosclerotia** (Gr. *pseûdos*, false + NL. *sclerotium* < Gr. *sklerotés*, hardness < *sklerós*, hard + L. dim. suf. *-ium*): a compact earthy mass of humus or siliceous particles conglomerated by mycelium. *Armillaria* (=*Armillariella*) *mellea* (Agaricales), *Xylaria polymorpha* (Xylariales), and *Monilinia fructicola* (Helotiales) are examples of fungi that form pseudosclerotia. Among the entomogenous fungi, *Cordyceps* (Hypocreales) also can be considered as forming pseudosclerotia, in this case produced from tissues of the host insect and of the mycelium of the parasitic fungus; from these pseudosclerotia are developed the perithecial stromata.

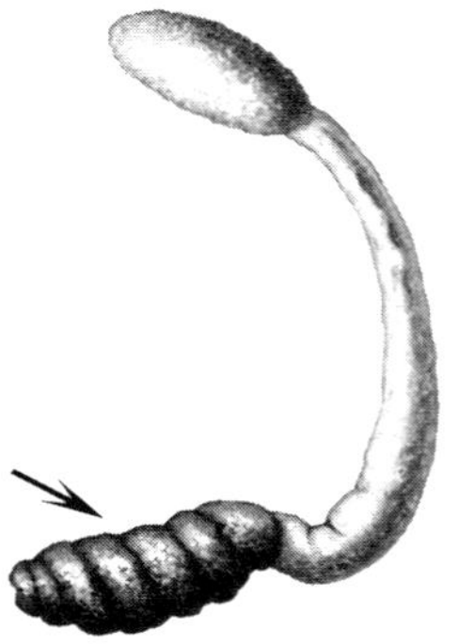

Pseudosclerotium (in the insect host) of *Cordyceps militaris*, x 1.5.

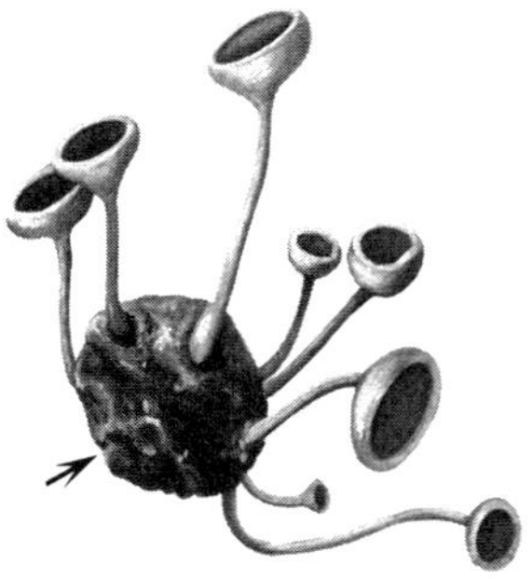

Pseudosclerotium (in the parasitized peach fruit) of *Monilinia fructicola*, x 0.6.

pseudoseptum, pl. **pseudosepta** (NL. *pseudoseptum* < Gr. *pseûdos*, false + L. *septum*, barrier, partition): **1**. A protoplasmic membrane or vacuolar membrane similar to a septum, as in *Corynespora* (dematiaceous asexual fungi) that has spores with up to 50 or more pseudosepta; also called **distoseptum**. **2**. Each of the membranes with pores that are present in the thallus of the Blastocladiales, and the celluline plugs that are found in the constricted hyphae of the aquatic fungi of the order Leptomitales (such as *Apodachlya* and *Leptomitus*). See **cellulin**.

pseudospore (Gr. *pseûdos*, false + *sporá*, spore): a spore lacking a cell wall; occasionally found in the sporocarps or fruiting bodies of some Dictyosteliomycota (e.g., in *Dictyostelium*).

pseudostem (Gr. *pseûdos*, false + ME. < OE. *stemn* < *standan*, to stand): see **pseudostipe**.

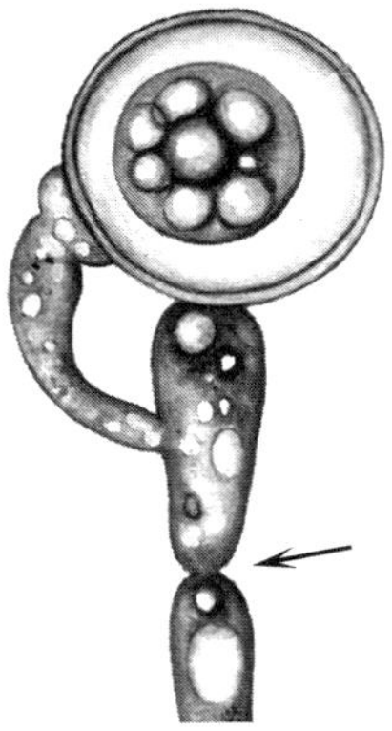

Pseudosepta of the hyphae of *Apodachlya* sp., x 750.

pseudostipe (Gr. *pseûdos*, false + L. *stipes*, pedicel, stipe): *Gasteromycetes*. A stem-like mass of spongy tissue (subgleba), which supports the gleba, in which the hyphae are not oriented parallel to the axis of the stipe; e.g., as occurs in *Lycoperdon* (Lycoperdales) and *Phallus* (Phallales). In Gasteromycetes with a true stipe (such as *Tulostoma*, of the Tulostomatales), the hyphae that compose it are arranged in a more or less parallel form.

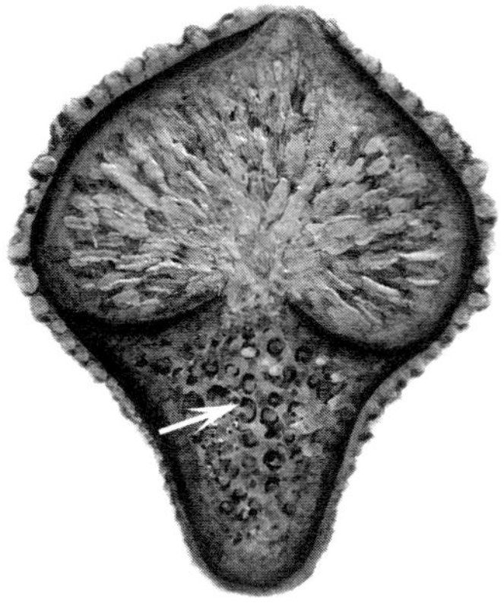

Pseudostipe of the fruiting body of *Lycoperdon perlatum*, x 1.

pseudostolon (Gr. *pseûdos*, false + NL. *stolon-*, *stolo*, branch, shoot, sucker): an unbranched vegetative hypha that connects two groups of sporangiophores, as in *Mucor racemosus* (Mucorales); it differs from a stolon in that it does not connect two groups or fascicles of rhizoids with sporangiophores opposite them, as occurs in *Rhizopus* and other Mucorales with typical stolons. Cf. **stolon**.

pseudostroma, pl. **pseudostromata** (Gr. *pseûdos*, false + *strõma*, bed): a complex stroma composed of tissues of the fungus and of the host plant; e.g., like that of *Polystigma rubrum* (Phyllachorales). Cf. **eustroma**.

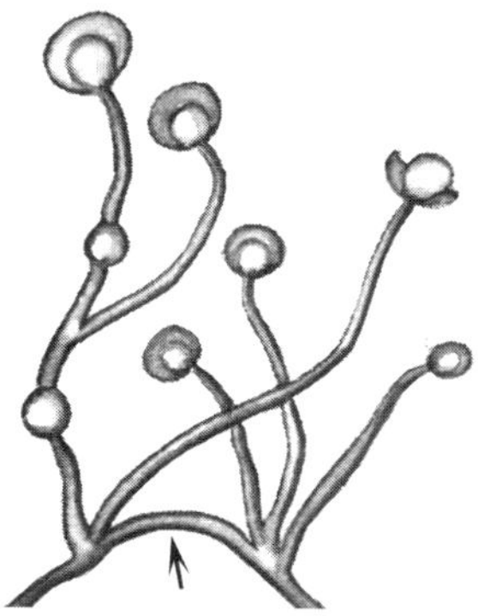

Pseudostolon of *Mucor racemosus*, x 300.

pseudothecium, pl. **pseudothecia** (NL. *pseudothecium* < Gr. *pseûdos*, false + *thekíon*, dim. of *thêke*, case, box; here, of the asci): an ascostroma bearing asci in unwalled locules (Loculoascomycetes); restricted by some to perithecioid unilocular ascostromata, such as *Leptosphaerulina*, *Munkiella* and *Mycosphaerella* (Dothideales), *Sporormiella* (Melanommatales) and *Venturia* (Pleosporales). Cf. **euthecium**.

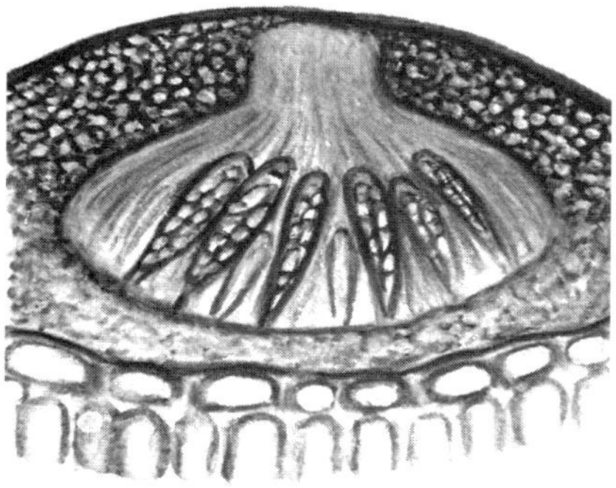

Pseudothecium of *Munkiella caa-guazu*, seen in longitudinal section, x 500.

psychrophile, **psychrophilic** (Gr. *psychrós*, cold + *phílos*, have an affinity for; or + suf. *-íkos* > L. *-icus* > E. *-ic*, belonging to, relating to): an organism that develops well at low temperatures, often near 0ºC. In the fungi, the species *Eupenicillium crustaceum* (Eurotiales) and *Penicillium spinulosum* (moniliaceous asexual fungi), among other examples, are psychrophiles, since their conidia germinate well and develop colonies of appreciable size at 5ºC. These molds, together with some species of *Fusarium* (tuberculariaceous asexual fungi) and yeasts (cryptococcaceous asexual yeasts and Saccharomycetales) cause biodeterioration of milk products and other foods maintained under refrigeration. Cf. **mesophile** and **thermophile**.

pterate (NL. *pteratus*, winged < Gr. *pterón*, wing + suf. *-atus* > E. *-ate*, provided with or likeness): winged. Also called **alate**.

ptyophagous (Gr. *ptýo* < *ptyein*, to throw out, spit + *phágos*, to eat + L. *-osus* > OF. *-ous*, *-eus* > E. *-ous*, possessing the qualities of): *Endomycorrhizae*. Any form of mycetization in which the hyphae of the fungus suffer **plasmoptysis** and empty their plasmatic contents (the so-called **ptyosomes**) into the cell of the host plant that it has penetrated; afterwards the host cell absorbs and assimilates the extruded materials. Cf. **thamnisophagous** and **tolypophagous**.

ptyosome (Gr. *ptýo*, to throw out + *sõma*, body): see **ptyophagous**.

pubescent (L. *pubescens*, genit. *pubescentis*, with hairs of puberty, downy < *puber*, *pubes*, *pubis*, downy, with hairiness + *-escens*, genit. *-escentis*, that which turns, beginning to, slightly > E. *-escent*): covered with fine, soft hair, like down. For example, the surface of the pileus of some species of *Xerocomus* (Agaricales) and of *Auricularia polytricha* (Auriculariales) are pubescent. Cf. **glabrous**.

Pubescent lower surface of the basidiocarps of *Auricularia polytricha*, x 0.8

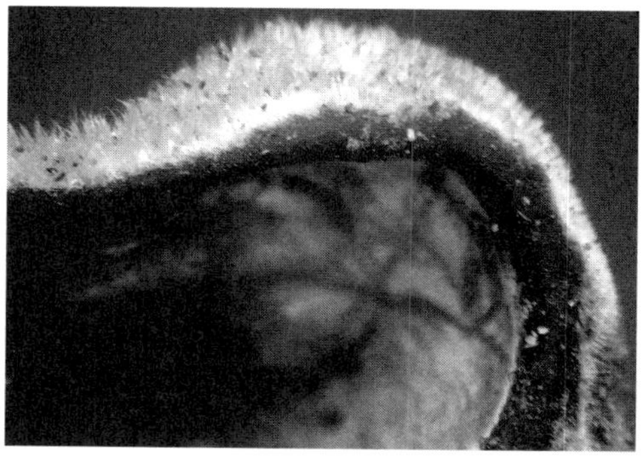

Pubescent lower surface of the basidiocarp of *Auricularia polytricha*, x 3 (*CB*).

puffball (ME. *puf*, *puffe* < *puffen*, to send forth + *bal*, *balle* < OF. *balle*, ball): the name given to species of the Lycoperdales which discharge clouds of spores when the outer peridium of the fruiting body (usually the endoperidium at maturity) is compressed rapidly; e.g., as in *Lycoperdon* (Lycoperdales).

puffing (E. < *puff* < ME. *puffen* < OE. *pyffan*, to blow in short gusts; to exhale forcibly): a phenomenon in which thousands of asci in an apothecium forcibly discharge their ascospores simultaneously, producing a perceptible smokelike cloud and a barely appreciable hissing sound. The basidiomata of Lycoperdales are called **puff-balls** for their puffing action, discharging their spores when they are pressed or struck, e.g., by rain drops.

pulque (Azt. *poliuhqui*, spoiled): the fermented sap (mead) of various *Agave* spp., consumed as an alcoholic and nutritious beverage in many regions of Mexico for several centuries. The microorganisms involved in the fermentation of the sap includes a wide range of bacteria, mainly lactic acid and ethanol producers (*Lactobacillus*, *Leuconostoc*, *Zymomonas*), and yeasts, especially *Saccharomyces*, *Pichia* (Saccharomycetales) and *Candida* (cryptococcaceous asexual yeasts).

pulvillus, pl. **pulvilli** (L. *pulvillus*, little cushion): a gelatinous swelling, turgescent, pulvinate, that is formed in the ostiolar canal of the perithecium of *Turgidosculum ulvae*, a marine ascomycete of uncertain affinities. The pulvillus functions as a plug that prevents the entrance of water to the perithecial cavity, which contains a gelatinous matrix that covers the asci, protecting them from dessication when the perithecia are exposed to the air, along with their algal host (*Blidingia minima*=*Ulva vexata*), during low tide. The pulvillus closes the ostiole of wet perithecia by swelling, but when dry it dehydrates and shrinks, putting pressure on the asci, which liberate the ascospores through the ostiole. Rewetting causes the pulvillus to swell and close the ostiolar canal, protecting the immature asci against exposure to water.

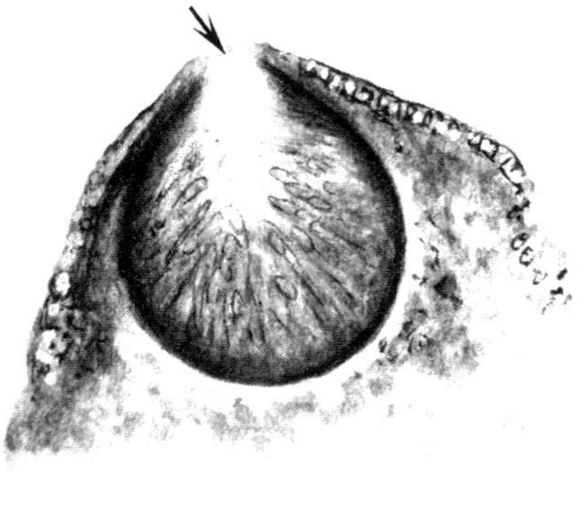

Pulvillus closing the ostiolar canal and covering the periphyses of the perithecium of *Turgidosculum ulvae*, which is parasitizing the alga *Blidingia minima*, x 660.

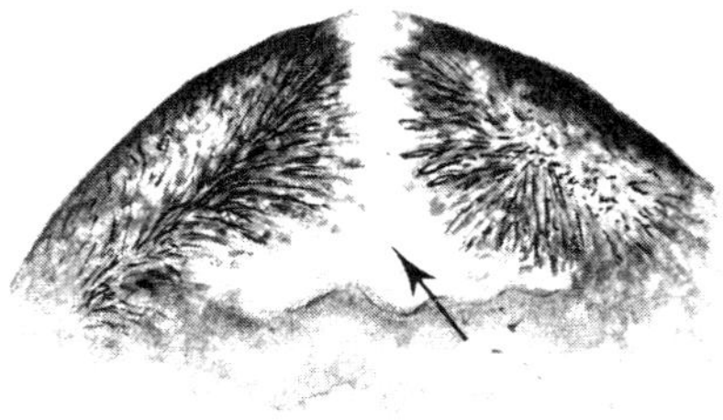

Pulvillus (a close-up of the previous figure), x 1 500.

pulvinate, **pulvinated**, **pulviniform** (L. *pulvinus*, cushion + suf. *-atus* > E. *-ate*, provided with or likeness; *pulvinus* + L. *-formis* < *forma*, shape): refers to any structure in the shape of a small cushion, such as the stroma of *Tubercularia vulgaris* (tuberculariaceous asexual fungi) or that of *Hypoxylon thouarsianum* (Xylariales).

Pulvinate perithecial stromata of *Hypoxylon thouarsianum*, x 2 (*EPS*).

punctate (NL. *punctatus*, dotted or spotted as with punctures < L. *punctum*, point, dot, puncture + suf. *-atus* > E. *-ate*, provided with or likeness): applied to surfaces (membranes, cell walls, etc.) that have more or less equidistant projections or prominences that are microscopic in size, appearing as punctures. For example, the perithecial stroma of *Poronia punctata* (Xylariales), the teliospores of *Sphacelotheca destruens* (Ustilaginales), and the basidiospores of *Cortinarius punctatus* and of *Panaeolina foenisecii* (Agaricales) are punctate.

punctiform (NL. *punctiformis* < *punctum*, point + *-formis* < *forma*, shape): like a point; small colonies (of yeasts and bacteria), less than 1 mm in diam., are called punctiform, although they can be seen without a magnifying lens.

purse (ME. < OE. *purs* < *pursa*, bag < ML. *bursa*, bag < Gr. *býrsa*, hide, leather): *Gasteromycetes*. The upper part of the funiculus which contains the coiled funicular cord in some species of Nidulariales, e.g., *Cyathus striatus*.

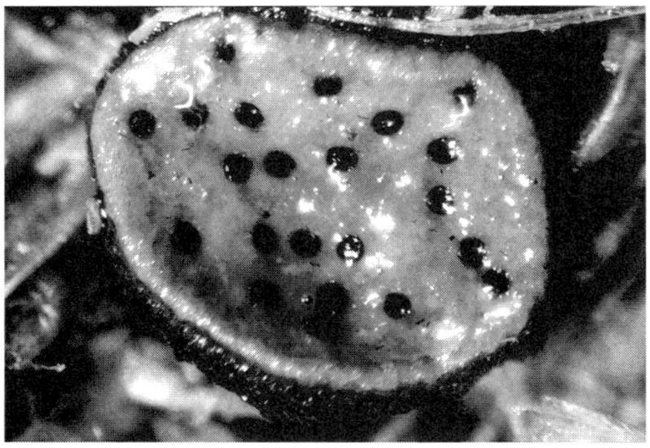

Punctate perithecial stroma of *Poronia punctata*, x 10 (*MU*).

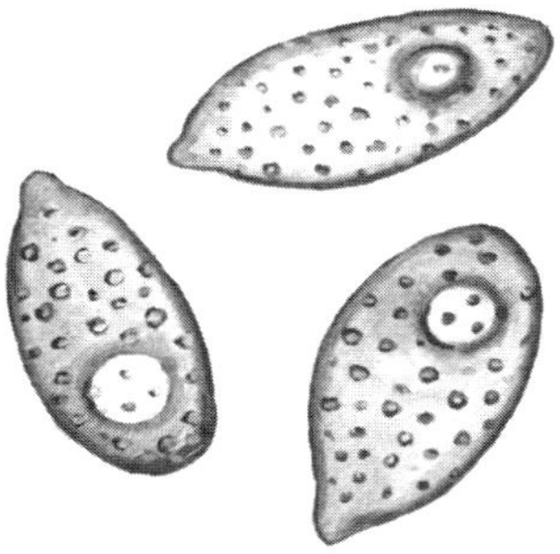

Punctate wall of the basidiospores of *Panaeolina foenisecii*, x 1 200.

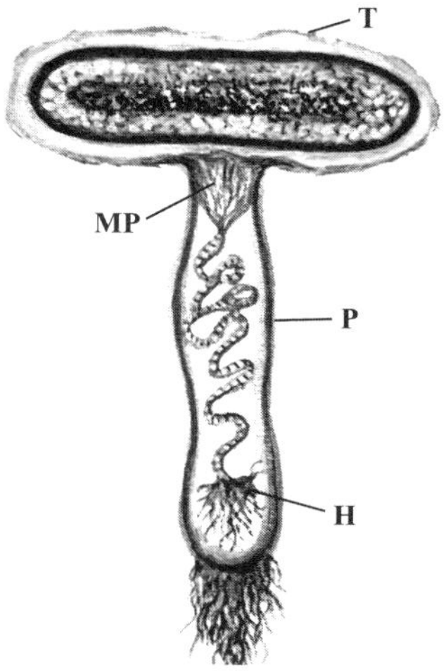

Purse (P) of a peridiolum of *Cyathus striatus*. In this diagram of a longitudinal section, the hapteron (H), middle piece (MP) and tunica (T) also are shown, x 20.

pustulate, **pustulose** (L. *pustulatus*, blistered < *pustula*, blister + suf. *-atus* > E. *-ate*, provided with or likeness; L. *pustulosus*, full of pimples < *pustula* + *-osus*, full of, augmented, prone to > ME. *-ose*): covered with blisters or pustules, like the lesions on stems and leaves that various plant pathogenic fungi cause, e.g., those called the white rusts (*Albugo*, of the Peronosporales) and rusts (*Puccinia*, of the Uredinales).

Pustulate leaf of mallow; the pustules are uredinia of *Puccinia malvacearum*, x 1.5 (*MU*).

pustule (L. *pustula*, blister, pimple, with the E. ending *-ule*): a small ampulliform elevation formed by fructifications of plant pathogenic fungi, or by lesions that they originate in the epidermal tissues of the host plant, such as the uredinia and telia of the rust fungi (e.g., *Puccinia malvacearum*, of the Uredinales). In the lichens, the small ampulliform elevations that are present on the surface of the thallus are called pustules.

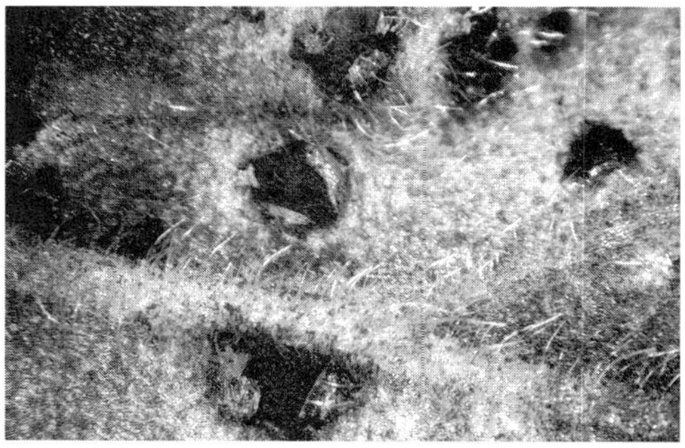

Pustules (uredinia) of *Puccinia malvacearum* on a leaf of mallow, x 3 (*MU*).

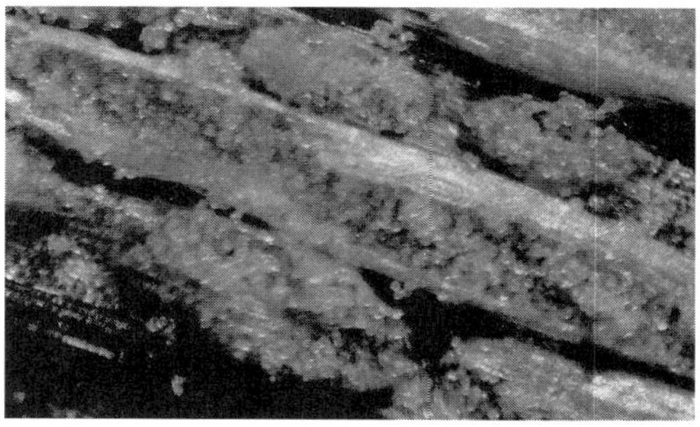

Pustules (uredinia) of *Puccinia graminis tritici* on a leaf of wheat, x 12 (*MU & CB*).

putrescent (L. *putrescens*, genit. *putrescentis*, to undergo putrefaction, becoming rotten, prp. of *putrescere*, to become rotten + *-escens*, genit. *-escentis*, that which turns, beginning to, slightly > E. *-escent*): rotting or decomposing rapidly, becoming soft and pulpy.

pycnidiospore (Gr. *pyknós*, dense, packed, concentrated + dim. suf. *-ídion* > L. *-idium* + Gr. *sporá*, spore): a spore produced in a pycnidium; e.g., as in *Phoma* and *Endothiella gyrosa* (sphaeropsidaceous asexual fungi).

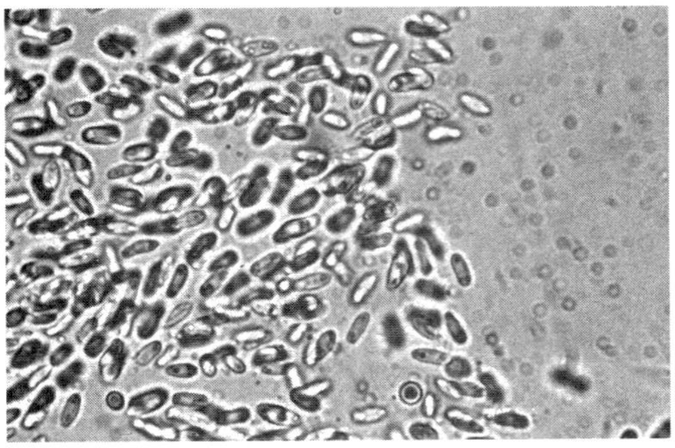

Pycnidiospores from a pycnidial cirrhus of *Phoma* sp., x 600 (*MU*).

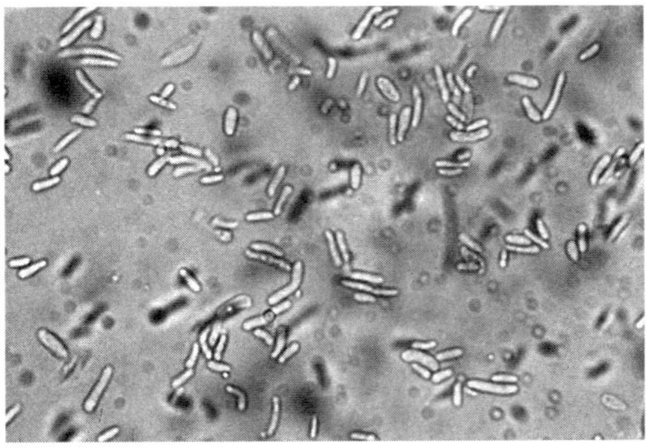

Pycnidiospores from a pycnidial cirrhus of *Endothiella gyrosa*, x 440 (*MU*).

pycnidium, pl. **pycnidia** (NL. *pycnidium* < Gr. *pyknós*, dense, packed, concentrated + dim. suf. *-ídion* > L. *-idium*): an asexual fruiting body, generally spherical or obpyriform, with an internal cavity lined with conidiophores; characteristic of the sphaeropsidaceous asexual fungi, such as *Phoma*, *Phomopsis*, *Pyrenochaeta* and *Botryodiplodia*.

pycniospore (Gr. *pyknós*, dense, packed, concentrated + *sporá*, spore): a spore formed in a pycnium. See **spermatium**.

pycnoascocarp (Gr. *pyknós*, dense, compact, alluding to the pycnidium + *askós*, wine-skin; here, ascus + *karpós*, fruit): *Lichens*. A peculiar type of apothecium that is formed from a pycnidium; it begins with the formation of ascogonia beneath a pycnidium, followed by the development of paraphyses and asci among the conidiophores of the pycnidium, and the transformation of the pycnidial wall into the excipulum of the mature apothecium. Characteristic of the family Lichiniaceae, such as *Ephebe* and *Zahlbrucknerella* (Lichinales). Cf. **thallinocarp**.

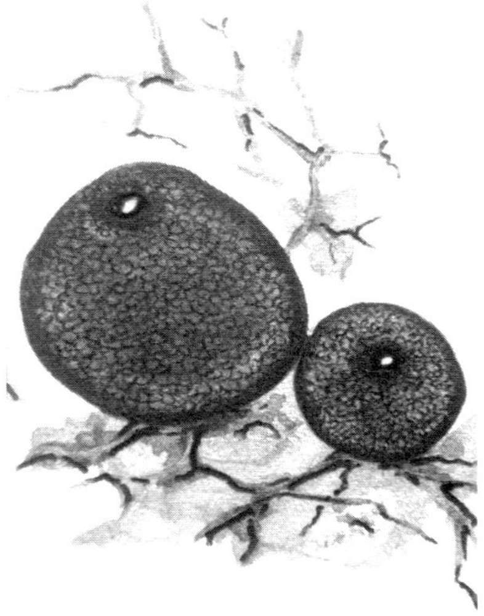

Pycnidia of *Phoma* sp., x 280.

pycnosclerotium, pl. **pycnosclerotia** (Gr. *pyknós*, dense, compact, concentrated + *sklerotés*, hardness < *sklerós*, hard + L. dim. suf. *-ium*): a plectenchymatous body with an external layer of cells with a thick, hard wall, similar to a pycnidium, since its interior becomes hollow, but lacking spores; e.g., *Phyllosticta* (=*Phyllostictina*) *carpogena* (sphaeropsidaceous asexual fungi) forms pycnosclerotia. Apparently, this term has fallen into disuse.

pycnosis, pl. **pycnoses**, **pyknosis**, pl. **pyknoses** (Gr. *pyknós*, dense, compact, concentrated + *-osis*, state or condition): *Microthyriaceae* (Dothideales). A process in which a rounded or elliptical portion of the thallus arches up and thickens, while the ascigerous hymenium is formed below the raised portion.

pycnostroma, pl. **pycnostromata** (Gr. *pyknós*, dense, compact, concentrated, solid, thick, crowded + *strõma*, cushion, bed): a stroma in which pycnidia are embedded. An asexual fructification shaped like a cushion or a flattened pycnidium, in which the conidiophores grow from the roof or upper wall downward; characteristic of pycnidial fungi, such as *Munkia* (=*Pycnostroma*) and *Peltasterella*, gen. that some classify in the sphaeropsidaceous or in the pycnothyriaceous asexual fungi. Also called **pycnothyrium**.

pycnothyrium, pl. **pycnothyria** (Gr. *pyknós*, dense, compact, concentrated + *thirís*, window + L. dim. suf. *-ium*): an asexual fructification, shaped like a cushion or a flat pycnidium, in which the conidiophores grow from the roof or upper wall downward; characteristic of *Pycnothyrium* and *Peltasterella* (sphaeropsidaceous or pycnothyriaceous asexual fungi). Also called **pycnostroma**.

pyramidal (ML. *pyramidalis* < L. *pyramis* < Gr. *pyramís*, pyramid + L. suf. *-alis* > E. *-al*, relating to or belonging to): shaped like a pyramid, i.e., a structure with a polygonal base and three or four triangular sides that meet in a point; e.g., like the scales of the pileus of *Amanita cokeri* (Agaricales), and of the fruiting body of *Lycoperdon candidum* and *L. echinatum* (Lycoperdales).

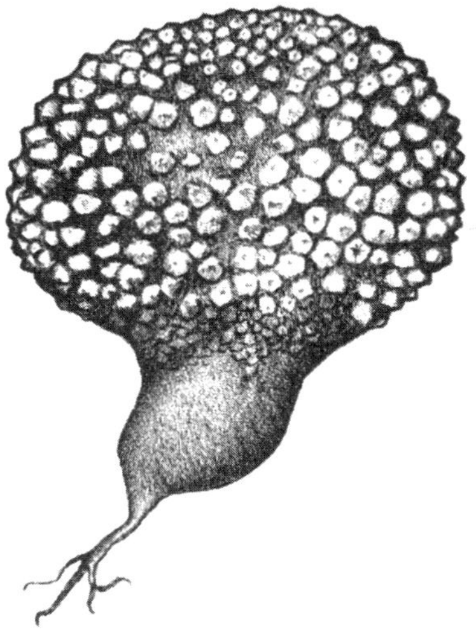

Pyramidal peridial scales of the fruiting body of *Lycoperdon candidum*, x 2.

Pyramidal peridial scales of the fruiting body of *Lycoperdon echinatum*, x 10 (*MU*).

pyramid-shaped wart (L. *pyramis* < Gr. *pyramís*, pyramid + ME. *shappe*, *shipe* < OE. *sceppen*, to make; ME. < OE. *wearte*, wart): a wart with a square base and three or four triangular sides, pointed or nearly so, often with zone lines, e.g., the warts on the peridium of *Calvatia* (Lycoperdales).

pyrenocarp (Gr. *pyrén*, *pyrênos*, stone of the drupe fruits + *karpós*, fruit): syn. of **perithecium**; sometimes used informally for any fungus that has an ascoma (ascocarp) similar to a perithecium. The lichens with perithecia (such as *Dermatocarpon*, Verrucariales) are called pyrenocarpic.

Pyrenomycetes (Gr. *pyrén*, *pyrênos*, stone of the drupe fruits + L. *-mycetes*, ending of class < Gr. *mýkes*, genit. *mýketos*, fungus): a class of ascomycetes with asci in a hymenium or fascicle contained in a perithecium-type ascocarp. According to the classification adopted in this dictionary (Alexopoulos *et al.*, 1996), the Pyrenomycetes include the orders Hypocreales, Melanosporales, Microascales, Phyllachorales, Ophiostomatales, Diaporthales, Xylariales, Sordariales, Meliolales and Halosphaeriales. Together with the classes Archiascomycetes, Plectomycetes, Discomycetes, and Loculoascomycetes, along with the ascosporogenous yeasts (Saccharomycetales), and other filamentous ascomycetes (such as the Erysiphales, Laboulbeniales, and Spathulosporales, which are not included in any of the above mentioned classes) constitute the phylum Ascomycota of the kingdom Fungi.

pyriform (NL. *pyriformis* < ML. *pyrus* < L. *pirus*, pear + *-formis* < *forma*, shape): pear-shaped; e.g., like the basidiospores of *Cortinarius pseudosalor* (Agaricales), and the sporangia of *Trichia floriformis* (Trichiales), *Helicostylum piriforme* and *Pirella circinans* (Mucorales). Cf. **obpyriform**.

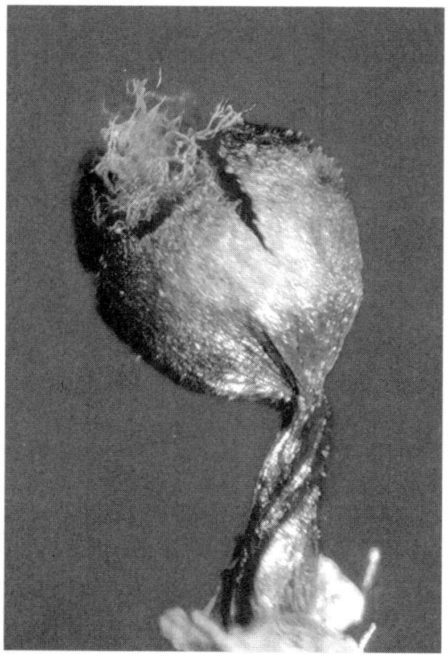

Pyriform sporangium of *Trichia floriformis*, x 80 (*MU*).

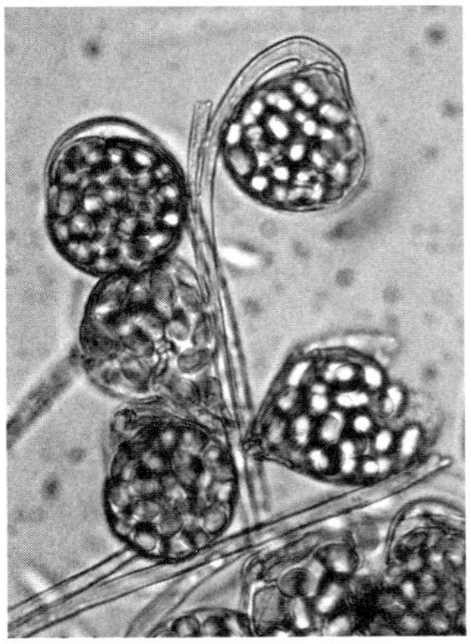

Pyriform sporangia of *Helicostylum piriforme*, x 1 300 (*MU*).

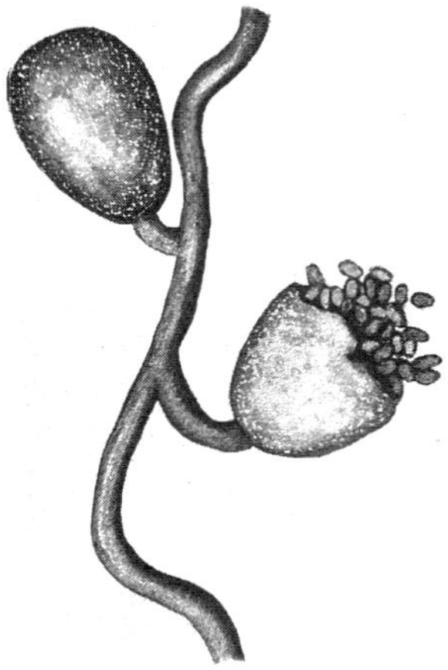

Pyriform sporangia of *Pirella circinans*, x 200.

pyrophilous (Gr. *pýr*, genit. *pyrós*, fire + *phílos*, have an affinity for + L. *-osus* > OF. *-ous*, *-eus* > E. *-ous*, possessing the qualities of): growing on burned or steam-sterilized substrates, such as soil or wood; e.g., like *Pyronema* (Pezizales). See **anthracobiontic**, **carbonicole** and **phoenicoid fungus**.

pyroxylophilous (Gr. *pýr*, genit. *pyrós*, fire + *xýlon*, wood + *phílos*, have an affinity for + L. *-osus* > OF. *-ous*, *-eus* > E. *-ous*, possessing the qualities of): growing on burned wood; e.g., like *Pyronema* (Pezizales) and certain species of *Xylaria* (Xylariales).

pyxidate (L. *pyxidatus*, box-like, cubical < Gr. *pyxís*, wooden box + suf. *-atus* > E. *-ate*, provided with or likeness): box-like; in the form of a box; provided with a **pyxidium**, which is a box.

pyxidium, pl. **pyxidia** (NL. *pyxidium* < Gr. *pyxídion* < *pyxís*, box with a lid + dim. suf. *-ídion* > L. *-idium*): a small box, or container, usually cylindrical, with a lid; e.g., the ascocarp of *Pyxidiophora* (Laboulbeniales).

q

quadrate arm (L. *quadratus*, squared < *quadrus*, fourfold + suf. *-atus* > E. *-ate*, provided with or likeness; ME. < OE. *earm*, arm): a receptaculum or arm in the Phallales which is 4-sided in cross section, as in *Clathrus columnatus*.

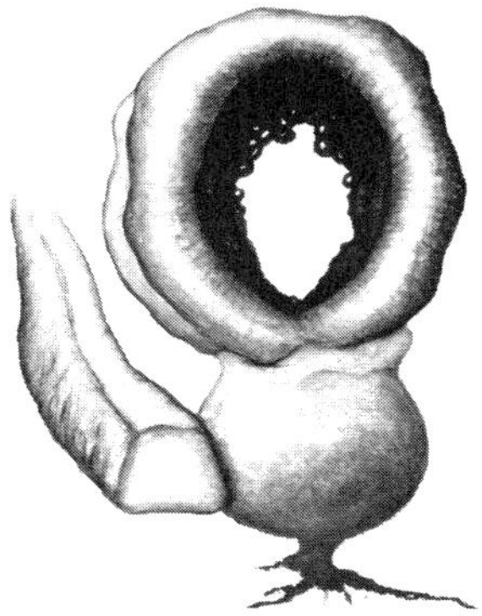

Quadrate arms of the clathrate fruiting body of *Clathrus columnatus*, lined with the gleba, x 0.5.

quadriverticillate (L. *quatour*, four > *quadri* + NL. *verticillatus* < *verticillus*, verticil + suf. *-atus* > E. *-ate*, provided with or likeness): branched on four levels, i.e., with branches (rami), branchlets (ramuli), metulae and phialides; this occurs in the conidiophore of *Penicillium chrysogenum* and *P. viridicatum* (moniliaceous asexual fungi). Cf. **monoverticillate**, **biverticillate**, **triverticillate** and **polyverticillate**.

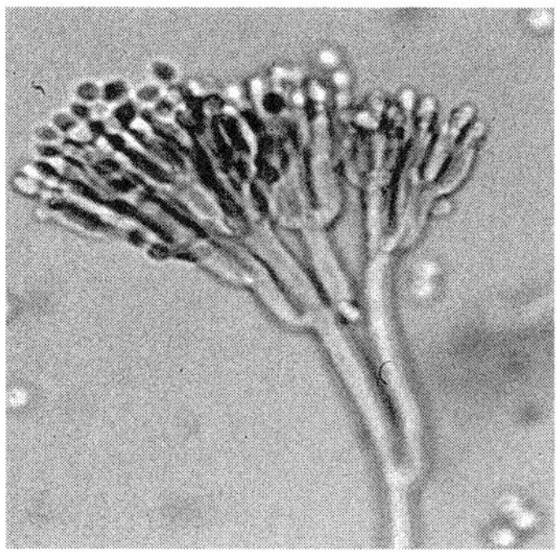

Quadriverticillate conidiophore of *Penicillium viridicatum*, x 770 (*CB*).

Quellkörper (G. *quellen*, to swell + *Körper*, body): *Ascomycetes*. A mass of gelatinized and swollen cells at the apex of the ascocarp of certain pyrenomycetous fungi that disintegrates at maturity, allowing release of the ascospores. Characteristic of *Bertia moriformis* (Sordariales).

r

race (MF. *race* < OIt. *razza*, generation; an actually or potentially interbreeding group within a species) : a group of **biotypes** with a similar virulence-avirulence pattern on a particular group of plants. Races, like **formae speciales**, are defined on the basis of physiological differences and may be capable of attacking only one particular cultivar of a host. For example, there have been identified about 350 races of *Puccinia graminis* forma specialis *tritici* (Uredinales), the rust of wheat, that are either very similar or identical morphologically but differ in their abilities to parasitize a host species or a group of host species.

racemose, **racemous** (L. *racemus*, raceme of grapes + *-osus*, full of, augmented, prone to > ME. *-ose*; or + *-osus* > OF. *-ous*, *-eus* > E. *-ous*, possessing the qualities of): with the shape of a raceme, like a cluster of grapes; e.g., like the stipe of *Collybia racemosa* (Agaricales), which has lateral branches, or the sporangiophore of *Mucor racemosus* (Mucorales).

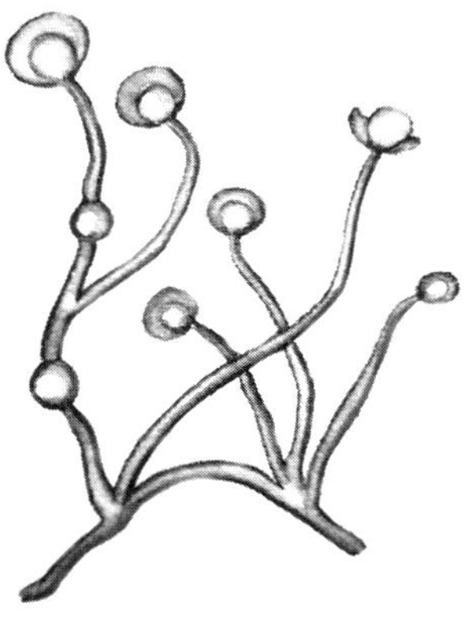

Racemose sporangiophores of *Mucor racemosus*, x 300.

rachiform (Gr. *ráchis*, axis, spine + L. *-formis* < *forma*, shape): a narrow conidiogenous axis (**rachis**), in comparison with the width of the portion on which each conidium is attached to the axis, and of sympodial growth that gives rise to a geniculate (zigzag) conidiophore, which produces conidia in basipetal succession, as occurs in some gen. of moniliaceous (*Tritirachium*, *Beauveria*, *Pyricularia*) and dematiaceous asexual fungi (*Racodium cellare*). Cf. **raduliform** and **ampulliform**.

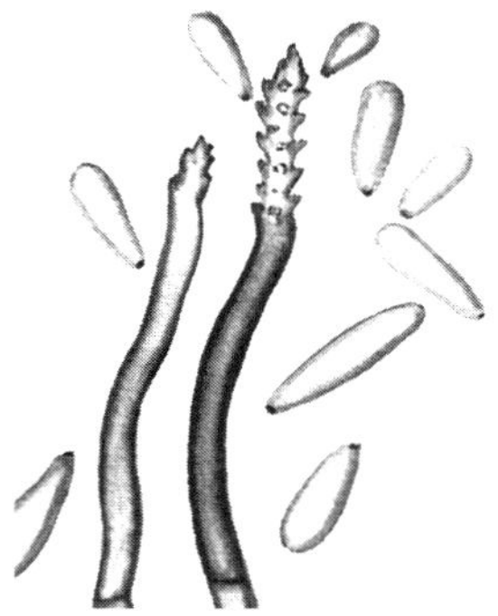

Rachiform conidiophores of *Racodium cellare*. Note the **rachis** or conidiogenous axis with sympodially arranged conidium-bearing denticles of the *Pleurophragmium* (=*Acrotheca*) type, and conidia of the *Cladosporium* type, x 600.

rachis, pl. **rachises** (Gr. *ráchis*, axis, spine): a geniculate or zigzag extension of a conidiogenous cell, as in *Tritirachium* and *Beauveria* (moniliaceous asexual fungi), and *Rhinocladiella* (dematiaceous asexual fungi), that results from the sympodial development of the cell and which produces spores sympodially along the major axis. See **rachiform**.

racket cell, **racquet cell**, **racket hypha**, **racquet hypha**, pl. **hyphae** (MF. *raquette*, racket < Ar. *rnhet*, palm of the hand; L. *cellula*, dim. of *cella*, storeroom, the cell of honeycombs; cell; Gr. *hyphé*, tissue, spider web; hypha): having the shape of a racket or with segments of such a shape, like the hyphae of the dermatophytic fungi (*Microsporum* and *Trichophyton*, of the moniliaceous asexual fungi).

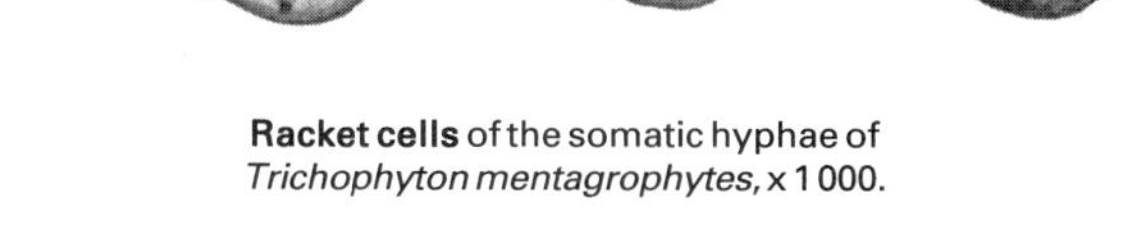

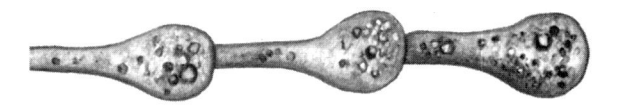

Racket cells of the somatic hyphae of *Trichophyton mentagrophytes*, x 1 000.

radial (ML. *radialis*, arranged or having similar parts arranged like rays < L. *radius*, ray + suf. *-alis* > E. *-al*, relating to or belonging to): developing uniformly around a central point or axis. **1**. *Lichens*. In some fruticulose thalli (such as *Usnea*, *Alectoria* and other Lecanorales) the symmetry is radial in cross section. **2**. *Agaricales*. The arrangement of the gills in the pileus of the agarics is radial; e.g., in *Coprinus* and *Xeromphalina*. Also called **radiate**.

Radial arrangement of the lamellae in the hymenophore of an agaric, x 2 (*MU*).

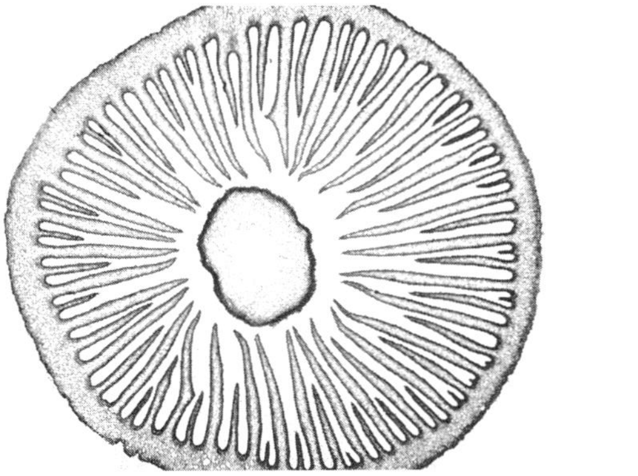

Radial arrangement of the lamellae in the hymenophore of *Coprinus* sp., seen in a transverse section of the pileus, x 7 (*MU*).

radiate (L. *radiatus*, rayed, surrounded by rays < *radius*, ray + suf. *-atus* > E. *-ate*, provided with or likeness): extending outward or arranged around a common center, like the gills of the Agaricales. Also called **radial**.

radiating plate (E. *radiating*, ptp. of *radiate*, rayed, surrounded by rays < L. *radius*, ray; ME. *plate* < OF. *plate*, flat object < VL. *plattus* < Gr. *platýs*, broad, flat): *Gasteromycetes*. Each of the veins or plates of sterile tissue that proliferate throughout the gleba of the fructification, originating radially from a gelatinous-watery columella; occurs in members of the Phallales, family Protophallaceae, such as *Kobayasia*. In the fruiting bodies of *Truncocolumella* (Hymenogastrales), the radial plates proliferate from a thick, fleshy pedicel-columella, forming locules or irregular chambers in the gleba.

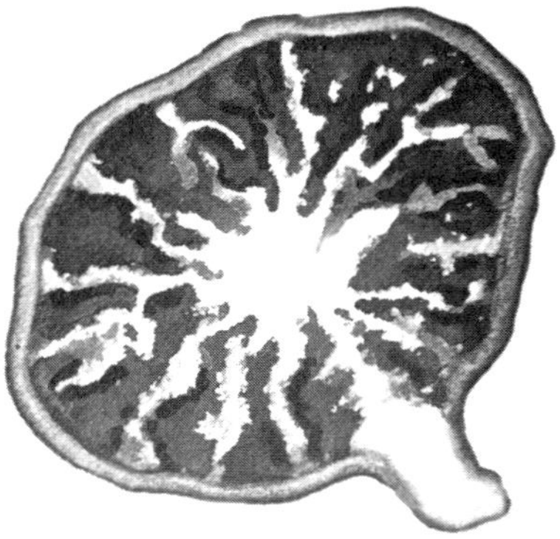

Radiating plates throughout the gleba, radiating from the columella of a fruiting body of *Kobayasia nipponica*, seen in longitudinal section, x 2.

radicate (L. *radicatus*, having roots < *radix*, genit. *radicis*, root + suf. *-atus* > E. *-ate*, provided with or likeness): provided with structures with the shape of roots. Generally applied to a stipe that has projections similar to roots, such as the rhizomorphs of *Armillaria* (=*Armillariella*) *mellea* and the pseudorhiza of *Oudemansiella radicata* and *Phaeocollybia attenuata* (Agaricales), as well as to certain types of cystidia, like those of *Tubulicium* (=*Tubulixenasma*) *vermiferum* (Aphyllophorales).

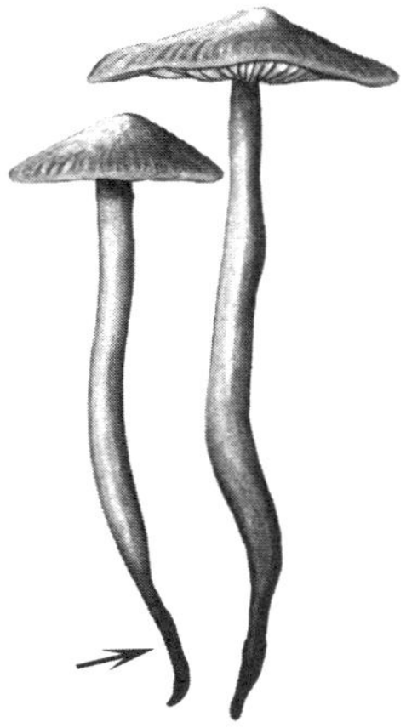

Radicate stipe of the basidiocarps of *Oudemansiella radicata*, x 0.2.

radicicolous (L. *radix*, genit. *radicis*, root + *-cola*, inhabitant + L. *-osus* > OF. *-ous*, *-eus* > E. *-ous*, possessing the qualities of): living in roots, usually as a parasite.

radula, pl. **radulae** (L. *radula*, rasp, scraper): a structure (hypha, conidiophore or ascospore) that becomes rough on its surface, due to the formation of small denticles on which are borne conidia known as radulaspores, e.g., *Arthrobotrys* (moniliaceous asexual fungi). In *Nectria coryli* (Hypocreales) the ascospores, still contained in the asci, become radulae, since on their surface originate viscous spores, supported by small denticles. See **raduliform**.

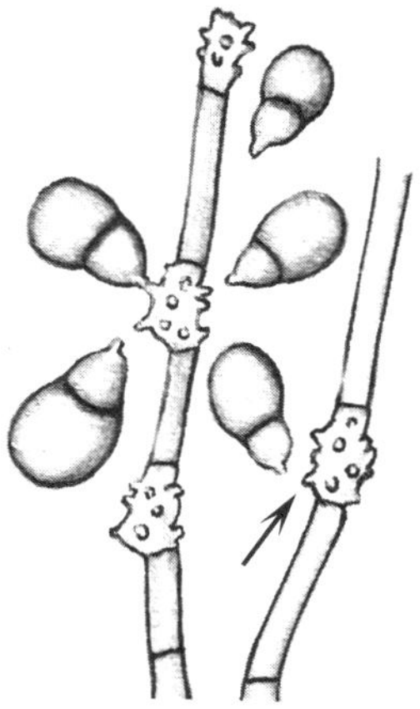

Radulae (conidiogenous cells) at the nodes of a **raduliform** conidiophore of *Arthrobotrys oligospora*, x 800.

raduliform (L. *radula*, rasp + *-formis* < *forma*, shape): a conidiogenous axis, relatively wide in comparison with the width of the portion on which is attached each conidium, and which tends to become clavate or somewhat inflated, instead of adopting a sympodial (zigzag) growth, as is observed in the gen. *Arthrobotrys* (moniliaceous asexual fungi), *Cordana* and *Gonatobotryum* (dematiaceous asexual fungi). Cf. **ampulliform** and **rachiform**.

ragged (ME. *ragget*, tattered < *ragge*, worn cloth): see **erose**.

ragi (Hind. *ragi*): small balls or roundish tablets of rice flour, containing viable spores and cells of amylolytic microorganisms: *Rhizopus*, *Mucor*, *Amylomyces* (Mucorales), *Saccharomyces fibuligera*, *Saccharomycopsis malanga* (Saccharomycetales), and cocci (*Pediococcus* sp.). Ragi is used as a starter culture for the elaboration of several oriental fermented foods, such as *tapé ketan* (from rice) and *tapé ketella* (from cassava tubers).

ramicole, **ramicolous** (L. *rami*, pl. of *ramus*, branch + *-cola*, inhabitant; or + *-osus* > OF. *-ous*, *-eus* > E. *-ous*, possessing the qualities of): living on the branches of plants.

ramiferous (L. *ramifer* < *rami*, pl. of *ramus*, branch + *-ferous*, bearer < *ferre*, to bear, carry + *-osus* > OF. *-ous*, *-eus* > E. *-ous*, possessing the qualities of, having): with branches; branched. Also called **ramose**.

ramoconidium, pl. **ramoconidia** (L. *ramus*, branch + *conidium* < Gr. *kónis*, dust + dim. suf. *-ídion* > L. *-idium*): an intercalary or apical branch of a conidiophore, which on detaching functions as a conidium, such as occurs in species of the gen. *Cladosporium* (dematiaceous asexual fungi).

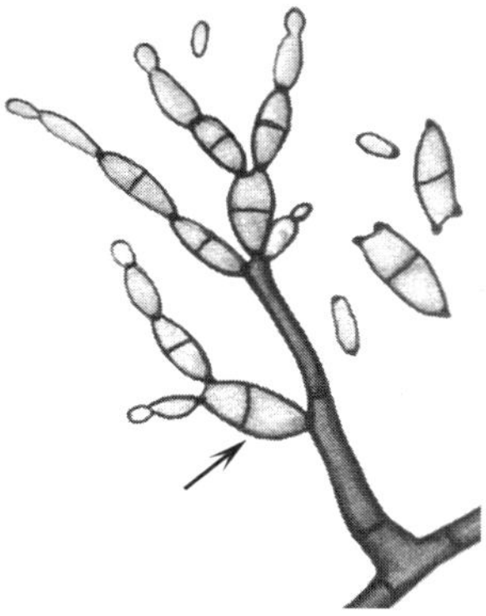

Ramoconidia of the conidiophores of *Cladosporium* sp., x 560.

ramose (L. *ramosus*, branched < *ramus*, branch + *-osus*, full of, augmented, prone to > ME. *-ose*): branched; applied to sporangiophores, conidiophores, pedicles and other structures. Syn. of **ramiferous**.

ramulus, pl. **ramuli** (L. *ramulus*, dim. of *ramus*, branch): a small branch; e.g., like the basidiocarps of the clavariaceous fungi.

rangiferoid (< *Rangifer*, the gen. of the reindeer + L. suf. *-oide* < Gr. *-oeídes*, similar to): branched like the antlers of a reindeer; e.g., as in the thallus of the lichen *Cladonia rangiferina* (Lecanorales).

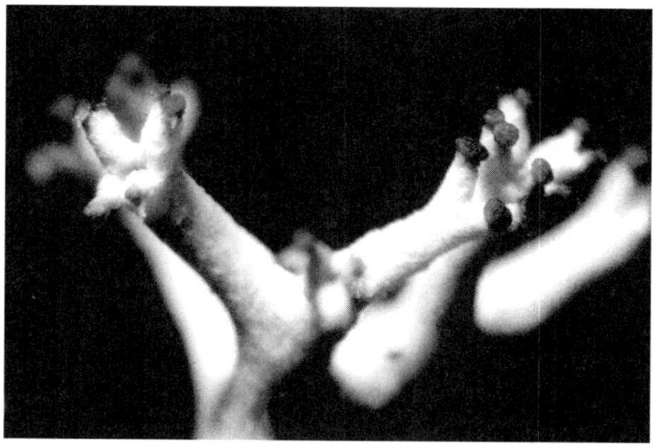

Rangiferoid thallus of *Cladonia rangiferina*, x 20 (*MU*).

raphe (Gr. *raphé*, seam, suture): a longitudinal line of dehiscence that is present in the pycnidial conidiomata of *Chaetomella* (sphaeropsidaceous asexual fungi);

also used for the longitudinal line in the conidial wall of some species of the gen. *Coniella*, such as *C. fragariae* (sphaeropsidaceous asexual fungi).

Raphe of the conidia of *Coniella fragariae*, x 2 000.

raphidium, pl. **raphidia** (NL. *raphidium* < Gr. *raphídion*, a small needle < *ráphis*, needle + dim. suf. *-ídion* > L. *-idium*): an acicular crystal, sharp-pointed at both ends and composed of calcium oxalate, like those present in some lichen thalli.

ray (ME. *raie*, *raye* < OF. *rai* < L. *radius*, radius, ray): see **involute ray** and **saccate ray.**

ray fungus, pl. **fungi** (ME. < MF. *rai* < L. *radius*, rod, ray; L. *fungus*, fungus): a member of the Actinomycetes, which are not fungi but filamentous bacteria.

receptacle, **receptaculum**, pl. **receptacula** (L. *receptaculum*, a receptacle, reservoir < *receptor*, a receiver + dim. suf. *-culum*): a structure that bears one or more reproductive organs, whether on a pedicel or on a hollow body. Applied to any structure that supports a hymenium. In the Phallales (such as *Phallus* and *Dictyophora*, e.g.), it consists of a pedicel (pseudostipe) and the pileus; in Phallales of the gen. *Clathrus* it consists of the clathrate body. In the Nidulariales (such as *Cyathus*) it refers to the goblet or funnel-shaped structure that contains the peridioles. In the Laboulbeniales (such as *Laboulbenia*, *Rhachomyces* and *Herpomyces*) the receptacle includes the perithecium, its appendages, and basal supporting cells. In the lichens it refers to the goblet of the thallus that contains soredia.

receptive hypha, pl. **hyphae** (L. *receptum*, supine of *recipere*, to receive + *-ivus* > E. *-ive*, quality or tendency, fitness; Gr. *hyphé*, tissue, spider web; hypha): a hypha that receives a gametic nucleus during sexual reproduction; the fertilization of a receptive hypha (considered as female) occurs, e.g., in the union of a protoperithecium with a microconidium (as in *Neurospora*, of the Sordariales), or in the union of a hypha of a spermogonium with a spermatium (as in *Puccinia*, of the Uredinales).

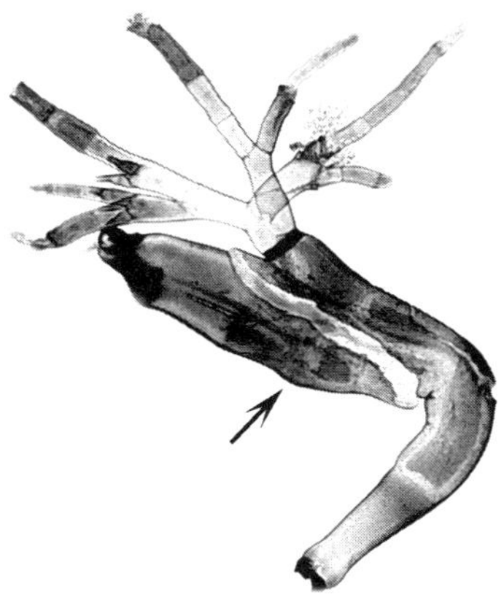

Receptacle of the thallus of *Laboulbenia flagellata*, x 125 (*EPS*).

Receptacle of the thallus of *Rhacomyces quetzalcoatl*, x 125 (*EPS*).

Receptacle (basidiocarp) of *Phallus ravenellii*, x 0.8 (*TH*).

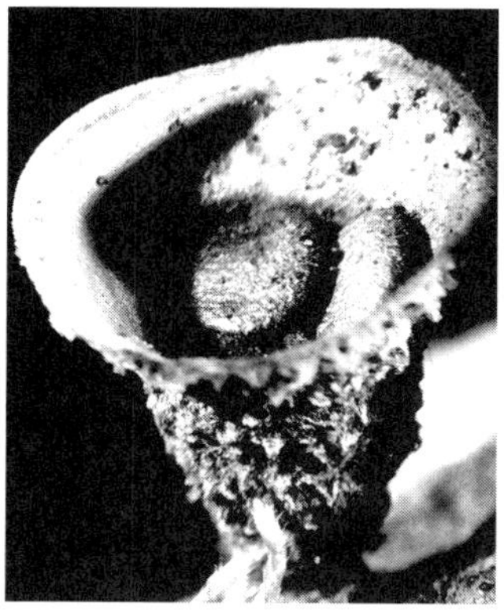

Receptacle (basidiocarp) of *Cyathus* sp., x 9 (*MU*).

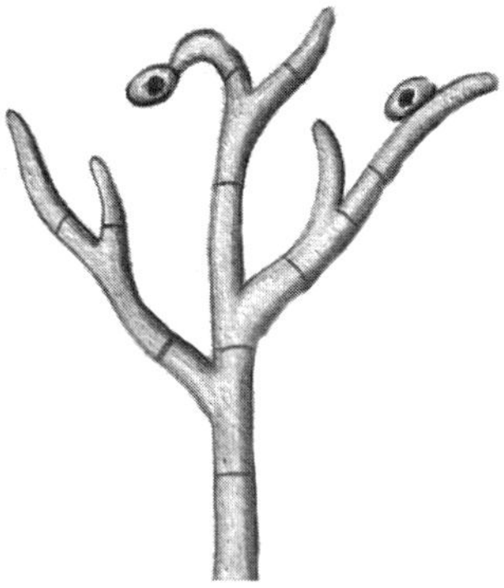

Receptive hyphae of *Neurospora sitophila*, x 1 000.

reclinate (L. *reclinatus*, leaning < pref. *re-*, back, again, against + *clinare*, to bend + suf. *-atus* > E. *-ate*, provided with or likeness): turned or bent downward.

recurvate, **recurved** (L. *recurvatus*, bent backward < pref. *re-*, back, again, against + *curvatus*, curved < *curvus*, curve + suf. *-atus* > E. *-ate*, provided with or likeness): curved in a manner that the concavity is on the outer or upper side, like the pileus of *Hygrophorus fornicatus* and *Inocybe patouillardi* (Agaricales). Cf. **incurvate** and **inflexed**.

Recurvate pileus of the basidiocarp of *Hygrophorus fornicatus*, x 1.

recurved ray (L. *recurvatus*, bent backward; ME. *raie*, *raye* < OF. *rai* < L. *radius*, radius, ray): *Gasteromycetes*. A ray in which the outer peridium is convex in profile. Cf. **revolute ray**.

reflexed (L. *reflexus* < *reflectere*, to curve backward): see **recurvate**. Cf. **inflexed**.

refracted (L. *refractus*, broken < *refringere*, to break): bent backward from the base; changed in direction, like a light ray when it passes obliquely through a medium of different density.

reindeer lichen, **reindeer moss** (ME. *reindere* < ON. *hreinn*, reindeer + ME. *deer*, animal, deer; L. *lichen* < Gr. *leichén*, lichen; ME. *mos*, moss): mainly *Cladonia rangiferina and C. stellaris*, fruticose lichens that form extensive patches in arctic and north-temperate regions, which are grazed by reindeer and caribou; these lichens are sometimes eaten by man.

reliquiae (L. *reliquiae*, remains): the remains of the pseudoparenchymatous centrum tissue of *Magnaporthe* (Diaporthales) after elongation of the asci.

reniform (L. *reniformis* < *renes*, kidney + *-formis* < *forma*, shape): in the shape of a kidney, or with a contour similar to that of a kidney. Also called **fabiform**, i.e., in the shape of a bean (*Vicia faba*), like the secondary zoospores of *Saprolegnia* (Saprolegniales) and other Oomycota, the ascospores of *Kluyveromyces marxianus* var. *marxianus* (Saccharomycetales) and *Petriella* (Microascales), or like the basidiospores of *Inocybe fastigiata* (Agaricales).

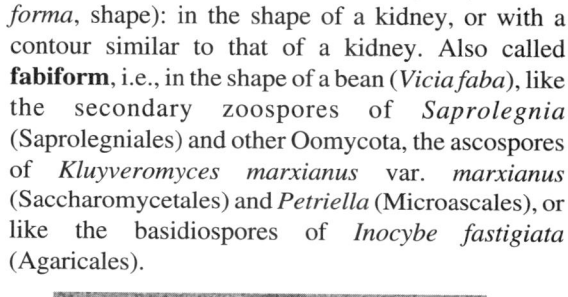

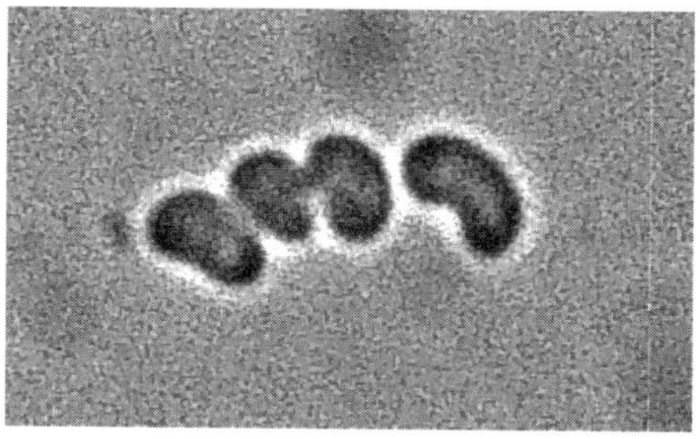

Reniform ascospores of *Kluyveromyces marxianus* var. *marxianus*, x 2 000 (*PL*).

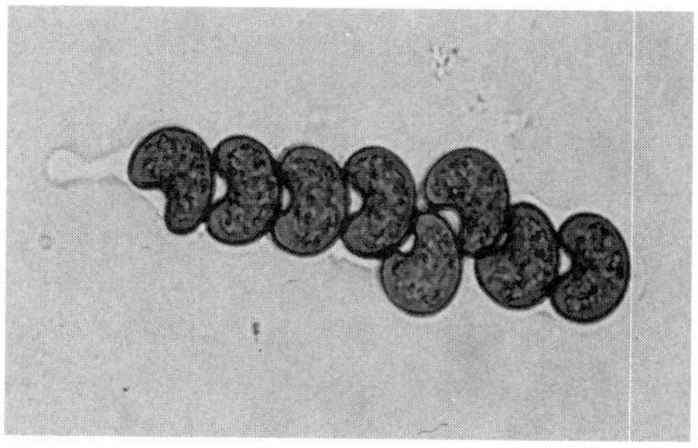

Reniform ascospores within an ascus of *Petriella* sp., x 1 000 (*CB*).

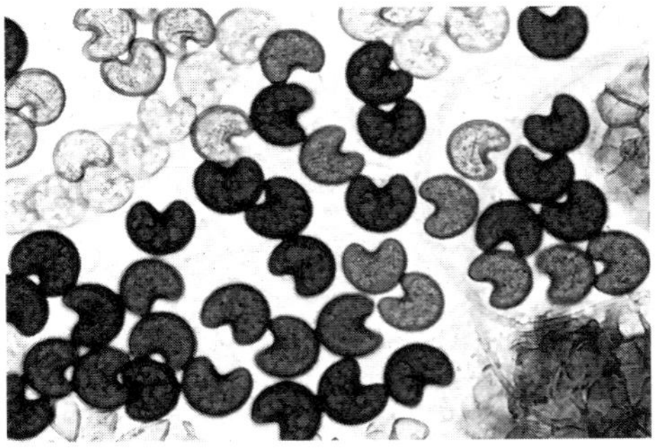

Reniform ascospores of *Petriella* sp., x 770 (*CB*).

repand, **repandous** (L. *repandus*, curved; or + L. *-osus* > OF. *-ous*, *-eus* > E. *-ous*, having, possessing the qualities of): with the margin wavy and elevated or reflexed (turned backward), like the pileus of *Collybia driophylla* and *Lepiota clypeolaria* (Agaricales).

Repand pileus of a basidiocarp of *Lepiota* sp., x 0.6.

resting spore or **latent spore** (E. *resting*, dormant < ME. *resten* < OE. *restan*, to rest; NL. *spora* < Gr. *sporá*, spore; L. *latens*, genit. *latentis* < *latere*, to lurk, lie hid, hidden + *-entem* > E. *-ent*, being; NL. *spora*): *Aquatic fungi*. The latent spore or sporangium, which results from the transformation of the zygote, and which remains dormant for a time before germination; as happens, e.g., in *Rozella* (Spizellomycetales), *Allomyces* (Blastocladiales) and *Chytriomyces* (Chytridiales). In the Plasmodiophorales, such as *Plasmodiophora*, *Spongospora* and *Woronina*, the zoosporangia become resting spores inside the host cells. Resting spores are released into the environment following the death and disintegration of host cells, where some are thought to lie dormant for a few years before germinating to form primary zoospores. The supposed zygospore of the Trichomycetes is also called a resting spore. The resting spore is also sometimes called a resistant spore.

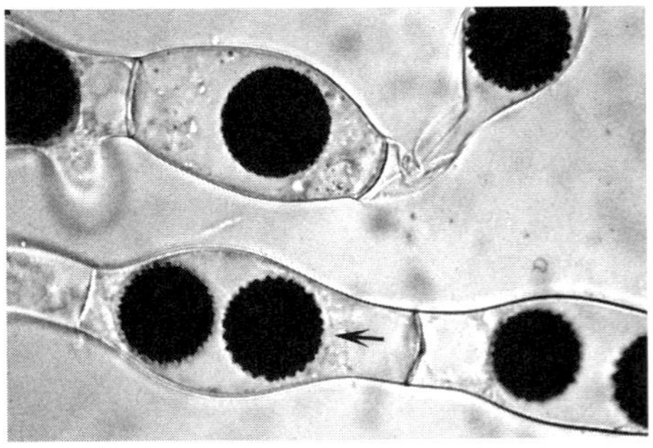

Resting spores of *Rozella allomycis* within hyphal cells of *Allomyces javanicus*, x 500 (*RTH*).

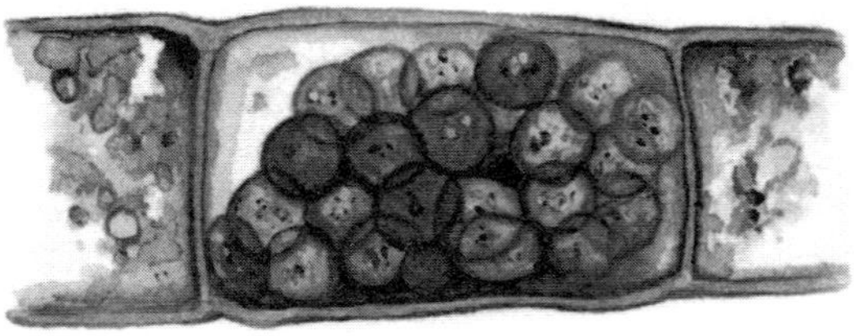

Resting spores (zoosporangia) of *Woronina polycystis* within a cell of the host (*Achlya*), x 530.

resupinate, **resupinated** (L. *resupinatus*, to bend back to a supine position, inverted < pref. *re-*, back, again, against < *supinus*, supine + suf. *-atus* > E. *-ate*, provided with or likeness): refers to any part or organ inverted with respect to the normal position; the basidiocarps of *Megasporoporia mexicana* and *Poria* (Aphyllophorales) are resupinate, with the poroid hymenium upward, on the side opposite that which is attached to the substrate.

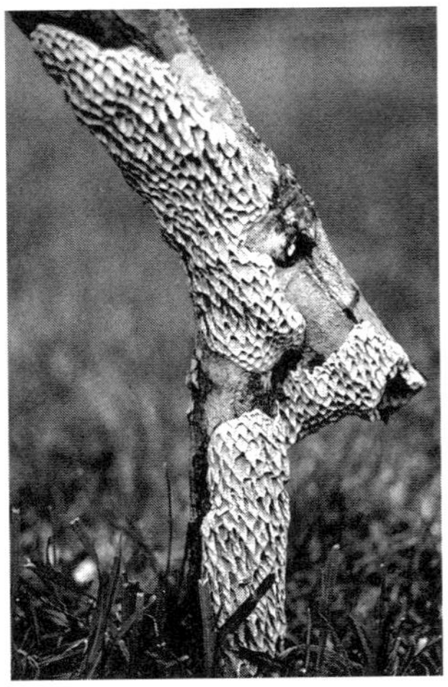

Resupinate basidiocarp of *Megasporoporia mexicana*, x 0.5 (*RV*).

Resupinate basidiocarps of *Poria*, on dead wood, x 0.8 (*MU*).

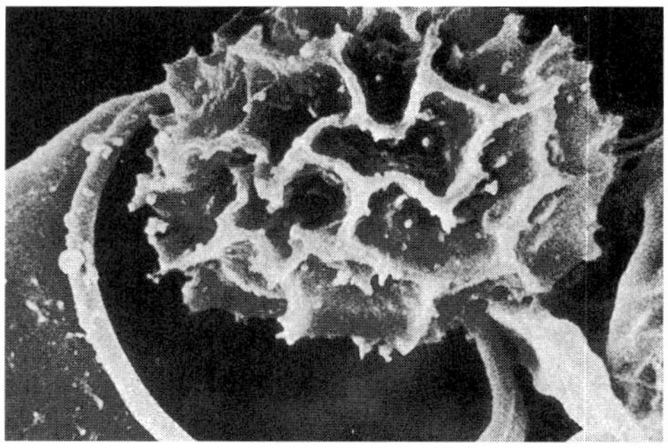

Reticulate wall of an ascospore of *Ascodesmis* sp., x 4 700 (*KO*).

reticulate (L. *reticulatus*, made like a net < *reticulum*, dim. of *rete*, net + suf. *-atus* > E. *-ate*, provided with or likeness): in the shape of a **reticulum**, i.e., with the veins, lines or borders of low relief intercrossed like a net (surface of a pileus, stipe, spores, etc.). For example, the pedicel of the sorocarp of *Dictyostelium mucoroides* (Dictyosteliomycota), the ascospore wall of *Ascodesmis* (Pezizales), the basidiocarp of *Auricularia delicata* (Auriculariales), and the basidiospores of *Lycoperdon spadiceum* (Lycoperdales) are reticulate.

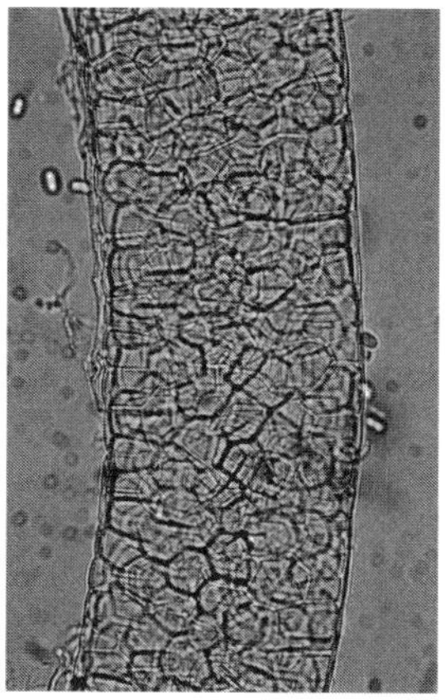

Reticulate pedicel of the sorocarp of *Dictyostelium mucoroides*, x 750 (*EAA*).

Reticulate lower surface of the basidiocarps of *Auricularia delicata*, x 0.8 (*TH*).

reticulum, pl. **reticula** (L. *reticulum*, a little net < *rete*, net + dim. suf. *-culum*): a network or mesh composed of filaments, veins, nerves, ridges, etc.

retrocurved (L. *retro*, backward + *curvus*, curved): curved or bent backward. Equal to **retroflexed** and **retrorse**. Cf. **antrorse**.

retroflex, **retroflexed**, **retroflexous** (L. *retroflexus*, bent backward < *retro*, backward + *flexus*, bent, turned, curved; or + L. *-osus* > OF. *-ous*, *-eus* > E. *-ous*, having, possessing the qualities of): same as **retrocurved**.

retrorse (L. *retrorsus*, toward the back): same as **retrocurved**. Cf. **antrorse**.

revolute (L. *revolutus*, rolled back, ptp. of *revolvere*, to turn or roll back): *Agaricales*. Applied to a pileus with borders rolled toward the upper surface; revolute is like reflexed, but rolled up. Cf. **involute**.

revolute ray (L. *revolutus*, rolled back, ptp. of *revolvere*, to turn or roll back; ME. *raie*, *raye* < OF. *rai* < L. *radius*, radius, ray): *Gasteromycetes.* A ray in which the outer peridium is recurved and also inrolled at the edge. Cf. **recurved ray**.

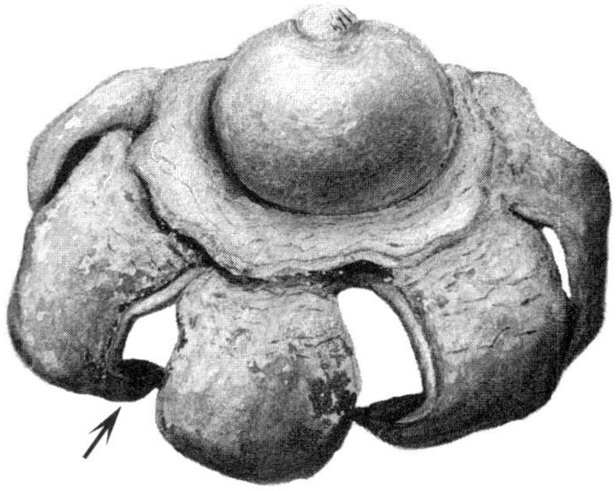

Revolute exoperidial **rays** of the fruiting body of *Geastrum saccatum*, x 2.

rhagadiose (L. *rhagadiosus*, fissured, cracked < *rhagádos*, genit. of *rhagás*, clink, break + L. *-osus*, full of, augmented, prone to > ME. *-ose*): with deep fissures, like the surface of the pileus of *Boletus castaneus* (Agaricales) and the exoperidial rays of *Astraeus hygrometricus* (Sclerodermatales). Also used for the surface of certain lichens, such as *Acarospora rhagadiosa* (=*A. scabra*) of the Lecanorales. Cf. **rimose**.

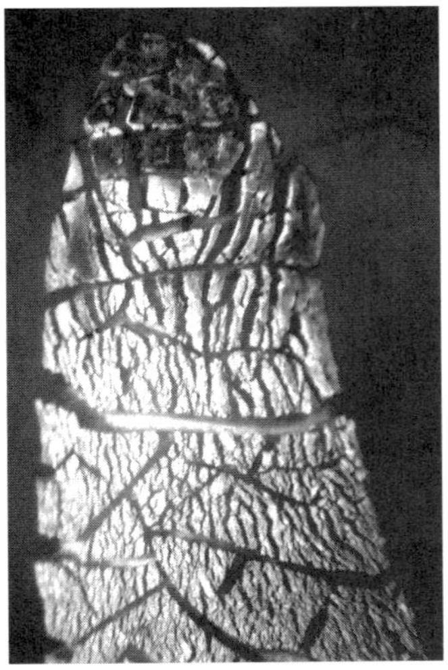

Rhagadiose surface of an exoperidial ray of the fruiting body of *Astraeus hygrometricus*, x 5 (*MU*).

rhexigenesis, pl. **rhexigeneses** (Gr. *rhêxis*, tear + *génesis*, engendering): see **rhexolytic**.

rhexolysis, pl. **rhexolyses** (Gr. *rhêxis*, tear + *lýsis*, dissolution, disintegration): see **rhexolytic**.

rhexolytic (Gr. *rhêxis*, tear + *lytikós*, able to loosen < *lýtos*, dissolvable, broken + suf. *-íkos* > L. *-icus* > E. *-ic*, belonging to, relating to): a spore that separates by the tearing of a conjunctive cell, as happens, e.g., in the detachment of thallic conidia that some fungi form (such as *Microsporum gypseum*, of the moniliaceous asexual fungi). This process is called **rhexigenesis** or **rhexolysis**. Cf. **schizolytic**, **schizolysis**, and **schizogenesis**.

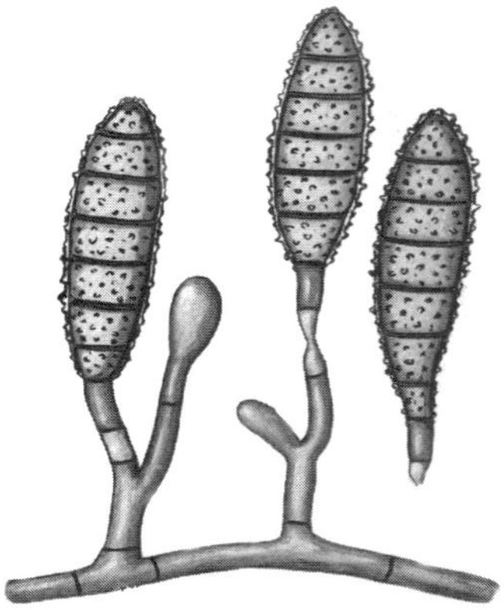

Rhexolytic conidia of *Microsporum gypseum*, x 600.

rhinophycomycosis, pl. **rhinophycomycoses**, or **rhinoentomophthoramycosis**, pl. **rhinoentomophthoramycoses** (NL. < Gr. *rhín*, *rhinós*, nose + *phýkos*, alga + NL. *mycosis* < Gr. *mýkes*, fungus + *-osis*, state or condition; < gen. *Entomophthora* < Gr. *éntomos*, insect + *phthorá*, death, destruction + NL. *mycosis* < Gr. *mýkes*, fungus + *-osis*): an inflammatory or chronic granulomatous disease, generally circumscribed to the subcutaneous tissues or nasal submucose, characterized by swellings and deformations of the face of humans. The etiological agent is *Conidiobolus* (=*Entomophthora*) *coronatus* (Entomophthorales), a soil inhabitant that attacks spiders and termites and other insects.

rhinosporidiosis, pl. **rhinosporidioses** (< gen. *Rhinosporidium* < Gr. *rhín*, *rhinós*, nose + *sporá*, spore + L. dim. suf. *-ium* + Gr. *-osis*, state or condition): an infection of the mucocutaneous tissue, principally of the nose and conjunctive, caused by *Rhinosporidium seeberi* (classified as an *incertae sedis* taxon in the Hyphochytriales), marked by the formation of penduculated polyps, tumors, papilomas and hyperplastic verrucous lessions.

rhizina, pl. **rhizinae** (L. *rhizina*, rootlet, a rootlike hair or strand < *rhiza* < Gr. *rhíza*, root + L. *-ina*, likeness): each one of the root-like structures that the mycobiont of a lichen thallus forms and which serves to attach it to the substrate; they are formed, e.g., in *Peltigera canina* (Peltigerales) and *Umbilicaria vellea* (Lecanorales).

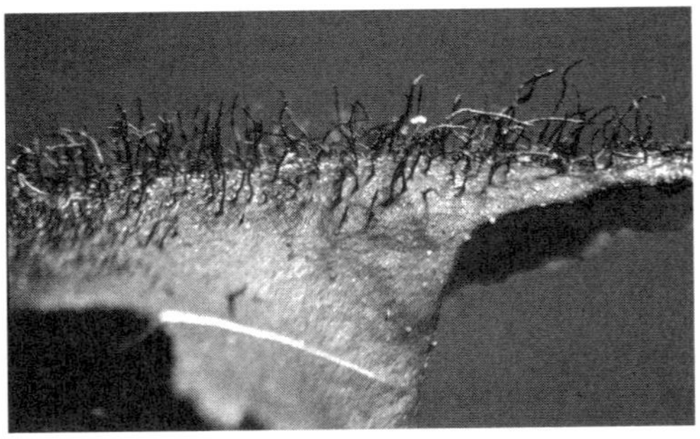

Rhizinae of the lower surface of the thallus of *Peltigera* sp., x 5 (*MU*).

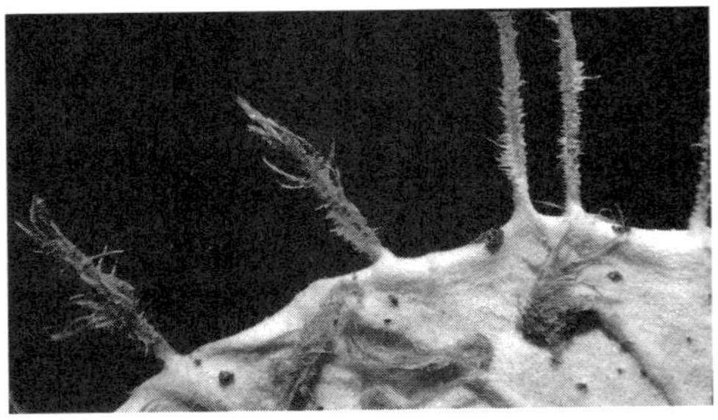

Rhizinae of the lower cortex of the thallus of *Peltigera canina*, x 10 (*MU*).

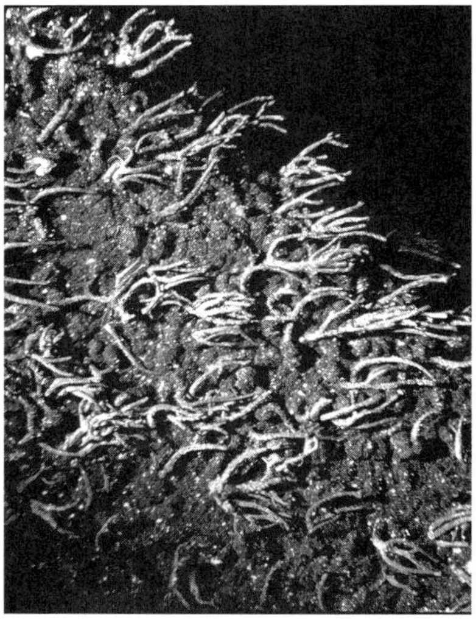

Rhizinae of the lower cortex of the thallus of *Umbilicaria vellea*, x 10 (*MU*).

rhizinose strand (L. *rhizina*, rootlet, a rootlike hair or strand < *rhiza*, root + *-ina*, likeness + L. *-osus*, full of, augmented, prone to > ME. *-ose*; ME. *strond*, fibers or filaments twisted, plaited, or laid parallel to form a unit for further twisting or plaiting into yarn, thread, rope, or cordage): *Lichens*. An attachment organ composed of strong, irregularly branched hyphae that grow among the soil particles; found in *Buellia pulchella* (Lecanorales) and other squamose lichens.

rhizoid (Gr. *rhíza*, root + *-oeídes*, similar to): a slender, tapered, often branched hypha, superficially similar to a plant root, that serves to attach a thallus or mycelium to the substrate and absorb nutritive substances from it. Rhizoids are characteristic of aquatic fungi, such as the Chytridiales and Hyphochytriales (e.g., *Rhizophydium pollinis-pini* and *Hyphochytrium*) and of some terrestrial species, such as the Mucorales (e.g., *Rhizopus*, *Absidia*, and *Actinomucor*).

Rhizoids of *Rhizopus nigricans*, x 400 (*PL*).

rhizomorph (Gr. *rhíza*, root + *morphé*, shape): a thick, resistant mycelial cord, composed of plectenchyma and with a growing tip similar to that of a plant root, which penetrates the substrate; they are capable of reaching more than a meter in length, with a width of up to 1 cm, generally dark in color, but also white or yellow to pink or purple, simple or branched. Rhizomorphs function as organs of absorption and conduction of nutritive substances, and on occasion also are capable of having latent life; they are typical of *Armillaria* (=*Armillariella*) *mellea* (Agaricales) and of some Gasteromycetes, such as some species of *Lycoperdon* (Lycoperdales) and *Scleroderma* (Sclerodermatales). Cf. **mycelial strand**.

Rhizomorphs of the fruiting body of *Lycoperdon* sp., x 0.8.

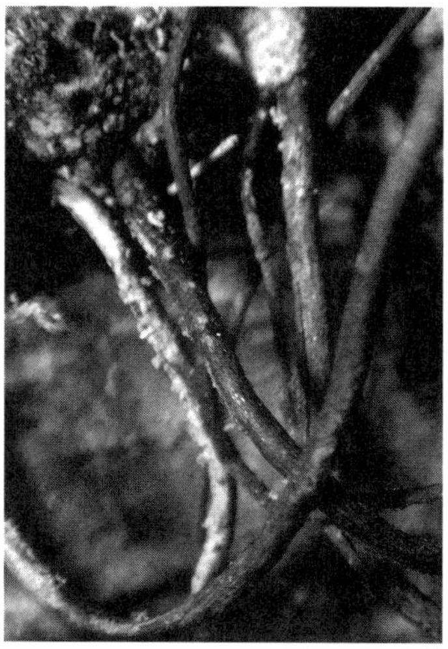

Rhizomorphs of an agaricaceous fungus, x 5 (*MU*).

rhizomycelium, pl. **rhizomycelia** (Gr. *rhíza*, root + NL. *mycelium* < Gr. *mýkes*, fungus + L. *-elis*, pertaining to + dim. suf. *-ium*): a rudimentary rhizoidal system, more or less developed and similar to a mycelium, but with the rhizoidal branches lacking nuclei.

Characteristic of aquatic fungi of the order Chytridiales and of the family Megachytriaceae, such as *Nowakowskiella ramosa*.

Rhizomycelium of *Nowakowskiella ramosa*, x 550.

rhizoplane (Gr. *rhíza*, root + L. *planus*, plain, flat surface): the surface of a root on which are established the biological relationships (such as those of the mycorrhizae) with the medium of the soil. Cf. **rhizosphere**.

rhizoplast (Gr. *rhíza*, root + *plastós*, formed, organized particle or body): an endoplasmatic chromatoid filament that connects the **blepharoplast** (an organelle related to the kinetic functions of flagellate cells) at the base of the flagellum with the nucleus. It is present in Chytridiomycota (such as *Rhizophlyctis rosea*), Hyphochytriomycota and Oomycota.

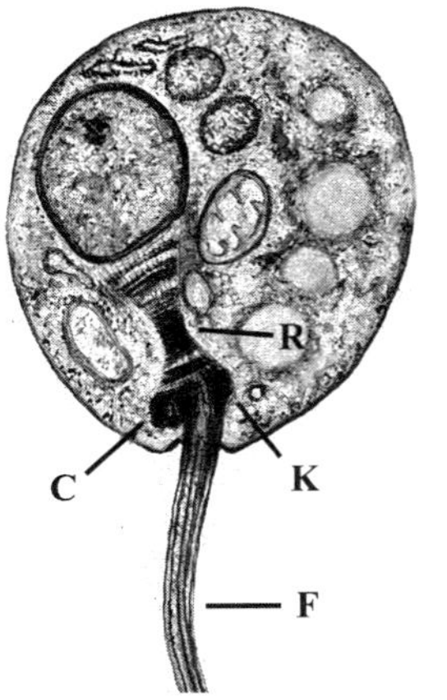

Rhizoplast (R) of the zoospore of *Rhizophlyctis rosea*, seen in longitudinal section. The kinetosome (K), centriole (C), and flagellum (F), also are shown, x 15 000.

rhizosphere (Gr. *rhíza*, root + *sphaîra*, sphere): the area of the soil near the living roots of a plant, with a microbiota frequently richer in species than that of the soil farther from the roots, and with certain relationships with the mycorrhizae. Cf. **edaphosphere** and **rhizoplane**.

rhodosporous (Gr. *rhódon*, rose, red + *sporá*, spore + L. *-osus* > OF. *-ous*, *-eus* > E. *-ous*, having, possessing the qualities of): refers to fungi that have rose, red or reddish spores, as seen in mass in a spore print; e.g., as in the basidiocarp of *Rhodophyllus* (Agaricales).

rhombic, **rhomboid**, **rhomboidal** (Gr. *rhómbos*, rhombus < *rhémbein*, to revolve + suf. *-íkos* > L. *-icus* > E. *-ic*, belonging to, relating to; or + Gr. *-oeídes*, similar to; or + L. suf. *-alis* > E. *-al*, relating to or belonging to): rhombus- or diamond-shaped, or quadrangular, like the cells of the yeasts of the gen. *Kloeckera* (cryptococcaceous asexual yeasts), the conidia of *Beltrania rhombica* (dematiaceous asexual fungi), or the ascospores of *Melanospora zamiae* (Melanosporales).

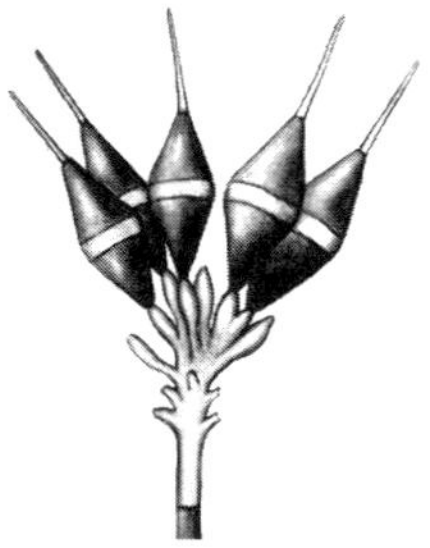

Rhombic conidia of *Beltrania rhombica*, x 700.

Rhombic conidia of *Beltrania rhombica*, x 450 (*GH*).

rhynchosporous (Gr. *rhýnchos*, peak + *sporá*, spore + L. *-osus* > OF. *-ous*, *-eus* > E. *-ous*, having, possessing the qualities of): having spores with beaks at the distal end, like those of *Rhynchosporium secalis* (tuberculariaceous asexual fungi).

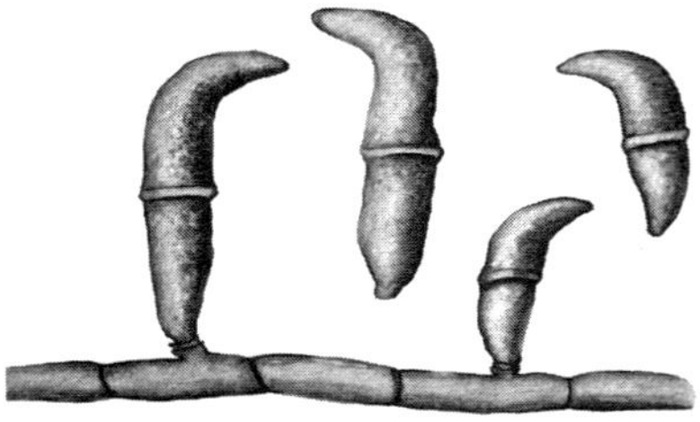

Rhynchosporous conidia of *Rhynchosporium secalis*, x 1 600.

ribbed (ME. < OE. *ribb*, rib): having ribs. See **costate**.

ribosome (E. *ribo-* < L. *ribose* + Gr. *sõma*, body): a nucleoprotein mass, 15-25 nm diam., that is found in the cytoplasm or associated with the endoplasmic reticulum and the nuclear envelope. Ribosomes can aggregate into clusters called polyribosomes, which function in protein synthesis.

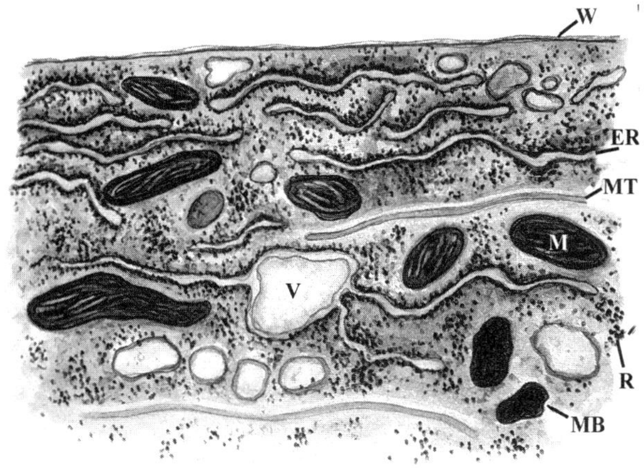

Ribosomes (R) of a young hypha of *Armillaria mellea*, seen in longitudinal section; these organelles are associated with the cisternae of the endoplasmic reticulum (ER). The microbodies (MB), mitochondria (M), microtubules (MT), vesicles (V), and cell wall (W) are also evident. Drawing composed from various transmission electron micrographs published by Beckett *et al.* in their *Atlas of Fungal Ultrastructure*, 1974, x 15 800.

rimose, **rimous** (L. *rimosus*, cracked, split < *rima*, fissure, crack, slit + *-osus*, full of, augmented, prone to > ME. *-ose*; or + *-osus* > OF. *-ous*, *-eus* > E. *-ous*, having, possessing the qualities of): with the surface cracked or split, like the pileus of *Amanita solitaria*, *Boletus rubellus* and *Inocybe fastigiata* (Agaricales) or the peridium of *Astraeus* (Sclerodermatales). Cf. **rhagadiose**.

Rimose pileus of the basidiocarp of *Inocybe fastigiata*, x 5 (*CB*).

rimula, pl. **rimulae** (L. *rimula*, dim. of *rima*, fissure, crack, slit): a small crack or slit. See **rimulose**.

rimulose, **rimulous** (L. *rimula*, dim. of *rima*, fissure, crack, slit + L. *-osus*, full of, augmented, prone to > ME. *-ose*; or + *-osus* > OF. *-ous*, *-eus* > E. *-ous*, having, possessing the qualities of): provided with small cracks or slits, like the pileus of some Boletaceae and the surface of the thallus of certain lichens, such as *Graphis rimulosa* (Ostropales).

ringworm (E. *ring* < ME. < OE. *hring*, an encircling arrangement + *worm* < ME. < OE. *wyrm*, serpent, worm): any of several infectious diseases of the skin, hair or nails of man and domestic animals caused by dermatophytes (fungi of the gen. *Epidermophyton*, *Microsporum* and *Trichophyton*, of the moniliaceous asexual fungi) and producing skin lessions characterized by ring-shaped discolored patches that are covered with vesicles and scales. See **tinea**.

rivose (L. *rivosus*, channeled, grooved < *rivus*, canal, a channel, groove, arroyo + *-osus*, full of, augmented, prone to > ME. *-ose*): having sinuate canals on the surface, like the thallus of some crustaceous lichens.

rivulose (L. *rivulosus*, channeled, grooved < *rivulus*, dim. of *rivus*, canal, a channel, groove, arroyo + *-osus*, full of, augmented, prone to > ME. *-ose*): with the surface marked by lines that resemble a system of small rivers or canals on a map, such as one sees in the pileus of the basidiocarp of *Clitocybe angustifolia* (Agaricales), and in the crustaceous thallus of the lichen *Lecidea rivulosa* (Lecanorales).

Rivulose pileus of the basidiocarp of *Clitocybe angustifolia* (= *C. rivulosa*), x 1.

rodlet (ME. < OE. *rodd*, rod, club, bar): structural unit of conidial and some hyphal walls composed of particles (ca. 5 nm in diam.) arranged in linear series.

roestelioid (< gen. *Roestelia* < L. *rostellum*, dim. of *rostrum*, bill, snout, beak + L. *-ia*, quality of or state of being + L. suf. *-oide* < Gr. *-oeídes*, similar to): the long, tubiform, cornute aecium characteristic of *Roestelia* and *Gymnosporangium* (Uredinales).

Rohr (G. *Rohr*, tube): an elongated, tubular cavity that forms in the interior of the encysted zoospores of *Plasmodiophora brassicae* (Plasmodiophorales) when these zoospores encyst on the outer surface of the epidermal cells of cabbage roots. The end of the Rohr or tube, which is oriented toward the cell wall of the host, has a kind of plug that stains lightly. Inside the tube is a cylindrical, sharp-pointed body (which stains intensely) called a sting or **Stachel**, which is what perforates the host cell wall so that the protoplast of the encysted zoospore can enter the host and initiate infection.

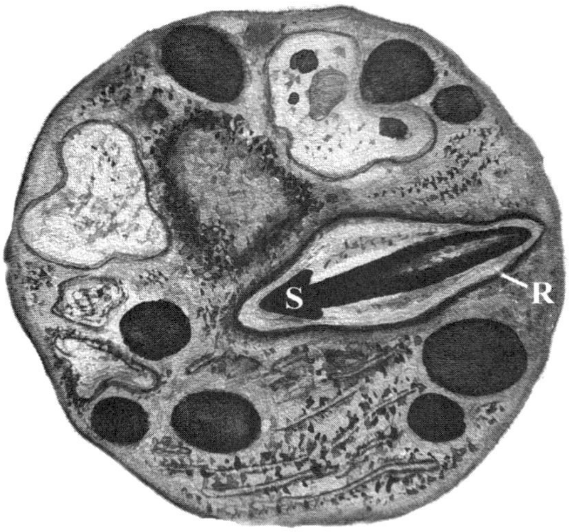

Rohr (R) of an encysted zoospore of *Plasmodiophora brassicae*, seen in longitudinal section. The Stachel (S) is contained in the Rohr. Drawing based on a transmission electron micrograph published by Beckett *et al.* in their *Atlas of Fungal Ultrastructure*, 1974, x 25 700.

roridous (L. *roridus*, dewy < *ros*, genit. *roris*, dew + *-osus* > OF. *-ous*, *-eus* > E. *-ous*, having, possessing the qualities of): refers to mycelium, sporiferous apparatus, etc., which has droplets of moisture on the surface, in the manner of dew. The colonies of many species of molds have this characteristic. Also, the sclerotia of *Sclerotinia* (Helotiales) are roridous in the presence of humidity; e.g., when they are developing on agar.

rostrate, **rostrated** (E. *rostrate*, having a rostrum < L. *rostratus*, beaked, hooked < *rostrum*, beak, bill, snout + suf. *-atus* > E. *-ate*, provided with or likeness): having a **rostrum**.

rostrate dehiscence (L. *rostratus*, beaked, hooked < *rostrum*, beak, bill, snout + suf. *-atus* > E. *-ate*, provided with or likeness; NL. *dehiscentia* < L. *dehiscere*, to open itself, to part + suf. *-entia* > F. *-ence* < E. *-ence*, state, quality or action): a type of ascus dehiscence present in some lichen-forming Discomycetes (Lecanorales), in which the asci have walls that may be thickened at the apex with the inner layer elongating and breaking through the outer wall layers in spore discharge. Cf. **fissitunicate** and **semifissitunicate**.

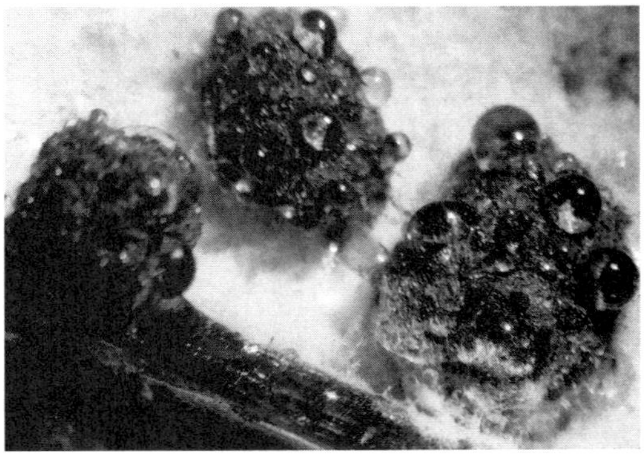

Roridous sclerotia of *Sclerotinia* sp. on agar, x 10 (*MU & CB*).

rostrum, pl. **rostra** (L. *rostrum*, beak, snout): a beak-like extension or process, like that present in various fungal structures, e.g., the spores of *Exserohilum* (=*Helminthosporium*) *rostratum* (dematiaceous asexual fungi) and the perithecia of *Guanomyces* (Sordariales); such structures are said to be **rostrate**.

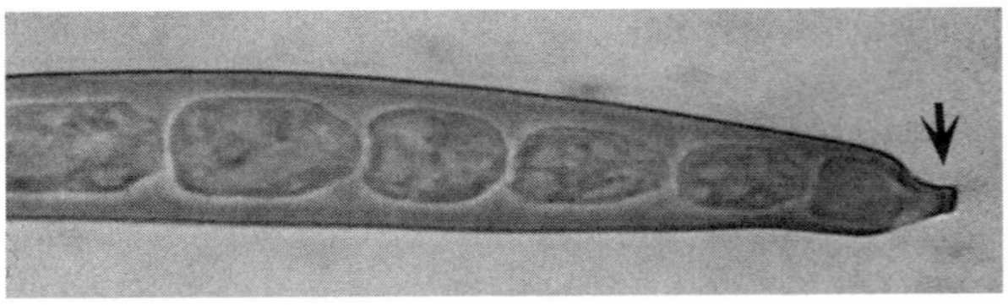

Rostrum of a conidium of *Exserohilum* (=*Helminthosporium*) *rostratum*, x 1 200 (*MU*).

Rostrum of the perithecium of *Guanomyces polythrix*, x 10 (*MU*).

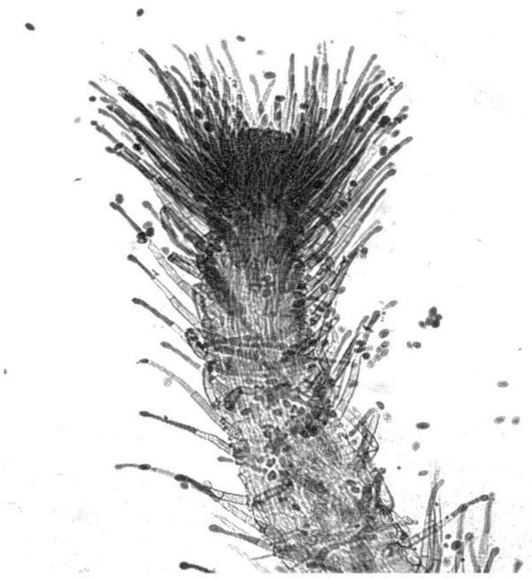

Rostrum of the perithecium of *Guanomyces polythrix*, x 500 (*MU*).

rosulate, **rosette-like** (L. *rosulatus*, like a small rose < *rosula*, a little rose, rosette + suf. *-atus* > E. *-ate*, provided with or likeness): arranged in the manner of a rosette; like a small rose.

rot (ME. *roten* < OE. *rotian*, to rot): to undergo decomposition from the action of microorganisms; a plant disease marked by breakdown of tissues and caused especially by fungi or bacteria. Based on symptoms, several rot types caused by fungi are recognized: *Brown rot* - the cellulose is utilized (not the lignin), brick-shaped cracks develop, the wood becomes brown and crumbles (caused by many species of Aphyllophorales, e.g., *Fomitopsis pinicola*). *Dry rot* - a brown rot caused by *Serpula lacrymans* (Aphyllophorales). This name is misleading as damp conditions are needed for initation of growth. *Soft rot* - caused by many fungal species, including hyphomycetous asexual fungi and Chaetomiaceae (Sordariales), among others. *Wet rot* - a brown rot caused by *Coniophora cerebella* (the cellar fungus, Aphyllophorales). *White rot* - both cellulose and lignin are decomposed (or mainly the latter), and the wood bleaches and becomes stringy (also caused by Aphyllophorales, e.g., *Trametes versicolor*).

rotaceous (L. *rota*, wheel + *-aceus* > E. *-aceous*, of or pertaining to, with the nature of): with the shape of a wheel, flat and circular, like the apothecia of *Rotula* and *Rotularia* (=*Mazosia*), lichens in the order Arthoniales.

rotund (L. *rotundus*, round): round, circular, orbicular, but not perfectly spherical, like the pileus of *Morchella rotunda* (Pezizales).

rubiginose, **rubiginous** (L. *rubiginosus*, of a reddish-brown color < *rubigo* or *robigo*, rust + *-osus*, full of, augmented, prone to > ME. *-ose*; or + *-osus* > OF. *-ous*, *-eus* > E. *-ous*, having, possessing the qualities of): rust-colored; ferruginous.

ruderal (L. *ruderalus*, growing in waste places < *rudus*, genit. *ruderis*, old rubbish, rubble, ruins, debris, wastes + L. suf. *-alis* > E. *-al*, relating to or belonging to): growing on ruins or debris, on media with rubbish, removed earth and other similar materials that are created by human habitation and old construction, as happens in many urban environments.

rugose, **rugous** (L. *rugosus*, wrinkled, creased < *ruga*, wrinkle + *-osus*, full of, augmented, prone to > ME. *-ose*; or + *-osus* > OF. *-ous*, *-eus* > E. *-ous*, having, possessing the qualities of): wrinkled or creased. The ornamentation of the wall of various spores is said to be rugose when the wrinkles or irregularities are abundant and confluent, as is observed in the basidiospores of *Panaeolina foenisecii* (Agaricales) and in the conidia of *Clasterosporium cocoicola* (dematiaceous asexual fungi). Also applied to the macrocolony which some yeasts develop in culture, such as *Pichia membranaefaciens* (Saccharomycetales).

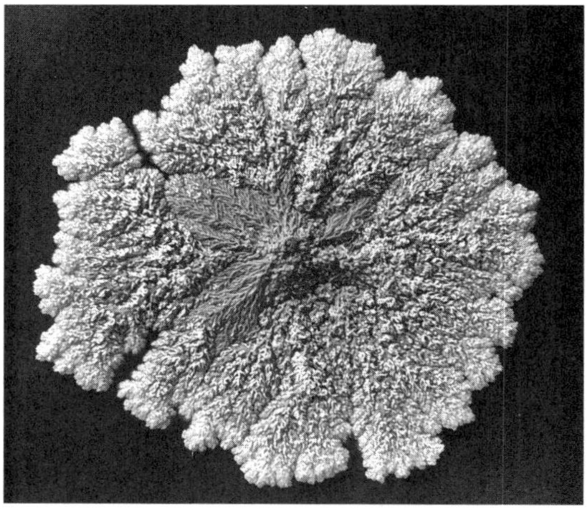

Rugose giant colony of *Pichia membranaefaciens* on agar, x 2 (*MU*).

rugulate, **rugulose** (L. *rugulatus*, having small wrinkles or creases < *rugula*, dim. of *ruga*, wrinkle + suf. *-atus* > E. *-ate*, provided with or likeness; *rugulosus* < *rugula*, small wrinkle + *-osus*, full of, augmented, prone to > ME. *-ose*): with slight creases or wrinkles, like the surface of the basidiospores of *Galerina clavata* (Agaricales).

rumposome (ME. *rump*, buttocks < MHG. *rumph*, torso + Gr. *sõma*, body): an intracellular organelle at the rear of the zoospores of *Chytridium*, *Rhizophydium* and other Chytridiales, as well as *Monoblepharella* (Monoblepharidales), composed of tubules arranged like a honeycomb, which covers part of the surface of the microbody-lipid body complex. The nucleus, the ribosomes, the single mitochondrion (or several mitochondria, depending upon the species), and the microbody-lipid body complex, form compartments that are interconnected by a system of double membrane of smooth endoplasmic reticulum. This system of membranes is continuous with the outer membrane of the nucleus and extends like a network around the lipid body and is connected with the rumposome, which is situated near the plasma membrane of the zoospore. The rumposome is also united with the kinetosome by means of microtubules, which suggests that all of this mechanism serves to anchor the lipid globule in a certain position within the zoospore, generally in the lateral and posterior part, near the flagellar apparatus.

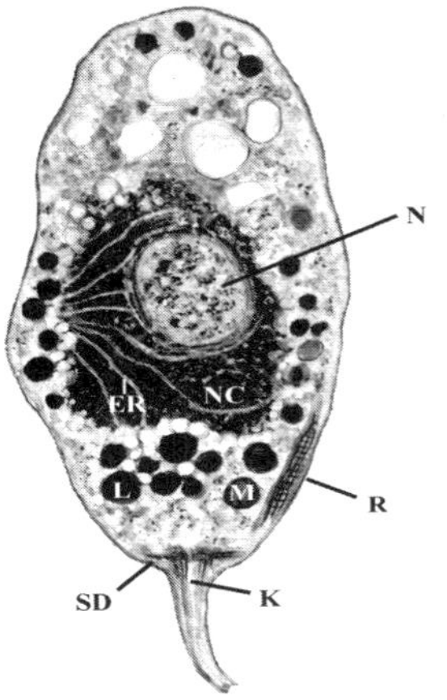

Rumposome (R) of a zoospore of *Monoblepharella* sp., seen in longitudinal section. The ribosomes are arranged in a nuclear cap (NC) which is not bounded by an envelope but is permeated by endoplasmic reticulum (ER). The nucleus (N) lies centrally in the zoospore while the kinetosome (K) is embedded in the striated disk (SD) at the posterior end of the spore. Microtubules radiate into the cytoplasm from the disk. Lipid droplets (L) and mitochondria (M) are dispersed through the spore. The rumposome is seen at a side in the posterior end of the spore Drawing based on a transmission electron micrograph published by Beckett *et al.* in their *Atlas of Fungal Ultrastructure,* 1974, x 4 700.

rupestral, **rupestrine** (NL. *rupestris*, growing among rocks < *rupes*, rock, stone + L. suf. *-alis* > E. *-al*, relating to or belonging to; or + L. suf. *-inus* > E. *-ine*, of or pertaining to): living on walls or rocks, e.g., *Caloplaca saxicola* (Teloschistales). See **saxicolous**.

Rupestral thallus of *Caloplaca saxicola*, x 10 (*CB*).

rust (ME. < OE. *rust*, red): *Uredinales*. 1. Any of numerous destructive plant diseases caused by a species of this order. 2. One of the Uredinales. *Black stem rust* - of cereals, *Puccinia graminis*. *Blister rust* - of *Pinus* and *Ribes*, *Cronartium ribicola*. *Brown leaf rust* - of barley, *Puccinia hordei*; of rye and wheat, *P. recondita*. *Crown rust* - of oats, *P. coronata*. *Red rust* - urediniospore state of cereal rusts, especially *P. graminis*. *White rust* (*white blister*) - of crucifers, *Albugo candida* (Peronosporales).

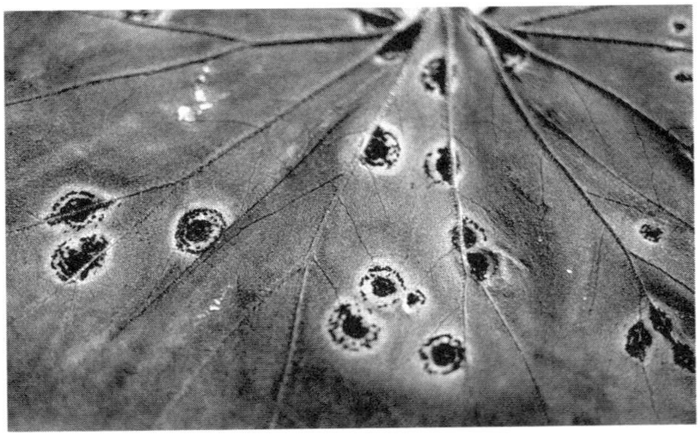

Rust (*Puccinia malvacearum*) of mallow. The pustules on the lower surface of the leaf correspond to uredinia, x 2 (*MU*).

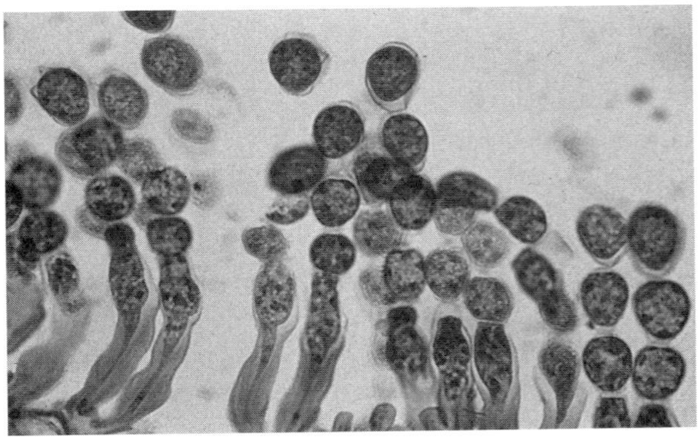

White **rust** (*Albugo candida*) of a cruciferous plant. The pustule below the host epidermis contains a palisade of sporangiophores with conidiosporangia, seen in longitudinal section, x 500 (*MU*).

S

sabulicole, **sabulicolous**, **sabulose**, **sabuline** (L. *sabulicola* < *sabulum*, sand + *-cola*, inhabitant; or + L. *-osus* > OF. *-ous*, *-eus* > E. *-ous*, having, possessing the qualities of; or + *-osus*, full of, augmented, prone to > ME. *-ose*; or + L. suf. *-inus* > E. *-ine*, of or pertaining to): living in sand. Syn. of **arenicolous**.

saccate, **saclike** (NL. *saccatus*, provided with little sacks < L. *saccus*, sack + suf. *-atus* > E. *-ate*, provided with or likeness; *saccus* + E. adj. *like*, similar to or having the characteristics of): **1**. Shaped like a small sack; e.g., many ascomycetes have sack-shaped asci. **2**. Provided with small sacks or bags; e.g., like the surface of the thallus of the lichen *Solorina saccata* (Peltigerales), which has the apothecia sunken in the thallus and with their excipulum concolorous with the thallus, which gives the appearance of small bottles or sacks.

saccate ray (NL. *saccatus*, provided with little sacks < *saccus*, sack, purse + suf. *-atus* > E. *-ate*, provided with or likeness; ME. *raie*, *raye* < OF. *rai* < L. *radius*, radius): *Gasteromycetes*. In *Geastrum saccatum* (Lycoperdales) it refers to rays of the mesoperidium that are turned upside down and split, forming a shallow basin in which the endoperidium of the sporiferous sack is directly seated; in *G. triplex* the mesoperidium breaks and remains completely separate from the saccate rays and from the endoperidium, forming a third layer in the base of the sporiferous sack.

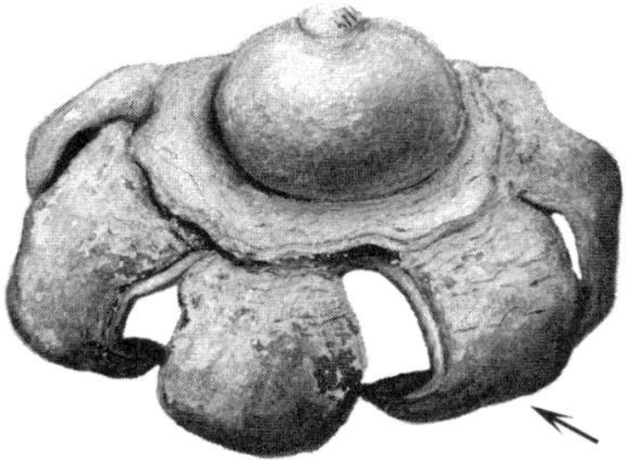

Saccate rays of the exoperidium in the fruiting body of *Geastrum saccatum*. These rays become revolute and form a shallow bowl in which the endoperidium is seated directly on the mesoperidium, x 1.

saccule (L. *sacculus*, small sack < *saccus*, sack + dim. suf. *-ulus* > E. *-ule*): a sacklike, swollen structure, referred to as a sporiferous saccule, at the base of a hypha, which gives rise to the chlamydospore in vesicular-arbuscular mycorrhizal fungi, such as *Acaulospora* and *Entrophospora* (Glomales).

sacculiform (L. *sacculiformis*, with the form of a small sack < *sacculus*, dim. of *saccus*, sack, purse + *-formis* < *forma*, shape): shaped like a sack or small purse; e.g., like the volva of the basidiocarp of *Amanita caesarea* (Agaricales).

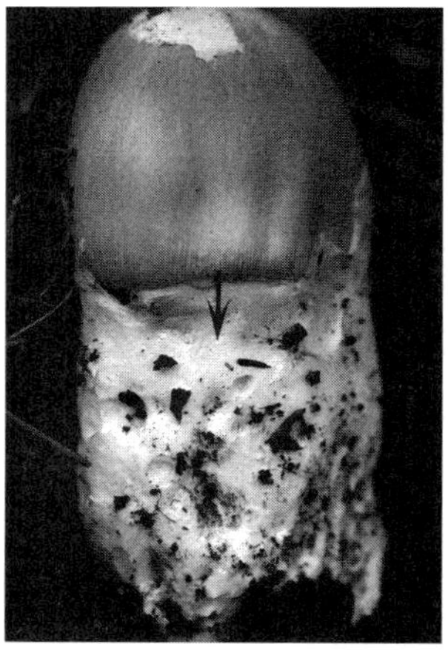

Sacculiform volva of the basidiocarp of *Amanita caesarea*, x 0.6 (*EPS*).

Saccharomycetes (Gr. *sáccharon*, sugar + L. *-mycetes*, ending of class < Gr. *mýkes*, genit. *mýketos*, fungus): a name assigned to the yeasts that have the capacity to ferment sugars, such as *Saccharomyces* (Saccharomycetales), among other gen.

saddle fungus, pl. **fungi** (ME. *sadel* < OE. *sadol*, saddle; L. *fungus*, fungus): the ascoma of *Helvella* spp. (Pezizales) whose pileus is saddle-shaped.

sagenogen (L. *sagena* < Gr. *sagéne*, a seine + Gr. *génos*, engenderment): each one of the organelles from which the ectoplasmic nets of Labyrinthulales are produced. See **bothrosome**.

sagittate (L. *sagitta*, arrow + suf. *-atus* > E. *-ate*, provided with or likeness): having the shape of an arrow, like the Stachel of the encysted zoospore of *Plasmodiophora brassicae* (Plasmodiophorales).

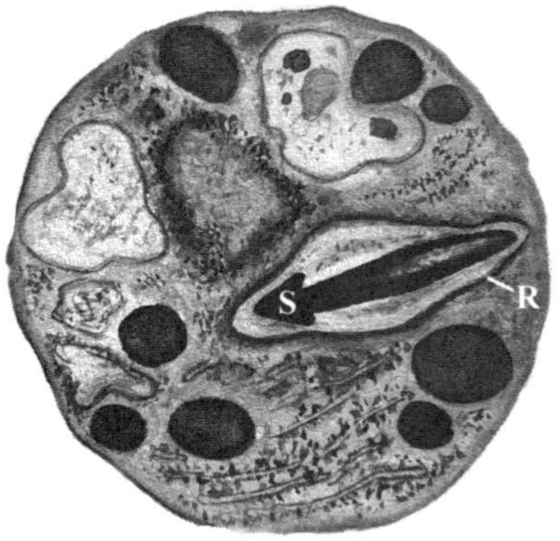

Sagittate organelle (the Stachel, S) lying in an invagination of the cell membrane (the Rohr, R) of an encysted zoospore of *Plasmodiophora brassicae*, seen in longitudinal section. Drawing based on a transmission electron micrograph published by Beckett *et al.* in their *Atlas of Fungal Ultrastructure*, 1974, x 23 800.

sand-case (ME. < OE. *sand*, sand; ME. *cas* < AF. *casse* < OF. *chasse* < L. *capsa*, case, box): applied to the exoperidium of *Disciseda candida* and *Gastrosporium* (Hymenogastrales), which in the mature fructification is composed of loosely intertwined and interwoven hyphae with small particles of sand and gravel; this sandy covering remains at the base, on the substrate, and the endoperidium and the ostiole are exposed in the upper part of the fructification.

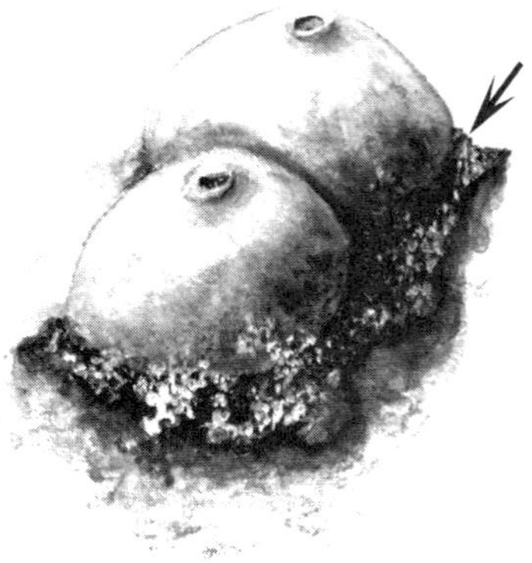

Sand case of the exoperidium in the fruiting bodies of *Disciseda candida*, x 1.2.

saprobe, **saprobiont** (Gr. *saprós*, putrid, rotten + *bíos*, life; *saprós* + *bíos* + *óntos*, genit. of *ón*, a being): an organism that develops on another dead organic being or on organic substances, and utilizes these substrates as food. The majority of the fungi are saprobes. There are **facultative** and **obligate saprobes**.

saprotrophic (Gr. *saprós*, putrid, rotten + *trophós*, something that nourishes, serves as food + suf. *-íkos* > L. *-icus* > E. *-ic*, belonging to, relating to): living and obtaining its food from dead organic material. This term is preferable to saprophyte when referring to a fungus, considering that fungi are not plants.

sarciniform (L. *sarcina*, raceme + *-formis* < *forma*, shape): having the shape of a raceme, like the dictyospores of *Sarcinella* (dematiaceous asexual fungi), asexual state of *Schiffnerula pulchra* (Dothideales).

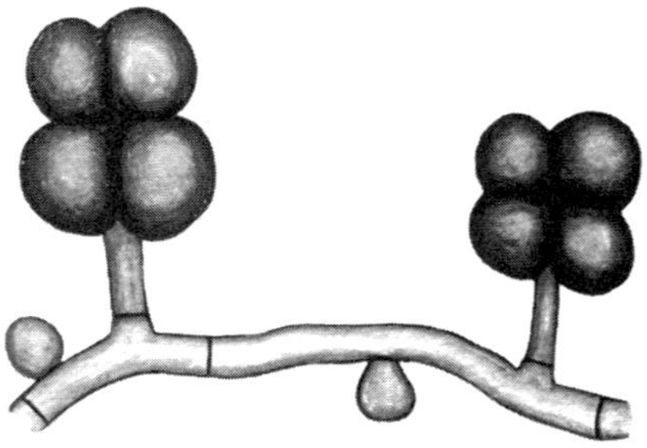

Sarciniform conidia of *Sarcinella*, asexual state of *Schiffnerula pulchra*, x 800.

saturnine (< planet Saturn + L. suf. *-inus* > E. *-ine*, of or pertaining to): with an equatorial flange, like the ascospores of *Debaryomyces* (=*Schwanniomyces*) *occidentalis* (Saccharomycetales).

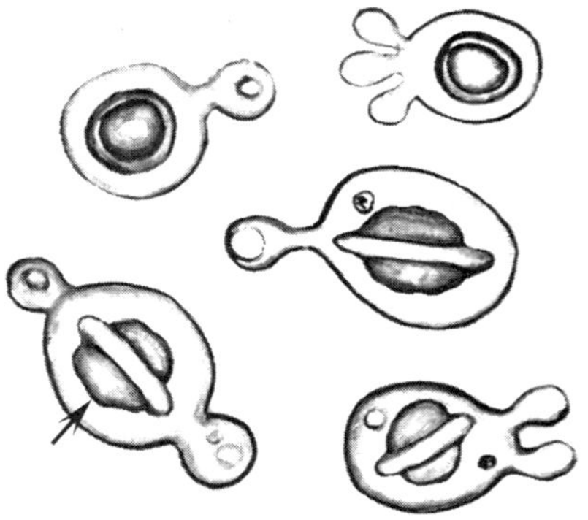

Saturnine ascospores of *Debaryomyces* (=*Schwanniomyces*) *occidentalis*, x 3 700.

saxicoline, **saxicolous** (L. *saxum*, rock, stone + *-cola*, inhabitant + suf. *-inus* > E. *-ine*, of or pertaining to; or + L. *-osus* > OF. *-ous*, *-eus* > E. *-ous*, having, possessing the qualities of): an organism that lives attached to the surface of rocks, as do some lichens (e.g., *Caloplaca saxicola*, of the Teloschistales, and *Rhizocarpon geographicum*, various species of *Parmelia*, *Lecanora calcarea* and *L. frustulosa*, of the Lecanorales, among many other species). Also called **epilithic**, **petricolous** and **rupestral**.

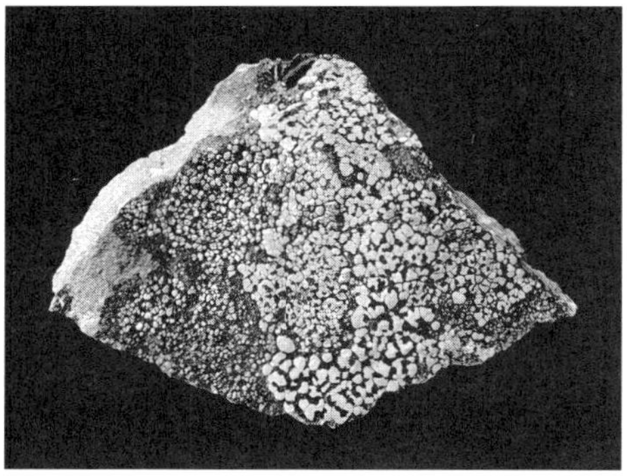

Saxicoline thallus of *Rhizocarpon geographicum*, x 1 (*MU*).

scab (ME., of Scand. origin *skabbr*, scab; akin to OE. *sceabb* and L. *scabies*, mange < *scabere*, to scratch): any of various bacterial or fungal diseases of plants characterized by crustaceous spots (hyperplastic scab-like lesions). *Apple-*, *pear-*, and *cherry-scab* (respectively caused by *Venturia inaequalis*, *V. cerasi* and *V. pirina*, of the Pleosporales). *Scab of cereals* (caused by *Gibberella zeae*, usually *G. saubinetti*, of the Hypocreales, and *Fusarium* spp. of the tuberculariaceous asexual fungi). *Citrus scab* (caused by *Elsinoë fawcettii*, of the Myriangiales). *Peach scab* (caused by *Fusicladium carpophilum*, of the dematiaceous asexual fungi). *Potato scab* (caused by *Spongospora subterranea*, of the Plasmodiophorales).

scabrid, **scabridous** (L. *scabridus*, rough, rugged < *scaber*, fem. *scabra*, rough, scurfy; or + L. *-osus* > OF. *-ous*, *-eus* > E. *-ous*, having, possessing the qualities of): with a rough, ragged, or rugose surface, with irregular, rigid projections. Same as **scabrose**.

scabrose, **scabrous** (L. *scaber*, fem. *scabra*, rough, scurfy + *-osus*, full of, augmented, prone to > ME. *-ose*, or + *-osus* > OF. *-ous*, *-eus* > E. *-ous*, having, possessing the qualities of): rough, with short, rigid projections, which are appreciated well on touch, like the surface of the stipe of *Leccinum scabrum* (=*Boletus scaber*) and *L. aurantiacum* (Agaricales). Also called **scabrid**.

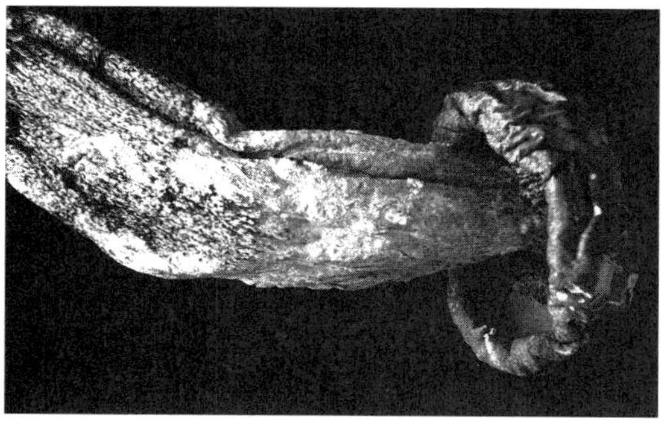

Scabrose stipe of the basidiocarp of *Leccinum scabrum*, x 0.8 (*MU*).

scaly (ME. *scale* < MF. < OF. *escale* < OHG. *skala*, scale, thin platelike structure + E. suf. *-y*, abounding in, full of, having the quality of): having scales on the surface; e.g., like the sporangial peridium of *Lepidoderma tigrinum* (Physarales) and the pileus of the basidiocarp of *Macrolepiota procera* and *Amanita muscaria* (Agaricales).

Scaly peridium of the sporangia of *Lepidoderma tigrinum*, x 35 (*MU*).

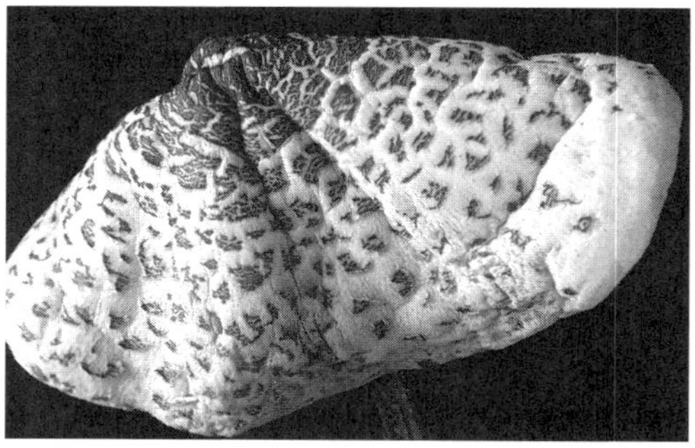

Scaly pileus of the basidiocarp of *Macrolepiota procera*, x 0.7 (*MU*).

scaphoid (Gr. *skáphe*, skiff, sloop + L. suf. *-oide* < Gr. *-oeídes*, similar to): equal to **carinate**, **cymbiform** and **navicular**.

scar (ME. *escare*, *scar* < MF. *escare*, scab < LL. *eschara*, hearth, scab): a mark in the parent cell of a yeast left when the bud (daughter cell) separates from it (*bud scar*); the mark on the daughter cell is the *birth scar*. Also a mark at the conidiogenous locus and conidial base/apex left after secession of conidium.

scariose, **scarious** (ML. *scariosus*, thin, dry, membranous < *escaria*, a spiny shrub < Gr. *schára*, crust, scale + L. *-osus*, full of, augmented, prone to > ME. *-ose*; or + *-osus* > OF. *-ous*, *-eus* > E. *-ous*, having, possessing the qualities of): applied to sheet-like organs that have a membranous consistency and are more or less stiff and dry, generally translucent, like paper; e.g., like the pileus scales of *Naucoria escaroides* (Agaricales).

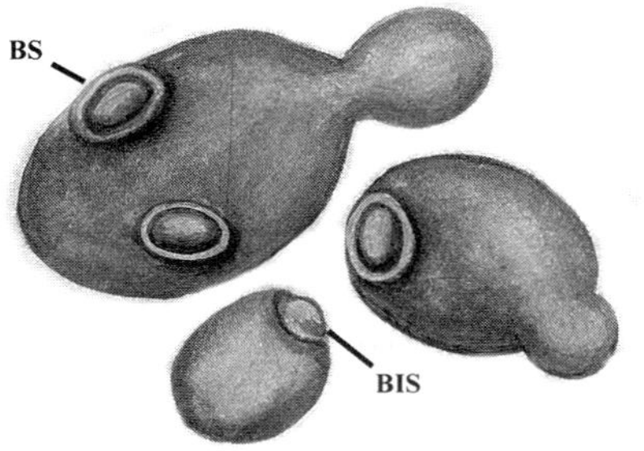

Somatic cells of *Saccharomyces cerevisiae*, showing bud **scars** (BS) and developing buds (B); a daughter cell shows a birth **scar** (BIS), x 3 000.

scarlet cup (ME. *scarlat* < *scarlet* < OF. *escarlate* < ML. *scarlata*, any of various bright reds; ME. *cuppe* < OE. < LL. *cuppa* < L. *cupa*, tub, cup): the ascoma of *Sarcoscypha coccinea* (Pezizales).

Schaeffer reaction (< mycologist J.C. Schaeffer; E. *reaction* < L. *re-*, again, movement in reverse + *acta*, something done < *agere*, to do + *-io*, *-ionis* > E. suf. *-ion*, result of an action, state of): *Agaricales*. The bright orange-red reaction that occurs when tissue is streaked with aniline and crossed with a streak of nitric acid; occurs in *Longula* and Section Arvensis of the gen. *Agaricus*.

schizidium, pl. **schizidia** (NL. *schizidium* < Gr. *schízo*, to split, separate + dim. suf. *-ídion* > L. *-idium*): *Lichens*. A vegetative propagule formed by excision from the upper layers of the lichen thallus, which detaches from the main lobes as segments similar to scales. For example, in *Fulgensia bracteata* subsp. *deformis* (Teloschistales) schizidia are produced as little lobes.

schizogenesis, pl. **schizogeneses** (Gr. *schízo*, to split, divide + suf. *-gén esis*, origin, engenderment, beginning): syn. of **fission** or **bipartition**; cell multiplication by means of simple division; typical of the bacteria and some yeasts, such as *Schizosaccharomyces* (Schizosaccharomycetales).

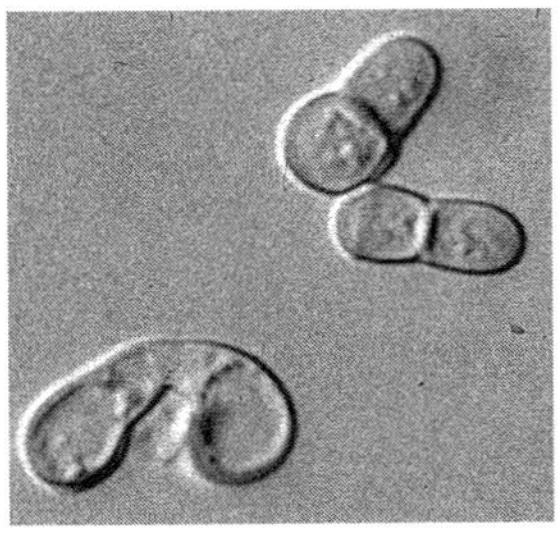

Schizogenesis of the vegetative cells of *Schizosaccharomyces octosporus*, x 850 (*RTH*).

schizogenous (Gr. *schízo*, to split, divide + *génos*, origin < *gennáo*, to engender, produce + L. *-osus* > OF. *-ous*, *-eus* > E. *-ous*, having, possessing the qualities of): see **schizolytic**. Cf. **lysigenous**.

schizolytic (Gr. *schízo*, to split, divide + *lytikós*, able to loosen < *lýtos*, dissolvable, broken + suf. *-íkos* > L. *-icus* > E. *-ic*, belonging to, relating to): syn. of **schizogenous**; a propagule that separates or detaches from the adjacent cell by dissolution of the septum which separates the two. This process is termed **schizolysis** or **schizogenesis**. The blastic conidia of some fungi are examples of schizolytic propagules, which detach by schizolysis. Ostiole formation in ascocarps and pycnidia also is schizogenous. Cf. **rhexolytic**, **rhexolysis**, and **rhexigenesis**.

schizoont (Gr. *schízōon*, *schízōontos* < *schízo*, to split, separate + *óntos*, genit. of *ón*, a being): any structure that divides or separates into two or more portions, but it refers especially to the wall-less, endobiotic, multinucleate, assimilative thallus of the Plasmodiophorales, which undergoes a simple or multiple division to give rise to the sporangia with spores within the cells of the host plant.

scissile (L. *scissilis*, capable of being cut smoothly or split easily < *scinoere*, a cleaving + suf. *-ilis* > E. *-ile*, capable, of the character of): splitting easily; e.g., the flesh of the pileus of some agarics can separate or divide itself into horizontal layers.

sclerobasidium, pl. **sclerobasidia** (Gr. *sklerós*, hard + *basídion* < *básis*, base + suf. dim. *-ídion* > L. *-idium*): *Agaricales*. A basidium with a thick wall (0.5 μm or more); sclerobasidia can be fertile but are more commonly sterile. An example of a fertile sclerobasidium can be found in some species of *Fayodia*, whereas sterile sclerobasidia occur in *Armillaria* (=*Armillariella*) *mellea* and certain species of *Hygrotrama*, all of the Agaricales.

scleroplectenchyma, pl. **scleroplectenchymata** (Gr. *sklerós*, hard + *plektós*, intertwined, braided + *énchyma*, filled, stuffed): a plectenchyma composed of very thick-walled, conglomerated cells, which constitute the **stereoma** or principal supporting tissue of the thallus of certain lichens, such as *Cladonia* and *Alectoria* (Lecanorales). Also applied to hard fungal tissue, like the peridium of the ascocarps of many species of *Leptosphaeria* (Pleosporales), which is composed of large, globose cells with hard, thick walls.

sclerotium, pl. **sclerotia** (NL. *sclerotium* < Gr. *sklerotés*, hardness, stiffness < *sklerós*, hard + L. dim. suf. *-ium*): a hardened structure composed of fungal tissue that resists to unfavorable conditions; sclerotia are composed of plectenchyma, with a firm cortex, frequently chitinized and brown or blackish in color, although they can be other colors; they are capable of germinating and reinitiating vegetative growth. Among the agonomycetaceous asexual fungi, sclerotia can be macroscopic (like those of *Cenococcum graniforme* and *Sclerotium rolfsii*) or microscopic (as in *Rhizoctonia solani*). Among the ascomycetes, the sclerotia of *Sclerotinia* (Helotiales) are also macroscopic; the large sclerotia of *Claviceps gigantea* (Hypocreales) are notable, as are the enormous sclerotia of *Wolfiporia cocos* (Aphyllophorales) in the basidiomycetes. Another example of a macroscopic sclerotium is that of *Coprinus sclerotigenous* (Agaricales).

Sclerotia of *Cenococcum graniforme*, x 20 (*CB & MU*).

Sclerotia of *Sclerotium rolfsii* on agar, x 10 (*MU*).

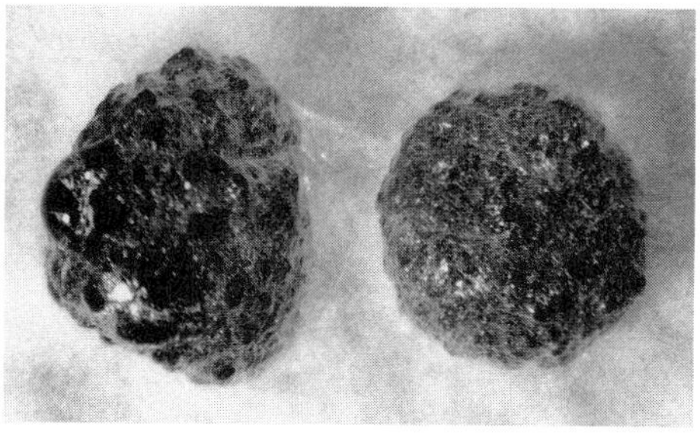

Sclerotia of *Sclerotinia* sp. on agar, x 1.6 (*MU*).

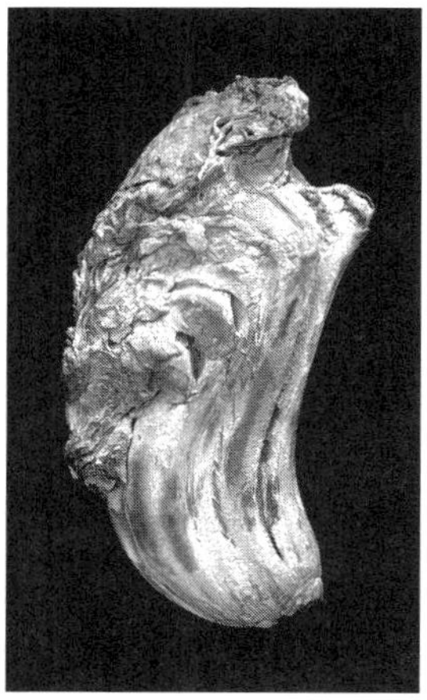

Sclerotium of *Claviceps gigantea*, x 2 (*MU*).

Sclerotium of *Wolfiporia cocos*, x 2 (*RTH*).

scobiform (NL. *scobiformis* < L. *scobs*, genit. *scobis*, filing, sawdust, scrapings, dust + *-formis* < *forma*, shape): having the appearance of filings or small grains, like those of sawdust; e.g., as is seen on the surface of the peridium of some Lycoperdaceae.

scolecospore (Gr. *skolékos*, genit. of *skólex*, worm + *sporá*, spore): an elongate, vermiform or filiform spore, such as the ß-conidia of *Phomopsis* (sphaeropsidaceous asexual fungi). Used in Saccardoan terminology.

Scolecospore of *Cercospora coffeicola*, x 1 250 (*CB*).

scrobiculate (L. *scrobiculatus*, having holes or ditches < *scrobiculus*, dim. of *scrobs*, hole, trench + suf. *-atus* > E. *-ate*, provided with or likeness): syn. of **foveolate**; having pits or tapers, arranged densely and regularly so that the projecting parts constitute a kind of reticulum; e.g., like the thallus of some lichens (*Lobaria scrobiculata*, of the Peltigerales) and the stipe of *Lactarius scrobiculatus* and *L. pallidus* (Agaricales).

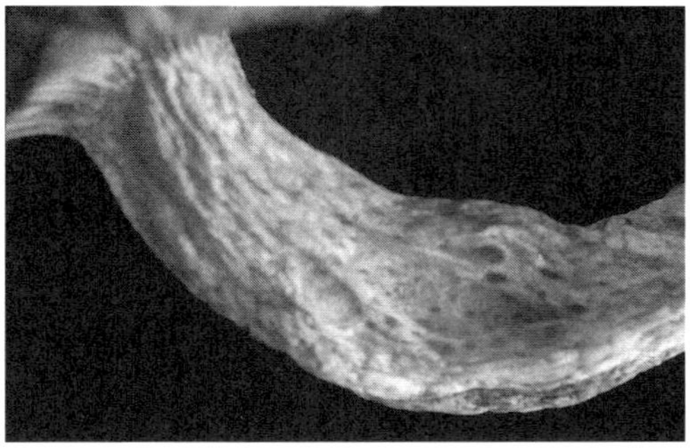

Scrobiculate stipe of the basidiocarp of *Lactarius scrobiculatus*, x 1 (*MU*).

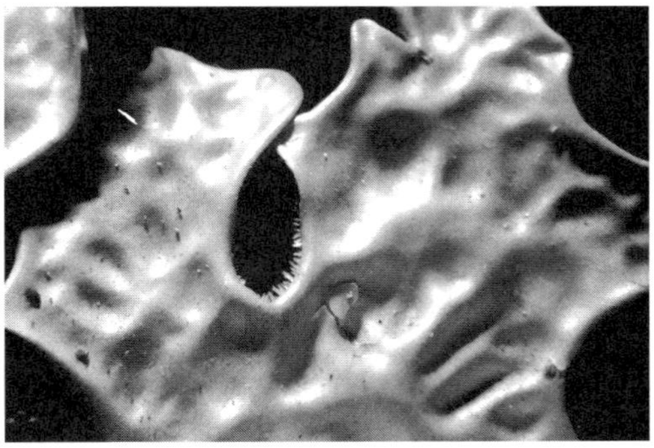

Scrobiculate thallus of *Lobaria scrobiculata*, x 15 (*MU*).

scrupose (L. *scruposus*, jagged, rough < *scrupus*, a small sharp stone + *-osus*, full of, augmented, prone to > ME. *-ose*): covered with small, sharp, hard roughenings; e.g., like the pileus of *Lepiota histrix* (Agaricales), whose scales are bristly.

scutate, **scutated** (L. *scutate*, shielded < *scutum*, an oblong shield + suf. *-atus* > E. *-ate*, provided with or likeness): syn. of **clypeate**, **peltate** and **scutiform**.

scutellum, pl. **scutella** (L. *scutellum* < *scutum*, an oblong shield + dim. suf. *-ellum*): the shield-shaped upper portion of a dimidiate ascostroma, which at maturity covers the ascigerous locule or locules; as in Loculoascomycetes of the order Dothideales, e.g., *Microthyrium microscopicum*.

scutiform (L. *scutiformis* < *scutum*, an oblong shield + *-formis* < *forma*, shape): having the shape of a shield. See **clypeate**, **peltate** and **scutate**.

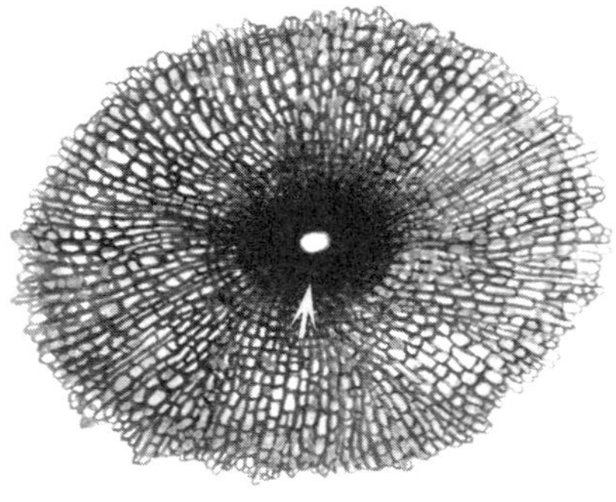

Scutellum (dark area) around the ostiole of a thyriothecium of *Microthyrium microscopicum*, drawn in a surface view, x 350.

scutulum, pl. **scutula** (NL. *scutulum*, dim. of L. *scutum*, an oblong shield + dim. suf. *-ulum*): a cup-like crust or patch of hyphae produced in the follicles of the scalp or the body in the infections caused by *Trichophyton schoenleinii* (moniliaceous asexual fungi).

scyphus, pl. **scyphi** (L. *scyphus* < Gr. *skýphos*, drinking vessel, cup): widening in the manner of a glass or goblet; e.g., like the apex of the podetia of certain lichens, such as *Cladonia fimbriata* and *C. chlorophaea* (Lecanorales).

Scyphi of the podetia of the thallus of *Cladonia chlorophaea*, x 10.

secondary (L. *secundarius* < *secundus*, second in order or grade, following + suf. *-arius*, belonging to + E. suf. *-y*, having the quality of): of or relating to the second order or stage in a series. **Secondary hyphae**: another term for the setose hyphae formed on the mycelium of *Blumeria* (=*Erysiphe*), of the Erysiphales. **Secondary mycelium**: *Basidiomycetes.* Dikaryotic mycelium that results from the plasmogamy of the primary mycelium, as *in Coprinus lagopus* (Agaricales). It is characterized by having clamp connections and by its assimilative function; it does not give rise to a fruiting body, as happens in the ascomycetes. See **primary mycelium** and **tertiary mycelium**. **Secondary spore**: *Basidiomycetes.* A spore that is not formed on basidia, but is a conidium, chlamydospore, etc., which is produced directly on the assimilative mycelium or on a hypha of the fruiting body. Also refers to a conidium formed on a previous spore. **Secondary thallus**: *Lichens*. The podetia that are developed on the squamulose primary thallus (Lecanorales).

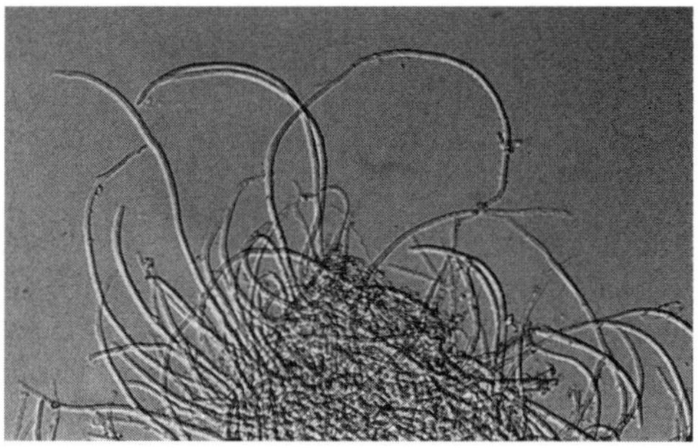

Secondary hyphae of the ascocarp of *Blumeria* (=*Erysiphe*) sp., x 100 (*RTH*).

secondary appendage (L. *secundarius* < *secundus*, second, following + suf. *-arius*, belonging to; *appendage* < *appendere*, to append): *Laboulbeniales*. An outgrowth that develops from a receptacle cell, which in turn derives from the basal cell of the ascospore; e.g., *Laboulbenia*. Cf. **primary appendage**.

secondary hypha, pl. **hyphae** (L. *secundarius* < *secundus*, second, following + suf. *-arius*, belonging to; NL. *hypha* < Gr. *hyphé*, tissue, spider web; hypha): see **secondary**.

secondary marine fungus, pl. **fungi** (L. *secundarius* < *secundus*, second, following + suf. *-arius*, belonging to + E. suf. *-y*, having the quality of; L. *marinus*, of the sea < *mare*, sea + suf. *-inus* > E. *-ine*, of or pertaining to; L. *fungus*, fungus < Gr. *spóngos*, *sphóngos*, sponge): a species derived from a terrestrial ancestor and which secondarily emigrated to the marine environment, as some Loculoascomycetes have done, among others. Cf. **primary marine fungus**.

secondary mycelium, pl. **mycelia** (L. *secundarius* < *secundus*, second, following + suf. *-arius*, belonging to + E. suf. *-y*, having the quality of; NL. *mycelium* < Gr. *mýkes*, fungus + L. *-elis*, pertaining to + dim. suf. *-ium*): see **secondary**.

secondary spore: (L. *secundarius* < *secundus*, second, following + suf. *-arius*, belonging to + E. suf. *-y*, having the quality of; Gr. *sporá*, spore): see **secondary**.

secondary thallus, pl. **thalli** (L. *secundarius* < *secundus*, second, following + suf. *-arius*, belonging to + E. suf. *-y*, having the quality of; L. *thallus* < Gr. *thallós*, sprout, thallus): see **secondary**.

secotioid (< gen. *Secotium* < L. *secare*, to cut + dim. suf. *-ium* + L. suf. *-oide* < Gr. *-oeídes*, similar to): a term applied to an epigeous fruiting body with the appearance of an unopened agaric or bolete, generally with a stipe-columella and a gleba often composed of contorted lamellae, as in *Secotium*, *Truncocolumella* and *Podaxis* (Hymenogastrales); strongly correlated with orthotropic spore development, but not necessarily with obvious affinities to known agarics.

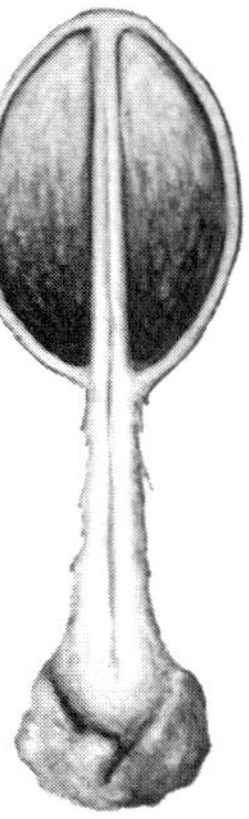

Secotioid type of basidiocarp, represented by *Podaxis pistillaris*, seen in longitudinal section, x 0.5.

sectoring (E. *sectoring*, to divide into or furnish with sections < L. *section* < *sectus*, ptp. of *secare*, to cut) : the formation of morphologically differentiated regions of a fungus colony, known as sectors, which contain different nuclear populations as a result of heterokaryosis during parasexual recombination of genes.

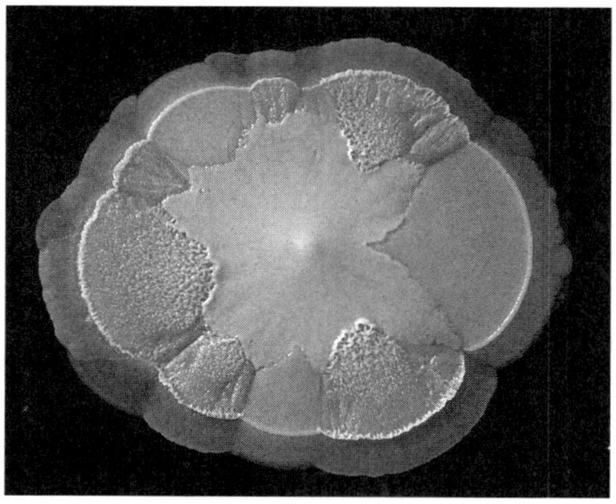

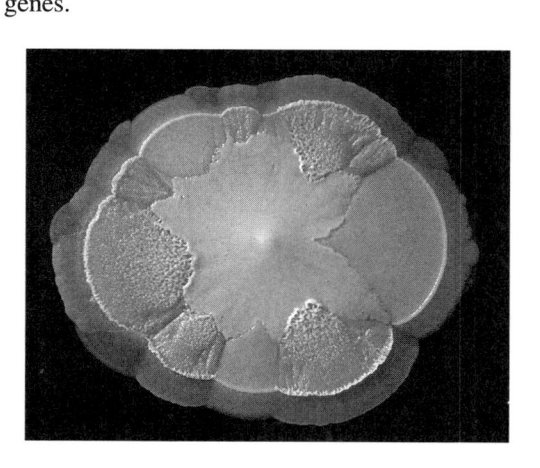

Sectoring of a giant colony of a yeast on agar, x 1 (*MU*).

self-inhibitor (AS. *self*, same, of or pertaining to one's self; one's own; ME. *inhibiten* < L. *inhibitus* pp. of *inhibere*, to prohibit from doing something; restrain):

a fungal metabolite which imposes dormancy to spores whenever unfavorable conditions to germinate are present. High concentrations of spores fail to germinate; this phenomenon is thought to be a direct effect of the presence of fungal self-inhibitors. Dilution of spore suspensions is one of the strategies used to overcome self-inhibition. See **mycosporine**.

semifissitunicate (L. *semis*, half + *fissio*, partition, cleaving + *tunicatus*, covered with a tunic < *tunica*, tunica, covering + suf. *-atus* > E. *-ate*, provided with or likeness): a type of ascus dehiscence, characteristic of the Arthoniales, that is distinguished by lack of complete separation of the two ascus walls during discharge; only the upper part of the inner wall separates from the outer wall to become everted in dehiscence. This and other variants may occur in members of Loculoascomycetes. Cf. **fissitunicate** and **rostrate dehiscence**.

seminicole, **seminicolous** (L. *semen*, genit. *seminis*, seed + *-cola*, inhabitant; or + L. *-osus* > OF. *-ous*, *-eus* > E. *-ous*, having, possessing the qualities of): living in seeds.

senescense (L. *senescentia* < *senescere*, to grow old): the action and effect of growing old; in certain fungi there is a degeneration that makes its propagation difficult or impossible, as is observed, e.g., in *Neurospora* (Sordariales).

sensus, pl. **sensi** (L. *sensus*, meaning of a concept): *sensu lato* (L. *latus*, extended), *lato sensu* or *sensu amplo* (L. *amplus*, broad), *amplo sensu*: in a broad sense, i.e., having a broad circumscription of a taxon, such as a genus or species. *Sensu stricto* (L. *strictus*, strict, restricted) or *stricto sensu*: circumscribing a taxon narrowly, in a strict sense. For example, *Melanospora* (Melanosporales) sensu lato includes species with several different types of ascospores, whereas *Melanospora* sensu stricto is limited to species with smooth ascospores with depressed germ pores; species with other types of ascospores are placed in other gen.

separable (L. *separ*, separate, different, disjoined + *-abilis* > E. *-able*, tendency toward, able to be, that may be, fit to be): *Agaricales*. Refers to a gill that at first is adnate to the stipe, but later separates from it; or a hymenophore that separates from the context in some boletes.

septal pore (E. *septal*, adj. of L. *septum*, barrier, partition; L. *porus* < Gr. *póros*, passage): a pore in the septum that separates one cell from another, either in a somatic or sporogenous hypha, which allows the free passage of cytoplasm and organelles, including nuclei. See **septum**. Cf. **dolipore**.

septal pore cap (L. *septum*, barrier, partition + L. suf. *-alis* > E. *-al*, relating to or belonging to; L. *porus* < Gr. *póros*, passage; LL. *cappa*, hooded cloak, cap): see **parenthesome**.

septal pore organelle (L. *septum*, barrier, partition + L. suf. *-alis* > E. *-al*, relating to or belonging to; L. *porus* < Gr. *póros*, passage; NL. *organella* < L. *organum*, a specialized cellular part that is analogous to an organ + dim. suf. *-ella*): a membrane-bound structure, often shaped like a pulley wheel, that may plug a septal pore. Septal pore organelles are more complex than Woronin bodies, which may also plug septal pores, occur in some filamentous ascomycetes, and are distributed in parts of the mycelium so that structures involved in sexual reproduction are isolated from other regions of the mycelium, e.g., *Barssia oregonensis* and *Acervus episparticus* (Pezizales). Cf. **Woronin body**.

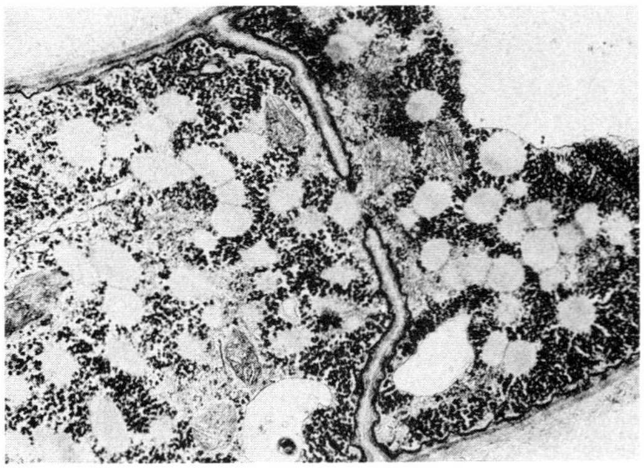

Septal pore in a hypha of an ascomycete, seen in longitudinal section, x 10 000 (transmission electron micrograph of *RTH*).

septate (L. *septatus*, divided by or having a septum < *septum*, barrier, partition + suf. *-atus* > E. *-ate*, provided with or likeness): with septa or partitions, as in hyphae of the majority of the fungi. Cf. **coenocytic** or **aseptate**.

septic, **septicous** (L. *septicus* < Gr. *septikós*, causing putrefaction < *sépsis*, decay + *-tikós* > L. *-ticus* > E. *-tic*, relation, fitness, inclination or ability + L. *-osus* > OF. *-ous*, *-eus* > E. *-ous*, having, possessing the qualities of): pertaining to or relative to sepsis. Cf. **aseptic**.

septum, pl. **septa** (L. *septum*, barrier, partition): a transverse wall in a cell or a hypha. The septa or partitions are formed by centripetal growth of the cell wall and are present with a certain spatial regularity in a septate mycelium. **Primary septum**: a septum that is formed in direct association with nuclear division (by constriction, mitosis or meiosis), which separates the daughter cells and is provided with a pore, which can be modified like a dolipore (in basidiomycetes) or be associated with Woronin bodies (in ascomycetes and their asexual states). **Adventitious septum**: a septum that is not formed in association with nuclear division, e.g., as when a hypha is broken naturally or artifically, and the formation of the septum impedes the loss of protoplasm.

sequestrate (LL. *sequestratus*, ptp. of *sequestrare*, sequester < L. *sequester*, agent, depositary, bailee; to set apart, segregate + suf. *-atus* > E. *-ate*, provided with or likeness): taxa of fungi which have evolved from having exposed hymenia and forcibly discharged spores to a closed or even hypogeous fruiting body habit. Many cases are recognized as being derived from specific spore-discharging ancestors, e.g., *Rhizopogon* (Hymenogastrales) from *Suillus* (Agaricales), whose spores are retained in the fruiting bodies until they decay or are eaten by an animal vector.

sericeous (L. *sericeus*, silken < *ser*, genit. *seris*, silk + *-aceus*, of or pertaining to, with the nature of): silken; with thin, white, shiny filaments that are arranged in one direction; e.g., like those on the base of the stipe of *Rhodophyllus sericeus* (Agaricales) which is sericeous when dry.

serrate (LL. *serratus* < ptp. of *serrare*, to saw + L. suf. *-atus* > E. *-ate*, provided with or likeness): with small, sharp teeth near the edge; generally applied to foliaceous or laminar structures, such as the gills of *Lentinellus* species (Agaricales).

Serrate lamellae of the basidiocarp of *Lentinellus* sp., x 3.

serrulate, **serrulated** (LL. *serrulatus*, dim. of *serratus*, toothed < ptp. of *serrare*, to saw + L. suf. *-atus* > E. *-ate*, provided with or likeness): toothed, but with smaller teeth, like the border of the gills of *Leptonia serrulata* (Agaricales).

sessile (L. *sessilis*, of or belonging to sitting < *sessor*, seated + suf. *-ilis* > E. *-ile*, capable, of the character of): refers to a fruiting body that is attached directly to the surface of the substrate, and lacking a foot or stalk (stipe), e.g., the sporangia of *Perichaena* (Trichiales). Same as **apodal**. Cf. **pedicellate** and **stipitate**.

Sessile sporangia of *Perichaena* sp., x 15 (*MU*).

seta, pl. **setae** (L. *seta*, bristle): refers to stiff hairs or bristles, usually thick-walled and tapered to a point, that occur on the fruiting bodies of various fungi, such as the acervuli of *Colletotrichum* (melanconiaceous asexual fungi), the pycnidia of *Pyrenochaeta*, *Chaetoseptoria* and *Chaetodiplodia caulina* (sphaeropsidaceous asexual fungi), the mycelium of *Meliola* (Meliolales), or the perithecia of *Podospora* (Sordariales). Also called setae are the bristles that originate in the hymenium of *Dennisiella babingtonii* (Dothideales), as well as those produced in the context and hymenium (hymenial setae) of many species of Aphyllophorales, e.g., *Phellinus ferruginosus*.

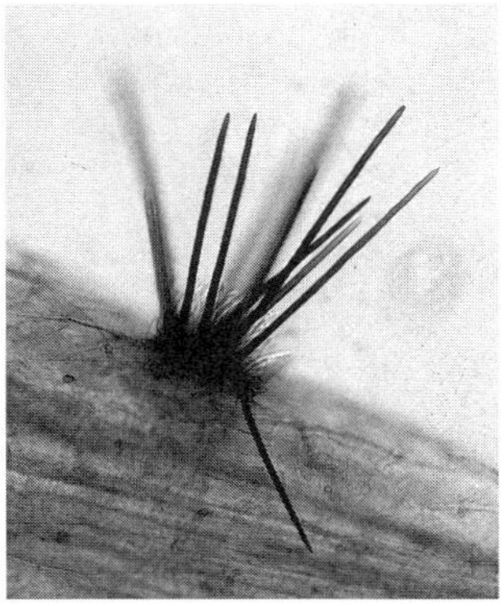

Setae of the acervulus of *Colletotrichum circinans*, x 125 (*MU*).

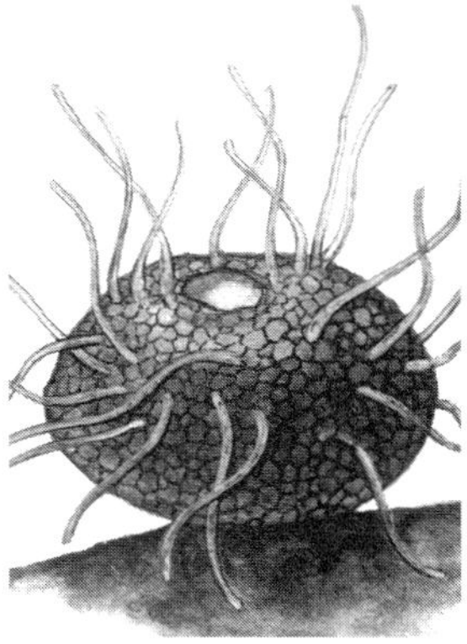

Setae of the pycnidium of *Chaetodiplodia caulina*, x 320.

Setae originating from pigmented cells in the hymenium of *Dennisiella babingtonii*, x 175.

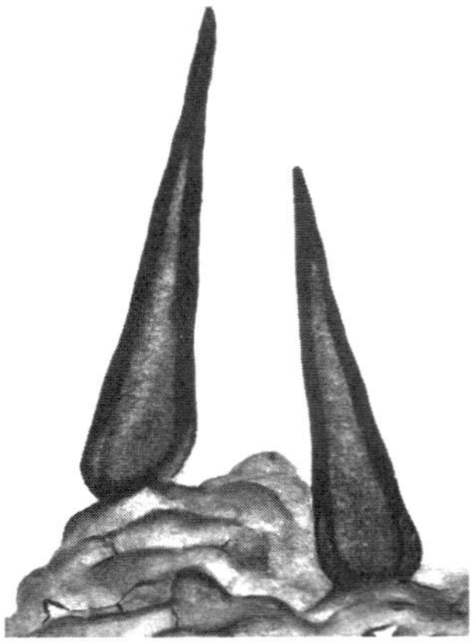

Setae of the hymenium of *Phellinus ferruginosus*, x 370.

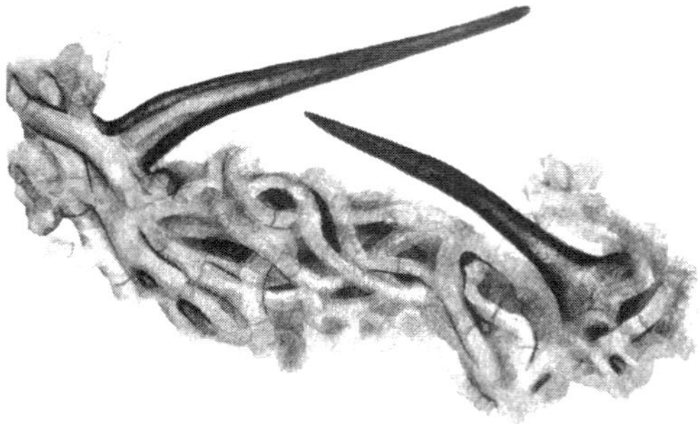

Setae of the context in the margin of the pileus of *Phellinus ferruginosus*, x 350.

setaceous (L. *setaceus*, like a seta < *seta*, bristle + suf. *-aceus* > E. *-aceous*, of or pertaining to, with the nature of): having stout, pointed, usually thick-walled hairs, such as those around the edge of the cup in *Cyathus* (Nidulariales).

seticole, **seticolous** (L. *seta*, bristle, mane + *-cola*, inhabitant < *colere*, to inhabit; or + L. *-osus* > OF. *-ous*, *-eus* > E. *-ous*, having, possessing the qualities of): living or developing on bristles or setae of the host organism; e.g., like the thalli of *Rickia passalina* (Laboulbeniales) which live attached to the setae of *Chondrocephalus debilis*, a coleopteran in the passalid family.

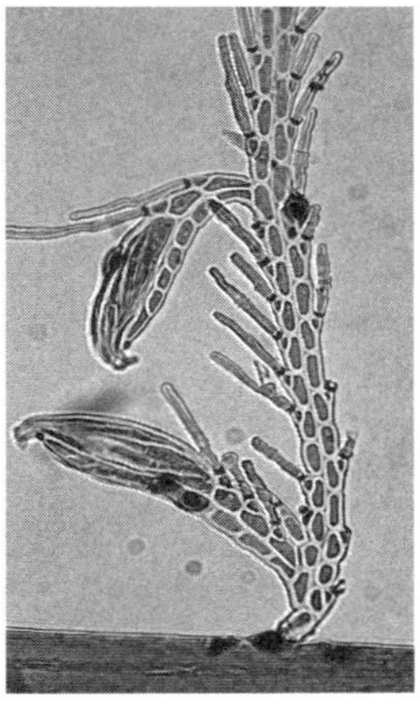

Seticole thallus of *Rickia passalina* on a host (coleopteran) body seta, x 140 (*HL*).

setiform (L. *setiformis* < *seta*, bristle, mane + *-formis* < *forma*, shape): having the shape of a seta; bristle-like.

setigerous (L. *seta*, bristle, mane + E. suf. *-gerous* < L. *ger*, to bear, carry + *-osus* > OF. *-ous*, *-eus* > E. *-ous*, having, possessing the qualities of): bearing setae.

setose, **setous** (L. *setosus*, having setae < *seta*, bristle, mane + *-osus*, full of, augmented, prone to > ME. *-ose*; or + *-osus* > OF. *-ous*, *-eus* > E. *-ous*, having, possessing the qualities of): **1**. Provided with setae or stiff hairs; e.g., like the acervuli of *Colletotrichum* (melanconiaceous asexual fungi) and the perithecia of *Chaetomidium heterotrichum* (Sordariales). **2**. Having the shape of a seta or bristle (**setiform**), i.e., a setose hair; e.g., the lamprocystidia of *Hohenbuehelia niger* (Agaricales) are setiform.

Setose perithecium of *Chaetomidium heterotrichum*, x 650.

setose exoperidium, pl. **exoperidia** (L. *setosus*, having setae < *seta*, bristle, mane + *-osus*, full of, augmented, prone to > ME. *-ose*; Gr. *éxo*, outside + *péridion*, small leather purse): an exoperidium in which some cells are transformed into setae; typical of some species of *Bovista* and *Lycoperdon* (Lycoperdales).

setula, pl. **setulae** (L. *setula*, dim. of *seta*, bristle, mane): a very fine seta or appendage, like those of the conidia of *Pestalotia* (melanconiaceous asexual fungi). Also called setulae are the bristles with a thick wall and dark brown color that some cystidia (of the trama) have in certain Hymenomycetes, in particular Aphyllophorales (they do not exist in agarics).

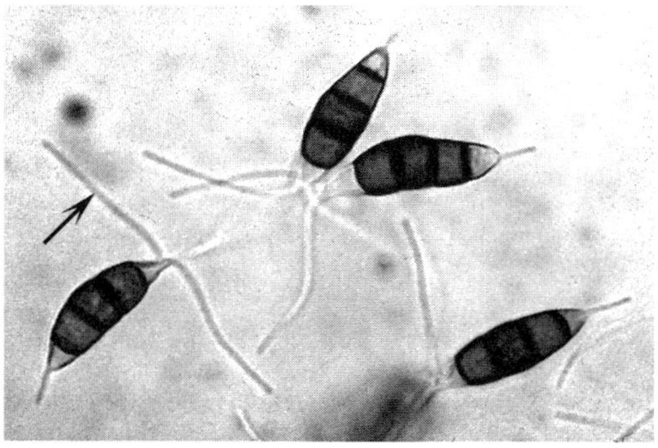

Setulae of the conidia of *Pestalotia* sp., x 1 000 (*MU*).

setuloid (L. *setula*, dim. of *seta*, bristle + suf. *-oide* < Gr. *-oeídes*, similar to): see **setuloid cystidium**.

setuloid cystidium, pl. **cystidia** (L. *setula*, dim. of *seta*, very fine bristle + suf. *-oide* < Gr. *-oeídes*, similar to; NL. *cystidium* < Gr. *kystídion* < *kýstis*, bag, bladder, cell + dim. suf. *ídion* > L. *-idium*): a **leptocystidium** with a thick, pigmented wall, present in the hymenium of some Agaricales.

sewage fungus, pl. **fungi** (E. *sewage*, refuse liquids or waste matter carried off by sewers < ME. < MF. *esseweur*, *seweur* < *esewer*, to drain < VL. *exaquare* < *ex*, out of + *aqua*, water; L. *fungus*, fungus): *Leptomitus lacteus* (Leptomitales), commonly found in polluted water, sometimes clogging sewage filters.

sexual (L. *sexualis*, pertaining or relative to sex < *sexus*, sex + suf. *-alis* > E. *-al*, relating to or belonging to): applied to a nucleus, cell, organ, process or phase pertaining to sex. For example, sexual reproduction is that involving plasmogamy, karyogamy, and meiosis; and a sexual phase is the portion of a fungal life cycle during which sexual reproduction occurs. Cf. **asexual**.

sexual phase (L. *sexualis*, relative to sex < *sexus*, sex + suf. *-alis* > E. *-al*, relating to or belonging to; Gr. *phásis* < *phaíno*, to shine): the phase or stage in the life cycle in which are formed sexually reproductive cells or organs. In the fungi it is more appropriate to speak of a sexual phase than of sex because fungi do not always have true sex, as occurs in other organisms, i.e., with clearly differentiated male and female cells or organs.

sheath (ME. *scheth* < OE. *sceath*, sheath): **1**. The hyaline, gelatinous covering that surrounds the ascospores of some ascomycetes, e.g., *Sordaria* (Sordariales); also called **capsule**. **2**. *Uredinales*. The structure that separates the host cell cytoplasm from the cell wall of the haustorium; this envelope or covering is electron transparent and surrounds the haustorial body, but not the haustorial neck. It is observed in plant pathogenic fungi, such as *Melampsora*. **3**. *Nidulariales*. The structure on the fruiting body to which the middle piece is attached, as in species of *Cyathus*. Also called **basal piece**.

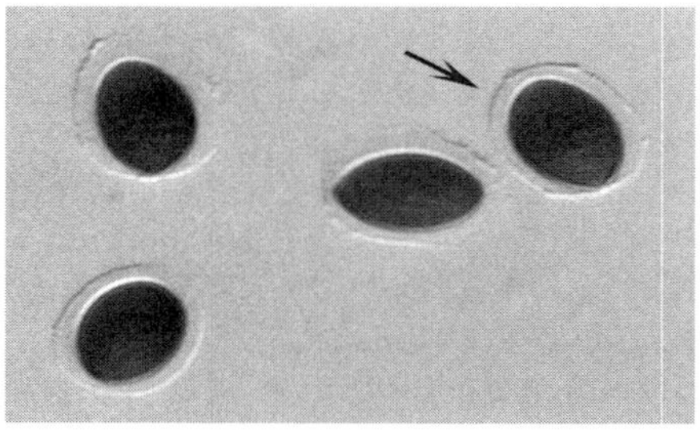

Sheath of the ascospores of *Sordaria fimicola*, x 680 (*RTH*).

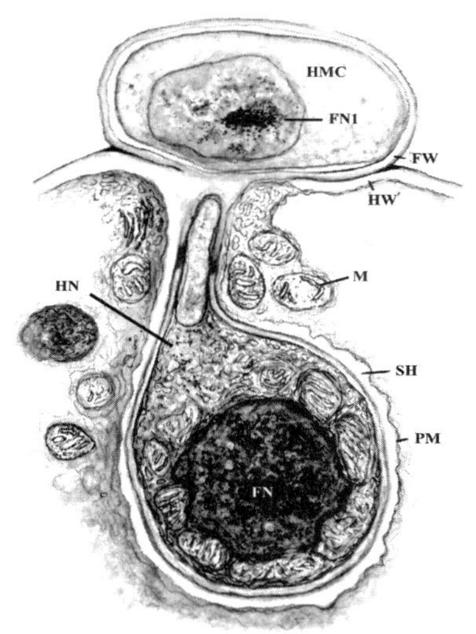

Sheath (SH) of *Melampsora lini*. Longitudinal non-median section through the haustorial mother cell (HMC), at the point where it penetrates the cell wall of its host (flax, *Linum usitatissimum*). The fungus wall (FW) is in contact with the host cell wall (HW). The haustorium and haustorial neck (HN) are surrounded by the invaginated host plasma membrane (PM). One fungal nucleus (FN) is in the haustorium, and the other fungal nucleus (FN1) appears to be partly in the haustorial neck and partly in the haustorial mother cell. The host cell mitochondria (M) are also shown. Drawing based on a transmission electron micrograph published by Beckett *et al.* in their *Atlas of Fungal Ultrastructure*, 1974, x 9 800.

shield-shaped (ME. *shelde* < OE. *sceld* < G. *Schild*, a protective piece of armor, shield; E. adj. *shaped*, ptp. of *to shape* < ME. *shap* < OE. *gesceap* < *sceapen*, ptp. of *sceppan* < G. *shaffen*, to make): having the shape of a shield. See **peltate**, **scutate** and **scutiform**.

shoestring fungus, pl. **fungi** (ME. *shoo* < OE. *schoh*, shoe + ME. < OE. *streng*, to bind tight; shoelace + L. *fungus*, fungus): the honey agaric, *Armillaria mellea* (Agaricales), especially refering to its rhizomorphs which resemble shoestrings.

shot-hole (ME. < OE. *scot*; akin to ON. *skot*, shot; ME. < OE. *hol*, hollow, hole): a plant disease characterized by leaf spots and holes made by the dead parts dropping out, e.g., shot-hole of peach leaves, caused by *Otthia* (=*Stigmina*) *carpophila*, of the Dothideales.

shoyu (Japanese *shoyu* < Chin. *shi-yau*, soybean oil): an oriental sauce of soybeans and wheat obtained by a double fermentation, one that depends on the enzymatic activities (proteolytic and lipolytic) of *Aspergillus oryzae* and *A. soyae* (moniliaceous asexual fungi), and another carried out by salt-tolerant yeasts (*Zygosaccharomyces soya* and *Z. major*, of the Saccharomycetales), and salt-tolerant lactic acid bacteria (principally *Pediococcus cerevisiae* and *Lactobacillus delbrueckii*).

sicyoid, **sicyodic** (Gr. *síkyos*, gourd + L. suf. *-oide* < Gr. *-oeídes*, similar to; or + Gr. suf. *-íkos* > L. *-icus* > E. *-ic*, belonging to, relating to): see **lageniform**, **utriculose** and **ventricose-rostrate**.

side body complex (ME. < OE. *side*, side; ME. < OE. *bodig*, body; LL. *complexus*, totality, complex): *Blastocladiales*. A complex structure, found in zoosporic fungi such as *Blastocladia*, *Blastocladiella* and *Allomyces*, which consists of a double membrane system associated with microbodies and lipid globules, and whose function is unknown. The complex is located just below the plasma membrane of the zoospore and planogamete, near the posterior end of the cell; it is reminiscent of the microbody-lipid globule complex found in zoospores of the Chytridiales and Monoblepharidales, although it lacks the associated rumposome.

siderophore (MF. < L. < Gr. *síderos* < *sidéreos*, iron + *-phóros*, bearer): a metabolic product of a fungus (or other microorganism) which binds iron and allows its translocation from the environment into the microbial cell.

sigmoid (Gr. *sigmoeidés* < *sígma*, or ζ, 18th letter of the Gr. alphabet corresponding to "s" + L. suf. *-oide* < Gr. *-oeídes*, similar to): curved twice; shaped like an S; e.g., such as the basidiospores of *Clavaria* species (Aphyllophorales) and some of the Hülle cells of *Aspergillus puniceus* (moniliaceous asexual fungi).

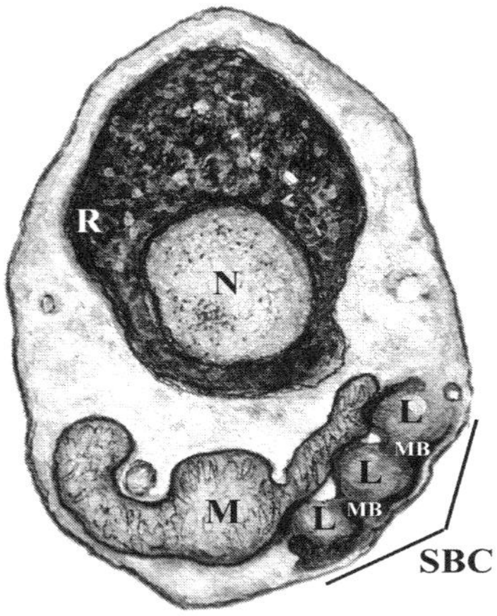

Side body complex (SBC) of a mature zoospore of *Blastocladiella emersonii*, seen in longitudinal section. The ribosomes (R) are enclosed by a pair of membranes. One large, lobed mitochondrion (M) is present at the posterior end of the zoospore, and the lipid droplets (L) and microbodies (MB) are aggregated into a side body complex. Drawing based on a transmission electron micrograph published by Beckett *et al.* in their *Atlas of Fungal Ultrastructure*, 1974, x 10 280.

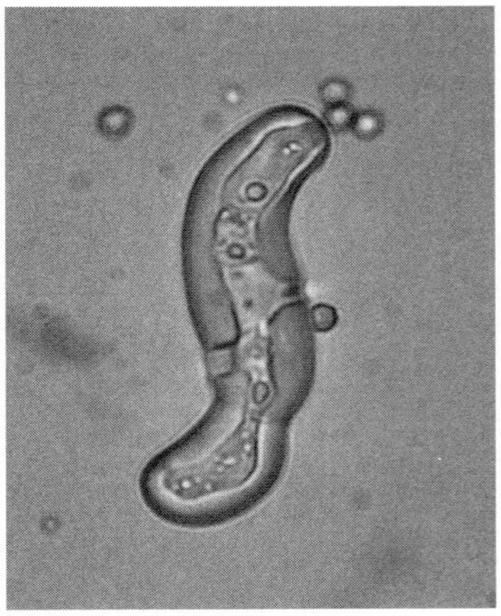

Sigmoid Hülle cell of *Aspergillus puniceus*, x 340 (*MU*).

sikospore or **sikyotic spore** (Gr. *síkyos*, gourd + *sporá*, spore; or *síkyos* + *-tikós* > L. *-ticus* > E. *-tic*, relation, fitness, inclination or ability): the spore formed from the **sikyotic cell** of *Parasitella simplex* (Mucorales), which serves to anchor its hyphal terminal cell to the host hypha (*Absidia glauca*, of the Mucorales).

sikyotic cell (Gr. *síkyos*, gourd + Gr. *-tikós* > L. *-ticus* > E. *-tic*, relation, fitness, inclination or ability; NL. *cellula*, cell, living cell < dim. of L. *cell*, small room): the terminal cell of a *Parasitella simplex* (Mucorales) hypha by which this parasite anchors itself to its host (*Absidia glauca*, also of the Mucorales). After some period of time, this organ differentiates into a **sikospore**.

siliquiform (L. *siliquiformis* < *siliqua*, an elongate seed pod + *-formis* < *forma*, shape): similar to a silique, long and slender; such as the sporangia of the aquatic fungus *Blastocladia pringsheimii* (Blastocladiales).

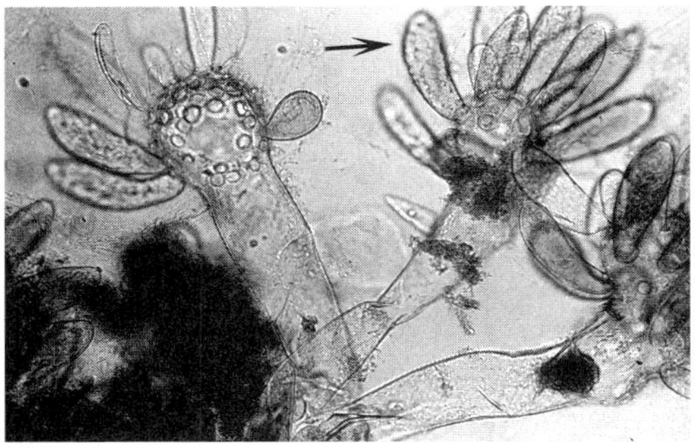

Siliquiform zoosporangia of *Blastocladia pringsheimii*, x 140 (*MU*).

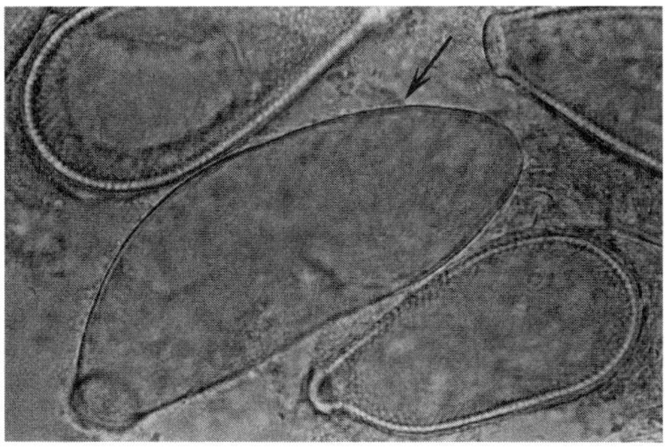

Siliquiform zoosporangium of *Blastocladia pringsheimii*, x 660 (*MU*).

simblospore (Gr. *símblos*, swarm of bees + *sporá*, spore): a flagellated, swimming spore; same as **zoospore**.

sinuate (L. *sinuatus*, bent, curved < *sinnus*, curve + suf. *-atus* > E. *-ate*, provided with or likeness): having a sinus; wavy, undulating or with indentations or furrows. Generally applied to the type of gill insertion on the stipe, when a notch or sinus is present in the area next to the point of insertion, such as occurs in the gen. *Agrocybe* (Agaricales). Cf. **adnate**, **decurrent** and **free**. Also applied to the edge of the giant colony of some yeasts when grown on solid media.

sinuose, **sinuous** (L. *sinuosus*, tortuous, serpentine, twisted < *sinus*, curve + *-osus*, full of, augmented, prone to > ME. *-ose*; or + *-osus* > OF. *-ous*, *-eus* > E. *-ous*, having, possessing the qualities of): with many undulations, like the plasmodiocarp of *Physarum sinuosum* (Physarales).

sirenin (LL. *Sirena* < L. *Siren* < Gr. *Seirén*, a seductive sea nymph, siren + NL. *-in*, suf. used in chemistry to denote an activator or compound): a hormone secreted by the female gametes of the aquatic fungus *Allomyces* (Blastocladiales), which attracts the male gametes prior to planogametic copulation.

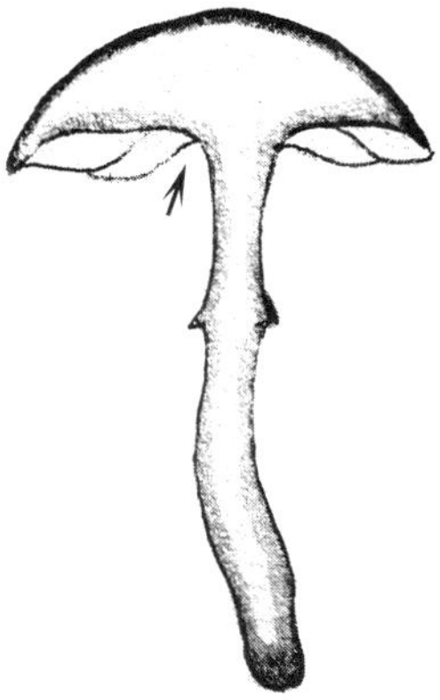

Sinuate lamellae of the basidiocarp of *Agrocybe* sp., seen in longitudinal section, x 1.

skeletal hypha, pl. **hyphae** (Gr. *skeletós*, dried up < *skéllo*, to dry, dry up + L. suf. *-alis* > E. *-al*, relating to or belonging to; Gr. *hyphé*, tissue, spider web; hypha): *Basidiomycetes*. A thick-walled, unbranched, aseptate hypha, straight or slightly flexuous, with little or no lumen, that provides rigidity to sporophores; e.g., as in *Ramaria gracilis*, *Lentinus tigrinus*, *L. lepideus* and *Schizophyllum commune* (Aphyllophorales). See **binding hypha** and **generative hypha**.

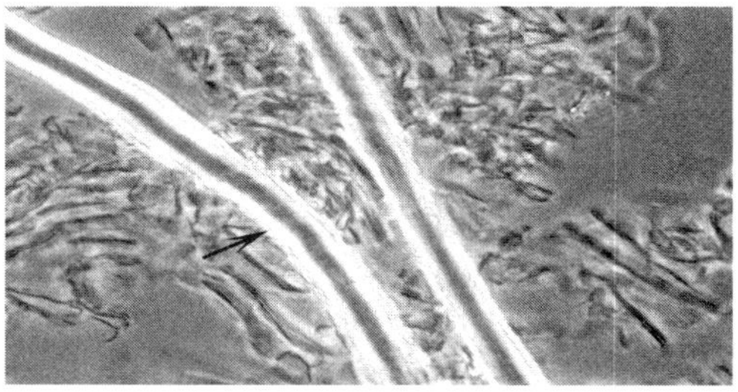

Skeletal hyphae of the basidiocarp of *Ramaria gracilis*, x 450 (*MV*).

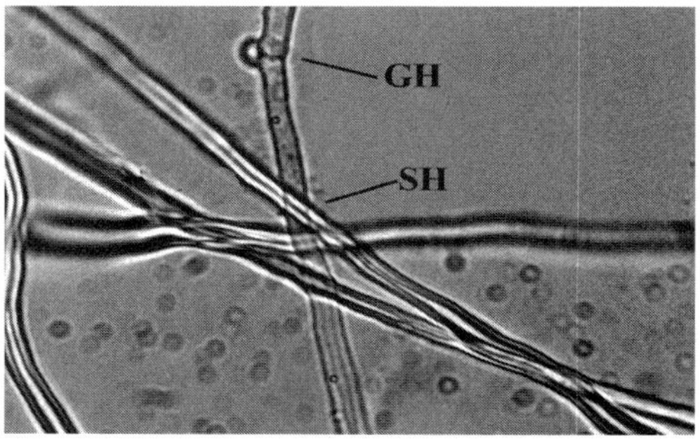

Skeletal hyphae (SH) and generative hypha (GH) of the basidiocarp of *Schizophyllum commune*, x 800 (*MU*).

slime mold (ME. < OE. *slim*, to smooth < L. *lima*, file; a soft, mucous or mucoid substance; ME. *mowlde* < *mowled*, ptp. of *moulen*, *malwen*, to grow moldy): one of the Acrasiomycota, Dictyosteliomycota (cellular slime molds) or Myxomycota (plasmodial slime molds).

slime net (ME. < OE. *slim*, to smooth < L. *lima*, file; a soft, mucous or mucoid substance; ME. *nett*, net < OE. < L. *nodus*, knot): the filoplasmodium or net plasmodium of the Labyrinthulales.

slimy (E. *slime* < ME. *slyme* < OE. *slim*, slime + E. suf. *-y*, abounding in, full of, having the quality of): viscous liquid, e.g., the colonies of *Cryptococcus albidus* (cryptococcaceous asexual yeasts) and *Aureobasidium pullulans* (dematiaceous asexual fungi). See **mucilaginous**.

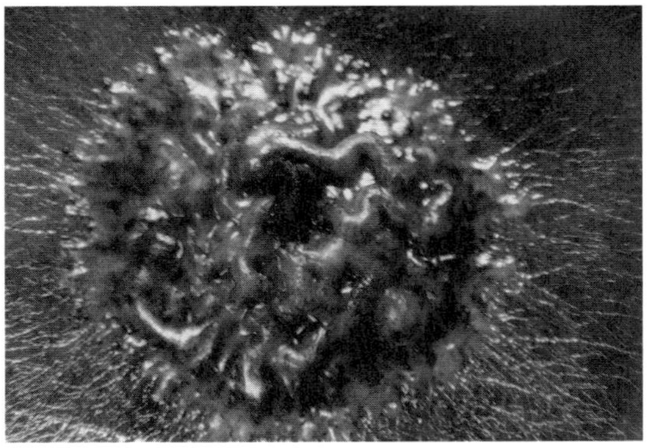

Slimy colony of *Aureobasidium pullulans* on agar, x 2.5 (*MU*).

sloughing exoperidium, pl. **exoperidia** (E. *sloughing*, ptp. of *slough*, to become shed or cast off < ME. *slughe*, *slouh*, snake skin; Gr. *éxo*, outside + *péridion*, small leather purse): an exoperidium which becomes detached and falls away, e.g., in *Lycoperdon marginatum* Lycoperdales).

smoke-colored (ME. < OE. *smoca*, smoke + E. *colored*, ptp. of *color* < ME. *colur* < AF. *coleur* < L. *color*, hue): dark gray.

smoky (E. *smoky*, resembling smoke < ME. < OE. *smoca* < ME. *smoken* < OE. *smocian*, to smoke + E. suf. *-y*, having the quality of): see **fumaceous**.

smut (prob. < *smot*, to stain < ME. *smotten*, to stain): any of various destructive diseases, especially of cereal grasses, caused by parasitic fungi (order Ustilaginales) and distinguished by the transformation of plant organs permeated by hyphae into dark masses of spores. *Covered smut*: a smut in which the mature spore mass remains for a time within a covering of host (or fungal) tissue (e.g., of barley and oats, caused by *Ustilago segetum*). *Stinking smut* or *bunt*: caused by *Tilletia caries* and *T. laevis* (=*T. foetida*). *Loose smut*: a smut in which the spores form an uncovered mass (e.g., of barley and wheat, caused by *Ustilago segetum* var. *tritici* (=*U. nuda*). *Stripe smut* of grasses (*U. striiformis*), of wheat (*U. agropyri*), of rye (*U. occulta*).

soleiform (E. *sole* < ME. < MF. < L. *solea*, sandal + L. *-formis* < *forma*, form): shaped like the sole of a shoe, e.g., elongate-ellipsoid.

solid (L. *solidus*, not hollow): solid, not hollow; in general, applied to the stipe of the agaricaceous fungi (such as *Agaricus* and *Boletus*) that lack a medulla or hollow, cavernose or fistulose central part. Also called **farctate**. Cf. **cavernose** and **fistular**.

solitary (L. *solitarius*, occurring singly < *solitas*, aloneness < *solus*, alone + suf. *-arius*, pertaining to + E. suf. *-y*, having the quality of): occurring singly, by itself; not in clusters. Applied to the growth habit of various fungi whose fruiting bodies occur scattered and separate; such as certain Agaricales, like *Amanita solitaria*. Cf. **caespitose**, **connate** and **gregarious**.

soma, pl. **somata** (NL. < Gr. *sõma*, body): the vegetative body of an organism, which carries out functions of assimilation and growth; it is distinguished by its physiology and morphology from the reproductive organs (or reproductive phase).

Soma (Hind. *Soma* ?): in Wasson´s theory, one of the gods of ancient India that supposedly corresponded to *Amanita muscaria* (Agaricales). The exact cause of the psychotropic activity of *A. muscaria* is unknown, but it may relate to the presence of various chemicals, including ibotenic acid and its derivative muscimol, as well as muscarine and other toxins, such as bufotenine.

somatic (Gr. *somatikós*, of the body < *sõma*, body + *-tikós* > L. *-ticus* > E. *-tic*, relation, fitness, inclination or ability): pertaining or relative to the body, whether in structure or function, but not the reproductive phases or parts.

somatogamy (Gr. *sõma*, body + *gamía* < *gámos*, sexual union + *-y*, E. suf. of concrete nouns): the fusion of somatic cells during plasmogamy, i.e., without the intervention of differentiated sexual organs, as happens in some chytrids, such as *Chytriomyces*, the majority of basidiomycetes and in many species of yeasts, such as *Saccharomyces* and *Zygosaccharomyces* (Saccharomycetales).

sooty mold (E. *sooty*, of, relating to, or producing soot < ME. < OE. *sot*, soot + E. suf. *-y*, having the quality of; ME. *mowlde* < *mowled*, ptp. of *moulen*, *malwen*, to grow moldy): a dark mat of fungus mycelium growing in insect honeydew on living leaves of tropical plants (such as lemon and other citrus plants), especially Capnodiales and their anamorphs.

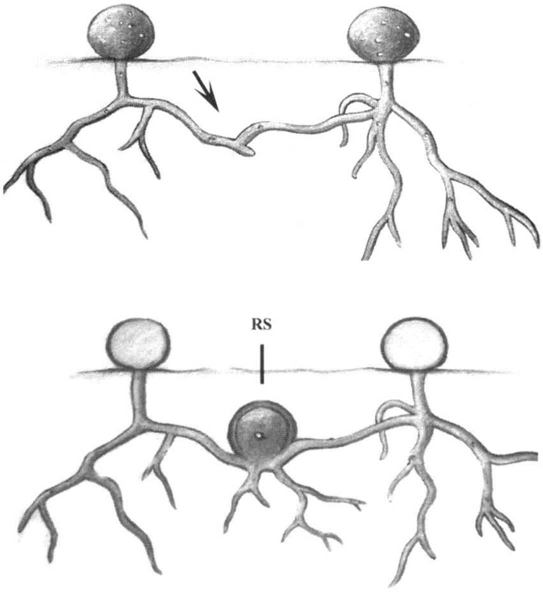

Somatogamy between the rhizoidal filaments of two thalli of *Chytriomyces hyalinus*, which results in the formation of a resting spore (RS), x 300.

soralium, pl. **soralia** (L. *soralium* < Gr. *sorós*, pile, heap + L. *-alis*, pertaining to + dim. suf. *-ium*): *Lichens.* A cortical pustule of various types in the thallus through which the soredia are released to the exterior; as in *Parmelia flaventior* and *P. sulcata* (Lecanorales),

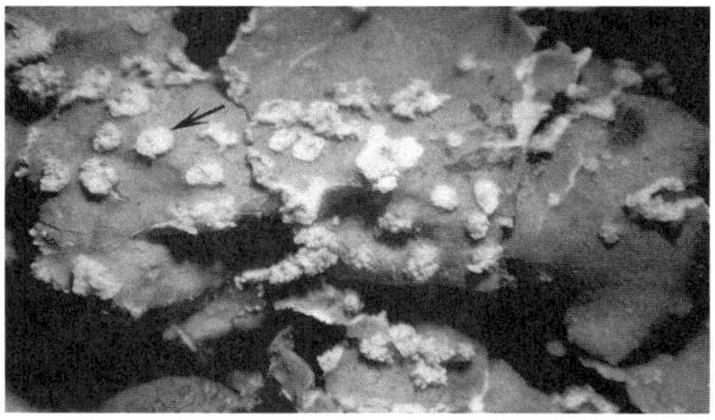

Soralium with soredia of the thallus of *Parmelia flaventior*, x 15 (*MU*).

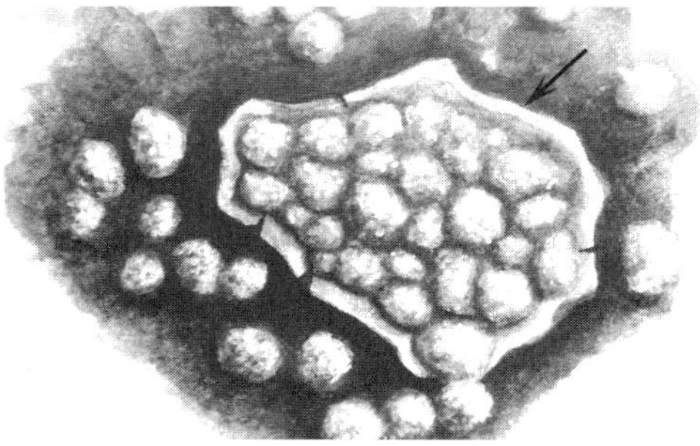

Soralium with soredia of *Parmelia sulcata*, x 20.

sorediate (ML. *soredium*, dim. of *sorus* < Gr. *sorós*, heap + suf. *-atus* > E. *-ate*, provided with or likeness): having **soredia**.

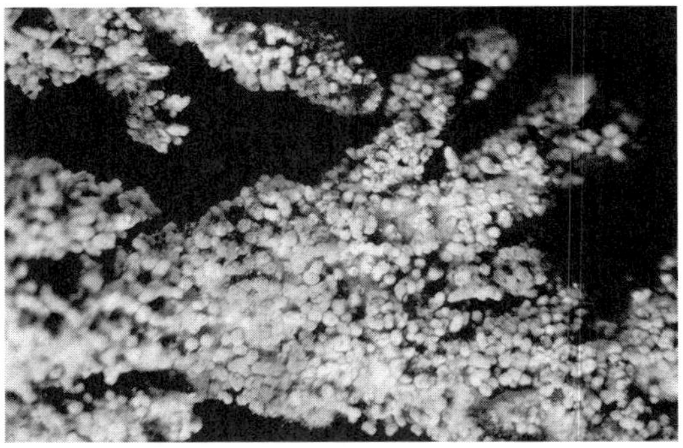

Sorediate thallus of *Stereocaulon saxatile*, x 15 (*MU*).

soredium, pl. **soredia** (ML. *soredium* < Gr. *sorídion* < *sorós*, pile + dim. suf. *-ídion* > L. *-idium*): *Lichens.* A more or less spherical microscopic body, formed by groups of algae surrounded by hyphae, that forms in pustules (**soralia**) on the surface of the thallus of many lichens (such as *Letharia vulpina*, *Parmelia perlata*, *P. subrudecta*, *Parmotrema* and *Stereocaulon saxatile*, among other Lecanorales); the soredia function as buds or propagules that permit vegetative reproduction. They differ from isidia in lacking a cortical layer. Cf. **isidium**.

Soredia of the thallus of *Parmotrema* sp., x 10 (*MU*).

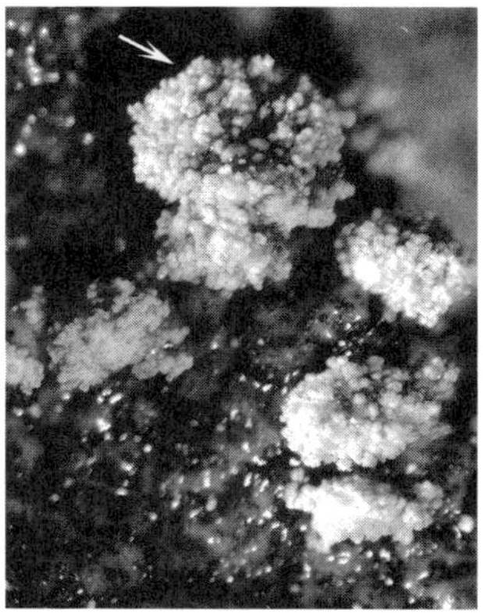

Soredia of the thallus of *Parmelia subrudecta*, x 20 (*MU*).

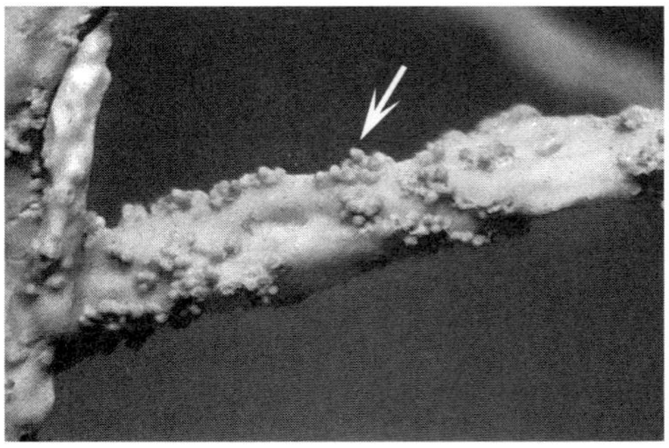

Soredia of the thallus of *Letharia vulpina*, x 40 (*MU*).

sorocarp (Gr. *sorós*, pile, heap + *karpós*, fruit): the fructification or sporophore of the Acrasiomycota, e.g., *Dictyostelium mucoroides* and *Polysphondylium violaceum* (Dictyosteliales).

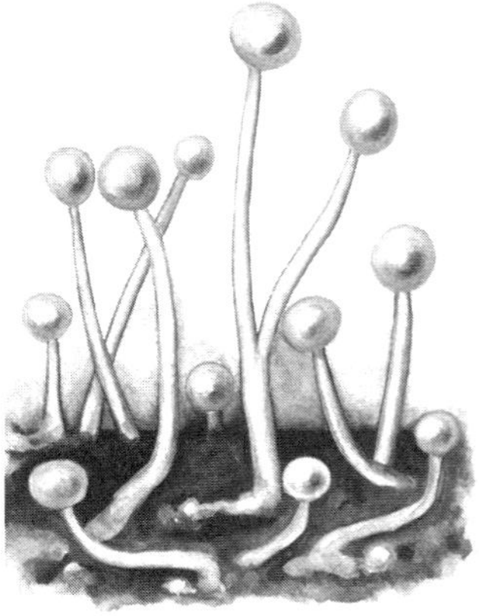

Sorocarps, each with a terminal sorus, of *Dictyostelium mucoroides*, x 30.

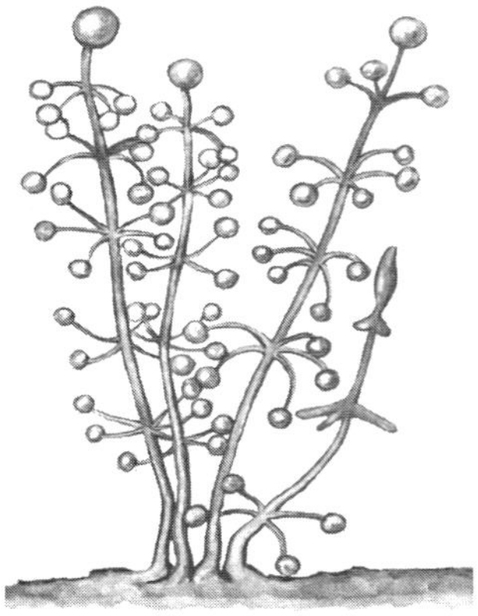

Sorocarps, each with verticills of sori and a terminal sorus, of *Polysphondylium violaceum*, x 12.

sorocyst (Gr. *sorós*, pile, heap + *kýstis*, vesicle, cell): each of the encysted cells of the sorocarp of *Copromyxa arborescens* (Acrasiomycota), considered as the simplest type of sorocarp because it is not differentiated into cells that give rise to spores, and cells that form a pedicel; the sorocyst lacks a pedicel.

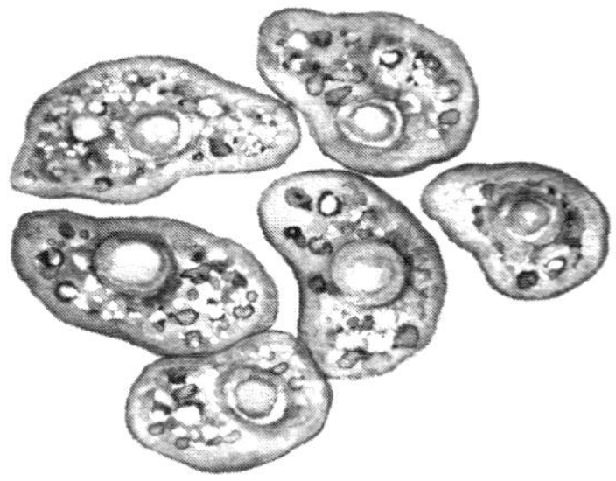

Sorocysts of *Copromyxa arborescens*, x 1 200.

sorogen (Gr. *sorós*, pile, heap + *génos*, origin < *gennáo*, to engender, produce): part of the pseudoplasmodium which is differentiated into a sorus or group of spores, and in the multicellular pedicel, of the sorocarp of the Dictyosteliomycota and Acrasiomycota. It is observed clearly during sorogenesis of *Polysphondylium* and *Dictyostelium* (Dictyosteliomycota).

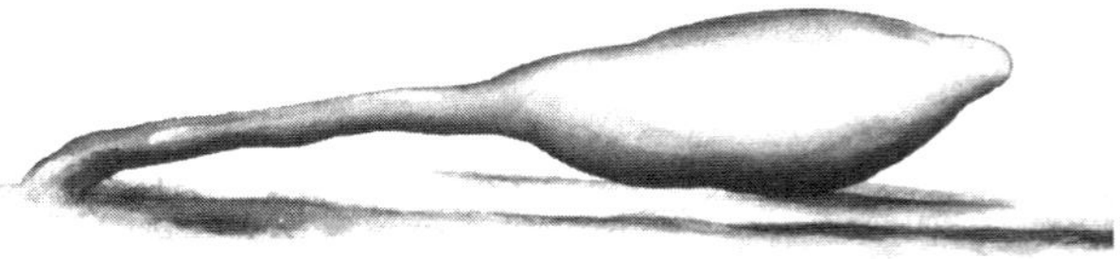

Sorogen of a young sorocarp of *Dictyostelium mucoroides*, x 65.

sorophore (Gr. *sorós*, pile, heap + *-phóros*, bearer < *phéro*, to carry, support): a pedicel (multicellular or acellular according to the species) that supports the sorus or mass of spores of the sorocarp of the Acrasiomycota and Dictyosteliomycota, such as *Acrasis* and *Dictyostelium*, respectively. The sorophore originates from the sorogen during sorogenesis or fructification of the pseudoplasmodium.

sorus, pl. **sori** (Gr. *sorós*, pile, heap): a spore mass formed in certain fungi, such as the rusts (Uredinales) and smuts (Ustilaginales), the sorocarp or fruiting body of the cellular slime molds or Dictyosteliomycota (such as *Dictyostelium* and *Polysphondylium*), or the groups of sporangia in fungi like *Synchytrium* (Chytridiales).

spathulate, **spatulate** (L. *spathulatus*, shaped like a spatula < *spathula*, dim. of *spatha*, spatula + suf. *-atus* > E. *-ate*, provided with or likeness): having the shape of a spatula or spoon; e.g., like the fruiting body of *Geoglossum difforme*, *Spathularia flavida* (Helotiales) and *Dacryopinax elegans* (Dacrymycetales).

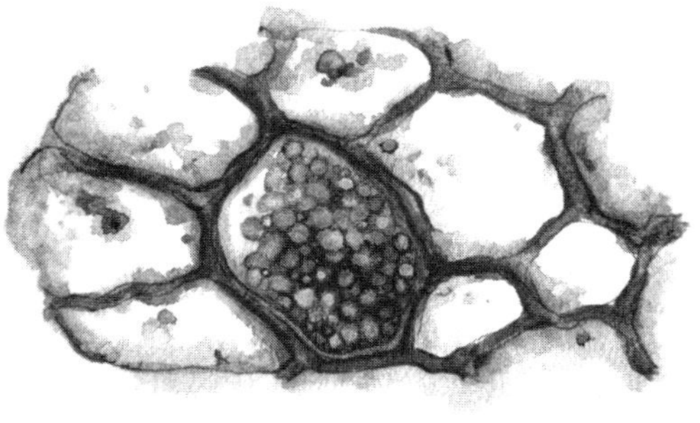

Sorus with zoosporangia of *Synchytrium endobioticum* within a cell of potato tuber, seen in transverse section, x 900.

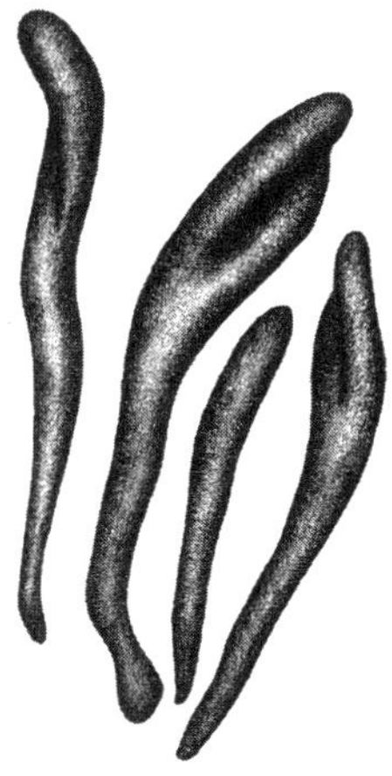

Spathulate apothecia of *Geoglossum difforme*, x 1.

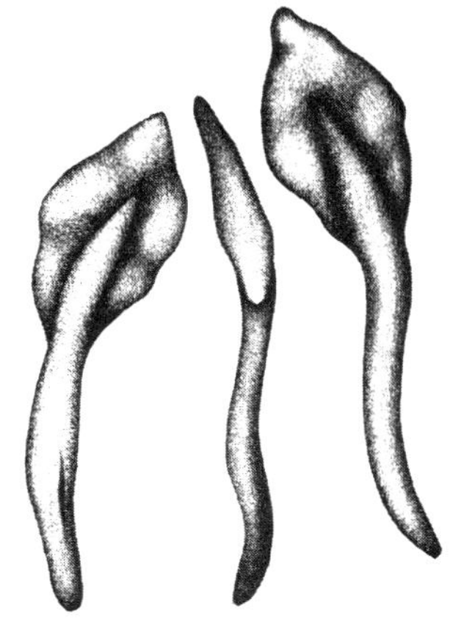

Spathulate apothecia of *Spathularia flavida*, x 1.3.

spawn (ME. *spawnen*, prob. < AF. *espaundre* < OF. *espandre*, to spread out, expand): the inoculum used in the cultivation of edible mushrooms, such as *Agaricus* and *Pleurotus* (Agaricales), which consists of grains of wheat (or a similar cereal) that have been invaded by the mycelium of the fungus; the spawn is used to inoculate a suitable substrate, which later will produce the mushroom fructifications to be harvested. To put inoculum (spawn) into a mushroom bed (compost) or other substrate.

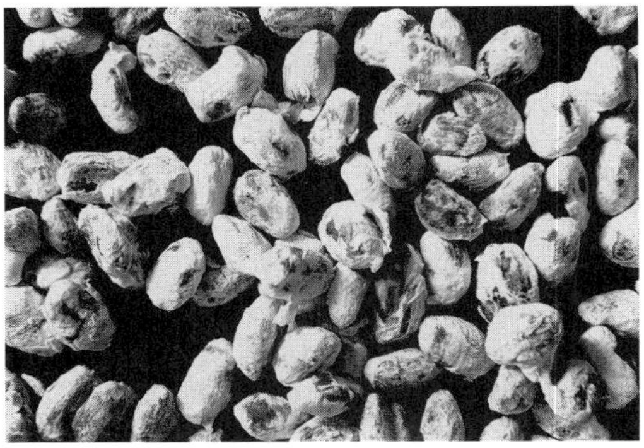

Spawn prepared with wheat grains invaded by the mycelium of *Pleurotus ostreatus*, x 1.5 (*MU*).

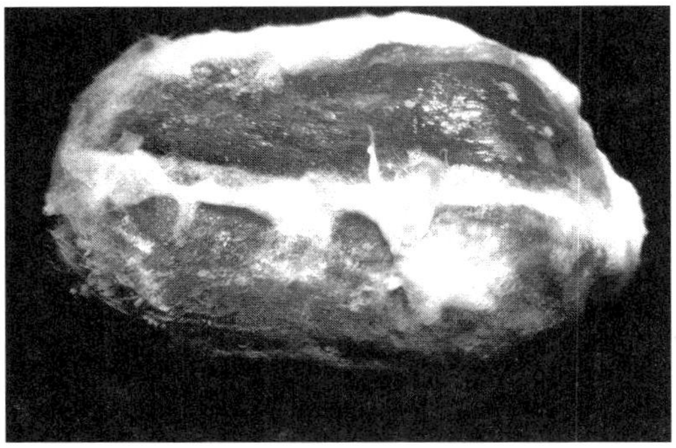

Spawn (a close up) consisting of a wheat grain overgrown with the mycelium of *Pleurotus ostreatus*, x 12 (*MU*).

spermagonium, pl. **spermagonia**, **spermogonium**, pl. **spermogonia**, **spermatogonium**, pl. **spermatogonia** (NL. *spermogonium* < Gr. *spérma*, genit. *spérmatos*, seed, sperm + *gónos*, that engendered, offspring + L. dim. suf. *-ium*): a pycnidium-like structure in which spermatia are produced, e.g., in *Cyttaria* (Cyttariales). Spermogonia in the rust fungi, such as *Aecidium* and *Puccinia* (Uredinales) have often been called pycnia, sing. pycnium; they function in spermatization in sexual reproduction.

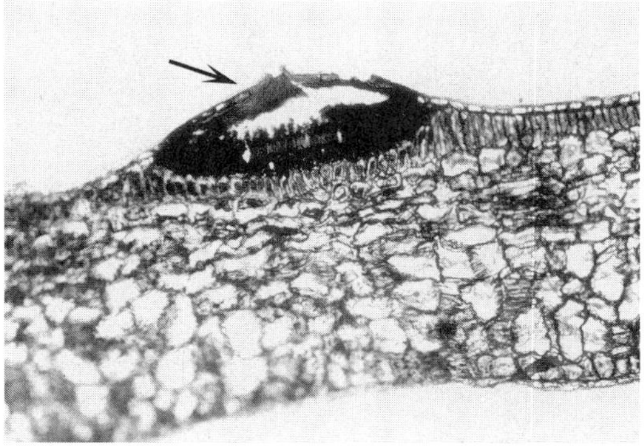

Spermogonium of *Aecidium* sp. in the upper surface of a leaf of *Diospyros digyna*, seen in longitudinal section, x 100 (*MU*).

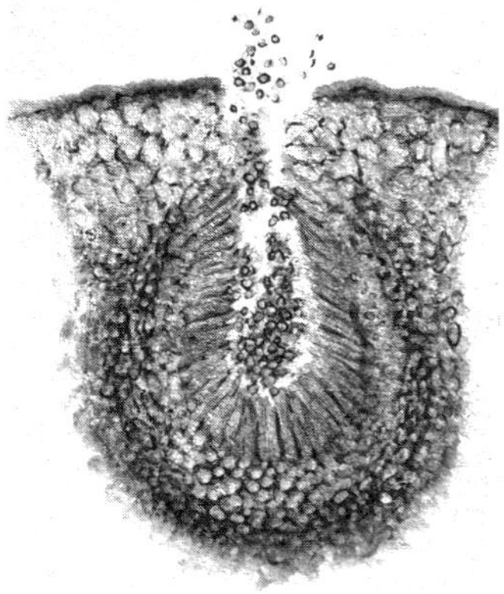

Spermogonium of *Cyttaria deformans* embedded in the branch tissue of *Nothofagus pumilio*, seen in longitudinal section, x 200.

spermatiophore (Gr. *spermátion*, small seed < *spérma*, seed, sperm + *-phóros*, bearer): a specialized hypha that produces spermatia; e.g., as happens in the Laboulbeniales and in some Uredinales.

spermatium, pl. **spermatia** (NL. *spermatium* < Gr. *spermátion*, small seed < *spérma*, genit. *spérmatos*, seed, sperm + dim. suf. *-íon* > L. *-ium*): a small, hyaline, uninucleate, nonmotile cell, similar to a spore, that is believed to function as a male gamete, and which fertilizes a receptive female hypha; e.g., as happens in *Phyllachora fusicarpa* (Phyllachorales), *Neurospora sitophila* (Sordariales), in the Laboulbeniales, and in the rust *Puccinia graminis tritici* (Uredinales); the spermatia of the rust fungi often have been called **pycniospores**.

Spermatia produced in a spermogonium of *Phyllachora fusicarpa* parasitizing a leaf of *Duranta* sp., seen in longitudinal section, x 1 000 (*RTH*).

spermatization (Gr. *spérma*, genit. *spérmatos*, seed, sperm + *-ationem*, action, state or condition, or result > E. suf. *-ation*): a type of plasmogamy in which a spermatium unites with a receptive hypha, as happens in *Neurospora* (Sordariales) or in *Puccinia* (Uredinales).

spermatozoid (Gr. *spérma*, genit. *spérmatos*, seed, and in animals and man, sperm or germ, semen + *zõon*, of animals + L. suf. *-oide* < Gr. *-oeídes*, similar to): see **antherozoid**.

spermidium, pl. **spermidia** (Gr. *spérma*, seed, sperm + dim. suf. *-ídion* > L. *-idium*): a fructification that produces spermatia.

spermodermium, pl. **spermodermia** (Gr. *spérma*, seed, sperm + *dérma*, skin + L. dim. suf. *-ium*): a hymenium of spermatiophores formed beneath the cuticle in necrotic areas of infected host leaves.

spermodochium, pl. **spermodochia** (Gr. *spérma*, seed, sperm + *docheíon*, recipient): a fasciculate or tuberculate aggregation of branched spermatiophores, usually arising from a single cell, and borne free on aerial mycelium.

spermoplane (Gr. *spérma*, seed, sperm + L. *planus*, flat, flat surface): the surface of a seed, on which fungi may occur.

spermosphere (Gr. *spérma*, seed, sperm + *sphaîra*, sphere, enclosure): the microhabitat around a seed in the soil. The spermosphere as well as the spermoplane have a characteristic mycobiota.

sphaeridium, pl. **sphaeridia** (Gr. *sphairidion* < *sphaîra*, sphere + dim. suf. *-ídion* > L. *-idium*): a stalked globose apical apothecium, as in the Caliciales. A syn. of **capitulum**.

Sphaeridia of the thallus of *Calicium farietinum*, x 10 (*MU*).

sphaerocyst (Gr. *sphaîra*, sphere + *kýstis*, bladder, cell): **1**. A spherical cell that forms part of the heteromerous trama of the gills (hymenophore) in basidiocarps of *Russula* and *Lactarius* (Agaricales). **2**. An inflated or variously shaped cell found in the exoperidium of *Bovista*, *Calvatia* and *Lycoperdon* (Lycoperdales).

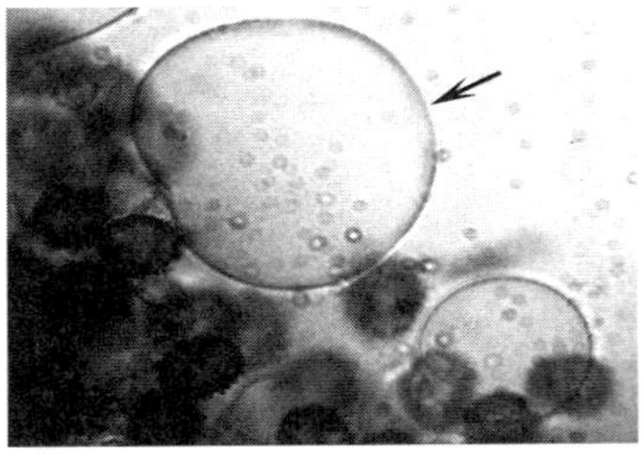

Sphaerocysts from the trama of a basidiocarp of *Russula* sp., x 500 (*MU*).

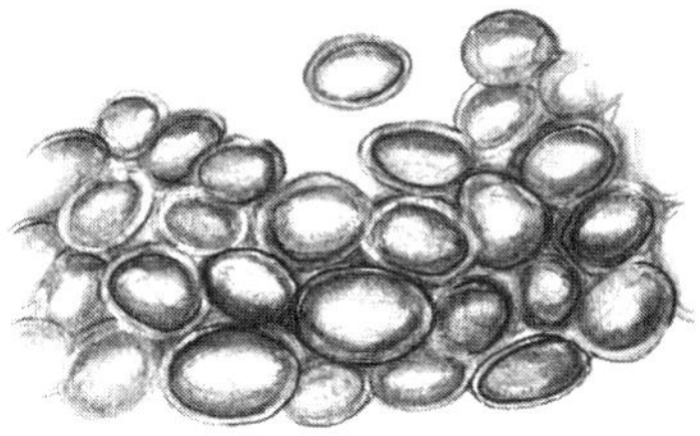

Sphaerocysts from the polycystoderm of an exoperidial fibril of *Calvatia cretacea*, x 500.

sphaeropedunculate (Gr. *sphaîra*, sphere + L. *pedunculatus*, furnished with a foot < *pedunculus*, dim. of *pes*, genit. *pedis*, foot, support + suf. *-atus* > E. *-ate*, provided with or likeness): a spherical or globose structure with a prolongation that tapers toward the basal end; e.g., like the hyphal cells that are found in the volva of the basidiocarp of *Amanita herrerae* (Agaricales). The so-called sphaeropedunculate cell is a cystidium with a nearly globose apex and a narrow basal stalk.

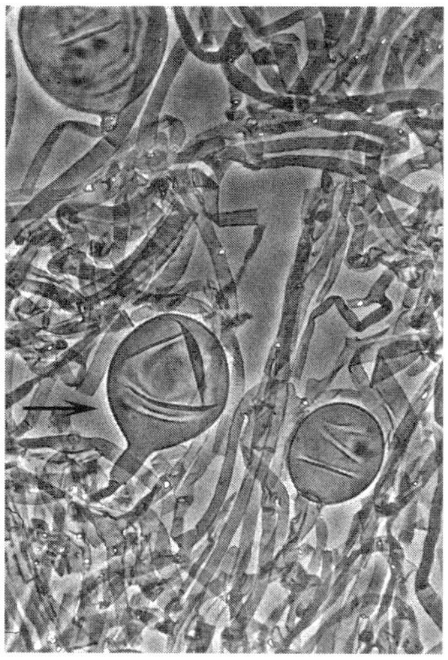

Sphaeropedunculate hyphal cells from the volva of a basidiocarp of *Amanita herrerae*, x 570 (*RMA*).

spherule (L. *spherula*, small sphere < *sphaera* < Gr. *sphaîra*, sphere, globe + dim. suf. *-ula* > E. *-ule*): **1**. A sporangium-like, globose structure (the parasitic phase) in *Coccidioides* (moniliaceous asexual fungi) that originates from the swelling and enlargement of the inhaled arthrospores (the saprobic phase). **2**. A globose or subglobose structure, subsessile or with a short pedicel, pearl white in color, 0.5-1.5 (2.5) mm diam., that functions as a sporodochial former of conidia in *Termitomyces* (Agaricales), a fungus symbiotic with termites. The spherules are originated on the surface of the termite nests as erect clusters of hyphae that are intertwined, catenulate or moniliform, with multiple verticillate buds on whose tips are borne the conidia.

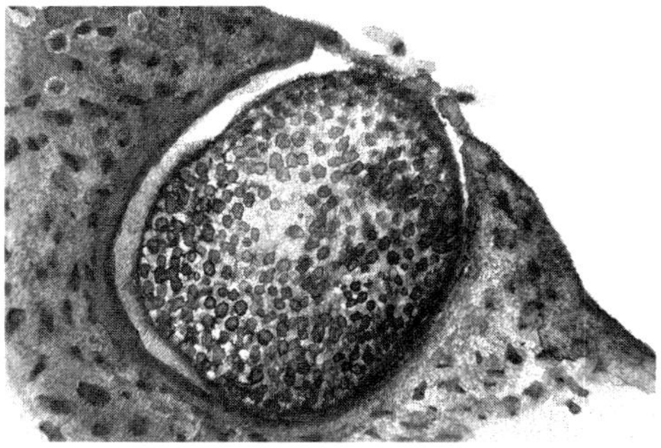

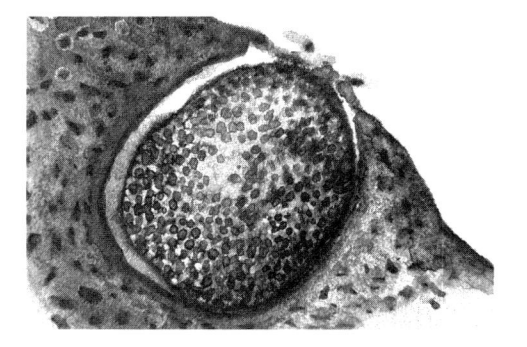

Spherule of *Coccidioides immitis* in human tissue, x 500.

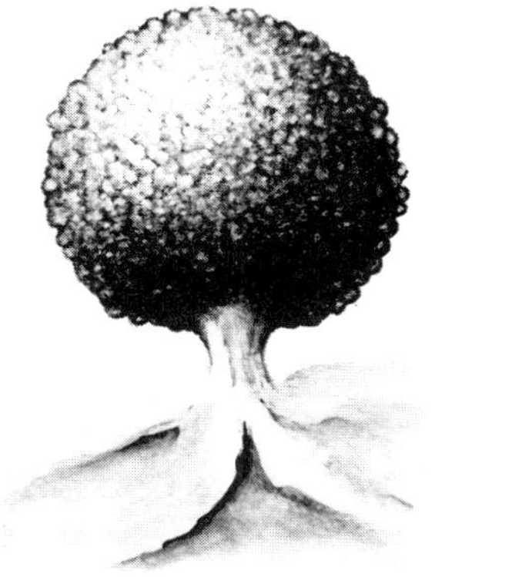

Spherule of *Termitomyces albuminosus*, x 50.

spiculate (L. *spiculatus*, furnished with spikes < *spiculum*, dim. of *spica*, dart, arrow, point, spike + suf. *-atus* > E. *-ate*, provided with or likeness): having **spicules**.

spicule (L. *spiculum* < *spica*, dart, arrow, point, spike + dim. suf. *-ulum* > E. *-ule*): a prong or spine-like appendage.

spiculospore (NL. < L. *spiculum*, spike, dart, prick + Gr. *sporá*, spore): a spore formed at the apex of a pointed, often elongate (spike-like) structure, as in *Hirsutella* and *Akanthomyces* (stilbellaceous asexual fungi).

spilodium, pl. **spilodia** (Gr. *spílos*, freckle + L. *-odium*, resembling < Gr. *ode*, like < *-oeídes*, similar to): *Lichens*. A small, round, blackish structure, composed of dark, compact hyphae, that is present in the thallus of the lichen *Dirina stenhammari* (Arthoniales). It is called a spilodium because of its similarity to a spilus or navel of the grains of wheat and other grasses.

spindle pole body (ME. *spindel*, to spin, in relation to the spindle used to form and twist the thread in a spinning wheel; L. *polus*, pole; ME. < OE. *bodig*, body): a small, electron-dense cytoplasmic structure, of unknown chemical composition, that lies adjacent to the nuclear envelope in most true fungi, i.e., nonflagellate. Evidence indicates that these organelles function as microtubule-organizing centers during mitosis and meiosis. Microtubules emanating from the spindle pole bodies develop into the spindle apparatus. A spindle-shaped achromatic figure along which the chromosomes are distributed during mitosis and meiosis. Species of fungi that produce flagellate cells lack these spindle pole bodies, possessing instead a pair of centrioles that are associated with the nuclear envelope. Cf. **centriole**.

spine (ME. *spine*, thorn < L. *spina*): a narrow, sharply pointed process. *Spiny*, having spines.

spinose (L. *spine*, thorn, spine + *-osus*, full of, augmented, prone to > ME. *-ose*): spiny.

spinule (L. *spinula*, small spine < *spina*, thorn, spine + dim. suf. *-ula* > E. *-ule*): a small spine. In the lichens it is applied to the small prominence on the thallus, with a narrow base, which on separating functions as a vegetative propagule; e.g., in *Bryoria bicolor* and other species of fruticose lichens of the order Lecanorales.

spinulose, **spinulous** (L. *spinulosus*, spiny < *spinula*, dim. of *spina*, thorn, spine + *-osus*, full of, augmented, prone to > ME. *-ose*; or + *-osus* > OF. *-ous*, *-eus* > E. *-ous*, having, possessing the qualities of): with small spines, at times microscopic; e.g., like the sporangial wall of *Rhizopus arrhizus* (Mucorales).

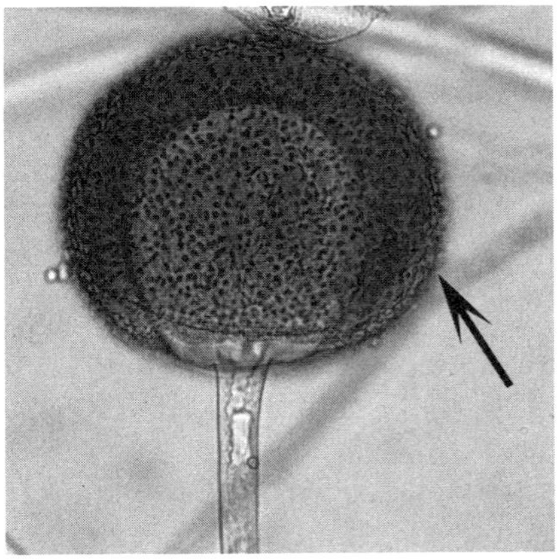

Spinulose sporangium of *Rhizopus arrhizus*, x 500 (*MU*).

spiral hypha, pl. **hyphae** (ML. *spiralis* < L. *spira*, coil, a three-dimensional curve, as a helix, with one or more turns about an axis + suf. *-alis* > E. *-al*, relating to or belonging to; Gr. *hyphé*, tissue, spider web, hypha): a hypha ending in a spiral or helical coil, as in *Trichophyton* (moniliaceous asexual fungi).

Spitzenkörper (G. *Spitzen*, tip, apex + *Körper*, body): a granular and/or vesicular region of cytoplasm found in the apex of hyphae of the Ascomycota and Basidiomycota, in association with vesicles of the cell wall, which are formed from the cisternae of the Golgi apparatus or dictyosomes and which participate in the synthesis of the cell wall of growing hyphae. The hyphae of the lower fungi (the phyla Oomycota, Hyphochytriomycota and Labyrinthulomycota of the kingdom Stramenopila) lack a Spitzenkörper. The term Spitzenkörper was originally described in German and subsequently incorporated into English unchanged, as is commonly done in English.

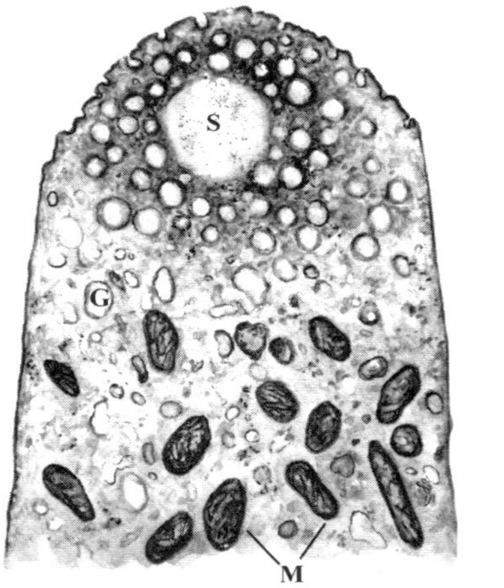

Spitzenkörper (S) in the sub-apical zone of a hypha of *Armillaria mellea*, seen in longitudinal section. Note the absence of vesicles in this spherical central region. The clusters of vesicles are associated with Golgi cisternae or tubules (G) in zones of cytoplasm which contain few ribosomes compared to the surrounding protoplasm. Mitochondria (M) and endoplasmic reticulum occur throughout the region. Drawing based on a transmission electron micrograph published by Beckett *et al.* in their *Atlas of Fungal Ultrastructure*, 1974, x 17 000.

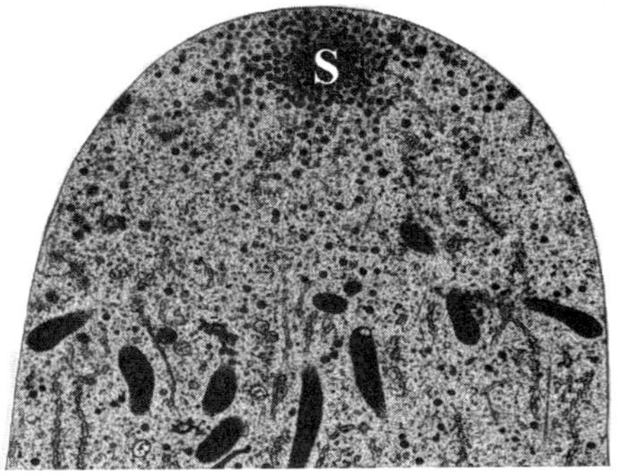

Spitzenkörper (S) in a hypha of a basidiomycete. A transmission electron micrograph, x 17 000 (*CM*).

splash cup (perhaps ME. *plasch*, pool, puddle < OE. *plæsc*; L. *cuppa* < *cupa*, tub, cask): an open cup-like structure, often with flaring side walls, in which raindrops land and forcibly discharge the peridioles; typically found in the Nidulariales (*Crucibulum*, *Cyathus*, *Nidula*, *Nidularia*).

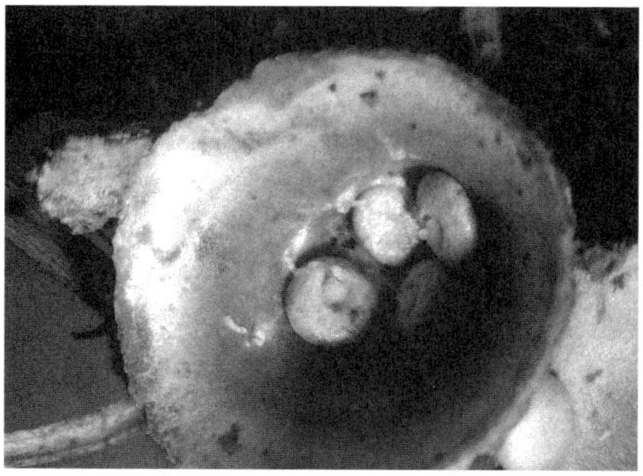

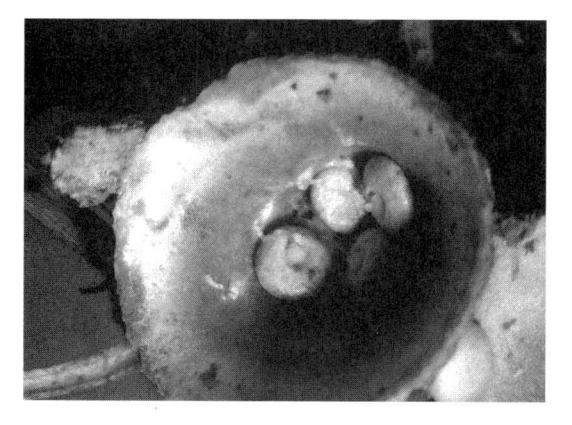

Splash cup of the fruiting body of *Crucibulum laeve*, x 6 (*CB*).

split lamella, pl. **lamellae** (Dutch *splitten*, to split; to divide lengthwise; L. *lamella*, dim. of *lamina*, plate): split gill-like structures on the underside of the basidiocarp of *Schizophyllum* (Aphyllophorales). The marginal proliferation of the basidiocarp forms the "split gills" that may fold over the hymenium upon drying. In some instances the split may be shallow and resemble only a groove. The lamellae in *Schizophyllum* are not believed to be homologous with the gills of the agarics.

Split lamellae of the basidiocarp of *Schizophyllum commune*, x 2 (*MU*).

spongy (L. *spongia*, sponge + E. suf. *-y*, having the quality of): having a soft consistency and a tendency to soak up water; e.g., like the fruiting body of *Boletus*, *Suillus* and other boletaceous Agaricales.

sporabola, pl. **sporabolae** (Gr. *sporabolá* < *sporá*, spore + *bállo*, to throw): in mycology it refers to the curve that the basidiospore makes after it is discharged from the sterigma of the basidium.

sporangiogenous (NL. *sporangium* < Gr. *sporá*, spore + *angeîon*, vessel, receptacle + *génos*, origin < *gennáo*, to engender, produce + L. *-osus* > OF. *-ous*, *-eus* > E. *-ous*, having, possessing the qualities of): *Plasmodiophorales*. The plasmodium of the asexual phase, which is transformed into a sporangium and gives rise to zoospores. Cf. **cystogenous**.

sporangiole, **sporangiolum**, pl. **sporangiola** (NL. *sporangium* < Gr. *sporá*, spore + *angeîon*, receptacle + L. suf. *-olum*, small > E. *-ole*): a small sporangium that contains only a few spores, and sometimes a single spore ("conidium"); sporangiola are present in *Thamnidium* and other Mucorales, such as *Blakeslea trispora*, *Cokeromyces* and *Choanephora*. In the Entomophthorales, such as *Conidiobolus*, sporangiola are also formed and are called conidia because they are unispored; on a sporangiophore (or conidiophore, if the sporangiolum is considered a conidium) a primary sporangiolum is developed that is forcibly discharged; this can germinate repeatedly by producing secondary conidia.

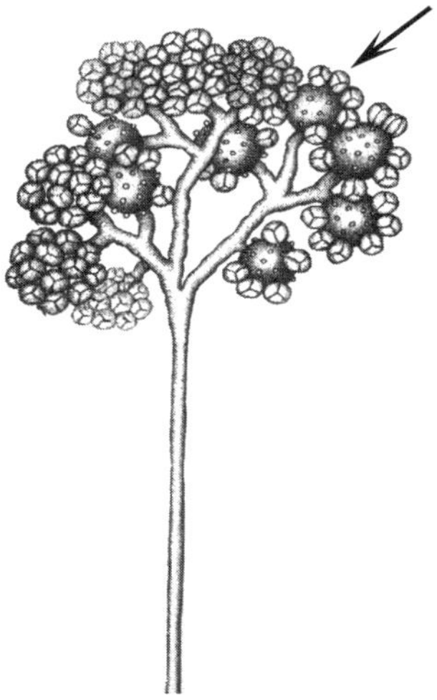

Sporangiola of *Blakeslea trispora*, x 300.

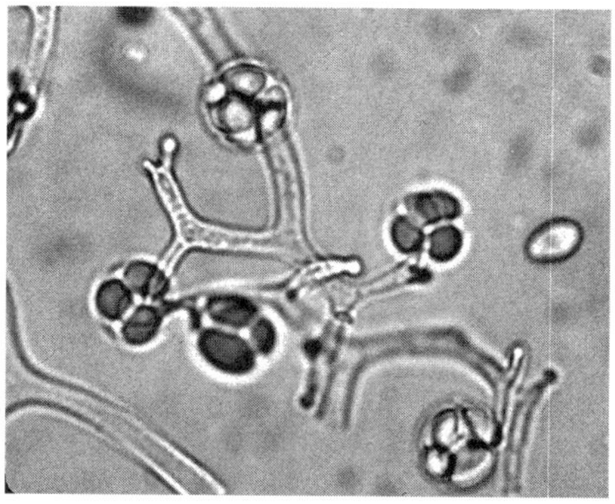

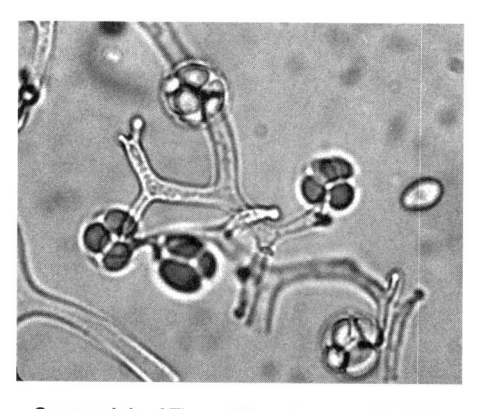

Sporangiola of *Thamnidium elegans*, x 690 (*MU*).

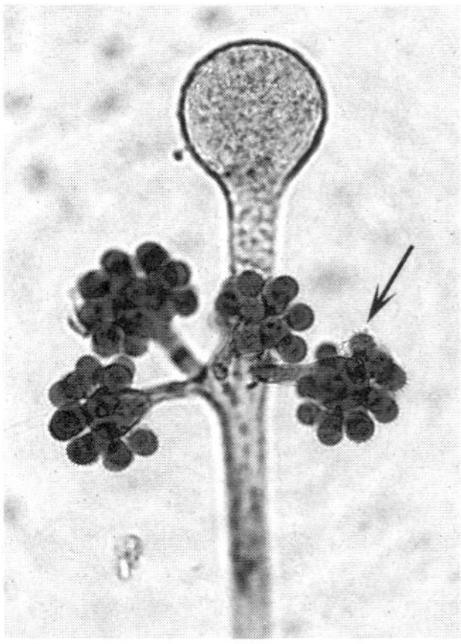

Sporangiola of *Cunninghamella echinulata*, x 550 (*MU*).

sporangiophore (Gr, *sporá*, spore + *angeîon*, vessel, receptacle + *-phóros*, bearer): a specialized hypha that produces and supports one or more sporangia, e.g., *Phycomyces nitens* (Mucorales).

Sporangiophores of *Phycomyces nitens* on agar, x 0.4 (*MU*).

sporangiospore (Gr. *sporá*, spore + *angeîon*, vessel, receptacle + *sporá*, spore): a spore produced in a sporangium by any of the modes; the sporangiospores can be immotile (aplanospores) or flagellate and motile (zoospores), and they are present in various groups, e.g., in the Chytridiomycota, the Oomycota, and the Zygomycota. Of the latter phylum, some illustrated examples of sporangiospores are *Rhizopus* and *Parasitella* (Mucorales).

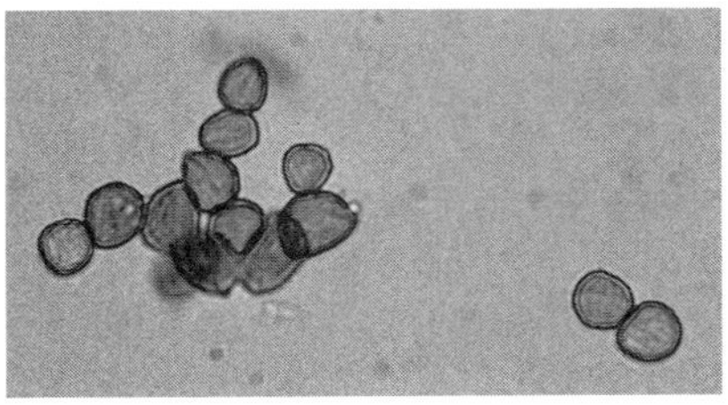

Sporangiospores of *Rhizopus nigricans*, x 1 300 (*MU*).

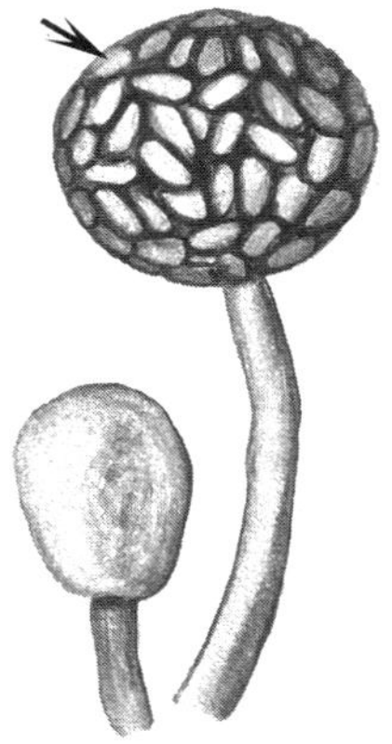

Sporangiospores of the sporangia of *Parasitella parasitica*, x 1 150.

sporangium, pl. **sporangia** (Gr. *sporá*, spore + *angeîon*, vessel, receptacle): a structure, of various shapes according to the species, which produces endogenous spores of asexual origin; all the protoplasmic contents of a sporangium is converted into an indefinite number of spores, whether they be zoospores or aplanospores, depending upon the species. Sporangia are present, for example, in Myxomycetes, such as *Cribraria* (=*Dictydium*) *cancellata* (Liceales), in aquatic fungi, such as *Phlyctochytrium* (Chytridiales), and in common molds, like *Mucor* and *Rhizopus* (Mucorales).

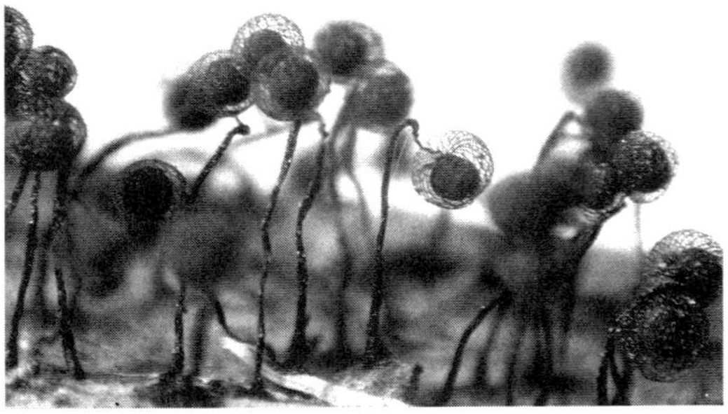

Sporangia of *Cribraria* (=*Dictydium*) *cancellata*, x 18 (*MU*).

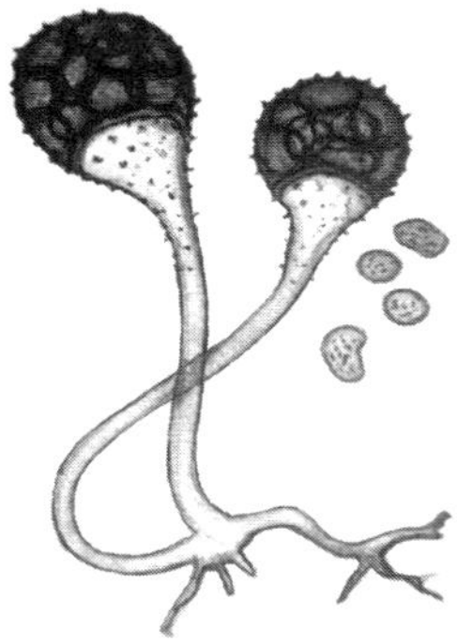

Sporangia of *Rhizopus oligosporus*, x 350.

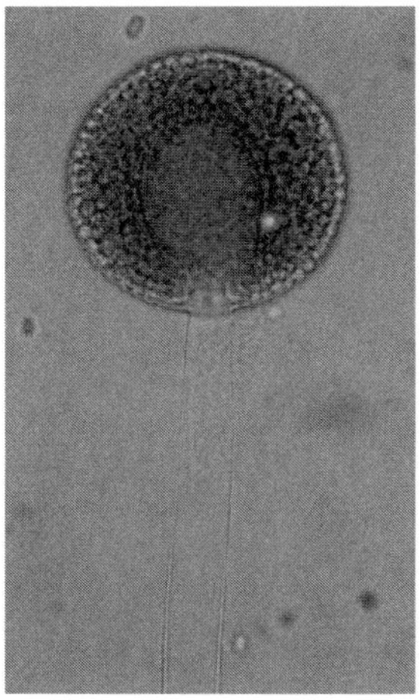

Sporangium of *Mucor hiemalis*, x 350 (*CB*).

spore (NL. *spora* < Gr. *sporá*, spore): a small unit of propagation, unicellular or multicellular, asexual or sexual, motile or immotile, that is capable of giving rise to a new individual. A spore functions like a seed, although it differs from the latter in that the spore does not contain a preformed embryo. Biologically, spores can be divided into various groups according to whether they are disseminated by air, water, insects and other animals, etc. Spores can survive during periods unfavorable for germination and growth, although a single type of spore can have various functions. The morphology and development of the spores serve as basic taxonomic criteria. In the description of spores the following characteristics are taken into consideration: *type of development* (ascospore, basidiospore, sporangiospore, etc.); *motile* or *immotile* (zoospore, aplanospore); if motile, the number and type of flagella, or amoeboid; *individual* or *aggregated; shape* (although spores are tridimensional, the shape is frequently described as seen in optical section, as a bidimensional figure); *contents* (hyaline appearance, annular, etc.; and the presence of oil droplets or air bubbles); *septation* (aseptate or with one or several septa); *cell wall* (dry or viscous, and its thickness); *ornamentation* (warts, spines, appendages, etc., and the reaction to dyes or reagents, such as cyanophilic, amyloid, non-amyloid, etc.); *color* (light or dark); and size in µm.

spore ball (Gr. *sporá*, spore; E. *ball* < ME. *bal*, *balle*, ball): a compact, round group of spores, or of spores and sterile cells, that function as a unit of dispersal; e.g., *Ascosphaera apis* (Ascosphaerales) and *Sorosporium* and *Tolyposporium* (Ustilaginales) form balls composed only of spores, whereas *Urocystis* (Ustilaginales) forms an aggregation of spores and sterile cells.

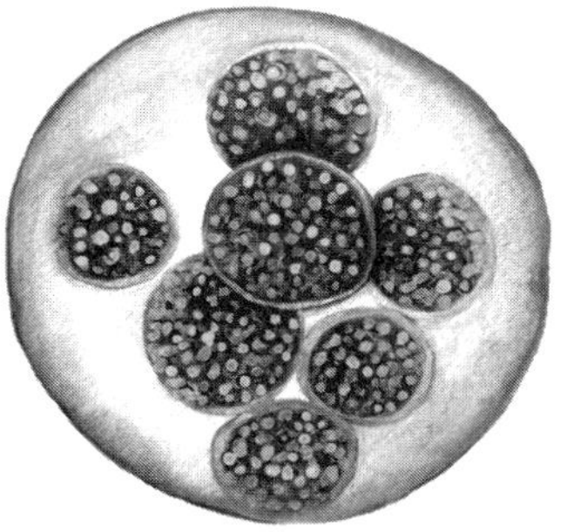

Spore balls within a sporocyst of *Ascosphaera apis*, x 1 200.

spore case (Gr. *sporá*, spore; ME. *cas* < AF. *casse* < OF. *chasse* < L. *capsa*, cylindrical scroll case): generally applied to the spore receptacle of some Gasteromycetes, such as the Lycoperdales, Sclerodermatales, and Nidulariales. Of the first, one can cite, e.g., *Geastrum*, whose spore case corresponds to the endoperidium, which develops an ostiole and a characteristic peristome or disk. In *Broomeia* (of the second order mentioned), a large, fleshy stroma is formed, which bears numerous sporiferous sacks corresponding to the endoperidium, each with an ostiole delimited by a peristome. In *Sphaerobolus* (Nidulariales) a single sporiferous sack is present, which in this case is called a peridiole (which is delimited by its own peridium), that remains exposed when the mature exoperidium splits into segments or coronate lobes. The violent ejection of the peridiole leaves the endoperidium evaginated, like a white sphere on the split exoperidium.

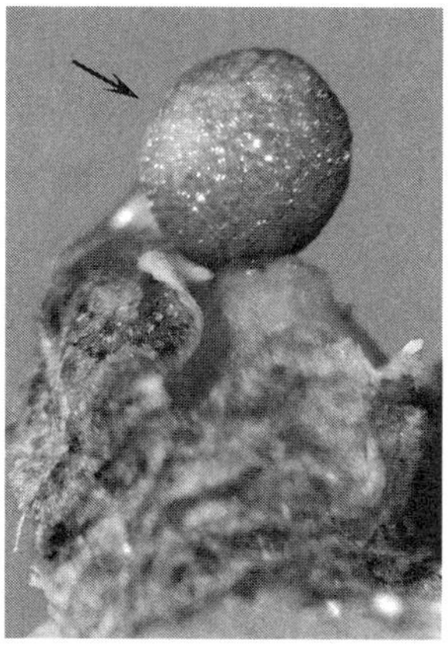

Spore case of the fruiting body of *Sphaerobolus stellatus*, x 25 (*MU*).

spore collar (Gr. *sporá*, spore; L. *collare*, neckband): a distinctive cylinder of tissue, derived from the upper part of the sterigma, which remains on many detached orthotropic spores; typically seen in *Hysterangium* (Phallales). See **collar**.

spore print, spore deposit (Gr. *sporá*, spore; ME. *prente*, *printe* < OF. *priente*, impression, print; L. *depositus*, laid down): the mass of spores of a fungus discharged from the sporophore onto a surface. In the agarics, the spore print is obtained by placing a mature pileus on white or black paper, with the gills downward; at the end of several hours, generally at night, the spore print is deposited, and its color characteristics can be determined. These are of taxonomic importance; e.g., in *Agaricus* (Agaricales) the spore print is dark brown.

Spore print from a basidiocarp of *Agaricus* sp., x 1.

sporidiole, sporidiolum, pl. **sporidiola** (Gr. *sporídion*, dim. of *sporá*, spore + L. dim. suf. *-olum* > E. *-ole*): spore-like structures produced inside **sporidiomata**, perithecium-like bodies, in *Kathistes* (an ascomycetous gen. of uncertain affinity).

sporidioma, pl. **sporidiomata** (Gr. *sporídion*, dim. of *sporá*, spore + suf. *-oma*, which implies entirety): a unique spore-producing body, associated with the submerged ascocarp base, characteristic of *Kathistes* (an ascomycetous gen. of uncertain affinity), a fungus that has been found only in moist chamber cultures of dung. At maturity masses of ascospores (**sporidiola**) collect in long incurved extensions of the wall cells of the perithecial neck.

sporidium, pl. **sporidia** (Gr. *sporídion*, dim. of *sporá*, spore): the name applied to the spore produced on the promycelium or basidium of the Uredinales and Ustilaginales, i.e., a basidiospore that is borne on an epibasidium. In the gen. *Tilletia* (Ustilaginales), e.g., the epibasidium forms eight primary sporidia which fuse in pairs (forming four H-shaped structures), and from these the secondary sporidia (secondary basidiospores) originate.

sporocarp (Gr. *sporá*, spore + *karpós*, fruit): a fruiting body that bears spores (a sporophore). Among the structures called sporocarps are the sporophores of the Protosteliales, such as *Nematostelium*, *Schizoplasmodium* and *Protostelium*, as well as the special structures of the endomycorhizogenous Zygomycetes (e.g., *Endogone*, of the Endogonales, and *Sclerocystis*, of the Glomales) which consist of a group of zygospores enveloped in a common membrane.

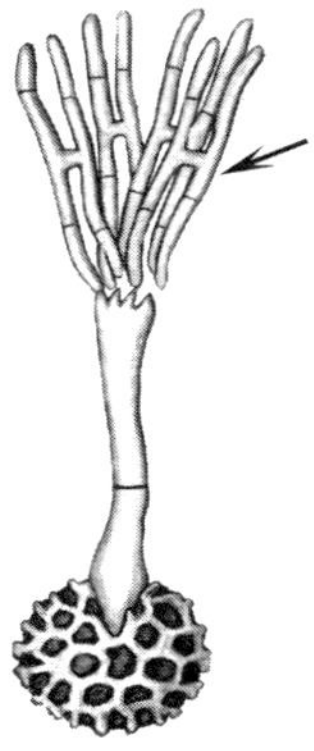

Sporidia of *Tilletia caries*, x 610.

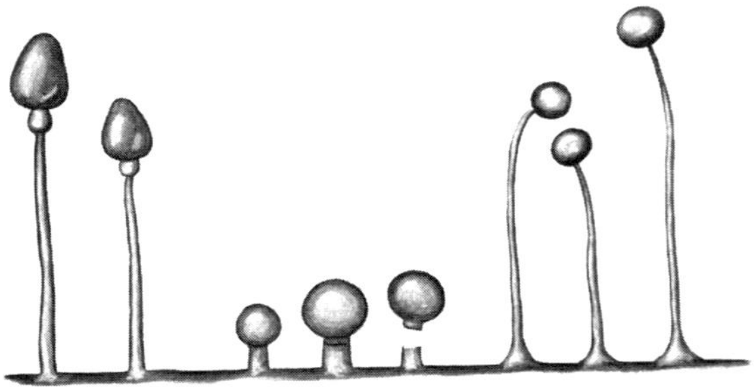

Sporocarps (left to right) of *Nematostelium ovatum*, *Schizoplasmodium cavostelioides* and *Protostelium mycophaga*, x 110.

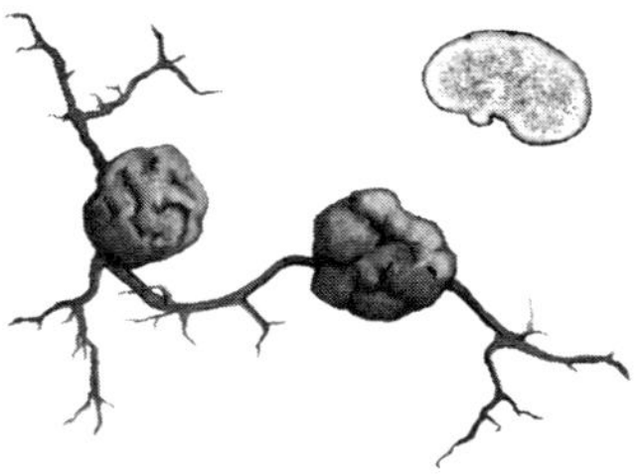

Sporocarps of *Endogone* sp., associated with the roots of *Pseudotsuga*, x 1.

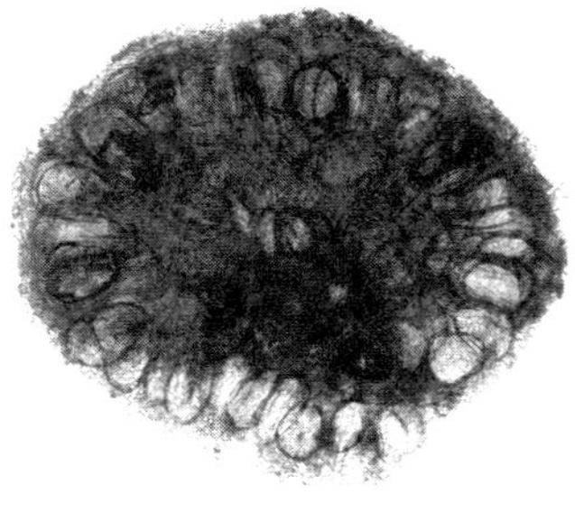

Sporocarp of *Sclerocystis* sp., x 80 (*CB*).

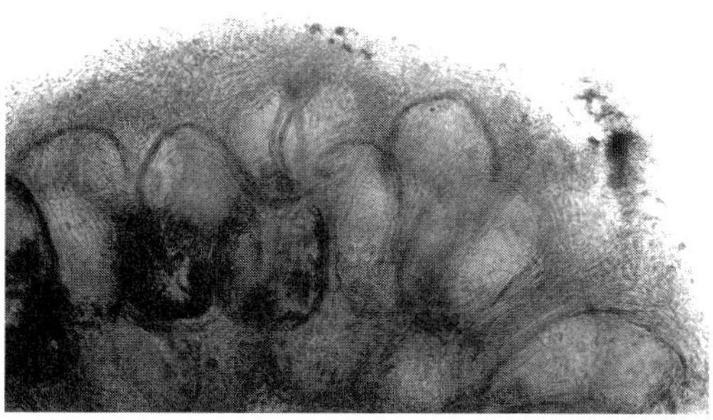

Sporocarp of *Sclerocystis* sp.
The spores inside are radially arranged, x 200 (*CB*).

sporocladium, pl. **sporocladia** (Gr. *sporá*, spore + *kládos*, little branch + L. dim. suf. *-ium*): a special type of hyphal branch, septate or aseptate, which produces unispored merosporangiola in the Zygomycetes of the order Kickxellales, e.g., in *Linderina*, *Coemansia* and *Spirodactylon*, among other gen.

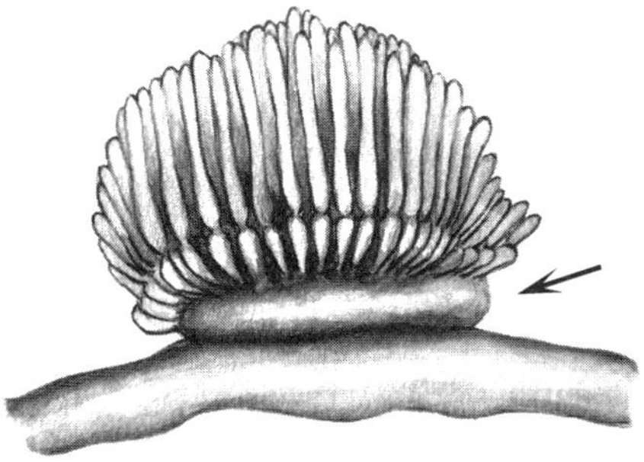

Sporocladium bearing pseudophialides with merosporangia of *Linderina pennispora*, x 850.

sporocyst (Gr. *sporá*, spore + *kýstis*, vesicle, cell): **1**. A cyst in which are produced asexual spores. **2**. The structure that develops from the nutriocyte in the Plectomycete *Ascosphaera apis* (Ascosphaerales), and which contains spore balls composed of ascospores of sexual origin. See **nutriocyte**.

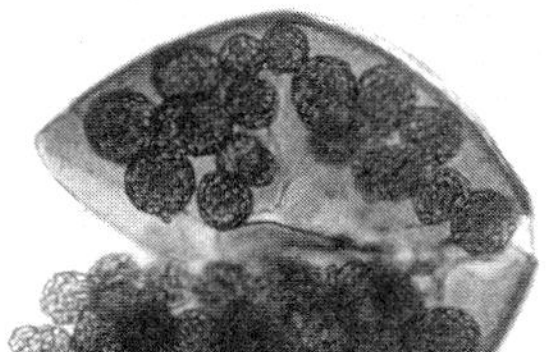

Sporocyst containing balls of ascospores of *Ascosphaera apis*, x 400 (*RTH*).

sporodochium, pl. **sporodochia** (Gr. *sporá*, spore + *docheíon*, recipient): a small, compact, cushion-shaped stroma on which are borne short conidiophores that produce the spores (conidia). This fructification, an asexual sporophore, is characteristic of the tuberculariaceous asexual fungi, such as *Epicoccum*, *Tubercularia* and *Volutella*, and of some melanconiaceous asexual fungi, such as *Pestalotia*.

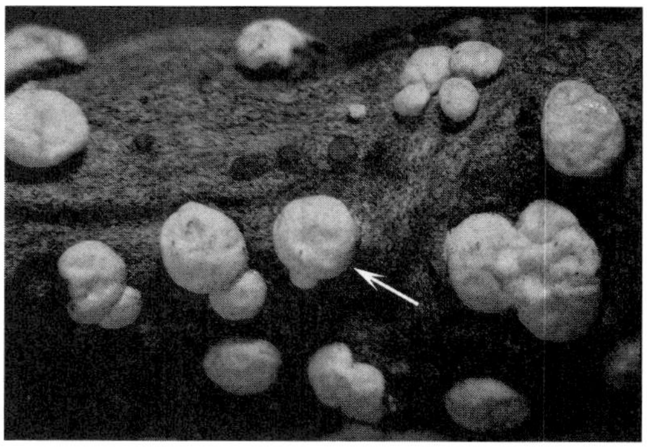

Sporodochia of *Tubercularia vulgaris* on a plum tree branch, x 3 (*MU*).

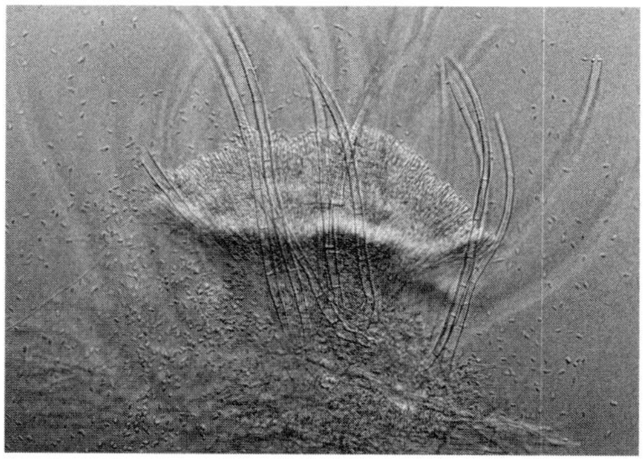

Sporodochium of *Volutella* sp., x 300 (*RTH*).

sporogenesis, pl. **sporogeneses** (Gr. *sporá*, spore + *génesis*, engenderment): the process by which spores are produced, whether asexual or sexual.

sporogenous (Gr. *sporá*, spore + *génos*, origin < *gennáo*, to engender, produce + L. *-osus* > OF. *-ous*, *-eus* > E. *-ous*, having, possessing the qualities of): a structure that produces spores or is capable of producing them. In the fungi, a sporogenous cell is the cell or hyphal branch of the sporangiophore of *Dimargaris crystalligena* (Dimargaritales) which produces the bispored merosporangia. The phialides of many imperfect fungi, such as *Aspergillus* (moniliaceous asexual fungi), are examples of sporogenous (conidiogenous) cells.

sporont (Gr. *sporá*, spore + *ón*, genit. *óntos*, a being): a thallus on which is produced any type of spore.

sporophagous (Gr. *sporá*, spore + *phágos*, one that eats + L. *-osus* > OF. *-ous*, *-eus* > E. *-ous*, having, possessing the qualities of): feeding on spores.

Certain thrips (insects of the order Thysanoptera) feed on spores of various species of moniliaceous and dematiaceous asexual fungi, and some beetles (Coleoptera) feed on myxomycete spores. Cf. **plasmodiophagous**.

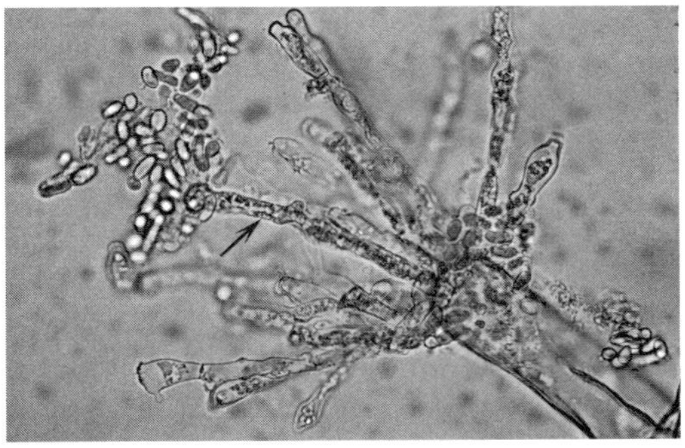

Sporogenous cells of *Dimargaris crystalligena*, x 660 (*PL & EAA*).

sporophore (Gr. *sporá*, spore + *-phóros*, bearer): any structure that bears spores, whether asexual or sexual.

sporoplasm (Gr. *sporá*, spore + *plásma*, formation, bland material with which a living being is formed): the portion of the protoplasm in sporangia and asci that is destined for the production of spores. Cf. **epiplasm**.

sporothallus, pl. **sporothalli** (Gr. *sporá*, spore + *thallós*, sprout, thallus): a thallus that produces sporangia with asexual spores; present in aquatic fungi in whose life cycle there occurs an alternation of generations, as in *Allomyces javanicus* (Blastocladiales). Cf. **gametothallus**.

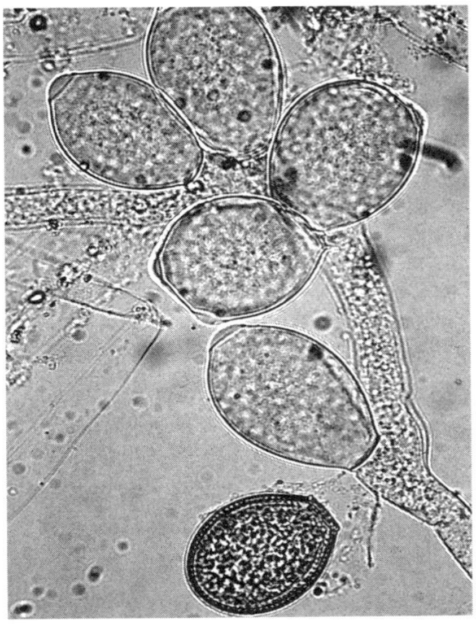

Sporothallus of *Allomyces javanicus* bearing thin-walled, hyaline zoosporangia (mitosporangia) and thick-walled, resting zoosporangia (meiosporangia), x 250 (*MU*).

sporotrichosis, pl. **sporotrichoses** (< gen. *Sporothrix* < Gr. *spóros*, spore + *thríx*, hair + *-osis*, state or condition): a subcutaneous, more or less localized, chronic mycosis of man and higher animals, acquired by the traumatic inoculation with plant splinters and thorns, or soil. Linfatic and linfocutaneous are the most common types of sporotrichosis but there are several clinical types, including the systemic or disseminated. The causal agent is the dimorphic fungus *Sporothrix schenckii* (moniliaceous asexual fungi).

squamule (L. *squamula*, small scale < *squama*, scale + dim. suf. *-ula* > E. *-ule*): a small shingle-like scale often present on the peridium, e.g., in *Endoptychum* (Hymenogastrales).

squamulose (L. *squamulosus*, scaly < *squamula*, dim. of *squama*, scale + *-osus*, full of, augmented, prone to > ME. *-ose*): *Lichens.* A thallus composed of numerous small lobes, scales or tiny scales; it can be considered intermediate in habit between the crustose and foliose lichens; it is present, e.g., in *Psora crenata*, *P. russellii*, *Cladonia didyma* and C. *pyxidata* (Lecanorales). Cf. **alepidote**.

Squamulose podetium of the thallus of *Cladonia didyma*, x 5 (*MU*).

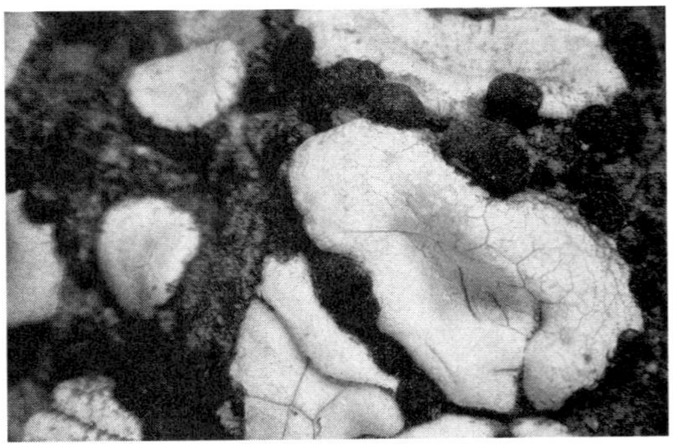

Squamulose thallus of *Psora russellii*, x 10 (*CB*).

squarrose (L. *squarrosus*, scaly, crusty, rough, scurfy < *squarra*, a gross scale + *-osus*, full of, augmented, prone to > ME. *-ose*): applied to a superficial covering of thick, rough, recurved scales; e.g., as in the basidiocarp of *Pholiota squarrosa* and *Ph. acutesquamosa* (Agaricales).

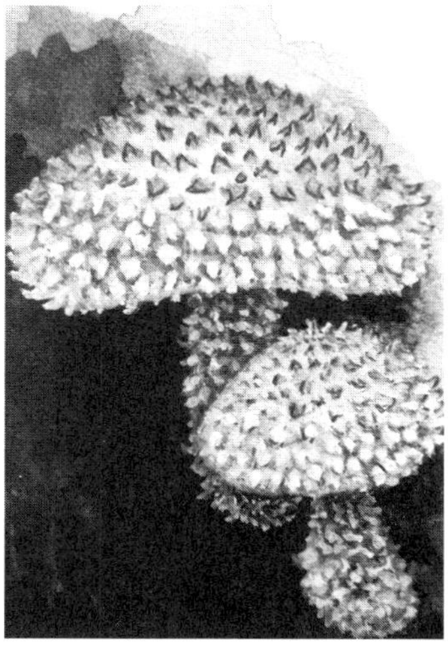

Squarrose basidiocarps of *Pholiota squarrosa*, x 0.7.

Stachel (G. *Stachel*, spine, thorn, sting): *Plasmodiophorales*. A cylindrical, sharp-pointed (aculeate) body that is formed in a specialized pouch or sheath (**Rohr**) in the encysted zoospores of *Plasmodiophora brassicae*. During the infection process, the zoospores encyst on the outer surface of the epidermal cells of cabbage root (the host) and the Rohr or sheath evaginates and forms a bulbous structure that adheres to the host cell wall. The Stachel perforates the host cell wall and the protoplast of the encysted zoospore penetrates the host cell, infecting it, to continue the life cycle of the fungus.

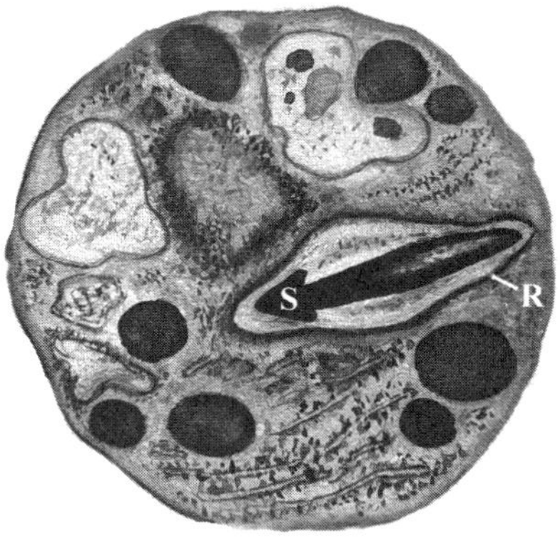

Stachel (S) contained in the Rohr (R) of an encysted zoospore of *Plasmodiophora brassicae*, seen in longitudinal section. Drawing based on a transmission electron micrograph published by Beckett *et al.* in their *Atlas of Fungal Ultrastructure*, 1974, x 24 000.

stalagmoid (Gr. *stalágmos*, drop, drip + L. suf. *-oide* < Gr. *-oeídes*, similar to): syn. of **dacryoid** and **lacrimoid**; an obconic spore with the elongated shape of a teardrop (stalagmospore); e.g., like the basidiospores of *Clitocybe gibba* (Agaricales).

stalk (ME. *stalke*, stem): see **stipe**.

staphylum, pl. **staphyla** (Gr. *stáphyle*, cluster + L. dim. suf. *-ium*): see **bromatium** and **gongylus**.

statismospore (Gr. *státis*, steadiness, immobility, or *statós*, standing, placed + *-ismós*, state or condition + *sporá*, spore): a spore that is not forcibly discharged from the sterigma of the basidium; the basidiospores of the Gasteromycetes (e.g., *Podaxis*, of the Hymenogastrales) are of this type, which distinguishes them from the basidiospores of the Hymenomycetes which are forcibly discharged from the basidia. Cf. **ballistospore**.

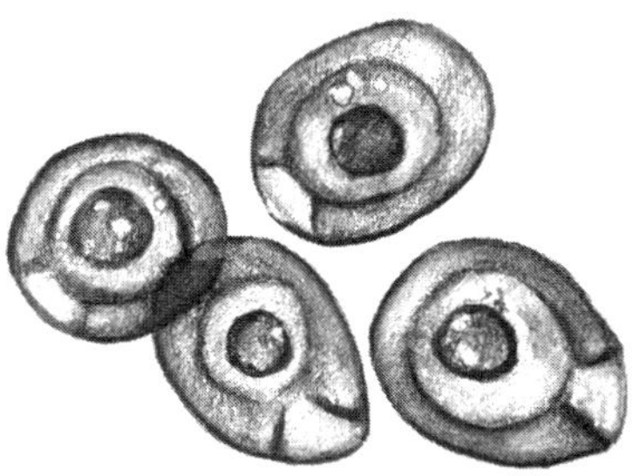

Statismospores of *Podaxis* sp., x 1 500.

statismosporic (Gr. *státis*, steadiness, immobility, or *statós*, standing, placed + *-ismós*, state or condition + *sporá*, spore + suf. *-íkos* > L. *-icus* > E. *-ic*, belonging to, relating to): having spores which cannot be forcibly discharged from the basidium (**statismospores**). Gasteromycetes, by definition, are statismosporic.

staurospore (Gr. *staurós*, cross + *sporá*, spore): a stellate asexual spore with three or more points. A Saccardoan term that is mainly applied to the conidial fungi, e.g., those of the gen. *Orbimyces* (dematiaceous asexual fungi), and *Riessia* and *Spegazzinia* (moniliaceous asexual fungi).

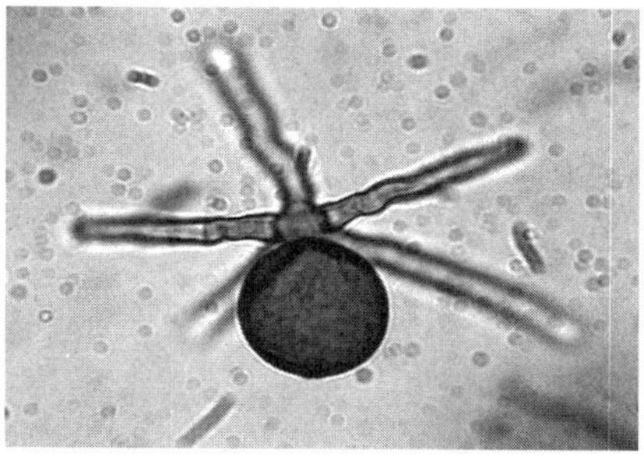

Staurospore of *Orbimyces spectabilis*, x 1 200 (*MU*).

steliogen (Gr. *stéle*, column + *géno*, origin < *gennáo*, to engender, produce): the finely granular portion of the cytoplasm of the prespore cell of the Protosteliales, which secretes the cellulosic tube of the sporocarp stalk during sporogenesis; it is present, e.g., in *Nematostelium ovatum*.

Steliogen of the sporocarp of *Nematostelium ovatum*, x 250.

stellate, **stelliform** (L. *stella*, star + suf. *-atus* > E. *-ate*, provided with or likeness; or + suf. *-formis* < *forma*, shape): having the shape of a star; e.g., like the sporangia of *Diderma radiatum* (Physarales), the fruiting body of *Geastrum saccatum* (Lycoperdales) or the ascospores of *Emericella variecolor* (Eurotiales).

Stellate peridial dehiscence of the sporangia of *Diderma radiatum*, x 10 (*MU*).

Stelliform basidiocarps of *Geastrum saccatum*, x 0.4 (*MRO*).

stenochora (Gr. *stenós*, narrow, limited + *chóra*, country): syn. of **endemic**; a species that has a more or less restricted geographic distribution. Cf. **eurichora**.

stenoecious (Gr. *stenós*, narrow, limited + *oîkos*, house + L. *-osus* > OF. *-ous*, *-eus* > E. *-ous*, having, possessing the qualities of): an organism that is limited to a narrow range of environmental conditions, such as the aquatic fungi of the gen. *Aqualinderella* (Rhipidiales), which only live in stagnant water with little or no oxygen. Cf. **euryoecious**.

stenospore (Gr. *stenós*, narrow + *sporá*, spore): a spore having the shape of a bullet or projectile; e.g., like the basidiospores of *Lepiota* (Agaricales).

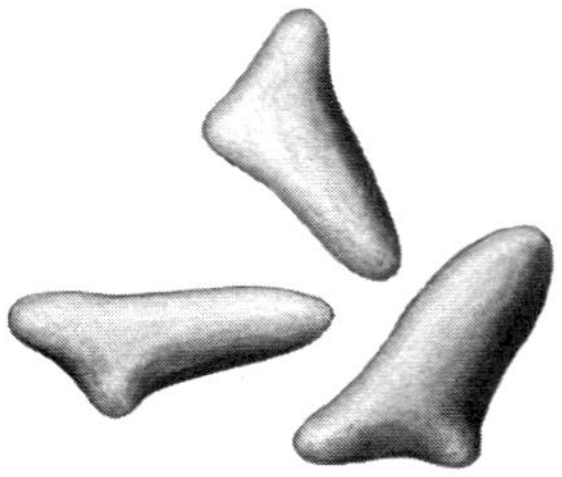

Stenospores of *Lepiota* sp., x 2 000.

stenoxenous (Gr. *stenós*, narrow + *xénos*, stranger; here, in the sense of host + L. *-osus* > OF. *-ous*, *-eus* > E. *-ous*, having, possessing the qualities of): used for parasitic organisms with a narrow host range; this may involve a few species, a single species, or only varieties of species susceptible to invasion by the parasite. Among the fungi, e.g., the cereal rusts are stenoxenous. Cf. **eurixenous**.

stephanocyst (Gr. *stéphanos*, crown + *kýstis*, bladder, cell): a bicellular structure in certain basidiomycetes, such as *Hyphoderma praetermissum* (Aphyllophorales), with the basal cell in the shape of a goblet and the terminal cell globose.

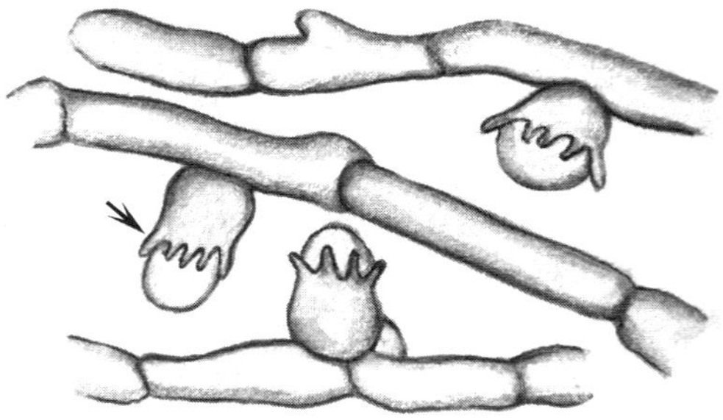

Stephanocysts of *Hyphoderma praetermissum*, x 1 000.

stereoma, pl. **stereomata** (Gr. *steréoma*, base or fundament, skeleton): *Lichens*. Tissues and cells of the thallus in certain lichens that provide mechanical support; e.g., in the Lecanorales (*Cladonia*, *Alectoria*). See **scleroplectenchyma**.

stereonema, pl. **stereonemata** (Gr. *stereós*, solid + *nêma*, genit. *nêmatos*, filament): a compact filament, free or anastomosed, which together with others make up the capillitium in the sporangia of many Myxomycetes, such as those of the gen. *Calomyxa* (Trichiales). Cf. **coelonema**.

sterigma, pl. **sterigmata** (Gr. *stérigma*, support): **1**. *Basidiomycetes*. Each of the small diverticula (usually 4) that form on the apex of each basidium, and which support the basidiospores. **2**. *Conidial fungi*. The small, tapered hyphal branch which bears a ballistospore (holoblastic conidium) of *Itersonilia* and *Sporobolomyces* (sporobolomycetaceous asexual yeasts). **3**. *Oomycota*. The small, sporangium-bearing structure of the sporangiophores of *Basidiophora entospora* (Peronosporales). **4**. *Lichens*. Elongated cells, simple or slightly branched, aseptate, situated in the interior of the spermogonium and which produce spermatia.

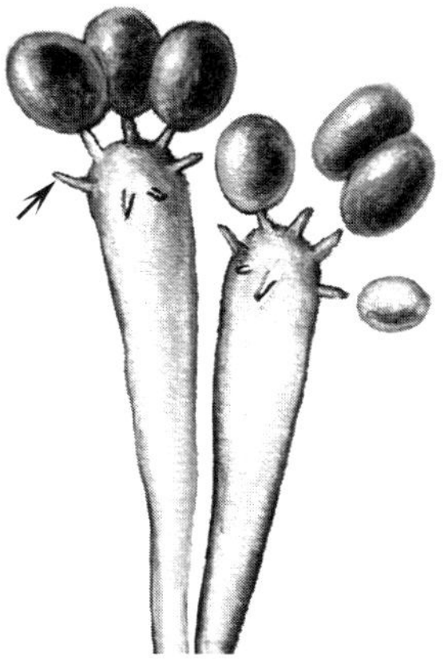

Sterigmata of the sporangiophores of *Basidiophora entospora*, x 115.

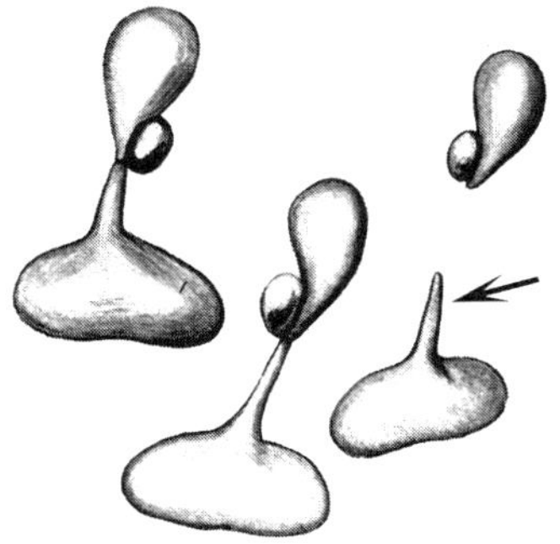

Sterigmata of the somatic cells of *Itersonilia perplexans*, x 2 100.

stichobasidium, pl. **stichobasidia** (NL. *stichobasidium* < Gr. *stíchos*, row, line + *basídion*, dim. of *básis*, base): the main type of basidium without septa (holobasidium or autobasidium), elongate in shape, cylindric, in which the achromatic spindle is arranged parallel to the main axis during the second meiotic division of the nucleus, as happens, e.g., in the gen. *Cantharellus* and *Craterellus* (Aphyllophorales). Cf. **chiastobasidium**.

stigmatocyst (Gr. *stígma*, genit. *stígmatos*, to sting, puncture + *kýstis*, vesicle, cell): the terminal cell of a capitate hyphopodium, from which develops the ascocarp of the Meliolales. Also known as nodal cell. Cf. **stigmatopod**.

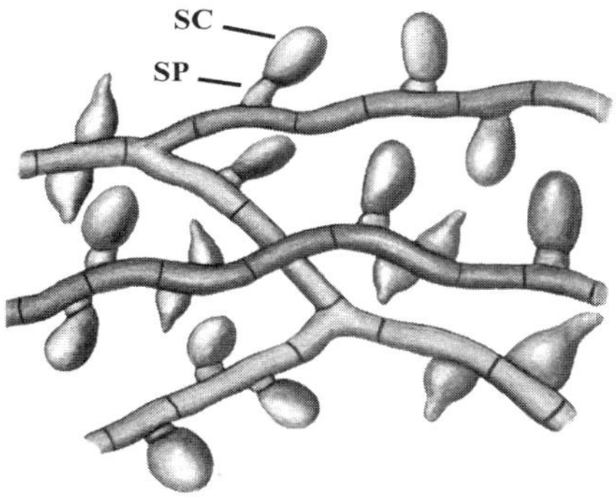

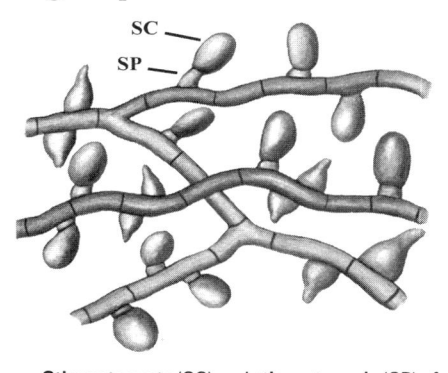

Stigmatocysts (SC) and **stigmatopods** (SP) of the hyphopodia of *Meliola caesariae*, x 350.

stigmatopod, **stigmatopodium**, pl. **stigmatopodia** (Gr. *stígma*, genit. *stígmatos*, to sting, puncture + NL. *podium* < Gr. *pódion*, dim. of *poús*, *podós*, foot): the basal or supporting cell of a capitate hyphopodium, which is considered as a reminiscence of the appressoria of the Erysiphales and possible relationship to the haustoria. The stigmatopodium (basal) and the stigmatocyst (terminal) are the two components of the capitate hyphopodium of the Meliolales, such as *Meliola caesariae*. Cf. **stigmatocyst**.

stilboid (< gen. *Stilbum*, of the stilbellaceous asexual fungi + L. suf. *-oide* < Gr. *-oeídes*, similar to): a structure without spores, similar to a capitate basidiocarp, that functions as a propagule, present in *Mycena citricolor* and other Agaricales. This structure was described as belonging to the gen. *Stilbella* (=*Stilbum*), as *S. flavida*, probably due to confusing the stilboid with the synnema that is formed by the stilbellaceous asexual fungus.

stinkhorn (ME. *stinken* < OE. *stinkan*, to emit a smell + ME. < OE. *horn*, horn < L. *cornu*, horn): a basidioma of certain Phallales, such as *Phallus* and *Dictyophora*, which emit a strong offensive odor.

stipe (L. *stipes*, genit. *stipitis*, stem, pedicel, foot): **1**. The part of a conidiophore that supports the conidiogenous cells (e.g., *Periconia byssoides*, a dematiaceous asexual fungus), **2**. A foot or stalk that supports the pileus of a basidiocarp (e.g., *Amanita frostiana* and *Marasmius siccus*, of the Agaricales) or the cap of a stipitate ascocarp, such as that of *Helvella lacunosa* (Pezizales). Another term for pseudopodetium. **3**. A more or less parallel tissue system which supports the gleba but has no direct structural connection with it, e.g., in *Tulostoma* (Tulostomatales).

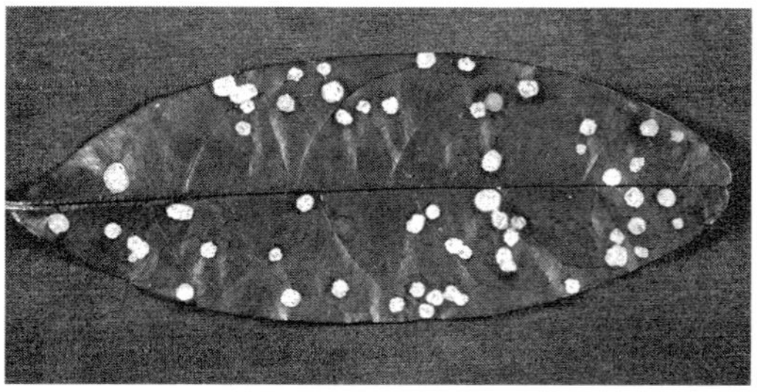

Stilboids of *Mycena citricolor* on the upper surface of a coffee leaf, x 1 (*RTH*).

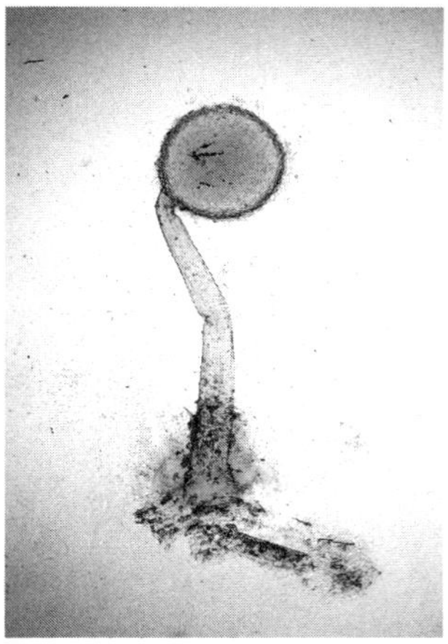

Stilboid of *Mycena citricolor*, x 60 (*MU*).

Stipe of the conidiophores of *Periconia byssoides*, x 500 (*CB*).

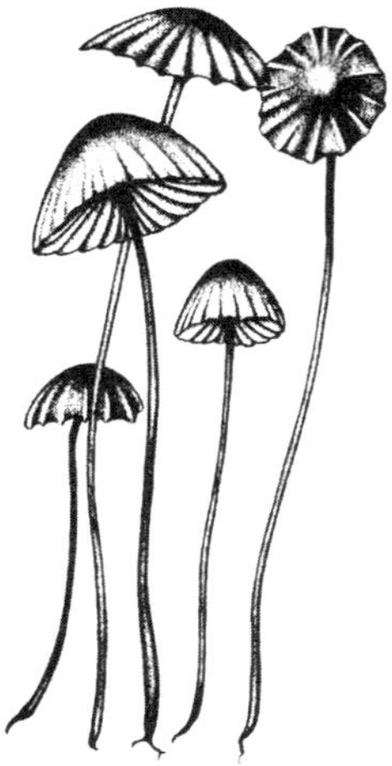

Stipe of the basidiocarps of *Marasmius siccus*, x 1.

stipe-columella, pl. **columellae** (L. *stipes*, stem, pedicel + *columella*, dim. of *columna*, column): a regular stipe or stalk which supports the receptaculum and, in addition, penetrates the gleba, forming a percurrent columella, e.g., in *Podaxis* (Hymenogastrales). Also called **dendritic columella** and **percurrent columella**.

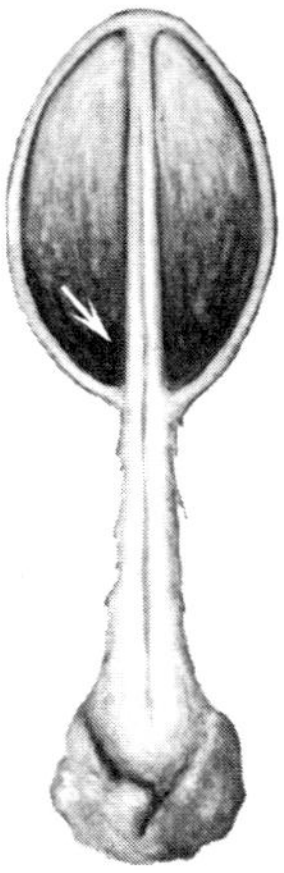

Stipe-columella of the fruiting body of *Podaxis pistillaris*, seen in longitudinal section, x 0.5.

stipitate (L. *stipitatus*, having or borne on a stipe < *stipes*, genit. *stipitis*, stem, pedicel, foot + suf. *-atus* > E. *-ate*, provided with or likeness): having a **stipe**. Cf. **apodal**, **pedicellate** and **sessile**.

stipitipellis (L. *stipes*, genit. *stipitis*, stem, pedicel, foot + *pellis*, skin): see **pellicule**.

stolon (L. *stolo*, genit. *stolonis*, shoot, sprout, branch): an aerial vegetative running hypha, unbranched, that connects two groups or fascicles of rhizoids; characteristic of *Rhizopus* and other gen. of Mucorales. Cf. **pseudostolon**.

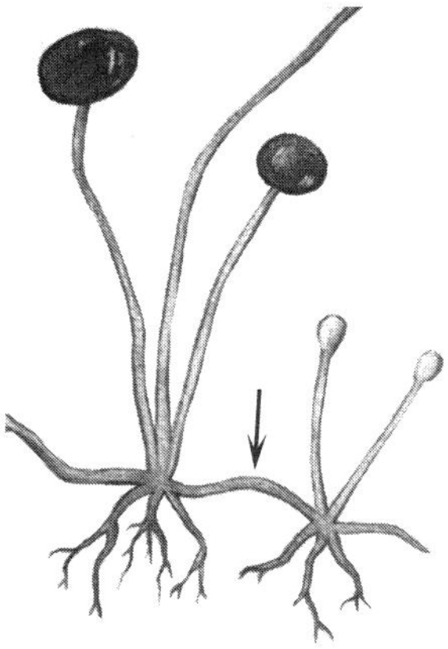

Stolon of *Rhizopus nigricans*, x 1 000.

stoma, pl. **stomata** (Gr. *stóma*, mouth): an opening or ostiole through which are liberated the spores from the sporiferous portion of a fructification; the term is usually restricted to the ostiole of *Calostoma* and *Tulostoma* (Tulostomatales), and to the multiple ostioles (**myriostomes**) of *Myriostoma* (Lycoperdales).

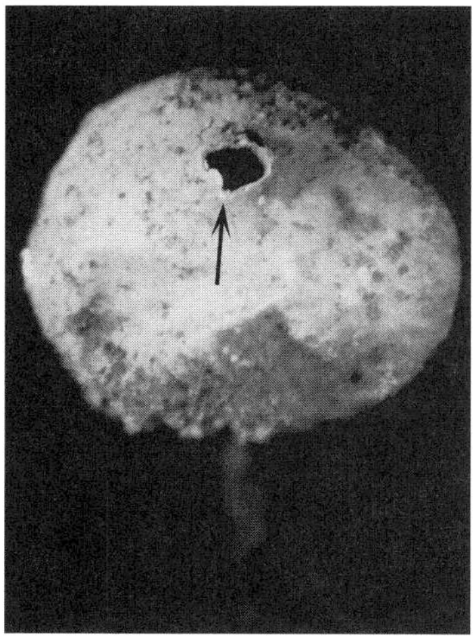

Stoma of the fruiting body of *Tulostoma* sp., x 4 (*MU & CB*).

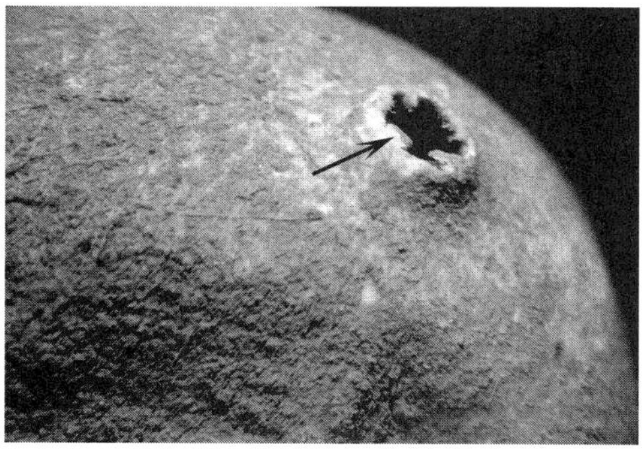

Stoma of the fruiting body of *Tulostoma* sp., x 6 (*MU & CB*).

stomapod, **stomapodium**, pl. **stomapodia**, **stomatopodium**, pl. **stomatopodia**, **stomopodium**, pl. **stomopodia** (Gr. *stóma*, genit. *stómatos*, mouth + NL. *podium* < Gr. *pódion*, dim. of *poús*, *podós*, foot): *Meliolales*. A hyphal branch that penetrates through the open stomata of a host plant to develop in the mesophyll and form numerous haustoria.

stone-fungus, pl. **fungi** (ME. < OE. *stan*, stone, akin to OHG. *stein*; L. *fungus*, fungus): the hard stone-like pseudosclerotium of *Polyporus tuberaster* (Aphyllophorales). It can be eaten after being watered. The Canadian tuckahoe is the same species.

strain (ME. *strene* < OE. *streon*, lineage, race): a fungus isolated in pure culture and characterized taxonomically, with one or more morphological, physiological or other properties that distinguishes it from other isolates of the same species. Individual strains of a fungus in a culture collection are distinguished by a code (of letters or numbers, or both), whether the collection is personal or internationally recognized, such as the American Type Culture Collection (ATCC) of the United States, the Centraalbureau voor Schimmelcultures (CBS) of the Netherlands, and other important national collections, in which are maintained numerous strains of fungi, yeasts, bacteria, viruses, algae, protozoans and other organisms. Fungi or other organisms maintained in culture can represent a **type culture** (i.e., the original culture of an organism on which a new name is based), or they can be strains of existing species that have a particular characteristic, such as the capacity to form antibiotics or toxins, or that are pathogenic; these characteristics of the original strain can be preserved and recovered in subsequent cultivations. Cf. **isolate**.

Stramenopila (L. *stramen*, straw + *pila* < *pilus*, hair): in the classification system of *Introductory Mycology* (Alexopoulos *et al.*, 1996) it corresponds to a monophyletic kingdom that includes the phyla Oomycota, Hyphochytriomycota, and Labyrinthulomycota as separate lineages among certain algal groups. The term Stramenopila was introduced by Patterson (1989) in recognition of the new monophyletic concept of these organisms based in part on flagellum hair structure. The members of these three phyla have zoospores with mastigonemate or tinsel flagella, that is, bearing hairs or mastigonemes.

straminipilous or **straminopilous** (L. *stramen*, genit. *straminis*, of, relating to, straw + *pilus*, hair + L. *-osus* > OF. *-ous*, *-eus* > E. *-ous*, having, possessing the qualities of): bearing tripartite tubular hairs, like the auxiliary cysts of *Saprolegnia* (Saprolegniales) which bear a tuft of tripartite tubular hairs (straminipilous cysts). The term is applied to cells (whether uniflagellate, multiflagellate or non-flagellate) and/or flagella, in the latter case called straminipilous flagella or **flimmergeissel**.

straw mushrooms (ME. < OE. *streaw*, straw; ME. *musseroum* < MF. *mousseron* < LL. *mussirion*, mushroom): the edible basidiomata of *Volvariella volvacea* and *V. diplasia* (Agaricales). Widely used in tropical regions of Asia (also called the paddy straw or Chinese mushrooms).

striate, **striated** (L. *striatus*, lined < *stria*, line, groove + suf. *-atus* > E. *-ate*, provided with or likeness): applied to surfaces that have nearly parallel ridges, grooves or furrows, as is observed in the wall of many spores, e.g., the ascospores of *Nectria haematococca* (Hypocreales), the hysterothecium of *Hysterium* (Dothideales), the teliospores of *Cintractia pachyderma* (Ustilaginales), and the basidiospores of *Boletus chrysenteroides* (Agaricales) and *Ramaria* sp. (Aphyllophorales). Other structures are also striate, such as the pileus of *Coprinus micaceus* (Agaricales) and the fruiting bodies of some Nidulariales and Lycoperdales.

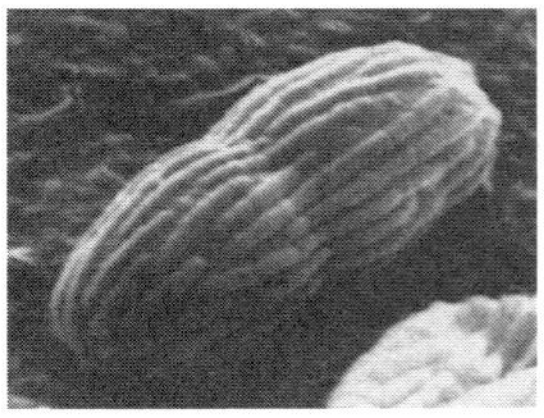

Striate wall of an ascospore of *Nectria haematococca*, x 2 800 (*RTH*).

Striate hysterothecium of an unidentified loculoascomycete, x 15 (*CB*).

Striate pileus of the basidiocarps of *Coprinus micaceus*, x 0.2.

strigose, **strigous** (NL. *strigosus*, bristled < *striga*, a rigid, sharp hair + *-osus*, full of, augmented, prone to > ME. *-ose*; or + *-osus* > OF. *-ous*, *-eus* > E. *-ous*, having, possessing the qualities of): syn. of **hirsute** and **hispid**; having long, stiff, sharp-pointed hairs; e.g., like the pileus of the basidiocarps of *Panus badius* and *Hexagonia* (=*Pogonomyces*) *hydnoides* (Aphyllophorales).

Strigose basidiocarp of *Panus badius*, x 5 (*MU*).

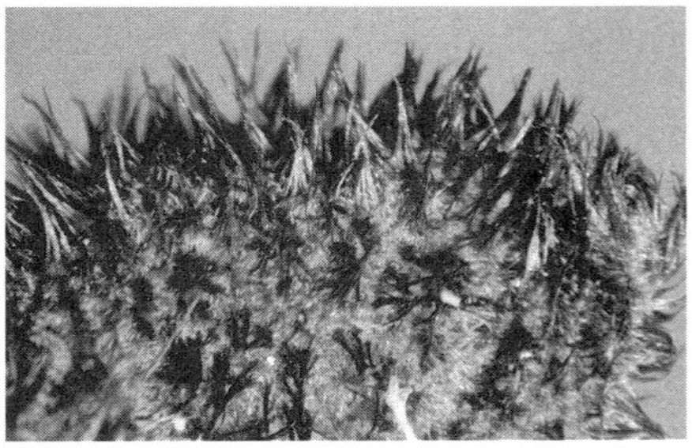

Strigose basidiocarp of *Panus badius* (a close up of the pileus), x 20 (*MU*).

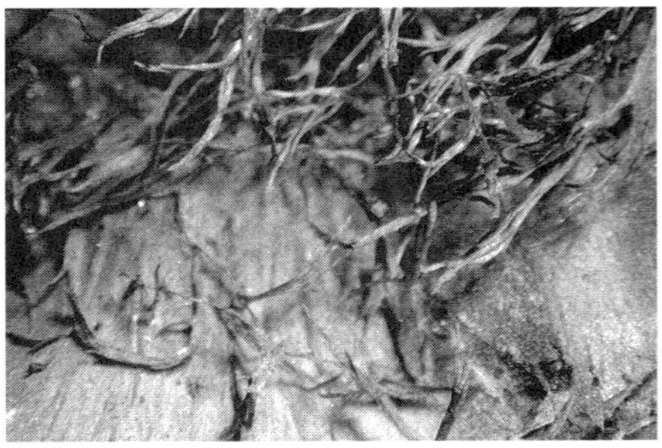

Strigose surface of a basidiocarp of *Hexagonia* (=*Pogonomyces*) sp., x 20 (*MU*).

stroma, pl. **stromata** (Gr. *strõma*, bed, cushion): a compact mass of somatic hyphae, made up of plectenchyma (prosenchyma, pseudoparenchyma, or both) on or in which are commonly produced fertile hyphae that generate asexual or sexual reproductive organs, such as sporodochia, pycnidia, perithecia, apothecia, etc. For example, in the Pyrenomycete *Daldinia concentrica* (Xylariales) the stroma is pulvinate and contains perithecia embedded in the peripheral layer. Other examples of stromata are *Rhytisma acerinum* (Rhytismatales) and *Broomeia congregata* (Lycoperdales). See **ectostroma** and **endostroma**.

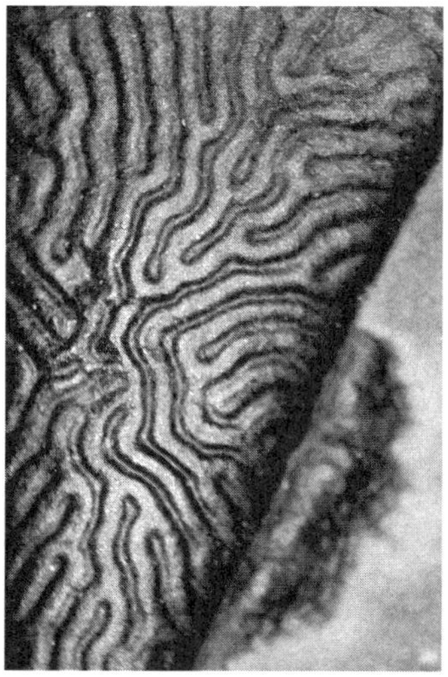

Stroma of *Rhytisma acerinum* on a leaf of maple, x 60 (*RTH*).

Stroma of *Broomeia congregata* bearing numerous spore sacs on the upper part, x 0.9.

stylospore (L. *stilus*, *stylus* < Gr. *stýlos*, column, pick, awl + *sporá*, spore): **1**. *Mucorales*. An asexual reproductive spore, united with the hypha that produces it by a pedicel or slender sterigma. Stylospores are characteristic of the family Mortierellaceae (Mucorales), e.g., *Mortierella indohii*, in which case they also are known as chlamydospores. **2**. *Ascomycetes*. A small filiform spore believed to function as a spermatium; they usually cannot be germinated.

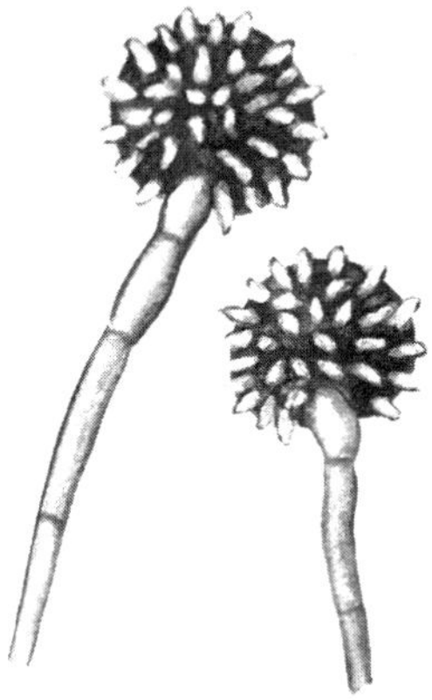

Stylospores of *Mortierella indohii*, x 750.

subalate (L. pref. *sub-*, low, attenuating the word that follows + *alatus*, winged < *ala*, wing + suf. *-atus* > E. *-ate*, provided with or likeness): generally applied to spores on whose wall the borders are of smaller proportions than those called winged (**alate**). Also called ribbed, for the ridges or ribs, as in the basidiospores of *Boletellus russellii* (Agaricales).

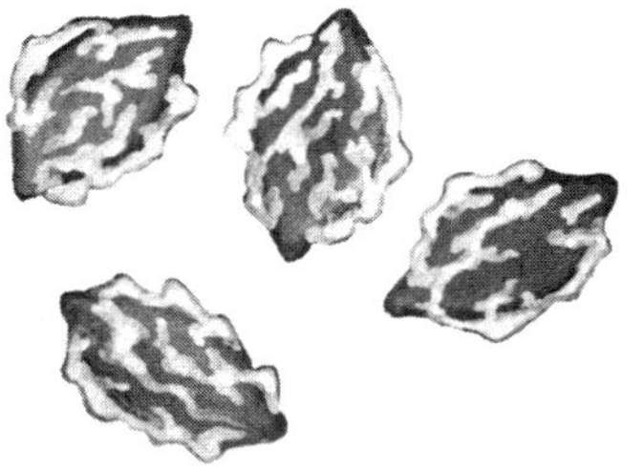

Subalate basidiospores of *Boletellus russellii*, x 1 980.

subcutis (L. pref. *sub-*, low + *cutis*, skin): the lower layer (stratum) of the cuticular layer of the pileus of the Agaricales; it is found beneath the **epicutis**. Also called **hypodermis**.

suberose (NL. *suberosus*, corky < L. *suber*, cork + *-osus*, full of, augmented, prone to > ME. *-ose*): having the consistency of cork; e.g., like the fruiting body of *Inonotus* and other Aphyllophorales.

subgleba, pl. **subglebae** (L. pref. *sub-*, below + *gleba*, lump, mass): *Gasteromycetes*. A sterile region located beneath the gleba in the fructification of some Lycoperdales. In *Lycoperdon*, the subgleba is delimited by a diaphragm, is composed of chambers and contains a pseudocolumella that goes from rudimentary to well developed. In other gen. of Lycoperdales the subgleba is lacking, as in

Mycenastrum, *Disciseda* and *Langermannia*, and in others in which it is present its composition is cellular or interwoven and it is separated from the gleba by a diaphragm, as in *Bovistella* and *Morganella*, respectively. See **pseudostem** and **pseudocolumella**.

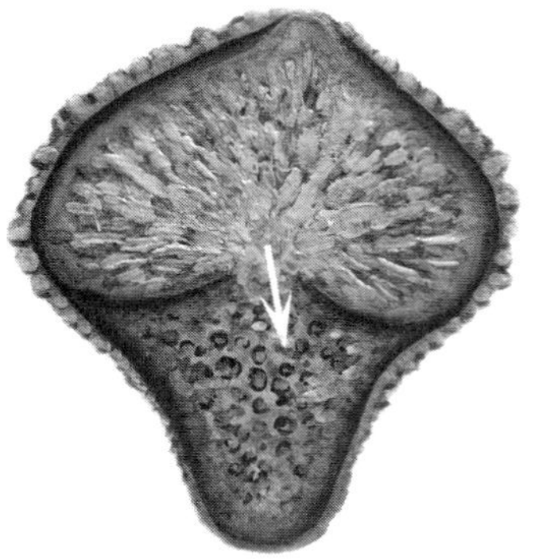

Subgleba of the fruiting body of *Lycoperdon perlatum*, seen in longitudinal section, x 1.

subglobose (L. pref. *sub-*, below, almost + *globosus* round as a ball < *globus*, ball + *-osus*, full of, augmented, prone to > ME. *-ose*): almost globose. Applied to structures (spores, sporocarps, etc.) with a length:width ratio of between 1:1.05 and 1:1.15. In polar axis it is sphaeroid prolate in shape and in equatorial axis it is sphaeroid oblate. For example, the sporangiospores of *Pilobolus longipes* (Mucorales), the asci of *Eurotium repens* (Eurotiales), and the basidiospores of *Amanita phalloides* (Agaricales) are subglobose.

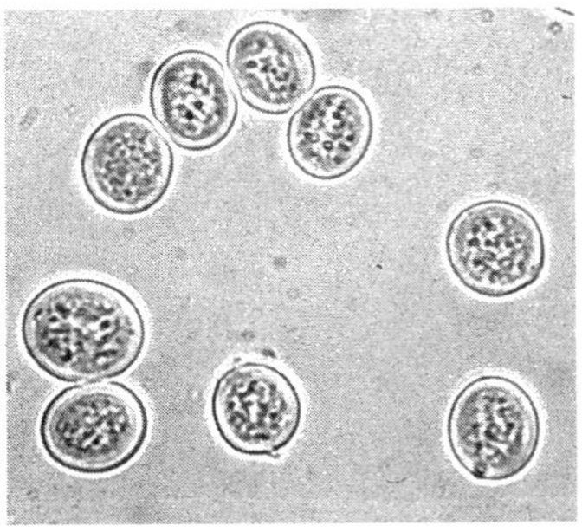

Subglobose sporangiospores of *Pilobolus longipes*, x 1 250 (*EAA*).

subhypothallic (L. *sub-*, beneath, under + Gr. *hypó*, under + *thallós*, shoot + suf. *-íkos* > L. *-icus* > E. *-ic*, belonging to, relating to): *Myxomycetes.* A type of sporocarp development in which the plasmodium is concentrated in various places to form hemispherical mounds that separate from one another by reabsorption of the protoplasmatic filaments that initially interconnected them. The calcareous layer of the plasmodium forms the hypothallus, when this is present, and the mounds continue elongating upward to form columnar structures called papillae, which are the primordia of the sporocarps. As these papillae elongate, their bases constrict to form the pedicel of the sporocarps; there is a protoplasmic flow toward each expanding primordium from the adjacent areas and from below on the substrate, so that the apex inflates while the pedicel contracts and forces the apex upward. The distinctive characteristic of this type of development is the continuity of the peridium with the pedicel and with the hypothallus of each sporocarp. It is called subhypothallic development because the sporocarp emerges from the lower part of the hypothallus, which is differentiated in the upper surface of the plasmodium. It is present in the orders Liceales, Trichiales, Echinosteliales and Physarales, and is typified by *Arcyria cinerea* (Trichiales). The other type of sporocarp development is the **epihypothallic**, which is characteristic of members of the order Stemonitales.

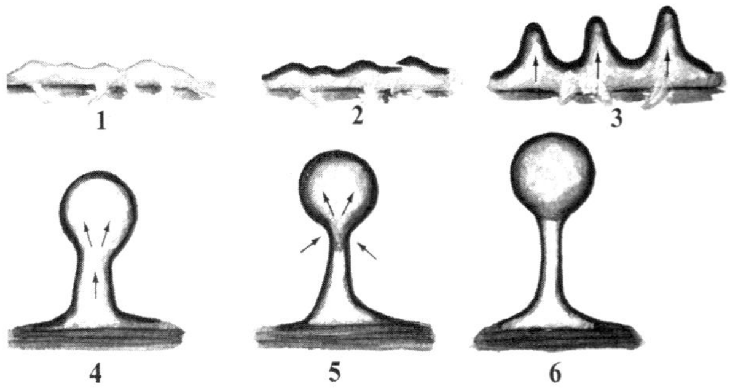

Subhypothallic type of sporangial development in nonstemonitid orders of Myxomycetes. 1. Emerging plasmodium. 2. Formation of supraplasmodial hypothallus. 3. Differentiation into primordia; protoplasmic flow into primordia. 4. Enlargement of primordia as protoplasmic flow continues; apex expansion. 5. Continuous expansion at apex as wall of stalk constricts. 6. Full-size sporangium prior to capillitial formation; exterior stalk is continuous with peridium and with supraplasmodial hypothallus. Protoplasm may remain in the stalk in some forms; stalk may be charged with debris in others. Diagram based on figures and information published by I. K. Ross in *Mycologia* *65*:477-485, 1973, x 20.

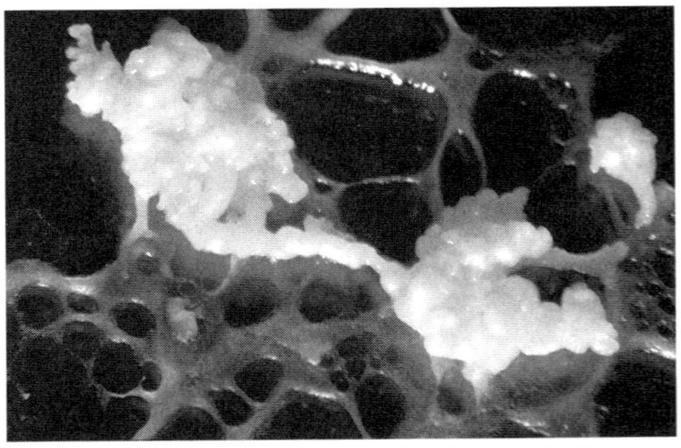

Subhypothallic type of sporangial development in *Physarum polycephalum*. Differentiation into primordia, x 6 (*MU*).

Subhypothallic type of sporangial development in *Physarum polycephalum*. Differentiation into primordia, x 10 (*MU*).

subiculum, pl. **subicula** (L. *subiculum* < *subex*, base, bottom, something located below + dim. suf. *-culum*): a felt-like layer or trama of loosely intertwined hyphae (hyphenchyma), which covers the substrate, and on which are seated fruiting bodies (perithecia, apothecia, etc.); the subiculum can be reticular, woolly or crustaceous. In the species of *Hypomyces*, such as *H. lactifluorum* (Hypocreales), e.g., many of them parasites of basidiomycete sporophores, the perithecia are sunken in a thick, cottony subiculum, whose trama approximates that of a stroma. *Nitschkia* (Sordariales) is a saprobe that forms its perithecia on a stroma-like subiculum that is embedded in dead wood.

Subiculum of *Nitschkia cupularis*, embedded in the branch tissues of *Acer*; the erumpent, gregarious, perithecia lie on the subiculum, x 58.

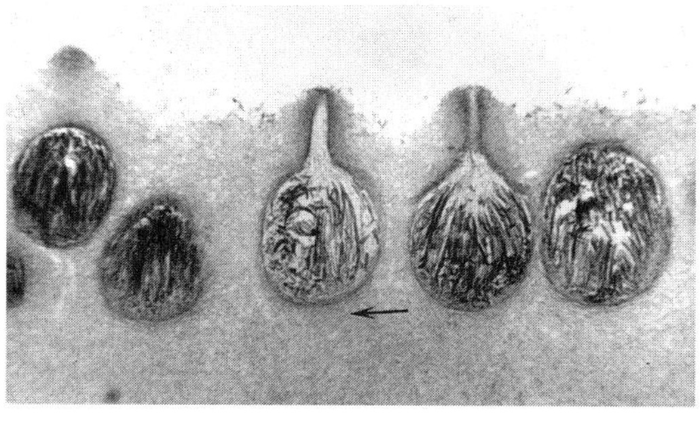

Subiculum with embedded perithecia of *Hypomyces lactifluorum* parasitizing a basidiocarp of *Lactarius* sp., seen in longitudinal section, x 47 (*RTH*).

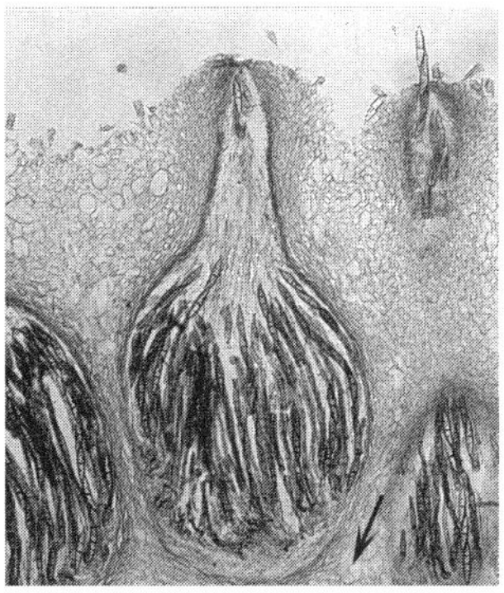

Subiculum with embedded perithecia of *Hypomyces lactifluorum* parasitizing a basidiocarp of *Lactarius* sp., seen in longitudinal section, x 75 (*RTH*).

subpellis (L. *sub-*, below, underneath + *pellis*, skin): see **pellicule**.

subprolate (L. *sub-*, low, to attenuate the following word + *prolatus*, bringing forward, elongated in the direction of the main axis): generally applied to spores broadly ellipsoidal in shape, not so elongate, with a width:length ratio of between 1:1.15 and 1:1.30.

subradicating (L. *sub-*, low, almost, less + *radicatus*, radicate, provided with a root): refers to a stipe provided with a small prolongation, but which does not attain a radiciform extension, as is seen in *Collybia maculata* (Agaricales).

substomatal vesicle (L. *sub*, below + NL. *stoma* < Gr. *stóma*, mouth + L. suf. *-alis* > E. *-al*, relating to or belonging to; L. *vesicula*, dim. of *vesica*, bladder): a structure formed by rust fungi in the substomatal cavity of the host leaf. For example, in the peanut rust fungus, *Puccinia arachidis* (Uredinales), a tiny penetration peg arises from the appressorium (which in turn was formed by the germinating urediniospore) and enters the host leaf by growing into the stomatal opening between the guard cells (a type of penetration referred to as indirect penetration). The infection peg then expands to form the substomatal vesicle into which the cytoplasm and nuclei of the appressorium move; these nuclei divide synchronously and an outgrowth termed the primary infection hypha develops from the substomatal vesicle, and from this hypha the haustorium formation is initiated.

substrate (ML. *substratum*, the base on which an organism lives): the substance on which an enzyme acts (enzymology); the substances (e.g. culture medium constituents) utilized by a microorganism for growth (microbiology). Cf. **substratum**.

substratum, pl. **substrata** (ML. *substratum*, the base on which an organism lives; an underlying support): the material on which an organism is growing or to which it is attached (ecology). Cf. **substrate**.

subulate, **subuliform** (NL. *subulatus*, awl-shaped < L. *subula*, awl + suf. *-atus* > E. *-ate*, provided with or likeness; or + L. *-formis* < *forma*, shape): narrowed toward the apex, ending in a broad point; also called awl-shaped; not as sharp-pointed as acicular. For example, the cheilocystidia of *Mycena atkinsoniana* and the pleurocystidia of *Boletus* (Agaricales) are subulate. Cf. **obsubulate**.

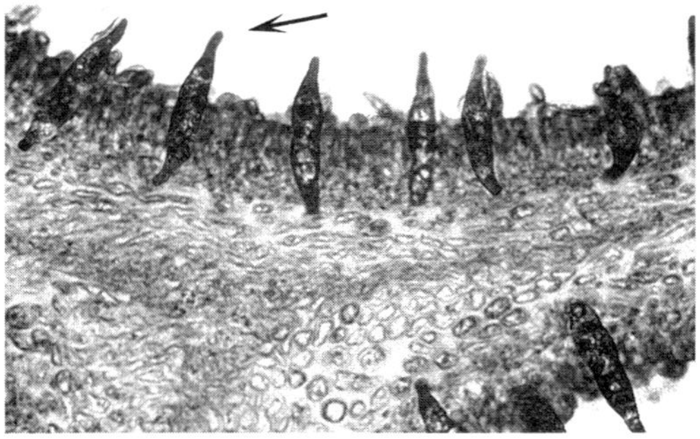

Subulate pleurocystidia of the hymenium of *Boletus* sp., x 246 (*MU*).

sulcate (L. *sulcatus*, furrowed < *sulcus*, furrow, canal + suf. *-atus* > E. *-ate*, provided with or likeness): provided with furrows, like the colonies of *Penicillium puberulum* (moniliaceous asexual fungi), the surface of the thallus of *Parmelia sulcata* (Lecanorales), the ostiole tips of many Diatrypaceae (Xylariales), and the peristome of the fruiting body of *Geastrum pectinatum* (Lycoperdales). Also called **plaited** or **plicate**.

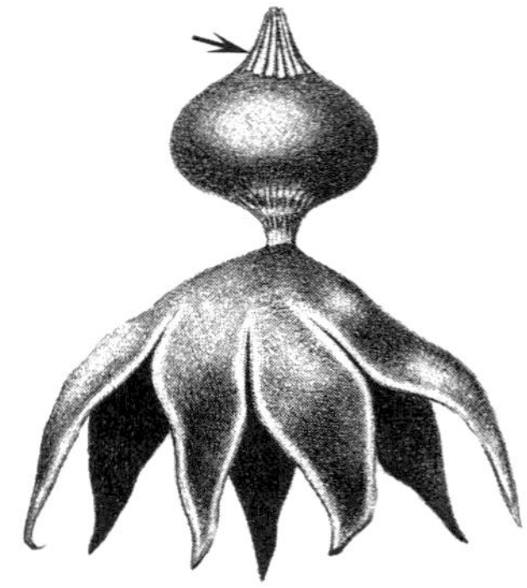

Sulcate peristome of the fruiting body of *Geastrum pectinatum*, x 1.3.

supine (L. *supinus*, lying on the back with the face upward, with the E. ending *-ine* < suf. *-inus*): refers to fructifications that are closely applied to the substratum.

suprahilar (L. *supra*, on + *hilum*, scar or notch of bean seeds; here, spore halo + suf. *-aris* > E. *-ar*, like, pertaining to): situated on or above the hilum. In certain basidiospores (of the Russulaceae, and of the gen. *Galerina*, Agaricales), the **suprahilar depression** corresponds to the rounded areole that is formed in the vicinity of the hilum, above it, and on its internal side, i.e., in the flat face or slightly concave face of the basidiospores. See **hilar appendix**.

suprahilar depression (L. *supra*, on, above + *hilum*, areola, halo + suf. *-aris* > E. *-ar*, like, pertaining to; ME. < ML. < LL. *depressio*, genit. *depressionis*, press down < L. *depressus*, pressed down + *-io*, *-ionis* > E. suf. *-ion*, result of an action, state of): *Basidiomycetes*. A smooth zone on the wall of certain basidiospores, near the **hilar appendix**, as in *Galerina* (Agaricales). Also called **suprahilar disc**, **suprahilar plage** and **suprahilar spot.**

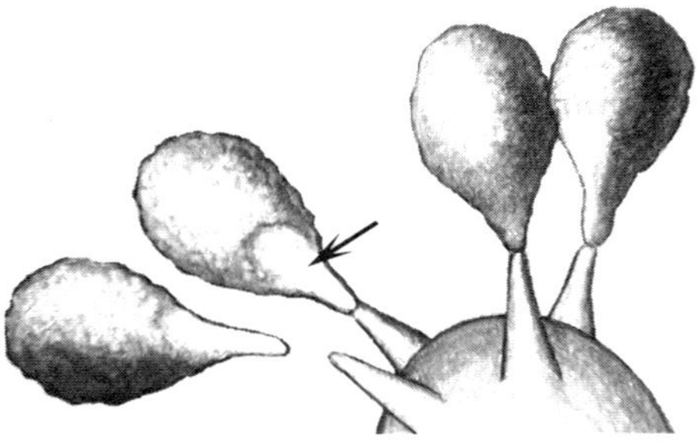

Suprahilar depression of the basidiospores of *Galerina phillipsi*, x 1 000.

suprahilar disc (L. *supra*, on, above + *hilum*, areola, halo + suf. *-aris* > E. *-ar*, like, pertaining to; L. *discus* < Gr. *dískos*, disc): see **suprahilar depression**.

suprahilar plage (L. *supra*, on, above + *hilum*, areola, halo + suf. *-aris* > E. *-ar*, like, pertaining to; Gr. *plágos*, side, plaque, beach): see **suprahilar depression**.

suprahilar spot (L. *supra*, on, above + *hilum*, areola, halo + suf. *-aris* > E. *-ar*, like, pertaining to; ME. *spotte*, stain, mark): see **suprahilar depression**.

suprapellis (L. *supra*, on + *pellis*, skin): see **cystoderm** and **pellicule**.

suscept (L. *susceptus*, open, subject, or unresistant to some stimulus, influence or attack, undertaken, ptp. of *suscipere*, to take up, to support, admit): a living organism which is susceptible to (able to be attacked by; not immune to) a given disease, pathogen, or toxin. Cf. **susceptible** and **tolerant**.

susceptible (LL. *susceptibilis* < L. *susceptus*, open, subject, or unresistant to some stimulus, influence or attack, undertaken, ptp. of *suscipere*, to take up, to support, admit + *-ilis*, suf. added to denote capability, of the character of): capable of submiting to an action or a process; open, subject, or unresistant to some stimulus, influence or agency (a pathogen, a toxin, cold, etc.). Cf. **suscept** and **tolerant**.

suspensor (NL. *suspensor* < *suspendere*, to suspend): a special hyphal branch that supports the zygosporangium in the Zygomycetes; the suspensors constitute the later stage of the zygophores. They can be more or less equal in size, as in *Mucor* and *Rhizopus*, or unequal, as in *Zygorhynchus* and *Absidia* (Mucorales).

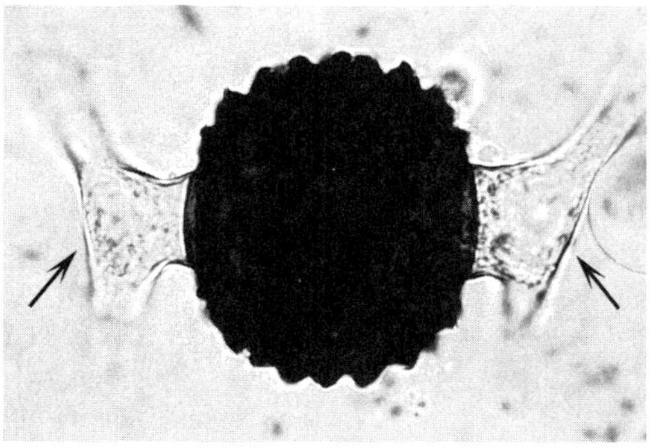

Suspensors of the zygosporangium of *Mucor hiemalis*, x 230 (*RTH*).

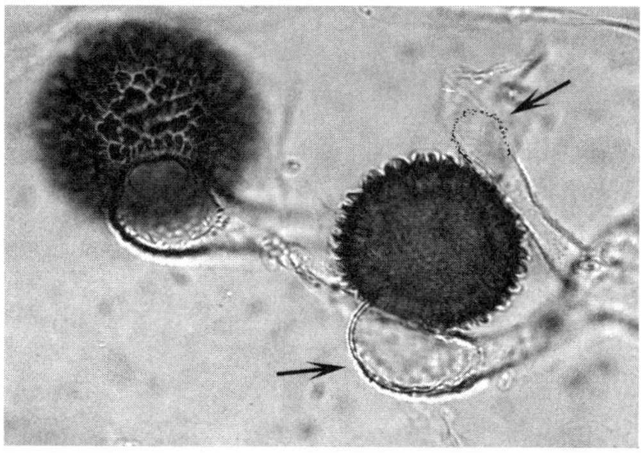

Suspensors of the zygosporangium of *Zygorhynchus vuilleminii*, x 500 (*MU*).

sycosis, pl. **sycoses** (NL. < Gr. *sykõsis* < *sýkon*, fig + *-osis*, state or condition): a fungus disease characterized by a chronic inflammatory disorder of the hair follicles, especially of the face (ringworm of the beard) marked by papules, pustules, and tubercles with crusting, caused by *Trichophyton rubrum* (moniliaceous asexual fungi).

symbiont (Gr. *symbiõntos* < *symbióo*, to live in common + *óntos*, genit. of *ón*, a being): each one of the organisms of a symbiosis; e.g., in the lichens the fungus and the alga both are symbionts. See **symbiosis**.

symbiosis, pl. **symbioses** (Gr. *symbíosis*, life in common): in biology the term is applied to life in common of two or more organisms, with mutual benefit for the participants or symbionts. There exists a whole series of gradations that includes symbiosis properly called (or **mutualistic symbiosis**) and **commensalism** to **antagonistic symbiosis** (or **parasitism**), in which the benefit is only in favor of the invador and against the host. Cf. **helotism**, **metabiosis**, **mutualism**, **parabiosis**, **parasymbiosis** and **synergy**.

symmetrical (Gr. *symmetría*, symmetry + suf. *-íkos* > L. *-icus* > E. *-ic*, belonging to, relating to + L. suf. *-alis* > E. *-al*, relating to or belonging to): refers to an organ or structure that has at least one plane of symmetry, which permits dividing it into two equal parts, as occurs in many fungal spores. Cf. **asymmetric**.

sympatric (Gr. *sým*, with + *pátra*, fatherland < *patér*, *patrós*, father + suf. *-íkos* > L. *-icus* > E. *-ic*, belonging to, relating to): occurring in the same geographical area. Cf. **allopatric**.

symphogenous, **symphyogenous** (Gr. *symphyós* < *symphýo*, grow together + *génos*, origin < *gennáo*, to engender, produce + L. *-osus* > OF. *-ous*, *-eus* > E. *-ous*, having, possessing the qualities of): applied to a type of pycnidium formation in which several different hyphae intertwine to form a small spherical initial that will develop into the pycnidium. *Zythia* (sphaeropsidaceous asexual fungi) has this type of pycnidial development. Also called **symphogenetic** (with the E. suf. *-ic* < L. *-icus* < Gr. *-íkos*, belonging to, relating to). Cf. **meristogenous**.

symphyogenesis, pl. **symphyogeneses** (Gr. *symphyós* < *symphýo*, grow together + *génesis*, origin): see **symphogenous**.

symplesiomorphy (Gr. pref. *sym-*, which entails the idea of concrescence, solidarity + *plésios*, near, recent + *morphé*, shape + *-y*, E. suf. of concrete nouns): a term used in cladistics to refer to a shared, ancestral character state. If the character is not shared then it is called plesiomorphy.

sympodial (NL. *sympodium* < Gr. pref. *sym-*, which entails the idea of concrescence, solidarity + NL. *podium* < Gr. *pódion*, dim. of *poús*, *podós*, foot, support + L. suf. *-alis* > E. *-al*, relating to or belonging to): having or involving the formation of an apparent main axis from succesive secondary axes, such as the conidiophores of *Helminthosporium* and *Stachybotrys* (dematiaceous asexual fungi). See **sympodium**.

sympodial branching (NL. *sympodium* < Gr. pref. *sym-*, which entails the idea of concrescence, solidarity + NL. *podium* < Gr. *pódion*, dim. of *poús*, *podós*, foot, support + L. suf. *-alis* > E. *-al*, relating to or belonging to; E. *branching*, ptp. of *to branch*, ramify < ME. *branche* < AF. < OF. *branche* < LL. *branca*, paw): having or involving the formation of an apparent main axis from succesive secondary axes. See **sympodium**.

Sympodial conidiophore of *Stachybotrys chartarum*, x 400 (*CB*).

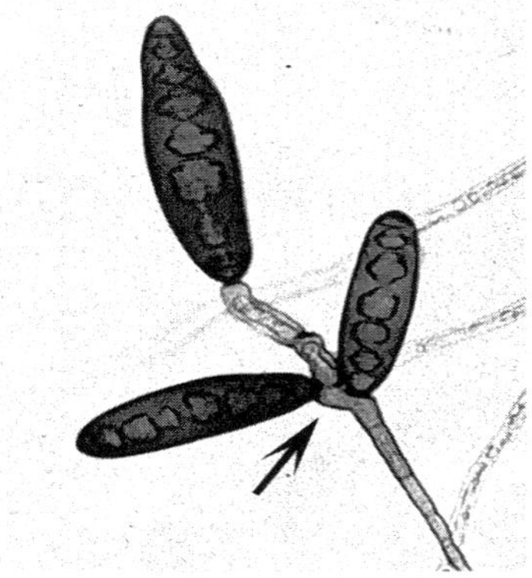

Sympodial conidiophore of *Helminthosporium* sp., x 300 (*MU*).

sympodioconidium, pl. **sympodioconidia** (NL. *sympodium* < Gr. pref. *sym-*, which entails the idea of concrescence, solidarity + NL. *podium* < Gr. *pódion*, dim. of *poús*, *podós*, foot, support + *kónis*, dust + dim. suf. *-ídion* > L. *-idium*): see **sympodulospore**.

sympodium; pl. **sympodia** (NL. *sympodium* < Gr. pref. *sym-*, which entails the idea of concrescence, solidarity + NL. *podium* < Gr. *pódion*, dim. of *poús*, *podós*, foot, support): a type of conidiophore which appears simple, but which terminates and branches repeatedly; since the successive branches form on alternate sides of the main axis, a structure with sympodial branching adopts a zigzag form; e.g., like the conidiophore of *Cercospora* and *Helminthosporium* (dematiaceous asexual fungi), or like the perithecial hairs of *Ascotricha guamensis* (Sordariales). Cf. **monopodium**.

sympodulospore (NL. *sympodium* < Gr. pref. *sym-*, which entails the idea of concrescence, solidarity + NL. *podium* < Gr. *pódion*, dim. of *poús*, *podós*, foot, support + *sporá*, spore): an asexual reproductive spore (conidium) which is formed on a conidiophore with sympodial branching. After formation of the terminal conidium, growth continues as a lateral branch that originates beneath the conidium; this branch soon produces another terminal conidium, and the branching process is repeated. Successive branches usually form on alternate sides of the conidiophore axis, resulting in a zigzag structure. This type of spore is also called a **sympodioconidium** or **terminal botryospore**; it is present, e.g., in *Tritirachium roseum* and *Arthrobotrys oligospora* (moniliaceous asexual fungi).

synanamorph (Gr. *sýn*, with, together + *anamorphóo*, to transform): one of the various morphs or conidial states associated with certain pleomorphic fungi, such as the interesting dematiaceous asexual fungus *Kionochaeta* (=*Phialicorona*) *pleomorpha*, which has, besides the characteristic phialidic state, three other different conidial states, or synanamorphs: *Sporidesmiella*, *Selenosporella* and *Heteroconium*, whose conidiophores originate from the same hyphae as the *Kionochaeta* (=*Phialicorona*). Cf. **anamorph**, **teleomorph** and **pleomorph**.

synapomorphy (Gr. *sýn*, with, together + *apó*, away from + *morphé*, shape + *-y*, E. suf. of concrete nouns): a term used in cladistics to refer to a shared, derived character state. When such a character is not shared it is called apomorphy.

synaptonemal complex (Gr. *synápto*, to bind, unite together + *nêma*, thread, filament + L. suf. *-alis* > E. *-al*, relating to or belonging to; LL. *complexus*, totality, complex): a structure that maintains the precise alignment of the two homologous chromosomes of a bivalent throughout its length. The synaptonemal complexes represent evidence of meiosis since crossing over of genes occurs in them. These complexes consist of two banded lateral components separated by a central region 10,000 to 12,000 nm in width, within which there is a central component. This structure can be seen in *Octospora* (=*Neottiella*) *rutilans* (Pezizales).

synascus, pl. **synasci** (Gr. *sýn*, with, together + *askós*, wine bag, sack; ascus): a cell containing several asci; applied to the female gametangium or ascogonium of *Ascosphaera* (Ascosphaerales), which, after plasmogamy, is converted into an inflated nutriocyte in which are developed clusters of asci. The asci form from a system of ascogenous hyphae, in a manner similar to the development of a cleistothecium, from which it differs in being unicellular and in lacking a hyphal peridium. Also called a synascus is the structure formed by *Mixia osmundae*, initially included in the Taphrinales but now considered as a basidiomycete, in the fern *Osmunda*.

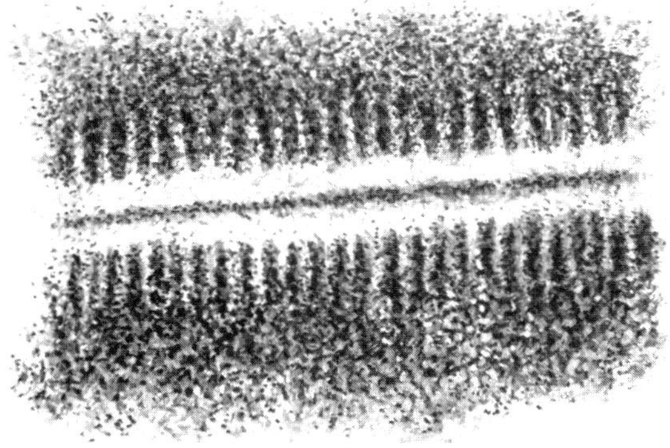

Synaptonemal complex of *Octospora* (=*Neottiella*) *rutilans*. Longitudinal section through a bivalent; the two paired homologous chromosomes (C) are held in register by the synaptonemal complex, which consists of two banded lateral components (L) separated by the central region (C). Drawing based on a transmission electron micrograph published by Beckett *et al.* in their *Atlas of Fungal Ultrastructure*, 1974, x 150 000.

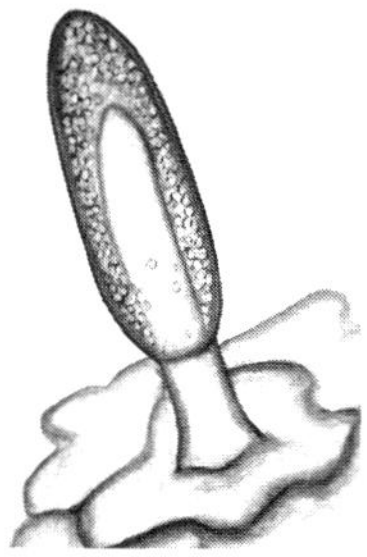

Synascus of *Mixia osmundae* on a host cell of *Osmunda cinnamomea*. The mature spore sac (synascus), which has a central columella surrounded by ascospores, is supported by a stalk cell that emerges from a chlamydospore contained in a host cell, x 520.

syncytium, pl. **syncytia** (Gr. *sýn*, with, together + *kýtos*, cavity; here, cell + L. dim. suf. *-ium*): the fusion plasmodium of the Myxomycetes, which results from the complete fusion of uninucleate amoeboid cells (myxamoebae) and the loss of their individual membranes. See **aphanoplasmodium**, **phaneroplasmodium**, **protoplasmodium** and **pseudoplasmodium**.

syndrome (NL. < Gr. *syndromé*, combination, syndrome < *sýn*, with + *drameín*, to run): a complex of signs and symptoms that occur together and characterize a particular abnormality or disease.

synergy, **synergism** (Gr. *synérgeia*, cooperation, mutual help + E. suf. *-y*, having the quality of; or + Gr. *-ismós* > L. *-ismus* > E. *-ism*, state, phase, tendency, action): in general, a combined action. Unlike metabiosis, in synergism two organisms coexist in a synergistic manner, in an associated or combined action on a substrate or particular host. It is common among the saprobic fungi, e.g., for a substrate to be altered specifically by the combined activities of two species that have complementary enzymatic capabilities. Also, among parasitic fungi there is a reciprocal enhancement of pathogenicity when two species exist on the same host. An example of synergy is the fermentative action that the bacteria and yeasts which make up the sugary kefir grains or tibi grains perform in sugary liquids. These tibi grains are utilized in Mexico and other places to produce fermented beverages, such as tepache and vinegar. **Antagonism** is the opposite phenomenon to synergy. Cf. **commensalism**, **helotism**, **metabiosis**, **mutualism**, **parasymbiosis** and **symbiosis**.

syngamy (Gr. *sýn*, with, together + *gamía* < *gámos*, sexual union + *-y*, E. suf. of concrete nouns): fertilization or fusion of male and female gametes to form a zygote.

synnema, pl. **synnemata** (Gr. *sýn*, with, together + *nêma*, filament): a compact, erect bundle of conidiophores that bear conidia at the apex and sometimes along the sides; the individual conidiophores may or may not be united laterally. Synnemata are characteristic of the stilbellaceous asexual fungi, such as *Doratomyces*, *Dendrostilbella* and *Graphium*, among other gen., and some moniliaceous asexual fungi, such as *Penicillium claviforme* and *Beauveria cretacea*. Cf. **coremium**.

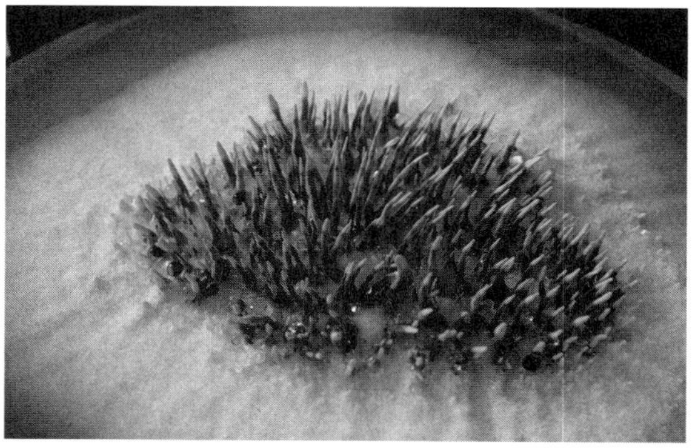

Synnemata of *Beauveria cretacea* on agar, x 1.5 (*MU*).

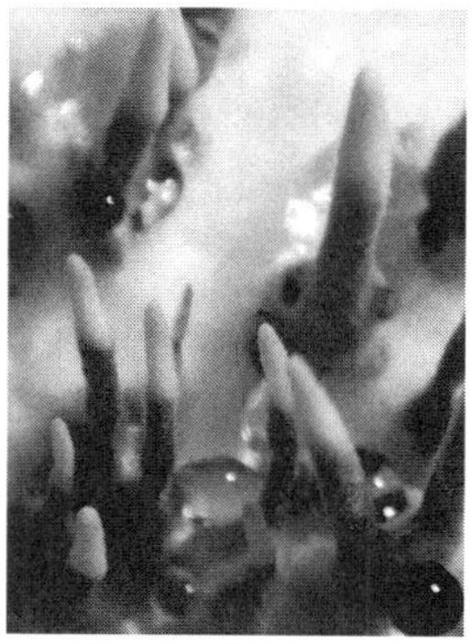

Synnemata of *Beauveria cretacea* on agar, x 7 (*MU*).

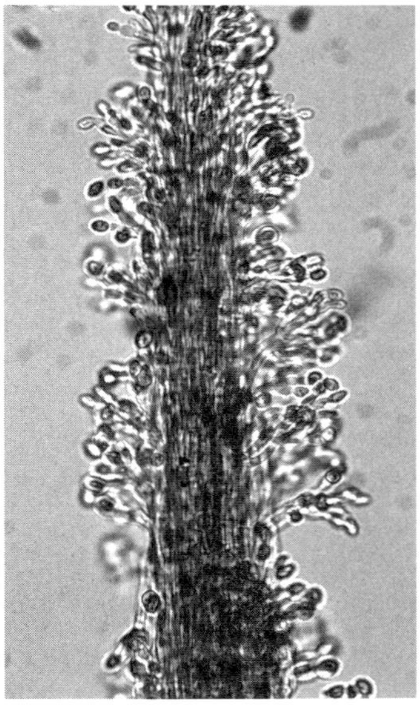

Synnema of *Doratomyces microsporus*, x 1 500 (*MU*).

synnematous (Gr. *sýn*, with, together + *nêma*, genit. *nêmatos*, filament + L. *-osus* > OF. *-ous*, *-eus* > E. *-ous*, having, possessing the qualities of): with **synnemata**. Cf. **mononematous**.

systematics (Gr. *systematikós* < *sýstema*, whole, aggregate, entirety, methodic compendium + *-tikós* > L. *-ticus* > E. *-tic*, relation, fitness, inclination or ability): in biology, *biosystematics* (Gr. *bíos*, life) is the study of the relationships and classification of the living world or biota in a system or conceptual hierarchy. It includes the subdisciplines **taxonomy** and **nomenclature**.

systemic (E. adj., of, relating to, or common to a system < LL. *systemat-*, *systema* < Gr. *sýstema*, whole, aggregate, entirely + suf. *-íkos* > L. *-icus* > E. *-ic*, belonging to, relating to): affecting the body generally. A systemic parasite spreads throughout the host. A systemic fungicide (or an antibiotic) is absorbed, especially by the roots, and translocated to other parts of the plant.

systole (Gr. *systolé*, contraction): a phase in the movement of a contractile vacuole, during which it contracts rapidly to expel its contents to the interior of the cell or a non-contractile reservoir, as occurs in certain plasmodia and zoospores. Cf. **diastole**.

t

tactiosensible, **tactiosensitive** (L. *tactus*, touch + *sensibilis*, sensible < *sensus*, sense + *-ilis*, suf. added to denote capability, of the character of; L. *tactus* + *sensitivus*, having sense or feeling < *sensus* + *-ivus* > E. *-ive*, quality or tendency, fitness): applied to Discomycetes in which ascospore discharge is stimulated by any contact, whether a solid body (such as an insect) or by light air currents.

tanninolytic (F. *tanin*, tannin < ML. *tannum*, tanbark, oak bark + Gr. *lytikós*, able to loosen < *lýtos*, dissolvable, broken + suf. *-íkos* > L. *-icus* > E. *-ic*, belonging to, relating to): capable of degrading tannins by means of the enzyme tanase; e.g., *Aspergillus carbonarius* (moniliaceous asexual fungi) has been utilized in solid state fermentation systems to degrade tannins from pods of the carob tree, making them tastier, with the object of utilizing them as cattle feed.

tar spot (ME. *terre*, *tarre*, a thick, brown to black, viscous liquid obtained by the destillation of wood, coal, peat, and other organic materials; ME. *spot*, mark, blot): the flat, circular, black, tarlike stromata, which bear apothecia within them, formed by *Rhytisma acerinum* (Rhytismatales) in fallen, dead maple leaves, where this fungus overwinters (tar spot of maple). The surface of each mature stroma is characterized by radiate fissures along which the stroma splits above the apothecia in the spring.

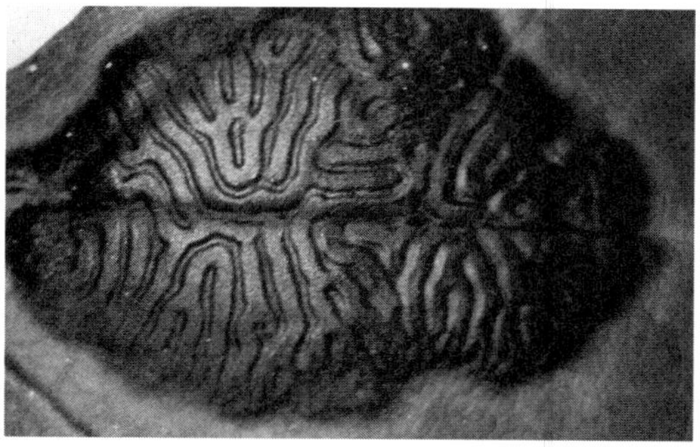

Tar spot of a maple leaf caused by *Rhytisma acerinum;* the mature, tarlike stroma bears apothecia within, and splits along the radiate fissures, x 16 (*RTH*).

tartarean, **tartareous** (L. *tartareus* < Tartar of Hell, frightening, horrible, infernal L. suf. *-anus* > E. *-an*, belonging to; or + L. *-osus* > OF. *-ous*, *-eus* > E. *-ous*, having, possessing the qualities of): with a surface thick, rough and crumbled, like that of certain crustaceous lichens; e.g., *Ochrolechia tartarea*, of the Pertusariales.

taxis, pl. **taxes** (Gr. *táxis*, arrangement, disposition): reflex translational or orientation movement by a freely motile organism in relation to a source of stimulation. It is positive when the movement is in the direction of the stimulus; negative when it is away from the stimulus. See **aerotaxis**, **chemotaxis**, **electrotaxis** and **gravitaxis**.

taxon, pl. **taxa** (Gr. *táxon*, to order, constitute): a group or category of any hierarchy within a taxonomic or classification system, from species to kingdom or superkingdom.

taxonomy (Gr. *táxon*, to order, constitute + *nómos*, legitimate according to law, normal + *-y*, E. suf. of concrete nouns): the part of natural history dealing with the classification of beings. Analysis of the characteristics of an organism in order to assign it to a category or taxon. See **nomenclature** and **systematics**.

tea fungus (Chin. *tea*, a shrub, *Camellia sinensis*, used to prepare an aromatic beverage, especially from the leaves; L. *fungus*, fungus): a symbiotic association of yeasts (*Saccharomycodes ludwigii*, of the Saccharomycetales) and bacteria (*Acetobacter xylinum*) grown in tea. This microbial association receives various names in different parts of the world: *mother of vinegar*, *kombucha*, *Wunderpilze*, and others. Cf. **kefir grains** and **tibi grains**. See **microbiogloea**.

teleblem, **teleoblem**, **teleoblema** (Gr. *téleos*, perfect, entire, complete + *blêma*, covering): syn. of **protoblem**.

telemorphosis, pl. **telemorphoses** (Gr. *téle*, far + *morphé*, shape + *-osis*, condition or state of being): a phenomenon in which a hypha stimulates another from a certain distance, provoking in it morphological and physiological changes, such as is observed in the process of sexual reproduction in *Achlya bisexualis* (Saprolegniales). In this case, it is a primordial hypha of the oogonium that secretes a hormone (antheridiol) that acts on a primordial hypha of the antheridium, which on reacting sets free the series of morphological and physiological changes that precede gametangial contact.

teleomorph (Gr. *téleos*, complete, perfect + *morphé*, shape): the sexual (perfect) state of a fungus (ascogenous or basidiogenous), whose spores are produced by meiosis. Cf. **anamorph** and **holomorph**.

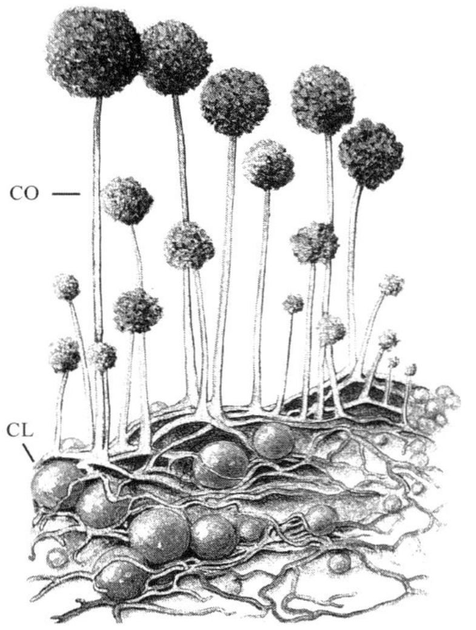

Teleomorph (*Eurotium rubrum*) of *Aspergillus rubrobrunneus* (the anamorph); the corresponding cleistothecia (CL) and conidiophores (CO) are shown, x 30.

teleutosorus, pl. **teleutosori** (Gr. *télos*, *téleos*, end + *sorós*, pile, heap): an old term for **telium**.

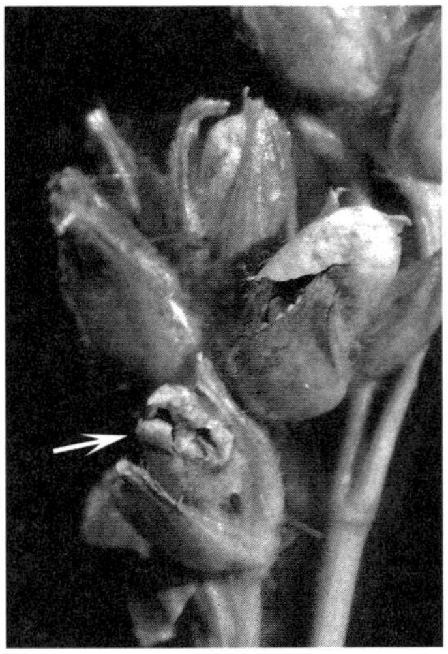

Teleutosori of *Sphacelotheca sorghii* in a spike of sorghum, x 9 (*MU*).

Teleutosori of *Ustilago maydis* on an ear of corn, x 0.2 (*MU*).

teliospore (Gr. *teleuté*, completion, finishing, end; or *télos*, *téleos*, end + *sporá*, spore): a thick-walled, unicellular or multicellular resting spore with binucleate (dikaryotic) cells, that is produced in a telium; characteristic of the rusts and smuts (Uredinales and Ustilaginales). The teliospore, also called teleutospore, is the site of karyogamy and meiosis, after which the basidial apparatus forms by germination.

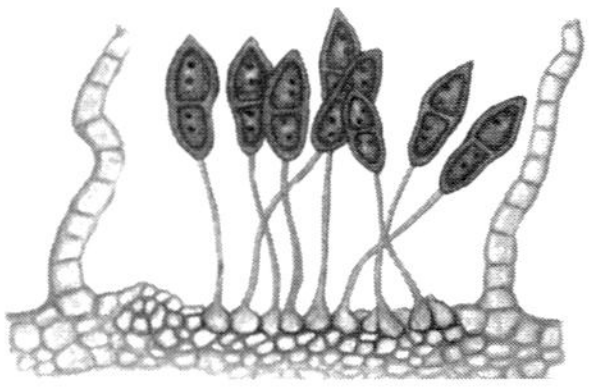

Teliospores of a telium of *Puccinia graminis tritici* on a wheat leaf, x 300.

telium, pl. **telia** (NL. *telium* < Gr. *télos*, *téleos*, end + L. dim. suf. *-ium*): the group of binucleate cells in a sorus that produces teliospores, which are resistant spores in the parasitic fungi of plants called rusts and smuts (Uredinales and Ustilaginales). The telia of the rusts are a stage in the life cycle that is always present, unlike the spermogonia, aecia and uredinia that can be lacking in said cycle, depending upon the species being treated. The morphology of the telia varies, as well as its position on the plant; in *Puccinia* (Uredinales), e.g., they are erumpent structures on leaves and stems. In *Ustilago* and *Sphacelotheca* (Ustilaginales) the sori of teliospores are produced mainly in the inflorescences of the host plants, mainly Gramineae.

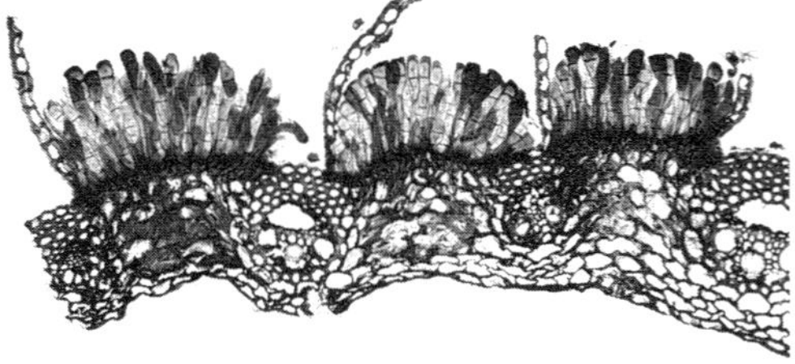

Telia with teliospores of *Puccinia graminis tritici* on a wheat leaf, seen in longitudinal section, x 100 (*MU*).

tenacle (L. *tenacula* < *tenere*, to hold fast, fasten): see **haerangium**.

terebrate (L. *terebratus*, a boring, perforation < *terebrare*, to bore, perforate + suf. *-atus* > E. *-ate*, provided with or likeness): bored or perforated. In lichenology, it refers to a thallus that has sparse perforations, like that of *Verrucaria terebrata* (Verrucariales).

terebrator (L. *terebratus*, a boring, perforation): a drilling apparatus; it has been utilized for the hyphal trichogyne of the lichens.

terete (L. *teres*, genit. *tereis*, round, cylindrical): round in cross section, cylindrical; often used in terms like tereticaule, teretifolia, etc., i.e., of stems, leaves, etc. For example, the perithecial stroma of *Camillea leprieurii* (Xylariales) is cylindrical and terete.

termitophilic (NL. *termites*, white ant < L. *tarmes*, wood-eating worm + Gr. *phílos*, have an affinity for + suf. *-íkos* > L. *-icus* > E. *-ic*, belonging to, relating to): growing on termite nests, like the fungus *Termitomyces* (Agaricales).

terricole, **terricoline**, **terricolous** (L. *terra*, earth + *-cola*, inhabitant < *colere*, to inhabit; or + L. suf. *-inus* > E. *-ine*, of or pertaining to; or + L. *-osus* > OF. *-ous*, *-eus* > E. *-ous*, having, possessing the qualities of): an organism that lives in the soil. Many species of micro as well as macromycetes are terricolous. See **epigeous** and **hypogeous**.

tertiary mycelium, pl. **mycelia** (L. *tertius*, third + suf. *-aris* > E. *-ar*, pertaining to + E. suf. *-y*, having the quality of; NL. *mycelium* < Gr. *mýkes*, fungus + L. *-elis*, pertaining to + dim. suf. *-ium*): *Basidiomycetes*. The mycelium that is differentiated from secondary (dikaryotic) mycelium in the formation of the fructifications. This term tends to be suppressed. See **primary mycelium** and **secondary mycelium**.

terverticillate (L. *ter*, three + NL. *verticillatus*, arranged in a verticil < *verticillus*, verticil, whorl + suf. *-atus* > E. *-ate*, provided with or likeness): see **triverticillate**.

tesgüino (Azt. *tecuin*, heart beat): an opaque maize beer consumed as a nutritious and ceremonial beverage by several ethnic groups of northern Mexico. The fermenting microbiota includes lactic acid bacteria (*Lactobacillus* spp.) and yeasts (especially *Saccharomyces* of the Saccharomycetales, and *Candida* spp., of the cryptococcaceous asexual yeasts).

tessellate (L. *tessellatus*, checkered < *tessella*, mosaic + suf. *-atus* > E. *-ate*, provided with or likeness): checkered, as if formed of small squares or mosaics, like the surface of the sporangium of *Perichaena tessellata* (Trichiales), and the thallus of *Rhizocarpon geographicum* (Lecanorales).

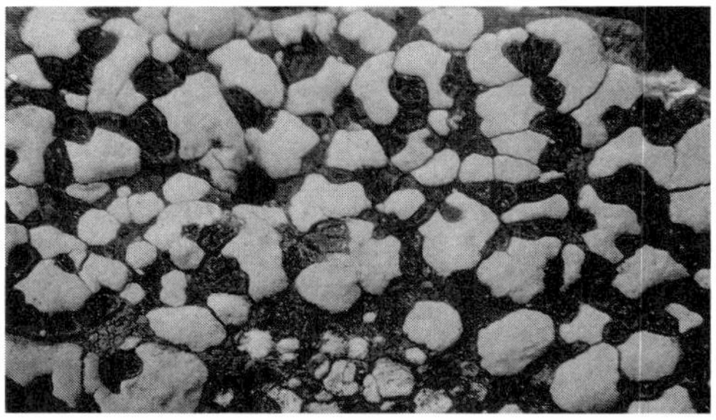

Tessellate thallus of *Rhizocarpon geographicum*, x 6.5 (*MU*).

testaceous (L. *testa*, vessel, glass, generally of mud, shell, covering or rind + *-aceus* > E. *-aceous*, of or pertaining to, with the nature of): having a shell or hard cover, generally of a light brick-red color; e.g., like the peridium of the sporangia of *Diderma testaceum* (Physarales).

tetrachotomous (Gr. *tetrás*, four + *scházo*, divide + *témeo*, to cut + L. *-osus* > OF. *-ous*, *-eus* > E. *-ous*, having, possessing the qualities of): with four branches arising from the same point.

tetrad (L. *tetras*, genit. *tetradis* < Gr. *tetrás*, genit. *tetrádos*, four): a group of four cells or spores, like the tetraspored basidia of many basidiomycetes, or an ascus with four ascospores, typical of *Saccharomyces* (Saccharomycetales).

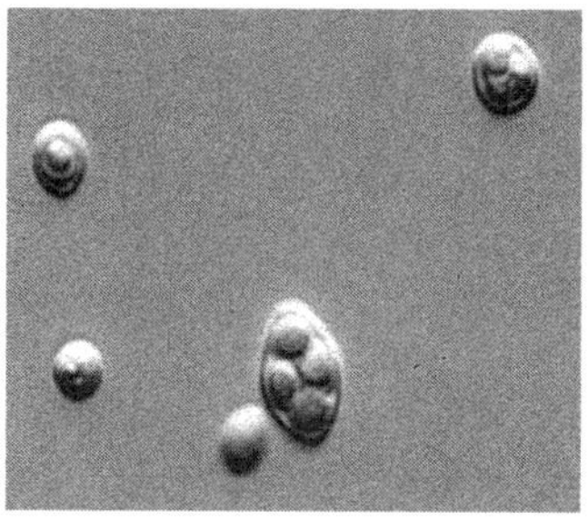

Tetrad (tetraspored ascus) and somatic cells of *Saccharomyces cerevisiae*, x 2 200 (*RTH*).

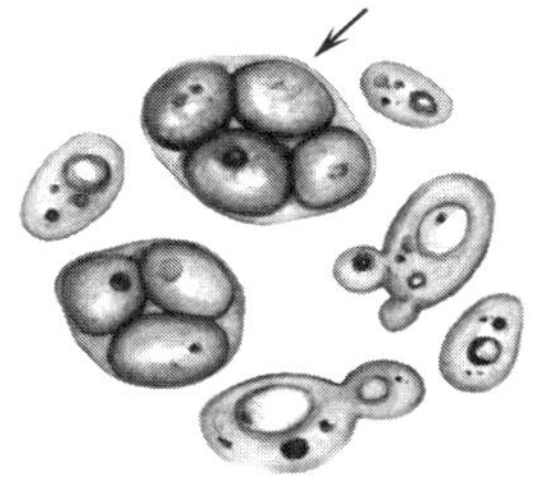

Tetrad (tetraspored ascus) and somatic cells of *Saccharomyces cerevisiae*; a trispored ascus is also shown, x 4 570.

tetrapolar (L. *tetras*, four + *polaris*, polar, relative or pertaining to the pole < *polus*, the end of an axis, pole + *-aris* > E. *-ar*, like, pertaining to): refers to the sexual compatibility of the basidiomycetes, which have sex factors in four groups (AB, Ab, aB, and ab). Fusions among monosporic mycelia are only possible when these mycelia carry the complementary pairs of these factors. Cf. **bipolar**.

texospore (Gr. *teixós*, wall + *sporá*, spore): an ascospore covered by a layer of cells of paraphysal origin, as in *Texosporium*, a lichen gen. of the Caliciales.

textura, pl. **texturae** (L. *textura*, tissue, link, connection): refers to the shapes of the cells that make up the different types of hyphal tissues found in conidiomata and ascomata; they are utilized in taxonomic descriptions. *Textura angularis* (L. *angularis*, angular): with cells polyhedral due to mutual compression, without intercellular spaces. *T. epidermoidea* (Gr. *epidermís*, membrane of pellicle that covers the skin + L. suf. *-oide* < Gr. *-oeídes*, similar to): a thin, membranous tissue composed of cells with very irregular outlines, with united walls, and lacking intercellular spaces (like pieces of a jigsaw puzzle). *T. globulosa* (L. *globulus*, small ball): with rounded-polyhedral cells, almost isodiametric, with intercellular spaces, but not resembling hyphae as such. *T. intricata* (L. *intricatus*, interwoven): composed of interwoven hyphae, with the walls not united, and generally with intercellular spaces. *T. oblita* (L. *oblitus*, forgotten, erased): composed of more or less parallel hyphae with long cells that have a narrow lumen and very thick walls, coherent. *T. porrecta* (L. *porrectus*, extended, long): hyphae with long cells, parallel, with wide lumen and thin walls, not coherent. *T. prismatica* (L. *prisma* < Gr. *prisma*, prism): with short cells, more or less rectangular, not isodiametric.

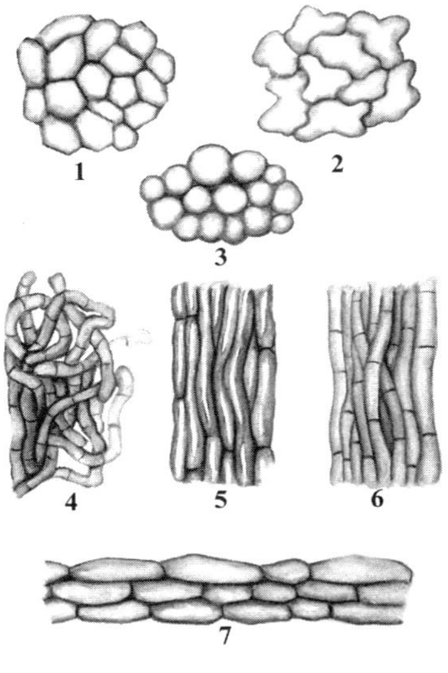

Textura *angularis* (1), *epidermoidea* (2), *globulosa* (3), *intricata* (4), *oblita* (5), *porrecta* (6), and *prismatica* (7), that may be present in different types of discomycete apothecia and deuteromycete pycnidia, x 500.

thallic (L. *thallus* < Gr. *thallós*, shoot, thallus + suf. *-íkos* > L. *-icus* > E. *-ic*, belonging to, relating to): *Conidiogenesis*. One of the two main modes of conidium development in the conidial fungi. The thallic conidium is characterized by being formed by the transformation of an existing cell in a hypha or conidiophore. If all the layers of the conidiogenous cell wall participate in the formation of the conidial wall, it is called **holothallic**; if the external wall of the conidiogenous cell is not involved in the formation of the conidial wall, it is termed **enterothallic**. **Arthrospores** and **meristem arthrospores** are types of thallic conidia. Cf. **blastic**.

thalliles, **thallyles** (L. *thallus* < Gr. *thallós*, shoot, thallus + suf. *-ilis* > E. *-ile*, capable, of the character of): minute thallus-like propagules produced on the underside of certain lichen thalli, as in *Umbilicaria* (Lecanorales).

thallinocarp (L. *thallus* < Gr. *thallós*, thallus + *karpós*, fruit): *Lichens*. A type of apothecium, similar to a lichen gall, that originates from ascogonia, ascogenous hyphae, and asci arranged in a swelling directly among the hyphae of the thallus, with the hymenium covered by groups of algal cells, as *in Lichinella (=Gonohymenia)*, of the family Lichinaceae (Lichinales). Cf. **pycnoascocarp**.

thallospore (NL. *thallospore* < Gr. *thallós*, shoot, thallus + *sporá*, spore): a spore produced directly by the thallus or mycelium, whether in an intercalary manner (alone or in chains) or terminally, from a preexisting cell, like the arthrospores and chlamydospores. Thallospores or thallic spores do not form *de novo*, like the other types of spores (which are **blastic**).

thallus, pl. **thalli** (L. *thallus* < Gr. *thallós*, shoot, thallus): the vegetative body or soma of a fungus. The thallus can be unicellular, multicellular, or dimorphic, depending upon the species or the phases of the life cycle and the conditions of the medium on which it develops, but it never has sap conducting vessels nor is it differentiated into root, stem, flowers, and fruits, like the vascular plants.

thamnisophagous (Gr. *thámnion*, dim. of *thámnos*, shrub + *phágos*, to eat + L. *-osus* > OF. *-ous*, *-eus* > E. *-ous*, having, possessing the qualities of): *Mycorrhizae*. One of the four types of endotrophic mycorrhizae of pteridophytes, characterized by the formation of intracellular haustorial arbuscules, finely branched, that finally are digested and assimilated by the host cells. Cf. **ptyophagous** and **tolypophagous**.

thamnoblast (Gr. *thámnos*, shrub + *blastós*, sprout, bud, germ, shoot): *Lichens*. A shrub-like, dendroid bud, as occurs in the fruticose lichens, e.g., *Teloschistes*

(Teloschistales) and *Ramalina* (Lecanorales).

thanatochresis, pl. **thanatochreses** (Gr. *thánatos*, death + *chrêsis*, enjoyment): the taking advantage of cadavers, secretions, excrement and other productions of one species by the live individuals of a second one as a means of sustenance. This phenomenon is observed in saprobic organisms that develop at the expense of other dead organisms or substrates coming from dead organisms. Among the fungi, there are innumerable examples of this phenomenon, and one of them is the growth and sporulation of *Harposporium anguillulae* (moniliaceous asexual fungi) from the remains of nematodes in soil.

thecium, pl. **thecia** (NL. *thecium* < Gr. *thekíon* < *thêke*, box, case; here, of the asci + L. dim. suf. *-ium*): refers to types of fructifications (cleistothecium, perithecium and apothecium) that are delimited by a proper wall that can remain closed during the major part of the development of the fructification and only open in some manner on reaching maturity (angiocarpic, like a cleistothecium), or the fructification has the hymenium or fertile layer partially exposed from the young stage (hemiangiocarpic, like a perithecium), or totally exposed (gymnocarpic, like an apothecium). At times thecium is used as the equivalent of hymenium.

thermophile, **thermophilic**, **thermophilous** (Gr. *thermós*, warm, hot + *phílos*, have an affinity for; or + suf. *-íkos* > L. *-icus* > E. *-ic*, belonging to, relating to; or + L. *-osus* > OF. *-ous*, *-eus* > E. *-ous*, having, possessing the qualities of): an organism that is capable of developing at elevated temperatures (from 50-60ºC or a little higher); the thermophiles or thermophilic fungi (such as *Chaetomium thermophile*, of the Sordariales) prosper at these temperatures and are incapable of growing at ordinary laboratory temperatures (20ºC); they have a maximum growth temperature of 50ºC or a little more, and a minimum growth temperature of 20ºC. Cf. **mesophile** and **psychrophile**.

thigmomorphosis, pl. **thigmomorphoses** (Gr. *thégma*, contact + *mórphosis*, the action of forming, of giving shape): development or change of form determined by a contact stimulus; equivalent to **haptomorphosis**: applied to the swelling at the ends of zygophores when they come into contact, as occurs, e.g., in *Rhizopus* and other Mucorales. The formation of an appresorium from a germ tube, as observed in many plant pathogenic fungi, is also considered as thigmomorphosis.

tholus, pl. **tholi** (L. *tholus* < Gr. *thólos*, rotunda): see **dome**.

thriptogen (Gr. *triptós* < *tríbo*, to hurt, debilitate + *génos*, origin < *gennáo*, to engender, produce): refers to an organism that attacks a host without killing it, although it debilitates it and makes it susceptible to other pathogens or physical agents (such as drought, freezing, etc.). For example, certain plant pathogenic nematodes are thriptogens that predispose the host plants to attack by vascular plant pathogenic fungi, such as *Verticillium* (moniliaceous asexual fungi) and *Fusarium* (tuberculariaceous asexual fungi).

thrush (prob. of Scand. origin): a disease caused by *Candida albicans* (cryptococcaceous asexual yeasts), especially of infants and children, characterized by white patches in the mucous membranes of the oral cavity and genital areas. A clinical type of **candidiasis**.

thyriothecium, pl. **thyriothecia** (NL. *thyriothecium* < Gr. *thyreós*, large, long shield + *thekíon* < *thêke*, case, box; here, of the asci + L. dim, suf. *-ium*): a lenticular ascoma (convex hemisphaeric) with its upper wall formed of plates of dark cells radially arranged (the upper plate corresponds to the **scutellum**). The asci are almost horizontal and grow inward with their apices pointing toward the ostiole; the cavity between the ostiole and the basal wall is occupied at first by pseudoparaphyses, which disappear when the asci mature. Characteristic of the families Microthyriaceae and Micropeltidaceae (Dothideales); *Microthyrium* and *Stomiopeltis* are the respective examples of fungi that form thyriothecia (also called **catathecia**, sing. **catathecium**) on the tissues of leaves and dead branches.

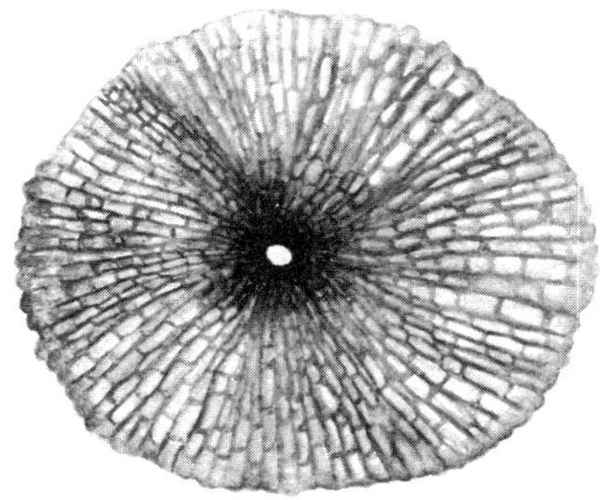

Thyriothecium of *Microthyrium microscopicum* in a surface view; the scutellum (dark region) surrounds the ostiole in the center, x 330.

thyrsus, pl. **thyrsi** (L. *thyrsus* < Gr. *thýrsos*, a stalk, wand; a type of inflorescence; a staff surmounted by a pinecone or by a bunch of vine or ivy leaves with grapes or berries that is carried by Bacchus and by satyrs and others engaging in bacchic rites): the densely branched apices of some lichens, e.g., *Cladonia stellaris* (Lecanorales).

tibi grains (E. *tibi* < Sp. *tibico* < unk.; ME. *gryne* < OF. *grein*, a seed, grain < L. *granum*, a seed, kernel): macroscopic (from a few mm to 1-2 cm), gelatinous grains (called *tibicos* in Mexico, and *sugary kefir grains* elsewhere), composed of dextran, that contain a symbiotic association of lactic bacteria and yeasts, and which are used as a starter culture for the production of fermented beverages (the so-called tepache and vinegar of tibicos) from sugary solutions. Cf. **kefir grains** and **tea fungus**. See **microbiogloea**.

tichus, pl. **tichi** (NL. *tichus* < Gr. *teychós*, instrument, armor, funeral urn): peripheral layer of cells of pseudothecial walls forming a dark protective layer, as in *Pleospora herbarum* (Pleosporales).

tinea (ME. < ML. < L. *tinea*, worm, moth): any of the several diseases of the skin, hair, and nails in humans and animals caused by parasitic fungi (especially the **dermatophytes**, of the gen. *Epidermophyton*, *Microsporum* and *Trichophyton*, moniliaceous asexual fungi). *Tinea barbae* (L. *barba*, beard): beard ringworm; see **sycosis**. *T. capitis* (L. *capit-*, *caput*, head): head or scalp ringworm; *tinea tonsurans* (ML. *tonsura*, act of shearing < L. *tonsus*, ptp. of *tondere*, to shear). *T. corporis* (L. *corpus*, *corporis*, body): body ringworm; *tinea circinata* (L. *circinatus*, ptp. of *circinare*, to make round, rolled up in the form of a flat coil with the apex at the center). *T. cruris* (L. *cruris*, groin): groin ringworm. *T. favosa*, *favus* or *tinea kerion* (L. *favus*, honeycomb; Gr. *kérion*, honeycomb): see **favus**. *T. imbricata* (L. *imbricatus*, ptp. of *imbricare*, to cover with pantiles): the so-called *tokelau*, an infection caused by *Trichophyton concentricum*. *T. nigra* (L. *nigra*, black): pigmented cutaneous infection by a dematiaceous asexual fungus (*Exophiala werneckii*). *T. nodosa* (L. *nodosus* < *nodus*, knot, protuberance): see **piedra**. *T. pedis* (L. *ped-*, *pes*, foot): athlete's foot, foot ringworm. *T. unguium* (L. *unguis*, finger nail): ringworm of nails. *T. versicolor* (L. *versicolorius*, *versicolorus*, of varied colors): same as *pityriasis versicolor* (< gen. *Pityrosporum*, of the cryptococcaceous asexual fungi < Gr. *pítyron*, dandruff, tinea + *spóron* > *sporum*, spore).

toadstool (ME. *tode* < OE. *tade*, *tadige*, toad + *stol*, stool, seat): an agaric or a bolete fruiting body, especially an inedible or poisonous one, usually having an umbrella-shaped cap. Cf. **mushroom**.

tolerant (E. adj., inclined to tolerate < L. *tolerare*, to endure + suf. *-ant* > E. *-ant*, one that performs, being): an organism exhibiting tolerance, i.e., giving little reaction to infection by a pathogen, or to the effect of other factors (e.g., tolerant of a toxic substance, of a virus, etc.). Cf. **suscept** and **susceptible**.

tolypophagous (Gr. *tolýpe*, ball of yarn, ball + *phágos*, to eat + L. *-osus* > OF. *-ous*, *-eus* > E. *-ous*, having, possessing the qualities of): in the endotrophic mycorrhizae, it is applied to forms of mycetization when the living cells of the host plant digest the dead hyphae of the fungus. Cf. **ptyophagous** and **thamnisophagous**.

tomentose, **tomentous** (L. *tomentum*, coarse interwoven wool + *-osus*, full of, augmented, prone to > ME. *-ose*; or + *-osus* > OF. *-ous*, *-eus* > E. *-ous*, having, possessing the qualities of): covered with densely matted hairs, such as the basidiocarps of *Cyathus* and *Nidula* (Nidulariales). See **tomentum**.

Tomentose basidiocarps of *Nidula niveo-tomentosa*, on dead wood, x 4.

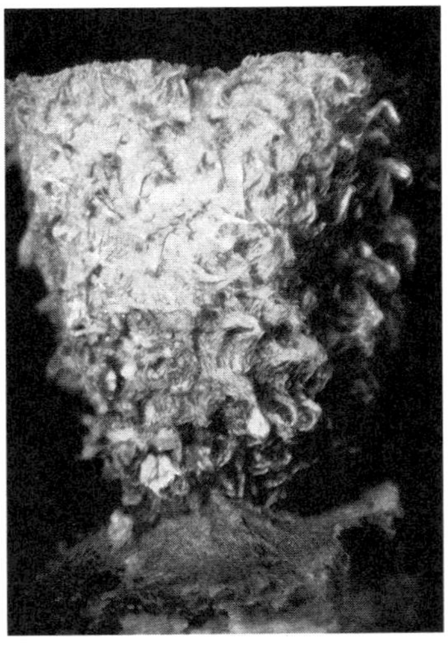

Tomentose peridium of the fruiting body of *Cyathus intermedius*, x 8 (*CB & MU*).

tomentum, pl. **tomenta** (L. *tomentum*, coarse interwoven wool): an aggregate of filaments or hairs, simple or branched, generally intertwined and very close, that resembles a flock of wool. A covering of dense, fine hairs over the surface of the stipe of some Hymenomycetes, such as *Boletus tomentosus* and *Laccaria laccata* (Agaricales). Also refers to hyphae, similar to hairs, that project in abundance from the lower cortex of crustaceous lichen thalli (heteromerous), and which serve to attach the thalli to the substratum. Structures that have this type of hyphae are called **tomentose**.

tonophily (L. *tonus* < Gr. *tónos*, tension, tone + NL. *philia* < Gr. *phílía*, tendency toward < *phílos*, having an affinity for + *-y*, E. suf. of concrete nouns): the ability to grow under conditions of high osmotic pressure. **Osmophilic** or **osmophilous** are the organisms that have this ability.

torn (E., ptp. of *tear* < ME. *teren* < OE. *teran*, pull apart): see **erose**.

torsive (LL. *torsus*, ptp. of L. *torquere*, to twist + *-ivus* > E. *-ive*, quality or tendency, fitness): spirally twisted.

tortuous (L. *tortuosus*, twisted < *tortus*, a twisting, winding + L. *-osus* > OF. *-ous*, *-eus* > E. *-ous*, having, possessing the qualities of): bent or twisted in different directions, like the cells of the cortex of the pileus of *Mycena tortuosa* (Agaricales) and the hyphae of many micromycetes.

torulose, **torulous** (L. *torulosus* < *torulus*, dim. of *torus*, a bulge or swelling, cushion, an elevation + *-osus*, full of, augmented, prone to > ME. *-ose*; or + *-osus* > OF. *-ous*, *-eus* > E. *-ous*, having, possessing the qualities of): elongate in shape, but with constrictions, like the filaments of the gen. *Torula* (dematiaceous asexual fungi); similar to **moniliform** (like the hyphae of *Monilia*, a moniliaceous asexual fungus), but with shallower constrictions. For example, the chains of microconidia of *Fusarium moniliforme* (tuberculariaceous asexual fungi), the cystidia of *Pholiota oedipus*, and the hyphae of the pileal cortex of *Leptonia subeuchroa* (Agaricales) are moniliform.

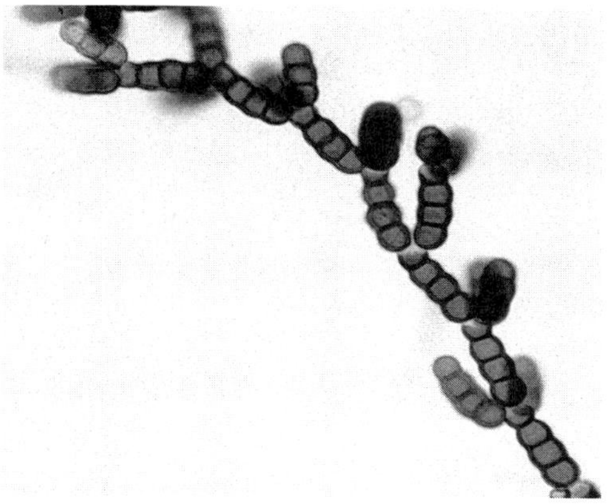

Torulose hyphae with blastic conidia of *Torula herbarum*, x 800 (*MU*).

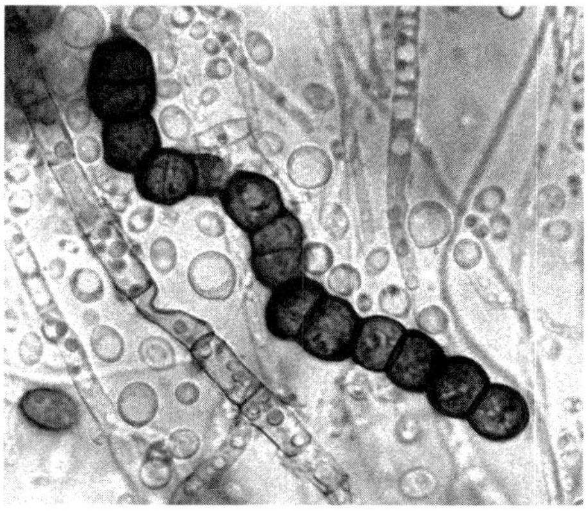

Torulose chain of blastospores of *Torula* sp., x 1 750 (*MU*).

toxiphilous (LL. *toxicus* < L. *toxicum*, poison < Gr. *toxikón*, arrow poison + Gr. *phílos*, having an affinity for + L. *-osus* > OF. *-ous*, *-eus* > E. *-ous*, having, possessing the qualities of): an organism favoring a polluted habitat (e.g., *Lecanora conizaeoides*, of the Lecanorales, in areas of high sulphur dioxide pollution in the air). Cf. **toxiphobous** and **toxitolerant**.

toxiphobous (E. *toxin* or *toxic* < LL. *toxicus* < L. *toxicum*, poison < Gr. *toxikón*, arrow poison + *phóbos*, fright, having an aversion for, lacking affinity for + L. *-osus* > OF. *-ous*, *-eus* > E. *-ous*, having, possessing the qualities of): an organism not tolerating a polluted habitat. Cf. **toxiphilous** and **toxitolerant**.

toxitolerant (LL. *toxicus* < L. *toxicum*, poison < Gr. *toxikón*, arrow poison + E. *tolerant* < L. *toleratus*, ptp. of *tolerare*, to endure + suf. *-ant* > E. *-ant*, one that performs, being): tolerant of toxins or a polluted habitat, with high sulphur dioxide concentrations in the atmosphere. Cf. **toxiphilous** and **toxiphobous**.

trabecula, pl. **trabeculae** (L. *trabecula*, dim. of *trabs*, genit. *trabis*, beam, girder): each of the solid columns that extends from the peridium toward the center of the fructification in *Mesophellia* (Lycoperdales); this center or heart also is solid and is supported by the trabeculae, with the gleba located between the center and the trabeculae. In *Laternea* (Phallales) the trabeculae support the **glebifer** or **lantern**, beneath the vault formed by the arms of the receptacle united at the apex.

tractellum, pl. **tractella** (L. *tractellum* < *trahere*, to drag, pull + dim. suf. *-ellum*): refers to the forward directed motor flagellum (tractor) of the zoospores of the Hyphochytriomycota and Oomycota.

trama, pl. **tramae** (L. *trama*, trama, warp, texture): a hyphal tissue that comprises the pileus or the layer which bears the hymenium of the basidiomycetes with holobasidia. The trama can be **homomerous** or

heteromerous. **Hymeniiferous trama**, also known as **context**, is the layer of tissue that is found beneath the hymenium, which is called **gill trama** in the agarics, or **tube trama** in the boletes. Depending on the disposition of the hyphae in the trama, it can be **convergent**, **divergent** or **parallel**. In *Coprinus* (Agaricales), e.g., it is parallel, with the hyphal elements arranged more or less regularly. In *Amanita muscaria* (Agaricales), it is divergent, with a central zone from which diverge two rows of cells.

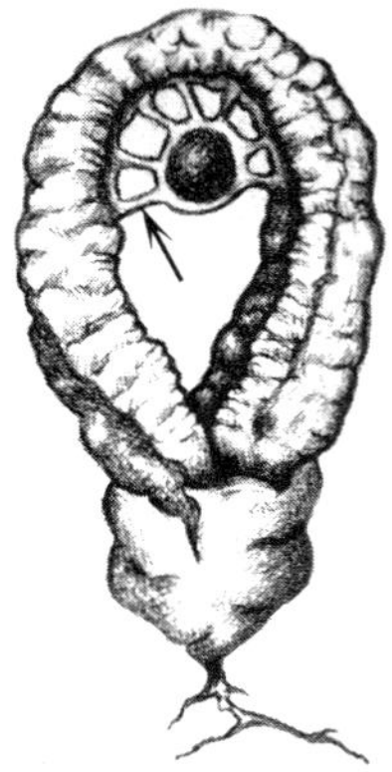

Trabeculae holding the gleba (lantern) of the fruiting body of *Laternea pusilla*, x 0.3.

tramal autolysis, pl. **autolyses** (L. *trama*, trama, warp, texture + L. suf. *-alis* > E. *-al*, relating to or belonging to; Gr. *autós*, same, self + *lýsis*, a loosening, releasing): a process in which the **tramal plates** collapse at maturity, forming a viscous or slimy mass of tissue and spores; typical of species in the Phallales.

tramal plate (L. *trama*, trama, warp, texture + L. suf. *-alis* > E. *-al*, relating to or belonging to; ME. *plate* < OF. *plate*, flat object < VL. *plattus* < Gr. *platýs*, broad, flat, plate): *Hymenogastrales*. A layer of tissue partially or totally devoted to the production of basidia. See **gusset**.

transgenic (L. pref. *trans-*, across + NL. *-genic*, adjectival combining form meaning giving rise to, originating): the condition of carrying genetic material from another organism which becomes incorporated into the genome of the host; e. g., hypovirulent strains of *Cryphonectria parasitica* (Diaporthales), the chestnut blight fungus, carry a virus responsible for the loss of virulence, and this virus can be spread to virulent strains which then disseminate it as they reproduce.

translucent (L. *translucent-*, *translucens*, prp. of *translucere*, to shine through < *trans-*, across + *lucere*, to shine < *lux*, genit. *lucis*, light + *-entem* > E. *-ent*, being): diaphanous, able to transmit light without being transparent, like the pileus of the basidiocarp of *Oudemansiella canarii* (Agaricales).

transverse (L. *transversus*, going or lying across): located or directed across. For example, the transverse planes are arranged at a right angle with respect to the axis of growth of the organ in question, as happens with the septa or partitions of many hyphae, and a transverse cut or section is one that passes perpendicularly to said axis. Cf. **longitudinal**.

trembling fungus, pl. **fungi** (ME. *tremblen* < MF. *trembler* < ML. *tremulare* < L. *tremulus*, tremulous < *tremere*, to tremble; L. *fungus*, fungus): the Tremellales, also called **jelly fungi**.

tremelloid (< gen. *Tremella* < L. *tremulus*, tremulous + L. dim. suf. *-ella* + L. suf. *-oide* < Gr. *-oeídes*, similar to): *Tremella*-like (Tremellales); like jelly or wet gelatin; gelatinous.

tremulous (L. *tremulus* < *tremere*, to tremble + L. *-osus* > OF. *-ous*, *-eus* > E. *-ous*, having, possessing the qualities of): something that trembles, not very firm, like the fruiting body of *Tremella* and other gelatinous fungi of the order Tremellales, which are also known as the trembling or jelly fungi.

trichiform (Gr. *thríx*, *trichós*, hair + L. *-formis* < *forma*, shape): see **aculeate**.

trichitomous (Gr. *trícha*, threefold < *treis*, three + L. *tomous* < Gr. *témeo*, to cut + L. *-osus* > OF. *-ous*, *-eus* > E. *-ous*, having, possessing the qualities of): with three branches arising from the same point.

trichocyst (Gr. *thríx*, *trichós*, hair + *kýstis*, bladder, vesicle): the name applied to the injective cell of *Haptoglossa mirabilis* (Myzocytiopsidales) endoparasitic on rotifers of the gen. *Adineta*. This cell originates in the encysted zoospore, which on germinating produces a harpoon-like apparatus, with a function similar to that of a hypodermic syringe, with a conical plug in the apex, which is introduced into the rotifer when the injective cell is fired, and from which is derived an infective sporidium, which becomes evident inside the attacked rotifer.

trichoderm (Gr. *thríx*, *trichós*, hair + *dérma*, skin): a cuticular layer of fruiting bodies of various Agaricales (such as *Leptonia perfusca* and *L. corvina*), formed by piliform or filiform hyphae, septate, and more or less perpendicular (anticlinal) to the respective organ surface. If said piliform hyphae reach more or less the same level, this type of dermis is called **trichoderm palisade** (as in *Leptonia aethiops*).

trichoderm palisade (Gr. *thríx*, *trichós*, hair + *dérma*, skin; F. *palissade* < OPr. *palissada* < *palissa*, paling < *pal*, stake < L. *palus*, stake): see **trichoderm**.

trichogyne (Gr. *thríx*, *trichós*, hair + *gyné*, female): a receptive hypha, or neck of the ascogonium in the shape of a filament, which is entrusted with capturing the spermatia during sexual reproduction in some ascomycetes, such as *Neurospora* (Sordariales) and *Laboulbenia* (Laboulbeniales). See **gynotrichous**, a term preferable to trichogyne, and **carpogonium**.

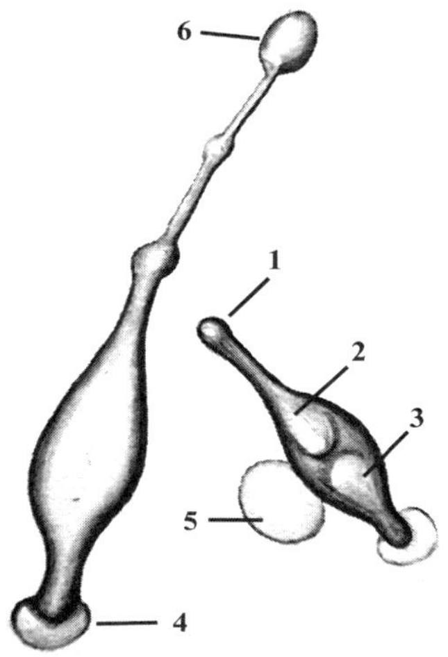

Trichocyst, injecting cell or gun cell, of an encysted zoospore of *Haptoglossa mirabilis*. Diagrammatic representation based on that of Barron, 1980, 1987. The structures shown are: apical cone (1), projectile chamber (2), basal vacuole (3), adhesive pad (4), empty zoospore cyst (5), and discharged harpoon-shaped projectile (6), x 2 000.

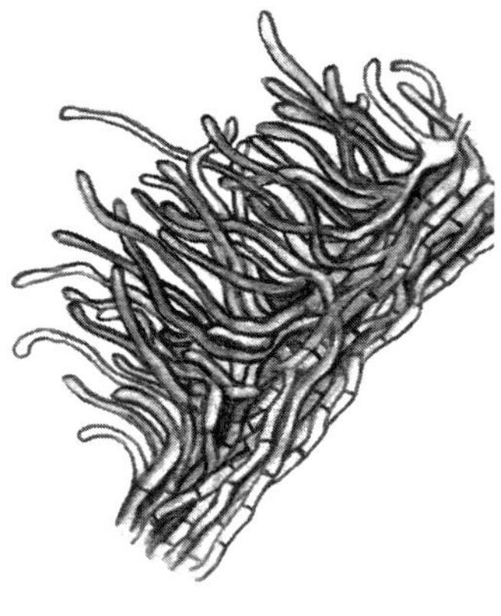

Trichoderm (stipitipellis) of the stipe of the basidiocarp of *Leptonia corvina*, x 80.

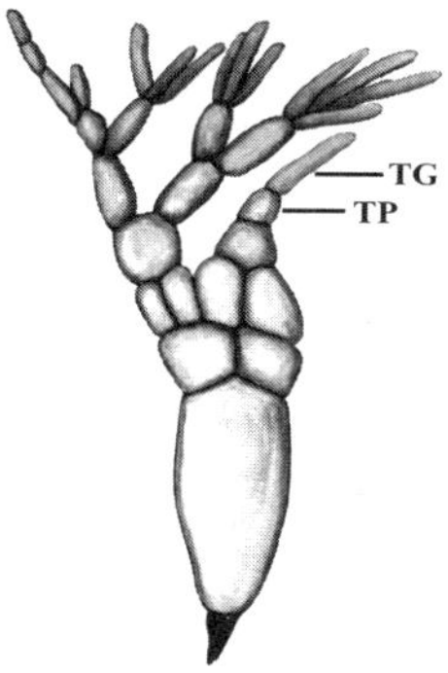

Trichogyne (TG) and trichophore (TP) of the thallus of *Laboulbenia flagellata*, x 100.

tricholomatoid (< gen. *Tricholoma* < Gr. *thríx, trichós*, hair + *lõma*, margin, border + L. suf. *-oide* < Gr. *-oeídes*, similar to): *Agaricology*. A type of fruiting body (13 principal types are recognized), with sinuate gills attached to the stipe, which is fibrous-fleshy or chalky in consistency, centrally implanted in the pileus and lacking a volva, as represented by *Tricholoma* and certain species of *Russula*, *Inocybe* and *Entoloma*, among other gen. of Agaricales.

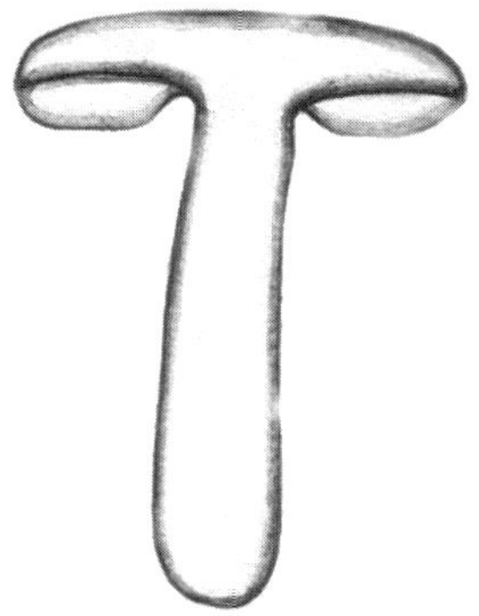

Tricholomatoid type of basidiocarp of *Tricholoma* sp., seen in longitudinal section, x 0.5.

Trichomycetes (Gr. *thríx, trichós*, hair + L. *-mycetes*, ending of class < Gr. *mýkes*, genit. *mýketos*, fungus): a class of fungi that live as commensals or as obligate symbionts in the intestinal tract of arthropods, or attached externally to their exoskeleton. Their thallus can be branched or not and coenocytic, or branched and septate. Their asexual reproduction is by means of trichospores, sporangiospores, arthrospores or amoeboid cells, depending upon the species. Their sexual reproduction is by conjugation of different thalli that form zygospores. The Trichomycetes (which includes the orders Harpellales, Amoebidiales, Asellariales, and Eccrinales), together with the Zygomycetes constitute the phylum Zygomycota of the kingdom Fungi.

trichophore (Gr. *thríx, trichós*, hair + *-phóros*, bearer < *phéro*, to carry, support): that which carries or supports a hair or filament; in the Laboulbeniales, the trichophore is the cell that supports the trichogyne. Both cells can be seen clearly in the thallus of *Laboulbenia flagellata*. See **carpogonium**.

trichospore (Gr. *thríx, trichós*, hair + *sporá*, spore): a unispored sporangium, provided with one or several filamentous or piliform appendages, characteristic of the Trichomycetes, as e.g., *Smittium* and *Stachylina* (Harpellales).

trimitic (Gr. *trís*, three times + *mítos*, filament; here, hypha + suf. *-íkos* > L. *-icus* > E. *-ic*, belonging to, relating to): a fructification with three types of hyphae: generative or fertile, skeletal or supportive, and binding, as is seen, e.g., in *Coriolus versicolor* (Aphyllophorales). Cf. **dimitic** and **monomitic**.

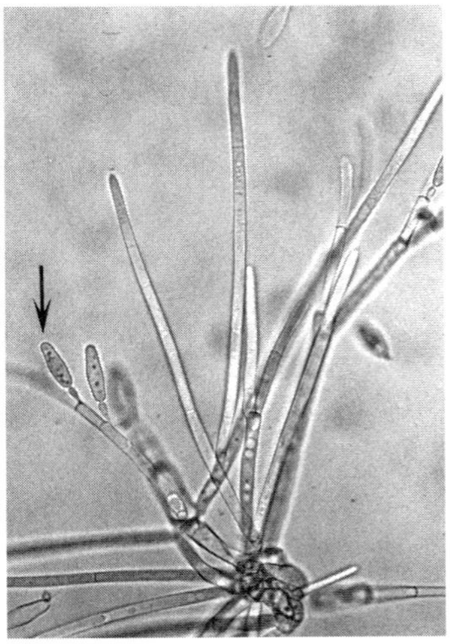

Trichospores of the thallus of *Smittium* sp., x 40 (*RTH*).

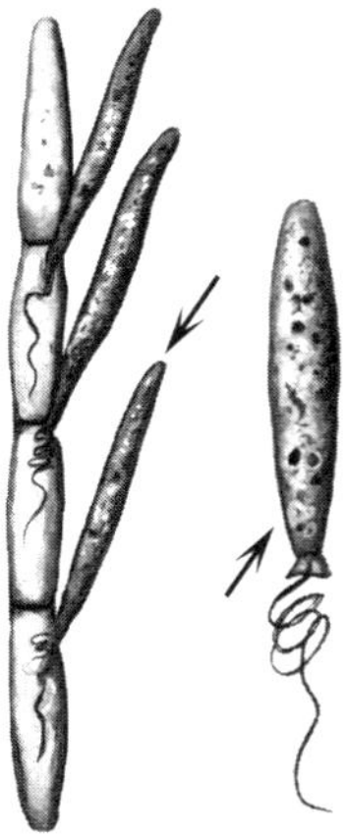

Trichospores of a thallus branch of *Stachylina grandispora*. Note the piliform appendage at the posterior end of the trichospores, x 320.

triquetrous arm (L. *triquetrus*, triangular; ME. < OE. *earm*, arm + L. *-osus* > OF. *-ous*, *-eus* > E. *-ous*, having, possessing the qualities of): *Phallales*. A receptaculum or arm which is three-cornered in cross section, as in *Pseudocolus*.

triverticillate (Gr. *trís*, three times > L. *tri*, *ter*, three + NL. *verticillatus*, arranged in a verticil < *verticillus*, verticil, whorl + suf. *-atus* > E. *-ate*, provided with or likeness): refers to a penicillus that has branching at three levels, i.e., that its phialides are borne on metulae, and the metulae generally borne on terminal branches of a well defined conidiophore, as happens, e.g., in *Penicillium chrysogenum*, *P. echinulatum* and *P. puberulum* (moniliaceous asexual fungi). Cf. **monoverticillate**, **biverticillate** and **polyverticillate**.

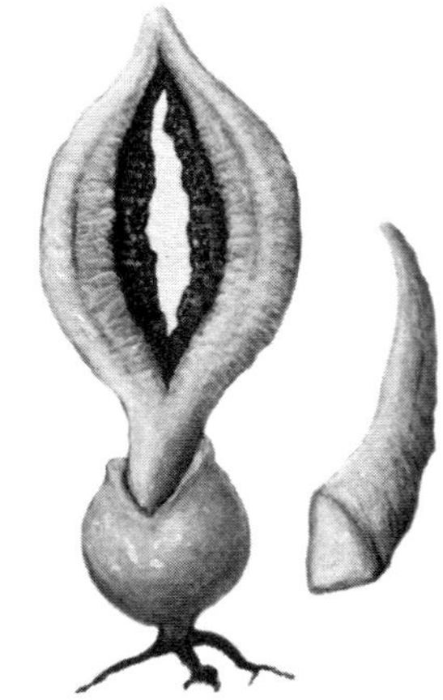

Triquetrous arms, lined with the gleba, of the clathrate fruiting body of *Pseudocolus fusiformis*. One arm is shown in transverse section, x 1.

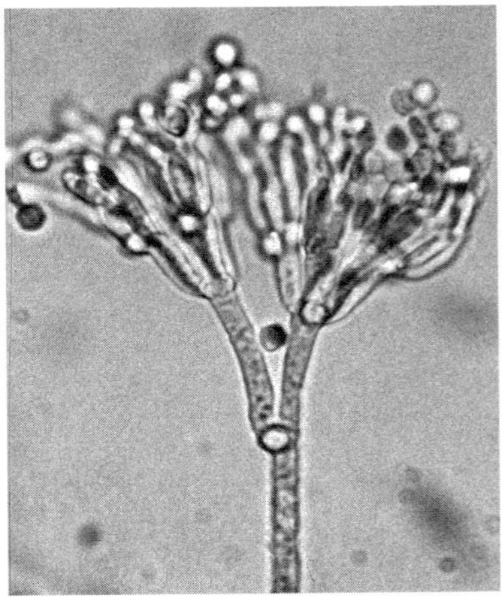

Triverticillate conidiophore of *Penicillium chrysogenum*, x 600 (*CB*).

trochleariform (L. *trochlea* < Gr. *trochilía*, pulley + *-formis* < *forma*, shape): with the shape of a pulley wheel, like the ascospores of *Eurotium* (Eurotiales).

troglobiotic (Gr. *tróglo*, hole, cave + *biotikós*, of or relating to life < *bíos*, life + *-tikós* > L. *-ticus* > E. *-tic*, relation, fitness, inclination or ability): living in caves, like *Histoplasma capsulatum* (moniliaceous asexual fungi), that along with other fungi lives in bat's dung and soil inside caves.

trophic (Gr. *trophós*, that which nourishes < *trépho*, to nourish + suf. *-íkos* > L. *-icus* > E. *-ic*, belonging to, relating to): generally applied to the vegetative phase of the Acrasiomycota, Dictyosteliomycota and Myxomycota (myxamoebae and plasmodia), which, like that of some protozooans, is characterized by having phagotrophic or holozoic nutrition, but can refer to the phase of any organism in which the latter mainly feeds.

trophocyst (Gr. *trophós*, that which nourishes, which serves as food + *kýstis*, bladder, cell): a swollen portion of a hypha, or an enlarged cell, immersed in the substrate from which it obtains its food, from which is formed the sporangiophore, as is observed in *Pilobolus crystallinus* and *P. longipes* (Mucorales).

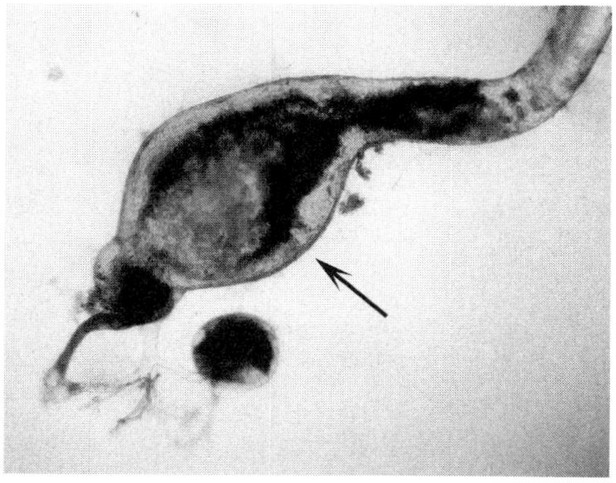

Trophocyst of *Pilobolus longipes*, x 21 (*MU*).

tropism (Gr. *trópos*, a turn, change in manner < *trépo*, to turn, revolve + *-ismós* > L. *-ismus* > E. *-ism*, state, phase, tendency, action): a turning, or curving growth, of an organism as a reaction to a source of stimulation. It is positive when the tropism is in the direction of the stimulus; negative when it is away from the stimulus. See **aerotropism**, **autotropism**, **chemotropism**, **diageotropism**, **galvanotropism**, **geotropism**, **gravitropism**, **heliotropism**, **hydrotropism**, **orthogeotropism**, **phototropism**, **plagiotropism**, **thigmotropism**, and **zygotropism**.

truffle (modification of MF. *truffe* < OPr. *trufa* < VL. *tufera*, akin to L. *tuber*, tuber): the edible, generally subterranean, ascoma of several European ascomycetous fungi of the gen. *Tuber* or other Pezizales or Elaphomycetales, or a basidioma of Melanogastrales. *Burgundy truffle*: *Tuber uncinatum*. *False truffle*: *Hymenogaster*. *Périgord* (*French*) *truffle*: *T. melanosporum*. *Red truffle*: *Melanogaster variegatus*. *Summer truffle*: *T. aestivum*. *White truffle*: *Choiromyces meandriformis*. *White Piedmont truffle*: *T. magnatum*. *White winter truffle*: *T. hiemalbum*. *Winter truffle*: *T. brumale*. *Yellow truffle*: *Terfezia* and *Tirmania*.

truncate (NL. *truncatus*, mutilated, cut off < *truncare*, to mutilate, cut off + suf. *-atus* > E. *-ate*, provided with or likeness): applied to structures that end abruptly, with a flat end, as if they had been cut off. For example, the meiosporangia of *Allomyces javanicus* (Blastocladiales) have a truncate base, as well as the conidia of *Scopulariopsis* (moniliaceous asexual fungi) and the basidiospores of *Coprinus angulatus* (Agaricales); truncate warts occur in the peridium of the fruiting bodies of *Calvatia* and *Lycoperdon* (Lycoperdales).

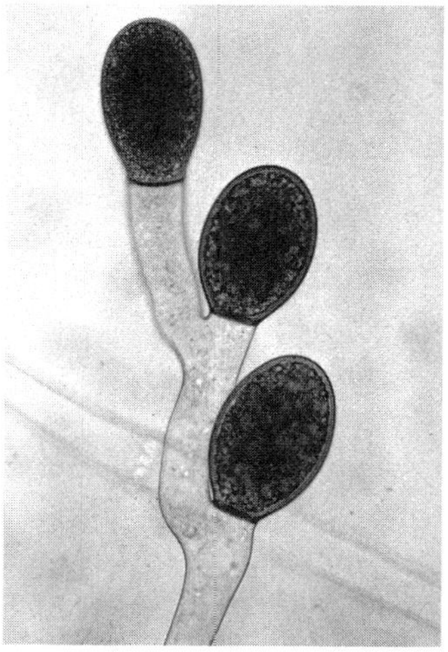

Truncate base of the resting zoosporangia in a thallus of *Allomyces javanicus*, x 400 (*RTH*).

tuba (Tagalog *tubá*): the fermented sap of various species of palms (especially the coconut palm), with alcoholic and lactic acid contents, as well as the microbiota responsible for the fermentation, very similar to those of the Mexican beverage known as **pulque**, which is prepared by fermentation of *Agave* sap. Tuba is consumed in the western coastal regions of Mexico, as well as in various parts of Asia, in particular the Phillipine Islands.

tubaeform, **tubiform** (L. *tuba*, trumpet + *-formis* < *forma*, shape): with the shape of a trumpet or of a tube expanded on one end, like the apothecium of *Sclerotinia sclerotiorum* (Helotiales), the podetia of the lichen *Cladonia* (Lecanorales), and the basidiocarps of *Omphalina chrysophylla* (Agaricales).

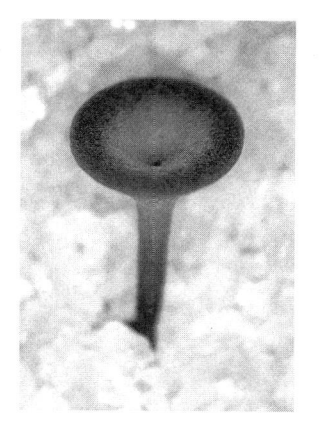

Tubaeform ascocarp arising from a germinating sclerotium of *Sclerotinia sclerotiorum*, x 2 (*RTH*).

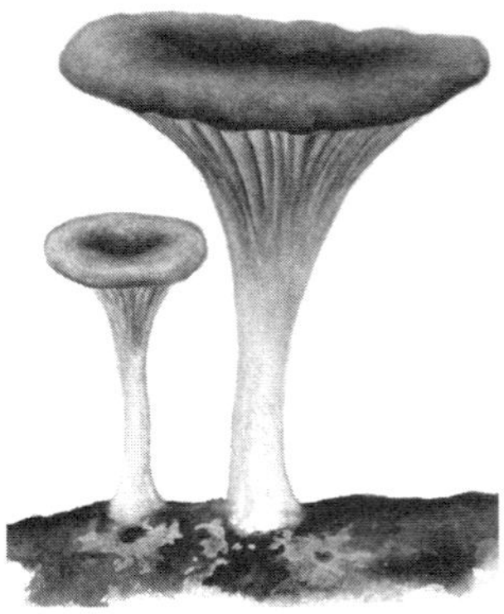

Tubaeform fruiting bodies of *Omphalina chrysophylla*, x 3.4.

tube trama, pl. **tramae** (L. *tubus*, tube, pipe; L. *trama*, trama, warp, texture): see **trama**.

tubercular, **tuberculate** (L. *tuberculum*, dim. of *tuber*, bump, swelling + suf. *-aris* > E. *-ar*, like, pertaining to; NL. *tuberculatus*, having tubercles < L. *tuberculum* + suf. *-atus* > E. *-ate*, provided with or likeness): having small rounded bumps or projections. For example, the conidia of *Histoplasma capsulatum*, *Scopulariopsis brevicaulis* (moniliaceous asexual fungi) and *Curvularia tuberculata* (dematiaceous asexual fungi), as well as the aeciospores of *Cronartium ribicola* (Uredinales), have a tuberculate wall.

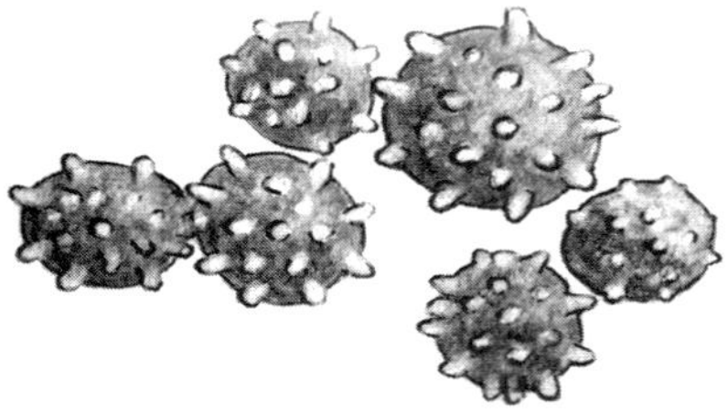

Tuberculate wall of the macroconidia of *Histoplasma capsulatum*, x 100.

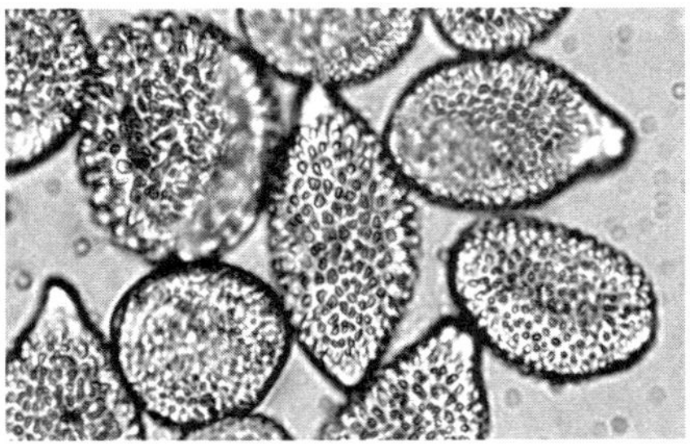

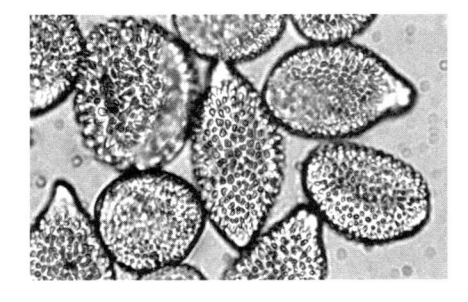

Tuberculate wall of the aeciospores of *Cronartium ribicola*, x 1 100 (*MU*).

tubular (NL. *tubularis* < L. *tubulus* < *tubus*, tube, pipe + suf. *-aris* > E. *-ar*, like, pertaining to): having a cylindrical, hollow shape, like the neck or rostrum of many perithecia, e.g., that of *Ceratocystis* (Microascales) and *Guanomyces* (Sordariales).

tubulus, pl. **tubuli** (L. *tubulus*, dim. of *tubus*, tube, pipe): refers to the neck or rostrum of many perithecia (Pyrenomycetes), as well as to any cylindrical, hollow and small structure.

tumid (L. *tumidus*, swollen): swollen, inflated.

tunica, pl. **tunicae** (L. *tunica*, tunic, covering): **1**. Any covering or investing layer. **2**. *Gasteromycetes*. Refers to the thin white membrane that covers the peridioles in the majority of species of nidulariaceous Gasteromycetes, such as *Cyathus* and *Nidularia* (Nidulariales).

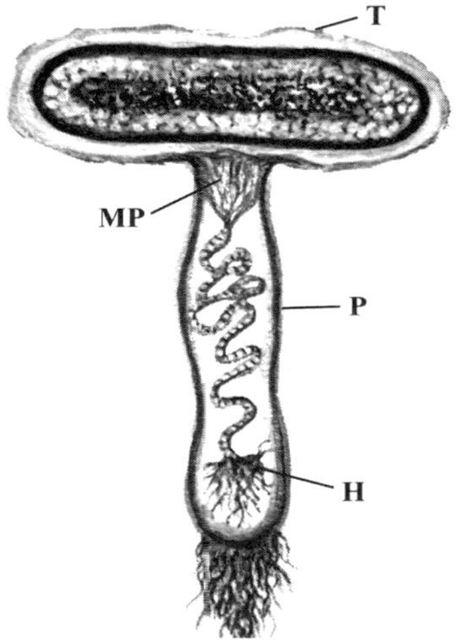

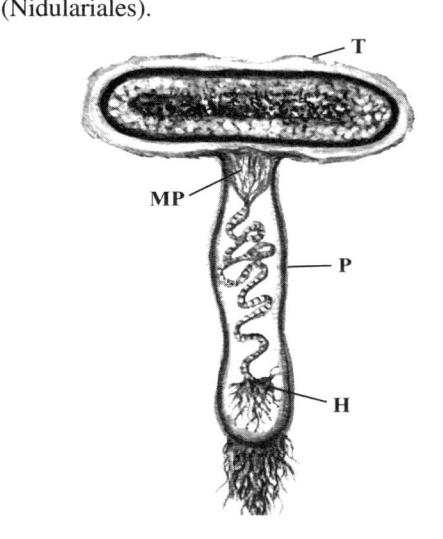

Tunica (T) covering a peridiole of *Cyathus striatus*, seen in longitudinal section. The purse (P) surrounding the funicular cord, and the hapteron (H), which is attached to the middle piece (MP), are also shown, x 100.

turbinate (L. *turbinatus*, shaped like a top < *turbin-*, *turbo*, top + suf. *-atus* > E. *-ate*, provided with or likeness): with the shape of an inverted cone or top, narrow at the base and wide at the apex; e.g., like the cystidia of certain agarics, the basal cells of the conidia of *Polyschema variabilis* (dematiaceous asexual fungi) and the perithecial stromata of *Camillea* and *Phylacia turbinata* (Xylariales). Similar to **obconic**.

Turbinate perithecial stromata of *Phylacia turbinata*, x 2.

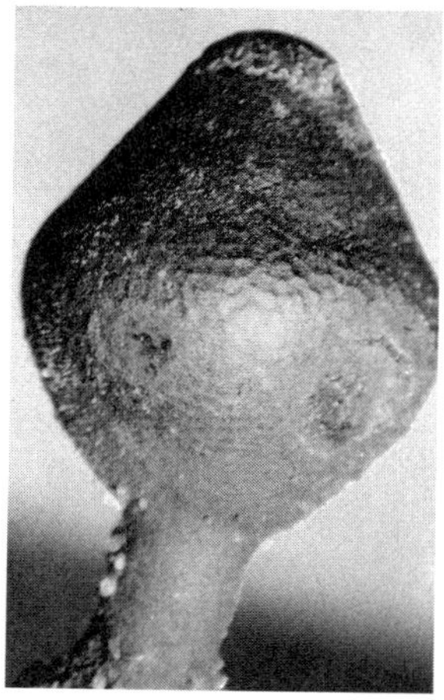

Turbinate perithecial stroma of *Camillea* sp., x 7 (*MU*).

u

ubiquitous (L. *ubique* > F. *ubiquite*, everywhere + L. *-osus* > OF. *-ous*, *-eus* > E. *-ous*, having, possessing the qualities of): refers to an organism that occurs in almost any type of habitat, like many species of fungi, e.g., the gen. *Aspergillus* and *Penicillium* (moniliaceous asexual fungi).

uliginose, **uliginous** (L. *uliginosus*, swampy, marshy, moist < *uligo*, genit. *uliginis*, moisture + *-osus*, full of, augmented, prone to > ME. *-ose*; or + *-osus* > OF. *-ous*, *-eus* > E. *-ous*, having, possessing the qualities of): applied to organisms that grow in wet places, such as swamps.

umbellate (L. *umbellatus*, bearing, consisting of, or arranged in umbels, resembling an umbel in form < *umbella*, parasol, umbel + suf. *-atus* > E. *-ate*, provided with or likeness): provided with umbels, like the thallus of *Rhipidium americanum* (Rhipidiales), and the sporangiophore of *Circinella umbellata* (Mucorales), whose sporangial branches start out at the same point and are nearly all the same length, like the ribs of an inverted parasol on the end of the main axis. Also umbellate is the arrangement of the sporocladia of *Kickxella alabastrina* (Kickxellales).

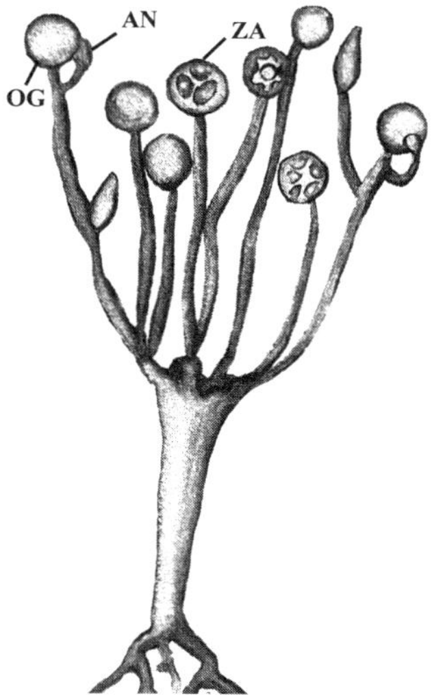

Umbellate thallus of *Rhipidium americanum*, with the branches bearing zoosporangia (ZA), oogonia (OG) and antheridia (AN), x 100.

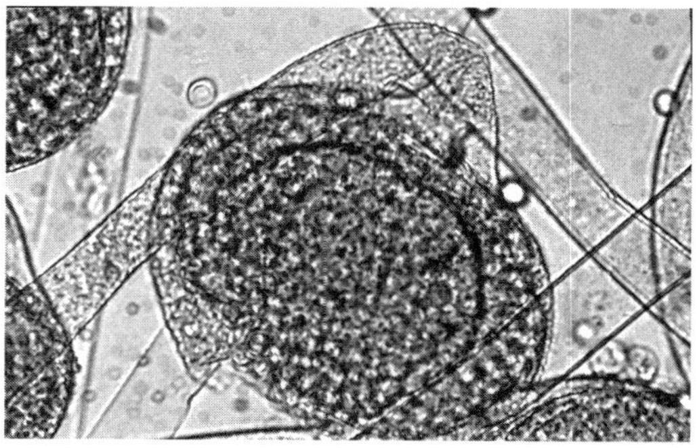

Umbellate sporangiophores of *Circinella umbellata*, x 780 (*EAA*).

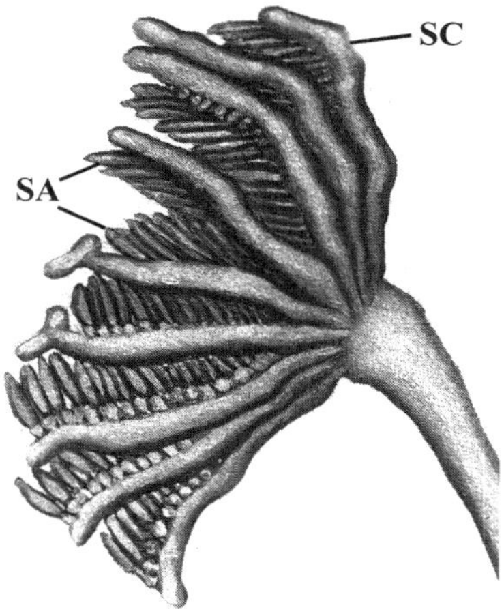

Umbellate sporangiophore of *Kickxella alabastrina*, bearing sporocladia (SC) with sporangiola (SA), x 570.

umbilicate (L. *umbilicatus*, navel-shaped < *umbilicus*, navel + suf. *-atus* > E. *-ate*, provided with or likeness): having a depression like a navel, like the pileus of the apothecium of *Leotia* (Helotiales) and the pileus of the basidiocarp of *Clitocybe* and *Omphalina* (=*Omphalia*), of the Agaricales. In the lichens, it is applied to a more or less concave foliaceous thallus that is united to the substrate only by its central part (the umbilical cord or funiculum), as in species of Umbilicariaceae, e.g., *Umbilicaria* (= *Gyrophora*) of the Lecanorales.

Umbilicate pileus of the apothecia of *Leotia lubrica*, x 0.8 (*RTH*).

Umbilicate pileus of the basidiocarps of *Clitocybe lazulina*, x 0.6 (*EPS*).

umbilicus, pl. **umbilici** (L. *umbilicus*, navel): see **funiculus**.

umbo (L. *umbo*, genit. *umbonis*, prominence, projection): **1**. A central, cone-shaped projection. **2**. *Agaricales*. It refers to the elevated mound or cone-shaped part that projects upward in the center of a certain type of pileus, whether concave or convex. See **umbonate**.

umbonate (NL. *umbonatus*, shielded < *umbo*, genit. *umbonis*, prominence, shield + suf. *-atus* > E. *-ate*, provided with or likeness): **1**. Having a central cone-shaped prominence or projection. **2**. Shaped like an umbo; e.g., like the pileus of *Psilocybe mazatecorum* and *P. caerulescens* (Agaricales).

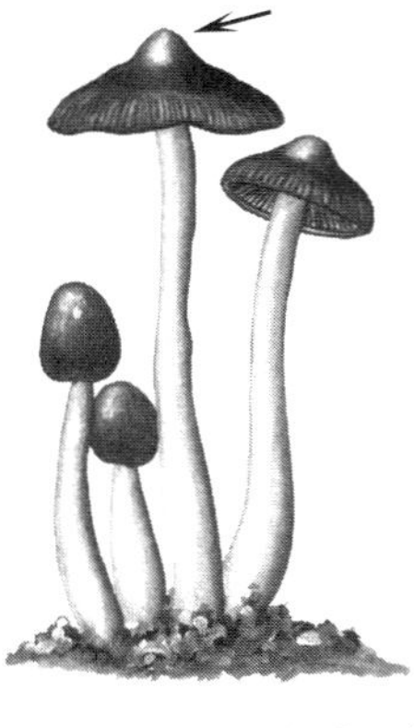

Umbonate pileus of the basidiocarps of *Psilocybe caerulescens*, x 0.5.

umbraticous (L. *umbraticus*, belonging to shade < *umbra*, a shaded area + *ticus* < Gr. *tikós*, belonging to + L. *-osus* > OF. *-ous*, *-eus* > E. *-ous*, having, possessing the qualities of): growing in shady places. See **photophobic**.

unciform (L. *unciformis* < *uncus*, hook + *-formis* < *forma*, shape): with the shape of a hook; also called **uncinate.** Forming a hook, like the ascocarp appendages of *Uncinula* (Erysiphales).

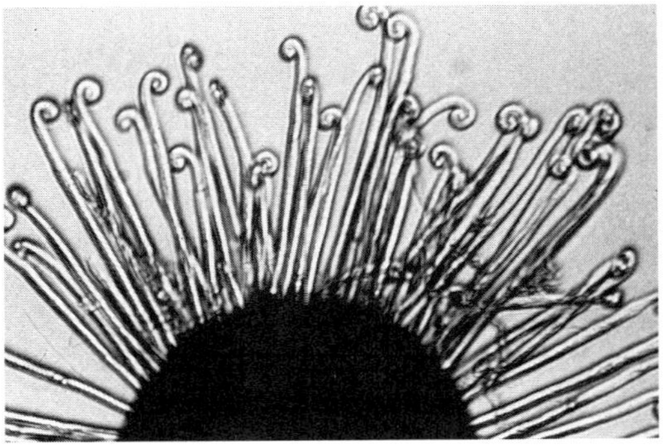

Unciform perithecial appendages of *Uncinula* sp., x 300 (*RTH*).

uncinate (L. *uncinatus*, bent at the tip like a hook < *uncinus*, dim. of *uncus*, hook + suf. *-atus* > E. *-ate*, provided with or likeness): see **unciform**.

uncinulum, pl. **uncinula** (L. *uncinulum* < *uncinus*, *uncus*, hook + dim. suf. *-ulum*): a hook-shaped branch that is produced at the apex of the ascogenous hyphae of many ascomycetes and which becomes the ascus mother cell, e.g., *Eurotium* (Eurotiales) and *Hypomyces* (Hypocreales). The simultaneous division of the two nuclei of each hook, the partition and delimitation of the cells are a typical process in the ascomycetes, which have their parallel in the clamp connections of the secondary mycelium of the basidiomycetes. See **crozier**.

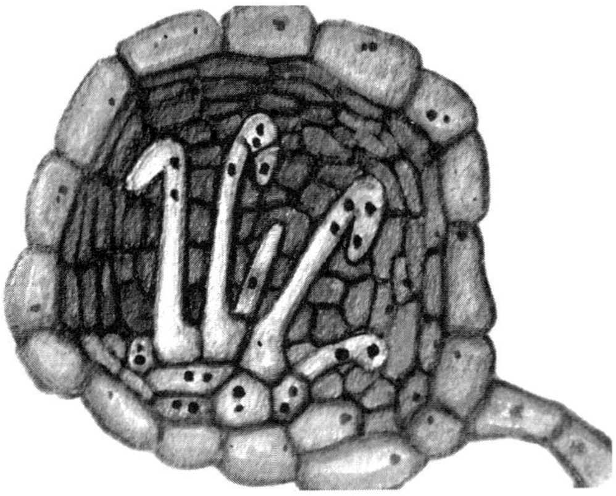

Uncinula or **croziers** of a cleistothecium of *Eurotium rubrum*, x 150.

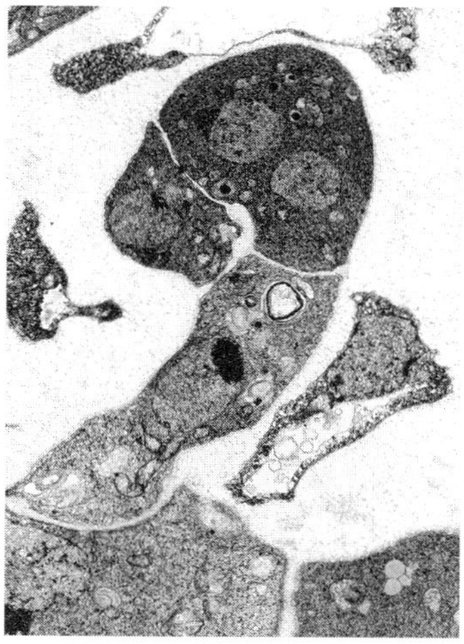

Uncinulum or **crozier** of *Hypomyces* sp. The dikaryotic penultimate cell of the hook will eventually undergo karyogamy and become the ascus mother cell, x 1 000 (transmission electron micrograph of *CM*).

undate (L. *undatus*, wavy, wave-like form < *unda*, wave + suf. *-atus* > E. *-ate*, provided with or likeness): with a wavy border; e.g., like the pileus of the basidiocarp of *Clitocybe phyllophila* (Agaricales). Cf. **undulate**.

undulate (L. *undula*, dim. of *unda*, wave + suf. *-atus* > E. *-ate*, provided with or likeness): like **undate** but with smaller waves, e.g., the border of the pileus of *Collybia fusipes* (Agaricales).

undulipode, **undulipodium**, pl. **undulipodia** (L. *undula*, dim. of *unda*, wave + NL. *podium* < Gr. *pódion* < *poús*, *podós*, foot, support + L. dim. suf. *-ium*): a flexible intracellular extension in the shape of a small whip (traditionally called a flagellum), that the motile cells of the majority of the eukaryotic organisms (many plants, the majority of the protoctists, or protists, and the majority of the animals) have. The undulipodia, which grow from the kinetosomes or basal bodies, are composed of packets of microtubules which, in cross section, show the characteristic 9 + 2 arrangement, and are constituted of more than 100 proteins. There are undulipodia with or without mastigonemes, depending upon the species. Cf. **flagellum**.

ungulate (LL. *ungulatus*, having claws or hoofs < L. *ungula*, claw, hoof + suf. *-atus* > E. *-ate*, provided with or likeness): applied to a fruiting body that has the shape of a horse's hoof, like the fruiting body of *Fomes igniarius* and *F. dochmius* (Aphyllophorales).

uniaxial (NL. *uniaxialis*, having a single axis < *unus*, one + *axis*, axis + L. suf. *-alis* > E. *-al*, relating to or belonging to): with a single central axis, around which are radially arranged its branches, like the majority of apothecia and stipitate basidiocarps. Also called **monaxial**. Cf. **multiaxial**.

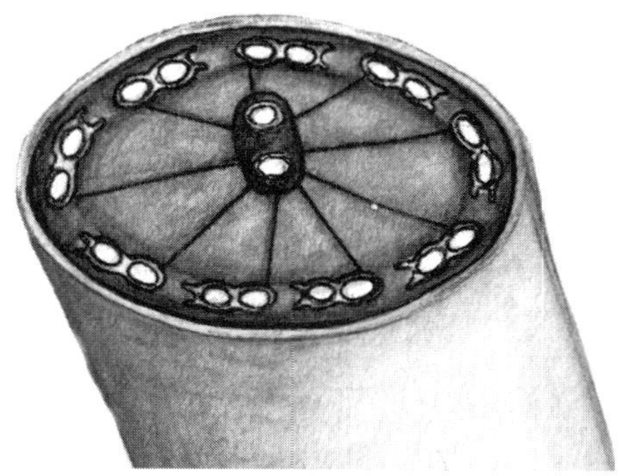

Undulipodium of a typical eukaryotic cell, showing the 9+2 arrangement of microtubules. Diagram of a transverse section based on figures and information published by Margulis and Schwartz in their book *Five Kingdoms*, 1988, x 300 000.

Ungulate fruiting body of *Fomes* (= *Ungulina*) sp., x 0.5 (*MU*).

unicellular (L. *unicellularis* < *unus*, one + *cellula*, cell, living cell, dim. of *cella*, small room, compartment, cell of a honeycomb + suf. *-aris* > E. *-ar*, like, pertaining to): consisting of a single cell, like the thallus of the yeasts, when the somatic cells have not budded, or like many types of asexual and sexual fungal spores. See **amerospore**.

uniflagellate (L. *unus*, one + *flagellatus*, provided with a flagellum < *flagellum*, whip + suf. *-atus* > E. *-ate*, provided with or likeness): having a single flagellum, like the zoospores of the Chytridiomycetes, which have a whiplash flagellum (without mastigonemes) on the posterior end, and those of the Hyphochytriomycota, which possess a single mastigonemate flagellum on the anterior end. The antherozoids or spermatozoids of the Monoblepharidales also are uniflagellate on the posterior end.

uniguttulate (L. *unus*, one + *guttulatus*, containing drops or drop-like masses < *guttula*, dim. of *gutta*, droplet + suf. *-atus* > E. *-ate*, provided with or likeness): refers to a structure that contains a single oil droplet, like the ascospores of *Podospora comata* and *Sordaria fimicola* (Sordariales). See **guttulate**.

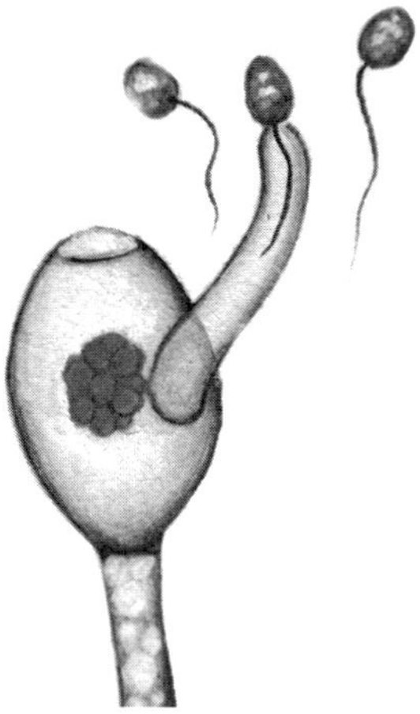

Uniflagellate antherozoids of *Monoblepharis polymorpha;* one antherozoid is escaping from the antheridium, x 650.

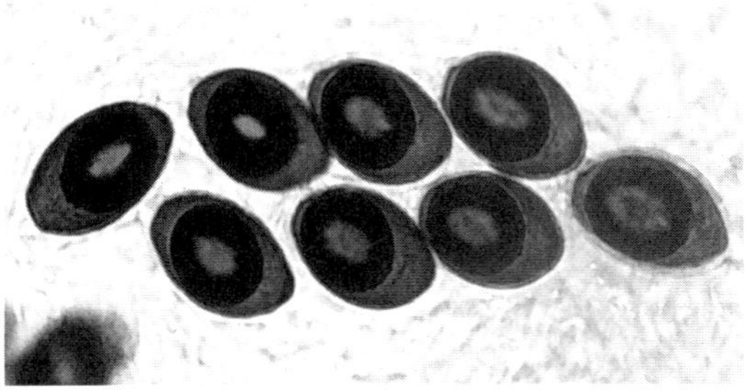

Uniguttulate ascospores of *Podospora comata*, x 530 (*CB*).

unilocular (L. *unilocularis* < *unus*, one + *loculus*, dim. of *locus*, cavity, locule + suf. *-aris* > E. *-ar*, like, pertaining to): with a single cavity or locule, like the ascostromata of *Mycosphaerella* (Dothideales) and *Sporormiella* (Melanommatales), which because of its similarity to a perithecium is called a **pseudothecium**. Cf. **multilocular**.

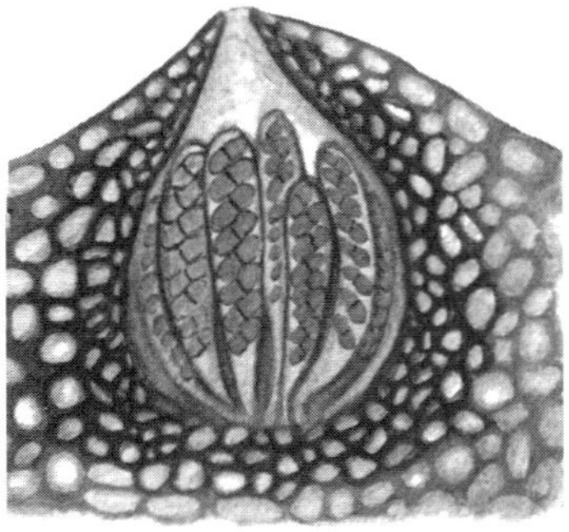

Unilocular ascostroma of *Mycosphaerella musicola*, x 150.

uniserial, **uniseriate** (L. *unus*, one + *series*, series, row + L. suf. *-alis* > E. *-al*, relating to or belonging to; or + L. suf. *-atus* > E. *-ate*, provided with or likeness): arranged in a single series, like the conidial head of some species of *Aspergillus* (moniliaceous asexual fungi), such as *A. fumigatus* and *A. repens*, among others, that have a series or row of phialides with conidia, borne directly on the vesicle of the conidiophore. Also called **monostichous**. The asci of the majority of species of ascomycetes have their ascospores in a single row. Cf. **biseriate** and **distichous**.

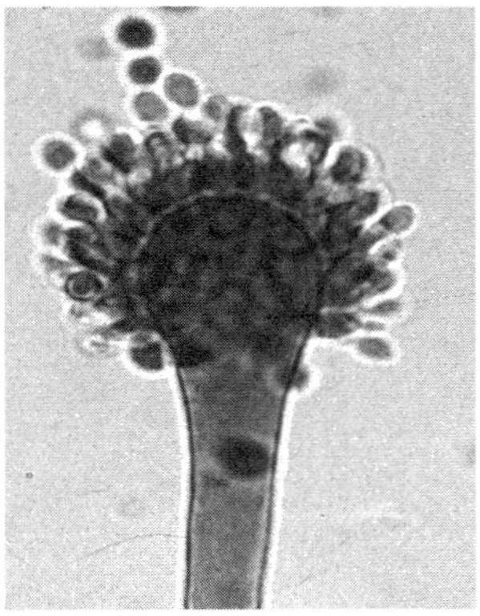

Uniseriate conidial head of *Aspergillus fumigatus*, x 600 (*CB*).

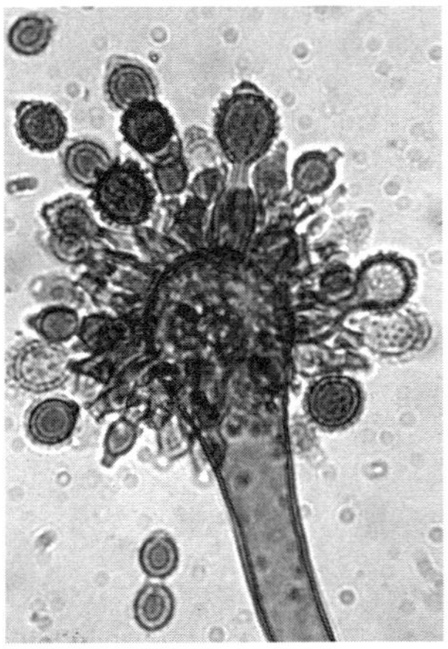

Uniseriate conidial head of *Aspergillus repens*, x 1 000 (*PL*).

unisexual (L. *unus*, one + *sexualis*, sexual, pertaining or relative to sex + suf. *-alis* > E. *-al*, relating to or belonging to): a thallus with one sex only. See **dioecious**. Cf. **hermaphrodite** and **monoecious**.

unispored (L. *unus*, one + Gr. *sporá*, spore): containing a single spore, like the sporangiola of various Mucorales and Entomophthorales, such as *Cunninghamella* and *Conidiobolus*, respectively. Since in these cases the spore wall is fused with the wall of the sporangium, it resembles a conidium of the mitosporic fungi, for which reason unispored sporangiola have been called conidia.

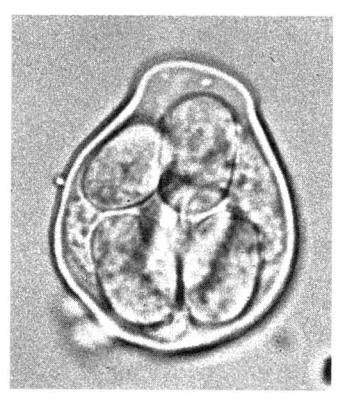

Unispored sporangiola of *Conidiobolus coronatus*, x 1 200 (*CB*).

unitunicate (L. *unus*, one + *tunica*, covering, blanket + suf. *-atus* > E. *-ate*, provided with or likeness): a type of ascus in which the internal and external layers of the wall are more or less rigid and do not separate during ascospore liberation, e.g., *Taphrina deformans* (Taphrinales) and *Brasiliomyces malachrae* (Erysiphales). Unitunicate asci are present in most ascomycetes; an exception are the Loculoascomycetes, which have bitunicate asci. Cf. **bitunicate** and **prototunicate**.

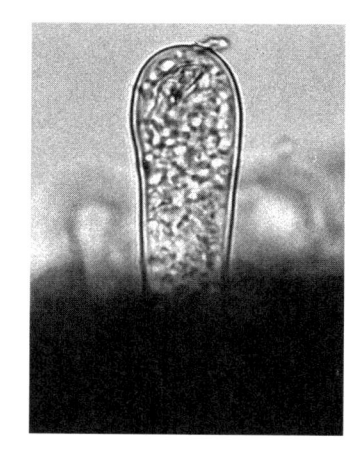

Unitunicate ascus of *Brasiliomyces malachrae*, with eight ascospores, x 800 (*RTH*).

Unitunicate ascus of *Taphrina caerulescens*, with numerous budding ascospores, x 1 560 (*RTH*).

universal veil (L. *universalis*, general < *universus*, universe + suf. *-alis* > E. *-al*, relating to or belonging to; L. *velum*, veil): see **veil**.

urban, urbanus (L. *urbanus*, urban, urbane < *urbs*, city + suf. *-anus* > E. *-an*, belonging to): developing in city habitats, like the basidiocarps of *Panaeolina foenisecii* and other Agaricales, such as *Bolbitius vitellinus*, common in lawns and parks.

urceolar, urceolate (L. *urceolus*, dim. of *urceus*, urn, pot, clay jar, pitcher + suf. *-aris* > E. *-ar*, like, pertaining to; NL. *urceolatus*, in the shape of an urn < *urceolus* + suf. *-atus* > E. *-ate*, provided with or likeness): having the shape of a kettle; in lichenology it is applied to the discoid apothecia of some lichens, with the hymenium sunken and the parathecium elevated, and with its converging borders forming a narrow mouth, as is observed in the species of *Aspicilia* (=*Urceolaria*) and in *Conotrema urceolatum* (Ostropales). Also called **orculiform**, and for the spores **polarilocular** and **polocellate.**

Urceolar apothecia of *Conotrema urceolatum*, x 6.

uredine (L. *uredo*, rust, a blight, a burning < *urere*, to burn + suf. *-inus* > E. *-ine*, of or pertaining to): an old name applied to uredinial rust fungi.

urediniospore (L. *uredo*, rust, genit. *uredinis*, a blight, a burning + Gr. *sporá*, spore): a binucleate spore, unicellular, sessile or pedicellate, that is formed in the uredinia or uredosori of the rust fungi, or Uredinales, so-called because of their rust-colored spores, which are the principal structures of propagation of these plant parasitic fungi, of great importance in epiphytology, because generally they produce several successive generations of uredinia on the host plant, before passing into the following stage in the life cycle, which is that of the telia.

uredinium, pl. **uredinia** (L. *uredo*, rust, genit. *uredinis*, a blight, a burning + dim. suf. *-ium*): also called **uredium** and **uredosorus**; each of the pustules, ochre or red-oxide in color, that produces urediniospores, and which the rust fungi (Uredinales) form in tissues of the infected plant. The uredinia lack a peridium or proper wall, since they push out through the torn epidermis of the host, i.e., they are erumpent, and are observed clearly in *Puccinia graminis tritici* (Uredinales), the wheat rust.

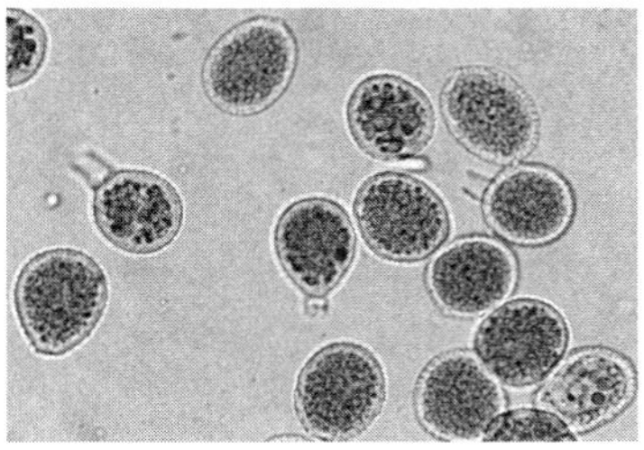

Urediniospores of *Puccinia* sp., x 780 (*CB*).

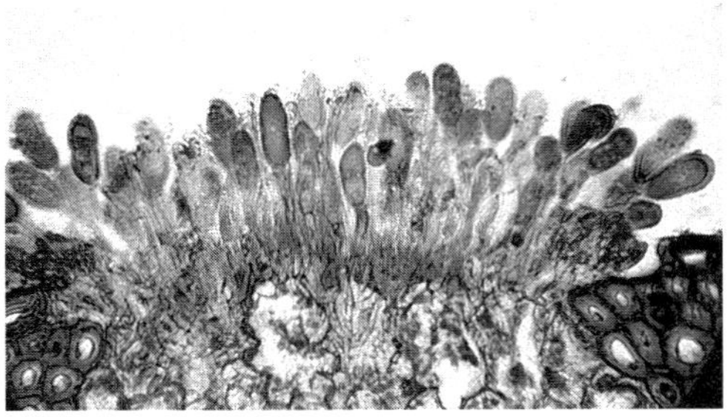

Uredinium with urediniospores of *Puccinia graminis tritici*, seen in longitudinal section of a wheat leaf, x 240 (*GGA*).

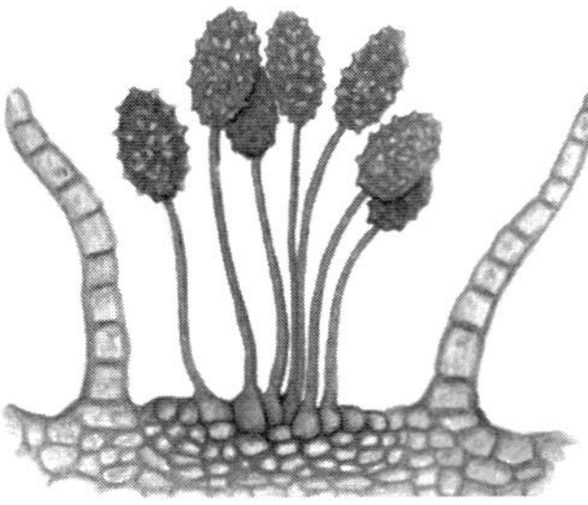

Uredinium with urediniospores of *Puccinia graminis tritici* on a wheat leaf, x 280.

uredium, pl. **uredia** (NL. *uredium* < *uredo*, rust + dim. suf. *-ium*): see **uredinium**.

uredosorus, pl. **uredosori** (L. *uredosorus* < *uredo*, rust + *sorus*, sorus, pile, heap): see **uredinium**.

uredospore (L. *uredo*, rust + Gr. *sporá*, spore): see **urediniospore**.

urinophilic (L. *urina*, urine + Gr. *phílos*, have an affinity for + suf. *-íkos* > L. *-icus* > E. *-ic*, belonging to, relating to): applied to operculate Discomycetes that appear predictably on substrates only after they have been soaked with urine, such as dung. Species of *Ascobolus*, *Saccobolus*, *Coprotus*, and *Iodophanus* (Pezizales) are prominent members of the dung mycota, which also are found in disturbed areas such as burns.

urnula, pl. **urnulae** (L. *urnula*, dim. of *urna*, urn, jar): a structure with the shape of a small urn, like the ascocarp of *Urnula* (Pezizales).

Urnula of *Urnula craterium*, x 0.4.

uroid (Gr. *ourá*, tail + L. suf. *-oide* < Gr. *-oeídes*, similar to): refers to the posterior region that myxamoebae of *Copromyxa arborescens* (Acrasiomycota) develop when they are in an aqueous medium; this region has the shape of a button or knob, often with extensions of hyaloplasm of a filose type (filopodia), and with a contractile vacuole.

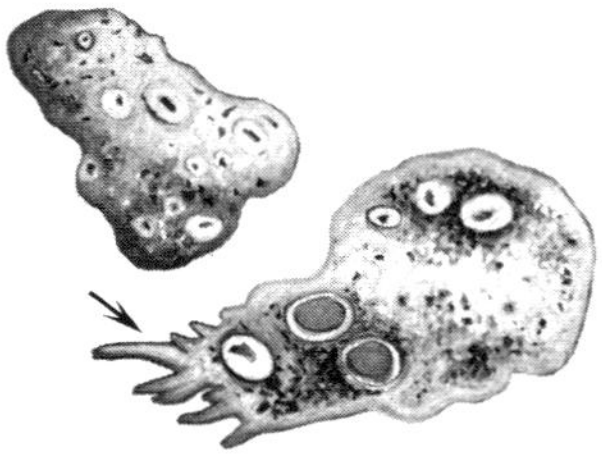

Uroid of the myxamoebae of *Copromyxa arborescens*, x 1 400.

ustilospore (L. *ustus*, *ustulatus*, burned < *ustulare*, to burn + Gr. *sporá*, spore): a smut spore. Also called **ustospore** (of the same etymology). See **bunt** and **smut**.

ustospore (L. *ustus*, burned + Gr. *sporá*, spore): see **ustilospore**.

utricle, **utriculus**, pl. **utriculi** (L. *utriculus* < *uter*, bag, skin purse or skin bottle, or < *uterus*, uterus, womb + dim. suf. *-ulus* > E. *-ule*): a cover, similar to a purse or bladder (the perispore), that surrounds the basidiospores of certain fungi, such as *Hymenogaster* (Hymenogastrales); such spores are said to be **utriculate**.

utriculate (L. *utriculatus* < *utriculus*, dim. of *uter*, bag, skin purse or skin bottle, or of *uterus*, uterus, womb + suf. *-atus* > E. *-ate*, provided with or likeness): provided with a pouch, like the basidiospores of *Hymenogaster thwaitesii* (Hymenogastrales), that are completely surrounded by a cover or purse consisting of the perispore. Also called **utriculose**. Cf. **calyptrate**.

utriculose (L. *utriculosus* < *utriculus*, dim. of *uter*, bag, skin purse or skin bottle, or of *uterus*, uterus, womb + *-osus*, full of, augmented, prone to > ME. *-ose*):

purse-shaped. Also called **lageniform**, **sicyoid**, **ventricose-rostrate** and **utriculate**.

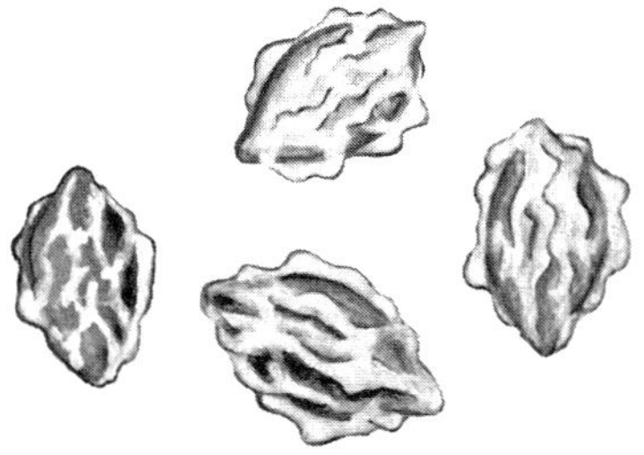

Utriculate basidiospores of *Hymenogaster thwaitesii*, x 1 400.

utriform (L. *utriformis* < *uter*, bag, skin purse or skin bottle + *-formis* < *forma*, shape): having the shape of a purse, bladder or skin bottle; e.g., like the phialides of *Stachybotrys chartarum* (dematiaceous asexual fungi), which are arranged in verticils on the apex of erect conidiophores, and the cheilocystidia of *Conocybe utriformis* (Agaricales), ventricose below and thick above, but almost isodiametric and typically with a slight constriction beneath a large, rounded head.

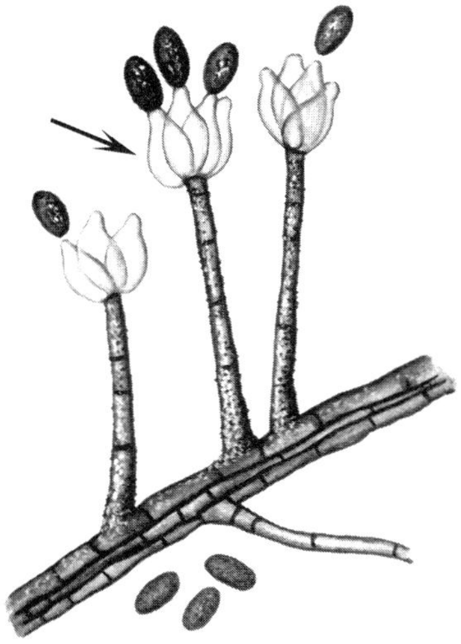

Utriform phialides of the conidiophores of *Stachybotrys chartarum*, x 700.

V

vacuolate (L. *vacuolum*, vacuole, dim. of *vaccum* < *vaccus*, empty + suf. *-atus* > E. *-ate*, provided with or likeness): containing one or more vacuoles. The hyphae of the aquatic fungi of the order Monoblepharidales are distinctively vacuolate, to the point of appearing foamy.

vacuole (L. *vacuolum*, vacuole, dim. of *vaccum* < *vaccus*, empty): a cavity or space, delimited by a membrane, that is formed in the cytoplasm of a cell and whose functions are for excretion and digestion. Vacuoles are most abundant in old cells.

vaginate (L. *vaginatus*, sheathed < *vagina*, sheath + suf. *-atus* > E. *-ate*, provided with or likeness): provided with a sheath or envelope, like the basidiocarp of *Amanita vaginata* (Agaricales), whose long volva covers the lower part of the stipe.

Vaginate stipe of the basidiocarp of *Amanita vaginata*, x 0.5.

vaginatoid (< old gen. *Vaginata*, syn. of *Amanitopsis*, syn. of *Amanita* + L. suf. *-oide* < Gr. *-oeídes*, similar to): *Agaricales*. A type of fruiting body (13 principal types are recognized), with free or finely attached gills, and with a volva but without a ring, like that of *Amanita vaginata*, of the species of *Volvariella*, and certain species of *Agaricus* and *Coprinus* (Agaricales).

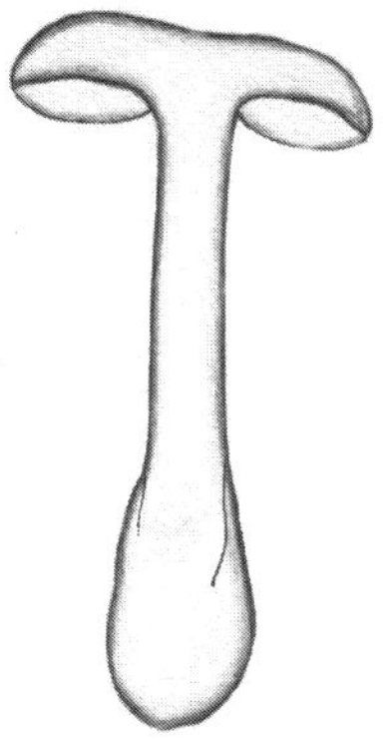

Vaginatoid type of basidiocarp, represented by *Amanita* sp., seen in longitudinal section, x 0.5.

valid (MF. *valide* < ML. *validus*, strong, valid): a name published in accordance with the Code Articles; such a name may be illegimate or legitimate. A *prevalid* (ME. < OF. < L. *prae*, before) name (or author) is the one published before 1753, the starting point for the nomenclature of fungi under the Code. See **nomen**.

valsoid (< gen. *Valsa* + L. suf. *-oide* < Gr. *-oeídes*, similar to): applied to stromatic ascomycetes in which the perithecia form in an endostroma, with the perithecial necks projecting through the ectostroma and converging toward a central point, as occurs in the Pyrenomycete gen. *Valsa* (Diaporthales), very common on decorticated branches. Cf. **eutypoid** and **diatrypoid**.

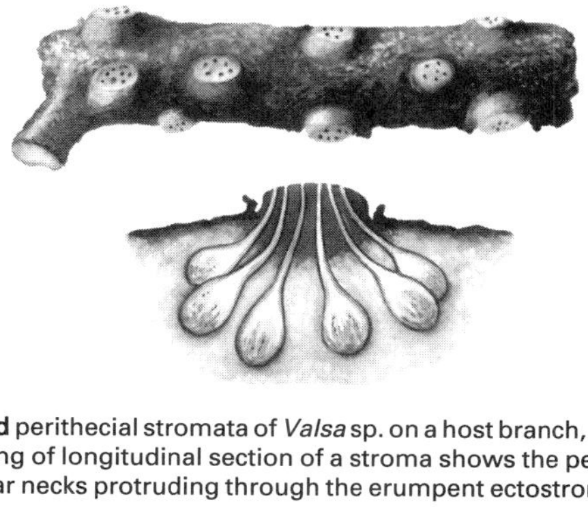

Valsoid perithecial stromata of *Valsa* sp. on a host branch, x 0.1. The drawing of longitudinal section of a stroma shows the perithecial ostiolar necks protruding through the erumpent ectostroma, x 10.

variegated (LL. *variegatus*, to look varied < *variegare* < *varius*, various + suf. *-egare*, to drive + suf. *-atus* > E. *-ate*, provided with or likeness): having a variety of colors and shades, or areas of diverse textures, intermixed, like the colonies of some molds, or the basidiocarps of certain Aphyllophorales, like *Stereum* and *Coriolus versicolor*.

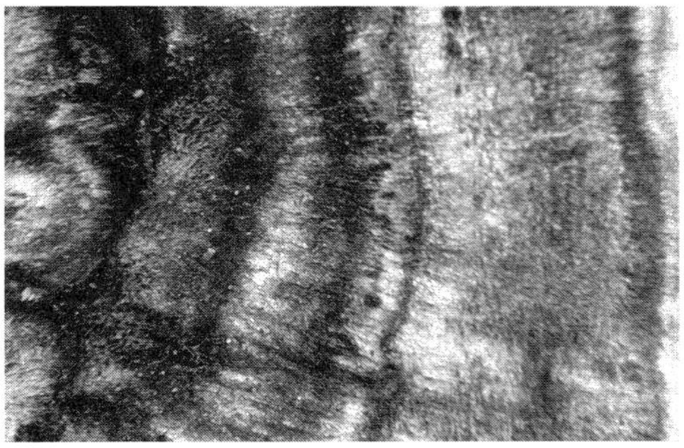

Variegated surface of the fruiting body of *Coriolus versicolor*, x 10 (*MU*).

variolate (L. *variolatus*, pustulated < *variola*, pustule, wart + suf. *-atus* > E. *-ate*, provided with or likeness): with pustules or large granular warts, like the thallus surface of the lichen *Lecanora subfusca* var. *variolosa* (Lecanorales).

vascelloid (< gen. *Vascellum* < L. *vas*, urn, vessel + dim. suf. *-cellum* + suf. *-oide* < Gr. *-oeídes*, similar to): *Gasteromycetes*. A type of fruiting body, such as in *Vascellum* (Lycoperdales), in which the gleba is separated from the sterile subgleba by a well developed diaphragm.

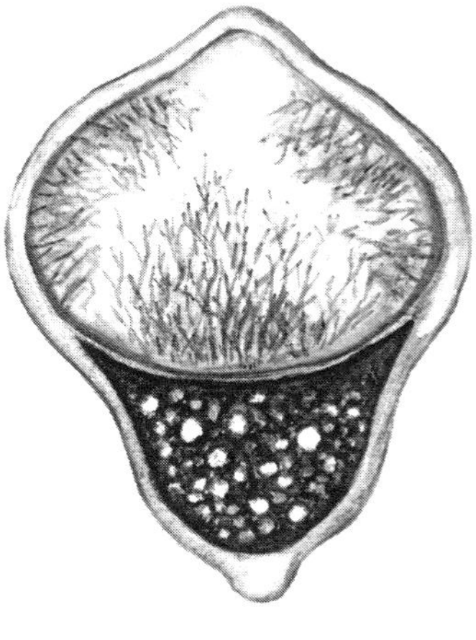

Vascelloid type of development, represented by the fruiting body of *Vascellum* sp., seen in longitudinal section, x 5.

vasiform (L. *vasiformis* < *vas*, genit. *vasis*, glass, vessel + *-formis* < *forma*, shape): shaped like a glass, like the fruiting body of *Vascellum* (Lycoperdales).

Vasiform basidiocarps of *Vascellum depressum*, x 2.

vaulted ray (E. *vaulted*, ptp. of *vault*, arch < ME. *voute* < OF. *voulte*, *volte*; L. *radius*, radius): a ray which arches over the exoperidium, forming a chamber, usually conic in profile, as seen in *Geastrum fornicatum* (Lycoperdales).

vector (L. *vector*, genit. *vectoris*, conductor, that which transports): applied to an organism that transmits or propagates another organism, the latter generally a pathogen. For example, *Olpidium viciae* (Spizellomycetales) is a chytrid vector of a virus that affects plants of *Vicia unijuga*. On the other hand, there exist many examples of species of fungi that are disseminated from a substrate or host to another by insect vectors, as occurs in *Claviceps purpurea* (Hypocreales), whose conidia are produced in the infected ovaries of rye flowers, mixed with a sweet, sticky secretion, similar to nectar; attracted by this nectar, the insects visit the flowers and disseminate the fungus to uninfected flowers.

vegetative (ML. *vegetativus* < L. *vegetar*, *vegetare*, to enliven + *-ivus* > E. *-ive*, quality or tendency, fitness): that which vegetates or is capable of vegetating; i.e., it undergoes various vital functions (germination, growth, development and asexual multiplication) but not true sexual reproduction. Present mycologists prefer to use the term **assimilative** (L. *assimilatio*, similarity of one thing with respect to another; the ensemble of processes by which the absorbed substances are converted into living material of the absorbing organism) in order to avoid the implication that the fungi are treated as plants. Thus, one says vegetative mycelium or assimilative mycelium, vegetative function or assimilative function, etc.

vegetative multiplication (L. *vegetare*, to vegetate + *-ivus* > E. *-ive*, quality or tendency, fitness; L. *multiplicatio*, genit. *multiplicationis* < *multiplicare*, to multiply + *-ationem*, action, state or condition, or result > E. suf. *-ation*): to grow, develop and multiply asexually. See **asexual**.

veil or **velum**, pl. **vela** (L. *velum*, veil, curtain): a curtain or cloth that covers something. The felt-like hyphenchymatous covering that surrounds the basidiocarp primordium of the angiocarpic Agaricales is called the **universal** (or **general**) **veil**; in certain gen. the veil is evanescent (as in *Coprinus*), in others it is persistent in the adult state (as in *Amanita* and *Volvariella*) and it constitutes the volva at the base of the stipe, and the plates or scales, generally white, on the upper surface of the pileus. The **partial veil** is the hyphenchymatous covering of the gills (in the hemiangiocarpic Agaricales), which extends from the edge of the pileus to the stipe, and which on first opening tears and its remains form the **ring** (as in *Agaricus*, *Lepiota*, etc.) and the **cortina** (as in *Cortinarius*). The ring also is called **hymenial veil** (and at times, and imprecisely, partial veil), which derives from the perpendicular primordial tissue, and differentiates later than the volva.

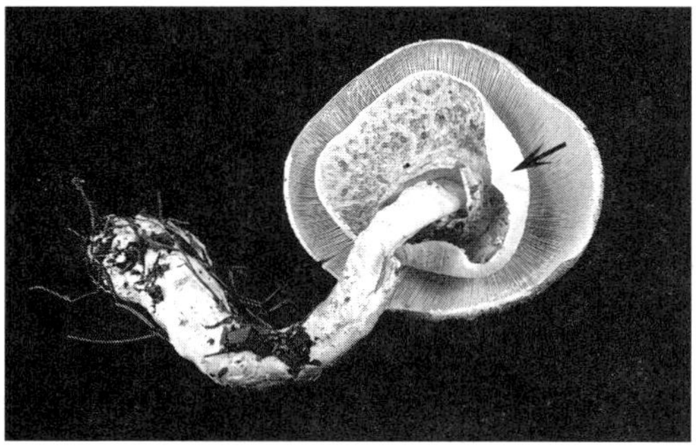

Veil (**hymenial**) of the basidiocarp of *Agaricus silvicola*, x 0.7 (*MRO*).

Veil (**universal**) of the basidiocarp of *Amanita* sp., x 1 (*RMA*).

velutinate, **velutinous** (NL. *velutinus*, velvet + suf. *-atus* > E. *-ate*, provided with or likeness; or + L. *-osus* > OF. *-ous*, *-eus* > E. *-ous*, having, possessing the qualities of): velvety, i.e., a surface in which the hairs are short, fine and soft, and are arranged compactly, like the colonies of some species of *Penicillium* (*P. chrysogenum*, of the moniliaceous asexual fungi) and *Cladosporium* (*C. cladosporioides*, of the dematiaceous asexual fungi), among many others, or like the surface of the pileus of *Psathyrella velutina* or the stipe of *Collybia velutipes* and *Xeromphalina tenuipes* (Agaricales). Same as **velvety**.

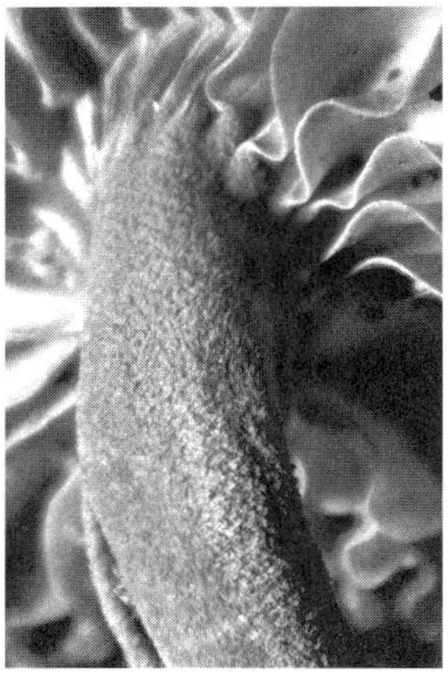

Velutinate stipe of the basidiocarp of *Xeromphalina tenuipes*, x 5 (*MU*).

velvety (ME. *velvet*, velvet < NL. *velutinus* < L. *villus*, shaggy nap + E. suf. *-y*, having the quality of): suggestive of or resembling velvet; smooth, soft; applied to surfaces that are covered with dense, short hair that is fine and brilliant or opaque, like the colony of *Penicillium chrysogenum* (moniliaceous asexual fungi). Same as **velutinate**.

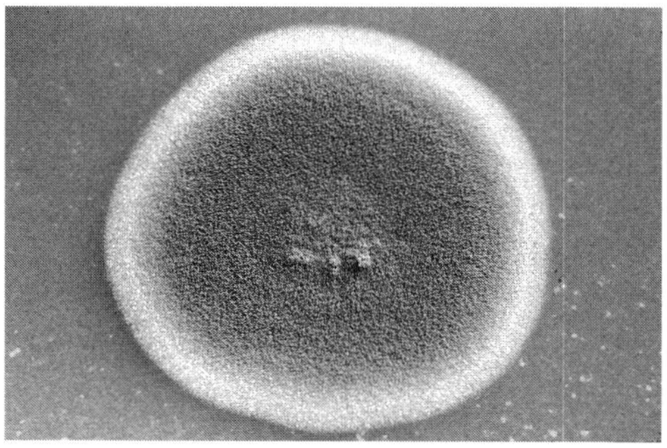

Velvety colony of *Penicillium chrysogenum* on agar, x 1 (*MU*).

ventricose (NL. *ventricosus*, swollen < L. *ventris*, genit. of *venter*, belly + *-osus*, full of, augmented, prone to > ME. *-ose*): swollen in the manner of a belly; also called **lageniform**, with the base wide and the apical part narrower, like the ascocarps of *Ceratocystis* (Microascales) and the cheilocystidia of *Psathyrella cyanescens* (Agaricales). When the apical part is extended, forming a prolongation similar to a beak or snout, it is called **ventricose-rostrate**, as is seen in the pleurocystidia of *Pholiota astragalina* (Agaricales). Also, the young ascospores and the perithecia of *Podospora comata* (Sordariales) are ventricose.

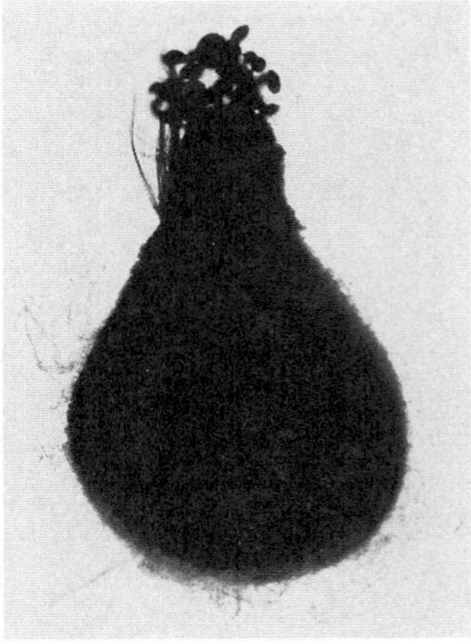

Ventricose perithecium of *Podospora comata*, x 87 (*RTH*).

ventricose-rostrate (NL. *ventricosus*, swollen < *ventris*, genit. of *venter*, belly + *-osus*, full of, augmented, prone to > ME. *-ose*; L. *rostratus*, beaked < *rostrum*, beak, snout + suf. *-atus* > E. *-ate*, provided with or likeness): see **ventricose**. Also called **lageniform**, **sicyoid** and **utriculose**.

vermicular, **vermiculate** (L. *vermicularis* < *vermiculus*, little worm < *vermis*, worm + dim. suf. *-culus* + suf. *-aris* > E. *-ar*, like, pertaining to; L. *vermiculatus*, wormy, worm-shaped < *vermiculus* + *-atus* > E. *-ate*, provided with or likeness): elongate and somewhat twisted, like a worm, or which moves like a worm; e.g., like the multinucleate plasmodium of *Protosporangium bisporum* (Protosteliales).

vermiform (L. *vermiformis* < *vermis*, worm + *-formis* < *-forma*, shape): shaped like a worm; e.g., like the ascospores of *Balansia henningsiana* (Hypocreales), *Ophiodothella vaccinii* (Phyllachorales) and *Geoglossum diforme* (Helotiales). See **anguilliform**.

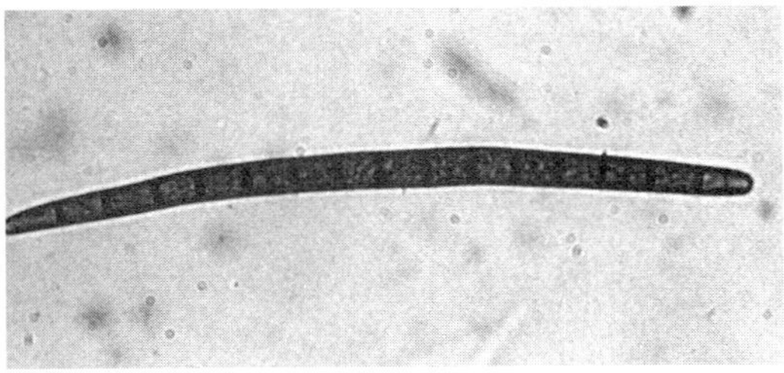

Vermiform ascospore of *Geoglossum difforme*, x 660 (*RTH*).

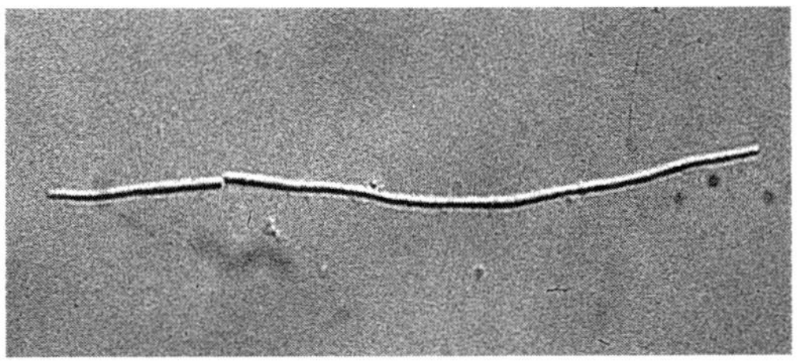

Vermiform ascospore of *Balansia henningsiana*, x 400 (*RTH*).

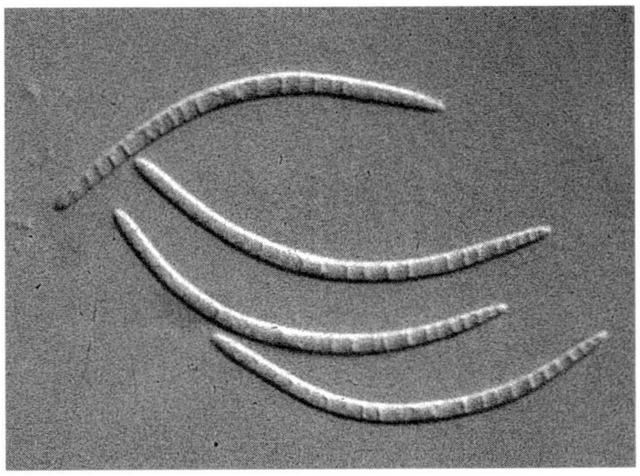

Vermiform ascospores of *Ophiodothella vaccinii*, x 580 (*RTH*).

vermivorous (L. *vermis*, worm + *vorare*, to devour + L. *-osus* > OF. *-ous*, *-eus* > E. *-ous*, having, possessing the qualities of): that which feeds on worms, in this case nematodes, like the moniliaceous asexual fungi of the gen. *Arthrobotrys*, *Dactylaria* and *Dactylella*, among others, which trap and consume nematodes in the soil. In *A. dactyloides*, e.g., constricting rings are formed that strangle their prey; from the rings are developed hyphae that invade and digest the dead body of the nematode. Syn. of **nematophagous**.

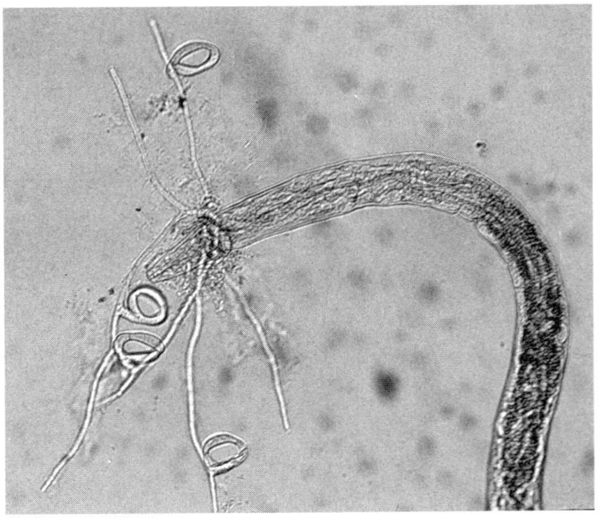

Vermivorous constricting rings of *Arthrobotrys dactyloides*, x 150 (*PL*).

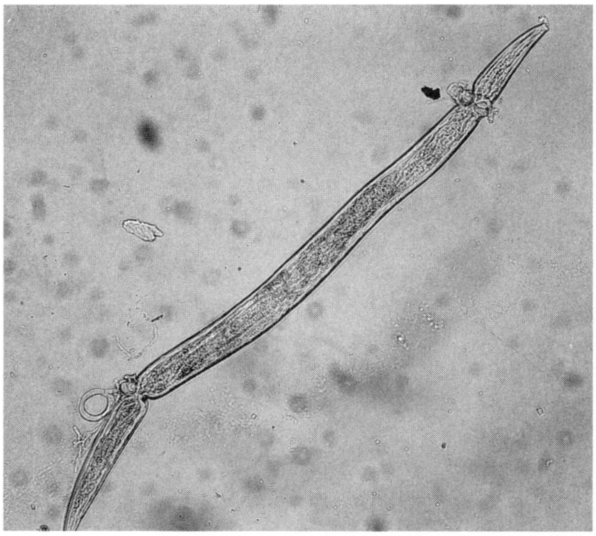

Vermivorous constricting rings of *Arthrobotrys dactyloides*, x 150 (*PL*).

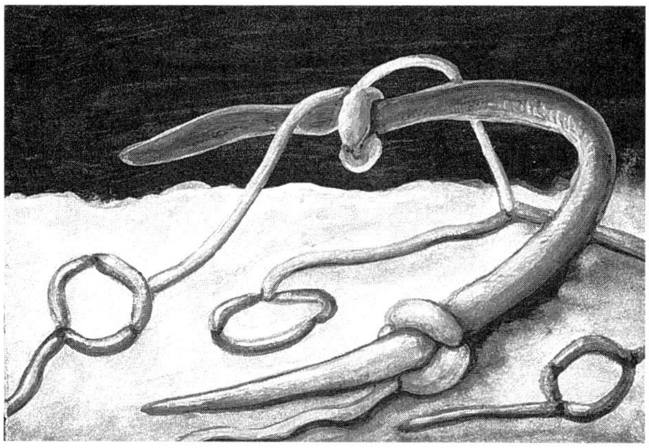

Vermivorous constricting rings of *Arthrobotrys dactyloides*, x 300.

verrucose (L. *verrucosus*, warted < *verruca*, wart, a height + *-osus*, full of, augmented, prone to > ME. *-ose*): with prominences in the manner of warts on the surface, like the conidiophore of *Penicillium roquefortii* (moniliaceous asexual fungi) and the conidia of *Stachybotrys chartarum* (dematiaceous asexual fungi).

verruculose (L. *verruculosus*, full of small warts < *verrucula*, dim. of *verruca*, wart + *-osus*, full of, augmented, prone to > ME. *-ose*): with a rough surface, ornamented with microscopic or minute prominences similar to warts, as is observed in spores of the gen. *Cortinarius* (Agaricales). If the prominences or warts are larger the term **verrucose** is applied.

versiform (L. *versiformis*, variable in form < *versus*, turned < *vertere*, to turn, to change + *-formis* < *forma*, shape): applied principally to fungal structures that change shape with age, like certain cystidia (cheilocystidia of *Bolbitius vitellinus*, Agaricales) that exhibit so many shapes that it is difficult to describe all the variations. See **pleomorphic**.

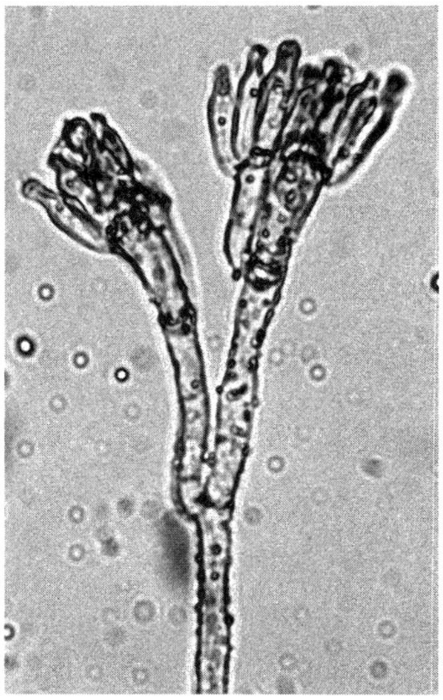

Verrucose wall of the conidiophore of *Penicillium roquefortii*, x 600 (*PL*).

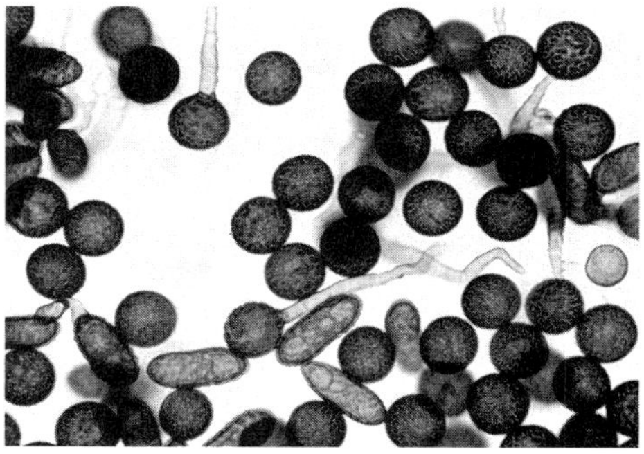

Verrucose wall of the conidia of *Stachybotrys chartarum*, x 800 (*MU*).

verticillate (NL. *verticillatus*, arranged in a verticil < L. *verticillus*, whorl + suf. *-atus* > E. *-ate*, provided with or likeness): a type of branching in which the branches (pedicels, metulae, phialides, etc.) are borne at the same level on the hypha or support (sporangiophore, conidiophore, etc.) and they grow obliquely upward with respect to the central axis, generally reaching the same length and surrounding the hypha or support; e.g., the sporangiophore of *Actinomucor* (Mucorales), and the conidiophores of *Scopulariopsis* and *Verticillium* (moniliaceous asexual fungi). The last one is a gen. that owes its name to this feature. Cf. **monoverticillate**, **biverticillate**, **triverticillate** and **polyverticillate**.

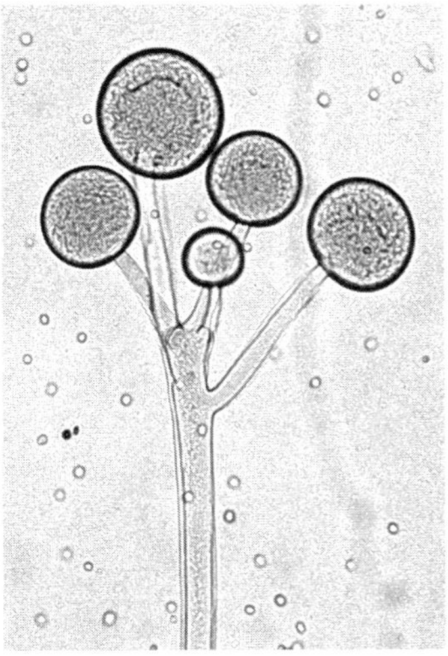

Verticillate sporangiophore of *Actinomucor* sp., x 175 (*CB*).

Verticillate phialides of a conidiophore of *Scopulariopsis brevicaulis*, x 750 (*CB*).

Verticillate phialides of a conidiophore of *Verticillium lecanii*, x 900 (*CB*).

vesicle (L. *vesicula*, dim. of *vesica*, bladder): **1**. A receptacle or purse shaped like a bladder or ampule; applied especially to the mucilaginous structure produced by the zoosporangia in some Chytridiales (*Chytriomyces*) and Peronosporales (*Pythium*), within which the zoospores mature before being liberated. **2**. The subsporangial swelling of species of *Pilobolus* (such as *P. longipes*, of the Mucorales). **3**. The vesicular structures of the hyphae, with a thick cell wall, that probably constitute reserve deposits or resistant structures, and which the mycorrhizogenous fungi form in the interior of the cells of the associated plant. The vesicles are one of the forms (the other constitute the arbuscles) present in endotrophic mycorrhizae (e.g., *Endogone*, of the Endogonales, in onion root). **4**. A swelling in the apical part of the conidiophores of the gen. *Aspergillus* (moniliaceous asexual fungi), which has various shapes depending upon the species (claviform, spheroid, etc.), and which gives rise to the phialides, producers of conidia (in species with a uniseriate conidial head), or to metulae and phialides (in those that have a biseriate conidial head).

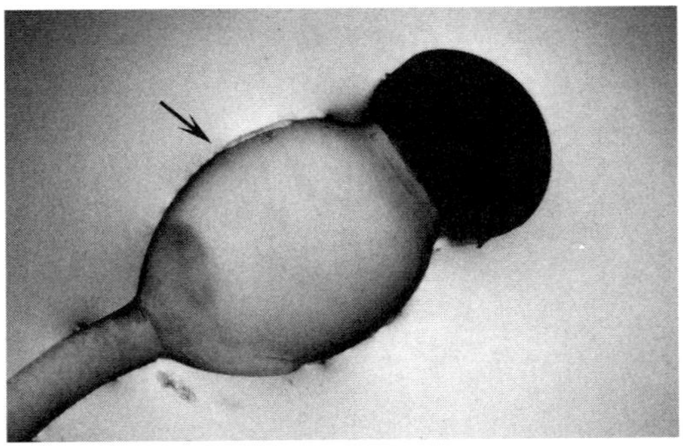

Vesicle (subsporangial) of the sporangiophore of *Pilobolus longipes*, x 430 (*MU*).

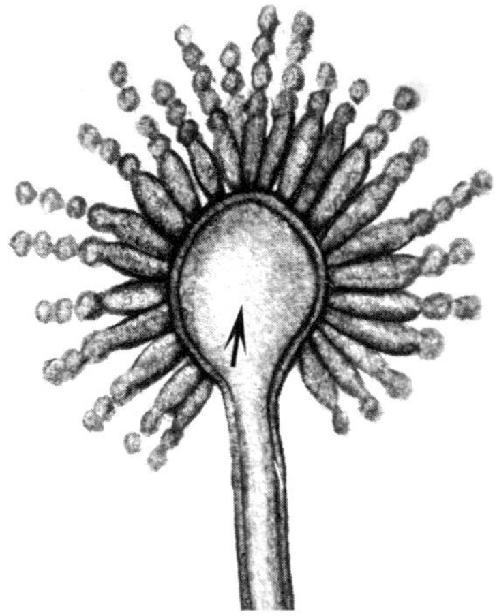

Vesicle of the conidial head of *Aspergillus rubrobrunneus*, x 400.

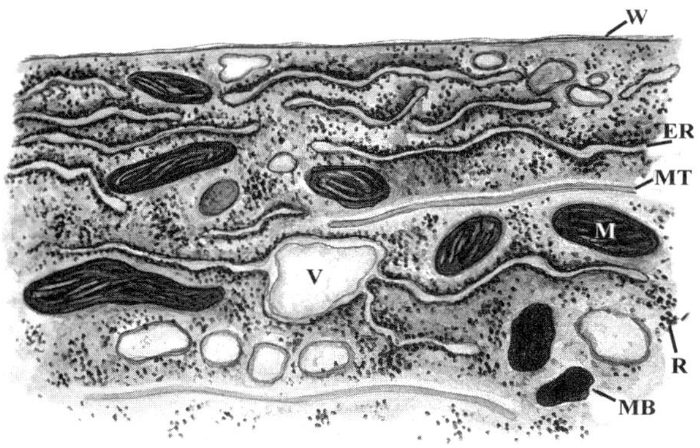

Vesicles of a young hypha of *Armillaria mellea*, seen in longitudinal section; the endoplasmic reticulum (ER), microbodies (MB), mitochondria (M), microtubules (MT), ribosomes (R), and cell wall (W) are also shown. Drawing composed from various transmission electron micrographs published by Beckett *et al.* in their *Atlas of Fungal Ultrastructure*, 1974, x 27 000.

vesicular, **vesiculate**, **vesiculiform**, **vesiculose** (L. *vesicularis* < *vesicula*, blister, bladder, vesicle + suf. *-aris* > E. *-ar*, like, pertaining to; L. *vesiculatus*, to become vesicular < *vesicula* + suf. *-atus* > E. *-ate*, provided with or likeness pertaining or relative to the vesicle; L. *vesiculiformis*, with the shape of a vesicle < *vesicula* + *-formis* < *forma*, shape; L. *vesiculose*, having vesicles < *vesicula* + ME. suf. *-ose* < L. *-osus*, full of, augmented, prone to): with the shape of a vesicle or having a vesicle; e.g., like the cystidia of *Leptonia fulva* and other Agaricales, which are swollen like a bladder or vesicle, with only its base abruptly narrowed.

vestigial (L. *vestigium*, track, trace + L. suf. *-alis* > E. *-al*, relating to or belonging to): not well developed, rudimentary; in mycology it is applied generally in the sense of residual or scanty. For example, there are vestigial rings and cortinas on mature basidiocarps of some Agaricales (various species of *Cortinarius* and *Stropharia*), that remain as remnants of the partial veil (the hyphenchymatous membrane that covers the gills of the young basidiocarp).

villose (L. *villosus*, hairy, shaggy < *villus*, fine hair, down + *-osus*, full of, augmented, prone to > ME. *-ose*): refers to structures that have down or hair, the latter not being too fine, in which case it is **pubescent**, not very short and dense, as then they are called **velvety** or **velutinous**; also they are not considered villose when the hairs are rough or rigid, because then they are called **hirsute** or **hispid**. The pileus of the basidiocarps of *Coprinus lagopus* and *Volvariella bombycina* (Agaricales), e.g., are villose.

virgate (L. *virgatus*, twiggy, made of twigs; also striped < *virga*, stick, cane; also ray or band + suf. *-atus* > E. *-ate*, provided with or likeness): long, straight and slender like a stick. Generally applied to the surface of the pileus of a basidiocarp, when the latter has rays or bands, frequently due to the presence of fibrils of different color, as in *Tricholoma virgatum* (Agaricales).

Villose pileus of the basidiocarps of *Coprinus lagopus*, x 0.3 (*MU*).

virose (L. *virosus*, poisoned, fetid, muddy, covered with slime < *virus*, poison, slime + *-osus*, full of, augmented, prone to > ME. *-ose*): poisonous, like the basidiocarp of *Amanita virosa* (Agaricales), cause of phalloidian mycetism.

virulence (L. *virulentus*, full of poison < *virus*, poison, slime + suf. *-entia* > F. *-ence* < E. *-ence*, state, quality or action): the quality or state of being **virulent**; the relative capacity of a pathogen to overcome host defenses; the degree or measure of pathogenicity.

virulent (ME. < L. *virulentus*, full of poison < *virus*, poison, slime + suf. *-ulentus*, that abounds in): marked by a rapid, severe, and malignant course (a virulent infection); extremely pathogenic, poisonous or venomous.

viscid, **viscosus** (L. *viscidus*, clammy, sticky like bird-lime < *viscum*, viscous, bird-lime; L. *viscosus*, sticky < *viscum* +): viscous or sticky, like the surface of the pileus of *Suillus* (Agaricales).

viticolous (L. *vitis*, grapevine + *-cola*, inhabitant + *-osus* > OF. *-ous*, *-eus* > E. *-ous*, having, possessing the qualities of): living on grapevines and fruits, like *Plasmopara viticola* (Peronosporales).

volutin (G. < NL. *volutans*, specific epithet of the bacterium *Spirillum volutans* < L. *voluta*, a spiral scroll, in which it was first found + NL. *-in*, suf. used in chemistry to denote an activator or compound): an electron-dense, basophilic granular substance, made out of polymetaphosphate, which serves as a reserve material of fungal cells, especially yeasts. Common in microorganisms.

volva, pl. **volvae** (L. *volva*, matrix, cover): *Basidiomycetes*. A remnant structure of the universal veil, which like a cyathiform sheath, surrounds the base of the stipe of the basidiocarp, in the gen. *Amanita* (Agaricales) and *Clathrus* (Phallales), among others.

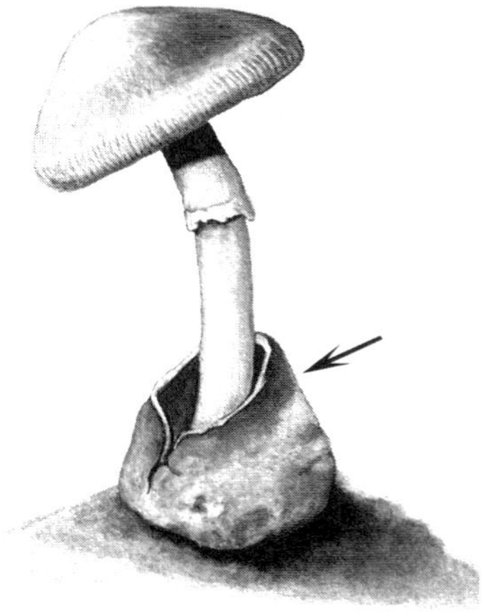

Volva of the basidiocarp of *Amanita* sp., x 0.3.

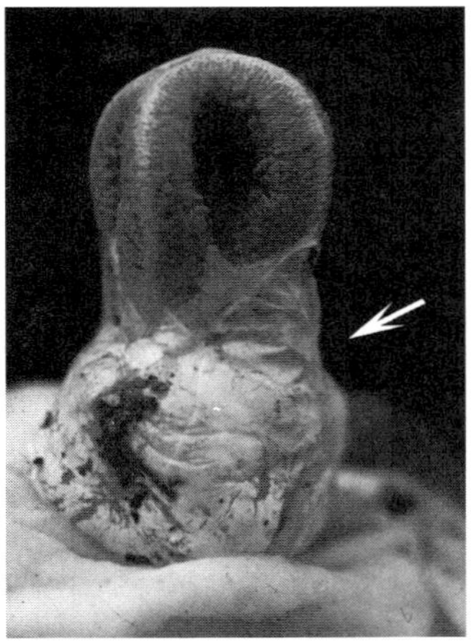

Volva of the basidiocarp of *Clathrus ruber*, x 1 (*MEV*).

W

wall (AS. *weal* < L. *vallum*, a wall set with palisades; the bounding side of a natural cavity or vessel): a dynamic structure that surrounds every fungal hypha or cell, that is subject to change and modification at different stages in the life of a fungus. The wall gives fungi most of their unique features. The wall's ability to safely contain turgor pressure appears to be a primordial reason for survival and evolution of fungi. In addition to containing turgor pressure, the cell wall also plays several other important roles in the life of a fungus. For example, the wall confers shape to the hypha, it acts as a filter controlling to some extent what enters the fungal protoplast, it protects the protoplast against environmental hazards, and it functions in the recognition of events associated not only with sexual reproduction but also with various interactions of the fungi with potential plant and animal symbionts. The chemical composition of the cell wall varies according to the group of species. It is composed basically of skeletal or microfibrillar components located on the inner side of the wall and usually embedded in an amorphous matrix material that extends to the outer surface of the wall. The skeletal component consists of chitin, which may be present in combination with diverse polysaccharides (e.g., hemicelluloses) and small amounts of lipids. In some groups, chitin may be absent and the wall is mainly made up of cellulose (e.g., the Oomycota), quitosan, B-glucan, manan, glycogen, galactan and galactosamine polymers. Fungal cell walls are stratified, usually made up of two or more layers of microfibrils embedded in an amorphous matrix, with the internal layer attached to the plasma membrane.

wart (ME. < OE. *wearte*, wart): a small, cone-shaped ornamentation; e.g., as on spores or the exoperidium of *Lycoperdon* (Lycoperdales).

water molds (ME. < OE. *waeter*, water; ME. *mowlde* < *mowled*, ptp. of *moulen*, *malwen*, to grow moldy): same as **aquatic fungi**.

wilt (alteration of earlier *welk* < ME. *welken*, to wilt; to lose freshness and become flaccid): a plant disease (especially caused by fungi, such as *Verticillium* [moniliaceous asexual fungi] and *Fusarium* (tuberculariaceous asexual fungi), marked by loss of turgidity in soft tissues and collapse of leaves and stems.

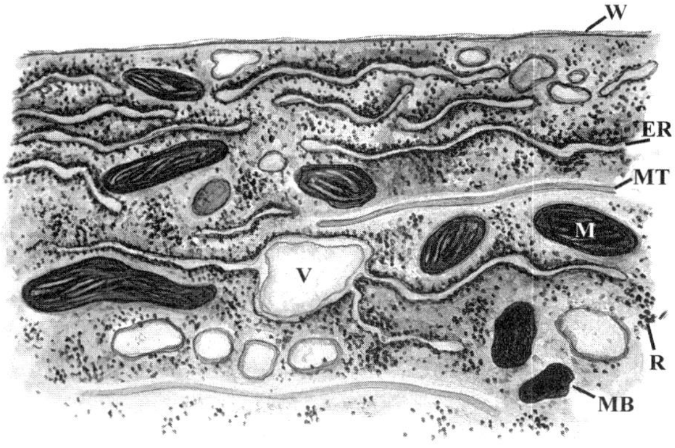

Wall of a young hypha of *Armillaria mellea*, seen in longitudinal section; the endoplasmic reticulum (ER), microbodies (MB), mitochondria (M), microtubules (MT), ribosomes (R), and vesicles (V) are also shown. Drawing composed from various transmission electron micrographs published by Beckett *et al.* in their *Atlas of Fungal Ultrastructure*, 1974, x 27 000.

Woronin body (named after its discoverer, Russian microbiologist M. Woronin; ME. < OE. *bodig*, body): a cytoplasmic organelle, round or oval in shape, crystalline in structure, and surrounded by a simple membrane (i.e., not double like that of other intracellular organelles), found near the septa in the cells of ascomycetes and their conidial states. Woronin bodies serve to plug the septal pore when the hyphal wall breaks for any reason, so to impede the loss of cytoplasm in the damaged mycelium; e.g., as in *Geotrichum candidum* (moniliaceous asexual fungi) and *Ascodesmis* (Pezizales). They may also function to temporarily plug the septal pore, for regulation of cytoplasmic flow and the distribution of nuclei during the formation and secession of conidia while they are maturing. Cf. **septal pore organelle**.

wrinkled peristome (ME. *wrinkled* < OE. *gewinclod*, ptp. of *gewinclian*, to wind around; Gr. *perí*, around + *stóma*, mouth): a peristome with uneven tissue, which is neither sulcate nor ridged and grooved.

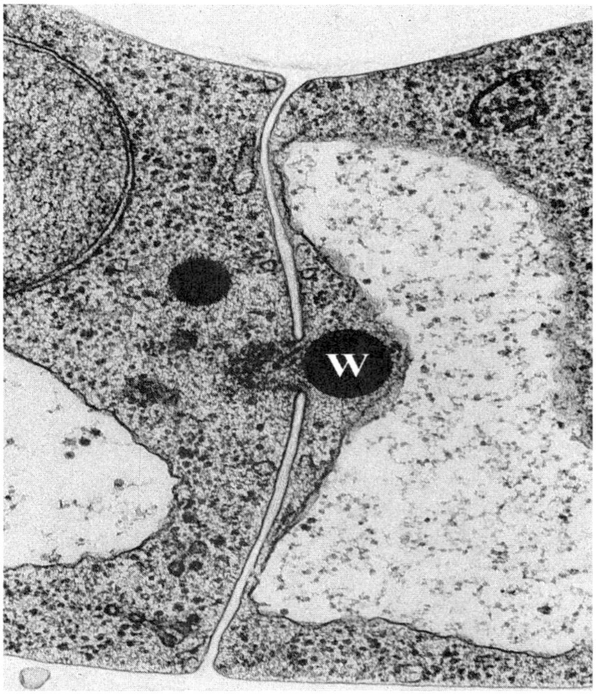

Woronin body (W) plugging the septal pore of a hyphal cell of *Ascodesmis* sp., seen in longitudinal section. Transmission electron micrograph, x 10 000 (*CM*).

X

xanthochroic (Gr. *xanthós*, the various shades of yellow + *chrõs*, color + suf. *-íkos* > L. *-icus* > E. *-ic*, belonging to, relating to): having a reddish- or yellowish-brown context which darkens on treatment with KOH, as in Hymenochaetaceae of Aphyllophorales.

xenospore (Gr. *xénos*, stranger, foreigner + *sporá*, spore): a term that on occasion is applied to a spore that functions in an important manner in the dispersion of the fungus that produces it. Conidia, sporangiospores and urediniospores are examples of xenospores. Cf. **hypnospore**.

xerophile, **xerophilic** (Gr. *xerós*, dry + *phílos*, have an affinity for; or + suf. *-íkos* > L. *-icus* > E. *-ic*, belonging to, relating to): living and developing on arid or dry substrates or places, or growing on media containing little free water. Among the fungi, some moniliaceous fungi are xerophiles, such as certain species of *Aspergillus* and *Penicillium*, and certain Gasteromycetes, such as *Scleroderma* (Sclerodermatales) and *Montagnea* (Hymenogastrales). Also considered xerophiles are species of *Schizophyllum*, *Lenzites*, *Daedalea*, *Polyporus*, *Stereum* and *Corticium*, among other gen. of Aphyllophorales, whose fruiting bodies are capable of resisting dryness without losing their vitality and the power to revive and liberate spores under wet conditions.

xerospore (Gr. *xerós*, dry + *sporá*, spore): a spore with a dry wall, not viscous, adapted to dissemination by means of air currents. This type of spore is present, e.g., in *Aspergillus*, *Penicillium* (moniliaceous asexual fungi), *Alternaria* (dematiaceous asexual fungi) and many others. Cf. **gleospore**.

xiphioid (Gr. *xíphos*, sword + L. suf. *-oide* < Gr. *-oeídes*, similar to): with the shape of a sword. Syn. of **ensiform**.

xylophagous (Gr. *xýlon*, wood + *phágos*, eater < *phágomai*, to eat + L. *-osus* > OF. *-ous*, *-eus* > E. *-ous*, having, possessing the qualities of): wood-eating. Cf. **lignicolous**.

xylosaprobic (Gr. *xýlon*, wood + *saprós*, rotten + *bíos*, life + suf. *-íkos* > L. *-icus* > E. *-ic*, belonging to, relating to): applied to species that live as saprobes on dead wood, as with most Aphyllophorales and many other fungi belonging to diverse groups. Cf. **lignicolous** and **xylophagous**.

xylostroma, pl. **xylostromata** (Gr. *xýlon*, wood + *strõma*, bed, cushion): a compact, hard, blackish stromalike structure extruded through longitudinal cracks in the bark of roots infected by *Armillaria* (=*Armillariella*) *tabescens* (Aphyllophorales). Also, the sheets of mycelium formed by *Daedalea* (=*Xylostroma*), of the Aphyllophorales.

y

yeast (ME. *yest* < OE. *gist*, yeast, akin to MHG. *jest*, foam, and Gr. *zeío*, to boil): unicellular, budding fungi that do not constitute a formal taxonomic group but a growth form exhibited by a range of unrelated fungi. Some primary filamentous species, including those pathogenic to animals, adopt yeast or yeast-like forms as a part of the life cycle or under cetain environmental conditions. See **dimorphic**. Yeasts are classified in different gen. depending on the authority and definition used. They constitute a phylogenetically heterogeneous grouping, including the ascosporogenous Saccharomycetaceae (Saccharomycetales), the anascosporogenous cryptococcaceous asexual yeasts, and also some tremellaceous forms (with heterobasidia), all consisting of single cells or a somewhat developed pseudomycelium and/or true mycelium, with multiplication by budding, by fission, or by a combination of the two processes. Yeasts are involved in many natural and man-controlled fermentation processes. See **fermentation**, **zymogen** and **zymosis**. *Apiculate yeasts* (NL. *apiculatus*, provided with an *apiculus* < L. *apex*, *apicis*, apex + dim. suf. *-ulus*): *Saccharomycodes*, *Nadsonia*, *Hanseniaspora* (ascosporogenous), and *Kloeckera* (anascosporogenous), having minute polar projections which are multiple scars (annellides). *Asporogenous yeasts* (L. *asporogenous* < Gr. *a-*, without + *sporá*, spore + *génos*, origin, birth): species that do not form ascospores (cryptococcaceous asexual yeasts). *Baker's yeast* (ME. *baken* < OE. *bacan*, akin to OHG. *bahhan*, to bake, and Gr. *phognýon*, to roast): *Saccharomyces cerevisiae* (Saccharomycetales), used or suitable for use as leaven, the same species used for making beer, therefore called *brewer's yeast* (ME. *brewen* < OE. *breowan*, akin to L. *fervere*, to boil) or *beer yeast* (ME. *ber* < OE. *beor*, beer, the alcoholic beverage usually made from a malted cereal grain, such as barley, flavored with hops, and brewed by slow fermentation). *Black yeast* (ME. *black* < OE. *blaec*, black): yeast-like states of *Aureobasidium*, *Cladosporium*, *Moniliella* (dematiaceous asexual yeasts), etc., and especially anamorphs of Herpotrichielleae, including *Exophiala*, *Ramichloridium* and *Rhinocladiella* (dematiaceous asexual fungi). *Bottom yeast* (ME. *botme* < OE. *botm*, bottom): one settling out at the bottom of a fermented liquid (the wort), e.g., *Saccharomyces uvarum*. *Top yeast* (ME. < OE. *top*): one aggregating at the surface of the fermented wort, e.g. *S. cerevisiae*. *Chinese yeast*: *Amylomyces rouxii* (Mucorales) and other fungi used for inoculating different substrates for preparation of fermented foods (see **ragi**). *Food yeast* (ME. *fode* < OE. *foda*, food): dry *Candida utilis* (cryptococcaceous asexual yeasts) and other yeasts used as food supplements for animals. *Petit yeast* (ME. < MF. *petit*, small, minor): a respiratory deficient mutant. *Scum yeast* (ME. < MD. *scum*, foam): one forming a surface scum or slime layer (*Trichosporon cutaneum*, of the cryptococcaceous asexual yeasts). *Shadow* or *mirror yeast* (ME. *shadwe* < OE. *sceaduw*, *sceadu*, shade, shadow; ME. *mirour* < OF. < *mirer*, to look at < L. *mirari*, to wonder at): *Bullera*, *Sporobolomyces*, *Itersonilia* (sporobolomycetaceous asexual yeasts), etc., producing ballistospores, thus forming mirror-images of the colonies growing on agar culture media. *Toddy yeast* (Hind. *tari*, juice of the palmyra palm < *tar*, palmyra palm): a mixture of yeasts which ferment the sap of the palmyra palm (*Borassus flabellifer*). *Wine yeast* (ME. < OE. *win* < L. *vinum*, wine; the fermented juice of fresh grapes used as a beverage): races of *S. cerevisiae*.

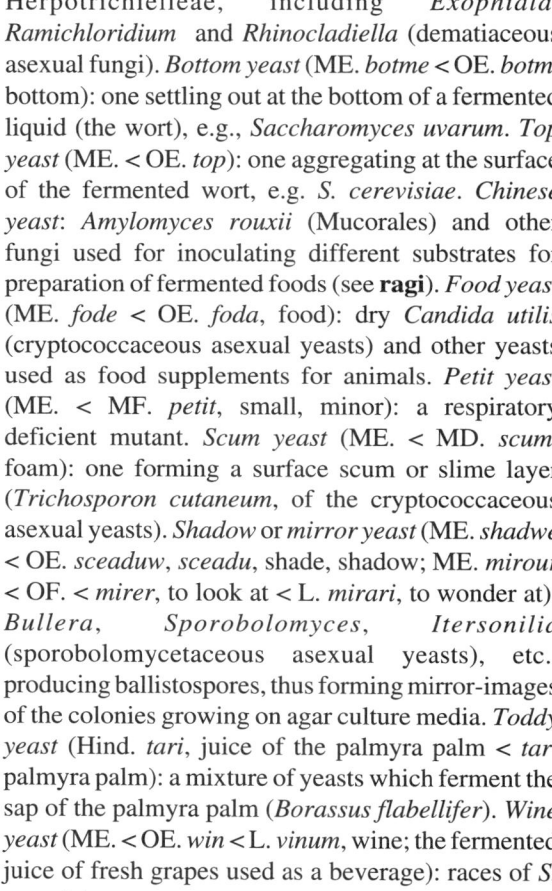

Black **yeast**. Colonies of *Aureobasidium pullulans* on agar, x 5 (*MU*).

yeast-like (ME. *yest* < OE. *gist*, yeast + ME. *lic*, *lik* < ON. *likr*, like, of the same form): refers to filamentous fungi that have the appearance of yeasts, both in macroscopic characters, such as colony growth, and in microscopic characters, such as the multiplication by budding of its somatic cells, e.g., *Aureobasidium pullulans* (dematiaceous asexual fungi). Many fungi are **dimorphic**; i.e., generally they form true hyphae and mycelium, but under certain conditions they become yeast-like. Some human pathogens form hyphae when grown in culture, but acquire the characteristics of yeasts while inside their host, as happens with *Histoplasma capsulatum* (moniliaceous asexual fungi) and other fungi causing mycoses in humans and vertebrate animals.

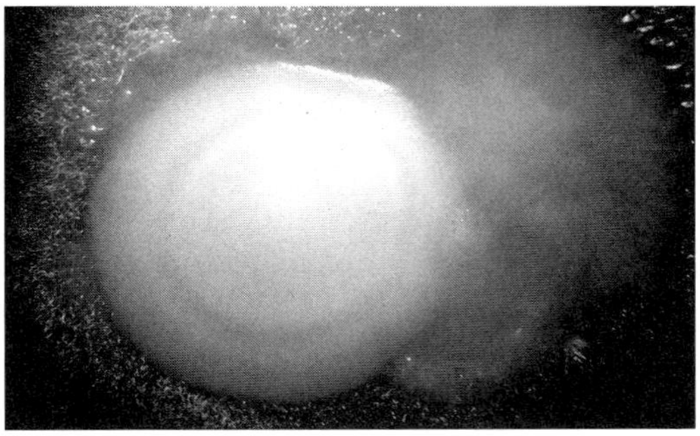

Yeast-like colony of *Aureobasidium pullulans* on agar, x 3 (*MU*).

ypsiliform (Ψ, *úpsilon*, *ýpsilon*, 20th letter of Gr. alphabet + L. suf. *-formis* < *forma*, shape): Y-shaped, like the pycnidiospores of *Ypsilonia* (cryptococcaceous asexual yeasts).

Z

zeugite (Gr. *zeugítes*, connected): the spore or cell of a fungus that results from the fusion of two nuclei when the process of fertilization is completed and the dikaryon ends, as occurs in the teliospore of Ustilaginales.

zonate, **zonated** (L. *zonatus*, banded < Gr. *zóne*, band, sash, strip + suf. *-atus* > E. *-ate*, provided with or likeness): arranged in bands or zones of distinct appearance, due to differences in color, texture or other characteristics. The colonies of many fungi cultivated on agar have bands that form concentric circles, with alternating areas of somatic mycelium and fertile areas of sporulation, such as is seen in *Aspergillus parasiticus* (moniliaceous asexual fungi); likewise, the surface of the basidiocarps of *Coriolus versicolor* (Aphyllophorales) is zonate.

Zonate colony of *Aspergillus parasiticus* on agar, x 0.7 (*MU*).

Zonate surface of the fruiting bodies of *Coriolus versicolor*, x 0.7 (*RV*).

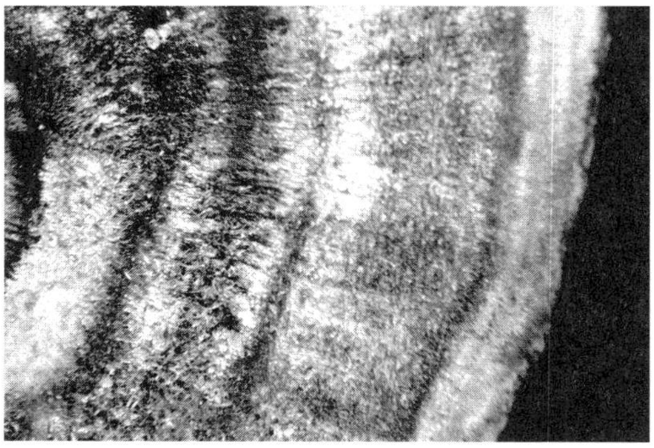

Zonate surface of the fruiting body of *Coriolus versicolor*, x 15 (*MU*).

zoochore, **zoochoric** (Gr. *zõon*, animal + *choréo*, to change place, move away; or + suf. *-íkos* > L. *-icus* > E. *-ic*, belonging to, relating to): it refers to an organism whose spores are normally disseminated by animals. There exist innumerable cases of fungi that are dispersed by animals (insects, rodents, cattle, etc.); some possess spiny, hooked, or viscous spores which attach themselves to the exterior of the animal vector (e.g., the mucilaginous conidia of *Graphium ulmi*, of the stilbellaceous asexual fungi, which are carried by beetles from infected elms to healthy ones, spreading the disease); the spores of other fungi are transported by insects that have special organs called **mycangia** (e.g., *Stereum*, of the Aphyllophorales, whose propagules are inoculated into the trunks of certain trees when the wood wasps oviposit on them), whereas other species of fungi, known as fimicolous, have spores that are ingested by various kinds of animals and pass through their digestive system unharmed, and germinate on being deposited on the media contained in the excreta of the animals (e.g., *Pilobolus*, of the Mucorales, common in cow and horse dung).

zoocyst (Gr. *zõon*, animal + *kýstis*, vesicle, cell): the encysted structure, with a thin wall and of short duration, that is derived from the multinucleate protoplast, that in turn came from a germinated spore, in *Ceratiomyxella tahitiensis* (Protosteliales). All of the nuclei of the zoocyst, except one, degenerate, and the persistent nucleus undergoes three successive divisions to give rise to eight nuclei. When the zoocyst germinates eight flagellate cells are produced.

zoogamete (Gr. *zõon*, animal + *gamétes*, husband, sexual cell): a motile, flagellate cell, like that of the aquatic fungi of the class Chytridiomycetes, such as *Allomyces* and *Blastocladia*, among other Blastocladiales.

zooglea, pl. **zoogleae**, **zoogloea**, pl. **zoogloeae** (Gr. *zõon*, animal + *gloiós*, viscous): refers to the mass of bacteria or lower algae conglutinated by the mucilage of their cell walls swollen by water. The so-called tibicos in Mexico (**tibi grains** or sugary kefir grains in other areas) are zoogloeae or macroscopic colonies composed of bacteria (*Lactobacillus brevis* and *Streptococcus lactis*) and yeasts (*Saccharomyces cerevisiae* and *Pichia membranaefaciens*, of the Saccharomycetales) in symbiosis, which are embedded in a dextran matrix. These zoogloeae are utilized in the popular elaboration of fermented products such as tepache and tibico vinegar. Nevertheless, since the microorganisms involved are not animals, perhaps it would be better to refer to these microbial associations as **microbiogloeae**.

Zoogloeae or microbiogloea known as tibi grains or sugary kefir grains, x 1.6 (*MU*).

zoophagous (Gr. *zõon*, animal + *phágos*, eater < *phágomai*, to eat + L. *-osus* > OF. *-ous*, *-eus* > E. *-ous*, having, possessing the qualities of): that which consumes animals or animal products, like the majority of the fungi included in the order Zoopagales (*Zoopage*, *Cochlonema* and others), which are predators of amoebae, rhizopods, and nematodes; another example, which is peculiar for having lateral adhesive hyphal branches, with which they trap and destroy rotifers, is the gen. *Zoophagus* (Zoopagales).

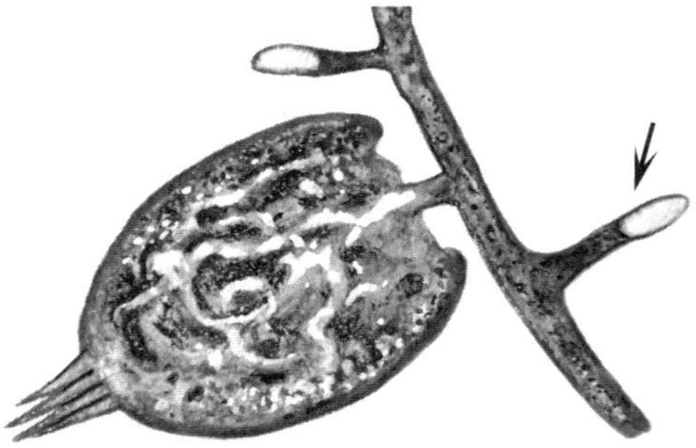

Zoophagous trapping hyphal pegs of *Zoophagus insidians*, x 250.

zoophilic, **zoophilous** (Gr. *zõon*, animal + *phílos*, have an affinity for + suf. *-íkos* > L. *-icus* > E. *-ic*, belonging to, relating to; or + L. *-osus* > OF. *-ous*, *-eus* > E. *-ous*, having, possessing the qualities of): *Med. Mycol.* Refers to a fungus that is preferentially pathogenic for animals instead of man, e.g., the dermatophyte *Trichophyton equinum* (moniliaceous asexual fungi), although on occasion man can be infected from diseased animals. Cf. **anthropophilic** and **geophilous**.

zoosporangiophore (Gr. *zõon*, animal + *sporá*, spore + *angeîon*, vessel, receptacle + *-phóros*, bearer): a sporangiophore that bears one or more zoosporangia.

zoosporangium, pl. **zoosporangia** (Gr. *zõon*, animal + *sporá*, spore + *angeîon*, vessel, receptacle): a sporangium that contains zoospores, as in the aquatic fungi belonging to the Chytridiomycota, Hyphochytriomycota and Oomycota, e.g., *Phlyctochytrium* (Chytridiales) and *Saprolegnia* (Saprolegniales).

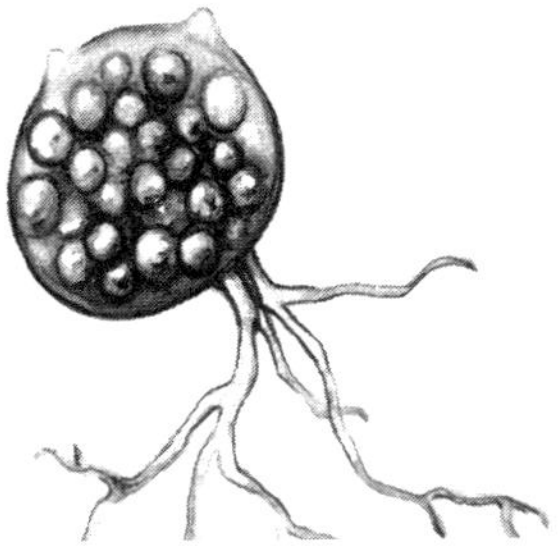

Zoosporangium of *Phlyctochytrium* sp., x 1 200.

zoosporic (Gr. *zõon*, animal + *sporá*, spore + suf. *-íkos* > L. *-icus* > E. *-ic*, belonging to, relating to): pertaining to or relative to the zoospores. Aquatic fungi are sometimes called zoosporic fungi (phyla Chytridiomycota, Hyphochytriomycota and Oomycota). See **aquatic fungi** and **water molds**.

zoosporiferous (Gr. *zõon*, animal + *sporá*, spore + L. *-ferous*, bearer < *ferre*, to bear, carry + *-osus* > OF. *-ous*, *-eus* > E. *-ous*, having, possessing the qualities of): that which produces and bears zoospores.

Zygomycetes (Gr. *zygón*, marriage tie, pair + L. *-mycetes*, ending of class < Gr. *mýkes*, genit. *mýketos*, fungus): a class of fungi that reproduce sexually by fusion of gametangia to form zygospores; asexual reproduction is by means of nonmotile (non-flagellate) sporangiospores, or by conidiola. The majority of the species have a well-developed mycelium that is coenocytic in most species but septate in certain groups. The Zygomycetes (which include the orders Mucorales, Dimargaritales, Kickxellales, Endogonales, Glomales, Entomophthorales and Zoopagales) together with the Trichomycetes constitute the phylum Zygomycota of the kingdom Fungi.

zygomycote (L. *zygomycete* < Gr. *zygón*, marriage tie, pair + *mýkes*, genit. *mýketos*, fungus): one of the Zygomycota, a phylum of the kingdom Fungi according to the classification system of Margulis and Schwartz, 1988, and Alexopoulos *et al.*, 1996. The latter is the one adopted in this dictionary.

zygophore (Gr. *zygón*, marriage tie, pair + *-phóros*, bearer): *Zygomycetes*. A specialized hyphal branch that gives rise to a progametangium. See **suspensor**.

zygosporangium, pl. **zygosporangia** (Gr. *zygón*, marriage tie, pair + NL. *sporangium* < Gr. *sporá*, spore + *angeîon*, vessel, receptacle): *Zygomycetes*. A sporangium that contains a zygospore; it develops after the fusion of the two gametangia (gametangial copulation) in gen. such as *Absidia*, *Mucor*, *Rhizopus*, *Phycomyces*, *Gilbertella*, *Syzygites* and other Mucorales. Cf. **azygosporangium**.

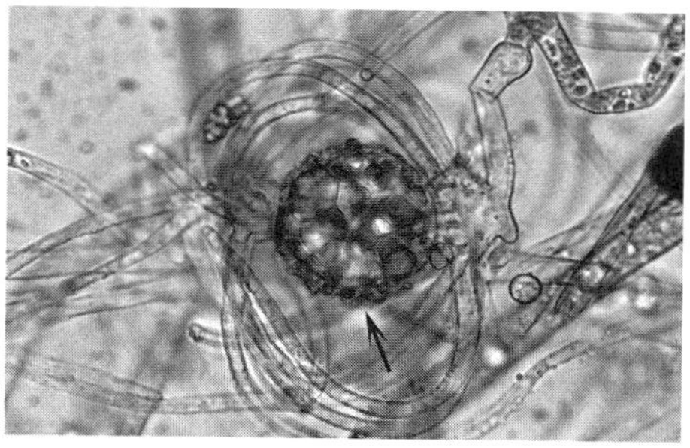

Zygosporangium between the suspensors of *Absidia spinosa*, x 125 (*EAA*).

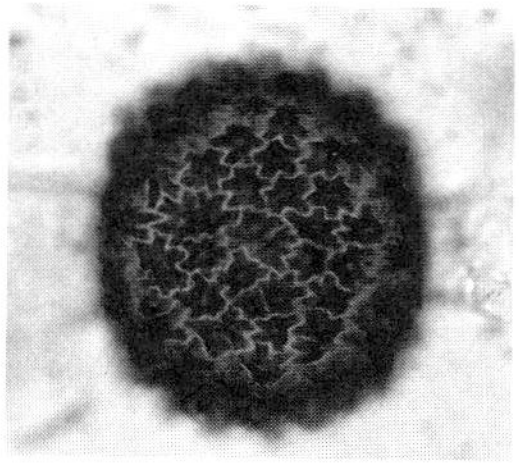

Zygosporangium between the suspensors of *Mucor hiemalis*, x 200 (*RTH*).

Zygosporangium between the suspensors of *Rhizopus nigricans*, x 200.

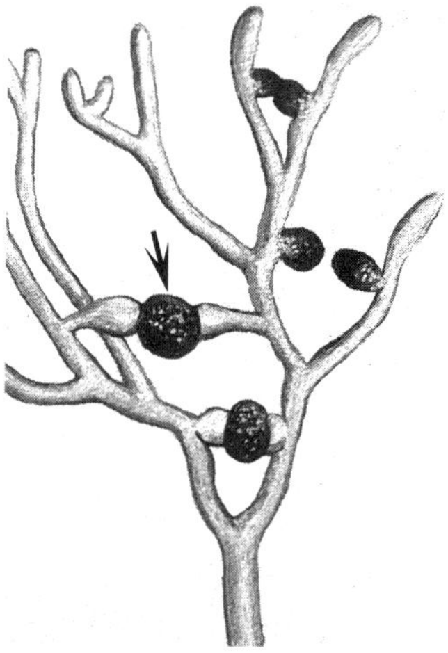

Zygosporangia between pairs of suspensors of *Syzygites megalocarpus*, x 80.

zygospore (Gr. *zygón*, marriage tie, pair + *sporá*, spore): *Zygomycetes*. A latent spore, contained in a zygosporangium, that results from the fusion of two gametangia (gametangial copulation). Cf. **azygospore**.

zygosporocarp (Gr. *zygón*, marriage tie, pair + *sporá*, spore + *karpós*, fruit): *Zygomycetes*. A fruiting body that contains zygosporangia with zygospores, e.g., as are present in *Mortierella* (Mucorales) and in *Endogone* (Endogonales), *Glomus*, *Gigaspora* (Glomales) and others.

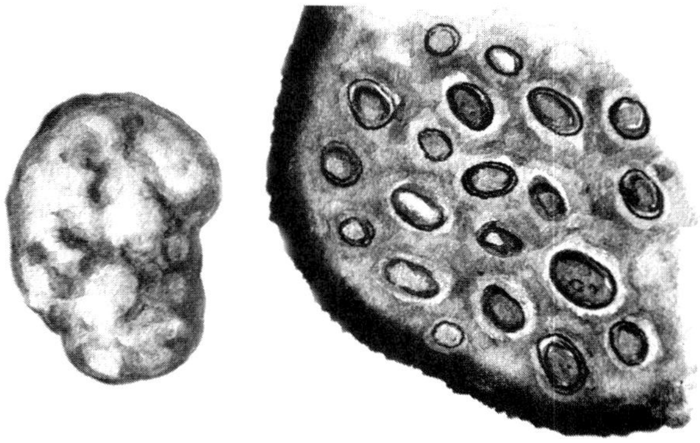

Zygosporocarps of *Endogone lactiflua*; in the transversally sectioned zygosporocarp, note the zygospores inside, x 4.

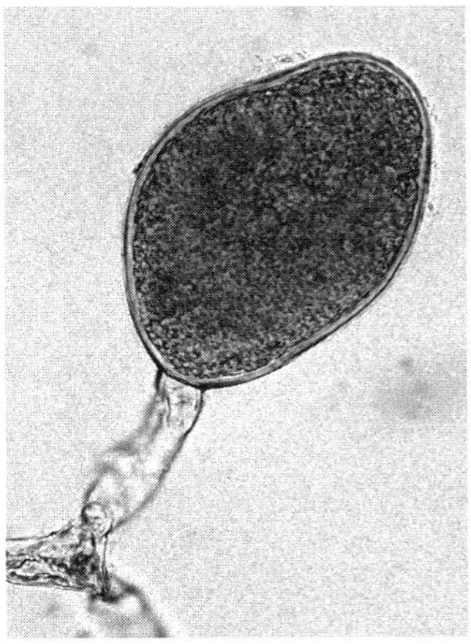

Zygospore of *Endogone* sp., x 50 (*MU*).

zygosporocarpic (Gr. *zygón*, marriage tie, pair + *sporá*, spore + *karpós*, fruit + suf. *-íkos* > L. *-icus* > E. *-ic*, belonging to, relating to): *Zygomycetes*. Having a fruiting body (sporocarp) in which are produced zygospores, as occurs, e.g., in some members of the order Endogonales (*Endogone*).

zygosporophore (Gr. *zygón*, marriage tie, pair + *sporá*, spore + *-phóros*, bearer): *Trichomycetes*. The supporting cell of the zygospore, derived from a bud that is formed by the union of two hyphae of different sexes; the zygosporophore enlarges in the upper part until it differentiates a pileate zygospore (plano-convex or biconcave), which separates from the zygosporophore that supports it by means of a septum, e.g., in *Stipella vigilans* (Harpellales).

Zygosporophore of *Stipella vigilans*, with medianly attached zygospore, x 800.

zygotactic (Gr. *zygón*, marriage tie, pair + *taktós*, a handling, touch + *-tikós* > L. *-ticus* > E. *-tic*, relation, fitness, inclination or ability): pertaining to **zygotaxis**; zygotactic gamete.

zygotaxis, pl. **zygotaxes** (Gr. *zygón*, marriage tie, pair + NL. *taxis* < Gr. *táxis*, arrangement < *tássein*, to arrange, put in order): the mutual attraction of two gametes or of two gametangia, as happens, e.g., in planogametic copulation (Chytridiales) and in gametangial copulation (Mucorales), among others.

zygote (Gr. *zygotós*, yoked, joined together < *zygo-* < *zygóun*, to join together < *zygón*, yoke): a diploid cell that results from the union of two haploid cells during sexual reproduction; also called the egg. Depending upon the group of fungi to which it refers, the zygote can be immotile or motile, and the latter amoeboid (amebozygote) or flagellate.

zygotropism (Gr. *zygón*, marriage tie, pair + *trópos*, a turn, change in manner < *tropé*, a turning < *trépo*, to revolve, turn towards + *-ismós* > L. *-ismus* > E. *-ism*, state, phase, tendency, action): *Zygomycetes*. A phenomenon in which two hyphae grow and curve toward one another in a reciprocal manner, as a result of mutual stimulation, until they touch and fuse, as happens with the + and - progametangia of many heterothallic Mucorales (*Rhizopus nigricans*, *Mucor hiemalis*, *Phycomyces nitens* and others).

zymocide (Gr. *zymé*, ferment > L. *zymo*, yeast + L. *-cida* < *caedere*, to kill): applied to the factor possessed by certain yeast strains, which in the presence of other strains sensitive to it, acts as an antagonist, causing the death of their cells. At present 11 different toxins, proteinaceous in nature, are recognized. Some strains of *Saccharomyces cerevisiae* (Saccharomycetales), e.g., are zymocides and cause the extermination of certain strains of *Candida glabrata* (cryptococcaceous asexual yeasts). In contrast, some strains of *S. cerevisiae* are sensitive to the zymocidal factor present in other strains of the same or of another species of yeast. The existence of this competitive phenomenon is of biological relevance and economic importance in the field of industrial fermentations, e.g., in those concerned with the production of yeast for baking or the production of distilled alcoholic beverages, such as rum, and others, nondistilled, such as beer.

zymogenous (Gr. *zýme*, ferment, yeast + *génos*, origin < *gennáo*, to engender, produce + L. *-osus* > OF. *-ous*, *-eus* > E. *-ous*, having, possessing the qualities of): an organism that produces ferments or enzymes that carry out fermentation.

zymogram (Gr. *zýme*, ferment, yeast + *grámma*, to describe): literally, a diagram of enzymes and their activity; it is a laboratory method that is applied to the culture of yeasts, bacteria and other microorganisms that develop on certain culture media in which the carbon source is varied, with the object of determining if the organism under study possesses the enzyme or the group of enzymes necessary to produce the fermentation of the various carbon compounds (pentoses, hexoses, etc.), which is manifested by the formation of organic acids and carbon dioxide in the culture medium. Together with an auxanogram and other methods of study, the zymogram is used routinely for the identification and classification of yeasts. Cf. **auxanogram**.

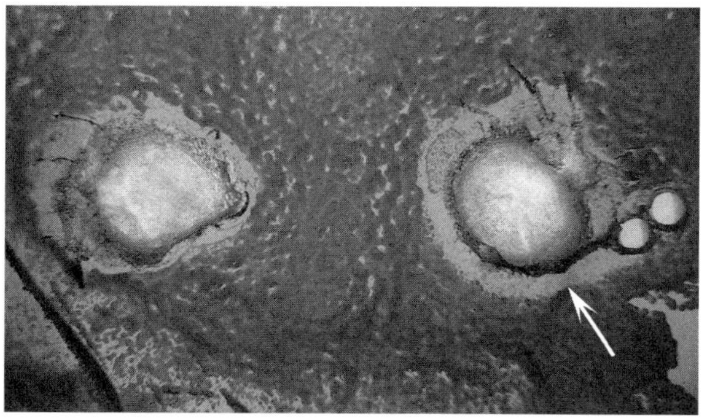

Zymocide effect of a wild strain of *Saccharomyces cerevisiae* (placed in the well of the agar) on a strain of baker's yeast, also of the species *Saccharomyces cerevisiae*, grown around the well. The halo that is surrounding the well indicates the zymocide activity, x 2 (*MU*).

zymologist (Gr. *zýme*, ferment > L. *zymo*, yeast + *lógos*, discourse, study): one who studies yeasts and their fermentations.

zymology (Gr. *zýme*, ferment > L. *zymo*, yeast + *lógos*, discourse, study + *-y*, E. suf. of concrete nouns): the science that treats of the study of yeasts.

zymosis, pl. **zymoses** (Gr. *zýmosis* < *zýme*, ferment > L. *zymo*, yeast + Gr. suf. *-osis*, condition, state of something, a process): see **fermentation.**

OUTLINE OF CLASSIFICATION OF INCLUDED TAXA (TO GENUS LEVEL)

Mainly based on the system followed in the 4th ed. of the book *Introductory Mycology* (Alexopoulos *et al.*, 1996), but with some modifications made after *Ainsworth & Bisby's Dictionary of the Fungi* (8th ed., 1995), for the taxa that are not considered by the former book. Families and genera are lacking for some orders included in the classification, as none were mentioned in the text.

KINGDOM FUNGI
PHYLUM CHYTRIDIOMYCOTA

Class	***Order***	***Family***	***Genus***
Chytridiomycetes	Spizellomycetales	Olpidiaceae	*Nucleophaga*
			Olpidium
			Rozella
		Spizellomycetaceae	*Rhizophlyctis*
	Neocallimasticales	Neocallimasticaceae	*Piromyces*
	Chytridiales	Synchytriaceae	*Synchytrium*
		Chytridiaceae	*Batrachochytrium*
			Chytridium
			Chytriomyces
			Phlyctidium
			Phlyctochytrium
			Polyphagus
			Rhizidium
			Rhizophydium
		Cladochytriaceae	*Megachytrium*
			Nowakowskiella
		Endochytriaceae	*Catenochytridium*
	Blastocladiales	Blastocladiaceae	*Allomyces*
			Blastocladia
			Blastocladiella
		Catenariaceae	*Catenaria*
	Monoblepharidales	Monoblepharidaceae	*Monoblepharella*
			Monoblepharis

PHYLUM ZYGOMYCOTA

Class	***Order***	***Family***	***Genus***
Zygomycetes	Mucorales	Mucoraceae	*Actinomucor*
			Amylomyces
			Chlamydoabsidia
			Gongronella
			Mucor
			Parasitella
			Rhizomucor
			Zygorhynchus
		Gilbertellaceae	*Gilbertella*
		Dicranophoraceae	*Dicranophora*
			Syzygites
		Saksenaeaceae	*Saksenaea*
		Phycomycetaceae	*Phycomyces*
		Absidiaceae	*Absidia*
			Circinella
			Rhizopus
		Pilobolaceae	*Pilaira*
			Pilobolus

		Choanephoraceae	*Blakeslea*
			Choanephora
		Thamnidiaceae	*Cokeromyces*
			Helicostylum
			Pirella
			Thamnidium
		Cunninghamellaceae	*Cunninghamella*
		Mortierellaceae	*Mortierella*
		Syncephalastraceae	*Syncephalastrum*
	Dimargaritales	Dimargaritaceae	*Dimargaris*
			Dispira
	Kickxellales	Kickxellaceae	*Coemansia*
			Kickxella
			Linderina
			Spirodactylon
	Endogonales	Endogonaceae	*Endogone*
	Glomales	Acaulosporaceae	*Acaulospora*
			Entrophospora
		Gigasporaceae	*Gigaspora*
		Glomaceae	*Glomus*
			Sclerocystis
	Entomophthorales	Entomophthoraceae	*Entomophthora*
		Basidiobolaceae	*Basidiobolus*
		Ancylistaceae	*Conidiobolus*
		Neozygitaceae	*Neozygites*
	Zoopagales	Zoopagaceae	*Acaulopage*
			Stylopage
			Zoopage
			Zoophagus
		Cochlonemataceae	*Cochlonema*
		Piptocephalidaceae	*Piptocephalis*
			Syncephalis
	Genera incertae sedis		*Geosiphon*
			Loboa
Trichomycetes	Harpellales	Harpellaceae	*Harpella*
			Stachylina
		Legeriomycetaceae	*Smittium*
			Stipella
	Amoebidiales	Amoebidiaceae	*Amoebidium*
	Asellariales		
	Eccrinales	Eccrinaceae	*Eccrinidus*
			Enterobryus

PHYLUM ASCOMYCOTA

Class	***Order***	***Family***	***Genus***
Archiascomycetes	Taphrinales	Taphrinaceae	*Taphrina*
		Protomycetaceae	*Protomyces*
	Schizosaccharomycetales	Schizosaccharomycetaceae	*Schizosaccharomyces*
Saccharomycetes	Saccharomycetales	Saccharomycetaceae	*Debaryomyces*
			(=*Schwanniomyces*)
			Hansenula

			Kluyveromyces
			Pichia
			Saccharomyces
			(=*Mycoderma*)
			Zygosaccharomyces
		Nadsoniaceae	*Nadsonia*
		Saccharomycodaceae	*Hanseniaspora*
			Saccharomycodes
		Saccharomycopsidaceae	*Saccharomycopsis*
		Eremotheciaceae	*Nematospora*
		Metschnikowiaceae	*Coccidiascus*
			Metschnikowia
		Cephaloascaceae	*Cephaloascus*
		Dipodascaceae	*Ascoidea*
			Galactomyces
			(=*Endomyces*)
		Lipomycetaceae	*Dipodascopsis*
Plectomycetes	Eurotiales	Cephalothecaceae	*Cephalotheca*
		Monascaceae	*Monascus*
		Eurotiaceae	*Emericella*
			Eupenicillium
			Eurotium
		Pseudeurotiaceae	*Pseudeurotium*
	Ascosphaerales	Ascosphaeraceae	*Ascosphaera*
	Onygenales	Onygenaceae	*Aphanoascus*
			Auxarthron
			Onygena
			Shanorella
		Arthrodermataceae	*Arthroderma*
			(=*Nannizzia*)
		Myxotrichaceae	*Byssoascus*
			Eidamella
			Myxotrichum
			Pseudogymnoascus
		Gymnoascaceae	*Gymnoascus*
Pyrenomycetes	Hypocreales	Hypocreaceae	*Calonectria*
			Hypocrea
			Hypomyces
		Nectriaceae	*Gibberella*
			Nectria
		Clavicipitaceae	*Balansia*
			Claviceps
			Cordyceps
			Epichloë
	Melanosporales	Melanosporaceae	*Melanospora*
			(=*Lithomyces*)
	Microascales	Microascaceae	*Microascus*
			Petriella
			Pseudallescheria
		Ceratocystidaceae	*Ceratocystis*
	Phyllachorales	Phyllachoraceae	*Glomerella*
			Haloguignardia

			Ophiodothella
			Phycomelaina
			Phyllachora
			Polystigma
	Ophiostomatales	Ophiostomataceae	*Ophiostoma*
	Diaporthales	Valsaceae	*Cryphonectria*
			Endothia
			Gnomonia
			Valsa
		Vialaeaceae	*Vialaea*
		Magnaporthaceae	*Magnaporthe*
	Calosphaeriales	Calosphaeriaceae	*Calosphaeria*
	Xylariales	Xylariaceae	*Camillea*
			Daldinia
			Hypoxylon
			Nummularia
			Phylacia
			Poronia
			Xylaria
		Diatrypaceae	*Diatrype*
			Eutypa
		Clypeosphaeriaceae	*Ceratostomella*
	Sordariales	Sordariaceae	*Gelasinospora*
			Neurospora
			Sordaria
		Tripterosporaceae	*Podospora*
		Chaetomiaceae	*Ascotricha*
			Chaetomidium
			Chaetomium
			Guanomyces
			Thielavia
		Lasiosphaeriaceae	*Cercophora*
			Lasiosphaeria
			Zopfiella
		Nitschkiaceae	*Bertia*
			Nitschkia
	Meliolales	Meliolaceae	*Asteridiella*
			Meliola
	Halosphaeriales	Halosphaeriaceae	*Abyssomyces*
			Aniptodera
			Bathyascus
			Carbosphaerella
			Chadefaudia
			Ceriosporopsis
			Corollospora
			Halosphaeria
			Lignincola
			Lulworthia
Discomycetes	Medeolariales		
	Rhytismatales	Rhytismataceae	*Lophodermella*
			Lophodermium
			Rhytisma
	Ostropales	Graphidaceae	*Graphis*

	Stictidaceae	*Conotrema*
		Schizoxylon
		(=*Agyriella*)
Cyttariales	Cyttariaceae	*Cyttaria*
Helotiales	Leotiaceae	*Cudoniella*
		(=*Helotium*)
		Leotia
	Sclerotiniaceae	*Ciboria*
		Mitrula
		Monilinia
		Sclerotinia
	Dermateaceae	*Blumeriella*
	Geoglossaceae	*Geoglossum*
		Spathularia
		Trichoglossum
	Hyaloscyphaceae	*Hyaloscypha*
	Vibrisseaceae	*Vibrissea*
Neolectales		
Gyalectales		
Lecanorales	Collemataceae	*Collema*
		Leptogium
	Lecideaceae	*Lecidea*
	Psoraceae	*Psora*
	Rhizocarpaceae	*Rhizocarpon*
	Lecanoraceae	*Lecanora*
	Acarosporaceae	*Acarospora*
	Umbilicariaceae	*Umbilicaria*
		(=*Gyrophora*)
		(=*Omphalodiscus*)
	Parmeliaceae	*Bryoria*
		Cetraria
		Chondropsis
		Evernia
		Letharia
		Parmelia
		Parmotrema
		Pseudevernia
		Usnea
	Physciaceae	*Anaptychia*
		Buellia
		Physcia
		Rinodina
	Ramalinaceae	*Ramalina*
	Cladoniaceae	*Cladina*
		Cladonia
	Stereocaulaceae	*Stereocaulon*
	Alectoriaceae	*Alectoria*
	Hymeneliaceae	*Aspicilia*
		(=*Sphaerothallia*)
		(=*Urceolaria*)
	Haematommataceae	*Haematomma*
Lichinales	Lichinaceae	*Ephebe*
		Lempholemma
		Lichinella

			(=*Gonohymenia*)
			Zahlbrucknerella
	Peltigerales	Lobariaceae	*Lobaria*
			Sticta
		Peltigeraceae	*Peltigera*
			Solorina
	Pertusariales	Pertusariaceae	*Ochrolechia*
			Pertusaria
	Teloschistales	Teloschistaceae	*Caloplaca*
			Fulgensia
			Teloschistes
	Caliciales	Caliciaceae	*Calicium*
			Texosporium
			Thelomma
	Pezizales	Pezizaceae	*Peziza*
			Tirmania
		Tuberaceae	*Tuber*
		Terfeziaceae	*Terfezia*
		Elaphomycetaceae	*Elaphomyces*
		Balsamiaceae	*Barssia*
		Otideaceae	*Acervus*
			Aleuria
			Geopora
			(=*Sepultaria*)
			Inermisia
			(=*Byssonectria*)
			Octospora
			(=*Neottiella*)
			Otidea
			Scutellinia
			(=*Patella*)
		Sarcoscyphaceae	*Cookeina*
			Pithya
			Sarcoscypha
			Wynnea
		Sarcosomataceae	*Urnula*
		Thelebolaceae	*Thelebolus*
			(=*Pezizella*)
		Ascobolaceae	*Ascobolus*
			Iodophanus
			Saccobolus
		Ascodesmidaceae	*Ascodesmis*
		Pyronemataceae	*Pyronema*
		Morchellaceae	*Morchella*
		Helvellaceae	*Choiromyces*
			Gyromitra
			Helvella
Loculoascomycetes	Coryneliales	Coryneliaceae	*Lagenulopsis*
	Dothideales	Coccodiniaceae	*Dennisiella*
		Dothideaceae	*Bagnisiella*
			Dothidea
		Englerulaceae	*Schiffnerula*
		Pseudosphaeriaceae	*Leptosphaerulina*

	Microthyriaceae	*Microthyrium*
		Muyocopron
		(=*Ellisiodothis*)
		Trichothyrium
	Parmulariaceae	*Coccodothis*
	Hypsostromataceae	*Manglicola*
	Hysteriaceae	*Glonium*
		Hysterium
		Hysterographium
	Lophiostomataceae	*Massarina*
	Micropeltidaceae	*Stomiopeltis*
	Mycosphaerellaceae	*Mycosphaerella*
		Stigmidium
		(=*Pharcidia*)
	Polystomellaceae	*Munkiella*
	Schizothyriaceae	*Schizothyrium*
Genera incertae sedis		*Didymella*
		Heliascus
		Otthia
		(=*Stigmina*)
Myriangiales	Myriangiaceae	*Myriangium*
	Elsinoaceae	*Elsinoë*
	Piedraiaceae	*Piedraia*
Arthoniales	Arthoniaceae	*Arthonia*
		Arthothelium
	Roccellaceae	*Chiodecton*
		Dirina
		Mazosia
		Opegrapha
		Roccella
		Schismatomma
Pyrenulales	Pyrenulaceae	*Pyrenula*
Asterinales	Asterinaceae	*Asterina*
Capnodiales	Capnodiaceae	*Capnodium*
Chaetothyriales	Strigulaceae	*Strigula*
	Herpotrichiellaceae	*Capronia*
		(=*Herpotrichiella*)
Patellariales	Patellariaceae	*Rhytidhysteron*
Pleosporales	Pleosporaceae	*Cochliobolus*
		Pleospora
		Pyrenophora
	Leptosphaeriaceae	*Leptosphaeria*
	Sporormiaceae	*Preussia*
		(=*Sporormia*)
		(=*Sporormiella*)
	Venturiaceae	*Apiosporina*
		Venturia
	Botryosphaeriaceae	*Botryosphaeria*
		Guignardia
Melanommatales	Mytilinidiaceae	*Lophium*
		Mytilinidion
Trichotheliales	Trichotheliaceae	*Porina*
Verrucariales	Verrucariaceae	*Dermatocarpon*
		Endocarpon

			Staurothele
			Verrucaria
Other filamentous Ascomycetes	Erysiphales	Erysiphaceae	*Blumeria*
			Brasiliomyces
			Erysiphe
			Microsphaera
			Phyllactinia
			Podosphaera
			Uncinula
	Laboulbeniales	Laboulbeniaceae	*Laboulbenia*
			Rhachomyces
			Rickia
		Herpomycetaceae	*Herpomyces*
		Pyxidiophoraceae	*Pyxidiophora*
	Spathulosporales	Spathulosporaceae	*Spathulospora*
	Familiae incertae sedis	Melaspileaceae	*Melaspilea*
		Thelenellaceae	*Thelenella* (=*Microglaena*)
	Genera incertae sedis		*Kathistes*
			Oceanitis
			Turgidosculum

PHYLUM BASIDIOMYCOTA

Class	***Order***	***Family***	***Genus***
Hymenomycetes	Agaricales	Boletaceae	*Boletus*
			Leccinum
			Suillus
		Gomphidiaceae	*Gomphidius*
		Gyrodontaceae	*Boletinus*
			Gyrodon
		Paxillaceae	*Omphalotus*
			Paxillus
		Strobilomycetaceae	*Strobilomyces*
		Xerocomaceae	*Boletellus*
			Xerocomus
		Hygrophoraceae	*Hygrocybe*
			Hygrophorus
			Hygrotrama
		Tricholomataceae	*Armillaria* (=*Armillariella*)
			Clitocybe
			Collybia
			Crinipellis
			Fayodia
			Hohenbuehelia
			Laccaria
			Lyophyllum
			Macrocystidia
			Marasmius
			Melanoleuca

	Mycena
	Nyctalis
	Omphalina
	(=*Omphalia*)
	Oudemansiella
	Pleurotus
	Resinomycena
	Tricholoma
	Trogia
	Xeromphalina
Amanitaceae	*Amanita*
	(=*Amanitopsis*)
	(=*Vaginata*)
	Termitomyces
Pluteaceae	*Pluteus*
	Volvariella
Agaricaceae	*Agaricus*
	Chlorophyllum
	Cystoderma
	Lepiota
	Leucocoprinus
	Macrolepiota
Coprinaceae	*Coprinus*
	Psathyrella
	(=*Drosophila*)
	(=*Lacrymaria*)
Bolbitiaceae	*Agrocybe*
	Bolbitius
	Conocybe
Strophariaceae	*Anellaria*
	Hypholoma
	(=*Naematoloma*)
	Panaeolus
	Panaeolina
	Pholiota
	(=*Flammula*)
	Psilocybe
	Stropharia
Cortinariaceae	*Cortinarius*
	(=*Phlegmacium*)
	Galerina
	Inocybe
	Naucoria
	Phaeocollybia
	Rozites
Galeropsidaceae	*Galeropsis*
	Weraroa
Crepidotaceae	*Crepidotus*
	Tubaria
Entolomataceae	*Entoloma*
	Leptonia
	Nolanea
	Pouzarella
	Rhodophyllus

	Russulaceae	*Lactarius*
		Russula
Aphyllophorales	Cantharellaceae	*Cantharellus*
		Craterellus
	Gomphaceae	*Gloeocantharellus*
		(=*Linderomyces*)
	Ramariaceae	*Ramaria*
	Clavariaceae	*Clavaria*
	Clavariadelphaceae	*Clavariadelphus*
	Pterulaceae	*Pterula*
	Sparassidaceae	*Sparassis*
	Clavicoronaceae	*Clavicorona*
	Auriscalpiaceae	*Auriscalpium*
	Hericiaceae	*Hericium*
		Lentinellus
	Hydnaceae	*Hydnum*
	Coniophoraceae	*Coniophora*
		Serpula
	Aleurodiscaceae	*Aleurodiscus*
		(=*Cyphella*)
	Corticiaceae	*Corticium*
	Hyphodermatacaee	*Cylindrobasidium*
		(=*Himantia*)
		Hyphoderma
	Meruliaceae	*Dictyonema*
		(=*Cora*)
	Peniophoraceae	*Peniophora*
	Podoscyphaceae	*Cotylidia*
	Steccherinaceae	*Irpex*
	Stereaceae	*Stereum*
	Tubulicrinaceae	*Tubulicium*
		(=*Tubulixenasma*)
	Thelephoraceae	*Pleurobasidium*
		Sarcodon
		Tomentella
		(=*Hypochnus*)
	Lachnocladiaceae	*Vararia*
		(=*Asterostromella*)
	Schizophyllaceae	*Schizophyllum*
	Fistulinaceae	*Fistulina*
	Ganodermataceae	*Amauroderma*
		Ganoderma
	Asterostromataceae	*Asterodon*
		Asterostroma
	Hymenochaetaceae	*Coltricia*
		(=*Polystictus*)
		Cyclomyces
		Hymenochaete
		Inonotus
		Phellinus
	Coriolaceae	*Coriolus*
		Daedalea
		(=*Xylostroma*)
		Fomes

			(=*Ungulina*)
			Fomitopsis
			Heterobasidion
			Hexagonia
			(=*Pogonomyces*)
			Laetiporus
			Lenzites
			Megasporoporia
			Meripilus
			Poria
			Pycnoporus
			Trametes
			Wolfiporia
		Lentinaceae	*Lentinus*
			Panus
		Polyporaceae	*Echinochaete*
			(=*Dendrochaete*)
			Polyporus
Gasteromycetes	Lycoperdales	Lycoperdaceae	*Bovista*
			Bovistella
			Calvatia
			Disciseda
			Langermannia
			Lycoperdon
			Morganella
			Vascellum
		Geastraceae	*Geastrum*
			Myriostoma
			Radiigera
		Broomeiaceae	*Broomeia*
		Mesophelliaceae	*Mesophellia*
		Mycenastraceae	*Mycenastrum*
	Tulostomatales	Battarreaceae	*Battarrea*
		Calostomataceae	*Calostoma*
		Tulostomataceae	*Tulostoma*
	Sclerodermatales	Astraeaceae	*Astraeus*
		Sclerodermataceae	*Pisolithus*
			Scleroderma
	Phallales	Clathraceae	*Aseroë*
			Clathrus
			Laternea
			Pseudocolus
		Hysterangiaceae	*Hysterangium*
			Phallogaster
		Phallaceae	*Dictyophora*
			Itajahya
			Mutinus
			Phallus
		Protophallaceae	*Kobayasia*
	Nidulariales	Nidulariaceae	*Crucibulum*
			Cyathus
			Nidula
			Nidularia

		Sphaerobolaceae	*Sphaerobolus*
	Hymenogastrales	Gastrosporiaceae	*Gastrosporium*
		Hymenogastraceae	*Hymenogaster*
		Octavianinaceae	*Sclerogaster*
		Podaxaceae	*Montagnea*
			Podaxis
		Rhizopogonaceae	*Rhizopogon*
		Secotiaceae	*Endoptychum*
			Secotium
	Gautieriales	Gautieriaceae	*Gautieria*
	Melanogastrales	Melanogastraceae	*Melanogaster*
		Niaceae	*Nia*
	Familiae incertae sedis	Mixiaceae	*Mixia*
Urediniomycetes	Uredinales	Cronartiaceae	*Cronartium*
		Melampsoraceae	*Melampsora*
		Phragmidiaceae	*Phragmidium*
		Pucciniaceae	*Gymnosporangium*
			Puccinia
			Uromyces
		Raveneliaceae	*Diorchidium*
		Uropyxidaceae	*Dasyspora*
	Genera incertae sedis		*Aecidium*
			Caeoma
			Hemileia
			Peridermium
			Roestelia
Ustilaginomycetes	Ustilaginales	Tilletiaceae	*Tilletia*
			Urocystis
		Ustilaginaceae	*Cintractia*
			Sorosporium
			Sphacelotheca
			Tolyposporium
			Ustilago
Other basidiomycetes			
	Tremellales	Tremellaceae	*Tremella*
	Auriculariales	Auriculariaceae	*Auricularia*
	Dacrymycetales	Dacrymycetaceae	*Calocera*
			Dacrymyces
			Dacryopinax
	Ceratobasidiales	Ceratobasidiaceae	*Ceratobasidium*
	Tulasnellales		
	Sporidiales	Sporidiobolaceae	*Aessosporon*
			Rhodosporidium
	Septobasidiales	Septobasidiaceae	*Septobasidium*
	Exobasidiales	Brachybasidiaceae	*Brachybasidium*
		Exobasidiaceae	*Exobasidium*

DEUTEROMYCETES : Asexual ascomycetes and other asexual fungi

	Genus
Sporobolomycetaceous asexual yeasts	*Bullera*
	Itersonilia
	Sporobolomyces
	Tilletiopsis
Cryptococcaceous asexual yeasts	*Candida*
	Cryptococcus
	Kloeckera
	Malassezia
	(=*Pityrosporum*)
	Rhodotorula
	Symbiotaphrina
	Torulopsis
	Trichosporon
	Ypsilonia
Agonomycetaceous asexual fungi	*Cenococcum*
	Papulaspora
	Rhizoctonia
	Sclerotium
Moniliaceous asexual fungi	*Acremonium*
	(=*Cephalosporium*)
	Alatospora
	Anguillospora
	Anthopsis
	Arachnophora
	Arthrobotrys
	Aspergillus
	Beauveria
	Blastomyces
	Botryosporium
	Botrytis
	Brachysporiella
	(=*Monotosporella*)
	Calceispora
	Candelabrella
	Ceratophorum
	Chloridium
	Chrysosporium
	(=*Emmonsia*)
	Coccidioides
	Cylindrocarpon
	Dactylaria
	Dactylella
	Dendrodochium
	Epidermophyton
	Flagellospora
	Geosmithia
	Geotrichum
	Gliocladium

	Harpagomyces
	Harposporium
	Helicoon
	Helicosporium
	Histoplasma
	Meria
	Metarhizium
	Microsporum
	Monilia
	Mycogone
	Myrioconium
	Oedocephalum
	Oidium
	Ostracoderma
	(=*Chromelosporium*)
	Paecilomyces
	Paracoccidioides
	Penicillium
	Phymatotrichopsis
	Pleurophragmium
	(=*Acrotheca*)
	Polyscytalum
	Pyricularia
	Riessia
	Rotiferophthora
	Scopulariopsis
	Spegazzinia
	Spiniger
	Sporendonema
	Sporothrix
	Trichoderma
	Trichophyton
	Trichosporiella
	Trichothecium
	Triramulispora
	Tritirachium
	Valdensia
	Varicosporina
	Verticillium
	Zygosporium
Dematiaceous asexual fungi	*Acrogenospora*
	Acremoniella
	(=*Monopodium*)
	Alternaria
	Amblyosporium
	Arthrinium
	Aureobasidium
	Beltrania
	Bipolaris
	Brachysporium
	Cercospora
	Cercosporidium
	Chalara

	(=*Thielaviopsis*) *Cheiromyces* *Cladosporium* *Clasterosporium* *Cordana* *Corynespora* *Curvularia* *Cyphellophora* *Dictyosporium* *Drechslera* *Exophiala* *Exserohilum* *Fusicladium* *Gonatobotryum* *Helicoma* *Helminthosporium* *Heteroconium* *Humicola* *Isthmospora* *Kionochaeta* (=*Phialicorona*) *Madurella* *Memnoniella* *Moniliella* *Nigrospora* *Oidiodendron* *Orbimyces* *Periconia* *Phialogeniculata* *Phialophora* *Pithomyces* *Polyschema* *Racodium* *Ramichloridium* *Rhinocladiella* (=*Fonsecaea*) *Sarcinella* *Scytalidium* *Selenosporella* *Spilocaea* *Sporidesmiella* *Stachybotrys* *Stemphylium* *Torula* *Ulocladium*
Stilbellaceous asexual fungi	*Akanthomyces* *Antromycopsis* *Dendrostilbella* *Didymostilbe* *Doratomyces* *Graphium* *Hirsutella* *Isaria*

	Stilbella
Tuberculariaceous asexual fungi	*Epicoccum* *Fusarium* *Myrothecium* *Rhynchosporium* *Sphacelia* *Tubercularia* *Volutella*
Melanconiaceous asexual fungi	*Colletotrichum* *Marssonina* (=*Gloeosporium*) *Pestalotia* *Pestalotiopsis* *Sphaceloma*
Sphaeropsidaceous asexual fungi	*Botryodiplodia* *Ceratophoma* *Chaetodiplodia* *Chaetomella* *Chaetoseptoria* *Coniella* *Coniothyrium* *Dichomera* *Endothiella* *Lasiodiplodia* *Microsphaeropsis* *Phoma* *Phomopsis* *Phyllosticta* (=*Phyllostictina*) *Pyrenochaeta* *Sphaeropsis* *Zythia*
Pycnothyriaceous asexual fungi	*Munkia* (=*Pycnostroma*) *Peltasterella* *Pycnothyrium*
Nomen confusum	*Dematium* *Fumago*
Deuterolichenes - Lichenized asexual fungi	*Lepraria*

KINGDOM STRAMENOPILA

PHYLUM OOMYCOTA

Order	***Family***	***Genus***
Saprolegniales	Saprolegniaceae	*Achlya*
		Aphanomyces
		Dictyuchus
		Pythiopsis
		Saprolegnia
		Sommerstorffia
Salilagenidiales	Haliphthoraceae	*Atkinsiella*
Lagenidiales	Lagenidiaceae	*Lagenidium*
Leptomitales	Apodachlyellaceae	*Apodachlyella*
	Leptomitaceae	*Apodachlya*
		Leptomitus
Myzocytiopsidales	Ectrogellaceae	*Ectrogella*
		Haptoglossa
	Pontismaceae	*Petersenia*
Rhipidiales	Rhipidiaceae	*Aqualinderella*
		Rhipidium
Peronosporales	Pythiaceae	*Phytophthora*
		Pythium
	Peronosporaceae	*Basidiophora*
		Peronospora
		Plasmopara
		Sclerospora
	Albuginaceae	*Albugo*
Familia incertae sedis	Lagenaceae	*Pythiella*

PHYLUM HYPHOCHYTRIOMYCOTA

Order	***Family***	***Genus***
Hyphochytriales	Hyphochytriaceae	*Hyphochytrium*
	Rhizidiomycetaceae	*Rhizidiomyces*
Genera incertae sedis		*Rhinosporidium*

PHYLUM LABYRINTHULOMYCOTA

Order	***Family***	***Genus***
Labyrinthulales	Labyrinthulaceae	*Labyrinthula*

KINGDOM PROTISTA

PHYLUM PLASMODIOPHOROMYCOTA

Class	***Order***	***Family***	***Genus***
Plasmodiophoromycetes	Plasmodiophorales	Plasmodiophoraceae	*Octomyxa*
			Plasmodiophora
			Polymyxa
			Sorosphaera
			Spongospora
			Tetramyxa
			Woronina

PHYLUM DICTYOSTELIOMYCOTA

Order	***Family***	***Genus***
Dictyosteliales	Dictyosteliaceae	*Dictyostelium*
		Polysphondylium

PHYLUM ACRASIOMYCOTA

Order	***Family***	***Genus***
Acrasiales	Acrasiaceae	*Acrasis*
	Copromyxaceae	*Copromyxa*

PHYLUM MYXOMYCOTA

Class	***Order***	***Family***	***Genus***
Myxomycetes	Liceales	Cribrariaceae	*Cribraria*
			(=*Dictydium*)
		Liceaceae	*Licea*
		Lycogalaceae	*Dictydiaethalium*
			Lycogala
	Echinosteliales	Echinosteliaceae	*Echinostelium*
	Trichiales	Arcyriaceae	*Arcyria*
		Dianemataceae	*Calomyxa*
		Trichiaceae	*Hemitrichia*
			Metatrichia
			Perichaena
			Trichia
	Physarales	Didymiaceae	*Diachea*
			Diderma
			Didymium
			Lepidoderma
			Mucilago
		Physaraceae	*Badhamia*
			Craterium
			Fuligo
			Leocarpus
			Physarum
	Stemonitales	Stemonitidaceae	*Comatricha*
			Enerthenema
			Lamproderma
			Stemonitis
	Ceratiomyxales		
Protosteliomycetes	Protosteliales	Cavosteliaceae	*Cavostelium*
			Ceratiomyxella
			Protosporangium
		Protosteliaceae	*Nematostelium*
			Protostelium
			Schizoplasmodium

PHYLUM ASCOMYCOTA: APPENDAGE TO THE OUTLINE OF CLASSIFICATION

For purposes of comparison, the families of Ascomycota listed in the outline of classification are placed in the new classification scheme of the phylum Ascomycota, that follows, proposed by Eriksson and Winka (1997), which is based primarily on molecular data.

PHYLUM ASCOMYCOTA

Subphylum Taphrinomycotina

Class	***Order***	***Family***
Taphrinomycetes	Taphrinales	Taphinaceae
	Protomycetales	Protomycetaceae
Schizosaccharomycetes	Schizosaccharomycetales	Schizosaccharomycetaceae
Neolectomycetes	Neolectales	

Subphylum Saccharomycotina

Class	***Order***	***Family***
Saccharomycetes	Saccharomycetales	Saccharomycetaceae
		Saccharomycodaceae
		Saccharomycopsidaceae
		Eremotheciaceae
		Metschnikowiaceae
		Cephaloascaceae
		Dipodascaceae
		Ascoideaceae
		Lipomycetaceae

Subphylum Pezizomycotina

Class	***Order***	***Family***
Eurotiomycetes	Eurotiales	Monascaceae
		Ascosphaeraceae
		Elaphomycetaceae
	Onygenales	Onygenaceae
		Arthrodermataceae
		Gymnoascaceae
Sordariomycetes		
Subclass Hypocreomycetidae	Hypocreales	Hypocreaceae
		Nectriaceae
		Clavicipitaceae
	Microascales	Microascaceae
	Halosphaeriales	Halosphaeriaceae
Subclass Sordariomycetidae	Ophiostomatales	Ophiostomataceae
		Kathistaceae
	Diaporthales	Valsaceae
		Vialaeaceae
	Sordariales	Sordariaceae
		Chaetomiaceae
		Lasiosphaeriaceae
		Nitschkiaceae

Subclass Xylariomycetidae	Xylariales	Xylariaceae
		Diatrypaceae
		Clypeosphaeriaceae
Leotimycetes	Rhytismatales	Rhytismataceae
	Cyttariales	Cyttariaceae
	Leotiales	Leotiaceae
		Sclerotiniaceae
		Dermateaceae
		Geoglossaceae
		Hyaloscyphaceae
		Vibrissiaceae
	Erysiphales	Erysiphaceae
		Thelebolaceae
Lecanoromycetes	Lecanorales	
	Suborder Lecanorineae	Collemataceae
		Lecideaceae
		Psoraceae
		Rhizocarpaceae
		Lecanoraceae
		Acarosporaceae
		Umbilicariaceae
		Parmeliaceae
		Physciaceae
		Ramalinaceae
		Cladoniaceae
		Stereocaulaceae
		Alectoriaceae
		Hymeneliaceae
		Caliciaceae
	Suborder Peltigerinae	Lobariaceae
		Peltigeraceae
Pezizomycetes	Pezizales	Pezizaceae
		Tuberaceae
		Terfeziaceae
		Otideaceae
		Sarcoscyphaceae
		Sarcosomataceae
		Ascobolaceae
		Ascodesmidaceae
		Pyronemataceae
		Morchellaceae
		Helvellaceae
Dothideomycetes	Dothideales	Coccodiniaceae
		Dothideaceae
		Microthyriaceae
		Hypsostromataceae
		Hysteriaceae
		Lophiostomataceae
		Micropeltidaceae

		Mycosphaerellaceae
		Polystomellaceae
		Schizothyriaceae
		Myriangiaceae
		Elsinoaceae
		Piedraiaceae
		Asterinaceae
		Capnodiaceae
		Pleosporaceae
		Leptosphaeriaceae
		Sporormiaceae
		Venturiaceae
		Botryosphaeriaceae
		Mytilinidiaceae
Arthoniomycetes	Arthoniales	Arthoniaceae
		Roccellaceae
Chaetothyriomycetes	Chaetothyriales	Herpotrichiellaceae
	Orders of uncertain position	
	Phyllachorales	Phyllachoraceae
	Meliolales	Meliolaceae
	Medeolariales	
	Ostropales	Graphidaceae
		Stictidaceae
	Gyalectales	
	Lichinales	Lichinaceae
	Pertusariales	Pertusariaceae
	Teloschistales	Teloschistaceae
	Coryneliales	Coryneliaceae
	Pyrenulales	Pyrenulaceae
	Patellariales	
	Trichoteliales	Trichoteliaceae
	Verrucariales	Verrucariaceae
	Laboulbeniales	Laboulbeniaceae
		Herpomycetaceae
		Pyxidiophoraceae
	Spathulosporales	Spathulosporaceae
		Families of uncertain position
		Cephalothecaceae
		Pseudeurotiaceae
		Myxotrichaceae
		Melanosporaceae
		Strigulaceae
		Thelenellaceae

Bibliography

Alexopoulos, C.J., C.W. Mims and M. Blackwell, 1996. *Introductory Mycology*, 4th ed. John Wiley, New York, 868 pp.

Batra, L.R. (Ed.), 1979. *Insect-Fungus Symbiosis*. John Wiley, New York, 276 pp.

Beckett, A., I.B. Heath and D.J. McLaughlin, 1974. *An Atlas of Fungal Ultrastructure*. Longman, London, 221 pp.

Blanco García, V., 1948. *Diccionario latino-español y español-latino*, 3th ed. Aguilar, Madrid, 506 pp.

Buller, A.H.R., 1909-1950. *Researches on Fungi*, 7 Vols.; Vols. 1-6, Longman, Green, London; Vol. 7, University of Toronto Press, Toronto.

Carmichael, J.W., W.B. Kendrick, I.L. Conners and L. Sigler, 1980. *Genera of Hyphomycetes*. The University of Alberta Press, Alberta, 386 pp.

Cifuentes Blanco, J., M. Villegas Ríos and L. Pérez, 1986. *Hongos*. In: Lot, A. and Chiang, F. (Comp.), *Manual de herbario: administración y manejo de colecciones técnicas de recolección y preparación de ejemplares botánicos*. Consejo Nacional de la Flora de México, A. C., pp 55-64.

Ellis, M.B., 1971. *Dematiaceous Hyphomycetes*. Commonwealth Mycological Institute, Kew, 608 pp.

Ellis, M.B., 1976. *More Dematiaceous Hyphomycetes*. Commonwealth Mycological Institute, Kew, 507 pp.

Eriksson, O.E. and K. Winka, 1997. Supraordinal taxa of Ascomycota. *Myconet* 1:1-16.

Font Quer, P., 1963. *Diccionario de botánica*. Labor, Barcelona, 1244 pp.

Fuller, M.S. (Ed.), 1978. *Lower Fungi in the Laboratory*. Department of Botany, University of Georgia, Athens, Ga., 212 pp.

Guzmán, G., 1977. *Identificación de los hongos comestibles, venenosos, alucinantes y destructores de la madera*. Limusa, México, D.F., 236 pp.

Gwynne-Vaughan, H.C.I. and B. Barnes, 1965. *The Structure and Development of the Fungi*, 2nd ed. Cambridge University Press, London, 449 pp.

Hanlin, R.T. and M. Ulloa, 1988. *Atlas of Introductory Mycology*, 2nd ed. Hunter Textbooks, Winston-Salem, 196 pp.

Hawksworth, D.L., 1974. *Mycologist's Handbook*. Commonwealth Mycological Institute, Kew, 231 pp.

Hawksworth, D.L. and D.J. Hill, 1984. *The Lichen-Forming Fungi*. Blackie, Glasgow, 158 pp.

Hawksworth, D.L., P.M. Kirk, B.C. Sutton and D.N. Pegler, 1995. *Ainsworth & Bisby's Dictionary of the Fungi*, 8th ed. International Mycological Institute, CAB International, Wallingford, 616 pp.

Herrera, T. and M. Ulloa, 1990. *El reino de los hongos: micología básica y aplicada*. UNAM-Fondo de Cultura Económica, México, D.F., 552 pp.

Holliday, O., 1989. *A Dictionary of Plant Pathology*. Cambridge University Press, Cambridge, 369 pp.

Jaeger, E., 1950. *A Source-book of Biological Names and Terms*. 2nd ed. Charles C. Thomas, Publisher, Springfield, 287 pp.

Kohlmeyer, J. and E. Kohlmeyer, 1979. *Marine Mycology. The Higher Fungi.* Academic Press, New York. 690 pp.

Largent, D.L., 1973. *How to Identify Mushrooms to Genus. I: Macroscopic Features.* Mad River Press, Eureka, 86 pp.

Largent, D. L., D. Johnson and R. Watling, 1977. *How to Identify Mushrooms to Genus. III: Microscopic Features.* Mad River Press, Eureka, 148 pp.

Margulis, L., H.I. McKhann and L. Olendzenski (Eds.), 1993. *Illustrated Glossary of Protoctista.* Jones and Bartlett Publishers, Boston, 288 pp.

Margulis, L. and K.V. Schwartz, 1988. *Five Kingdoms. An Illustrated Guide to the Phyla of Life on Earth.* 2nd ed. W.H. Freeman and Company, New York, 376 pp.

McVaugh, R., R. Ross and F.A. Stafleu, 1968. An annotated glossary of botanical nomenclature. *Regnum Vegetabile* 56:1-31.

Martin, G.W. and C.J. Alexopoulos, 1969. *The Myxomycetes.* University of Iowa Press, Iowa, 560 pp.

Moreno, N.P., 1984. *Glosario botánico ilustrado.* Compañía Editorial Continental, México, D.F., 300 pp.

O'Donnell, K.L., 1979. *Zygomycetes in Culture.* Department of Botany, University of Georgia, Athens, Ga., 257 pp.

Ortega Pedraza, E., 1980. *Etimologías. Lenguaje culto y científico.* Diana, México, D.F., 286 pp.

Pabón S. de Urbina, J.M. and E. Echauri Martínez, 1959. *Diccionario griego-español.* Publicaciones y Ediciones Spes, Barcelona, 633 pp.

Padres Escolapios, 1943. *Diccionario manual griego-latino-español,* 2th ed. Albatros, Buenos Aires, 966 pp.

Reynolds, D.F. (Ed.), 1981. *Ascomycete Systematics.* Springer-Verlag, New York, 242 pp.

Olive, L.S., 1975. *The Mycetozoans.* Academic Press, New York, 293 pp.

Snell, W.H. and E.A. Dick, 1957. *A Glossary of Mycology.* Harvard University Press, Cambridge, 171 pp.

Snell, W.H. and E.A. Dick, 1971. *A Glossary of Mycology.* Rev. ed. Harvard Univ. Press, Cambridge, MA., 181 pp.

Sparrow, F.K., 1960. *Aquatic Phycomycetes,* 2nd ed. The University of Michigan Press, Ann Arbor, 1187 pp.

Stearn, W.T., 1983. *Botanical Latin.* David & Charles, London, 566 pp.

Sutton, B.C., 1980. *The Coelomycetes.* Commonwealth Mycological Institute, Kew, 696 pp.

Sousa S., M. and S. Zárate P., 1983. *Flora Mesoamericana. Glosario para Spermatophyta, español-inglés.* Instituto de Biología, UNAM, México, D.F., 88 pp.

Swartz, D., 1971. *Collegiate Dictionary of Botany.* The Ronald Press Co., New York, 520 pp.

Ulloa, M., 1991. *Diccionario ilustrado de micología.* Instituto de Biología, UNAM, México, D.F., 310 pp.

Ulloa, M. and T. Herrera, 1994. *Etimología e iconografía de géneros de hongos.* Cuadernos del Instituto de Biología Núm. 21, UNAM, México, D.F., 304 pp.

Wanng, C.J.K. and R.A. Zabel (Eds.), 1990. *Identification Manual for Fungi from Utility Poles in the Eastern United States.* American Type Culture Collection. Allen Press, Lawrence, 356 pp.

Webster's Ninth New Collegiate Dictionary, 1989. Merriam-Webster Inc., Publishers, Springfield, 1562 pp.

Credits of photographs and drawings

Samuel Aguilar (*SA*), Departamento de Botánica, Instituto de Biología, Universidad Nacional Autónoma de México (UNAM), México (2)*

Elvira Aguirre Acosta (*EAA*), Departamento de Botánica, Instituto de Biología, UNAM, México (13)

Regla María Aroche (*RMA*), Universidad Autónoma de Hidalgo, Pachuca, México (11)

Víctor Bandala Muñoz (*VBM*), Instituto de Ecología, Xalapa, México (2)

Calixto Benavides (*CB*), México (115)

Carmen Calderón (*CC*), Departamento de Aerobiología, Centro de Ciencias de la Atmósfera, UNAM, México (6)

Angélica Calderón Villagómez (*ACV*), Universidad del Valle de México, México (2)

Joaquín Cifuentes (*JC*), Facultad de Ciencias, UNAM, México (3)

Gerald van Dyke (*GVD*), Department of Botany, North Carolina State University, Raleigh, North Carolina, USA (2)

Martín Esqueda Valle (*MEV*), Departamento de Tecnología de Alimentos de Origen Vegetal, CIAD, A.C. (1)

Jorge Galindo (*JG*), Centro de Fitopatología, Chapingo, México (1)

Genoveva García Aguirre (*GGA*), Departamento de Botánica, Instituto de Biología, UNAM, México (2)

María C. González (*MCG*), Departamento de Botánica, Instituto de Biología, UNAM, México (1)

Richard T. Hanlin (*RTH*), Department of Plant Pathology, University of Georgia, Athens, Georgia, USA (147)

Gabriela Heredia (*GH*), Instituto de Ecología, Xalapa, México (1)

Teófilo Herrera (*TH*), Departamento de Botánica, Instituto de Biología, UNAM, México (16)

Patricia Lappe (*PL*), Departamento de Botánica, Instituto de Biología, UNAM, México (14)

Rubén López Martínez (*RLM*), Departamento de Microbiología y Parasitología, Facultad de Medicina, UNAM, México (3)

Bernard Lowy † (*BL*), Department of Botany, Louisiana State University, Baton Rouge, Louisiana, USA (1)

Héctor Luna (*HL*), Centro de Investigaciones en Ciencias Biológicas, Universidad Autónoma de Tlaxcala, Tlaxcala,

México (1)

Don Marx (*DM*), Department of Forestry Sciences, University of Georgia, Athens, Georgia, USA (3)

Charles W. Mims (*CM*), Department of Plant Pathology, University of Georgia, Athens, Georgia, USA (6)

Kenny O'Donnell (*KO*), NRRL, ARS, USDA, 1815 N. University, Peoria, Illinois, USA (3)

Sergio Palacios (*SP*), Instituto de Geología, UNAM, México (1)

Evangelina Pérez Silva (*EPS*), Departamento de Botánica, Instituto de Biología, UNAM, México (16)

Manuel Ruiz Oronoz † (*MRO*), Departamento de Botánica, Instituto de Biología, UNAM, México (4)

José Alfredo Samaniego (*JAS*), Campo Experimental Agrícola de la Laguna, CIAN, Torreón, México (3)

Miguel Ulloa (*MU*), Departamento de Botánica, Instituto de Biología, UNAM, México (383)

Ricardo Valenzuela (*RV*), Departamento de Botánica, Escuela Nacional de Ciencias Biológicas, IPN, México (8)

Margarita Villegas (*MV*), Facultad de Ciencias, UNAM, México (1)

W.K.Wynn † (*WKW*), Department of Plant Pathology, University of Georgia, Athens, Georgia, USA (2)

* The figures in parenthesis correspond to the number of contributed photographs.

The captions with no author initials in parentheses represent drawings done by Miguel Ulloa.